Lecture Notes in Artificial Intelligence   7458

Subseries of Lecture Notes in Computer Science

LNAI Series Editors

Randy Goebel
  University of Alberta, Edmonton, Canada
Yuzuru Tanaka
  Hokkaido University, Sapporo, Japan
Wolfgang Wahlster
  DFKI and Saarland University, Saarbrücken, Germany

LNAI Founding Series Editor

Joerg Siekmann
  DFKI and Saarland University, Saarbrücken, Germany

W0235260

Patricia Anthony   Mitsuru Ishizuka
Dickson Lukose (Eds.)

PRICAI 2012:
Trends in
Artificial Intelligence

12th Pacific Rim International Conference
on Artificial Intelligence
Kuching, Malaysia, September 3-7, 2012
Proceedings

 Springer

Series Editors

Randy Goebel, University of Alberta, Edmonton, Canada
Jörg Siekmann, University of Saarland, Saarbrücken, Germany
Wolfgang Wahlster, DFKI and University of Saarland, Saarbrücken, Germany

Volume Editors

Patricia Anthony
Lincoln University, Faculty of Environment, Society and Design
Department of Applied Computing
P.O. Box 84
Christchurch 7647, New Zealand
E-mail: patricia.anthony@lincoln.ac.nz

Mitsuru Ishizuka
University of Tokyo, School of Information Science and Technology
7-3-1, Hongo, Bunkyo-ku
Tokyo 113-8656, Japan
E-mail: ishizuka@i.u-tokyo.ac.jp

Dickson Lukose
MIMOS Berhad, Knowledge Technology
Technology Park Malaysia
Kuala Lumpur 57000, Malaysia
E-mail: dickson.lukose@mimos.my

ISSN 0302-9743   ISSN 1611-3349 (eBook)
ISBN 978-3-642-32694-3   ISSN 978-3-642-32695-0 (eBook)
DOI 10.1007/978-3-642-32695-0
Springer Heidelberg Dordrecht London New York

Library of Congress Control Number: 2012944045

CR Subject Classification (1998): I.2, H.3, H.2.8, F.4, H.4, J.3

LNCS Sublibrary: SL 7 – Artificial Intelligence

Typesetting: Camera-ready by author, data conversion by Scientific Publishing Services, Chennai, India

Printed on acid-free paper

Springer is part of Springer Science+Business Media (www.springer.com)

Preface

The Pacific Rim International Conferences on Artificial Intelligence (PRICAI) is a biennial international event. The PRICAI series aims at stimulating research by promoting exchange and cross-fertilization of ideas among different branches of artificial intelligence (AI). It also provides a common forum for researchers and practitioners in various fields of AI to exchange new ideas and share their experience.

This volume contains the proceedings of the 12th Pacific Rim International Conference on Artificial Intelligence (PRICAI 2012) held in Kuching, Sarawak, Malaysia. PRICAI 2012 received 240 submissions from 43 countries. From these, 82 papers (34%) were accepted as regular and short papers and included in this volume. Eleven papers were accepted for poster presentations, and a four-page paper for each poster is included in this volume. All submitted papers were reviewed by three or more reviewers selected by the Program Committee members. The reviewers' evaluations and comments were carefully examined by the core members of the Program Committee to ensure fairness and consistency in the paper selection.

The papers in this volume give an indication of new movements in AI. The topics roughly include, in alphabetical order, AI foundations, applications of AI, cognition and intelligent interactions, computer-aided education, constraint and search, creativity support, decision theory, evolutionary computation, game playing, information retrieval and extraction, knowledge mining and acquisition, knowledge representation and logic, linked open data and semantic Web, machine learning and data mining, multimedia and AI, natural language processing, robotics, social intelligence, vision and perception, Web and text mining, Web and knowledge-based systems.

The technical program comprised two days of workshops and tutorials, followed by paper and poster sessions, invited talks, and six special sessions. The keynote speakers of PRICAI 2012 were Hiroaki Kitano (Sony CS Lab./System Biology Institute, Japan) and Francesca Rossi (University of Padova, Italy). In addition to these two talks, the participants of PRICAI 2012 could attend four invited talks of two other AI-related conferences co-located at the same venue: the 15th International Conference on Principles and Practice of Multi-Agent Systems (PRIMA 2012) and the International Conference on Dublin Core and Metadata Applications (DC 2012).

The success of a conference depends on the support and cooperation from many people and organizations; PRICAI 2012 was no exception. PRICAI 2012 was supported by MIMOS Berhad, Malaysia. MIMOS organizes a yearly event called the Knowledge Technology Week, and this year PRICAI 2012 operated as the main part of this umbrella event.

We would like to take this opportunity to thank the authors, the Program Committee members, reviewers, and fellow members of the Conference Committee for their time and effort spent in making PRICAI 2012 a successful and enjoyable conference. We also thank the PRICAI Steering Committee, especially its Chair, Abdul Sattar, for giving us helpful advice.

Finally, we thank Springer, its computer science editor, Alfred Hofmann, and Anna Kramer, for their assistance in publishing the PRICAI 2012 proceedings as a volume in its *Lecture Notes in Artificial Intelligence* series.

September 2012   Patricia Anthony
                                                        Mitsuru Ishizuka
                                                        Dickson Lukose

Organization

Steering Committee

Tru Hoang Cao	Ho Chi Minh City University of Technology, Vietnam
Tu-Bao Ho	Japan Advanced Institute of Science and Technology, Japan
Mitsuru Ishizuka	University of Tokyo, Japan
Hideyuki Nakashima	Future University Hakodate, Japan
Duc Nghia Pham	National ICT Australia Ltd., Australia
Abdul Sattar	Griffith University, Australia (Chair)
Byoung-Tak Zhang	Seoul National University, Korea
Chengqi Zhang	University of Technology Sydney, Australia
Zhi-Hua Zhou	Nanjing University, China (Secretary)

Organizing Committee

General Chairs
Dickson Lukose	MIMOS Berhad, Malaysia
Mitsuru Ishizuka	University of Tokyo, Japan

Program Chair
Patricia Anthony	Lincoln University, New Zealand

Local Arrangements Chairs
Edmund Ng Giap Weng	University Malaysia Sarawak, Malaysia
Abdul Rahim Ahmad	University Tenaga Nasional, Malaysia

Special Session Chairs
Bernhard Pfahringer	The University of Waikato, New Zealand
Shahrul Azman Mohd Noah	Universiti Kebangsaan Malaysia, Malaysia

Tutorials Chair
Duc Ngia Pham	Queensland Research Lab, NICTA Ltd., Australia

Workshops Chairs
Aditya Ghose	University of Wollongong, Australia
Ali Selamat	University Technology Malaysia, Malaysia

Treasury Chair
Lai Weng Kin   Tunku Abdul Rahman College, Malaysia

Publicity Chairs
Zili Zhang   Deakin University, Australia
Rayner Alfred   University Malaysia Sabah, Malaysia
Masayuki Numao   Osaka University, Japan

Program Committee

Junaidi Abdullah	Multimedia University
Nik Nailah Abdullah	MIMOS Berhad
Alessandro Agostini	Prince Mohammad Bin Fahd University
Abdul Rahim Ahmad	Universiti Tenaga Nasional
Mohd Sharifuddin Ahmad	Universiti Tenaga Nasional
Eriko Aiba	Japan Advanced Industrial Science and Technology
Akiko Aizawa	National Institute of Informatics
Umar Al-Turki	King Fahd University of Petroleum and Minerals
David Albrecht	Monash University
Rayner Alfred	University Malaysia Sabah
Zulaiha Ali Othman	Universiti Kebangsaan Malaysia
Arun Anand Sadanandan	MIMOS Berhad
Patricia Anthony	Lincoln University
Judith Azcarraga	De La Salle University
Daniel Bahls	Leibniz Information Centre for Economics
Quan Bai	Auckland University of Technology
Mike Bain	The University of New South Wales
Mike Barley	University of Auckland
Laxmidhar Behera	Indian Institute of Technology Kanpur
Khalil Ben Mohamed	MIMOS Berhad
Lee Beng Yong	Universiti Teknologi MARA
Elisabetta Bevacqua	Lab-STICC, CERV - ENIB
Ghassan Beydoun	University of Wollongong
Albert Bifet	University of Waikato
Patrice Boursier	AMA International University
Rafael Cabredo	Osaka University
Xiongcai Cai	The University of New South Wales
Longbing Cao	University of Technology Sydney
Tru Cao	Ho Chi Minh City University of Technology
Yu-N Cheah	Universiti Sains Malaysia
Ali Chekima	Universiti Malaysia Sabah
Shu-Ching Chen	Florida International University
Songcan Chen	Nanjing University of Aeronautics and Astronautics

Stefan Mandl	EXASOL AG
Ivan Marsa-Maestre	University of Alcala
Eric Martin	The University of New South Wales
Thomas Meyer	UKZN/CSIR Meraka Centre for AI Research
Benjamin Chu Min Xian	MIMOS Berhad
Riichiro Mizoguchi	Osaka University
Fitri Suraya Mohamad	Universiti Malaysia Sarawak
Shahrul Azman Mohd Noah	Universiti Kebangsaan Malaysia
Diego Molla	Macquarie University
James Mountstephens	Universiti Malaysia Sabah
Raja Kumar Murugesan	Taylor's University
Edmund Ng	Universiti Malaysia Sarawak
Radoslaw Niewiadomski	CNRS LTCI Telecom ParisTech
Masayuki Numao	Osaka University
Magalie Ochs	CNRS LTCI Télécom ParisTech
Hayato Ohwada	Tokyo University of Science
Manabu Okumura	Tokyo Institute of Technology
Isao Ono	Tokyo Institute of Technology
Mehmet Orgun	Macquarie University
Noriko Otani	Tokyo City University
Maurice Pagnucco	The University of New South Wales
Jeffrey Junfeng Pan	Google Inc.
Jeng-Shyang Pan Pan	National Kaohsiung University of Applied Sciences
Russel Pears	Auckland University of Technology
Jose-Maria Pena	Universidad Politecnica de Madrid
Bernhard Pfahringer	University of Waikato
Duc-Nghia Pham	NICTA/Griffith University
Jeremy Pitt	Imperial College London
Marc Plantevit	LIRIS - Université Claude Bernard Lyon 1
Mikhail Prokopenko	CSIRO Australia
Joel Quinqueton	LIRMM - University of Montpellier
David Rajaratnam	The University of New South Wales
Anca Ralescu	University of Cincinnati
Mukhtar Masood Rana	University of Hail
Dennis Reidsma	University of Twente
Fenghui Ren	University of Wollongong
Napoleon Reyes	Massey University
Debbie Richards	Macquarie University
Pedro Pereira Rodrigues	University of Porto
Vedran Sabol	Know-Center
Kazumi Saito	Univesity of Shizuoka
Rolf Schwitter	Macquarie University
Ali Selamat	Universiti Teknologi Malaysia
Nazha Selmaoui-Folcher	PPME - University of New Caledonia

Chee Seng Chan	University of Malaya
Rudy Setiono	National University of Singapore
Yi-Dong Shen	Chinese Academy of Sciences
Tan Sieow Yeek	MIMOS Berhad
Yashwant Prasad Singh	Multimedia University
Dominik Ślęzak	University of Warsaw
Tony Smith	University of Waikato
Tran Cao Son	New Mexico State University
Kalaiarasi Sonai Muthu	Multimedia University
Safeeullah Soomro	Institute of Business and Technology
Markus Stumptner	University of South Australia
Merlin Teodosia Suarez	De La Salle University
Wing-Kin Sung	National University of Singapore
Xijin Tang	CAS Academy of Mathematics and Systems Science
Yi Tang	Xi'an Institute of Optics and Precision Mechanics of CAS
David Taniar	Monash University
Jason Teo	Universiti Malaysia Sabah
Bui The Duy	Vietnam National University - Hanoi
Patrick Then	Swinburne University of Technology
Michael Thielscher	The University of New South Wales
John Thornton	Griffith University
Klaus Tochtermann	Leibniz Information Center for Economics
Satoshi Tojo	Japan Advanced Institute of Science and Technology
Kah Mein Tracey Lee	Singapore Polytechnic
Eric Tsui	The Hong Kong Polytechnic University
Kuniaki Uehara	Kobe University
Ventzeslav Valev	Bulgarian Academy of Sciences
Michael Cheng Wai Khuen	Universiti Tunku Abdul Rahman
Toby Walsh	NICTA and UNSW
Kewen Wang	Griffith University
Lipo Wang	Nanyang Technological University
Qi Wang	Xi'an Institute of Optics and Precision Mechanics of CAS
Ian Watson	University of Auckland
Chin Wei Bong	MIMOS Berhad
Goh Wei Wei	Taylor's University
Kow Weng Onn	MIMOS Berhad
Wayne Wobcke	The University of New South Wales
Kok Sheik Wong	University of Malaya
M.L. Dennis Wong	Xi'an Jiaotong-Liverpool University
Sheng-Chuan Wu	Franz Inc.
Guandong Xu	Victoria University
Ke Xu	Microsoft

Ming Xu	Xi'an Jiaotong-Liverpool University
Xiangyang Xue	Fudan University
Ikuko Yairi	Sophia University
Seiji Yamada	National Institute of Informatics
Jianmei Yang	South China University of Technology
Roland Yap	National University of Singapore
Dayong Ye	University of Wollongong
Dit-Yan Yeung	Hong Kong University of Science and Technology
Lim Ying Sean	MIMOS Berhad
Philip Yu	University of Illinois at Chicago
Pong C Yuen	Hong Kong Baptist University
Fadzly Zahari	MIMOS Berhad
Yi Zeng	CAS Institute of Automation
Bo Zhang	Tsinghua University
Byoung-Tak Zhang	Seoul National University
Chengqi Zhang	University of Technology Sydney
Daoqiang Zhang	Nanjing University of Aeronautics and Astronautics
Dongmo Zhang	University of Western Sydney
Du Zhang	California State University
Junping Zhang	Fudan University
Minjie Zhang	University of Wollongong
Peng Zhang	Chinese Academy of Sciences
Shichao Zhang	University of Technology Sydney
Wen Zhang	CAS Institute of Software
Xiatian Zhang	Business Intelligence Center Tencent
Xuguang Zhang	Yanshan University
Zhongfei Zhang	SUNY Binghamton
Zili Zhang	Deakin University
Yanchang Zhao	RDataMining.com
Zhi-Hua Zhou	Nanjing University
Xingquan Zhu	Florida Atlantic University
Indre Zliobaite	Bournemouth University
Jean Daniel Zucker	Institut de Recherche pour le Développement

Additional Reviewers

Abul Hasanat, Mozaherul Hoque	Chen, Feng	Du, Liang
Ali, Memariani	Chitsaz, Mahsa	El-Alfy, El-Sayed M.
Bardi, Yausefi	Chongjun, Wang	Fanaee Tork, Hadi
Cao, Huiping	Chua, Fang-Fang	Fatima, Shaheen
Chang, Meng	David, Kalhor	Feng, Yachuang
Chao, Wen-Han	Dehzangi, Abdollah	Ferreira, Carlos
	Dinh, Thien Anh	Fleites, Fausto

Editorial Board

Sponsoring Organizations

 Ministry of Science, Technology and Innovation, Malaysia

 MIMOS Berhad, Malaysia

 Sarawak Convention Bureau, Malaysia

 Leibniz Information Centre for Economics, Germany

 Japanese Society for Artificial Intelligence, Japan

 SIG-MACC, Japan Society for Software Science and Technology, Japan

 AGRO-KNOW Technologies, Greece

 agINFRA: A Data Infrastructure for Agriculture

 Franz Inc., USA

 Quintiq Sdn. Bhd., Malaysia

 Centre for Agricultural Bioscience International, UK

 NOTA ASIA Sdn. Bhd., Malaysia

 University of Tasmania, Australia

Supporting Institutions

Universiti Tenaga Nasional, Malaysia
Swinburne University of Technology, Malaysia
The University of Nottingham, Malaysia
Monash University, Malaysia
Multimedia University, Malaysia
Sunway University, Malaysia
University Tunku Abdul Rahman, Malaysia
Universiti Sains Malaysia, Malaysia
Universiti Malaya, Malaysia
The British University in Dubai, United Arab Emirates
The University of Tulsa, USA
The University of New South Wales, Australia
Nanyang Technological University, Singapore
The University of Waikato, New Zealand
Macquarie University, Australia
Waseda University, Japan
National ICT Australia Ltd., Australia
Universiti Malaysia Sarawak, Malaysia
Universiti Kebangsaan Malaysia, Malaysia
Universiti Malaysia Sabah, Malaysia

Table of Contents

Short Papers

Poster Papers

Systems Biology Powered by Artificial Intelligence

Hiroaki Kitano

The Systems Biology Institute,
Okinawa Institute of Science and Technology,
Sony Computer Science Laboratories, Inc.
kitano320@gmail.com

Abstract. Systems biology is an attempt to understand biological system as system thereby triggering innovations in medical practice, drug discovery, bio-engineering, and global sustainability problems. The fundamental difficulties lies in the complexity of biological systems that have evolved through billions of years. Nevertheless, there are fundamental principles governing biological systems as complex evolvable systems that has been optimized for certain environmental constraints. Broad range of AI technologies can be applied for systems biology such as text-mining, qualitative physics, marker-passing algorithms, statistical inference, machine learning, etc. In fact, systems biology is one of the best field that AI technologies can be best applied to make high impact research that can impact real-world. This talk addresses basic issues in systems biology, especially in systems drug discovery and coral reef systems biology, and discusses how AI can contribute to make difference.

P. Anthony, M. Ishizuka, and D. Lukose (Eds.): PRICAI 2012, LNAI 7458, p. 1, 2012.
© Springer-Verlag Berlin Heidelberg 2012

Preference Reasoning and Aggregation: Between AI and Social Choice

Francesca Rossi

Department of Mathematics,
University of Padova, Via Trieste 63
35121 Padova, Italy
frossi@math.unipd.it

Abstract. Preferences are ubiquitous in everyday decision making. They should therefore be an essential ingredient in every reasoning tool. Preferences are often used in collective decision making, where each agent expresses its preferences over a set of possible decisions, and a chair aggregates such preferences to come out with the "winning" decision. Indeed, preference reasoning and multi-agent preference aggregations are areas of growing interest within artificial intelligence.

Preferences have classically been the subject also of social choice studies, in particular those related to elections and voting theory. In this context, several voters express their preferences over the candidates and a voting rule is used to elect the winning candidate. Economists, political theorist, mathematicians, as well as philosophers, have made tremendous efforts to study this scenario and have obtained many theoretical results about the properties of the voting rules that one can use.

Since, after all, this scenario is not so different from multi-agent decision making, it is not surprising that in recent years the area of multi-agent systems has been invaded by interesting papers trying to adapt social choice results to multi-agent setting. An adaptation is indeed necessary, since, besides the apparent similarity, there are many issues in multi-agent settings that do not occur in a social choice context: a large set of candidates with a combinatorial structure, several formalisms to model preferences compactly, preference orderings including indifference and incomparability, uncertainty, as well as computational concerns.

The above considerations are the basis of a relatively new research area called computational social choice, which studies how social choice and AI can fruitfully cooperate to give innovative and improved solutions to aggregating preferences given by multiple agents. This talk will present this interdisciplinary area of research and will describe several recent results regarding some of the issues mentioned above, with special focus on robustness of multi-agent preference aggregation with respect to influences, manipulation, and bribery.

P. Anthony, M. Ishizuka, and D. Lukose (Eds.): PRICAI 2012, LNAI 7458, p. 2, 2012.

Economical Operation of Thermal Generating Units Integrated with Smart Houses

Shantanu Chakraborty[1], Takayuki Ito[1], and Tomonobu Senjyu[2]

[1] Department of CSE, Nagoya Institute of Technology
Nagoya, Japan
[2] Department of EEE, University of the Ryukyus
Okinawa, Japan

Abstract. This paper presents an economic optimal operation strategy for thermal power generation units integrated with smart houses. With the increased competition in retail and power sector reasoned by the deregulation and liberalization of power market make optimal economic operation extremely important. Moreover, the energy consumption is multiplying due to the proliferation of all-electric houses. Which is why, controllable loads such as electric water heater, heat pump (HP) and electric vehicles (EV) have great potentials to be introduced in a smart-grid oriented environment. The presented strategy models thermal power generators with controllable loads (HP and EV) in a coordinated manner in order to reduce the production cost as a measure of supply side optimization. As of demand side, the electricity cost is minimized by means of reducing the interconnection point (IP) power flow. Particle swarm optimization (PSO) is applied to solve both of the optimization problems in efficient way. A hypothetical power system (with practical constraints and configurations) is tested to validate the performance of the proposed method.

Keywords: Smart grid, thermal unit commitment, smart house, renewable energy sources, particle swarm optimization.

1 Introduction

The restructuring in the electric power industry leads the deregulation of electric utilities and the number of new power suppliers in electric power market has grown significantly due to this deregulation. Consequently such deregulation and liberalization in power market lead to the increased competition in retail power sector and electric sources operated by independent power producer. Such situation demands efficient thermal generation strategy while minimizing the production cost as well as maximizing the profit. On the other hand, CO_2 emission and energy consumption need to be reduced to compensate fossil fuel burning and global warming [1]. Therefore, renewable power plant such as photovoltaic (PV) facilities and wind turbine generators (WG) are integrated with conventional power system.

P. Anthony, M. Ishizuka, and D. Lukose (Eds.): PRICAI 2012, LNAI 7458, pp. 3–14, 2012.

The reduction of production cost by controlling the thermal units and storage system is essentially feasible and researchers have proposed optimization strategies for such economical operations [2]. Since the amount of energy consumption being increasing in domestic house holdings, controllable loads are thought to be very useful inclusion. The usage of smart grid is very effective considering the load leveling and cost reduction facilitating controllable loads and batteries. Earlier, a method [3] is proposed which provides power balance while coordinating thermal generators with controllable load. On that work, the authors introduced HP, water heater and EV as smart grid components. The effective demand side management was ensured by optimal operation of thermal units and smart grid components. Such research further enhanced in [4] where smart house concept was introduced comprising with HP, batteries, solar collector system and PV system. Moreover, as a measure of control signal, the IP power flow from supply side to demand side was also discussed.

This paper determines the reference of the IP power flow sent to the demand side by the integration of HP, EV as controllable loads while the thermal generating units are in supply side. Henceforth, an optimal economical operation methodology for smart grid is proposed which can reduce the cost for both supply and demand sides. The organization of the paper is briefed as follows: Section 2 presents the optimal operation of supply side where as the demand side operation is presented in Section 3. The operation of smart grid introducing bidirectional connection is presented in Section 4. The conduction simulation and conclusion are presented in Section 5 and 6, respectively. The simulation section shows the effectiveness of the proposed method by various numerical simulations and verify the capability and effect of the optimal operation of thermal units introduced by the new electricity tariff based on IP power flow.

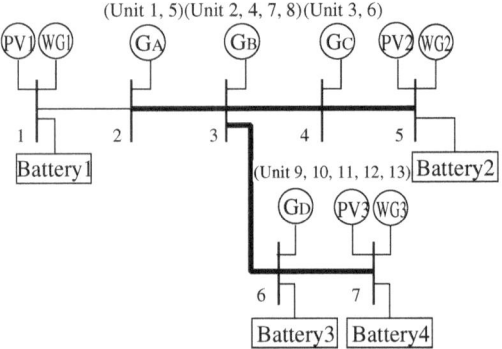

Fig. 1. Hypothetical power system model with 13-units

2   Power Supply Side Optimal Operation

The hypothetical power system model used in this paper is shown in Fig. 1. The model comprises with PV systems, wind turbines, batteries, controllable loads and thermal units. The each battery inverter ratings are 50 MW and 250 MWh. The state of charge (SOC) of each battery is ranged from 20% to 80%. The rated power of the thermal units Unit 1-6, Unit 7-9 and Unit 10-13 are 162 MW, 130 MW and 85 MW, respectively.

Table 1. Forecasted load demand for 24-hours

Node No.	1	2	3	4	5	6	7
Node Load P_{Li}	$\frac{P_L}{16}$	$\frac{P_L}{8}$	$\frac{P_L}{8}$	$\frac{P_L}{8}$	$\frac{P_L}{16}$	$\frac{P_L}{4}$	$\frac{P_L}{4}$

Table 1 shows the perspective load demand in each node of the system model. The assumed maximum rated power and the total thermal units rated power are 1,145 MW and 1,702 MW, respectively.

2.1   Formulation

The formulation of the optimization problem including the considered constraints will be shown in this section.

Objective Function. The objective function is formulated to minimize the operational cost which is a summation of fuel cost and start-up cost for the committed units.

$$min \; F = \sum_{t \epsilon T}[FC_i(P_{g,i}(t)).u_i(t) + cost_i.u_i(t).(1 - u_i(t-1))] \qquad (1)$$

where T is the total scheduling period [hour], $FC_i(P_{g,i}(t))$ is the fuel cost[\$] for unit i at t-th hour for generating $P_{g,i}$ power, $u_i(t)$ is the on/off status of unit i at t-th hour and $cost_i$ is the start-up cost [\$]. The fuel cost is a quadratic function comprises of various cost coefficients.

Constraints. The constraints considered in this paper for power supply side include power balance constraints, thermal unit constraints, the controllable loads constraints and battery constraints. The minimum up/down constraints and thermal unit output power constraints are considered as thermal unit constraints. The batteries are possessed in the main electricity power company. They can be controlled within the acceptable range to coordinate the thermal units optimal operation. The battery constraints include the battery output limit and the state of the charge.

As controllable loads constraints, it can be safely assumed that these loads' output are high in night time and low in daytime. Accordingly, the controllable loads output is limited in the acceptable range as follows:

$$-\frac{1.0 \times 10^5}{P_{Li}(t)} \le P_{co,i}(t) \le \frac{1.0 \times 10^5}{P_{Li}(t)} \tag{2}$$

Note that, the controllable load constraint of the node 1 described in Section 4 is determined by HP and EV in demand side.

2.2   Optimal Operation

Particle swarm optimization is used in this paper to solve the optimization problem. PSO was introduced as a swarm based computational algorithm, imitating the behavior of flocks of birds or school of fish coupled with their knowledge sharing mechanisms. Individual, in the context of PSO, referred as 'particle' which represents potential solution, utilizes two important kinds of information in decision process; 1) its own experience, which says the choices it has tried so far and the best one within them, 2) knowledge of the other particles, i.e. it has also the information of how the other agents performed. The movement of particle requires updating the velocity and position according following equations

$$v_i(t+1) = w.v_i(t) + c_1.r_1.[pbest_i(t) - x_i(t)] \tag{3}$$
$$+ c_2.r_2.[gbest - x_i(t)]$$

$$x_i(t+1) = x_i(t) + \Delta t.v_i(t+1) \tag{4}$$

where w is the weight; t represents the current iteration; c_1 and c_2 represent the acceleration constants of cognitive and social components, respectively and $r_{1,2} = U(0,1)$ are the uniform random numbers distributed within $[0,1]$.

Since the generation scheduling requires binary value optimization, the corresponding binary version of PSO is introduced. Henceforth, the position updating equation is changed as follows:

$$x_i(t+1) = \begin{cases} 1 & \text{if } rand < Sigmoid(v_i(t+1)) \\ 0 & \text{otherwise} \end{cases} \tag{5}$$

where $Sigmoid(\lambda)$ is defined as

$$Sigmoid(\lambda) = \frac{1}{1 + \exp(\lambda)} \tag{6}$$

The outline of the applied PSO method for supply side optimization is shown in Fig. 2.

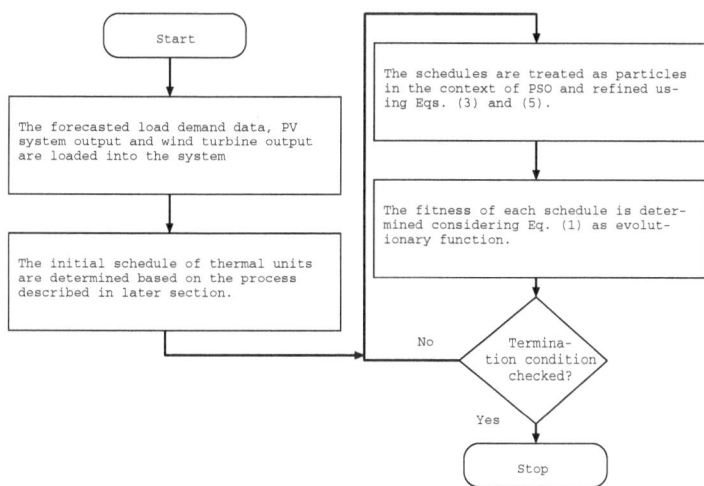

Fig. 2. Flow chart of the PSO based supply side optimization

2.3   Generating Initial Solution

All thermal units in this paper always operated at rated output level. Therefore, the thermal units initial schedule are produced based on the thermal units, the controllable and the batteries condition. The difference of the thermal units output and the load demand is controlled by the controllable loads and the batteries. For producing the priority list, the proposed method considers on/off frequency of thermal unit. Therefore, the largest rated thermal unit has the highest priority.

3   Optimal Operation in Demand Side

3.1   PV System

Fig. 3 shows the smart house model used in demand side. The system parameters used in this paper for PV are, the conversion efficiency of solar cell array η_{PVd} (= 14.4%), the panel number n_{PVd} (= 18), the array area S_{PVd} (= 1.3 m^2), the rated power (= 3.5 kw). The PV generated output P_{PVd} [kW] from the amount of solar radiation is calculated as following equation:

$$P_{PVd} = \eta_{PVd} n_{PVd} S_{PVd} I_{PVd}(1 - 0.005(T_{CR} - 25)) \qquad (7)$$

where, I_{PVd} and T_{CR} are the solar radiation[kW/m^2] and the out side air temperature[°C].

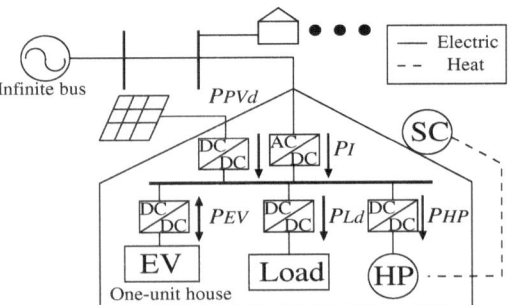

Fig. 3. Applied smart house model

3.2 Solar Collector System

This paper employs the solar collector system referred in [4]. This solar collector system contains the HP as an auxiliary heat source. The hot water heated up from the solar collector is adjusted by diluting the water flow. The water temperature lower than 60 °C at 18 hour is heated to 60 °C by the heat pump.

3.3 Objective Function and Constraints

$P_I(t)$, $P_{Ld}(t)$, $P_{PVd}(t)$, $P_{EV}(t)$ and $P_{HP}(t)$ in Fig. 3 are the IP power flow to the smart houses, the power consumption except the controllable loads, the battery output, the PV system output, and heat pump output, respectively. The power balance condition in demand side is represented as following equation:

$$P_{Ld}(t) = P_I(t) + P_{PVd}(t) + P_{EV}(t) - P_{HP}(t) \qquad (8)$$

The objective function in the demand side is formulated to minimize the IP power flow from the power supply side to the smart houses. The objective function and the constraints are as follows:

Objective Function

$$\min F = \sum_{t \in T} [B_{Icen}(t) - P_I(t)]^2 \qquad (9)$$

where, T, $B_{Icen}(t)$ and $P_I(t)$ are the total scheduling period [hour], the referenced IP power flow and injected the IP power flow to the smart house, respectively.

Constraint Conditions. The constraints considered in the smart house contain the IP power flow, the EV battery output and the capacity of EV battery. The IP power flow is represented as following equations

$$P_{Imin}(t) < \Delta P_I(t) < P_{Imax}(t) \qquad (10)$$

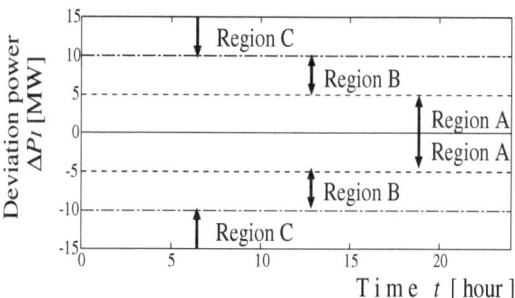

Fig. 4. Electricity tariff in different region

$$\begin{cases} \Delta P_I(t) = P_I(t) - B_{Icen} & (P_I(t) > B_{Icen}) \\ \Delta P_I(t) = B_{Icen} - P_I(t) & (B_{Icen} > P_I(t)) \end{cases} \quad (11)$$

where, $P_{Imin}(t)$, $P_{Imax}(t)$ and $\Delta P_I(t)$ are IP power flow minimum bandwidth, IP power flow maximum bandwidth and the violated interconnection point power flow, respectively.

Fig 4 shows the new electricity tariff system. The electricity tariff system is determined advantaging both power supply and demand sides, which agrees with the concept of future smart grid system. The electricity tariff to buy new electricity tariff is 10 Yen/kWh within the region/bandwidth A shown in Fig 4, and it is 20 and 30 Yen/kWh in the violated bandwidth B and C, respectively. The electricity tariff to sell is 10 Yen/kWh.

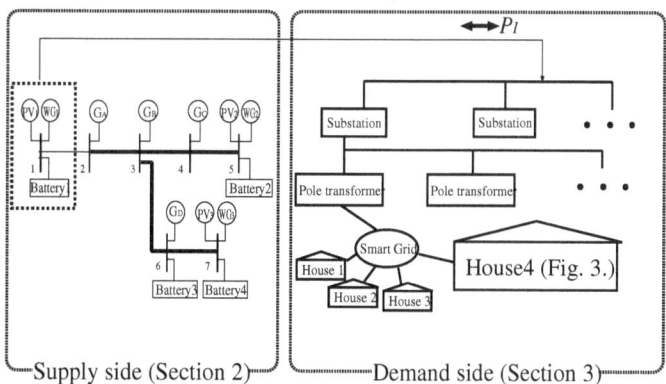

Fig. 5. Concept of smart grid

Optimization Method. PSO method is used here as the optimization algorithm. It is assumed that load demand and heat load in demand side are forecasted. Therefore,the operating time should be estimated by the solar radiation. Continuous PSO determines the operating time of the EV output and the heat pump output for each hour.

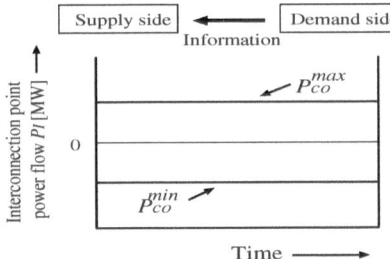

6(a) Producing the controllable load constraint (STEP1)

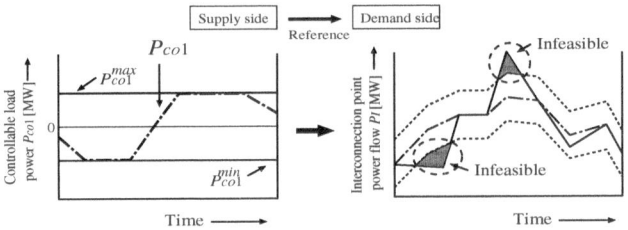

6(b) Optimal operation of supply side and demand side (STEP2)

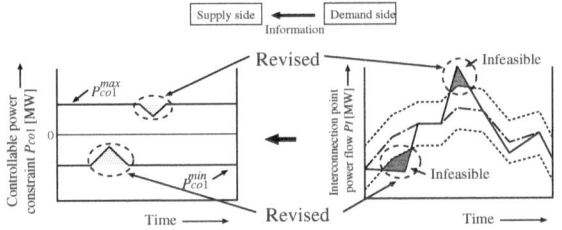

6(c) Revising the controllable load constraint (STEP3)

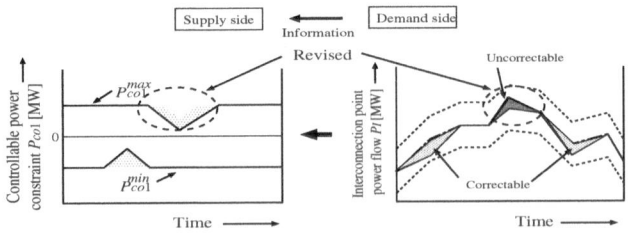

6(d) Revising the controllable load constraint (STEP4)

Fig. 6. Operation of smart grid

4   Optimal Operation Based on Smart Grid

The IP power flow is controlled by the controllable loads within the bandwidth in the new electricity tariff based on the IP power flow in demand side. Fig 5 shows the smart grid scheme used in this paper. Here are the outline of the optimization algorithm.

Step 1: The controllable loads information, shown in Fig. 6(a) is assembled from the demand side.

Step 2: The thermal units schedule shown in Fig. 6(b) is determined in the power supply side and the IP power flow reference sent to the node 1 (Section 2). Consequently, the demand side operation is performed (Section 3).

Step 3: The IP power flow is checked whether it can be controlled within the bandwidth. If not, the controllable loads constraint is revised (as shown in Fig. 6(c)) and the searching procedure back to Step 2.

Step 4: The difference of the IP power flow and the reference of that is controlled by the batteries in the power supply side. When the difference is minimized, the searching procedure backs to Step 1.

Table 2. Parameter in demand side (node 1)

	Node 1	Smart house
Maximum power (P_{Ld})	78 MW(25,000 houses)	3 kW
Inverter of EV (P_{EV})	19 MW(7,500 houses)	2.5 kW
Capacity of EV (S_{EV})	120 MWh(7,500 houses)	16 kWh
Heat pump (P_{HP})	11 MW(7,500 houses)	1.5 kW
Photovoltaic (P_{PVd})	26 MW(7,500 houses)	3.5 kW
Used hours of shower	18~21 hour	18~21 hour
Amount of Shower	750 kl(7,500 houses)	100 l
Area of SC	1.6 m^2	1.6 m^2
Number of SC	22,500(7,500 houses)	3
Tank capacity	2775 kl(7,500 houses)	370 l

5   Simulation

5.1   Simulation Condition

Supply Side System. The load demandP_L, the PV system output P_{PV} and the wind energy system output P_{WG} (shown in Fig. 1) are assumed to have no forecasting error. The 30,000 houses are cumulated in node 1 among 30 % of them are assumed to be smart house. These houses comprised with the PV system, the solar collector system, HP and EV. The conditions in node 1 are summarized in Table 3.

Demand Side System. Assuming no forecasting error of the load demand P_{Ld}, the PV system output P_{PVd} in smart house shown in Fig. 3. The simulation results contain the fair, cloudy and rainy environmental conditions. The PV system output in each weather condition is shown in Fig. 7(d). The load demands except the controllable loads are divided into each node as shown in Table 1. For the heat load, we assume the use of showers from 18th to 21st hour.

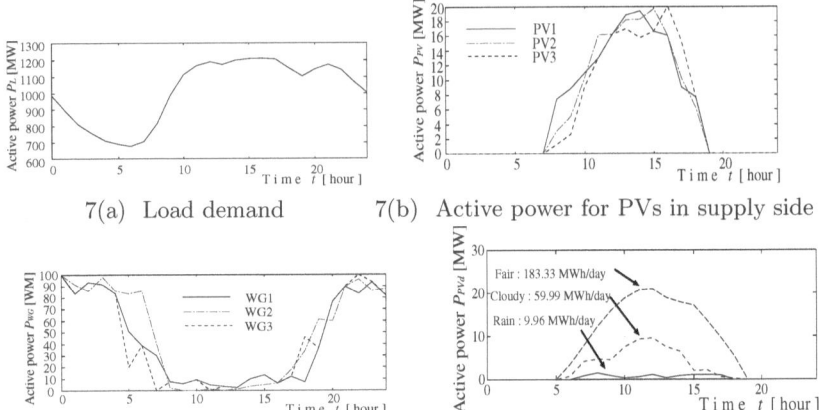

7(a) Load demand   7(b) Active power for PVs in supply side

7(c) Active power for WGs in supply side 7(d) PV output power in demand side

Fig. 7. Simulation conditions

5.2   Simulation Results

The simulation results of the optimal operation considering the smart house is shown Fig. 8 and Table 4. Table 4 shows result in cloudy condition. The "controllable load" and "inverter" in supply side are represented (in Table 4) as the sum of the each controllable loads P_{co} and the each battery output P_{inv} in 0~8, 9~17 and 18~24 hour, respectively. The load demand is increased in 0~8 hour, decreased in 9~17 by the controllable loads in demand side, hence smoothing the demand. "state of charge" in Table 4 indicates the state of charge of charge of battery in 0~25 hour. The "Output of EV" in "Demand" are represented (in Table 4) as the sum of the each battery output P_{EV} in 0~8, 9~17 and 18~24 hour. Fig. 8(a) shows the revised controllable loads output constraint in node 1. In this controllable loads output constraint, the controllable loads in demand side can control the IP power flow within the bandwidth. Fig. 8(b) shows the thermal units output. Fig. 8(c) shows the water temperature in demand side. The water temperature lower than 60 °C is heated by the heat pump. Fig. 8(d) shows the IP power flow and the bandwidth and the reference of that. The search direction of the IP power flow reference is determined to increase the load demand in night time and to decrease the load demand in day time in order to reduce the output from thermal units. The IP power flow is controlled within the region A shown in Fig. 4 by optimizing the heat pump and the electricity. The electricity tariff is expected to be reduced in demand side. Table 5 shows the calculated electricity tariff for each weather condition in demand side. The tariff shown in table 5 is the one that employed in Tokyo Electric Power Company. The tariff in 7~23 hour is 21.87 Yen/kWh, and tariff of the other time is 9.17 Yen/kWh.

Table 3. Summarized result for cloudy condition

			0~8	9~17	18~24	0	25
Supply	Inverter of battery	P_{inv1}[MW]	34.7	15.3	0		
		P_{inv2}[MW]	0	5.2	35.2		
		P_{inv3}[MW]	-3.0	37.8	11.4		
		P_{inv4}[MW]	0	0	0		
	State of charge of battery	S_{B1}[MWh]				100.0	59.6
		S_{B2}[MWh]				100.0	59.6
		S_{B3}[MWh]				100.0	53.8
		S_{B4}[MWh]				100.0	100.0
	Controllable load	P_{co1}[MW]	-42.1	61.8	8.4		
		P_{co2}[MW]	-50.7	88.9	49.3		
		P_{co3}[MW]	-64.0	60.8	46.9		
		P_{co4}[MW]	-64.0	30.4	38.4		
		P_{co5}[MW]	-25.7	9.9	18.8		
		P_{co6}[MW]	-48.2	24.8	75.4		
		P_{co7}[MW]	-27.1	24.8	59.7		
Demand	Output of EV	P_{EV}[MW]	-70.0	168.0	-21.0		
	State of charge of EV	S_{EV}[MWh]				80.0	91.0
	Heat pump	Time[hour]		6~8			
		P_{HP}[MW]		9.7			

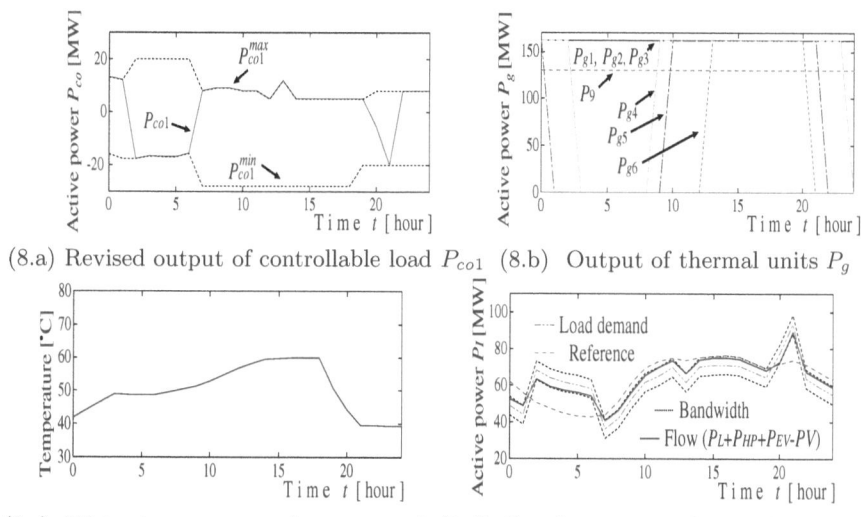

(8.a) Revised output of controllable load P_{co1} (8.b) Output of thermal units P_g

(8.c) Water temperature of storage tank (8.d) Supplying power from infinite bus

Fig. 8. Simulation results in cloudy condition

The IP power flow is controlled within the Region A in each weather condition, thereby reducing the price. Table 6 shows the total cost of thermal units in the supply side. Operating the thermal units in the rated power by controlling the controllable loads in demand side leads to the total cost reduction of thermal units. It is noted that, the operating cost in fair condition is lower than cloudy and rainy conditions because of the usage of HP facilitating solar collector system.

Table 4. Electric cost for smart houses (node 1)

	Fair	Cloudy	Rain
Conventional cost [$]	236,050	275,020	290,870
Proposed cost [$]	121,470	146,140	156,150

Table 5. Operating cost

	Conventional	Fair	Cloudy	Rain
Fuel cost[$]	591,630	481,190	484,940	484,940
Start cost[$]	2,200	4,800	4,800	4,800
Total energy[MWh]	25,469	24,775	24,937	24,937
Total cost[$]	593,830	485,990	489,740	489,740

6 Conclusion

With the increasing concerns regarding renewable energy sources and smart grid infrastructure, the applicability of smart houses is growing. Considering such scenario in mind, this paper presents an economical optimal operation of conventional thermal units integrated with smart houses. The heat pump, the solar collector system and electric vehicle are modeled into the smart house. The optimization procedure tries to minimize the cost and controlling IP power flow in supply and demand side, respectively. Both continuous and binary version of PSO are applied to carry out the procedure. This optimal operation shows that the electricity tariff in demand side can be reduced by controlling the IP power flow within the bandwidth in new electricity tariff based on the IP power flow injected to the demand side.

References

1. Chakraborty, S., Senjyu, T., Saber, A.Y., Yona, A., Funabashi, T.: Optimal Thermal Unit Commitment Integrated with Renewable Energy Sources Using Advanced Particle Swarm Optimization. IEEJ Trans. on Electrical and Electronics Eng. 4(5), 609–617 (2009)
2. Tokudome, M., Senjyu, T., Yona, A., Funabashi, T.: Frequency and Voltage Control of Isolated Island Power Systems by Decentralized Controllable Loads. In: IEEE T& D Conference. IEEE Press, Korea (2009)
3. Goya, T., Senjyu, T., Yona, A., Chakraborty, S., Urasaki, N., Funabashi, T., Kim, C.H.: Optimal Operation of Thermal Unit and Battery by Using Smart Grid. In: International Conference on Electrical Engineering, Korea (2010)
4. Tanaka, K., Uchida, K., Oshiro, M., Goya, T., Senjyu, T., Yona, A.: Optimal Operation for DC Smart-Houses Considering Forecasted Error. In: The 9th International Power and Energy Conference, no. P0375, pp. 722–727. IEEE Press, Singapore (2010)

Concept Learning for $\mathcal{EL}^{++}$ by Refinement and Reinforcement

Mahsa Chitsaz, Kewen Wang, Michael Blumenstein, and Guilin Qi

School of Information and Communication Technology, Griffith University, Australia
School of Computer Science and Engineering, Southeast University, China
mahsa.chitsaz@griffithuni.edu.au,
{kewen.wang,m.blumenstein}@griffith.edu.au,
gqi@seu.edu.cn

Abstract. Ontology construction in OWL is an important and yet time-consuming task even for knowledge engineers and thus a (semi-) automatic approach will greatly assist in constructing ontologies. In this paper, we propose a novel approach to learning concept definitions in $\mathcal{EL}^{++}$ from a collection of assertions. Our approach is based on both refinement operator in inductive logic programming and reinforcement learning algorithm. The use of reinforcement learning significantly reduces the search space of candidate concepts. Besides, we present an experimental evaluation of constructing a family ontology. The results show that our approach is competitive with an existing learning system for $\mathcal{EL}$.

Keywords: Concept Learning, Description Logic $\mathcal{EL}^{++}$, Reinforcement Learning, Refinement Operator.

1 Introduction

Description logics have become a formal foundation for ontology languages since the Web Ontology Language (OWL), which is adapted as the World Wide Web Consortium (W3C) standard for ontology languages. Recently, OWL has evolved into a new standard OWL 2 [1], which consists of three ontology language profiles: EL, QL and RL. OWL 2 standard provides different profiles that trade some expressive power for the efficiency of reasoning, and vice versa. Depending on the structure of the ontologies and the reasoning tasks, one can choose either of these profiles. OWL 2 EL, which is based on $\mathcal{EL}^{++}$ [1], is suitable for applications employing ontologies that contain very large numbers of properties and classes, because the basic reasoning problems can be performed in time that is polynomial with respect to the size of the ontology.

The ontology consists of a *terminology* box, *Tbox*, and an *assertion* box, *Abox*. Besides, concept learning concerns learning a general hypothesis from the given examples of the ontology that we want to learn; those examples are instances

---

[1] http://www.w3.org/TR/owl2-overview/

P. Anthony, M. Ishizuka, and D. Lukose (Eds.): PRICAI 2012, LNAI 7458, pp. 15–26, 2012.

of Abox. The problem arises when constructing ontologies in OWL, which is an onerous task even for knowledge engineers. Additionally, this construction may have diverse presentations with different engineers. For example, there are many ontologies for a particular subject like wine, which can be found in Falcons ontology search engine [2]. One of the applications of concept definition learning is to assist knowledge engineers construct an ontology harmoniously.

In the past few decades, research has been performed towards learning in various logics such as first order logic (FOL) and logic programs. Inductive logic programming has been intensively investigated; there are also different existing ILP approaches such as GOLEM [11], and FOIL [13]. Similar works have been less performed on description logic despite the recent attempts [3,5,10]. Currently, these approaches to concept learning for description logics are not scalable, not because of their methodology, but due to the fact that the underlying description logics such as $\mathcal{ALC}$ are inherently intractable. As a result, existing concept learning systems for expressive logics are not scalable.

One natural option for obtaining scalable systems for concept learning in ontologies is to use a tractable ontology language. A theory of learning concepts in $\mathcal{EL}$ ontologies have been proposed in [9]. In particular, $\mathcal{EL}^{++}$ extends $\mathcal{EL}$ by allowing concept disjointedness, nominals and concrete domains. $\mathcal{EL}^{++}$ is attractive due to at least three key facts: (1)it is a lightweight description logic, in particular, the subsumption problem is tractable even in the presence of general concept inclusion axioms (GCIs) [2], which is important for large scale ontology applications; (2) many practical ontologies can be represented in $\mathcal{EL}^{++}$ such as SNOMED CT [15], the Gene ontology [16], and large parts of GALEN [14], and (3)several efficient reasoners for $\mathcal{EL}^{++}$ are available, e.g. Snorocket system [7] and ELK reasoner [6].

In this paper we propose a novel approach to learn a concept in $\mathcal{EL}^{++}$ by employing techniques from reinforcement learning [17] and inductive logic programming [12]. Our method has been implemented and some preliminary experiments have been conducted. The experimental results show that the new system for concept learning in $\mathcal{EL}^{++}$ is competitive compared to the method proposed in [9]. The method of learning concepts in this paper is based on the *refinement operator* that is used in inductive logic programming. Our refinement operators for $\mathcal{EL}^{++}$ consists of a downward operator for concepts transformation. This operator is designed to specialise a concept to not satisfy negative examples in the class definition. A key issue encountered in concept learning is how to efficiently explore the search space and eventually find a correct hypothesis with respect to the given examples, reinforcement learning technique is a suitable candidate for this purpose. The RL agent can learn to efficiently build an hypothesis over time by systemical trial and error.

The rest of the paper is organized as follows. In Section 2, we introduce the preliminaries of the description logics, concept learning problem, and refinement operators. In Section 3, we comprehensively explain how to apply the reinforcement learning technique and refinement operators in solving the concept learning

---

[2] http://ws.nju.edu.cn/falcons/objectsearch/index.jsp

problem. We report the experimental results in Section 4 and discuss the related work in next section. Finally, some conclusions are drawn and further enhancements to this work are mentioned.

2 Preliminaries

2.1 The Description Logic $\mathcal{EL}^{++}$

Description logics is a family of knowledge representation language that represent knowledge in terms of concept, role, and individuals. Individuals represent constants, concepts correspond to classes of individuals in the domain of interest, while roles represent binary relations in the same domain. In description logics, information is represented as a *knowledge base* which is divided into *TBox*, the *terminology* axioms, and *ABox*, the *assertional* axioms. The description logic $\mathcal{EL}^{++}$ is a fragment of DL that is restricted to few concept and role constructors, which are shown in Table 1. In Table 2, the possible knowledge base axioms in DL $\mathcal{EL}^{++}$ are listed. Besides, an *interpretation* $\mathcal{I}$ is a *model* of a knowledge base $\mathcal{K}$ iff the restriction on the right-side of Table 2 are accomplished for all axioms in $\mathcal{K}$.

Definition 1. *Let $\mathcal{A}$ be an Abox, $\mathcal{T}$ be a Tbox and $\mathcal{K} = (\mathcal{T}, \mathcal{A})$ be an $\mathcal{EL}^{++}$ knowledge base. Let N_C be a set of concept names, N_I be a set of individuals, and N_R be a set of roles. $a \in N_I$ is an instantiation of a concept $\mathcal{C}$, denoted by $\mathcal{K} \models \mathcal{C}(a)$, iff in all models $\mathcal{I}$ of $\mathcal{K}$, we have $a^{\mathcal{I}} \in C^{\mathcal{I}}$. Furthermore, a and $b \in N_I$ are an instantiation of a role $\mathcal{R}$, denoted by $\mathcal{K} \models \mathcal{R}(a, b)$, iff in all models $\mathcal{I}$ of $\mathcal{K}$, we have $(a^{\mathcal{I}}, b^{\mathcal{I}}) \in R^{\mathcal{I}}$.*

Example 1. The following shows a sample *Tbox* and *Abox* for family knowledge base.
$N_C = \{$Person, Male, Female, Parent$\}$
$N_R = \{$hasChild, marriedTo, hasSibling$\}$
$N_I = \{$Christopher, Pennelope, Arthur, Victoria, Colin, Charlotte$\}$

$Tbox = \{$ Male $\sqsubseteq$ Person, Female $\sqsubseteq$ Person, Parent $\equiv \exists$ hasChild.Person,
Parent $\equiv$ hasChild$\top$
$range($hasChild$) \sqsubseteq$ Person, hasSibling is *symmetric*, marriedTo is *symmetric* $\}$

$Abox = \{$ Parent(Christopher), Parent(James),
Parent(Pennelope), Parent(Victoria),
hasChild(Christopher,Victoria), hasChild(Pennelope,Victoria),
hasChild(Victoria,Colin), hasChild(Victoria,Charlotte),
hasChild(James,Colin), hasChild(James,Charlotte),
marriedTo(Christopher,Pennelope), marriedTo(Victoria,James) $\}$

Table 1. $\mathcal{EL}^{++}$ syntax and semantics

Name	Syntax	Semantic
top	$\top$	$\Delta^{\mathcal{I}}$
nominal	$\{a\}$	$\{a^{\mathcal{I}}\}$
bottom	$\bot$	$\emptyset$
concept name	A	$A^{\mathcal{I}} \subseteq \Delta^{\mathcal{I}}$
role name	r	$r^{\mathcal{I}} \subseteq \Delta^{\mathcal{I}} \times \Delta^{\mathcal{I}}$
conjunction	$C \sqcap D$	$C^{\mathcal{I}} \cap D^{\mathcal{I}}$
existential restriction	$\exists r.C$	$\{x \in \Delta^{\mathcal{I}}\mid$ there is $y \in \Delta^{\mathcal{I}}$ with $(x,y) \in r^{\mathcal{I}} \wedge y \in C^{\mathcal{I}}\}$

Table 2. Knowledge Base Axioms

Name	Syntax	Semantic
concept inclusion	$C \sqsubseteq D$	$C^{\mathcal{I}} \subseteq D^{\mathcal{I}}$
concept equivalence	$C \equiv D$	$C^{\mathcal{I}} = D^{\mathcal{I}}$
role inclusion	$r_1 \circ \ldots \circ r_k \sqsubseteq r$	$r_1^{\mathcal{I}} \circ \ldots \circ r_k^{\mathcal{I}} \subseteq r^{\mathcal{I}}$
domain restriction	$dom(r) \sqsubseteq C$	$r^{\mathcal{I}} \subseteq C^{\mathcal{I}} \times \Delta^{\mathcal{I}}$
range restriction	$rang(r) \sqsubseteq C$	$r^{\mathcal{I}} \subseteq \Delta^{\mathcal{I}} \times C^{\mathcal{I}}$
disjointness	$C \sqcap D \sqsubseteq \bot$	$C^{\mathcal{I}} \cap D^{\mathcal{I}} = \emptyset$
concept assertion	$C(a)$	$a^{\mathcal{I}} \in C^{\mathcal{I}}$
role assertion	$r(a,b)$	$(a^{\mathcal{I}}, b^{\mathcal{I}}) \in r^{\mathcal{I}}$

2.2 Concept Learning Problem for $\mathcal{EL}^{++}$

Concept learning in $\mathcal{EL}^{++}$ concerns learning a general hypothesis from the given examples of the background knowledge that we want to learn. There are two kinds of examples: positive examples, which are true, and negative examples, which are false. The positive and negative examples are given as sets E^{+} and E^{-}, respectively, of *Abox* assertions. Literally, if one assumes a set $\mathcal{A}' \subseteq \mathcal{A}$, then $\mathcal{A}'$ can be viewed as a finite example set of the *goal* concept G (or the learned concept):

$$\mathcal{A}' = \{G(a_1), G(a_2), \ldots, G(a_p), \neg G(b_1), \neg G(b_2), \ldots, \neg G(b_n)\}$$
$$E^{+} = \{a_1, a_2, \ldots, a_p\} \quad E^{-} = \{b_1, b_2, \ldots, b_n\}.$$

Besides, let G be a concept, $\mathcal{C}_G$ be the definition of G, and $\mathcal{T}_G = \mathcal{T} \cup \{G \equiv \mathcal{C}_G\}$. G is complete *w.r.t.* E^{+}, if $(\mathcal{T}_G, \mathcal{A}) \models G(a)$ for every a $\in E^{+}$. G is consistent *w.r.t.* E^{-}, if $(\mathcal{T}_G, \mathcal{A}) \not\models G(b)$ for every b $\in E^{-}$. G is correct, if G is complete and consistent *w.r.t.* E^{+} and E^{-} respectively. In the following definition, we define the concept learning problem in general.

Definition 2 (Concept Learning Problem). *The concept learning problem consists in finding a concept definition, denoted by $\mathcal{C}_G$, for G, such that G is correct with respect to the examples [12].*

Example 2. If one wants to learn the concept of *Grandfather* for the ontology mentioned in the previous example with a positive example, $E^+ = \{Christopher\}$, and a negative example, $E^- = \{James\}$, the possible correct solution is:

$$G \equiv \text{Male} \sqcap \exists \text{hasChild.Parent}$$

2.3   Refinement Operators

The refinement operators were first defined in *inductive logic programming* (ILP). In the learning system, there are large space of hypothesis that would be traversed to reach an optimal hypothesis. This could not happen by a simple search algorithm, thus external heuristics are needed to traverse the search space flexibly. Refinement operators introduce this heuristic to be used in the search algorithm. Downward (upward) refinement operators construct specializations (generalization) of hypotheses [12].

The pair $\langle G, R \rangle$ is a *quasi-ordered set*, if a relation R on a set G is reflexive and transitive. If $\langle G, \leq \rangle$ is a quasi-ordered set, a *downward refinement operator* for $\langle G, \leq \rangle$ is a function ρ, such that $\rho(C) \subseteq \{D | C \leq D\}$.

An *upward refinement operator* for $\langle G, \leq \rangle$ is a function δ, such that $\delta(C) \subseteq \{D | D \leq C\}$.

A *refinement chain* from C to D is a finite sequence $C_0, C_1, ..., C_n$ of concepts such that $C = C_0, C_1 \in \rho(C_0), C_2 \in \rho(C_1), ..., C_n \in \rho(C_{n-1}), D = C_n$.

3   Concept Learning Using Reinforcement Learning

Although, the *reinforcement learning* technique has been used in other areas of machine learning, its application in concept learning has not to the best of our knowledge been explored. Concept learning exploits the RL due to its dynamic nature. The goal of the RL agent is to find an optimal way to reach the best concept definition. Nevertheless, we do not use the basic RL system in our approach, because the algorithm that generates a possible action among many ones is heuristics that may not converge. Therefore, we have modified the standard RL system to suit our approach.

In the standard reinforcement learning, a RL agent interacts with its environment via perception and action. On each step of interaction the agent receives an input, the current state of the environment. The RL agent then chooses an action to generate an output. The action changes the state of the environment and the value of this state transition is then received by the RL agent through a reinforcement signal. The RL agent's behavior should lead the RL agent to choose actions that tend to increase the overall sum of values of the reinforcement signals, except for occasional exploratory actions. The state, action and reward function should be defined when using RL method. The number of instances in E^+ and E^- identifies the states. Firstly, the RL agent starts with the top concept as an initiative, then finds the current state of the learned concept. Each state in the search space is associated with some actions; these are the refinement operators to explore the search space. The actions will make changes

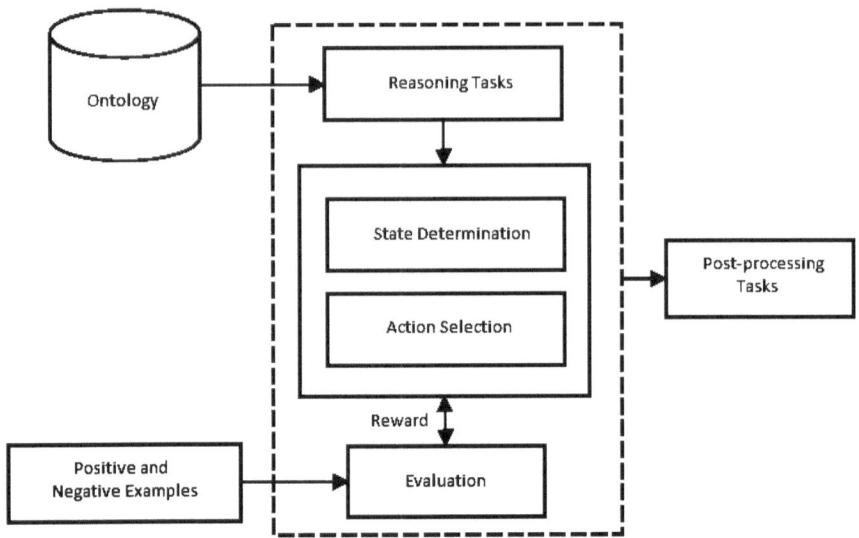

Fig. 1. The global view of the proposed method

in the hypothesis to specialize it. Therefore, we used only downward refinement operator in our method. Then, after each iteration the RL agent receives a signal as its evaluation value. Therefore, the RL agent tries to achieve the best concept definition while maximizing the given rewards.

In the proposed method, as shown in Figure 1, the input of the method is a knowledge base, $\mathcal{K}$, and as well a list of positive and negative examples. The system finds the state of hypothesis with the help of reasoning API such as finding instance of a concept, or subsumption properties. Then, the RL agent chooses one action among those possible actions of this undecided state. The evaluation part will determine the correctness of the hypothesis. It evaluates the changes that have been done by the RL agent, although, it is possible to ignore the chosen action and forces the RL agent to the previous state in order to reach the goal state more efficiently and rapidly. Consequently, we score the RL agent. After the RL agent has finished its work on the hypothesis, we use some of the reasoning features to polish the hypothesis, because it is possible that a few of its items are redundant. Besides, the possible actions for each state guide the RL agent to achieve the goal. These definition of actions are based on refinement operators. In the following, we define the state, action and reward function for concept learning in $\mathcal{EL}^{++}$.

Definition 3 (State). *Let p and n be the number of instances in E^+ and E^-, respectively. A state, s, is a pair (a, b) where $0 \leq a \leq p$ and $0 \leq b \leq n$. The goal state is $(p, 0)$.*

Table 3. The possible actions for each states

State	Action	Meaning
(p,n)	a1 or a2	$(\mathcal{T}_G, \mathcal{A}) \models G(a)$ for every a $\in E^+$ and $(\mathcal{T}_G, \mathcal{A}) \models G(b)$ for every b $\in E^-$
$(0,0)$	a6	$(\mathcal{T}_G, \mathcal{A}) \not\models G(a)$ for every a $\in E^+$ and $(\mathcal{T}_G, \mathcal{A}) \not\models G(b)$ for every b $\in E^-$
(p,i)	a3-a5	$(\mathcal{T}_G, \mathcal{A}) \models G(a)$ for every a $\in E^+$ and $(\mathcal{T}_G, \mathcal{A}) \models G(b)$ for some b $\in E^-$
$(p,0)$	N/A	The goal state:$(\mathcal{T}_G, \mathcal{A}) \models G(a)$ for every a $\in E^+$ and $(\mathcal{T}_G, \mathcal{A}) \not\models G(b)$ for every b $\in E^-$).

The number of state sets depends on the number of examples, because the number of pairs are dependent on $|\mathcal{A}'|$. Therefore, the total number of states is $(p+1) \times (n+1)$.

Example 3. The given positive and negative examples for learning *Grandfather* concept based on the ontology described in Example 1 are:
$E^+ = \{$Christopher$\}$ and $E^- = \{$James$\}$.
The states are: $\{ [(1,1),0], [(1,0),1], [(0,1),2], [(0,0),3] \}$.
The first state means that $(\mathcal{T}_G, \mathcal{A}) \models G(\text{Christopher})$ and $(\mathcal{T}_G, \mathcal{A}) \models G(\text{James})$.
The next state is the goal state that $(\mathcal{T}_G, \mathcal{A}) \models G(\text{Christopher})$ and $(\mathcal{T}_G, \mathcal{A}) \not\models G(\text{James})$.
The third state means that $(\mathcal{T}_G, \mathcal{A}) \not\models G(\text{Christopher})$ and $(\mathcal{T}_G, \mathcal{A}) \models G(\text{James})$.
The last state happens when $(\mathcal{T}_G, \mathcal{A}) \not\models G(\text{Christopher})$ and $(\mathcal{T}_G, \mathcal{A}) \not\models G(\text{James})$.

The actions are the refinement operators that we need for $\mathcal{EL}^{++}$. These actions try to change $\mathcal{C}_G$, in the way that the change improves the learned concept to be correct eventually.

Definition 4 (Action). *Let $\mathcal{C}_G$ be the definition of the goal concept, N_C be the set of concept names, and N_R be the set of role names. An action for the RL agent is one of the following refinement operators, for concepts C, C_1, C_2, and rules r, r_1 and r_2:*

a1. $\mathcal{C}_G = \mathcal{C}_G \sqcap C$, if $C \in N_C$;
a2. $\mathcal{C}_G = \mathcal{C}_G \sqcap \exists r.C$ if $r \in N_R$ and $range(r) \sqsubseteq C$;
a3. $\mathcal{C}_G = C_1 \rightsquigarrow \mathcal{C}_G = C_2$ if $C_1 \sqsubseteq C_2$;
a4. $\mathcal{C}_G = \exists r.C_1 \rightsquigarrow \mathcal{C}_G = \exists r.C_2$, if $C_1 \sqsubseteq C_2$;
a5. $\mathcal{C}_G = \exists r_1.C \rightsquigarrow \mathcal{C}_G = \exists r_2.C$, if $r_1 \sqsubseteq r_2$;
a6. $\mathcal{C}_G = \mathcal{C}_G \sqcap C \rightsquigarrow \mathcal{C}_G = \mathcal{C}_G$ or $\mathcal{C}_G = \mathcal{C}_G \sqcap \exists r.C \rightsquigarrow \mathcal{C}_G = \mathcal{C}_G$, if a1 or a2 was chosen respectively.

From the definition of actions, it is easily seen that a3 to a5 are downward refinement operators because every operation changes the current concept in the definition with one of its subsumers. Moreover, there is a function to produce the possible actions for each state. For example, when the program starts the top concept has to be changed by one of its subsumers, therefor the number of possible actions for this states are the number of subsumers of top concepts.

The reward function returns a reinforcement signal to the RL agent as an evaluation of the current concept definition.

Definition 5 (Reward). *The reward function is defined as follows:*

$$Reward(state \Leftarrow (a, b)) = \begin{cases} 1 & \text{if } a = |E^+| \text{ and } b < |E^-| \\ -1 & \text{if } a < |E^+| \\ 100 & \text{if } a = |E^+| \text{ and } b = 0 \end{cases}$$

Each state has its own meaning (as it is explained in Example 3), therefore it is possible to change G state when this change improves it and helps to reach the goal state. Consequently, for each state there are different actions to be considered. In Table 3, the possible actions for each state are shown. The first row shows that when the learned concept entails all of positive and negative examples, a role or a concept should be added to the concept definition. The next row illustrates the condition that the learned concept does not entail both positive and negative examples, in this case the last item that was added to the concept definition is removed. The third row happens when the learned concept entails all positive examples and some negative examples, then it is specialized to remove negative examples. The last row shows the goal state that no further action is needed.

However, it is not mentioned in Table 3 that for each state when the possible action is not applicable, the system will allow the RL agent to perform the action a_1 or a_2; this is respectively adding a concept or role to the concept definition. For instance, when the RL agent is in the state that should change the concept to one of its subsumer, and there is no sub-class for this concept, it is configured to leave the changes and add another item . This configuration allows the addition of some overlapping items to the concept definition, but it will be ignored after the RL agent has reached the goal with some post-processing tasks. Moreover, if the RL agent received a negative signal, it means that the concept definition does not entail some positive examples, therefore the system forces the RL agent to the previous state and return to the previous definition, although the Q-value of the chosen action is updated to make it less be chosen in the future.

For each state s, after performing an action on G, the next state s' should be an improved state. Otherwise, the action is dismissed and G is backtracked to state s. The improvement state means that as long as G is going to entail more instances in E^+ and not entail any instances from E^-, then the next state is an improvement state. The reinforcement signal is the way of communicating to the RL agent what we want it to achieve, not how we want it to achieve. Therefore, in the reward definition, the RL agent receives a reinforcement signal of 1, if the G state is an improvement state, and it receives a significant greater reward if the G state is a goal state.

In Algorithm 1, the whole procedure for the proposed method is shown. The inputs of the algorithm are the positive and negative examples, as well as the number of exploration times for the RL agent. Initially, the concept definition

---

Algorithm 1. Concept Learning in $\mathcal{EL}^{++}$

Input: E^+, E^- and iterations
Output: $\mathcal{C}_G$

    $state \longleftarrow 0$
    while $state$ is not the goal state and iterations > 0 **do**
        $\mathcal{C}_G \longleftarrow \top$
        $state \longleftarrow \text{FindState}(G \equiv \mathcal{C}_G, E^+, E^-)$
        $action \longleftarrow \text{GetAction}(state)$
        $\text{DoActionOn}(action, G \equiv \mathcal{C}_G)$
        $nextState \longleftarrow \text{FindState}(G \equiv \mathcal{C}_G, E^+, E^-)$
        $reward \longleftarrow \text{Reward}(nextState)$
        $Q(state, x) = (1-\alpha)Q(state, x) + \alpha(reward + \gamma max_{x'} Q(nextState, x'))$ where $x \in$
        all possible actions for $state$ and $x' \in$ all possible actions for $nextState$
        if $nextState$ is not an improvement state **then**
            $\text{Backtrack}(G \equiv \mathcal{C}_G, state)$
        else
            $state \longleftarrow nextState$
        end if
    end while

---

is the top concept, then *FindState* function finds the *state* of G, after that *GetAction* function will produce a list of possible actions for the *state* of G, then return a possible action for the current state by max-random rule which returns an action with maximum Q-Value with probability P_{max}. *DoActionON* function will perform the *action* on the G. Then the *nextState* is found by the *FindState* function. If the *nextState* is not an improvement state, G will revert to the previous sentence. Otherwise, the *state* will be replaced by *nextState*. This loop continues until, the RL agent reaches to the goal statE.

4   Empirical Evaluations

The experiment illustrates our result in learning concepts from a family ontology. The ontology used is a combination of *forte* ontology [10] and *BasicFamily* ontology [5]; it has 2 classes, 4 properties and 474 *Abox* assertions. The only existing method of concept learning for $\mathcal{EL}$ has been proposed by Lehmann and Haase [9]. We compared our result with the latest version of DL-Learner [3], which configure the program by ELTL algorithm. The tests were run on an Intel Core i5-2400 3.10GHz CPU machine with 4GB RAM. We used the Pellet 2.3.0 reasoner[4] which was connected to our test program via OWL API 3 interface[5].

    The program started with two concepts, *Female* and *Male*, then progressed to finding a definition for each concept. For learning the concepts from *Uncle* till *Grandmother* concepts, we added the concept of *Parent* in the ontology,

---

[3] http://sourceforge.net/projects/dl-learner/files/DL-Learner/
[4] http://clarkparsia.com/pellet
[5] http://owlapi.sourceforge.net/index.html

Table 4. The results of learning concept from the family ontology. The average and standard deviation of computation time is shown in milliseconds.

Concept	$\| E^+ \|$	$\| E^- \|$	Our approach		ELTL	
			average	SD	average	SD
Mother	31	89	23.7	0.67	27.67	0.87
Father	33	87	23.8	0.62	28.94	1.51
Brother	38	79	23.69	0.59	28.44	2.53
Sister	39	79	23.67	0.56	28.22	2.24
Child	77	40	23.53	0.91	26.4	1.95
Sibling	77	40	23.68	0.78	27.29	2.77
Son	40	77	24.0	1.66	27.43	2.54
Daughter	37	80	24.38	2.01	27.62	2.45
Parent	64	53	25.26	3.5	32.7	9.56
Uncle	17	100	24.79	2.48	26.29	4.48
Aunt	20	97	25.04	2.8	27.67	5.76
Nephew	26	91	25.29	3.05	29.13	7.04
Niece	31	86	25.32	3.08	30.42	7.87
Grandfather	16	101	25.25	3.12	31.69	8.62
Grandmother	15	102	25.5	3.41	32.96	9.82
Grandparent	31	86	25.36	3.58	33.36	9.57

otherwise the learned concept definition is not presented in DL $\mathcal{EL}^{++}$. Besides, for learning the concept *Niece* and *Nephew*, the concept of *Sibling* is added to the ontology because of the mentioned reason. Table 4 shows the results of this experiment. The first column is the learned concept. The next two columns contains the number of positive and negative examples. We run the system 10 times, subsequently, the next two columns of Table 4 show the average and the standard deviation (SD) of computation time for our approach, respectively. The last two columns are the average and standard deviation of [9] approach. In their system, there were some initializations of different components which are needed for learning the concept, we are included only the ELTL component taken times in Table 4.

The most interesting outcome is that the definition was always correct based on the given examples. It is observed from the result that the average computation time for our approach is between 23 to 25 ms, which is less than the average computation time of EL Tree approach. On the other hand, the DL-Learner system has an efficient built-in instance checker which decreases the total computation time. As a result of this experiment, our approach is a competitive approach with the current system for learning concept definition in $\mathcal{EL}^{++}$.

5   Related Work

The only existing method of concept learning for DL $\mathcal{EL}$ has been proposed by Lehmann and Haase [9]. They used minimal trees to construct DL $\mathcal{EL}$ axioms then refined these by refinement operators. The description logics were converted to trees and four different operators were defined to amend these trees. We compare our approach with this system which is explaned in Section 4.

Currently, the concept learning approaches in DLs are an extension of *inductive logic programming* (ILP) methods. Additionally, these approaches consider the *Open World Assumption* (OWA) of DLs rather than *Close World Assumption* (CWA) of LP. In the area of concept learning in description logic promising research has been investigated and described in [5,3,10]. All these approaches have been proposed for expressive description logics like $\mathcal{ALC}$, although we propose an approach for less expressive logic, $\mathcal{EL}^{++}$, in the hope of developing a scalable concept learning system. The most significant concept learning system for DL is DL-Learner. Besides, there are few works in concept learning in DLs that transferred DL axioms to *logic programs* (LP), then applied the ILP method in order to learn a concept [4]. This approach is too expensive in terms of computation time. Furthermore, it is not always guaranteed that this conversion is possible. Additionally, another approach to tackle the concept learning problem in DL is by employing a *Machine Learning* approach such as *Genetic Programming* [8]. They performed *genetic programming* (GP) in concept learning in two different ways. First, standard GP was used in concept learning; however, the results were not satisfactory because the subsumption properties of concepts could not be employed in the model. Then, a hybrid system that uses refinement operators in GP was proposed, it was DL-Learner GP. The results for the hybrid system outperformed the standard GP, although it generated longer hypotheses and the performance is not the best.

6   Conclusion and Further Work

In summary, we propose a novel approach to solve the concept learning problem in DL $\mathcal{EL}^{++}$. Our approach takes into account refinement operator then applies it to our reinforcement learning algorithm. Subsequently, we compare it with the current approach, which our approach is competitive. It is also possible to use this approach to construct an ontology, the average computation time for the family ontology was less than 25ms. However, our approach has few shortcomings. In future, we will improve the optimization process of our approach to reduce the redundant items in the concept definition. Moreover, we will work on the implementation of our approach that can be used for a very large ontology in order to show its scalability.

Acknowledgement. This work was supported by the Australia Research Council (ARC) Discovery grants DP1093652 and DP110101042.

References

1. Baader, F., Brandt, S., Lutz, C.: Pushing the $\mathcal{EL}$ Envelope. In: Proceedings of the 19th International Joint Conference on Artificial Intelligence, pp. 364–369 (2005)
2. Baader, F., Lutz, C., Suntisrivaraporn, B.: Efficient Reasoning in $\mathcal{EL}^{++}$. In: Proceedings of the 2006 International Workshop on Description Logics, vol. 189 (2006)

3. Fanizzi, N., d'Amato, C., Esposito, F.: DL-FOIL Concept Learning in Description Logics. In: Železný, F., Lavrač, N. (eds.) ILP 2008. LNCS (LNAI), vol. 5194, pp. 107–121. Springer, Heidelberg (2008)
4. Fanizzi, N., Ferilli, S., Iannone, L., Palmisano, I., Semeraro, G.: Downward Refinement in the $\mathcal{ALN}$ Description Logic. In: Proceedings of the 4th International Conference on Hybrid Intelligent Systems, pp. 68–73. IEEE Computer Society Press (2004)
5. Iannone, L., Palmisano, I., Fanizzi, N.: An Algorithm Based on Counterfactuals for Concept Learning in the Semantic Web. Applied Intelligence 26(2), 139–159 (2007)
6. Kazakov, Y., Krötzsch, M., Simancik, F.: Unchain my $\mathcal{EL}$ Reasoner. In: Proceedings of the 24th International Workshop on Description Logics. Description Logics, vol. 745 (2011)
7. Lawley, M., Bousquet, C.: Fast Classification in Protege: Snorocket as an OWL 2 EL Reasoner. In: Proceedings of the Australasian Ontology Workshop 2010: Advances in Ontologies, vol. 122, pp. 45–50 (2010)
8. Lehmann, J.: Hybrid Learning of Ontology Classes. In: Perner, P. (ed.) MLDM 2007. LNCS (LNAI), vol. 4571, pp. 883–898. Springer, Heidelberg (2007)
9. Lehmann, J., Haase, C.: Ideal Downward Refinement in the $\mathcal{EL}$ Description Logic. In: De Raedt, L. (ed.) ILP 2009. LNCS, vol. 5989, pp. 73–87. Springer, Heidelberg (2010)
10. Lehmann, J., Hitzler, P.: Concept Learning in Description Logics using Refinement Operators. Machine Learning 78(1-2), 203–250 (2010)
11. Muggleton, S., Feng, C.: Efficient Induction of Logic Programs. In: Proceedings of the New Generation Computing, pp. 368–381 (1990)
12. Nienhuys-Cheng, S.-H., de Wolf, R.: Foundations of Inductive Logic Programming. LNCS, vol. 1228. Springer, Heidelberg (1997)
13. Quinlan, J.R., Cameron-Jones, R.M.: FOIL: A Midterm Report. In: Brazdil, P.B. (ed.) ECML 1993. LNCS, vol. 667, pp. 3–20. Springer, Heidelberg (1993)
14. Rector, A., Horrocks, I.: Experience Building a Large, Re-usable Medical Ontology using a Description Logic with Transitivity and Concept Inclusions. In: Proceedings of the Workshop on Ontological Engineering. AAAI Press (1997)
15. Stearns, M.Q., Price, C., Spackman, K.A., Wang, A.Y.: SNOMED Clinical Terms: Overview of the Development Process and Project Status. In: Proceedings of the Annual Symposium AMIA, pp. 662–666 (2001)
16. The Gene Ontology Consortium: The Gene Ontology Project in 2008. Nucleic Acids Research 36, 440–444 (2008)
17. Watkins, C.: Learning from Delayed Rewards. Ph.D. thesis. University of Cambridge, England (1989)

Hierarchical Training of Multiple SVMs for Personalized Web Filtering

Maike Erdmann[1], Duc Dung Nguyen[2], Tomoya Takeyoshi[1], Gen Hattori[1], Kazunori Matsumoto[1], and Chihiro Ono[1]

[1] KDDI R&D Laboratories, Saitama, Japan
{ma-erdmann,to-takeyoshi,gen,matsu,ono}@kddilabs.jp
[2] Vietnam Academy of Science and Technology, Hanoi, Vietnam
nddung@ioit.ac.vn

Abstract. The abundance of information published on the Internet makes filtering of hazardous Web pages a difficult yet important task. Supervised learning methods such as Support Vector Machines can be used to identify hazardous Web content. However, scalability is a big challenge, especially if we have to train multiple classifiers, since different policies exist on what kind of information is hazardous. We therefore propose a transfer learning approach called Hierarchical Training for Multiple SVMs. HTMSVM identifies common data among similar training sets and trains the common data sets first, in order to obtain initial solutions. These initial solutions then reduce the time for training the individual training sets without influencing classification accuracy. In an experiment, in which we trained five Web content filters with 80% of common and 20% of inconsistently labeled training examples, HTMSVM was able to predict hazardous Web pages with a training time of only 26% to 41% compared to LibSVM, but the same classification accuracy (more than 91%).

Keywords: Hazardous Web content, Hierarchical training, Transfer learning, SVM, Machine learning.

1 Introduction

The abundance of information published on the Internet makes filtering of hazardous Web pages a difficult yet important task. Supervised learning methods such as Support Vector Machines (SVM) can be used as an accurate way of identifying hazardous Web content [1], but scalability is a big challenge due to the fact that SVM needs to solve the quadratic programming (QP) problem.

With the help of an Internet monitoring company, we have collected a set of 3.1 million training examples with 12,000 features for automatically detecting hazardous Web pages. Training the whole data set is expected to take at least one month. If we assume that the training process has to be conducted only once, such a long training time might be acceptable. However, we will certainly encounter different policies on what kind of information is hazardous, depending

P. Anthony, M. Ishizuka, and D. Lukose (Eds.): PRICAI 2012, LNAI 7458, pp. 27–39, 2012.
© Springer-Verlag Berlin Heidelberg 2012

on which organization is requesting the Web content filter and for what purpose. For instance, parents might want a Web filter that protects their children from adult content such as gambling and legal drugs. A company might want to set up a filter that prevents employees from visiting Web sites that are not work related. Moreover, laws and ethical standards vary widely among countries.

Under the circumstances explained above, a training example might have to be labeled hazardous in one Web content filter and harmless in another, but the majority of instances is still labeled equally. We therefore propose a new approach called Hierarchical Training for Multiple SVMs (HTMSVM), which can identify common data among similar training sets and train the common data sets first in order to obtain initial solutions. These initial solutions can then reduce the time for training the individual training sets, without influencing classification accuracy.

The rest of the paper is structured as follows. In Section 2, we introduce the HTMSVM algorithm. In Section 3, we present and discuss an experiment, which shows that our personalized Web content filters can predict hazardous Web pages with an accuracy of more than 91%, and that HTMSVM can reduce training time by more than half without affecting classification accuracy. In Section 4, we give an overview on related research in training time reduction for SVM as well as on the training of similar tasks. Finally, we draw a conclusion and outline future work in Section 5.

2   Hierarchical Training for Multiple SVMs

In this section, we give a broad overview on Support Vector Machines and describe how Hierarchical Training for Multiple SVMs (HTMSVM) can be used to improve training time without affecting classification accuracy.

2.1   Support Vector Machines

The basic principle of Support Vector Machines can be expressed as follows. We are given a data set of l training examples $x_i \in R^d, 1 \leq i \leq l$ with labels $y_i \in \{-1, +1\}$, and want to solve the following quadratic problem:

$$\min L(\alpha) = \frac{1}{2} \sum_{i,j=1}^{l} y_i y_j \alpha_i \alpha_j K(x_i, x_j) - \sum_{i=1}^{l} \alpha_i \qquad (1)$$

$$\text{subject to } \sum_{i=1}^{l} y_i \alpha_i = 0, \ 0 \leq \alpha_i \leq C, \ 1 \leq i \leq l$$

where $K(x_i, x_j)$ is a kernel function calculating the dot product between two vectors x_i and x_j in a feature space. C is a parameter penalizing each "noisy" example in the given training data. The optimal coefficients $\{\alpha_i | 1 \leq i \leq l\}$ form the decision function:

$$y = \text{sign}\left(f(x) = \sum_{\alpha_i \neq 0} y_i \alpha_i K(x_i, x) + b \right), \tag{2}$$

$$b = \frac{1}{|\text{SV}|} \sum_{i \in \text{SV}} \left(y_i - \sum_{j=1}^{N} \alpha_i y_i K(x_i, x) \right)$$

for predicting the target class of an example, where $\text{SV} = \{i | \alpha_i \in (0, C]\}$ is the set of true support vectors. Initially, the values of all parameters α_i are zero. In each iteration of the training process, new values for the parameters are calculated. If the new values provide an optimal solution, the training stops.

2.2   Learning Problem

If we need to produce several classifiers, e.g. due to different policies in the filtering of hazardous Web pages, each classifier is usually trained independently in standard SVM, even if they share a significant number of common training examples. HTMSVM, however, detects common training examples and trains them first, after which it uses the obtained alpha parameters for training each training set individually. Given a number of M training sets, HTMSVM constructs SVM classifiers with the decision function y.

$$T^u = \left\{ (x_i, y_i^u) \in R^d \times \{-1, +1\} | i = 1, \ldots, l \right\}, u = 1, \ldots, M \tag{3}$$

$$y = \text{sign}\left(f^u(x) = \sum_{\alpha_i^u \neq 0} y_i^u \alpha_i^u K(x_i, x) + b^u \right), u = 1, \ldots, M \tag{4}$$

2.3   Training Order

In order to decide the training order, we identify clusters of overlapping training data. The algorithm is specified in Figure 1.

We detect suitable clusters of overlapping training data in a bottom-up approach. In Steps 1-3 of the algorithm, one cluster is created for each training set. The index set of each cluster $(c \rightarrow Idx)$ has one element, which is the index of the data, and the two children $(c \rightarrow Child)$ are set to zero. The α parameter $(c \rightarrow \alpha)$ is initialized to be zero. In Step 4, $ClusterSet$ is defined as the initial set of M clusters.

In Steps 5-10, the bottom-up hierarchical clustering is applied to $ClusterSet$. In each iteration, the two closest clusters c_i and c_j are selected (Step 6). They are then combined into a new cluster c_{ij} (Step 7). The two child clusters c_i and c_j are removed from $ClusterSet$ (Step 8) and the newly created cluster c_{ij} is added instead (Step 9). The distance between the two clusters c_i and c_j is

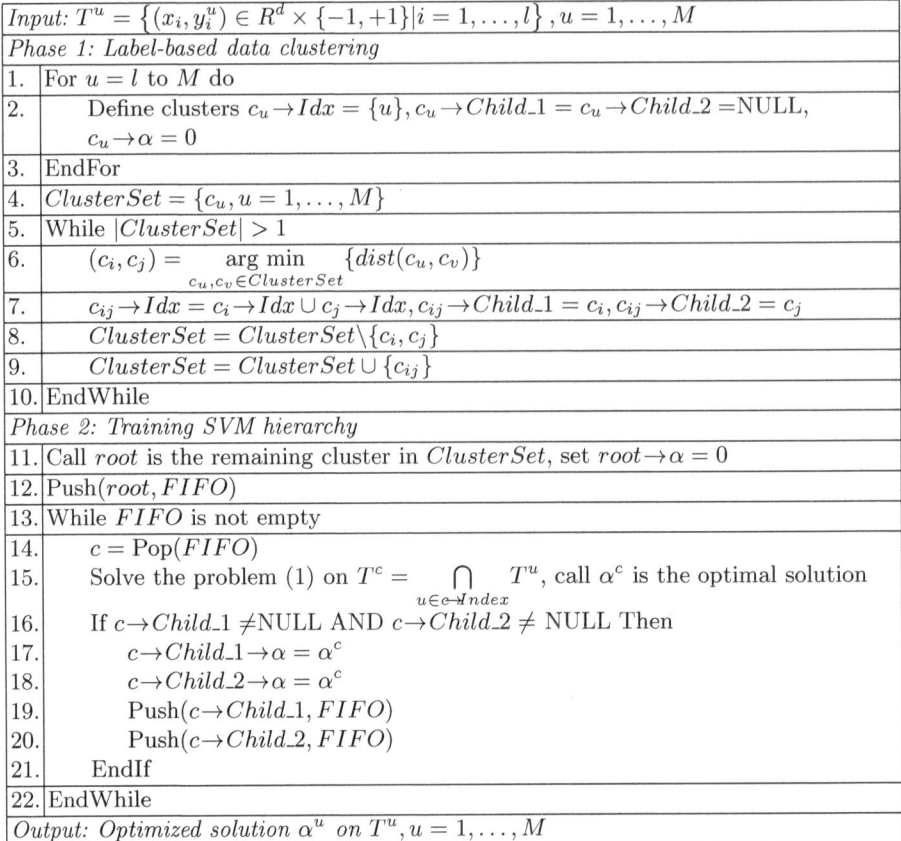

Fig. 1. Hierarchical Training Algorithm

calculated from the total number of training examples without the number of common training examples in the two clusters:

$$dist(c_i, c_j) = l - \left| \bigcap_{u \in c_{ij} \to Index} T^u \right|, c_{ij} \to Index = c_i \to Index \cup c_j \to Index \quad (5)$$

The cluster structure derived from Step 5-10 is visualized in Figure 2.

In Steps 11-12, the remaining cluster in $ClusterSet$ becomes the root of the hierarchy. Its coefficient vector α is initialized to be zero, and it is pushed into a FIFO structure (First-in-First-out).

In Steps 13-22, the SVM training is performed in the order determined by the cluster hierarchy. In each iteration, one data cluster c is picked from $ClusterSet$ in the FIFO structure (Step 14). The SVM training algorithm finds the optimal solution α^c on that cluster. If c has two child clusters, α^c will be used as an initial solution for training the child clusters. The training order follows the

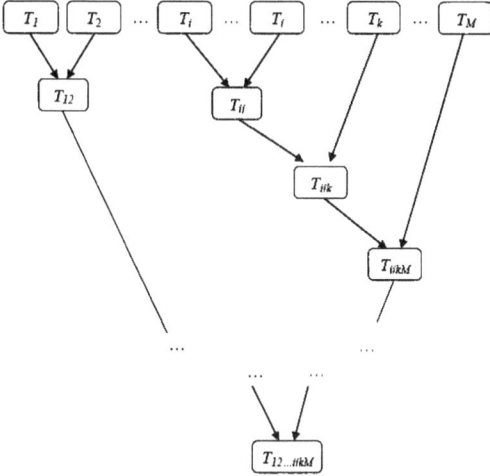

Fig. 2. Visualization of Common Data Selection

FIFO structure (Steps 16-21) as visualized in Figure 3. When all child clusters have been trained, the training process stops. The output of the training are the classifiers for each child cluster.

2.4   Finding an Initial Solution

In the following, we describe how HTMSVM finds a first feasible solution for a cluster, as described in Steps 17-18 of the algorithm (Figure 1).

For the Sequential Minimal Optimization (SMO) algorithm, the following margin plays a crucial role:

$$E_i = \sum_{k=1}^{l} y_k \alpha_k K(x_k, x_i) - y_i, i = 0, \dots, l \tag{6}$$

More specifically, the maximum gain selection heuristic for a pair of vectors to be optimized is

$$\begin{cases} i = \arg \max_k \{E_k | k \in I_{up}(\alpha)\}, \\ j = \arg \max_k \{|\Delta_{ik}| | k \in I_{low}(\alpha)\} \end{cases} \tag{7}$$

where I_{up}, I_{low} and Δ_{ij} are defined as

$$I_{up}(\alpha) = \{k | \alpha_k < C, y_k = +1 \text{ or } \alpha_k > 0, y_k = -1\} \tag{8}$$

$$I_{low}(\alpha) = \{k | \alpha_k < C, y_k = -1 \text{ or } \alpha_k > 0, y_k = +1\} \tag{9}$$

$$\Delta_{ij} = \frac{(E_i - E_j)^2}{2(K_{ii} + K_{jj} - 2K_{ij})} \tag{10}$$

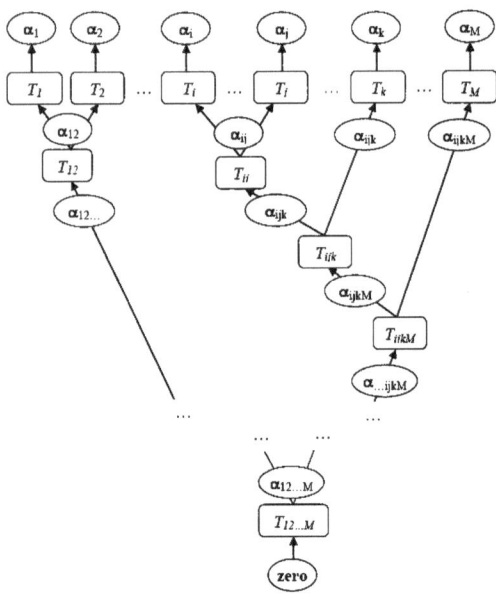

Fig. 3. Visualization of Training Order

The stopping condition of the SMO is also based on the degree of improvement in objective function Δ_{ij}. There are two ways for calculating $E_i, i = 1, \ldots, l$. Firstly we could use the formula (6) directly. However, this direct method requires $K(x_i, x_j)$, which could be very expensive.

Alternatively, we can assume that $\alpha^1 = \{\alpha_1^1, \alpha_2^1, \ldots, \alpha_{|T|}^1\}$ is the optimal solution on the common data $T = \{(x_i, y_i^u) | y_i^u = y_i^1, \forall u = 2, \ldots, M\}$, then the coefficient vector

$$\alpha_0^u = \{\alpha_1^u = \alpha_1^1, \alpha_2^u = \alpha_2^1, \ldots, \alpha_{|T|}^u = \alpha_{|T|}^1, \alpha_{|T|+1}^u = 0, \ldots, \alpha_l^u = 0\} \qquad (11)$$

is a feasible solution on $T^u, u = 2, \ldots, M$. The margins E_i^u in the u-th training data can be calculated efficiently from E_i^1, for $i = 2, \ldots, l$

$$E_i^u = \begin{cases} E_i^1 \text{ if } y_i^u = y_i^1, \\ (E_i - 2) \text{ if } y_i^u y_i^1 = -1 \end{cases} \qquad (12)$$

In (12), there is no need to re-calculate the margin information of the new learning problem. Instead, they are transferred from training a parent cluster to sub-clusters.

3   Web Content Filter

In this section, we discuss an experiment, for which we constructed personalized Web content filters using our proposed method and compared their performance to standard SVM (LibSVM).

Table 1. Web Content Categorization Examples

Class	Category	Subcategory
Harmless	Shopping	Auctions
		Shopping, general
		Real estate
		IT related shopping
	Hobby	Music
		Celebrities
		Food
		Recreation, general
Hazardous	Illegal	Terror, extremism
		Weapons
		Defamation
		Suicide, runaway
	Adult	Sex
		Nudity
		Prostitution
		Adult search

3.1 Construction of Training Sets

For the experiment, we used a training corpus of about 3,1 million manually labeled examples of hazardous and harmless Web pages with 12,000 features. The training data is categorized into into 27 categories with 103 subcategories. Some examples of hazardous and harmless categories are shown in Table 1.

From the training corpus, we constructed five training sets representing five different data labeling policies ranging between conservative (e.g. parental control) Web filters to very liberal Web filters. For the first set, we used the original labels. In order to construct the other four sets, we changed the labels of selected categories and subcategories from harmless to hazardous or vice versa. As a result, 80% of the training examples were identical in all five sets, 10% were different in only one of the training sets and another 10% were different in two training sets. Although the training sets were created solely for the experiment, we made sure that they represented realistic Web filtering policies.

The training order for our corpus is visualized in Figure 4. The first training is conducted on $T_{1,2,3,4,5}$, i.e. the training data which is identical in all five sets (80% of the data). After that, the set $T_{1,4,5}$ is trained, which consists of the training data that is identical in three of the training sets (90% of the data). Subsequently, the sets $T_{1,4}$ and $T_{2,3}$ are trained, both containing 95% of the training examples. Finally, each training set is trained individually.

In our experiment, we compared the training time of HTMSVM with the training time of LibSVM for different amounts of training data. The smallest set contained 50,000 training examples per training set whereas the largest contained 250,000 training examples. We selected the RBF kernel (for both HTMSVM and LibSVM), since we want to ensure a high classification accuracy. Our proposed

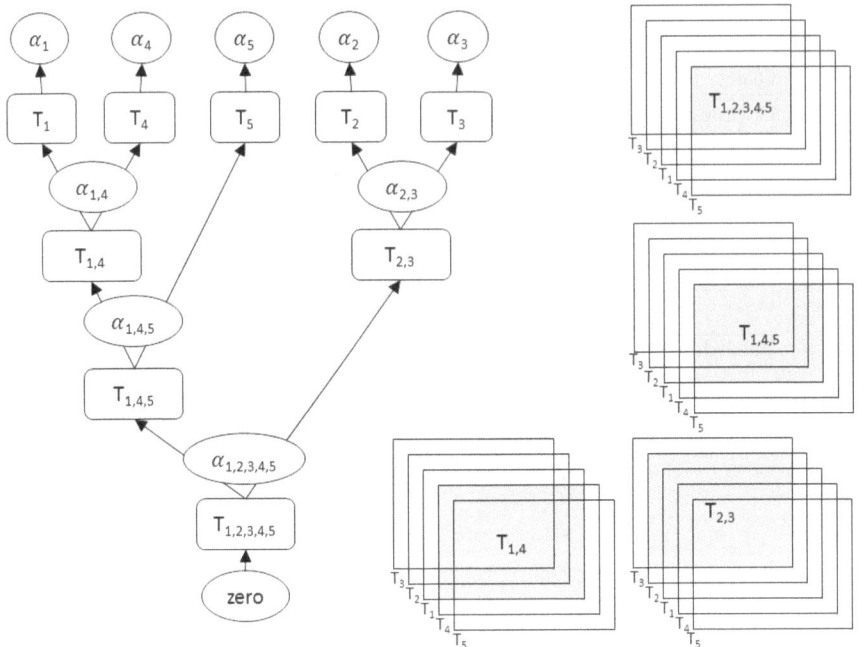

Fig. 4. Training Order for Web Content Filter

method can be applied to any kernel, but due to the complexity of our classification problem, especially due to the large number of features, the classification accuracy of the linear kernel is not sufficient.

3.2   Training Time

The training time for HTMSVM and LibSVM is given in Table 2. The experiment was conducted on a server with 8 CPUs and 60GB of memory. The column "common" shows the training time for training the data that is identical in multiple training sets. The common training time is zero in the case of LibSVM. The column "private" shows the training time for training all five training sets individually.

As the results show, the training time for the private training data is noticeably shorter for HTMSVM than for LibSVM. As a consequence, the overall training time of HTMSVM lies between 26% and 41% of that of LibSVM, even though the training time for the common data sets has to be added.

Noticable is that the training time for the common data in HTMSVM is a lot shorter than the training time for private data in LibSVM, even though the size of both data sets is similar. The main reason for that phenomenon is that the common data contains much fewer support vectors than the private data. Since the complexity of SVM training algorithms depends on the number of support vectors, it is not unusual for training data sets of the same size to have different training times.

Table 2. Training Time of Web Content Filter

Training Size	Method	Common (hh:mm:ss)	Private (hh:mm:ss)	Overall (hh:mm:ss)	Ratio (HTMSVM/LibSVM)
50,000	LibSVM		2:11:58	2:11:58	
	HTMSVM	0:31:49	0:22:55	0:54:44	41.48%
100,000	LibSVM		15:14:17	15:14:17	
	HTMSVM	2:07:21	1:51:35	3:58:56	26.13%
150,000	LibSVM		25:38:48	25:38:48	
	HTMSVM	5:21:52	3:30:54	8:52:46	34.62%
200,000	LibSVM		59:06:13	59:06:13	
	HTMSVM	8:41:48	8:16:17	16:58:05	28.71%
250,000	LibSVM		83:10:33	83:10:33	
	HTMSVM	14:20:44	19:53:11	34:13:55	41.16%

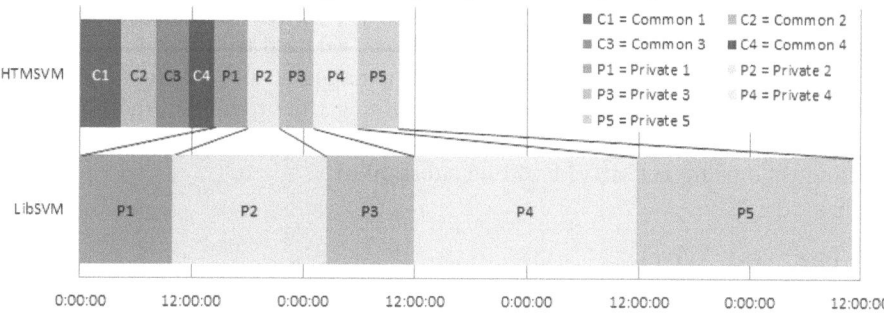

Fig. 5. Detailed Training Time of Web Content Filter

The large discrepancy in training time ratio for different training sets seems to be caused by the interference of the memory swapping process. In our experiments, the number of support vectors in the initial solutions was unusually large and therefore did not fit into the main memory. Thus, memory swapping was necessary, but the number of memory swappings depends on the actual data and can differ significantly. Nevertheless, HTMSVM reduced training time by more than half in each of the experiments.

Figure 5 shows details of the training process for 250,000 training examples. Training time for each "private" data set varies, but the training time of HTMSVM is at all times much shorter than that of LibSVM.

3.3 Classification Accuracy

After training the classifiers, we tested their classification accuracy on a test set of 50,000 examples. We were able to confirm our assumption that the shorter

Table 3. Classification Accuracy of Web Content Filter

Method	Training Size				
	50,000	100,000	150,000	200,000	250,000
LibSVM (linear)	82.1988%	83.7752%	84.6660%	85.3120%	85.7216%
LibSVM (RBF)	89.0704%	90.0500%	90.7504%	91.0444%	91.2540%
HTMSVM (RBF)	89.0696%	90.0500%	90.7528%	91.0432%	91.2208%

training time of HTMSVM does not affect classification accuracy. As the results in Table 3 show, the classification accuracy of LibSVM (RBF) and HTMSVM (RBF) is not identical, yet the difference in accuracy ($< 0.04\%$) is neglectible. Furthermore, the classification accuracy of the linear kernel was noticably lower than that of the RBF kernel, which shows that saving training time by using the linear kernel is not an option.

For the classifier trained on 250,000 examples, a classification accuracy of about 91% was achieved. Since a minimum accuracy of 95% is required to ensure applicability of the classifier to real applications, we need to train the whole data set of 3.1 million training examples, which is expected to take more than one month, even assuming the usage of other training time reduction methods such as Condensed SVM [2]. For that reason, we can expect that the reduction of training time using HTMSVM will be substantial.

4   Related Work

Support Vector Machines [3] have proved to be an effective tool for a wide range of classification problems including Web content filtering. One of the most popular translations of SVM into practice is Sequential Minimal Optimization (SMO) [4], which is implemented in e.g. the LibSVM software [5].

Many attempts have been made to optimize SVM for large-scale training data by e.g. online and offline-style learning algorithms, parallelization of the training process, or data sampling. Impressive results have been achieved for linear classification problems [6], but linear classification is not suitable for complex tasks that require a large numbers of features. Only few attempts have been made to optimize training time for RBF kernel classification [2, 7], which is more suitable for complex tasks such as Web content filtering. However, our proposed method can easily be combined with those training time reduction approaches.

In order to differentiate our contribution, we will introduce research on the training of similar tasks, which focusses on improving classification accuracy rather than reducing training time.

In multitask learning [8–11], related tasks are learned simultaneously or sequentially. The approach is often used when the number of training examples is insufficient for a single task. One typical example for multitask learning is the simultaneous recognition of attributes such as age, sex and facial expression of a person in a photograph.

Multitask learning is a form of transfer learning, but transfer learning [12, 13] does not necessarily require the feature space of the related tasks to be identical. Transfer learning can be applied, for instance, if the knowledge obtained from training a classifier for customer reviews should be transfered to customer reviews on a different product. The previous knowledge can be transfered in form of e.g. reusing features or training examples that are present in both tasks. Several proposals have been made to apply transfer learning to the task of spam filtering, but based on the assumption that the personalization of the classifiers has to be undertaken using unlabeled data [14].

Yet another related research area is incremental learning [15–17], which is suitable particularly for dynamic applications, where new training data is added or existing training data needs to be revised frequently. Thereby, a first classifier is built on the training data available at a given time, and a second classifier is built on the updated training set, i.e. new and modified examples, by reusing the results of the first training process. Incremental learning can also be used to train subsets of a training corpus in cases where training of the whole data at once would take too much time.

Our task is not to be confused with multi-label classification [18], in which more than one label can be assigned to each training example. Multi-label classification often occurs in e.g. text classification. For instance, a newspaper article might be assigned both the label "politics" and the label "economy". In our classification problem, however, each training example is assigned only one label per training set.

5   Conclusion

In this paper, we introduced a transfer learning approach called Hierarchical Training for Multiple SVMs (HTMSVM), that can identify common data among similar training sets and train the common data sets first in order to obtain initial solutions. These initial solutions can then reduce the time for training the individual training sets, without influencing classification accuracy.

Furthermore, we tested our proposed method in an experiment in which we trained five personalized Web content filters from an identical training corpus. In the experiment, 20% of the training examples were labeled inconsistently, assuming that different policies exist on which Web pages should be labeled as hazardous. HTMSVM was able to predict hazardous Web pages with an accuracy of more than 91%, but only 26% to 41% of the training time of LibSVM.

In the next step, we want to further reduce training time by optimizing memory usage. If the number of support vectors in the initial solutions is too large, the time required for memory swapping undermines the training time advantage of HTMSVM.

HTMSVM can be applied to other kinds of transfer learning problems, where we have several classification tasks with overlapping training sets. For that reason, we also want to apply HTMSVM to other applications, such as recommendation systems. In order to avoid negative transfer, i.e. causing an increase in training time through a bad initial solution, it is necessary to develop an algorithm to estimate for which applications our proposed method is useful.

References

1. Ikeda, K., Yanagihara, T., Hattori, G., Matsumoto, K., Takisima, Y.: Hazardous Document Detection Based on Dependency Relations and Thesaurus. In: Li, J. (ed.) AI 2010. LNCS, vol. 6464, pp. 455–465. Springer, Heidelberg (2010)
2. Nguyen, D.D., Matsumoto, K., Takishima, Y., Hashimoto, K.: Condensed vector machines: Learning fast machine for large data. IEEE Transactions on Neural Networks 21(12), 1903–1914 (2010)
3. Cortes, C., Vapnik, V.: Support-vector networks. Machine Learning 20(3), 273–297 (1995)
4. Platt, J.C.: Sequential minimal optimization: A fast algorithm for training support vector machines. Technical report, Advances in Kernel Methods - Support Vector Learning (1998)
5. Chang, C.C., Lin, C.J.: LIBSVM: A library for support vector machines. ACM Transactions on Intelligent Systems and Technology 2, 27:1–27:27 (2011), Software, http://www.csie.ntu.edu.tw/~cjlin/libsvm
6. Menon, A.K.: Large-scale support vector machines: Algorithms and theory, research exam. Technical report, University of California San Diego (2009)
7. Cervantes, J., Li, X., Yu, W.: Svm classification for large data sets by considering models of classes distribution. In: Mexican International Conference on Artificial Intelligence (MIKAI), pp. 51–60 (2007)
8. Abu-Mostafa, Y.S.: Learning from hints in neural networks. Journal of Complexity 6(2), 192–198 (1990)
9. Caruana, R.: Multitask learning: A knowledge-based source of inductive bias. In: Proceedings of the Tenth International Conference on Machine Learning, pp. 41–48 (1993)
10. Thrun, S.: Is learning the n-th thing any easier than learning the first? In: Advances in Neural Information Processing Systems, pp. 640–646 (1996)
11. Baxter, J.: A model of inductive bias learning. Journal of Artificial Intelligence Research 12, 149–198 (2000)
12. Arnold, A., Nallapati, R., Cohen, W.W.: A comparative study of methods for transductive transfer learning. In: Proceedings of the Seventh IEEE International Conference on Data Mining Workshops, pp. 77–82 (2007)
13. Pan, S.J., Yang, Q.: A survey on transfer learning. IEEE Transactions on Knowledge and Data Engineering 22(10), 1345–1359 (2010)
14. Bickel, S.: Ecml-pkdd discovery challenge 2006 overview. In: ECML-PKDD Discovery Challenge Workshop, pp. 1–9 (2008)

15. Cauwenberghs, G., Poggio, T.: Incremental and decremental support vector machine learning. In: Advances in Neuronal Information Processing Systems, vol. 13, pp. 409–415 (2000)
16. Ruping, S.: Incremental learning with support vector machines. In: IEEE International Conference on Data Mining, pp. 641–642 (2001)
17. Shilton, A., Palaniswami, M., Ralph, D., Tsoi, A.C.: Incremental training of support vector machines. IEEE Transactions on Neural Networks 16(1), 114–131 (2005)
18. Tsoumakas, G., Katakis, I.: Multi-label classification: An overview. International Journal of Data Warehousing and Mining, 1–13 (2007)

Citation Based Summarisation of Legal Texts

Filippo Galgani, Paul Compton, and Achim Hoffmann

School of Computer Science and Engineering,
The University of New South Wales, Sydney, Australia
{galganif,compton,achim}@cse.unsw.edu.au

Abstract. This paper presents an approach towards using both incoming and outgoing citation information for document summarisation. Our work aims at generating automatically catchphrases for legal case reports, using, beside the full text, also the text of cited cases and cases that cite the current case. We propose methods to use catchphrases and sentences of cited/citing cases to extract catchphrases from the text of the target case. We created a corpus of cases, catchphrases and citations, and performed a ROUGE based evaluation, which shows the superiority of our citation-based methods over full-text-only methods.

1 Introduction

Citations have been used for summarisation of scientific articles: sets of citations to a target article (sentences about the cited paper) are believed to explain its important contributions, and thus can be used to form a summary [1,8,12,9]. In this paper we explore a different direction: we consider not only documents that cite the target document, but also documents cited by the target one and how they can be used for summarisation. Furthermore we consider not only sentences about the document, but also catchphrases of related documents, and how they can be used both as summary candidates and to identify important fragments in the full text of the target document.

We apply our approach to a particular summarisation problem: creating catchphrases for legal case reports. The field of law is one where automatic summarisation can greatly enhance access to legal repositories: Legal cases, rather than summaries, often contain a list of catchphrases: phrases that present the important legal points of a case, giving a quick impression on what the case is about: *"the function of catchwords is to give a summary classification of the matters dealt with in a case. [...] Their purpose is to tell the researcher whether there is likely to be anything in the case relevant to the research topic"* [11]. As citations are a peculiar feature of legal documents (decisions always cite precedent cases to support their arguments), this work focuses on using a citation-based approach to generate catchphrases. Examples of catchphrases and citations for a case report are shown in Table 1.

We created a large corpus of seed cases, and downloaded cases that cite or are cited by them (Section 2). We investigate different novel methods to use citing sentences and catchphrases of related cases to generate catchphrases for the

P. Anthony, M. Ishizuka, and D. Lukose (Eds.): PRICAI 2012, LNAI 7458, pp. 40–52, 2012.

Table 1. Examples of (1) the catchphrases of a case (Re Read), and (2) two citations to the case

CORPORATIONS - winding up - court-appointed liquidators - entry into agreement - able to subsist more than three months - no prior approval under s 477(2B) of Corporations Act 2001 (Cth) - application to extend "period" for approval under s 1322(4)(d) - no relevant period - s 1322(4)(d) not applicable - power of Court under s 479(3) to direct liquidator - liquidator directed to act on agreement as though approved - implied incidental powers of Court - prior to approve agreement - power under s 1322(4)(a) to declare entry into agreement and agreement not invalid - COURTS AND JUDGES - Federal Court - implied incidental power - inherent jurisdiction
However, in Re Read and Another [2007] FCA 1985 ; (2007) 164 FCR 237 (Read), French J (as the Chief Justice then was) expressed disagreement with the approach taken in those two cases. His Honour was of the opinion (at [32]-[39]) that s 1322(4)(d) of the Act could not be relied upon to extend time under s 477(2B). French J's reason was that s 477(2B) did not specify a "period for" the making of an application for approval under s 477(2B).
In Re Read [2007] FCA 1985 ; (2007) 164 FCR 237 French J (as his Honour then was) held that s 1322(4)(d) of the Corporations Act could not be relied upon to extend time under s 477(2B) because the latter provision did not fix a period for making application which could be extended under s 1322(4)(d) (at [32]-[39]).

current documents: using only the citation text or using the citation to extract sentences from the target document. These methods are explained in Section 4. We use an evaluation procedure based on ROUGE to match generated candidates with author-given catchphrases, described in Section 3, to compare the different methods. In Section 5 we show that using catchphrases of related cases to identify important sentences in the full text gives the best results. Section 6 explains how this work differs from other citation-based summarisation approaches, and in Section 7 we draw the conclusions and indicate directions of future work.

2   Corpus of Legal Catchphrases and Citations

In Australia documents that record court decisions are made publicly available by AustLII[1], the Australasian Legal Information Institute [4]. AustLII is one of the largest sources of legal material on the net, with over four million documents.

We created an initial corpus of 2816 cases accessing case reports from the Federal Court of Australia (FCA), for the years 2007 to 2009, for which author-made catchphrases are given, extracting the full text and the catchphrases of every document. Each document contains on average 221 sentences and 8.3 catchphrases. In total we collected 23230 catchphrases, of which 15359 (92.7%) are unique, appearing only in one document in the corpus.

In order to investigate if we could automatically generate catchphrases for those cases, we downloaded citation data from LawCite[2]. LawCite is a service

---

[1] http://www.austlii.edu.au
[2] http://www.lawcite.org

provided by AustLII that, for a given case, presents a list of cited cases and a list of more recent cases that cite it. We build a script that for each case queries LawCite, obtains the two lists, and downloads the full texts and the catchphrases (where available) from AustLII, of both cited (previous) cases and more recent cases that cite the current one (citing cases). Of the 2816 cases, 1904 are cited at least by one other case (on average by 4.82 other cases). We collected the catchphrases of these citing cases, and searched the full texts to extract the location where a citation is explicitly made, and extracted the containing paragraph(s). For each of the 1904 cases we collected on average 21.17 citing sentences and 35.36 catchphrases (from one or more other documents). We also extracted catchphrases from previous cases cited by the judge, obtaining on average 67.41 catchphrases for each case (all cases cite at least one other case).

The corpus thus contains the initial 2816 cases with given catchphrases, and all cases related to them by incoming or outgoing citations, with catchphrases and citing sentences explicitly identified. We plan to release the whole corpus for researchers interested in citations analysis. For the experiments described in the remainder of this paper, of the 2816 initial documents we selected all those which have at least 10 citing sentences and at least 10 catchphrases of related cases, resulting in a total of 926 cases.

We refer to the document from our FCA corpus for which we want to create catchphrases as the **target document**, to sentences collected from cases that cite it as its **citances** [16] and to the catchphrases of the citing and cited cases as **citphrases**, the latter can be divided into citing and cited, but for most of the discussion we will consider the union of them (see Figure 1 for a schematic representation). We use the term **citations** to refer to both citphrases and citances together.

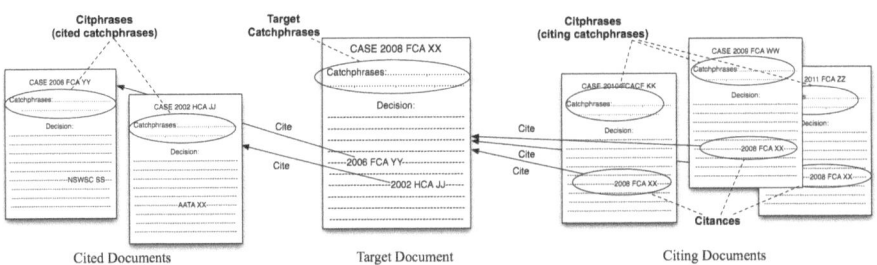

Fig. 1. Citances and Citphrases for a given case

3   Evaluation Methodology

We propose a simple method to evaluate candidate catchphrases automatically by comparing them with the author-made catchphrases from our AustLII corpus (considered as our "gold standard"), in order to quickly estimate the relative performance of a number of methods on a large number of documents. As our system extracts sentences from text as candidate catchphrases, we propose an evaluation method which is based on ROUGE scores between extracted sentences and given catchphrases. ROUGE [7] includes several measures to quantitatively compare system-generated summaries to human-generated summaries, counting the number of overlapping n-grams of various lengths, word pairs and word sequences between two or more summaries. Among the various scores in the ROUGE family we used ROUGE-1, ROUGE-SU and ROUGE-W.

If we follow the standard ROUGE evaluation, we would compare the whole block of catchphrases to the whole block of extracted sentences. However when evaluating catchphrases, we do not have a single block of text, but rather several catchphrase candidates, these should thus be evaluated individually: the utility of any one catchphrase is minimally affected by the others, or by their particular order. On the other hand we want to extract sentences that contain an entire individual catchphrase, while a sentence that contains small pieces of different catchphrases is not as useful: see the example in Figure 2.

Sentences:

1 The *Tribunal* should have *procedures* to guard against such a *possibility*.
2 The two grounds of this application are that, first, the Tribunal's decision was affected by reasonably apprehended bias and, secondly, **denial** of **procedural fairness**

Catchphrases

1 **denial** of *procedural* **fairness**
2 decision to issue warrant required forming a view about a *possible* criminal offence by applicant
3 *Tribunal's* decision to be set aside

Fig. 2. In this example both sentences 1 and 2 have three words in common with the catchphrases, thus they have the same ROUGE score. Using our evaluation methods, however, only sentence 2 is considered a match, as it covers all three terms of catchphrase 1, while sentence 1 has terms from different catchphrases, and thus is not considered a match. This corresponds to the intuition that sentence 2 is a better catchphrase candidate than sentence 1.

We therefore devised the following evaluation procedure: we compare each extracted sentence with each catchphrase individually, using ROUGE. If the recall (on the catchphrase) is higher than a threshold, the catchphrase-sentence pair is considered a **match**, and the sentence is considered **relevant**. For example if we have a 10-words catchphrase, and a 15-words candidate sentence, if they have 6 words in common we consider this as a match using ROUGE-1 with threshold 0.5, but not a match with a threshold of 0.7 (requiring at least 7/10 words from the catchphrase to appear in the sentence). Using other ROUGE scores (ROUGE-SU or ROUGE-W), the order and sequence of tokens are also considered in defining a match. Once defined the matches between single sentences and catchphrases, for one document and a set of extracted (candidate) sentences, we can compute precision and recall as:

$$Recall = \frac{MatchedCatchphrases}{TotalCatchphrases} \qquad Precision = \frac{RelevantSentences}{ExtractedSentences}$$

The recall is the number of catchphrases matched by at least one extracted sentence, divided by the total number of catchphrases, the precision is the number of sentences extracted which match at least one catchphrase, divided by the number of extracted sentences.

This evaluation method lets us compare the performance of different extraction systems automatically, by giving a simple but reasonable measure of how many of the desired catchphrases are generated by the systems, and how many of the sentences extracted are useful. This is different from the use of standard ROUGE overall scores, where precision and recall do not relate to the number of catchphrases or sentences, but to the number of smaller units such as n-grams, skip-bigrams or sequences, which makes it more difficult to interpret the results.

4   Methods for Catchphrase Generation Using Citations

Given a target document, and the set of its **citances** and **citphrases** extracted from connected documents, we use two kinds of methods to generate candidate catchphrases for the target document: (1) directly using the text from citances and citphrases as catchphrase candidates for the target document, or (2) using the text from citances and citphrases to identify relevant sentences in the target documents, and use those sentences as candidate catchphrases.

4.1   Using Citations Text

We experimented with using citphrases and citances directly as candidate catchphrases for the target documents. To choose the best citations as candidates for extraction, we rank all citations to find the most "central" ones, hoping to identify "concepts" that are repeated across different citations. The proposed methods belong thus to the class of centroid or centrality-based summarisation (see for example [2,14], which are used as baselines in our evaluation). For one target case, we take all citphrases and citances, both for citing and cited cases, and we compute scores to measure group similarity and extract only the most "central" citations (either citphrases or citances). We propose two types of centrality scores:

- AVG: for each citance or citphrase, we compute the average value of similarity with all the other citances/citphrases, where the similarity is measured using ROUGE-1 or ROUGE-SU recall.
- THR: we establish a link between two citances/citphrases if their similarity (ROUGE recall) exceed a certain threshold (which we set to 0.5), and then we score the citations based on the number of such links.

Because we use a method based on similarity, we also applied a re-ranker to avoid selecting sentences very similar to those already selected.

The results of these methods applied to the citances and citphrases of every case are presented in Figure 3. In this plot we used ROUGE-1 with threshold 0.5 as matching criteria (as defined in Section 3). The plot clearly shows that citphrases give results significantly better than citances in terms of precision, while the recall is comparable. The use of different ROUGE scores (ROUGE-1 or ROUGE-SU) and of different similarity measures (AVG or THR) does not influence the results substantially. The plot shows that we can find "good" citances and citphrases, that are similar to the target catchphrases, however the recall fails to grow over a certain limit (around 0.6) even when increasing the number of extracted citations, thus suggesting that we cannot cover all aspects of a case relying only on citances or citphrases.

4.2   Using Citations to Rank Sentences

To increase the number of recalled catchphrases, we experimented with using citation text to select candidates from the sentences of the target case, as opposed to using citations directly as candidates (this is in some way similar to finding implicit citing sentences, i.e. [6], but used both for citing and cited documents). We use again a centrality-based approach. For each sentence in the target document, we measure similarity with all the citations (either citphrases or citances), and rank the sentences in order of decreasing similarity: a sentence which has high similarity with many citations is preferred. We rely on the idea that citations indicate the main issues of the case, and that they can be used to identify the sentences that describe these relevant issues. We measure similarity between a sentence and the citphrases or citances in the following ways:

- AVG: for each sentence, compute ROUGE-1 or ROUGE-SU6 recall with each citphrase/citance, and take the average of such scores.
- THR: for each sentence, compute ROUGE-1 or ROUGE-SU6 recall with each citation, and count how many exceed a certain threshold (set to 0.5).

The results for this class of methods are shown in Figure 4, where again we use ROUGE-1, with threshold 0.5 as matching criteria. As before citphrases give better results than citances both in terms of precision and recall, although the difference is less pronounced, and the use of different similarity measures (ROUGE-1/ROUGE-SU6 and AVG/THR) does not bring significant differences.

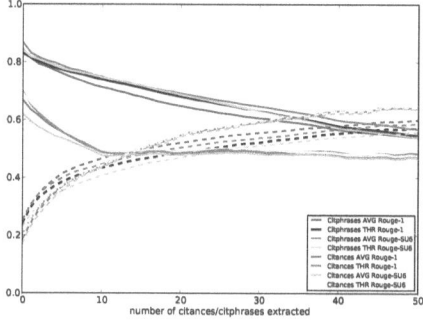

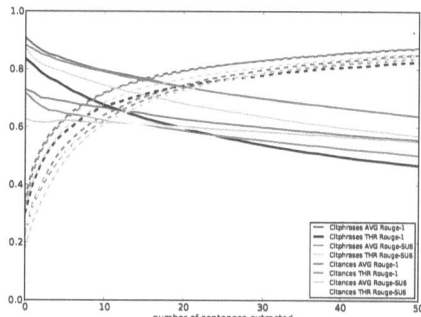

Fig. 3. Precision and Recall (dashed) of different ways of ranking citances and citphrases from related cases

Fig. 4. Precision and Recall (dashed) of different ways of ordering sentences using citances and citphrases

Compared to the first class of methods, these give better performances: they cover a larger number of catchphrases and at the same time fewer irrelevant sentences are extracted. An example of sentences extracted by this class of methods (using AVG on citphrases) is given in Figure 5. Results are further examined and discussed in the next section.

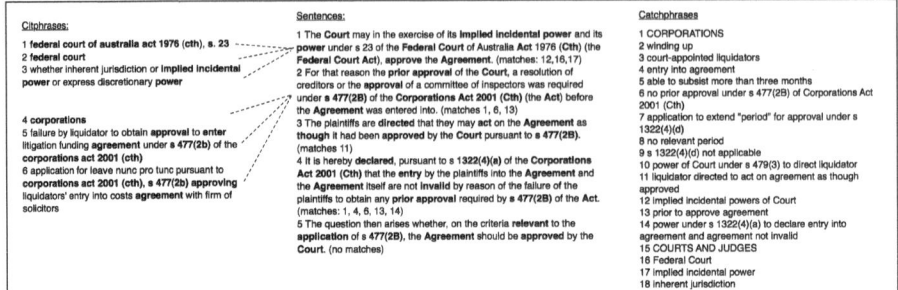

Fig. 5. The first 5 sentences (centre) as extracted using citphrases for a case. Words in bold appear also in the catchphrases (right). For each sentence the matching catchphrases (if any) are indicated in brackets. In this case the recall is 9/18=0.5 and the precision 4/5=0.8. For the first two sentences we also show (on the left) some of the citphrases which contribute to their rank, words in bold occur in the corresponding sentence.

5   Experimental Results

To better characterize the performances of citation based methods, in this section we compare our methods to other approaches which do not use citation data. We show that our citation based methods outperform both general purpose summarisation approaches, and domain specific methods based solely on the full text.

Baselines

The first baseline we use is the **FcFound** method, an approach developed for legal catchphrases extraction. FcFound was found to be the best method for legal catchphrase extraction when compared to a number of other frequency based approaches [3]. FcFound uses a database of known catchphrases to identify relevant words: words that, if they appear in the text, are more likely to appear also in the catchphrases of the case. The FcFound score of a term t (a term is a single token) is defined as the ratio between how many times (that is in how many documents) the term appears both in the catchphrases and in the text of the case, and how many times in the text (of the case):

$$FcFound(t) = \frac{NDocs_{text\&catchp.}(t)}{NDocs_{text}(t)}$$

We computed FcFound scores for every term, using our original corpus of 2816 case reports and corresponding given catchphrases. Then for each document we gave a score to each sentence, computing the average FcFound score of the terms in the sentence.

We used four other baselines: Random, Lead, Mead Centroid and Mead Lexrank. Random is a random selection of sentences from the document, Lead (a commonly used baseline in summarisation, see for example [8]) takes the sentences of the document in their order. As a more competitive baseline, we used also Mead, a state-of-the-art general purpose summariser [14]. In Mead we can set different policies: Mead Centroid (the original Mead score) builds a vector representation of the sentences, and extracts the sentences most similar to the centroid of the document in the vector space. A variation of Mead is LexRank [2], which first builds a network in which nodes are sentences and a weighted edge between two nodes shows lexical cosine similarity. Then it performs a random walk to find the most central nodes in the graphs and takes them as the summary. We downloaded the Mead toolkit[3] and applied both methods, obtaining a score for each sentence.

Results and Discussion

An overall evaluation of all the methods is given in Figure 6, which shows precision and recall of the different methods against the number of extracted sentences, averaged over all the documents. The matches are computed using ROUGE-1 with threshold=0.5. In the plot, we have the four citation based methods we developed and five baselines:

- **CpOnly**: citphrases are used directly as candidates.
- **CsOnly**: citances are used directly as candidates.
- **CpSent**: citphrases are used to rank sentences of the target document.
- **CsSent**: citances are used to rank sentences of the target document.
- Fcfound, Mead (LexRank and Centroid), Random and Lead, which do not use citation information in selecting sentences from the document.

The four citation methods use ROUGE-1 AVG as similarity score to rank text fragments (it was shown in Section 4 that varying the similarity score does not significantly impact the performances).

We can see from Figure 6 that the best methods are those using citations to select sentences: CpSent is the method which consistently gives the highest performances both for precision and recall, CsSent gives the second best recall, while CpOnly the second best precision. These results suggest that using citation data to select sentences can improve significantly the performance over using only citation text or only the full text of the case. While CpSent and CsSent give high precision and recall, CpOnly and CsOnly have high precision but lower recall, confirming the hypothesis that citation data alone, while useful in identifying some key issues of the target case, is not enough to cover all the main aspects of

---

[3] `www.summarization.com/mead/`

the case, and should be used in combination with the full text to obtain better summaries (see Figure 9).

We performed other evaluations, varying the matching criterion: using different ROUGE scores (ROUGE-1, ROUGE-SU and ROUGE-W) with different thresholds (0.5 and 0.7, we tried also other values and the results were comparable) to define a match between a sentence and a catchphrase. More "strict" match conditions give lower values for recall and precision, and vice-versa. As a comparison Figure 7 shows the results using ROUGE-SU6. We can see that, for any criteria used, the shape of the curves and the performance of the methods relative to each other are still consistent.

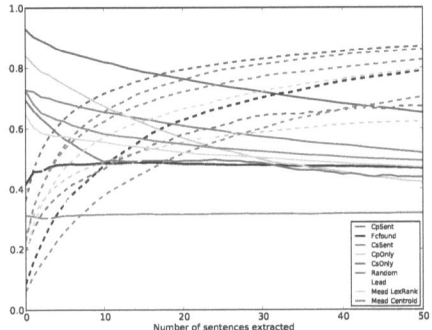 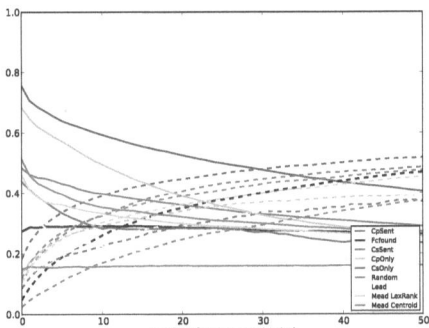

Fig. 6. Precision and Recall (dashed) of different methods. The matching criteria is ROUGE-1 with threshold 0.5.

Fig. 7. Precision and Recall (dashed) of different methods. The matching criteria is ROUGE-SU6 with threshold 0.5.

Regarding the use of different types of citations, generally citphrases give higher performances than citances: it seems that citances match more irrelevant sentences, while citphrases select better candidates. A reason may be that citphrases use a language at the right level of abstraction to describe the relevant issues, while citances may examine those issues in a more specific way. The difference in performance may also be influenced by the fact that citphrases are generally present in larger numbers than citances. Thus for many cases citances may be just not enough to cover all the main issues. Another difference is that citphrases are taken from both previous cited cases and following citing cases, while citances can be extracted only from the latter. Both citphrases and citances, however, outperform methods based only on the full text of the case.

All the methods based on citphrases were evaluated using two kind of citphrases, those from citing documents (cases that cite the target case) and cited documents (cases cited by the target case). In Figure 8 we evaluate the two kinds of citations separately: plotting separate results from the two sources as well as the union of them. The difference between using citing citphrases or cited citphrases is minimal in terms of recall and precision, while using both kinds of citation together improves the results, especially in the CpOnly case. This has important implications for the general applicability of these methods: contrary to most citation-based summarisation approaches, we can apply our methods also for new cases that have not yet been cited, using cited cases to extract catchphrases. Thus we show that it is possible to use citation to summarise new documents not yet cited, using only outgoing citations (not considering incoming citations) is applicable to all cases and it still outperforms text-only methods.

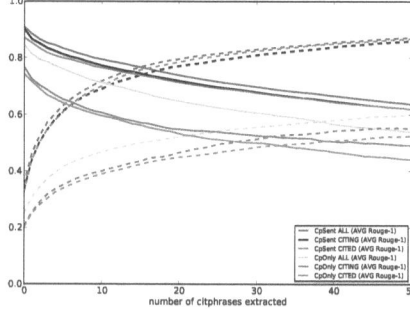

	Random	FcFound	Mead-C	CpSent
Precision	0.314	0.484	0.589	**0.827**
Recall	0.359	0.454	0.609	**0.702**
F-Measure	0.335	0.469	0.599	**0.759**

Fig. 8. Precision and Recall (dashed) of candidates for different citation types

Fig. 9. Evaluation for 10 sentences extracts

Finally, to compare our evaluation method with a more traditional method, we also ran a standard ROUGE evaluation in which we compare the whole block of extracted sentences with the block of given catchphrases, without distinguishing among single catchphrases or single sentences (using all the gold standard). Table 2 presents the results of such evaluation.. CpSent still emerges as the best method outperforming all baselines. However the ranking between other system differs, as FcFound is the second best, followed by CpOnly. The main difference between the two evaluation methods is that standard ROUGE rewards any matching token, while our evaluation method only considers matches between two units with a

Table 2. ROUGE evaluation for extracts (10 sentences)

	ROUGE-1			ROUGE-SU6			ROUGE-W-1.2		
	Pre	Rec	Fm	Pre	Rec	Fm	Pre	Rec	Fm
CpSent	**0.1876**	**0.4660**	**0.2469**	**0.0674**	**0.1850**	**0.0895**	**0.1230**	**0.2264**	**0.1426**
FcFound	0.1733	0.4389	0.2293	0.0598	0.1672	0.0797	0.1154	0.2168	0.1346
CpOnly	0.1724	0.4034	0.2216	0.0599	0.1537	0.0774	0.1165	0.1996	0.1308
Mead LexRank	0.1629	0.4071	0.2145	0.0569	0.1559	0.0753	0.1092	0.2013	0.1263
CsSent	0.1524	0.3871	0.2015	0.0502	0.1420	0.0668	0.1029	0.1934	0.1198
Lead	0.1432	0.3606	0.1890	0.0487	0.1332	0.0644	0.0985	0.1822	0.1141
Mead Centroid	0.1343	0.3405	0.1777	0.0439	0.1225	0.0584	0.0897	0.1679	0.1043
Random	0.1224	0.3164	0.1624	0.0326	0.0948	0.0436	0.0812	0.1567	0.0952
CsOnly	0.1043	0.2416	0.1319	0.0291	0.0753	0.0371	0.0716	0.1218	0.0790

minimum number of common elements (as discussed in Section 3, see Figure 2). Another difference is that our evaluation method does not take into account the length of the extracted sentences. Methods like FcFound and CpOnly extract shorter sentences, and thus obtain an higher score in this kind of evaluation. While on the one hand it is correct to penalize long sentences in the evaluation (if two sentences contain the same information - having the same units in common with citations- we prefer to select the shorter one), on the other hand methods that penalize longer sentences, having a bias towards selecting short sentences, have also problems, as many catchphrases are actually quite long and they are covered only by long sentences.

6   Related Work

The use of citations for summarisation has been mainly applied to scientific articles. In 2004 Nakov et.al. [10] pointed out the possibility of using citation contexts directly for text summarisation, as they provide information on the important facts contained in a paper. An application of the idea can be found in the work of Qazvinian and Radev [12,13], where they propose different methods to create a summary by extracting a subset of the sentences that constitute the citation context. Mohammad et.al. [9] apply this approach to multi-document summarisation, building on the claim by Elkiss et.al. [1] about the difference of information given by the abstract and the citation summary of a paper. Mei and Zhai [8] use citation data to summarise the impact of a research paper. The use of citation contexts to improve information retrieval is analysed in [15].

Rather than scientific papers, our work investigates the application of using citations for summarisation of legal text. While most approaches generate summaries directly from citation text only, we also show how to use citations to extract a summary ranking sentences from the full text (analogous to what was done by [8]). Another point of distinction is the use of catchphrases of citing/cited cases for summarisation. This is not possible for scientific articles as catchphrases are not given; at best a few keywords are given. A way to apply a similar idea to articles might be to use the abstracts of cited papers, normally provided, for extractive summarisation. Our work is also novel in the sense that it considers for each document both incoming and outgoing citations, while work on scientific articles is usually confined to incoming citations. This is particularly important if we want to summarise new documents, which have not yet been cited.

In the legal field, researchers have investigated extractive summarisation of legal cases, but without using any citation information: examples include the work of Hachey and Grover [5], ProdSum [17] and [3].

7   Summary and Future Work

In this paper, we proposed an innovative approach to using citations for summarisation that (1) uses also older documents cited by the target document for summarisation (as opposed to using only newer documents citing the target one, as done in scientific papers), this allows us to summarise a new case which has not been cited yet with no significant loss of performances; (2) uses catchphrases taken from cited and citing cases and show that they improve performances compared to using citing sentences (which is what is usually done in citation summarisation); (3) uses ROUGE scores to match citation text with the full text of the target document to find important sentences, rather than extracting the summary directly from the citation text.

We applied our methods to catchphrase extraction for legal case reports and presented the results of our validation experiments. Catchphrases are considered to be a significant help to lawyers searching through cases to identify relevant precedents and are routinely used when browsing documents. We created a large corpus of case reports, corresponding catchphrases and both incoming and outgoing citations. We set an evaluation framework, based on ROUGE, that let us automatically evaluate catchphrase candidates on a large number of documents. Our results show that extracting the sentences which are most similar to citation catchphrases gives the best results, outperforming several baselines that represent both general purpose and legal-oriented summarisation methods. The method can also be applied to recent cases that do not have incoming citations, using cited cases. It is worth noting that legal case texts use citations differently to most scientific papers which, e.g. , also need to cite papers that presented competing work.

In future work we will explore the different ways of combining information from citation data with other features of the target document, such as frequency or centrality measures. We suspect that different combinations of methods depending on the documents being analysed will give the best results. Another direction for future development is to study how these techniques (using both incoming and outgoing citations, and use of citphrases) can be applied in other domains such as scientific articles.

References

1. Elkiss, A., Shen, S., Fader, A., Erkan, G., States, D., Radev, D.: Blind men and elephants: What do citation summaries tell us about a research article? J. Am. Soc. Inf. Sci. Technol. 59(1), 51–62 (2008)
2. Erkan, G., Radev, D.: LexRank: Graph-based lexical centrality as salience in text summarization. Journal of Artificial Intelligence Research 22, 457–479 (2004)
3. Galgani, F., Compton, P., Hoffmann, A.: Towards Automatic Generation of Catchphrases for Legal Case Reports. In: Gelbukh, A. (ed.) CICLing 2012, Part II. LNCS, vol. 7182, pp. 414–425. Springer, Heidelberg (2012)
4. Greenleaf, G., Mowbray, A., King, G., Van Dijk, P.: Public Access to Law via Internet: The Australian Legal Information Institute. Journal of Law and Information Science 6, 49 (1995)
5. Hachey, B., Grover, C.: Extractive summarisation of legal texts. Artif. Intell. Law 14(4), 305–345 (2006)

6. Kaplan, D., Iida, R., Tokunaga, T.: Automatic extraction of citation contexts for research paper summarization: a coreference-chain based approach. In: Proceedings of the 2009 Workshop on Text and Citation Analysis for Scholarly Digital Libraries, pp. 88–95. Association for Computational Linguistics, Morristown (2009)
7. Lin, C.Y.: Rouge: A package for automatic evaluation of summaries. In: Text Summarization Branches Out: Proceedings of the ACL 2004 Workshop, pp. 74–81. Association for Computational Linguistics, Barcelona (2004)
8. Mei, Q., Zhai, C.: Generating impact-based summaries for scientific literature. In: Proceedings of ACL 2008: HLT, pp. 816–824 (2008)
9. Mohammad, S., Dorr, B., Egan, M., Hassan, A., Muthukrishan, P., Qazvinian, V., Radev, D., Zajic, D.: Using citations to generate surveys of scientific paradigms. In: Proceedings of Human Language Technologies: The 2009 Annual Conference of the North American Chapter of the Association for Computational Linguistics, Boulder, Colorado, pp. 584–592 (2009)
10. Nakov, P.I., Schwartz, A.S., Hearst, M.A.: Citances: Citation sentences for semantic analysis of bioscience text. In: Proceedings of the SIGIR 2004 Workshop on Search and Discovery in Bioinformatics (2004)
11. Olsson, J.L.T.: Guide To Uniform Production of Judgments, 2nd edn. Australian Institute of Judicial Administration, Carlton South (1999)
12. Qazvinian, V., Radev, D.R.: Scientific Paper Summarization Using Citation Summary Networks. In: Proceedings of the 22nd International Conference on Computational Linguistics (Coling 2008), pp. 689–696 (2008)
13. Qazvinian, V., Radev, D.R., Ozgur, A.: Citation summarization through keyphrase extraction. In: Proceedings of the 23rd International Conference on Computational Linguistics (Coling 2010), Beijing, China, pp. 895–903 (August 2010)
14. Radev, D., Allison, T., Blair-goldensohn, S., Blitzer, J., Çelebi, A., Dimitrov, S., Drabek, E., Hakim, A., Lam, W., Liu, D., Otterbacher, J., Qi, H., Saggion, H., Teufel, S., Winkel, A., Zhang, Z.: Mead - a platform for multidocument multilingual text summarization. In: LREC 2004 (2004)
15. Ritchie, A., Robertson, S., Teufel, S.: Comparing citation contexts for information retrieval. In: CIKM 2008: Proceeding of the 17th ACM Conference on Information and Knowledge Management, pp. 213–222. ACM, New York (2008)
16. Schwartz, A.S., Hearst, M.: Summarizing key concepts using citation sentences. In: Proceedings of the HLT-NAACL BioNLP Workshop on Linking Natural Language and Biology, pp. 134–135. Association for Computational Linguistics, New York (2006)
17. Yousfi-Monod, M., Farzindar, A., Lapalme, G.: Supervised Machine Learning for Summarizing Legal Documents. In: Farzindar, A., Kešelj, V. (eds.) Canadian AI 2010. LNCS (LNAI), vol. 6085, pp. 51–62. Springer, Heidelberg (2010)

A Novel Video Face Clustering Algorithm
Based on Divide and Conquer Strategy

Gaopeng Gou, Di Huang, and Yunhong Wang

Laboratory of Intelligent Recognition and Image Processing,
Beijing Key Laboratory of Digital Media,
Beihang University, Beijing, China
gougp@cse.buaa.edu.cn,
{dhuang,yhwang}@buaa.edu.cn

Abstract. Video-based face clustering is a very important issue in face analysis. In this paper, a framework for video-based face clustering is proposed. The framework contains two steps. First, faces are detected from videos and divided into subgroups using temporal continuity information and SIFT (Scale Invariant Feature Transform) features. Second, the typical samples are selected from these subgroups in order to remove some non-typical faces. A similarity matrix is then constructed using these typical samples in the subgroups. The similarity matrix is further processed by our method to generate the face clustering results of that video. Our algorithm is validated using the SPEVI datasets. The experiments demonstrated promising results from our algorithm.

Keywords: face clustering, similarity measure, subgroup.

1   Introduction

Considering that video sequences contain much more information than images and text, increasing attention has been paid to video-based content understanding, searching and analysis in recent years. Among these applications, human face is one of the most important features. Face clustering and indexing in videos is a general method, which could enhance computational efficiency and decrease memory usage in searching a specified person from a clip of video or making personal digital albums. The performance of most existing methods [1] [2] [3] [4] [5] is easily influenced by variations of pose, expression, illumination, and also partial occlusion and low resolution.

Some methods have been proposed to solve these difficulties mentioned. Ji *et al.* [1] propose a constraint propagation algorithm to get face clusters in videos. However, their method only solves the influence of pose in clustering. Based on SIFT (Scale Invariant Feature Transform) features [6], Li *et al.* [2] create subgroups of a video and incorporate these subgroups into the final face clusters. Although their method alleviates these difficulties using SIFT, the strategies of creating and incorporating subgroups can be further improved. Frey *et al.* [3] introduce an affinity propagation method to generate cluster results without defining the number of clusters and the initial cluster centers. But this method produces duplicated cluster results for the same person in a video.

P. Anthony, M. Ishizuka, and D. Lukose (Eds.): PRICAI 2012, LNAI 7458, pp. 53–63, 2012.
© Springer-Verlag Berlin Heidelberg 2012

In this paper, a novel approach of creating and incorporating subgroups is proposed. Our method is similar to the SIFT-based method proposed by Li *et al.* [2]. However, there are major differences, and our contributions lie in the following two aspects. First, a new face similarity measurement method named SIFT_fusion is proposed to divide the detected faces into different subgroups based on SIFT features of faces. Unlike [2] that uses partial matching points, SIFT_fusion integrates all matching SIFT key points information and temporal information to measure the similarity between two faces. This enhancement makes the similarity measurement more accurate and robust. Second, a new method of combining subgroups is proposed. In this method, the typical samples of subgroups are selected and the similarity matrix of these subgroups is generated. Then the similarity matrix is analyzed to generate the final clustering results of the detected faces in a video. The experimental results show that our method improves the robustness and accuracy of face clustering in video surveillance.

The remainder of the paper is organized as follows. In Section 2, the framework of video-based face clustering using SIFT features is introduced. In Section 3, we explain the proposed face similarity measurement method to generate the subgroups of the detected faces in video surveillance. The proposed method for merging face sub-groups into the final clusters is introduced in Section 4. Experiments are conducted and the results are discussed in Section 5. Concluding remarks are given in Section 6.

2   The Framework of Face Clustering Algorithm

This section introduces the framework of video-based face clustering using both SIFT features and temporal continuity information. In Figure 1, the process of the framework is described. In construction step of the face subgroups, we use AdaBoost [7] to detect faces from a video, and then we divide them into subgroups using the proposed SIFT_fusion method. In the combination step of the subgroups, we construct a similarity matrix of the subgroups that only contain typical face samples. Then, the similarity matrix is further processed to get the final clustering results.

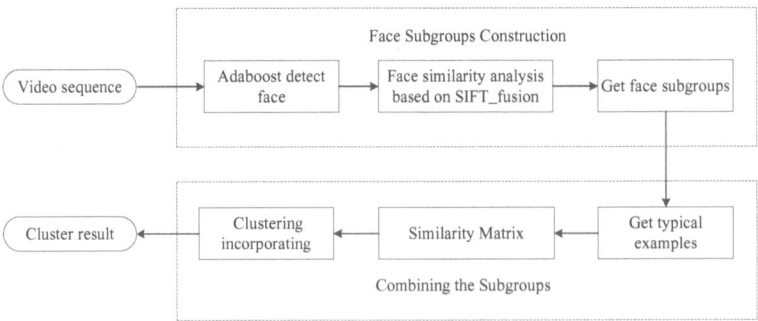

Fig. 1. Framework of the video-based face clustering algorithm

3   Face Subgroups Construction

Lowe's SIFT [6], which has been proved to be effective for object recognition, is used in our method. SIFT features are invariant to image scaling, transformation and rotation, and partially invariant to illumination and viewpoint changes. There are mainly four steps to extract SIFT features from a given face:

1. select the peak of scale-space;
2. localize the accurate positions of the key points;
3. assign the orientation;
4. get key point descriptors $(1 \times 128 vectors)$.

In [2], a face similarity measurement method is proposed based on SIFT. We name the method SIFT_max. The details of SIFT_max are: First, the SIFT key point descriptors $\{(scl_i, pos_i, rot_i, vct_i), i \in A\}$ is extracted from face image A using [6]. Let $scl_i, pos_i, rot_i, vct_i$ denote respectively the scale-space, position, orientation and descriptor of SIFT key point i. $\{(scl_j, pos_j, rot_j, vct_j), j \in B\}$ is extracted from face B also using [6]. Second, the similarity value between key point i and j is defined as:

$$S_{ij} = exp(-\frac{1}{\sigma^2} |\text{pos}_i - pos_j|) \cdot \|vct_i, vct_j\| \tag{1}$$

Assume S_{i1} and S_{i2} are the two largest similarity scores. If S_{i1}/S_{i2} is larger than a $ratio$ (taken as 1.16 in [2]), the similarity value between A and B is defined as S_{i1}.

3.1 $SIFT\_fusion$ Method

$SIFT\_max$ only considers the largest similarity scores of the matching pairs and discards the surplus matching pairs between two faces. However, the surplus matching pairs contain much more information that can be utilized to get more exact similarity measure between two faces. Therefore, a new SIFT-based face similarity measurement method is proposed. We name it $SIFT\_fusion$, which fuses the similarity scores of all matching pairs between two face A and B. $\{(pos_i, rot_i, scl_i, vct_i), i \in A\}$ and $\{(scl_j, pos_j, rot_j, vct_j), j \in B\}$ are the SIFT descriptors. $dis(\bullet, \bullet)$ is the correlation coefficient of two vectors. j is the matching key point in face B with key point i in face A. k is another key point in image B. If i and j satisfy the following 4 conditions:

$$dis(vct_i, vct_j) < dis(vct_i, vct_k) \times R, \quad k \neq j \tag{2}$$

$$\begin{cases} If\ scl_i/scl_j{<}1,\ satisfy\ scl_j/scl_i < scl_{thres} \\ else\ satisfy\ scl_i/scl_j < scl_{thres} \end{cases} \tag{3}$$

$$|pos_i - pos_j| < pos_{thres} \tag{4}$$

$$|rot_i - rot_j| < rot_{thres} \tag{5}$$

then S_{ij} is the similarity value of the key point i in image A and the key point j in image B. S_{ij} can be calculated by Equation 6. The more similar the two key points i and j, the smaller the similarity value will be.

$$S_{ij} = \begin{cases} 1 & \text{if i and j are not the matching key points} \\ \min_{j \in B} \{arccos(vct_i \times vct_j^T)\} & else \end{cases} \tag{6}$$

Finally the similarity of image A and B, DIS is defined as:

$$DIS = \prod_{i=1}^{M} \prod_{j=1}^{N} S_{ij} \tag{7}$$

where M is the number of the key points in image A, N is the number of the key points in image B. Using Equation 7, an accurate measurement of the similarity value between A and B could be obtained because the information of all matching pairs is considered and fused. The more similar two faces are, the smaller DIS is. If A and B are faces in adjacent frames, the temporal information can be used to improve the similarity between A and B using:

$$DISNEW = \begin{cases} exp(-\frac{1}{\sigma^2}|p_A - p_B|) \times DIS, \text{if A and B are adjacent} \\ DIS, \ else \end{cases} \tag{8}$$

The parameters p_A and p_B are the face positions of A and B in the frame if the faces are adjacent faces.

Because of using $SIFT\_fusion$ method, $DISNEW$ will be very small, so the absolute value of the similarity value's natural logarithm is introduced.

$$newsim = |log(DISNEW)| \tag{9}$$

Using Equation 9 the improved similarity value of two faces is $newsim$. It can be found that the more similar the two images are, the bigger the similarity value will be.

3.2   Obtaining the Face Subgroups

Faces in a video are detected by AdaBoost. Then $SIFT\_fusion$ is implemented to compute the similarity between these faces and the faces are divided into different subgroups by a threshold δ_1.

Firstly, the faces in the first frame of the video sequence are detected by AdaBoost. Once a human face is detected, a new subgroup is created to label that face. The face is the index face of this subgroup. The process lasts until all the faces are detected from the first frame in the video.

Secondly, detect the faces along the time sequence of the frames in the video by AdaBoost. If a face is detected in one frame, the similarity between that face and the index face of every subgroup is calculated using Equation 9. If the largest similarity value is larger than the given threshold δ_1, the face is added to that subgroup whose index face is most similar to it. Otherwise, a new subgroup is generated which contains that face, and it is the index face of the new subgroup.

Finally, do the second step until every frame in the video is processed and every detected face is labeled.

4   Combining the Subgroups

In this section, the method of combining the subgroups into several clusters is introduced. It is a matrix integrating method using the similarity matrix which is constructed based on the typical samples from the subgroups.

4.1   Getting the Typical Samples

In this step, the first face in every subgroup is kept and used to be the index face, and then we compute the similarity value between the index face and every remaining face in the subgroup using Equation 9. If the similarity value is smaller than a pre-determined threshold δ_2, it will be deleted from this subgroup. The deleted samples are the redundant faces that are easily affected by pose, expression, illumination, partial occlusion and low resolution. Because the remaining faces are much more similar to the index face than the ones deleted in the same subgroup, every subgroup keeping similar samples will decrease the within-class scatter and increase the between-class scatter of subgroups. This will improve clustering efficiency in the next step.

4.2   Similarity Matrix Construction

Now a set $G = \{G_1, \cdots, G_i, \cdots, G_n\}$ is defined, G_i is the ensemble of the typical examples in the ith. A similarity matrix of the different subgroups is constructed in this step. The similarity between different subgroups G_u and G_p is defined as:

$$Sim_{u,p} = \min_{i \in G_u, j \in G_p} (sim_{i,j}) \tag{10}$$

The value of $sim_{i,j}$ can be calculated by Equation 7, now a $n \times n$ similarity matrix M can be constructed. The element $M_{u,p}$ in M equals to $Sim_{u,p}$, the diagonal elements in M equal to zero. The similarity value of two subgroups is defined as the minimize similarity values in the between-class of them.

4.3   Clustering Incorporating

In this part, the APC (affinity propagation clustering) is introduced to incorporate the subgroups to the final clustering. APC is an efficient clustering algorithm published on Science in 2007 [3]. This method could automatically determine the cluster number, so it can be used to incorporate our subgroups. Now we should input the parameters $similarity$ and $preference$ to the cluster method, the $similarity$ is generated by the similarity matrix, and $preference$ equals to the mean of the similarity matrix.

In this part, an incorporating method named $SIFT\_fusion + APC$ is proposed, which utilizes the similarity matrix generated by Equation 10. Then the parameters similarity and preference could be generated by the similarity matrix. Inputting the two parameters to APC the finally clustering results will be generated. A novel subgroups incorporation method is also proposed. There are the details:

1. Initialize an empty set $R = \{\}$, then assign $R = \{R_1, \cdots, R_n\}$, n is the number of subgroups. $R_i = \{M_{i,i+1}, \cdots, M_{i,n}, \}$, M is the similarity matrix of the subgroups;
2. Check every element in R_i and see whether $M_{i,j} > Threshold$. Then delete the element $M_{i,j}, j \in \{i+1, \cdots, n\}$ from R_i and update the set R. The $Threshold$ can be calculated by Equation 11;
3. Initialize an empty set $C = \{\}$, in the beginning, $C = \{C_1\}$, where $C_1 = \{R_1\}$;

4. Update set C and its element C_i by the relationship between C_i and R_j. If $C_i \cap R_j \neq \emptyset$, update $C_i = C_i \cup R_j$. Otherwise, generate a new element in set C, and then assign $C = \{C_1, \cdots, C_n, C_{n+1}\}$, assign $C_{n+1} = R_j$;
5. Repeat the step (4) until all elements R_i in set R has been compared with the element C_i in set C;
6. If the number of elements in set C is not equal to the number of elements in set R, we calculate the similarity matrix M of C by the method mentioned by Equation 10, repeat the step (1) to (5) until the number of elements in set C is equal to the number of elements in set R.

$$Threshold = \min_{G_u \in G, G_p \in G, G_u \neq G_p} \left\{ \max_{i \in G_u, j \in G_p} (sim_{i,j}) \right\} \tag{11}$$

The similarity matrix is generated by the minimum similarity value of between-class of two subgroups. The $threshold$ is the minimum value of between-class matrix, which is composed of the maximum similarity value of between-class. The $threshold$ generating in subgroups incorporation method and the similarity matrix generating method determine the wrongly incorporated rate will be very small because the get-ting typical samples step decreases the within-class scatter and increases the between-class scatter of subgroups. After several iterations, the final clustering results will be generated.

5   Experimental Results and Discussion

5.1   Database Introduction

The multiple faces dataset in SPEVI datasets [8] is used for evaluating the proposed method. The dataset has three sequences. People in these videos repeatedly occluded each other while appearing and disappearing along with the fast changes of rotation and position. Figure 2 shows the three multiple face datasets in SPEVI datasets and their face detection results ,these three face videos in SPEVI named $motinas\_multi\_face\_frontal$, $motinas\_multi\_face\_turning$, $motinas\_multi\_face\_fast$ from left to right in Figure 2.

Fig. 2. SPEVI datasets and their face detection results

We select 22 different persons from FERET face database [9] to compare our algorithm $SIFT\_fusion$ with $SIFT\_max$, PCA and LBP. Figure 3 shows the 22 persons.

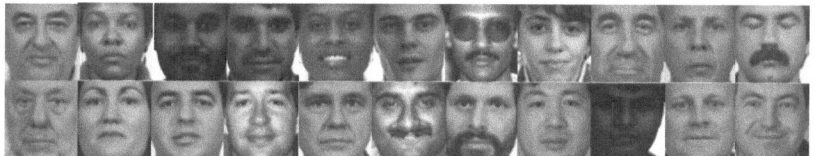

Fig. 3. 22 persons selected from FERET database

5.2   The Comparison of Similarity Measure Algorithms

In this experiment, $SIFT\_fusion$, $SIFT\_max$, PCA and LBP are tested on the FERET face database. 22 people are selected from FERET. The 22 persons can be found in Figure 3. 10 different pictures are selected from the datasets of FERET face database, $R = 0.8, \sigma = 9, pos_{thres} = 7, rot_{thres} = 0.27, scl_{thres} = 0.77$ are the parameters in $SIFT\_fusion$. $ratio = 1.16, \sigma = 9$ are the parameters in $SIFT\_max$. The Eigenvector number of PCA is 219. LBP operator is $(8, 1)$. We randomly select n faces from the 10 faces of every person as the training temples, then find the most similar face of everyone in the 220 faces by using $SIFT\_fusion$, $SIFT\_max$, LBP and PCA from the $22 \times n$ training temples. If the testing face and its most similar template face are from the same person, the testing face is correctly recognized. When the template number is n, the cumulative recognition rate equals to the number of the correctly recognized faces divide the total number of faces. It can be found that our algorithm $SIFT\_fusion$ get the best cumulative recognition rate at different template numbers from Figure 4.

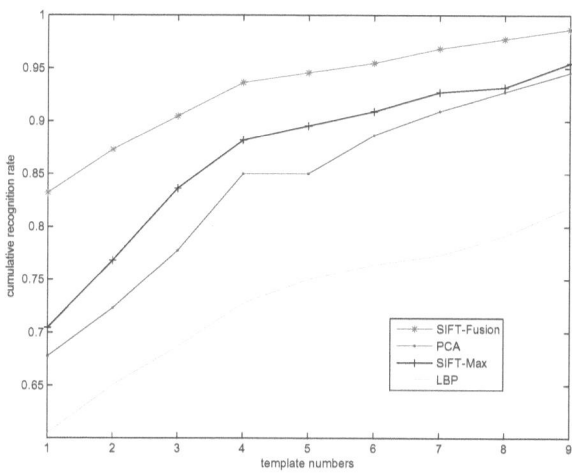

Fig. 4. The experiment of $SIFT\_fusion$, $SIFT\_max$, PCA and LDA

5.3   Selecting the Optimal Threshold to Obtain Subgroups

The video named $motinas\_multi\_face\_frontal$, which has 4 targets in SPEVI [8], is used as the training sample, 3396 faces including 35 non-faces is detected by AdaBoost. In this experiment, the relationships among the wrongly divided rate of faces, the value of parameter δ_1 used in the $SIFT\_fusion$, and the number of the divided sub-groups are tested. Equation 12 is applied to calculate the wrongly divided rate. Set A is the subgroups we have divided, K is the number of subgroups. Set M is manual divided the detected faces by the person occludes in the test video, B_j is the set of the faces of the jth person. Set B is used as the ground truth.

$$wronglydividedrate = \frac{\sum_{i=1}^{K}\left\{|A_i| - \underset{B_j \in B}{argmax}\,|A_i \cap B_j|\right\}}{\sum_{i=1}^{K}|A_i|} \qquad (12)$$

In Figure 5, the data on the curve is the number of the subgroups. It can be found that when the threshold δ_1 increases the wrongly divided rate decreases. The wrongly divided rate and the subgroups number are moderated when the threshold is bigger than 4.5, so finally we assign the threshold as 4.9 for testing. Different surveillance video will have different, but for face clustering, the value near to 5.

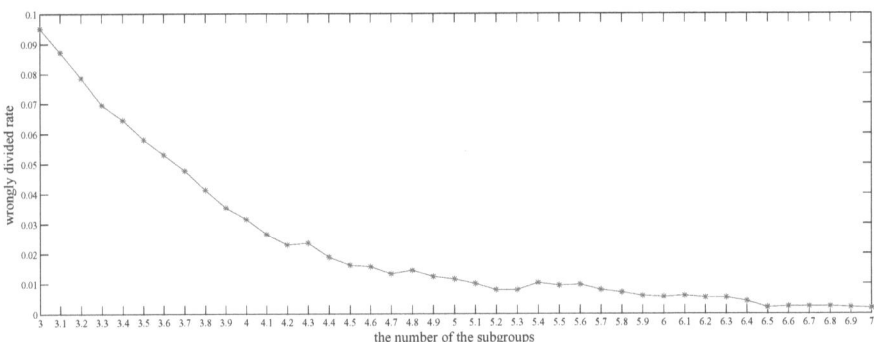

Fig. 5. The relationships between wrongly divided rate, subgroups number and threshold δ_1

5.4   The Comparison of $SIFT\_fusion$ and $SIFT\_max$ on the Efficiency of Getting Subgroups

In this experiment, the number of subgroups between $SIFT\_fusion$ and $SIFT\_max$ are compared when the wrongly divided rate of faces is approximately equal. From Table 1 and Table 2, it is easy to find that using $SIFT\_fusion$ method only needs about 10 percent of the subgroups that $SIFT\_max$ method needs to get the same wrongly

Table 1. The relationship between subgroup and wrongly divided rate by using $SIFT\_fusion$

subgroup number	180	201	214	223
wrongly divided rate	0.0188	0.0133	0.0124	0.0115

divided rate. Therefore, less subgroup numbers means less time to get typical samples, construct similarity matrix, generate final clustering, and decrease the final wrongly divided rate in incorporating subgroups.

Table 2. The relationship between subgroup and wrongly divided rate by using $SIFT\_max$

subgroup number	2079	1583	1221	685
wrongly divided rate	0.0035	0.0121	0.02	0.0439

5.5   Choose the Optimum Threshold to Get Cluster Results

In this experiment, the typical samples are obtained from the subgroups by modify-ing the threshold δ_2, which determines the within-class scatter and the between-class scatter of subgroups and influences the final clustering result. Now we change δ_2 to get the typical samples and then get its minimum similarity matrix, and then use the proposed incorporating method to get the final clustering result. It can be found from Table 3 that when δ_2 changes from 13 to 14.5, the clustering results are optimal. Because there are 4 persons in the testing video named $motinas\_multi\_face\_frontal$.

Table 3. The relationship between cluster number and threshold δ_2

cluster number	2	2	4	4	4	5	5
δ_2	10	12	13	14	14.5	14.8	15

5.6   The Testing of Our Algorithm

The rest two multiple faces dataset in SPEVI [8] are used as the testing samples. 1112 faces including 12 non-faces are detected from $motinas\_multi\_face\_fast$, which has 3 targets, while from $motinas\_multi\_face\_turning$, 2511 faces including 53 non-faces are detected, which has 4 targets. The testing results of our algorithm, method in [2] [3] and $SIFT\_fusion + APC$ are listed in Table 4. The results show that our algorithm could get nearly exact clustering results and lower wrongly divided rate com-pared to the method in [2] [3]. Figure 6 gives the face clustering results of $motinas\_multi\_face\_fast$ using our method.

In Figure 6, it can be found that the top 3 clusters has get the satisfied result, in the bottom, there has some wrongly results. That is because some non-typical samples reside in the subgroups because of the threshold δ_1 . Through adjusting the threshold, we could get more perfect clustering results.

Table 4. The testing results of our algorithm, method in [2] [3] and $SIFT\_fusion + APC$

Method	Testing video	Cluster number	Wrongly divided rate
$SIFT\_fusion + APC$	$motinas\_multi\_face\_fast$	4	0.3052
Method in[2]	$motinas\_multi\_face\_fast$	8	0.3902
Method in[3]	$motinas\_multi\_face\_fast$	12	0.2275
our algorithm	$motinas\_multi\_face\_fast$	4	0.1359
$SIFT\_fusion + APC$	$motinas\_multi\_face\_turning$	4	0.3658
Method in[2]	$motinas\_multi\_face\_turning$	8	0.1072
Method in[3]	$motinas\_multi\_face\_turning$	13	0.0848
our algorithm	$motinas\_multi\_face\_turning$	5	0.1215

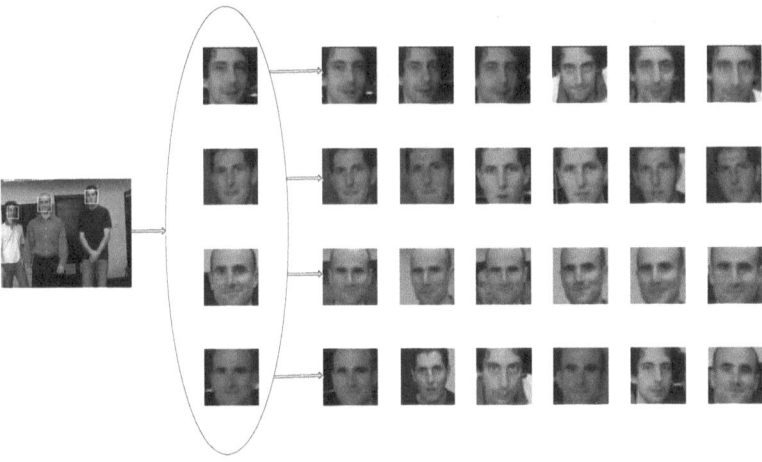

Fig. 6. Face clustering results of $motinas\_multi\_face\_fast$ using our method (only show parts of the results)

5.7   The Face Searching from Different Video

Using our proposed method we get the cluster results of $motinas\_multi\_face\_fast$, in this part, then form the video $motinas\_multi\_face\_turning$, we randomly select a detected face, we use the Equation 9 to get the most similarity cluster from the cluster results of $motinas\_multi\_face\_fast$. Figure 7 shows the searching results.

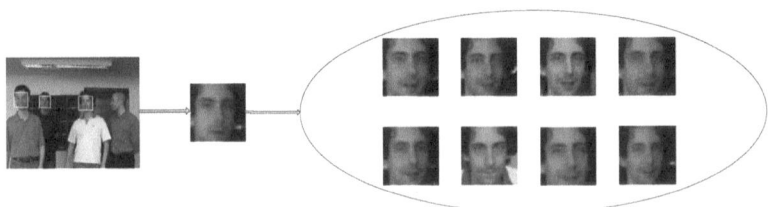

Fig. 7. Face searching results from $motinas\_multi\_face\_fast$(only show parts of the results)

6   Conclusion and Future Work

This paper proposed an effective face clustering algorithm based on SIFT for video surveillance. $SIFT\_fusion$, a new SIFT-based similarity method is proposed, which divides the detected faces into more efficient subgroups with lower wrongly divided rate. Then we propose a novel similarity matrix incorporation method, which integrates the subgroups into the final clustering results. Our algorithm has better performance over general APC [3], Li's method [2] and the proposed $SIFT\_fusion+APC$ method on SPEVI [8]. Our algorithm can be used in video and surveillance face indexing, searching and clustering. Our algorithm is robust to the variations of pose, expression, illumination, partial occlusion and low resolution.

Acknowledgments. This work is funded by the National Basic Research Program of China (No. 2010CB327902), the National High Technology Research and Development Program of China (No. 2011AA010502), the National Natural Science Foundation of China (No. 61005016, No. 61061130560), the Fundamental Research Funds of Beihang University and the Central Universities.

References

1. Tao, J., Tan, Y.P.: Face clustering in videos using constraint propagation. In: Proc. ISCAS, pp. 3246–3249 (2008)
2. Li, J., Wang, Y.: Face indexing and searching from videos. In: Proc. ICIP, pp. 1932–1935 (2008)
3. Frey, B.J., Dueck, D.: Clustering by passing messages between data points. Science, 972–976 (2007)
4. Eickeler, S., Wallhoff, F., Lurgel, U., Rigoll, G.: Content based indexing of images and video using face detection and recognition methods. In: Proc. ICASSP, vol. 3, pp. 1505–1508 (2001)
5. Chan, Y., Lin, S.H., Tan, Y.P., Kung, S.: Video shot classification using human faces. In: Proc. ICIP, pp. 843–846 (1996)
6. Lowe, D.G.: Distinctive image features from scale-invariant keypoints. IJCV, 91–110 (2004)
7. Viola, P., Jones, M.: Rapid object detection using a boosted cascade of simple features. In: Proc. CVPR, vol. 1, pp. 511–518 (2001)
8. Surveillance performance evaluation initiative (spevi),
   `http://www.elec.qmul.ac.uk/staffinfo/andrea/spevi.html`
9. Jonathon Phillips, P., Wechsler, H., Huangand, J., Rauss, A.J.: The feret database and evaluation procedure for face-recognition algorithms. Image Vision Comput. 16(5), 295–306 (1998)

A Hybrid Local Feature for Face Recognition

Gaopeng Gou, Di Huang, and Yunhong Wang

Laboratory of Intelligent Recognition and Image Processing,
Beijing Key Laboratory of Digital Media, Beihang University, Beijing, China
gougp@cse.buaa.edu.cn,
{dhuang,yhwang}@buaa.edu.cn

Abstract. Efficient face encoding is an important issue in the area of face recognition. Compared to holistic features, local features have received increasing attention due to their good robustness to pose and illumination changes. In this paper, based on the histogram-based interest points and the speeded up robust features, we propose a hybrid local face feature, which provides a proper balance between the computational speed and discriminative power. Experiments on three databases demonstrate the effectiveness of the proposed method as well as its robustness to the main challenges of face recognition and even in practical environment.

Keywords: Local feature detection, feature descriptor, face recognition.

1 Introduction

The human face is one of the most important cues to identify a person, therefore face recognition has been the key component in biometric recognition systems for many decades. However, the performance of a face recognition algorithm tends to be impaired by poor face quality, pose variation, illumination changes, and partial occlusions. According to [1], the success of a face recognition algorithm relies on effective facial features which enlarges inter-class variations while decreases intra-class ones, especially in a practical environment.

Many approaches have been proposed to extract facial features from texture images. Both holistic and local ones can be used to represent faces and measure their similarities. For holistic based, the whole facial images are directly projected and compared in a relatively low dimensional subspace in order to avoid the curse of dimensionality. Holistic features can be extracted using methods like principal component analysis (PCA) [2] and linear discriminant analysis (LDA) [3]*etc*. While local based face recognition ones, the facial images are partitioned into local blocks to extract features and then classify faces by combining and matching with corresponding features extracted from local blocks. Local features can be extracted using methods like scale-invariant feature transform (SIFT) [4], local binary patterns (LBP) [5], and speeded up robust features (SURF) [6] *etc*. But the limitation of holistic face recognition is that it requires accurate face normalization according to pose, illumination, scale, facial expression and occlusions. Variations in these factors can affect the holistic features extracted from the faces leading to inaccuracies of the recognition rate of system. Compared to holistic features, local features based face recognition algorithms have an advantage over holistic ones because they are more robust to pose and illumination changes [5], also insensitive

P. Anthony, M. Ishizuka, and D. Lukose (Eds.): PRICAI 2012, LNAI 7458, pp. 64–75, 2012.

to clutter and occlusion [7]. Therefore, in this paper, we aim to construct a novel and effective local feature which could be utilized in face recognition.

When local features are extracted, the next step is matching these features for face recognition. The matching method could be divided to the strategy they use *i.e.,*the dense as well as the sparse. Dense matching is used to find matches for all points in the image, when the local features are extracted from all the pixels of the facial image one by one, have to choose dense matching method. The typical a good representative of dense matching is LBP [5], LBP is an efficient texture operator which labels the pixels of an image by thresholding the neighborhood of each pixel and considers the result as a binary number. The dense matching method may cause the probably induces curse of dimensionality when concatenating all local features in a whole vector for classification or high computational cost since too many pixels are to be matched. Sparse matching method is used to establish a set of robust matches between the interest points of an image pair. Extracting sparse local features can be divided to be two phases [8]: 1)interest point detection, 2)local feature description. The typical sparse features are SIFT [4], and SURF [6] in face recognition domain. SIFT utilizes difference-of-Gaussian (DoG) detector to extract keypoints from image, then assign the orientation and scale, finally generate a 1×128 vector to describe the keypoint. SURF feature is extracted by the similarity framework to SIFT, but using hessian matrix to detect keypoints and based on the neighborhood information of keypoints to describe them.

In the past decade, many interest point detectors and local feature descriptors have been proposed [9]. Current detecting methods normally obtain interest points from images by measuring pixel-wise differences in image intensity. For instance, the Harris corner detector [10] uses the trace and the determinant of the second moment matrix, which is obtained from the image intensity function, to explore the locations of interest points. Mikolajczyk and Schmid [11] developed the Harris-Laplace and the Harris-Affine detectors, which are based on affine normalization around Harris and Hessian points. Lowe [4] proposed the DoG detector that approximates the scale-space Laplacian by DoG. Kadir *et al.*[12] proposed a salient point detector that takes into account local intensity histograms. Maver [13] divided local regions by radial, tangential, and residual saliency measures and then located the local extreme in the scale space of each saliency measure. SIFT descriptors are computed for normalized image patches with the code provided by Lowe [4]. Gradient location-orientation histogram (GLOH) [14] is an extension of the SIFT descriptor designed to increase its robustness and distinctiveness. SURF descriptor describes the distribution of the intensity content within the neighborhood of interest point, this descriptor is robust and fast than SIFT descriptor.

In recent years, some new local sparse features are proposed in general object recognition domain may also be applied to face recognition. Laptev and Lindeberg [15] used space-time interest points to interpret events recorded in videos. Dalal and Triggs [16] introduced the histograms of oriented gradients (HOG) for human detection. Lee and Chen [17] proposed the histogram-based interest points (HIP) to match images and showed that the matching results are invariant to scale and illumination changes.

In this paper, a hybrid local feature is proposed for face recognition based on HIP and SURF. Existing interest point detectors generally use pixel-based (intensity/color) representations to characterize local features. However, for face images consisting of highly textured regions, the pixel-based information may generate too many unstable corners during interest point detection. Even though the variations in the distribution of texture patterns may be insignificant, the sum of squared differences will increase dramatically. In contrast, the histogram-based interest point detectors are able to iden-tify interest points that exhibit a distinctive distribution of low-level features in a local area of a face image. The widely-used histogram-based detectors are HOG and SIFT. The HIP detector proposed by Lee and Chen [17] is relatively new and have not been applied to face recognition. HIP is able to capture large-scale structures and distinctive textured patterns. In addition, HIP exhibits strong invariance to rotation, illumination variation, and blur. Therefore, we employ HIP in the interest point detection step of our local feature extraction method. For the description step, a robust descriptor is required in practical environments. Currently, the most popular descriptors are SIFT descriptor and SURF descriptor. SIFT descriptor is robust to lighting variations and small posi-tion shift. SURF descriptor was designed as an efficient alternative to SIFT descriptor, and therefore it has smaller time and space complexity compared to SIFT. Hence, we employ SURF to describe the interest points detected by HIP in the first step. Exper-imental results demonstrate that the proposed face feature, *i.e.*, HIP+SURF performs well on different databases even under noisy condition.

The remainder of the paper is organized as follows. In Section 2, the proposed HIP+SURF face feature extraction method is described in detail. The experimental re-sults are discussed in Section 3. Conclusion and future work are given in Section 4.

2   Methodology

Figure 1 illustrates the general process of face feature extraction. The face recognition framework used in this paper is as follows:

1. Candidate face segmented images are detected by Adaboost [18] from the input images.
2. Interest points are detected from the face images and then described using descrip-tors. These descriptors are used for face training and recognition in the next step.
3. The detected face images from step 1 are divided into gallery set and probe set. The nearest neighbor (NN) classifier is used to recognize faces by calculating the similarity between the images in gallery and probe sets using the local features extracted in step 2.

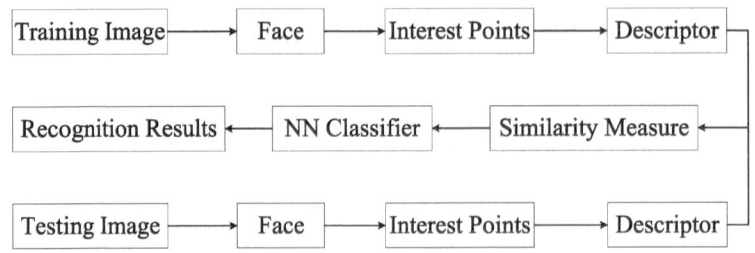

Fig. 1. The process of face feature extracting and the framework of face recognition

The three main phases in our face feature extraction method are: 1) Extract interest points using HIP; 2) Generate descriptors of the interest points using SURF; 3) Calculate similarity measure of face images.

2.1 Histogram-Based Interest Points

Given a face sequence/image, HIP [17] is employed to extract robust interest points from each frame/image. A face image is first divided into different patches. A color histogram or oriented gradient histogram [16] $\bar{h}$ with L bins is then built for each patch. Based on this histogram $\bar{h}$, a weighted histogram $h(x, y)$ is constructed from pixels in the neighborhood $\Omega(x, y)$ of each pixel location (x, y) in each patch. More specifically, the kth bin of $h(x, y)$, denoted by $h_k(x, y)$, is computed using Equation 1. Note that only those neighboring pixels in $\Omega(x, y)$, which take values equivalent to $\bar{h}_k j$, contribute to $h_k(x, y)$, where $\bar{h}_k$ denotes the value (*i.e.*, color or oriented gradient) associated with the kth bin of $\bar{h}$.

$$h_k(x, y) = \frac{1}{Z} \sum_{\substack{(x_i, y_i) \\ \in \Omega(x,y) \\ b(x_i,y_i)=k}} w(x_i - x, y_i - y) \tag{1}$$

where Z is a normalizing parameter; $w(x_i - x, y_i - y)$ is a Gaussian weighting function, which takes the form $w(x, y) = e^{-\frac{(x^2+y^2)}{\sigma^2}}$; and $b(x_i, y_i)$ is the discrete quantity derived from the color or oriented gradient of image.

In order to estimate whether (x, y) is an interest point, the Bhattacharyya coefficient ρ [19], a measurement of the amount of overlap between two samples, is used to evaluate the similarity between $h(x, y)$ and the weighted histogram $h(x + \Delta x, y + \Delta y)$ of its shifted pixel. The smaller ρ is, the more different $h_k(x, y)$ and $h_k(x + \Delta x, y + \Delta y)$ are. Therefore, the pixel location which projects to the local minimum of ρ are selected as the interest points in a patch. However, these interest points obtained from HIP cannot be directly applied to measure the similarity of different faces because the information associated with them is insufficient for effective comparison. Hence, based on the location information of the interest points, we further describe them using SURF descriptor.

Color histogram and oriented gradient histogram are used to generate the histogram-based interest points. Here, we give a brief description.

Color Histogram. For a RGB color face image, each color channel (256 levels assumed) is quantized into 8 bins, and a histogram with $8^3 = 512$ bins can be obtained. The quantization function is given by:

$$b(x, y) = \lfloor R_{x,y}/32 \rfloor \times 8^2 + \lfloor G_{x,y}/32 \rfloor \times 8 + \lfloor B_{x,y}/32 \rfloor + 1, \tag{2}$$

where 32 is calculated as the 256 levels divided by the 8 bins; $R_{x,y}$, $G_{x,y}$, $B_{x,y}$ are the RGB values of pixel (x, y) in face images. Then, $b(x, y)$ is substituted into the constraint condition of Equation 1 where $b(x, y) = k$ in the set $\Omega(x, y)$.

Oriented Gradient Histogram. The intensity gradients can also be used to construct the histograms for face images. We quantize the orientation of the gradient into 8 bins, and each bin covers a 45° angle (360° ÷ 8). The magnitude of the gradient is also divided into 8 bins, thus the resulting histogram contains 64 bins. The magnitude of the gradient provides useful information for interest point detection. Equation 1 can be changed to Equation 3. $\|g(x_i, y_i)\|^\alpha$ is the magnitude of the gradient at pixel (x_i, y_i), and α is a scaling parameter.

$$h_k(x, y) = \frac{1}{Z} \sum_{\substack{(x_i, y_i) \\ \in \Omega(x, y) \\ b(x_i, y_i) = k}} w(x_i - x, y_i - y) \|g(x_i, y_i)\|^\alpha \tag{3}$$

2.2 Interest Point Descriptors

In this phase, a descriptor is built for each interest point based on its neighborhood information using SURF [6]. SURF is chosen because it is more efficient than other invariant local descriptors like SIFT and it is robust to the variation of scales, rotations, and poses [6]. In order to generate the descriptors, a square area around each interest point is selected as the interest region, whose size is 20 times of the interest point's scale. This interest region is then split equally into 16 square subregions with 5×5 regularly spaced sample points in each subregion. Haar wavelet responses d_x and d_y for the x- and y-directions are calculated in every subregion. The generation of descriptors is shown in Figure 2. The responses in each subregion are weighted using a Gaussian function and then summed up in the x- and y-directions respectively to generate the feature descriptor $(\sum d_x, \sum d_y)$, which is a 32-dimensional vector ($4 \times 4 \times 2$). When the absolute responses are considered, the descriptor becomes a 64-dimensional vector $(\sum d_x, \sum d_y, \sum |d_x|, \sum |d_y|)$. Furthermore, when $\sum d_x$ and $\sum |d_x|$ are computed according to the signs of d_y's, similarly $\sum d_y$ and $\sum |d_y|$ according to the signs of d_x's, the descriptor becomes a 128-dimensional vector. In this paper, the normalized 128-dimensional descriptors are used in order to provide more information for the face similarity measure and recognition in the next phase.

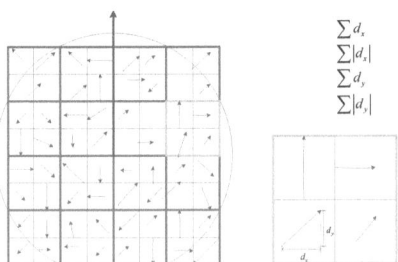

Fig. 2. The generation of descriptors [6]

2.3 Face Similarity Measure

Given two face images I^i and I^j, we respectively generate the SURF descriptors $D^i = \{d_1^i, \cdots, d_M^i\}$ and $D^j = \{d_1^j, \cdots, d_N^j\}$ of HIPs extracted from I^i and I^j, where M and

N are the numbers of HIPs in I^i and I^j respectively. First, we find the two matching descriptors between D^i and D^j using the strategy proposed in [4,6]. For each descriptor d_m^i in D^i, a similarity score $s_{m,n}$ is computed between it and each descriptor d_n^j in D^j using Equation 4.

$$s_{m,n} = [d_m^i - d_n^j][d_m^i - d_n^j]^T. \tag{4}$$

The smaller $s_{m,n}$ is, the more similar d_m^i and d_n^j are. Because d_m^i and d_n^j are normalized, $s_{m,n} \leqslant 2$. In practical situations, normally $s_{m,n} \leqslant 1$. After all the similarity scores $\{s_{m,1}, \cdots, s_{m,N}\}$ are calculated for a descriptor d_m^i, the ratio between the smallest score $s_{m,p}$ and the second smallest score $s_{m,q}$ can be calculated. If this ratio is less than 0.8, d_m^i and d_p^j are considered as a good match; otherwise, no good match is found and we set $s_{m,p} = 1$.

The more matching descriptor pairs are found, the more similar the two faces are. Therefore, the similarity between I^i and I^j is defined by the product rule:

$$S^{i,j} = \prod_{p=1}^{M} \prod_{q=1}^{N} s_{p,q}. \tag{5}$$

The smaller $S^{i,j}$ is, the more similar I_i and I_j are. Because $S^{i,j}$ may get extremely small, we take the absolute value of the logarithm of $S^{i,j}$ to measure the similarity of I_i and I_j, i.e.,

$$\bar{S}^{i,j} = |\log(S^{i,j})|. \tag{6}$$

It could be found that a larger $\bar{S}^{i,j}$ indicates the face pair is more similar.

3   Experimental Results

We evaluated the performance of the proposed color-HIP+SURF (using color histogram) and gradient-HIP+SURF (using oriented gradient histogram) against SIFT, SURF and LBP on three databases.

3.1   Databases

Three databases are used to test the effectiveness of the proposed local features, i.e., color-HIP+SURF and gradient-HIP+SURF. The first database was collected by ourselves in an indoor environment and contains 17 subjects. All subjects have head movements and face expression changes. Two videos were collected for each subject. One is used for training, and the other one for testing. Face sequences were automatically generated using Adaboost and scaled to 100×100 pixels. We use the first 25 and 100 frames respectively from every subject's training and testing face sequences for performance evaluation. Some face images of different subjects can be found in Figure 3. It could be found that the face conditions vary according to different poses, lighting conditions, and expressions.

The second database is the multi-modal VidTIMIT database [20], which was also collected in an indoor environment. The VidTIMIT database contains 43 subjects and 13 sequences were recorded for each subject. Each subject rotates the head in four

Fig. 3. Faces of some subjects in our database

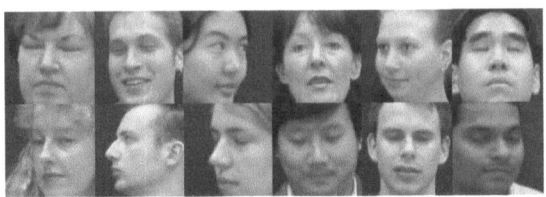

Fig. 4. Faces of some subjects in the VidTIMIT database

directions and recites short sentences. Their head poses and expressions also vary when they are reciting sentences. Same as in our database, face sequences were automatically generated using Adaboost and scaled to 100×100 pixels. We use the first 25 and 100 frames respectively from every subject's training and testing face sequences for performance evaluation. Figure 4 shows some face images from different subjects in this database.

The third database is the Computer Vision Laboratory (CVL) face database [21]. This database has 114 subjects. For 111 subjects, 7 images were collected for each person. The 7 images are of seven different poses: far left face pose, angle $45°$ face pose, face with serious expression, angle $135°$ face pose, far right face pose, smile (showing no teeth), and smile (showing teeth). Figure 5 shows 7 images of one person detected by Adaboost. For the other 3 subjects, only 6 images were collected for each person. Because every subject in this database has only 7 (or 6) images, we choose the leave-one-out cross-validation method to test the performance of different feature extraction methods.

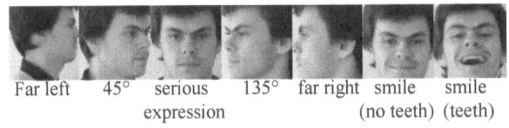

Far left   45°   serious   135°   far right   smile   smile
                   expression   (no teeth) (teeth)

Fig. 5. Seven face images of a subject in the CVL face database

3.2   Experiment I - Face Recognition on Clean Data

In this experiment, we compared the performance of the proposed color-HIP+SURF and gradient-HIP+SURF with other local features, *i.e.,* SIFT, SURF and LBP. The face recognition rates are reported in Table 1. For our database, gradient-HIP+SURF gets the

Table 1. Face recognition rates on CVL database, our database and VidTIMIT database

Database		color-HIP+SURF	gradient-HIP+SURF	SIFT	SURF	LBP
CVL	far left	6.14%	7.02%	2.65%	5.26%	6.14%
	45°	4.39%	8.77%	7.08%	5.26%	5.26%
	serious expression	91.23%	82.46%	84.96%	64.91%	78.95%
	135°	6.14%	5.26%	7.08%	5.26%	7.02%
	far right	3.51%	7.02%	10.61%	4.39%	8.77%
	smile no teeth	88.50%	90.27%	89.29%	68.14%	85.84%
	smile teeth	93.52%	95.37%	92.59%	65.09%	84.26%
	All	41.26%	41.64%	41.13%	30.57%	38.97%
Our		82.47 %	87.53%	81.00 %	81.47 %	86.47 %
VidTIMIT		67.21%	70.84%	60.00%	53.84%	55.81%

Fig. 6. Clean face samples (first row) and the same faces corrupted with AGWN (second row)

best recognition rate. color-HIP+SURF shows similar performance to SIFT and SURF. For VidTIMIT, color-HIP+SURF and gradient-HIP+SURF perform better than all the other three features. On the uniform test framework (Section 2), the results suggest that color-HIP+SURF and gradient HIP+SURF encode face images more effectively than SIFT, SURF, and LBP. Color-HIP+SURF and gradient-HIP+SURF extract more useful information from the face images. Gradient-HIP+SURF is better than color-HIP+SURF on our database and VidTIMIT database because the magnitude of gradient could provide useful information for interest point detection.

Different from the tests on our database and VidTIMIT database. We divide the CVL database to be 7 different probe sets according to faces' poses or expressions. From Table 1, it can be seen that all the five local features perform relatively well on 3 expression probe sets and very poor on the other 4 pose probe sets. This is due to the leave-one-out cross-validation method used in testing the performance of these face features. When a pose of one subject is in the probe set, there is no similar poses of that subject in the gallery set. Therefore, the recognition result is very poor on the 4 pose probe sets. On the 3 expression probe sets, color-HIP+SURF and gradient-HIP+SURF show better performance than SURF and LBP, and similar or slightly better performance than SIFT. Therefore, color-HIP+SURF and gradient-HIP+SURF are robust to expression variation.

From the recognition results in Table 1, we can find that when the head pose of the subject does not exist in the gallery sets, our system will recognize the subject to be another subject which has the similar head pose in the gallery set. Because our goal is to explore a robust face feature to represent faces, we compare these features on the simple NN classifier without preprocessing. The face recognition framework cannot process profile faces with small samples efficiently. The algorithm of recognizing the profile faces with small samples could be found in [22].

3.3   Experiment II - Face Recognition on Noisy Data

In this experiment, we evaluated color-HIP+SURF and gradient-HIP+SURF against SIFT, SURF and LBP on our database, VidTIMIT database and CVL database corrupted with additive Gaussian white noise (AGWN) with 0 mean and 0.002 variance. Some corrupted face images are shown against clean images in Figure 6. Table 2 gives the test results of the five face features on our database, VidTIMIT database and CVL database. Under noisy conditions, the face recognition rates of color-HIP+SURF and gradient-HIP+SURF are decreased, but they still perform better than SIFT, SURF, and LBP on the average. Comparing Table 2 and Table 1, we can see that the face recognition rates using color-HIP+SURF and gradient-HIP+SURF decrease much slower than SIFT, SURF and LBP, when noise is introduced. Hence, color-HIP+SURF and gradient-HIP+SURF are more robust under noisy conditions than SIFT, SURF, and LBP.

Table 2. Face recognition rates on CVL database, our database and VidTIMIT database corrupted with AGWN

Database		color-HIP+SURF	gradient-HIP+SURF	SIFT	SURF	LBP
CVL	far left	4.39%	8.77%	7.08%	5.26%	5.26%
	45°	4.39%	5.26%	1.75%	7.89%	7.89%
	serious expression	76.32%	71.05%	69.30%	51.75%	71.93%
	135°	4.39%	6.14%	2.63%	6.14%	8.77%
	far right	4.39%	3.51%	7.89%	5.26%	7.89%
	smile no teeth	80.53%	76.99%	74.34%	59.29%	76.99%
	smile teeth	81.48%	76.85%	75.93%	57.01%	76.85%
	All	35.72%	33.96%	32.87%	26.67%	35.81%
Our		81.00 %	85.29%	83.82 %	70.59 %	83.94 %
VidTIMIT		65.51%	69.53%	50.14%	50.14%	53.81%

3.4   Experiment III - Computational Complexity

The average computation times of color-HIP+SURF, gradient-HIP+SURF, SIFT, SURF, and LBP are compared in Table 3 on the databases. Note that the proposed feature extraction algorithm is not optimized. LBP is the most efficient feature extraction method, but LBP has worse recognition results compared to color-HIP+SURF or gradient-HIP +SURF. Although color-HIP+SURF and gradient-HIP+SURF take twice the times needed by SURF, the recognition results of SURF are the worst and easily effected by noise. SIFT is quite time consuming even though it gets close performance to color-HIP+SURF and gradient-HIP+SURF. Considering both recognition rates and computation speed, color-HIP+SURF and gradient-HIP+SURF are the best local features compared to SIFT, SURF, and LBP.

Table 3. Comparison of average computation times (seconds) on the three test database

Features	color-HIP+SURF	gradient-HIP+SURF	SIFT	SURF	LBP
Our	0.3729	0.3092	3.9753	0.1681	0.0214
VidTIMIT	0.3017	0.2524	5.3559	0.1389	0.0143
CVL	0.4670	0.3447	4.9103	0.1550	0.0123

Table 4. Average numbers of interest points detected by different interest point detector form our database

Interest point detector	color-HIP	gradient-HIP	DoG(SIFT)	Hessian Matrix(SURF)
Avg. #	28.09	18.99	68.73	10.32

3.5   Experiment IV - Impact of the Number and Location of Interest Points in the Recognition Results

In this section, we discuss the impact of the number and location of interest points extracted from face images in our database. LBP is not discussed because it is a dense matching local feature. Average numbers of interest points detected by color-HIP+SURF, gradient-HIP+SURF, SIFT, and SURF in a 100×100 face image are reported in Table 4. Our proposed two local features using HIP to detect interest points, color-HIP+SURF and gradient-HIP+SURF get more interest points than SURF but fewer than SIFT. Although the color-HIP+SURF, gradient-HIP+SURF use SURF as the descriptor to describe the detected interest points, more interest points are detected than the original SURF, therefore, color-HIP+SURF, gradient-HIP+SURF get better recognition rates than SURF. Although color-HIP+SURF and gradient-HIP+SURF features detect fewer interest points than SIFT, the interest points are mainly located on the eye and mouth areas as shown in Figure 7, which are the most important recognition areas in face recognition [5]. Therefore, the color-HIP+SURF, gradient-HIP+SURF lead to better face recognition rates than SIFT. The face recognition results in Table 1 and Table 2 validate our above conclusion and analysis.

We could also find the below results. SIFT extracts more interest points and some of them have no contribution for face recognition, this is why extracting SIFT feature

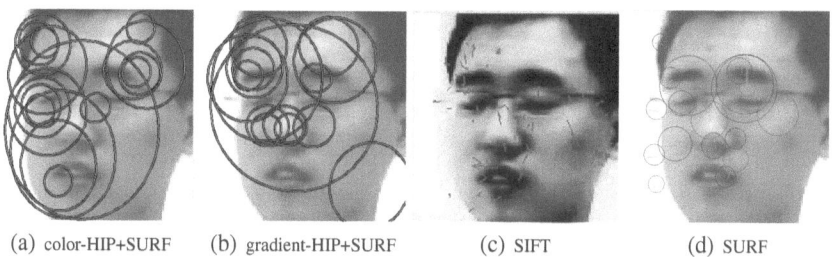

(a) color-HIP+SURF   (b) gradient-HIP+SURF   (c) SIFT   (d) SURF

Fig. 7. The interest points detected by color-HIP, gradient-HIP, SIFT, and SURF from one face in our database

needs more time consuming in Experiment III and SIFT feature could not perform better than color-HIP+SURF and gradient-HIP+SURF in Experiment I and II. Because SURF gets fewer interest points than color-HIP+SURF, gradient-HIP+SURF and SIFT, SURF features will fail when the face images are low resolution. But color-HIP+SURF, gradient-HIP+SURF and SIFT work well on the low resolution face images.

4   Conclusion and Future Work

In this paper, we proposed a novel face feature HIP+SURF to encode face images. HIP is used to extract interest point locations from face images, and SURF is used to generate descriptors of these interest points for a robust feature representation. The hybrid HIP+SURF face feature was tested on our database, VidTIMIT database and CVL database. Experimental results suggest that compared with other local face features, HIP+SURF produces good face recognition accuracy. HIP+SURF feature also gets robust results under noisy conditions. In future work, we will extend the proposed HIP+SURF face feature to low resolution faces taken under more challenging conditions, *e.g.,* outdoor scenes and surveillance.

Acknowledgments.   This work is funded by the National Basic Research Program of China (No. 2010CB327902), the National High Technology Research and Development Program of China (No. 2011AA010502), the National Natural Science Foundation of China (No. 61005016, No. 61061130560), the Fundamental Research Funds of Beihang University and the Central Universities.

References

1. Yan, S., Wang, H., Tang, X., Huang, T.: Exploring feature descritors for face recognition. In: Proc. ICASSP, pp. 629–632 (2007)
2. Liu, X., Chen, T., Thornton, S.M.: Eigenspace updating for non-stationary process and its application to face recognition. Pattern Recognition, 1945–1959 (2003)
3. Lu, J., Plataniotis, K.N., Venetsanopoulos, A.N.: Regularization studies on lda for face recognition. In: Proc. ICIP, pp. 63–66 (2004)
4. Lowe, D.G.: Distinctive image features from scale-invariant keypoints. IJCV, 91–110 (2004)
5. Ahonen, T., Hadid, A., Pietikainen, M.: Face description with local binary patterns: Application to face recognition. IEEE Trans. PAMI 28(12), 297–301 (2006)
6. Bay, H., Ess, A., Tuytelaars, T., Gool, L.V.: Surf: Speeded up robust features. CVIU 110(3), 346–359 (2008)
7. van de Sande, K.E.A., Gevers, T., Snoek, C.G.M.: Evaluating color descriptors for object and scene recognition. IEEE Trans. PAMI 32(9), 1582–1596 (2009)
8. Cai, J., Zha, Z., Zhao, Y., Wang, Z.: Evaluation of histogram based interest point detector in web image classification and search. In: Proc. ICME, pp. 613–618 (2010)
9. Mikolajczyk, K., Tuytelaars, T., Schmid, C., Zisserman, A., Matas, J., Schaffalitzky, F., Kadir, T., Gool, L.V.: A comparison of affine region detectors. IJCV 65(30), 43–72 (2005)
10. Harris, C., Stephens, M.: A combined corner and edge detection. IEEE Trans. PAMI, 147–151 (1988)
11. Mikolajczyk, K., Schmid, C.: Scale and affine invariant interest point detectors. IJCV 60(1), 63–86 (2005)

12. Kadir, T., Zisserman, A., Brady, M.: An Affine Invariant Salient Region Detector. In: Pajdla, T., Matas, J(G.) (eds.) ECCV 2004. LNCS, vol. 3021, pp. 228–241. Springer, Heidelberg (2004)
13. Maver, J.: Self-similarity and points of interest. IEEE Trans. PAMI 32(7), 1211–1226 (2010)
14. Mikolajczyk, K., Schmid, C.: A performance evaluation of local descriptors. IEEE Trans. PAMI 10(27), 1615–1630 (2005)
15. Quelhas, P., Monay, F., Odobez, J.M., Gatica-Perez, D., Tuytelaars, T., Gool, L.V.: Modeling scenes with local descriptors and latent aspects. In: Proc. ICCV, pp. 883–890 (2005)
16. Dalal, N., Triggs, B.: Histograms of oriented gradients for human detection. In: Proc. CVPR, pp. 886–893 (2005)
17. Lee, W., Chen, H.: Histogram-based interest point detectors. In: Proc. CVPR, pp. 1590–1596 (2009)
18. Viola, P., Jones, M.: Robust real-time face detection. In: Proc. ICCV, pp. 590–595 (2001)
19. Comaniciu, D., Ramesh, V., Meer, P.: Real-time tracking of non-rigid objects using mean shift. In: Proc. CVPR, pp. 142–149 (2000)
20. Sanderson, C., Paliwal, K.K.: Polynomial features for robust face authentication. In: Proc. ICIP, pp. 997–1000 (2002)
21. Kovac, J., Peer, P., Solina, F.: Illumination independent color-based face detection. In: Proc. ISPA, pp. 510–515 (2003)
22. Li, S.Z., Jain, A.K.: Handbook of Face Recognition. Springer (2005)

Ontology Querying Support in Semantic Annotation Process

Miha Grčar, Vid Podpečan, Borut Sluban, and Igor Mozetič

Jožef Stefan Institute, Jamova c. 39, 1000 Ljubljana, Slovenia
{miha.grcar,vid.podpecan,borut.sluban,igor.mozetic}@ijs.si

Abstract. In this paper, we present an approach to ontology querying for the purpose of supporting the semantic annotation process. We present and evaluate two algorithms, (i) a baseline algorithm and (ii) a graph-based algorithm based on the bag-of-words text representation and PageRank. We evaluate the two approaches on a set of semantically annotated geospatial Web services. We show that the graph-based algorithm significantly outperforms the baseline algorithm. The devised solution is implemented in Visual OntoBridge, a tool that provides an interface and functionality for supporting the user in the semantic annotation task. The improvement over the baseline is also reflected in practice.

1 Introduction and Motivation

Semantic annotations are formal, machine-readable descriptions that enable efficient search and browse through resources and efficient composition and execution of Web services. It this work, the semantic annotation is defined as a set of interlinked domain-ontology elements associated with the resource being annotated. For example, let us assume that our resource is a database table. We want to annotate its fields in order to provide compatibility with databases from other systems. Further on, let us assume that this table has a field called "employee_name" that contains employee names (as given in Fig. 1, left side). On the other hand, we have a domain ontology containing knowledge and vocabulary about companies (an excerpt is given in Fig. 1, right side). In order to state that our table field in fact contains employee names, we first create a variable for the domain-ontology concept *Name* and associate it with the field. We then create a variable for an instance of *Person* and link it to the variable for *Name* via the *hasName* relation. Finally, we create a variable for an instance of *Company* and link it to the variable for *Person* via the *hasEmployee* relation. Such annotation (shown in the middle in Fig. 1) indeed holds the desired semantics: the annotated field contains names of people which some company employs (i.e., names of employees).

Note that it is possible to instantiate any of the variables with an actual instance representing a real-world entity. For example, the variable *?c* could be replaced with an instance representing an actual company such as, for example, *Microsoft ∈ Company*. The annotation would then refer to "names of people employed at Microsoft".

P. Anthony, M. Ishizuka, and D. Lukose (Eds.): PRICAI 2012, LNAI 7458, pp. 76–87, 2012.

The annotation of a resource is a process in which the user (i.e., the domain expert) creates and interlinks domain-ontology instances and variables in order to create a semantic description for the resource in question. Formulating annotations in one of the ontology-description languages (e.g., WSML [1]) is not a trivial task and requires specific expertise.

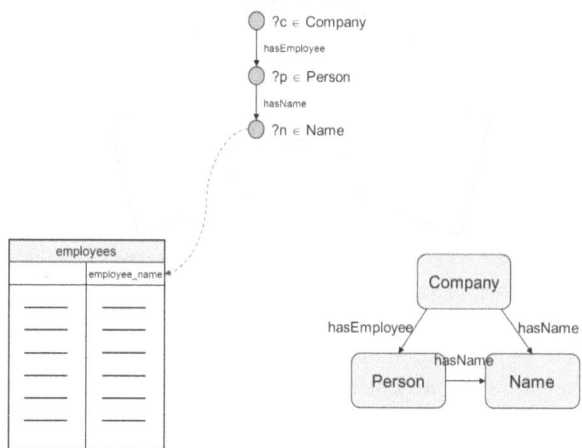

Fig. 1. Annotation as a "bridge" between a resource and the domain ontology

In the European projects SWING[1] and ENVISION[2], we have for this reason developed Visual OntoBridge (VOB) [2], a system that provides a graphical user interface and a set of machine learning algorithms that support the user in the annotation task. VOB allows the user to (i) visualize the resource and the domain ontology (much like this is done in Fig. 1), (ii) create variables by clicking on the domain-ontology concepts, and (iii) interlink the variables and/or instances by "drawing" relations between them.

In addition, the user is able to enter a set of natural-language (Google-like) queries, according to which the system provides a set of "building blocks" that can be used for defining the annotation.

The main purpose of this paper is to present and evaluate the developed ontology querying facilities implemented in VOB. The paper is organized as follows. In Section 2, we present two approaches to ontology querying, the baseline approach and the proposed graph-based approach. We evaluate these two approaches in Section 3 and present our conclusions in Section 4. We discuss some related work in Section 5.

---

[1]   Semantic Web Services Interoperability for Geospatial Decision Making (FP6-26514), http://138.232.65.156/swing/
[2]   Environmental Services Infrastructure with Ontologies (FP7-249120), http://www.envision-project.eu/

2   Ontology Querying Approach

Establishing annotations manually is not a trivial task, especially if the domain ontology contains a large number of entities and/or the user is not fully familiar with the conceptualizations in the ontology. VOB provides functionality for querying the domain ontology with the purpose of finding the appropriate concepts and triples. A triple in this context represents two interlinked instance variables (e.g., *?Company* hasEmployee *?Person*) and serves as a more complex building block for defining semantic annotations.

The process of querying the domain ontology is as follows. The user enters a set of Google-like natural-language queries. The system then provides the user with two lists—the list of proposed concepts and the list of proposed triples. The user inspects the two lists, from top to bottom, and selects the relevant entities. If the required concepts/triples are not found at the top of the list, the user should consider reformulating the queries. The selected concepts and triples are transferred to the graphical annotation editor, where the user is able to revise and extend the annotation as required.

VOB employs text mining techniques, the PageRank algorithm, and consults a Web search engine to populate the two lists of recommended building blocks. In the following sections, we first present two important technical aspects of our ontology querying approach and later on discuss the developed ontology querying algorithms in more details.

2.1   Text Preprocessing and PageRank

In this section, we present two important aspects of our ontology querying procedure, i.e., the bag-of-words vector representation of documents and the (Personalized) PageRank algorithm.

Bag-of-Words Vectors and Cosine Similarity

Most text mining approaches rely on the bag-of-words vector representation of documents. To convert text documents into their bag-of-words representation, the documents are first tokenized, stop words are removed, and the word tokens are stemmed [4, 5]. N-grams (i.e., word sequences of up to a particular length) can be considered in addition to unigrams [17]. Infrequent words and terms (n-grams) are removed from the vocabulary (regularization). The remaining words and terms represent the dimensions in a high-dimensional bag-of-words space in which every document is represented with a (sparse) vector. The words and terms in this vector are weighted according to the TF-IDF weighting scheme (TF-IDF stands for "term frequency—inverse document frequency"; see http://en.wikipedia.org/wiki/Tf*idf). A TF-IDF weight increases proportionally to the number of times a word appears in the document, but is decreased according to the frequency of the word in the entire document corpus.

Cosine similarity is normally used to compare bag-of-words vectors. It measures the cosine of the angle between two vectors. Note that cosine similarity is equal to dot product, provided that the two vectors are normalized in the Euclidean sense.

PageRank and Personalized PageRank

PageRank [6] is a link analysis algorithm that assigns a numerical weighting to each element of a set of interlinked documents, such as the World Wide Web, with the purpose of "measuring" its relative importance within the set. The algorithm can be applied to any collection of entities with reciprocal references.

PageRank effectively computes the importance of a vertex in the graph with respect to a set of source vertices. The way PageRank is typically used, every vertex is considered a source vertex. This means that we are interested in the importance of vertices in general (i.e., with respect to the entire set of vertices). Note that it is possible to modify the algorithm to limit the set of source vertices to several selected vertices or only one vertex. This variant of PageRank is called Personalized PageRank (P-PR). "Personalized" in this context refers to using a predefined set of nodes as the starting nodes for random walks. At each node, the random walker decides whether to teleport back to the source node (this is done with the probability $1 - d$ where d is the so-called damping factor) or to continue the walk along one of the edges. The probability of choosing a certain edge is proportional to the edge's weight compared to the weights of the other edges connected to the node. In effect, for a selected source node i in a given graph, P-PR computes a vector of probabilities with components $PR_i(j)$, where j is one of the nodes in the graph. $PR_i(j)$ is the probability that a random walker starting from node i will be observed at node j at an arbitrary point in time.

2.2   Baseline Ontology Querying Algorithm

In [3], we presented and evaluated several term matching techniques that serve as the basis for automating the annotation process. To produce the two lists of recommendations as discussed in the previous section, it is possible to directly apply the term matching techniques. The algorithm is as follows:

1. Each concept and each possible domain-relation-range triple in the domain ontology is grounded through a Web search engine [3]. Grounding a term means collecting a set of documents and assigning them to the term. In our case, the terms are the concept and relation labels in the domain ontology. With the ontology being grounded, it is possible to compare a natural-language query to the grounded domain-ontology entities. To ground a concept, the search engine[3] is queried with the corresponding concept label. To ground a triple, on the other hand, the search engine is queried with the search term created by concatenating the label of the relation domain, the label of the relation, and the label of the relation range, respectively.
2. The groundings are converted into TF-IDF bag-of-words vectors [5]. Each vector is labeled with the corresponding domain ontology entity (either the concept or the triple label). These vectors constitute the training set (i.e., the set of labeled examples).

---

[3]   We use the Yahoo search engine through their API.

3. The training set is used to train the centroid classifier [7]. Each centroid is computed as the l^2-normalized sum of the corresponding TF-IDF vectors.
4. The set of queries, provided by the user, is first grounded through a Web search engine. For each query, the corresponding centroid TF-IDF vector is computed. These TF-IDF vectors constitute the test set (i.e., the set of unlabeled examples).
5. Given a bag-of-words vector from the test set, the centroid classifier is employed to assign a classification score to each target class, that is, to each ontology entity. These scores are aggregated over the entire set of query vectors.

Given the set of bag-of-words vectors representing the user's queries, the classifier is thus able to sort the domain ontology concepts and triples according to the relevance to the queries. This gives us the two required lists of annotation building blocks: the list of concepts and the list of triples.

2.3   Incorporating Ontology Structure: The OntoBridge Approach

To establish the baseline discussed in the previous section, we treated the domain ontology as a flat list of entities. What we did not take into account is that these entities are in fact interlinked. This means that the domain ontology can be represented as a graph in which vertices are entities and edges represent links. In this section, we show how we can couple text similarity assessments with PageRank to exploit the ontology structure for determining relevant ontology entities.

To employ PageRank, the domain ontology is first represented as a graph. This can be done in numerous different ways. Naively, we could represent each concept with a vertex and interconnect two vertices with an undirected edge if there would exist at least one domain-relation-range definition involving the two concepts. However, since only the concepts would then be represented with vertices and would thus be the only entities able to "accumulate" rank, we would not be able to rank the triples. With a slightly more sophisticated transformation approach, it is possible to include the triples as well. This type of transformation is illustrated in Fig. 2. We create additional vertices (i.e., vertices representing triples; drawn as squares and triangles in the figure) to "characterize" all possible relations between the two concepts. We also include vertices representing triples based on inverse relations (drawn as triangles in the figure) even though they are not explicitly defined in the domain ontology. The reason for this is that we do not want the random walker to reach a triple vertex and then head back again; we want it to reach the other concept through a pair of directed edges.

In more details, the proposed ontology-to-graph transformation process is as follows:

1. Represent each concept with a vertex.
2. Represent each triple c_1-r-$c_2 \in \mathbf{T}$, where $\mathbf{T}$ is the set of triples in the domain ontology, with two vertices: one representing c_1-r-c_2 and one representing the corresponding inverse relation, c_2-r^{-1}-c_1.
3. For each pair of concepts c_1, c_2 and for each relation r such that c_1-r-$c_2 \in \mathbf{T}$, do the following:

- Connect the vertex representing c_1 to the vertex representing $c_1\text{-}r\text{-}c_2$ with a directed edge and weight it with $C(\mathbf{Q}, c_1\text{-}r\text{-}c_2)$. Here, $\mathbf{Q} = \{q_1, q_2, q_3...\}$ is a set of natural-language queries and $C(\mathbf{Q}, c_1\text{-}r\text{-}c_2)$ is computed as $\Sigma_{q \in \mathbf{Q}} C(q, c_1\text{-}r\text{-}c_2)$. $C(a, b)$ refers to cosine similarity between the centroid of groundings of concept/relation a and the centroid of groundings of concept/relation b. A centroid is computed by first converting the corresponding groundings to TF-IDF feature vectors and then computing the l^2-normalized sum of these feature vectors.
- Connect the vertex representing $c_1\text{-}r\text{-}c_2$ to the vertex representing c_2 with a directed edge and weight it with 1.
- Connect the vertex representing c_2 to the vertex representing $c_2\text{-}r^{-1}\text{-}c_1$ with a directed edge and weight it with $C(\mathbf{Q}, c_1\text{-}r\text{-}c_2)$.
- Connect the vertex representing $c_2\text{-}r^{-1}\text{-}c_1$ to the vertex representing c_1 with a directed edge and weight it with 1. Note that since this is the only edge going out of this vertex, its weight can in fact be an arbitrary positive value. This is because PageRank normalizes the weights of the outgoing edges at each vertex so that they sum up to 1.

4. Represent each bag-of-words vector q_i, representing the test set $\mathbf{Q} = \{q_1, q_2, q_3...\}$, with a vertex. Note that the test set represents the user queries.
5. For each bag-of-words vector q_i representing the query and each concept c_j, if $C(q_i, c_j) > 0$, create a directed edge from q_i to c_j and weight it with $C(q_i, c_j)$.

This process is illustrated in Fig. 2 where w_i represent the weights computed in Step 3 of the presented ontology-to-graph transformation process.

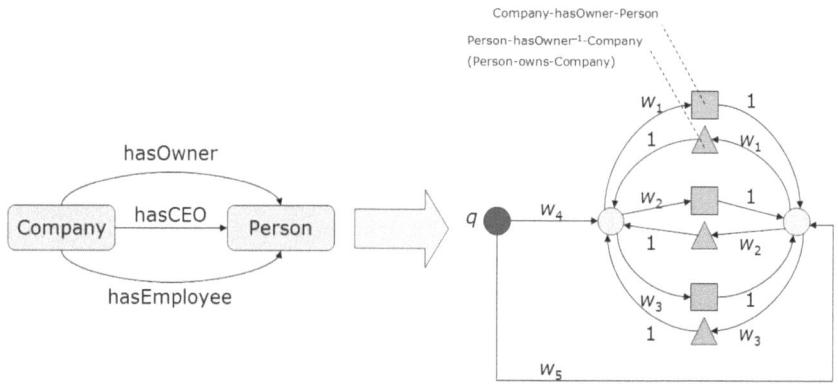

Fig. 2. Representing ontologies as graphs

When the graph is created and properly weighted, we run PageRank to rank vertices (i.e., concepts and triples) according to the relevance to the query. The vertices representing the query are therefore used as the source vertices for PageRank. Note that a triple $c_1\text{-}r\text{-}c_2 \in \mathbf{T}$ "accumulates" the ranking score in two different vertices: in the vertex representing $c_1\text{-}r\text{-}c_2$ and in the vertex representing $c_2\text{-}r^{-1}\text{-}c_1$. It is thus necessary to sum the ranking scores of these two vertices to obtain the ranking score of the corresponding triple.

With the discussed procedure, every concept and every triple is ranked by Page-Rank. We can therefore populate the two lists of annotation building blocks and present these to the user.

3   Evaluation of the Ontology Querying Algorithms

We evaluated our approach to ontology querying in the context of a project commit-ted to develop an infrastructure for handling geospatial Web services. In the devised system, geo-data is served by a database of spatial-information objects through stan-dardized interfaces. One of such standard interfaces, defined by the Open Geospatial Consortium (OGC), is the Web Feature Service (WFS) [16]. WFS are required to describe the objects that they provide (e.g., rivers, roads, mountains...). These objects are also termed "features". Each feature is described with a set of attributes (e.g., water bodies can have depth, temperature, water flow...).

We used a set of WFS schemas, enriched with the golden-standard annotations and user queries, to evaluate the developed ontology querying algorithms. The evaluation process and the evaluation results are presented in the following sections.

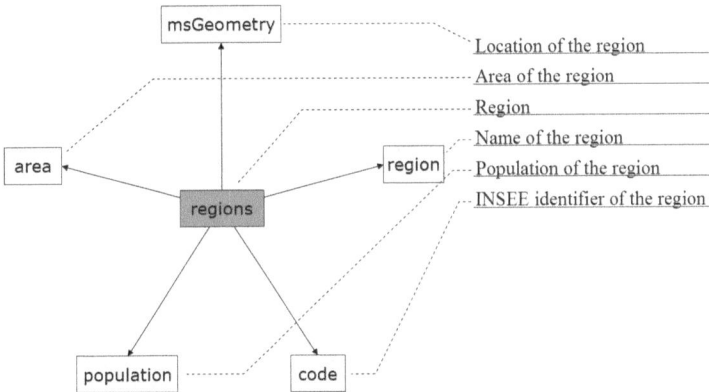

Fig. 3. The golden-standard acquisition form for the feature type "regions". The feature type (green box) with all its attributes (yellow boxes) is visualized in the left-hand side, the corres-ponding queries, provided by one of the participants, can be seen in the right-hand side.

3.1   Golden Standard

For the experiments, we acquired a set of Web Feature Services (WFS). Each WFS was accompanied with the corresponding semantic annotation and several sets of user queries. The service schemas were annotated with the SWING ontology (available at http://first-vm2.ijs.si/envision/res/swing.n3). It contains 332 concepts, 141 relations, and 4,362 domain-relation-range triples (taking the basic triple-inference rules into account). We asked the domain experts at Bureau of Geological and Mining Research (BRGM, France) to provide us with natural-language queries with which they would hope to retrieve relevant building blocks for the annotations. For this purpose, we gave each of the participating domain experts a set of forms presenting the WFS schemas. A participant had to describe each feature type with a set of English queries, one query per attribute and one additional query for the feature type itself. Fig. 3 shows one of such golden-standard acquisition forms.

We received input from 3 domain experts, each assigning queries to 7 feature types (41 queries altogether by each of the participants). We have identified 114 concepts and 96 triples (unique in the context of the same feature type) relevant for annotating the feature types involved in the golden-standard acquisition process. Since the acquired golden standard thus contained both, the queries and the corresponding building blocks, we were able to assess the quality of the ontology querying algorithms by "measuring" the amount of golden-standard building blocks discovered in the domain ontology, given a particular set of queries. We measured the area under the Receiver Operating Characteristic (ROC) curve to evaluate the lists produced by the algorithm. We discuss the evaluation process and present the results in the following section.

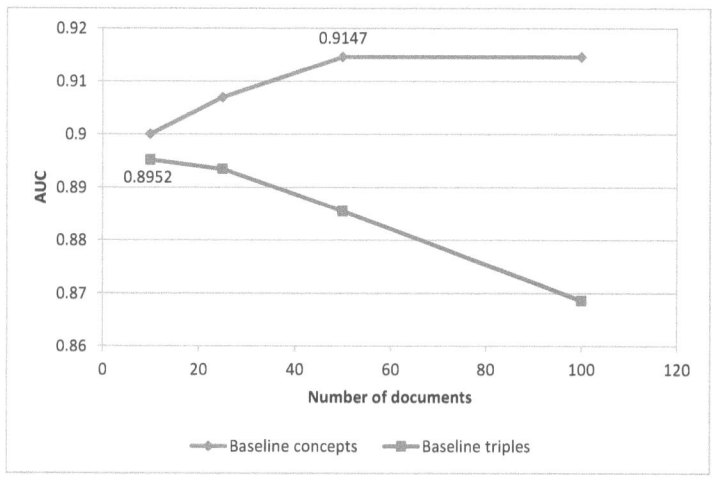

Fig. 4. Evaluation results for the baseline methods

3.2   Evaluation of the Baseline Algorithm

In this section, we establish the baselines and determine the setting in which the baseline algorithm, presented in Section 2.2, performs best. Through the evaluation, we determined the number of search-result snippets used for grounding domain ontology entities and user queries. We experimented with 10, 25, 50, and 100 documents per grounding.

The results are shown in Fig. 4. The chart in the figure presents the evaluation results for the list of proposed concepts and the list of proposed triples. Both series show the average area under the ROC curve (y axis) with respect to the number of documents per grounding (x axis).

From the results, we can conclude that the concepts—as well as the queries when used for ranking the concepts—should be grounded with at least 50 documents

(91.47% AUC). As we can see from the chart, at around 50 documents, all available useful information is already contained in the collected documents. On the other hand, the triples—as well as the queries when used for ranking the triples—should be grounded with only around 10 documents (89.52% AUC). We believe this is because the triples are more precisely defined than the concepts (i.e., the corresponding search terms contain more words), which results in a smaller number of high-quality search results.

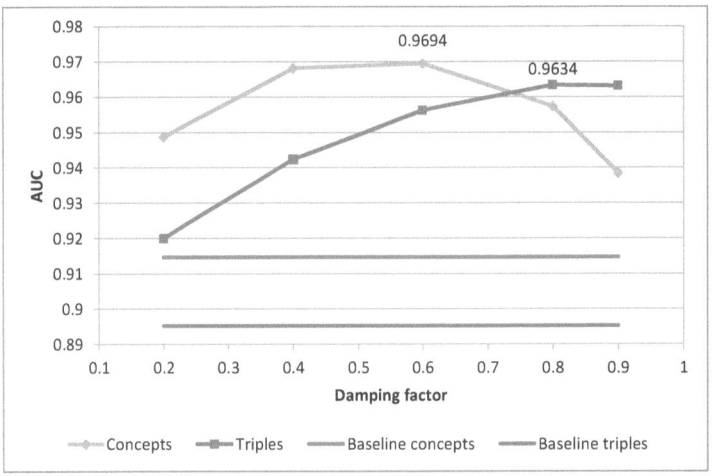

Fig. 5. Evaluation results for the graph-based methods

3.3   Evaluation of the Graph-Based Algorithm

When evaluating the graph-based algorithm, the most important parameter to tune is the PageRank damping factor. We experimented with the damping factor values of 0.2, 0.4, 0.6, 0.8, and 0.9. The results are presented in Fig. 5. The chart in the figure presents the evaluations result for the list of proposed concepts and the list of proposed triples. Both series represent the average area under the ROC curve (y axis) with respect to the value of the damping factor (x axis). The chart also shows the baselines (see the previous section).

When evaluating the baseline algorithm, we learned that the concepts should be grounded with 50 documents each, so should the queries when used to rank the concepts. On the other hand, the triples should be grounded with only 10 documents each, so should the queries when used to rank the triples. The evaluation of the graph-based algorithms fully confirms these findings at low damping factor values. This is expected because low damping factor values mean putting less emphasis on the structure—the random walker "gets tired" after only a few steps and "jumps" back to a source vertex. However, as we increase the damping factor towards the values at which the graph-based algorithms perform best, we achieve better results when simply grounding concepts, triples, and queries with 50 documents each. Note that this grounding size setting was also used for computing the results in Fig. 5.

The damping factor should be set to 0.6 for the concepts and 0.8 for the triples. This means that we can either run PageRank twice or set the damping factor to 0.7 to increase the speed at the slight expense of quality on both sides. The rewarding fact is that we managed to significantly beat the baselines. We have increased the average AUC for 5.48% on the concepts and for 6.82% on the triples. This presents a big difference. For example, if the correct triples were distributed amongst the top 597 of 4,362 suggestions with the baseline algorithm, they will now be distributed amongst the top 319 suggestions (almost half less). Also, we believe that the user will have to inspect far less than 319 items to find the required building blocks as he will be able to interact with the system (i.e., reformulate queries to direct the search). To support this claim, we computed the average AUC by taking, for each annotation, only the most successful annotator into account (i.e., the annotator that formulated the query resulting in the highest AUC). In this modified setting, the average AUC on the triples rose to 98.15%. This reduces the number of items that need to be inspected from 319 to 161 (of 4,362). The improvement over the baseline method is also reflected in practice. Several user queries and the corresponding retrieved concepts (top 10 according to the relevance) from the SWING ontology are shown in Table 1.

Table 1. Several user queries and the corresponding retrieved concepts

User query	Retrieved concepts
"rare birds, sequoia forests, natural parks"	ImportantBirdArea, AviFauna, Bird, NationalProgram, QuarrySite, Lake, NationalCoordinator, ProtectedArea, NaturalSite, Mammal…
"protected fauna and flora in France"	WildFauna, WildFlora, Fauna, Flora, ProtectedArea, AviFauna, Znieff [4], ZnieffTypeI, ZnieffTypeII, Natural-Site…
"locaiton[5] of open-pit mine"	AllowedMiningDepth, QuarrySite, Legislation, QuarryAdministration, QuarrySiteManagement, MineralResource, QuarryLocation, MineralProperty, ConstructionApplication, Location…

4   Conclusions

Semantic annotations are formal, machine-readable descriptions that enable efficient search and browse through resources and efficient composition and execution of Web services. Formulating annotations in one of the ontology-description languages is not

---

[4] ZNIEFF stands for "Zone naturelle d'intérêt écologique, faunistique et floristique" and denotes protected natural areas in France. This example demonstrates that we are able to use descriptive queries to discover concepts labeled with acronyms.

[5] This example shows that our ontology-querying approach is even resilient to typos.

a trivial task and requires specific expertise. In this paper, we presented an approach to ontology querying for the purpose of supporting the semantic annotation process. In the approach, we use the bag-of-words text representation, employ a Web search engine to ground ontology objects and user queries with documents, and run Page-Rank to take the ontology structure into account.

We evaluated the approach in the context of geospatial Web services. In the evaluation process, we used a set of Web Feature Service (WFS) schemas, enriched with the golden-standard annotations and user queries. In the experiments, we varied the number of grounding documents and PageRank damping factor. We concluded that it is best to ground the concepts, triples, and user queries with 50 documents each and to set the damping factor to 0.7. The achieved AUC for retrieving concepts was 96.94% and for retrieving triples, 96.34%. With these results, we managed to achieve a significant improvement over the baselines.

5   Related Work

The main contribution of this paper is a novel ontology querying algorithm. In this section, we overview several techniques that can be used to assess the relevance of an object (with respect to another object or a query) or the similarity between two objects in a network. Some of these techniques are: spreading activation [8], hubs and authorities (HITS) [9], PageRank and Personalized PageRank [6], SimRank [10], and diffusion kernels [11]. These methods are extensively used in information-retrieval systems. The general idea is to propagate "authority" from "query nodes" into the rest of the graph or heterogeneous network, assigning higher ranks to more relevant objects.

ObjectRank [12] employs global PageRank (importance) and Personalized Page-Rank (relevance) to enhance keyword search in databases. Specifically, the authors convert a relational database of scientific papers into a graph by constructing the data graph (interrelated instances) and the schema graph (concepts and relations). To speed up the querying process, they precompute Personalized PageRank vectors for all possible query words. HubRank [13] is an improvement of ObjectRank in terms of space and time complexity without compromising the accuracy. It examines query logs to compute several hubs for which PPVs are precomputed. In addition, instead of precomputing full-blown PPVs, they compute fingerprints [14] which are a set of Monte Carlo random walks associated with a node.

Stoyanovich et al. [15] present a ranking method called EntityAuthority which defines a graph-based data model that combines Web pages, extracted (named) entities, and ontological structure in order to improve the quality of keyword-based retrieval of either pages or entities. The authors evaluate three conceptually different methods for determining relevant pages and/or entities in such graphs. One of the methods is based on mutual reinforcement between pages and entities, while the other two approaches are based on PageRank and HITS, respectively.

Acknowledgements. This work has been partially funded by the European Commission in the context of the FP7 project ENVISION (Environmental Services Infrastructure with Ontologies) under the Grant Agreement No. 249120.

References

1. Roman, D., de Bruijn, J., Mocan, A., Lausen, H., Domingue, J., Bussler, C., Fensel, D.: WWW: WSMO, WSML, and WSMX in a Nutshell. In: Mizoguchi, R., Shi, Z.-Z., Giunchiglia, F. (eds.) ASWC 2006. LNCS, vol. 4185, pp. 516–522. Springer, Heidelberg (2006)
2. Grcar, M., Mladenic, D.: Visual OntoBridge: Semi-automatic Semantic Annotation Software. In: Buntine, W., Grobelnik, M., Mladenić, D., Shawe-Taylor, J. (eds.) ECML PKDD 2009. LNCS (LNAI), vol. 5782, pp. 726–729. Springer, Heidelberg (2009)
3. Grcar, M., Klien, E., Novak, B.: Using Term-Matching Algorithms for the Annotation of Geo-Services. Knowledge Discovery Enhanced with Semantic and Social Information 220, 127–143 (2009)
4. Feldman, R., Sanger, J.: The Text Mining Handbook: Advanced Approaches in Analyzing Unstructured Data. Cambridge University Press, Cambridge (2006)
5. Salton, G.: Automatic Text Processing: The Transformation, Analysis, and Retrieval of Information by Computer. Addison-Wesley Longman Publishing, Boston (1989)
6. Page, L., Brin, S., Motwani, R., Winograd, T.: The PageRank Citation Ranking: Bringing Order to the Web. Technical Report 1999-66. Stanford InfoLab, Stanford, USA (1999)
7. Cardoso-Cachopo, A., Oliveira, A.L.: Empirical Evaluation of Centroid-Based Models for Single-Label Text Categorization. Technical Report 7/2006. Instituto Superior Tecnico, Lisbon, Portugal (2006)
8. Crestani, F.: Application of Spreading Activation Techniques in Information Retrieval. Artificial Intelligence Review 11, 453–482 (1997)
9. Kleinberg, J.M.: Authoritative Sources in a Hyperlinked Environment. Journal of the Association for Computing Machinery 46, 604–632 (1999)
10. Jeh, G., Widom, J.: SimRank: A Measure of Structural Context Similarity. In: Proceedings of KDD 2002, pp. 538–543 (2002)
11. Kondor, R.I., Lafferty, J.: Diffusion Kernels on Graphs and Other Discrete Structures. In: Proceedings of ICML 2002, pp. 315–322 (2002)
12. Balmin, A., Hristidis, V., Papakonstantinou, Y.: ObjectRank: Authority-based Keyword Search in Databases. In: Proceedings of VLDB 2004, pp. 564–575 (2004)
13. Chakrabarti, S.: Dynamic Personalized PageRank in Entity-Relation Graphs. In: Proceedings of WWW 2007, pp. 571–580 (2007)
14. Fogaras, D., Rácz, B.: Towards Scaling Fully Personalized PageRank. In: Leonardi, S. (ed.) WAW 2004. LNCS, vol. 3243, pp. 105–117. Springer, Heidelberg (2004)
15. Stoyanovich, J., Bedathur, S., Berberich, K., Weikum, G.: EntityAuthority: Semantically Enriched Graph-based Authority Propagation. In: Proceedings of the 10th International Workshop on Web and Databases (2007)
16. Open Geospatial Consortium: Web Feature Service Implementation Specification, Version 1.0.0 (OGC Implementation Specification 02-058) (2002)
17. Mladenic, D.: Feature Subset Selection in Text Learning. In: Nédellec, C., Rouveirol, C. (eds.) ECML 1998. LNCS, vol. 1398, pp. 95–100. Springer, Heidelberg (1998)

Learning to Achieve Socially Optimal Solutions in General-Sum Games

Jianye Hao and Ho-fung Leung

Department of Computer Science and Engineering,
The Chinese University of Hong Kong
{jyhao,lhf}@cse.cuhk.edu.hk

Abstract. During multi-agent interactions, robust strategies are needed to help the agents to coordinate their actions on efficient outcomes. A large body of previous work focuses on designing strategies towards the goal of Nash equilibrium under self-play, which can be extremely inefficient in many situations. On the other hand, apart from performing well under self-play, a good strategy should also be able to well respond against those opponents adopting different strategies as much as possible. In this paper, we consider a particular class of opponents whose strategies are based on best-response policy and also we target at achieving the goal of social optimality. We propose a novel learning strategy TaFSO which can effectively influence the opponent's behavior towards socially optimal outcomes by utilizing the characteristic of best-response learners. Extensive simulations show that our strategy TaFSO achieves better performance than previous work under both self-play and against the class of best-response learners.

1 Introduction

Multi-agent learning has been studied extensively in the literature and lots of learning strategies [9,17,2,6,8] are proposed to coordinate the interactions among agents. As multiple agents from different parties may coexist in the same environment, we need robust strategies for efficient coordination among the agents and also effectively responding to other external agents.

The multi-agent learning criteria proposed in [2] require that an agent should be able to converge to a stationary policy against some class of opponent (convergence) and the best-response policy against any stationary opponent (rationality). If both agents adopt rational learning strategies in the context of repeated games and also their strategies converge, then they will converge to the Nash equilibrium of the stage game. Unfortunately, in many cases Nash equilibrium does not guarantee that the agents receive their best payoffs. One typical example is the prisoner's dilemma game shown in Fig. 1(a). By converging to the Nash equilibrium (D, D), both agents obtain a low payoff of 1, and also the sum of the agents' payoffs is minimized. On the other hand, both agents could have received better payoffs by coordinating on the non-equilibrium outcome of (C, C) in which the sum of the players' payoffs is maximized.

P. Anthony, M. Ishizuka, and D. Lukose (Eds.): PRICAI 2012, LNAI 7458, pp. 88–99, 2012.

One major reason for resulting in inefficient Nash equilibria is that the rationality mentioned above is myopic, i.e., the strategies are designed for best response towards stationary opponents only. They do not explicitly leverage the fact that their opponents' behaviors are adaptive and can be affected by the behaviors of their interacting partners as well. To see how the failure to consider this can decrease the agents' payoffs, let us consider the stacklberg game shown in Fig. 1(b). We know that the outcome will converge to the Nash equilibrium (D, L) of the stage game if two agents play this game repeatedly by adopting a best-response strategy against each other. On the other hand, if the left-side agent can realize that its opponent is learning its past behaviors and always making the best response, then it would be better for the left-side agent to take the role of a teacher by choosing action U deliberately. After a while, it is reasonable to expect that its opponent will start responding with action R. Therefore both agents' payoffs are increased by 1 compared with their Nash equilibrium payoffs and also the achieved outcome (U, R) is socially optimal, i.e., the sum of their payoffs is maximized.

From the above example, we can see that the role of teaching can be particularly important for improving the learning performance. When choosing a course of actions, an agent should take into account not only what it learns from its opponent's past behaviors, but also how it can better influence its opponent's future behaviors towards its targeted outcomes. We refer to the learning strategies employing the role of a teacher as *teacher* strategies, and those learnig strategies always making best response to their opponents (e.g., fictitious play [6], Q-learning [17]) as *follower* strategies [4,5]. A number of different teaching strategies [10,11,5] have been proposed aiming at achieving better payoffs for the agents. However, those strategies suffer from at least one of the following drawbacks.

- The punishment mechanism is usually implemented as choosing an action to minimize the payoff of the opponent in the following rounds after its deviation from the target action. The resulting problem is that the opponent may still be punished when it has already been back to choose the target action, thus making it difficult for a follower agent to determine which of its actions is being punished. Therefore the teaching performance can be degraded due to mistaken punishments.
- The agent's own payoff is not taken into consideration when the opponent is being punished, which makes the punishment too costly.
- The agents using those strategies may fail to coordinate their actions (e.g., under self-play) if there exist multiple optimal target solutions.

To solve the above problems, in this paper, we propose a novel learning strategy TaFSO combining the characteristics of both teacher and follower strategies. We are interested in how the opponent's behavior can be influenced towards the coordination on socially optimal outcomes through repeated interactions in the context of repeated games. We consider an interesting variant of sequential play here. In addition to choose from their original action spaces, the agents are given a free option of deciding whether to entrust its opponent to make decision for itself or not (denoted as action F). If agent i asks its opponent j to make decision for itself, then the agents will execute the action pair assigned by agent j in this round. The motivation behind is to determine whether the introduction of the additional action F can improve the coordination between agents towards socially optimal outcomes. Similar idea has been adopted in investigating multi-agent learning in multi-agent resource selection problems [13], which has been shown to be effective for agents to coordinate on optimal utilizations of the resources. Similar with previous work [11,5], the TaFSO strategy is also based on punishment and reward mechanisms, however, all the drawbacks existing in previous work as previously mentioned are remedied. Firstly, in TaFOS, it can be guaranteed that the opponent never perceives the wrong punishment signal and thus the coordination efficiency among agents is greatly improved. Secondly, The agent adopting TaFSO always picks an action in the best response to the opponent's strategy from the set of candidate actions suitable for punishment instead of adopting the minmax strategy, thus the punishment cost is reduced. Thirdly, due to the introduction of action F, the agents are guaranteed to always coordinate on the same optimal outcome even if multiple optimal outcomes coexist. The performance of TaFSO is evaluated under self-play and also against a number of best-response learners. Simulation results show that better performance can be achieved when the agents adopt TaFSO strategy compared with previous work.

The remainder of the paper is structured as follows. The related work is given in Section 2. In Section 3, the learning environment and the learning goal are introduced. In Section 4, we present the learning approach TaFSO in the context of two-player repeated games. Experimental simulations and performance comparisons with previous work are presented in Section 5. In the last section, we conclude our paper and present the future work.

2   Related Work

A number of approaches [16,12] have been proposed targeting at the prisoner's dilemma game only, and the learning goal is to achieve pareto-optimal solution of mutual cooperation instead of Nash equilibrium solution of mutual defection. Besides, there also exists some work [15,1] which address the problem of achieving Pareto-optimal solution in the context of general-sum games as following. Sen et al. [15] propose an innovative expected utility probabilistic learning strategy by incorporating action revelation mechanism. Simulation results show that agents using action revelation strategy under self-play can achieve pareto-optimal outcomes which dominate Nash Equilibrium in certain games, and also the average performance with action revelation is significantly better than Nash Equilibrium solution over a large number of randomly generated game matrices. Banerjee et al. [1] propose the conditional joint action learning strategy (CJAL) under which each agent takes into consideration the probability of an action taken by its opponent given its own action, and utilize this information to make its own decision. Simulation results show that agents adopting this strategy under self-play can learn to converge to the pareto-optimal solution of mutual cooperation in prisoner's dilemma game when the game structure satisfies certain condition. However, all previous strategies are based on the assumption of self-play, and there is no guarantee of the performance against the opponents using different strategies.

There also exists a number of work [10,11,5,14] which assumes that the opponent may adopt different strategies. One representative work is Folk Theorem [14] in the literature of Game Theory. The basic idea of Folk Theorem is that there are some strategies based on punishment mechanism which can enforce desirable outcomes and are also in Nash equilibrium, assuming that all players are perfectly rational. Our focus, however, is to utilize the ideas in Folk theorem to design efficient strategy against adaptive best-response opponents from the learning perspective. Also the strategies we explore need not be in equilibrium in the strict sense, since it is very difficult to construct a strategy which is the best response to a particular learning strategy such as Q-learning. Besides, a number of teacher strategies [10,11] have been proposed to induce better performance from the opponents via punishment mechanism, assuming that the opponents adopt best-response strategies such as Q-learning. Based on the teacher strategy Goldfather++ [11], Crandall and Goodrich [5] propose the strategy SPaM employing both teaching and following strategies and show its better performance in the context of two-player games against a number of best-response learners.

3   Learning Environment and Goal

In this paper, we focus on the class of two-player repeated normal-form games with average payoffs. Formally a two-player normal-form game G is a tuple $\langle N, (A_i), (u_i) \rangle$ where

- $N = \{1, 2\}$ is the set of players.
- A_i is the set of actions available to player $i \in N$.
- u_i is the utility function of each player $i \in N$, where $u_i(a_i, a_j)$ corresponds to the payoff player i receives when the outcome (a_i, a_j) is achieved.

At the end of each round, each agent receives its own payoff based on the outcome and also observes the action of its opponent. Two example normal-form games are already shown in Fig. 1(a) and Fig. 1(b).

We know that every two-player normal-form game has at least one pure/mixed strategy Nash equilibrium. Each agent's strategy in a Nash equilibrium is always its best response to the strategy of the other agent. Though this is the best any rational agent can do myopically, the equilibrium solution can be extremely inefficient in terms of the payoffs the agents receive. In this paper, we target at achieving a more efficient solution from the system's perspective, social optimality, in which the sum of both agents' payoffs is maximized. We are interested in designing a learning strategy such that the agent adopting this strategy can effectively influence its opponent towards cooperative behaviors through repeated

1's payoff, 2's payoff		Player 2's action	
		C	D
Player 1's action	C	3, 3	0, 5
	D	5, 0	1, 1

1's payoff, 2's payoff		Player 2's action	
		L	R
Player 1's action	U	1, 0	3, 2
	D	2, 1	4, 0

(a)   (b)

Fig. 1. Payoff matrices for (a) prisoner's dilemma game, and (b) stackelberg game

interactions, and thus reach socially optimal outcomes finally. We adopt the same setting as [10,5] assuming that the opponents are adaptive best-response learners, which cover a wide range of existing strategies. Specifically in this paper we consider the opponent may adopt one of the following best-response strategies: Q-learning [17], WoLF-PHC [2], and Fictitious play [6]. For our purpose, we do not take into consideration the task of learning the games and assume that the game structure is known to both agents beforehand.

4 TaFSO:A Learning Approach towards Socially Optimal Outcomes

In this section, we present the learning approach TaFSO aiming at achieving socially optimal outcomes. The strategy TaFSO combines the properties of both *teacher* and *follower* strategies. On one hand, its *teacher* component is used to influence its opponent to cooperate and behave in the expected way based on punishment and reward mechanisms; on the other hand, its *follower* component guarantees that the TaFSO agent can obtain as much payoff as possible against its opponent at the same time.

4.1 *Teacher* Strategy of TaFSO

To enable the *teacher* strategy to exert effective influence on the opponent's behavior, we consider an interesting variation of sequential play by allowing entrusting decision to others. During each round, apart from choosing an action from its original action space, every agent is also given an additional option of asking its opponent to make the decision for itself. Whenever the opponent j decides to entrust TaFSO agent i to make decisions (denoted as choosing action F), TaFSO agent i will select an optimal joint action pair (s_1, s_2) and both agents will execute their corresponding actions accordingly. Similar to previous work [11,5], in TaFSO, the optimal joint action pair (s_1, s_2) is defined as the joint action maximizing the product of both player's positive advantages (i.e., the difference between their payoffs in (s_1, s_2) and their minimax payoffs). We assume that every agent will honestly execute the action assigned by its opponent whenever it asks its opponent to do so. If both agents choose action F

simultaneously, then one of them will be randomly picked to make the joint decision.

We can see that the opponent j may obtain higher payoff by deviating from action F to some action from A_j. Therefore an TaFSO agent i needs to enable its opponent j to learn that entrusting the TaFSO agent to make decisions is always its best choice. To achieve this, the TaFSO agent i will teach its opponent by punishment if the opponent j choose actions from A_j. To make it effective, the punishment must exceed the profit of deviating. In other words, the opponent j's any possible gain from its deviation has to be wiped out. The TaFSO agent i keeps the record of the opponent j's accumulated gains G_j^t from deviation by each round t and updates G_j^t as follows (Fig. 2).

- If the opponent j entrusts the TaFSO agent i to make decision for itself (i.e., $a_j^t = c$), and also its current gain $G_j^t \geq 0$, then its gain remains the same in the next round $t+1$.
- If the opponent j asks the TaFSO agent i to make decision for itself (i.e., $a_j^t = c$), and also its current gain $G_j^t < 0$, this indicates it suffers from previous deviations. In this case, we forgive its previous deviations and set $G_j^{t+1} = 0$.
- If the opponent j chooses its action independently and also $G_j^t > 0$, then its gain is updated as $G_j^{t+1} = \max\{G_j^t + u_j(a_i^t, a_j^t) - u_j(s_i, s_j), \epsilon\}$. If it obtains higher payoff than that in the optimal outcome (s_i, s_j) from this deviation, then its gain will be increased, and vice versa. Besides, its total gain by round $t+1$ cannot become smaller than zero since it deviates in the current round.
- If the opponent j chooses its action independently and also $G_j^t \leq 0$, its gain is updated as $G_j^{t+1} = u_j(a_i^t, a_j^t) - u_j(s_i, s_j) + \epsilon$. That is, we forgive the opponent j if since it learns the lesson by itself by suffering from previous deviations ($G_j^t \leq 0$), and its previous gain is counted only as ϵ.

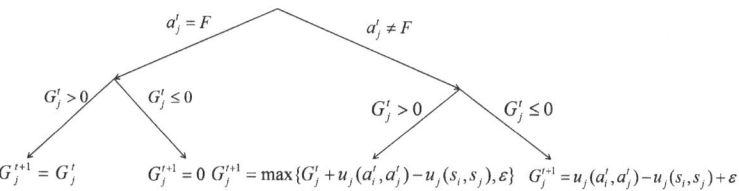

Fig. 2. The rules of calculating the opponent j's accumulated gain G_j^t by each round t. Each path in the tree represents one case for calculating G_j^t.

Based on the above updating rules of G_j^t, the TaFSO agent needs to determine which action is chosen to punish the opponent j. To do this, the TaFSO agent i keeps a teaching function $T_i^t(a)$ for each action $a \in A_i \cup \{F\}$ in each round t, indicating the action's punishment degree. From the Folk theorem [14] we know that the minimum payoff the TaFSO agent can guarantee that the opponent j

receives via punishment is its minimax payoff $minimax_j$. If $G_j^t > minimax_j$, we can only expect to exert $minimax_j$ amount of punishment on the opponent j since the opponent j can always obtain at least $minimax_j$ by playing its maxmin strategy; if $G_j^t \leq minimax_j$, then punishing the opponent j by the amount of G_j^t is already enough. Therefore the punishment degree of action a in round t is evaluated as the difference between the expected punishment on the opponent j by choosing a and the minimum value between G_j^t and the punishment degree under the minimax strategy. Formally the teaching function $T_i^t(a)$ is defined as follows.

$$T_i^t(a) = u_j(s_1, s_2) - E[u_j(., a)] - \min\{G_j^t, u_j(s_1, s_2) - minimax_j\} \quad (1)$$

where $E[u_j(., a)]$ is the expected payoff that the opponent j will obtain if the TaFSO agent i chooses action a based on the past history, and $minimax_j$ is the minimax payoff of the opponent j.

If $T_i^t(a) \geq 0$, it means it is sufficient to choose action a to punish the opponent j. We can see that there can exist multiple candidate actions for punishing the opponent j. If $T_i^t(a) < 0$, $\forall a \in A_i$, then we only choose the action with the highest $T_i^t(a)$ as the candidate action. Overall the set C_i^t of candidate actions for punishment is obtained based on TaFSO's *teacher* strategy in each round t. Based on this information, the TaFSO agent i chooses an action from this set C_i^t to punish its opponent according to its *follower* strategy, which will be introduced next.

4.2 *Follower* Strategy of TaFSO

The *follower* strategy of TaFSO is used to determine the best response to the strategy of the opponent if the opponent chooses its action from its original action space. Here we adopt the Q-learning algorithm [17] as the basis of the *follower* strategy. Specifically the TaFSO agent i holds a Q-value $Q_i^t(a)$ for each action $a \in A_i \cup \{F\}$, and gradually updates its Q-value $Q_i^t(a)$ for each action a based its own payoff and action in each round. The Q-value update rule for each action a is as follows:

$$Q_i^{t+1}(a) = \begin{cases} Q_i^t(a) + \alpha_i(u_i^t(O) - Q_i^t(a)) & \text{if } a \text{ is chosen in round } t \\ Q_i^t(a) & \text{otherwise} \end{cases} \quad (2)$$

where $u_i^t(O)$ is the payoff agent i obtains in round t under current outcome O by taking action a. Besides, α_i is the learning rate of agent i, which determines how much weight we give to the newly acquired payoff $u_i^t(O)$, as opposed to the old Q-value $Q_i^t(a)$. If $\alpha_i = 0$, agent i will learn nothing and the Q-value will be constant; if $\alpha_i = 1$, agent i will only consider the newly acquired information $u_i^t(O)$.

In each round t, the TaFSO agent i chooses its action based on the ϵ-greedy exploration mechanism as follows. With probability $1 - \epsilon$, it chooses the action with the highest Q-value from the set C_i^t of candidate actions, and chooses one action randomly with probability ϵ from the original action set $A_i \cup F$. The value of ϵ controls the exploration degree during learning. It initially starts at a high value and decreased gradually to nothing as time goes on. The reason is that initially the approximations of both the teaching function and the Q-value function are inaccurate and the agent has no idea of which action is optimal, thus the value of ϵ is set to a relatively high value to allow the agent to explore potential optimal actions. After enough explorations, the exploration has to be stopped so that the agent will focus on only exploiting the action that has shown to be optimal before.

4.3 Overall Algorithm of TaFSO

The overall algorithm of TaFSO is sketched in Algorithm 1, and it combines the *teacher* and *follower* elements we previously described. One difference is that a special rule (line 5 to 9) is added to identify whether the opponent is adopting TaFSO or not. If the opponent also adopts TaFSO, it is equivalent to the reduced case that both agents alternatively decide the joint action and thus the pre-calculated optimal outcome (s_1, s_2) is always achieved. Next we make the following observations. First, in TaFSO, the teaching goal is to let the opponent be aware that entrusting the TaFSO agent to make decisions for itself is in its best interest. The opponent is always rewarded by the payoff in the optimal outcome (s_1, s_2) when it chooses action F, and punished to wipe out its gain whenever it deviates by choosing action from its original action space. In this way, it prevents the occurrence of mistaken punishment and the opponent never wrongly perceives the punishment signal, thus making the teaching process more efficient. Second, the TaFSO agent always chooses the action which is in its best interest among the candidate actions through exploration and exploitation according to the *follower* strategy. Thus the TaFSO agent can reduce its own cost as much as possible when it exerts punishment on its opponent.

5 Experiments

In this section, we present the experimental results in two parts. The first part focuses on the case of self-play. We compare the performance of TaFSO with previous work [1,15] using the testbed proposed in [3] based on a number of evaluation criteria. In the second part, we evaluate the performance of TaFSO against a variety of best-response learners and also make comparisons with previous strategy SPaM [5].

5.1 Under Self-play

In this section we compare the performance of TaFSO with CJAL [1], Action Revelation [15] and WOLF-PHC [1][2] in two-player's games under self-play. Both

---

[1] WOLF-PHC is an algorithm which has been theoretically proved to be converge to a Nash equilibrium for any two-action, two-player general-sum games, and we consider it here for comparison purpose to show the inefficiency of always achieving Nash equilibrium solution.

Algorithm 1. Overall Algorithm of TaFSO

1: Initialize G_j^t, $Q(a)$, $\forall a \in A_i \cup F$
2: Observe the game G, calculate the optimal outcome (s_1, s_2).
3: **for** each round t **do**
4:     Compute the set C_j^t of candidate actions.
5:     **if** t = 1 **then**
6:         Choose action F.
7:     **else**
8:         **if** $a_j^{t-1} = F$ **then**
9:             Choose action F.
10:         **else**
11:             Choose an action a_i^t according to the *follower* strategy in Sec 4.2.
12:         **end if**
13:     **end if**
14:     **if** it becomes the joint decision-maker **then**
15:         Choose the pre-computed optimal outcome (s_1, s_2) as the joint decision.
16:         Update G_j^t based on the update rule in Sec 4.1.
17:     **else**
18:         Update $Q(a)$, $\forall a \in A_i \cup F$ following Equation 2 after receiving the reward of the joint outcome (a_i^t, a_j^t).
19:         Update G_j^t based on the update rule in Sec 4.1.
20:     **end if**
21: **end for**

---

players play each game repeatedly for 2000 time steps with learning rate of 0.6. The exploration rate starts at 0.3 and gradually decreases by 0.0002 each time step. For all previous strategies the same parameter settings as those in their original papers are adopted.

Here we use the 57 conflicting-interest game matrices with strict ordinal payoffs proposed by Brams in [3] as the testbed for evaluation. Conflicting interest games are the games in which the players disagree on their most-preferred outcomes. These 57 game matrices cover all the structurally distinct two-player conflicting interest games and we simply use the rank of each outcome as its payoff for each agent. For the non-conflicting interest games, it is trivial since there always exists a Nash equilibrium that both players prefer most and also is socially optimal, and thus we do not consider here.

The performance of each approach is evaluated in self-play on these 57 conflicting interest games, and we compare their performance based on the the following three criteria [1]. The comparison results are obtained by averaging over 50 runs across all the 57 conflicting interest games.

Utilitarian Social Welfare. The utilitarian collective utility function $sw_U(P)$ for calculating utilitarian social welfare is defined as $sw_U(P) = \sum_i^n p_i$, where $P = \{p_i\}_i^n$ and p_i is the actual payoff agent i obtains when the outcome is converged. Since the primary learning goal is to achieve socially optimal outcomes, utilitarian social welfare is the most desirable criterion for performance evaluation.

Nash Social Welfare. Nash social welfare is also an important evaluation metrics in that it strikes a balance between maximizing utilitarian social welfare and achieving fairness. Its corresponding utility function $sw_N(P)$ is defined as $sw_N(P) = \prod_i^n p_i$, where $P = \{p_i\}_i^n$ and p_i is the actual payoff agent i obtains when the outcome is converged. One one hand, Nash social welfare reflects utilitarian social welfare. If any individual agent's payoff decreases, the Nash social welfare also decreases. On the other hand, it also reflects the fairness degree between individual agents. If the total payoffs is a constant, then Nash social welfare is maximized only if the payoffs is shared equally among agents.

Success Rate. Success rate is defined as the percentage of times that the socially optimal outcome (maximizing utilitarian social welfare) is converged at last.

The comparison results based on these three criteria are shown in Table 1. We can see that TaFSO outperforms all these three strategies in terms of the above criteria. Players using ToFSO can obtain utilitarian social welfare of 6.45 and Nash social welfare of 10.08 with success rate of 0.96, which are higher than all the other three approaches. Note that the performance of WOLF-PHC approach is worst since this approach is designed for achieving Nash equilibrium only which often does not coincide with socially optimal solution. For Action Revelation and CJAL, they both fail in certain types of games, e.g., the prisoner's dilemma game. For Action Revelation, self-interested agent can always exploit the action revelation mechanism and have the incentive to choose defection D, thus leading the outcome to converge to mutual defection; for CJAL, it requires the agents to randomly explore for a finite number of rounds N first and the probability of converging to mutual cooperation tends to 1 only if the value of N approaches infinity.

Table 1. Performance comparison with CJAL, action revelation and WOLF-PHC using the testbed in [3]

	Utilitarian Social Welfare	Nash Social Welfare	Success Rate
TaFSO (our strategy)	6.45	10.08	0.96
CJAL [1]	6.14	9.25	0.86
Action Revelation [15]	6.17	9.30	0.81
WOLF-PHC [2]	6.03	9.01	0.75

5.2 Against Best-Response Learners

In this part, we evaluate the performance of TaFSO against the opponents adopting a varitey of different best-response strategies [2]. Specifically here we consider the opponent may adopt one of the following strategies: Q-learning [17], WoLF-PHC [2], and Fictitious play (FP) [6]. We compare the performance of TaFSO with SPaM [5] against the same set of opponents. The testbed we adopt here is the same as the one in [5] by using the following three representative games: prisoner's dilemma game (Fig. 1(a)), game of chicken (Fig. 3(a)), and tricky game (Fig. 3(b)).

---

[2] In our current learning context, we assume that the agents using best-response strategies will simply choose the joint action pair with the highest payoff for itself when it becomes the decision-maker for both agents.

	Player 2's action	
1's payoff, 2's payoff	C	D
Player 1's action C	4, 4	2, 5
Player 1's action D	5, 2	0, 0

	Player 2's action	
1's payoff, 2's payoff	C	D
Player 1's action C	0, 3	3, 2
Player 1's action D	1, 0	2, 1

(a)   (b)

Fig. 3. Payoff matrices for (a) game of chicken, and (b)tricky game

For the prisoner's dilemma game, the socially optimal outcome is (C, C), in which both agents receive a payoff of 3. For the game of chicken, the target solution is also (C, C), in which both agents obtain a payoff of 4. In the tricky game, the socially optimal solution is (C, D), and the agents' average payoffs are 2.5. Table 2 shows the agents' average payoffs when both TaFSO and SPaM strategies are adopted to repeatedly play the above representative games against different opponents.[3] We can see that the agents adopting TaFSO can always receive the average payoffs corresponding to the socially optimal outcomes for different games. For comparison, for those cases when SPaM is adopted, the agents' average payoffs are relatively lower than the maximum value of the socially optimal outcomes. The main reason is that the opponent agents adopting the best-response strategies may wrongly perceive the punishment signals of the SPaM agent, and thus result in mis-coordination occasionally.

Table 2. The agents's average payoffs using TaFSO and SPaM strategies against a number of best response learners in three representative games

Average Payoffs	Prisoner's Dilemma Game	Game of Chicken	Tricky Game
TaFSO vs. TaFSO	3.0	4.0	2.5
SPaM vs. SPaM	2.85	3.86	2.31
TaFSO vs. Q-learning	3.0	4.0	2.5
SPaM vs. Q-learning	2.46	3.46	2.15
TaFSO vs. WoLF-PHC	3.0	4.0	2.5
SPaM vs. WoLF-PHC	2.45	3.48	2.1
TaFSO vs. FP	3.0	4.0	2.5
SPaM vs. FP	2.5	3.6	2.17

6   Conclusion and Future Work

In this paper, we propose a learning strategy TaFSO consisting of both *teacher* and *follower* strategies' characteristics to achieve socially optimal outcomes. We consider an interesting variation of sequential play by introducing an additional action F for each agent. The TaFSO agent rewards its opponent by choosing its optimal joint action if its opponent chooses action F, and punish its opponent

---

[3] Note that only the payoffs obtained after 500 rounds are counted here since at the beginning the agents may achieve very low payoffs due to initial explorations. The results are averaged over 50 runs.

based on its *follower* strategy otherwise. Simulation results show that TaFSO can effectively influence the best-response opponents towards socially optimal outcomes and better performance can be achieved than previous work. As future work, we are going to further explore how to utilize the characteristics of the opponents' strategies towards the targeted goal when interacting with non best-response learners. Besides, we are also interested in investigating how to handel the cases when the targeted solution consists of a sequence of joint actions such as achieving fairness [7].

Acknowledgement. The work presented in this paper was partially supported by a CUHK Research Committee Direct Grant for Research (Ref. No. EE09423).

References

1. Banerjee, D., Sen, S.: Reaching pareto optimality in prisoner's dilemma using conditional joint action learning. In: AAMAS 2007, pp. 91–108 (2007)
2. Bowling, M.H., Veloso, M.M.: Multiagent learning using a variable learning rate. In: Artificial Intelligence, pp. 215–250 (2003)
3. Brams, S.J.: Theory of Moves. Cambridge University Press, Cambridge (1994)
4. Camerer, C.F., Ho, T.H., Chong, J.K.: Sophisticated ewa learning and strategic teaching in repeated games. Journal of Economic Theory 104, 137–188 (2002)
5. Crandall, J.W., Goodrich, M.A.: Learning to teach and follow in repeated games. In: AAAI Workshop on Multiagent Learning (2005)
6. Fudenberg, D., Levine, D.K.: The Theory of Learning in Games. MIT Press (1998)
7. Hao, J.Y., Leung, H.F.: Strategy and fairness in repeated two-agent interaction. In: ICTAI 2010, pp. 3–6. IEEE Computer Society (2010)
8. Jafari, A., Greenwald, A., Gondek, D., Ercal, G.: On no-regret learning, fictitious play, and nash equilibrium. In: ICML 2001, pp. 226–233 (2001)
9. Littman, M.: Markov games as a framework for multi-agent reinforcement learning. In: ICML 1994, pp. 322–328 (1994)
10. Littman, M.L., Stone, P.: Leading best-response strategies in repeated games. In: IJCAI Workshop on Economic Agents, Models, and Mechanisms (2001)
11. Littman, M.L., Stone, P.: A polynomial time nash equilibrium algorithm for repeated games. Decision Support Systems 39, 55–66 (2005)
12. Moriyama, K.: Learning-rate adjusting q-learning for prisoner's dilemma games. In: WI-IAT 2008. pp. 322–325 (2008)
13. oH, J., Smith, S.F.: A few good agents: multi-agent social learning. In: AAMAS 2008, pp. 339–346 (2008)
14. Osborne, M.J., Rubinstein, A.: A Course in Game Theory. MIT Press, Cambridge (1994)
15. Sen, S., Airiau, S., Mukherjee, R.: Towards a pareto-optimal solution in general-sum games. In: AAMAS 2003, pp. 153–160 (2003)
16. Stimpson, J.L., Goodrich, M.A., Walters, L.C.: Satisficing and learning cooperation in the prisoner's dilemma. In: IJCAI 2001, pp. 535–540 (2001)
17. Watkins, C.J.C.H., Dayan, P.D.: Q-learning. In: Machine Learning, pp. 279–292 (1992)

Incorporating Fairness into Agent Interactions Modeled as Two-Player Normal-Form Games

Jianye Hao and Ho-fung Leung

Department of Computer Science and Engineering
The Chinese University of Hong Kong
{jyhao,lhf}@cse.cuhk.edu.hk

Abstract. Many multi-agent interaction problems, like auctions and negotiations, can be modeled as games. Game theory, a formal tool for game analysis, thus can be used to analyse the strategic interactions among agents and facilitate us to design intelligent agents. Typically the agents are assumed individually rational consistent with the principle of classical game theory. However, lots of evidences suggest that fairness emotions play an important role in people's decision-making process. To align with human behaviors, we need to take the effects of fairness motivation into account when analysing agents' strategic interactions. In this paper, we propose a fairness model which incorporates two important aspects of fairness motivations, and the solution concept of fairness equilibrium is defined. We show that the predictions of our model successfully reflect the intuitions from both aspects of fairness motivations. Besides, some general results for identifying which outcomes are likely to be fairness equilibria are presented.

1 Introduction

In a typical multi-agent system, multiple autonomous entities, called agents, interact with each other to achieve a common or individual goal. The interacting agents can be software, robots or even humans. When designing intelligent agents for multi-agent systems, it is usually beneficial to model the multi-agent interaction problems as games. The tools and language of game theory [1] provide us with an elegant way to analyse how the agents should behave in such strategic interaction environments. The analysis under the game-theoretic framework can serve as the guidance for us to design effective strategies for the agents.

The typical assumption of individual agents in multi-agent system is self-rationality, according to the principle of classical game theory. However, substantial evidences show that humans are not purely self-interested and thus this strong assumption of self-rationality has been recently relaxed in various ways [2–4]. One important aspect is that people strongly care about fairness. Since many multi-agent systems are designed to interact with human or act on behalf of them, it is important that an agent's behavior should be (at least partially) aligned with human expectations, and failure to consider this may incur significant cost during the interactions. To assist the design of intelligent agents

P. Anthony, M. Ishizuka, and D. Lukose (Eds.): PRICAI 2012, LNAI 7458, pp. 100–111, 2012.

towards fairness, it is important for us to have some game-theoretic frameworks which can formally model fairness. Within these frameworks, we can formally analyse the strategic interactions among agents and help the agents predict the behaviors of their interacting partners (especially with human) under the concern of fairness.

When people make decisions, apart from their material payoffs, they also care about the motive behind others' actions [5]. People are willing to sacrifice their own material interests to help those who are being kind and also punish those who are being unkind [6, 7]. Besides, people also care about the relative material payoffs with others and are willing to sacrifice their own material payoffs to move in the direction of decreasing the material payoff inequity [8]. However, in previous fairness models [5, 8], only either of the previous factors is considered, which thus only reflects a partial view of the motivations behind fairness. In this paper, we propose a fairness model within the game-theoretic framework, which combines both aspects of fairness motivations, to provide better predictions on the agents' behaviors.

In our fairness model, following the inequity-averse theory proposed in [8], each agent is equipped with a pair of inequity-averse factors to reflect his fairness concern of the relative material payoff with others. Furthermore, the fairness motion that people care about the motive behind others' actions is modeled by introducing kindness functions. To combine these two aspects of fairness motivations together, these kindness functions are defined based on the agents' inequity-averse factors and also their higher-order beliefs on those factors. In this context, we define the game-theoretic solution concept of fairness equilibrium similar to the equilibrium in psychological games. In general, fairness equilibria do not constitute either a subset or a superset of Nash equilibria, and we show that it can both add new predictions and rule out conventional predictions. We use some examples to show that our fairness model is able to predict all behaviors consistent with every aspect of fairness we considered. We also present some general results about the conditions for an outcome to be a fairness equilibrium.

The remainder of the paper is organized as follows. In Section 2, we give a brief overview of the related work. In Section 3, we introduce the fairness model and the concept of fairness equilibrium we propose. Section 4 presents some general results about fairness equilibrium. Lastly Section 5 concludes our paper and presents some future work.

2   Related Work

One main line of research for modeling fairness is based on the theory of inequity aversion with the premise that people not only care about their own material payoffs, but also the relative material payoffs with other people's. Fehr and Schmidt [8] propose an inequity aversion model, which is based on the assumption that people show a weak aversion towards advantageous inequity, and a strong aversion towards disadvantageous inequity. The authors have shown that this model can accurately predict human behavior in various games such as ultimatum game, pubic good game with punishments and so on. Similar model based on the same assumptions in the literature includes the ERC model in [9]. However, these work are purely outcome-oriented in the sense that they only model the effects resulting from the inequitable material payoffs, but not explicitly account for the role of intentions. Accordingly their models fail to predict certain important phenomenons which can be obtained only by considering the effect of intentions.

Another popular model for fairness is based on the theory of reciprocity [5], which is modeled as the behavioral response to actions which are perceived as being kind or not. The distinction with the inequity-averse model is that it takes the fairness intention into account, and fairness is achieved by reciprocal behavior which is a response to kindness instead of a desire for reducing inequity. Other similar models are developed for sequential games using the same framework [10, 11]. However, the role of inequity aversion is ignored in previous models, and thus they fail to predict certain behaviors caused by the effect of inequity aversion.

A number of prescriptive, computational models are proposed to assist the agents to achieve fairness in multi-agent systems [4, 12, 13]. Nowé et al. [4] propose a periodical policy for achieving fair outcomes among multiple agents in a distributed way. This policy can be divided into two periods: reinforcement learning period and communication period. Simulation results show that the average material payoff each agent obtains is approximately equal. de Jong et al. [12] propose a prescriptive fairness model in multi-agent system to obtain the best possible alignment with human behavior. The model combines the Homo Egualis utility function [8]with Continuous Action Learning Automata together. They analyze two forms of games: Ultimatum Game and Nash Bargaining Game and show that agents' actions using this model are consistent with the empirical results observed from human subjects in behavioral economics. However, in our work, we focus on proposing a descriptive fairness model within the game-theoretic framework instead of a prescriptive one, which is fundamental for designing prescriptive models towards the goal of fairness.

3   Fairness Model and Fairness Equilibrium

3.1   Motivation

Consider the battle-of-the-sexes game shown in Fig. 1, where $X > 0$. In this game, both agents prefer to take the same action together, but each prefers a different action choice. Specifically, agent 1 prefers that both agents choose action Opera together, while agent 2 prefers that both agents jointly take action Boxing.

In this game there exist two pure strategy Nash equilibria, i.e., (Opera, Opera) and (Boxing, Boxing). The outcomes (Opera, Boxing) and (Boxing, Opera) are not Nash equilibria, in which each agent always has the incentive to deviate from his current strategy to increase his material payoff. However, in reality, people usually do not make their decisions based on the material payoffs only. Apart from material payoff, people also take their fairness emotions into their decision-making processes. People not only care about their material payoff, but also the way that they are being treated by other people [6, 7, 14]. In reality, if agent 1 thinks that agent 2 is doing him a favor, then he will have the desire to do agent

1's payoff, 2's payoff		Agent 2's action	
		Opera	Boxing
Agent 1's action	Opera	2X, X	0, 0
	Boxing	0, 0	X, 2X

Fig. 1. Example 1: Battle of the Sexes

1's payoff, 2's payoff		Agent 2's action	
		C	D
Agent 1's action	C	4X, 10X	2X, 2X
	D	2X, 2X	4X, X

Fig. 2. Example 2

2 a favor in return; if agent 2 seems to be hurting agent 1 intentionally, agent 1 will be motivated to hurt agent 2 as well.

Based on the above theories, let us look at the outcome (Opera, Boxing) again. If agent 1 believes that agent 2 deliberately chooses Boxing instead of Opera which decreases his material payoff, he will have the incentive to hurt agent 2 in return in reality by sticking with Opera. If agent 2 is thinking in the same way, then he will also have the motivation to sacrifice his own interest and stick with Boxing to hurt agent 1. In this way, both agents are hurting each other to satisfy their emotional desires. If this kind of emotions is strong enough, then no agent is willing to deviate from the current strategy, and thus the outcome (Opera, Boxing) becomes an equilibrium.

Another important aspect of fairness emotions that can influence people's decisions is the fact that people not only care about their own material payoffs, but also the relative material payoffs with others. It has been revealed that people show a weak aversion towards advantageous inequity, and a strong aversion towards disadvantageous inequity [8, 15–17]. People may be willing to give up their material payoffs to move in the direction of more equitable outcomes. For example, consider the game shown in Fig. 2, where $X > 0$. It is easy to check that a Nash equilibrium is (C, C). However, if we also consider the factor that people are inequity-averse, then agent 1 may be very unhappy about the current situation that his material payoff is much lower than that of agent 2. Thus agent 1 may have the incentive to switch to action D to achieve an equitable outcome instead, and (C, C) is not an equilibrium any more.

Furthermore, consider the outcome (C, D), which is not a Nash equilibrium. If agent 1 believes that agent 2 intentionally chooses D to avoid obtaining a higher material payoff than agent 1, agent 1 may wish to reciprocate by sticking with C, even though he could have chosen action D to obtain a higher material payoff of $4X$ (if he chooses action D and obtains $4X$, he may feel guilty for taking advantage of agent 2.). Similarly, if agent 2 also believes that agent 1 is doing a favor to him by choosing C instead of D, he will also have the motivation to reciprocate by sticking with action D. Thus the outcome (C, D) becomes an equilibrium if this kind of emotions between the agents is taken into consideration.

From the above examples, we can see that apart from the material payoff, each agent's decision also depends on his fairness emotions resulting from both his beliefs about his opponent's motive and his own inequity-averse degree towards unequal material payoffs. Is it possible to incorporate all these fairness motivations into the agents' decision-making processes by transforming the material payoffs to some kind of *perceived payoffs*, so that we can analyze the transformed game in the conventional way? It turns out that the answer is "No". In Example 2, we have analyzed that both (C, C) and (C, D) can be strict equilibria if emotion is taken into account. In the equilibrium (C, C), agent 2 strictly prefers C to D; while in the equilibrium (C, D), agent 2 strictly prefers D to C. We can see that there will always be contradictions no matter how perceived payoffs are defined in this game, if they depend on the action profile only. Similarly, in Example 1, both (Opera, Opera) and (Opera, Boxing) can be strict equilibria, in which we can also get the contradictory conclusion no matter how perceived payoffs are defined. To directly model these fairness emotions, therefore, it is necessary to explicitly take the agents' thoughts into consideration. In this paper, we propose a game-theoretic fairness model which incorporates fairness motivations in the context of two-player finite games.

3.2   Fairness Model

Consider a two-player normal form game, and let A_1, A_2 be the action spaces of agents 1 and 2. For each action profile $(a_1, b_2) \in A_1 \times A_2$, let $p_1(a_1, b_2)$ and $p_2(a_1, b_2)$ be the material payoff for agent 1 and 2 respectively.

Following the work of Fehr and Schmidt [8], to represent the inequity-averse degree of the agents, we assume that each agent i has two inequity-averse factors α_i and β_i. These two factors represent agent i's suffering degree when he receives relatively lower and higher material payoffs than his opponent respectively. We use α_i' and β_i' to denote agent i's belief on agent j's $(i \neq j)$ inequity-averse factors; and α_i'' and β_i'' to denote his belief on agent j's belief on agent i's inequity-averse factors. The overall utility of agent i thus depends on the following factors:[1] (1) the strategy profile played by the agents (2) his inequity-averse factors (3) his belief about agent j's inequity factors (4) his belief about agent j's belief about his own inequity-averse factors.

Each agent's inequity-averse factors and his higher-order beliefs model how he perceives the kindness from his opponent and also how kind he is to his opponent. Before formally defining these kindness functions, we adopt the inequity-averse model in [8] to define the *emotional utility* of agent i as follows.

$$u_i(a_i, b_j) = p_i - \alpha_i \max\{p_j - p_i, 0\} - \beta_i \max\{p_i - p_j, 0\}, i \neq j \qquad (1)$$

where p_i and p_j are the material payoffs of the agents i and j under strategy profile (a_i, b_j) and α_i and β_i are agent i's inequity-averse factors.

Given that agent j chooses strategy b_j, how kind is agent i being to agent j if agent i chooses strategy a_i? Based on Equation 1, agent i believes that agent j's emotional utility is

---

[1] Note that it is possible to perform higher-level modeling here. However, as we will show, modeling at the second level is already enough and also can keep the model analysis tractable.

$$u'_j(a_i, b_j) = p_j - \alpha'_i \max\{p_j - p_i, 0\} - \beta'_i \max\{p_i - p_j, 0\}. \tag{2}$$

Let $\prod(b_j)$ be the set of all possible emotional utility profiles if agent j chooses action b_j from agent i's viewpoint, i.e., $\prod(b_j) = \{(u_i(a_i, b_j), u'_j(a_i, b_j) \mid a_i \in A_i\}$. We define the reference emotional utility $u^e_j(b_j)$ as the expected value of agent j's highest emotional utility $u^h_j(b_j)$ and his lowest one $u^l_j(b_j)$ among those Pareto-optimal points in $\prod(b_j)$. Besides, let $u^{min}_j(b_j)$ be the minimum emotional utility of agent j in $\prod(b_j)$.

Based on the above definitions, now we are ready to define the *kindness* of agent i to agent j. This kindness function reflects how much agent i believes that he is giving to agent j with respect to the reference emotional utility.

Definition 1. *The* kindness *of agent i to agent j under strategy profile (a_i, b_j) is given by*

$$K_i(a_i, b_j, \alpha'_i, \beta'_i) = \frac{u'_j(a_i, b_j) - u^e_j(b_j)}{u^h_j(b_j) - u^{min}_j(b_j)}. \tag{3}$$

If $u^h_j(b_j) - u^{min}_j(b_j) = 0$, *then* $K_i(a_i, b_j, \alpha'_i, \beta'_i) = 0$.

Similarly we can define how kind agent i believes that agent j is being to him. One major difference is that now agent i's emotional utility is calculated based on his belief about agent j's belief about his inequity-averse factors. Given a strategy profile (a_i, b_j), agent i's believes that agent j wants him to obtain emotional utility $u'_i(a_i, b_j)$, which is defined as follows.

$$u'_i(a_i, b_j) = p_i - \alpha''_i \max\{p_j - p_i, 0\} - \beta''_i \max\{p_i - p_j, 0\}. \tag{4}$$

Besides, we can similarly define $\prod(a_i) = \{(u'_i(a_i, b_j), u'_j(a_i, b_j) \mid b_j \in A_j\}$, which is the set of all possible emotional utility profiles if agent i chooses a_i from agent i's viewpoint. The reference utility $u^e_i(a_i)$ is the expected value of agent i's highest emotional utility $u^h_i(a_i)$ and his lowest one $u^l_i(a_i)$ among those Pareto-optimal points in $\prod(a_i)$. Besides, let $u^{min}_i(a_i)$ be the minimum emotional utility of agent i in $\prod(a_i)$.

Finally, agent i's *perceived kindness* from agent j is defined as how much agent i believes that agent j is giving to him.

Definition 2. *Player i's* perceived kindness *from agent j under strategy profile (a_i, b_j) is given by*

$$\delta_i(a_i, b_j, \alpha''_i, \beta''_i) = \frac{u'_i(a_i, b_j) - u^e_i(a_i)}{u^h_i(a_i) - u^{min}_i(a_i)}. \tag{5}$$

If $u^h_i(a_i) - u^{min}_i(a_i) = 0$, *then* $\delta_i(a_i, b_j, \alpha''_i, \beta''_i) = 0$.

Each agent's *overall utility* over an outcome (a_i, b_j) is jointly determined by his emotional utility, his kindness and perceived kindness from his opponent. Each agent i always chooses action a_i to maximize his own overall utility when he makes his decisions. Player i's overall utility function is defined as follows.

$$U_i(a_i, b_j, \alpha_i, \beta_i, \alpha_i', \beta_i', \alpha_i'', \beta_i'') = u_i(a_i, b_j) + \\ K_i(a_i, b_j, \alpha_i', \beta_i') \times \delta_i(a_i, b_j, \alpha_i'', \beta_i'') \tag{6}$$

In this way, we have incorporated all the fairness motivations considered in previous discussions into the agents' decision-making processes. If agent i believes that agent j is kind to him, then $\delta_i(a_i, b_j, \alpha_i'', \beta_i'') > 0$. Thus agent i will be emotionally motivated to show kindness to agent j and choose an action a_i such that $K_i(a_i, b_j, \alpha_i', \beta_i')$ is high to increase his overall utility. Besides, agents' kindness emotions reflect the agents' inequity-averse degrees. For example, assuming that the value of α_i is large, then agent i will believe that agent j is unkind to him if his material payoff is much lower than that of agent j even if his current material payoff is already the highest among all his possible material payoffs. Thus agent i will reciprocate by choosing an action a_i' to decrease agent j's emotional utility based on his own beliefs. Note that both kindness functions are normalized, and thus they are insensitive to the positive affine transformations of the material payoffs. However, the overall utility function is sensitive to such transformations. If the material payoffs are extremely large, the effect of $u_i(a_i, b_j)$ may dominate the agents' decisions and the influences of the agents' reciprocity emotions can be neglected, which is consistent with the evidences in human experiments [6, 18].

3.3   Fairness Equilibrium

We define the concept of equilibrium using the concept of psychological Nash equilibrium defined by [19]. This is an analog of Nash equilibrium for psychological games with the additional requirement that all higher-order beliefs must match the actual ones.

Definition 3. *The strategy profile* $(a_1, a_2) \in A_1 \times A_2$ *is a fairness equilibrium iff the following conditions hold for* $i = 1, 2, j \neq i$

- $a_i \in arg \max_{a_i \in A_i} U_i(a_i, a_j, \alpha_i, \beta_i, \alpha_i', \beta_i', \alpha_i'', \beta_i'')$
- $\alpha_i' = \alpha_j = \alpha_j''$
- $\beta_i' = \beta_j = \beta_j''$

In this fairness equilibrium definition, the first condition simply represents that each agent is maximizing his overall utility, while the second and the third conditions state that all higher-order beliefs are consistent with the actual ones. This solution concept is consistent with the discussions in previous examples. Considering Example 1, suppose that $\alpha_1 = \alpha_2 = \beta_1 = \beta_2 = 0.5$, and let the higher order beliefs be consistent with these actual ones. For the outcome (Opera, Boxing), we have $K_1(Opera, Boxing, 0.5, 0.5) = -1$ and $\delta_i(Opera, Boxing, 0.5, 0.5) = -1$, and thus the overall utility of agent 1 is 1. If agent 1 chooses Boxing, then his overall utility will be $0.5X$. Thus if $X < 2$, agent 1 prefers Opera to Boxing given these beliefs. For the sake of symmetry, we can have that agent 2 would prefer Boxing to Opera in the same situation. For $X < 2$, therefore, (Opera, Boxing) is a fairness equilibrium. Intuitively, in this equilibrium, both agents are hostile towards each other and unwilling to concede to the other. If $X > 2$, the agents' desires for pursuing emotional utility will override their concerns for fairness, thus (Opera, Boxing) is not a fairness equilibrium. On the other hand, it is not difficult to check that both (Opera, Opera) and (Boxing, Boxing) are always equilibria. In these equilibria, both agents feel the kindness from the other and thus are willing to coordinate with the other for reciprocation, which also maximizes their emotional utilities.

Next let us consider Example 2, and the same set of values for the agents' inequity-averse factors is adopted as Example 1. For the outcome (C, D), we have $K_1(C, D, 0.5, 0.5) = \frac{1}{2}$ and $\delta_i(C, D, 0.5, 0.5) = \frac{1}{2}$, and thus agent 1's overall utility is $2X + \frac{1}{4}$. If agent 1 switches to D, his kindness to agent j is changed to $K_1(D, D, 0.5, 0.5) = -\frac{1}{2}$, and thus his overall utility becomes $2.5X - \frac{1}{4}$. Therefore, if $2X + \frac{1}{4} > 2.5X - \frac{1}{4}$, i.e., $X < 1$, he will prefer action C to action D. Similarly, we can have the conclusion that if $X < 0.1$, agent 2 will prefer action D to action C. Thus (C, D) is a fairness equilibrium if $X < 0.1$. In this equilibrium, both agents are kind to the other by sacrificing their own personal interest. However, if the emotional utility becomes large enough such that the effect of the emotional utility dominates the overall utility, (C, D) is not an equilibrium any more. For example, if $X > 1$, agent 1 will have the incentive to choose action D to maximize his overall utility. Similarly, we can also check that (C, C) is no longer an equilibrium any more for any $X > 0$, since agent 1's strong inequity-averse emotion always induces him to deviate to action D to obtain an equal payoff.

Overall, we can see that in both examples, new equilibria can be introduced based on our fairness model, apart from the traditional Nash equilibria. Besides, some Nash equilibria are ruled out in certain situations (e.g., the Nash equilibrium (C, C) is not a fairness equilibrium in Example 2). In next section, we are going to present some general conclusions about fairness equilibrium explaining why this is the case.

4   General Theorems

For any game G, we can obtain its *emotional extension* G' by replacing every material payoff profile with its corresponding emotional utility profile defined in Equation 1. Considering the battle-of-the-sexes game, its emotional extension using the same set of parameters as in the previous section is shown in Fig. 3. If both agents play strategy Opera, then both of them are maximizing their opponents' emotional utilities simultaneously. Similarly, if agent 1 chooses strategy Opera while agent 2 chooses Boxing, then both agents are minimizing their opponents's emotional utilities at the same time. These strategy profiles are examples of *mutual-max* and *mutual-min* outcomes, which are formally defined as follows.

1's payoff, 2's payoff		Agent 2's action	
		Opera	Boxing
Agent 1's action	Opera	1.5X, 0.5X	0, 0
	Boxing	0, 0	0.5X, 1.5X

Fig. 3. Emotional Extension of Battle of the Sexes

Definition 4. *A strategy profile* $(a_1, a_2) \in A_1 \times A_2$ *is a* mutual-max *outcome if, for* $i \in \{1, 2\}$, $j \neq i$, $a_i \in \arg\max_{a \in A_i} u_j(a, a_j)$.

Definition 5. *A strategy profile* $(a_1, a_2) \in A_1 \times A_2$ *is a* mutual-min *outcome if, for* $i = \{1, 2\}$, $j \neq i$, $a_i \in \arg\min_{a \in A_i} u_j(a, a_j)$.

For any game G, we can also define Nash equilibrium in its emotional extension G' as follows.

Definition 6. *For any two-player normal form game* G, *a strategy profile* $(a_1, a_2) \in A_1 \times A_2$ *is a Nash equilibrium in its emotional extension* G' *if, for* $i = 1, 2$, $j \neq i$, $a_i \in \arg\max_{a \in A_i} u_i(a, a_j)$.

We also characterize an outcome of a game based on the kindness function of the agents as follows.

Definition 7. *An outcome* (a_i, b_j) *is*

- strictly (or weakly) positive *if, for* $i = 1, 2$, $K_i(a_i, b_j, \alpha'_i, \beta'_i) > 0$
  (or $K_i(a_i, b_j, \alpha'_i, \beta'_i) \geq 0$*);*
- strictly (or weakly) negative *if, for* $i = 1, 2$, $K_i(a_i, b_j, \alpha'_i, \beta'_i) < 0$
  (or $K_i(a_i, b_j, \alpha'_i, \beta'_i) \leq 0$*);*
- neutral *if for* $i = 1, 2$, $K_i(a_i, b_j, \alpha'_i, \beta'_i) = 0$;
- mixed *if for* $i = 1, 2$, $i \neq j$, $K_i(a_i, b_j, \alpha'_i, \beta'_i) \times K_j(a_i, b_j, \alpha'_j, \beta'_j) < 0$.

Based on previous definitions, we are ready to state the sufficient conditions for a Nash equilibrium in game G's emotional extension G' to be a fairness equilibrium in the original game G.

Theorem 1. *Suppose that the outcome* (a_1, b_2) *is a Nash equilibrium in* G', *if it is also either a mutual-max outcome or mutual-min outcome, then it is a fairness equilibrium in* G.

Proof. Since (a_1, b_2) *is a Nash equilibrium in* G', *both agents must maximize their emotional utilities in this outcome. If* (a_1, b_2) *is also a mutual-max outcome, then both* $K_1(a_1, b_2, \alpha'_1, \beta'_1)$ *and* $K_2(a_1, b_2, \alpha'_2, \beta'_2)$ *must be non-negative, which implies that both agents perceive kindness from the other. Therefore, for each agent, there is no incentive for him to deviate from the current strategy in terms of maximizing both his emotional utility and the satisfaction of his emotional* *reactions to the other agent (maximizing the product of his kindness factors). Thus the outcome (a_1, b_2) is a fairness equilibrium. Similarly, if (a_1, b_2) is a mutual-min outcome, then both $K_1(a_1, b_2, \alpha_1', \beta_1')$ and $K_2(a_1, b_2, \alpha_2', \beta_2')$ must be non-positive. Therefore both agents have the incentive to stick with the current strategy to maximize the product of their kindness factors. Also their emotional utilities are maximized in (a_1, b_2). Therefore their overall utilities are maximized which implies (a_1, b_2) is a fairness equilibrium.*

Next we characterize the necessary conditions for an outcome to be a fairness equilibrium.

Theorem 2. *If an outcome (a_1, b_2) is a fairness equilibrium, then it must be either strictly positive or weakly negative.*

Proof. This theorem can be proved by contradiction. Suppose that (a_1, b_2) is a fairness equilibrium, and also we have $K_i(a_i, b_j, \alpha_i', \beta_i') > 0$, $K_j(a_i, b_j, \alpha_j', \beta_j') \le 0$. Since $K_i(a_i, b_j, \alpha_i', \beta_i') > 0$, it indicates that agent i is kind to agent j. Based on Definition 1, we know that there must exist another strategy a_i' such that by choosing it agent i can increase his own emotional utility and also decrease agent j's emotional utility at the same time. Besides, since $K_j(a_i, b_j, \alpha_j', \beta_j') \le 0$, it implies that agent j is treating agent i in a bad or neutral way at least. Thus agent i also emotionally have the incentive to choose strategy a_i' rather than being nice to agent j. Overall, agent i would have the incentive to deviate from strategy a_i to maximize his overall utility. Thus (a_1, b_2) is not an equilibrium, which contradicts with our initial assumption. The only outcomes consistent with the definition of fairness equilibrium, therefore, are those outcomes that are either strictly positive or weakly negative.

Previous theorems state the sufficient and necessary conditions for an outcome to be a fairness equilibrium in the general case. Next we present several results which hold when the emotional utilities are either arbitrarily large or small. For the sake of convenient representation, given a game G, for each strategy profile $(a_1, b_2) \in A_1 \times A_2$, let us denote the corresponding pair of emotional utilities as $(X \times u_1^o(a_1, b_2), X \times u_2^o(a_1, b_2))$, where X is a variable used as a scaling factor of the emotional utility. By setting the value of X appropriately, therefore, we can get the corresponding game $G'(X)$ (or $G(X)$) of any scale.

Let us consider the emotional extension of the battle-of-the-sexes game in Fig. 3. Notice that (Opera, Boxing) is not a Nash equilibrium in the sense of the emotional utilities, but it is a strictly negative mutual-min outcome. If the value of X is arbitrarily small ($X < 2$), the emotional utilities become unimportant, and the agents' emotional reactions begins to take control. As we have previously shown, in this condition, the outcome (Opera, Boxing) can be a fairness equilibrium. Similarly, we can check that, in Example 2, the outcome (C, D) which is not a Nash equilibrium, is strictly positive mutual-max. Besides, we have also shown that if the value of X is small enough, (C, D) becomes a fairness equilibrium. In general, we can have the following theorem, and we omit the proof due to space limitation.

Theorem 3. *For any outcome* (a_1, b_2) *that is either strictly positive mutual-max or strictly negative mutual-min, there always exists an* X' *such that for all* $X \in (0, X')$, *the outcome* (a_1, b_2) *is a fairness equilibrium in the game* $G(X)$.

Now let us consider another case when the value of X is arbitrarily large. Consider the battle-of-the-sexes example again. As we have shown, if the value of X is large enough, i.e., $X > 2$, then the emotional utility dominates, and thus the outcome (Opera, Boxing) is not a fairness equilibrium any more. This can be generalized to the following theorem, and the proof is omitted due to space constraints.

Theorem 4. *Given a game* $G(X)$, *if the outcome* (a_1, b_2) *is not a Nash equilibrium in* $G'(X)$, *then there always exists an* X' *such that for all* $X > X'$, (a_1, b_2) *is not a fairness equilibrium in* $G(X)$.

Similarly, we can have the symmetric theorem as following.

Theorem 5. *Given a game* $G(X)$, *if the outcome* (a_1, b_2) *is a strict Nash equilibrium in* $G'(X)$, *then there always exists an* X' *such that for all* $X > X'$, (a_1, b_2) *is a fairness equilibrium in* $G(X)$.

For example, consider Example 2, it can be checked that the outcome (D, C), which is a Nash equilibrium in Example 2's emotional extension, is always a fairness equilibrium when the value of X is large enough. However, note that if an outcome is weak Nash equilibrium in $G'(X)$, there may not exist an X such that it is a fairness equilibrium in $G(X)$.

Overall, we have theoretically analyzed and proved why some Nash equilibria can be ruled out and some non-Nash equilibria become fairness equilibria in different cases. Based on the above theorems, it can provide us with valuable instructions to identify fairness equilibria in games of different scales.

5   Conclusion and Future Work

In this paper, we propose a descriptive fairness model within the game-theoretic framework which incorporates two important aspects of fairness: reciprocity and inequity-aversion. The game-theoretic solution concept of fairness equilibrium is defined using the concept of psychological Nash equilibrium. Besides some general results about which outcomes can be fairness equilibria in games of different scales are presented.

As future work, more experimental evaluations of the model will be conducted to show that our model can better fit the empirical results found in human behaviors, comparing with the existing fairness models in the literature. Another important direction is to apply this fairness model into some practical multi-agent applications such as multi-agent negotiation scenarios. We believe that by utilizing the implications resulting from this model, more intelligent agents can be designed to better align with human expectations, especially for the cases involving interactions with humans or software agents with human-like emotions. It would also be an interesting direction to extend this model to the n-player case and sequential games.

Acknowledgement. The work presented in this paper was partially supported by a CUHK Research Committee Direct Grant for Research (Ref. No. EE09423).

References

1. Osborne, M.J., Rubinstein, A.: A Course in Game Theory. MIT Press (1994)
2. Simon, H.: Theories of bounded rationality. Decision and Organization 1, 161–176 (1972)
3. Chevaleyre, Y., Endriss, U., Lang, J., Maudet, N.: A Short Introduction to Computational Social Choice. In: van Leeuwen, J., Italiano, G.F., van der Hoek, W., Meinel, C., Sack, H., Plášil, F. (eds.) SOFSEM 2007. LNCS, vol. 4362, pp. 51–69. Springer, Heidelberg (2007)
4. Verbeeck, K., Nowé, A., Parent, J., Tuyls, K.: Exploring selfish reinforcement learning in repeated games with stochastic rewards. In: AAMAS, vol. 14, pp. 239–269 (2006)
5. Rabin, M.: Incorporating fairness into game theory and economics. American Economic Review 83, 1281–1302 (1993)
6. Dawes, R.M., Thaleri, R.H.: Anomalies: Cooperation. Journal of Economic Perspectives 2, 187–198 (1988)
7. Thaler, R.H.: Mental accounting and consumer choice. Marketing Science 4, 199–214 (1985)
8. Fehr, E., Schmidt, K.M.: A theory of fairness, competition and cooperation. Quarterly Journal of Economics 114, 817–868 (1999)
9. Bolton, G., Ockenfels, A.: Erc-a theory of equity, reciprocity and competition. American Economic Review 90, 166–193 (2000)
10. Dufwenberg, M., Kirchsteiger, G.: A theory of sequential reciprocity. Games and Economic Behavior 47, 268–298 (1998)
11. Falka, A., Fischbache, U.: A theory of reciprocity. Games and Economic Behavior 54, 293–315 (2006)
12. de Jong, S., Tuyls, K., Verbeeck, K.: Artificial agents learning human fairness. In: AAMAS 2008, pp. 863–870. ACM Press (2008)
13. Hao, J.Y., Leung, H.F.: Strategy and fairness in repeated two-agent interaction. In: ICTAI 2010, pp. 3–6 (2010)
14. Kahneman, D., Knetsch, J.L., Thaler, R.H.: Fairness as a constraint on profit seeking: Entitlements in the market. American Economic Review 76, 728–741 (1986)
15. Camerer, C., Thaler, R.H.: Ultimatums, dictators, and manners. Journal of Economic Perspectives, 209–219 (1995)
16. Agell, J., Lundberg, P.: Theories of pay and unemployment: Survery evidence from swedish manufacturing firms. Scandinavian Journal of Economics XCVII, 295–308 (1995)
17. Bewley, T.: A depressed labor market as explained by participants. In: American Economic Review, Papers and Proceedings vol. LXXXV, pp. 250–254 (1995)
18. Leventhal, G., Anderson, D.: Self-interest and the maintenance of equity. Journal of Personality and Social Psychology 15, 57–62 (1970)
19. Geanakoplos, J., Pearce, D., Stacchetti, E.: Psychological games and sequential rationality. Games and Economic Behavior 1, 60–79 (1989)

Semi-Supervised Discriminatively Regularized Classifier with Pairwise Constraints

Jijian Huang, Hui Xue*, and Yuqing Zhai

School of Computer Science and Engineering
Southeast University, Nanjing, 211189, P.R. China
{huangjijian2008,hxue,yqzhai}@seu.edu.cn

Abstract. In many real-world classifications such as video surveillance, web retrieval and image segmentation, we often encounter that class information is reflected by the pairwise constraints between data pairs rather than the usual labels for each data, which indicate whether the pairs belong to the same class or not. A common solution is combining the pairs into some new samples labeled by the constraints and then designing a smoothness-driven regularized classifier based on these samples. However, it still utilizes the limited discriminative information involved in the constraints insufficiently. In this paper, we propose a novel semi-supervised discriminatively regularized classifier (SSDRC). By introducing a new discriminative regularization term into the classifier instead of the usual smoothness-driven term, SSDRC can not only use the discriminative information more fully but also explore the local geometry of the new samples further to improve the classification performance. Experiments demonstrate the superiority of our SSDRC.

Keywords: Discriminative information, Structural information, Pairwise constraints, Semi-supervised classification.

1 Introduction

Semi-supervised learning is a class of machine learning techniques that makes use of both labeled and unlabeled data, which has achieved considerable development in theory and application [1-4]. According to different actual circumstances, semi-supervised learning usually involves two categories of class information, that is, the class label and the pairwise constraint. The class label specifies the concrete label for each datum, which is common in the traditional classification, while the pairwise constraint is defined on the data pair, which indicates that whether the pair belongs to the same class (must-link) or not (cannot-link). In many applications, the pairwise constraint is actually more general than the class label, because sometimes the true label may not be known prior, while it is easier for a user to specify whether the data pair belong to the same class or different class. Moreover, the pairwise constraints can be derived from the class label but not

---

* Corresponding author.

P. Anthony, M. Ishizuka, and D. Lukose (Eds.): PRICAI 2012, LNAI 7458, pp. 112–123, 2012.

vice versa. Furthermore, different from the class label, the pairwise constraints can sometimes be obtained automatically [5].

Recently, the research on semi-supervised classification with pairwise constraints has attracted more and more interests in machine learning [6,7]. However, compared to the common models, such classification problem is much harder due to the difficulties in extracting the discriminative information from the constraints. Generally, common classifiers mostly contain loss functions and regularization terms, where loss functions measure the difference between the predictions and the initial class labels. However, pairwise constraints just represent the relationship between the data pairs rather than certain labels. Consequently, the constraints cannot be incorporated into the loss functions directly, which leads to the inapplicability of most existing classifiers. Due to the particularity of pairwise constraints, Zhang and Yan proposed a method dependent on the value of pairwise constraints (OVPC), which combines the data pairs into some new samples by outer product [8] and then designs a smoothness-driven regularized classifier based on these samples. OVPC can deal with a large number of pairwise constraints and avoid local minimum problem simultaneously. Yan et al. [9] presented a unified classification framework which consists of two loss functions for labeled data and pairwise constraints respectively and a common Tikhonov regularization term to penalize the smoothness of the classifier.

Although these methods have shown much better classification performance, they still extract the prior information from the pairwise constraints insufficiently. Firstly, they use the Tikhonov term as the regularization term which only emphasizes on the smoothness of the classifier but ignores the limited discriminative information inside the constraints. Xue et al. [10] have presented that relatively speaking, the discriminability of the classifier is more important than the usual smoothness. Hence, such regularization term is obviously insufficient for classification. Secondly, they also neglect the structural information in the data. Yeung et al. [11] have indicated that a classifier should be sensitive to the structure of the data distribution. Much related research has further validated the effectivity of the structural information for classification [12-17]. The absence of such vital information in these methods undoubtedly influences the corresponding classification performance.

In this paper, we propose a new classifier with pairwise constraints called SSDRC which stands for semi-supervised discriminatively regularized classifier for binary classification problems . Inspired by OVPC, SSDRC firstly combines the pairs into new samples. Then it applies a discriminative regularization term into the classifier instead of the traditional smoothness-driven term, which directly emphasizes on the discriminability of these new samples through using two terms to measure the intra-class compactness and inter-class separability respectively. Moreover, SSDRC also introduces the local sample geometries into the construction of the two terms in order to further fuse the structural information.

The rest of the paper is organized as follows. Section 2 briefly reviews the OVPC methods. Section 3 presents the proposed SSDRC. In Section 4, the experiment analysis is given. Conclusions are drawn in Section5.

2   Classifier Dependent On the Value of Pairwise Constraints(OVPC)

Given the training data which consist of two parts, $(x_1, y_1),...,(x_m, y_m)$ are labeled data with y_i representing the class label, and $(x_{11}, x_{12}, y'_1),...,(x_{n1}, x_{n2}, y'_n)$ are pairwise constraints with y'_i indicating the relationship between pairs. 1 represents must-link constraint and -1 represents cannot-link constraint. That is,

$$y'_i = \begin{cases} +1, \ y_{i1} = y_{i2} \\ -1, \ y_{i1} \neq y_{i2} \end{cases}$$

OVPC firstly combines the pair (x_{i1}, x_{i2}) into a new sample by outer product [5], which equals a transformation from the original space X to a new space $\tilde{X}$. Let z_i denote the corresponding transformed sample in $\tilde{X}$. Therefore, (z_i, y'_i) is the new sample whose class label is y'_i. Then OVPC constructs a common regularized least-squares classifier in $\tilde{X}$. That is,

$$\hat{\varphi}(n, \lambda_n) = \arg\min\{\frac{1}{n}\sum_{i=1}^{n}(y'_i - \varphi^T z_i)^2 + \lambda_n \parallel \varphi \parallel^2\} \qquad (1)$$

where λ_n is the regularization parameter. $\|\varphi\|^2$ is the Tikhonov regularization term which penalizes the smoothness of the classifier. Finally, for a testing datum x, OVPC applies some simple inverse transformations on the estimator $\hat{\varphi}(n, \lambda_n)$ to get the final classification result.

Though OVPC has been shown much better performance in applications such as video object classification [8,9], it still has some limitations. On the one hand, the regularization term in OVPC is still the common Tikhonov term which cannot fully mine the underlying discriminative information inside the pairwise constraints. On the other hand, OVPC also ignores the structural information of the data distribution which can be used to enhance the classification performance.

3   Semi-Supervised Discriminatively Regularized Classifier(SSDRC)

In this section, we present SSDRC which introduces a new discriminative regularization term into the classifier instead of the smoothness-driven Tikhonov term. As a result, SSDRC can not only mine the discriminative information in the pairwise constraints more sufficiently but also preserve the local geometry of the new samples (z_i, y'_i) derived from pairwise constraints.

3.1   Data Pair Transformation

Inspired by OVPC, SSDRC firstly projects the data pair in each pairwise constraint into a new space $\tilde{X}$ as a single sample. Given the pairwise constraints

$(x_{11}, x_{12}, y_1'), ..., (x_{n1}, x_{n2}, y_n')$, where $x_i \in R^p$. Let A denote the result of outer product between the data pair in each pairwise constraint, that is,

$$A = x_{i1} \circ x_{i2} \qquad (2)$$

Here, we also use the operator *vech* which returns the upper triangular elements of a symmetric matrix in order of row as the new sample z, whose length is $(p + 1) \times p/2$. That is,

$$z = vech(A + A^T - diag(A)) \qquad (3)$$

For example, given $A = [a_{i,j}] \in R^{3 \times 3}$, $z = [a_{11}, a_{12} + a_{21}, a_{13} + a_{31}, a_{22}, a_{23} + a_{32}, a_{33}]$. Consequently, the pairwise constraints $(x_{11}, x_{12}, y_1'), ..., (x_{n1}, x_{n2}, y_n')$ in the original space X are transformed into new samples $(z_1, y_1'), ..., (z_n, y_n')$ in the new transformed space $\tilde{X}$.

3.2   Classifier Design in the Transformed Space

In view of the limitations in OVPC, here we aim to further fuse the discriminative and structural information hidden in the new samples into the classifier. Obviously, through the data pair transformation, the semi-supervised classification problem in the original space X has been transformed to a supervised binary-class classification task in the transformed space $\tilde{X}$, which can be solved by some state-of-the-art supervised classifiers. In terms of the least-squares loss function, Xue et al. [10] have proposed a new discriminatively regularized least-squares classifier(DRLSC). Instead of the common Tikhonov regularization term, DRLSC defines a discriminative regularization term

$$R_{disreg}(f, \eta) = \eta A(f) - (1 - \eta)B(f) \qquad (4)$$

where $A(f)$ and $B(f)$ are the matrices which measure the intra-class compactness and inter-class separability of the data respectively. η is the regularization parameter which controls the relative significance of $A(f)$ and $B(f)$.

Following the line of the research in DRLSC, we further introduce the discriminative regularization term into the classifier design in $\tilde{X}$. Based on the spectral theory[14], we also use two graphs, intra-class graph G_w and inter-class graph G_b with the weight matrices W_w and W_b respectively to define $A(f)$ and $B(f)$, which can characterize the local geometry of the sample distribution in order to utilize the structural information of the new samples better.

Concretely, for each sample z_i, let $ne(z_i)$ denote its k nearest neighborhood and divide $ne(z_i)$ into two non-overlapping subsets $ne_w(z_i)$ and $ne_b(z_i)$. That is,

$$ne_w(z_i) = \{z_i^j | \text{ if } y_i' = y_j', 1 \leq j \leq k\}$$

$$ne_b(z_i) = \{z_i^j | \text{ if } y_i' \neq y_j', 1 \leq j \leq k\}$$

Then we put edges between z_i and its neighbors, and thus obtain the intra-class graph and inter-class graph respectively. The corresponding weights are defined as follows:

$$W_{w,ij} = \begin{cases} 1, & \text{if } z_j \in ne_w(z_i) \text{ or } z_i \in ne_w(z_j); \\ 0, & \text{otherwise.} \end{cases}$$

$$W_{b,ij} = \begin{cases} 1, & \text{if } z_j \in ne_b(z_i) \text{ or } z_i \in ne_b(z_j); \\ 0, & \text{otherwise.} \end{cases}$$

The goal of SSDRC is to keep the neighboring samples of G_w as close as possible while separate the connected samples of G_b as far as possible. Thus,

$$A(f) = \frac{1}{2} \sum_{i=1}^{n} \sum_{j=1}^{n} W_{w,ij} \parallel f(z_i) - f(z_j) \parallel^2$$

Similarly,

$$B(f) = \frac{1}{2} \sum_{i=1}^{n} \sum_{j=1}^{n} W_{b,ij} \parallel f(z_i) - f(z_j) \parallel^2$$

Assume that the classifier has a linear form, that is,

$$f(z) = w^T z \tag{5}$$

Substitute the equation (5) into $A(f)$ and $B(f)$ and then obtain

$$A(f) = \frac{1}{2} \sum_{i=1}^{n} \sum_{j=1}^{n} W_{w,ij} \parallel f(z_i) - f(z_j) \parallel^2$$

$$= w^T Z (D_w - W_w) Z^T w$$

$$= w^T Z L_w Z^T w$$

where D_w is a diagonal matrix and its entries $D_{w,ij} = \sum_j W_{w,ij}$, $L_w = D_w - W_w$ is the laplacian matrix of G_w.

$$B(f) = \frac{1}{2} \sum_{i=1}^{n} \sum_{j=1}^{n} W_{b,ij} \parallel f(z_i) - f(z_j) \parallel^2$$

$$= w^T Z (D_b - W_b) Z^T w$$

$$= w^T Z L_b Z^T w$$

where D_b is a diagonal matrix and its entries $D_{b,ij} = \sum_j W_{b,ij}$, $L_b = D_b - W_b$ is the laplacian matrix of G_b.

The final optimization function can be formulated as

$$\min\{\frac{1}{n} \sum_{i=1}^{n} (y_i' - f(z_i))^2 + \eta A(f) - (1-\eta)B(f)\} \tag{6}$$

that is,

$$\min\{\frac{1}{n} \sum_{i=1}^{n} (y_i' - w^T z_i)^2 + w^T Z[\eta L_w - (1-\eta)L_b] Z^T w\} \tag{7}$$

The solution of the optimization function can follow from solving a set of linear equations by embedding equality type constraints in the formulation. Interested reader can refer the literature [10] for more details.

It is worthy to point out that, although the classifier design in SSDRC is similar to DRLSC, the corresponding classifier is defined in the transformed space $\tilde{X}$ rather than the original space as that in DRLSC. As a result, for a new testing sample, SSDRC should firstly conduct the classifier in the transformed space and then get the final classifier through some additional inverse transformations rather than DRLSC that predicts the class label in the original space directly. The particular process of the inverse transformation will be given in the next subsection.

3.3   Classification in the Original Space

Here we apply the inverse operation of *vech* to the discriminative vector w obtained in the transformed space $\tilde{X}$, resulting in a $p \times p$ symmetric matrix Θ. That is,

$$\Theta = vech^{-1}(w) \tag{8}$$

For example, given $w = \{a_{11}, a_{12}, a_{13}, a_{22}, a_{23}, a_{33}\}$, $\Theta = [a_{11}, a_{12}, a_{13}; a_{12}, a_{22}, a_{23}; a_{13}, a_{23}, a_{33}] \in R^{3 \times 3}$, . Then we perform the eigen-decomposition to the symmetric matrix Θ, and obtain the largest eigenvalue s_1 and its corresponding eigenvector u_1. We select

$$\hat{\theta} = \sqrt{s_1} u_1$$

as the sign-insensitive estimator of $\hat{\beta}$, which is the discriminative vector in the original space. Here sign-insensitive means that the real sign of $\hat{\beta}$ is still unknown. The labeled sample $(x_1, y_1), ..., (x_m, y_m)$ are used to determine the correct sign of $\hat{\beta}$.

To be more specific, the real sign of $\hat{\beta}$ can be computed as

$$s(\hat{\theta}) \begin{cases} +1, \sum_{i=1}^{m} I(y_i \hat{\theta}^T x_i) \geq \lceil \frac{m}{2} \rceil; \\ -1, \text{otherwise.} \end{cases} \tag{9}$$

where $\lceil t \rceil$ is the ceil function which returns the smallest integer value that is no less than t [5]. So the real estimator of the discriminative vector in the original space is

$$\hat{\beta} = s(\hat{\theta})\hat{\theta} \tag{10}$$

For a new testing datum x, the predicted class label is

$$\tilde{y} = \hat{\beta}^T x \tag{11}$$

3.4   The Pseudo-code for SSDRC

Based on the previous analysis, we present the SSDRC method. The corresponding pseudo-code is summarized in Algorithms 1.

input : Labeled Samples$\{(x_i, y_i)\}_{i=1}^m$;
         Pairwise Constraints $\{(x_{j1}, x_{j2}, y'_j)_{j=1}^n\}$;
         The number k of the nearest neighbors of new sample z_i derived
         from the pairwise constraints;
         The regularization parameters η $(0 \leq \eta \leq 1)$
output: the estimator $\hat{\beta}$ of discriminative vector

for $i \leftarrow 1$ **to** n **do**
 $\quad$ $A = x_{i1} \circ x_{i2}$;
 $\quad$ $z_i = vech(A + A^T - diag(A))$;
end
for $j \leftarrow 1$ **to** n **do**
 $\quad$ $z_j^k \leftarrow$ kth nearest neighbor of z_j among $(z_i)_{i=1}^n$;
end
for $i \leftarrow 1$ **to** n **do**
 $\quad$ **for** $j \leftarrow 1$ **to** k **do**
 $\quad\quad$ **if** $y'_i = y'_j$ **then** $W_{w,ij} \leftarrow 1$;
 $\quad\quad$ **else** $W_{b,ij} \leftarrow 1$;
 $\quad$ **end**
end
$D_{w,ij} = \sum_j W_{w,ij}$; $D_{b,ij} = \sum_j W_{b,ij}$;
$L_w = D_w - W_w$; $L_b = D_b - W_b$;
$w \leftarrow$ solve the optimization function:

$$\arg\min\{\frac{1}{n}\sum_{i=1}^n (y'_i - w^T z_i)^2 + w^T Z[\eta L_w - (1-\eta)L_b]Z^T w\}$$

$\Theta = vech^{-1}(w)$
Compute the largest eigenvalue s_1 and its corresponding eigenvector u_1 of Θ
$\hat{\theta} = \sqrt{s_1} u_1$
if $\sum_{i=1}^m I(y_i \hat{\theta}^T x_i \geq 0) \geq \lceil \frac{m}{2} \rceil$ **then** $s(\hat{\theta}) = +1$;
else $s(\hat{\theta}) = -1$;

$$\hat{\beta} = s(\hat{\theta})\hat{\theta}$$

Algorithm 1. Pseudo-code for SSDRC

4   Experiments

In this section, we evaluate the performance of our SSDRC algorithms on the real-word classification datasets: six datasets in UCI[1] and IDA datasets[2] in comparison to some state-of-the-art algorithms shown in the Table 1. We select the supervised method RLSC as the baseline. OVPC and PKLR are two popular semi-supervised classifiers with pairwise constraints.

---

[1] The dataset is available from
   http://www.ics.uci.edu/ mlearn/MLRepository.html
[2] The database is available from
   http://ida.first.fraunhofer.de/projects/bench/benchmarks.htm

Table 1. The acronyms, full names and citations algorithms compared with the SS-DRC in the experiments

Acronym	Full name	Citation
RLSC	Regularized Least Square Classifier	[18]
OVPC	On the Value of Pairwise Constraints	[8]
PKLR	Pairwise Kernel Logistic Regression	[9]

4.1 UCI Dataset

In this section we compare the relative performance of SSDRC with other three classification algorithms on six datasets in UCI, namely *Water(39, 116), Sonar(60, 208), Ionosphere(34, 351), Wdbc*(31,569), *Pima*(9,768), *Spambase*(58,4600). These datasets are typical binary-classification datasets in UCI. To be more specific, the first element in the brackets represents the dimension while the second means the number of samples in each dataset. Notice that the scale of the dataset is increment, and we divide each dataset into two equal parts. One is for training set and the other is for testing set. In our experiment, the pairwise constraints are obtained randomly selecting pairs of instances from the training set, and creating must-link and cannot-link constraints. The number of constraints is changed from 10 to 50 at a rate of 10 increment. Moreover, In SSDRC, the number of the k nearest neighbors is selected from $\{5, 10, 15, 20\}$. Especially, when the number of pairwise constraints is 10 which is relatively less to select the large number of nearest neighbors, the value of k is selected from $\{5, 10\}$. Moreover, in PKLR, we select the liner kernel as the kernel function. The regularization parameters λ in OVPC and PKLR are selected from $\{2^{-10}, 2^{-9}, ..., 2^9, 2^{10}\}$, and the regularization parameter η in SSDRC is chosen in $[0, 0.1, ..., 0.9, 1]$. All the parameter selections are done by cross-validation. Since labeled samples are only used to determine the real sign of the estimator, so we only select one sample for each class. The whole process is repeated 100 runs and the average results are reported.

Figure 1 shows the corresponding average classification accuracies of the algorithms in the six datasets. From the figure, we can see that the accuracies of OVPC, PKLR and SSDRC are basically improved with the increase of the number of the pairwise constraints step by step, which validates the "No Free Lunch" Theorem [14], that is, with more prior information incorporated, the better classification performance we can get. In the comparison of the four algorithms, the performance of RLSC is always the worst as a straight line, since it only uses the limited labeled data, which justifies the significance of the pairwise constraints data in semi-supervised learning. Furthermore, SSDRC outperforms PKLR and OVPC at each same number of pairwise constraints in all the six datasets, especially in Water, Wdbc and Pima, with more than 10% improvement in average. Besides, the variance of experimental result in SSDRC is much less than the ones in other three algorithms on most datasets. This also demonstrates that the utilization of pairwise constraints and structural information in SSDRC is much better than PKLR and OVPC.

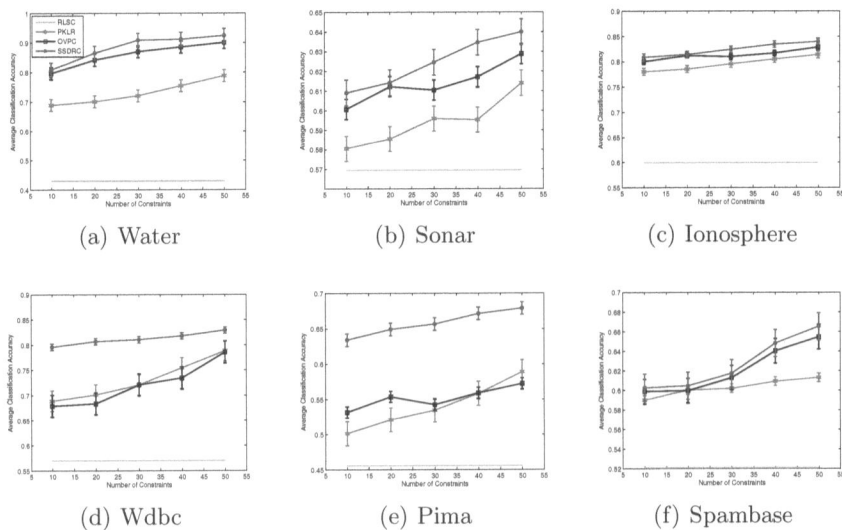

Fig. 1. Classification accuracy on UCI database with different number of constraints

Table 2. The attributes of the thirteen datasets in the IDA database

Dataset	Dimension	Training Set Size	Testing Set Size
Heart	13	170	100
Banana	2	400	4900
Breast-cancer	9	200	77
Diabetis	8	468	300
Flare-solar	9	666	400
German	20	700	300
Ringnorm	20	400	7000
Thyroid	5	140	75
Titanic	3	150	2051
Twonorm	20	400	7000
Waveform	21	400	4600
Image	18	1300	1010
Splice	60	1000	2175

4.2   IDA Database

In this subsection, we further evaluate the performance of the SSDRC algorithm
on the IDA database, which consists of thirteen datasets, and all of them has
two classes. The training and testing sets have been offered in each dataset already. Table 2 shows the attributes of the thirteen datasets in the IDA database:
the number of dataset's dimension, the size of training set and testing set respectively. The experimental settings are the same as those in the previous UCI
datasets.

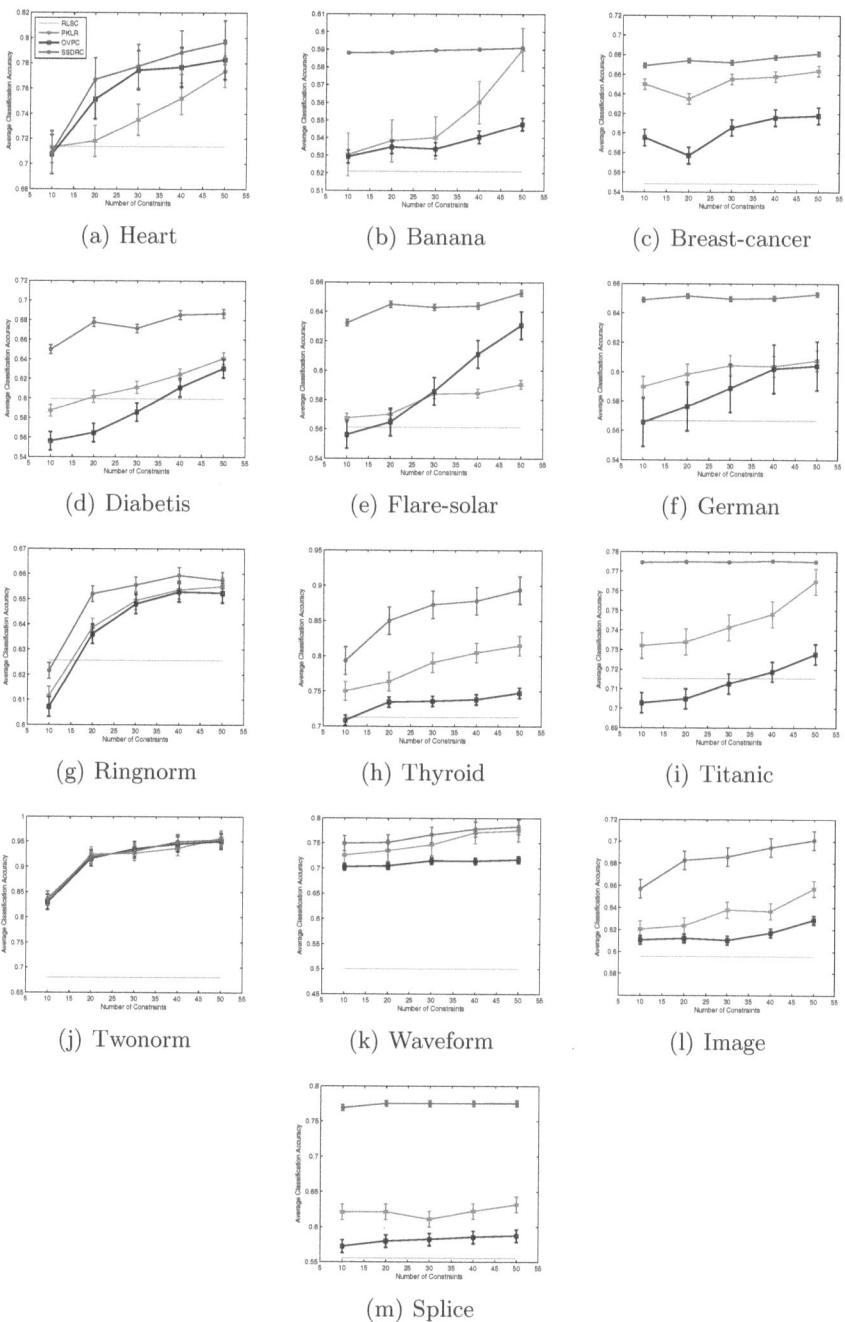

Fig. 2. Classification accuracy on IDA database with different number of constraints

Figure 2 shows the corresponding average classification accuracies of the four algorithms in the IDA datasets. From the figure, we can see that the SSDRC outperforms other three algorithms obviously in most datasets, especially in Banana, Splice, Titanic and German. The reason more likely lies in that different from the other algorithms, SSDRC embeds the local structure involved in data and makes use of the discriminative information in constraints more sufficiently, which results in its superior performance in the real-world classification tasks.

5   Conclusion

In this paper, we propose a novel classification method with pairwise constraints SSDRC. Different form many existing classifiers, SSDRC firstly transforms the data pairs in pairwise constraints into some new samples and then designs a discriminability-driven regularized classifier in the transformed space, which can not only fully capture the discriminative information in the constraints but also preserve the local structure of these new samples. Experimental results demonstrates that SSDRC is much better than the popular related classifiers OVPC and PKLR.

Throughout the paper, SSDRC focuses on the binary classification problems. How to extend SSDRC to the multi-class problems deserves our further work. Furthermore, the kernelization of SSDRC also needs more study.

Acknowledgements. This work was supported by National Natural Science Foundation of China (Grant No. 60905002) and the Scientific Research Startup Project for New Doctoral Faculties of Southeast University. Furthermore, the work was also supported by the Key Laboratory of Computer Network and Information Integration (Southeast University), Ministry of Education.

References

1. Zhu, X.: Semi-supervised learning literature survey. Technical Report, 1530, Madison: Department of Computer Sciences, University of Wisconsin (2008)
2. Chapelle, O., Schölkopf, B., Zien, A.: Semi-Supervised Learning. MIT Press, Cambridge (2006)
3. Han, X., Chen, Y., Xiang, R.: Semi-supervised and Interactive Semantic Concept Learning for Scene Recognition. In: 20th International Conference on Pattern Recognition, pp. 3045–3048 (2010)
4. Martnez-Us, A., Sotoca, J.M.: A Semi-supervised Gaussian Mixture Model for Image Segmentation. In: 20th International Conference on Pattern Recognition, pp. 2941–2944 (2010)
5. Tang, W., Zhong, S.: Pairwise constraints-guided dimensinality reduction. In: SDM 2006 Workshop on Feature Selection for Data Mining, Bethesda, MD (2006)
6. Zhang, D., Zhou, Z., Chen, S.: Semi-supervised dimensionality reduction. In: SIAM Conference on Data Mining, ICDM (2007)

7. Chen, S., Zhang, D.: Semi-supervised dimensionality reduction with pairwise constraints for hyper spectral image classification. IEEE Geoscience and Remote Semsing Letters 8(2), 369–373 (2011)
8. Zhang, J., Yan, R.: On the value of pairwise constraints in classification and consistency. In: Proceedings of the 24th International Conference on Machine Learning, pp. 1111–1118 (2007)
9. Yan, R., Zhang, J., Yang, J., Hauptmann, A.: A discriminative learning framework with pairwise constraints for video object classification. IEEE Transactions on Pattern Analysis and Machine Intelligence 28(4), 578–593 (2006)
10. Xue, H., Chen, S., Yang, Q.: Discriminatively regularized least-squares classification. Pattern Recognition 42(1), 93–104 (2009)
11. Yeung, D.S., Wang, D., Ng, W.W.Y., Tsang, E.C.C., Zhao, X.: Structured large margin machines: sensitive to data distributions. Machine Learning 68, 171–200 (2007)
12. Xue, H., Chen, S., Yang, Q.: Structural regularized support vector machine: a framework for structural large margin classifier. IEEE Trans. on Neural Networks 22(4), 573–587 (2011)
13. Lanckriet, G.R.G., Ghaoui, L.E., Bhattacharyya, C., Jordan, M.I.: A robust minimax approach to classfication. J. Machine Learning Research 3, 555–582 (2002)
14. Huang, K., Yang, H., King, I., Lyu, M.R.: Learning large margin classifiers locally and globally. In: The 21th International Conference on Machine Learning, ICML (2004)
15. Belkin, M., Niyogi, P., Sindhwani, V.: Manifold regularization: A geometric framework for learning from examples. Department of Computer Science, University of Chicago, Tech. Rep: TR-2004-06 (2004)
16. Duda, R.O., Hart, P.E., Stork, D.G.: Pattern Classification. Wiley, New York (2001)
17. Pan, J.J., Pan, S.J., Yin, J., Ni, L.M., Yang, Q.: Tracking mobile users in wireless networks via semi-supervised colocalization. IEEE Trans. on Pattern Analysis and Machine Intelligence 34(3), 587–600 (2012)
18. Haykin, S.: Neural Networks: A Comprehensive Foundation. Tsinghua University

Outdoor Situation Recognition Using Support Vector Machine for the Blind and the Visually Impaired

Jihye Hwang, Kyung-tai Kim, and Eun Yi Kim

Visual Information Processing Labratory, Department of Advanced Technology Fusion,
Konkuk University, Hwayang-dong, Gwangjin-gu, Seoul, South Korea
{hjh881120,eykim04}@gmail.com, kkt1341@konkuk.ac.kr

Abstract. Traffic intersections are most dangerous situations for the pedestrian, in particular the blind or the visually impaired person. In this paper, we present a novel method for automatically recognizing the situations where a user stands on, to help safe mobility of the visually impaired in their travels. Here, the situation means the place type where a user is standing on, which is classified as sidewalk, roadway and intersection. The proposed method is performed by three steps: ROIs extraction, feature extraction and classification. The ROIs corresponding to the boundaries between sidewalks and roadways are first extracted using Canny edge detector and Hough transform. From those regions, features are extracted using Fourier transform, and they fed into two SVMs. One SVM is trained to learn the textural properties of sidewalk and the other is for intersection. On online stage, these two SVMs are hierarchically performed; the current situation is first categorized as sidewalks and others, then it is re-categorized as intersections and others. The proposed method was tested with about 500 outdoor images, then it showed the accuracy of 93.9 %.

Keywords: Traffic intersection, Situation recognition, Support vector machine, Fourier transform.

1 Introduction

Traffic intersections are most dangerous situations for the pedestrian, in particular the blind or the visually impaired person. Practically, the average rate of 22% among total accidents is occurred on intersections [1-3].

So far various solutions to safely cross intersections have been proposed and implemented. The accessible pedestrian signals (APS) [5] and Talking Signs [4] have been developed, however their adoption is not spread in many countries and some additional devices should be involved.

As alternative to these approaches, the vision-based methods such as "Crosswatch" [6] and "Bionic Eyeglasses" [7] have been recently proposed. These systems are hand-held devices and automatically find the location and orientation of crosswalk, thereby guiding the user to safely cross intersection.

However, such systems are missing the essential fact that the people require the different guidance solutions according to the situation where a user stands on. For

P. Anthony, M. Ishizuka, and D. Lukose (Eds.): PRICAI 2012, LNAI 7458, pp. 124–132, 2012.
© Springer-Verlag Berlin Heidelberg 2012

example, if a user is on intersection, he (or she) wants to find crosswalk. On the other hand, when a user is on the sidewalk, the system leads the user to walk along opposite side to the road. Accordingly, recognizing the outdoor situations where a user stands on is essentially performed.

In this paper, a novel method for automatically recognizing user's current situations is proposed to help safe mobility of the visually impaired in their travels. Here, the situation means the place type where a user is standing on, which is classified as sidewalk, roadway and intersection. As a key clue to recognize the situation, it uses the alignment of boundaries between sidewalks and roadways, which is the ROIs. Thus, it is performed by three steps: ROIs extraction, feature extraction and classification. It first extracts ROIs using Canny edge detector and Hough transform. From those regions, features are extracted using Fourier transform, and they fed into two SVMs. One SVM is trained to learn the textural properties of sidewalk and the other is for intersection. On online stage, these two SVMs are hierarchically performed; the current situation is first categorized as sidewalks and others, then it is re-categorized as intersections and others. The proposed method was tested with about 300 outdoor images, then it showed the accuracy of 93.9 %.

2   Outline of Proposed Method

The goal of this study is to develop the method to automatically recognize user's current situation, thereby providing user-centered guidance to the disabled people.

In this work, a *situation means the type of place the user is located*, which is categorized into three types: {roadway, sidewalk, and intersection}. Among these situations, the discrimination of sidewalk and intersection is more important.

For this, specific characteristics need to be identified according to the situation types, that is, visual properties associated with the respective situations. However, the diverse ground patterns were used to represent sidewalk, and the colors of ground are

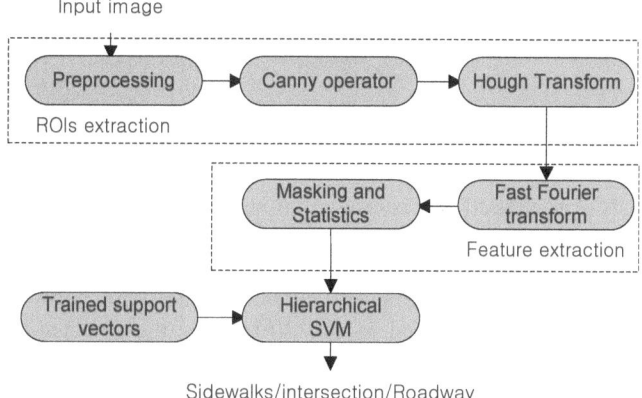

Fig. 1. System overview

likely to be distorted by shadows and time-varying illumination, so it is difficult to find robust characteristics. Accordingly, we use the support vector machines (SVMs) to understand the features of the sidewalk and intersection.

Figure 1 shows the outline of the proposed method, which is performed by three steps: ROIs extraction, feature extraction and classification. The ROIs corresponding to the boundaries between sidewalks and roadways are first extracted using Canny edge detector and Hough transform. From those regions, features are extracted and they fed into two SVMs. One SVM is trained to learn the textural properties of sidewalk and the other is for intersection. On online stage, these two SVMs are hierarchically performed; the current situation is first categorized as sidewalks and others, then it is re-categorized as intersections and others.

3   ROIs Extraction

We assume the pixels around boundaries of sidewalks and roadways have distinctive textures. At such boundaries, we can easily observe the vertical or horizontal lines, as shown in Figure 2(a). Then, the boundaries oriented to the horizontal were found in the images corresponding to intersection, whereas boundaries close to the vertical were observed in the images corresponding to sidewalks.

Therefore, the proposed method first detects the boundaries in between roadways and sidewalks. The boundary detection is performed by three steps: preprocessing, Canny operator and Hough transform, as shown in Figure. 1. As a preprocessing, histogram equalization is first applied to an input image for higher contrast and then Gaussian smoothing filter is used to remove noise. Thereafter, the input image is binarized using Canny operator, then followed by Hough transform.

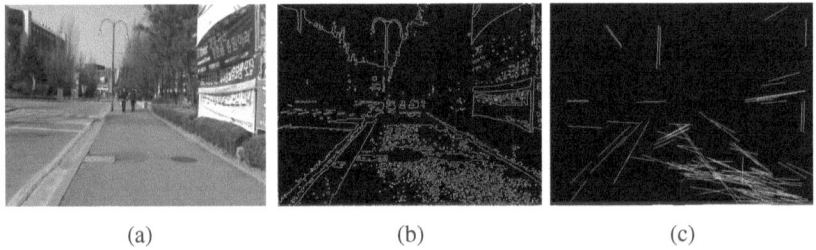

(a)   (b)   (c)

Fig. 2. Regions of interests (ROI) extraction: (a) input image, (b) edges detected by Canny operator (c) lines detected by Hough transform

Figure 2 illustrate how the ROIs are extracted. Figure 2(b) shows the edge images produced by Canny operator, and Figure 2(c) shows the detected lines. However, although some boundaries are correctly extracted, there are too many false alarms to be removed. For this, the textural properties are used.

4   Feature Extraction

To classify the outdoor situation, textural properties of ROIs are considered. To describe textural properties of intersection and sidewalk, various visual information have been considered such as histogram of oriented gradient (HOG), fast Fourier transform (FFT) coefficients, and so on. Through experiments, FFT coefficients were proven to show the better performance in discriminating the situations of three types.

The FFT is applied to the gray scale values of pixels within ROIs and its neighboring pixels. Through the experiments, 128×128 sized window was selected to define the neighborhood.

Figure 3 shows sample images corresponding to sidewalks and intersections, respectively. Figures 3(a) shows the original images, where the images have diverse ground patterns even if they belong to the same category. Then, Figure 3(b) shows the transformed images to frequency domain using FFT. As shown in Figure 3(b), FFT coefficients are similarly distributed within the same situation, despite of the different patterns and viewing angles, whereas they are distinctively distributed between different situations.

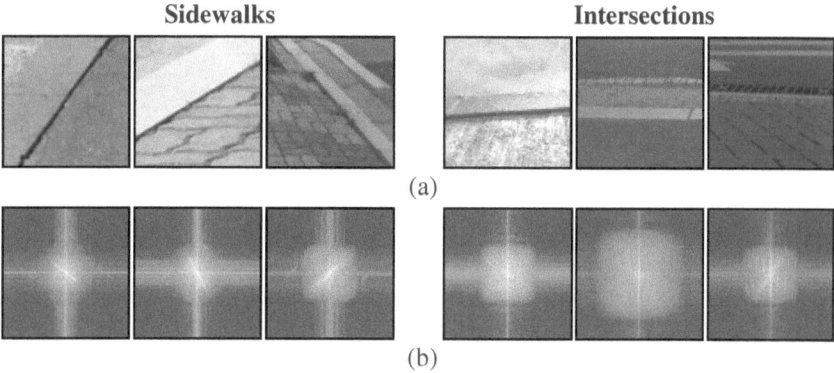

Fig. 3. Visual feature extraction using FFT: (a) original images, (b) Fourier transformed image

As shown in Figure 3(b), the prominent difference in between sidewalks and intersections is found on the center of images and nearby the cross: 1) the frequencies on the images corresponding to sidewalks are less concentrated on the center than ones on the interaction, and 2) any specific orientations were found from the images of sidewalks.

To capture such distinctive features and filter non-necessary ones, a flower-shaped mask is newly designed as shown in Figure 4. It is composed of five leafs. Then using the whole of FFT coefficients within the mask as an input of SVM is too time-consuming. Thus, we calculate the statistics of FFT coefficients, using moments.

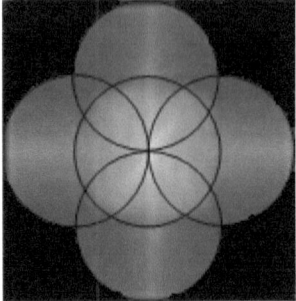

Fig. 4. A flower-shaped mask to filter unnecessary FFT coefficients: it composed of five leafs, then we calculate the statists of FFT coefficients within a leaf

Given an 128×128 sub-block, I, $M(I)$ is the average value, and $\mu_2(I)$ and $\mu_3(I)$ are the second-order and third-order central moments, respectively.

$$M(I) = \frac{1}{N^2} \sum_{i=0}^{N-1} \sum_{j=0}^{N-1} I(i,j)$$

$$\mu_2(I) = \frac{1}{N^2} \sum_{i=0}^{N-1} \sum_{j=0}^{N-1} \big(I(i,j) - M(I)\big)^2$$

$$\mu_3(I) = \frac{1}{N^2} \sum_{i=0}^{N-1} \sum_{j=0}^{N-1} \big(I(i,j) - M(I)\big)^3$$

Thus, as the mask is composed of five leaves, a 15-D feature vector is extracted from every pixel corresponding to ROIs. The vector is fed into SVM-based texture classifier to determine which situation is assigned to the pixel.

5   Classification

To determine the current situation out of three possible situations, {sidewalk, roadway, intersection}, two SVMs are hierarchically used, as follows:

$$O(x,y) = \begin{cases} 2 & \text{if} \quad SVM_s(x,y) > 0 \\ 1 & \text{else if} \quad SVM_i(x,y) > 0 \\ 0 & \text{else} \end{cases},$$

where $SVM_s(x,y)$ classifies a pixel's class as sidewalk and others, by learning the textural properties between sidewalk and others. Similarly, $SVM_i(x,y)$ discriminates a pixel's class as intersection and roadway, by learning the textures between intersection and roadway.

On online stage, the current situation is first categorized as sidewalk and others, then it is re-categorized as intersection and roadway.

Figure 5 shows the classification results. For the input image of Figure 2(a), the classification result is shown in Figure 5(a), where a red-color pixel and a blue-color are used to denote pixels' class as sidewalk and intersection, respectively.

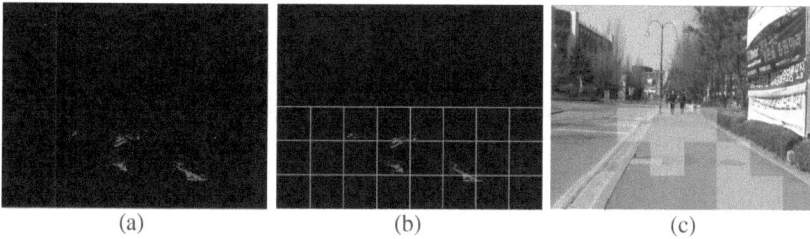

| (a) | (b) | (c) |

Fig. 5. Sample of situation recognition result: (a) classification results for pixels corresponding to ROIs, (b) classification result overlapped to a grid map, (c) cell classification result

As shown in Figure 5(a), the classification result has some errors (noises), as the decision is locally performed on each pixel. Thus, the smoothing is performed on the texture classification results to globally combine the individual decisions on a whole image. For this, we design the grid map, where each cell is sized at 80×80 (See Figure 5(b)). This grid map is overlaid onto the classification result, then the decision is made on each cell, instead of on every pixel. The situation of a cell is determined by the majority rule. For example, if most pixels of a cell are assigned as intersections, the cell is also considered as intersection. Figure 5(c) shows the final classification result, where the red-block and the blue block denote the sidewalk and intersection classes, respectively.

6   Experimental Results

To assess the effectiveness of the proposed recognition method, experiments were performed on the images obtained from outdoors. A total of 500 images were collected: 300 images were captured from real outdoors using iPhone4 and the others were downloaded from Google street view at San Francisco. In case of the former data, each place was captured while pedestrian is walking, so they were captured on a variety of viewpoints. Among 300 images, 100 images were used for training SVMs and the others were used for testing.

Figure 6 shows some images that are captured at various viewpoints, where the images have the diverse ground patterns and colors even if they belong to the same categories. These properties make the classification process is difficult.

| (a) | (b) |

Fig. 6. Examples of collected data: (a) images downloaded from Google street view (b) images captured from real outdoors using iPhone4

Figure 7 shows some classification results, where the original images are shown in Figure 7(a). Then, extracted ROIs and texture classification results on every cell are shown in Figures. 7(b) and (c), respectively. For the first two images, the proposed method accurately classified all pixels and all cells, however some errors are occurred for the last image. As shown in the bottom of Figure 7(c), some cells are colored by pink color, which means it has similar probabilities to be assigned as intersection and sidewalk, that is, the major voting is not found on the cells. We guess these errors were caused by crosswalk which has the texture characteristics of both sidewalk and intersection, and by shadows of surrounding buildings. Although some cells are mis-classified, a majority of cells is classified as intersection, thus its situation is considered as intersection.

The results showed that the proposed method have a robust performance to the pattern of ground and viewing angle.

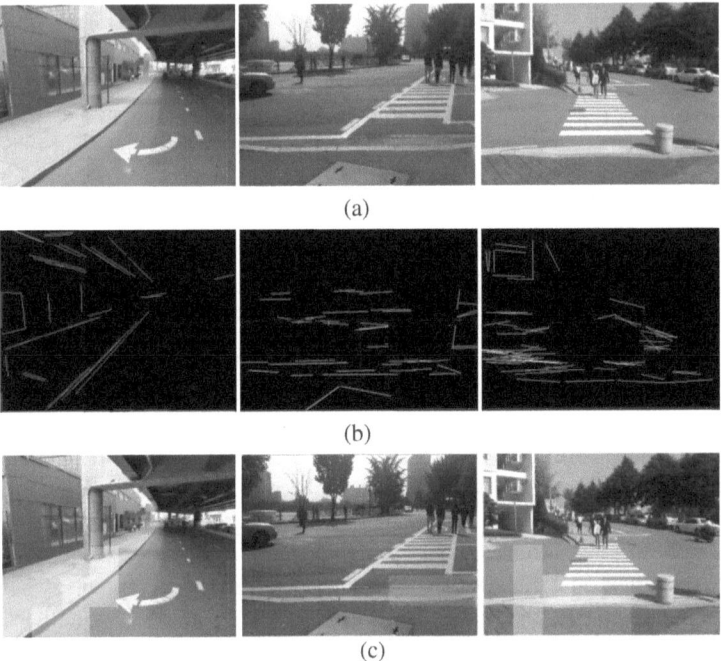

(a)

(b)

(c)

Fig. 7. Results of outdoor situation recognition: (a) origin images (b) results of ROIs Extraction (c) SVM results (d) cell classification results

Table 1. Confusion matrix of situation recognition (%)

	Sidewalks	Intersection	Roads
Sidewalks	91.0	0	0
Intersections	0	90.8	0
Roads	9.0	10.2	100

Table 1 summarizes the performance of the situation recognition under various outdoor environments. The average accuracy was about 93.9%. For the roadway, the proposed method showed the perfect recognition, which is very important as mis-recognizing the roadways as sidewalks or intersections can cause the collisions of vehicles. On the other hand, it has relatively low precisions for the sidewalks and intersections, which seems to be caused by camera's viewing angle and various patterns of sidewalks.

Currently, we considered only textural properties to discriminate outdoor situations, however, the distinctive features in accordance with geometric shape are found between intersection and sidewalk. So, we expect that using both textural and shape properties can improve the classification results, which is under working.

Table 2. Processing Time

Modules	Processing time (.ms)
ROIs Extraction	141.2716
Feature Extraction	1.38734
Texture Classification	6.93
Total	149.58894

The proposed method aims at providing the most appropriate guidance solution to users by recognizing their current situation, thus it should be performed in real-time.

Table 2 shows the processing time in proposed method, when it was performed on an Intel Core2 CPU 2.40 GHz. As shown in Table 2, the average time taken to process a frame through all the stages was about 150ms, thereby an average processing of up to 7 frames/second.

The experiment results showed that the proposed system could provide essential guidance information for people with visual and cognitive impairments.

7   Conclusions

Individuals with visual impairments face a daily challenge to their ability to move about alone. In particular, urban intersections are known as the most dangerous parts of the disabled people's travel. Accordingly, this paper presents a new method to recognize the outdoor situations. It was performed by three steps: ROIs extraction, feature extraction, texture classification using hierarchical SVMs. Then, for learning the texture properties of ROIs on the respective situations, FFT coefficients were extracted from ROIs, and their statics were used as the input of SVMs. To demonstrate the validity of the proposed method, it was tested with 500 images, then it showed the accuracy of 93.9%.

Acknowledgement. "This research was supported by Basic Science Research Program through the National Research Foundation of Korea(NRF) funded by the Ministry of Education, Science and Technology(grant number)" (20110002565) and (20110005900).

References

1. Ju, J., Shin, Y., Kim, E.: Vision based interface system for hands free control of an intelligent wheelchair. Journal of NeuroEngineering and Rehabilitation (2009)
2. Ko, E., Ju, J., Kim, E.: Situation-based indoor wayfinding system for the visually impaired. In: ACM ASSETS 2011, pp. 35–42 (2011)
3. Ju, J., Ko, E., Kim, E.: EYECane: Navigating with camera embedded white cane for visually impaired person. In: ACM ASSETS 2009, pp. 237–238 (2009)
4. Brabyn, J., Crandall, W., Gerrey, W.: Talking signs: a remote signage, solution for the blind, visually impaired and reading disabled. Engineering in Medicine and Biology Society (1993)
5. Barlow, J.M., Bentzen, B.L., Tabor, L.: Accessible pedestrian signals: Synthesis and guide to best practice. National Cooperative Highway Research Program (2003)
6. Ivanchenko, V., Coughlan, J.M., Shen, H.: Crosswatch: A Camera Phone System for Orienting Visually Impaired Pedestrians at Traffic Intersections. In: Miesenberger, K., Klaus, J., Zagler, W.L., Karshmer, A.I. (eds.) ICCHP 2008. LNCS, vol. 5105, pp. 1122–1128. Springer, Heidelberg (2008)
7. Karacs, K., Lazar, A., Wagner, R., Balya, D., Roska, T., Szuhaj, M.: Bionic Eyeglass: an Audio Guide for Visually Impaired. In: Biomedical Circuits and Systems, BioCAS 2006, pp. 190–193 (2006)

Mining Rules for Rewriting States in a Transition-Based Dependency Parser

Akihiro Inokuchi[1], Ayumu Yamaoka[1], Takashi Washio[1], Yuji Matsumoto[2], Masayuki Asahara[3], Masakazu Iwatate[4], and Hideto Kazawa[5]

[1] Institute of Scientific and Industrial Research, Osaka University
[2] Nara Institute of Science and Technology
[3] National Institute for Japanese Language and Linguistics
[4] HDE, Inc.
[5] Google, Inc.

Abstract. Methods for mining graph sequences have recently attracted considerable interest from researchers in the data-mining field. A graph sequence is one of the data structures that represent changing networks. The objective of graph sequence mining is to enumerate common changing patterns appearing more frequently than a given threshold from graph sequences. Syntactic dependency analysis has been recognized as a basic process in natural language processing. In a transition-based parser for dependency analysis, a transition sequence can be represented by a graph sequence where each graph, vertex, and edge respectively correspond to a state, word, and dependency. In this paper, we propose a method for mining rules for rewriting states reaching incorrect final states to states reaching the correct final state, and propose a dependency parser that uses rewriting rules. The proposed parser is comparable to conventional dependency parsers in terms of computational complexity.

1   Introduction

Data mining is used to mine useful knowledge from large amounts of data. Recently, methods for mining graph sequences (dynamic graphs [4] or evolving graphs [3]) have attracted considerable interest from researchers in the data-mining field [9]. For example, human networks can be represented by a graph where each vertex and edge respectively correspond to a human and relationship in the network. If a human joins or leaves the network, the numbers of vertices and edges in the graph increase or decrease. A graph sequence is one of the data structures used to represent a changing network. Figure 1(a) shows a graph sequence that consists of four steps, five vertices, and edges among the vertices. The objective of graph sequence mining is to enumerate subgraph subsequence patterns, one of which is shown in Fig. 1(b), appearing more frequently than a given threshold from graph sequences.

Syntactic dependency parsing has been recognized as a basic process in natural language processing, and a number of studies have been reported [12,14,16,8].

P. Anthony, M. Ishizuka, and D. Lukose (Eds.): PRICAI 2012, LNAI 7458, pp. 133–145, 2012.
© Springer-Verlag Berlin Heidelberg 2012

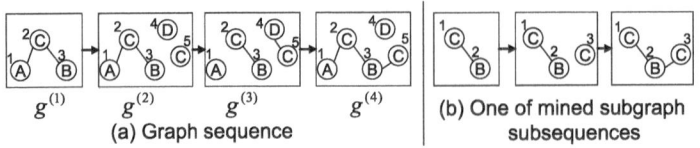

Fig. 1. Examples of a graph sequence and one of mined frequent patterns

One reason for its increasing popularity is the fact that dependency-based syntactic representations seem to be useful in many applications of language technology [11], such as machine translation [5] and information extraction [6]. In a transition-based dependency parser, a transition sequence can be represented by a graph sequence where each graph, vertex, and edge respectively correspond to a state, word, and dependency. Because of the nature of the algorithm where transition actions are selected deterministically, an incorrect selection of an action may adversely affect the remaining parsing actions. If characteristic patterns are mined from transition sequences for sentences analyzed incorrectly by a parser, it is possible to design new parsers and to generate better features in the machine learner in the parser to avoid incorrect dependency structures.

The first and main objective of this study is to demonstrate the usefulness of graph sequence mining in dependency analysis. Since methods for mining graph sequences were developed, they have been applied to social networks in Web services [3], article-citation networks [2], e-mail networks [4], and so on. In this paper, we demonstrate a novel application of graph sequence mining to dependency parsing in natural language processing. The second objective is to propose a method for mining rewriting rules that can shed light on why incorrect dependency structures are returned by transition-based dependency parsers. To mine such rules, the rules should be human-readable. If we identify the reason for incorrect dependency structures, it is possible to design new parsers and to generate better features in the machine learner in the parser to avoid incorrect dependency structures. The third objective is to propose a dependency parser that uses rewriting rules, where the method is comparable to conventional methods whose time complexity is linear with respect to the number of segments in a sentence. The fourth, but not a main, objective is to improve the attachment score, which is a measure of the percentage of segments that have the correct head, and the exact match score for measuring the percentage of completely and correctly parsed sentences.

2   Transition-Based Dependency Parsing

In this paper, we focus on dependency analysis using an "arc-standard parser" [14], which is a parser based on a transition system, for "Japanese sentences", for the sake of simplicity, because constraints in Japanese dependency structures are stronger than those in other languages. Japanese dependency structures have

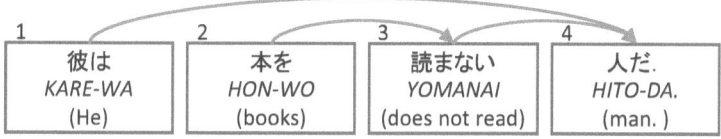

Fig. 2. Example of a Japanese sentence and its dependency structure

Parse$(x = \langle w_1, w_2, \cdots, w_n \rangle)$
1) $c \leftarrow c_s(x)$
2) while $c \notin C_F$
3) $c \leftarrow [o(c)](c)$
4) return $c_= (N, A)$

Transitions
Arc $(N, A) \Rightarrow (N, A \cup \{(i, j)\})$
       where $\{i, j\} = roots((N, A))$
Shift $(N, A) \Rightarrow (N \cup \{|N| + 1\}, A)$
Preconditions
Arc c is not a tree, but a forest.
Shift $|N| \neq n$

Fig. 3. Dependency parser based on   **Fig. 4.** Transitions of an arc-standard parser
a transition system

strictly head-final, single-head, single-rooted, connected, acyclic, and projective constraints [10]. However, the principle of the method proposed in this paper can basically be applied to any parser based on a transition system for sentences in any language.

Most Japanese dependency parsers are based on bunsetsu segments (hereafter segments), which are a similar in concept to English base phrases.

Definition 1. *A dependency structure for a sentence $x = \langle w_1, \cdots, w_n \rangle$ consisting of n segments is represented as a directed rooted tree $g = (V, E)$, where $V = \{1, \cdots, n\}$, $E \subset V \times V$, and n is the root of the tree.* ∎

Example 1. A dependency structure for a sentence $x = \langle KARE\text{-}WA, HON\text{-}WO, YOMANAI, HITO\text{-}DA.\rangle$ is represented by a directed graph without edge crossings, as shown in Fig. 2.

We define a transition-based dependency parser whose input is $x = \langle w_1, \cdots, w_n \rangle$ and output is $g = (V, E)$.

Definition 2. *A transition-based parser consists of $S = (C, T, c_s, C_F)$, where*
 − *$C = \{(N, A)\}$ is a set of states, where N and A are subsets of $V = \{1, \cdots, n\}$ and $N \times N$, respectively,*
 − *T is a set of transitions, where $t \in T$ is a partial function s.t. $t : C \rightarrow C$,*
 − *c_s is an initial function satisfying $c_s(x) = (\{1\}, \emptyset)$, and*
 − *$C_F \subseteq C$ is a set of final states, and $c_F \in C_F$ is a tree where n is the root.* ∎

A transition sequence for $x = \langle w_1, \cdots, w_n \rangle$ on $S = (C, T, c_s, C_F)$ is represented as $C_{1,m} = \langle c_1, \cdots, c_m \rangle$, satisfying (1) $c_1 = c_s(x)$, (2) $c_m \in C_F$, and (3) $\exists t \in T$ for c_i $(1 < i \leq m)$, $c_i = t(c_{i-1})$. We denote sets of vertices and edges for a state c as N_c and A_c, respectively.

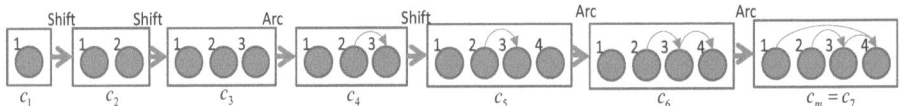

Fig. 5. Transition sequence for the sentence in Example 1

Definition 3. *A transition-based parser* $S = (C, T, c_s, C_F)$ *is incremental if* $N_c \subseteq N_{t(c)}$ *and* $A_c \subseteq A_{t(c)}$. ∎

If a dependency parser is incremental, the numbers of vertices and edges in a state $c = (N, C)$ increase monotonically. In addition, the state $c = (N_c, A_c)$ is a forest that is a set of ordered trees, and a graph (N_c, A_c) is a subgraph of $c_m = (N_{c_m}, A_{c_m})$ that S returns.

Figure 3 shows the algorithm of a transition-based dependency parser. In Fig. 3, o is an oracle for selecting t to transit to the next state in a deterministic way. In particular, the arc-standard parser, which is a transition-based parser, selects either Arc or Shift to analyze Japanese sentences, as shown in Fig. 4, where *roots* returns a pair of the largest roots $\{i, j\}$ ($i < j$) from a forest $c = (N, A)^1$. If o selects Arc, then an edge (i, j) is added to transit from c to $t(c)$. Otherwise, the smallest vertex that does not exist in $c = (N, A)$ is added to c to transit from c to $t(c)$. Since o is a function for determining whether the i-th segment is the dependent of the j-th segment, it is implemented with a binary classifier, such as a support vector machine (SVM), for feature vectors that characterize the i-th and j-th segments [11].

Since the arc-standard parser is incremental, Arc is selected $n - 1$ times and Shift is also selected $n - 1$ times to reach the final state. The time complexity of the parser for a sentence with n segments is therefore $O(n\theta)$, where we assume o, which is a binary classifier, returns its output in at most θ time.

Example 2. Figure 5 shows a transition sequence from the initial state to the final state for the dependency structure shown in Fig. 2. The words are omitted because of a lack of space. In the sequence, Shift, Shift, Arc, Shift, Arc, and Arc are selected by o, in that order.

Figure 6 shows the search space for the sentence in Example 1. Since a search space consisting of states is a tree, there is only one transition sequence from the initial state to the correct final state. In addition, the branching factor of the tree is at most two. Since the function o selects a transition between two branches, if the function o selects an incorrect transition once, the parser never reaches the correct final state. A straightforward approach to avoiding this mistake is to integrate backtracking or a probabilistic algorithm with the parser. However, this impairs the advantages of a parser whose time complexity is linear with respect to the number of segments in a sentence.

---

1 Although the arc-standard parser is defined using a stack and queue in many books and articles, in this paper, we define it using graphs to link dependency parsing to graph sequence mining.

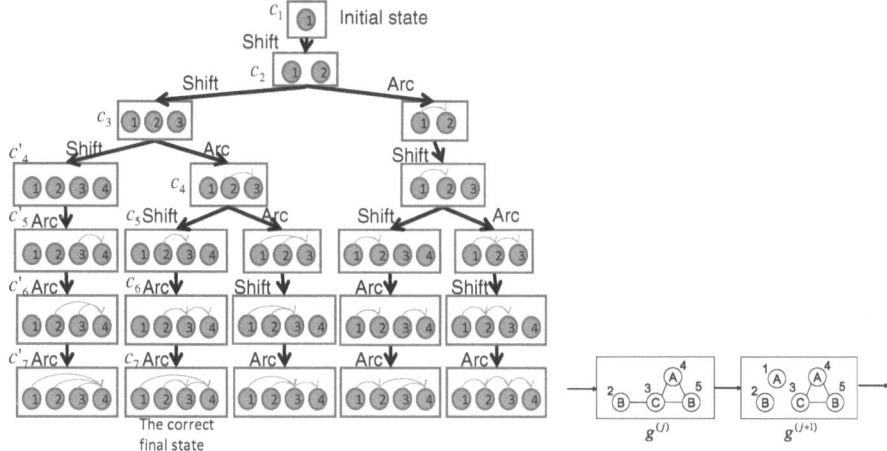

Fig. 6. Search space for the sentence in Example 1   **Fig. 7.** Change between two successive graphs

In this paper, we propose a method for mining rules for rewriting from states reaching incorrect final states to states reaching the correct final state, and propose a dependency parser that uses rewriting rules. The rewriting rules correspond to bypasses among states in the search tree shown in Fig. 6, and the proposed parser is comparable to a conventional dependency parser in terms of computational complexity. To describe the proposed method, we explain another method, called GTRACE, for mining graph sequences corresponding to transition sequences in the next section.

3   Graph Sequence Mining

Figure 1(a) shows an example of a graph sequence. The graph $g^{(j)}$ is the j-th labeled graph in the sequence. The problem we address in this section is how to mine patterns that appear more frequently than a given threshold from a set of graph sequences. We have proposed transformation rules for representing graph sequences compactly under the assumption that "the change is gradual" [9]. In other words, only a small part of the structure changes, while the other part remains unchanged between two successive graphs $g^{(j)}$ and $g^{(j+1)}$ in a graph sequence. For example, the change between two successive graphs $g^{(j)}$ and $g^{(j+1)}$ in the graph sequence shown in Fig. 7 is represented as an ordered list of two transformation rules $\langle vi^{(j)}_{[1,A]}, ed^{(j)}_{[(2,3),\bullet]} \rangle$. This list implies that a vertex with ID 1 and label A is inserted (vi), and then an edge between vertices with IDs 2 and 3 is deleted (ed). By assuming the change in each graph to be gradual, we can represent a graph sequence compactly, even if the graph in the graph sequence has many vertices and edges. We have also proposed a method, called GTRACE, for efficiently mining all frequent patterns from ordered lists of transformation

Table 1. TRs for representing graph sequence

Vertex Insertion $vi^{(j,k)}_{[u,l]}$	Insert a vertex u with label l into $g^{(j,k)}$ to transform to $g^{(j,k+1)}$.
Vertex Deletion $vd^{(j,k)}_{[u,\bullet]}$	Delete an isolated vertex u in $g^{(j,k)}$ to transform to $g^{(j,k+1)}$.
Vertex Relabeling $vr^{(j,k)}_{[u,l]}$	Relabel a label of a vertex u in $g^{(j,k)}$ as l to transform to $g^{(j,k+1)}$.
Edge Insertion $ei^{(j,k)}_{[(u_1,u_2),l]}$	Insert an edge with label l between vertices u_1 and u_2 into $g^{(j,k)}$ to transform to $g^{(j,k+1)}$.
Edge Deletion $ed^{(j,k)}_{[(u_1,u_2),\bullet]}$	Delete an edge between vertices u_1 and u_2 in $g^{(j,k)}$ to transform to $g^{(j,k+1)}$.
Edge Relabeling $er^{(j,k)}_{[(u_1,u_2),l]}$	Relabel a label of an edge between vertices u_1 and u_2 in $g^{(j,k)}$ as l to transform to $g^{(j,k+1)}$.

rules. A transition sequence in the dependency parser is represented as a graph sequence. In addition, since the change between two successive graphs in the graph sequence is an addition of a vertex (Shift) or of an edge (Arc), the assumption holds.

A labeled graph g is represented as $g = (V, E, L, l)$, where $V = \{1, \cdots, n\}$ is a set of vertices, $E \subseteq V \times V$ is a set of edges, and L is a set of labels such that $l : V \cup E \to L$. In addition, a graph sequence is an ordered list of labeled graphs and is represented as $d = \langle g^{(1)}, \cdots, g^{(z)} \rangle$.

To represent a graph sequence compactly, we focus on differences between two successive graphs $g^{(j)}$ and $g^{(j+1)}$ in the sequence.

Definition 4. *The differences between the graphs $g^{(j)}$ and $g^{(j+1)}$ in d are interpolated by a virtual sequence $d^{(j)} = \langle g^{(j,1)}, \cdots, g^{(j,m_j)} \rangle$, where $g^{(j,1)} = g^{(j)}$ and $g^{(j,m_j)} = g^{(j+1)}$. The graph sequence d is represented by the interpolations as $d = \langle d^{(1)}, \cdots, d^{(z-1)} \rangle$.* ■

The order of graphs $g^{(j)}$ represents the order of graphs in an observed sequence. On the other hand, the order of graphs $g^{(j,k)}$ is the order of graphs in the artificial interpolation, and there can be various interpolations between the graphs $g^{(j)}$ and $g^{(j+1)}$. We limit the interpolations to be compact and unambiguous by taking one having the shortest length in terms of the graph edit distance to reduce both the computational and spatial costs.

Definition 5. *Let a transformation of a graph by either insertion, deletion, or relabeling of a vertex or an edge be a unit, and let each unit have edit distance 1. A graph sequence $d^{(j)} = \langle g^{(j,1)}, \cdots, g^{(j,m_j)} \rangle$ is defined as an interpolation in which the edit distance between any two successive graphs is 1 and the edit distance between any two graphs is minimum.* ■

Transformations are represented in this paper by the following "transformation rule (TR)".

Definition 6. *A TR transforming $g^{(j,k)}$ to $g^{(j,k+1)}$ is represented by $tr^{(j,k)}_{[o_{jk},l_{jk}]}$, where*

- *tr is a transformation type that is either insertion, deletion, or relabeling of a vertex or an edge,*
- *o_{jk} is a vertex or edge to which the transformation is applied, and*
- *$l_{jk} \in L$ is a label to be assigned to the vertex or edge in the transformation.* ∎

For the sake of simplicity, we simplify $tr^{(j,k)}_{[o_{jk},l_{jk}]}$ to $tr^{(j,k)}_{[o,l]}$. We use six TRs in Table 1. In summary, we define a transformation sequence as follows.

Definition 7. *A graph sequence $d^{(j)} = \langle g^{(j,1)}, \cdots, g^{(j,m_j)} \rangle$ is represented by $s_d^{(j)} = \langle tr^{(j,1)}_{[o,l]}, \cdots, tr^{(j,m_j-1)}_{[o,l]} \rangle$. Moreover, a graph sequence $d = \langle g^{(1)}, \cdots, g^{(z)} \rangle$ is represented by a transformation sequence $s_d = \langle s_d^{(0)}, \cdots, s_d^{(z-1)} \rangle$.* ∎

The notation of transformation sequences is far more compact than the original graph-based representation since only differences between two successive graphs in d are kept in the sequence. In addition, any graph sequence can be represented by the six TRs in Table 1.

When a transformation sequence s_d' is a subsequence of a transformation sequence s_d, there is a mapping ϕ from vertex IDs in s_d' to those in s_d, and it is denoted as $s_d' \sqsubseteq s_d$. We omit its detailed definition because of a lack of space (see [9] for a detailed definition). Given a set of graph sequences $DB = \{\langle g^{(1)}, \cdots, g^{(z)} \rangle\}$, we define a support $\sigma(s_p)$ of a transformation sequence s_p as $\sigma(s_p) = |\{d \mid d \in DB, s_p \sqsubseteq s_d\}|/|DB|$, where s_d is a transformation sequence of d. We call a transformation sequence whose support is no less than the minimum support σ' a frequent transformation subsequence (FTS). Given a set of graph sequences, GTRACE efficiently enumerates a set of all FTSs from the set.

4   Mining Rules for Rewriting States

As mentioned in Section 2, if the parser shown in Fig. 3 selects the incorrect transition once, it never reaches the correct state. In this paper, we aim to discover rules for rewriting from states reaching incorrect final states to states reaching the correct final state. To discover these rewriting rules, we mine FTSs from graph sequences $\langle c_1, \cdots, c_m, g \rangle$, each of which consists of a transition sequence $\langle c_1, \cdots, c_m \rangle$ traversed by the parser and its correct dependency structure g. If $c_m = g$, then the final state c_m is correct and there are no TRs for transforming c_m into g. Otherwise, c_m is an incorrect final state and the TRs for transforming c_m into g are either

- transformation rules for inserting edges in g and not in c_m, or
- transformation rules for deleting edges in c_m and not in g.

As mentioned above, the rewriting rules to be mined are rules for transforming graphs in states that do not reach the correct dependency structure into

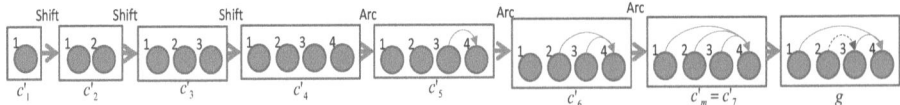

Fig. 8. Graph sequence for a transition sequence

graphs in other states that reach the correct structure for many sentences. The rewriting rules are therefore FTSs containing TRs for transforming c_m into g. To distinguish TRs for transforming c_m to g from other rules, we assign a label l_2 to edges in g and not in c_m, and a label l_1 to the other edges.

Example 3. Figure 8 shows a graph sequence d_A generated by appending the correct dependency structure g to the transition sequence for the sentence in Example 1, where the function o in a parser selects an incorrect transition from c'_3 to c'_4. Since the edge $(2,3)$ is not in c'_m and is in g, the label l_2 is assigned to the edge. The transformation sequence of the graph sequence is given as
$$s_{d_A} = \langle vi_{[1]}^{(0,1)} vi_{[2]}^{(1,1)} vi_{[3]}^{(2,1)} vi_{[4]}^{(3,1)} ei_{[(3,4),l_1]}^{(4,1)} ei_{[(2,4),l_1]}^{(5,1)} ei_{[(1,4),l_1]}^{(6,1)} ei_{[(2,3),l_2]}^{(7,1)} \rangle^{23}.$$

We select an FTS whose confidence is the highest among mined FTSs whose last TR is the edge insertion of label l_2. The definition of the confidence of an FTS is similar to basket analysis [1], as described in the following. We call the selected FTS a rewriting rule.

Definition 8. *Given an FTS s, let s' be the prefix of s, obtained by removing the last TR in s. The confidence of s is defined as $\sigma(s')/\sigma(s)$, and s' is called a body of s. In addition, a function for returning an edge to which the last TR in s is applied is defined as $lastEdge(s)$.* ∎

Example 4. When $r = \langle vi_{[2]}^{(0,1)} vi_{[3]}^{(1,1)} vi_{[4]}^{(2,1)} ei_{[(2,4),l_1]}^{(3,1)} ei_{[(2,3),l_2]}^{(4,2)} \rangle$ is a rewriting rule, its confidence is $\sigma(\langle vi_{[2]}^{(0,1)} vi_{[3]}^{(1,1)} vi_{[4]}^{(2,1)} ei_{[(2,4),l_1]}^{(3,1)} \rangle)/\sigma(r)$, and $lastEdge(r) = (2,3)$.

If a parser has the rewriting rule r of Example 4 and is in the state c'_6 of Example 3, the method proposed in this paper adds an edge $(2,3)$ to c'_6, and deletes an edge $(2,4)$ from c'_6, by applying r to transit another state c_6 in Fig. 5 that can reach the correct final state, since the transformation sequence of a transition sequence $\langle c'_1, \cdots, c'_6 \rangle$ contains the body of r as a subsequence. Therefore, the rewriting rule corresponds to a bypass from c'_6 to c_6 in the search tree shown in Fig. 6.

---

[2] Although each vertex is labeled by information such as words and parts-of-speech (POSs), the labeling depends on the features generated for a binary classifier in a parser. The details of labeling vertices are discussed in Section 5.

[3] We have *a priori* knowledge that each vertex in a state has at most one parent. Therefore, the fact that a TR t for inserting an edge with l_2 exists in a transformation sequence s indicates that another TR for deleting an edge whose dependent is identical to t must exist in s. For this reason, we do not include TRs for deleting edges in s to reduce the computation time of GTRACE, which exponentially increases with the average length of the transformation sequences in its input.

RuleMiner$(D, \sigma')\{$

1) $R \leftarrow \emptyset$
2) $r \leftarrow null$
3) while
4) $DB \leftarrow \emptyset$
5)   for sentence $(x = \langle w_1, \cdots, w_n \rangle, g) \in D$
6) $d \leftarrow$ ParseWithRules $(x, R \cup \{r\})$
7) $DB \leftarrow DB \cup \{d \Diamond g\}$
8) $evaluete(c_m, g)$, where $d = \langle c_1, \cdots, c_m \rangle$
9) if $R \neq \emptyset$ and the attachment
     score is saturated.
10)   return R
11)   if $r \neq null$
12) $R \leftarrow R \cup \{r\}$
13) $r \leftarrow MineRewritingRule(DB, \sigma')$
14)   if $r = null$
15)   return R

ParseWithRules$(x = \langle w_1, \cdots, w_n \rangle, R)$

1) $c \leftarrow c_s(x)$
2) $d \leftarrow \langle c \rangle$
3) while $c \notin C_F$
4) $c \leftarrow [o(c)](c)$
5) $d \leftarrow d \Diamond c$
6) s_d—the transformation sequence of d
7)   for $r \in R$
8) $(a, b) \leftarrow lastEdge(r)$
9)   if $body(r) \sqsubseteq s_d$,
     where $\phi : ID(body(r)) \rightarrow ID(s_d)$
10) $(i, j) \leftarrow (\phi(a), \phi(b))$
11) $c \leftarrow (N_c, A_c \cup \{(i, j)\})$
12)   if $\exists j'$ s.t. $(i, j') \in A_c \wedge j' \neq j$
13) $c \leftarrow (N_c, A_c \setminus \{(i, j')\})$
14) $d \leftarrow d \Diamond c$
15) return d

Fig. 9. Algorithms for mining rewriting rules and for parsing with the rules

Another way to generate a graph sequence from a transition sequence is to append to the correct state to the transition sequence immediately after the oracle in the parser selects the incorrect transition. In the case of Example 3, the graph sequence generated in this way is $d_B = \langle c'_1, \cdots, c'_4, c_4 \rangle$, where c_4 is of Fig. 5. Any subsequence of the transformation sequence s_{d_B} of d_B is always a subsequence of s_{d_A} in Example 3. In addition, d_A contains the information about vertices and edges that are not contained in d_B. Therefore, we use the approach to generate graph sequences in the form of d_A. Similarly, if r is a subsequence of s_{d_B}, r is a subsequence of s_{d_A}. Therefore, we apply r to c'_6 that is not the final state.

We propose a method for mining rewriting rules from transition sequences traversed by a dependency parser. The left part of Fig. 9 shows the pseudo-code for mining a set of rewriting rules R from the transition sequences. Let D be a corpus $D = \{(x, g)\}$ consisting of tokenized sentences $x = \langle w_1, \cdots, w_n \rangle$ and their dependency structures g. In Line 6, ParseWithRules returns a transition sequence d by parsing a sentence x using rewriting rules R. Next, in Line 7, after appending g to the tail of d, which is denoted by $d \Diamond g$, $d \Diamond g$ is added to DB. Subsequently, in Line 8, the attachment score is updated after comparing the final state c_m with the correct dependency g of the sentence x. In Line 9, if the attachment score for $R \cup \{r\}$ is no greater than that for R, then R is returned. Otherwise, r is added to R. In Line 13, a rewriting rule r with the highest confidence is mined among the FTSs enumerated by GTRACE from DB under the minimum support threshold σ'.

The right part of Fig. 9 shows the pseudo-code for parsing a sentence x using rewriting rules R to return a transition sequence for x. The procedures from Line 1 to Line 5 are similar to those in Fig. 3. If there is a rewriting rule whose body is contained in s_d and its mapping ϕ from vertex IDs in the body of r to vertex IDs in s_d, the state c is rewritten in Line 11 or 13 and is transited to

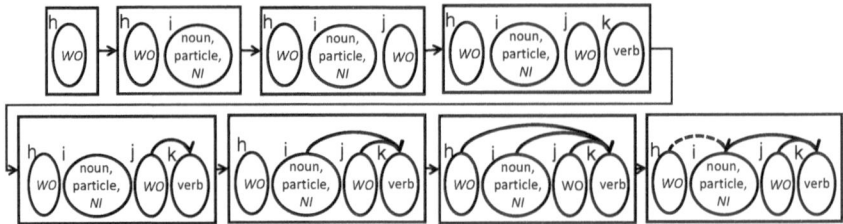

Fig. 10. One of the mined rules in the first loop

another state. In Line 11, an edge (i, j), corresponding to (a, b), is added to A_c. In addition, if the i-th segment has another parent j' rather than j, an edge (i, j') is deleted from A_c in Line 13. The parser in Fig. 9 is not incremental, but $|N_c| \leq |N_{t(c)}|$ and $|A_c| \leq |A_{t(c)}|$ hold.

In the remainder of this section, we discuss the time complexity of ParseWith-Rules. Loops of Lines 3 and 7 in ParseWithRules are repeated $2n - 2$ times, as discussed in Section 2, and $|R|$ times, respectively. If a graph sequence consists of general graphs, the computation time needed to check whether $body(r) \sqsubseteq s_d$ is identical to the subgraph isomorphism is known to be NP-complete [7]. However, since a graph sequence in this paper consists of ordered forests, the computation time to check it is linear with respect to the number of vertices in a forest, which corresponds to the number of segments in a sentence [13]. The complexity of ParseWithRules is therefore $O(n(\theta + |R|n))$. Additionally, in our implementation, for a transformation sequence s'_d of $d\Diamond c$, the computation to check whether $body(r) \sqsubseteq s'_d$ is solvable in constant time by storing mappings between vertices in $body(r)$ and vertices in s_d,; that is, the complexity of ParseWithRules is $O(n(\theta + |R|))$. The complexity of parsing a sentence x is therefore linear with respect to the number of segments n in a sentence, and is equivalent to that of the conventional method.

5   Experiments

We evaluated the proposed method using Kyoto Text Corpus v4.0, which consists of newspaper articles. In the implementation, CaboCha-0.60, which is a representative transition-based parser for Japanese [12], was integrated into the proposed method. We used the period from January 1 to 8 (7635 sentences) to train the SVM. In addition, we used data for eight days between January 9 to 17 (12054 sentences) to mine the rewriting rules and data for one day that is not used to mine the rules in evaluating the proposed method. We repeated this process nine times, which corresponds to nine-fold cross-validation.

We assigned a feature name with value of 1 to each vertex as a vertex label, since a feature vector that characterizes a segment w_i and is processed in the SVM of CaboCha-0.60 is a binary feature vector. In addition, we assigned feature names to each vertex as labels, although the original GTRACE assumes

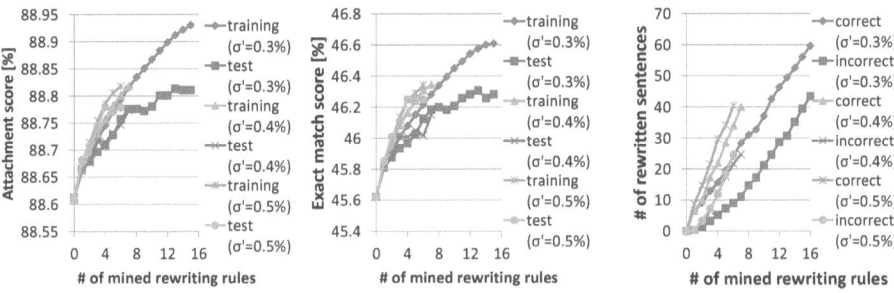

Fig. 11. Attachment score   **Fig. 12.** Exact match score   **Fig. 13.**   # of sentences rewritten by rules

that a label is assigned to each vertex in a graph sequence. Therefore, in our experiments, the insertion of a vertex with labels $\{lv_1, \cdots, lv_n\}$ and vertex ID 1 is naturally represented as $\langle tr_{[1,lv_1]}^{(j,1)}, \cdots, tr_{[1,lv_n]}^{(j,n)} \rangle$. For example, the second segment shown in Example 1 is characterized by a set of features $\{HON, WO, \text{noun}, \text{common\_noun}, \text{particle}\}$, and a vertex for the segment is labeled by the features.

Figure 10 shows one of the rewriting rules mined using the proposed method under the minimum support threshold 0.5%, where h, i, j, and k are segment IDs satisfying $h < i < j < k$, and the terms in each circle are labels. The rule was mined in the first loop in Line 13 in Fig. 9. The support and confidence of the rule are 0.58% and 89.7%, respectively. Japanese native speakers know that a segment containing WO usually modifies the first transitive verb appearing after the segment, and that the segment is the object of the verb in the sentence. However, a segment containing a transitive verb has only one dependent that contains WO and appears as the nearest before the verb. In the case that there are two segments containing WO before a verb, the former segment modifies a segment containing NI appearing after the former segment.

The rewriting rule shown in Fig. 10 mentions that the oracle selects Shift when determining whether the h-th segment is the dependent of the i-th segment in the second state, because a segment containing WO usually modifies a transitive verb after the segment, as mentioned above. At this point, the parser does not know that another segment containing WO appears between the h-th segment and a segment containing a verb, because the parser is the arc-standard parser. Subsequently, the oracle selects Arc to transit from the fourth state, for the same reason. In the sixth state, the arc-standard parser cannot add an edge (h, i), and it adds an edge (h, k), although the segment containing the verb already has a dependent containing WO. This rule rewrites the seventh state by deleting the edge (h, k), and adds an edge (h, i). The rule is therefore valid grammatically.

As shown above, the proposed method has the benefits that the rules mined by the method are human-readable and easily understandable. In addition, the rewriting rules contain context that is more complex and detailed than a set of features of the conventional parser, because of the use of the graph representation. Furthermore, if the mined rules are valid grammatically, and a dependency structure obtained by the proposed method, after being rewritten by the rules, is different from a dependency structure in the corpus made by humans, the latter dependency structure may contain incorrect dependencies. The proposed method is therefore also useful for rectifying human errors in the corpus.

Figures 11 and 12 show the attachment score and the exact match score of the proposed method for the number of mined rewriting rules for nine-fold cross-validation under the minimum support thresholds 0.3%, 0.4%, and 0.5%[4]. The number of mined rewriting rules $|R|$ increases one by one in each loop in Line 3 of Fig. 9. The number of mined rules is not very large, because some of the rules are unlexicalized and the others are lexicalized not by content words but by functional words (case markers such as WO or NI and auxiliary verbs). The attachment and exact match scores at $|R| = 16$ were improved by 0.20% and 0.66% under the minimum support threshold 0.3%, respectively. The number of mined rules differs for each trial in the nine-fold cross-validation, and the scores of the proposed method were finally improved by 0.22% and 0.90% under the minimum support threshold 0.3%, respectively. Since the number of mined rules is small, the improvements in the scores are not high. However, we can conclude that the mined rewriting rules are valid because improvements in the scores were obtained. In addition, Fig. 13 shows the numbers of sentences rewritten correctly and incorrectly under the various minimum support thresholds for the test datasets. It shows that when the number of rewriting rules mined by the proposed method increases, the numbers of sentences rewritten correctly and incorrectly decrease and increase, respectively, because confidence in the rewriting rules decreases with a progressive increase in the number of rules.

6   Discussion and Conclusion

In this paper, we proposed a method for mining rules for rewriting states reaching incorrect final states, and proposed a dependency parser with rules maintaining time complexity linear with respect to the number of segments in a sentence. The rewriting rules mined by the proposed method are human-readable, and it is possible for us to design new parsers and to generate features in the machine learner in the parser to avoid obtaining incorrect dependency structures. In this paper, we used GTRACE to analyze transition sequences, although there are other data structures for representing graph sequences, such as dynamic graphs and evolving graphs, and algorithms for mining the graphs. Since insertions of vertices cannot be represented by dynamic graphs, and a vertex in an evolving graph always comes with an edge connected to the vertex, these data structures cannot be used to analyze transition sequences in transition-based parsers to mine rewriting rules. The class of graph sequences is therefore general enough to apply to the analysis of transition sequences, compared with dynamic graphs and evolving graphs. The principle of the method proposed in this paper can basically be applied to any parsers based on a transition system, including parsers

---

[4] The attachment and exact match scores of the conventional method trained using data for 15 days were 89.4% and 47.7%, respectively.

employing the beam search [16,8], for sentences in any language. We plan to apply the method proposed in this paper to other transition-based parsers and to corpora of other languages in the future.

References

1. Agrawal, R., Srikant, R.: Fast Algorithms for Mining Association Rules in Large Databases. In: Proc. of Int'l Conf. on Very Large Data Bases (VLDB), pp. 487–499 (1994)
2. Ahmed, R., Karypis, G.: Algorithms for Mining the Evolution of Conserved Relational States in Dynamic Networks. In: Proc. of IEEE Int'l Conf. on Data Mining (ICDM) (2011)
3. Berlingerio, M., Bonchi, F., Bringmann, B., Gionis, A.: Mining Graph Evolution Rules. In: Buntine, W., Grobelnik, M., Mladenić, D., Shawe-Taylor, J. (eds.) ECML PKDD 2009. LNCS, vol. 5781, pp. 115–130. Springer, Heidelberg (2009)
4. Borgwardt, K.M., et al.: Pattern Mining in Frequent Dynamic Subgraphs. In: Proc. of IEEE Int'l Conf. on Data Mining (ICDM), pp. 818–822 (2006)
5. Culotta, A., Sorensen, J.: Dependency Tree Kernels for Relation Extraction. In: Proc. of Annual Meeting of Association for Comp. Linguistics (ACL), pp. 423–429 (2004)
6. Ding, Y., Palmer, M.: Automatic Learning of Parallel Dependency Treelet Pairs. In: Su, K.-Y., Tsujii, J., Lee, J.-H., Kwong, O.Y. (eds.) IJCNLP 2004. LNCS (LNAI), vol. 3248, pp. 233–243. Springer, Heidelberg (2005)
7. Garey, M., Johnson, D.: Computers and Intractability: A Guide to the Theory of NP-Completeness. W. H. Freeman and Company, New York (1979)
8. Huang, L., Sagae, K.: Dynamic Programming for Linear-Time Incremental Parsing. In: Proc. of Annual Meeting of the Association for Computational Linguistics (ACL), pp. 1077–1086 (2010)
9. Inokuchi, A., Washio, T.: A Fast Method to Mine Frequent Subsequences from Graph Sequence Data. In: Proc. of IEEE Int'l Conf. on Data Mining (ICDM), pp. 303–312 (2008)
10. Iwatate, M., et al.: Japanese Dependency Parsing Using a Tournament Model. In: Proc. of Int'l Conf. on Comp. Linguistics (COLING), pp. 361–368 (2008)
11. Kubler, S., et al.: Dependency Parsing. Morgan and Claypool Publishers (2009)
12. Kudo, T., Matsumoto, Y.: Japanese Dependency Analysis using Cascaded Chunking. In: Proc. of Conf. on Comp. Natural Language Learning (CoNLL), pp. 63–69 (2002)
13. Makinen, E.: On the Subtree Isomorphism Problem for Ordered Trees. Information Processing Letters 32(5), 271–273 (1989)
14. Nivre, J.: Algorithms for Deterministic Incremental Dependency Parsing. Comp. Linguistics 34(4), 513–553 (2008)
15. Pei, J., et al.: PrefixSpan: Mining Sequential Patterns by Prefix-Projected Growth. In: Proc. of Int'l Conf. on Data Engineering (ICDE), pp. 2–6 (2001)
16. Zhang, Y., Clark, S.: A Tale of Two Parsers: Investigating and Combining Graph-based and Transition-based Dependency Parsing. In: Proc. of Conf. on Empirical Methods in Natural Language Processing (EMNLP), pp. 562–571 (2008)

The Comparison of Stigmergy Strategies for Decentralized Traffic Congestion Control: Preliminary Results

Takayuki Ito[1], Ryo Kanamori[1], Jun Takahashi[1], Iván Marsa Maestre[2], and Enrique de la Hoz[2]

[1] Nagoya Institute of Technology, Gokiso, Showa-ku, Nagoya 466-8555, Japan
{ito.takayuki,kanamori.ryo}@nitech.ac.jp,
takahashi.jun@itolab.nitech.ac.jp
[2] Computer Engineering Department, Universidad de Alcal, 28803 Alcala de Henares, Madrid, Spain
{ivan.marsa,enrique.delahoz}@uah.es

Abstract. We investigate several stigmergies models for decentralized traffic congestion control. For realizing a smart city, one of the main problems that should be handled is traffic congestion. There have been a lot of works on managing traffic congestion with information technology. There is a relatively long history on observing traffic flow and then providing **stochastic estimation** on traffic congestion. Recently, more dynamic coordination methods are becoming possible by using more short term traffic information. Short term traffic information can be provided by car navigation systems with GPS (Global Positioning System)s and probe-vehicle information. There are several approaches to handle short term traffic information, in which stigmergy-based approach is a popular. Stigmergy is employed for indirect communication for cooperation among distributed agents. We can imagine several types of stigmergies : long term memory, short term memory, and anticipatory memory. However, there have been no discussion what kind of stigmergies can work well for managing traffic congestion. We conducted several simulations to compare the different kind of stigmergies. Our preliminary results demonstrate that if the traffic network is static, the combination of long term and short term stigmergies overcome the other stigmergies.

1 Introduction

In this paper we investigate several stigmergies models for decentralized traffic congestion control. The concept of smart city has been focused recently because of a lot of consideration on environments and the rapid development of information technologies. For realizing a smart city, one of the main problems that should be handled is traffic congestion.

There have been a lot of works on managing traffic congestion with information technology. There is a relatively long history on observing traffic flow and then providing information on traffic congestion. This is usually done by observing and counting the number of vehicles that pass certain locations by using

P. Anthony, M. Ishizuka, and D. Lukose (Eds.): PRICAI 2012, LNAI 7458, pp. 146–156, 2012.
© Springer-Verlag Berlin Heidelberg 2012

sensor equipped gates. These gates are usually equipped on the main trunk roads. In real world these information are rarely stored, and can not work as stigmergy. Rather they just provide the near current information to the vehicles.

Recently, more meaningful coordination methods are becoming possible by gathering and storing traffic information. More precise traffic information can be provided by car navigation systems with GPS (Global Positioning System)s and probe-vehicle information. These data are being stored into central servers like as a **long term memory**, which can provide stochastic traffic congestion information to the vehicles. This type of information technologies has been already applied to real world.

In the research field on traffic information and multi-agent systems, the dynamic **short term memory** has been focused very much. Vehicles are sharing these dynamic information, and drivers can choose their routes more dynamically based on the real time information. As we show in Section 5, there are several approaches to handle short term traffic information, in which stigmergy-based approach is a popular. Stigmergy is employed for indirect communication for cooperation among agents. For example, ants' pheromone is a kind of stigmergy for cooperation among them. Ants can been seen as agents in multiagent model, and also seen as vehicles in traffic situation. Vehicles can estimate nearest future situation based these stigmergies.

In addition, we can imagine stigmergies for future. For example, the vehicles can declare where they intend to move. We call it **anticipatory stigmergy**. Some papers focused on anticipation mechanisms for car routings, traveling salesman problems, etc. Such anticipation should be applied to stigmergies as well.

In this paper, as we discussed above, we assume the following types of stigmergies : long term, short term, and anticipatory. We conducted several simulations to compare the different kind of stigmergies. The results demonstrate that if the traffic network is static, the combination of long term and short term stigmergies overcome the other stigmergies.

In the rest of this paper, 2nd section present our basic simulation model for traffic simulation. In 3rd section, we propose anticipatory stigmergy model, and 4th section, we show preliminary experiments and their results. In 5th section, we review the related works, and finally in 6th section, we show conclusion.

2   A Smart Trafic Simulation Model

In this paper, we model the road network as a graph. Let the graph $G = (N, E, f_{cap}, f_{max})$ serve as a model of the road network, where N is the finite set of nodes, modeling intersections, and $E \subseteq N \times N$ is the set of links, modeling one-way roads between intersections. The link $l = (n, n') \in E$ if and only if there is a road that permits traffic to flow from intersection n to intersection n'. Function $f_{cap}(l)$ defines the capacity of a road l. Function $f_{max}(l)$ defines the maximum speed of a road l. Each vehicle i has its origin node n_i^o and its destination node n_i^d. $|l|$ is the length of l.

In the experiments in this paper, the simulated road network has 25 nodes and 80 directed links. The distance of each link is 2.5km. We assume two classifications on roads: trunk roads and ordinal roads. Trunk roads are represented by bold lines and have a ring shape in the Figure. The other roads are represented by normal lines. The traffic capacity and the maximum speed of the trunk roads are 7.5 - 10 vehicles/min and 20 - 25 km/h under the uniform distribution, respectively. Those of the ordinal roads are 5 - 7.5 vehicles/min and 15 - 20 km/h under the uniform distribution.

We assume each road (link) has 10 cells. While 1 cell in a trunk road can have 2 vehicles, 1 cell in an ordinal road can have 1 vehicle. 1 vehicle can move 1 cell for 1 step-time. 1 step-time is assumed as 1 minute. Thus, 1 road need 10 minutes to pass. A vehicle can move from the current cell $c_{current}$ to the next cell c_{next} at time $t + 1$ if there is no other vehicle at that cell c_{net} at current time t. If there is a vehicle at that cell c_{next} at the current time t, then he stop at the current cell $c_{current}$.

3   Trafic Simulation

We assume 6 cases for traffic simulation to see the effect of stigmergies.

(Case 0) No Information: Any information on transportation are not provided and gathered. Each vehicle finds the best path by the Dijkstra search only before it departs. We assume several different original(start) points and goal points as like people have different home and different destinations. The cost of a road l is $c = |l|/f_{max}(l)$. $f_{max}(l)$ defines the maximum speed of a road l. $|l|$ is the length of l.

(Case 1) Current Information: Each link l has a counter to measure the volume of traffic for every 5 minutes. Each link sends the current time required for passing it to the server. Each vehicle replans by the Dijkstra search its route for every 5 minutes based on the information sent from all links. We assume all vehicle can gather all information from all links. The information is not stored. The equation (1) shows the cost of a road l in case 1. This function follows the standard BPR functions.

$$c = \frac{|l|}{f_{max}}(1 + \alpha * (\frac{CA}{f_{cap}(l)/20})^{\beta}) \tag{1}$$

CA is the current amount of traffic. Function $f_{cap}(l)$ defines the capacity of a road l. Function $f_{max}(l)$ defines the maximum speed of a road l. $|l|$ is the length of l. $\alpha = 0.48$. $\beta = 2.48$. This cost function is a heuristic, where if there are many vehicles, then the c will be increased briefly.

(Case 2) Current Information of Only Trunk Roads: A truck road has a count to measure the volume of traffic for every 5 minutes. Each link sends the current time required for passing it to the server. Each vehicle replans by the Dijkstra search its route for every 5 minutes based on the information sent from all trunk roads. We assume all vehicle can gather all information from all trunk roads. Vehicles estimate the time to pass the ordinal roads by their length divided by the maximum speed. The cost function is same as Case 1.

(Case 3) Long Term Stigmergy: A road (link) stores and manages long term stigmergy information forever. Concretely, the load keeps storing data about the time required for passing it, and provide the long term stigmergy value $v_l = a + s * 0.1$, where a is the average and s is the standard deviation of the entire stored data, as the long term stigmergy information. Each vehicle utilizes this long term stigmergy information to make a new plan by Dijkstra search for every 5 minutes. Long term stigmergy updates for every x hours, where x is 3, 12, or 24 in our simulations.

(Case 4) Short Term Stigmergy: A road (link) stores and manages stigmergy information for only recent 5 minutes. Concretely, the load keeps storing data about the time required for passing it for only recent 5 minutes, and provide the short term stigmergy value $v_s = a + s * 0.1$, where a is the average and s is the standard deviation of the recent 5-min stored data, as the long term stigmergy information. Each vehicle utilizes this long term stigmergy information to make a new plan by Dijkstra search for every 5 minutes.

(Case 5) Long Term and Short Term Stigmergy: A road (link) stores and manages long term and short term stigmergy. Long term stigmergy information is the value of $a + s * 0.1$, where a is the average and s is the standard deviation of the entire stored data. Short term stigmergy information is the value of $a + s * 0.1$, where a is the average and s is the standard deviation of the recent 5-min stored data. Each vehicle utilizes this both long term and short term stigmergies information to make a new plan by Dijkstra search for every 5 minutes. The equation 2 shows how to integrate long term and short term stigmergies. v_{ls} is the combined stigmergy information.

$$v_{ls} = v_s * (1 - w) + v_l * w \tag{2}$$

where v_s is the short term stigmergy value, v_l is the long term stigmergy value, and w is the weight of the long term stigmergy.

(Case 6) Anticipatory Stigmergy: For each 5 min, each vehicle find the best pass to its objective node based on long term stigmergy and short term stigmergy as same as case 5. Then, they submit where (which a link) they will be after 5 min. This is anticipatory stigmergy in this paper. Then, they retry to find the best pass based on the anticipatory stigmergies.

The equation (3) shows the heuristic cost of a road l by using anticipatory stigmeriges. Actually anticipatory stigmergies are assumed to work as same as the current information in case 1 or case 2.

$$c = \frac{|l|}{f_{max}}(1 + \alpha * (\frac{CA}{f_{cap}(l)/40})^{\beta}) \tag{3}$$

CA is the current amount of traffic by anticipatory stigmergy. Function $f_{cap}(l)$ defines the capacity of a road l. Function $f_{max}(l)$ defines the maximum speed of a road l. $|l|$ is the length of l. $\alpha = 0.48$. $\beta = 2.48$. This cost function is a heuristic, where if there are many vehicles, then the c will be increased briefly.

We varied the trust rate about anticipatory stigmergies among 1.0, 0.75, 0.5, and 0.25. The trust rate means the possibility that each vehicle believe stigmergy information to re-plan his/her route.

4   Comparable Experiments

4.1   Setting

First, we use a simple road network shown in Figure (1), where bold liks are the trunk roads. For make congestion artificially, the shape is not a complete square. The links from nodes 10 to 15 and from nodes 14 to 19 are missing.

There are 1000 vehicles. 500 vehicles starts from node 0 to node 24. The other 500 vehicles starts from node 4 to node 20. For every one minute, each vehicle starts from node 0 or node 24. Information gathering take place in every 5 minutes.

Also, we assume the ratio of having equipment to send and receive information (like a car navigation system). In case 0 to 5, 0.75 percent of the vehicles have it. In case 6, we varied the ratio among 0.25, 0.5, and 0.75.

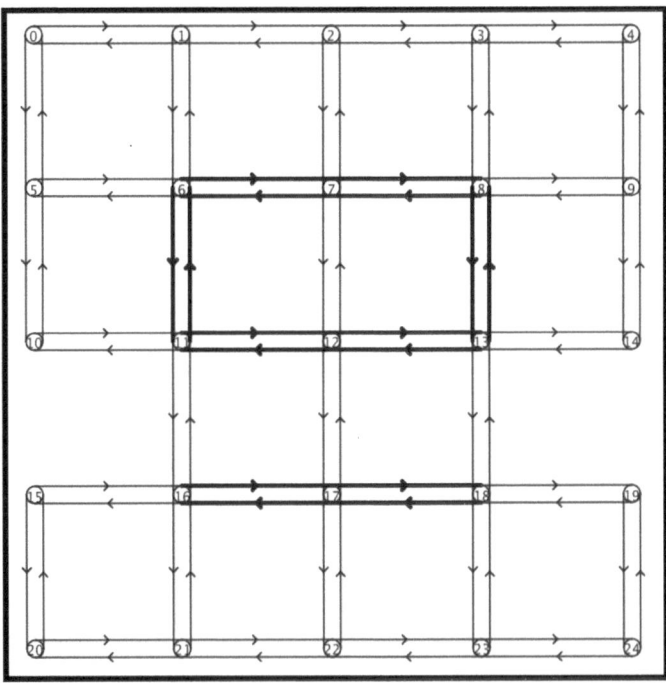

Fig. 1. Simulated road network (Bold links are the trunk roads)

Our simulator keep the map information based on XML format that is almost equivalent to the format of the OpenStreetMap[1]. Future work includes simulations on the real city or town map. Figure 2 shows an interface of this simulator. Each small circle that has the number (like 1, 2, ...) is a vehicle.

5   Comparable Experiments

5.1   Setting

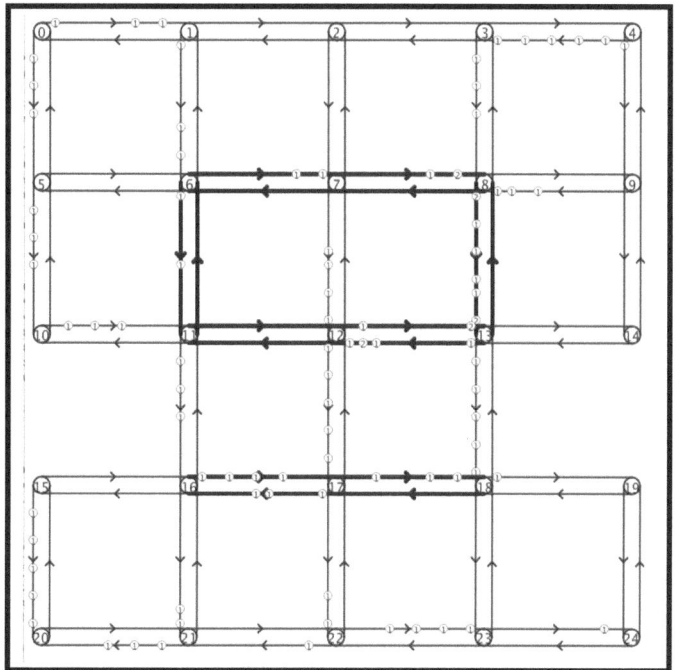

Fig. 2. Simulated road network interface

5.2   Preliminary Results

Figure 3 shows the total required time for all iterations.

Firstly, interestingly, case 3, case 4, and case 5 outperformed case 1, and case 2. While case 1 and case 2 use traffic counting systems for each road, case 3, case 4, and case 5 use stigmergy information from each vehicle. This means that if each vehicle that has an equipments to communicate (like a car-navigation system) properly, then the congestion is reduced compared with the road counting gates that is a kind of huge social infrastructure.

Let us see the detailes. Case 0 performs badly because they do not know any information. Case 1 works better than Case 0 and Case 2 because all vehicles

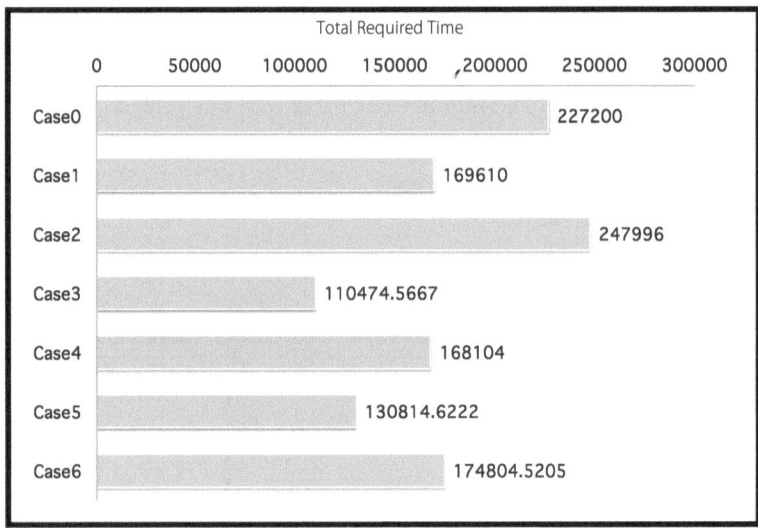

Fig. 3. Total Required Time for Each Case

know all information of all links. About Case 2, because the number of trunk roads is small and trunk roads are located at the region that might cause congestions, Case 2 is worse than Case 1 and Case 0. We can guess that congestion information from trunk roads made more congestion on the narrow area of our map.

Case 3 shows the best performance in total required time. Case 3 employed only long term stigmergy by probe-vehicle data. Case 4 utilizes short term stigmergy information. Case 5 utilizes the combination of short term and long term stigmergies. The results shows natural because if we only use short term memory, it is little bit difficult to find the better way compared with long term memory. Also, case 5 is in the middle of them.

In terms of Case 6, we expected anticipatory stigmergy works better than the result. However, it did not work well as we expected. There could be many reasons but we list up some of the reasons:

1) The road network was static. Because it was static, the long term memory works very well for finding the best path for each vehicle. This means that the area where there are less accidents and less constructions, it might not so good idea to adopt real-time anticipatory stigmergy-like equipments.

2) Our anticipatory stigmergy used only anticipatory intentions. Each vehcile tried to find the best path based on anticipatory intentions from the other vehciles after find the best path based on long term and short term stigmergy. This means that each vehcile first find the best way from the history, and then secondly try to find another path. Here, there is a large possibility that the best ways found is collapsed by this second trials.

We will investigate Case 5 and Case 6 more in details.

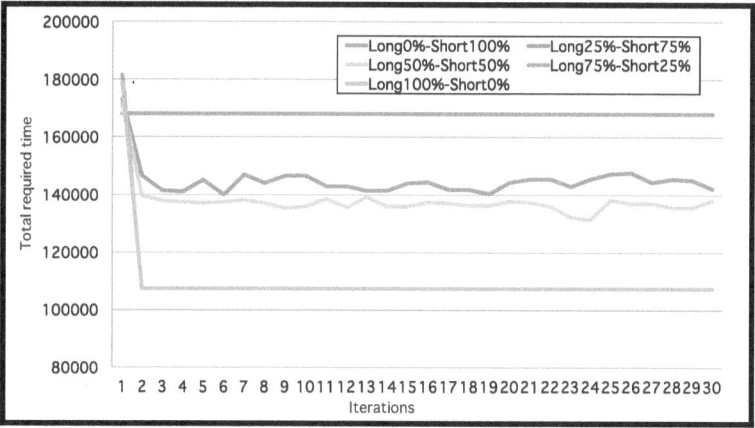

Fig. 4. Case 5: Long term and Short term Stigmergy

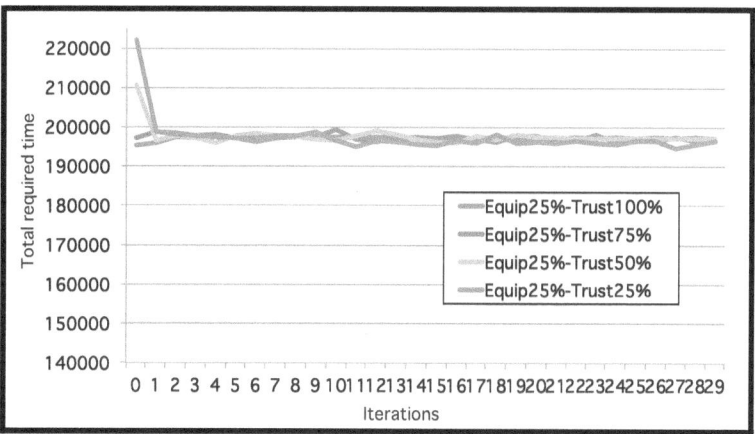

Fig. 5. Case6 : Anticipatory Stigmergy (Equip Rate 25% and Several Trust Rates)

Figure 4 shows a sensitivity analysis of Case 5 about the weight w of Equation (4)(= Equation (2)). Here, in Figure 4, Long0% means $w = 0$, Long25% means $w = 25$, etc. The best performance is Long 100% and Short 0%. Actually this almost equals to Case 3. Also, unfortunately we can not show it on the Figure 4, but Long 75% and Short 25% performed as same as Long 100% and Short 0%. They are overlapped in Figure, but their performance are equivalent. Long 25% and Short 75%, and Long 50% and Short 50% performed in middle.

$$v_{ls} = v_s * (1 - w) + v_l * w \qquad (4)$$

Figure 5, Figure 6, and Figure 7 show a sensitive analysis of equipped rates and trust rates of anticipatory stigmergy.

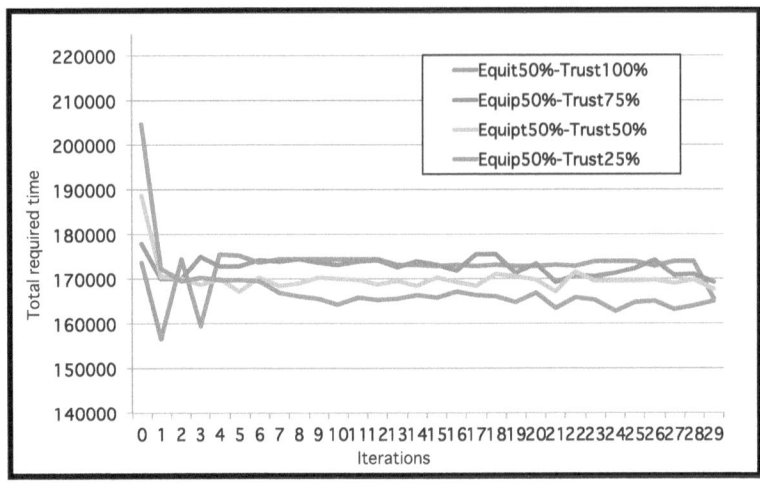

Fig. 6. Case6 : Anticipatory Stigmergy (Equip Rate 50% and Several Trust Rates)

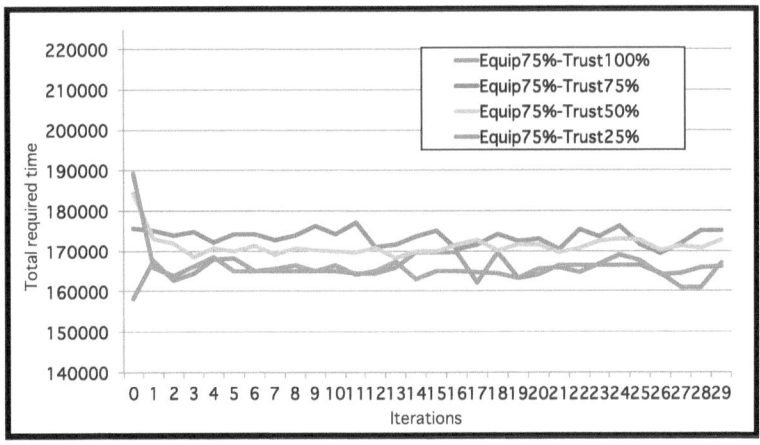

Fig. 7. Case6 : Anticipatory Stigmergy (Equip Rate 75% and Several Trust Rates)

Figure 5 shows the situation where only 25% vehicle equipped the system. In this case, all cases of the trust rates show almost same total required time. This is understandable because many vehicle do not have the equipments.

In Figure 6 and Figure 7, there is some distribution between the cases of the trust rates. Interestingly, the full trust case (Trust 100%) varied from worst in Figure 6 to best in Figure7. This means that if the number of equipped vehicles increases, and also the number of people who trust anticipatory stigmergy, there is a tendency that anticipatory stigmergy works better.

The results discussed about is largely depending on the map. However, we think we succeeded a certain level understanding of the difference of the effects of stigmergies.

6   Related Works

There have been many approaches to handle traffic congestions from real world approaches to theoretical approaches.

Stochastic congestion estimation from long term stored data has been used in order to provide an estimation of congestion in the real world(For example, [2]). As we showed in this paper, the long term stochastic congestion estimation is working well in a pure experiment. However, as real world shows, congestions are still happening. The reasons might be incentive issues, dynamic nature of traffic (accidents, constructions, human unpredictable behaviors, etc.) or drivers's habitual activities (like bounded rationality).

Congestion pricing[3] is one of the popular topics about avoiding congestion. Congestion pricing is a system of surcharging drivers of a traffic network in periods of peak demand to reduce traffic congestion. Some toll-like road pricing fees, or car pool lanes are real examples. Congestion pricing is not new. Credit Based Congestion Pricing[4,5] is one of the next directions for congestion pricing. Here, points-based mechanism is adopted for exchanging the rights to pass a congested road in periods of peak demand. Congestion pricing is a kind of centralized mechanism. However, our mechanism is rather able to be implemented as distributed mechanism with indirect communication stigmergy. Paper [6] proposes evolutionary game-theoretic model for dynamic congestion pricing in traffic networks. Their learning model improves the dynamic congestion pricing for some of real world road networks.

Distributed approaches have been widely studied in the field of multi-agent sytems and artificial intelligences. Very large scale survey paper [7] investigate the whole area of the transportation field with multi-agent systems from traffic congestion to railway transportation. In particular, it seems people are very focusing that ACO (Ant Colony Optimization)[8,9] can fit into the congestion avoiding problems. We also think swarm intelligence approach can fit well about traffic congestion control. Paper[10] proposes anticipatory vehicle routing using multi-agent systems. Here, their multi-agent systems is actually more like ACO. For anticipation purpose, they use ACO like optimization mechanism before actual routing. Rather paper [11] models a vehicle as an ant. This approach is close to our approach. The difference is that paper [11] adopts only pheromone mechanisms without probe information that are being used in real world. There is some possibility to improve their methodology by integrating with the current real world approach to reduce traffic congestions. Another approach [12] is to use cloud computing with mobile agent technologies for managing traffic congestion. This is rather implementation issue.

The paper[13] aims at providing some knowledge on drivers 簀 dynamic route choice behavior using probe-vehicle data. Namely, modeling drivers' route choice from the real probe-vehicle data is essential because real human's route choice could be biased on its habitual activities. Also, the papers [14,15] are also approaching to model drivers' dynamic routing modeling.

7   Conclusion

In this paper we investigate several stigmergies models for decentralized traffic congestion control. In this paper, as we discussed above, we assume the following types of stigmergies : long term, short term, and anticipatory. We conducted several simulations to compare the different kind of stigmergies. The results demonstrate that if the traffic network is static, the combination of long term stigmergies overcome the other stigmergies. Also, the stigmergy approach, where each driver has a car-navigation like system that can communicate to outside servers about stigmergy, overcome the situation where many vehcile counting gates are located for all the roads. We conducted some sensitivity analysis about the integration weights of long term and short term stigmergies in Case 5, and, about the equip rates and the trust rates in Case 6.

Future work includes to test more variety of types of the stigmergies, larger maps, dynamic environments (including accidents, constructions, etc.). In our expectation, in some settings, the anticipatory stigmegy could make better effect to avoid traffic congestion. We are now investigating such settings.

References

1. OSM: Openstreetmap (2012)
2. IBM: Ibm and singapore's land transport authority pilot innovative traffic prediction tool (2007)
3. Wikipedia: Congestion pricing (2012) (Online; accessed March 31, 2012)
4. Kockelman, K.M., Kalmanje, S.: Credit-based congestion pricing: a policy proposal and the public's response. Transportation Research Part A, 671–690 (2005)
5. Gulipallia, P.K., Kockelman, K.M.: Credit-based congestion pricing: A dallas-fort worth application. Transport Policy (2008)
6. Dimitriou, L., Tsekeris, T.: Evolutionary game-theoretic model for dynamic congestion pricing in multi-class traffic networks. Netnomics (2009)
7. Chen, B., Cheng, H.H.: A review of the applications of agent technology in traffic and transportation systems. IEEE Transactions on Intelligent Transportation Systems (2010)
8. Dorigo, M., Sttzle, T.: Ant Colony Optimization. MIT Press (2004)
9. Dorigo, M., Sttzle, T.: Ant Colony Optimization: Overview and Recent Advances. Springer (2010)
10. Claes, R., Holvoet, T., Weyns, D.: A decentralized approach for anticipatory vehicle routing using delegate multiagent systems. IEEE Transactions on Intelligent Transportation Systems 12(2), 364–373 (2011)
11. Narzt, W., Wilflingseder, U., Pomberger, G., Kolb, D., Hortner, H.: Self-organising congestion evasion strategies using ant-based pheromones. Intelligent Transport Systems, IET 4(1), 93–102 (2010)
12. Li, Z., Chen, C., Wang, K.: Cloud computing for agent-based urban transportation systems. IEEE Intelligent Systems 26(1), 73–79 (2011)
13. Morikawa, T., Miwa, T.: Preliminary analysis on dynamic route choice behavior using probe-vehicle data. Journal of Advanced Transportation (2006)
14. Pillac, V., Gendreau, M., Gueret, G., Medaglia, A.L.: A review of dynamic vehicle routing problems. Technical Report CIRRELT-2011-62 (2011)
15. Thomas, B.W., White III, C.C.: Anticipatory route selection. Transportation Science (2004)

Life-Logging of Wheelchair Driving on Web Maps for Visualizing Potential Accidents and Incidents

Yusuke Iwasawa and Ikuko Eguchi Yairi

Graduate School of Science and Technology, Sophia University,
7-1, Kioicho, Chiyodaku, Tokyo, Japan. 102-8554
{yuusuke.0519,i.e.yairi}@sophia.ac.jp
http://www.yairilab.net

Abstract. Life-logging has attracted rising attention as the most fundamental elements for developing every rich software today. This paper presents computational estimation and mapping of potential accidents and incidents of wheelchairs from life-logs with a single cheap and mini-sized three-axis accelerometer mounted on a wheelchair. Wheelchair driving data was obtained by real wheelchair users driving with their wheelchair on real roads, but has the sampling time delay and noises. As a first step of computational estimation, wheelchair driving behavior was classified into moving and static action, and the moving action was divided into tough and smooth status of the ground surface. We employed Support Vector Machine for classification, and made the precise supervised data from the video of wheelchair driving. As the result of classification, estimation of moving/static was achieved 98.2% accuracy rate and estimation of tough/smooth surface was achieved 82.6% accuracy rate. From the surface estimation result, wheelchair-driving difficulty was mapped and evaluated.

Keywords: Life-Log, SVM, time-series classification, wheelchair.

1 Introduction

Recently life-loggers no longer wear heavy computers and large devices in order to capture their entire lives, or large portions of their lives. In accordance with the expanding smartphone sales, life-logging has become popular application and attracted rising attention as the most fundamental elements for developing every rich software, from the end-user applications to the big data management tools in the era of ubiquitous and cloud computing. So we have been developing life-logging application of wheelchair users[1].

This paper introduces our approach to the time series data analysis of wheelchair user's life-log on driving with three-axis accelerometers. Three-axis accelerometers are mounted on most recent popular smart phones, and their time-series data includes useful human behavior patterns. If wheelchair users sense and

P. Anthony, M. Ishizuka, and D. Lukose (Eds.): PRICAI 2012, LNAI 7458, pp. 157–169, 2012.

record their driving behavior with three-axis accelerometers as life-logs, their motion such as stopping, moving and near-falling accidents, and the status of the ground surfaces such as smooth and bumpy will be estimated from time-series patterns of three-axis accelerometers. The human behavior information where near-falling accidents occurred is very important for all wheelchair users to prevent accidents on driving. The information of the environment surrounding wheelchairs such as where bumpy roads are is also necessary for them to choose maneuverable routes on driving. Of course, simple trails of wheelchairs are practical information for wheelchair users as the evidential fact where wheelchairs were able to access. But a simple trail provides no information about whether a user took a lot of trouble on driving on the route or not. The information about this driving difficulty can be extracted by the estimation of human behavior and environmental information from wheelchair driving logs with time-series data of accelerometers. If wheelchair trails and the extracted information from driving logs such as near-falling accidents and bumpy roads are mapped on web maps, essential support will be provided to expand the mobility of each wheelchair user.

We would like to classify time-series of acceleration values into some driving action patterns for estimating and visualizing the information about driving difficulty in this paper. Although there are some application to evaluate the ground surface condition [2][3], these application have focused on improving the wheelchair driving environment, it is not enough to visualize the information about human action. In order to estimate a information about driving difficulty for a wheelchair user from life-logs, classifying action pattern is necessary as conducting at a research which purpose to achieve the health management system fitting for individuals from human behavior 'life-log'[4][5]. There are a few studies which focus on classifying wheelchair driving behavior. However, these studies have only dealt with the data of non-handicapped person [6], so classification using wheelchair users data at moving on actual environment have not been discussed. Our approach to classify and visualize the mobility barrier is different with another studies in terms of using wheelchair users driving data and classifying it, and novelty.

According to Zhengzheng's survey [7], sequence classification can be classified into three main groups, 'Feature Based Classification', 'Sequence Distance Based Classification' such as K nearest neighbor classifier(KNN) or SVM, and 'Model Based Classification' such as Hidden Markov Model(HMM). Especially, regarding time series data like acceleration values, 'Sequence Distance Based Classification' are widely adopted to classify. We adopt SVM to classify time series of acceleration values into driving actions. Recently, SVM has proved to be an effective method and some studies use SVM for sequence classification [8][9][10][11]. Before to make clear what kind of classification method is optimal and how is its accuracy with wheelchair driving data, this paper focuses on the SVM as an ordinary classification method to precisely analyze the obtained wheelchair driving data on the experiments.

In this paper, we introduce our collected data of wheelchair driving by real users and our first step analysis of the data with SVM for the estimation of the human behavior and the status of the ground surface. The collected data is not clean because of sampling time delay and is noisy, containing outlier. The estimation from the data may require some creative algorithm which is optimum for the data characteristics. However, we never mentioned about an optimum algorithm in this paper. Our purpose is to clarify the characteristics of our collected data of wheelchair driving from the classification analysis using SVM. The remainder of this paper is organized as follows. The classification analysis framework is proposed in Section 2. A classification tree to digitize and visualize the potential accidents and incidents is discussed. Section 3 is devoted to a brief introduction to the collected data on wheelchair driving experiments and our employed classification method for data analysis. Results of the data analysis are presented in Section 4. The conclusion and future works are showed in Section 5.

2   Visualization of Potential Accidents and Incidents of Wheelchairs Driving

Typical serious accidents of wheelchairs were reported, for examples, as follows: wheelchair falls because of sidewalk curbs, collision accidents of wheelchairs because of loss of control at long or steep slopes, and car accidents with wheelchairs which were reluctantly moving on driving road to avoid uneven sidewalks. There is no question that these accidents were caused by the physical barriers to wheelchair mobility on the ground surface such as roughness and terrains. Behind these serious accidents, the physical barriers on the ground surface also cause so many incidents, as non-injury accidents [12][13][14].

Wheelchair users can not access the information where these accidents and incidents tend to occur because these information are difficult to digitize without manpower and special skills in wheelchair mobility barrier assessment. So each wheelchair user faces a daily challenge to manipulate his/her wheelchair on the ground surface with undiscovered dangers. These challenges under undiscovered dangers lead to the mental and physical stress for wheelchair users on outdoor activities. To lessen the mental and physical stress of wheelchair users, it is indispensable to digitize the information about physical barriers for wheelchair mobility on the ground surface and to visualize the possibility where accidents and incidents tend to occur. To digitize and visualize this information is a important mobility support method which enables wheelchair users to judge the risk of driving on a certain route in advance and to plan a safer route.

As a first step of digitization and visualization of mobility barriers of wheelchairs on the ground surface, we attached single three-axis accelerometer to wheelchairs of seven mobility-disabled persons, and collected the driving data on several routes in Akihabara area. The data was mapped and color-coded according to the Vibration Acceleration Level (abbr. VAL). But this color-coded VAL map shows little information of the ground surface about accidents and incidents risk. Because the simple acceleration value indicates the degree of vibration which a wheelchair user felt on driving and which was influenced by a wheelchair user's driving such as speed, sudden starting and stopping, and so on. Required solution for wheelchair users' needs is not mapping the degree of vibration from acceleration value on web, but mapping the ground surface conditions which are estimated from time series of acceleration values with machine learning techniques.

Consequently, this paper proposes the estimation of mobility barriers on the ground surface from time series of acceleration values on wheelchair driving with machine learning techniques. To clarify the estimation objectives, we describe wheelchair driving behavior as the classification model shown in Figure 1. As the first level classification, wheelchair-driving behavior is divided into moving and static actions. As the second level classification, moving action is divided into moving on the ground surface with/without mobility barriers. Through this two step classification, this paper investigates the possibility of the estimation of mobility barriers on the ground surface from time series of acceleration values on wheelchair driving with machine learning techniques.

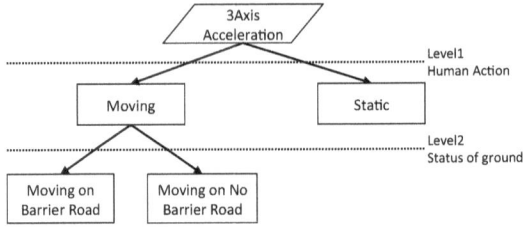

Fig. 1. The classification model of wheelchair driving behavior

3   Estimation of Mobility Barriers on the Ground Surface from Life-Log

3.1   Wheelchair Diving Data

Seven mobility disabled persons participated in wheelchair driving experiments of our laboratory and then wheelchair driving movement data was measured by a Sun SPOT, three-axis accelerometers attached under the wheelchair seat, as figure 2. Sun SPOT is cheap like smartphone and is easier to attach to the narrow space under the wheelchair. In order to obtain natural driving data, each person was asked to drive his/her own wheelchair which is used in his/her everyday life. The acceleration value was measured about 20 minutes per person continuously, and the routes was selected as including some mobility barriers of wheelchairs, such as sidewalk curbs or tactile indicators for dealing with a lot of and various wheelchair driving action patterns.

In terms of classifying wheelchair driving behavior, it is important to prepare the precise supervised data, which are tuples of three-axis acceleration values and video data taken from the back of wheelchair at the experiment in our case. Thus, we use the single person's data whose acceleration values and video data were mostly correct. The data has comparatively little missing of acceleration values, and wheelchair-driving action could be judged using video data at almost all time.

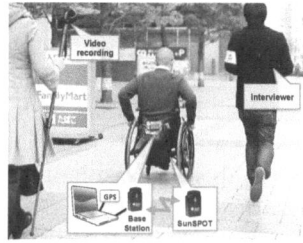

Fig. 2. System of life-logging wheelchair driving data

Figure 3 shows the histogram of sampling time of acceleration values. Horizontal axis indicates sampling time and vertical axis indicates the number of frames. As a result, the sampling time varied widely and the mean of sampling rate was about 15 Hz instead of the theoretical sampling rate of Sun Spot, 50Hz. Naturally, it is desirable that sensor data is taken as precisely as possible. However, such a precise data is not always available in actual environment. Accordingly, we investigate the available classification and accuracy of the classification from widely dispersion of sampling rate.

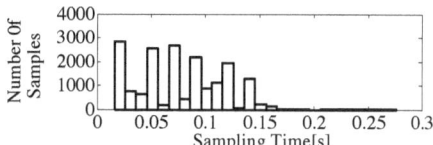

Fig. 3. Histogram of sampling time

3.2   Classification Method

Having cleared the features of using data, next I would like to explain about classification method of acceleration time-series data. SVM, a novel machine learning technique, was used to classify the time-series of acceleration values. Concretely, we divide time-series of acceleration values into some windows by the number of frames and then classify the window using SVM. Dividing time-series data using above method causes the difference of window size because of various sampling time. Although the difference of window size can decrease the classification accuracy, however, classifying the window was conducted with accepting it.

In addition, window sizes affect the classification accuracy. For instance, classification must be difficult when the continued movement, such as climbing over the sidewalk curbs, was divided into some windows Also, window sizes affect the classification finesse because the smaller window size lead to higher time resolution. Thus, it is important to optimize the window size for improvement of performances, such as classification accuracy and fineness. Result of searching the optimized window size will be shown at chapter 4.

SVM is a supervised learning method; therefore, we need information of correct class, in which what kind of driving action is taken in each window. Correct class was confirmed using video taken from the back of wheelchair at experiment. Figure 4 shows the picture captured from the video. As seen from the figure 4, it is possible to confirm the kind of ground surface which wheelchair moves on, for instance wheelchair moving on paved asphalt at (a) and moving on rough ground surface at (b). The status of road was judged like the above example.

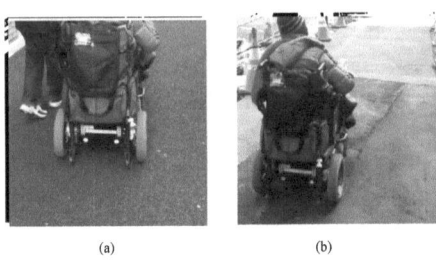

(a)   (b)

Fig. 4. Video taken from the back of wheelchair at the experiment (a) moving on paved asphalt, (b) moving on rough ground surface

4   Result of Estimation and Visualization

4.1   Estimation of Human Action

Figure 5 shows the comparison between the moving action and the static action of acceleration values taken for five seconds each. Horizontal axis indicates time and vertical axis indicates acceleration value. (a), (b), and (c) represent the acceleration value of horizontal, traveling, and vertical direction of wheelchair movement. As seen from figure 5, there is big difference in two curves, that is, the curves of moving action expressed by x-mark has bigger amplitude of acceleration value than that of static action expressed by heavy line. So, it seems to be easy that we classify the driving action into moving and stopping.

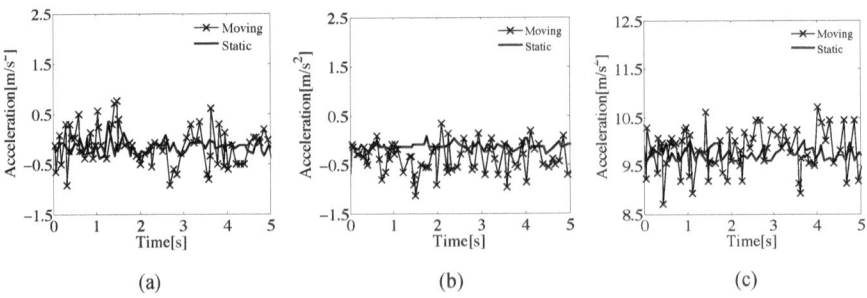

(a)   (b)   (c)

Fig. 5. Comparison the acceleration values of moving action with static action (a) horizontal axis (b) traveling axis (c) vertical axis

Three methods were conducted in order to classify human action. The first method is simple classification that uses raw data as feature value. The second method also uses raw data as feature value but add a preprocessing to reduce the difference of window size before we divide into some windows. The preprocessing cut off frames which sampling time over the average plus 2 times standard deviation. The third method use three-axis statistics as feature value. Using statistics as feature value make the classification more tolerant for noise, though the temporal feature is rounded. As it compresses the dimension of feature value, improving the generalization ability is also expected.

Figure 6 shows F values of the above three classification methods as for each action class. Horizontal axis indicates the number of frames in a window and vertical axis indicates average of 100 F value which was calculated by 10-fold cross validation. Although F value was calculated from 5 frames to 120 frames step by 1 frame, figure 6 shows the result of 5 frames to 120 frames step by 5 frames because of the restriction of space. The missing value in figure is not a number because of few windows of static class.

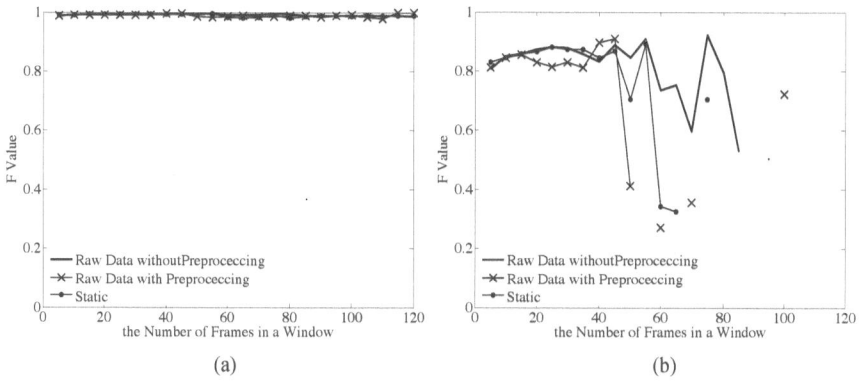

(a)   (b)

Fig. 6. F value comparison between three classification method (a) moving action class (b) static action class

As the figure 6 indicates, F value of moving action class was very high. To put it more concretely, the value was greater than 0.978 in each classification method and window size. On the other hand, F value of static action class was low than that of moving action class and tended to remarkably decrease when the number of frames is bigger than 55 frames. Comparing the result of each method, there was no big difference in both classes, however, when we focused on the area of the number of Frame which achieve high F value, classification using raw data without preprocessing (heavy line in figure 5) was better than the other methods. This result suggests that the preprocessing and using statistic value were of no effect to improve the classification accuracy. Hence, it seems reasonable to use raw data for classifying the time-series acceleration values into moving action class and static action class.

Table 1. Concrete classification accuracy at 25, 35, 45, 55 frame

the Number of Frames	Accuracy	Recall Ratio (Moving)	Recall Ratio (Static)	Precision Ratio (Moving)	Precision Ratio (Static)
25 frames	98.85%	99.30%	89.66%	99.49%	86.47%
35 frames	98.80%	98.98%	94.12%	99.76%	78.75%
45 franes	99.04%	**99.31%**	92.86%	99.70%	**85.10%**
55 frames	**99.30%**	99.27%	**100%**	**100%**	83.33%

Let us examine the classification using raw data without preprocessing in more detail. Then F value of moving class tended to be high when the number of frames was from 25 to 75, and that of static class tended to be high when the number of frames was from 25 to 55. Thus, optimized frames seem to be from 25 frames (about 1.75 sec) to 55 frames (about 3.85 sec) as for the classifying. Table 1 shows the concrete accuracy, recall ratio of each class and precision ratio of each class at 25, 35, 45, 55 frame.

4.2   Estimation of Status of the Ground Surface

Figure 7 shows ground surfaces chosen physical barrier for wheelchair mobility. These ground surface can cause falling down, paralyzing the limbs, or discomfort. The details are as follows; (a) rough road surface causing discomfort, (b) sidewalk curbs causing falling down, (c) tactile surface indicators and (d) tile paving with rough joint causing discomfort and paralyzing the limbs.

(a)   (b)   (c)   (d)

Fig. 7. Examples of ground surface with mobility barriers (a) rough road surface, (b) sidewalk curbs, (c) tactile surface indicator, (d) tile paving with rough joint

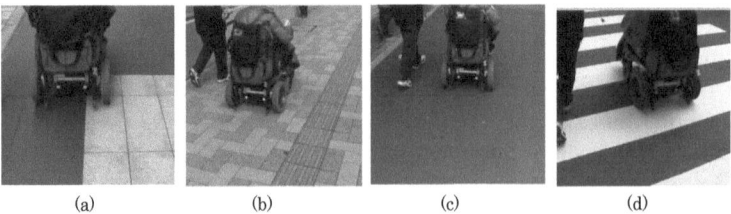

(a)   (b)   (c)   (d)

Fig. 8. Examples of road surface without mobility barriers (a) pavement, (b) tile paving with smooth joint, (c) paved Asphalt, (d) pedestrian crossing (paved Asphalt)

Figure 8 shows the other ground surfaces included in moving path, which were judged not having physical barrier. The details are as follows; (a) paved concrete, (b) tile paving with smooth joint (c) paved asphalt, (d) pedestrian crossing; paved asphalt. These ground surface cause only small risk for wheelchair mobility because it is comparatively flat.

We conducted three methods to classify the moving action into the state of moving on the ground surface having physical barrier (with barrier class) and not having physical barrier (without barrier class). Each method uses different feature value, raw data, statistic of each axis, and frequency component. It is expected that using frequency component help to classify more concretely.

F values of each method are shown in figure 9 as well as figure 6. Figure 9 (a) shows the result of with barrier class and Figure 9 (b) shows that of without barrier class. As a result, next two things were seen. Firstly, as the window size gets bigger, F value of with barrier class tended to increase, whereas that of without barrier class tended to decrease. It is inferred from this result that there was the domain which is difficult to divide into two classes in feature space. Secondly, method using statistics expressed by x-mark in figure 9 showed the highest performance in each class. Especially concerning without barrier class, the reduction of F value was gradually. In consequence, it is reasonable to use statistics as feature value in terms of maximize the accuracy of classification.

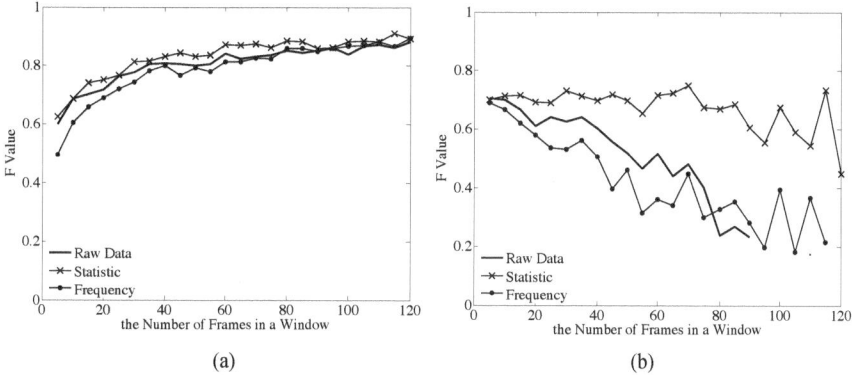

Fig. 9. F value of each classification whose feature value is raw data, statistics, and frequency component (a) moving on ground surface with mobility barrier, (b) moving on ground surfaced without mobility barrier

Figure 10 shows the accuracy curve of classification using statistic as feature value. Notation of Figure 10 similar to that of Figure 6 with the exception of that the number of frames is 5 frames to 120 frames step by 1 frame. The accuracy curve tended to increase until in front and behind of 60 frames, and to flat after 60 frames. Thus, optimized window length is bigger than 60 frames (about 4.2 sec) to classify moving class into with barrier class and without barrier class. To put it more concretely, recall ratio of with barrier class and without barrier class, precision ratio of with barrier class and without barrier class, and accuracy at the 60 frames were 89.2%, 67.9%, 85.1%, 75.5%, 82.6% respectively.

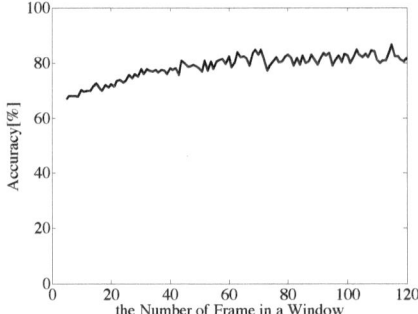

Fig. 10. The accuracy curve of classification using statistic as feature value (from 5 to 120 frames step by 1 frame)

4.3   Visualization

Figure 11 shows the result of visualizing the driving difficulty estimated through the above two classifications. High difficulty in wheelchair mobility is visualized as light shade color marker and low difficulty in wheelchair mobility is visualized as dark shade color marker. Let us look at correspondence of ground surface to visualizing result at place A to D. At the place A and C where wheelchair drove on tile paving with small joint which caused only small risk to move, visualizing result tended to be light shade color marker. At the place B having strong rough road surface and the place D being tile paving with rough joint, a lot of deep shade color marker was seen on the visualizing result. Like the above cases, visualizing result correspond to driving difficulty at least in rough point of view. Thus, to visualize driving difficulty estimated through classifications, which was shown in this paper, enables wheelchair users to judge the trends of the difficulty of terrain caused by ground surfaces.

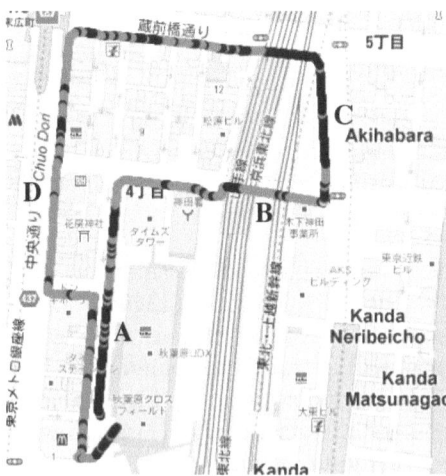

Fig. 11. Result of visualizing driving difficulty on the map

Next, we will compare our proposing approach of visualizing the difficulty with our previous approach which simply visualizes the acceleration values on trails of wheelchair. Figure 12 shows the comparison between two visualization. Each visualization use the same wheelchair driving movement data and the wheelchair drove from right to left as the heavy arrow in figure 12. The dark shade color marker means that there is no physical barrier and light shade color means that there is physical barriers in figure 12 (a), and the color is deeper as acceleration value is bigger in figure 12 (b).

Comparing two visualization at the places where wheelchair climbed over the sidewalk curbs pointed by the solid line in figure 12, there are many light shade color marker in both visualization, then we could find there was physically barrier, but, as for the places where wheelchair climbed down sidewalk curbs pointed by the dotted line, as shown in figure 12 (c), both visualization was totally different. That is, these places were estimated as physical barrier correctly by our proposing approach, and had small vibration at visualizing acceleration values, which lead the users to erroneous judge. The reason why the place caused only small vibration was the habit of the user which is decreasing the speed of wheelchair when climbed down the sidewalk curbs for avoiding the impactprobably for avoiding impact. On the other hand, such movement habit was de-noised by estimating ground surface conditions from acceleration value and that enable to judge correctly. As shown above discussion, our proposing approach was effective to digitize the potentially dangerous place and to support for wheelchair mobility.

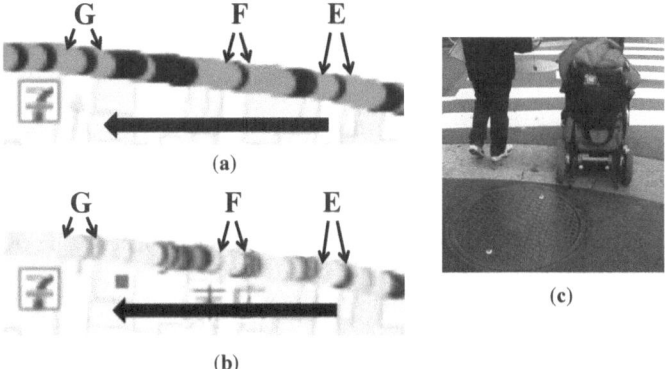

Fig. 12. The estimation result of each visualization at sidewalk curbs (a) our proposing approach, (b) visualizing VAL, (c) the sidewalk curbs

5 Conclusion

Although there is no question that physical barriers on the ground surface cause some wheelchair accidents, users can not access the information about where these accidents and incidents tend to occur because of difficulty of digitizing these information. This paper proposed to estimate the mobility barriers on the ground surface from wheelchair driving life-log measured by single cheap and mini-sized accelerometer in accordance with the classification model in Figure 1. As the results of classification, human action level classification was achieved 98.2 % accuracy rate by using raw data as feature value and setting the window size as 30 frames, and the status of the ground level classification was achieved 82.1 % accuracy rate by using statistic data as feature value and setting the window size as 60 frames, from the life-log of wheelchair driving using SVM. To visualize the estimated ground surface conditions was effective to digitize more correct information than simply mapping the VAL on Map. Open problems for developing more useful system are as follows; (1)analyzing wheelchair driving data of various properties and a large number of people, (2)reviewing another mobile and cheap sensor, such as smartphone or quasi-zenith satellites which enable to use high precision location information, to improve the accuracy of classification and to classify more complex action, (3)reviewing algorithms specialize to classify the time-series of acceleration values of wheelchair driving action into some detailed action. To clear these open problems enables us to digitize the potential dangerous place corresponding to the properties of various users and to support wheelchair mobility.

References

1. Yusuke, F., et al.: Sensing human movement of mobility and visually impaired people. In: ASSETS 2011 The Proceedings of the 13th International ACM SICACCESS Conference on Computers and Accessibility, pp. 279–280. ASSETS Press, NY (2011)
2. Coyle, E., et al.: Vibration-based terrain classification for electric powered wheelchairs. In: Telehealth/AT 2008 Proceedings of the IASTED International Conference on Telehealth/Assistive Technologies, pp. 139–144. ACTA Press, CA (2008)
3. Hashizume, T., et al.: Study on the wheelchair user's body vibration and wheelchair driving torque when wheelchair is ascending/descending the boundary curb between pavement and roadway. In: SICE Annual Conference, pp. 1273–1276. IEEE Press, Tokyo (2008)
4. Mathie, M.J., et al.: Classification of basic daily movements using a triaxial accelerometer. Med. Biol. Eng. Comput. 42, 679–687 (2004)
5. Mathie, M.J., et al.: Detection of daily physical activities using a triaxial accelerometer. Med. Biol. Eng. Comput. 41, 296–301 (2003)
6. Noriaki, K., et al.: A study on ubiquitous system for collecting barrier-free information for wheelchair users. In: CASAME 2010 Proceedings of the 4th ACM International Workshop on Context-Awareness for Self-Managing Systems, Article No. 5. CASEMANS Press, NY (2010)
7. Zhenzheng, X., Jian, P., Eamonn, K.: A Brief Survey on Sequence Classification. ACM SIGKDD Explorations Newsletter 12(1), 40–48 (2010)
8. Leslie, C.S., Eskin, E., Noble, W.S.: The spectrum kernel: A string kernel for SVM protein classification. In: Pacific Symposium on Biocomputing, NY, pp. 566–575 (2002)
9. Ming, L., Sleep, R.: A robust approach to sequence classification. In: 17th IEEE International Conference on Tools with Artificial Intelligence, ICTAI 2005, vol. 5, p. 201. IEEE Press, Hong Kong (2005)

10. Sonnenburg, S., Rätsch, G., Schäfer, C.: Learning Interpretable SVMs for Biological Sequence Classification. In: Miyano, S., Mesirov, J., Kasif, S., Istrail, S., Pevzner, P.A., Waterman, M. (eds.) RECOMB 2005. LNCS (LNBI), vol. 3500, pp. 389–407. Springer, Heidelberg (2005)

11. Kampouraki, A., Manis, G., Nikou, C.: Heartbeat Time Series Classification With Support Vector Machines. IEEE Transactions on Information Technology in Biomedicine 13, 512–518 (2009)

12. Chen, W.-Y., et al.: Wheelchair-Related Accidents: Relationship With Wheelchair-Using Behavior in Active Community Wheelchair Users. Archives of Physical Medicine and Rehabilitation 92, 892–898 (2011)

13. Benjamin, S., et al.: A Survey of Outdoor Electric Powered Wheelchair Driving. In: RESNA Annual Conference, Las Vegas (2010)

14. Xiang, H., et al.: Wheelchair related injuries treated in US emergency departments. Inj. Prev. 12, 8–11 (2006)

Texture Feature Extraction Based on Fractional Mask Convolution with Cesáro Means for Content-Based Image Retrieval

Hamid A. Jalab[1] and Rabha W. Ibrahim[2]

[1] Faculty of Computer Science & Information Technology
University Malaya, 50603.Kuala Lumpur, Malaysia
hamidjalab@um.edu.my
[2] Institute of Mathematical Sciences
University Malaya, 50603.Kuala Lumpur, Malaysia
rabhaibrahim@yahoo.com

Abstract. This paper introduces a texture features extraction technique for content-based image retrieval using fractional differential operator mask convolution with Cesáro means. We propose one general fractional differential mask on eight directions for texture features extraction. Image retrieval based on texture features is getting unusual concentration because texture is an important feature of natural images. Experiments show that, the capability of texture features extraction by fractional differential-based approach appears efficient to find the best combination of relevant retrieved images for different resolutions. To compare the performance of image retrieval method, average precision and recall are computed for query image. The results showed an improved performance (higher precision and recall values) compared with the performance using other methods of texture extraction.

Keywords: Fractional calculus, fractional differential, Cesáro means, fractional mask, texture segmentation, content based image retrieval.

1 Introduction

A typical content-based image retrieval (CBIR) system consists of two main tasks, feature extraction and similarity measurement. The key to a successful retrieval system is choosing the right features to accurately represent the images and the size of the feature vector. Commonly the features used in CBIR are include color, shape, texture or any combination of them.

Texture is an important feature of natural images a variety of image texture applications and has been a subject of intense study by many researchers [1]. Image texture, defined as a function of the spatial variation in pixel intensities (gray values). Therefore, texture retrieval is relevant to CBIR since texture characteristics are powerful in discriminating between images [2]. A wide variety of techniques for texture have been proposed. Few of the techniques used global color and texture features [3-5].

P. Anthony, M. Ishizuka, and D. Lukose (Eds.): PRICAI 2012, LNAI 7458, pp. 170–179, 2012.

The objective of this paper is to develop a technique which captures texture features in an image using fractional differential mask convolution with Cesáro means, which is the main contribusion of this work. Each pixel of the image is convolved with the fractional differential mask on eight directions. The features computed on these masks serve as local descriptors of texture. Compare to co-occurrence matrice and Gabor filters for texture features extraction methods, the proposed method performs well with very less computational time. The outline of the paper is as follows: Fractional calculus is presented in Section 2 . The construction of fractional differential mask scheme is presented in Section 3. Experimental results and conclusion are are shown in Sections 4 and 5, respectively.

2   Fractional Calculus

Fractional calculus and its applications (that is the theory of derivatives and integrals of any arbitrary real or complex order) has importance in several widely diverse areas of mathematical physical and engineering sciences. It generalized the ideas of integer order differentiation and n-fold integration. Fractional derivatives introduce an excellent instrument for the description of general properties of various materials and processes. This is the main advantage of fractional derivatives in comparison with classical integer-order models, in which such effects are in fact neglected. The advantages of fractional derivatives become apparent in modeling mechanical and electrical properties of real materials, as well as in the description of properties of gases, liquids and rocks, and in many other fields.

Nowadays, fractional calculus (integral and differential operators) arises in signal and image possessing. The fractional calculation is able to enhance the quality of images, with interesting possibilities in edge detection and image restoration, to reveal faint objects in astronomical images and devoted to astronomical images analysis [6,7]. Furthermore, fractional calculus is employed in texture segmentation [8], design problems of variables [9] and image denoising [10]. Finally, the fractional calculus (differential operators) is used in different applications in engineering [11].

3   Construction of Fractional Differential Mask

This section briefly describes the mathematical background for the fractional differen-tial mask (φ) that has been used by the proposed algorithm. We have proceeded to construct the generalized fractional mask using the following generalized fractional differential operator [12] :

$$D_z^{\alpha,\mu} f(z) := \frac{(\mu+1)^\alpha}{\Gamma(1-\alpha)} \frac{d}{dz} \int_0^z \frac{\zeta^\mu f(\zeta)}{(z^{\mu+1}-\zeta^{\mu+1})^\alpha} d\zeta; 0 < \alpha \le 1, \tag{1}$$

where the function $f(z)$ is analytic function.

Proposition 3.1. The generalized derivative of the function $f(z) = z^\nu, \nu \in \mathsf{R}$ is given by

$$D_z^{\alpha,\mu} f(z) = \frac{(\mu+1)^{\alpha-1}\Gamma(\frac{\nu}{\mu+1}+1)}{\Gamma(\frac{\nu}{\mu+1}+1-\alpha)} z^{(1-\alpha)(\mu+1)+\nu-1}.$$

Proof. We let $\eta := (\frac{\zeta}{z})^{\mu+1}$ then we have

$$
\begin{aligned}
D_z^{\alpha,\mu} z^\nu &= \frac{(\mu+1)^\alpha}{\Gamma(1-\alpha)} \frac{d}{dz} \int_0^z \frac{\zeta^{\mu+\nu}}{(z^{\mu+1}-\zeta^{\mu+1})^\alpha} d\zeta \\
&= \frac{(\mu+1)^{\alpha-1}}{\Gamma(1-\alpha)} \frac{d}{dz} z^{(1-\alpha)(\mu+1)+\nu} \int_0^1 \eta^{\frac{\nu+\mu+1}{\mu+1}-1}(1-\eta)^{(1-\alpha)-1} d\eta \\
&= \frac{(\mu+1)^{\alpha-1}\Gamma(\frac{\nu}{\mu+1}+1)}{\Gamma(\frac{\nu}{\mu+1}+1-\alpha)} z^{(1-\alpha)(\mu+1)+\nu-1}.
\end{aligned}
$$

Assume that the uniformly sampled signal satisfying the $n-$degree polynomial

$$s_n(z) = \sum_{k=0}^n z^k$$

which is used to fit the given signal $z = 1,2,...,I.$ Now in view of proposition 3.1, we have

$$D_z^{\alpha,\mu} s_n(z) = \sum_{k=0}^n D^{\alpha,\mu} z^k = \sum_{k=0}^n \phi_k z^{(1-\alpha)(\mu+1)+k-1}. \tag{2}$$

with the following condiments

$$\phi_0 = \frac{(\mu+1)^{\alpha-1}}{\Gamma(1-\alpha)}$$

$$\phi_1 = \frac{(\mu+1)^{\alpha-1}\Gamma(\frac{1}{\mu+1}+1)}{\Gamma(\frac{1}{\mu+1}+1-\alpha)}$$

$$\vdots$$

$$\tag{3}$$

$$\phi_{n-1} = \frac{(\mu+1)^{\alpha-1}\Gamma(\frac{n-1}{\mu+1}+1)}{\Gamma(\frac{n-1}{\mu+1}+1-\alpha)}.$$

Note that when $\mu = 0$, (3) reduces to the case of the Riemann-Liouville fractional operator. However, in the context of image processing Eq.(3) applies uniformly in the whole digital image and therefore should be in two directions of z and w . Now for two variables function like images, the negative direction of z and w coordinates, can be expressed as

$$
\begin{aligned}
D_z^{\alpha,\mu} s(z_-, w) &= \phi_0 s(z, w) \\
&+ \sum_{k=1}^{n} \phi_k s(z - k, w)
\end{aligned}
\tag{4}
$$

and

$$
\begin{aligned}
D_z^{\alpha,\mu} s(z, w_-) &= \phi_0 s(z, w) \\
&+ \sum_{k=1}^{n} \phi_k s(z, w - k).
\end{aligned}
\tag{5}
$$

While for two variables on the positive direction of z and w coordinates, we have

$$
\begin{aligned}
D_z^{\alpha,\mu} s(z_+, w) &= \phi_0 s(z, w) \\
&- \sum_{k=1}^{n} \phi_k s(z + k, w)
\end{aligned}
\tag{6}
$$

and

$$
\begin{aligned}
D_z^{\alpha,\mu} s(z, w_+) &= \phi_0 s(z, w) \\
&- \sum_{k=1}^{n} \phi_k s(z, w + k).
\end{aligned}
\tag{7}
$$

For digital images, two dimensional fractional differential mask coefficients can be obtained in eight directions of 180°, 90°, 0°, 270°, 45°, 135°, 315°, 225°, as shown in Fig.1.

The output of each image block is eight values, which are representing the texture information in each image. Then the final texture value for each image block is calculated by using Cesáro means of the following equation [13]:

$$
g_k = \sum_{n=1}^{k} \left(\frac{k - n + 1}{k} \right) z^n
\tag{8}
$$

All texture image block values are combined to produce one texture feature for each image. In order to create the texture features vector, in the beginning each image is divided into non- overlapped image blocks. The size of each image block is equal to the size of the fractional differential mask (4x4). The number of image block is chosen to achieve the requirements of the image detail. The main objective of applying Cesáro means is to reduce the size of the texture feature vector to one value instead of 16 for each image block without compromising their discriminating ability. The similarity measurement of a new texture feature vector will be possible by searching the most similar vector into the database, where the retrieved images are ranked by their Euclidean distances D(i,j) to their respective query image which is calculated as follows[1].

$$D_{(i,j)} = \sqrt{\sum_{n=1}^{m}(x_i - x_j)^2 + (y_i - y_j)^2}$$ (9)

where Di,j is the distance of two images xi,j, yi,j and n, m is the image size.

The proposed texture features extraction includes the following steps:

i.   Resize the images to 128 × 128 × 3 and converted to grayscale.
ii.   Divide each image into a specific number of blocks (32x32). The size of each image block is equal to the size of the fractional differential mask (4x4).
iii. Set the mask window size and the values of the fractional powers α and μ.
iv.   Apply fractional differential mask convolution on eight directions with the gray value of corresponding image pixels, and adding all product terms to obtain weighting sum on eight directions.
v.   Find the arithmetic mean of the weighting sum value on the eight directions as approximate value of fractional differential for image pixel.
vi.   Apply Cesáro means to reduce the size of the texture feature vector.
vii. Repeat steps iii to vi for whole image pixels.
viii. Store the complete texture vector for each image.

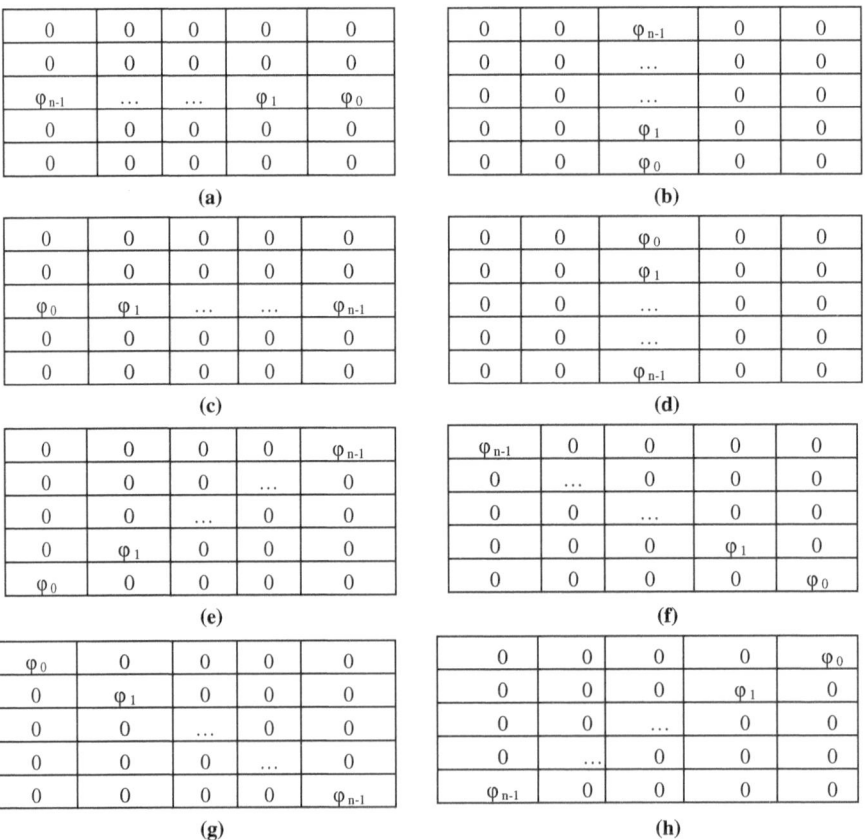

Fig. 1. Fractional differential masks on directions of 180°, 90°, 0°, 270°, 45°, 135°, 315°, 225°

4   Experimental Results

Image retrieval is the process of finding similar images from a large image database with the help of some key attributes related to the images or features contained in the images [1]. Texture features for all images are extracted and stored in a database for comparison with the texture feature value of query image. Each row in the feature vectors represents the feature vector of an image, and each column represents the result vector for each mask windows (1x1024) as the complete texture vector for each image to be used for image retrieval. The main objective of applying Cesáro means is to reduce the size of the texture feature vector to one value instead of 16 for each mask windows(4x4).The measurement of a new feature vector will be possible by searching the most similar vector into the database, where the retrieved images are ranked by their Euclidean distances to their respective query image.

Performance tests for the system proposed by this paper   were implemented using Matlab 2010a on Intel(R) Core i7 at 2.2GHz, 4GB DDR3 Memory, system type 64-bit, Window 7. The method is tested using the VisTex database which is a collection of texture images with images spread across classes containing 16 images each. Fig. 2, illustrates some of these images.

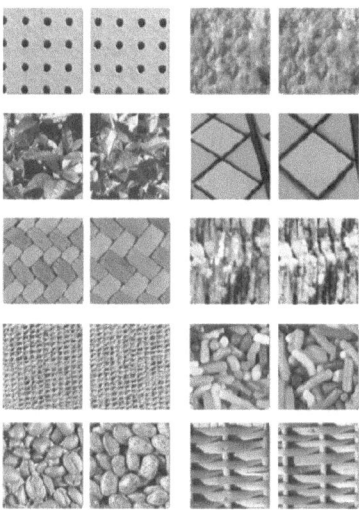

Fig. 2. Images from the database

To evaluate the retrieval performance, we have used the precision-recall crossover point. Precision–Recall, are calculated as follows:

$$\text{Precision} = \frac{\text{Number of relevant images retrieved}}{\text{Total number of images retrieved}}$$

$$\text{Recall} = \frac{\text{Number of relevant images retrieved}}{\text{Total number of relevant images in database}}$$

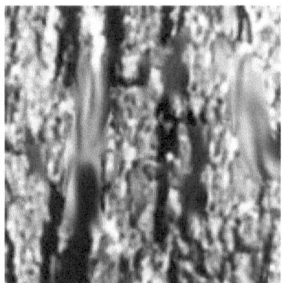

Fig. 3. Query Image

To discuss the performance of the algorithm we have used one class image, which is shown in Fig. 3 considered as an example. The algorithm is applied on the database of 304 images to generate texture feature vector for each image in the database and the query image and then calculate the Euclidian distance to find the relevant images. The proposed algorithm exhibited good results as shown Fig. 4, where the first 9 retrieved images are shown. The number of relevant images from same class are 7, and a few images from different classes retrieved. The precision-recall crossover plot for the same is shown in the Fig. 5, where both the precision and recall curves intersect is called crossover point in precision and recall. Crossover point can be used in a way to measure how correct the proposed algorithm is, higher the crossover point, better is the performance of the method [14]. The average precision-recall crossover of the CBIR method acts as one of the important parameters to judge its performance. A comparison with co-occurrence matrix and Gabor filters as standard methods for texture extraction is presented in this paper.

Co-occurrence matrix is a statistical method to describe textures in an image is achieved primarily by modeling texture as a two-dimensional gray level variation[15]. Gabor filter (or Gabor wavelet), widely adopted to extract texture features from images for image retrieval. Many proposed retrieval techniques adopt the Gabor wavelet as a useful texture descriptor[1].

As the comparison is done on the bases of these two algorithms, the relevancy of image decreases are shown in Figs. 6 - 8 for co-occurrence matrix and Gabor filter respectively. The average precision-recall crossover for co-occurrence matrix and Gabor filter are shown in Fig.7 and 9 respectively. Table 1, shows the comparison of experimental results of proposed method with co-occurrence matrix and Gabor filter.

The proposed technique for CBIR system provides satisfactory results, extracting quite relevant images from the same class. The higher crossover point of the proposed algorithm, acts as one of the important parameters to judge its performance.

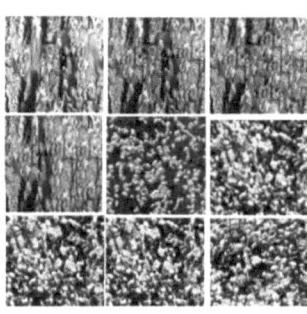

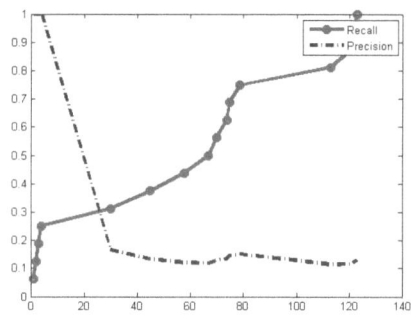

Fig. 4. Retrieved images for query image of Fig. 2 (proposed algorithm)

Fig. 5. Precision - recall for query image of Fig. 2 with the number of images retrieved(proposed algorithm)

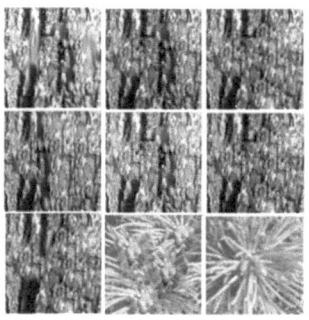

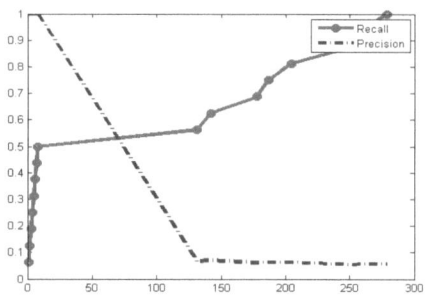

Fig. 6. Retrieved images for query image of Fig. 2 (co-occurrence matrix)

Fig. 7. Precision - recall for query image of Fig.2 with the number of images retrieved (co-occurrence matrix)

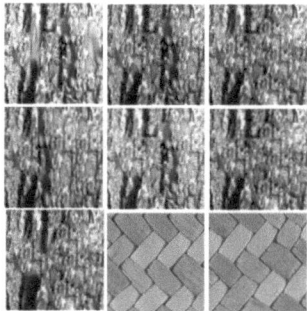

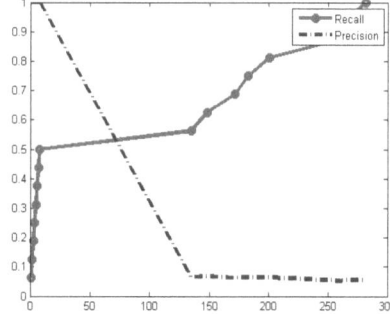

Fig. 8. Retrieved images for query image of Fig. 2 (Gabor Filters)

Fig. 9. Precision - recall for query image of Fig.2 with the number of images retrieved (Gabor Filters)

Table 1. Comparison of the experimental results with other standard methods

Algorithm	Average Precession	Average Recall	Cross over point
Proposed	0.5432	0.5401	0.53
Co-occurrence matrix	0.5207	0.5113	0.51
Gabor Filters	0.3491	0.5313	0.32

5   Conclusion

In the current paper, a texture features extraction technique, using fractional differential mask convolution with Cesáro means was used to retrieve desired images from their databases. Fractional differential mask convolution on eight directions with the gray value had been applied on eight directions. The experiment results means demonstrate the efficacy of this algorithm in comparison with the existing standard methods for texture extraction. Beside, the proposed algorithm exhibited better retrieval precision than the co-occurrence matrix and Gabor filters as standard methods for texture extraction.

Acknowledgement. This research has been funded by the University of Malaya, under the Grant no. RG066-11ICT.

References

1. Jalab, H.A.: Image Retrieval System Based on Color Layout Descriptor and Gabor Filters. In: IEEE Conference on Open System (ICOS 2011), pp. 32–36 (2011)
2. Baaziz, N., Abahmane, O., Missaoui, R.: Texture feature extraction in the spatial-frequency domain for content-based image retrieval (2010), Arxiv preprint ar-Xiv:1012.5208
3. Yue, J., Li, Z., Liu, L., Fu, Z.: Content-based image retrieval using color and texture fused features. Mathematical and Computer Modelling 54, 1121–1127 (2011)
4. Lin, C.H., Lin, W.C.: Image retrieval system based on adaptive color histogram and texture features. The Computer Journal 54(7), 1136–1147 (2010)
5. Lin, C.: A smart content-based image retrieval system based on color and texture feature. In: Image and Vision Computing, vol. 27, pp. 658–665 (2009)
6. Sparavigna, A.C.: Using fractional differentiation in astronomy. Computer Vision and Pattern Recognition (2010), arXiv.org cs arXiv:0910.2381
7. Marazzato, R., Sparavigna, A.C.: Astronomical image processing based on fractional calculus: the AstroFracTool. Instrumentation and Methods for Astrophysics (2009), arXiv.org astro-ph - arXiv:0910.4637
8. Kekre, H.B., Thepade, S.D., Maloo, A.: Image retrieval using fractional coefficients of transformed image using DCT and Walsh transform. International Journal of Engineering Science and Technology 2, 362–371 (2010)

9. Tseng, C.C.: Design of variables and adaptive fractional order FIR differentiators. Signal Processing 86, 2554–2566 (2006)
10. Jalab, H.A., Ibrahim, R.W.: Denoising algorithm based on generalized fractional integral operator with two parameters. Discrete Dynamics in Nature and Society, 1–14 (2012)
11. Tenreiro Machado, J.A., Silva, M.F., Barbosa, R.S., Jesus, I.S., Reis, C.M., Marcos, M.G., Galhano, A.F.: Some applications of fractional calculus in engineering. Mathematical Problems in Engineering, Article ID 639801, 34 Pages (2010)
12. Ibrahim, R.W.: On generalized Srivastava-Owa fractional operators in the unit disk. Advances in Difference Equations 55, 1–10 (2011)
13. Srivastava, H.M., Darus, M., Ibrahim, R.W.: Classes of analytic functions with fractional powers defined by means of a certain linear operator. Integ. Tranc. Special Funct. 22, 17–28 (2011)
14. Kekre, H.B., Thepade, S.D., Sarode, T.K., Suryawanshi, V.: Color feature extraction for CBIR. International Journal of Engineering Science and Technology 3(12), 8357–8365 (2011)
15. Kekre, H.B., Thepade, S.D., Sarode, T.K., Suryawanshi, V.: Image Retrieval using Texture Features extracted from GLCM, LBG and KPE. International Journal of Computer Theory and Engineering 2(5), 560–600 (2010)

Possibilistic Reasoning in Multi-Context Systems: Preliminary Report

Yifan Jin, Kewen Wang, and Lian Wen

School of Information and Communication Technology
Griffith University, Brisbane, QLD 4116, Australia
yifan.jin@griffithuni.edu.au,
{k.wang,l.wen}@griffith.edu.au

Abstract. This paper makes the first attempt to establish a framework for possibilistic reasoning in (nonmonotonic) multi-context systems, called possibilistic MCS. We first introduce the syntax for possibilistic MCS and then define its equilibrium semantics based on Brewka and Eiter's nonmonotonic multi-context systems. Then we investigate several properties and develop a fixoint theory for possibilistic MCS.

1 Introduction

Sharing and reasoning about information in a distributed and heterogeneous environment is becoming more important than ever with the advent of the web and of ubiquitous connectivity. In many cases, such information is not organized as a unique, homogeneous and coherent knowledge base, but is scattered in a large set of local and inter-related contexts. As a result, advanced information systems for the web should be able to deal with such heterogeneity. Moreover, this kind of information is usually incomplete and the information flow between different sources can be quite diverse. During the last decade, there have been extensive efforts in resolving this challenge and in particular, multi-context systems are regarded a promising tool for formalizing and processing heterogeneous and incomplete information [2; 3]. In artificial intelligence, a context is either a situation in the general sense of the term or a part of knowledge or both. Informally, a multi-context system is a formal description of the information available in a number of contexts and specifies the information flow between those contexts. Several logical approaches to context systems have been proposed, most notably McCarthy's propositional logic of context [10] and the multi-context systems devised by Giunchiglia and Serafini [7]. We note that multi-context systems are different from multi-agent systems in that, unlike an agent, a context is not autonomous in general while there is information flow between contexts.

Several different logic-based approahces to MCS have been proposed, e.g. in [10] the contexts are based on classical monotonic reasoning and in [12] and [5] the contexts allow for reasoning based on the absence of information from a context, and in [4] a formalism of heterogeneous nonmonotonic multi-context systems is introduced, which is capable of combining arbitrary monotonic and nonmonotonic logics.

On the other hand, possibility logic, which is developed from Zadeh's possibility theory [14], provides a useful framework for representing states of partial ignorance

P. Anthony, M. Ishizuka, and D. Lukose (Eds.): PRICAI 2012, LNAI 7458, pp. 180–193, 2012.

owing to the use of a dual pair of possibility and necessity measures [6]. We note that some efforts have been made to merge possibilistic reasoning in multiple-source information, e. g. [1], but MCS is different from the frameworks for merging multiple-source information as MCS aims to provide a suitable framework for performing distributed reasoning across multiple information sources.

To our best knowledge, *the problem of incorporating possibilistic reasoning into MCS has not been studied yet*. This paper makes the first attempt to establish a framework for combining nonmonotonic MCS and possibilistic reasoning. We first introduce the syntax for our possibilistic MCS and then define the equilibrium semantics based on Brewka and Eiter's nonmonotonic MCS in [4]. In our framework, each context is represented as a possibilistic logic program [11]. Then we investigate several properties and develop algorithms for the possibilistic MCS.

We proceed, in section 2, with a brief review of possibilistic normal logic program with answer set semantic. In section 3 we introduce the poss-MCS and deal with a part of it, then we extend the result in section 4. Finally, we conclude the work in section 5.

2   Preliminary

We first recall some basics of possibilistic logic and then introduce the syntax and semantics for possibilistic logic programs proposed in [11]. We deal with propositional logic and logic programs. Throughout the paper, a possibilistic concept is denoted $\overline{X}$ while its classical counterpart is denoted X.

We assume that Σ is a set of atoms. A (classical) interpretation I is a subset of Σ. An atom a is true under I if $a \in I$; otherwise, a is false under I. By 2^{Σ} we denote the set of all interpretations on Σ, i. e. the power set of Σ.

A *possibilistic formula* $\overline{\phi}$ on Σ is a pair $(\phi, [\alpha])$ where ϕ is a propositional formula on Σ and $\alpha \in [0, 1]$. Informally, $(\phi, [\alpha])$ expresses that the formula ϕ is certain at least to the level α. This degree α is evaluated by a necessity measure but it is not a probability. The higher is the level, the more certain is the formula. In particular, a possibilistic formula $\overline{\phi}$ is called a *possibilistic atom* if ϕ is an atom. A possibilistic knowledge base (poss-KB) $\overline{K}$ on Σ is a finite set of possibilistic formula on Σ. If $\overline{K} = \{(\phi_1, [\alpha_1]), \ldots, (\phi_n, [\alpha_n])\}$ $(n \geq 0)$, then the classical part of $\overline{K}$ is denoted $K = \{\phi_1, \ldots, \phi_n\}$.

The basic part of the semantics for possibilistic logic is the *possibility distributions*, each of which is a mapping from 2^{Σ} to the interval $[0, 1]$.

Given a possibility distribution π, for each interpretation ω, $\pi(\omega)$ represents the degree of compatibility of the interpretation ω with the available information (or beliefs) about the real world.

A possibility distribution π defines two different weights for propositional formulas. For each propositional formula ϕ, we define

- Possibility degree: $\Pi(\phi) = max\{\pi(\omega) \mid \omega \models \phi\}$.
- Necessity degree: $N(\phi) = 1 - \Pi(\neg\phi)$.

The possibility degree $\Pi(\phi)$ evaluates the extent to which ϕ is consistent with the available beliefs expressed by π. Thus the possibility degree is also referred to as the consistent degree. The necessity degree $N(\phi)$, also called certainty degree evaluates the extent to which ϕ is entailed by the available beliefs expressed by π.

We say a possibility distribution is *compatible* with a poss-KB $\overline{K}$ if, $N(\phi) \geq \alpha$ for every $(\phi, [\alpha]) \in \overline{K}$. Generally, there may exist several possibility distributions compatible with $\overline{K}$. The most desirable distribution is usually selected by the minimum specificity principle [13]. A possibility distribution π is said to be the *least specific distribution* among all compatible distributions if there is no possibility distribution π' such that it is compatible with $\overline{K}$, $\pi' \neq \pi$, and $\forall w$, $\pi'(w) \geq \pi(w)$.

Definition 1. *Let Σ be a finite set of atoms. A possibilistic atom is $\overline{a} = (a, [\alpha])$, where $a \in \Sigma$ and $\alpha \in [0, 1]$.*

The classical projection of $\overline{a}$ is the atom a and $n(a) = \alpha$ is the necessity degree of the possibilistic atom $\overline{a}$.

Definition 2. *A possibilistic normal logic program (or poss-program) is a set of possibilistic rules of the form:*

$$\overline{r} = a \leftarrow a_1, \ldots, a_m, not\ b_1, \ldots, not\ b_n, [\alpha]. \qquad (1)$$

where $m \geq 0$, $n \geq 0$, $\{a_1, \ldots, a_m, b_1, \ldots, b_n, a\} \subseteq \Sigma$, and $n(\overline{r}) = \alpha \in [0, 1]$.

The symbol "*not*" denotes the *default negation* and for each atom b_i, *not* b_i is a negative literal.

Similar to possibilistic propositional logic, the classical projection r of a possibilistic rule $\overline{r}$ is the classical rule $a \leftarrow a_1, \ldots, a_m, not\ b_1, \ldots, not\ b_n$. Also, α represents the certainty level of the information described by the rule $\overline{r}$.

Given a rule $\overline{r}$ of the form (1), its head is defined as $head(\overline{r}) = a$ and its body is $body(\overline{r}) = body^+(\overline{r}) \cup not\ body^-(\overline{r})$ where $body^+(\overline{r}) = \{a_1, \ldots, a_m\}$, $body^-(r) = \{b_1, \ldots, b_n\}$.

The positive projection of $\overline{r}$ is $\overline{r}^+ = head(\overline{r}) \leftarrow body^+(\overline{r}), [\alpha]$.

The set of all rules of $\overline{P}$ with the head a is $H(\overline{P}, a) = \{\overline{r} \in \overline{P} \mid head(\overline{r}) = a\}$.

If a poss-program $\overline{P}$ does not contain any default negation (i. e. $body^-(\overline{P}) = \emptyset$), then $\overline{P}$ is called a *definite poss-program*.

We first introduce the semantics for definite poss-programs.

The reduct of a poss-program $\overline{P}$ w. r. t. a set A of atoms is the definite poss-program defined by:

$$\overline{P}^A = \{\overline{r}^+ \mid \overline{r} \in \overline{P}, body^-(\overline{r}) \cap A = \emptyset\} \qquad (2)$$

We note that the rule $\overline{r}^+$ is actually the possibilistic rule formed by the classical reduct r^+ together with the certainty level of $\overline{r}$.

For a set of atoms $A \subseteq \Sigma$ and a rule $\overline{r}$ in $\overline{P}$, we say $\overline{r}$ is applicable in A if $body^+(\overline{r}) \subseteq A$ and $body^-(\overline{r}) \cap A = \emptyset$. $App(\overline{P}, A)$ denotes the set of rules in poss-program $\overline{P}$ that are applicable in A.

$\overline{P}$ is said to be *grounded* if it can be ordered as a sequence $\langle \overline{r}_1, \ldots, \overline{r}_n \rangle$ such that

$$\forall i, 1 \leq i \leq n, \overline{r}_i \in App(\overline{P}, head(\{\overline{r}_1, \ldots, \overline{r}_{i-1}\})) \qquad (3)$$

Given a poss-program $\overline{P}$ over a set Σ of atoms, similar to the case of propositional possibilistic logic, the semantics of $\overline{P}$ is also defined through possibility distributions on Σ.

The compatibility of a possibility distribution with definite poss-program $\overline{P}$ is defined in [11] (Definition 4). There may exist several different possibility distributions that are compatible with a given definite poss-program. Among these compatible distributions, we are particularly interested in the least specific one, which is given in the next result.

Proposition 1. *Let $\overline{P}$ be a definite poss-program. We define a possibilistic distribution $\pi_{\overline{P}}$ for $\overline{P}$ as, for each $A \in 2^{\Sigma}$,*

$$\pi_{\overline{P}}(A) = \begin{cases} 0, & \text{if } A \nsubseteq head(App(P, A)) \\ 0, & \text{if } App(P, A) \text{ is not grounded} \\ 1, & \text{if } A \text{ is a model of } P \\ 1 - max\{n(\overline{r}) \mid A \nvDash r\}, & \text{otherwise.} \end{cases} \tag{4}$$

Then $\pi_{\overline{P}}$ is the least specific distribution compatible with $\overline{P}$.

The least specific distribution for $\overline{P}$ determines its possibilistic measures.

Definition 3. *Let $\overline{P}$ be a definite poss-program and $\pi_{\overline{P}}$ the least specific distribution compatible with $\overline{P}$. Then the possibility and necessity degrees for an atom a is defined by*

$$\Pi_{\overline{P}}(a) = max\{\pi_{\overline{P}}(A) \mid a \in A\}.$$
$$N_{\overline{P}}(a) = 1 - max\{\pi_{\overline{P}}(A) \mid a \notin A\}.$$

$\Pi_{\overline{P}}(a)$ gives the level of consistency of a w. r. t. the definite poss-program $\overline{P}$ and $N_{\overline{P}}(a)$ evaluates the level at which a is inferred from $\overline{P}$. For instance, whenever an atom a belongs to the model of the classical program, its possibility is equal to 1.

The necessity measure allows us to introduce the following definition of the possibilistic model of a definite poss-program.

Definition 4. *Let $\overline{P}$ be a definite poss-program. Then the set*

$$\overline{M}(\overline{P}) = \{(a, N_{\overline{P}}(a)) \mid a \in \Sigma, N_{\overline{P}}(a) > 0\} \tag{5}$$

is referred to as its possibilistic model.

So far we have introduced the semantics for definite poss-programs. Now we turn to study the computation of the possibility distribution and possibilistic model for a given poss-program. First we define β-applicability of a rule $\overline{r}$ to capture the certainty of an conclusion that the rule can derive w. r. t. a set $\overline{A}$ of possibilistic atoms.

Definition 5. *Let $\overline{r}$ be a possibilistic rule of the form $c \leftarrow a_1, \ldots, a_n, [\alpha]$ and $\overline{A}$ be a set of possibilistic atoms.*

1. *$\overline{r}$ is β-applicable in $\overline{A}$ with possibility $\beta = min\{\alpha, \alpha_1, \ldots, \alpha_n\}$ if $\{(a_1, \alpha_1), \ldots, (a_n, \alpha_n)\} \subseteq \overline{A}$.*
2. *$\overline{r}$ is 0-applicable otherwise.*

If the rule body is empty, then the rule is applicable with its own certainty degree and if the body is not satisfied by $\overline{A}$, then the rule is 0-applicable and it is actually not at all applicable w. r. t. $\overline{A}$. So the applicability level of the rule depends on the certainty level of atoms in its body and its own certainty degree.

The set of rules in $\overline{P}$ that have the head a and are applicable w. r. t. $\overline{A}$ is denoted $App(\overline{P}, \overline{A}, a)$:

$$App(\overline{P}, \overline{A}, a) = \{\overline{r} \in H(\overline{P}, a), \overline{r} \text{ is } \beta\text{-applicable in } \overline{A}, \beta > 0\} \tag{6}$$

Having defined the applicability of possibilistic rules, we can generalise the consequence operator of classical logic programs to poss-programs.

Definition 6. *Let $\overline{P}$ be a poss-program, a be an atom and $\overline{A}$ be a set of possibilistic atoms. Then we define the consequence operator for $\overline{P}$ by*

$$\overline{T}_{\overline{P}}(\overline{A}) = \{(a, \delta) \mid a \in head(P), App(\overline{P}, \overline{A}, a) \neq \emptyset, \delta = max\{\beta \mid \overline{r} \text{ is } \beta\text{-applicable in } \overline{A}\}\}$$

Then the iterated operator $\overline{T}_{\overline{P}}^k$ is defined by

$$\overline{T}_{\overline{P}}^0 = \emptyset \text{ and } \overline{T}_{\overline{P}}^{n+1} = \overline{T}_{\overline{P}}(\overline{T}_{\overline{P}}^n), \forall n \geq 0. \tag{7}$$

$\overline{T}_{\overline{P}}$ has a least fixpoint that is the possibilistic consequences of $\overline{P}$ and it is denoted by $\overline{Cn}(\overline{P})$. We have $\overline{Cn}(\overline{P}) = \overline{M}(\overline{P})$ (see [11] for more details).

For poss-programs, it is easy to formalize the notion of stable models by a generalized reduct [8].

Definition 7. *Let $\overline{P}$ be a poss-program and A a set of atoms.*
We say $\overline{A}$ is a stable model of poss-program $\overline{P}$ if $\overline{A} = \overline{Cn}(\overline{P}^A)$.

The possibility distribution for $\overline{P}$ is defined in terms of its reduct's possibility distribution as follows.

Definition 8. *Let $\overline{P}$ be a possibilistic logic program and A be an atom set, then $\tilde{\pi}_{\overline{P}}$ is the possibility distribution defined by:*

$$\forall A \in 2^\Sigma, \tilde{\pi}_{\overline{P}}(A) = \pi_{\overline{P}^A}(A) \tag{8}$$

With these two definitions we can also define the possibility and necessity measures to each atom by Definition 3.

3   Possibilistic Multi-Context Systems

In this section we will first incorporate possibilistic reasoning into multi-context systems (MCS) and then discuss their properties.

3.1 Syntax of Poss-MCS

A possibilistic multi-context system (or poss-MCS) is a collection of contexts where each context is a poss-program with its own knowledge base and bridge rules. In this paper, a possibilistic context $\overline{C}$ is a triple $(\Sigma, \overline{K}, \overline{B})$ where Σ is a set of atoms, $\overline{K}$ is a poss-program, and $\overline{B}$ is a set of possibilistic bridge rules for the context $\overline{C}$. Before formally introducing poss-MCS, we first give the definition of possibilistic bridge rules. Intuitively, a possibilistic bridge rule make it possible to infer new knowledge for a context based on some other contexts. So possibilistic bridge rules provide an effective way for the information flow between related contexts.

Definition 9. *Let* $\overline{C}_1, \ldots, \overline{C}_n$ *be* n *possibilistic contexts. A possibilistic bridge rule* $\overline{br}_i$ *for a context* $\overline{C}_i$ *(*$1 \le i \le n$*) is of the form*

$$a \leftarrow (C_1 : a_1), \ldots, (C_k : a_k), not\,(C_{k+1} : a_{k+1}), \ldots, not\,(C_n : a_n), [\alpha] \qquad (9)$$

where a *is an atom in* $\overline{C}_i$, *each* a_j *is an atom in context* $\overline{C}_j$ *for* $j = 1, \ldots, n$.

Intuitively, a rule of the form (9) states that the information a is added to context $\overline{C}_i$ with necessity degree α if, for $1 \le i \le k$, a_j is present in context $\overline{C}_j$ and for $k+1 \le j \le n$, a_j is not provable in $\overline{C}_j$.

By br_i we denote the classical projection of $\overline{br}_i$:

$$a \leftarrow (C_1 : a_1), \ldots, (C_k : a_k), not\,(C_{k+1} : a_{k+1}), \ldots, not\,(C_n : a_n). \qquad (10)$$

The necessity degree α of the bridge rule $\overline{br}_i$ is written $n(\overline{br}_i)$.

Definition 10. *A possibilistic multi-context system, or just* poss-MCS, $\overline{M} = (\overline{C}_1, \ldots, \overline{C}_n)$ *is a collection of contexts* $\overline{C}_i = (\Sigma_i, \overline{K}_i, \overline{B}_i)$, $1 \le i \le n$, *where each* Σ_i *is the set of atoms used in context* $\overline{C}_i$, $\overline{K}_i$ *is a poss-program on* Σ_i, *and* $\overline{B}_i$ *is a set of possibilistic bridge rule over atom sets* $(\Sigma_1, \ldots, \Sigma_n)$.

A poss-MCS *is* definite *if the poss-program and possibilistic bridge rules of each context is definite.*

Definition 11. *A possibilistic belief set* $\overline{S} = (\overline{S}_1, \ldots, \overline{S}_n)$ *is a collection of possibilistic atom sets* $\overline{S}_i$ *where each* $\overline{S}_i$ *is a collection of possibilistic atoms* $\overline{a_i}$ *and* $a_i \in \Sigma_i$

In the next two subsections we will study the semantics for possibilistic definite MCS.

3.2 Model Theory for Definite Poss-MCS

Like poss-programs we will first specify the semantics for definite poss-MCS (i. e. without default negation) and then define the semantics for poss-MCS with default negation by reducing the given poss-MCS to a definite poss-MCS.

For convenience, by a *classical MCS* we mean a multi-context system (MCS) without possibility degrees as in [4]. The semantics of a classical MCS is defined by the set of its equilibria, which characterize acceptable belief sets that an agent may adopt based on the knowledge represented in a knowledge base. Let us first recall the semantics of classical MCS. Let $M = (C_1, \ldots, C_n)$ be a classical MCS with each (classical) context $C_i = (\Sigma_i, K_i, B_i)$, where Σ_i is a set of atoms, K_i is a logic program, and B_i is a set of (classical) bridge rules.

A *belief state* of a MCS $M = (C_1, \ldots, C_n)$ is a collection $S = (S_1, \ldots, S_n)$ where each S_i is a set of atoms that $S_i \subseteq \Sigma_i$. A (classical) bridge rule (10) is *applicable* in a belief state S iff for $1 \leq j \leq k$, $a_j \in S_j$ and for $k + 1 \leq j \leq n$, $a_j \notin S_j$.

In general, not every belief state is acceptable for an MCS. Usually, the equilibrium semantics selects certain belief states for a given MCS as acceptable belief states. Intuitively, an equilibrium is a belief state $S = (S_1, \ldots, S_n)$ where each context C_i respects all bridge rules applicable in S and accepts S_i. The definition of equilibrium can be found in [4]. There may exist several different equilibrium, among these equilibrium, we are particularly interested in the minimal one. Formally, S is a grounded equilibrium of an MCS $M = (C_1, \ldots, C_n)$ iff for each i $(1 \leq i \leq n)$, S_i is an answer set of logic program $P = K_i \cup \{head(r) \mid r \in B_i$ is applicable in $S\}$. Then if the logic program P is grounded, we can use $Cn(P)$ to obtain its (unique) answer set, which is the smallest set of atoms closed under P and alternatively, it can be computed as the least fixpoint of the consequence operator T_P: $T_P(A) = head(App(P, A))$ (see [9]).

So for a classical definite MCS, its unique grounded equilibrium is the collection consisting of the least model of each context. The grounded equilibrium of a definite MCS M is denoted $GE(M)$.

Then we clarify the links between the grounded equilibrium S of a definite MCS M and the rules producing it. We see that for each context C_i, S_i is underpinned by a set of applicable rules $App_i(M, S)$, that satisfies a stability condition and that is grounded.

Proposition 2. *Let M be a definite MCS and S be a belief state,*

$$S \text{ is the grounded equilibrium of } M \Leftrightarrow \begin{cases} S_i = head(App_i(M, S)) \\ \bigcup_i App_i(M, S) \text{ is grounded} \end{cases} \qquad (11)$$

Now let us turn to the semantics of definite poss-MCS. Thus, we will specify the possibility distribution of belief states for a given definite poss-MCS. As we know that the satisfiability of a rule r is based on its applicability w. r. t. an belief state S and $S \not\models r$ iff $body^+(r) \subseteq S \wedge head(r) \notin S$. But this is not enough to determine the possibility degree of a belief state. For example, if we have a definite MCS $M = (C_1, C_2)$ consisting of two definite MCS. Assume K_1 and K_2 are empty, and B_1 consists of the single bridge rule $p \leftarrow (2 : q)$ and B_2 of the single bridge rule $q \leftarrow (1 : p)$. Now $S = (\{p\}, \{q\})$ satisfies every rule in M. But it is not an equilibrium because the groundedness is not satisfied. Besides, assume that an definite MCS $M = (C_1)$ with a single context, K_1 is $\{a\}$ and B_1 consists of the bridge rule $b \leftarrow (1 : c)$. Now $S = (\{a, b\})$ satisfies every rules in M but it is not an equilibrium because b cannot be produced by any rule from C_1 applicable in S. In these two cases, the possibility of S must be 0 since they cannot be an equilibrium at all, even if they satisfy every rule in their MCS.

Definition 12. *Let $\overline{M} = (\overline{C}_1, \ldots, \overline{C}_n)$ be a definite poss-MCS and $S = (S_1, \ldots, S_n)$ be a belief state. The possibility distribution $\pi_{\overline{M}} : 2^{\Sigma} \to [0, 1]$ for $\overline{M}$ is defined as, for $S \in 2^{\Sigma}$,*

$$\pi_{\overline{M}}(S) = \begin{cases} 0 & \text{if } S \nsubseteq head(\bigcup_i App_i(M, S)) \\ 0 & \text{if } \bigcup_i App_i(M, S) \text{ is not grounded} \\ 1 & \text{if } S \text{ is an equilibrium of } M \\ \pi_{\overline{M}}(S) = 1 - max\{n(\overline{r}) \mid S \nvDash r, \overline{r} \in \overline{B}_i \text{ or } \overline{r} \in \overline{K}_i\}, & \text{otherwise.} \end{cases}$$
(12)

The possibility distribution specifies the degree of compatibility of each belief set S with poss-MCS $\overline{M}$.

Recall that $GE(M)$ denotes the grounded equilibrium of a (classical) definite MCS M. Then the possibility distribution for definite poss-MCS has the following useful properties.

Proposition 3. *Let $\overline{M} = (\overline{C}_1, \ldots, \overline{C}_n)$ be a definite poss-MCS, $S = (S_1, \ldots, S_n)$ be a belief state and $GE(M)$ the grounded equilibrium of M, then*

1. $\pi_{\overline{M}}(S) = 1$ iff $S = GE(M)$.
2. If $S \supset GE(M)$, then $\pi_{\overline{M}}(S) = 0$.
3. If $GE(M) \neq \emptyset$, then $\pi_{\overline{M}}(\emptyset) = 1 - max\{n(\overline{r}) \mid body^+(\overline{r}) = \emptyset, \overline{r} \in \overline{B}_i \text{ or } \overline{r} \in \overline{K}_i\}$.

Proof. 1. $\Rightarrow$: Let $\pi_{\overline{M}}(S) = 1$. On the contrary, assume that $S \neq GE(M)$. Then by Equation (12), $\pi_{\overline{M}}(S) = 1$ would only be obtained from the last case:

$$\pi_{\overline{M}}(S) = 1 - max\{n(\overline{r}) \mid S \nvDash r, \overline{r} \in \overline{B}_i \text{ or } \overline{r} \in \overline{K}_i\}.$$

This implies that $S \models r$.

By the first case in Equation (12), we have that $S \subseteq head(\bigcup_i App_i(M, S))$.

By the second case in Equation (12), $\bigcup_i App_i(M, S)$ is grounded.

Therefore, S must be the least equilibrium of M.

$\Leftarrow$: If $S = GE(M)$ by the definition we have $\pi_{\overline{M}}(S) = 1$.

2. Because $S \supset GE(M)$ we have for each $i : S_i \supset head(App_i(M, S)) \vee App(M, S)$ is not grounded by properties 2. So by definition $\pi_{\overline{M}}(S) = 0$.

3. It is obvious that $\emptyset \subseteq head(App_i(M, \emptyset))$ and $App_i(M, \emptyset)$ is grounded. So it can only apply to the forth case of Equation 12.

Definition 13. *Let $\overline{M}$ be a definite poss-MCS and $\pi_{\overline{M}}$ be the possibilistic distribution for $\overline{M}$. The possibility and necessity of an atom in a belief state S is defined by:*

$$\Pi_{\overline{M}}(a_i) = max\{\pi_{\overline{M}}(S) \mid a_i \in S_i\}$$
(13)

$$N_{\overline{M}}(a_i) = 1 - max\{\pi_{\overline{M}}(S) \mid a_i \notin S_i\}$$
(14)

Proposition 4. *Let $\overline{M}$ be a definite poss-MCS and $S = (S_1, \ldots, S_n)$ is a belief state. Then*

1. $a_i \notin S_i$ *iff* $N_{\overline{M}}(a_i) = 0$.
2. *If* $a_i \in S_i$, *then* $N_{\overline{M}}(a_i) = min\{max\{n(\overline{r}) \mid S_i \nvdash r, \overline{r} \in \overline{B}_i \text{ or } \overline{r} \in \overline{K}_i\} \mid a_i \notin S_i, S_i \subset GE(M)\}$.

Proof. 1. Because $\pi_{\overline{M}}(S) = 1$, $N_{\overline{M}}(a_i) = 0$ is obvious when $a_i \notin S_i$. And when $N_{\overline{M}}(a_i) = 0$ it means there is an S such that $a_i \notin S_i$ and $\pi_{\overline{M}}(S) = 1$, So such an S must be the equilibrium of M. Thus, $a \notin GE(M)$

   2.

$$
\begin{aligned}
N_{\overline{M}}(a_i) =& 1 - max\{\pi_{\overline{M}}(S) \mid a_i \notin S_i\} \\
=& 1 - max\{\pi_{\overline{M}}(S) \mid a_i \notin S_i, S_i \subset GE(M)\} \\
& \text{since by Proposition 3} \\
=& 1 - max\{1 - max\{n(\overline{r}) \mid S_i \nvdash r, \overline{r} \in \overline{B}_i \text{ or } \overline{r} \in \overline{K}_i\} \mid a_i \notin S_i, S_i \subset GE(M)\} \\
& \text{since } \pi_{\overline{M}}(S) = 1 - max\{n(\overline{r}) \mid S_i \nvdash r, \overline{r} \in \overline{B}_i \text{ or } \overline{r} \in \overline{K}_i\} \\
=& min\{max\{n(\overline{r}) \mid S_i \nvdash r, \overline{r} \in \overline{B}_i \text{ or } \overline{r} \in \overline{K}_i\} \mid a_i \notin S_i, S_i \subset GE(M)\}.
\end{aligned}
$$

The semantics for definite poss-MCS is determined by its unique possibilistic grounded equilibrium.

Definition 14. *Let $\overline{M}$ be a definite poss-MCS. Then the following set of possibilistic atoms is referred to as the possibilistic grounded equilibrium.*

$$\overline{MD}(\overline{M}) = (\overline{S}_1, \ldots, \overline{S}_n)$$

where $\overline{S}_i = \{(a_i, N_{\overline{M}}(a_i)) \mid a_i \in \Sigma_i, N_{\overline{M}}(a_i) > 0\}$ *for* $i = 1, \ldots, n$.

By the first statement of Proposition 4, it is easy to see the following result holds.

Proposition 5. *Let $\overline{M}$ be a definite poss-MCS and M be the classical projection of $\overline{M}$. Then the classical projection of $\overline{MD}(\overline{M})$ is the grounded equilibrium of the definite MCS M.*

Example 1. *Let $\overline{M} = (\overline{C}_1, \overline{C}_2)$ be a definite poss-MCS where $\Sigma_1 = \{a\}$, $\Sigma_2 = \{b, c\}$, $\overline{K}_1 = \{(a, [0.9]), (c \leftarrow b, [0.8])\}$, $\overline{K}_2 = \overline{B}_1 = \emptyset$, $\overline{B}_2 = \{(b \leftarrow 1 : a, [0.7])\}$.*

By Definition 12, $\pi_{\overline{M}}(\{\emptyset\}, \{\emptyset\})=1 - max\{0.9\}=0.1$, $\pi_{\overline{M}}(\{\emptyset\}, \{b\})=1 - max\{0.9, 0.8\}=0.1$, $\pi_{\overline{M}}(\{\emptyset\}, \{c\}) = 1 - max\{0.9\} = 0.1$, $\pi_{\overline{M}}(\{\emptyset\}, \{b, c\}) = 0$ (not inclusion), $\pi_{\overline{M}}(\{a\}, \{\emptyset\}) = 1 - max\{0.7\} = 0.3$, $\pi_{\overline{M}}(\{a\}, \{b\}) = 1 - max\{0.8, 0.7\} = 0.2$, $\pi_{\overline{M}}(\{a\}, \{c\}) = 0$ (not inclusion), and $\pi_{\overline{M}}(\{a\}, \{b, c\}) = 1$ (the grounded equilibrium).

And thus, by Definition 13, we can get the necessity value for each atom: $N_{\overline{M}}(a) = 1 - max\{0.1\} = 0.9$, $N_{\overline{M}}(b) = 1 - max\{0.1, 0.3, 0\} = 0.7$, and $N_{\overline{M}}(c) = 1 - max\{0.1, 0.3, 0.2\} = 0.7$.

Then by Definition 14 we can get the possibilistic grounded equilibrium $\overline{S} = (\{(a, [0.9])\}, \{(b, [0.7]), (c, [0.7])\})$.

3.3   Fixpoint Theory for Definite Poss-MCS

In the last subsection we introduced the possibilistic grounded equilibrium and the possibilistic distribution of the belief states. In this subsection we will develop a fixpoint theory for the possibilistic grounded equilibrium and thus provide a way for computing the equilibrium.

Similar to Definition 5 for poss-programs, we can define the applicability of possibilistic rules and thus, for an atom $a_i \in \Sigma_i$ and a possibilistic belief state $\overline{S}$ we define

$$App(\overline{M}, \overline{S}, a_i) = \{\overline{r} \in H(\overline{M}, a_i), \overline{r} \text{ is } \beta\text{-applicable in } \overline{S}, \beta > 0\} \qquad (15)$$

The above set is the collection of rules that have the head a_i and are β-applicable in $\overline{S}$.

By modifying the approach in [4], we introduce the following consequence operator for a definite poss-MCS. As we already know from Definition 6, the possibilistic consequences of a poss-program $\overline{P}$ is denoted by $\overline{Cn}(\overline{P})$.

Definition 15. *For each context* $\overline{C}_i = (\Sigma_i, \overline{K}_i, \overline{B}_i)$ *in a definite poss-MCS* $\overline{M} = (\overline{C}_1, \ldots, \overline{C}_n)$, *we define* $\overline{K}_i^{t+1} = \overline{K}_i^t \cup \{(head(\overline{r}), [\beta]) \mid \overline{r} \in \overline{B}_i$ *and is* β-applicable in $\overline{E}^t, \beta > 0\}$, *where* $\overline{K}_i^0 = \overline{K}_i$ *for* $1 \le i \le n$, $\overline{E}^t = (\overline{E}_1^t, \ldots, \overline{E}_n^t)$, $\overline{E}_i^t = \overline{Cn}(\overline{K}_i^t)$ *for* $t > 0$.

Since K_i and B_i of each context are finite, the iteration based on $\overline{K}_i^t$ will reach a fixpoint, which is denoted $\overline{K}_i^\infty$. Then we have the following proposition.

Proposition 6. *Let* $\overline{M} = (\overline{C}_1, \ldots, \overline{C}_n)$ *be a definite poss-MCS with* $\overline{C}_i = (\Sigma_i, \overline{K}_i, \overline{B}_i)$ *for* $1 \le i \le n$ *and* $\overline{S} = (\overline{S}_1, \ldots, \overline{S}_n)$ *be the grounded equilibrium for* $\overline{M}$. *Then*

$$\overline{Cn}(\overline{K}_i^\infty) = \overline{S}_i.$$

The key idea above is that, for each knowledge base $\overline{K}_i^t$, we use the operator $\overline{Cn}$ to obtain its possibilistic answer set, and then by the first item in Definition 15, add atoms from bridge rules that are derivale from $\overline{K}_i^t$ to get $\overline{K}_i^{t+1}$. Then, we apply the operator $\overline{Cn}$ again. Repeat this process until we reach the fixpoint.

Let us consider an example.

Example 2. *Let* $\overline{M} = (\overline{C}_1, \overline{C}_2, \overline{C}_3)$ *be a definite poss-MCS, where*

- $\overline{K}_1 = \{(a, [0.9])\}$, $\overline{B}_1 = \emptyset$;
- $\overline{K}_2 = \emptyset$, $\overline{B}_2 = \{(b \leftarrow (1:a), [0.8])\}$;
- $\overline{K}_3 = \{(c, [0.7]), (d \leftarrow c, [0.6]), (f \leftarrow e, [0.5])\}$, $\overline{B}_3 = \{(e \leftarrow (2:b), [0.4])\}$.

At the beginning we will start with $\overline{K}_i^0$ for each context.
For context $\overline{C}_1$ with $\overline{K}_1^0 = \overline{K}_1 = \{(a, [0.9])\}$:
$\overline{T}_{1,0}^0 = \emptyset, \overline{T}_{1,0}^1 = \overline{T}_{1,0}(\emptyset) = \{(a, [0.9])\}, \overline{T}_{1,0}^2 = \{(a, [0.9])\}$.
For context $\overline{C}_2$ with $\overline{K}_2^0 = \overline{K}_2 = \emptyset$:
$\overline{T}_{2,0}^0 = \emptyset, \overline{T}_{2,0}^1 = \overline{T}_{2,0}^0 = \emptyset$.
For context $\overline{C}_3$ with $\overline{K}_3^0 = \overline{K}_3 = \{(c, [0.7]), (d \leftarrow c, [0.6]), (f \leftarrow e, [0.5])\}$:

$\overline{T}_{3,0}^0 = \emptyset$, $\overline{T}_{3,0}^1 = \overline{T}_{3,0}(\emptyset) = \{(c, [0.7])\}$, $\overline{T}_{3,0}^2 = \overline{T}_{3,0}(\{(c, [0.7])\}) = \{(c, [0.7]), (d, [0.6])\}$, $\overline{T}_{3,0}^3 = \overline{T}_{3,0}^2 = \{(c, [0.7]), (d, [0.6])\}$.

Thus, $\overline{E}^0 = \{(a, [0.9]), (c, [0.7]), (d, [0.6])\}$.

Then starting from the fixpoint of $T_{S_i,0}$, for each context $\overline{C}_i$ we have:

For context 1 with $\overline{K}_1^1 = \{(a, [0.9])\}$: $\overline{T}_{1,1}^0 = \overline{T}_{1,0}^2 = \{(a, [0.9])\}$.

For context 2 with $\overline{K}_2^1 = \{(b, [0.8])\}$: $\overline{T}_{2,1}^0 = \overline{T}_{2,0}^1 = \emptyset$, $\overline{T}_{2,1}^1 = T_{2,1}(\emptyset) = \{(b, [0.8])\}$, $\overline{T}_{2,1}^2 = \overline{T}_{2,1}(\{(b, [0.8])\}) = \{(b, [0.8])\}$.

For context 3 with $\overline{K}_3^1 = \{(c, [0.7]), (d \leftarrow c, [0.6]), (f \leftarrow e, [0.5])\}$: $\overline{T}_{3,1}^0 = \overline{T}_{3,0}^2 = \{(c, [0.7]), (d, [0.6])\}$.

Thus, $\overline{E}^1 = \{(a, [0.9]), (b, [0.8]), (c, [0.7]), (d, [0.6])\}$.

Then from the fixpoint of $T_{S_i,1}$, for each context $\overline{C}_i$:

For context 1 with $\overline{K}_1^2 = \{(a, [0.9])\}$: $\overline{T}_{1,2}^0 = \overline{T}_{1,1}^1 = \{(a, [0.9])\}$

For context 2 with $\overline{K}_2^2 = \{(b, [0.8])\}$:
$\overline{T}_{2,2}^0 = \overline{T}_{2,1}^2 = \{(b, [0.8])\}$, $\overline{T}_{2,2}^1 = \overline{T}_{2,2}(\{(b, [0.8])\}) = \{(b, [0.8])\}$.

For context 3 with $\overline{K}_3^2 = \{(c, [0.7]), (d \leftarrow c, [0.6]), (f \leftarrow e, [0.5]), (e, [0.4])\}$:
$\overline{T}_{3,2}^0 = \overline{T}_{3,1}^2 = \{(c, [0.7]), (d, [0.6])\}$, $\overline{T}_{3,2}^1 = \overline{T}_{3,2}(\{(c, [0.7]), (d, [0.6])\}) = \{(c, [0.7]), (d, [0.6]), (e, [0.4])\}$,
$\overline{T}_{3,2}^2 = \overline{T}_{3,2}(\{(c, [0.7]), (d, [0.6]), (e, [0.4])\}) = \{(c, [0.7]), (d, [0.6]), (e, [0.4]), (f, [0.4])\}$,
$\overline{T}_{3,2}^3 = \overline{T}_{3,2}(\{(c, [0.7]), (d, [0.6]), (e, [0.4]), (f, [0.4])\}) = \{(c, [0.7]), (d, [0.6]), (e, [0.4]), (f, [0.4])\}$.

Thus, $E^2 = \{(a, [0.9]), (b, [0.8]), (c, [0.7]), (d, [0.6]), (e, [0.4]), (f, [0.4])\}$.

Therefore, the possibilistic grounded equilibrium of $\overline{M}$ is
$\overline{S} = (\{(a, [0.9])\}, \{(b, [0.8])\}, \{(c, [0.7]), (d, [0.6]), (e, [0.4]), (f, [0.4])\})$.

Proposition 7. Let $M = (\overline{C}_1, \dots, \overline{C}_n)$ be a definite poss-MCS and $S = (S_1, \dots, S_n)$ be a belief state. Then for each i $(1 \leq i \leq n)$ and a cardinal t, $\overline{T}_{S_i,t}$ is monotonic, i. e. for all sets $\overline{A}$ and $\overline{B}$ of possibilistic atoms with $\overline{A} \sqsubseteq \overline{B}$, it holds that

$$\overline{T}_{S_i,t}(A) \sqsubseteq \overline{T}_{S_i,t}(B).$$

Proof. For any $\overline{A} \sqsubseteq \overline{B}, \forall a \in head(\overline{K})$, $App(\overline{K}, \overline{A}, t) \subseteq App(\overline{K}, \overline{B}, t)$. And relies on the *max* operator, $A \sqsubseteq B \Rightarrow \overline{T}_{S_i,t}(A) \sqsubseteq \overline{T}_{S_i,t}(B)$. Thus $\overline{T}_{S_i,t}$ is monotonic.

By Taski's fixpoint theorem, we can state the following result.

Proposition 8. The operator $\overline{T}_{S_i,t}$ has a least fixedpoint when $\overline{S}_i$ is a definite poss-program. We denote $\overline{T}_{S_i,t}^\infty = \overline{S}_i$ then the $\overline{S} = (\overline{S}_1, \dots, \overline{S}_n)$ is the equilibrium of $\overline{M}$ and we denote it as $\Pi GE(\overline{M})$.

We can now show the relationship between the semantical approach and fixed point approach:

Theorem 1. Let $\overline{M}$ be an definite poss-MCS, then $\overline{GE}(\overline{M}) = \overline{MD}(\overline{M})$.

The proof is similar to that in [11].

4   Normal Poss-MCS

Having introduced the semantics for definite poss-MCS, we are ready to define the semantics for normal poss-MCS with default negation. The idea is similar to the definition of answer sets, we will reduce a poss-MCS with default negation to a definite poss-MCS. Based on the definition of rule reduct in Equation (2), we can define the reduct for normal poss-MCS.

Definition 16. *Let $\overline{M} = (\overline{C}_1, \ldots, \overline{C}_n)$ be a normal poss-MCS and $S = (S_1, \ldots, S_n)$ be a belief state. The possibilistic reduct of $\overline{M}$ w. r. t. S is the poss-MCS*

$$\overline{M}^S = (\overline{C}_1^S, \ldots, \overline{C}_N^S). \tag{16}$$

where $\overline{C}_i^S = (\Sigma_i, \overline{K}_i^S, \overline{B}_i^S).$

We note that the reduct of $\overline{K}_i$ relies only on S_i while the reduct of $\overline{B}_i$ depends on the whole belief state S. This is another role difference of $\overline{K}_i$ from $\overline{B}_i$.

Given the notion of reduct for normal poss-MCS, the equilibrium semantics of normal poss-MCS can be defined easily.

Definition 17. *Let $\overline{M}$ be a normal poss-MCS and $\overline{S}$ be a possibilistic belief state. $\overline{S}$ is a possibilistic equilibrium of $\overline{M}$ if $\overline{S} = \overline{GE}(\overline{M}^S).$*

Example 3. *Let $\overline{M} = (\overline{C}_1, \overline{C}_2, \overline{C}_3)$ be a definite poss-MCS with 3 contexts:*
- $\overline{K}_1 = \{(a, [0.9])\},$ $\overline{B}_1 = \{(b \leftarrow not\ 3 : e, [0.8])\};$
- $\overline{K}_2 = \{(d \leftarrow not\ c, [0.7])\},$ $\overline{B}_2 = \{(c \leftarrow 1 : a, [0.6])\};$
- $\overline{K}_3 = \emptyset,$ $\overline{B}_3 = \{(e \leftarrow not\ 1 : b, [0.5])\}$

We have $S_1 = (\{a\}, \{c\}, \{e\})$ and thus $\overline{M}^{S_1}$ is obtained as

$$\overline{M}^{S_1} = \begin{cases} \overline{K}_1 = \{(a, [0.9])\}, & \overline{B}_1 = \emptyset \\ \overline{K}_2 = \emptyset, & \overline{B}_2 = \{(c \leftarrow 1 : a, [0.6])\} \\ \overline{K}_3 = \emptyset, & \overline{B}_3 = \{(e, [0.5])\} \end{cases} \tag{17}$$

Following Definition 15, we can get $\overline{S}_1 = \{(a, [0.9]), \{c, [0.6]\}, \{e, [0.5]\}).$
And also $S_2 = (\{a, b\}, \{c\}, \{\emptyset\})$, then $\overline{M}^{S_2}$ is as follows:

$$\overline{M}^{S_2} = \begin{cases} \overline{K}_1 = \{(a, [0.9])\}, & \overline{B}_1 = \{(b, [0.8])\} \\ \overline{K}_2 = \emptyset, & \overline{B}_2 = \{(c \leftarrow 1 : a, [0.6])\} \\ \overline{K}_3 = \emptyset, & \overline{B}_3 = \emptyset \end{cases} \tag{18}$$

So we have $\overline{S}_2 = (\{(a, [0.9]), (b, [0.8])\}, \{(c, 0.6)\}, (\{\emptyset\})).$
The following proposition shows that a possibilistic equilibrium is actually determined by its classical counterpart and the necessity function, and vice versa.

Proposition 9. *Let $\overline{M}$ be a poss-MCS.*

1. If $\overline{S}$ is a possibilistic equilibrium of $\overline{M}$ and $a_i \in \Sigma_i$, then $(a_i, \alpha) \in \overline{S}_i$ iff $\alpha = N_{M^S}(a_i)$.

2. If S is an equilibrium of M, then $\overline{S} = (\overline{S}_1, \ldots, \overline{S}_n)$ where $\overline{S}_i = \{(a_i, N_{M^S}(a_i)) \mid N_{M^S}(a_i) > 0 \text{ and } a_i \in \Sigma_i\}$ $(i = 1, \ldots, n)$.

3. If $\overline{S}$ be a possibilistic equilibrium of $\overline{M}$, then S is an equilibrium of M.

Let us introduce the possibility distribution for normal poss-MCS.

Definition 18. *Let $\overline{M}$ be a normal poss-MCS and S be an belief state. Then the possibility distribution, denoted $\tilde{\pi}_{\overline{M}}$, is defined by:*

$$\forall S, \tilde{\pi}_{\overline{M}}(S) = \pi_{\overline{M}^S}(S). \tag{19}$$

The possibility degree for a normal poss-MCS $\overline{M}$ and an equilibrium of the classical projection M of $\overline{M}$ has the following connection.

Proposition 10. *Let $\overline{M}$ be a poss-MCS and S be a belief state. Then $\tilde{\pi}_{\overline{M}}(S) = 1$ iff S is an equilibrium of M.*

Proof. If $\tilde{\pi}_{\overline{M}}(S) = 1$ then $\pi_{\overline{M}^S}(S) = 1$, thus $S = GE(M^S)$. So S is an equilibrium of M. And if S is an equilibrium of M, then $S = GE(M^S)$, thus $\pi_{\overline{M}^S}(S) = 1$ then $\tilde{\pi}_{\overline{M}}(S) = 1$.

The possibility distribution for normal poss-MCS defines two measures.

Definition 19. *The two dual possibility and necessity measures for each atom in a normal poss-MCS are defined by*

- $\tilde{\Pi}_{\overline{M}}(a_i) = max\{\tilde{\pi}_{\overline{M}}(S) \mid a_i \in S_i\}$
- $\tilde{N}_{\overline{M}}(a_i) = 1 - max\{\tilde{\pi}_{\overline{M}}(S) \mid a_i \notin S_i\}$

5   Conclusion

In this paper we have established the first framework for possibilistic reasoning and nonmonotonic reasoning in multi-context systems, called possibilistic multi-context systems (poss-MCS). In our framework, a context is represented as a possibilistic logic programs and the semantics for a poss-MCS is defined by its equilibria that are based on the concepts of possibilistic answer sets and possibility distributions. We have studied several properties of poss-MCS and in particular, developed a fixpoint theory for poss-MCS, which provides a natural connection between the declarative semantics and the computation of the equilibria. As a result, algorithms for poss-MCS are also provided. Needless to say, this is just the first and preliminary attempt in this direction. There are several interesting issues for future study. First, as we have seen in the last two sections, the possibilistic equilibriua of a poss-MCS are computed using a procedure based double iterations. Such an algorithm can be inefficient in some cases. So it would be useful to develop efficient algorithm for computing possibilistic equilibriua. Another important issue is to apply poss-MCS in some semantic web applications.

Acknowledgement. This work was supported by the Australia Research Council (ARC) Discovery grants DP1093652 and DP110101042.

References

1. Benferhat, S., Sossai, C.: Reasoning with multiple-source information in a possibilistic logic framework. Information Fusion 7, 80–96 (2006)
2. Bettini, C., Brdiczka, O., Henricksen, K., Indulska, J., Nicklas, D., Ranganathan, A., Riboni, D.: A survey of context modelling and reasoning techniques. Pervasive and Mobile Computing 6(2), 161–180 (2010)
3. Bikakis, A., Patkos, T., Antoniou, G., Plexousakis, D.: A Survey of Semantics-Based Approaches for Context Reasoning in Ambient Intelligence. In: Mühlhäuser, M., Ferscha, A., Aitenbichler, E. (eds.) AmI 2007 Workshops. CCIS, vol. 11, pp. 14–23. Springer, Heidelberg (2008)
4. Brewka, G., Eiter, T.: Equilibria in heterogeneous nonmonotonic multi-context systems. In: Proc. AAAI, pp. 385–390 (2007)
5. Brewka, G., Roelofsen, F., Serafini, L.: Contextual default reasoning. In: Proc. IJCAI, pp. 268–273 (2007)
6. Dubois, D., Lang, J., Prade, H.: Possibilistic logic. In: Handbook of Logic in Artificial Intelligence and Logic Programming, vol. 3, pp. 439–513 (1995)
7. Serafini, L., Giunchiglia, F.: Multilanguage hierarchical logics, or: how we can do without modal logics. In: Artificial Intelligence, pp. 29–70 (1994)
8. Gelfond, M., Lifschitz, V.: The stable model semantics for logic programming. In: Proc. 5th ICLP, pp. 1070–1080 (1988)
9. Lloyd, J.W.: Foundations of Logic Programming, 2nd edn. Springer, New York (1987)
10. McCarthy, J.: Notes on formalizing context. In: Proc. IJCAI, pp. 555–560 (1993)
11. Nicolas, P., Garcia, L., Stéphan, I., Lefèvre, C.: Possibilistic uncertainty handling for answer set programming. In: Annals of Mathematics and Artificial Intelligence, pp. 139–181 (2006)
12. Roelofsen, F., Serafini, L.: Minimal and absent information in contexts. In: Proc. IJCAI, pp. 558–563 (2005)
13. Yager, R.R.: An introduction to applications of possibility theory. Human Syst. Manag., 246–269 (1983)
14. Zadeh L.A.: Fuzzy sets as a basis for a theory of possibility. Fuzzy Sets and Systems, 3–28 (1978)

A Two-Step Zero Pronoun Resolution by Reducing Candidate Cardinality

Kye-Sung Kim, Su-Jeong Choi, Seong-Bae Park, and Sang-Jo Lee

Department of Computer Engineering
Kyungpook National University 702-701 Daegu, Korea
{kskim,sjchoi,sbpark}@sejong.knu.ac.kr,
sjlee@knu.ac.kr

Abstract. The high cardinality of antecedent candidates is one of the major reasons which make zero pronoun resolution difficult. To improve performance, it is necessary to reduce this cardinality before defining the features to choose the most plausible antecedent. This paper proposes a two-step method for intra-sentential zero pronoun resolution. First, the clause which contain the antecedent for a given zero pronoun is determined using structural relationships between clauses. Then, the antecedent of the zero pronoun is chosen from the noun phrases within the identified clause. The cardinality of candidates reduces to the number of antecedent candidates present in clauses. Our experimental results show that the proposed method outperforms other methods without the first step, no matter what features are used to identify antecedents.

Keywords: zero pronoun, candidate cardinality, hierarchy of linguistic units, relationship between clauses.

1 Introduction

In pro-drop languages such as Chinese, Japanese, and Korean, repetitive elements are easily omitted when they are inferable from the context. This phenomenon is commonly referred to as zero anaphora, and the omitted element is called as a zero pronoun (or zero anaphor). The resolution of zero pronouns is essential for natural language understanding, and thus is important for many NLP applications such as machine translation, information extraction, question answering, and so on.

One way to resolve zero pronouns is to pair each zero pronoun with the most plausible antecedent chosen from the antecedent candidates [1]. That is, if a number of noun phrases precede a zero pronoun, the most plausible antecedent should be chosen from these noun phrases. Various kinds of features have been proposed to choose the most plausible antecedent. Ferrandez and Peral [2] used surface features such as person and number agreement and some heuristic preferences. Zhao and Ng [3] adopted syntactic information including parts of speech, grammatical role of the antecedents. Some recent studies employed parse trees directly as structured features [4,5].

P. Anthony, M. Ishizuka, and D. Lukose (Eds.): PRICAI 2012, LNAI 7458, pp. 194–205, 2012.
© Springer-Verlag Berlin Heidelberg 2012

Zero pronoun resolution is still difficult, no matter how nice are the features defined to choose the most plausible antecedent. This is because the number of antecedent candidates is generally very large, since a sentence has many noun phrases and all noun phrases preceding a zero pronoun can be its antecedent. This high cardinality of antecedent candidates is one of the major reasons which make zero pronoun resolution difficult. Therefore, a method to reduce this cardinality should be considered before defining the features to choose the most plausible antecedent.

Languages are believed to have a hierarchy of linguistic units. On the basis of the hierarchy structure, candidate antecedents can be grouped into a larger unit such as phrase, clause, sentence, and paragraph. Since the number of clauses in a sentence is far less than that of phrases, it is easier, from the standpoint of the number of targets, to choose the clause with the antecedent of a zero pronoun than to find the antecedent directly. Therefore, this paper proposes a two-step method for intra-sentential zero pronoun resolution. The proposed method first identifies the clause which may contain the antecedent for a given zero pronoun using structural information of clauses. A parse tree kernel [6] is adopted for modeling the structural information. Then, the antecedent of the zero pronoun is chosen from the noun phrases within the identified clause. If the first step is trustworthy enough, the proposed method outperforms other methods without the first step since the second step chooses the antecedent from far less number of candidates. According to our experiments on STEP 2000 data set, the accuracy of the first step is 91.1 % which is high enough. As a result, the proposed method improves other methods without the first step up to 65.9 of F-score no matter what features are used in the second step.

The remainder of this paper is organized as follows. Section 2 surveys previous work on zero pronouns. Section 3 describes the problem and intra-sentential zero pronoun resolution, and Section 4 presents our proposed method in detail. Experimental results are presented in Section 5, and finally Section 6 provides conclusions and future work.

2   Related Work

Most studies on anaphora resolution including zero anaphora resolution are widely classified into two groups. One group is the one based on heuristic rules and Centering theory. Centering theory provides a model of local coherence in discourse, and has been usually used to resolve overt pronouns in English. However, it is not easy to apply the centering framework to zero anaphora resolution. First, the presence of non-anaphoric zero pronouns is not considered in the framework of centering. In addition, notations such as utterance, realization and ranking can be considered as parameters of centering [7]. Several studies have examined the impact of setting centering parameters in different ways. Especially, in languages other than English, some researchers have attempted to treat complex sentences as pieces of discourse [8,9]. Most of these studies focused on how to analyze complex sentences in existing centering models.

The other is the one based on machine learning methods. Various kinds of features have been proposed to improve the resolution performance. Features of agreement on person and number are simple but widely used for anaphora resolution [2,1]. However, in languages like Korean and Japanese which do not have verb agreement, there is little evidence to guide the choice of antecedent candidates by this kind of features. Thus, the use of agreement features is not promising in these languages. Iida et al. [4] used syntactic pattern features to resolve intra-sentential zero pronouns, but the representation of syntactic patterns increases the risk of information loss, since the paths between a zero pronoun and its antecedent in parse trees should be encoded as syntactic patterns. In the work of Hu [10], zero anaphora is categorized into intra- and inter-clausal anaphora. They focused on finding paths between zero pronouns and antecedents, but this study is limited to examining the paths which can be resolved by the centering algorithm.

Several researchers have tried to filter out non-anaphoric cases before identifying antecedents [11]. This is because zero pronouns can appear without explicit antecedents in some languages. However, determining the anaphoricity of zero pronouns is not trivial. Most of the previous methods are based on the use of heuristics or parameters which have to be set by the user [4], and the performance is not good enough for antecedent identification. More recently, some researchers worked on the joint determination of anaphoricity and coreference model [1]. However, the performance is still far from satisfactory, and efforts to improve it are still ongoing.

Many attempts have been made to choose the most plausible antecedent for a given zero pronoun. However, the importance of the cardinality of candidate antecedents for a zero pronoun has not been investigated sufficiently, even if the high cardinality of candidates does harm on the performance of antecedent identification.

3   Problem Setting

Zero pronoun resolution is a process of determining the antecedent of a zero pronoun. Figure 1 shows an example of a sentence in Korean. In this example, zero pronoun zp_1 refers to noun phrase 'sun' (NP_5) in the same sentence. That is, the antecedent of zp_1 is NP_5, and zero pronoun zp_1 in this case is anaphoric, since its antecedent precedes the zero pronoun. However, all zero pronouns do not refer to antecedents in the preceding context. Especially, in languages like Japanese and Korean, antecedents may be found in the following context. These zero pronouns are known as cataphoric pronouns. However, the resolution of both anaphoric and cataphoric zero pronouns is identical except the direction of reference. Thus, we do not distinguish anaphoric and cataphoric zero pronouns. For a given zero pronoun zp, intra-sentential zero pronoun resolution can be defined as

$$a^* = \arg\max_{a \in \mathcal{A}} f(a, zp; w), \tag{1}$$

낮-도$_{NP1}$	짧아지-어$_{VP1}$	9시$_{NP2}$	40$_{NP3}$	,	50 분-에$_{NP4}$	태양-이$_{NP5}$	나타나-아$_{VP2}$
the days-TOP	get short-CAUSE	nine o'clock	forty	PUNC	fifty minutes-COMP	sun-NOM	show-CONJ
(∅₁-이)$_{NP6}$	2시$_{NP7}$	30$_{NP8}$	,	40 분-에는$_{NP9}$	사라진-다$_{VP3}$	.	
(∅₁-NOM)	two o'clock	thirty	PUNC	forty minutes-COMP	go down- PRED	PUNC	

Since the days get short, the sun shows about around 9:40, 50 am. and (∅₁-NOM) goes down at around 2:30, 40 pm.

Fig. 1. Example of a sentence in Korean

where $\mathcal{A}$ is a set of antecedent candidates composed of all noun phrases in a sentence and $f(a, zp; w)$ is an arbitrary decision function specified by the model parameter w. One of the key problems which affect the performance of zero pronoun resolution is the large norm of the candidate set, $|\mathcal{A}|$. This is because the antecedent of zp is just one out of $\mathcal{A}$. If $|\mathcal{A}|$ is large, it will be difficult to choose the optimal antecedent from $\mathcal{A}$.

4   Zero Pronoun Resolution with Multi-class Classification

The task addressed in Equation (1) can be regarded as a multiclass classification in the viewpoint of machine learning. In multiclass classification, large scale target classes have been raised as one of the critical obstacles to improve classification performances. An efficient way to overcome this problem is to reduce the size of target classes based on some underlying shared characteristics among classes [12]. Normally shared characteristics are implicitly appeared, and thus they are extracted by learning from data. However, shared characteristics among antecedent candidates are explicitly specified by a hierarchy of linguistic units. This paper adopts these explicit shared characteristics to reduce the size of $|\mathcal{A}|$.

A clause is a higher-level linguistic unit that is composed of various phrases. Since antecedent candidates are all noun phrases of zp, a clause can be regarded as a shared characteristic among antecedent candidates. That is, the antecedent candidates in a clause are said to share the same characteristic. Assuming a function $\phi(a)$ to map an antecedent candidate a to a clause-level feature space, the mapped antecedent candidates $\phi(a)$ and $\phi(a')$ are same if they are contained in the same clause c. That is, $\phi(a) = \phi(a')$ if both $a \in c$ and $a' \in c$ are true. The number of clauses is normally less than that of noun phrases. That is,

$$|\mathcal{C}| \leq |\mathcal{A}|, \tag{2}$$

where $\mathcal{C}$ is a set of clause-level antecedent candidates. $|\mathcal{C}| = |\mathcal{A}|$ is met only when all antecedent candidates are obtained from different clauses. Note that Equation (2) becomes $|\mathcal{C}| \ll |\mathcal{A}|$ in general cases.

In this paper, two-step zero pronoun resolution is proposed by adopting $\mathcal{C}$. First, the most plausible clause candidate c^* is detected which contains the antecedent of the given zp. That is, c^* is obtained by

$$c^* = \underset{c \in \mathcal{C}}{\arg\max}\ g(c, \phi(zp); w_c), \tag{3}$$

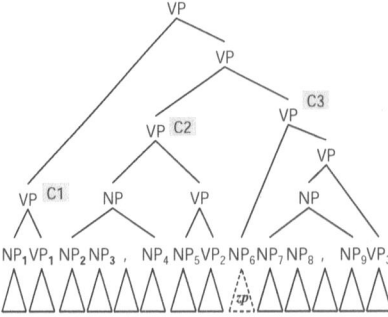

Fig. 2. An example of a parse tree

where $g(c, \phi(zp); w_c)$ is a decision function specified by w_c. Then, the antecedent of zp is detected within antecedent candidates $\mathcal{A}'$ which is

$$\mathcal{A}' = \{a | a \in \mathcal{A} \text{ and } \phi(a) = c^*\}. \tag{4}$$

Then, antecedent detection is defined as

$$a^* = \arg\max_{a \in \mathcal{A}'} f(a, zp; w_p), \tag{5}$$

where w_p is a model parameter to be learned.

4.1   Clause-Level Antecedent Detection

Shared characteristic among antecedent candidates is defined as clauses which contain antecedent candidates. Thus, in the first step of the proposed method, the most plausible clause is detected. When a sentence and its parse tree are given as shown in Figure 2, the clauses in the parse tree are easily obtained. Figure 3 shows the clauses which constitute the sentence in Figure 2. As shown in Figure 3, all clauses are specified with their tree structures. In the first step, clause candidates and a zero pronoun are represented as their clause-level trees shown in Figure 3. It implies that $c \in \mathcal{C}$ and $\phi(zp)$ are all clause-level trees which contain antecedent candidates or a zero pronoun.

When a pair of trees c and $\phi(zp)$ is given, the decision function $g(c, \phi(zp); w_g)$ determines a score which is the possibility that the clause tree c contains the antecedent of zp. There could exist a lot of scoring functions for this purpose. In this paper, Support Vector Machines (SVMs) is adopted to construct the decision function. SVMs is a well-known classifier that is applied to various problems and it achieves state-of-the-art performances in many cases. When the clause candidate c is regarded as a class in SVMs, there are a large number of classes and these classes make it difficult to train a decision model.

In order to adopt SVMs for the decision function, it is needed to introduce a Path-Enclosed Tree (PET) generated by using the shortest path linking two

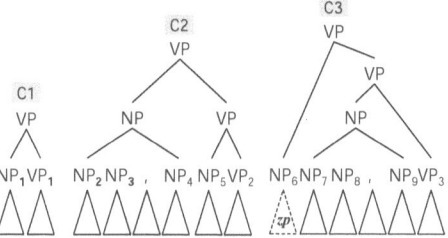

Fig. 3. An example of clause trees extracted from the tree in Figure 2

trees [13]. Figure 4 shows a simple example of a PET which generated from a pair of clauses in Figure 3. As shown in this figure, the path-enclosed tree of the clause c_1 and $c_3 (= \phi(zp))$ is generated by combining them with all the edges and nodes on their shortest path. Given a PET of c and $\phi(zp)$, the decision function $g(c, \phi(zp); w_g)$ is substituted by

$$g(c, \phi(zp); w_g) = g'(PET_{c,zp}; w_g), \tag{6}$$

where $PET_{c,zp}$ is the PET generated from both c and $\phi(zp)$. By using $g'(PET_{c,zp}; w_g)$ instead of $g(c, \phi(zp); w_g)$, SVMs can be trained to determines the margin value for $PET_{c,zp}$ which denotes how possibly c contains the antecedent of zp.

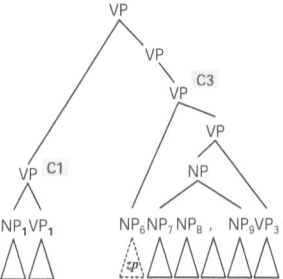

Fig. 4. An example of a path-enclosed tree of clauses, $C1$ and $C3$

Given $PET_{c,zp}$, the decision function $g'(PET_{c,zp}; w_g)$, constructed by SVMs determines where $PET_{c,zp}$ is valid or not. The result of this determination is measured by a margin value obtained from SVMs. Thus, if $PET_{c,zp}$ bring about higher margin value than others, it implies that $PET_{c,zp}$ contains the most plausible clause candidate. Therefore, based on PETs and the decision function $g'(PET_{c,zp}; w_g)$, clause-level antecedent detection is done by

$$c^* = \arg\max_{c \in \mathcal{C}} g'(PET_{c,zp}; w_g).$$

SVMs estimates model parameter w_g by minimizing the hinge loss with regularizer $||w_g||^2$ to control model complexity. For this purpose, $PET_{c,zp}$ should be

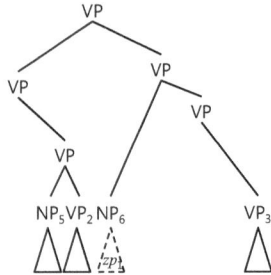

Fig. 5. An example of a simple-expansion tree

mapped onto a feature space. A parse tree kernel is one of the efficient ways to do this. In the parse tree kernel, a parse tree is mapped onto the space spanned by all subtrees which possibly appear in the parse tree. The explicit enumeration of all subtrees is computationally problematic since the number of subtrees in a tree increases exponentially as the size of tree grows. Collins and Duffy proposed a method to compute the inner product of two trees without having to enumerate all subtrees [6].

4.2 Phrase-Level Antecedent Detection

A phrase-level antecedent is determined by the decision function, $f(a, zp; w_p)$ in Equation (5). Given an antecedent candidate a and the zero pronoun zp, the decision function $f(a, zp; w_p)$ returns the score of a as the antecedent for zp. Without considering the first step to reduce the number of antecedent candidates, $f(a, zp; w_p)$ has been already studied in contemporary zero pronoun resolution techniques. Thus, we can use any existing zero pronoun resolution method for the second step.

In this paper, phrase-level antecedents are represented with their simple-expansion trees introduced in [14]. When a and zp are given, the simple-expansion tree that is the shortest path which includes a, zp, and their predicates is extracted from the parse tree of a sentence. An example of a simple-expansion tree is given in Figure 5. Given zp_1 and candidate antecedent NP_5 in Figure 2, the simple-expansion tree which is extracted from the parse tree of the sentence is as shown in Figure 5.

The decision function $f(a, zp; w_p)$ is constructed in the same manner of the first step based on SVMs and the parse tree kernel by using these simple-expansion trees of all candidates. Note that, even though the second step is similar with contemporary detection methods, the size of antecedent candidates considered in this step is significantly reduced by using the clause-level detection. This is why our two step antecedent detection method can improve the performance of zero pronoun resolution.

Table 1. Simple statistics on the dataset used in our experiments

Dataset	Number
Sentences	5,221
Clauses	20,748
Clauses containing zero pronouns	13,171

Table 2. Distribution of zero pronouns (ZPs) in Korean

Intra-sentential	Inter-sentential	Extra-sentential
10,371	666	2,134
(78.74%)	(5.06%)	(16.20%)

5 Experimentation and Discussion

5.1 Dataset

In our experiments, the parsed corpus which is a product of STEP 2000 project supported by Korean government is used. Recent studies indicate that subjects in a sentence are often dropped highlighting the importance of dealing with zero subjects in zero pronoun resolution [1]. Especially in Korean, zero subjects are more frequent in that over 90% of zero pronouns are zero subjects [15]. Therefore only zero pronouns in subject positions are considered in this study to investigate the effect of reducing candidate cardinality.

We first manually identified zero subjects in the parsed corpus, and then the complex sentences with one or more zero subjects were extracted from the parsed corpus. A simple statistics on the dataset is given in Table 1. The number of selected sentences is 5,221, and the sentences are segmented into 20,748 clauses (on average, 3.97 clauses/sentence and 7.67 words/clause).

Table 2 shows the distribution of zero pronouns in Korean. In this paper, non-anaphoric uses of zero pronouns such as deictic, indefinite personal and expletive [15] are classified as extra-sentential ones. As shown in Table 2, a large number of zero pronouns are intra-sentential, and it implies that most zero pronouns appear within complex sentences.

5.2 The Experimental Results and Analysis

Our experiments are performed in five-fold cross validation and SVM_{light} [16] is used as classifiers. Three metrics of accuracy, precision, and recall are used to investigate the effect of the proposed method in zero anaphora resolution.

$$\text{Accuracy} = \frac{\text{\# of correctly classified pairs}}{\text{total \# of pairs of ZPs and clauses}}$$

$$\text{Precision} = \frac{\text{\# of correctly identified pairs}}{\text{\# of identified pairs of ZPs and clauses}}$$

Table 3. Result of the identification of clauses having antecedents

	Accuracy	Precision	Recall	F1
Baseline	0.6234	0.4585	0.6168	0.5260
Our first step	0.9144	0.8736	0.8737	0.8736

Table 4. Performance of intra-sentential zero pronoun resolution

	(a) Without the first step			(b) With the first step		
Features used	P_{intra}	R_{intra}	F1	P_{intra}	R_{intra}	F1
Syntactic	0.3447	0.4238	0.3802	0.4738	0.5789	0.5211
Simple-expansion	0.6023	0.7374	0.6631	0.6246	0.7631	0.6870
Dynamic-expansion	0.5310	0.6501	0.5846	0.6188	0.7560	0.6806

$$\text{Recall} = \frac{\text{\# of correctly identified pairs}}{\text{\# of true pairs of ZPs and clauses}}$$

Table 3 shows the result of the first step that is the identification of clauses containing an antecedent. In this table, the baseline method assumes that the antecedent of a given zero pronoun is located in the adjacent clauses [17], and its accuracy is nearly 62.0%. On the other hand, the accuracy of the proposed method is significantly higher than that of the baseline. This result indicates that syntactic structures of sentences are useful in determining clauses with an antecedent. As a result of the identification of clauses with an antecedent, the cardinality of candidates reduces to the number of noun phrases in the identified clause.

Table 4 shows the performance of intra-sentential zero pronoun resolution. Three models with different features are compared to evaluate the effect of the first step of clause-level antecedent detection. These features are used in the second step of our method. 'Syntactic' model is the one based on 26 features proposed by Zhao and Ng [3]. However, two headline features ("A_In_Headline" and "Z_In_Headline") are excluded in this study, since we focus on intra-sentential zero pronoun resolution. 'Simple-expansion' model denotes a simple-expansion tree mentioned in Section 4.2 to capture syntactic structured information in parse trees. 'Dynamic-expansion' model adopts a tree expanded from the simple-expansion tree by using additional antecedent competitor-related information. [18,5]. A parse tree kernel [19] is used to train both 'Simple-' and 'Dynamic-expansion' model. To investigate the effect of reducing candidate cardinality for antecedent identification, the results of antecedent identification with the first step are compared with ones without the first step. The evaluation is measured by precision and recall by

$$P_{intra} = \frac{\text{\# of correctly detected antecedents}}{\text{\# of ZPs classified as intra-sentential}}$$

$$R_{intra} = \frac{\text{\# of correctly detected antecedents}}{\text{total \# of intra-sentential ZPs}}$$

According to this table, two-step methods with the first step always outperform the ones without the first step regardless of features used in the second step. Also, the performance of the proposed method which consists of the first step of clause-level antecedent detection and the second step using simple-expansion tree structure is better than other methods. This implies that the cardinality of candidates influences the performance of antecedent identification, and thus our two-step approach is more effective in resolving zero pronouns.

An interesting finding is that the performance of 'simple-expansion' model is higher than that of 'dynamic-expansion' model. Recently, Zhou et al. [18] reported that the dynamic-expansion model is better than the simple-expansion model in pronoun resolution. However, our experimental results show that the proposed method which contains two-step detection with simple-expansion model is more suitable for the resolution of zero pronouns in Korean than the two-step detection with the dynamic-expansion model. This is because the word order in Korean is freer than the languages such as English and Chinese. Furthermore, additional information about candidate competitors can reduce the discrimination performance of the dynamic-expansion trees by making trees to be similar, and the f value of the dynamic-expansion model which does not include the first step was 0.5846 which which is about 12% lower than the simple-expansion model. To improve the performance drop of dynamic-expansion model, it is essentially demanded to select informative candidate competitors by considering agreements. However, by determining clauses with antecedents, the proposed two-step method helps to avoid these problems and to resolve zero pronouns in languages such as Korean. The quality of 'Syntactic' model without the first step is very low. It seems that the features to resolve Chinese zero pronouns are not good enough for the resolution of anaphoric and cataphoric zero pronouns in Korean. However, the result is better when clauses with antecedents are first identified, although the features are mostly binary. This means that candidate sets of small cardinality should be obtained before features are designed to choose the antecedent.

In this paper, the average cardinality of candidates of each zero pronoun is reduced from 7.18 to 2.63 when the first step is done. That is, the ratio of the cardinality of candidates compared with corresponding baseline methods is reduced to approximately 63% in our dataset while improving performance. In particular, the proposed method attempts to assign a similar cardinality to each zero pronoun even though the sentences are quite long. This is because the cardinality corresponds to the number of candidate antecedents in clauses.

Another advantage of the proposed methods in terms of intra-sentential resolution is that non-intrasentential zero pronouns can be filtered out in the first step. Table 5 shows this effect. For comparison, anaphoricity determination in Kong and Zhou [5] is adopted. The evaluation is measured by

$$P_{ana} = \frac{\text{\# of correctly identified ZPs}}{\text{\# of ZPs classified as not having intra-sentential antecedents}}$$

$$R_{ana} = \frac{\text{\# of correctly identified ZPs}}{\text{\# of ZPs which do not have intra-sentential antecedents}}$$

Table 5. Observing the effect of anaphoricity determination in the first step

	P_{ana}	R_{ana}	F1
Anaphoricity	0.7397	0.2063	0.3225
Clause-level detection	0.5169	0.7364	0.6074

In this experiment, the proposed method determines the number of zero pronouns which do not have intra-sentential antecedents by using margin values of candidates. When there is no candidate whose margin value is over 1.0 (that means significantly positive), the proposed method regards the target zero pronoun as non-intra sentential one. As shown in this table, even though the proposed method is not designed to identify non-intra sentential zero pronouns, the clause-level antecedent detection achieves about 47% improvement compared with anaphoricity determination.

6 Conclusion

Zero pronoun resolution is a central task in natural language understanding. This paper proposed a two-step approach for resolving intra-sentential zero pronouns in Korean. In order to reduce the cardinality of candidate antecedents, the clause containing an antecedent is first identified. Then, the antecedent of a zero pronoun is searched within the identified clause. Our experimental results show that the proposed method outperforms other methods without the first step no matter what features are used in antecedent identification. This indicates that the reduction of the cardinality of candidates is an important factor for zero pronoun resolution.

In order to achieve better performance in zero pronoun resolution, it is necessary to continuously improve the first step. In the future, the performance of zero anaphora resolution will be much better, if additional semantic information or a combination of structural and semantic features is used to identify the clause with antecedent. We also plan to extend this approach to resolve zero pronouns which occur in various grammatical positions as well as to resolve inter-sentential zero pronouns.

Acknowledgement. This work was supported by the Industrial Strategic Technology Development Program (10035348, Development of a Cognitive Planning and Learning Model for Mobile Platforms) funded by the Ministry of Knowledge Economy (MKE, Korea).

References

1. Iida, R., Poesio, M.: A cross-lingual ilp solution to zero anaphora resolution. In: Proceedings of the 49th Annual Meeting of the Association for Computational Linguistics: Human Language Technologies, pp. 804–813 (2011)

2. Ferrández, A., Peral, J.: A computational approach to zero-pronouns in Spanish. In: Proceedings of the 38th Annual Meeting of the Association for Computational Linguistics, pp. 166–172 (2000)
3. Zhao, S., Ng, H.T.: Identification and resolution of Chinese zero pronouns: A machine learning approach. In: Proceedings of the 2007 Joint Conference on EMNLP and CNLL, pp. 541–550 (2007)
4. Iida, R., Inui, K., Matsumoto, Y.: Zero-anaphora resolution by learning rich syntactic pattern features. ACM Transactions on Asian Language Information Processing 6(4), 1–22 (2007)
5. Kong, F., Zhou, G.: A tree kernel-based unified framework for Chinese zero anaphora resolution. In: Proceedings of EMNLP, pp. 882–891 (October 2010)
6. Collins, M., Duffy, N.: Convolution kernels for natural language. In: Proceedings of Neural Information Processing Systems, pp. 625–632 (2001)
7. Poesio, M., Stevenson, R., Eugenio, B.D., Hitzeman, J.: Centering: A parametric theory and its instantiations. Computational Linguistics 30(3), 309–363 (2004)
8. Sakurai, Y.: Centering inJapanese: How to rank forward-looking centers in a complex sentence. In: Proceedings of the 22nd Northwest Linguistic Conference, pp. 243–256 (2006)
9. Wiesemann, L.M.: The function of Spanish and English relative clauses in discourse and their segmentation in Centering Theory. PhD thesis, Department of Linguistics, Simon Fraser University (2009)
10. Hu, Q.: A Corpus-based Study on Zero Anaphora Resolution in Chinese Discours. PhD thesis, Department of Chinese, Translation and Linguistics, City University of Hong Kong (2008)
11. Abdul-Mageed, M.: Automatic detection of Arabic non-anaphoric pronouns for improving anaphora resolution. ACM Transactions on Asian Language Information Processing 10(1) (2011)
12. Amit, Y., Fink, M., Srebro, N., Ulman, S.: Uncovering shared structures in multi-class classification. In: Proceedings of the 24th ICML (2007)
13. Zhang, M., Zhang, J., Su, J.: Exploring syntactic features for relation extraction using a convolution tree kernel. In: Proceedings of the HLT/NAACL, pp. 288–295 (2006)
14. Yang, X., Su, J., Tan, C.: Kernel-based pronoun resolution with structured syntactic knowledge. In: Proceedings of COLING-ACL, pp. 41–48 (2006)
15. Han, N.R.: Korean Zero Pronous: Analysis and Resolution. PhD thesis, Department of Linguistics at the University of Pennsylvania (2006)
16. Joachims, T.: Making large-Scale SVM Learning Practical. In: Advances in Kernel Methods - Support Vector Learning. MIT Press (1999)
17. Leffa, V.J.: Clause processing in complex sentences. In: Proceedings of the 1st International Conference on Language Resources and Evaluation, pp. 937–943 (1998)
18. Zhou, G., Kong, F., Zhu, Q.: Context-sensitive convolution tree kernel for pronoun resolution. In: Proceedings of the IJCNLP, pp. 25–31 (2008)
19. Moschitti, A.: Making tree kernels practical for natural language learning. In: Proceedings of the 11th International Conference on European Association for Computational Linguistics, pp. 113–120 (2006)

Hybrid Techniques to Address Cold Start Problems for People to People Recommendation in Social Networks

Yang Sok Kim, Alfred Krzywicki, Wayne Wobcke, Ashesh Mahidadia,
Paul Compton, Xiongcai Cai, and Michael Bain

School of Computer Science and Engineering,
University of New South Wales Sydney NSW 2052, Australia
{yskim,alfredk,wobcke,ashesh,compton,xcai,mike}@cse.unsw.edu.au

Abstract. We investigate several hybrid approaches to suggesting matches in people to people social recommender systems, paying particular attention to *cold start* problems, problems of generating recommendations for new users or users without successful interactions. In previous work we showed that interaction-based collaborative filtering (IBCF) works well in this domain, although this approach cannot generate recommendations for new users, whereas a system based on rules constructed using subgroup interaction patterns can generate recommendations for new users, but does not perform as effectively for existing users. We propose three hybrid recommenders based on user similarity and two content-boosted recommenders used in conjunction with interaction-based collaborative filtering, and show experimentally that the best hybrid and content-boosted recommenders improve on the IBCF method (when considering user success rates) yet cover almost the whole user base, including new and previously unsuccessful users, thus addressing cold start problems in this domain. The best content-boosted method improves user success rates more than the best hybrid method over various "cold start" subgroups, but is less computationally efficient overall.

1   Introduction

In recent work we have investigated people to people recommendation with particular application to online dating sites. This problem is particularly interesting since it requires a *reciprocal recommender* [8], one where "items" are users and hence have their own preferences that must be taken into account when recommending them to others. Reciprocal recommenders must consider user "taste" (how they view potential matches) and "attractiveness" (how they are viewed by others), Cai *et al.* [3]. In particular, in contrast to typical product recommender systems, it is not generally appropriate to recommend popular users, since popular users are typically highly attractive, hence less likely to respond positively to an expression of interest from the average user.

Cold start problems are particularly acute in people to people recommendation in online dating since the user base is highly dynamic, so there are many new users at any point in time. In addition, our previous work has shown that the best information to use for recommendation is the set of positive interactions between users, those interactions initiated by one user where the other replies positively (whether a reply is positive is

P. Anthony, M. Ishizuka, and D. Lukose (Eds.): PRICAI 2012, LNAI 7458, pp. 206–217, 2012.
© Springer-Verlag Berlin Heidelberg 2012

fixed by the system, hence is reliably known). However, many existing users have no experience of positive interactions, either as senders, i.e. they have not sent a message to another user who has replied positively though they may be sending messages, or as receivers, i.e. they may have received some messages but never replied positively to any of them. In effect, these users are part of the cold start problem for people to people recommender systems. On this basis, roughly following Park and Chu [7], we consider separately the cold start problems of recommending existing users to new users who are either non-senders, non-receivers or both (this last group is the most difficult as least is known about them).

Previously we have developed several approaches to people to people recommendation. Our best approach, interaction-based collaborative filtering (IBCF), Krzywicki *et al.* [5], uses two types of collaborative filtering based on positive interactions, one based on similarity of users as senders (so they have similar taste) and another based on similarity of users as recipients (so they have similar attractiveness). A combined method using both types of similarity provides high quality recommendations, but only for a subset of users, those who have had positive interactions. In contrast, a method based on statistical properties of interactions between user subgroups, Kim *et al.* [4], constructs rules that can generate recommendations for all users, but the method does not perform as well as IBCF. This motivates combining these methods (and investigating other hybrid methods) in order to address cold start problems.

Key to this is to observe that the IBCF methods use interactions *both* to generate and rank recommendations from user similarity *and* to define user similarity. We conjectured that IBCF derives its power from the use of interactions to generate and rank recommendations, not from the particular similarity relation based on interactions, and hence that other types of similarity used with IBCF would perform well. This motivates the investigation of various hybrid recommender systems based on using collaborative filtering with content-based methods for determining similarity, which, as observed by Adomavicious and Tuzhilin [1], are traditionally used to address cold start problems. In this paper, we investigate such hybrid methods based on user similarity and also content-boosted recommender systems [6] used in conjunction with our IBCF methods. In a content-boosted recommender, the recommendations provided by one system (in this case, that based on compatible subgroup rules) are used to bootstrap a collaborative filtering recommender. We show that, surprisingly, a simple method based on user similarity provides high quality recommendations for almost all users. However, in line with earlier results [6], we show that our best content-boosted recommender outperforms the original IBCF recommender and those hybrids based on user similarity. With particular attention to cold start problems, we show that the content-boosted recommender performs especially well for both non-senders and non-receivers (both of which include the class of new users). We also examine a breakdown by gender (males and females behave very differently in this domain) and show that the performance of the recommenders is consistent across these subgroups.

The remainder of this paper is organised as follows. In Section 2 we summarize our previous methods in online dating recommenders, then in Sections 3 and 4 introduce hybrids based on user similarity and content-boosted recommenders based on compatible subgroup rules. Section 5 provides experimental evaluation and discussion.

2   Recommendation Approaches

2.1   Compatible Subgroup Rules

The recommender based on compatible subgroup rules (Rules) works by constructing rules for each user of the form: *if* $u_1, \cdots, u_n$ (condition) *then* $c_1, \cdots, c_n$ (conclusion), where the u_i are profile features of the user and c_i are corresponding profile features of the candidate, Kim *et al.* [4]. If the user satisfies the condition of such a rule, any candidate satisfying the conclusion can be recommended to the user. We now summarize how the rules are constructed and candidates ranked.

Each profile feature is an attribute with a specific value, e.g. age = 30–34, location = Sydney (each attribute with a discrete set of possible values). An initial statistical analysis determines, for each possible attribute a and each of its values v (taken as a sender feature), the "best" matching value for the *same* attribute a (taken as a receiver feature), here treating male and female sender subgroups separately. So, for example, for males the best matching value for senders with feature age = 30–34 might be females with age = 25–29.

The best matching value is determined from historical data that records interactions between users, and is essentially a harmonic mean of the "interest" between the two subgroups of users defined by the rule's condition and conclusion. The *interest* $I_{s,r}$ of one group of users s (senders) in another r (receivers) is defined as the proportion of messages sent from members of the group s that are sent to members of the group r. Thus "compatibility" is defined as follows.

Definition 1. *The compatibility $c_{s,r}$ between a sender subgroup s and a receiver subgroup r is defined by:*

$$c_{s,r} = \frac{2I_{s,r}I_{r,s}}{I_{s,r} + I_{r,s}} \tag{1}$$

Rule construction for each user starts with an empty rule (no condition and no conclusion) and produces a sequence of rules $R_1, \cdots, R_n$. At each step, starting with a rule R, a feature u_i ($a = v_i$) of the user is chosen that has the highest subgroup success rate; u_i is added to the condition, and the corresponding feature c_i ($a = v_i^*$) is added to the conclusion, where v_i^* is the best matching value for v_i of a for the gender (male/female) of the sender. These conditions are added to the rule only if it improves the overall rule success rate ($SR_{s,r}$) defined as follows, and this improvement is statistically significant:

$$SR_{s,r} = \frac{n(s \overset{+}{\leftarrow} r)}{n(s \to r)} \tag{2}$$

where s is the sender subgroup defined by the current rule condition and r is the receiver subgroup defined by the current rule conclusion, $n(s \to r)$ is the number of messages sent from s to r and $n(s \overset{+}{\leftarrow} r)$ is the number of positive replies from r to s. Attribute addition stops when the number of interactions between rule subgroups in the historical data is less than a minimum threshold.

Finally, for recommendation, the candidates are any users who satisfy the conclusion of any of the rules $R_1, \cdots, R_n$. The method ranks candidates that satisfy the more specific conclusions of rules R_i higher than those satisfying the less specific conclusions of rules R_j where $j < i$, and otherwise breaks ties randomly.

2.2   Interaction-Based Collaborative Filtering

In Krzywicki *et al.* [5], we introduced the application of interaction-based collaborative filtering (IBCF) to people to people recommendation, and defined three IBCF methods that are variants of collaborative filtering. An *interaction* consists of a user sending a message to a receiver, which may have a positive, negative or null reply. We defined notions of user similarity also based on interactions. In particular, two users are *similar senders* to the extent they have contacted other users in common, and are *similar recipients* to the extent they have been contacted by other users in common. By considering the links in the network of interactions, we defined various collaborative filtering methods based on the two notions of similarity, as follows.

- Basic CF: Recommend similar recipients to the user's contacts (equivalently, recommend the contacts of similar senders)
- Inverted CF Recipient: Recommend the contacts of similar recipients

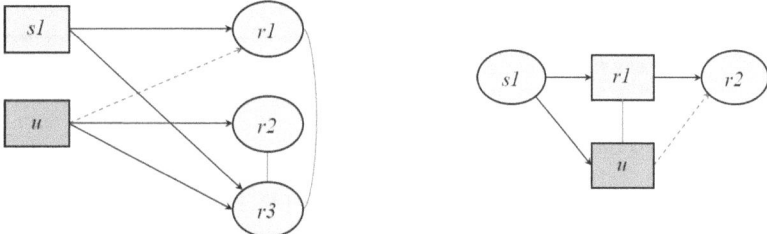

Fig. 1. Basic CF+ and Inverted CF+ Recipient

It was found that these IBCF methods provide much better recommendations if they are based only on positive interactions, denoted 'CF+' instead of 'CF'. The best results were obtained by combining Basic CF+ and Inverted CF+ Recipient, called Best CF+ here (Figure 1). In this and all subsequent figures, an arrow from s to r indicates a message sent from s to r that receives a positive reply, lines represent similarity, and the convention is that users of the same gender are in nodes of the same type (boxes or ovals), ignoring same-sex interactions for simplicity. The left half of the figure shows Basic CF+, where u is the active user to whom recommendations are to be given. Since $r1$ and $r3$ have both been contacted by $s1$, they are similar recipients, hence $r1$, who has not been contacted by u, is recommended to u, shown as a dashed line in the figure. The right half of the figure shows Inverted CF+ Recipient, where $r1$ and u are similar recipients since they have both been contacted by $s1$. The contacts of $r1$, in this case just $r2$, are recommended to u. The two methods complement each other very well, with Inverted CF+ Recipient able to recommend candidates to users who have received no positive replies, while Basic CF+ can recommend candidates to users who have received positive replies.

The ranking of a candidate for each method is given by the number of "votes" of similar users for that candidate. For the combined method Best CF+, the ranking is obtained by adding together the votes of Basic CF+ and Inverted CF+ Recipient.

3   Hybrids Based on Similarity

In this section, we define various hybrids of content-based methods for determining user similarity and collaborative filtering. We examine two similarity measures, a simple method based on profile similarity restricted by age and location that we apply to senders and to receivers, and a rule-based approach based on assigning the user to a preferred sender subgroup that we apply only to senders (the analogous measure applied to receivers is described in Section 4 as an example of a content-boosted method).

The basic picture is similar to the left half of Figure 1, except that instead of the similarity of u and $s1$ being determined by their interactions, their similarity is determined by their profiles. As in the figure, the contacts of those similar senders to the user u who have not already been contacted by u are recommended to u. The ranking of a candidate is determined by the number of similar senders who initiated a successful interaction with the candidate.

We use the following (very simple) definition of profile similarity.

Definition 2. *For a given user u, the class of similar users consists of those users of the same gender and sexuality who are either in the same age band as u or one age band either side of u, and who have the same location as u.*

3.1   Profile-Based Similar Senders CF (SIM-CF)

Profile-Based Similar Senders CF (SIM-CF), user-based similarity followed by collaborative filtering, uses the above definition of similarity applied to senders. Thus SIM-CF works by finding users similar in age and location to a user u and recommending the contacts of those users. This basic idea is also used in the work of Akehurst *et al.* [2].

3.2   Profile-Based Similar Recipients CF (CF-SIM)

Where SIM-CF computes user-based profile similarity, the analogous method Profile-Based Similar Recipients CF (CF-SIM), collaborative filtering followed by similarity over candidates, uses candidate-based (item-based) similarity. More precisely, candidates with similar profiles to the contacts of u are recommended to u. Note that, whereas SIM-CF can provide candidates for almost any given user u, CF-SIM requires that u have at least one positive interaction. This severely limits the applicability of CF-SIM.

3.3   Rule-Based Similar Senders CF (RBSS-CF)

The profile-based similarity measure used by SIM-CF and CF-SIM generates a large number of similar users for any given user u. Rule-Based Similar Senders CF (RBSS-CF) uses compatible subgroup rules to define the similarity relation over users. More precisely, the users similar to a given user u are those satisfying the condition of the most specific compatible subgroup rule defined for u. Since all compatible subgroup rules include age and location, a subset of the users with similar profiles as defined above (but those with profiles even more similar to the user) will be found using this approach.

4   Content-Boosted Methods

Hybrid recommendation methods based on profile similarity can provide recommendations for many users, however the measures used for similarity are simple. In this section, we define methods that exploit knowledge about the domain expressed as compatible subgroup rules, rules that relate the attributes of senders and receivers to define a statistically significant increased chance of a successful interaction. Since compatible subgroup rules can also be used to provide recommendations, the methods described here that combine this method with interaction-based collaborative filtering are examples of content-boosted methods as defined in Melville, Mooney and Nagarajan [6]. We define two content-boosted methods that are analogous to Basic CF+ and Inverted CF+ Recipient described above (Figure 1).

4.1   Rule-Based Similar Recipients CF (RBSR-CF)

Rule-Based Similar Recipients CF (RBSR-CF) uses a compatible subgroup rule's conclusion, which defines the features of preferred receivers, to indirectly define similar senders for a given user u. This assumes that two senders are similar if they have had interactions with similar receivers, where those similar receivers are defined by a compatible subgroup rule generated for u as a sender. This is illustrated in the left half of Figure 2, where u satisfies the condition of a compatible subgroup rule and so is similar (by rule) to $s1$ and $s2$. No further use is made of $s1$ and $s2$, except to define the compatible recipient subgroup that here contains $r1$ and $r2$. In general, this compatible subgroup consists of the top 100 potential candidates for u ranked according to their age and location compatibility using Equation 1. The Rules method would recommend $r1$ and $r2$ to u. However in the content-boosted method RBSR-CF, these potential candidates are used to "seed" interaction-based collaborative filtering. The method finds users who actually contacted $r1$ and $r2$, in this case $s3$ and $s4$, then applies Basic CF+ to generate and rank candidates being the contacts of those users whom u has not yet contacted, in this case $r1$ and $r3$, as shown by dashed lines in the figure. Note that the left part of this figure corresponds to Basic CF+ as shown in Figure 1, and $s3$ and $s4$ are similar senders according to their interactions.

4.2   Rule-Based Similar Recipients Inverted CF (RBSR-ICF)

Analogous to RBSR-CF, which corresponds to Basic CF+, we define Rule-Based Similar Recipients Inverted CF (RBSR-ICF) which corresponds to Inverted CF+ Recipient, the right half of Figure 1. RBSR-ICF uses a subgroup rule's conclusion to indirectly define similar receivers for a given user u. Under the inverted scheme, this method assumes that two such users are similar if they have been contacted by users considered similar as defined by a compatible subgroup rule generated for u as a sender, again based on the top 100 potential candidates determined by the Rules method. Thus in the right half of Figure 2, u satisfies the condition of a rule so is similar (by rule) to $s1$ and $s2$. Again $s1$ and $s2$ play no further role, but serve to define the compatible recipient subgroup, here containing $r1$ and $r2$. Now under the inverted scheme, the contacts of $r1$ and $r2$ are similar as recipients, so their contacts can in turn be recommended to u

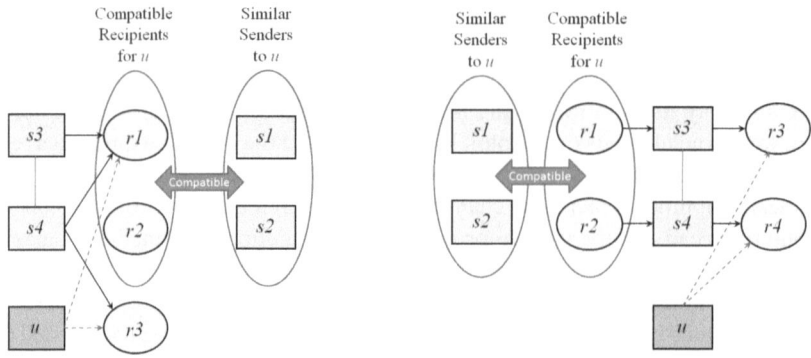

Fig. 2. Rule-Based Similar Recipients CF and Inverted CF

if u has not already contacted them. In the diagram, $s3$ and $s4$ are similar recipients according to their interactions, so their contacts $r3$ and $r4$ can be recommended to u. Note that the right part of this figure corresponds to Inverted CF+ Recipient as shown in Figure 1.

5   Evaluation

In this section, we discuss the experimental methodology and evaluation of the hybrid and content-boosted methods used with interaction-based collaborative filtering, with special emphasis on the cold start problems of recommending candidates to users who have no positive interactions as a sender (non-senders), no positive interactions as a receiver (non-receivers), and the intersection of these two groups, who have had no positive interactions at all.

For evaluation we used a historical dataset from a commercial dating site, which records both profile information about each user and interactions between users. Profile information consists of basic data about the user, such as age, location, marital and family status, plus attributes such as smoking and drinking habits, education and occupation. Each interaction is recorded with a date/time stamp, the type of message and the response message type, which is classified as either positive or negative, and is null if no reply has been received. Null replies are highly significant in this domain, with around a third of all messages going without a reply. In our evaluations, null replies are counted as negative interactions, since they would correspond to an unhelpful recommendation. The test set consists of around 130,000 users with around 650,000 interactions, of which roughly 15% are positive. Thus another characteristic of this dataset is the high imbalance between positive and non-positive (negative and null) interactions.

5.1   Experimental Setup and Metrics

All methods were tested on the historical dataset using interactions from the first three weeks of March 2010. This contrasts with the six day testing period used in Krzywicki *et al.* [5] so as to capture more interactions involving "cold start" users, making this analysis more reliable. The candidate set for each recommender was constructed once for a notional date of March 1, 2010, using only interactions initiated prior to the start of the test period. The purpose of the testing was to find out how well the recommendations conform to actual user interactions occurring in the test period and how well the methods cover various subgroups of users.

The metrics used are: user coverage, success rate (SR) and success rate improvement (SRI), and are measured for the top N recommendations, where $N = 10, 20, \cdots, 100$. Success rate is the same as precision over a set of recommendations.

The active users of interest in the study implicitly define the user pool U, the users to whom we aim to provide recommendations *and* the candidate pool from which these recommendations will be drawn.

Definition 3. *An* active user *is a user who (i) sent at least one message in the 28 days before the start of the test period, or (ii) received at least one message in the 28 days before the start of the test period and read the message, or (iii) is a new user who joined the site in the 28 days before the start of the test period. Call the set of all active users the* user pool, *denoted* U.

Definition 4. *Let* R *be a set of ranked recommendations* (u_1, u_2) *where* $u_1, u_2 \in U$. *Let* U_R *be the projection of* R *on the first argument, i.e. the set of* users *who are given recommendations in* R. *Let* $U' \subseteq U$ *be any subgroup of active users. Then the* user coverage *of* R *(and implicitly of the method used to generate* R*) with respect to the set* U' *is the proportion of users in* U' *who are in* U_R *(i.e. receive recommendations in* R*).*

  – *user coverage* $= |U_R \cap U'|/|U'|$

Definition 5. *Let* T *be a test set consisting of user pairs* (u_1, u_2) *where* $u_1, u_2 \in U$ *and* u_1 *has initiated a sequence of one or more interactions between* u_1 *and* u_2. *Let* $T^+ \subseteq T$ *be the set of user pairs in* T *where at least one of those interactions is positive. Let* $U' \subseteq U$ *be a subset of active users and let* $T_{U'} \subseteq T$ *and* $T_{U'}^+ \subseteq T^+$ *be the subsets of* T *and* T^+ *consisting of all user pairs* (u_1, u_2) *where* $u_1 \in U'$, *i.e.* $T_{U'}$ *and* $T_{U'}^+$ *are the test sets relative to the senders in* U'. *Let* $R_N \subseteq R$ *be the set of user-candidate pairs* (u_1, u_2) *where the candidate* u_2 *is ranked in* R *in the top* N *candidates for* u_1. *Then relative to* U', *the baseline success rate (BSR), the top* N *success rate* (SR_N) *and the top* N *success rate improvement* (SRI_N) *are defined as follows.*

  – $BSR(U') = |T_{U'}^+|/|T_{U'}|$
  – $SR_N(U') = |R_N \cap T_{U'}^+|/|R_N \cap T_{U'}|$
  – $SRI_N(U') = SR_N(U')/BSR(U' \cap U_R)$

The baseline success rate (BSR) for U' is the proportion of user pairs with successful interactions (in the test set) initiated by users in U', where U' in our analysis can be sets such as the users U_R receiving recommendations from a method R, males and females, or the "cold start" subgroups of non-senders and non-receivers; thus the baseline success rate varies with U'. The success rate (SR) at N is the proportion of user pairs with successful interactions (in the test set) from the top N recommendations for the users in U' receiving recommendations. The main metric is success rate improvement (SRI), which gives a ratio representing how much more likely users are to receive a positive reply from a recommended candidate compared to their baseline success rate.

5.2   Results and Discussion

Table 1 shows the user coverage (the proportion of users for whom recommendations can be provided) for each of the methods. The first column shows the user coverage for the whole user pool and the other columns for the different "cold start" user groups, the non-senders, the non-receivers and their intersection (defined based on positive interactions). The results for Best CF+ are consistent with those reported earlier [5], though here popular users are also included. As expected, SIM-CF and the content-boosted methods have very high coverage of the user pool, while it is surprising that RBSS-CF has such a low user coverage. To understand why, we calculated the number of similar users generated across the user pool for each of these methods. RBSS-CF generates similar users for only around 40% of all users: the compatible subgroup rules are so highly personalized that they fail to generate any similar users for a large number of users. The size of the user similarity relation (not in the table) also shows that there are computational advantages to the content-boosted methods RBSR-CF and RBSR-ICF since they have almost as high user coverage as SIM-CF whilst generating far fewer similar users on average (220 and 218 per user as compared to 1545).

Table 1. User Coverage

	User Pool	Non-Send (35%)	Non-Rec (16%)	Non-Send & Non-Rec (8%)
Best CF+	0.630	0.400	0.068	0.0
SIM-CF	0.997	0.997	0.998	0.496
CF-SIM	0.405	0.0	0.17	0.0
Rules	0.980	0.971	0.951	0.941
RBSS-CF	0.373	0.143	0.187	0.044
RBSR-CF	0.978	0.968	0.946	0.934
RBSR-ICF	0.973	0.961	0.933	0.916

Regarding the cold start user groups (Table 1), as was known, CF-SIM can cover only the non-receiver group, and this coverage is quite low (17%). But for the other methods, user coverage on the subgroups is consistent with user coverage on the whole user pool (again with the possible exception of RBSS-CF), showing that SIM-CF and the content-boosted methods can provide recommendations to the majority of users in these subgroups, though the user coverage of SIM-CF is markedly lower than for the content-boosted methods on the intersection of non-senders and non-receivers.

To examine the quality of recommendations, we calculated success rate improvement (SRI) measured over the whole user pool. As shown in Figure 3, both SIM-CF and the content-boosted method RBSR-CF exceed the performance of Best CF+, whilst covering almost all users. For example, for the top 10 recommendations, users are around 2.54 times more likely with RBSR-CF to have a positive interaction. This represents a substantial improvement over Best CF+ because that method provides no recommendations at all for cold start users. The improvements for RBSS-CF and RBSR-ICF are roughly in line with results for Basic CF+ and Inverted CF+ Recipient [5]. What we consider remarkable in this figure is that SIM-CF performs so well given the simplicity of the definition of profile similarity, providing further evidence that the power of the IBCF methods derives from the use of interactions to rank candidates rather than from the determination of similar users.

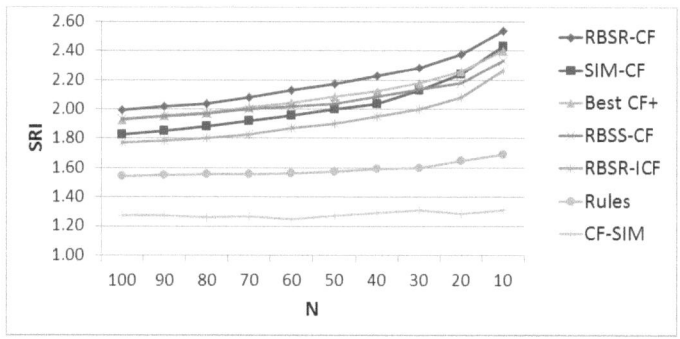

Fig. 3. SRI for User Pool

We now focus on the cold start user groups, the non-senders and non-receivers, as defined using positive interactions (absolute numbers for their intersection are too low for meaningful analysis). For the non-senders, the general pattern shown in Figure 4 follows that for the whole user pool, RBSR-CF and SIM-CF performing the best with RBSR-CF achieving an SRI of around 2.26 for the top 10 recommendations (Best CF+ cannot be shown since it cannot generate any recommendations for users in this class). The difference between the two methods is more pronounced than for all users, suggesting that RBSR-CF is the superior method for this class of users. Note also that the baseline success rate for this class of users is *higher* than for users in general, so such an SRI for this class of users is a good result.

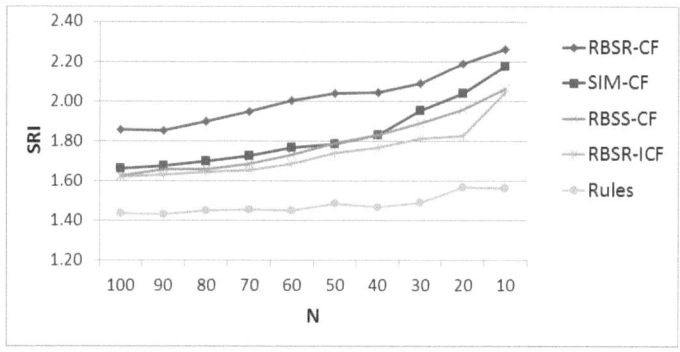

Fig. 4. SRI for Non-Senders

Figure 5 shows the analogous results for non-receiving users. Again the pattern is the same as in Figure 3, but this time the SRI is much higher, around 2.85 for the top 10 recommendations for RBSR-CF. To help understand this difference from Figure 4, we examined a breakdown of top 100 SRI by gender, since as indicated by the data, non-receivers are dominated by males while non-senders by females (with the split around 60/40 for each group). The results are shown in Table 2. The SRI for the male subgroups is consistently higher than for females (this is because males send far more messages and have a far lower baseline success rate). Since the non-receivers contain proportionally more males, the SRI for both SIM-CF and RBSR-CF is higher for this subgroup than for the non-senders.

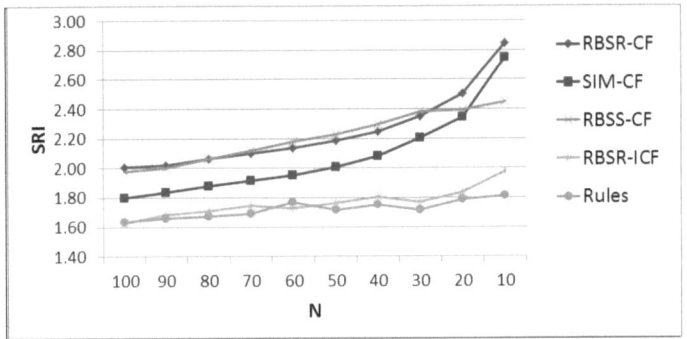

Fig. 5. SRI for Non-Receivers

Table 2. Top 100 SRI for Male/Female Subgroups

	Non-Send (M,F)	Non-Rec (M,F)
SIM-CF	1.750,1.500	1.808,1.547
RBSR-CF	1.922,1.686	2.018,1.752
RBSR-ICF	1.613,1.452	1.593,1.416

In summary, the hybrid method SIM-CF achieves a high SRI over the whole user pool and over the cold start subgroups, and performs surprisingly well in this domain. However, the content-boosted method RBSR-CF has the highest SRI and performs much better than SIM-CF over all the cold start subgroups, including over the male and female subgroups within these groups. Moreover, RBSR-CF achieves this high SRI by generating candidates from far fewer similar users on average than SIM-CF (220 per user as compared to 1545), confirming the effectiveness of content-boosted methods based on compatible subgroup rules. SIM-CF, however, is computationally more efficient overall, as it does not require the calculation of compatible recipients.

6   Conclusion

In this paper we have shown how hybrid and content-boosted recommender systems can address cold start problems in people to people recommender systems, a reciprocal recommendation framework where "items" are also users and hence have their own preferences that must be taken into account in recommendation. We presented three hybrid methods based on user similarity, and two content-boosted methods based on compatible subgroup rules. Using data obtained from a commercial online dating site, we examined all methods on the whole user pool and on the cold start subgroups of non-senders and non-receivers. Both the best hybrid and the best content-boosted method enable interaction-based collaborative filtering (IBCF) to generate candidates for almost all users, exceeding the quality of recommendations generated by the best IBCF method, that applies only to users with prior positive interactions. The content-boosted method gives a greater improvement in success rate than the hybrid method over the cold start subgroups, but the hybrid method is more computationally efficient overall.

Acknowledgements. This work was funded by Smart Services Cooperative Research Centre. We would also like to thank our industry partners for providing the datasets.

References

1. Adomavicius, G., Tuzhilin, A.: Toward the Next Generation of Recommender Systems: A Survey of the State-of-the-Art and Possible Extensions. IEEE Transactions on Knowledge and Data Engineering 17, 734–749 (2005)
2. Akehurst, J., Koprinska, I., Yacef, K., Pizzato, L., Kay, J., Rej, T.: CCR - A Content-Collaborative Reciprocal Recommender for Online Dating. In: Proceedings of the Twenty-Second International Joint Conference on Artificial Intelligence, pp. 2199–2204 (2011)
3. Cai, X., Bain, M., Krzywicki, A., Wobcke, W., Kim, Y.S., Compton, P., Mahidadia, A.: Collaborative Filtering for People to People Recommendation in Social Networks. In: Li, J. (ed.) AI 2010. LNCS, vol. 6464, pp. 476–485. Springer, Heidelberg (2010)
4. Kim, Y.S., Mahidadia, A., Compton, P., Cai, X., Bain, M., Krzywicki, A., Wobcke, W.: People Recommendation Based on Aggregated Bidirectional Intentions in Social Network Site. In: Kang, B.-H., Richards, D. (eds.) PKAW 2010. LNCS, vol. 6232, pp. 247–260. Springer, Heidelberg (2010)
5. Krzywicki, A., Wobcke, W., Cai, X., Mahidadia, A., Bain, M., Compton, P., Kim, Y.S.: Interaction-Based Collaborative Filtering Methods for Recommendation in Online Dating. In: Chen, L., Triantafillou, P., Suel, T. (eds.) WISE 2010. LNCS, vol. 6488, pp. 342–356. Springer, Heidelberg (2010)
6. Melville, P., Mooney, R.J., Nagarajan, R.: Content-Boosted Collaborative Filtering for Improved Recommendations. In: Proceedings of the Eighteenth National Conference on Artificial Intelligentce (AAAI 2002), pp. 187–192 (2002)
7. Park, S.T., Chu, W.: Pairwise Preference Regression for Cold-Start Recommendation. In: Proceedings of the Third ACM Conference on Recommender Systems, pp. 21–28 (2009)
8. Pizzato, L., Rej, T., Chung, T., Koprinska, I., Kay, J.: RECON: A Reciprocal Recommender for Online Dating. In: Proceedings of the Fourth ACM Conference on Recommender Systems, pp. 207–214 (2010)

An Improved Particle Swarm Optimisation for Image Segmentation of Homogeneous Images

Weng Kin Lai[1] and Imran M. Khan[2]

[1] School of Technology, TARC, 53300 Kuala Lumpur, Malaysia
[2] Dept. of Electrical and Computer Engineering, IIUM, 53100 Kuala Lumpur, Malaysia
laiwk@mail.tarc.edu.my

Abstract. Image segmentation is one of the fundamental and important steps that is needed to prepare an image for further processing in many computer vision applications. Over the last few decades, many image segmentation methods have been proposed as accurate image segmentation is vitally important for many image, video and computer vision applications. A common approach is to look at the grey level intensities of the image to perform multi-level-thresholding. In our approach we treat image segmentation as an optimization problem to identify the most appropriate segments for a given image where a two-stage population based stochastic optimization with a final refinement stage has been adopted.

Nevertheless, the ability to quantify and compare the resulting segmented images can be a major challenge. Information theoretic measures will be used to provide a quantifiable measure of the segmented images. These measures would also be compared with the total distances of the pixels to its centroid for each region.

Keywords: PSO, Image Segmentation, Data Clustering.

1 Introduction

As computer power improves and their prices drop, the use of computers to automatically process digital images becomes increasingly attractive. Application areas include the processing of scene data for autonomous machine perception, which may be for automatic character recognition, industrial machine vision for product assembly and inspection, military recognizance, automatic processing of fingerprints etc. Nevertheless, there are several fundamental steps involved in processing these images and one very important one is image segmentation.

Good image segmentation partitions an image into the appropriate regions based on some appropriate homogeneity measure so that the segmented sections/images are homogeneous. Hence, the segments of an image which shows similar characteristics should be put together with those which exhibit similarities based on these measures. On the other hand, those which are significantly different should then be labeled as belonging to different or separate segments.

P. Anthony, M. Ishizuka, and D. Lukose (Eds.): PRICAI 2012, LNAI 7458, pp. 218–228, 2012.
© Springer-Verlag Berlin Heidelberg 2012

The problem of segmentation is well-studied and there are a variety of approaches that may be used. Nevertheless, different approaches are suited to different types of images and the quality of output of a particular algorithm is difficult to measure quantitatively due to the fact that there may be many "*correct*" segmentations for a single image. Here we treat image segmentation as a data clustering problem and propose an improved population based stochastic optimization, namely particle swarm optimization [1] (*PSO*), to compute the optimal clusters. The experimental results indicate that the proposed *PSO* is capable of delivering improved segments as compared to the regular *PSO* approach.

The rest of the paper is organized as follows. Section 2 reviews prior work on image segmentation with PSO while section 3 introduces the experimental setup. Section 4 describes and briefly discusses the results and finally section 5 gives the conclusion of the entire study.

2   Prior Work

Accurate image segmentation is important for many image, video and computer vision applications. Over the last few decades, many image segmentation methods have been proposed. A common approach is to look at the grey level luminance of the image to perform multi-level-thresholding.

Many image segmentation techniques are available in the literature [2],[3],[4],[5],[6]. Some of these techniques use only the grey level histogram, some use spatial details while others use fuzzy set theoretic approaches. Some may be used to perform bilevel thresholding in which only the foreground and background images are identified. Others were used to perform multilevel image segmentations instead. Liping *et al* [7] proposed an approach based on a 2D entropy to segment the image based on the thresholds identified.

Particle Swarm Optimization (PSO) was originally modeled by Kennedy and Eberhart in 1995 after they were inspired by the social behavior of a bird flock. PSO had been used to solve a myriad of various problems. Kiran *et al* [8] have investigated using PSO for Human Posture recognition. And in recent years, segmentation with PSO has been investigated whereby the segmentation problem is now viewed as an optimization problem. Mahamed G. H. Omran first used Particle Swarm Optimization to perform image segmentation where the problem was modeled as a data clustering problem[9]. CC Lai [10] adopted a different approach in using the PSO approach for image segmentation. He used the PSO to compute the parameters for the curve that had been determined to give the best fit for the given image histogram. Subsequently, this was used to identify the appropriate thresholds for the various segments of the given image. Kaiping *et al* [11] proposed an effective threshold selection method of image segmentation which was then refined with the PSO embedded into a two-dimensional Otsu algorithm[4]. Meanwhile, Tang, Wu, Han and Wang [12] adopted a different approach in using PSO whereby they used it to search for the regions that returned the maximum entropy.

In our approach, we adopted the PSO to firstly identify the best thresholds and subsequently refined with pivotal alignment to optimize the regions. The segmented images were measured based on three different approaches. Two are based on information theoretic measure to quantify the quality of the segmentation. Finally, the last one measures how well the segments are organized within each region.

3   Methodology

An image is assumed to have L grey levels [1, ..., L] spatially distributed across the image. Nonetheless, the individual distinctive grey levels can be summed up to form a histogram $h(g)$.

The PSO clustering algorithm which mimicked the social behaviour of a flock of birds can be summarised into two main stages; a global search stage with pivotal alignment as a local refining stage. Within the PSO context, a swarm refers to a set of particles or solutions which could optimize the problem of clustering data. In a typical PSO system, the set of particles will fly through the search space with the aim of finding the particle position which results in the best evaluation of the given fitness function. Each particle will have its own memory which is effectively the local best (*pbest*) and the best fitness value out of all the *pbest* which is then labeled as the global best (*gbest*). The *gbest* will be the reference point where the other particles will strive to achieve while searching for better solutions. The PSO may be represented by the following equations:

$$V_{id}^{new} = WV_{id}^{old} + \left(C_1 rand_1 \left(p_{best} - X_{id}^{old}\right)\right) + \left(C_2 rand_2 \left(g_{best} - X_{id}^{old}\right)\right) \quad (1)$$

where

$$X_{id}^{new} = X_{id}^{old} + V_{id}^{old} \quad (2)$$

V_{id} represents the velocity which is involved in updating the movement and magnitude of the particles, X_{id}^{new} represents the new position of the particles after updating, W the inertia weight, C_1 and C_2 are acceleration coefficients while $rand_1$ and $rand_2$ are random values that varies between 0 to 1. These parameters provide the necessary diversity to the particle swarm by changing the momentum of the particles to avoid the stagnation of particles at the local optima. Nevertheless, PSO can still get trapped in local minima or it may take a longer time to converge to optimum results. Thus, we propose a pivotal alignment to improve the local optimum results.

(a) Select initial set of thresholds for n segments randomly.
(b) Repeat the following until $gbest_{t+1}$ does not change
  i)   Compute new thresholds with PSO
  ii)   If $gbest_{t+1} < gbest_t$ then $gbest_{t+1} = gbest_t$
(c) Pivotal Alignment
  -    identify centroids of current segments,
  -    for each pixel, compute the distance to centroid of current and the neighboring regions,
  -    realign the pixel to nearest region.

Fig. 1. Pivotal *Alignment PSO* for image segmentation

The thresholds for the n-segments, B_i are randomly selected but ensuring that there will not be any overlaps with its neighbours, i.e.

$$B_i = B_i + \text{rand}(b_1, b_2) \tag{3}$$

where $\text{rand}(b_1, b_2)$ are randomly generated numbers on the interval $[b_1, b_2]$ with a uniform distribution, $b_1 = \dfrac{-L}{2n}$ and $b_2 = \dfrac{L}{2n}$.

The *Brodatz* set of images is a well-known benchmark database for evaluating texture recognition algorithms[13] and the set of homogeneous *Brodatz* images used for testing the improved PSO is shown in Fig. 2. With a homogeneous set of images, it is easy for visual verification of the segmentation. Each image consists of 370×370 pixels and the number of distinctive regions range from 2 to 7.

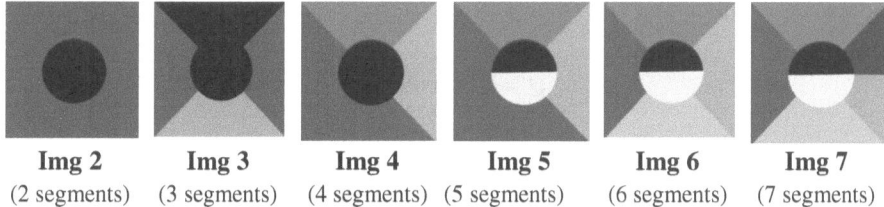

| Img 2 | Img 3 | Img 4 | Img 5 | Img 6 | Img 7 |
| (2 segments) | (3 segments) | (4 segments) | (5 segments) | (6 segments) | (7 segments) |

Fig. 2. Six images(colour) from the *Brodatz* set

4   Experimental Results

For the PSO with Pivotal Alignment (PSOWPA) used in this experiment, 20 particles are trained for a maximum of 200 iterations. The PSOWPA parameters are set as follows: $W = 0.001$ and $C_1 = C_2 = 1.49$. PSO starts by segmenting the target image using the grey level luminance values of each particle and computing the fitness of each one using the following evaluation function.

The error for each pixel is computed as the difference of the luminance from the mean of the associated luminance for segment j, i.e.

$$err_j = | mean_j - p_{j,i}| \tag{4}$$

where $p_{j,i}$ is the luminance of the ith pixel in segment j,
   $mean_j$ is the mean value of the luminance for segment j.

4.1   Segmentation Quality Measure

It is not uncommon to perform the evaluation visually, qualitatively, or indirectly by the effectiveness of the segmentation on the subsequent processing steps. Such methods are either subjective or tied to particular applications and they do not judge the performance of a segmentation method objectively. To overcome this problem, a few quantitative evaluation methods have been proposed[14].

Liu and Yang proposed an evaluation function that is based on empirical studies[15]. However, their approach suffers from the fact that unless the image has very well defined regions with very little variation in luminance and chrominance, the resultant segmentation score would tend to lean towards results with very few segments. Subsequently, an improved approach proposed using a modified quantitative evaluation with a new measure[16] to minimise such undesirable effects. Nevertheless, both these approaches were criticized for their lack of theoretical grounding of information theory as they were merely based on empirical analysis. *Zhang et al* [17] then came up with a measure that combines two measures - expected region entropy $\mathbf{H}_{ere}$ and layout entropy $\mathbf{H}_{lay}$. Given a segmented image I with n regions, V_j is the set of all possible values for the luminance in region j while $L_m(R_j)$ represents the number of pixels in region R_j that have luminance of m, the entropy Q'' would then be,

$$Q'' = H_{lay}(I) + H_{ere}(I) \tag{5}$$

$$= -\sum_{j}^{n} \frac{L(R_j)}{S_I} \log_2 \frac{L(R_j)}{S_I} + \left(-\sum_{j}^{n} \frac{L(R_j)}{S_I} \left(\sum_{m \in V_j} \frac{L_m(R_j)}{L(R_j)} \log \frac{L_m(R_j)}{L(R_j)} \right) \right) \tag{6}$$

S_I defines the total number of pixels for image I.

One measures the lack of uniformity whereas the other calculates the layout of the segmentation itself. This is a good approach as it maximises the uniformity of the pixels in each segmented region $\mathbf{H}_{ere}$ while maximising the differences between different regions with $\mathbf{H}_{lay}$. When the segmented region is uniform, the entropy $\mathbf{H}_{ere}$ would be small. On the other hand, such a development would result in an increase in the layout entropy $\mathbf{H}_{lay}$ instead.

Unlike Zhang's entropy-based measure, Mohsen *et al* [18] recommended a weighted measure Q' that gives a slightly larger bias to the layout entropy, i.e.

$$Q' = w_1 H_{lay}(I) + w_2 H_{err}(I) \tag{7}$$

where,

$H_{lay}(I)$: Layout entropy of image I i.e.,

$$H_{lay}(I) = -\sum_{j}^{n} \frac{L_j(R_e')}{S_I} \log_2 \frac{L_j(R_e')}{S_I} \tag{8}$$

$H_{cee}(I)$: Colour error entropy of image I, i.e.

$$H_{cee}(I) = -\sum_{j=\min_e}^{\max_e} \frac{L_j(R_e'')}{S_I} \log_2 \frac{L_j(R_e'')}{S_I} \tag{9}$$

n : total number of segments

$L_j(R_e')$: area of the jth region for *near-similar* pixels

min_e : minimum colour errors in image I"

w_1, w_2 : weighting parameters

S_I :total number of pixels for image I

$L_j(R_e'')$: area of *non-similar* pixels

max_e : maximum colour errors in image I"

$w_1 + w_2 = 1.0$

Identifying which pixels are near-similar or otherwise is quite simple. Any pixel with an error less than the threshold value of d are deemed to be near-similar. Anything else would then be non-similar. Near-similar pixels would contribute towards the layout entropy H_{lay} whereas those non-similar ones, the colour error entropy H_{err}. This is illustrated in Fig. 3.

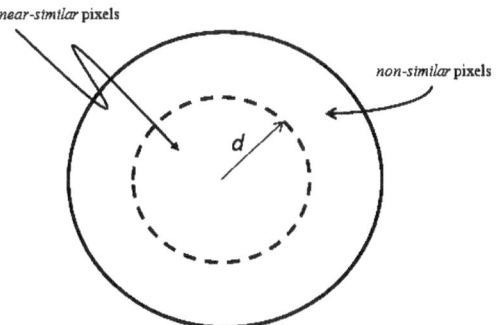

Fig. 3. Near-similar and non-similar pixels

A value of $d = 0.15$ with $w_1 = 0.55$ and $w_2 = 0.45$ were used to provide a measure of the segmentation result. Furthermore, we would normalized the results against the computed maximum values.

4.1.1   Normalised EntropyMeasure

Rewriting Equations (8) and (9), i.e.

$$H_{lay} = -\sum_{j}^{n} \frac{\hat{S}_j}{S_I} \log_2 \frac{\hat{S}_j}{S_I}$$ (10)

$$H_{cee} = -\sum_{j}^{n} \frac{S_j''}{S_I} \log_2 \frac{S_j''}{S_I}$$ (11)

where ,

$\hat{S}_j$: area of the jth region for near-similar pixels

S_j'' : area of the jth region for non-similar pixels,

S_j : area of jth region, i.e. $S_j = S_j'' + \hat{S}_j$

a) $Q' = w_1 H_{lay}(I)$ when $H_{cee} = 0$, or

$$\frac{S_j - \hat{S}_j}{S_I} \log_2 \frac{S_j - \hat{S}_j}{S_I} = 0$$

$$S_j = \hat{S}_j$$

i.e. all the pixels in the jth segment are near or are near-similar pixels. Hence,

$$Q' = -w_1 \sum_{j}^{n} \frac{\hat{S}_j}{S_I} \log_2 \frac{\hat{S}_j}{S_I}$$ (12)

It may then be shown that,

$$Q'_{max} = w_1 \frac{\log_2 e}{e}$$ (13)

b)   Similarly, $Q'' = w_2 H_{err}(I)$ when $H_{lay}(I) = 0$

Similarly, it can be shown that

$$Q''_{max} = w_2 \frac{\log_2 e}{e}$$ (14)

Hence we may then scale the entropy values by normalising it with $\frac{\log_2(e)}{e} = 0.5039$.

To investigate and compare the accuracy and effectiveness of *PSO with* Pivotal Alignment with the regular PSO, 7 different images shown in fig. 2 were used. 5 different runs $(a - e)$ were tested for each image and the results of the segmentation were measured using both of the entropy measures discussed in the previous sections as well as the sum of the distances from the mean of each region.   The results are summarized in table 1.

Table 1. Comparison of segmentation results

Image	Distance to centroids			Zhang's Entropy, Q"		Mohsen's Entropy, Q'	
	PSO	*PSOWPA*	*%*	*PSO*	*PSOWPA*	*PSO*	*PSOWPA*
$Img2_a$	11,554	11,123	3.73	0.4269	0.4187	0.1329	0.1327
$Img2_b$	12,500	11,123	11.02	0.4327	0.4187	0.1330	0.1327
$Img2_c$	12,638	11,123	11.99	0.4342	0.4187	0.1330	0.1327
$Img2_d$	13,368	11,123	16.79	0.4369	0.4187	0.1331	0.1327
$Img2_e$	14,249	11,123	21.94	0.4404	0.4187	0.1331	0.1327
$Img3_a$	31,434	31,081	1.12	0.7710	0.7712	0.2044	0.2053
$Img3_b$	31,483	31,081	1.28	0.7721	0.7712	0.1936	0.2053
$Img3_c$	32,519	31,081	4.42	0.7704	0.7712	0.1939	0.2053
$Img3_d$	32,832	31,081	5.33	0.7716	0.7712	0.1925	0.2053
$Img3_e$	46,178	31,081	32.69	0.7845	0.7712	0.1927	0.2053
$Img4_a$	847,779	847,779	0.0000	0.9164	0.9165	0.1743	0.1744
$Img4_b$	847,803	847,779	0.0028	0.9179	0.9164	0.1741	0.1745
$Img4_c$	847,843	847,779	0.0075	0.9195	0.9165	0.1667	0.1745
$Img4_d$	847,925	847,779	0.0172	0.9130	0.9164	0.1739	0.1744
$Img4_e$	848,172	847,779	0.0463	0.9232	0.9164	0.1656	0.1745
$Img5_a$	24,287	24,010	1.14	1.0252	1.0230	0.1501	0.1559
$Img5_b$	25,184	24,010	4.66	1.0237	1.0230	0.1448	0.1559
$Img5_c$	26,672	24,010	9.98	1.0248	1.0230	0.1451	0.1559
$Img5_d$	26,897	24,010	10.73	1.0341	1.0230	0.1412	0.1559
$Img5_e$	30,073	24,010	20.16	1.0443	1.0230	0.1339	0.1559
$Img6_a$	24,286	24,141	0.60	1.3346	1.3332	0.1497	0.1494

Table 1. (*Continued*)

Img6$_b$	24,940	24,141	3.20	1.3272	1.3332	0.1441	0.1494
Img6$_c$	25,340	24,141	4.73	1.3334	1.3332	0.1448	0.1494
Img6$_d$	26,424	24,141	8.64	1.3447	1.3332	0.1369	0.1494
Img6$_e$	26,754	24,141	9.77	1.3317	1.3332	0.1452	0.1494
Img7$_a$	337,147	336,006	0.34	1.6057	1.6023	0.1212	0.1254
Img7$_b$	338,543	336,099	0.72	1.6090	1.6021	0.1264	0.1253
Img7$_c$	23,232	22,383	3.65	1.4994	1.4956	0.1309	0.1351
Img7$_d$	27,705	22,383	19.21	1.5149	1.4956	0.1236	0.1351
Img7$_e$	33,234	22,383	32.65	1.5007	1.4956	0.1252	0.1351

Improvements in the segmentation with Pivotal Alignment can be quite significant as illustrated in fig. 4. While the regular PSO got stuck with a local optimum configuration (Fig.4 (a)), PSO with Pivotal Alignment was able to correct this to arrive at a significantly improved results. (Fig.4(b)).

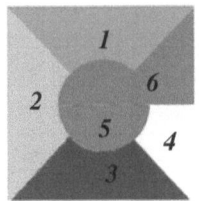   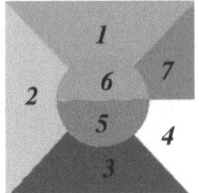

(a) Regular PSO   (b) *PSO with Pivotal Alignment*

Fig. 4. Segmented images for the 7-segment *Brodatz* image

Nonetheless, on most occasions as may be seen from the results in table 1, the regular PSO was able to identify the good segmentations. However, Pivotal Alignment was able to enhance upon the local optimal configurations found by the regular PSO. In such cases, the improvements are typically less than 10% (column 4).

Entropy-based measures are expected to show lower values for better segmentation[17]. Zhang's entropy Q'' showed smaller values for all the segmented images tested here. Even though the normalized entropy measure showed decreased values for the images with smaller segments, nonetheless there was a noticeable increase when it comes to those images with a larger number of segments, e.g. 7 segments. For such cases, there was a significant increase of the near-similar pixels as compared with those segmented with the regular PSO as illustrated in Fig. 5.

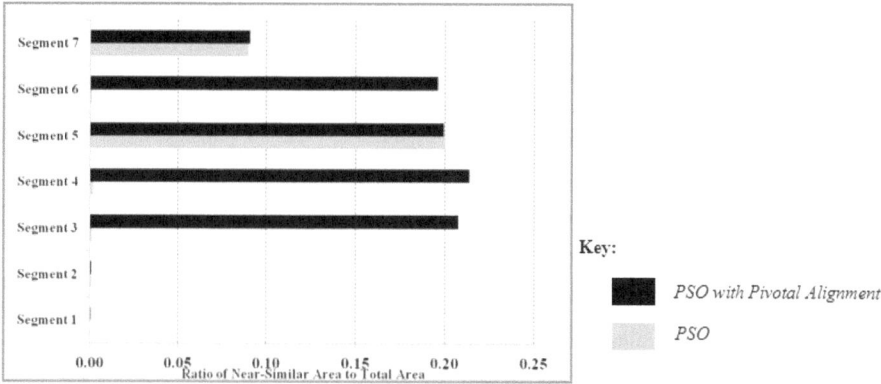

Fig. 5. Layout Entropy: Near-Similar pixels

Clearly it may be seen that the total number of near-similar pixels segmented (as measured by the ratio of the number of near-similar pixels to the total number of pixels) with the PSO were significantly lesser, especially when it comes to segments 1, 2, and 6.

5   Discussions and Conclusions

We have demonstrated via our preliminary set of experiments on homogeneous images, the effectiveness of an improved PSO with Pivotal Alignment for image segmentation. We have also compared the segmentation results with 3 different approaches. In general, all three measures were able to capture the improvements of the PSOWPA. In this investigation, it has been noticed that the more sophisticated segmentation measures using the concept of near-similar pixels and non-similar pixels increased when it comes to images with a higher number of regions. Further analysis showed that this was because of the improvements in the segmentation due to the better regions found in the near-similar regions.

One immediate area for further investigation is to explore PSO with Pivotal Alignment for non-homogeneous images. Nevertheless, the ability to accurately quantify the amount of improvements can be another challenge even though some of the measures investigated here may be used.

References

1. Kennedy, J., Eberhart, R.: Particle swarm optimization. In: Proceedings of IEEE International Conference on Neural Networks, Perth, Australia, vol. 4, pp. 1942–1948 (1995)
2. Guo, R., Pandit, S.M.: Automatic threshold selection based on histogram modes and discriminant criterion. Mach. Vis. Appl. 10, 331–338 (1998)
3. Pal, N.R., Pal, S.K.: A review of image segmentation techniques. Pattern Recognition 26, 1277–1294 (1993)

4. Otsu, N.: A threshold selection method from grey-level histograms. IEEE Trans. System. Man & Cybernetics, SMC 9, 62–66 (1979)
5. Shaoo, P.K., Soltani, S., Wong, A.K.C., Chen, Y.C.: Survey: A survey of thresholding techniques. Computer Vis. Graph. Image Process 41, 233–260 (1988)
6. Sydner, W., Bilbro, G., Logenthiran, A., Rajala, S.: Optimal thresholding: A new approach. Pattern Recognition Letters 11, 803–810 (1990)
7. Zheng, L., Pan, Q., Li, G., Liang, J.: Improvement of Grayscale Image Segmentation Based On PSO Algorithm. In: Proceedings of the Fourth International Conference on Computer Sciences and Convergence Information Technology, pp. 442–446 (2009)
8. Kiran, M., Teng, S.L., Seng, C.C., Kin, L.W.: Human Posture Classification Using Hybrid Particle Swarm Optimization. In: Proceedings of the Tenth International Conference on Information Sciences, Signal Processing and Their Application (ISSPA 2010), Kuala Lumpur, Malaysia, May 10-13 (2010)
9. Omran, M.G.H.: Particle Swarm Optimization Methods for Pattern Recognition and Image Processing. PhD Thesis, University of Pretoria (2005)
10. Lai, C.-C.: A Novel Image Segmentation Approach Based on Particle Swarm Optimization. IEICE Trans. Fundamentals E89(1), 324–327 (2006)
11. Wei, K., Zhang, T., Shen, X., Jingnan: An Improved Threshold Se-lection Algorithm Based on Particle Swarm Optimization for Image Segmentation. In: Proceedings of the Third International Conference on Natural Computation, ICNC 2007, pp. 591–594 (2007)
12. Tang, H., Wu, C., Han, L., Wang, X.: Image Segmentation Based on Improved PSO. In: Proceedings of the International Conference on Computer and Communications Technologies in Agriculture Engineering, Chengdu, China, pp. 191–194 (June 2010)
13. Brodatz, P.: Textures: A Photographic Album for Artists and Designers. Dover Publications, New York (1966)
14. Zhang, H., Fritts, J.E., Goldman, S.A.: Image Segmentation Evaluation: A Survey of Unsupervised Methods. In: Computer Vision and Image Understanding (CVIU), vol. 110(2), pp. 260–280 (2008)
15. Liu, J., Yang, Y.-H.: Multi-resolution Color Image Segmenta-tion. IEEE Transactions on Pattern Analysis and Machine Intelligence 16(7), 689–700 (1994)
16. Borsotti, M., Campadelli, P., Schettini, R.: Quantitative evaluation of color image segmentation results. Pattern Recognition Letters 19, 741–747 (1998)
17. Zhang, H., Fritts, J.E., Goldman, S.A.: An entropy-based objective segmentation evaluation method for image segmentation. In: Proceedings of SPIE Electronic Imaging - Storage and Retrieval Methods and Applications for Multimedia, pp. 38–49 (January 2004)
18. Mohsen, F.M.A., Hadhoud, M.M., Amin, K.: A new Optimization-Based Image Segmentation method By Particle Swarm Optimization. International Journal of Advanced Computer Science and Applications, Special Issue on Image Processing and Analysis 10–18 (Online) ISSN 2156-5570

A Regression-Based Approach for Improving the Association Rule Mining through Predicting the Number of Rules on General Datasets

Dien Tuan Le, Fenghui Ren, and Minjie Zhang

School of Computer Science and Software Engineering,
University of Wollongong,Wollongong, 2500, NSW, Australia
dtl844@uowmail.edu.au, {fren,minjie}@uow.edu.au

Abstract. Association rule mining is one of the useful techniques in data mining and knowledge discovery that extracts interesting relationships between items in datasets. Generally, the number of association rules in a particular dataset mainly depends on the measures of *'support'* and *'confidence'*. To choose the number of useful rules, normally, the measures of *'support'* and *'confidence'* need to be tried many times. In some cases, the measures of *'support'* and *'confidence'* are chosen by experience. Thus, it is a time consuming to find the optimal measure of *'support'* and *'confidence'* for the process of association rule mining in large datasets. This paper proposes a regression based approach to improve the association rule mining process through predicting the number of rules on datasets. The approach includes a regression model in a generic level for general domains and an instantiation scheme to create concrete models in particular domains for predicting the potential number of association rules on a dataset before mining. The proposed approach can be used in broad domains with different types of datasets to improve the association rule mining process. A case study to build a concrete regression model based on a real dataset is demostrated and the result shows the good performance of the proposed approach.

Keywords: Association rules mining, regression model, multiple correlation coefficient.

1 Introduction

Data mining has emerged as an important method to discover useful information, hidden patterns or rules from different types of datasets. Association rule is one of the most popular techniques and an important research issue in the area of data mining and knowledge discovery for many different purposes such as data analysis, decision support, patterns or correlation discovery on different types of datasets. In the past two decades, many approaches have been developed with the objective of solving the obstacles presented in the generation of association rules [1–3]. Agrawal et al. [4] proposed two algorithms, called Apriori and Apri-oriTid, to discover significant association rules between items in large datasets. Shi et al. [5]

P. Anthony, M. Ishizuka, and D. Lukose (Eds.): PRICAI 2012, LNAI 7458, pp. 229–240, 2012.
© Springer-Verlag Berlin Heidelberg 2012

used association rules for credit evaluation in the domain of farmers'credit history. Li et al. [6] specifically analyzed out-bound tourism datasets among Hong Kong residents using both positive and negative targeted rules. However, current approaches are facing two main challenges: (1) Each approach might be limited to special types of domains; (2) A huge amount of time needs to be consumed to find out the right association rules from large datasets.

In order to address these two challenges, a new approach is proposed in this paper to predict the potential number of association rules so as to improve the process of association rule mining. The approach consists of a regression model for general domains and an instantiation scheme to instantiate the general model to a concrete model based on a particular dataset in a domain for association rule prediction. By this way, the proposed approach can be applied to different domains with different types of datasets for the prediction before applying association rule mining. After ascertaining optimal values of *support* coefficient and *confidence* coefficient which exist in a dataset from our approach, association rule mining can be carried out effectively by using data mining software such as SPSS, R, and Tanagra, so as to reduce time consumption and quickly find useful rules. The major contributions of our approach are that (1) we designed a prediction model in a generic level by consideration of general domains so it can be used in broad applications; (2) the instantiation scheme of the generic model is carefully designed and evaluated by two statistical methods and can create concrete models in different domains on different types and sizes of datasets. In this paper, we also provide a case study to show the procedure of creating a concrete model by the use of our generic model, based on a real dataset, which is surveyed from customers of a supermarket in Wollongong City in 2011, to demonstrate the good performance of proposed approach.

The rest of this paper is organized as follows. Section 2 introduces the principle of our approach in detail, which includes the detail introduction of the generic model and the instantiation scheme. A concrete model generated by our approach on a case study and an experiment in a real dataset are demonstrated in Section 3. Section 4 compares our approach with some related work. Section 5 concludes this paper and points out our future work.

2 A Regression-Based Approach to Predicting the Potential Number of Associate Rules in General Domains

In this section, the proposed approach is introduced in detail. Our approach includes two parts, (1) generic model construction and (2) concrete model instantiation. These two parts are introduced in details in the following two subsections, respectively.

2.1 Generic Model Construction

In this part, a regression model is designed in a generic level between data items in general domains to express the relationships between the number of association rules, and *support* coefficient and *confidence* coefficient. In principle, the number of association rules in a dataset mainly depend on the *support* coefficient and the *confidence* coefficient. In general, the *support* coefficient and *confident* coefficient can be utilized to the measurement of association rules.

The meaning of *support* and *confidence* of the association rules can be formally described as follows. Let $I = k_1, k_2, ... k_m$ be a set of items in a dataset. Let D be a set of transactions, where each transaction T is a set of items such that $T \subseteq I$. Let A be a set of items in I, a transaction T is said to contain A iff $A \subseteq T$. An association rule is an implication of the form $A \Rightarrow B$, where $A \subset I$, $B \subset I$, and $A \cap B = \varnothing$. The rule $A \Rightarrow B$ holds in the transaction set D with **support** s, where $s\%$ of transactions in D contains $A \cup B$. The rule $A \Rightarrow B$ holds in the transaction set D with **confidence** c, where $c\%$ of transactions in D contains A and contains B.

The formula of *support* and *confidence* for the form $A \Rightarrow B$ can be formally represented [8] by Equation 1 and 2.

$$support(A \Rightarrow B) = P(A \cup B) \tag{1}$$

$$confidence(A \Rightarrow B) = P(A \setminus B) \tag{2}$$

where the values of *support* and *confidence* are in-between 0 and 1.

Based on the analysis of relationships between the number of association rules and the *support* and the *confidence*, we propose a non-linear regression model to predict the potential number of association rules on datasets in general domains. Our regression model is formally defined by Equation 3.

$$y = \beta_0 + \beta_1 \frac{1}{x_1^2} + \beta_2 \frac{1}{x_2^3} \tag{3}$$

where y is the number of association rules, x_1 represents the *support* variable, and x_2 represents the *confidence* variable, β_0, β_1 and β_2 are coefficients of the regression model. The coefficient β_1 measures the partial effect of x_1 on y while the coefficient β_2 measures the partial effect of x_2 on y. The coefficient β_0 is a constant and represents the mean of y when each independent variable equals to 0. In general, the coefficient β_0 is rarely explained and evaluated in a regression model because it includes all other factors except *support* and *confidence*. From Equation 3 we can see that the number of association rules mainly depend on values of *support* and *confidence*, as well as β_1 and β_2 in a particular domain.

After we make simple variable transformation, the Equation 3 can be rewritten as:

$$y = \beta_0 + \beta_1 X_1 + \beta_2 X_2 \tag{4}$$

where $X_1 = 1/x_1^2$ and $X_2 = 1/x_2^3$.

After defining the regression model, we need to determine the values of three coefficients in terms of result in minimizing the totally residual error of the model. The residual error Z could be defined by using Equation 5

$$Z = \sum_{i=1}^{n} (y_i - \hat{y}_i)^2 = \sum_{i=1}^{n} (y_i - \beta_0 - \beta_1 X_1 - \beta_2 X_2)^2 \tag{5}$$

where y_i is the real number of association rules in a data sample, $\hat{y}_i$ is the predicted number of association rules in a data sample. The values of coefficients in Equation 4 can be derived by Equation 5 and the results are represented as follows:

$$\begin{cases} \frac{\partial Z}{\partial \beta_0} = 0 \Leftrightarrow \sum (y_i - \beta_0 - \beta_1 X_1 - \beta_2 X_2) = 0 \\ \frac{\partial Z}{\partial \beta_1} = 0 \Leftrightarrow \sum (y_i - \beta_0 - \beta_1 X_1 - \beta_2 X_2) X_1 = 0 \\ \frac{\partial Z}{\partial \beta_2} = 0 \Leftrightarrow \sum (y_i - \beta_0 - \beta_1 X_1 - \beta_2 X_2) X_2 = 0 \end{cases} \tag{6}$$

After carrying out the partial derivatives, we have an equation group as follows:

$$\begin{cases} \sum y_i = n\beta_0 + \beta_1 \sum X_{1i} + \beta_2 \sum X_{2i} \\ \sum y_i X_{1i} = \beta_0 \sum X_{1i} + \beta_1 \sum X_{1i}^2 + \beta_2 \sum X_{1i} X_{2i} \\ \sum y_i X_{2i} = \beta_0 \sum X_{1i}^2 + \beta_1 \sum X_{1i} X_{2i} + \beta_2 \sum X_{2i}^2 \end{cases} \tag{7}$$

By solving Equation 7, coefficients β_0, β_1 *and* β_2 are represented by Equation 8.

$$\begin{cases} \beta_0 = \bar{y} - \beta_1 \bar{X}_1 - \beta_2 \bar{X}_2 \\ \beta_1 = \frac{\sum y_i X_{1i} \sum X_{2i}^2 - \sum y_i X_{2i} \sum X_{1i} X_{2i}}{\sum X_{1i}^2 \sum X_{2i}^2 - (\sum X_{1i} X_{2i})^2} \\ \beta_2 = \frac{\sum y_i X_{2i} \sum X_{2i}^2 - \sum y_i X_{1i} \sum X_{1i} X_{2i}}{\sum X_{1i}^2 \sum X_{2i}^2 - (\sum X_{1i} X_{2i})^2} \end{cases} \tag{8}$$

where $\bar{y} = \frac{\sum y_i}{n}$, $\bar{X}_1 = \frac{\sum X_{1i}}{n}$, $\bar{X}_2 = \frac{\sum X_{2i}}{n}$, and n is the number of observations.

2.2   Concrete Model Instantiation

In Subsection 2.1, we represented a generic regression model (refer to Equation 3) and proposed a calculation method to obtain all coefficients of generic model (refer to Equation 8). In order to apply the proposed model to predict the potential number of association rules on a particular dataset, we need to instantiate the generic model to a concrete model, which can fix the features of the domain in terms of confirmation of coefficients of the generic model.

In this subsection, the instantiation scheme of our approach is introduced. This scheme consists of two steps, which are (1) correlation evaluation between the number of association rules and *support* and *confidence* and (2) efficiency evaluation on coefficients.

Step 1: *Correlation Evaluation*
The purpose of correlation evaluation is to evaluate correlation between the dependent variable y and two independent variables x_1 and x_2. The multiple correlation efficient method is employed to achieve this purpose. The multiple correlation coefficient method is a common approach in statistics to measure the proportion of the total variation in a variable y, which is explained by the predictive power of independent variables, (only x_1 and x_2 in this paper).

The multiple correlation coefficient is defined as follows:

$$R = \sqrt{1 - \frac{\sum (y_i - \hat{y}_i)^2}{\sum (y_i - \bar{y}_i)^2}} \tag{9}$$

where R is the multiple correlation coefficient and $0 \leq R \leq 1$, $\bar{y}_i$ is the real association rule average in a data sample. The closer the R to 1, the smaller the residual error is, and the higher the validity of the model is.

Step 2: *Coefficient Evaluation*
This evaluation is used to test whether coefficients β_1 and β_2 exist in the regression model. We utilize two categories of tests, which are the entire model test and each coefficient test.

(1) Entire Model Test
We need to test whether the two independent variables x_1 and x_2 have a statistically significant effect on dependent variable y. Thus, a pair of hypothesis including the null hypothesis H_0 and the alternative hypothesis H_1 are created as follows.

$H_0 : \beta_1 = \beta_2 = 0$
$H_1 : (\beta_1 \neq 0)$ *or* $(\beta_2 \neq 0)$ *or* $(\beta_1 \neq 0$ *and* $\beta_2 \neq 0)$
The test of H_0 is carried out using the following statistical calculation:

$$F = \frac{R^2/2}{(1 - R^2)/(n - 3)} \tag{10}$$

where n indicates the number of observations and R^2 is the multiple determination coefficient. R^2 can be calculated by using the following equation:

$$R^2 = \frac{\sum (y_i - \bar{y}_i)^2 - \sum (y_i - \hat{y}_i)^2}{\sum (y_i - \bar{y}_i)^2} \tag{11}$$

Before carrying out hypothesis test, we need to calculate the critical value $F_{\alpha(2,n-3)}$ for this test, where α indicates the significant level and n is the number of observations. According to statistics, the chosen value of α is usually 5%. The critical value $F_{\alpha(2,n-3)}$ can be found in Fisher - Snerecor Distribution Table or it can calculated by FINV function in MS Excel.

In Equation 10, If $F > F_{\alpha(2,n-3)}$, then H_0 is rejected. It states that there is at least one independent variable x_1 or x_2 related to y. In other words, the dependent variable y can be effected by the independent variable x_1 or x_2 or both, then we need to test each coefficient to determine which variable exists in the model. In other cases, while $F \leq F_{\alpha(2,n-3)}$, we accept the null hypothesis. It states that little evidence exists to support the relationship between the dependent variable y and two the independent variables x_1 and x_2.

(2) Coefficient test
This test includes to test *support* coefficient x_1 and *confidence* coefficient x_2 separately. We need to build a pair of hypothesis for each coefficient.

Case 1: In this test, the null hypothesis H_0 and the alternative hypothesis H_1 for β_1 are created as: $H_0 : \beta_1 = 0$, $H_1 : \beta_1 \neq 0$

The test of H_0 is calculated by using the following statistics: $t = \frac{\hat{\beta}_1 - \beta_0}{Se(\hat{\beta}_1)}$, where $Se(\hat{\beta}_1)$ indicates a standard error of β_1. Before carrying out hypothesis test, we need to calculate the critical value $-t_{(\alpha/2,n-3)}$ and $t_{(\alpha/2,n-3)}$ for this test, where α indicates the significant level. According to statistics, the chosen value of α is usually 5%. The critical values $-t_{(\alpha/2,n-3)}$ and $t_{(\alpha/2,n-3)}$ can be found in Student Distribution Table or it can be calculated by TINV function in MS Excel. If $-t_{(\alpha/2,n-3)} \leq t \leq t_{(\alpha/2,n-3)}$ then H_0 is accepted. It states that little evidence exists of interaction between the dependent variable y and the independent variable x_1. In other cases, we accept the alternative hypothesis H_1. It states that there exists some evidence about interaction between the dependent variable y and the independent variable x_1.

Case 2: In this test, the null hypothesis H_0 and the alternative hypothesis H_1 for β_2 are created as: $H_0 : \beta_2 = 0$, $H_1 : \beta_2 \neq 0$.

The test of H_0 is calculated by using the following statistics: $t = \frac{\hat{\beta_2} - \beta_0}{Se(\hat{\beta_2})}$, where $Se(\hat{\beta_2})$ is a standard error of β_2. If $-t_{(\alpha/2,n-3)} \leq t \leq t_{(\alpha/2,n-3)}$ then we accept H_0. It states that little evidence exists about interaction between the dependent variable y and the independent variable x_2. In other cases, we accept the alternative hypothesis H_1. It states that there exists interaction between dependent variable y and the independent variable x_2.

3   A Case Study

This section presents a case study to show how to instantiate a concrete model from the proposed approach for a real world dataset which is collected from a customer survey in Woolworths Supermarket in Wollongong City, Australia. This case study not only shows the procedure of using our approach in a particular domain, but also demonstrates the performance of our approach in the real life situations. The background of the case study is introduced in Subsection 3.1. The procedure of generating a concrete model applied on the dataset of the case study is described step by step in Subsection 3.2. The prediction results and analysis are given by Subsection 3.3.

3.1   Background of the Case Study

The dataset used in this study is collected from Woolworths supermarket through survey of customers in November 2011 at Wollongong City, Australia. The survey's data was collected from different time and different types of customers arriving to buy goods at the Supermarket. The questionnaires in the survey consisted of some common groups such as vegetable, bread, fruit, meat, milk, cheese and seafood. Thus, the dataset created by the survey includes 48 attributes and 51 data samples. The dataset is created through SPSS software for preprocess to analyze data. The results of data summary in Vegetable Group and Bread Group are illustrated in Figure 1.

Among 51 interviewed customers, it has been observed from Fig 1.(a) that in the graph of Vegetable Group, most customers tend to buy the items including cauliflower, lettuce and broccoli because more than 50% of fifty one people

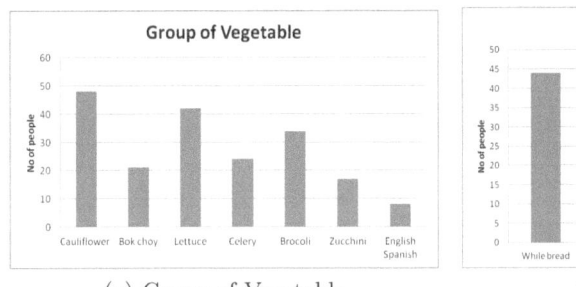
(a) Group of Vegetable

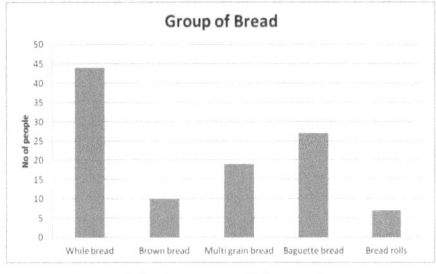
(b) Group of Bread

Fig. 1. Data Analysis of the survey

arrived Supermarket tend to buy these items. In particular, the proportion of cauliflower, lettuce, broccoli is 95%, 82% and 66%, a respectively. On the other hand, the proportion of other items is very slow. In particular, the proportion of boy choy, celery, zucchini and english spanish is 41%, 47%, 33% and 15%, a respectively.

From Fig 1.(b), it can be seen that the proportion of white bread and baguette purchased by customers is very high. In particular, the proportion of white bread and baguette bread is 86% and 53%, respectively. In the other hand, the proportion of other items is very slow.

The graphs of other groups including meat, seafood, milk, cheese, and fruit are in the similar format. Due to the space limit, we do not show them in this paper.

3.2 Generation of a Concrete Model for the Dataset of the Case Study

This subsection demonstrates the procedure of creating a concrete model for this case study. In our case study, we apply our approach to groups of the survey dataset. In this paper, we use the Vegetable Group to demonstrate our case study.

Step 1: *Generating a Regression Model for Vegetable Group*

This step illustrates a concrete regression model of the sample data of Vegetable Group from dataset of survey's customers. The sample data is created by using the algorithm of Apriori [4] for association rule mining to get the real number of rules in the conditions of different *support* thresholds and different *confidence* thresholds. The results are shown in Table 1.

From the results of Table 1, we can build a regression model by using Equation 3 to present the relationships between the number of association rules, *support* and *confidence*. In the case study, we use software of SPSS to calculate coefficients β_0, β_1, and β_2 of Equation 3. The results of coefficients β_0, β_1, and β_2 are shown in Table 2. Based on the results in Table 2, a concrete model for the dataset of Vegetable Group can be generated as

$$y = -70.408 + 4.000\frac{1}{x_1^2} + 3.774\frac{1}{x_2^3} \tag{12}$$

Table 1. The sample data of vegetable group

Observation	Support (%)	Confidence (%)	No of real rules
1	20	60	51
2	20	65	42
3	20	70	39
4	23	60	21
5	23	65	17
6	23	70	17
7	25	60	10
8	25	65	8
9	25	70	7

Table 2. Regression Statistics

		Standard Error
β_0	-70.408	6.054
β_1	4.000	0.212
β_2	3.774	1.136
Multiple correlation coefficient(R)	0.991	
Multiple determination coefficient(R^2)	0.983	

From Equation 12, we can figure out that if the value of *support* or *confidence* increases, the number of association rules decrease because the regression model is a none-linear model. The number of association rules can be estimated from Equation 12 if the values of support and confidence for this particular dataset are given.

Step 2: *Evaluating the Regression Model for Group of Vegetable*
Before predicting the number of association rules for Vegetable Group, we evaluate whether the regression model satisfies the standard evaluation (refer to subsection 2.2).
- **Correlation evaluation**
  The value of multiple correlation coefficient R shown in Table 2 was calculated by using Equation 9. The detail calculation of R is shown in Equation 12.

$$R = \sqrt{1 - \frac{\sum (y_i - \hat{y}_i)^2}{\sum (y_i - \bar{y}_i)^2}} = \sqrt{1 - \frac{34.4116}{2124.222}} = 0.991 \qquad (13)$$

$R = 0.991$ means that there is a strong relationship between the number of rules and *support* and *confidence* in the dataset of Vegetable Group.
- **Model test**
  Now we need to carry out test for the significance of the regression model. If the model is satisfied by test, we need to carry out the second test. In other cases, the data sample is considered by increasing the number of observations. The second test is to test individual regression coefficients β_1 and β_2. Generally,

coefficient β_0 needs not to be evaluated because it is a constant. The test data is illustrated as follows

The null hypothesis for the regression model is

$H_0 : \beta_1 = \beta_2 = 0$

$H_1 : (\beta_1 \neq 0)$ or $(\beta_2 \neq 0)$ or $(\beta_1 \neq 0$ and $\beta_2 \neq 0)$

R^2 is 0.983 from Table 2, the statistic to test H_0 is

$$F = \frac{R^2/2}{(1 - R^2)/(n - 3)} = \frac{0.983/2}{(1 - 0.983)/(9 - 3)} = 173.47 \tag{14}$$

The critical value for this test, corresponding to a significance level of 0.05 is $F_{\alpha(2,n-3)} = F_{0.05,2,6} = 5.143$ (refer to subsection 2.2). Since $F > F_{\alpha(2,n-3)}$, H_0 is rejected and it is concluded that at least one coefficient among β_1 and β_2 is significant. In other words, there exists the relationship between the number of association rules and either *support* factor or *confidence* factor or both of them. To consider which factor exists in the regression model, we need to carry out coefficient test.

• **Coefficient test**

Coefficient test is carried out to consider which factors (*support, confidence*) exist in the regression model.

(i) The null hypothesis to test β_1 is

$H_0 : \beta_1 = 0$

$H_1 : \beta_1 \neq 0$

Based on the Table 2, the statistic to test H_0 is $t = \frac{\hat{\beta_1} - \beta_0}{Se(\hat{\beta_1})} = \frac{4-0}{0.212} = 18.79$. The critical values of $t_{(\alpha/2,n-3)} = t_{(0.025,6)}$ and $-t_{(\alpha/2,n-3)} = -t_{(0.025,6)}$ with a significance of 0.05 are 2.968 and -2.968 respectively (refer to subsection 2.2). Since $t = 18.79 > t_{(0.025,6)} = 2.968$, the null hypothesis H_0 is rejected and it states that the number of association rule is effected by support.

(ii) Similarly, considering β_2, it can be seen that $t = \frac{3.744-0}{1.136} = 3.295$ is more than $t_{(\alpha/2,n-3)} = t_{(0.025,6)} = 2.968$. So, the null hypothesis H_0 is rejected and it is concluded that the number of association rule is effected by the confidence factor.

3.3   Prediction Results and Analysis for Vegetable Group

In this Subsection, we use Equation 12 to predict the number of association rules and compare the real number of association rules and the predicted number of association rules. Furthermore, association rule mining is carried out by data mining software for analysis.

• **Prediction results**

After the regression model is generated, we can compare the predicted number of rules with the actual number of rules in the condition of given *support* coefficient and *confidence* coefficient. It means that we replace the pair of values of x_1 and x_2 from Table 1 into Equation 12 to estimate the number of association rules. The result is presented in Table 3.

From the residuals and the rate of error in Table 3, we can see that the difference between the actual numbers of rules and the predicted numbers of rules in 9 different cases is acceptable and at the average error of 8.1%. It means that

Table 3. The predicted result of vegetable group

No	Support (%)	Confidence (%)	No of real rules	No of predicted rules	Residuals	Rate of error(%)
1	20	60	51	47	4	7.8
2	20	65	42	43	1	2.3
3	20	70	39	40	1	2.5
4	23	60	21	22	1	4.7
5	23	65	17	19	2	11.7
6	23	70	17	16	1	5.8
7	25	60	10	11	1	10
8	25	65	8	8	0	0
9	25	70	7	5	2	28.5

the performance of the prediction model is relatively good. Even if the forecast values and actual values are biased, the error is acceptable. Furthermore, the prediction results on the data groups of fruit, milk, meat, seafood, cheese and bread group are also good and a similar level to the result of Vegetable Group.

• **Analysis**

The number of estimated rules with the optimal values of *support* and *confidence* is estimated from the regression model. After we confirm the values of *support* x_1 and *confidence* x_2, the potential number of association rules is obtained for Group of Vegetable by using Equation 12.

For example, if the minimum *support* and *confidence* is 24% and 70%, respectively, the number of predicted association rules for Group of Vegetable coming from Equation 12 is

$$y = -70.408 + 4.000\frac{1}{0.24^2} + 3.774\frac{1}{0.7^3} = 10$$

After estimating the potential number of association rules with the optimal values of *support* and *confidence*, we utilize the optimal values of *support* and *confidence* to effectively find out the detailed association rules by using data mining software such as SPSS, R, Tanagra. The time consumption of association rule mining process is reduced and useful rules can be quickly found. For instance, we choose minimum *support* of 24% and minimum *confidence* of 70% randomly, we conduct the association rule mining by using Tanagra software for Vegetable Group and the results of detailed association rules are represented in Table 4. From Table 4, we can conclude that the level of difference between the actual number of rules and the predicted number of rules is very small because the prediction error between the number of real association rules (8) and the number of estimated association rules (10) is 10 - 8 =2. Our prediction result is 80% correct for this case. Such a prediction can reduce mining time significantly for the process of association rule mining in large datasets because the optimal values of *support* coefficient and *confidence* coefficient are achieved from our regression model. These values can provide useful information to setup best values for *support* and *confidence* coefficients so as to lead an effective mining

Table 4. Detailed Association Rules for Group of Vegetable

Rules	Association rules	Support (%)	Confidence (%)
1	Zucchini⇒ Lecture and Celery	25.49	76.47
2	Lecture and Zucchini⇒ Celery	25.49	81.25
3	Zucchini⇒ Celery	25.49	76.47
4	Celery and Zucchini⇒ Lecture	25.49	100.0
5	Zucchini⇒ Lecture	31.373	94.11
6	Cauliflower and Zucchini ⇒ Lecture	29.412	93.75
7	Zucchini⇒ Cauliflower and Lecture	29.412	88.23
8	Cauliflower and Bokchoy⇒ Brocolli	29.412	75.00

process. Furthermore, the proposed approach has applied to the groups of meat, seafood, milk, cheese, bread and fruit, and the performance of the regression model on this groups are relatively good.

4 Related Work

There has been a lot of previous work on improving the association rule mining. Yi et al [7] studied regression anaysis of the number of association rules. The authors argued that if dataset is large, it is difficult to effectively extract the useful rules. To solve this difficulty, the authors analyzed the meanings of the *support* and *confidence* and designed a variety of equations between the number of rules and the parameters. They only showed the preliminary state and did not provide the way about how to use it in different domains. Our approach is generic and can be instantiate to any domains through the rigorous statistical evaluations. Furthermore, the chosen values of *support* and *confidence* with the suitable number of association rules from the concrete models are used to extract the detailed association rules.

Shaharanee et al. [9] worked toward the combination of data mining and statistical techniques in ascertaining the extracted rules/patterns. The combination of the approaches used in their method showed a number of ways for ascertaining the significant patterns obtained using association rule mining approaches. They used statistical analysis approaches to determine the usefulness of rules obtained from a association rule mining method. The difference between Shaharanee's work and our work is that we focuss on building a generic regression model to predict the number of association rules in general domains and our model can be instantiated in different domains while Shaharanee's work pays more attention to ascertain the useful rules.

Brin et al. [10] developed the notion of mining rules that identifies correlations of rules, and considered both the absence and presence of items as a basis for generating rules. They also proposed a method to measure significance of associations via the Chi-squared test for correlation from classical statistics. Their approach is only suitable for small scale datasets while our approach can be applied on any scales of datasets.

5   Conclusion and Future Work

This paper proposed a new approach to improve the association rule mining process through the prediction of the potential number of association rules on datasets. The proposed approach is novel because the design of the regression model in the approach is generic so it can be applied into broad domains for predicting the potential number of association rules after instantiations to particular domains. The instantiation scheme in our approach is designed and evaluated by two statistical tests in terms of correlation and significance of confidence and support in the model. The case study demonstrates the good performance of our approach on a real life dataset. The future work will pay attention to apply our approach to datasets in other areas such as finance, telecommunication, and so on.

References

1. Fukuda, T., Morimoto, S., Morishita, S., Tokuyama, T.: Mining Optimized Association Rules for Numeric Attributes. In: Proceedings of the Fifteenth ACM Symposium on Principles of Database Systems, pp. 182–191 (1996)
2. Klemettinen, M., Mannila, H., Ronkainen, P., Toivonen, H., Verkamo, A.I.: Finding Interesting Rules from Large Sets of Discovered Association Rules. In: Proceedings of the 3rd International Conference on Information and Knowledge Management, pp. 401–407 (1994)
3. Mannila, H., Toivonen, H., Inkeri Verkamo, A.: Efficient Algorithms for Discovering Association Rules. In: Proceedings of the AAAI Workshop on Knowledge Discovery in Databases, pp. 181–192 (1994)
4. Agrawal, R., Srikant, R.: Fast Algorithms for Mining Association Rules. In: Proceedings of the 20th International Conference on Very Large Data Bases, pp. 487–499 (1994)
5. Shi, L., Jing, X., Xie, X., Yan, J.: Association Rules Applied to Credit Evaluation. In: 2010 International Conference on Computational Intelligence and Software Engineering (CiSE), pp. 1–4 (2010)
6. Li, G., Law, R., Rong, J., Vu, H.Q.: Incorporating both Positive and Negative Association Rules into the Analysis of Outbound Tourism in Hong Kong. Journal of Travel and Tourism Marketing 27, 812–828 (2010)
7. Yi, W.G., Lu, M.Y., Liu, Z.: Regression Analysis of the Number of Association Rules. International Journal of Automation and Computing 8, 78–82 (2011)
8. Han, J., Kamber, M.: Data mining: Concepts and Techniques, 2nd edn. Elsevier Inc. (2006)
9. Shaharanee, I.N.M., Dillon, T.S., Hadzic, F.: Assertaining Data Mining Rules Using Statistical Approaches. In: 2011 International Symposium on Computing, Communication, and Control, vol. 1, pp. 181–189 (2011)
10. Brin, S., Motwani, R., Silverstein, C.: Beyond Market Baskets: Generalizing Association Rules to Correlations. In: Pro. of the ACM SIGMOD Conference on Management of Data, pp. 265–276 (1997)

A Real-Time, Multimodal, and Dimensional Affect Recognition System

Nicole Nielsen Lee, Jocelynn Cu, and Merlin Teodosia Suarez

Center for Empathic Human-Computer Interactions, De La Salle University, Philippines
Nicole_lee@dlsu.ph, {jiji.cu,merlin.suarez}@delasalle.ph

Abstract. This study focuses on the development of a real-time automatic affect recognition system. It adapts a multimodal approach, where affect information taken from two modalities are combined to arrive at an emotion label that is represented in a valence-arousal space. The SEMAINE Database was used to build the affect model. Prosodic and spectral features were used to predict affect from the voice. Temporal templates called Motion History Images (MHI) were used to predict affect from the face. Prediction results from the face and voice models were combined using decision-level fusion. Using support vector machine for regression (SVR), the system was able to correctly identify affect label with a root mean square error (RMSE) of 0.2899 for arousal, and 0.2889 for valence.

Keywords: Emotion models, regression, multimodal, dimensional labels.

1   Introduction

The TALA [4] empathic space is an ubiquitous computing environment capable of automatically identifying its human occupant, model his/her affective state and activities, and provide empathic responses by controlling the ambient settings. One major component of TALA is an automatic affect recognition system (AARS). The AARS should be able to determine the affect of the user, even if only one modality (voice and/or facial expression) is available, and it should be able to do so in real-time (or with minimum amount of delay).

There are several design issues involved in building an AARS and these are: affect database and annotation; input modality; and affect model for automatic prediction.

An affect database is a collection of audio-video (AV) clips showing a subject in an affective state. These expressions can be posed or spontaneous. Posed affect is usually obtained within a controlled laboratory setting, with a subject acting out a specific emotion. It is usually exaggerated and more pronounced. Spontaneous affect, on the other hand, can be obtained by inducing a specific emotion from the subject, or it can be captured in a natural setting. Spontaneous affect is more subtle and varied. Coders annotate each AV clip with information, such as context, transcription of conversation, and affect label among others. The label used to describe the affective state of the subject is very important. There are three approaches in labelling the affective

P. Anthony, M. Ishizuka, and D. Lukose (Eds.): PRICAI 2012, LNAI 7458, pp. 241–249, 2012.

state of the subject and these are: the categorical approach, the dimensional approach, and the appraisal-based approach. The categorical approach believes in the universality of basic emotions. With this approach, affective states are described using either coarse-grained descriptors (e.g., positive-affect, negative affect), fine-grained descriptors (e.g., happiness, sadness, etc.), or a combination of both affect labels. The dimensional approach believes that discrete labels are not enough to describe the wide spectrum of emotions that one can experience and express. Thus, affective states are described using multi-dimensional parameters. An example is Russell's Circumplex Model of Affect, where an affect label is represented using the valence (pleasantness and unpleasantness) and arousal (relaxed or calm) pair. Others use PAD, for pleasure-arousal-dominance; or PAD plus expectation (degree of anticipation) and intensity (state of rationality). The appraisal-based approach believes that affect is the result of continuous, recursive, and subjective evaluation of both a person's own internal state and the state of his/her environment, such that affect is described in terms of a combination of various components. An example of this is the OCC Emotion Model, that describes affect in terms of events, actions and objects.

Affect can be expressed in various modalities. It can be expressed through voice, face, posture and body movement. Voice is a combination of linguistic, acoustic and prosodic cues. In the studies focusing on affective speech like that of [1], it was found that speech features such as pitch, mean intensity, pitch range, short pauses, speech rate, blaring timbre, and high frequency energy are usually associated with high arousal. While a fast speaking rate, large pitch range, less occurrences of high-frequency energy, and low pitch are usually associated with positive valence. The face is the most common modality monitored for affective expression. Approaches such as Ekman and Friesen's Facial Action Coding System (FACS) and Active Appearance Model (AAM) [3] are commonly used to extract facial features that can be used for affect recognition. Another useful modality is the body. Various body movement and postures, either standing or sitting, can give a lot of information regarding the affective state of a person. An important issue one must address in multimodal approach is how to merge the results to arrive at a single prediction. This can be achieve by performing feature-level fusion or decision-level fusion. In feature-level fusion, features extracted from various modalities are merged into a single vector prior to modelling and classification. Decision-level fusion, on the other hand, merge the prediction results after modelling and classification.

Affect recognition is successful if one has a reliable affect model. The type of classifier and machine learning technique used to build an affect model is dependent on the type of features extracted from the user and the annotations used to label the affect. Many studies have attempted to tackle the problem of affect modelling using dimensional labels and audio cues. The study of [16] used Long Short-Term Memory Recurrent Neural Networks (LSTM-RNN) for continuous affect recognition.

There were several studies that tried to address the issue of multimodality and dimensional data, but they may not work in real-time. There are also works that focus on real-time systems, but not necessary work with multimodal and/or dimensional data. EmoVoice [15] is a real-time emotion recognition system that focuses on speech signals. It used Naïve-Bayes and support vector machines (SVM) to classify data into the four quadrants of the valence-arousal space. A similar study done by [6] also focused on speech-based continuous emotion recognition. It uses acoustic Low-Level Descriptors (LLD) and LSTM-RNN to classify data into a 3D valence-activation-time space. The works of [8] and [14] focused on facial features. [8] extracted facial points that were mapped to facial animation parameters (FAPs). These FAPs are then classified using a neurofuzzy rule-based system. [14] worked on elaborating affect in the valence-arousal space using the locally linear embedding algorithm. NetProm [2] is an affect recognition system that can adapt to specific users, along with contextual information. The system uses facial expression most of the time and uses speech prosody or gestures as fall-back solutions. It was evident in these studies that the more successful classifiers that work well with multimodal and dimensional data, are not necessarily applicable for real-time implementation.

This paper aims to provide a methodology for a real-time implementation of automatic affect recognition system that works well with multimodal input and dimensional labels. Section 2 of this paper describes the framework for the real-time, multimodal, dimensional automatic affect recognition system. Section 3 presents and discusses the results of the experiments. Section 4 concludes the paper.

2   Methodology

The entire process of affect recognition can be divided into two stages. The first stage is to prepare the data to build an affect model and the second stage is to predict the affect automatically. This section discusses how these were implemented in this study.

Figure 1 shows the framework for the real-time, multimodal, and dimensional affect recognition system. This is the framework followed by the system when it already has an affect model built inside each predictor. The SEMAINE database [9] was used to build and train the affect model. Prosodic and facial features were extracted from the clips to build the affect model. The prediction of affect is separated into valence and arousal for each modality. Decision-level fusion is used to combine the results to arrive at a single prediction for valence and arousal label.

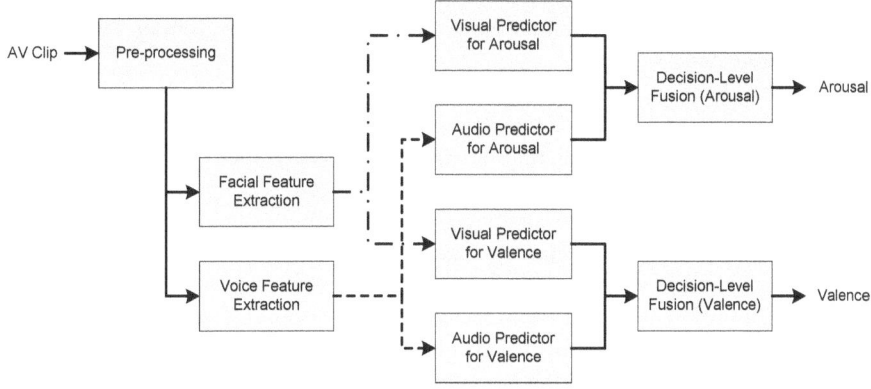

Fig. 1. Framework for the real-time, multimodal, dimensional affect recognition system

2.1   Data Preparation

The SEMAINE database [9] is a collection of AV clips that shows the spontaneous interaction and conversation between a participant and a virtual character with specific personality. The recorded sessions were fully transcribed and annotation in five affective dimensions (i.e., arousal, valence, power, anticipation and intensity), and partially transcribed in 27 other dimensions. This database is used to build the affect model.

Each clip in the database was annotated by 2 to 8 coders, who may or may not have the same perception on the affect displayed by the participant. To combined these annotations and arrive at a single label, an inter-coder agreement is computed. The inter-coder agreement is computed by getting the average correlation per coder [12] using the formula below:

$$cor'_{S,cj} = \frac{1}{|S| - 1} \sum_{i \in S, ci \neq cj} cor(ci, cj) \tag{1}$$

Where S is the relevant session annotated by |S| number of coders, and each coder annotating S is defined as $ci \in S$. This specific approach measures how similar a coder's annotations are with the rest and determined the contribution of each coder to the ground truth. The coder whose annotation achieved the highest average correlation is chosen as the ground truth.

2.2   Feature Extraction

Each AV clip is separated into its audio and video components. Features from the voice are extracted separately from the face.

From the audio clip, a sliding window is used to partition the whole recording into smaller frames. An overlap of 30% between frames is used to ensure signal continuity. From each frame, the following prosodic and spectral features are extracted: pitch (F0), energy, formants F1 – F3, Harmonics-to-noise ratio (HNR), 12 Linear Predictive Coding (LPC) coefficients, and 12 Mel-Frequency Cepstral Coefficients (MFCC). These main features, including their individual statistics such as mean, standard deviation, minimum and maximum values, were all extracted to form an audio feature vector of 76 features.

From the video clip, facial features are extracted in the form of temporal templates. Unlike the usual approach used by similar studies, e.g. [8] [12], which employ tracking the facial points and calculating the distances between these points from the neutral face, the temporal template motion feature provides a compact representation of the whole motion sequence into one image only. The most popular temporal template motion feature is the Motion History Image (MHI). An MHI is the weighted sum of past images, with the weights decaying through time. Therefore, an MHI image contains the past images within itself, in which the most recent image is brighter that past ones [10]. The strength of this approach is the use of a compact, yet descriptive, real-time representation capturing a sequence of motions in a single static image. The

MHI is constructed successively by layering selected image regions over time using a simple update rule [5]:

$$MHI_\delta(x,y) = \begin{cases} \tau, & if \ \Psi\big(I(x,y)\big) \neq 0 \\ 0, & else \ if \ MHI_\delta(x,y) < (\tau - \delta) \end{cases} \tag{2}$$

Where each pixel *(x,y)* in the MHI is marked with a current timestamp if the function signals object presence or motion in the current video image. The remaining time-stamps in the MHI are removed if they are older than the decay value. This study uses the edge orientation histograms (EOH) of MHI based on the [11]. Once MHI is obtained, the edges of the MHI were detected using Sobel operators. After which, the edges, magnitude and orientation of the edges were computed. The orientation interval is then divided into *N*-bins. In this study, 6 bins were used. The EOH is formed by summing the magnitude of the pixels belonging to the corresponding orientation bin. The MHI is then divided into 8x8 cells, with 2x2 cells forming a block. EOH is extracted from each cell and normalized according to the values within a block using the L2-norm block normalization technique. The EOH is then built from an 8-pixel overlapping blocks that are shifted both in the vertical and horizontal directions. Since there are 16 block shifts and 24 EOH in each block, the final feature vector has a dimension of 384.

2.3   Modelling and Classification

Since data is labelled using dimensional labels (i.e., valence-arousal), regression techniques are applied to build the affect model. Algorithms tested for modelling and classification in this research include Linear Regression (LR), Support Vector Machine for Regression (SVR), and Gaussian Process (GP). Results of these various classifiers are compared based on the root-mean square error (RMSE).

Linear Regression (LR) is probably the most widely used and useful statistical technique for prediction problems. It provides the relationship of a dependent variable, y, with observed values of one or more independent variables $x_1...x_n$. LR assumes that there is a linear relationship between two variables, x and y, and that relationship is additive.

Support Vector Machine for Regression (SVR) is under the discriminative paradigm, which tackles the prediction problem by choosing a loss function $g(t,y)$, being an approximation to the misclassification loss $l_{t_y} \leq 0$ and then searching for a discriminant $y(.)$, which minimizes $E_g\big(t_*, y(x_*)\big)$ for the points x_* of interest. Moreover, discriminative models are also called nongenerative models because they focus on selecting a discrimant between classes without the need to specify a probabilistic generative model for the data [13]. While SVMs are mostly used to perform classification by determining the maximum margin separation hyperplane between two classes, SVR attempts the reverse, that is, to find the optimal regression hyperplane such that most training samples lie within an -margin around this hyperplane [7].

Gaussian Process (GP), on the contrary, is a generative Bayesian model that makes use of the concept of Bayesian inference. It is a collection of random variable, indexed by X, such that each joint distribution of infinitely many variables is Gaussian. Such a process $y(.)$ is completely determined by the mean function $x \rightarrow E[y(x)]$ and the covariance kernel $K(x, x') = E[y(x)y(x')]$ [13]. The reliability of a GP is its dependence on the covariance function. Some examples of these are squared exponential covariance function and the radial basis function.

2.4   Decision-Level Data Fusion

Decision-level data fusion is the most common way of integrating asynchronous but temporally corrected modalities. With this method, each modality is first classified independently and the final classification is based on the fusion of the outputs from the different modalities [2]. Various approaches can be used to merge the outputs, such as the sum rule, product rule, weighted-sum rule, max/mix/median rule, majority voting, etc. For this study, the sum rule and weighted-sum rule are used. The latter multiplies a certain weight with the result from each modality and sums them up. It is simple and provides a way for allowing the best modality in specific scenarios to have a bigger impact on the final result.

3   Results and Discussions

40 recorded sessions from the SEMAINE database are used to build the affect model for each modality. For the face, a valence-based and arousal-based affect models are built. Corresponding models are also built for the voice. Data are segmented into various frame sizes (i.e., 1-, 3-, and 5-seconds frame sizes) to identify the optimal size for real-time implementation without sacrificing prediction accuracy. Accuracy of prediction is measured in terms of the root mean square error (RMSE). RMSE quantifies the difference between the values implied by the predictor and the true values of the item being estimated. All results were verified using the 10-fold cross validation approach.

3.1   Affect Recognition Based on a Single Modality

The first part of the experiment is focused on real-time affect recognition of dimensional data using a single modality only. The test separates the results for the voice and the face.

The result of affect recognition based on voice data only is presented in Table 1. Based on experiment results for both valence and arousal voice models, LR yielded the least RMSE, followed by SVR and GP, respectively. Due to hardware constraint, GP which is computationally expensive, was not run for the 1 second frame segmentation, which eliminates it for real-time implementation. The results also showed that the shorter frame length does not necessarily result to a more accurate prediction result. Based Table 1, it was evident that the longer frame length allow the system ample data to make a better prediction, in addition to being able to produce a result with minimum amount of delay. This result is consistent with [12], which states that a 2-6 seconds window or frame length provides better accuracy in prediction. However, contrary to literature, the results showed that valence has a higher prediction accuracy compared to arousal.

Examination of the feature extraction process also yielded the following findings: although MFCCs and HNR are useful and compact features, they take longer time to extract; LPC coefficients, which are also compact and useful features, contain redundant information with formants. However, since MFCC is already removed from the original feature set, formats are removed to retain the LPC coefficients. These moves shortened the overall voice feature set to 34 features and made GP-based affect model the best classifier model.

Table 1. RMSE of voice affect model

Dimension	Algorithm	Without feature selection Frame length			With feature selection Frame length		
		1 sec	3 sec	5 sec	1 sec	3 sec	5 sec
	LR	0.3142	0.2988	**0.2928**	0.3217	0.3069	0.2989
Arousal	SVR	0.3147	0.3009	0.2945	0.3205	0.3063	0.2983
	GP	-	0.2987	0.2942	-	**0.2878**	0.2983
	LR	0.3144	0.2949	**0.2841**	0.3282	0.3114	0.3010
Valence	SVR	0.3137	0.2972	0.2866	0.3243	0.3121	0.3052
	GP	-	0.2954	0.2846	-	**0.2830**	0.2846

Table 2 presented the result for the same experiment using face features. The results show that SVR consistently yielded the least RMSE across different window or frame length. Inversely, unlike the results presented in the voice model, it was evident in the results that smaller window or frame size for face feature analysis is better than using large window or frame size. This is expected since facial features are based on MHI, which show more significant information when analyzed in shorter time frame as opposed to being distorted by noisy data when analyzed in a longer time frame. In contrast to voice, LR performed the worst in this case, while SVR seemed to be the best predictor. Again, based on the results, valence has higher accuracy compared to arousal.

Reducing the number of facial features resulted to a slight increase in prediction error; however, in view of real-time prediction, the increase in error is tolerable in terms of prediction accuracy. From 384 facial features, only 96 features are used for the final set.

Table 2. RMSE of face affect model

Dimension	Algorithm	Without feature selection Frame length			With feature selection Frame length		
		1 sec	3 sec	5 sec	1 sec	3 sec	5 sec
	LR	0.3009	0.3250	0.3400	0.3204	0.3184	0.3259
Arousal	SVR	**0.2970**	0.3134	0.3161	**0.3141**	0.3145	0.3211
	GP	-	0.3218	0.3281	-	0.3219	0.3291
	LR	0.2984	0.3146	0.3321	0.3201	0.3038	0.3222
Valence	SVR	**0.2928**	0.3065	0.2116	0.3125	**0.2942**	0.3153
	GP	-	0.3096	0.3186	-	0.3027	0.3218

3.2   Affect Recognition with Decision-Level Data Fusion

For single modality, the face affect model performs best for short window size while voice affect model performs best for longer window size. To eliminate the issue of synchronization, a 3-second window size is used to analyze both modalities. Based on the comparison results in the previous section, SVR is used to build the face and voice affect models. Compared to LR and GP, SVR is more robust, has excellent generalization characteristic, and has a fast implementation in LIBSVM.

Table 3 shows the result of decision-level fusion to merge the results from both modalities. As expected, RMSE decreased when both modalities are used. For valence, using the sum rule provided the best result. This may be due to the almost equal performance of face (RMSE = 0.2942) and voice (RMSE = 0.3121) in predicting valence values. On the other hand, a weight of 0.6 for voice and 0.4 for face yielded the best result for weighted-sum rule. This is due to the fact that voice (RMSE = 0.3063) is better at predicting arousal than the face (RMSE = 0.3145) does.

Table 3. RMSE results of decision-level fusion

Fusion	Arousal	Valence
Sum Rule	0.2900	0.2884
Weighted (Voice = 0.2)	0.2964	0.2932
Weighted (Voice = 0.4)	0.2911	0.2890
Weighted (Voice = 0.6)	**0.2899**	0.2889
Weighted (Voice = 0.8)	0.2930	0.2928

4   Concluding Remarks

A straightforward approach to real-time automatic affect recognition is presented in this study. The system works well with multimodal inputs and dimensional data. Consistent with other studies, visual features perform better in predicting valence. In the fusion result, audio cues perform better in predicting arousal. Moreover, as the frame length is increased, voice performs better. On the contrary, face performs better in shorter frame length. This is due to the fact that voice rely more on the used-global statistics and face representation in MHI rely on motion details. Even with these differences in analysis requirements, using more than one modality for affect recognition reduces its prediction error. The ability of this system to automatically recognize affect, either from the face and/or voice of a person, with minimum delay makes it a useful system for the TALA empathic space. Timely and reliable prediction of a person's affect allows TALA to provide appropriate empathic support to its occupant. The next step for this project is to integrate with the TALA empathic space and verify how well it performs in actual deployment.

Acknowledgement. This study is supported by the Philippine Department of Science and Technology, under the Philippine Council for Industry, Energy and Emerging Technology Research and Development (PCIEERD), and the Engineering Research and Development for Technology (ERDT) programs. The authors would also like to thank the students and staff of the Center for Empathic Human-Computer Interactions laboratory of De La Salle University College of Computer Studies.

References

1. Calvo, R., D'Mello, S.: Affect detection – an interdisciplinary review of models, methods and their applications. IEEE Trans. on Affective Computing 1(1), 16–37 (2010)
2. Caridakis, G., Karpouzis, Z., Kollias, S.: User and context adaptive neural networks for emotion recognition. Neurocomputing 71, 2553–2562 (2008)
3. Cootes, T.F., Edwards, G.J., Taylor, C.J.: Active Appearance Models. In: Burkhardt, H., Neumann, B. (eds.) ECCV 1998, Part II. LNCS, vol. 1407, p. 484. Springer, Heidelberg (1998)
4. Cu, J., Cabredo, R., Cu, G., Legaspi, R., Inventado, P., Trogo, R., Suarez, M.: The TALA empathic space – integrating affect and activity recognition into a smart space. In: HumanCom (2010)
5. Davis, J.W.: Hierarchical motion history images for recognizing human motion. In: IEEE Workshop on Detection and Recognition of Events in Video, pp. 39–46 (2001)
6. Eyben, F., Wollmer, M., Graves, A., Schuller, B., Douglas-Cowie, E., Cowie, R.: Online emotion recognition in a 3-D activation-valence time continuum using acoustic and linguistic cues. JMUI (2009)
7. Grimm, M., Kroschel, K., Narayanan, S.: Support vector regression for automatic recognition of spontaneous emotions in speech. In: ICASSP, vol. 4, pp. 1085–1088 (2007)
8. Ioannou, S., Raouzaiou, A., Tzouvaras, V., Mailis, T.P., Karpouzis, K., Kollias, S.: Emotion recognition through facial expression analysis based on a neurofuzzy network. Neural Networks 18(4), 423–435 (2005)
9. McKeown, G., Valstar, M., Cowie, R., Pantic, M.: The SEMAINE corpus of emotionally coloured character interactions. In: ICME, pp. 1079–1084 (2010)
10. Meng, H., Pears, N.: Descriptive temporal template features for visual motion recognition. Pattern Recognition Letters 30(12), 1049–1058 (2009)
11. Meng, H., Romera-Paredes, B., Bianchi-Berthouze, N.: Emotion recognition by two-view svm_2k classifier on dynamic facial expression features. In: FG, pp. 854–859 (2011)
12. Nicolaou, M., Gunes, H., Pantic, M.: Continuous prediction of spontaneous affect from multiple cues and modalities in valence-arousal space. IEEE Trans. on Affective Computing 2(2), 92–105 (2011)
13. Seeger, M.: Relationships between Gaussian processes, support vector machines and smoothing splines. Technical Report (1999)
14. Shin, Y.S.: Facial expression recognition of various internal states via manifold learning. J. Computing Science Technology 24(4), 745–752 (2009)
15. Vogt, T., Andre, E., Bee, N.: Emovoice – a framework for online recognition of emotions from voice. In: Workshop on Perception and Interactive Technologies for Speech-Based Systems. Springer (2008)
16. Wollmer, M., Eyben, F., Reiter, S., Schuller, B., Cox, C., Douglas-Cowie, E., Cowie, R.: Abandoning emotion classes – towards continuous emotion recognition with modeling of long-range dependencies. In: INTERSPEECH, pp. 597–600 (2008)

Representing Reach-to-Grasp Trajectories Using Perturbed Goal Motor States

Jeremy Lee-Hand, Tim Neumegen, and Alistair Knott

Dept. of Computer Science, University of Otago, New Zealand

Abstract. In the biological system which controls movements of the hand and arm, there is no clear distinction between movement planning and movement execution: the details of the hand's trajectory towards a target are computed 'online', while the movement is under way. At the same time, human agents can reach for a target object in several discretely different ways, which have their own distinctive trajectories. In this paper we present a method for representing different reach movements to a target without reference to full trajectories: movements are defined through learned perturbations of the hand's ultimate goal motor state, creating distinctive deviations in the hand's trajectory when the movement is under way. We implement the method in a newly developed computational platform for simulating hand/arm actions.

Keywords: reach/grasp actions, touch receptors, reinforcement learning.

1 Introduction: Biological Models of Reaching, and the Problem of Action Representation

In this paper, we consider how human infants learn to reach and grasp target objects in their immediate environment. Performing this task is complex, as the human hand/arm system has many degrees of freedom. For any given peripersonal target, the infant must issue a sequence of commands to the muscle groups of the shoulder, elbow, wrist and fingers which result in her hand achieving a stable grasp on the target. Theorists term the function which computes a sequence of commands the **motor controller**. The research question for us is how infants learn a motor controller which lets them reach and grasp target objects.

Traditionally, control theorists have construed the motor controller as two separate functions: one which plans a trajectory along which the effector must move, and one which computes the appropriate motor commands to issue at each point to cause it to describe this trajectory (see e.g. Jordan and Wolpert, 2000). However, there is a growing consensus that the biological motor controller does not work in this way; in the biological system, it appears that a detailed trajectory is only computed while the movement is actually under way (Cisek, 2005). Before the movement is initiated, the agent only computes a rough trajectory representation, which specifies little more than the basic direction the movement will be in. The main evidence for this suggestion is that the regions of the brain in which reach-to-grasp movements are prepared overlap extensively with those

P. Anthony, M. Ishizuka, and D. Lukose (Eds.): PRICAI 2012, LNAI 7458, pp. 250–261, 2012.

involved in actually executing movements. The preparation of a visually guided reach-to-grasp movement involves activity in a neural pathway running from primary visual cortex and somatosensory cortex through posterior parietal cortex to premotor cortex (see e.g. Burnod *et al.*, 1999). When a prepared action is executed, these same pathways are involved in delivering a real-time signal to primary motor cortex, which sends torque commands to individual joints (see e.g. Cisek *et al.*, 2003; Cisek, 2005), bringing about motor movements. These movements create new reafferent sensory states, which provide updated inputs to the pathway, which in turn result in new motor commands, and so on, in a loop. In this conception of motor control, movements of the arm/hand are described by a complex dynamical system whose components are partly physical and partly neural: the task of the neural motor controller is to set the parameters of this system so that the attractor state it moves towards is one in which the agent's hand achieves the intended goal state. This model of motor control was first posited by Bullock and Grossberg (1988), and is now quite well established (see e.g. Hersch and Billard, 2006).

While there is some consensus that a detailed hand trajectory is not precomputed by the agent, at some level of abstraction human agents are clearly able to represent a range of different trajectories of the hand onto a target. For instance, when confronted with a target object, we can choose different ways to grasp it, which require the hand to approach from different angles. We can also choose to perform actions other than reaching-to-grasp, which have their own idiosyncratic trajectories. For instance, to 'squash' the target our hand must approach it from above, while 'slapping' the target requires an approach from the side; 'snatch', 'punch' and 'stroke' likewise have distinctive trajectories. If we do not compute detailed hand trajectories, how do we represent the discrete alternative grasps afforded by target objects, and how do we represent discretely different transitive actions like 'squash' and 'slap'? In this paper we propose an answer to this question, and investigate its feasibility in some initial experiments.

2   Representing Reach-to-Grasp Actions Using Perturbed Goal Motor States

Any model of reaching-to-grasp needs to make reference to the concept of a **goal motor state**: the state in which the agent has a stable grasp on the target object, and which must function as an attractor for the hand-arm system as a whole. We first envisage the existence of a simple feedback controller, which works to minimise the difference between the current and goal motor states of the hand and arm by generating a force vector in the direction of the difference. This controller defines a 'default' trajectory of the hand towards the target, which will be suboptimal in many ways, but which provides a framework for specifying more sophisticated controllers.

Our main proposal is that on top of the basic feedback controller there is a more abstract motor controller which is able to generate arbitrary *perturbations* of the goal motor state provided to the feedback controller. A given perturbation, applied for a specified amount of time, will cause a characteristic deviation in the hand's trajectory to the target. For instance, if the higher-level controller initially perturbs the goal motor state some distance to the right, the feedback controller will generate a hand trajectory which is deviated to the right.

Of course, appropriate perturbations, or perturbation sequences, must be learned. We suggest that infants learn useful trajectories bringing the hand into contact with a target object by exploring the space of possible perturbations of the goal hand state in which the hand is touching the object, and reinforcing those perturbation sequences which have successful consequences. The model we propose is a learning model, intended to simulate certain aspects of the motor learning done by infants.

Our perturbation model echoes a number of existing proposals. Oztop *et al.* (2004) present a model of infant reach-to-grasp learning in which infants learn 'via-points' for the hand to pass through on its way to the target. Their scheme conforms to the classical motor control model in which trajectories are first planned and then executed; they use a biologically plausible neural network to learn suitable via-points for reaches to a range of object locations, and then use techniques from robotics to compute the kinematics of a reach which brings the hand first to the appropriate via-point and then to the target. Via-points are specified in a motor coordinate system centred on the goal motor state, as our perturbed goal states are. However, in Oztop *et al.*'s model, the agent's hand actually reaches the learned intermediate state on its way to the target; in our model, the learned perturbation pulls the hand's trajectory in a particular direction, but the hand would not typically reach the perturbed goal state. Our model allows less fine-grained control over hand trajectory than Oztop *et al.*'s— of necessity, because it does not allow the precomputation of detailed trajectories.

Another notion of intermediate goal states is implemented in Fagg and Arbib's (1998) model of reaching-to-grasp. This model focusses on the grasp component of the reach, which maps a visually derived representation of the shape of the target onto a goal configuration of the fingers. It is important that the fingers are initially opened wider than this goal configuration, so that the hand can move to its own goal position close to the target. Fagg and Arbib model a circuit within the grasp pathway which operates simultaneously with the reach movement, which first drives the fingers to their maximum aperture, and then brings them into their goal configuration. In our model, we can think of the larger grasp aperture achieved during the reach movement as generated by a learned perturbation of the goal state of the fingers, rather than through a specialised circuit.

The idea that biological motor control involves a combination of a simple feedback controller and a more complex learned controller is fairly uncontroversial (see e.g. Kawato *et al.*, 1987). The learned controller is often modelled as a feedforward controller, which takes the current motor state and the goal motor state at the next moment and returns the command which produces this state. But this way of modelling it assumes a fully precomputed trajectory: since this assumption cannot be sustained, we have to find another way of representing the learned aspect of motor control. Our model provides a way of describing how learning supervenes on basic feedback control which does not depend on the assumption of a precomputed trajectory.

3   A Developmental Methodology

It remains to be seen whether learned perturbations to goal motor states provide a rich and accurate enough way of representing motor movements. To answer this question, we need to propose mechanisms which learn perturbations, and examine their effectiveness. Our approach is to model an agent learning simple perturbations, which support the action of reaching-to-grasp. The agent is assumed to be an infant, learning its earliest reach-to-touch and reach-to-grasp actions. We assume a reinforcement learning paradigm: the infant learns motor behaviours which maximise a reward term, which is derived from tactile sensations (what Oztop *et al.*, 2004 call 'the joy of grasping'). The assumption is that early in infant development, touch sensations are intrinsically rewarding, so that the infant's behaviours are geared towards achieving particular kinds of touch.

Roughly speaking, infants first learn actions which achieve simple touches on target objects, and later learn actions more reliably achieve stable grasps. Before five months of age, infants' fingers do not move as they reach for objects, and hand trajectories (in Cartesian space) are relatively convoluted; at around five months, reaches reliably touch their targets, but grasps are only reliably achieved from around nine months (see e.g. Gordon, 1994; Konczak and Dichgans, 1997).

Our agent learns in two phases, which roughly replicate these two developmental stages. During Stage 1, the agent learns a simple function in the reach sensorimotor pathway, which transforms a retinal representation of target location into a goal motor state. This function is trained whenever the agent achieves a touch sensation anywhere on the hand. In this stage, the agent's movements are generated by a simple feedback controller, with no perturbations; when learning at this stage is complete, the agent can generate an accurate goal motor state for targets presented at a range of retinal locations, and can reliably reach to touch these objects (through somewhat suboptimal trajectories). During Stage 2, the agent learns a function which perturbs the goal motor state generated from vision during the early reach. This function is trained whenever the agent achieves a particular kind of touch—an 'opposition touch' where contact is felt simultaneously on the inner surfaces of the thumb and opposing fingers, or (better still) a stable grasp.

One point to mention about infant motor development is that infants' reach-to-grasp actions become *reliable* some time before they become *straight*. A hallmark of adult point-to-point hand movements is that they are straight in Cartesian space (see e.g. Morasso, 1996). While infant grasps are reliable from around 9 months, the hand trajectories produced during grasping only become straight at around 15 months (see again Konczak and Dichgans, 1997). Until recently, the straightness of adult reach trajectories was taken as evidence that agents explicitly precompute trajectories—and moreover, that they do so using Cartesian (or at least, retinal) coordinates. However, there are ways of generating straight Cartesian trajectories without precomputing them: as Todorov and Jordan (2002) have shown, if reach movements are optimised to minimise error, and movement errors are represented in visual coordinates, this suffices to produce straight reach trajectories. In our current simulations, we want to model the developmental stage when infants reliably reach, but have not yet begun to optimise their reaches using visual error criteria. We will briefly discuss how this optimisation might happen in Section 7.

4   A Platform for Modelling Reach-to-Grasp Actions

We developed a new suite of tools for simulating hand/arm actions to investigate the model of reaching-to-grasp outlined in Section 2. There were two motivations for this. One was a desire to use a general-purpose physics engine to model the physics of the hand/arm and its interactions with the target. Grasp simulation packages often implement special-purpose definitions of 'a stable grasp'; we wanted to define a concept of stable grasp using pre-existing routines for collision detection and force calculation, to ensure that no unrealistic assumptions are built in. The second motivation was a desire to model the tactile system in more detail than most simulation environments allow. Our learning methodology assumes a gradation of tactile reward signals: we wanted to ensure that the signals we use correspond to signals which are obtained by the human touch system. These two motivations are in fact linked; the notion of a stable grasp should make reference to general-purpose physics routines, but must also make reference to the tactile system, because the agent's perception that a stable grasp is achieved occurs mainly through tactile representations.

4.1   Software Components

Our simulation package makes use of a Java game engine called JMonkey (jme2, http://jmonkeyengine.com). JMonkey combines a physics engine called Bullet with the OpenGL graphics environment. The basic unit of representation in JMonkey is the **node**: a rigid body with a defined shape and mass; JMonkey calculates the forces on nodes, and OpenGL renders them graphically. The user can connect nodes together using various types of joint; Bullet uses rigid body dynamics to compute their movements, interpreting joints as constraints on movement. On top of this, our own simulation defines routines for constructing a simple hand/arm system, a simple model of the visual field from a fixed point above the hand/arm, a basic feedback controller which moves the hand/arm from its current state to an arbitrary goal state, a manual controller which allows a user to select goal states, and a collection of neural networks which support motor learning. All these will be described in more detail in the remainder of the paper. The software implementing the physics, graphics and motor controller is available at `http://graspproject.wikispaces.com`.

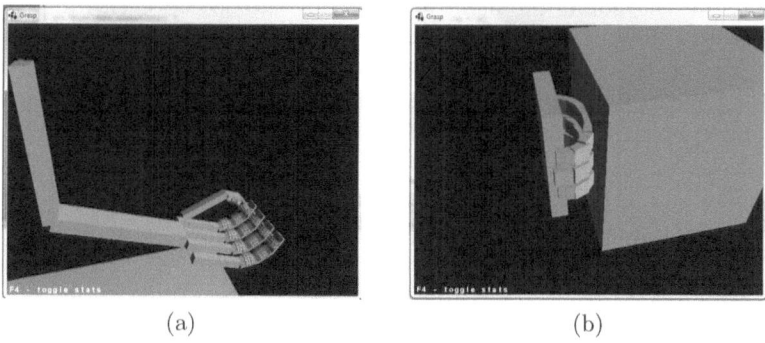

(a)   (b)

Fig. 1. (a) The hand/arm system. (b) detail of a single finger pad.

The hand/arm system we describe in this paper is illustrated in Figure 1a. It has five degrees of freedom: the shoulder can rotate in two dimensions, the elbow can bend, the forearm can twist, and the fingers of the hand can open and close. (A single parameter controls the angles of both thumb joints and all finger joints, so that fingers and thumb are straight when open and curled when closed.)

4.2   A Model of Soft Fingers

An important physical property of fingers is that they are not rigid: they deform if they make (forceful) contact with an object. There are many computational simulations of 'soft fingers'. One common solution is to model fingertips as spheres, and compute the deformations on these spheres which contacting objects would produce, using the result to calculate forces at the fingertips (see e.g. Barbagli *et al.*, 2004). But we implemented an alternative strategy inspired by a model of deformable objects from computer graphics, which represents a deformable object as a lattice of rigid cubes connected by springs (Rivers and James, 2007). This model allows us to obtain fine-grained tactile information from the fingertips: each object in the lattice delivers information about the forces applied to it. The structure of a single finger pad is shown in Figure 1b.

4.3   The Touch System

The inputs to the human touch system come from neurons called **mechanoreceptors** in the top two layers of the skin (the epidermis and the dermis). There are several types of mechanoreceptor: the most relevant ones for grasping are **Meissner's corpuscles** which sense light touches which do not deform the skin, **Pacinian corpuscles** which sense firmertouches which deform the skin surface, **Merkel's cells** which detect the texture of objects slipping over the skin, and **Ruffini endings** which detect stretches in the skin. The latter two detectors combine to provide information about the slippage of objects the agent is attempting to grasp: roughly speaking, a stable grasp is one where no slippage is detected at the points the hand is contacting the object, even when the arm is moved.

Our current implementation of touch sensors simulates Meissner's corpuscles by reading information about contacting objects from each link in each finger pad into an array of real-valued units (see Fig. 2a) and Pacinian corpuscles by reading information about the deviation of each link from its resting position (Fig. 2b). We simulate slip sensors by consulting the physics engine, and computing the relative motion of each fingerpad link with any object contacting it.

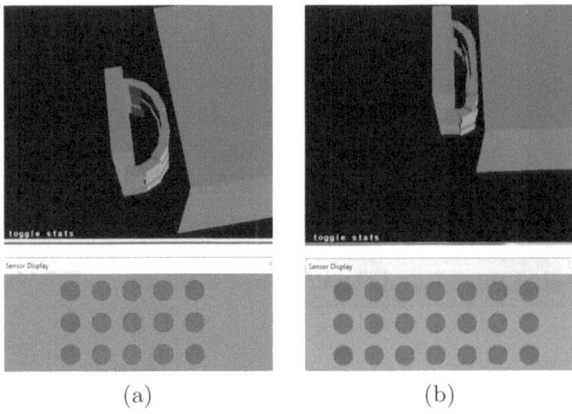

(a)   (b)

Fig. 2. Sensor outputs for a single fingerpad light touch (a) and firm touch (b). Strength of contacting force for links in the pad is shown by the shade of circles; degree of deformation of a link is shown by the shade of a circle's background.

4.4   A Basic Feedback Motor Controller

We assume a hardwired feedback motor controller. The one we implement is a proportional-integral-derivative (**PID**) controller (see Araki, 2006), which computes an angular force vector $u(t)$ based not only on the current difference $e(t)$ between the actual and goal motor states (given in joint angles) but also on the sum of this difference over time, and on its current rate of change.

$$u(t) = K_p e(t) + K_i \int_0^t e(\tau)d\tau + K_d \frac{d}{dt}e(t)$$

The integral term is to ensure the hand does not stop short of the goal state. The derivative term is to slow the hand's acceleration as it approaches the goal state, to minimise overshoot and help suppress oscillations around the goal state. The parameters K_p, K_i and K_d were optimised for a combined measure of reach speed and accuracy on a range of randomly selected target locations.

5   Stage 1: Reaching-to-Touch

The first developmental stage in our simulation involves learning a function which maps a retinal representation of target location onto a goal hand state. In this stage, the fingers of the hand are not moved; a 'goal hand state' is any state in which a touch sensation is generated anywhere on the hand.

The model proceeds through a series of trial reaches, each from the same starting position. In each trial, a horizontally oriented cylindrical target object is presented at a random location in reachable space (on average 40cm away from the hand), and its projection onto a simulated retina is calculated. The centroid of this projection is computed, to provide approximate x and y retinal coordinates of the object, and the physics engine is consulted to provide a retina-centred z (depth) coordinate (which in a real agent would be computed from depth cues, in particular stereopsis). These three coordinates are provided as input to a **reach network**, whose structure and training are described below.

Network Structure. The reach network is a feedforward neural network which takes a 3D retina-centred location as input and produces a 3D goal motor state as output. The network's output layer has three units, which represent goal angles for the two shoulder joints and the elbow joint, encoded using a localist scheme. Its structure is shown in Figure 3a.

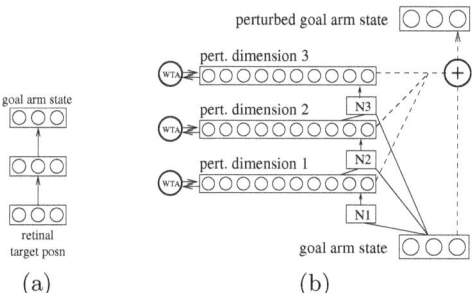

(a)   (b)

Fig. 3. (b) The reach network. (b) The arm perturbation network.

Training. In each trial, the goal arm state computed by the reach network is combined with a noise term, and the resulting vector is given to the feedback controller, which executes a reach to this state. The noise term is initialised to a high value, and gradually annealed over trials, so the arm begins by reaching to random locations, and over time comes to reach accurately to the location computed by the network.

The reach network is trained every time a touch sensation is generated. During each frame of each trial, a touch-based reward value is calculated, which sums touch inputs across all parts of the hand, including the back of the fingers and thumb as well as the pads. Each hand part contributes a touch value, and these values are summed to generate the reward value. The frame with the highest (non-zero) reward value is used to log a training item, pairing the visually-derived location of the target with the current motor state (the joint angles of the elbow and shoulder in that particular frame). The most recently logged training items are retained in a buffer (of size 175 in our experiments); the network is trained on all the items in this buffer after each trial in which a positive reward value is generated. The training algorithm is back-propagation (Rumelhart *et al.*, 1986). Each training item is also tagged with the magnitude of the reward, which is used to set the learning constant for that item, so that more learning happens for higher rewards.

Results. After training for 300 trials using the above scheme, the agent learns to reach for objects at any location quite reliably. It achieves a successful touch on objects at unseen locations in 76% of tests, and in the remainder of tests, gets on average within 3.5cm of the target (SD=1.3cm). A plot showing the areas of motor space where successful touches are achieved is shown in Figure 4a, and a learning curve showing error performance is shown in Figure 4b.

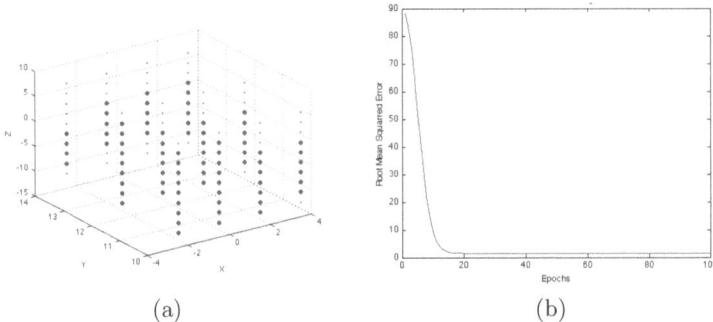

(a)   (b)

Fig. 4. (a) Coverage of the trained reach network, in motor coordinates. (Large dots denote touches, small dots denote misses.) (b) Root mean squared error (in cm) on unseen objects during training.

6   Stage 2: Reaching-to-Grasp

During the second developmental stage, the agent learns to generate a single perturbation to the visually computed goal motor state, which helps the hand achieve a more grasp-like touch on the target, and eventually a reliable stable grasp. The perturbation is computed by a second network, the **arm perturbation network**; it is applied at the start of the reach, and is removed when the hand reaches a certain threshold distance from the target.

Network Structure. The arm perturbation network takes a goal arm state as in-put, and computes a perturbation of this state as output. Its structure is shown in Figure 3b. To allow the representation of multiple alternative perturbations (and therefore multiple alternative trajectories), the network uses population coding to represent perturbations. Each of the three motor dimensions in which perturbations are applied is represented by a bank of units, with Gaussian response profiles tuned to different values in the range of possible perturbations. The network can generate more than one value for a given dimension; in order to return a single value, each bank of units also supports a winner-take-all (WTA) operation, which chooses the most active population in the layer.

As shown in Figure 3b, the arm perturbation network comprises three separate subnetworks, N1–N3, which compute the three dimensions of the perturbation separately, in series. Each subnetwork takes as input the goal motor state, plus the dimensions of the perturbation which have been computed so far. Subnetworks apply the WTA operation before passing their results on, to ensure that decisions in subsequent networks reflect a single selected value. When all three dimensions have been computed, they can be added to the goal arm state. (Note that although perturbations are defined in relation to the goal arm state, it is still important to provide this goal state as input to the network. This is a matter of fine-tuning: the perturbations which should be applied when grasping an object at different points in space will be similar, but not identical.)

Training Regime. As in Stage 1, training happens in a series of reach trials to targets at different locations. In each trial, a target is presented and the reach network generates a goal arm state from visual information; the arm perturbation network then uses this goal state to compute a perturbation. The arm perturbation network's outputs are annealed with noise over the course of Stage 2, so that early trials explore the space of possible perturbations, and later trials exploit learning in the network.

As before, training data for the network is only logged to the training buffer when a tactile reward is obtained. But now rewards have to be more grasp-like touches: either opposition touches, with contact on both the thumb and opposing fingers, or (better still) a fully stable grasp, detected through the hand's slip sensors. Tactile rewards again vary on a continuum, and training items are tagged with their associated reward, so that the learning constant used by the training algorithm (again backprop) can be adjusted to reflect reward magnitudes.

Results. During Stage 2 training, the arm perturbation network learns to produce trajectories which bring the target object into the open hand, since these states are those with the highest tactile rewards. The fingers of the hand are hardwired to close when contact is felt on the fingerpads. After training, the network was presented with objects at a grid of locations unseen during training. Figure 5a shows the perturbations applied for objects at each point in the grid, in motor coordinates. As can be seen, they vary continuously over the reachable space. Figure 5b shows the coverage of the network. There are regions of space where grasps are reliably achieved, but also regions where grasps are not achieved. Note that the reach network trained in Stage 1 continues to learn during Stage 2, so that by the end of this stage the goal arm state computed from vision is a state achieving a stable grasp on the target, rather than just a touch. As the goal arm state changes, so do the optimal perturbations which must be applied to it, so learning happens in parallel in the two networks during Stage 2.

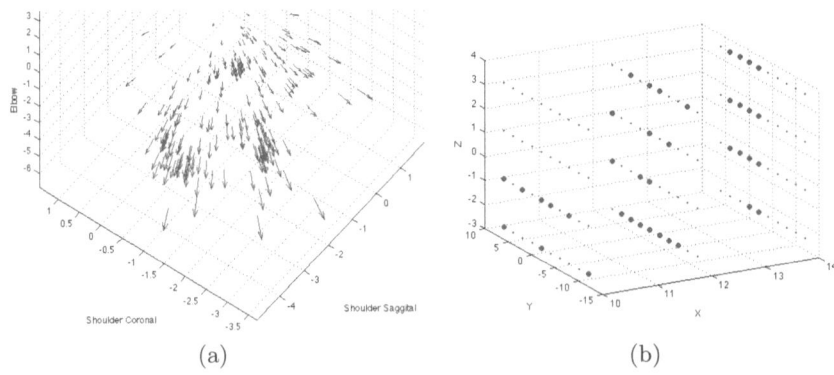

Fig. 5. (a) Perturbations for each point in reachable space (in motor coordinates). (b) Coverage of the trained perturbation network, in motor coordinates. (Large dots denote successful grasps, small dots denote misses.)

7   Conclusions and Further Work

From the experiments reported above, there is some indication that learned goal state perturbations can support the learning of useful hand trajectories. But these results are quite preliminary; there are many questions which remain to be explored. For one thing, our current model does not simulate the grasp component of reaching-to-grasp in any detail. The human grasp visuomotor pathway maps the visually perceived shape of an attended object onto a goal hand motor state; we need to add functionality to Stage 2 to implement this. For another thing, there are gaps in our current network's coverage of reachable space, especially for the Stage 2 network. We need to investigate the interpolation and generalisation potential of the combined networks, to determine if the perturbation model is a practical possibility. We know that population codes are helpful in this regard, but we do not have a good idea of the magnitude of the learning task. Finally, as mentioned at the outset, our main goal is to represent the characteristic trajectories associated with high-level hand/arm motor programmes such as 'squash', 'slap' and 'snatch'. It remains to be seen whether perturbations can be learned which implement high-level motor actions like these.

References

Araki, M.: PID control. In: Unbehauen, H. (ed.) Control Systems, Robotics and Automation, vol. II (2006)

Barbagli, F., Frisoli, A., Salisbury, K., Bergamasco, M.: Simulating human fingers: A soft finger proxy model and algorithm. In: Proceedings of the International Symposium on Haptic Interfaces, pp. 9–17 (2004)

Bullock, D., Grossberg, S.: Neural dynamics of planned arm movements: Emergent invariants and speed-accuracy properties during trajectory formation. Psychological Review 95(1), 49–90 (1988)

Burnod, Y., Baraduc, P., Battaglia-Mayer, A., Guigon, E., Koechlin, E., Ferraina, S., Laquaniti, F., Caminiti, R.: Parieto-frontal coding of reaching: An integrated framework. Experimental Brain Research 129, 325–346 (1999)

Cisek, P.: Neural representations of motor plans, desired trajectories, and controlled objects. Cognitive Processes 6, 15–24 (2005)

Cisek, P., Kalaska, J.: Neural correlates of reaching decisions in dorsal premotor cortex: Specification of multiple direction choices and final selection of action. Neuron 45, 801–814 (2005)

Cisek, P., Cramond, D., Kalaska, J.: Neural activity in primary motor and dorsal premotor cortex in reaching tasks with the contralateral versus ipsilateral arm. Journal of Neurophysiology 89, 922–942 (2003)

Fagg, A., Arbib, M.: Modeling parietal-premotor interactions in primate control of grasping. Neural Networks 11(7/8), 1277–1303 (1998)

Gordon, A.: Development of the reach to grasp movement. In: Bennett, K., Castiello, U. (eds.) Insights into the Reach to Grasp Movement, pp. 37–58. Elsevier, Amsterdam (1994)

Hersch, M., Billard, A.: A biologically-inspired controller for reaching movements. In: Proceedings of the 1st IEEE/RAS-EMBS Intl. Conference on Biomedical Robotics and Biomechatronics, Pisa, pp. 1067–1072 (2006)

Jordan, M., Wolpert, D.: Computational motor control. In: Gazzaniga, M. (ed.) The New Cognitive Neurosciences, pp. 71–118. MIT Press (2000)

Kawato, M., Furawaka, K., Suzuki, R.: A hierarchical neural network model for the control and learning of voluntary movements. Biological Cybernetics 56, 1–17 (1987)

Konczak, J., Dichgans, J.: The development toward stereotypic arm kinematics during reaching in the first 3 years of life. Experimental Brain Research 117(2), 346–354 (1997)

Morasso, P.: Spatial control of arm movements. Experimental Brain Research 42, 223–227 (1996)

Oztop, E., Bradley, N., Arbib, M.: Infant grasp learning: a computational model. Experimental Brain Research 158, 480–503 (2004)

Rivers, A., James, D.: FastLSM: Fast lattice shape matching for robust real-time deformation. ACM Trans. Graphics 26(3), Article 82 (2007)

Rumelhart, D.E., Hinton, G.E., Williams, R.J.: Learning internal representations by error propagation. In: PDP vol. 1: Foundations, pp. 318–362. MIT Press, Cambridge (1986)

Todorov, E., Jordan, M.: Optimal feedback control as a theory of motor coordination. Nature Neuroscience 5, 1226–1235 (2002)

TSX: A Novel Symbolic Representation for Financial Time Series

Guiling Li[1], Liping Zhang[1], and Linquan Yang[2]

[1] School of Computer Science, China University of Geosciences, Wuhan, China
guiling@cug.edu.cn, carolynzhang321@gmail.com
[2] Faculty of Information Engineering, China University of Geosciences, Wuhan, China
yanglinquan@gmail.com

Abstract. Existing symbolic approaches for time series suffer from the flaw of missing important trend feature, especially in financial area. To solve this problem, we present Trend-based Symbolic approximation (TSX), based on Symbolic Aggregate approximation (SAX). First, utilize Piecewise Aggregate Approximation (PAA) approach to reduce dimensionality and discretize the mean value of each segment by SAX. Second, extract trend feature in each segment by recognizing key points. Then, design multiresolution symbolic mapping rules to discretize trend information into symbols. Experimental results show that, compared with traditional symbol approach, our approach not only represents the key feature of time series, but also supports the similarity search effectively and has lower false positives rate.

Keywords: Data mining, Time series, Dimensionality reduction, Symbolic.

1 Introduction

Time series is a sequence of data changing with time order, which is increasingly important in many domains, such as nature science, engineering technology and social economics. Daily close prices of the stock are typical time series data. Approximation representation is a basic problem in time series data mining. It can support different mining tasks such as similarity search, clustering, classification, etc. Approximation representation should consider the following: (1) effectiveness on dimensionality reduction; (2) good effect of feature preservation; (3) effective support on time series data mining tasks. In recent years, there have emerged numerous approximation representation methods by extracting useful patterns from time series. Among them, symbolic representation is a popular approximation approach, due to its simplicity, efficiency and discretization. However, many methods miss important trend feature, which is important in some time series data, especially in financial time series.

Symbolic Aggregate approximation (SAX) [1] is a classical symbolic approach for time series. Since SAX is put forward, it attracts many researchers' interest in data mining community. SAX is based on Piecewise Aggregate Approximation (PAA) [2]. As PAA realizes dimensionality reduction by extracting segment mean value feature, SAX only reflects the segment mean value feature and misses the important trend

P. Anthony, M. Ishizuka, and D. Lukose (Eds.): PRICAI 2012, LNAI 7458, pp. 262–273, 2012.

feature. We present a novel symbolic representation, which can not only reflect the segment mean value feature, but also capture trend feature with good resolution, further support time series data mining tasks.

Our work can be simply summarized as follows:

- We extract the trend feature from time series, according to the Most Peak point (MP) and the Most Dip point (MD). Then we design multiresolution discretization method to transform the trend feature into symbols.
- We propose a novel symbolic dimensionality reduction approach, and call it as Trend-based Symbolic approximation (TSX). TSX extracts both mean value and trend information of segments in time series. Further, we put forward distance function TSX_DIST based on TSX.
- We demonstrate the effectiveness of the proposed approach by experiments on real financial datasets. The experiments validate the utility of our proposed approach.

The rest of the paper is organized as follows. In section 2 the related work is discussed. In section 3 we briefly review SAX. Section 4 presents our proposed TSX approach for financial time series. Section 5 is the experimental evaluation. The last section summarizes our work and points out the future work.

2   Related Work

Agrawal et al. put forward Discrete Fourier Transform (DFT), which is the pioneering work on approximation representation for time series [3]. Later, numerous researches on dimensionality reduction techniques for time series have spawned. Keogh summarizes the main techniques of time series representation in a hierarchy [4]. There are four categories in the hierarchy, i.e., model based, data adaptive, non-data adaptive and data dictated, respectively. Symbolic representation belongs to data adaptive category and has become popular, because such representation not only has the merits of simplicity, efficiency and discretization, but also allows researchers to utilize the prolific data structures and algorithms from text processing and bioinformatics communities [5]. Due to these merits, many researchers have proposed different symbolic methods.

Agrawal et al. define eight symbols according to segment trends, and define Shape Definition Language (SDL) to implement shape based query [6]. However, the query depends on SDL and does not have the generality, and the scalability of symbolization is not enough. Xia segments hierarchically trend interval (-90°, 90°), then defines Shape Definition Table and Shape Definition Hierarchy Tree to perform subsequence search and whole sequence search [7]. This method does not standardize the symbol sets in hierarchy tree.

SAX is the first proposed symbolic approach which allows dimensionality reduction and supports lower bounding distance measure [1]. Based on PAA, SAX discretizes the PAA segments into symbols according to breakpoint intervals with Gaussian distribution. When SAX is put forward, it is very popular in time series data mining community. Various kinds of work emerge flourishingly based on SAX and apply SAX into pattern extraction and visualization. SAX is simple and efficient, but it misses some important patterns.

To solve the problem of missing extreme points in SAX, Extend SAX (ESAX) adds two new values in each segment based on SAX, i.e., max value and min value, respectively [8]. ESAX captures more information than SAX from extreme point of view, without the view of segment trend. Combing SAX and Piecewise Linear Approximation (PLA) can make up the flaw of SAX [9], but PLA is only the post-processing step and this method does not symbolize the trends of PLA. Multiresolution symbolic approach, indexable Symbolic Aggregate approximation (iSAX), represents time series by string with 0 and 1, which is the extension of SAX [10]. Adaptive SAX (aSAX) and indexable adaptive SAX (iaSAX) improve the requirement of Gaussian distribution in SAX and iSAX, they adopt training method and k-means clustering algorithm to determine adaptively breakpoints intervals [11], but they still do not capture the trend feature.

Although various symbolic approaches have been proposed in the past decades, the symbolization on trend feature is still an open problem and has not been solved. Due to the classic and popularity of SAX, we focus on extracting trend feature and discretizing them into symbols in time series data based on SAX.

3   Background: SAX

Since our work is based on SAX, first we briefly review SAX. For more details, interested readers are recommended to [1]. The notations used in this paper are listed in Table 1.

Table 1. A summarization of notation used in this paper

Notation	Description
T	raw time series of length n, $T=\{t_1,t_2,...,t_n\}$
T'	normalized time series, $T'=\{t_1',t_2',...,t_n'\}$
$\overline{T}$	PAA representation, $\overline{T}=\{\overline{t_1},\overline{t_2},...,\overline{t_w}\}$
$\hat{T}$	SAX representation, $\hat{T}=\{\hat{t_1},\hat{t_2},...,\hat{t_w}\}$
$\tilde{T}$	TSX representation, $\tilde{T}=\{\tilde{t_1},\tilde{t_2}...,\tilde{t_w}\}$
w	the length of $\overline{T}$
a	alphabet cardinality of SAX(eg. for {**a,b,c,d**}, a=4)
t	alphabet cardinality of trend(eg. for {**a,b,c,d**}, t=4)

Given a raw time series T of length n, the SAX of T consists of the following steps:

Step 1: Normalization. Transform raw time series T into normalized time series T' with mean of 0 and standard deviation of 1.

Step 2: Dimensionality reduction via PAA. Represent normalized series T' by PAA approach and obtain $\overline{T}$ of length w with compression ratio n/w.

Step 3: Discretization. According to SAX breakpoints search table, choose alphabet cardinality, discretize $\overline{T}$ into symbols and obtain SAX representation $\hat{T}$.

Breakpoints in Table 2 are an ordered numerical set, denoted by $B=\beta_{1,\ldots,}\ \beta_{a-1}$. In N(0,1) Gaussian curve, the probabilities of region from β_i to β_{i+1} are equal with $1/a$, where $\beta_0=-\infty$, $\beta_a=+\infty$. Comparing the segment mean value of $\overline{T}$ with breakpoints, if the segment mean value is smaller than the smallest breakpoint β_1, the segment is mapped to symbol '**a**'; if the segment mean value is larger than the smallest break-point β_1 and smaller than β_2, the segment is mapped to symbol '**b**'; and so on. An example of SAX is illustrated as Fig. 1.

Table 2. SAX breakpoints search table with Gaussian distribution

β_i	SAX alphabet cardinality a										
	2	3	4	5	6	7	8	9	10	11	12
β_1	0.00	-0/43	-0.67	-0.84	-0.97	-1.07	-1.15	-1.22	-1.28	-1.34	-1.38
β_2		0.43	0	-0.25	-0.43	-0.57	-0.67	-0.76	-0.84	-0.91	-0.97
β_3			0.67	0.25	0.00	-0.18	-0.32	-0.43	-0.52	-0.6	-0.67
β_4				0.84	0.43	0.18	0	-0.14	-0.25	-0.35	-0.43
β_5					0.97	0.57	0.32	0.14	0	-0.11	-0.21
β_6						1.07	0.67	0.43	0.25	0.11	0
β_7							1.15	0.76	0.52	0.35	0.21
β_8								1.22	0.84	0.6	0.43
β_9									1.28	0.91	0.67
β_{10}										1.34	0.97
β_{11}											1.38

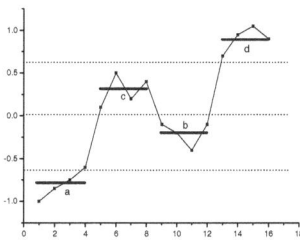

Fig. 1. A time series is discretized by first obtaining PAA representation and then using SAX breakpoints to map the PAA segments to symbols. Here, $n=16$, $w=4$ and $a=4$, the time series is mapped to the word **acbd**.

Suppose two raw time series $S=\{s_1, s_2,\ldots, s_n\}$ and $T=\{t_1, t_2,\ldots, t_n\}$, the correspond-ing SAX are $\hat{S}$ and $\hat{T}$. In order to measure the similarity based on SAX representa-tion, MINDIST is defined as Equation (1), where $dist()$ function can be calculated by search table in Table 3.

$$MINDIST(\hat{S},\hat{T}) = \sqrt{\frac{n}{w}}\sqrt{\sum_{i=1}^{w}(dist(\hat{s}_i,\hat{t}_i))^2} \tag{1}$$

Table 3. *dist*() search table used by *MINDIST*() (cardinality *a*=4)

	a	b	c	d
a	0	0	0.67	1.34
b	0	0	0	0.67
c	0.67	0	0	0
d	1.34	0.67	0	0

In Table 3, the distance between two symbols can be obtained by checking the corresponding row and column number in *dist*() search table. For example, *dist*(**a**, **b**)=0, *dist*(**a**, **c**)=0.67. If the cardinality is arbitrary, the *cell* in dist() search table can be calculated by Equation (2).

$$cell_{r,c} = \begin{cases} 0 & if \mid r-c \mid \leq 1 \\ \beta_{\max(r,c)-1} - \beta_{\min(r,c)} & otherwise \end{cases} \qquad (2)$$

Since SAX extracts the mean value information, and misses the trend information, Consequently, MINDIST only measures the similarity of mean value of segments through dimensionality reduction, it can not evaluate the trend similarity.

4   Proposed TSX Approach

In financial time series, tread feature is important. Financial traders pay much attention to typical peak and dip feature in financial series. Moreover, both long-term and short-term analysis should consider multiresolution.

The main idea of our proposed TSX approach is to first reduce dimensionality by PAA approach, then segment each PAA subsegment by trend feature, thus extract both mean and trend information, subsequently discretize mean and trend information into symbols respectively.

4.1   Trend Feature Extraction

To extract the trend information of segments, we need to detect and collect the peak points and dip points.

Definition 1. *Trend Line*: The line by connecting the start and the end of segment is called as trend line.

Definition 2. *The Most Peak point (MP)*: The point which is above the trend line and has the largest distance to the trend line is called as the Most Peak point.

Definition 3. *The Most Dip point (MD)*: The point which is below the trend line and has the largest distance to the trend line is called as the Most Dip point.

There are different methods to measure distance, such as Euclidean Distance (ED), Perpendicular Distance (PD), Vertical Distance (VD) [12]. Suppose p_1 is the start of segment and p_3 is the end of segment, we get the trend line by connecting p_1 and p_3. Three distances from point p_2 to the trend line are shown as Fig. 2. Because Vertical Distance can get better result in catching highly fluctuant points, we choose VD as distance measure. We evaluate VD from the points to trend line and determine whether the points are MP or MD.

The MP and MD in the segment are shown in Fig.3(a). By identifying MP and MD, there are four key points in the segment, the start, MP, MD and the end, respectively. According to the occurred order, connecting the four key points can extract three trend segments and thus obtain the slopes of three trend segments. However, not all occasions are identical. Affected by compression ratio parameter of PAA approach, not all segments have both MP and MD. If the segment only has MP or MD, it should be recognized and treated individually, the strategy we adopt is scaling processing. If the segment only has MD as Fig.3(b), now there are only three key points, the start point, MD, the end point. First connect three points successively and obtain two trend segments and slopes, then compare the length of two trend segments and bisect the longer one into two segments with the same trend, thus get three main trend slopes. If the segment only has MP as Fig.3(c), we utilize the similar strategy. The processing effects corresponding to three occasions in Fig.3 are shown as Fig.4.

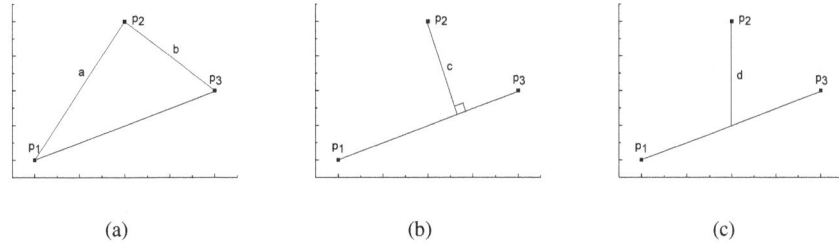

Fig. 2. Three distances from p_2 to the trend line p_1p_3 (a) ED is a+b (b) PD is c (c) VD is d

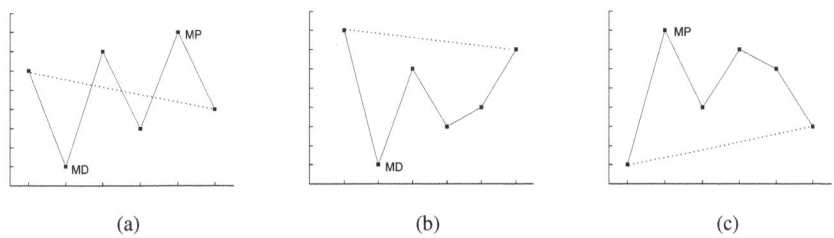

Fig. 3. Three occasions of MP and MD in segment (a) MP and MD both exist (b) only MD (c)only MP

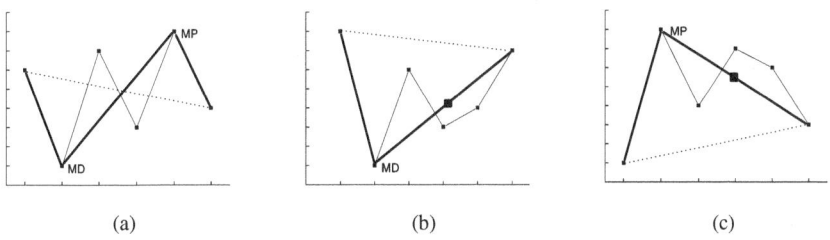

Fig. 4. Three trend segments (a) MP and MD both exist (b) trend scaling when only MD exists (c) trend scaling when only MP exists

4.2   Symbolic Representation

We design multiresolution angle breakpoint search table to symbolize the trend slopes. The angle space corresponding to the slopes of trend segment is (-90°, 90°), which can be divided into certain numbers of nonoverlapping intervals, each interval corresponds to a symbol. The trends of segment include three kinds: the falling trend, the horizontal trend and the rising trend. The falling trend can further comprise slow falling, fast falling and sharp falling; the rising trend can further comprise slow rising, fast rising and sharp rising. Based on the extent of trend change from judging whether it is high or low and whether it is slow or sharp, design multiresolution angle breakpoint intervals search table to map trends to symbols, which is shown as Table 4.

Definition 4. *Angle Breakpoints (AB)*: Angle breakpoints are sorted list of angles, denoted by $AB=\theta_{1,...,}\ \theta_{t-1}$. Suppose each interval from θ_i to θ_{i+1} has the equal probability of $1/t$, $\theta_0= -\infty$, $\theta_t= +\infty$.

Table 4. Angle breakpoint interval search table

θ_i	trend alphabet cardinality t										
	2	3	4	5	6	7	8	9	10	11	12
θ_1	0°	-5°	-30°	-30°	-30°	-45°	-45°	-60°	-60°	-75°	-75°
θ_2		5°	0°	-5°	-5°	-30°	-30°	-45°	-45°	-60°	-60°
θ_3			30°	5°	0°	-5°	-5°	-30°	-30°	-45°	-45°
θ_4				30°	5°	5°	0°	-5°	-5°	-30°	-30°
θ_5					30°	30°	5°	5°	0°	-5°	-5°
θ_6						45°	30°	30°	5°	5°	0°
θ_7							45°	45°	30°	30°	5°
θ_8								60°	45°	45°	30°
θ_9									60°	60°	45°
θ_{10}										75°	60°
θ_{11}											75°

The corresponding relationship between trend symbol and angle is determined by search table shown as Table 4, for each slope value in the segment, use symbol to represent trend feature according to Table 4. Comparing the angle reflected by slope with a series of angle breakpoints, if the angle of trend segment is smaller than the smallest angle breakpoint θ_1, this trend segment is mapped to symbol '**a**'; if the angle of trend segment is larger than the smallest angle breakpoint θ_1 and smaller than θ_2, this trend segment is mapped to symbol '**b**'; and so on.

We discretize the mean value of each segment to symbol according to the SAX breakpoint search table in Table 2, and discretize the slopes of three trend segments to symbols according to the angle breakpoint search table in Table 4, thus get the trend symbolic tuple of segment.

Definition 5. *Trend Symbolic Tuple of Segment*: After dimensionality reduction via PAA, We discretize the mean and three trend segment slopes in the ith segment to symbols, thus get the trend symbolic representation of the segment, which is denoted by a symbol tuple $\tilde{t}_i =< \tilde{t}_i^1, \tilde{t}_i^2, \tilde{t}_i^3, \tilde{t}_i^4 >$. In this tuple, $\tilde{t}_i^1$ is corresponding to the mean of the ith segment and symbolized by SAX approach; $\tilde{t}_i^2, \tilde{t}_i^3, \tilde{t}_i^4$ are the symbols which are corresponding to the three slopes in the ith segment and discretized based on search table in Table 4.

Definition 6. *Trend Symbolic Representation of Raw Time Series*: Given a raw time series T, its trend symbolic representation is denoted by $\tilde{T} = \{\tilde{t}_1, \tilde{t}_2 ..., \tilde{t}_w\}$, where $\tilde{t}_i$ stands for the ith trend symbolic tuple of segment as definition 5.

Hence, our proposed Trend-based Symbolic approXimation(TSX) approach includes the following steps:

Step 1: Normalization. Transform the raw time series T into the normalized time series T' with mean of 0 and standard deviation of 1.

Step 2: Use PAA approach to reduce dimensionality on T', get the mean of each segment; then break up each segment into three trend segments based on trend feature and extract the slopes of three trend segments.

Step 3: Choose SAX alphabet cardinality a and discretize each mean value in the segment into symbol by Table 2; Choose trend alphabet cardinality t and discretize three slopes in the segment to symbols by Table 4; thus obtain the trend symbolic representation of the segment; and finally obtain Trend Symbolic Representation of raw time series, denoted by $\tilde{T}$.

For example, given the time series T of length 16 in Fig.1, suppose compression ratio of PAA is 8, both a and t are 4, then the TSX of T is $\tilde{T} = \{\mathbf{a,c,c,b,c,b,c,b}\}$.

4.3   Distance Function

TSX reflects two parts of connotation, one is the mean information, and the other is the trend information. Distance between two symbols reflecting mean information can be calculated by *dist*() search table as Table 3 and Equation (2); distance between two symbols reflecting trend information can be calculated by *trendist*() as Table 5 and Equation (3).

By checking the corresponding row and column of two symbols in Table 5, the distance between trend symbols can be obtained. For example, *trendist*(**a,b**)=0, *trendist*(**a,c**)=tan25°, *trendist*(**a,d**)=tan35°. When trend cardinality t is arbitrary, the *cell'* in *trendist*() search table can be calculated by Equation (3).

Table 5. trendist() search table (t=5)

	a	b	c	d	e
a	0	0	tan25°	tan35°	tan60°
b	0	0	0	tan10°	tan35°
c	tan25°	0	0	0	tan25°
d	tan35°	tan10°	0	0	0
e	tan60°	tan35°	tan25°	0	0

$$cell'_{r,c} = \begin{cases} 0 & if \ |r-c| \le 1 \\ \tan(\theta_{\max(r,c)-1} - \theta_{\min(r,c)}) & otherwise \end{cases} \tag{3}$$

For economic traders, trend features are very important in financial time series. Therefore, to evaluate the similarity of TSX representation, the distance between trend symbols should have priority weight.

Suppose there are two raw time series, $S=\{s_1, s_2, \cdots, s_n\}$, $T=\{t_1, t_2, \cdots, t_n\}$, the length of PAA is w, $\tilde{S} = \{\tilde{s}_1, \tilde{s}_2 \dots, \tilde{s}_w\}$ and $\tilde{T} = \{\tilde{t}_1, \tilde{t}_2 \dots, \tilde{t}_w\}$ denote their TSX, where $\tilde{s}_i = <\tilde{s}_i^1, \tilde{s}_i^2, \tilde{s}_i^3, \tilde{s}_i^4>$, $\tilde{t}_i = <\tilde{t}_i^1, \tilde{t}_i^2, \tilde{t}_i^3, \tilde{t}_i^4>$. The distance function TSX_DIST based on TSX is defined as Equation (4).

$$TSX\_DIST(\tilde{S}, \tilde{T}) = \sqrt{\frac{n}{w} \sum_{i=1}^{w} (dist(\tilde{s}_i^1, \tilde{t}_i^1))^2 + (\frac{n}{w})^2 \sum_{i=1}^{w} \sum_{j=2}^{4} (trendist(\tilde{s}_i^j, \tilde{t}_i^j))^2} \tag{4}$$

When compression ratio n/w increases, the measure of trend similarity will be more important and significant, the factor before $trendist()$ will be larger, so TSX_DIST can achieve better measure effect.

5   Experimental Results and Evaluation

5.1   Experimental Settings

We demonstrate the utility of TSX with a comprehensive set of experiments. For these experiments, we use notebook with Pentium Dual 1.73GHz CPU, 1GB RAM, 120GB disk space. The source code is written in C++ language.

The experimental data include four economic datasets, i.e., EMC, Ford, Morgan Stanley and Pepsico [13]. The corresponding data sizes are 5653, 8677, 4096 and 8676, respectively. Our goal is to evaluate our TSX approach and show its improvement on supporting similarity search compared with SAX. We perform subsequence similarity search on datasets by brute force algorithm. We measure the effect of different approximation approaches by false positives rate as follows:

$$False\_Positives\_Rate = \frac{|R| - |R \cap C|}{|R|} \times 100\% \tag{5}$$

where R is the subsequence result set reported by embedding the representation and distance function into similarity search algorithm, C is the real correct subsequence result set.

5.2   Comparison of False Positives Rate

As both SAX and TSX are dimensionality reduction techniques for time series, we first explore how false positives rate changes with compression ratio. The comparison of false positives rate of two methods is shown in Fig.5. As the result indicates, when compression ratio changes, the false positives rate based on TSX is lower than that based on SAX.

SAX and TSX are both multiresolution symbolic representation, another common parameter is SAX alphabet cardinality. We observe how false positives rate changes with SAX alphabet cardinality. The comparison is shown in Fig.6. With SAX alphabet cardinality changes, the false positives rate based on TSX is lower than that based on SAX.

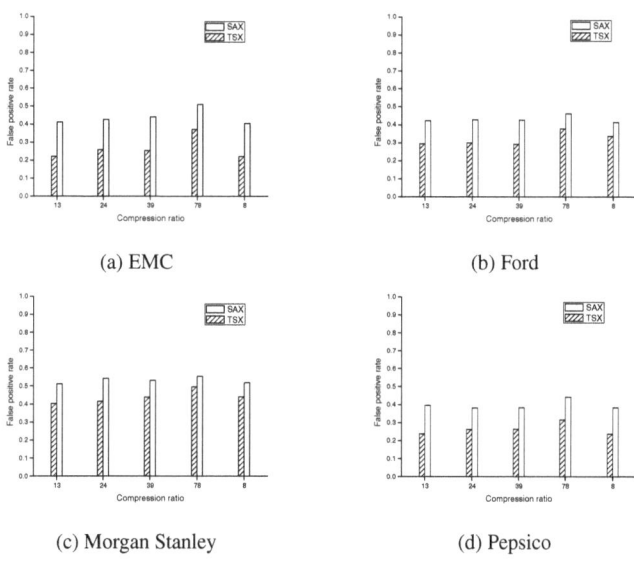

(a) EMC

(b) Ford

(c) Morgan Stanley

(d) Pepsico

Fig. 5. Comparison of false positives rate changing with compression ratio

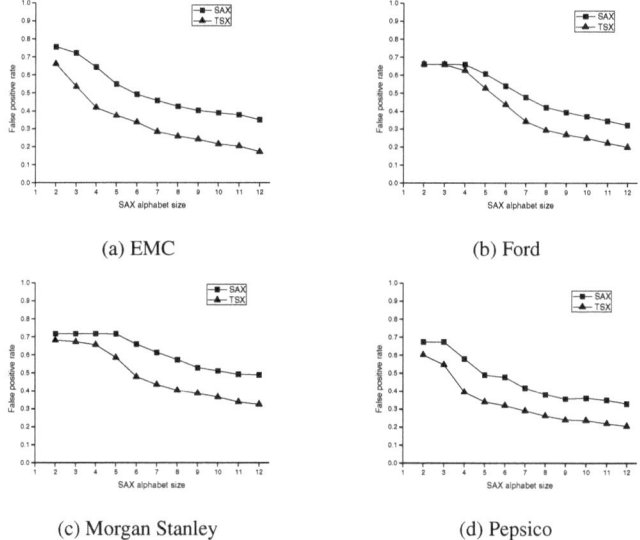

(a) EMC

(b) Ford

(c) Morgan Stanley

(d) Pepsico

Fig. 6. Comparison of false positives changing with SAX alphabet cardinality

5.3   Parameter Analysis

TSX belongs to multiresolution symbolic representation for time series, the effect of multiresolution is influenced by both SAX alphabet cardinality and trend alphabet cardinality. We investigate how these two parameters affect the similarity search. The result is shown in Fig.7. From Fig.7, we can observe, with two cardinalities increase synchronously, the total change trend of false positives rate is to reduce gradually.

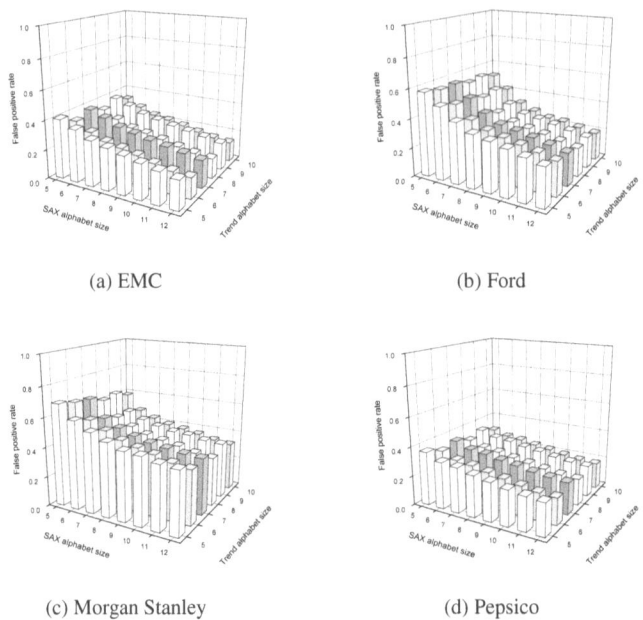

(a) EMC   (b) Ford

(c) Morgan Stanley   (d) Pepsico

Fig. 7. The effect how two cardinalities of TSX influence false positives rate

6   Conclusions

In this paper, we propose a novel symbolic representation for financial time series, namely, Trend-based Symbolic approximation (TSX). This symbolic approximation extracts both the mean information and trend feature, achieves multiresolution symbolization by designing angle breakpoints interval search table and symbolization mapping rule. Compared with classic SAX approach, the experimental results validate our approach is meaningful and effective. For further research, we will apply our TSX approach to discord discovery in time series to further explore its applicability.

Acknowledgements. The research is supported by the Fundamental Research Funds for the Central Universities, China University of Geosciences (Wuhan) under grants No. CUGL120281 and CUGL100243. The authors thank anonymous reviewers for their very useful comments and suggestions.

References

1. Lin, J., Keogh, E., Lonardi, S., et al.: A Symbolic Representation of Time Series, with Implications for Streaming Algorithms. In: Mohammed, J.Z., Charu, A. (eds.) Proc. of the 8th ACM SIGMOD Workshop on Research Issues in Data Mining and Knowledge Discovery, San Diego, CA, USA, pp. 2–11. ACM Press, New York (2003)
2. Keogh, E., Chakrabarti, K., Pazzani, M., et al.: Dimensionality Reduction for Fast Similarity Search in Large Time Series Databases. Knowledge and Information Systems 3(3), 263–286 (2000)
3. Agrawal, R., Faloutsos, C., Swami, A.: Efficient Similarity Search in Sequence Databases. In: Lomet, D.B. (ed.) FODO 1993. LNCS, vol. 730, pp. 69–84. Springer, Heidelberg (1993)
4. Keogh, E.: A Decade of Progress in Indexing and Mining Large Time Series Databases. In: VLDB, p. 1268 (2006)
5. Lin, J., Keogh, E., Wei, L., et al.: Experiencing SAX: A Novel Symbolic Representation of Time Series. In: Data Mining Knowledge Discovery, pp. 107–144 (2007)
6. Agrawal, R., Psaila, G., Wimmers, E.L., et al.: Querying Shapes of Histories. In: Umeshwar, D., Peter, G.M.D., Shojiro, N. (eds.) Proc. of the 21st Int'l Conf. on VLDB, Zurich, Switzerland, pp. 502–514. Morgan Kaufmann Press (1995)
7. Xia, B.B.: Similarity Search in Time Series Data Sets (Master thesis). Simon Fraser University (1997)
8. Lkhagva, B., Suzuki, Y., Kawagoe, K.: New Time Series Data Representation ESAX for Financial Applications. In: Liu, L., Reuter, A., Whang, K., et al. (eds.) Proc. of the 22nd Int'l Conf. on Data Engineering Workshops, Atlanta, Georgia, USA, pp. 17–22. IEEE Computer Society, Washington, DC (2006)
9. Huang, N.Q.V., Anh, D.T.: Combining SAX and Piecewise Linear Approximation to Improve Similarity Search on Financial Time Series. In: Int'l Symposium on Information Technology Convergence, pp. 58–62. IEEE Computer Society, Washington, DC (2007)
10. Shieh, J., Keogh, E.: iSAX: Indexing and Mining Terabyte Sized Time Series. In: Li, Y., Liu, B., Sarawagi, S. (eds.) Proc. of ACM SIGKDD, Las Vegas, Nevada, USA, pp. 623–631. ACM Press, New York (2008)
11. Pham, N.D., Le, Q.L., Dang, T.K.: Two Novel Adaptive Symbolic Representations for Similarity Search in Time Series Databases. In: Han, W., Srivastava, D., Yu, G., et al. (eds.) Proc. of the 12th Int'l Asia-Pacific Web Conference, Busan, Korea, pp. 181–187. IEEE Computer Society, Washington, DC (2010)
12. Fu, T., Chung, F., Luk, R., et al.: Representing Financial Time Series Based on Data Point Importance. Engineering Application of Artificial Intelligence 21(2), 277–300 (2008)
13. http://finance.yahoo.com

Genetic-Optimized Classifier Ensemble for Cortisol Salivary Measurement Mapping to Electrocardiogram Features for Stress Evaluation

Chu Kiong Loo[1], Soon Fatt Cheong[2], Margaret A. Seldon[2], Ali Afzalian Mand[2], Kalaiarasi Sonai Muthu[2], Wei Shiung Liew[1], and Einly Lim[3]

[1] Faculty of Computer Science and Information Technology
University of Malaya
Kuala Lumpur, Malaysia
[2] Faculty of Information Science and Technology
Multimedia University
Melaka, Malaysia
[3] Faculty of Biomedical Engineering
University of Malaya
Kuala Lumpur, Malaysia
{ckloo.um,einly_lim}@um.edu.my,
{sfcheong,margaret.seldon,ali.afzalian,kalaiarasi}@mmu.edu.my,
liew.wei.shiung@gmail.com

Abstract. This work presents our findings to map salivary cortisol measurements to electrocardiogram (ECG) features to create a physiological stress identification system. An experiment modelled on the Trier Social Stress Test (TSST) was used to simulate stress and control conditions, whereby salivary measurements and ECG measurements were obtained from student volunteers. The salivary measurements of stress biomarkers were used as objective stress measures to assign a three-class labelling (Low-Medium-High stress) to the extracted ECG features. The labelled features were then used for training and classification using a genetic-ordered ARTMAP with probabilistic voting for analysis on the efficacy of the ECG features used for physiological stress recognition. The ECG features include time-domain features of the heart rate variability and the ECG signal, and frequency-domain analysis of specific frequency bands related to the autonomic nervous activity. The resulting classification method scored approximately 60-69% success rate for predicting the three stress classes.

Keywords: stress classification, genetic algorithm, ARTMAP, electrocardiogram, salivary cortisol.

1   Introduction

Stress is a general term encompassing biological and psychological impact of external influences on human physiology. Repeated and chronic exposure to high stress levels can cause adverse health effects such as hypertension, heart disease, and depression. It

P. Anthony, M. Ishizuka, and D. Lukose (Eds.): PRICAI 2012, LNAI 7458, pp. 274–284, 2012.
© Springer-Verlag Berlin Heidelberg 2012

has been reported that workload stress has become a major contributor towards productivity loss and medical costs [1] in recent times, especially in highly competitive and/or high risk work environments.

Continuous and accurate monitoring of stress, whether at work or at home may allow timely intervention for stress mitigation. Several existing studies have attempted to discriminate induced stress from a resting state from physiological signals under laboratory conditions. A study by Soga et.al [2] induced mental workload using a mental arithmetic task to investigate the recovery patterns of physiological responses as indicators of stress. Nater et.al [3] discovered marked differences in alpha-amylase, cortisol, and heart rate activity before and after stressors were applied via the Trier Social Stress Test (TSST) as compared with a similar control condition.

TSST was a well-known lab stressor for inducing mental stress through psychosocial means. A meta-analysis of over 200 stress studies [14] suggested that by including a social evaluative threat, unpredictability, and by using a combination of speech tasks and cognitive tasks with the presence of evaluative observation, the most robust stress response could be achieved. Therefore for this study, TSST was chosen for its inherent combination of each of these factors and offers a comprehensive activation of various psychological stress systems.

Stress in the human body is largely regulated by the hypothalamus-pituitary-adrenal (HPA) axis and the autonomic nervous system (ANS). Both systems interact to maintain homeostasis of the human body in response to applied stressors. Cortisol is one of the biomarkers tied to the HPA axis [15] and can be reliably measured from salivary excretions with minimal invasive procedures. The TSST procedure was designed to elicit stress reactions and to measure stress indirectly through cortisol measurements for the HPA axis, and ECG measurements for the ANS. This study was conceived as an exercise to reconcile these two measures of stress by using an ARTMAP pattern learning and classification system. The features extracted from the ECG recording was labeled using the cortisol measurements as an objective stress classification, and used for training the ARTMAP system. The trained system will then act as a substitute for salivary cortisol stress measurement using only ECG inputs to predict the corresponding cortisol-based stress classification. If successful, the system can be used in conjunction with only an ECG sensor to replace the salivary cortisol stress assessment method.

The Augsburg Biosignal Toolbox [4] was used for reducing the dimensionality of the ECG recording into time-domain and frequency-domain features. The salivary cortisol measurements were divided into one of three stress classifications (Low, Medium, or High Stress). An ARTMAP pattern learning and classification system was used to map the ECG features to the stress classes, with the viability of the method determined by the resultant ARTMAP's ability to correctly predict the stress class given an unlabeled ECG feature input. A genetic permutation method was employed to optimize the training sequence of the ARTMAP. Multiple ARTMAPs were trained and their prediction outputs used for a probabilistic voting strategy to determine the best stress classification. In comparison, several other pattern classification methods will be used as benchmarks to the genetic-optimized ARTMAP system.

The following section details the experimental procedure used for data collection of salivary and ECG data. Section 3 explains our data processing and analysis methods, as well as the algorithm used for stress classification. Section 4 shows the results of our analyses, and Section 5 discusses our experimental findings.

2   Data Collection

2.1   Mobile Electrocardiogram Measurement

Measurement and recording of the ECG was performed using g.Tec's portable bio-signal acquisition device, g.Mobilab [19]. Three bipolar silver-chloride electrodes were used in a modified lead-2 configuration: an electrode was attached to the inside of both wrists, and the ground electrode attached to one ankle. The ECG data was recorded digitally on the host computer using g.Tec's proprietary software with 256Hz sampling rate.

2.2   Trier Social Stress Test

A total of 22 subjects participated in the data collection experiment, consisting of 17 male and 5 female, ranging from age 19 to 24 years old. Each participant was screened for exclusion criteria to decrease the likelihood of potential factors that might affect the results of the experiment. Criteria include not being on medication or drug use, non-smoking, and have had no strenuous physical activity, meals, alcohol or low pH soft drinks for at least two hours before testing.

Each subject was required to undergo one session for stress induction and one session for control (placebo session), each session on a different day. Each subject was randomly started with either control or stress session to eliminate confounding factors associated with habituation. The purpose of this arrangement was to create two distinct sets of salivary measurements and ECG recordings for each subject in order to discriminate between a stress state and a resting state. The double-track study design will help determine the baseline value of stress biomarkers of the subject during the placebo session, with the stressor replaced by routine activity. The experimental design was based partially on the works of Kirschbaum et al. [7], Cinaz et al. [8], and Nater et al. [3].

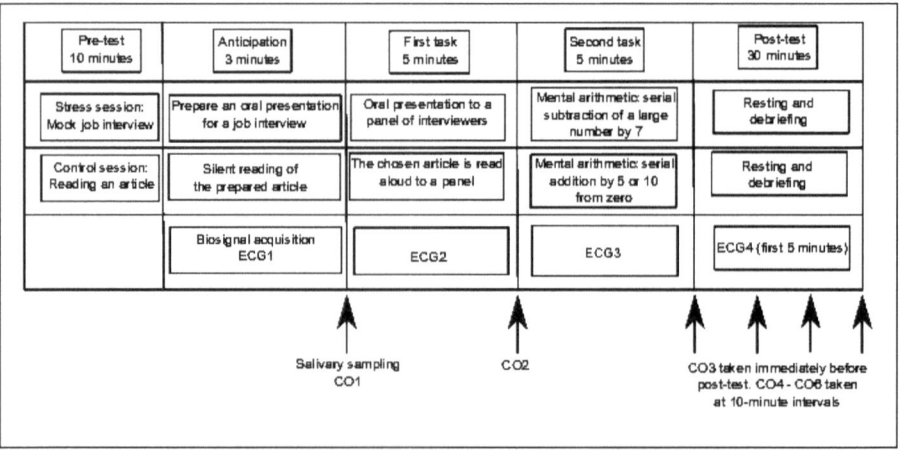

Fig. 1. Overview of the TSST experiment design

The experiment consisted of five parts for each session:

- Pre-test. During this period, ECG sensors were attached to the subject while the examiner summarized the experiment's methodology for that particular session.
- Anticipation. The subject was given three minutes to prepare talking points for the mock job interview in the case of the stress session, or reading a chosen article in the case of the control session. For the entire duration, the subject's heart rate was recorded (ECG1). At the end of the duration, the subject was given a saliva swab to obtain the first salivary measurement (CO1).
- First task. For the stress session, the subject was required to present their qualifications in a mock job interview. A panel of two examiners perform the roles of interviewers by interrogating the subject with standard job interview questions. For the control session, the subject was only required to read aloud the chosen article with no feedback from the interview panel. ECG2 was recorded for the entire duration, and CO2 was obtained at the end of the duration.
- Second task. For the stress session, the subject was instructed to serially subtract 7 from an arbitrarily chosen number as quickly and as accurately as possible. Whenever a mistake was made or when the subject slowed down, they were asked to start again with a new starting number. For the control session, the subject was required to serially add 5 or 10 from zero. ECG3 was recorded for the entire duration, and CO3 was obtained at the end of the duration.
- Post-test. ECG4 was recorded for the first five minutes while saliva swabbing was performed at ten-minute intervals (CO4-CO6). During this time, the subject was given a visual analog scale (VAS) questionnaire to assess the relative discomfort levels of the current session.

2.3   Biochemical Assay for Cortisol Measurement

Measurements for salivary cortisol were performed using Salivary Immunoassay Kits (Salimetric Inc. USA) [20] using the standard guidelines outlined in the manual provided by the manufacturer. Optical density reading was performed using an automated microtitre plate reader. Each salivary sample was measured twice and the mean value was reported. The microtitre plate spectrophotometer used for the reading was the Powerwave X5 (Biotek, USA) used along with the GEN5 Microplate Data Collection and Analysis Software [21].

3   Data Analysis

3.1   Feature Extraction

The experiment yielded a total of 176 ECG recordings and 264 salivary cortisol measurements. Unfortunately, 7 of the salivary measurements were omitted from analysis due to insufficient saliva volume. For analysis of the ECG data, Wagner's Augsburg Biosignal Toolbox [4] was used to extract numerical features representative of the signal as a whole. Features include time-domain analysis of the PQRST-wave and HRV features, and frequency-domain features such as power spectrum densities of specific frequency bands. Other features were added to complement the existing feature set, such as the sympathetic and parasympathetic activity derived using principal dynamic modes (PDM) method [5] [6] from HRV analysis.

3.2   Data Pre-processing

Analysis of variance was performed to determine the statistical difference of saliva measurements between control session and stress session (F=18.08, p=3.04 x 10^{-5}) and among different times (F=2.605, p=0.026). Based on the statistical analyses, the baseline for the Low Stress class was determined by taking the value at the longest time plus one standard deviation of the normalized cortisol measurement, while the lower cut-off for the High Stress class was taken as the average value plus one standard deviation. The lower and upper bounds for each of the three stress classes were thus derived as:

$$0.00 <= \text{Low Stress} < 0.35 <= \text{Medium Stress} < 0.53 <= \text{High Stress} <= 1.00$$

As this experiment involved physiological measurements of electrical activity and biochemical activity related to stress, it would be informative to analyse the effect of time delay between the two measures. To that end, a number of label sets were created to assign lagged saliva measurements to ECG recordings, as demonstrated in Table 1. A total of three sets of labels and their corresponding modified data sets were created from the originals.

3.3   Genetic Ordering for ARTMAP Training Optimization

The effectiveness of the ARTMAP classification system is highly dependent on the presentation sequence of training data. Thus, a genetic permutation algorithm was proposed to efficiently search for an optimal training sequence in order for the ARTMAP to yield the best possible classification accuracy. The genetic permutation algorithm as used in this experiment is a search heuristic mimicking natural evolution to rearrange the genetic sequence to find an optimal solution for the selected fitness function. The algorithm used was based on Haupt and Haupt's continuous genetic algorithm [22] for minimizing a cost function, adapted for predictive error in this experiment.

Table 1. Assigning time-delayed salivary measurements to corresponding ECG recording

$\text{Label}_{t=t+0}$		$\text{Label}_{t=t+1}$		$\text{Label}_{t=t+2}$	
ECG1	CO1	ECG1	CO2	ECG1	CO3
ECG2	CO2	ECG2	CO3	ECG2	CO4
ECG3	CO3	ECG3	CO4	ECG3	CO5
ECG4	CO4	ECG4	CO5	ECG4	CO6

The steps involved were:

3.3.1   Population Generation

A large population of candidate training sequences was generated random-ly. Each candidate, or chromosome, consisted of a number of genes each representing a single unique sequence in which the entire ECG data set will be presented to the ARTMAP for training. In this case, the chromosomes contain 176 genes to be optimally resequenced.

3.3.2   Fitness Testing and Selection

Each chromosome was fitness tested by training an ARTMAP system using 5-fold cross-validation. Fitness of each chromosome was computed as the percentage of correctly classified patterns. The top 50% of the fittest solu-tions were retained for the next generation's population, while the rest were discarded.

3.3.3   Reproduction and Mutation

The survivors were paired at random to generate new offspring to replace the discarded candidates. Common genetic traits shared by the parent chromosomes were carried over to the offspring, while uncommon traits were randomized, as demonstrated in Figure 2. A mutation operator was also introduced to limit genetic convergence among the generated offspring.

3.3.4   Iteration

Fitness testing, selection, reproduction, and mutation were iterated succes-sively for the initial population of candidate solutions over twenty genera-tions. At the end of the optimization exercise, the population ideally consists of candidate solutions which have a higher average fitness than the initial population. The training sequences which yielded the best predictive accuracy from the fitness test were then selected to determine the sequence of the training data presentation to the Biased ARTMAP.

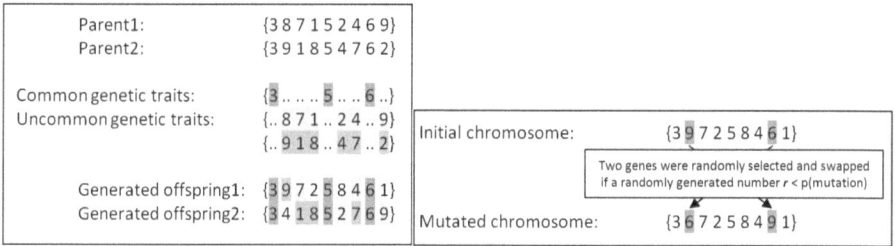

Fig. 2. Hereditary reproduction and simple mutation methods, using 9-gene chromosomes as example

3.4   Biased ARTmap

Adaptive resonance theory (ART) was developed as a theory of human cognitive information processing [9], the design principles of which, led to the development of real-time neural network models that perform supervised and unsupervised learning, pattern recognition and prediction. The ARTMAP model [10] is a hierarchical network architecture that can organize stable categorical mappings between M-dimensional input vectors and N-dimensional output vectors. An ARTMAP-based system was used as the system's pattern learning and classification system due to its fast and stable ability to incorporate new learning data without repeating the entire training process. The prototype stress identification system will be trained using an existing database of stress data, while the ARTMAP-based system allows for continuous improvement of the system's stress classification ability.

Under online fast learning conditions, the ARTMAP's critical features attention may be distorted by certain sequences of input presentation, causing less-suitable critical features to be overemphasized in future searching procedures. The Biased ARTMAP variant [11] introduces a new medium-term memory that would enable the network to selectively bias input features whenever a predictive error was generated. Whenever a predictive error was generated for a given input, the features which were currently active in the ARTMAP memory will be modified by a biasing vector in the medium-term memory for all future predictions. The strength of the biasing was controlled by an attention parameter λ, where the optimal value of λ can be determined by validation.

The Biased ARTMAP system was chosen for our base classifier for its suitability for fast and stable online learning. This experiment was designed as a proof-of-concept system for offline training of the Genetic Ensemble Biased ARTMAP system, which then can be implemented as a basic ECG-based stress predictor. Its main strengths will be its ability to learn and adapt to each user's unique physiological pattern on an individual basis, thus improving its stress prediction capability over time.

3.5   Probabilistic Ensemble Voting

The probabilistic ensemble voting strategy used here was based on research by Lin et.al [12] and Loo and Rao [13]. The probabilistic voting strategy is essentially a series of classifiers each with its own weights, classifying a pattern recognition task with multiple classes, each with a single constant representing the class's *a priori* probability. The probabilistic voting decision employed the Bayes' rule to select the class with the largest *a posteriori* probability.

The probabilistic ensemble voting strategy also used a method for enforcing a measure of reliability, which is the probability of a voting decision to be the correct decision given the voting results of all voting members. This allowed predictions with conflicting outcomes from each of the voting members to be set aside for expert review. The user-defined reliability threshold can be set to zero to implement a simple majority probabilistic voting decision, or set to one to reject all predictions that are not unanimous.

4   Experimental Results

A total of three labelled data sets were used for training and testing using the Genetic Ensemble Biased ARTMAP system. Twenty of the best training sequences were generated from the genetic optimization process, ordered by predictive accuracy. The first N training sequences were used to train the first N number of Biased ARTMAP voters. The trained ARTMAPs were then used to create a voting ensemble, and to derive the final class prediction through probabilistic voting. Each method was rated by the percentage of correctly identified data samples over the total number of tested data samples using leave-one-out training and testing strategy.

In addition, each data set was evaluated, with and without genetic optimization, using bootstrapped mean of 1000 resampling with 95% confidence interval. The data set was randomly ordered five times, and the randomized data sets were used for training a five-voter Biased ARTMAP. This was repeated for each value of attention parameter from 0 to 10. The resultant eleven results were used to generate a bootstrapped mean, to represent the data set's classification using Biased ARTMAP without genetic ordering. For the genetic-ordered bootstrapped mean, all training sequences generated by the genetic permutation algorithm for the data set were used.

As a comparison of methods, other classification methods were used for pattern classification using the same data sets. The methods used were linear discriminant analysis (LDA), K^{th} nearest neighbour (KNN), and multilayer perceptron (MLP), all of which were obtained via Wagner's Augsburg Biosignal Toolbox [4]. For KNN, the number of nearest-neighbours was set to 10% of the data set's total number of samples. For MLP, the initialized network was created with three hidden layers, learning rate set to 0.2, and 50 learning iterations. The results were the percentage of correct stress class prediction using leave-one-out method of training and testing.

Table 2. Comparison of pattern classification and feature selection methods

	Cortisol$_{t=t+2}$			
	Entire data set	ANOVA	Hamming	Conical corr.
LDA	27.01	42.08	38.74	30.82
kNN	38.07	41.86	36.66	37.86
MLP	36.14	37.79	41.97	44.83
Bootstrapped mean: no genetic optimization	48.23	47.05	47.05	47.05
Bootstrapped mean: genetic optimized	63.11	63.80	62.17	62.17
1-voter genetic Biased ARTMAP	68.82	69.41	69.41	69.41
3-voters	65.88	66.47	64.70	64.70
5-voters	63.52	65.29	64.11	64.11
7-voters	63.52	65.29	62.94	62.94
10-voters	62.94	64.70	61.76	61.76

Of the three data sets, only **Cortisol**$_{t=t+2}$ showed a number of features with significantly low p-values (20 features with p-value at least 0.05 or lower). The dataset **Cortisol**$_{t=t+2}$ was tested again using ANOVA feature reduction to first reduce the size of the data set from 106 features to 20 of the most correlated features.

Two alternative methods of feature selection was performed on the **Cortisol**$_{t=t+2}$ data set, using Hamming distance calculation and canonical correlation analysis. For Hamming distance calculations, all ECG features and the cortisol measurements were normalized to [0, 1] range. Then, the Hamming distance for each feature was calculated as the difference of values between the ECG feature and its assigned cortisol measurement, calculated as the mean across all data samples. The features with the shortest Hamming distance were selected for further analysis.

Canonical correlation analysis is an analysis method for cross-covariance matrices. The method is able to compute the linear combinations of features between two feature vectors that possess maximum correlations with each other. For this study, canonical correlation analysis was performed to derive the best combinations of ECG features and their respective cortisol measurements. Features with the best correlations were selected for further analysis.

5   Discussion

The analysis of results from the experiment yielded several interesting conclusions. Firstly, the low classification accuracy indicates an ineffective mapping between ECG features and salivary measurements. This may be caused by any number of reasons, such as flaws in the experimental data selection design, or stress class thresholding, or the effectiveness of the selected ECG features for stress classification. A more viable method for correlating salivary biomarkers to cardiovascular activity may be derived by using HRV features which are more relevant to stress studies.

This study introduces an innovation in comparison with previous stress recognition studies by attempting to verify objectively the stress outcome using measurements of body biomarker changes. In comparison, other studies [16] [17] [18] base their stress classification on the stressor time stamp or using a subjective self-report verification. The use of biomarkers in stress evaluation was common in behavioural science study. However, it was not very common among technology researchers due to the need for specialized protocol and equipment to obtain the measurements.

Biomarker measurements as quantitative data facilitates the multi-level classification of subject stress strength. This ensures that the stress recognition ability is realistic and precise in relation to the individual stress response, due to variations in periods of transient stress response for each individual throughout the day. The subjective nature of self-assessment and intensity of the stressor cannot be a reliable quantitative surrogate of stress response. A more realistic model should be able to perform multiple-level analysis or continuous measure of stress. Thus, objective measurement of stress-based biomarker will be a viable model for researchers who wish to design a pragmatic system that can be used in real environments.

Results of the experiment were not satisfactory, as multiple analyses show insufficient correlation to map the selected ECG features to the biomarker measurements. A hypothesis is that physiological responses and body chemistry responses to applied stressors vary from subject to subject, making it less straightforward for a simple ECG-to-biomarker mapping process. This speculation may be taken into account for future studies in this area.

The overall Genetic Ensemble Biased ARTMAP method was shown to be effective in classifying the data sets as compared with other pattern classification methods. An alternative voter selection method is required, as the current method degrades the system's classification accuracy when additional voters are added. A potential solution is to expand the genetic optimization to include voter selection.

Acknowledgement. This study was funded by the Ministry of Higher Education (MOHE) Malaysia under the Fundamental Research Grant Scheme for project code FRGS-1-10-TK-MMU-03-10.

References

1. Kiecolt-Glaser, J.K., McGuire, L., Robles, T.F., Glaser, R.: Psychoneuroimmunology: Psychological Influences on Immune Function and Health. Journal on Consulting and Clinical Psychology 70(3), 537–547 (2002)
2. Soga, C., Miyake, S., Wada, C.: Recovery patterns in the physiological responses of the autonomic nervous system induced by mental workload. In: Annual Conference of the Society of Instrument and Control Engineers (SICE), pp. 1366–1371 (2007)
3. Nater, U.M., Rohleder, N., Gaab, J., Berger, S., Jud, A., Kirschbaum, C., Ehlert, U.: Human salivary alpha-amylase reactivity in a psychosocial stress paradigm. International Journal of Psychophysiology 55, 333–342 (2005)
4. Wagner, J.: The Augsburg Biosignal Toolbox (2009), http://www.informatik.uni-augsburg.de/en/chairs/hcm/projects/aubt/ (retrieved June 29, 2011)
5. Zhong, Y., Wang, H., Ju, K.H., Jan, K.M., Chon, K.H.: Nonlinear analysis of the separate contributions of autonomic nervous systems to heart rate variability us-ing principal dynamic modes. IEEE Transactions on Biomedical Engineering 51(2), 255–262 (2004)
6. Choi, J., Gutierrez-Osuna, R.: Using heart rate monitors to detect mental stress. In: 6th International Workshop on Wearable and Implantable Body Sensor Networks, pp. 219–223 (2009)
7. Kirschbaum, C., Pirke, K.M., Hellhammer, D.H.: The 'Trier Social Stress Test' – A tool for investigating psychobiological stress responses in a laboratory setting. Neuropsychobiology 28, 76–81 (1993)
8. Cinaz, B., Arnrich, B., La Marca, R., Troster, G.: Monitoring of mental workload levels during an everyday life office-work scenario. Personal and Ubiquitous Computing Journal (2001), doi:10.1007/s00779-011-0466-1 (retrieved December 2011)
9. Grossberg, S.: Adaptive pattern classification and universal recoding; 1: Parallel development and coding of neural feature detectors; 2: Feedback, expectation, olfaction, and illusions. Biological Cybernetics 23, 121–134, 187–202 (1976)
10. Carpenter, G.A., Grossberg, S.: ARTMAP: a self-organizing neural network archi-tecture for fast supervised learning and pattern recognition. Neural Networks 4(5), 565–588 (1991)

11. Carpenter, G.A., Gaddam, S.C.: Biased ART: a neural architecture that shifts at-tention toward previously disregarded features following an incorrect prediction. Neural Networks 23(3), 435–451 (2010)
12. Lin, X., Yacoub, S., Burns, J., Simske, S.: Performance analysis of pattern classifier combination by plurality voting. Pattern Recognition Letters 24, 1959–1969 (2003)
13. Loo, C.K., Rao, M.V.C.: Accurate and reliable diagnosis and classification using probabilistic ensemble simplified fuzzy ARTMAP. IEEE Transactions on Knowledge and Data Engineering 17(11), 1589–1593 (2005)
14. Dickerson, S.S., Kemeny, M.E.: Acute stressors and cortisol responses: a theoreti-cal integration and synthesis of laboratory research. Psychological Bulletin 130(3), 355–391 (2004)
15. Piazza, J.R., Almeida, D.M., Dmitrieva, N.O., Klein, L.C.: Frontiers in the use of biomarkers of health in research on stress and aging. Journal of Gerontology: Psychological Sciences, 513–525 (2010)
16. Sun, F.T., Kuo, C., Cheng, H.T., Buthpitiya, S., Collins, P., Griss, M.L.: Activity-aware mental stress detection using physiological sensors. Silicon Valley Campus (2010), http://repository.cmu.edu/silicon_valley/23/ (retrieved February 1, 2012)
17. Healey, J., Nachman, L., Subramanian, S., Shahabdeen, J., Morris, M.: Out of the Lab and into the Fray: Towards Modeling Emotion in Everyday Life. In: Floréen, P., Krüger, A., Spasojevic, M. (eds.) Pervasive Computing. LNCS, vol. 6030, pp. 156–173. Springer, Heidelberg (2010), doi:10.1007/978-3-642-12654-3_10 (retrieved July 5, 2011)
18. van den Broek, E.L., Lisy, V., Janssen, J.H., Westerink, J.H.D.M., Schut, M.H., Tuinenbreijer, K.: Affective man-machine interface: Unveiling human emotions through biosignals. Communications in Computer and Information Science 52(1), 21–47 (2010), doi:10.1007/978-3-642-11721-3_2 (retrieved September 23, 2011)
19. g.tec Medical Engineering GmbH: g.Mobilab+ Mobile Laboratory (2012), http://www.gtec.at/Products/Hardware-and-Accessories/ g.MOBIlab-Specs-Features (retrieved May 16, 2012)
20. Salimetrics, L.L.C.: Salivary Assay Kits (2012), http://www.salimetrics.com/salivary-assay-kits/ (retrieved May 16, 2012)
21. Biotek Instruments Inc.: Powerwave HT Microplate Spectrophotometer (2012), http://www.biotek.com/products/microplate_detection/ powerwave_microplate_spectrophotometer.html (retrieved May 16, 2012)
22. Haupt, R.L., Haupt, S.E.: Continuous genetic algorithm. Practical Genetic Algorithms, 2nd edn., pp. 215–219 (2004)

Information Bottleneck with Local Consistency

Zhengzheng Lou, Yangdong Ye, and Zhenfeng Zhu

School of Information Engineering, Zhengzhou University, Zhengzhou, China
`iezzlou@gmail.com, {yeyd,iezfzhu}@zzu.edu.cn`

Abstract. Given the joint distribution $p(X, Y)$ of the original variable X and relevant variable Y, the Information Bottleneck (IB) method aims to extract an informative representation of the variable X by compressing it into a "bottleneck" variable T, while maximally preserving the relevant information about the variable Y. In practical applications, when the variable X is compressed into its representation T, however, this method does not take into account the local geometrical property hidden in data spaces, therefore, it is not appropriate to deal with non-linearly separable data. To solve this problem, in this study, we construct an information theoretic framework by integrating local geometrical structures into the IB methods, and propose Locally-Consistent Information Bottleneck (LCIB) method. The LCIB method uses k-nearest neighbor graph to model the local structure, and employs mutual information to measure and guarantee the local consistency of data representations. To find the optimal solution of LCIB algorithm, we adopt a sequential "draw-and-merge" procedure to achieve the converge of our proposed objective function. Experimental results on real data sets demonstrate the effectiveness of the proposed approach.

Keywords: Information Bottleneck, Local Consistency, Mutual Information, Entropy.

1 Introduction

In the information age, more and more data are collected. How to extract useful information or analyze the hidden patterns residing in the data has attracted many researchers. Consequently, many approaches have been proposed, among them, the *Information Bottleneck (IB)* method [1] is one of the popular and powerful unsupervised data analysis techniques, which is especially appropriate to deal with "co-occurrence" or "dyadic" data. Such kind of data is represented by a matrix which elements characterize the mutual relationships between the row and column objects. The co-occurrence data exist in many real world applications such as document classification [2] and object category recognition [5].

In order to discover the underlying patterns hidden in the data, based on the co-occurrence relationship between the variables X and Y, the IB method [1] first transforms the numerical matrix about the original variable X and the relevant variable Y into their joint distribution, denoted by $p(X, Y)$. Then, the IB method aims to extract an informative representation of X by compressing it

P. Anthony, M. Ishizuka, and D. Lukose (Eds.): PRICAI 2012, LNAI 7458, pp. 285–296, 2012.
© Springer-Verlag Berlin Heidelberg 2012

into a "bottleneck" variable T, while maximally preserving the relevant information about the variable Y. In general, the compressed representation from X to T is denoted by $p(t|x)$. Based on this idea, many IB algorithms are designed to optimize their objective functions and obtain impressive results on co-occurrence data analysis [4]. However, the corresponding representation $p(t|x)$ only characterizes solutions supported on the Euclidean spaces, and the local geometrical property hidden in the data spaces is ignored. This property is essential to reveal the global structures embedded in the original data matrix. In fact, the co-occurrence data are usually sampled from a nonlinear low-dimensional manifold which is embedded in the high-dimensional ambient space [6,7]. In many manifold methods, such as Laplacian Probabilistic Latent Semantic Indexing [6] and Locally-consistent Topic Modeling [7], the local geometrical structures are considered to achieve efficient nonlinear clustering methods.

To integrate the local geometric structure into the mapping representation $p(t|x)$, two fundamental challenges include (1) how to model the underlying local geometric structure, and (2) how to measure and guarantee the consistency of local structure in the compressed representation. In this study, the former is solved by using k-nearest neighbor graph, and for the latter, we use the mutual information as measure criterion. Then, we proposed the Locally-Consistent Information Bottleneck (LCIB) method. The mutual information can guarantee each sample and its local k-nearest neighbors tend to have similar representations during the novel IB optimization process. To find the optimal solution of LCIB algorithm, we adopt a sequential "draw-and-merge" procedure which can guarantee to converge to a local maximum of our proposed IB-functional. Experimental comparisons with some baseline methods, demonstrate that the proposed method can obtain better prediction accuracies by considering the local geometrical structures.

The rest of the paper is organized as follows. Section 2 provides a background of IB method. The proposed LCIB method is introduced in Section 3. Experimental results are presented in Section 4. Finally, we give conclusion and remarks in Section 5.

2   Information Bottleneck Method

The IB algorithms have shown impressive results in various applications [1,4]. In this section, we briefly introduce the IB principle, and describe one of IB algorithms, *sequential Information Bottleneck (sIB)* [2] algorithm, which is the base of our optimization procedure.

2.1   Information Bottleneck Principle

Given a joint distribution $p(X, Y)$, the IB method constructs a new representation variable T that defines partitions over the elements of X which are informative about Y. The compactness of the representation is calculated by the mutual information $I(T;X)$, while the accuracy of the representation is measured by mutual information $I(T;Y)$ which means how much useful information is preserved. Tishby et al. [1] suggest the following IB-functional:

$$\mathcal{L}_{min} = I(T;X) - \beta \cdot I(T;Y), \tag{1}$$

where β is the Lagrange multiplier controlling the trade-off between the compression from X to T and the preservation about T and Y. The mutual information between X and Y is defined in [3]:

$$I(X;Y) = \sum_{x \in X} \sum_{y \in Y} p(x,y) \log \frac{p(x,y)}{p(x)p(y)}. \tag{2}$$

Tishby et al. formulize an optimal solution of the IB-functional as follows:

$$\begin{cases} p(t|x) = \frac{p(t)}{Z(x,\beta)} e^{-\beta D_{KL}[p(y|x)||p(y|t)]} \\ p(y|t) = \frac{1}{p(t)} \sum_x p(x,y,t) = \frac{1}{p(t)} \sum_x p(x,y)p(t|x) \\ p(t) = \sum_{x,y} p(x,y,t) = \sum_x p(x)p(t|x) \end{cases} \tag{3}$$

where $D_{KL}[p(y|x)||p(y|t)] = \sum_y p(y|x) \log \frac{p(y|x)}{p(y|t)}$ is the *Kullback-Leibler(KL) divergence* between the conditional distributions $p(y|x)$ and $p(y|t)$, $Z(x,\beta)$ is a normalization function. Obviously, the variables $p(t)$ and $p(y|t)$ are determined through $p(t|x)$. To cluster data, the number of clusters, denoted as $|T|$, is much less than the original data size $|X|$ (i.e. $|T| \ll |X|$), which implies a significant compression. Therefore, we may concentrate only on maximally preserving the relevant information $I(T;Y)$, without considering the compression information $I(T;X)$. To make it, we can simply set $\beta = \infty$. Then, the probability $p(t|x)$ becomes deterministic: $p(t|x)$ takes values of zero or one, and we can get "hard" clustering result. In paper [4], N. Slonim rewrites the IB-functional Eq. (1) in alternative form $\mathcal{L}_{max} = I(T;Y) - \beta^{-1}I(T;X)$ via dividing Eq. (1) by $-\beta$. When β is set to be ∞, β^{-1} is 0. So, the IB-functional becomes

$$\mathcal{L}_{max} = I(T;Y). \tag{4}$$

2.2   The SIB Algorithm

Constructing the optimal solution is an NP-hard problem [4]. By now, several IB algorithms have been proposed to approximate to the optimal solution. The sIB algorithm is one of the popular IB algorithms, which generates a 'flat' partition of the original data. From the IB-functional Eq. (4), we observe that among all the possible partitions of $X = \{x_1, x_2, \ldots, x_{|X|}\}$ into groups $T = \{t_1, t_2, \ldots, t_{|T|}\}$, the desired clustering is the one that preserves the maximum mutual information between T and relevant variable Y. The sIB algorithm [2] takes a "draw-and-merge" procedure to find the optimal solution of Eq. (4). It starts with a random partition of X into $|T|$ clusters. At each step, a single $x_i \in X$ is drawn from its current clusters $t(x_i)$ as a new singleton cluster, and is merged into the cluster t^{new} such that $t^{new} = \arg\min_{t \in T} d(x_i, t)$. The merge criterion $d(x_i, t)$ is computed by

$$d(x_i, t) = (p(x_i) + p(t)) \cdot JS_{\Pi}[p(y|x_i), p(y|t)], \tag{5}$$

where $JS_{\Pi}[p(y|x_i), p(y|t)] = \pi_1 D_{KL}[p(y|x_i)||\bar{p}] + \pi_2 D_{KL}[p(y|t)||\bar{p}]$ is the *Jensen-Shannon* divergence, $\Pi = \{\pi_1, \pi_2\} = \{\frac{p(x_i)}{p(x_i)+p(t)}, \frac{p(t)}{p(x_i)+p(t)}\}, \bar{p} = \pi_1 p(y|x_i) + \pi_2 p(y|t)$. After a sequential "draw-and-merge" procedure for each element x, sIB algorithm can get a stable solution, that is, no more assignment updates can further improve the value $\mathcal{L}_{max}$.

3   Locally-Consistent Information Bottleneck

In the IB framework, the compression of variable X to T is measured by mutual information $I(T; X)$, and the preservation of relevant information is estimated by $I(T; Y)$. Our goal is to embed the constrain that similar data points tend to have similar compressed representation $p(t|x)$, into the original IB framework. To achieve this goal, we adopt the k-nearest neighbor graph to model the underlying local geometric structure and use the mutual information to measure the local consistency.

3.1   Formulation

For data point x_i, we use $\mathcal{K}_i$ to denote the set of x_i's k-nearest neighbors (in our work, the 1-nearest neighbor of x_i is itself, i. e. $x_i \in \mathcal{K}_i$). The mutual information $I_{\mathcal{K}_i}(T; X)$ is used to measure the local consistency of data representations in $\mathcal{K}_i$, which is defined as follows:

$$I_{\mathcal{K}_i}(T; X) = \sum_{t \in T} \sum_{x \in \mathcal{K}_i} p(x|\mathcal{K}_i) p(t|x) \log \frac{p(t|x)}{p(t|\mathcal{K}_i)}, \tag{6}$$

where $p(x|\mathcal{K}_i) = \frac{p(x)}{\sum_{x \in \mathcal{K}_i} p(x)}$, $p(t|\mathcal{K}_i) = \sum_{x \in \mathcal{K}_i} p(x|\mathcal{K}_i) p(t|x)$.

   The mutual information $I_{\mathcal{K}_i}(T; X)$ can be used to measure the average number of bits that data points $x \in \mathcal{K}_i$ contain about the representation T [10]. If the data and its k-nearest neighbors have the same compressed representations $p(t|x)$, the mutual information will be zero. So, we can interpret mutual information as a measure of consistency of the local geometric structure. The objective functional of Locally-Consistent Information Bottleneck is defined as follows:

$$\mathcal{L}[p(t|x)] = I(T; Y) - \mu \cdot \sum_{i=1}^{|X|} I_{\mathcal{K}_i}(T; X), \tag{7}$$

In this objective functional, the mutual information $I(T; Y)$ makes the compressed representation T preserve the global structure resided in relevant variable Y, and $\sum_{i=1}^{|X|} I_{\mathcal{K}_i}(T; X)$ will drive T to preserve the local geometric structure.

The balance parameter μ is used to control the trade-off between the preservation of global structure and local structure.

In this work, we only consider the "hard" clustering, where the value of $p(t|x)$ is either 0 or 1. Now we reconsider the mutual information $I_{\mathcal{K}_i}(T; X)$.

$$I_{\mathcal{K}_i}(T; X) = \sum_{t \in T} \sum_{x \in \mathcal{K}_i} p(x|\mathcal{K}_i)p(t|x) \log \frac{p(t|x)}{p(t|\mathcal{K}_i)}$$

$$= \sum_{t \in T} \sum_{x \in \mathcal{K}_i} p(x|\mathcal{K}_i)p(t|x) \log p(t|x) - \sum_{t \in T} \sum_{x \in \mathcal{K}_i} p(x|\mathcal{K}_i)p(t|x) \log p(t|\mathcal{K}_i)$$

In the above function, $p(t|x)$ is either 0 or 1, and $0 \log 0 = 0$, $1 \log 1 = 0$. So the first term of above function is equal to 0. Then $I_{\mathcal{K}_i}(T; X)$ can be abbreviate to

$$I_{\mathcal{K}_i}(T; X) = -\sum_{t \in T} \sum_{x \in \mathcal{K}_i} p(x|\mathcal{K}_i)p(t|x) \log p(t|\mathcal{K}_i) = -\sum_{t \in T} p(t|\mathcal{K}_i) \log p(t|\mathcal{K}_i) = H_{\mathcal{K}_i}(T).$$

Thus, the objective function of our work can be rewritten as below:

$$\mathcal{L}[p(t|x)] = I(T; Y) - \mu \cdot \sum_{i=1}^{|X|} H_{\mathcal{K}_i}(T). \tag{8}$$

3.2 Objective Functional Optimization

In this work, we adopt the sequential "draw-and-merge" procedure to optimize the proposed objective function Eq. (8). For the original IB function Eq. (4), we use the merge criterion described in Eq. (5). Now, we give the merge criterion of the last term in our objective function, i.e. $\mu \cdot \sum_{i=1}^{|X|} H_{\mathcal{K}_i}(T)$.

When the instance x_i is drawn from its current cluster and taken as a new singleton cluster, the entropy of $H_{\mathcal{K}_i}(T)$ is calculated as $H_{\mathcal{K}_i}[p(t_1), \cdots, p(t_{|T|}), p(x_i)]$. If x_i is merged into cluster t_l, $1 \leq l \leq |T|$, the merge cost with respect to $H_{\mathcal{K}_i}(T)$ is

$$d_{\mathcal{K}_i}(x_i, t_l) = H_{\mathcal{K}_i}[p(t_1), \cdots, p(t_{|T|}), p(x_i)] - H_{\mathcal{K}_i}[p(t_1), \cdots, p(t_l) + p(x_i), \cdots, p(t_{|T|})]$$

$$= -p(t_l|\mathcal{K}_i) \log p(t_l|\mathcal{K}_i) - p(x_i|\mathcal{K}_i) \log p(x_i|\mathcal{K}_i)$$

$$+ [p(t_l|\mathcal{K}_i) + p(x_i|\mathcal{K}_i)] \log[p(t_l|\mathcal{K}_i) + p(x_i|\mathcal{K}_i)]$$

$$= -p(t_l|\mathcal{K}_i) \log \frac{p(t_l|\mathcal{K}_i)}{p(t_l|\mathcal{K}_i) + p(x_i|\mathcal{K}_i)} - p(x_i|\mathcal{K}_i) \log \frac{p(x_i|\mathcal{K}_i)}{p(t_l|\mathcal{K}_i) + p(x_i|\mathcal{K}_i)}$$

$$= [p(t_l|\mathcal{K}_i) + p(x_i|\mathcal{K}_i)]H\left(\frac{p(t_l|\mathcal{K}_i)}{p(t_l|\mathcal{K}_i) + p(x_i|\mathcal{K}_i)}, \frac{p(x_i|\mathcal{K}_i)}{p(t_l|\mathcal{K}_i) + p(x_i|\mathcal{K}_i)}\right).$$

Let $p_{\mathcal{K}_i}(\bar{t}) = p(t_l|\mathcal{K}_i) + p(x_i|\mathcal{K}_i), \Pi_{\mathcal{K}_i} = \{\pi_{\mathcal{K}_i}^1, \pi_{\mathcal{K}_i}^2\} = \{\frac{p(t_l|\mathcal{K}_i)}{p(t_l|\mathcal{K}_i) + p(x_i|\mathcal{K}_i)}, \frac{p(x_i|\mathcal{K}_i)}{p(t_l|\mathcal{K}_i) + p(x_i|\mathcal{K}_i)}\}$, then we rewrite $d_{\mathcal{K}_i}(x_i, t_l)$ as $p_{\mathcal{K}_i}(\bar{t}) \cdot H(\Pi_{\mathcal{K}_i})$.

Let $\mathcal{N}_i = \{j\}, \mathring{x}_i \in \mathcal{K}_j$, be the set of data point's indexes whose k-nearest neighbors include x_i. When x_i is merged into t_l, the entropy values of $\mathcal{K}_j (j \in \mathcal{N}_i)$ will

---

Algorithm 1. sLCIB algorithm

Input: Joint distribution $p(X, Y)$, parameters M, k, μ.
Output: A partition T of X into M clusters.

Construct k-nearest neighbor graph:
for $i = 1$ **to** $|X|$ **do**
   Construct k-nearest neighbors $\mathcal{K}_i$ for x_i;
end for
Initialize:
$T \leftarrow$ random partition of X into M clusters;
Main Loop:
repeat
   $noChange = true$;
   for $i = 1$ **to** $|X|$ **do**
      draw x_i out of $t(x_i)$;
      for $j = 1$ **to** M **do**
         Calculate merge costs $d(x_i, t_j)$ based on Eq. (9);
      end for
      $t^{new}(x_i) = \arg\min_{t_j} d(x_i, t_j)$;
      if $t^{new}(x_i) \neq t(x_i)$ **then**
         Merge x_i into $t^{new}(x_i)$;
         $noChange = false$;
      end if
   end for
until $noChange$ is $true$

---

be changed. So, while x_i is merged into t_l, the total merger cost with respect to $\sum_{i=1}^{|X|} H_{\mathcal{K}_i}(T)$ is $\sum_{j \in \mathcal{N}_i} d_{\mathcal{K}_j}(x_i, t_l)$. Finally, we get the following merger criterion with respect to our objective functional.

$$d(x_i, t_l) = (p(x_i) + p(t_l)) \cdot JS_{\Pi}[p(y|x_i), p(y|t_l)] - \mu \cdot \sum_{j \in \mathcal{N}_i} p_{\mathcal{K}_j}(\bar{t}) \cdot H(\Pi_{\mathcal{K}_j}) \quad (9)$$

In the sLCIB algorithm (Algorithm 1), each "draw-and-merge" step either improves the score of Eq. (8), or leaves the current partition unchanged. Note that, $I(T; Y) \leq I(X; Y)$, and given joint distribute $p(X, Y)$, the value of $I(X; Y)$ is constant. The minimum value of $H_{\mathcal{K}_i}(T)$ is zero. Thus, the maximum value is upper bounded, which is $I(X; Y)$. Therefore, the sLCIB algorithm is guaranteed to converge to a local maximum while no assignment changes are further performed.

The sLCIB algorithm needs to construct the k-nearest neighbor graph to capture the local geometric structure, and the corresponding time complexity is $O(|X|^2|Y|)$. The complexity of sLCIB algorithm $O(LM|X||Y|)$ is the same as sIB [2], where L is the number of loops. So, compared to sIB algorithm's computational complexity, the sLCIB will take some more time to construct and obtain the local geometric structure. The following section shows that the additional time cost is valuable since sLCIB algorithm can get better accuracy than its rival methods.

4   Experiments

4.1   Data Sets

In this paper, we use three document corpus, which are widely used as benchmark data sets in clustering literature [6,7], and three image data sets, which are designed for object category discovery [8]. The six data sets are described as follows:

CSTR. This data set contains 476 abstracts which are technical reports published in the Department of Computer Science at a university from 1991 to 2007. There are four research areas in this data set, which are Natural Language Processing, Robotics/Vision, Systems and Theory.

TDT2. The TDT2 corpus (Nist Topic Detection and Tracking corpus) consists of data collected from 6 sources, including 2 newswires (APW, NYT), 2 radio programs (VOA, PRI) and 2 television programs (CNN, ABC). After removing the documents with multiple category labels, we select the documents appearing in the largest 10 categories in our experiments, and the final data set contains $7,456$ documents.

Reuters-21578. In this paper, we use the ModApte version of Reuters-21578. The documents appearing in two or more categories are discarded, and $8,293$ documents in 65 categories are left. From 65 categories, we select the largest 10 categories as our experiment data set, which includes $7,285$ documents.

amazon. This data set consists of products images downloaded from (www.amazon.com). There are 31 categories (backpack, bike, notebook, stapler etc) with total $2,813$ images in this data set.

dslr. The images in dslr are captured with a digital SLR camera in realistic environments with natural lighting conditions, which also contain 31 object categories. The size of this data set is 498.

webcam. The webcam image set, with 795 images in total, contains the same 31 object categories appeared in the above two image data sets. These images are collected by a simple webcam.

On the three document data sets, we select the top 1000 words according to their contribution to the mutual information about the documents [2]. For the image data preprocessing, we use the "Bag-of-Words" method to transform the image data set into a co-occurrence matrix. Local scale-invariant interest points in images are extracted by SURF detector [9]. The 64-dimensional SURF features are collected from the images, and a codebook with the size of 800 is generated by vector quantization via the k-means algorithm on a random subset of database. Using this codebook, the image database can be transformed into a co-occurrence matrix in the subsequent learning process.

4.2   Experimental Design

To evaluate the performance of the proposed method, we compare sLCIB algorithm with the following five methods:

- sIB algorithm [2]
- k-means
- Probabilistic latent semantic analysis (PLSA) [11]
- Locally-consistent Topic Modeling (LTM) [7]
- Laplacian Probabilistic Latent Semantic Indexing (LapPLSI) [6]

The sIB algorithm is a classical IB algorithm. PLSA is a topic modeling clustering algorithm. LTM and LapPLSI are topic modeling on manifold methods.

4.3   The Evaluation Method

In this paper, we use the clustering accuracy and the normalized mutual information as the performance measures, which are the standard measures widely used for clustering [6,7].

Clustering Accuracy (AC). AC evaluates the clustering performance by comparing the obtained clustering assignment of each data point with the ground truth label of that point. The AC is defined as:

$$AC = \frac{\sum_{i=1}^{n} \delta(l_i, \mathrm{map}(t_i))}{n}, \tag{10}$$

where t_i denotes the cluster assignment of x_i, l_i is the ground truth label of x_i, and n is size of the data. The delta function $\delta(x, y)$ equals 1 if $x = y$ and equals 0 otherwise. The permutation function $\mathrm{map}(t_i)$ maps each cluster assignment t_i to the equivalent label provided by the data corpus.

Normalized Mutual Information (NMI). After obtaining the clustering assignment T, we can get the confusion matrix between T and ground truth label L, from which we can produce the joint distribution $p(T, L)$. The mutual information $I(T; L)$ can be calculated by Eq. 2. The NMI evaluation metric is defined as:

$$NMI = \frac{I(T; L)}{\max(H(T), H(L))}. \tag{11}$$

4.4   Experimental Results

Parameter Selection. To evaluate the influence of the parameters, Figure 1 and Figure 2 give the performance of sLCIB varies with k and μ on document data set and image data set respectively. As we can see, sLCIB is stable with respect to μ and k. LCIB adopts a k-nearest graph to model the underlying local geometric structure. It achieves consistent better performance than sIB algorithm with k varying from 4 to 10. The sLCIB algorithm get better clustering accuracy when μ varying from 10^{-4} to 10^{-6}. When μ is larger, the sLCIB will concentrate on the local geometric structure of the data. While $\mu \to 0$, the local structure is ignored and the results of sLCIB algorithm will approximate to sIB's results.

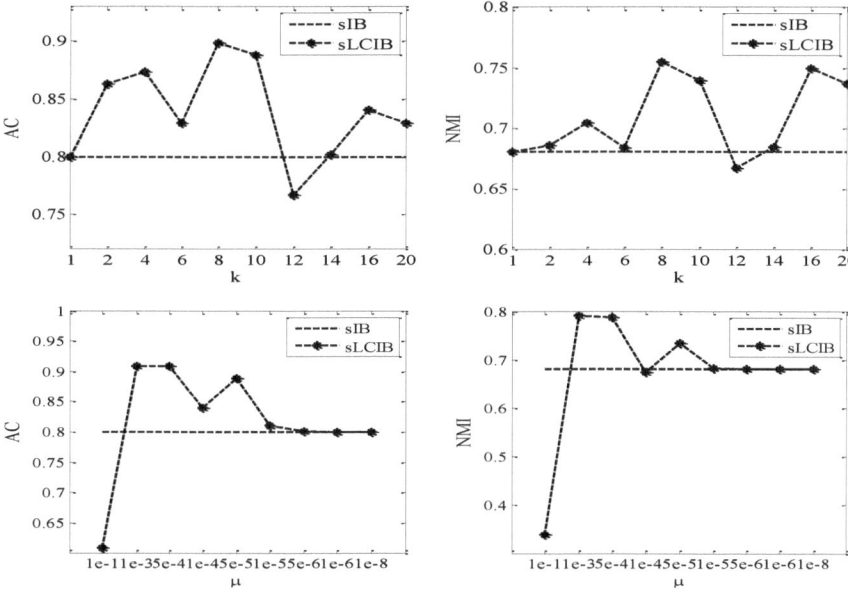

Fig. 1. The performance of sLCIB and sIB vary with the k and μ on CSTR data

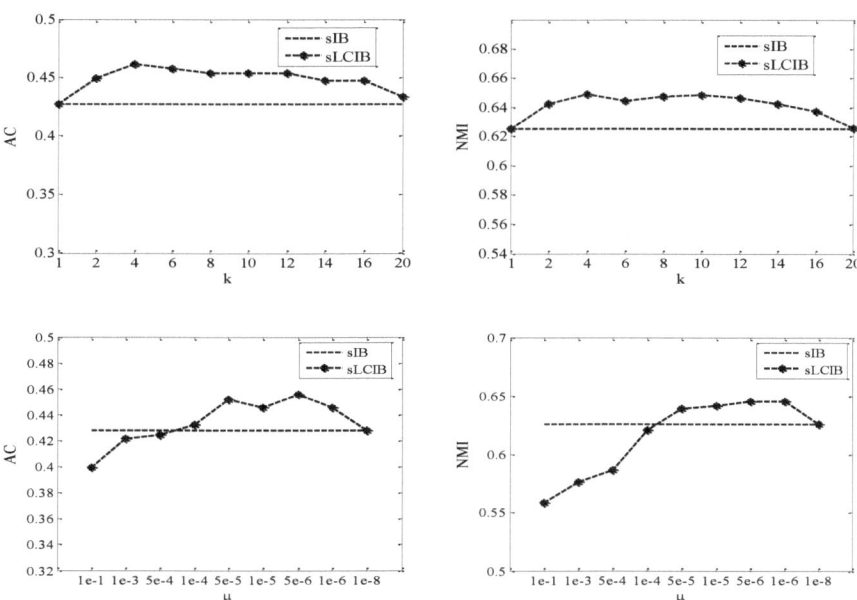

Fig. 2. The performance of sLCIB and sIB vary with the k and μ on dslr data

Performance Comparisons. Table 1 and Table 2 show the comparing results of the six clustering methods. In all experiments, the number of learned clusters is taken to be identical with the number of real categories on each data set. For the proposed sLCIB algorithm, the k-nearest neighbor graph $\mathcal{K}$ is constructed via JS divergence. The number of nearest neighbors and the value of balance parameter μ are set to be 8 and 0.00001 respectively according to Figure 1 and Figure 2. The parameters for LapPLSI and LTM is set to their default values [6,7]: for LapPLSI, the number of nearest neighbors and the regularization parameters are set to be 7 and 1000 respectively, and for LTM, these two parameters are set to be 5 and 1000. To alleviate the influence caused by random initialization, we run each algorithm 20 times, each with a new random initialization, and report the average evaluation score and standard deviation.

Observing the these experiments, several interesting points should be noted.

- The sIB and sLCIB outperforms all the other rival methods, typically with an impressive gap. This suggests that the IB method is a powerful data analysis approach.
- By integrating the local geometric structure into the IB framework, the sLCIB algorithm can get better performance than sIB algorithm. This demonstrates that the local geometrical property is essential to reveal the global structures hidden in the data set.
- In some data sets, such as TDT2, the improvements of sLCIB over sIB algorithm is more distinguishing than some other data sets. One possible reason is that the hidden patterns in some data sets are more compact and focused than some other data. Thus, the nearest neighbor graphs, which captures local geometrical structure of the data, are more consistent with the hidden structure.

Table 1. Clustering Accuracy on the six data sets (%)

Data Sets	PLSA	k-means	LapPLSI	LTM	sIB	sLCIB
CSTR	41.1±6.2	41.6±4.6	62.7±5.2	60.6±10.8	84.5±9.3	**86.8±8.1**
TDT2	39.6±3.2	39.0±2.3	60.3±4.1	75.3±4.2	81.0±3.7	**85.8±4.2**
Reuters	42.0±3.2	42.3±2.7	42.3±4.2	45.2±4.9	49.0±4.0	**49.1±4.8**
amazon	17.6±1.3	17.7±0.7	12.7±0.2	20.5±0.6	24.8±1.0	**26.9±0.8**
dslr	29.8±1.7	34.6±1.3	21.8±0.5	41.1±1.8	43.9±1.6	**44.6±1.8**
webcam	28.6±1.9	31.0±1.3	16.7±0.5	36.6±1.6	41.7±2.2	**43.6±2.6**
Avg.	33.1	34.4	36.1	46.5	54.2	**56.1**

Table 2. Clustering Normalized Mutual Information on the six data sets (%)

Data Sets	PLSA	k-means	LapPLSI	LTM	sIB	sLCIB
CSTR	10.2±5.8	11.5±4.2	56.3±2.2	42.7±12.1	72.2±6.0	**74.6±4.6**
TDT2	32.0±3.6	32.4±2.9	68.1±2.4	74.3±1.4	80.0±1.2	**83.2±2.0**
Reuters	30.0±3.6	29.5±2.9	33.2±3.0	40.1±3.7	46.2±2.2	**46.5±3.5**
amazon	19.5±1.5	21.3±0.8	14.2±0.1	24.5±0.4	27.7±0.6	**29.8±0.7**
dslr	42.0±1.6	50.8±1.2	37.1±0.6	56.5±1.4	62.0±1.2	**63.3±1.4**
webcam	40.3±1.8	48.8±1.2	28.7±0.4	53.0±1.4	60.1±1.3	**61.2±1.5**
Avg.	29.0	32.4	39.6	48.5	58.0	**59.8**

5   Conclusion

We have integrated the local geometrical structure into the IB methods, and proposed the Locally-Consistent Information Bottleneck, which takes a k-nearest neighbor graph to model the local geometric structure hidden in the data set and takes the mutual information as the consistency measure of data representations. To find the solution of LCIB functional, we have proposed the sLCIB algorithm, which adopts the sequential "draw-and-merge" procedure for guaranteeing to converge to a local optimal of the function. Experimental results on document and image clustering show that, the sLCIB algorithm can get better performance than the powerful sIB algorithm [2] and two topical modeling on manifold methods (LapPLSI [6] and LTM [7]).

As we can see from the experiments, the construction of local geometric structure influences the performance of sLCIB algorithm, and bad k-nearest neighbor graph will add worthless information into the IB framework. In the future work, we will try different ways to construct the graph, such as incorporating prior knowledge. In addition, we also plan to extend the LCIB to semi-supervised clustering, which will add constrains, must-link and cannot-link, to the IB framework. How to measure the constrains and design the corresponding algorithm will be the key points.

Acknowledgement. This work is supported by the National Natural Science Foundation of China under grant No. 61170223, China Postdoctoral Science Foundation under grant No. 2011M501189.

References

1. Tishby, N., Pereira, F.C., Bialek, W.: The Information Bottleneck Method. In: Proccedings of the 37th Allerton Conference on Communication and Computation, Illinois, USA, pp. 368–377 (1999)
2. Slonim, N., Friedman, N., Tishby, N.: Unsupervised Document Classification using Sequential Information Maximization. In: Proccedings of the 25th ACM International Conference on Research and Development of Information Retireval, pp. 129–136. ACM Press, Tampere (2002)
3. Cover, T.M., Thomas, J.A.: Elements of Information Theory. John Wiley, New York (1991)
4. Slonim, N.: The Information Bottleneck: Theory and Applications. PhD thesis, the Senate of the Hebrew University (2002)
5. Lou, Z., Ye, Y., Liu, D.: Unsupervised object category discovery via information bottleneck method. In: Proceedings of ACM International Conference on Multimedia, pp. 863–866 (2010)
6. Cai, D., Mei, Q., Han, J., Zhai, C.: Modeling Hidden Topics on Document Manifold. In: Proceeding of the 17th ACM Conference on Information and Knowledge Management, pp. 911–920 (2008)
7. Cai, D., Wang, X., He, X.: Probabilistic Dyadic Data Analysis with Local and Global Consistency. In: Proceedings of the 26th Annual International Conference on Machine Learning, pp. 105–112 (2009)

8. Saenko, K., Kulis, B., Fritz, M., Darrell, T.: Adapting Visual Category Models to New Domains. In: Daniilidis, K., Maragos, P., Paragios, N. (eds.) ECCV 2010. LNCS, vol. 6314, pp. 213–226. Springer, Heidelberg (2010)
9. Bay, H., Ess, A., Tuytelaars, T., Gool, L.V.: Speeded-Up Robust Features (SURF). Computer Vision and Image Understanding 110(3), 346–359 (2008)
10. Szummer, M., Jaakkola, T.: Partially labeled classification with markov random walks. In: Advances in Neural Information Processing Systems, pp. 1025–1032 (2001)
11. Hofmann, T.: Unsupervised Learning by Probabilistic Latent Semantic Analysis. Machine Learning 42(1), 177–196 (2001)

A D-S Theory Based AHP Decision Making Approach with Ambiguous Evaluations of Multiple Criteria

Wenjun Ma, Wei Xiong, and Xudong Luo*

Institute of Logic and Cognition,
Sun Yat-Sen University, Guangzhou, 510275, China
luoxd3@mail.sysu.edu.cn

Abstract. This paper proposes an ambiguity aversion principle of minimax regret to extend DS/AHP approach of multi-criteria decision making (MCDM). This extension can analyze the MCDM problems with ambiguous evaluations of multiple criteria. Such evaluations cannot be avoided in real life, but that existing MCDM theories and models cannot handle it well. We also give an example of real estate investment to illustrate our approach.

1 Introduction

Generally speaking, decision making can be regarded as a process that results in the selection of a course of action among several alternatives. In real-life, often the decisions have to be made in the case that only limited or ambiguous information is available. In particular, for multi-criteria decision making (MCDM) problems, (i) the consequence of an action chosen regarding some criteria is often unknown; (ii) the consequence of an action chosen regarding some criteria might take an imprecise interval-value; and (iii) the consequence of an action chosen, which is determined by a number of criteria, has multiple possibilities, even including a possible loss (i.e., risk) [13]. For example, the problem of real estate investment is such a MCDM problem under uncertainty and risk. In the problem, the investor is affected by different criteria such as house price, the wealth of the decision maker, spillover effects of the house, the policy of the government, the economic environment, investment profit, and so on. Moreover, some criteria such as the economic environment usually are uncertain for the decision maker. As a result, in order to make decisions based on ambiguous information provided by several different sources, we need a model that can assist the decision-maker to rank decision alternatives regarding different criteria [13], and properly make use of ambiguous information in decision analyzing. To construct such a model, this paper incorporates the D-S theory of evidence [7,18] with Analytic Hierarchy Process (AHP) [17] to deal with MCDM problems under uncertainty of ambiguity.

---

* Corresponding author.

P. Anthony, M. Ishizuka, and D. Lukose (Eds.): PRICAI 2012, LNAI 7458, pp. 297–311, 2012.

In the literature, AHP [17] is one of the most well-known MCDM techniques. Recently, Beynon et al. [2,3,5,6] extend the AHP model to a DS/AHP method with which a decision maker can make preference judgements on groups of options rather on individual options or through pairwise comparisons of options. After that, many researchers have used DS/AHP to analyzed different problems in specific domains. For example, Beynon *et al.* [6] employ the DS/AHP technique in the consumer choice analysis and Awasthi *et al.* [1] use AHP and D-S theory for evaluating sustainable transport solutions. Although Beynon et al. [2,3,5,6] claimed that ignorance in the judgements is allowed in DS/AHP, they did not provide the method for solving MCDM problems with some kind of ambiguous information, interval-valued evaluations for a decision alternative regarding some criteria, and imprecise probabilities over possible outcomes of a decision alternative against some criteria.

To tackle the problem, this paper extends the DS/AHP to handle ambiguous evaluations of of multiple criteria. More specifically, (i) this paper will introduces the ambiguity-avoided regret model that employs the cognitive factors to set a preference ordering over interval values; and (ii) this paper will apply the interval-valued DS/AHP to get the evaluation of a choice for making the best choice.

This paper advances the state of art in the field of MCDM in the following aspects. (i) We extend the DS/AHP model to deal with ambiguous information, which means that more practical application problems can be solved by our model. (ii) We relax the assumption of the precise point-valued decision matrix to interval-valued one, which is more feasible in real-life. (iii) We use the ambiguity-degree to revise minimax regret degrees so that interval-values can be compared properly. And (iv) our method can deal with three kinds of ambiguity in decision matrix.

The rest of this paper is organized as follows. Section 2 briefs some basics of D-S theory and DS/AHP theory. Section 3 gives a formal definition of the problems we are going to solve in this paper. Section 4 discusses how to set a proper preference ordering over interval-valued expected utilities. Section 5 proposes a method to obtain the optimal option by belief interval. Section 6 illustrates the applicability of our decision model by a real estate investment problem. Finally, Section 7 summarizes the paper and points out some possibilities of the future work.

2   Preliminaries

This section recaps D-S theory [7,18]and DS/AHP theory [2,3,5].

2.1   Basics of D-S Theory

Definition 1. *Let an exhaustive set of mutually disjoint atomic Θ be a frame of discernment.*

(i) *Function* $m : 2^{\Theta} \to [0,1]$ *is called a basic probability assignment or a mass function if* $m(\phi) = 0$ *and* $\sum_{A \subseteq \Theta} m(A) = 1$.

(ii) *Function* $Bel : 2^{\Theta} \to [0,1]$, *defined as follows, is a belief function over* Θ:

$$Bel(A) = \sum_{B \subseteq A} m(B). \tag{1}$$

If $m(B) > 0$, *then* B *is said to be a focal element of Bel.*

(iii) *Function* $Pl : 2^{\Theta} \to [0,1]$, *defined as follows, is called a plausibility function over* Θ:

$$Pl(A) = \sum_{B \cap A \neq \phi} m(B). \tag{2}$$

D-S theory also provides a rule to combine the mass functions evidenced by independent sources:

Definition 2 (Dempster's rule of combination). *Let* m_1 *and* m_2 *be two basic probability assignment over the discernment frame* Θ. *Then function* $m_{12} = m_1 \oplus m_2$ *is given by:*

$$m_{12}(\{x\}) = \begin{cases} 0 & \text{if } x = \emptyset \\ \dfrac{\sum\limits_{A_i \cap B_j = x} m_1(A)m_2(B)}{k} & \text{if } x \neq \emptyset \end{cases} \tag{3}$$

where normalization factor

$$k = 1 - \sum_{A_i \cap B_j = \emptyset} m_1(A)m_2(B), \tag{4}$$

which is the measure of the conflict between the pieces of evidence.

The Dempster's rule of combination satisfies commutativity and associativity. So, the pieces of evidence can be combined in any order [18].

Now we recall the concept of a mass function's ambiguity degree that is a normalized one of the generalized Hartley measure for nonspecificity [8]:

Definition 3. *Let* m *be a mass function over discernment frame* Θ, *and* $|A|$ *be the cardinality of set* A. *Then the ambiguity degree of* m, *denoted as* δ, *is given by*

$$\delta = \frac{\sum\limits_{A \subseteq \Theta} m(A) \log_2 |A|}{\log_2 |\Theta|}. \tag{5}$$

Given a frame of discernment $\Theta = \{x_1, ..., x_n\}$ where $x_1 < ... < x_n$. Let a be a choice, which corresponds to a mass function m over Θ.[1] Based on the concept of mass function, the point-valued expected utility formula can be extended to the expected utility interval [19]:

---

[1] If this mass function is a probability function, the specified choice is actually a lottery, defined by von Neumann and Morgenstern [22].

Definition 4. *For choice a specified by mass function m over Θ, its expected utility interval is $EUI(a) = [\underline{E}(a), \overline{E}(a)]$, where*

$$\underline{E}(a) = \sum_{A \subseteq \Theta} \min(A)m(A), \tag{6}$$

$$\overline{E}(a) = \sum_{A \subseteq \Theta} \max(A)m(A). \tag{7}$$

In the above definition, if each $A \subseteq \Theta$ has only one element, $m(A)$ degenerates to probability and formulas (6) and (7) degenerate to the point-valued expected utility. In other words, the interval-value of an expected utility is caused by $m(A) > 0$ where $A \subseteq \Theta$ has at least two elements. That is, the interval-value of expected utility is due to the decision maker's ambiguity about which consequence will be caused.

2.2 DS/AHP Theory

Following the idea of the Analytic Hierarchy Process (AHP) [17], the DS/AHP method [2,3,5] integrates D-S theory into AHP so that a preference on groups of decision alternatives can be set regarding different criteria. More specifically, with DS/AHP, a decision maker needs to make two sets of judgments: (i) the judgment on the weights of the criteria; and (ii) the judgment on the preference level of groups of decision alternatives regarding different criteria. Then, based on these two judgments, the decision maker needs: (i) set the discernment frame Θ for the decision alternatives regarding the criteria and the goal of the decision problem; (ii) assign a preference scale values on groups of decision alternatives (focal elements according to the evidence) to discern the preferences of the decision maker on an individual criterion; (iii) assign the weight to each criterion based on the judgment on their relative importance; (iv) set the mass function for the focal elements of each criteria by

$$m(s_i) = \frac{a_i\omega}{\sum_{j=1}^{d} a_j\omega + \sqrt{d}}, \quad (i = 1, 2, \ldots, d), \tag{8}$$

$$m(\Theta) = \frac{\sqrt{d}}{\sum_{j=1}^{d} a_j\omega + \sqrt{d}}, \tag{9}$$

where s_i is one of d focal elements of a criterion that assigned the scale values a_i, Θ is the frame of discernment, and ω is the weight for that criterion; and (v) use Dempster's rule of combination to combine the mass function of each criterion to get the overall evaluation of each alternative.

3   Problem Definition

This section gives the formal definition of the problems that we are going to solve in this paper.

In real-life, the decisions often have to be made in the situation where there are only very limited precise or only ambiguous information available from several different sources because of time pressure, lack of data, disturbance of unknown factors, randomness outcome of some attributes, and so on. Actually, we can distinguish the three kinds of ambiguity based on their performance in the decision matrix. (i) The decision maker is completely lack of information about the performance of a decision alternative regarding a criterion. As a result, the values of the decision alternatives regarding this criterion in the decision matrix are unknown (denoted by "∗"). (ii) The information the decision maker has about the performance of a decision alternative regarding a criterion is not a precise point-value but an interval-value or more than one qualitative evaluation. As a result, the values of the decision alternatives regarding this criterion in the decision matrix are not precise point-valued or unique. (iii) The information the decision maker has reveals that the performance of a decision alternative regarding a criterion will cause a set of possible outcomes, but the probabilities of these outcomes are unknown or not even meaningful.

Table 1. Decision matrix of the real estate investment problem

	Price ($/m^2$)	Quality (grade)	Service Level (%)	External Environment (grade)	Investment Profit($/m^2$) growth	recession	depression	Style (grade)
A	2200-3200	G	85	E,G[0.3];G,A[0.5]	3800	2800	2000	E,G
B	2200-3200	E	∗	E,G	3200	3000	2800	G
C	2800	∗	81-85	G,A	3000	2800	2500	G,A
D	3200	A,E	91	E	4200	3200	2500	E
E	2800-3000	A,G	81-95	A	3800	3000	1800	∗

For example, suppose a person wants to buy a house for investment and family both. There are five alternatives A, B, C, D, and E, and six criteria for evaluating these alternatives: Price, Quality, Service Level, External Environment, Investment Profit (after five years), and Style. Suppose the criteria of Quality, External Environment and Style can be assessed by using the assessment grades: Excellent (E), Good (G), Average (A) and Poor (P) as showed in Table 1. For the criterion of Investment Profit, each decision alternative has one of the three different outcomes of growth, recession, and depression and according to the market survey result of the trend of economic environment we can set $m(\{growth\}) = 0.4$, $m(\{growth, recession\}) = 0.3$, $m(\{recession, depression\}) = 0.3$. For alternative A, probably there is a noisy manufactory or a beautiful park will be built nearby and so the external environment of alternative A is ambiguously uncertain. Also, some values in the decision matrix of the real estate investment problem are unknown. For example, the quality of house C is unknown. Moreover, some style of the alternatives is uncertain for the buyer because it is a pre-sale of uncompleted houses and the price of an alternative could be estimated only by an interval-value because of some random factors. Then, which house should he choose?

In these cases, we might be not able to assign a unique preference scale value to each group of identified decision alternatives, which is assumed in DS/AHP. Moreover, all of these cases are caused by the ambiguous information: (i) Complete ignorance: The decision maker has no information about the consequence the decision alternatives will cause regarding to some criteria. (ii) Interval-value: The decision maker only knows the result of selecting the decision alternative regarding a criterion, but has no idea exactly where this result comes from. For example, when people are buying a second-hand house, the real estate agency can tell him the house owner might sell the house at a price between $300,000 and $350,000, but normally they would not tell the buyer how this price comes from. (iii) Traditional ambiguity: The decision makers only know the mass (belief or plausibility) function over the set of all possibilities. The following is the formal definition of the MCDM problem with ambiguous information:

Definition 5. *Give a multi-criteria decision making problem* (A, C, U, f) *where*

(i) $A = \{a_i \mid i = 1, \ldots, n\}$ *is a non-empty finite decision alternatives sets;*

(ii) $C = \{c_j \mid j = 1, \ldots, m\}$ *is a non-empty finite criteria sets, for each criterion* c_j; *and*

(iii) $f : A \times C \longrightarrow U$ *and* $U_{ij} = f(a_i, c_j)$ *is the evaluation of the decision alternative* a_i *regarding criterion* c_j,

If $\forall a_i \in A$, $\exists c_j \in C$, *the value of* U_{ij} *is one of the three cases: (i) missing, (ii) interval-value, and (iii) a mass function, then the problem is called an ambiguous MCDM problem.*

Actually, although the DS/AHP method [4] does not mention its applicability to (A,C,U, f) with missing evaluations, we find that in [4], Beynon has applied this method to handle an example with missing evaluations. So, this paper focuses on the other two cases of ambiguity.

To solve the ambiguous MCDM problem, we have to answer three questions, which are not discussed in DS/AHP. (i) How to calculate the expected evaluation intervals from ambiguous information about the decision alternatives regarding criteria in the third case? (ii) How to obtain a preference value of decision alternative in order to apply the DS/AHP method? And (iii) how to apply the belief function and plausibility function obtained by the DS/AHP method to get the optimal decision alternative? We will discuss the them in the following section.

4   Preference Setting

This section will set a proper preference ordering over interval-valued expected utilities.

4.1   Ambiguity Aversion Principle of Minimax Regret

Our approach is to extend the minimax regret principle [20], which is based on the observation that people often compare the outcome of what they chose with

what they could choose [10]. Further, we consider one more cognitive factor—
ambiguity aversion, which has been observed in both laboratory experiments [9]
and real-world problem of health care [21], to revise the definition of regret under
uncertainty of ambiguity. Formally, we have:

Definition 6 (The ambiguity aversion principle of minimax regret). *Let
m be an initial mass function over a set of utilities $\Theta = \{x_1, ..., x_n\}$, $EUI(x) = [\underline{E}(x), \overline{E}(x)]$ be the expected utility interval of choice x, $\varepsilon(b)$ be the ambiguity-
avoided upper expected utility of choice b, and $\delta(b)$ be the ambiguity degree of
choice b. Then the ambiguity-avoiding maximum regret of the choice a against
choice b is defined as*

$$\Re_a^b = \varepsilon(b) - \underline{E}(a), \tag{10}$$

where

$$\varepsilon(b) = \underline{E}(b) + (1 - \delta(b))(\overline{E}(b) - \underline{E}(b)). \tag{11}$$

For any two choices a and b, the strict preference ordering $\succ$ is defined as follow:

$$a \succ b \Leftrightarrow \Re_a^b < \Re_b^a. \tag{12}$$

Moreover, from formula (11), we can see that $\underline{E}(a) \leq \varepsilon(a) \leq \overline{E}(a)$ and ambiguity
degree δ actually works as a discounting factor: the higher it is, the more the
upper utility of a choice is discounted. In particular, for point-value expected
utility a, by formula (12), we can obtain that

$$\Theta = \{a\}, m(\{a\}) = 1, \delta = 0, \varepsilon = a. \tag{13}$$

The following theorem confirms the preference ordering $\succ$ defined in the above
enables us to compare any choices properly.

Theorem 1. *Let C be a finite choice set and the interval-valued expected utility
of choice $c_i \in C$ be $EUI(c_i) = [\underline{E}(c_i), \overline{E}(c_i)]$, and its ambiguity degree be $\delta(c_i)$.
A binary relation, denoted as $\succ$, over C is a preference ordering over C if it
satisfies:*

(i) if $\underline{E}(c_1) > \overline{E}(c_2)$, then $c_1 \succ c_2$;
(ii) if $\underline{E}(c_1) > \underline{E}(c_2)$, $\overline{E}(c_1) > \overline{E}(c_2)$, and $\delta(c_1) \leq \delta(c_2)$, then $c_1 \succ c_2$;
*(iii) if $\frac{\underline{E}(c_1) + \overline{E}(c_1)}{2} = \frac{\underline{E}(c_2) + \overline{E}(c_2)}{2}$ and $\underline{E}(c_1) \geq \underline{E}(c_2)$, $0 < \delta(c_1) \leq \delta(c_2)$, then
$c_1 \succeq c_2$ (in particular, $c_1 \sim c_2$ when $\underline{E}(c_1) = \underline{E}(c_2)$, $\delta(c_1) = \delta(c_2)$); and*
(iv) if $\underline{E}(c_1) > \underline{E}(c_2)$, and $\delta(c_1) = \delta(c_2) = 1$, then $c_1 \succ c_2$.

Proof. (i) By Definition 3 and $\underline{E}(c_1) > \overline{E}(c_2)$, we have

$$\varepsilon(x) = \underline{E}(x) + (1 - \delta_x)(\overline{E}(x) - \underline{E}(x)) \ \& \ 0 \leq \delta(c_1), \ \delta(c_2) \leq 1$$
$$\Rightarrow \overline{E}(c_1) \geq \varepsilon(c_1) \geq \underline{E}(c_1) > \overline{E}(c_2) \geq \varepsilon(c_2) \geq \underline{E}(c_2) \ (When \ \underline{E}(c_1) > \overline{E}(c_2))$$
$$\Rightarrow \underline{E}(c_1) + \varepsilon(c_1) > \underline{E}(c_2) + \varepsilon(c_2)$$
$$\Rightarrow c_1 \succ c_2.$$

(ii) When $\underline{E}(c_1) > \underline{E}(c_2)$, $\overline{E}(c_1) > \overline{E}(c_2)$ and $0 \leq \delta(c_1) \leq \delta(c_2) \leq 1$, we have

$$(\underline{E}(c_1) + \varepsilon(c_1)) - (\underline{E}(c_2) + \varepsilon(c_2))$$
$$= (2\underline{E}(c_1) + (1 - \delta(c_1))(\overline{E}(c_1) - \underline{E}(c_1))) - (2\underline{E}(c_2) + (1 - \delta(c_2))(\overline{E}(c_2) - \underline{E}(c_2)))$$
$$\geq (2\underline{E}(c_1) + (1 - \delta(c_2))(\overline{E}(c_1) - \underline{E}(c_1))) - (2\underline{E}(c_2) + (1 - \delta(c_2))(\overline{E}(c_2) - \underline{E}(c_2)))$$
$$= (1 - \delta(c_2))(\overline{E}(c_1) - \overline{E}(c_2)) + (1 + \delta(c_2))(\underline{E}(c_1) - \underline{E}(c_2))$$
$$> 0$$

That is, $c_1 \succ c_2$.

(iii) When $\underline{E}(c_1) \geq \underline{E}(c_2)$, $\frac{\underline{E}(c_1) + \overline{E}(c_1)}{2} = \frac{\underline{E}(c_2) + \overline{E}(c_2)}{2}$, and $0 < \delta(c_1) \leq \delta(c_2)$, we have

$$(\underline{E}(c_1) + \varepsilon(c_1)) - (\underline{E}(c_2) + \varepsilon(c_2))$$
$$= (2\underline{E}(c_1) + (1 - \delta(c_1))(\overline{E}(c_1) - \underline{E}(c_1))) - (2\underline{E}(c_2) + (1 - \delta(c_2))(\overline{E}(c_2) - \underline{E}(c_2)))$$
$$= (\underline{E}(c_1) + \overline{E}(c_1)) - (\underline{E}(c_2) + \overline{E}(c_2)) - \delta(c_1)(\overline{E}(c_1) - \underline{E}(c_1)) + \delta(c_2)(\overline{E}(c_2) - \underline{E}(c_2))$$
$$\geq -\delta(c_2)(\overline{E}(c_1) - \overline{E}(c_2)) + \delta(c_2)(\underline{E}(c_1) - \underline{E}(c_2)).$$

Also $\underline{E}(c_1) \geq \underline{E}(c_2)$ and $\frac{\underline{E}(c_1) + \overline{E}(c_1)}{2} = \frac{\underline{E}(c_2) + \overline{E}(c_2)}{2}$ imply that $\overline{E}(c_1) \leq \overline{E}(c_2)$. Thus, since $0 < \delta(c_1) \leq \delta(c_2)$, we have $(\underline{E}(c_1) + \varepsilon(c_1)) - (\underline{E}(c_2) + \varepsilon(c_2)) \geq 0$. When $\underline{E}(c_1) = \underline{E}(c_2)$, $\delta(c_1) = \delta(c_2)$, $(\underline{E}(c_1) + \varepsilon(c_1)) - (\underline{E}(c_2) + \varepsilon(c_2)) = 0$. That is, $c_1 \succeq c_2$.

(iv) When $\delta(c_1) = \delta(c_2) = 1$, we have $\varepsilon(c_1) = \underline{E}(c_1) + (1 - \delta(c_1))(\overline{E}(c_1) - \underline{E}(c_1)) = \underline{E}(c_1)$, and $\varepsilon(c_2) = \underline{E}(c_2) + (1 - \delta(c_2))(\overline{E}(c_2) - \underline{E}(c_2)) = \underline{E}(c_2)$. Thus, since $\underline{E}(c_1) > \underline{E}(c_2)$, we have $c_1 \succ c_2$.

To apply the AHP approach to conduct decision analysis for MCDM problems with ambiguous evaluations, all the evaluations need to be a unified range of assessment because of the four reasons as follows. (i) The criteria in an MCDM problem might be classified into different categories. For example, in the problem of house investment, against the *benefit* criterion, the higher a decision alternative is evaluated, the more the decision maker prefer this alternative; for the *cost* criterion, the higher the cost of a decision alternative, the less the decision maker prefer this alternative. (ii) The criteria in the MCDM problems might be measured in different ways, which cannot be combined directly. For example, the price of a house is measured by the price per square meter, the service level is measured by the satisfaction degree. (iii) The criteria might be measured in different levels. For example, it might cost million dollars to build a house, but the number of new houses built by a company per year might be not more than 100. (iv) Decision makers cannot combine the evaluations of the qualitative criterion and the quantitative criteria directly. So, we need a mutually exclusive and collectively exhaustive set of numeral assessment grade H_n ($n = 1, \ldots, 8$), which presents an eight-scale unit preference scale value on groups of decision alternatives (focal elements according to the evidence). The meanings of the scale values from 1 to 8 are from *"extremely dislike"* to *"extremely preferred"*. More specifically, for the real estate investment problem mention in Section 3, without

Table 2. Decision matrix of the real estate investment problem

	Price	Quality	Service Level	External Environment	Investment Profit($/m^2$)			Style
	($/m^2$)	(grade)	(%)	(grade)	growth	recession	depression	(grade)
A	2-8	6	4	6,8[0.3];4,6[0.5]	7	4	2	6,8
B	2-8	8	*	6,8	6	5	4	6
C	6	*	2-4	4,6	5	4	3	4,6
D	2	4,8	6	8	8	6	3	8
E	4-6	4,6	2-8	4	7	5	1	*

loss of generality we can assume the preference scale for each criterion as shown in Table 2.

Then, for the real estate problem, how to represent ambiguous evaluation of the decision alternatives regarding some criteria in the third case? In this case, we use a mass function to represent the evaluation of a decision alternative. Since all the consequences of the decision alternatives with ambiguous evaluations of criteria can be represented by utilities, we can assume the frame of discernment of the mass function is given by a set of utilities.

Now we will show how expected utility interval is calculated regarding the *investment profit* criterion of the real estate problem. Firstly, the frame of discernment is set as follows:

$$\Theta = \{x_1, x_2, x_3, x_4, x_5, x_6, x_7, x_8\} = \{1, 2, 3, 4, 5, 6, 7, 8\}.$$

And according to a survey of real estate, we can set mass function m over Θ as follows:

$$m_A(\{x_2\}) = m_B(\{x_3\}) = m_C(\{x_4\}) = m_D(\{x_1\}) = m_E(\{x_2\}) = 0.4,$$
$$m_A(\{x_2, x_5\}) = m_B(\{x_3, x_4\}) = m_C(\{x_4, x_5\}) = m_D(\{x_1, x_3\}) = m_E(\{x_2, x_4\}) = 0.3,$$
$$m_A(\{x_5, x_7\}) = m_B(\{x_4, x_5\}) = m_C(\{x_5, x_6\}) = m_D(\{x_3, x_6\}) = m_E(\{x_4, x_8\}) = 0.3.$$

Thus, by formulas (6) and (7), we can obtain the expected utility intervals of each decision alternatives regarding the *investment profit* criterion as follows:

$EUI(A) = [4.6, 6.1]$, $EUI(B) = [5.1, 5.7]$, $EUI(C) = [4.1, 4.7]$, $EUI(D) = [5.9, 7.4]$, $EUI(E) = [4.6, 6.4]$.

4.2 Preference Degrees of Choices with Expected Utility Interval

This subsection will discuss how to obtain a unique preference degree over a group of identified decision alternatives with interval-valued expected utility. This is very important for the decision makers. For example, the decision making of real estate investment, which will cost most of money for a normal person in China; the policy decision of a government, which will causes the depression of economic environment; and the banker decides whether or not to loan a lot of money to an investor. In all of these situations, the decision maker should be more cautious than a daily decision situation in order to avoid the loss that he cannot afford. As a result, we suggest that the decision maker should follow the ambiguous aversion principle of minimax regret, which we give in Definition 6.

In order to apply the DS/AHP model, we need to give a point-valued preference degree over a group of identified decision alternatives with the expected utility interval, which can induce a preference ordering equivalent to that in Definition 6:

Definition 7. *Let $EUI(x) = [\underline{E}(x), \overline{E}(x)]$ $(x = a, b)$ be the expected utility interval of x and $\delta(x)$ be the ambiguous degree of $EUI(x)$. Then the preference degree of x is given by*

$$k(x) = \frac{2\underline{E}(x) + (1 - \delta(x))(\overline{E}(x) - \underline{E}(x))}{2}. \tag{14}$$

Moreover, we can have the following theorem:

Theorem 2. *Let $EUI(x) = [\underline{E}(x), \overline{E}(x)]$ $(x = a, b)$ be the expected utility of x and $\delta(x)$ be the ambiguous degree of $EUI(x)$. Then*

(i) $a \succ b \Leftrightarrow k(a) > k(b)$; and
(ii) $\underline{E}(a) < k(a) < \overline{E}(a)$, $\underline{E}(b) < k(b) < \overline{E}(b)$.

Proof. By $a \succ b \Leftrightarrow \Re_a^b < \Re_b^a$, $\Re_a^b = \varepsilon(b) - \underline{E}(a)$, we can obtain that $a \succ b \Leftrightarrow \frac{\varepsilon(a) + \underline{E}(a)}{2} > \frac{\varepsilon(b) + \underline{E}(b)}{2}$. Moreover $2\underline{E}(a) < \varepsilon(a) + \underline{E}(a) < 2\varepsilon(a) \Leftrightarrow \underline{E}(a) < \frac{\varepsilon(a) + \underline{E}(a)}{2} < \varepsilon(a)$.

For the *investment profit* (ip) criterion of Example 1, the ambiguous degrees of the choices are the same and equal to $\frac{0.4 \log_2 1 + 0.3 \log_2 2 + 0.3 \log_2 2}{\log_2 8} = 0.2$ by Definition 3. By Definition 7, we have

$$k_{ip}(A) = \frac{\underline{E}(A) + \varepsilon(A)}{2} = 5.2, \ k_{ip}(B) = 5.34, \ k_{ip}(C) = 4.34, \ k_{ip}(D) = 6.5, \ k_{ip}(E) = 5.32.$$

It means that the preference ordering of the decision alternatives regarding the *investment profit* criterion is $D \succ B \succ E \succ A \succ C$ and we can obtain the unique preference value over a group of identified decision alternatives for the *investment profit* criterion.

Now we will apply the ambiguous aversion principle of minimax regret to deal with the second type of ambiguity: we just know the interval-valued evaluation of selecting the decision alternatives regarding a criterion, but we have no ideas where this result came from. As a result, suppose the evaluation of decision alternative a_l regarding a criterion is $[x_i, x_j]$ and the discernment frame of the criterion is $\{x_1, \ldots, x_i, \ldots, x_j, \ldots, x_n\}$, then we can express the complete ignorance of the decision maker for the interval-valued evaluation $[x_i, x_j]$ of alternative a_l as follows:

$$m_l(\{x_i, \ldots, x_j\}) = 1,$$

which means that the decision maker only knows that the evaluation value that he can obtain could be any between x_i and x_j, but which one cannot be known. Then, by formulas (6) and (7), we have

$$EUI_l = [x_i \times 1, x_j \times 1] = [x_i, x_j].$$

And by formula (5), we have the ambiguous degree as follows:

$$\delta_l = \frac{m_i(\{x_i, \ldots, x_j\}) \log_2(j - i + 1)}{\log_2 n} = \log_n(j - i + 1).$$

Thus, by formula (11) we have

$$\varepsilon_l = \underline{E_l} + (1 - \delta_l)(\overline{E_l} - \underline{E_l}) = x_i + (1 - \log_n(j - i + 1))(x_j - x_i). \quad (15)$$

Regarding the *style* (s) criterion in the real estate investment problem mentioned in Section 3, by formulas (13) and (15), we can obtain the preference degrees of all alternatives as follows:

$$k_s(A) = 6.47, \ k_s(B) = 6, \ k_s(C) = 4.47, \ k_s(D) = 8.$$

So, the preference ordering over the decision alternatives regarding the *style* criterion is $D \succ A \succ B \succ C$, and the preference relation of decision alternative E is unknown.

5 Preference Ordering over Belief Intervals

In this section, we offer a method to obtain the optimal decision alternative by the idea behind that of the ambiguous aversion principle of minimax regret.

As the belief (Bel) function of a decision alternative represents the confidence of the decision maker towards all the evidence and the plausibility (Pl) function of a decision alternative represents the extent to which the decision maker fails to disbelieve the choice is the best [18], the decision maker's Bel function of a choice can be taken as the lower bound of a belief interval and its Pl function can be taken as the upper bound of the belief interval. Moreover, as we find that the more elements the subset is, the less confident the decision maker believes the Pl function turns out to be the probability of the choice. So, we can consider the impact of the subset ambiguous for the choice in in finding the optimal one. For example, suppose choices A and B have the same Bel function, but the non-zero value of their Pl functions is only over subset $\{A, D\}$ and $\{B, C, E\}$, respectively. Although the Bel and Pl functions are both the same for these two choices, choice A still is better than choice B because the elements in subset $\{A, D\}$ is less than that in $\{B, C, E\}$. Formally, we can give the definition of ambiguous degree of such a belief interval and the degree of preference between two decision alternatives as follow:

Definition 8. *Let $Bel(A)$ be the belief function of $m(A)$ and $Pl(A)$ be its plausibility function Then the ambiguous degree for uncertain interval $[Bel(A), Pl(A)]$ over choice A, denoted as $\delta_U(A)$, is given by*

$$\delta_U(A) = \frac{\sum_{A \cap B \neq \emptyset} m(B) \log_2 |B|}{\log_2 |\Theta|}. \quad (16)$$

Definition 9. *The degree of preference of a over b, denoted by $P_b^a) \in [0,1]$, is given by:*

$$P_b^a = \frac{2Bel(\{a\}) + (1 - \delta_U(a))(Pl(\{a\}) - Bel(\{a\}))}{2Bel(\{b\}) + (1 - \delta_U(b))(Pl(\{b\}) - Bel(\{b\}))}. \tag{17}$$

By Definition 9, the preference relation between two choices can be defined as follows:

Definition 10. *The preference relation between two decision alternatives a and b is defined as follows: (i) $a \succ b$ if $P_b^a > 1$; (ii) $a \prec b$ if $P_b^a < 1$; and (iii) $a \sim b$ if $P_b^a = 1$.*

Theorem 3. *Let C be a finite choice set, $Bel(c)$ and $Pl(c)$ be the belief function and the plausibility function of choice $c \in C$, and the ambiguous degree for uncertain interval $[Bel(c), Pl(c)]$ over choice c be $\delta_U(c)$. Then the preference ordering $\succ$ over C satisfies:*

(i) *if $Bel(c_1) > Pl(c_2)$, then $c_1 \succ c_2$;*
(ii) *if $Bel(c_1) > Bel(c_2)$, $Pl(c_1) > Pl(c_2)$, and $\delta_U(c_1) \leq \delta_U(c_2)$, then $c_1 \succ c_2$;*
(iii) *if $\frac{Bel(c_1)+Pl(c_1)}{2} = \frac{Bel(c_2)+Pl(c_2)}{2}$ and $Bel(c_1) \geq Bel(c_2)$, $0 < \delta_U(c_1) \leq \delta_U(c_2)$, then $c_1 \succeq c_2$. ($c_1 \sim c_2$ when $Bel(c_1) = Bel(c_2)$, $\delta_U(c_1) = \delta_U(c_2)$); and*
(iv) *if $Bel(c_1) > Bel(c_2)$, and $\delta_U(c_1) = \delta_U(c_2) = 1$, then $c_1 \succ c_2$.*

The proof of the above theorem is similar to that of Theorem 1. According to Definition 10 and Theorem 3, we can set the preference ordering over the decision alternatives regarding all the criteria and thus obtain the optimal choice.

6   The Scenario of Real Estate Investment

Now we will illustrate our model by the real estate investment problem mentioned in Section 3. According to Tables 1, by the expected utility interval formulas (6) and (7) and the ambiguous aversion principle of minimax regret, we can obtain the preference degree for all groups of decision alternatives regarding each criterion.

Without loss of generality, suppose the subjective preference scale regarding each criterion is the same as that in Table 2 and the preference scale of the decision maker based on the four assessment grades of external environment criterion and style criterion are $u(P) = 2$, $u(A) = 4$, $u(G) = 6$, $u(E) = 8$.

By Definitions 6 and 7, formulas (13) and (15), we can obtain the point-value preference degrees for all groups of decision alternatives regarding each criterion as follow:

– for the *price* criterion $k_p(\{A, B\}) = 2.19, k_p(\{C\}) = 6, k_p(\{D\}) = 2, k_p(\{E\}) = 4.47$;

- for the *quality* criterion $k_q(\{A\}) = 6, k_q(\{B\}) = 8, k_q(\{D\}) = 4.45, k_q(\{E\}) = 4.47$;
- for the *service level* criterion $k_l(\{A\}) = 4, k_l(\{C\}) = 2.47, k_l(\{D\}) = 6, k_l(\{E\}) = 2.19$;
- for the *external environment* criterion $k_e(\{A\}) = 4.8, k_e(\{B\}) = 4.45, k_e(\{C\}) = 4.47, k_e(\{D\}) = 8, k_e(\{E\}) = 4$;
- for the *investment profit* criterion $k_{ip}(\{A\}) = 5.2, k_{ip}(\{B\}) = 5.34, k_{ip}(\{C\}) = 4.34, k_{ip}(\{D\}) = 6.5, k_{ip}(\{E\}) = 5.32$; and
- for the *style* criterion $k_s(\{A\}) = 6.47, k_s(\{B\}) = 6, k_s(\{C\}) = 4.47, k_s(\{D\}) = 8$.

Table 3. The mass function of each criterion in the house investment example

criterion	mass function
Price	$m_1(\{A, B\}) = 0.089$ $m_1(\{C\}) = 0.243$ $m_1(\{D\}) = 0.081$ $m_1(\{E\}) = 0.181$ $m_1(\Theta) = 0.406$
Quality	$m_2(\{A\}) = 0.203$ $m_2(\{B\}) = 0.271$ $m_2(\{D\}) = 0.15$ $m_2(\{E\}) = 0.151$ $m_2(\Theta) = 0.225$
Service Level	$m_3(\{A\}) = 0.116$ $m_3(\{C\}) = 0.071$ $m_3(\{D\}) = 0.173$ $m_3(\{E\}) = 0.063$ $m_3(\Theta) = 0.577$
External Environment	$m_4(\{A\}) = 0.1$ $m_4(\{B\}) = 0.093$ $m_4(\{C\}) = 0.093$ $m_4(\{D\}) = 0.166$ $m_4(\{E\}) = 0.083$ $m_4(\Theta) = 0.465$
Investment Profit	$m_5(\{A\}) = 0.137$ $m_5(\{B\}) = 0.141$ $m_5(\{C\}) = 0.115$ $m_5(\{D\}) = 0.172$ $m_5(\{E\}) = 0.14$ $m_5(\Theta) = 0.295$
Style	$m_6(\{A\}) = 0.144$ $m_6(\{B\}) = 0.133$ $m_6(\{C\}) = 0.1$ $m_6(\{D\}) = 0.178$ $m_6(\Theta) = 0.445$

Now without loss generality, we assign the criteria weights based on the judgement on importance of each criterion as follows: $\omega(Price) = 0.2$, $\omega(Quality) = 0.3$, $\omega(Service\ Level) = 0.1$, $\omega(External\ Environment) = 0.1$, $\omega(Investment Profit) = 0.2$, $\omega(Style) = 0.1$. Then the mass function for the focal elements of each criteria can be obtained by formulas (8) and (9) as shown in Table 3. Hence, the overall mass function for the houses after combining all evidence from the criteria by the Dempster combination rule in Definition 2 is:

$$m_{house}(\{A\}) = 0.216, m_{house}(\{B\}) = 0.202, m_{house}(\{C\}) = 0.109, m_{house}(\{D\}) = 0.276,$$
$$m_{house}(\{E\}) = 0.148, m_{house}(\{A, B\}) = 0.009, m_{house}(\Theta) = 0.04.$$

Further, a value of each mass function m_{house} represents the level of exact belief for all associated group of decision alternatives containing the best one. From the overall mass function, we can obtain the measures of belief (Bel) and plausibility (Pl) by formulas (1) and (2) as follows:

$$Bel(A) = 0.216, Bel(B) = 0.202, Bel(C) = 0.109, Bel(D) = 0.276, Bel(E) = 0.148;$$
$$Pl(A) = 0.265, Pl(B) = 0.251, Pl(C) = 0.149, Pl(D) = 0.316, Pl(E) = 0.188.$$

Hence, by Definitions 8 and 9 we can obtain

$$\delta_U(A) = \delta_U(B) = \frac{0.009 \times \log_2 2 + 0.04 \log_2 5}{\log_2 5} = 0.044, \ \delta_U(C) = \delta_U(D) = \delta_U(E) = 0.04.$$

Thus, by Theorem 3 and Definition 10, we can obtain the overall preference ordering as follows:

$$D \succ A \succ B \succ E \succ C.$$

So, D is the optimal one.

7   Summary

In this paper we firstly review D-S theory and the DS/AHP method. Then we propose an ambiguous aversion principle of minimax regret in order to set a proper preference ordering over choices with expected utility intervals. Further, we incorporate the expected utility interval and the ambiguous aversion principle of minimax regret into the DS/AHP theory so that ambiguous evaluations of criteria can be handled well in AHP. That is, we identify three types of ambiguous information and point out that the two among three types cannot be solved by the DS/AHP method but can be done by our model. Finally, we illustrate the applicability of our model by a problem of real estate investment in the domains of business.

In the future, it is interesting to do lots of psychological experiments to refine and validate our model, integrate more factors of emotions into our model, and apply our model into a wider range of domains, especially intelligent agent systems such as multi-attribute negotiating agents [11,14,15]. It is also interesting to apply our idea of this paper into other MCDM techniques such as fuzzy contraints [12] and fuzzy aggregation operators [13,16,23].

Acknowledgements. This paper is partially supported by National Natural Science Foundation of China (No. 61173019), Bairen plan of Sun Yat-sen University, National Fund of Philosophy and Social Science (No. 11BZX060), Office of Philosophy and Social Science of Guangdong Province of China (No. 08YC-02), Fundamental Research Funds for the Central Universities in China, and major projects of the Ministry of Education (No. 10JZD0006).

References

1. Awasthi, A., Chauhan, S.S.: Using AHP and Dempstere-Shafer Theory for Evaluating Sustainable Transport Solutions. Environmental Modelling & Software 26(6), 787–796 (2011)
2. Beynon, M.J., Curry, B., Morgan, P.H.: The Dempster-Shafer Theory of Evidence: An Alternative Approach to Multicriteria Decision Modelling. Omega 28(1), 37–50 (2000)
3. Beynon, M.J.: DS/AHP Method: A Mathematical Analysis, Including an Understanding of Uncertainty. European Journal of Operational Research 140(1), 149–165 (2002)
4. Beynon, M.J.: Understanding Local Ignorance and Non-Specificity within the DS/AHP Method of Multi-Criteria Decision Making. European Journal of Operational Research 163(2), 403–417 (2005)

5. Beynon, M.J.: The Role of the DS/AHP in Identifying Inter-Group Alliances and Majority Rule within Group Decision Making. Group Decision and Negotiation 15(1), 21–42 (2006)
6. Beynon, M.J., Moutinho, L., Veloutsou, C.: A Dempster-Shafer Theory Based Exposition of Probabilistic Reasoning in Consumer Choice. In: Casillas, J., Martínez-López, F.J. (eds.) Marketing Intelligent Systems Using Soft Computing. STUDFUZZ, vol. 258, pp. 365–387. Springer, Heidelberg (2010)
7. Dempster, A.P.: Upper and Lower Probabilities Induced by a Multivalued Mapping. The Annals of Mathematical Statistics 38(2), 325–339 (1967)
8. Dubois, D., Prade, H.: A Note on Measures of Specificity for Fuzzy Sets. International Journal of General Systems 10(4), 279–283 (1985)
9. Ellsberg, D.: Risk, Ambiguous, and the Savage Axioms. Quarterly Journal of Economics 75(4), 643–669 (1961)
10. Landman, J.: Regret: the Persistence of the Possible. Oxford University Press, Oxford (1993)
11. Luo, X., Jennings, N.R., Shadbolt, N., Leung, H.-f., Lee, J.H.-m.: A Fuzzy Constraint-based Knowledge Model for Bilateral, Multi-issue Negotiations in Semicompetitive Environments. Artificial Intelligence 148(1-2), 53–102 (2003)
12. Luo, X., Lee, J.H.-m., Leung, H.-f., Jennings, N.R.: Prioritised Fuzzy Constraint Satisfaction Problems: Axioms, Instantiation and Validation. Fuzzy Sets and Systems 136(2), 155–188 (2003)
13. Luo, X., Jennings, N.R.: A Spectrum of Compromise Aggregation Operators for Multi-Attribute Decision Making. Artificial Intelligence 171(2-3), 161–184 (2007)
14. Luo, X., Miao, C., Jennings, N.R., He, M., Shen, Z., Zhang, M.: KEMNAD: A Knowledge Engineering Methodology for Negotiating Agent Development. Computational Intelligence 28(1), 51–105 (2012)
15. Pan, L., Luo, X., Meng, X., Miao, C., He, M., Guo, X.: A Two-Stage Win-Win Multi-Attribute Negotiation Model. In: Intelligence Optimization and Then Concession (2012), doi: 10.1111/j.1467-8640.2012.00434.x
16. Ricci, R.G., Mesiar, R.: Multi-Attribute Aggregation Operators. Fuzzy Sets and Systems 181(1), 1–13 (2011)
17. Saaty, T.L.: The Analytic Hierarchy Process. McGraw Hill, New York (1980)
18. Shafer, G.: A Mathematical Theory of Evidence. Princeton University Press, Princeton (1976)
19. Strat, T.M.: Decision Analysis Using Belief Functions. International Journal of Approximate Reasoning 4(5-6), 391–418 (1990)
20. Savage, L.: The Theory of Statistical Decision. Journal of the American Statistical Association 46(253), 54–67 (1951)
21. Treich, N.: The Value of a Statistical Life under Ambiguous Aversion. Journal of Environmental Economics and Management 59(1), 15–26 (2010)
22. von Neumann, J., Morgenstern, O.: Theory of Games and Economic Behavior, Commemorative edition. Pricetion University Press, Pricetion (2004)
23. Yu, X., Xu, Z., Liu, S., Chen, Q.: Multicriteria Decision Making with 2-Dimension Linguistic Aggregation Techniques. International Journal of Intelligent Systems 27(6), 539–562 (2012)

Synthesizing Image Representations of Linguistic and Topological Features for Predicting Areas of Attention

Pascual Martínez-Gómez[1,2], Tadayoshi Hara[1], Chen Chen[1,2], Kyohei Tomita[1,2], Yoshinobu Kano[1,3], and Akiko Aizawa[1]

[1] National Institute of Informatics
{pascual,harasan,chen,kyohei,kano,aizawa}@nii.ac.jp
[2] The University of Tokyo
[3] PRESTO, Japan Science and Technology Agency

Abstract. Depending on the reading objective or task, text portions with certain linguistic features require more user attention to maximize the level of understanding. The goal is to build a predictor of these text areas. Our strategy consists in synthesizing image representations of linguistic features, that allows us to use natural language processing techniques while preserving the topology of the text. Eye-tracking technology allows us to precisely observe the identity of fixated words on a screen and their fixation duration. Then, we estimate the scaling factors of a linear combination of image representations of linguistic features that best explain certain gaze evidence, which leads us to a quantification of the influence of linguistic features in reading behavior. Finally, we can compute saliency maps that contain a prediction of the most interesting or cognitive demanding areas along the text. We achieve an important prediction accuracy of the text areas that require more attention for users to maximize their understanding in certain reading tasks, suggesting that linguistic features are good signals for prediction.

1 Introduction

Reading is an important method for humans to receive information. While skilled readers have powerful strategies to move fast and optimize their reading effort, average readers might be less efficient. When producing a text, the author may or may not be aware of the text areas that require more attention from users. Moreover, depending on the reading objective or strategy, there might be different areas that catch user's attention for longer periods of time. Ideally, readers would know *a priori* what are the pieces of text with the most interesting linguistic characteristics to attain his/her reading objectives, and proceed to the reading act consequently. However, due to the uncertainty on the distribution of time-demanding portions of text, users may incur in an inefficient use of cognitive effort when reading.

The goal is then, given a user and a text, provide a map with the most interesting text areas according to user's reading objective. We work under the assumption that on-line cognitive processing influences on eye-movements [10], and that people with different reading strategies and objectives fixate on words and phrases with different linguistic

P. Anthony, M. Ishizuka, and D. Lukose (Eds.): PRICAI 2012, LNAI 7458, pp. 312–323, 2012.

features. Thus, given the same text, different maps might be displayed when users have different objectives or reading strategies.

Traditionally, in the field of computational linguistics, features and models have been developed to explain observations of natural language, but not to explain the cognitive effort required to process those observations by humans. In the present work, the first step is to quantify the influence of linguistic features when users are performing different reading tasks. We will use an eye-tracker device to capture gaze samples when users read texts. Then, we will synthesize image representations of these gaze samples and image representations of several linguistic features. By assigning a relevance weight to the image representations of linguistic features, we can find the best configuration of the value of these scaling factors that best explain the image representation of the reference gaze samples. After obtaining the influence of each linguistic feature on every reading objective, maps showing the attention requirements of new texts could be automatically obtained and displayed to users before they start their reading act, or documents could be conveniently formatted according to these maps.

As far as we know, our approach is the first attempt to use natural language processing (NLP) techniques while preserving the topology of the words appearing on the screen. The necessity to develop this framework arises from the integrated use of gaze information that consists in spatial coordinates of gaze and the linguistic information that can be extracted using traditional NLP techniques [8]. We believe that by synthesizing image representations of linguistic features, image processing techniques become available to perform natural language processing that inherently incorporates the geometric information of the text and gaze.

This paper is organized as follows. Next section describes previous work related to the present investigation. Then, we briefly introduce a recent text-gaze alignment method using image registration techniques that we borrow from [9] for completeness. In section 4 we introduce the technique to build image representations of reference gaze data. We describe in detail in section 5 how image representations of linguistic features are synthesized and how their influence on reading behavior is estimated. A description of the reading objectives, experimental settings and empirical results are in section 6. Some conclusions and future directions are left for the final section.

2   Related Work

Gaze data and natural languages are different modalities that suffer from an important ambiguity. In order to use these sources of information successfully, meaningful features and consistent patterns have to be extracted. Under this perspective, there exist two types of approaches.

In the first approach, there is a search for linguistic features that activate cognitive processes. A recent example following this idea can be found in the field of *active learning*. Supervised machine learning NLP approaches require an increasing amount of annotated corpus and active learning has proved to be an interesting strategy to optimize human and economical efforts. Active learning assumes the same cost to annotate different samples, which is not accurate. With the purpose to unveil the real cost of annotation, [15] propose to empirically extract features that influence cognitive effort by

Raw gaze data on word bounding boxes   Linearly transformed gaze data on word bounding boxes

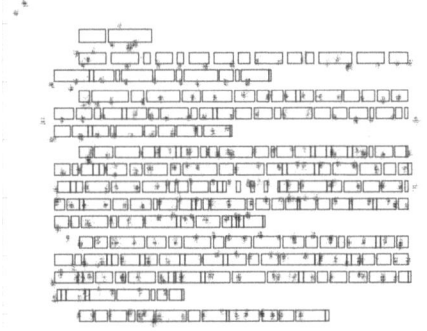

Fig. 1. On the left, raw gaze data superimposed on text bounding boxes. On the right, linearly transformed gaze data (translation = −38 pixels, scaling = 1.13) on text bounding boxes. Most gaze samples in the figure on the right are mapped onto a reasonable bounding box.

means of an eye-tracker. They affirm that their model built using these features explains better the real annotation efforts.

The objective of the second approach is to find reading behavior characteristics[1] that reflect certain linguistic features. Authors in [3] found that certain characteristics of reading behavior correlate well with traditional measures to evaluate machine translation quality. Thus, they believe that eye-tracking might be used to semi-automatically evaluate machine translation quality using the data obtained from users reading machine translation output.

In the present work, we make use of gaze data reflecting cognitive processing to extract the importance of linguistic features with the objective of predicting the attention that users will have when reading text. This will be implemented using techniques from image recognition for error-correction, function optimization and synthesis.

3 Text-Gaze Alignment

Eye-tracking technology is improving fast and modern devices are becoming more affordable for laboratories and companies. Despite of this rapid development, eye-trackers still introduce measurement errors that need to be corrected at a preprocessing stage. [4] describe two types of errors in a typical eye-tracking session. The first type is variable errors that can be easily corrected by aggregating the gaze samples into fixations. The second type is systematic errors. Systematic errors often come in the form of vertical drifts [5] and they are more difficult to correct.

Researchers and practitioners of eye-tracking have developed their own error correction methods, but they are either too task-specific or introduce constraints on reading behavior. [9] model the error-correction problem as an image registration task that is

---

[1] Under the assumption that reading behavior is an indicator of cognitive processing. This is also called the *eye-mind assumption*.

briefly described in this section. Image registration is a well studied technique to align two or more images captured by different sensors. In our context, it can be used as a general error correction method in unconstrained tasks where the objective is to align the image representation of gaze samples to the image representation of words appearing on the screen. This method works reasonably well under the assumption that users solely fixate on text. The key is to define a spatial transformation to map gaze coordinates into the space of words. Addressing the vertical drift reported in [5], a linear transformation is defined as $g_{a,b} : (x,y) \rightarrow (x, y \cdot b + a)$ where a (translation) and b (scaling) are the transformation parameters of the y-coordinates that have to be estimated by means of a non-linear optimization strategy. An easy objective function measures how well gaze samples are mapped *into* word bounding boxes. Let $\mathbf{G}_{a,b}$ be the image representation of the mapped gaze samples, where pixel $\mathbf{G}_{a,b}(i,j)$ has a high value if there is a gaze sample in coordinates (i,j). And let $\mathbf{W}$ the image representation of word bounding boxes in a text, where pixel $\mathbf{W}(i,j)$ has a high value if it falls inside a word bounding box. A measure of alignment between the two image representations can be defined as the sum of absolute differences of pixels (i,j):

$$f(\mathbf{G}_{a,b}, \mathbf{W}) = \sum_i \sum_j |\mathbf{G}_{a,b}(i,j) - \mathbf{W}(i,j)| \tag{1}$$

The intention is then to estimate the values $(\hat{a}, \hat{b})$ of the transformation parameters that minimize the objective function f:

$$(\hat{a}, \hat{b}) = \underset{a,b}{\text{argmin}} \, f(\mathbf{G}_{a,b}, \mathbf{W}) \tag{2}$$

$$= \underset{a,b}{\text{argmin}} \sum_i \sum_j |\mathbf{G}_{a,b}(i,j) - \mathbf{W}(i,j)| \tag{3}$$

Due to the non-convexity nature of the solution space, this optimization is iteratively performed using different levels of blurs in what is called multi-blur optimization and hill-climbing at every iteration. Typical results of the error correction of gaze samples using this method can be found in fig. 1. Then, by using the information on the structure of the text, gaze samples can be collapsed into fixations according to their closest word bounding box.

4   Gaze Evidence

Psycholinguistic studies have long noted that eye-movements reveal many interesting characteristics [14]. For example, fixation locations are usually strongly correlated with current focus of attention and, when users read text, they indicate the identity of the word or phrase that the subject is currently processing. Another variable, fixation durations, is useful in quantifying other hidden processes such as user's familiarity with the text or with specific terms, whether or not the text is written in the user's native language, etc. Among saccadic movements, length and direction of regressions are also interesting features of the reading act that occur in diverse situations as when the subject reads about a fact that gets in contradiction with prior knowledge. Although forward and backward saccadic movements provide relevant information on subjects and texts, they might be ambiguous and difficult to interpret by using automatic methods in unconstrained tasks. For this reason, in this work we will only focus on the interpretation of fixation locations and their duration.

Fixation locations and their durations depend on many variables that mainly come from two sources. The first source is from subject's personal characteristics, namely prior background knowledge, native language, cultural identity, interests or reading objectives. Examples of reading objectives are precise reading, question answering, writing a review or preparing a presentation. The second source of variables are related to the linguistic characteristics of the text.

As we have previously stated in the introduction, we are interested in finding the importance of individual linguistic features to explain certain reading behaviors when users are reading texts with different objectives in mind. When pursuing these objectives, different users may have different levels of success. There are many types of reading objectives that can be set to guide subject's reading strategy, where writing reviews or preparing presentations tasks are among them. The level of success of these reading tasks can be evaluated by assessing the performance of the actions that subjects have to carry out after reading, but it might be difficult since there are other variables that may influence subject's performance, such as personal (in)ability to prepare presentations or prior prejudices about the topic the subject writes the review about. For this reason, we limit ourselves to reading objectives whose attainment degree can be easily evaluated as a function of the level of understanding achieved, as measured by an interview with the subject to quantify the accuracy and completeness of his/her answers. Examples of this type of reading objectives are precise reading, question answering or obtaining general information in a very limited amount of time.

Our intention is to predict the text areas where subjects should fixate longer in order to maximize their level of understanding. Thus, we have to obtain gaze evidence that serves as a reference of effective reading behavior. One may be tempted to sample the population of subjects and select the most effective reader. There might be, however, other subjects that follow a different reading strategy and achieve other effectivity levels that we should take into consideration. In order to include this uncertainty in our system, we weight the gaze evidence obtained by all users according to their level of understanding. Let $\mathbf{G}_u$ be the image representation of gaze evidence obtained from user u when reading a certain text, and let λ_u his/her level of understanding on that text. By considering the image representation of gaze evidence as a matrix whose (i, j) positions denote pixels with a gray value between 0 and 1, we can obtain a reference gaze evidence by scaling the evidence of every subject with his/her level of understanding:

$$\mathbf{G}^\tau = \sum_u \lambda_u \cdot \mathbf{G}_u \qquad (4)$$

An schema representing the idea of scaling gaze evidence by user's level of understanding can be found in fig. 2. The linear combination of image representations of gaze evidence is carried out using scaling factors λ_u and it is essential to preserve the sense of uncertainty in our system.

Linear combination of gaze evidence from U users with different level of understanding λ_u

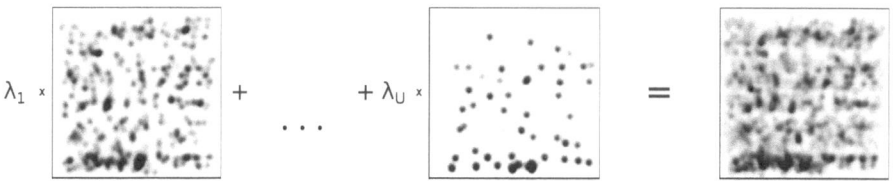

Fig. 2. In order to obtain a reference gaze evidence $\mathbf{G}^\tau$ that also accounts for the uncertainty of different reading strategies, image representations of gaze samples are scaled by the level of understanding λ_u of every user.

5   Quantifying Influence of Linguistic Features

Eye-actions can be roughly classified into two important categories. The first one consists in fixations, where eyes remain still, gazing on a word or phrase for lexical processing. The second category is saccades, consisting in abrupt eye-movements that are used to place the gaze on different text locations. Syntactic and semantic integration of lexical information is believed to influence these eye-movements [10]. While both eye-actions provide much information about on-line cognitive processing, in this work we will only use information about fixation locations and their durations.

The identity of fixated words and their fixation duration depend on many factors. Some of these factors are related to personal characteristics, e.g. reading objective, reading skill, prior knowledge that serves as background, user's interests, etc. Other factors depend on the linguistic features of the text [14], e.g. lexical properties of words, syntactic or semantic features, etc. The study on the impact of each of these factors is interesting, but it will be limited in this work to linguistic features and reading objectives. The factor of reading objectives influences on data collection and it will be discussed in its own section.

Within the image recognition field, saliency maps [6] represent visual salience of a source image. Similarly, we can synthesize image representations of a text that describe it, while preserving the topological arrangement of words. There are multiple possibilities to describe text according to its linguistic features. Many interesting linguistic features can be numerically described. For instance, we can think of word unpredictability as a probability, as given by an N-gram language model. Other examples are word length, or semantic ambiguity, according to the number of senses in WordNet [11].

To synthesize an image representation of a certain linguistic feature, we filled the word bounding boxes with a gray level between 0 and 1, proportional to the numerical value of the linguistic feature. Fig. 3 shows two examples of image representations of linguistic features. In order to account for the uncertainty introduced by measurement and user errors, images are slightly blurred by convolving them with an isotropic gaussian filter with spread $\sigma = 10$ pixels. Finally, to normalize the intensity when comparing different image representations of linguistic features, the intensity of the images are adjusted so that only the upper and lower 1% of the pixels are saturated.

Semantic ambiguity   Nouns

Fig. 3. Two examples of image representations of linguistic features. On the left, image representation of semantic ambiguity as given by the number of senses of every word in WordNet. On the right, image representation of Nouns, where bounding boxes of words that are nouns are filled with high pixel values. Note that the pixel values of these images are complemented for clarity.

The final step is to find the weight ω_f of the image representations of every linguistic feature. Let's consider again images as matrices whose pixels are elements with a value between 0 and 1 and denote by $\mathbf{G}^\tau$ the image representation of the error-corrected gaze evidence. We denote by $\mathbf{W}_1^F = \{\mathbf{W}_1, \ldots, \mathbf{W}_f, \ldots, \mathbf{W}_F\}$ the list of image representations of F linguistic features. A dissimilarity function between the gaze evidence and the linguistic features can be defined as the absolute pixel-wise (i, j) difference between the images:

$$g(\mathbf{G}^\tau, \mathbf{W}_1^F) = \sum_i \sum_j \left| \mathbf{G}^\tau(i, j) - \sum_{f=1}^F \omega_f \cdot \mathbf{W}_f(i, j) \right| \tag{5}$$

where the image representations of the linguistic features are linearly combined by scaling factors ω_f. Then, a standard algorithm can be used to perform a non-linear optimization to minimize the dissimilarity. Formally,

$$\hat{\omega} = \underset{\boldsymbol{\omega}}{\operatorname{argmin}} \, g(\mathbf{G}^\tau, \mathbf{W}_1^F) \tag{6}$$

$$= \underset{\boldsymbol{\omega}}{\operatorname{argmin}} \sum_i \sum_j \left| \mathbf{G}^\tau(i, j) - \sum_{f=1}^F \omega_f \cdot \mathbf{W}_f(i, j) \right| \tag{7}$$

A graphical schema of the combination process can be found in fig. 4. The objective of the optimization is to estimate the importance of different linguistic features so that the linear combination best explains the gaze evidence used as a reference.

6   Experimental Results

6.1   Experimental Settings

We believe that the importance of different linguistic features to explain certain reading behavior depends on the reading objective. Following this hypothesis, we designed three

Estimation of importance of several linguistic features

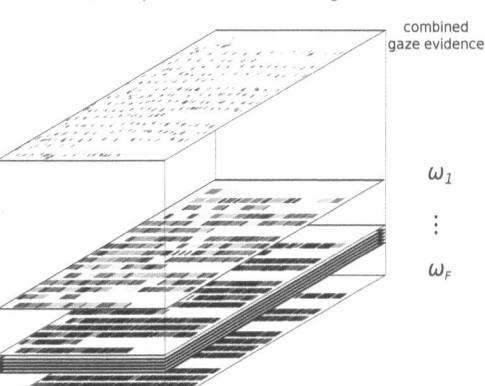

Fig. 4. Schema representing the estimation of several linguistic features (on the bottom) that best explain certain gaze evidence (on the top). Image representations of different linguistic features are linearly combined with scaling factors ω_f and their values are estimated by means of a standard non-linear optimization method. For clarity, the value of the pixels in the images are complemented and the image representations have not been blurred nor their intensity adjusted.

tasks with different reading objectives. There were two documents with a different topic for every task. In the first task, subjects were asked to carefully read a text and that some questions about the text will be asked after. Subjects were also told that they could spend as much time as they need to maximize their understanding about the text, but such a text would not be available during the evaluation of their understanding. In order to check the level of understanding, open questions, true-false and select-correct-answer questions were asked. In the second task, subjects were given questions before the reading act and asked them to find the answers in the text. We asked one and two short questions (for each document, respectively) for users to easily remember them and not to cause extra cognitive load. In the third task, we asked subjects to obtain as much information as possible from a text in only 10 seconds. We told them that they would be required to speak out as much information as they obtained after the reading and that their reading did not have to be necessarily sequential.

For every task, there were two documents in English presented to the subjects. These documents consisted in short stories extracted from interviews, fiction and news. The average number of sentences per document was 13.50 and the average number of words per sentence was 20.44.

While subjects were reading the documents on the screen, Tobii TX300 was used to capture gaze samples at a rate of 300 Hz in a 23" screen at a resolution of 1920 x 1080, resulting in more than 800MBs of gaze coordinates. Text was justified in the center of the screen, from the top-most left position located at pixels (600px, 10px) to the bottom-most right position located at (1300px, 960px). Words were displayed in a web browser using Text2.0 framework [1] allowing to easily format the text using CSS style-sheets

Table 1. List of linguistic features divided in three categories: lexical, syntactic and semantic. Instances of $tag are "Noun", "Adjective", "Verb", etc. Word unpredictability is computed as the perplexity by a 5-gram language model. Heads and parse trees are extracted using Enju [12], an English HPSG parser.

Category	Linguistic feature	type
	word length	Integer
	contains digit	Binary
Lexical	word unpredictability	Real
	contains uppercase	Binary
	is all uppercase	Binary
	is head	Binary
	is POS $tag (23 features)	Binary
Syntactic	height of parse tree of its sentence	Integer
	depth of the word in the parse tree	Integer
	word position in sentence	Integer
Semantic	is named entity	Binary
	ambiguity: number of senses from WordNet	Integer

and to recover the geometric boundaries of the words. The font family was Courier New of size 16px, text indentation of 50px and line height 30px, in black on a light gray background. A chin rest was used to reduce errors introduced by readers. Text on the screen was short enough not to require any action to entirely visualize it. There were 10 subjects participating in three tasks consisting in two documents each. The subjects were undergraduate, master, Ph.D. and post-doc students from China, Indonesia, Japan, Spain, Sweden and Vietnam with a background in Computer Science.

The eye-tracker was calibrated once per subject and document. Then, an unsupervised text-gaze alignment using the image registration technique [9] was used to automatically correct vertical measurement errors. There were two sessions[2] (out of 60) that needed manual correction of horizontal errors and another session had to be corrected using better vertical scaling and translation than the one automatically obtained by the unsupervised method.

There is a huge amount of linguistic features that could be considered to explain certain reading behaviors. Intuitively, some features might be more relevant than others, but ideally all of them should be included in the model with different scaling factors according to the gaze evidence. We divided the type of linguistic features into three classes: lexical, syntactic and semantic features. Although the list could be bigger, Table 1 contains the linguistic features that were used in this work. Examples of these features are the part-of-speech (POS) tag, or the word unpredictability as measured by the perplexity computed using a 5-gram language model estimated using the EuroParl corpus [7] and smoothed using modified Kneser-Ney smoothing [2]. In order to find the best estimates $\hat{\omega}$ of the scaling factors ω_f in eq. 7, we used Powell's dogleg trust region algorithm [13] as a standard non-linear optimization method.

---

[2] A session is defined as a subject reading a document.

Table 2. Average values and standard deviation of dissimilarity between the image representation of the test gaze evidence and the linear combination of weighted image representations of linguistic features in a cross-validation. The scale of the values can be found in the last column.

	precise reading		question answering		10-second reading		
	doc. 1	doc. 2	doc. 3	doc. 4	doc. 5	doc. 6	scale
baseline	403 ± 4.1	456 ± 5.9	463 ± 9.7	444 ± 4.5	431 ± 0.2	$\mathbf{415 \pm 0.1}$	$\times 10^3$
scaled feats.	$\mathbf{288 \pm 3.1}$	$\mathbf{325 \pm 2.6}$	$\mathbf{349 \pm 8.5}$	$\mathbf{371 \pm 3.6}$	$\mathbf{419 \pm 4.1}$	562 ± 5.7	$\times 10^3$

Table 3. Average correlation between the vectors of scaling factors obtained from the cross-validation. A high intra-document correlation can be appreciated between the scaling factors within the same document. A low inter-document correlation can be appreciated between the scaling factors estimated for different documents within the same task.

	precise reading		question answering		10-second reading	
	doc. 1	doc. 2	doc. 3	doc. 4	doc. 5	doc. 6
doc. 1	**0.94**	0.34	–	–	–	–
doc. 2	0.34	**0.96**	–	–	–	–
doc. 3	–	–	**0.93**	0.37	–	–
doc. 4	–	–	0.37	**0.96**	–	–
doc. 5	–	–	–	–	**0.96**	−0.19
doc. 6	–	–	–	–	−0.19	**0.94**

6.2 Results

In order to evaluate the predictive power of the linear combination of image representations of linguistic features, leaving-one-out cross-validation was used. Cross-validation was carried out among all subjects within the same document, and every observation consisted in a single session (per document) containing gaze evidence of a subject. Using the training set, scaling factors of the linguistic features were estimated and an image representation of the weighted linguistic features was synthesized and compared to the gaze evidence in the test data. The average results of such comparison can be found in Table 2. As a baseline, we used uniform weights to scale the image representations of the linguistic features in Table 1.

It can be observed that in the precise reading and question answering tasks, there is a consistent and significant reduction in the dissimilarity of the image representations of the test gaze observations and the linear combination of image representations of linguistic features, when compared to the baseline. However, in the 10-second task, the model fails at predicting the distribution of the gaze evidence since the dissimilarity is not consistently reduced. We have two hypotheses to explain such a fact. The first one is that subject's personal characteristics (e.g. background knowledge, native language, etc.) are essential features to explain the reading behavior in the latter task. The second hypothesis is that we have left out important linguistic features.

As we have seen, for the precise reading and the question answering task, the linear combination of image representations of linguistic features helps to explain the gaze evidence of readers within the same document. The remaining question is whether the scaling factors of the linguistic features are good predictors for different documents that are being read using the same reading strategy. To answer this question, we have computed the average correlation between the value of the estimated scaling factors for observations between different documents of the same reading task, together with the correlations within the same document. The results can be observed in Table 3. It can be appreciated that the inter-document correlation is low, suggesting that the estimated weights of the linear combination from one document are not good predictors for other documents. This contrasts with the high intra-document correlation, reinforcing the consistency of the estimations to explain gaze evidence from different subjects within the same document. Since the intra-document precision is considerable, the most plausible explanation is that more documents of the same reading objective are needed to robustly estimate the scaling factors of such amount of linguistic features.

7   Conclusions and Future Work

In the first stage, we have collected gaze evidence from subjects reading documents using three different reading objectives and measured their level of understanding. Well-known systematic errors were corrected using image-registration techniques. Then, a reference gaze evidence has been obtained by linearly combining the image representations of the gaze evidence of every subject scaled by their level of understanding. In the second stage, image representations of several linguistic features have been synthesized and the importance of every linguistic feature has been estimated to explain the reference gaze evidence. The predictive power of the linear combination of image representations of linguistic features have been assessed on held-out data. Our model obtains higher recognition accuracy than a non-informed system in the precision reading and question answering task. However, our model fails at predicting reading behavior in the 10-second reading task.

In order to evidence the generalization power of our model using the estimated scaling factors, we computed the correlation between the scaling factors trained on different documents of the same reading objective task. We found a high intra-document correlation but a low inter-document correlation within the same task.

The results of this work find an immediate application to collaborative filtering and recommendation using gaze data as implicit feedback, since using gaze data from different users within the same document proves to be useful for prediction. For user personalization, however, systems may need gaze data captured from a larger amount of documents.

For the future work, it might be interesting to include linguistic features that are related to the discourse of the document and that we believe may play a significant role in academic learning. Another interesting research direction is related to the study of the personal characteristics that help to explain certain gaze evidence beyond the linguistic features of the text. We acknowledge, however, the intrinsic difficulty of obtaining an accurate description of user's personal characteristics in a large scale real application. In such scenario where we are constrained to a low intrusion into personal characteristics, user's personal features can be included into the model as latent variables and they can be estimated, together with the patent variables (e.g. linguistic features), formulating the optimization as an incomplete data problem.

Acknowledgments. Work partially supported by Kakenhi, MEXT Japan [23650076] and JST PRESTO.

References

1. Biedert, R., Buscher, G., Schwarz, S., Möller, M., Dengel, A., Lottermann, T.: The Text 2.0 framework - writing web-based gaze-controlled realtime applications quickly and easily. In: Proc. of the International Workshop on Eye Gaze in Intelligent Human Machine Interaction, EGIHMI (2010)
2. Chen, S.F., Goodman, J.: An empirical study of smoothing techniques for language modeling. Computer Speech and Language 4(13), 359–393 (1999)
3. Doherty, S., O'Brien, S., Carl, M.: Eye tracking as an MT evaluation technique. Machine Translation 24, 1–13 (2010),
   http://dx.doi.org/10.1007/s10590-010-9070-9,
   doi:10.1007/s10590-010-9070-9
4. Hornof, A., Halverson, T.: Cleaning up systematic error in eye-tracking data by using required fixation locations. Behavior Research Methods 34, 592–604 (2002),
   http://dx.doi.org/10.3758/BF03195487, doi:10.3758/BF03195487
5. Hyrskykari, A.: Utilizing eye movements: Overcoming inaccuracy while tracking the focus of attention during reading. Computers in Human Behavior 22, 657–671 (2005),
   http://dx.doi.org/10.1016/j.chb.2005.12.013
6. Itti, L., Koch, C., Niebur, E.: A model of saliency-based visual attention for rapid scene analysis. IEEE Transactions on Pattern Analysis and Machine Intelligence 20(11), 1254–1259 (1998)
7. Koehn, P.: Europarl: A parallel corpus for statistical machine translation. In: Proc. of the 10th Machine Translation Summit, September 12-15, pp. 79–86 (2005)
8. Martinez-Gomez, P.: Quantitative analysis and inference on gaze data using natural language processing techniques. In: Proceedings of the 2012 ACM International Conference on Intelligent User Interfaces, IUI 2012, pp. 389–392. ACM, New York (2012),
   http://doi.acm.org/10.1145/2166966.2167055
9. Martinez-Gomez, P., Chen, C., Hara, T., Kano, Y., Aizawa, A.: Image registration for text-gaze alignment. In: Proceedings of the 2012 ACM International Conference on Intelligent User Interfaces, IUI 2012, pp. 257–260. ACM, New York (2012),
   http://doi.acm.org/10.1145/2166966.2167012
10. McDonald, S.A., Shillcock, R.C.: Eye movements reveal the on-line computation of lexical probabilities during reading. Psychological Science 14(6), 648–652 (2003)
11. Miller, G.A.: Wordnet: A lexical database for English. Communications of the ACM 38, 39–41 (1995)
12. Miyao, Y., Tsujii, J.: Feature forest models for probabilistic HPSG parsing. Computational Linguistics 34, 35–80 (2008),
    http://dx.doi.org/10.1162/coli.2008.34.1.35
13. Powell, M.: A new algorithm for unconstrained optimization. Nonlinear Programming, 31–65 (1970)
14. Rayner, K.: Eye movements in reading and information processing: 20 years of research. Psychological Bulletin 124, 372–422 (1998),
    http://dx.doi.org/10.1037/0033-2909.124.3.372
15. Tomanek, K., Hahn, U., Lohmann, S., Ziegler, J.: A cognitive cost model of annotations based on eye-tracking data. In: Proceedings of the 48th Annual Meeting of the Association for Computational Linguistics, ACL 2010, pp. 1158–1167. Association for Computational Linguistics, Stroudsburg (2010),
    http://dl.acm.org/citation.cfm?id=1858681.1858799

Stigmergic Modeling for Web Service Composition and Adaptation

Ahmed Moustafa[1], Minjie Zhang[1], and Quan Bai[2]

[1] School of Computer Science and Software Engineering,
University of Wollongong, Wollongong, NSW 2522, Australia
{aase995,minjie}@uow.edu.au
[2] School of Computing and Mathematical Sciences,
Auckland University of Technology, Auckland, New Zealand
quan.bai@aut.ac.nz

Abstract. As Web services become widespread, many complex applications require service composition to cope with high scalability and heterogeneity. Centralized Web service composition approaches are not sufficient as they always limit the scalability and stability of the systems. How to efficiently compose and adapt Web services under decentralized environments has become a critical issue, and important research question in Web service composition. In this paper, a stigmergic-based approach is proposed to model dynamic interactions among Web services, and handle some issues in service composition and adaptation. In the proposed approach, Web services and resources are considered as multiple agents. Stigmergic-based self-organization among agents are adopted to evolve and adapt Web service compositions. Experimental results indicate that by using this approach, service composition can be efficiently achieved, despite dealing with incomplete information and dynamic factors in decentralized environments.

Keywords: Web Services, Decentralized Composition, Dynamic Adaptation.

1 Introduction

In recent years, service composition and adaptation has drawn the interests of researchers in a number of fields. Generally, service composition is to fulfil users' requirements by combining services in different repositories. Those requirements can be functional or non-functional. The functional requirements are in relation to targeting the overall outcome concerning the the underway business process, while the non-functional requirements pertain to the quality of the composition as a whole in terms of availability, reliability, cost and response time. Service adaptation is to adjust service "behaviors" and/or composition strategies to adapt to the changes of users' requirements, QoS fluctuations and other prospective runtime dynamics.

Although in the Web service domain, many standards have been developed and adopted for facilitating service composition, lack of supports for decentralized working environments is still a standing challenge and an obvious drawback

P. Anthony, M. Ishizuka, and D. Lukose (Eds.): PRICAI 2012, LNAI 7458, pp. 324–334, 2012.

of many existing composition approaches. Hence, efficient Web service applications need not only to be managed from a centralized perspective, but also to incorporate decentralized scenarios.

Stigmergic interaction, exhibited by social insects to coordinate their activities, has recently inspired a vast number of computer science applications [7,9]. In these works, intelligent components (i.e., agents) interact by leaving and sensing artificial pheromones (i.e., markers) in a virtual environment. Such pheromones can encode and describe application-specific information to be used to achieve specific tasks. From a general perspective, stigmergic interactions have two major advantages: (i) Stigmergic interactions are mediated by pheromones and are completely decoupled. This feature is suitable for open and dynamic environments. (ii) Stigmergic interaction naturally supports application-specific context awareness, in that pheromones provide agents with an application-specific representation of their operational environment (e.g., in ant foraging, pheromones provide a representation of the environment in terms of paths leading to food sources).

In this paper, we propose a stigmergic-based approach for modeling Web service compositions and adaptations. The proposed approach exhibits the concept of decentralization, and allows Web services to compose and adapt in open decentralized environments. The rest of this paper is organized as follows. The principle and introduction of the stigmergic-based service composition and adaptation approach is introduced in Section 2. In Section 3, some experimental results are presented for evaluating the proposed approach. Section 4 gives a brief review of related work and discussions. Finally, the paper is concluded in Section 5.

2   Stigmergic Service Composition and Adaptation

In this section, an approach named Stigmergic Service Composition and Adaptation is proposed to address the problem of Web service composition and adaptation in decentralized environments. The proposed approach is inspired by the concept of *Digital Pheromone* [12], which is a stigmergic-based mechanism for coordinating and controlling swarming objects. Examples from natural systems show that stigmergic systems can generate robust, complex, intelligent behaviors at the system level even when the individual agents only possess very limited or even no intelligence. In these systems, intelligence resides not in a single distinguished agent (as in centralized control) nor in each individual agent, but in the whole group of agents distributed in the environment [3,10].

2.1   Web Service Composition/Adaptation Modeling

In our approach, each Web service is envisaged as a service agent, and a Web service system can then be considered as a multi-agent system with a number of interactive service agents. These agents achieve collaborations and self-organizations via exchanging "pheromone" and performing several pheromone operations. A service is composed if a group of service agents can form an organization to collaborate.

Definition 1: Service Agent. A *Service Agent* p_i is defined as a tuple $p_i =<$ $id, F >$, where id is the identifier of the service agent, F is the pheromone store for facilitating composition and decentralized self-coordination (as detailed in Definition 2).

Definition 2: Pheromone Store. A *Pheromone Store* F is a set of pheromone, i.e., $F = \{f_1, f_2, ..., f_n\}$. Each pheromone flavor f_j holds a scalar value which represents the trail of a certain service agent.

Based upon user requirements, a service agent p_i requests the composition of an abstract workflow r_j such that $r_j = \{r_{j1}, r_{j2}, ..., r_{jn}\}$, where r_{ji} represents a concrete service or a resource needs to composed for satisfying a single unit of functionality or a subtask. To this end, the requester service agent traverses a network of directly linked service agents, and look for suitable services and resources for constructing workflow r_j. Once constructed, the requester service agent deposits/withdraws certain amount (i.e., Q_{f_j}) of pheromone flavor f_j into/from the digital pheromone stores of all provider service agents, which took part in this composition round. The decision of depositing or withdrawing depends on the quality of the provided service. Based upon the strength of the trail of this pheromone flavor f_j, in future rounds, similar service requests will be guided towards the highest quality workflow quickly without the need to traverse the entire network again. Namely, even in open decentralized environments, a requester service agent can immediately make decisions based on the pheromone trail left by other requester service agents.

2.2 Pheromone Operations

In this approach, we define three primary pheromone operations for coordinating service agents, and enabling them to achieve self-organizations.

Pheromone Aggregation: A *Digital Pheromone* Q_{f_j} can be deposited and withdrawn from any service agent p_i. Deposits of a certain flavor f_j are added to the current amount of that flavor of pheromone $s(Q_{f_j}, p_i, t)$ located at that service agent p_i and time t.

Pheromone Evaporation: A *Pheromone Evaporation* is the operation wherein a pheromone flavor f_j evaporates over time t. This operation will generate an Evaporation Factor E_{f_j} ($0 < E_{f_j} \leq 1$) to weaken obsolete information. E_{f_j} can be obtained via Equation 1:

$$E_{f_j} = e^{\frac{-\triangle t(i)}{\lambda}} \tag{1}$$

where $trianglet(i)$ is the time difference between the current time and when f_i has been left, and λ is the parameter to control the evaporation speed.

Pheromone Propagation: A *Pheromone Propagation* G_{f_j} is the operation wherein a pheromone flavor f_j propagates from a service agent p_i to service agent p_k based upon a neighborhood relation N where $p_k \in N(p_i)$. The act of propagation causes pheromone gradients $g(Q_{f_j}, p_i, t)$ to be formed.

2.3   Dynamic Coordination and Adaptation

The three operations introduced in the above subsection serve as a basis for the coordination mechanism. By using these operations, pheromone can be exchanged and circulated among distributed service agents. Furthermore, a service composition and adaptation algorithm (i.e., Algorithm 1) is designed to enable agents to learn the best composition path, and to adapt to the changing environment. The underlying proposed algorithm rests on two fundamental equations to enhance the self-organization behavior, i.e., Equation 2 and Equation 3.

Let t represents an update cycle time, which is the time interval between propagation, pump auto-deposits, and evaporation of a certain flavor f_j; $P = \{p_1, p_2, ..., p_n\}$ represent a set of service agents; $N : P \longrightarrow P$ represents a neighbor relation between service agents; $s(Q_{f_j}, p_i, t)$ represents the strength of pheromone flavor f_j at service agent p_i and time t; $d(Q_{f_j}, p_i, t)$ represents the sum of external deposits of pheromone flavor f_j within the interval $(t-1, t]$ at service agent p_i; $g(Q_{f_j}, p_i, t)$ represents the propagated input of pheromone flavor f_j at time t to service agent p_i; $E_{f_j} \in (0,1)$ represents the evaporation factor for flavor f_j; $G_{f_j} \in (0,1)$ represents the propagation factor for flavor f_j; T_{f_j} represents the threshold below which $S(Q_{f_j}, p_i, t)$ is set to zero.

The evolution of the strength of a single pheromone flavor at a given service agent is defined in Equation 2:

$$s(Q_{f_j}, p_i, t) = E_{f_j} * [(1 - G_{f_j}) * (s(Q_{f_j}, p_i, t-1) + d(Q_{f_j}, p_i, t)) + g(Q_{f_j}, p_i, t)] \quad (2)$$

where E_{f_j} is the evaporation factor pheromone flavor f_j (refer to Equation 1), $1 - G_{f_j}$ calculates the amount remaining after propagation to neighboring service agents, $s(Q_{f_j}, p_i, t-1)$ represents the amount of pheromone flavor f_j from the previous cycle, $d(Q_{f_j}, p_i, t)$ represents the total deposits made since the last update cycle (including pump auto-deposits) and $g(Q_{f_j}, p_i, t)$ represents the total pheromone flavor f_j propagated in from all the neighbors of p_i. Every service agent p_i applies this equation to every pheromone flavor f_j once during every update cycle.

The propagation received from the neighboring service agents is described in Equation 3:

$$g(Q_{f_j}, p_i, t) = \sum_{p_k \in N(p_i)} \frac{G_{f_j}}{N(p_k)} (s(Q_{f_j}, p_k, t-1) + d(Q_{f_j}, p_k, t)) \quad (3)$$

In Equation 3, it can be found that each neighbor service agent p_k $(p_k \in N(p_i))$ propagates a portion of its pheromone to p_i in each update cycle t. This portion depends on the parameter G_{f_j} and the total number of p_k neighbors $N(p_k)$.

Using Equations 2 and 3, we can demonstrate several critical stability and convergence rules, including Local Stability, Propagated Stability and Global Stability.

Rule 1: Local Stability. A *Local Stability* S_l is the rule in which the strength of the output propagated from any set of service agents $M \subset P$ to their neighbors

$N \subset P$ at $t+1$ is strictly less than the strength of the aggregate input (external plus propagated) to those service agents at t.

Rule 2: Propagated Stability. A *Propagated Stability* S_p is the rule in which there exists a fixed upper limit α to the aggregated sum of all propagated inputs at an arbitrary service agent p_i if one-update cycle t and one-service agent p_i external input is assumed.

Rule 3: Global Stability. A *Global Stability* S_g is the rule in which the pheromone strength $s(Q_{f_j}, p_i, t)$ in any place agent is bounded. If the pheromone strength drops below threshold T_{f_j}, it disappears from the pheromone store, $s(Q_{f_j}, p_i, t) \geq T_{f_j}$.

The Stigmergic Service Composition and Adaptation can be represented by Algorithm 1.

Algorithm 1. Stigmergic Service Composition and Adaptation Algorithm

---

  for each $p_i \in P$ **do**
    while $s(Q_{f_j}, p_i, t) \geq T_{f_j}$ **do**
      for $t = 1 : n$ **do**
        $g(Q_{f_j}, p_i, t) = \sum_{p_k \in N(p_i)} \frac{G_{f_j}}{N(p_k)} (s(Q_{f_j}, p_k, t-1) + d(Q_{f_j}, p_k, t));$
        if $g(Q_{f_j}, p_i, t) < \alpha$ **then**
          $g(Q_{f_j}, p_i, t) = g(Q_{f_j}, p_i, t);$
        else
          $g(Q_{f_j}, p_i, t) = \alpha;$
        end if
        $p_i^{in} = g(Q_{f_j}, p_i, t);$
        $p_i^{out} = (G_{f_j}) * (s(Q_{f_j}, p_i, t-1) + d(Q_{f_j}, p_i, t));$
        while $p_i^{out} < p_i^{in}$ **do**
          $s(Q_{f_j}, p_i, t) = E_{f_j} * [(1-G_{f_j}) * (s(Q_{f_j}, p_i, t-1) + d(Q_{f_j}, p_i, t)) + g(Q_{f_j}, p_i, t)];$
        end while
      end for
    end while
  end for

Using Algorithm 1, the Stigmergic Service Composition and Adaptation approach can account for the decline and emergence of a set of service agents having the same pheromone flavor. By ever choosing the trail with the strongest flavor, the proposed approach can adapt to the dynamics of open runtime environments by employing new emerging service agents with stronger flavors and forgetting other service agents with fading flavors.

3 Experimentation and Results

3.1 Experiment Setup

Three experiments have been conducted to assess the proposed Stigmergic Service Composition and Adaptation approach. The proposed approach runs in successive iterations till reaching a convergence point. The proposed approach converges to a near optimal solution once it receives the same approximate value of accumulative QoS for a number of successive iterations. Those accumulated QoS values are compared iteration by iteration and the difference is projected against a threshold. For our experiments, this threshold value is set to 0.0001, and the number of successive iterations is set to 100. The first experiment examines the effectiveness of the proposed approach in composing Web services under a decentralized environment. The second experiment explores the speed of composing Web services using the proposed approach under a decentralized environment, whereas the third experiment inspects the performance and the efficiency of the proposed approach related to the scale of the environment. All the three experiments have been conducted on 3.33 GHz Intel core 2 Duo PC with 3 GB of RAM.

3.2   Result Analysis

The results of the three experiments are demonstrated and analyzed in details in the following three subsections, respectively.

Experiment 1: Decentralized Environment. To explore the effectiveness of the proposed Stigmergic Service Composition and Adaptation approach under decentralized environments, a composition request has been submitted to the proposed approach and the QoS values of the composed workflow is contrasted to those values obtained by a centralized service composition approach. The proposed approach assumes no prior knowledge of the environment including available service agents and their QoS values, while other centralized approaches assume the existence of a central service registry in which all the the functional and QoS qualities of Web services are recorded.

Table 1 shows the progression of the Qos values of the composed workflow using the proposed approach against those values obtained by centrally composing the same workflow using the centralized approaches, e.g., Critical Path Method (CPM). Those values are recorded on a 100-iteration basis.

Figure 1 depicts the contrast between the QoS values obtained form the proposed approach against those values obtained from a centralized approach (y axis) and the corresponding number of iterations (x axis). As shown in Figure 1, the centralized service composition approach shows a steady line with a QoS value of 177 regardless of the number of iterations. This is attributed to the assumption of a prior knowledge of the available Web services and their qualities, which is not the case in real world. On the other hand, the proposed Stigmergic Service Composition and Adaptation approach shows a progressive line representing an upward trend in the QoS values obtained as the number of iterations increases. Due to its decentralized nature, the proposed approach needs to run for a number of iterations before showing stability in results, which happens in our experiment when the number of iterations goes beyond 300. The best stable QoS obtained is 145, which shows 82% rate of effectiveness compared to the optimal value of 177 in a fully centralized approach. This result demonstrates a

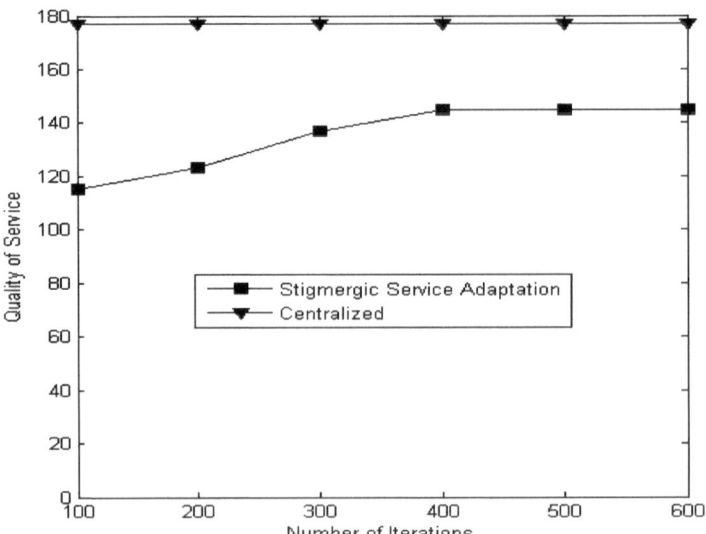

Fig. 1. Decentralized Environment

Table 1. Decentralized Environment

Iterations	Stigmergic Service Composition and Adaptation	Centralized(CPM)
100	115	177
200	123	177
300	137	177
400	145	177
500	145	177
600	145	177

good performance considering a decentralized open environment with no prior knowledge of available Web services and their QoS values.

Experiment 2: Composition Speed. To examine the composition speed of the Stigmergic Service Composition and Adaptation approach, a varying number of composition requests having the same composition goal have been submitted against an environment of 10000 service agents. Those composition requests supposedly have different starting points and each time the number of iterations to converge to the composition goal is measured.

Table 2 shows a range of composition requests and the number of iterations to converge.

Figure 2 depicts the relationship between the number of composition requests (x axis) and the corresponding number of iterations to converge (y axis). In Figure 2, the trend line shows a downward movement describing a linear relationship. As the number of running composition requests increases, the number of iterations to converge to the composition goal decreases. This explains the

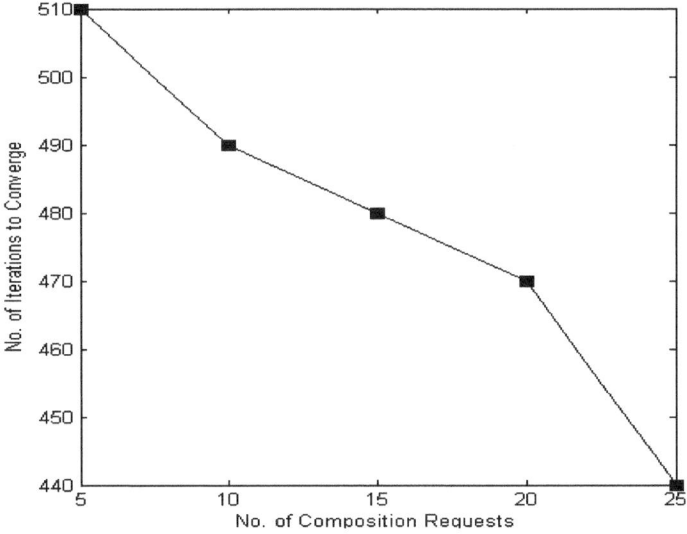

Fig. 2. Composition Speed

Table 2. Composition Speed

No. of Composition Requests	No. of Iterations to Converge
5	510
10	490
15	480
20	470
25	440

benefit of the accumulated pheromone trails left by previous service agents working for the same composition goal. The service agents in new iterations have higher probabilities to make benefit of the pheromone flavor trails left by previous service agents used by other composition requests for the same composition goal, thus to follow their trail to fulfill that goal faster. As the number of running composition requests increases, the pheromone trails left by service agents employed by those composition requests become more strengthened and thus the potentiality to guide new service agents working towards the same composition goal increases.

Experiment 3: Environment Scale. To examine the efficiency of the proposed approach under large scale open environments, twenty composition requests sharing the same composition goal have been submitted against a range of environment scales, expressed in terms of varying numbers of service agents. The composition requests supposedly have different starting points and each time the number of iterations to converge to the composition goal is measured.

Table 3 shows a range of environment scales with the number of iterations to converge.

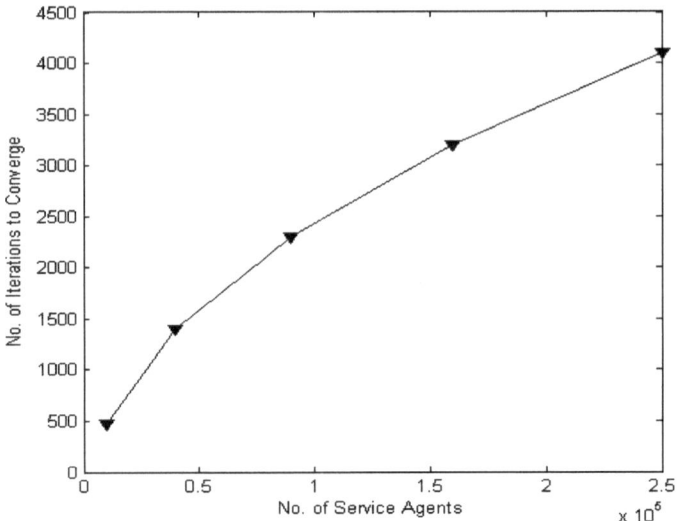

Fig. 3. Environment Scale

Table 3. Environment Scale

No. of Service Agents	No. of Iterations to Converge
10000	470
40000	1400
90000	2300
160000	3200
250000	4100

Figure 3 depicts the relationship between the scale of the environment in terms of the number of service agents (x axis) and the corresponding number of iterations needed to converge (y axis). The trend line shows an upward movement describing a linear relationship. This linear relationship proves the efficiency of the proposed approach in composing and adapting Web services in open large scale environments, e.g., the Internet. As the scale of the environment increases, the number of iterations to converge increases linearly not exponentially accordingly. This shows the robustness of the proposed approach under large scale service environments, e.g., the Internet.

4   Related Work and Discussions

Various approaches have been proposed to handle the problem of dynamic composition of Web services [8,13,4]. However, most of these approaches are based on a centralized architecture. In these models, a service requester submits requests to a centralized directory, and the directory decides which service on the list should be returned to the requester. This assumption on centralized control becomes impractical in scenarios with dynamic arrivals and departures of service providers, which require frequent updates of the central entities, resulting in a large system overhead. In contrast, our proposed approach can work under a fully decentralized environment where no central directory exists. The service agent has to find other matching service agents during run time based on neighborhood relationships and track for their QoS by exchanging pheromone flavors. This enables our approach to adapt effectively under dynamic environments accounting for the arrival and departure of service providers at any time.

Some approaches exploiting distributed techniques have already been proposed in the literature [6,11,5]. As the first step towards decentralization of service composition, these approaches introduced a distributed directory of services. For example, Kalasapur et al. [5] proposed a hierarchical directory in which the resource-poor devices depend on the resource-rich devices to support service discovery and composition. However, this approach is not fully decentralized. There exist nodes that perform the task of service discovery and composition for other nodes. In contrast, our approach is fully decentralized as a set of service agents cooperate together coherently to fulfill the composition request.

Another important approach was proposed by Basu et al. [1] in which a composite service is represented as a task-graph and subtrees of the graph are computed in a distributed manner. However, this approach assumes that a service requestor is one of the services and the approach relies on this node to coordinate service composition. A similar approach was proposed by Drogoul [2], whose work is different from Basu et al. [1] in terms of the way of electing the coordinator. For each composite request, a coordinator is selected from within a set of nodes. The service requestor delegates the responsibility of composition to the elected coordinator. Both approaches [2,1] exhibit some limitations. Although service discovery is performed in a distributed manner, service composition still relies on a coordinator assigned for performing the task of combining and invoking services. Also, both of the two approaches assume the direct interaction between nodes responsible for service discovery and service composition. In contrast, our approach overcomes those limitations as follows. (1) In our approach, there is no special entity to manage service composition process; (2) Service providers communicate only with their local neighbors where no service provider knows the full global information; and (3) Our approach adopts an indirect interaction scheme where the set of service agents responsible for service composition and adaptation can interact with each other through the environment by exchanging pheromone trails.

5   Conclusion

This paper focuses on modeling Web service composition and adaptation under decentralized dynamic environments. The paper presented a service composition and adaptation approach using stigmergic interactions to actively adapt to dynamic changes in complex decentralized Web service composition environments.

The proposed approach is able to adapt actively to QoS fluctuations considering both potential degradation and emergence of QoS values. The experimental results have shown the effectiveness of the proposed approach in highly complex and dynamic Web service composition environments. The future work is set to investigate the performance of the proposed approach in real world applications and to study the potentiality of building a web service trust model using the proposed approach.

References

1. Basu, P., Ke, W., Little, T.D.C.: A novel approach for execution of distributed tasks on mobile ad hoc networks. In: 2002 IEEE Wireless Communications and Networking Conference, WCNC 2002, vol. 2, pp. 579–585 (March 2002)
2. Chakraborty, D., Joshi, A., Finin, T., Yesha, Y.: Service composition for mobile environments. Mob. Netw. Appl. 10(4), 435–451 (2005)
3. Haack, J.N., Fink, G.A., Maiden, W.M., McKinnon, A.D., Templeton, S.J., Fulp, E.W.: Ant-based cyber security. In: 2011 Eighth International Conference on Information Technology: New Generations (ITNG), pp. 918–926 (April 2011)
4. Hatzi, O., Vrakas, D., Nikolaidou, M., Bassiliades, N., Anagnostopoulos, D., Vlahavas, I.: An integrated approach to automated semantic web service composition through planning. IEEE Transactions on Services Computing (99), 1 (2011)
5. Kalasapur, S., Kumar, M., Shirazi, B.A.: Dynamic service composition in pervasive computing. IEEE Transactions on Parallel and Distributed Systems 18(7), 907–918 (2007)
6. Liao, C.-F., Jong, Y.-W., Fu, L.-C.: Toward reliable service management in message-oriented pervasive systems. IEEE Transactions on Services Computing 4(3), 183–195 (2011)
7. Ren, C., Huang, H., Jin, S.: Stigmergic collaboration: A sign-based perspective of stigmergy. In: 2010 International Conference on Intelligent Computation Technology and Automation (ICICTA), vol. 3, pp. 390–393 (May 2010)
8. Ren, X., Hou, R.: Extend uddi using ontology for automated service composition. In: 2011 Second International Conference on Mechanic Automation and Control Engineering (MACE 2009), pp. 298–301 (July 2011)
9. Rosen, D., Suthers, D.D.: Stigmergy and collaboration: Tracing the contingencies of mediated interaction. In: 2011 44th Hawaii International Conference on System Sciences (HICSS), pp. 1–10 (January 2011)
10. Simonin, O., Huraux, T., Charpillet, F.: Interactive surface for bio-inspired robotics, re-examining foraging models. In: 2011 23rd IEEE International Conference on Tools with Artificial Intelligence (ICTAI), pp. 361–368 (November 2011)
11. Sirbu, A., Marconi, A., Pistore, M., Eberle, H., Leymann, F., Unger, T.: Dynamic composition of pervasive process fragments. In: 2011 IEEE International Conference on Web Services (ICWS), pp. 73–80 (July 2011)
12. Van Dyke Parunak, H.: Interpreting digital pheromones as probability fields. In: Proceedings of the 2009 Winter Simulation Conference (WSC), pp. 1059–1068 (December 2009)
13. Zahoor, E., Perrin, O., Godart, C.: Rule-based semi automatic web services composition. In: Proceedings of the 2009 Congress on Services - I, pp. 805–812. IEEE Computer Society, Washington, DC (2009)

Recognizing Human Gender in Computer Vision: A Survey

Choon Boon Ng, Yong Haur Tay, and Bok-Min Goi

Universiti Tunku Abdul Rahman, Kuala Lumpur, Malaysia
{ngcb,tayyh,goibm}@utar.edu.my

Abstract. Gender is an important demographic attribute of people. This paper provides a survey of human gender recognition in computer vision. A review of approaches exploiting information from face and whole body (either from a still image or gait sequence) is presented. We highlight the challenges faced and survey the representative methods of these approaches. Based on the results, good performance have been achieved for datasets captured under controlled environments, but there is still much work that can be done to improve the robustness of gender recognition under real-life environments.

Keywords: Gender recognition, gender classification, sex identification, survey, face, gait, body.

1   Introduction

Identifying demographic attributes of humans such as age, gender and ethnicity using computer vision has been given increased attention in recent years. Such attributes can play an important role in many applications such as human-computer interaction, surveillance, content-based indexing and searching, biometrics, demographic studies and targeted advertising. For example, in face recognition systems, the time for searching the face database can be reduced and separate face recognizers can be trained for each gender to improve accuracy [1]. It can be used for automating tedious tasks such as photograph annotation or customer statistics collection.

While a human can easily differentiate between genders, it is a challenging task for computer vision. In this paper, we survey the methods of human gender recognition in images and videos. We focus our attention on easily observable characteristics of a human which would not require the subject's cooperation or physical contact. Most researchers have relied on facial analysis, while some work have been reported on using the whole body, either from a still image or using gait sequences. We concentrate on approaches using 2-D (rather than the more costly 3-D) data in the form of still image or videos. Audio cues such as voice are not included.

In general, a pattern recognition problem such as gender recognition, when tackled with a supervised learning technique, can be broken down into several steps which are object detection, preprocessing, feature extraction and classification. In detection, the human subject or face region is detected and cropped from the image. This is

P. Anthony, M. Ishizuka, and D. Lukose (Eds.): PRICAI 2012, LNAI 7458, pp. 335–346, 2012.
© Springer-Verlag Berlin Heidelberg 2012

followed by some preprocessing, for example geometric alignment, histogram equalization or resizing. In feature extraction, representative descriptors of the image are found, after which selection of the most discriminative features may be made or dimension reduction is applied. As this step is perhaps the most important to achieve high recognition accuracy, we will provide a more detailed review in later sections.

Lastly, the classifier is trained and validated using a dataset. As the subject is to be classified as either male or female, a binary classifier is used, for example, support vector machine (SVM), Adaboost, neural networks and Bayesian classifier.

The rest of this paper is organized as follows: Section 2, 3, and 4 review aspects (challenges, feature extraction, and performance) of gender recognition by face, gait and body, respectively, followed by concluding remarks in Section 5.

2   Gender Recognition by Face

The face region, which may include external features such as the hair and neck region, is used to make gender identification. The image of a person's face exhibits many variations which may affect the ability of a computer vision system to recognize the gender, which can be categorized as being caused by the image capture process or the human. Factors due to the former are the head pose or camera view [2], lighting and image quality. Head pose refers to the head orientation relative to the view of the image capturing device, as described by the pitch, roll and yaw angles. Human factors are age [3][4], ethnicity [5], facial expression [6] and accessories worn (e.g hat).

2.1   Facial Feature Extraction

We broadly categorize feature extraction methods for face gender classification into *geometric-based* and *appearance-based* methods [3][7]. The former is based on distance measurements of *fiducial points*, which are important points that mark features of the face, such as the nose, mouth, and eyes. Psychophysical studies using human subjects established the importance of these distances in discriminating gender. While the geometric relationships are maintained, other useful information may be discarded [3] and the points need to be accurately extracted [8]. Brunelli and Poggio [9] used 18 point-to-point distances to train a hyper basis function network classifier. Fellous [10] selected 40 manually extracted points to calculate 22 normalized fiducial distances.

Appearance-based methods are based on some operation or transformation performed on the image pixels, which can be done globally (holistic) or locally (patches). The geometric relationships are naturally maintained [3], which is advantageous when gender discriminative features are not exactly known. But they are sensitive to variations in appearance (view, illumination, etc.) [3] and the large number of features [8].

Pixel intensity values can be directly input to train a classifier such as neural network or support vector machine (SVM). Moghaddam and Yang [11] found that Gaussian RBF kernel gave the best performance for SVM. Baluja and Rowley [12] proposed a fast method that matched the accuracy of SVM using simple pixel comparison operations to find features for weak classifiers which were combined using AdaBoost.

Viola and Jones[13] introduced *rectangle or Haar-like features* for rapid face detection, and used for real-time gender and ethnicity classification of videos in [14].

Dimension reduction methods such as Principal Component Analysis (PCA), used in early studies [15][16], obtain an image representation in reduced dimension space, which would otherwise be proportionate to the image size. Sun et al. [17] used genetic algorithm to remove eigenvectors that did not seem to encode gender information. Other methods such as 2-D PCA, Independent Component Analysis (ICA) and Curvilinear Component Analysis (CCA) have also been studied [6] [18][19].

Ojala et al. [20] introduced *local binary patterns* (LBP) for grayscale and rotation invariant texture classification. Each pixel in an image is labeled by applying the LBP operator, which thresholds the pixel's local neighborhood at its grayscale value into a binary pattern. LBP detect microstructures such as edge, corners and spot. LBP has been used for multi-view gender classification [2], and combined with intensity and shape feature[21], or with contrast information [22]. Shan [23] used Adaboost to learn discriminative LBP histogram bins. Other variants inspired by LBP have been used for gender recognition, such as Local Gabor Binary Mapping Pattern [24][25][26], centralized Gabor gradient binary pattern [27], and Interlaced Derivative Pattern [28].

Scale Invariant Feature Transform (SIFT) features are invariant to image scaling, translation and rotation, and partially invariant to illumination changes and affine projection [29]. Using these descriptors, objects can be reliably recognized even from different views or under occlusion and eliminates the need for preprocessing, including accurate face alignment [30]. Demirkus et al. [31] exploited these characteristics, using a Markovian model to classify face gender in unconstrained videos.

Research in neurophysiology has shown that *Gabor filters* fit the spatial response profile of certain neurons in the visual cortex of the mammalian brain. Gabor wavelets were used to label the nodes of an elastic graph representing the face [32] or extracted for each image pixel and then selected using Adaboost [33]. Wang et al. extracted SIFT descriptors at regular image grid points and combined it with Gabor features [34]. Gabor filters have also been used to obtain the simple cell units in *biologically inspired features* (BIF) model. This model contains simple (S) and complex (C) cell units arranged in hierarchical layers of S1, C1, S2 and C2. For face gender recognition, the C2 and S2 layers were found to degrade performance [4].

Other facial representations that have been used include a generic patch-based representation [35], regression function [36], DCT [37], and wavelets of Radon transform [38]. Features external to the face region such as hair, neck region [39] and clothes [7] are also cues used by humans to identify gender. Social context information based on position of a person's face in a group of people was used in [40].

2.2   Evaluation and Results

A list of representative works on face gender recognition is compiled in Table 1. Because of the different datasets and parameters used for evaluation, a straight comparison is difficult. The datasets that have been used tend to be from face recognition or detection since no public datasets have been designed specifically for gender recognition evaluation. Evaluation metric is based on the accuracy or classification rate.

Table 1. Face gender recognition results

First Author, Year	Feature extraction	Classifier	Training data*	Test data*	Ave. Acc.%	Dataset variety[@]
Moghaddam, 2002 [11]	Pixel values	SVM-RBF	FERET 1044m 711f	5-CV	96.62	F
Shakhnarovich, 2002 [14]	Haar-like	Adaboost	Web images	5-CV Video seqs.	79 90	P (<30°), A,E,L
Buchala, 2005 [41]	PCA	SVM -RBF	Various mixes 200m 200f	5-CV	92.25	F
Jain, 2005 [18]	ICA	SVM	FERET 100m 100f	FERET 150m 150f	95.67	F,S
Baluja, 2006 [12]	Pixel comp.	Adaboost	FERET 1495m 914f	5-CV	94.3	F,S
Lapedriza, 2006 [42]	BIF multi scale filt.	Jointboost	FRGC 3440t FRGC 1886t	10-CV 10-CV	96.77 91.72	uniform background cluttered background
Lian, 2006 [2]	LBP histogram	SVM-polynomial	CAS-PEAL 1800m 1800f	CAS-PEAL 10784t	94.08	P (up to 30° yaw & pitch)
Leng, 2008 [33]	Gabor	Fuzzy SVM	FERET 160m 140f	5-CV	98	
Xu, 2008 [43]	Haar-like, fiducial	SVM-RBF	Various mixes 500m 500f	5-CV	92.38	F,E,A,L,S
Xia, 2008 [24]	LGBMP hist.	SVM-RBF	CAS-PEAL 1800m 1800f	CAS-PEAL 10784t	94.96	P (up to 30° yaw & pitch)
Aghajanian, 2009 [35]	Patch-based	Bayesian	Web images 16km 16kf	Web images 500m 500f	89	U
Li, 2009 [37]	DCT	SGMM	YGA 6096t	YGA 1524t	92.5	F,A,S
Lu, 2009 [6]	2D PCA	SVM-RBF	CAS PEAL 300m 300f	CAS-PEAL 1800t	95.33	F,X
Demirkus, 2010 [31]	SIFT	Bayesian	FERET 1780m 1780f	Video seqs. 15m 15f	90	U (P,X,O,L)
Wang, 2010 [34]	SIFT, Gabor	Adaboost	Various mixes 4659t	10-CV	~97	F,X,L,O
Lee, 2010 [36]	regression	SVM	FERET 1158m, 615f		98.8	F
Alexandre, 2010 [21]	Intensity, edge, LBP	SVM-linear	FERET 152m 152f UND set B 130m 130f	FERET 60m 47f UND set B 171m 56f	99.07 91.19	F,S F,S
Li, 2011 [7]	LBP, hair, clothing	SVM	FERET 227m 227f	FERET 114m 114f	95.8	F
Wu, 2011 [25]	LGBP	SVM-RBF	CAS-PEAL 2142m 2142f	CAS-PEAL 2023m 996f	~91-97 per set	P (up to 67° yaw), S
Zheng, 2011 [26]	LGBP-LDA	SVMAC	CAS-PEAL 2706m 2706f (of 9 sets) FERET 282m 282f	CAS-PEAL 2175m 1164f FERET 307m 121f	≥ 99.8 per set 99.1	P (up to 30° yaw & pitch), S F
Shan, 2012 [23]	LBP hist. bins	SVM-RBF	LFW 4500m 2943f	5-CV	94.81	F,U,S

*The number of male and female faces is given; e.g. 500m 500f refers to 500 male and female faces each. Where the number was not given, the total faces used are given (e.g. 1000t.)
When the accuracy is reported based on cross-validation result, this is indicated in the *test data* field; e.g. 5-CV refers to five-fold cross validation, and the average rate from validation results is given. If classification rate for a different test set is given, this result is used and the dataset is indicated.
[@] The variations controlled or the variety used,, as mentioned by the authors, are indicated as follows:
   F–frontal only   A–age   E–ethnicity   P–pose/view   L–lighting   X–expression   O–occlusion
   U – uncontrolled   S – indicates the same individual does not appear on both training and test set

It is noted that the FERET dataset is the most often used (although the subset of images taken varies.) The best accuracy is 99.1%, using frontal face only [26][21]. Zheng et al. [26] achieved near 100% for pose variations up to 30° yaw and pitch on the CAS-PEAL dataset, with separate classifiers trained for each pose (thus requiring prior pose detection). For images taken in uncontrolled environments, Shan [23] obtained 94.8% on the LFW dataset containing frontal and near frontal faces.

3   Gender Recognition by Gait

Gait is defined to be the coordinated, cyclic combination of movements that result in human locomotion [44], which includes walking, running, jogging and climbing stairs. In computer vision research, the gait of a walking person is often used. Exploiting gait information is helpful in some situations such as when the face is not visible. In a video sequence of a person walking, the gait cycle can be referred to as the time interval between two consecutive left/right mid-stances [45]. Many factors affect the gait of a person, such as load [46], footwear, walking surface, injury, mood, age [47] and change with time. Video-based analysis of gait would also need to contend with clothing, camera view [47][48][49][50][51], walking speed and background clutter.

3.1   Feature Extraction

Early work on gait analysis used point lights attached to the body's joints. Based on the motion of the point lights during walking, identity and gender of a person could be identified [52] (see [53] for a survey on these early works). Human gait representation can be divided into *model-based* or *appearance-based* (*model-free*)[54][55]. Yoo et al., guided by anatomical knowledge, obtained 2-D *stick figures* from the body contour [56], of which a sequence from one gait cycle composed a gait signature. Such model-based approaches rely on accurate estimation of joints [55][57] and require high-quality gait sequences [45] where the body parts need to be tracked in each frame, thus incurring higher computational costs. Moreover, they ignore body width information [57]. However, they are view and scale invariant [45].

In many appearance-based methods, the *silhouette* of the human is obtained, for example using background subtraction. Lee and Grimson [58] divided each silhouette into 7 regions and fitted *ellipses* into each region. The mean and standard deviation of the ellipse moments together with the silhouette centroid height formed the gait features, with robustness to silhouette noise. However, it will be affected by viewpoint, clothing and gait changes [58]. Felez et al. [59] improvised by using a different regionalization of 8 parts to obtain more realistic ellipses while Hu et al. [57] used equal partitions formed by grids. Zhang and Wang [60] used *frieze patterns* to study multiview gender classification. A *frieze* pattern is a two-dimensional pattern that repeats along one dimension. The gait representation is generated by projecting the silhouette along its columns and rows, then stacking these 1-D projections over time [61]. Shan et al. [62] showed that the *Gait Energy Image* (GEI) [63] was an effective representation for gender recognition by fusing gait and face features. A GEI represents human motion in a single image while preserving temporal information by averaging the silhouette images in one or more gait cycles, thus saving on storage and computational cost and is robust to silhouette noise in individual frames [63]. A similar representation, *Average Gait Image* (AGI) [64], averages over one gait cycle.

The GEI can also be estimated from a whole gait sequence, without the need to detect the gait cycle frequency [49]. Yu et al. [55] divided the GEI into 5 different components, with each given a weight based on the results from psychophysical experiments. Li et al. [65] partitioned the AGI into 7 components corresponding to body parts, while Chen et al. [50] used 8 components based on consideration of walking patterns. Lu and Tan [48] obtained the difference GEI from different views, using uncorrelated discriminant simplex analysis (USDA) for efficient projection into lower dimensional subspace.

Chen et al. [66] applied Radon transform on the human silhouettes in a gait cycle and used Relevant Component Analysis (RCA) for feature transformation. Oskuie and Faez [67] extracted Zernike moments from Radon-transformed Mean Gait Energy Image [68]. Frequency-domain features obtained from the silhouette using Discrete Fourier Transform (DFT) [47] and wavelet decomposition on the silhouette contour width [69] have been used. Instead of extracting the silhouette, DCT coefficients was obtained from the image to train embedded hidden Markov models [70]. Hu et al. [46] applied Gabor filters and used Maximization of Mutual Information (MMI) to learn the discriminative low dimensional representation.

3.2   Evaluation and Results

Table 2 shows representative works on gait-based gender recognition. For the CASIA Gait Database (Set B), state of the art performance is 98.39% using side view sequences only [57]. Slightly higher rate of 98.5% was reported by [67] using a gender imbalanced set. For real-time videos, 84.38% was achieved on a small set of four subjects [49]. 94% average accuracy was obtained for multiview sequences without requiring prior knowledge of the view angle [70]. For the IRIP Gait Database, Hu et al. [57] reported 98.33% using side view sequences. Chen et al. [50] achieved 93.3% by fusion of multiviews, requiring a camera per view, thus increasing complexity.

As a conclusion, gait-based gender recognition can achieve high classification rate in controlled datasets, especially with a single side view. There is a need for more investigation into generalization ability through cross database testing and performance for datasets with larger number of subjects in unconstrained environments.

Table 2. Gait-based gender recognition

First Author, Year	Feature Extraction	Classifier	Training data	Test data	Ave. Acc.%	Dataset variety [@]
Yoo, 2005 [56]	2D stick figures	SVM-polynomial	SOTON 84m 16f	10-CV	96%	N
Chen, 2009 [66]	Radon transform of silhouette	Mahalano-bis distance	IRIP 32m 28f (300)	LOO-CV	95.7	N
Chen, 2009 [50]	AGI	Euclidean distance	IRIP 32m 28f (300 per angle)	LOO-CV	93.3	M(0°-180)
Yu, 2009 [55]	GEI	SVM	CASIA B 31m 31f (372)	31-CV	95.97	N
Chang, 2009 [49]	GEI+ PCA+LDA	Fisher boost	CASIA B 93m 31f (8856)	124-CV Videos	96.79 84.38	M(0°-180°) M(U)
Chang, 2010 [70]	DCT	EHMM	CASIA B 25m 25f	5-CV	94	M(0°-180°)

Table 2. (*Continued*)

Lu, 2010 [48]	GEI + UDSA	Nearest neighbour	CASIA B 31m 31f (4092)	LOO-CV	83-93 (per view)	M(0°-180°)
Felez, 2010 [59]	Ellipse fittings	SVM-linear	CASIA B 93m 31f (744)	10-CV	94.7	N
Hu, 2010 [46]	Gabor + MMI	GMM-HMM	CASIA B 31m 31f (372)	31-CV	96.77	N
Hu, 2011 [57]	ellipse fittings & stance indexes	MRCF	CASIA B 31m 31f (372)	31-CV	98.39	N
			IRIP 32m 28f (300)	LOO-CV	98.33	N
Handri, 2011 [69]	silhoutte contour width	kNN	Private 29m 14f (>172)	LOO-CV	94.3	N, A
Makihara, 2011 [47]	DFT of sil-houette + SVD	kNN	OU-ISIR 20m 20f	20-CV	~70-80 (per view)	M (0° -360° +overhead)
Oskuie, 2011 [67]	RTMGEI + Zernike momts.	SVM	CASIA B 93m 31f CASIA B 93m 31f		98.5 98.94	N N, W, C

Under *Test data*, the figure in the bracket is the total number of sequences used.
LOO-CV refers to leave-one-out cross validation.
[@] N – side view only   M– multi-view (the range of angles are given)
A – various age   W – wearing overcoat   C– carrying bag

4   Gender Recognition by Body

Here, we refer to the use of the static human body (either partially or as a whole) in an image which, like gait, would be useful in situations where using the face is not poss-ible or preferred. However it is challenging in several aspects. To infer gender, hu-mans use not only body shape and hairstyle, but additional cues such as type of clothes and accessories [71], which may be the similar among different genders. The classifier should also be robust to variation in pose, articulation and occlusion of the person and deal with varying illumination and background clutter.

4.1   Feature Extraction

The first attempt to recognize gender from full body images partitioned the centered and height-normalized human image into patches corresponding to some parts of the body [72]. Each part was represented using *Histogram of Oriented Gradients* (HOG) feature, which was previously developed for human detection in images [73]. HOG features are able to capture local shape information from the gradient structure with easily controllable degree of invariance to translations or rotations [73]. Collins et al. [74] proposed PixelHOG, which are dense HOG features computed from a custom edge map. This was combined with color features obtained from a histogram com-puted based on the hue and saturation values.

Bourdev et al. [71] used a set of patches they called *poselets*, represented with HOG features, color histogram and skin features The poselets were used to train attribute classifiers which were combined to infer gender using context information. Their method relies on training dataset that is heavily annotated with keypoints.

Biologically-inspired features (BIF) model were used for human body gender rec-ognition by Guo et al. [75]. Only C1 features obtained from Gabor filters were used, as it was found that C2 features degraded performance (as in the case of face gender recognition). Various manifold learning techniques were applied on the features. Best results were obtained by first classifying the view (front, back, or mixed) using BIF with PCA, and followed by the gender classifier.

4.2   Evaluation and Results

Table 3 summarizes the results obtained. Bourdev et al. [71] achieved 82.4 % accuracy but with imbalanced gender dataset. Collins et al. [74] achieved 80.6 % accuracy on a more balanced but smaller dataset with frontal view only. From these results, there is still room for improvement.

Table 3. Body-based gender recognition

First Author, Year	Feature Extraction	Classifier	Training data	Test data	Ave. Acc.%	Dataset variety
Cao, 2008 [72]	HOG	Adaboost variant	MIT-CBCL 600m 288f	5-CV	75	View (frontal, back)
Collins, 2009 [74]	PixelHOG, color hist.	SVM-linear	VIPeR 292m 291f	5-CV	80.62	View (frontal)
Guo, 2009 [75]	BIF + PCA/LSDA	SVM-linear	MIT-CBCL 600m 288f	5-CV	80.6	View (frontal, back)
Bourdev, 2011 [71]	HOG, color, skin pixels	SVM	Attributes of People 3395m 2365f		82.4	Unconstrained

5   Conclusion

In this paper, we have presented a survey on human gender recognition using computer vision-based methods, focusing on 2-D approaches. We have highlighted the challenges and provided a review of the commonly-used features. Good performance has been achieved for frontal faces, whereas for images which include non-frontal poses, there is room for improvement, especially in uncontrolled conditions, as required in many practical applications. Current gait-based methods depend on the availability of one or more complete gait sequences. High classification rate have been achieved with controlled datasets, especially with side views. Investigation of the generalization ability of the methods (through cross database testing) is called for. Performance for datasets containing larger number of subjects with sequences taken under unconstrained environments is not yet established. Some work has also been done based on static human body, but there is scope for further improvement.

References

1. Mäkinen, E., Raisamo, R.: An experimental comparison of gender classification methods. Pattern Recognition Letters 29(10), 1544–1556 (2008)
2. Lian, H.C., Lu, B.L.: Multi-view gender classification using local binary patterns and support vector machines. In: Advances in Neural Networks-ISNN 2006, pp. 202–209 (2006)
3. Benabdelkader, C., Griffin, P.: A Local Region-based Approach to Gender Classification From Face Images. In: IEEE Computer Society Conference on Computer Vision and Pattern Recognition-Workshops, CVPR Workshops, p. 52 (2005)

4. Guo, G., Dyer, C.R., Fu, Y., Huang, T.S.: Is gender recognition affected by age? In: 2009 IEEE 12th International Conference on Computer Vision Workshops (ICCV Workshops), pp. 2032–2039 (2009)
5. Gao, W., Ai, H.: Face gender classification on consumer images in a multiethnic environment. In: Advances in Biometrics, pp. 169–178 (2009)
6. Lu, L., Shi, P.: A novel fusion-based method for expression-invariant gender classification. In: 2009 IEEE International Conference on Acoustics, Speech and Signal Processing, ICASSP 2009, pp. 1065–1068 (2009)
7. Li, B., Lian, X.-C., Lu, B.-L.: Gender classification by combining clothing, hair and facial component classifiers. Neurocomputing, 1–10 (2011)
8. Kim, H., Kim, D., Ghahramani, Z.: Appearance-based gender classification with Gaussian processes. Pattern Recognition Letters 27, 618–626 (2006)
9. Brunelli, R., Poggio, T.: Face recognition: features versus templates. IEEE Transactions on Pattern Analysis and Machine Intelligence 15(10), 1042–1052 (1993)
10. Fellous, J.M.: Gender discrimination and prediction on the basis of facial metric information. Vision Research 37(14), 1961–1973 (1997)
11. Moghaddam, B., Yang, M.H.: Learning gender with support faces. IEEE Transactions on Pattern Analysis and Machine Intelligence 24(5), 707–711 (2002)
12. Baluja, S., Rowley, H.A.: Boosting sex identification performance. International Journal of Computer Vision 71(1), 111–119 (2007)
13. Viola, P., Jones, M.: Rapid object detection using a boosted cascade of simple features. In: Proceedings of the 2001 IEEE Computer Society Conference on Computer Vision and Pattern Recognition, CVPR 2001, vol. 1, pp. I-511–I-518 (2001)
14. Shakhnarovich, G., Viola, P., Moghaddam, B.: A unified learning framework for real time face detection and classification. In: Proceedings of Fifth IEEE International Conference on Automatic Face Gesture Recognition, pp. 16–23 (2002)
15. Abdi, H., Valentin, D., Edelman, B.: More about the difference between men and women: evidence from linear neural network and the principal-component approach. Perception 24(1993), 539–539 (1995)
16. Golomb, B.A., Lawrence, D.T., Sejnowski, T.J.: Sexnet: A neural network identifies sex from human faces. In: Advances in Neural Information Processing Systems, vol. 3, pp. 572–577 (1991)
17. Sun, Z., Bebis, G., Yuan, X., Louis, S.J.: Genetic feature subset selection for gender classification: A comparison study. In: Proceedings of Sixth IEEE Workshop on Applications of Computer Vision (WACV 2002), pp. 165–170 (2002)
18. Jain, A., Huang, J., Fang, S.: Gender identification using frontal facial images. In: IEEE International Conference on Multimedia and Expo., ICME 2005, p. 4 (2005)
19. Buchala, S., Davey, N., Gale, T.M.: Analysis of linear and nonlinear dimensionality reduction methods for gender classification of face images. International Journal of Systems Science 36(14), 931–942 (2005)
20. Ojala, T., Pietikainen, M.: Multiresolution gray-scale and rotation invariant texture classification with local binary patterns. IEEE Transactions on Pattern Analysis and Machine Intelligence 24(7), 971–987 (2002)
21. Alexandre, L.A.: Gender recognition: A multiscale decision fusion approach. Pattern Recognition Letters 31(11), 1422–1427 (2010)
22. Ylioinas, J., Hadid, A., Pietikäinen, M.: Combining contrast information and local binary patterns for gender classification. Image Analysis, 676–686 (2011)
23. Shan, C.: Learning local binary patterns for gender classification on real-world face images. Pattern Recognition Letters 33(4), 431–437 (2012)

24. Xia, B., Sun, H., Lu, B.-L.: Multi-view Gender Classification based on Local Gabor Binary Mapping Pattern and Support Vector Machines. In: IEEE International Joint Conference on Neural Networks, IJCNN 2008 (IEEE World Congress on Computational Intelligence), pp. 3388–3395 (2008)
25. Wu, T.-X., Lian, X.-C., Lu, B.-L.: Multi-view gender classification using symmetry of facial images. Neural Computing and Applications, 1–9 (May 2011)
26. Zheng, J., Lu, B.-L.: A support vector machine classifier with automatic confidence and its application to gender classification. Neurocomputing 74(11), 1926–1935 (2011)
27. Fu, X., Dai, G., Wang, C., Zhang, L.: Centralized Gabor gradient histogram for facial gender recognition. In: 2010 Sixth International Conference on Natural Computation (ICNC), vol. 4, pp. 2070–2074 (2010)
28. Shobeirinejad, A., Gao, Y.: Gender Classification Using Interlaced Derivative Patterns. In: 2010 20th International Conference on Pattern Recognition, pp. 1509–1512 (2010)
29. Lowe, D.G.: Distinctive image features from scale-invariant keypoints. International Journal of Computer Vision 60(2), 91–110 (2004)
30. Rojas-Bello, R.N., Lago-Fernandez, L.F., Martinez-Munoz, G., Sdnchez-Montanes, M.A.: A comparison of techniques for robust gender recognition. In: 2011 18th IEEE International Conference on Image Processing, pp. 561–564 (2011)
31. Demirkus, M., Toews, M., Clark, J.J., Arbel, T.: Gender classification from unconstrained video sequences. In: 2010 IEEE Computer Society Conference on Computer Vision and Pattern Recognition Workshops (CVPRW), pp. 55–62 (2010)
32. Laurenz, W., Fellous, J.-M.F., Kruger, N., von der Malsburg, C.: Face recognition and gender determination. In: Proceedings of the International Workshop on Automatic Face and Gesture Recognition, pp. 92–97 (1995)
33. Leng, X.M., Wang, Y.D.: Improving generalization for gender classification. In: 15th IEEE International Conference on Image Processing, ICIP 2008, pp. 1656–1659 (2008)
34. Wang, J.G., Li, J., Lee, C.Y., Yau, W.Y.: Dense SIFT and Gabor descriptors-based face representation with applications to gender recognition. In: 2010 11th International Conference on Control Automation Robotics & Vision (ICARCV), pp. 1860–1864 (December 2010)
35. Aghajanian, J., Warrell, J., Prince, S.J.D., Rohn, J.L., Baum, B.: Patch-based within-object classification. In: 2009 IEEE 12th International Conference on Computer Vision, pp. 1125–1132 (2009)
36. Lee, P.H., Hung, J.Y., Hung, Y.P.: Automatic Gender Recognition Using Fusion of Facial Strips. In: 2010 20th International Conference on Pattern Recognition (ICPR), pp. 1140–1143 (2010)
37. Li, Z., Zhou, X.: Spatial gaussian mixture model for gender recognition. In: 2009 16th IEEE International Conference on Image Processing (ICIP), pp. 45–48 (2009)
38. Rai, P., Khanna, P.: Gender classification using Radon and Wavelet Transforms. In: 2010 International Conference on Industrial and Information Systems (ICIIS), pp. 448–451 (2010)
39. Ueki, K., Kobayashi, T.: Gender Classification Based on Integration of Multiple Classifiers Using Various Features of Facial and Neck Images. Information and Media Technologies 3(2), 479–485 (2008)
40. Gallagher, A.C., Chen, T.: Understanding images of groups of people. In: 2009 IEEE Conference on Computer Vision and Pattern Recognition, pp. 256–263 (2009)
41. Buchala, S., Loomes, M.J., Davey, N., Frank, R.J.: The role of global and feature based information in gender classification of faces: a comparison of human performance and computational models. International Journal of Neural Systems 15, 121–128 (2005)

42. Lapedriza, A., Marin-Jimenez, M.: Gender recognition in non controlled environments. In: 18th International Conference on Pattern Recognition, ICPR 2006, 2006, vol. 3, pp. 834–837 (2006)
43. Xu, Z., Lu, L., Shi, P.: A hybrid approach to gender classification from face images. In: 19th International Conference on Pattern Recognition, ICPR 2008, pp. 1–4 (2008)
44. Boyd, J.E., Little, J.J.: Biometric Gait Recognition. In: Biometrics, pp. 19–42 (2005)
45. Boulgouris, V., Hatzinakos, D., Plataniotis, K.N.: Gait Recognition: A challenging signal processing technology for biometric identification. IEEE Signal Processing Magazine, 78–90 (November 2005)
46. Hu, M., Wang, Y., Zhang, Z., Wang, Y.: Combining Spatial and Temporal Information for Gait Based Gender Classification. In: 2010 20th International Conference on Pattern Recognition, pp. 3679–3682 (2010)
47. Makihara, Y., Mannami, H., Yagi, Y.: Gait Analysis of Gender and Age Using a Large-Scale Multi-view Gait Database. In: Kimmel, R., Klette, R., Sugimoto, A. (eds.) ACCV 2010, Part II. LNCS, vol. 6493, pp. 440–451. Springer, Heidelberg (2011)
48. Lu, J., Tan, Y.-P.: Uncorrelated discriminant simplex analysis for view-invariant gait signal computing. Pattern Recognition Letters 31(5), 382–393 (2010)
49. Chang, P.-C., Tien, M.-C., Wu, J.-L., Hu, C.-S.: Real-time Gender Classification from Human Gait for Arbitrary View Angles. In: 2009 11th IEEE International Symposium on Multimedia, pp. 88–95 (2009)
50. Chen, L., Wang, Y., Wang, Y.: Gender Classification Based on Fusion of Weighted Multi-View Gait Component Distance. In: 2009 Chinese Conference on Pattern Recognition, pp. 1–5 (2009)
51. Huang, G., Wang, Y.: Gender Classification Based on Fusion of Multi-view Gait Sequences. In: Yagi, Y., Kang, S.B., Kweon, I.S., Zha, H. (eds.) ACCV 2007, Part I. LNCS, vol. 4843, pp. 462–471. Springer, Heidelberg (2007)
52. Kozlowski, L.T., Cutting, J.E.: Recognizing the sex of a walker from a dynamic point-light display. Attention, Perception, & Psychophysics 21(6), 575–580 (1977)
53. Davis, J.W., Gao, H.: An expressive three-mode principal components model for gender recognition. Journal of Vision 4(5), 362–377 (2004)
54. Nixon, M.S., Carter, J.N.: Automatic Recognition by Gait. Proceedings of the IEEE 94(11), 2013–2024 (2006)
55. Yu, S., Tan, T., Huang, K., Jia, K., Wu, X.: A study on gait-based gender classification. IEEE Transactions on Image Processing 18(8), 1905–1910 (2009)
56. Yoo, J.-H., Hwang, D., Nixon, M.S.: Gender Classification in Human Gait Using Support Vector Machine. In: Blanc-Talon, J., Philips, W., Popescu, D.C., Scheunders, P. (eds.) ACIVS 2005. LNCS, vol. 3708, pp. 138–145. Springer, Heidelberg (2005)
57. Hu, M., Wang, Y., Zhang, Z., Zhang, D.: Gait-based gender classification using mixed conditional random field. IEEE Transactions on Systems, Man, and Cybernetics, Part B, Cybernetics 41(5), 1429–1439 (2011)
58. Lee, L., Grimson, W.: Gait analysis for recognition and classification. In: Proceedings of Fifth IEEE International Conference on Automatic Face Gesture Recognition, pp. 155–162 (2002)
59. Martin-Felez, R., Mollineda, R.A., Sanchez, J.S.: Towards a More Realistic Appearance-Based Gait Representation for Gender Recognition. In: 2010 20th International Conference on Pattern Recognition, pp. 3810–3813 (2010)
60. Zhang, D., Wang, Y.: Investigating the separability of features from different views for gait based gender classification. In: 19th International Conference on Pattern Recognition, ICPR 2008, pp. 3–6 (2008)

61. Liu, Y., Collins, R., Tsin, Y.: Gait Sequence Analysis Using Frieze Patterns. In: Heyden, A., Sparr, G., Nielsen, M., Johansen, P. (eds.) ECCV 2002, Part II. LNCS, vol. 2351, pp. 657–671. Springer, Heidelberg (2002)

62. Shan, C., Gong, S., McOwan, P.W.: Fusing gait and face cues for human gender recognition. Neurocomputing 71(10-12), 1931–1938 (2008)

63. Han, J., Bhanu, B.: Individual recognition using gait energy image. IEEE Transactions on Pattern Analysis and Machine Intelligence 28(2), 316–322 (2006)

64. Liu, Z., Sarkar, S.: Simplest Representation Yet for Gait Recognition: Averaged Silhouette. Pattern Recognition, no. 130768

65. Li, X., Maybank, S.J., Yan, S., Tao, D., Xu, D.: Gait Components and Their Application to Gender Recognition. IEEE Transactions on Systems, Man, and Cybernetics, Part C (Applications and Reviews) 38(2), 145–155 (2008)

66. Chen, L., Wang, Y., Wang, Y., Zhang, D.: Gender Recognition from Gait Using Radon Transform and Relevant Component Analysis. In: Huang, D.-S., Jo, K.-H., Lee, H.-H., Kang, H.-J., Bevilacqua, V. (eds.) ICIC 2009. LNCS, vol. 5754, pp. 92–101. Springer, Heidelberg (2009)

67. Bagher Oskuie, F., Faez, K.: Gender Classification Using a Novel Gait Template: Radon Transform of Mean Gait Energy Image. In: Kamel, M., Campilho, A. (eds.) ICIAR 2011, Part II. LNCS, vol. 6754, pp. 161–169. Springer, Heidelberg (2011)

68. Chen, X.T., Fan, Z.H., Wang, H., Li, Z.Q.: Automatic Gait Recognition Using Kernel Principal Component Analysis. In: Science and Technology (2010)

69. Handri, S., Nomura, S., Nakamura, K.: Determination of Age and Gender Based on Features of Human Motion Using AdaBoost Algorithms. International Journal of Social Robotics 3(3), 233–241 (2011)

70. Chang, C.Y., Wu, T.H.: Using gait information for gender recognition. In: 2010 10th International Conference on Intelligent Systems Design and Applications (ISDA), pp. 1388–1393 (2010)

71. Bourdev, L., Maji, S., Malik, J.: Describing People: A Poselet-Based Approach to Attribute Classification. In: 2011 IEEE International Conference on Computer Vision (ICCV), pp. 1543–1550 (2011)

72. Cao, L., Dikmen, M., Fu, Y., Huang, T.S.: Gender recognition from body. In: Proceeding of the 16th ACM International Conference on Multimedia, pp. 725–728 (2008)

73. Dalal, N., Triggs, B.: Histograms of oriented gradients for human detection. In: IEEE Computer Society Conference on Computer Vision and Pattern Recognition, CVPR 2005, vol. 1, pp. 886–893 (2005)

74. Collins, M., Zhang, J., Miller, P.: Full body image feature representations for gender profiling. In: 2009 IEEE 12th International Conference on Computer Vision Workshops, ICCV Workshops, pp. 1235–1242 (2009)

75. Guo, G., Mu, G., Fu, Y.: Gender from Body: A Biologically-Inspired Approach with Manifold Learning. In: Zha, H., Taniguchi, R.-i., Maybank, S. (eds.) ACCV 2009, Part III. LNCS, vol. 5996, pp. 236–245. Springer, Heidelberg (2010)

Web-Based Mathematics Testing
with Automatic Assessment

Minh Luan Nguyen[1], Siu Cheung Hui[1], and Alvis C.M. Fong[2]

[1] Nanyang Technological University, Singapore
{NGUY0093,asschui}@ntu.edu.sg
[2] Auckland University of Technology, New Zealand
acmfong@gmail.com

Abstract. Web-based testing is an effective approach for self-assessment and in-telligent tutoring in an e-learning environment. In this paper, we propose a novel framework for Web-based testing and assessment, in particular, for mathematics testing. There are two major components in Web-based mathematics testing and assessment: automatic test paper generation and mathematical answer verification. The proposed framework consists of an efficient constraint-based Divide-and-Conquer approach for automatic test paper generation, and an effective Probabilistic Equivalence Verification algorithm for automatic mathematical answer verification. The performance results have shown that the proposed framework is effective for web-based mathematics testing and assessment. In this paper, we will discuss the proposed framework and its performance in comparison with other techniques.

Keywords: Intelligent tutoring system, web-based testing, test paper generation, multi-objective optimization, answer verification, probabilistic algorithm.

1 Introduction

Web-based testing [3,4,6] is an effective approach for self-assessment and intelligent tutoring in an e-learning environment. There are two major components in Web-based testing: automatic test paper generation and answer verification. Automatic test paper generation generates a test paper automatically according to user specification on multiple assessment criteria, and the generated test paper can then be attempted over the Web by users. Automatic answer verification automatically verifies users' answers online to evaluate their proficiency for assessment purposes. In this paper, we focus on automatic answer verification for mathematical expressions which are commonly required by most math questions.

Web-based test paper generation is a challenging problem. It is not only a NP-hard multi-objective optimization problem, but also required to be solved efficiently in runtime. The popular test paper generation techniques such as genetic algorithm (GA) [10], particle swarm optimization [7] and ant colony optimization [8] are based on traditional weighting parameters for multi-objective optimization. However, determining appropriate weighting parameters is computationally expensive [5]. Therefore, these current techniques generally require long runtime for generating good quality test papers.

P. Anthony, M. Ishizuka, and D. Lukose (Eds.): PRICAI 2012, LNAI 7458, pp. 347–358, 2012.

As equivalent mathematical expressions can be expressed in different forms (e.g. $x + \frac{1}{x}$ and $\frac{x^2+1}{x}$), automatic mathematical answer verification which checks the equivalence of the user answers and standard solutions is another challenging problem. Computer Algebra Systems (CAS) such as Maple [13] and Mathematica [14] provide the expression equivalence checking function. However, their techniques have generated many false negative errors if the provided correct answers are written in different forms from the given standard solutions.

In this paper, we propose a novel framework for Web-based testing and assessment, in particular, for mathematics testing. The proposed framework consists of an efficient constraint-based Divide-and-Conquer (DAC) approach [15] for automatic test paper generation, and an effective Probabilistic Equivalence Verification (PEV) algorithm for mathematical answer verification. DAC is based on constraint decomposition for multi-objective optimization, whereas PEV is a randomized algorithm which can avoid false negative errors completely.

2   Related Work

Automatic test paper generation aims to find an optimal subset of questions from a question database to form a test paper based on criteria such as total time, topic distribution, difficulty degree, discrimination degree, etc. Many techniques have been proposed for test paper generation. In [11], tabu search (TS) was proposed to construct test papers by defining an objective function based on multi-criteria constraints and weighting parameters for test paper quality. TS optimizes test paper quality by the evaluation of the objective function. In [9], dynamic programming was proposed to optimize an objective function incrementally based on recursive optimal relation of the objective function. In [10], a genetic algorithm (GA) was proposed to generate quality test papers by optimizing a fitness ranking function based on the principle of population evolution. In addition, swarm intelligence algorithms have also been investigated. In [7], particle swarm optimization (PSO) was proposed to generate multiple test papers by optimizing a fitness function which is defined based on multi-criteria constraints. In [8], ant colony optimization (ACO) was proposed to generate test papers by optimizing an objective function which is based on the simulation of the foraging behavior of real ants.

One of the challenging problems of the current techniques is that they require weighting parameters and some other parameters such as population size, tabu length, etc. for each test paper generation that is not easy to determine. Moreover, they generally take long runtime for generating good quality test papers especially for large datasets of questions. Therefore, the DAC approach [15] in the proposed framework aims to overcome these challenging problems.

3   Proposed Approach

In this paper, we propose a novel framework for Web-based mathematics testing and assessment, which is shown in Figure 1. The proposed framework consists of 2 main components: *Web-based Test Paper Generation* which generates a high quality test paper that satisfies the user specification, and *Automatic Mathematical Answer Verification* which checks the correctness of users' answers. In particular, it determines the equivalence of mathematical expressions in users' answers and the standard solutions stored in the database.

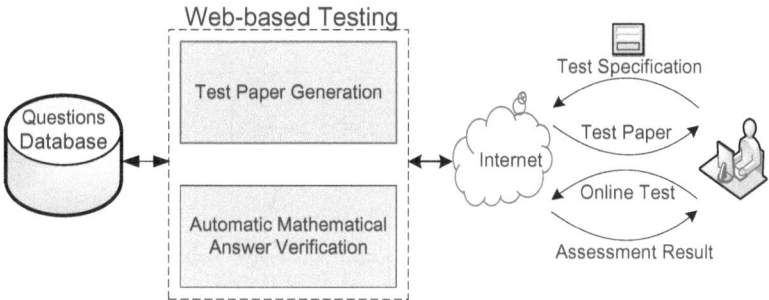

Fig. 1. The Proposed Framework for Web-based Testing

4   Web-Based Test Paper Generation

Let $\mathcal{Q} = \{q_1, q_2, .., q_n\}$ be a dataset consisting of n questions, $\mathcal{C} = \{c_1, c_2, .., c_m\}$ be a set of m topics, and $\mathcal{Y} = \{y_1, y_2, .., y_k\}$ be a set of k question types such as multiple choice questions, fill-in-the-blanks and long answers. Each question $q_i \in \mathcal{Q}$, $i \in \{1, 2, .., n\}$, has 8 attributes $\mathcal{A} = \{q, o, a, e, t, d, c, y\}$, where q is the question identity, o is the question content, a is the question answer, e is the discrimination degree, t is the question time, d is the difficulty degree, c is the related topic and y is the question type. A *test paper specification* $\mathcal{S} = \langle N, T, D, C, Y \rangle$ is a tuple of 5 attributes which are defined based on the attributes of the selected questions as follows: N is the number of questions, T is the total time, D is the average difficulty degree, $C = \{(c_1, pc_1), .., (c_M, pc_M)\}$ is the specified proportion for topic distribution and $Y = \{(y_1, py_1), .., ((y_K, py_K)\}$ is the specified proportion for question type distribution.

The test paper generation process aims to find a subset of questions from a question dataset $\mathcal{Q} = \{q_1, q_2, .., q_n\}$ to form a test paper P with specification $\mathcal{S}_P$ that maximizes the average discrimination degree and satisfies the test paper specification such that $\mathcal{S}_P = \mathcal{S}$. It is important to note that the test paper generation process occurs over the Web where user expects to generate a test paper within an acceptable response time. Therefore, the Web-based test paper generation process is as hard as other optimization problems due to its computational NP-hardness, and it is also required to be solved efficiently in runtime.

In the constraint-based Divide-And-Conquer (DAC) approach, we divide the set of constraints into two subsets of relevant constraints, namely *content constraints* and *assessment constraints*, which can then be solved separately and effectively. In the test paper specification $\mathcal{S} = \langle N, T, D, C, Y \rangle$, the content constraints include the constraints on topic distribution C and question type distribution Y, whereas the assessment constraints include the constraints on total time T and average difficulty degree D.

The DAC approach, which is shown in Figure 2, consists of 2 main processes: Offline Index Construction and Test Paper Generation. The Offline Index Construction process constructs an effective indexing structure for supporting local search in assessment constraint optimization to improve the generated paper quality. The Test Paper Generation process generates an optimal test paper that satisfies the specified content constraints and assessment constraints.

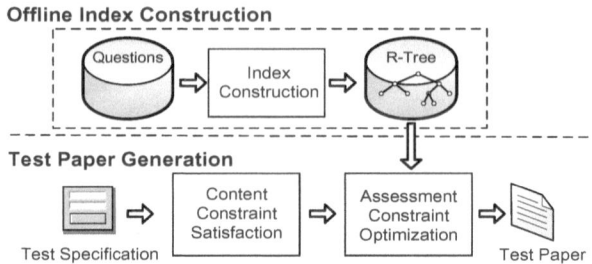

Fig. 2. The Proposed DAC Approach

Index Construction: We use an effective 2-dimensional data structure, called R-tree [2], to store questions based on the time and difficulty degree attributes. R-tree has been widely used for processing queries on spatial databases. The R-tree used here is similar to the R-tree version discussed in [2] with some modifications on operations such as insertion, subtree selection and overflow handling in order to enhance the efficiency.

Content Constraint Satisfaction: It is quite straightforward to generate an initial test paper that satisfies the content constraints based on the number of questions N. Specifically, the number of questions of each topic c_l is $pc_l * N$, $l = 1..M$. Similarly, the number of questions of each question type y_j is $py_j * N$, $j = 1..K$. There are several ways to assign the N pairs of topic-question type to satisfy the content constraints. Here, we have devised an approach which applies a heuristic to try to achieve the specified total time early. To satisfy the content constraints, the round-robin technique is used for question selection. More specifically, for each topic c_l, $l = 1..M$, we assign questions alternately with various question types y_j, $j = 1..K$, as much as possible according to the number of questions. Then, for each of the N pairs of topic-question type (c_l, y_j) obtained from the round-robin selection step, we assign a question q from the corresponding topic-question type (c_l, y_j) that has the highest question time in order to satisfy the total time early.

Assessment Constraint Optimization: Assessment constraint violations indicate the differences between the test paper specification and the generated test paper according to the total time constraint and the average difficulty degree constraint which are given as follows:

$$\triangle T(\mathcal{S}_P, \mathcal{S}) = \frac{|T_P - T|}{T} \quad \text{and} \quad \triangle D(\mathcal{S}_P, \mathcal{S}) = \frac{|D_P - D|}{D}$$

Based on the assessment constraint violations, the following objective function is defined for evaluating the quality of a test paper P:

$$f(P) = \triangle T(\mathcal{S}_P, \mathcal{S})^2 + \triangle D(\mathcal{S}_P, \mathcal{S})^2$$

To minimize assessment constraint violations, local search is used to find better questions to substitute the existing questions in the test paper. To form a new test paper, each question q_k in the original test paper P_0 is substituted by another better question q_m which has the same topic and question type such that the assessment constraint violations are minimized. Specifically, the choice of the neighboring solution $P_1 \in N(P_0)$, where $N(P_0)$ is a neighborhood region, is determined by minimizing the fitness function $f(P_1)$. The termination conditions for the local search are based on the criteria for quality satisfaction and the number of iterations. Algorithm 10 presents the local search algorithm.

Algorithm 1. Local_Search (DAC)

Input: $\mathcal{S} = (N, T, D, C, Y)$ - test paper specification;
$\qquad P_0 = \{q_1, q_2, .., q_N\}$ - original test paper;
$\qquad \mathcal{R}$ - R-Tree
Output: P_1 - Improved test paper
begin

1 $\quad \mathcal{P} \leftarrow \{P_0\}$;
2 $\quad$ **while** *Termination condition is not satisfied* **do**
3 $\qquad$ **foreach** q_i *in* P_0 **do**
4 $\qquad\quad$ Compute 2-dimensional region W ;
5 $\qquad\quad$ $q_m \leftarrow$ **Best_First_Search**$(q_i, W, \mathcal{R})$;
6 $\qquad\quad$ $P_1 \leftarrow \{P_0 - \{q_i\}\} \cup \{q_m\}$;
7 $\qquad\quad$ Compute fitness $f(P_1)$;
8 $\qquad\quad$ Insert new test paper P_1 into $\mathcal{P}$;
9 $\qquad$ $\mathcal{P} \leftarrow \{P_0\} \leftarrow \underset{P_1 \in \mathcal{P}}{\mathrm{argmin}}\ f(P_1)$ /* best move*/;
10 $\quad P_1 \leftarrow P_0$;

Pruning Search Space: The local search needs to scan through the entire question list several times to find the most suitable question for improvement. However, exhaustive searching on a long question list is computational expensive as it requires $O(N)$ time. To improve the local search process, we focus on substituting questions that help improve the assessment constraint violations such that the search process can converge faster. Hence, we prune the search space by finding a 2-dimensional region W that contains possible questions for substitution. Let $\mathcal{S}_{P_0} = \langle N, T_0, D_0, C_0, Y_0 \rangle$ be the specification of a test paper P_0 generated from a specification $\mathcal{S} = \langle N, T, D, C, Y \rangle$. Let P_1 be the test paper created after substituting a question q_k of P_0 by another question $q_m \in \mathcal{Q}$ with $\mathcal{S}_{P_1} = \langle N, T_1, D_1, C_1, Y_1 \rangle$. The relations of total time and average difficulty degree between P_1 and P_0 can be expressed as follows:

$$T_1 = T_0 + t_m - t_k \quad (1) \quad \text{and} \quad D_1 = D_0 + \frac{d_m}{N} - \frac{d_k}{N} \quad (2)$$

where t_k and t_m are the question time of q_k and q_m respectively; and d_k and d_m are the difficulty degree of q_k and q_m respectively.

A generated test paper P with specification $\mathcal{S}_P = \langle N, T_P, D_P, C_P, Y_P \rangle$ is said to satisfy the assessment constraints in $\mathcal{S}$ if $\triangle T(\mathcal{S}_P, \mathcal{S}) \leq \alpha$ and $\triangle D(\mathcal{S}_P, \mathcal{S}) \leq \beta$, where α and β are two predefined thresholds. Let's consider the total time violation of P_0.

If $\triangle T\,(\mathcal{S}_{P_0},\mathcal{S}) = \frac{|T_0-T|}{T} \geq \alpha$ and $T_0 \leq T$, to improve the total time satisfaction of P_1, q_m should have the question time value of $t_k + (T - T_0)$ such that $\triangle T(\mathcal{S}_{P_1},\mathcal{S})$ is minimized. Furthermore, as $\triangle T(\mathcal{S}_{P_1},\mathcal{S}) = \frac{|T_1-T|}{T} \leq \alpha$, q_m should have the total time t_m in the interval $t_k + (T - T_0) \pm \alpha T$. If $\triangle T(\mathcal{S}_{P_0},\mathcal{S}) = \frac{|T_0-T|}{T} \geq \alpha$ and $T_0 > T$, we can also derive the same result. Similarly, we can derive the difficulty degree d_m of q_m in the interval $d_k + N(D - D_0) \pm \beta ND$. As such, we can find the 2-dimensional region W based on t_m and d_m for question substitution.

Finding Best Question for Substitution: Among all the questions located in the 2-dimensional region W, the DAC approach finds the best question that minimizes the objective function in order to enhance the test paper quality. Consider question q_m as a pair of variables on its question time t and difficulty degree d. From Equations (1) and (2), the objective function $f(P_1)$ can be rewritten as a multivariate function $f(t,d)$:

$$
\begin{aligned}
f(t,d) &= (\frac{T_1 - T}{T})^2 + (\frac{D_1 - D}{D})^2 \\
&= \frac{(t - T + T_0 - t_k)^2}{T^2} + \frac{(d - ND + ND_0 - d_k)^2}{D^2} \\
&= \frac{(t - t^*)^2}{T^2} + \frac{(d - d^*)^2}{D^2} \\
&\geq \frac{(t - t^*)^2 + (d - d^*)^2}{T^2 + D^2} = \frac{distance^2(q_m, q^*)}{T^2 + D^2}
\end{aligned}
$$

where q^* is a question having question time $t^* = T - T_0 + t_k$ and difficulty degree $d^* = ND - ND_0 + d_k$.

As T and D are predefined constants and q^* is a fixed point in the 2-dimensional space, the good question q_m to replace question q_k in P_0 is the question point that is the nearest neighbor to the point q^* (i.e., the minimum value of the function $f(P_1)$) and located in the region W. To find the good question q_m efficiently, we perform the Best First Search (BFS) [16] with the R-Tree. BFS recursively visits the nearest question whose region is close to q^* with time complexity $O(\lg N)$.

5   Mathematical Answer Verification

To provide Web-based testing for mathematics, it is necessary to support automatic mathematical answer verification which verifies the correctness of user answer compared with the standard solution. In most mathematics questions, mathematical expressions are the most common form of the required answers. In this paper, we propose a novel Probabilistic Equivalence Verification (PEV) algorithm for equivalence verification of two mathematical expressions. PEV is a general and efficient randomized algorithm based on probabilistic testing methods [1].

In the proposed framework, mathematical expressions (or functions) are stored in Latex format in the question database. The mathematical answer verification problem is specified as follows. Given 2 mathematical functions $h(x_1, x_2, .., x_n)$ and $g(x_1, x_2, .., x_n)$, mathematical answer verification aims to verify the equivalence of two mathematical functions h and g. Mathematically, two functions h and g are said to be *equivalent* $(h \equiv g)$, if and only if they have the same value at every point in the domain

$\mathcal{D}$ (e.g. $\mathbb{Z}^n, \mathbb{R}^n$, etc.). Note that $h \equiv g$ iff $f = h - g \equiv 0$. Therefore, the proposed PEV algorithm aims to check whether the function f is equivalent to zero, i.e. $f(x_1, .., x_n) \equiv 0, \forall x_1, .., x_n \in \mathcal{D}$.

The PEV algorithm is given in Algorithm 6. It repeatedly evaluates the value of f at random points $r_1, r_2, .., r_n$ of the corresponding multivariate variables $x_1, x_2, .., x_n$, which are sampled uniformly from a sufficiently large range. If the evaluation result is equal to 0, it returns true. Otherwise, it returns false. Different from CASs such as Maple and Mathematica, PEV can avoid false negative errors completely while guaranteeing small possibility for false positive errors to occur. As such, the proposed PEV algorithm is effective for verifying the equivalence of 2 mathematical expressions which are equivalent but may be expressed in different forms.

Algorithm 2. Probabilistic_Equivalence_Verification (PEV)

Input: $f(x_1, x_2, .., x_n)$ - a multivariate function; $\mathcal{D}$ - domain value of f

Output: **true** if $f \equiv 0$ and **false** if otherwise

begin

1 | Transform the function f for evaluation;

2 | Choose set of trial points $V \subset \mathcal{D}$ randomly such that $|V|$ is sufficiently large;

3 | **repeat**

4 | | Select a random point $(r_1, .., r_n)$ from V uniformly and independently;

5 | | **if** $f(r_1, .., r_n) \neq 0$ **then return** *false*;

  | **until** $|V|$;

6 | **return** *true*;

Here, we first prove that the PEV algorithm has a guaranteed error bound for polynomial functions. For other mathematical functions, we transform them into equivalent polynomials, thereby guaranteeing the error bound for these functions.

Multivariate Polynomial Verification: The degree of a particular term in a polynomial P is defined as the sum of its variable exponents in that term and the degree of P is defined as the maximum degree of all the terms in P. For example, the degree of the term $x_1^6 x_2 x_3^5$ is 12, and the degree of $P = x_1^3 x_2^2 + x_1^6 x_2 x_3^5$ is 12.

Theorem 1. *If the polynomial function $P \equiv 0$, then the PEV algorithm shown in Algorithm 6 has no false negative error. If $P \not\equiv 0$, then the probability that the PEV algorithm has false positive error is bounded by $Pr[P(r_1, r_2, .., r_n) = 0] \leq \frac{d}{|V|}$, where V is the set of trial points $(r_1, r_2, .., r_n), r_i \in \mathcal{D}, \forall i = 1..n$, and d is the degree of P.*

Proof: It is straightforward that the PEV algorithm has no false negative error as it is a trial-and-error method. The false positive error bound can be proved using mathematical induction on the number of variables n of the polynomial.

For the case $n = 1$, if $P \not\equiv 0$, then P is a univariate variable polynomial with degree d. Thus, it has at most d roots. Hence, $Pr[P(r_1, r_2, .., r_n) = 0] \leq \frac{d}{|V|}$.

For the inductive step, we can solve the multivariate case by fixing $n - 1$ variables $x_2, .., x_n$ to $r_2, .., r_n$ and apply the result from the univariate case. Assume that $P \not\equiv 0$, we compute $Pr[P(r_1, .., r_n) = 0]$. First, let k be the highest power of x_1 in P. We can rewrite P as:

$$P(x_1, x_2, .., x_n) = x_1^k A(x_2, .., x_n) + B(x_1, x_2, .., x_n)$$

for some polynomials A and B. Let $\mathcal{X}_1$ be the case that $P(r_1, .., r_n)$ is evaluated to 0, and $\mathcal{X}_2$ be the case that $A(r_2, .., r_n)$ is evaluated to 0. Using Bayes rule, we can rewrite the probability that $P(r_1, .., r_n) = 0$ as:

$$\Pr[P(r) = 0] = \Pr[\mathcal{X}_1]$$
$$= \Pr[\mathcal{X}_1|\mathcal{X}_2]\Pr[\mathcal{X}_2] + \Pr[\mathcal{X}_1|\neg\mathcal{X}_2]\Pr[\neg\mathcal{X}_2]$$
$$\leq \Pr[\mathcal{X}_2] + \Pr[\mathcal{X}_1|\neg\mathcal{X}_2] \quad (3)$$

Consider the probability of $\mathcal{X}_2$ (or $A(r_2, .., r_n) = 0$), the polynomial A has degree $d-k$ with $n-1$ variables. So, by inductive hypothesis on the number of variables, we obtain:

$$\Pr[\mathcal{X}_2] = \Pr[A(r_2, .., r_n) = 0] \leq \frac{d-k}{|V|} \quad (4)$$

Similarly, if $\neg\mathcal{X}_2$ (or $A(r_2, .., r_n) \neq 0$), P is a univariate polynomial degree k. By inductive hypothesis, we have:

$$\Pr[\mathcal{X}_1|\neg\mathcal{X}_2] = \Pr[P(r_1, ., r_n) = 0|A(r_2, ., r_n) \neq 0] \leq \frac{k}{|V|} \quad (5)$$

Finally, we can substitute the results from substituting Equations (4) and (5) into Equation (3) to complete the inductive step. □

Based on Theorem 1 for the case of polynomials, the probability of false positive errors can be guaranteed to any desired small number ϵ (e.g. $\epsilon = 10^{-3}$) by conducting sufficiently large number of trials with $|V| \geq \frac{d}{\epsilon}$.

Mathematical Function Verification: We note that the PEV algorithm can work well with many function types such as trigonometric, logarithmic, exponential, fractional, root and rational functions. This can be done by using substitution techniques to transform a mathematical function into an equivalent polynomial. For instance, the function $f = xe^x + x^2 \sin x$ can be transformed into a multivariate polynomial $f^* = xy + x^2z$, where $y = e^x$ and $z = \sin x$.

For mathematical functions in continuous domains of variables, the PEV algorithm tackles the following 2 challenges, namely the selection of uniform and independent random points and the evaluation of functions (e.g. e^x) that produce very large numbers. Generally, complex pseudo-random generators are required to select random points uniformly and independently on continuous domain of variables than on discrete domain. In the proposed PEV algorithm, it overcomes these challenges and performs the computation efficiently as follows. We transform the original domain variable x into an integer domain variable $h_A(x)$ by using a hash function h_A. The hash function is based on arithmetic modular operator $mod\ p$, where p is a prime number. Similarly, we also transform a general domain function $f(x)$ into an integer domain $f(h_A(x))\ (mod\ p)$ using modular and other arithmetic operators. In doing so, we can evaluate a general function $f(x)$ by regularizing discrete values on an integer domain of variables. As such, we are able to overcome the 2 problems mentioned above. Let's consider the function $f = xe^x + x^2 \sin x$, we can transform f, using the formula $\sin x = \frac{e^{ix} - e^{-ix}}{2i}$, into $f^* = xe^x + x^2\frac{e^{ix} - e^{-ix}}{2i}$, where i is the imaginary axis. Then, the function f^* can be evaluated using arithmetic operators efficiently. By using the function substitution and transformation technique, we can ensure the probability that false positive errors occur for these function types is also bounded.

6   Performance Evaluation

Two datasets are used for performance evaluation: one small real dataset on undergraduate mathematics questions and one large synthetic dataset. In the undergraduate dataset, question attributes are labeled manually. In the synthetic dataset, the values of each attribute are generated automatically according to a normal distribution. Table 1 shows a summary of the 2 datasets. In the experiment, we have designed 12 test specifications with different parameters to generate online test papers. We evaluate the performance based on the 12 test specifications for each of the following 5 algorithms: GA, PSO, ACO, TS and DAC. The runtime and quality of the generated test papers are measured.

Table 1. Question Datasets

	Undergraduate	Synthetic
Number of Questions	3000	20000
Number of Topics	12	40
Number of Question Types	3	3

To evaluate the quality of k generated test papers on a dataset $\mathcal{D}$ w.r.t. any arbitrary test paper specification $\mathcal{S}$, we use Mean Discrimination Degree and Mean Constraint Violation. Let $P_1, P_2, ..., P_k$ be the generated test papers on a question dataset $\mathcal{D}$ w.r.t. different test paper specifications $\mathcal{S}_i$, $i = 1..k$. Let E_{P_i} be the average discrimination degree of P_i. The *Mean Discrimination Degree* $\mathcal{M}_d^{\mathcal{D}}$ is defined as:

$$\mathcal{M}_d^{\mathcal{D}} = \frac{\sum_{i=1}^k E_{P_i}}{k}$$

The Mean Constraint Violation consists of two components: Assessment Constraint Violation and Content Constraint Violation. The Assessment Constraint Violation consists of the total time constraint violation $\triangle T(\mathcal{S}_P, \mathcal{S})$ and the average difficulty degree constraint violation $\triangle D(\mathcal{S}_P, \mathcal{S})$. In Content Constraint Violation, Kullback-Leibler (KL) Divergence [12] is used to measure the topic distribution violation $\triangle C(\mathcal{S}_P, \mathcal{S})$ and question type distribution violation $\triangle Y(\mathcal{S}_P, \mathcal{S})$ between the generated test paper specification $\mathcal{S}_P$ and the test paper specification $\mathcal{S}$ as follows:

$$\triangle C(\mathcal{S}_P, \mathcal{S}) = D_{KL}(pc_p \| pc) = \sum_{i=1}^M pc_p(i) \log \frac{pc_p(i)}{pc(i)}$$

$$\triangle Y(\mathcal{S}_P, \mathcal{S}) = D_{KL}(py_p \| py) = \sum_{j=1}^K py_p(j) \log \frac{py_p(j)}{py(j)}$$

The Constraint Violation (CV) of a generated test paper P w.r.t. $\mathcal{S}$ is defined as:

$$CV(P, \mathcal{S}) = \frac{\lambda * \triangle T + \lambda * \triangle D + \log \triangle C + \log \triangle Y}{4}$$

where $\lambda = 100$ is a constant used to scale the value to a range between 0-100.

The *Mean Constraint Violation* $\mathcal{M}_c^{\mathcal{D}}$ of k generated test papers $P_1, P_2, ..., P_k$ on a question dataset $\mathcal{D}$ w.r.t different test paper specifications $\mathcal{S}_i$, $i = 1..k$, is defined as:

$$\mathcal{M}_c^{\mathcal{D}} = \frac{\sum_{i=1}^k CV(P_i, \mathcal{S}_i)}{k}$$

As such, we can determine the quality of the generated test papers. For high quality test papers, the Mean Discrimination Degree should be high and the Mean Constraint Violation should be low. We set a threshold $\mathcal{M}_c^{\mathcal{D}} \leq 10$, which is obtained experimentally, for high quality test papers.

Figure 3 gives the runtime performance of the DAC approach in comparison with other techniques on the 2 datasets. The results have shown that DAC outperforms other techniques in runtime. Moreover, it also shows that DAC satisfies the runtime requirement as it generally requires less than 1 minute to complete the paper generation process for the 2 different dataset sizes. In addition, DAC is scalable in runtime. Figure 4 shows the quality performance of DAC and other techniques based on Mean Discrimination Degree $\mathcal{M}_d^{\mathcal{D}}$ and Mean Constraint Violation $\mathcal{M}_c^{\mathcal{D}}$ for the 2 datasets. As can be seen, DAC has consistently outperformed other techniques in generating high quality papers with Mean Constraint Violation $\mathcal{M}_c^{\mathcal{D}} \leq 8$.

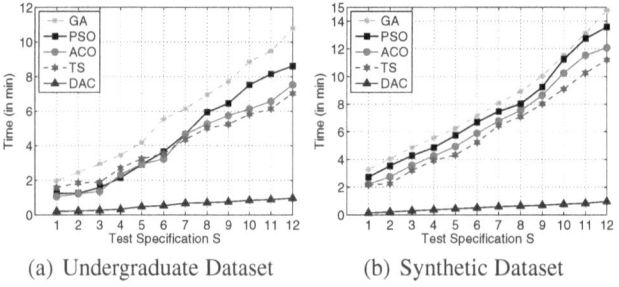

(a) Undergraduate Dataset   (b) Synthetic Dataset

Fig. 3. Performance Results Based on Runtime

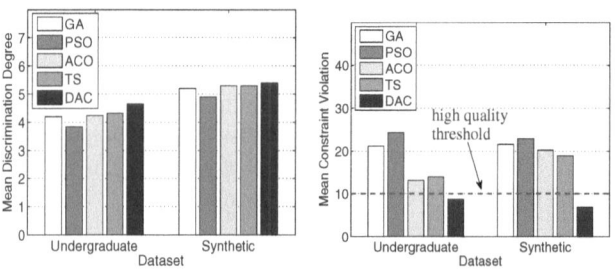

(a) Mean Discrimination Degree   (b) Mean Constraint Violation

Fig. 4. Performance Results based on Quality

To evaluate mathematical answer verification, we have devised 500 answer-solution pairs with 290 correct (positive) answers, and 210 incorrect (negative) answers for 500 test questions from the undergraduate dataset. The test questions are common questions selected according to some standard mathematical functions. Table 2 shows some sample answers and solutions.

We use the precision measure to evaluate the performance of the proposed PEV algorithm, which is defined as the percentage of total correct verifications from all verifications. In addition, we also measure the false negative errors and false positive

Table 2. Test Dataset

Function Type	# Qns	Example	
		Answer	Solution
Constant	50	$\frac{2}{\sqrt{5}}$	$\frac{2\sqrt{5}}{5}$
Polynomial	75	$(x+1)^2(1-x)$	$(1+x)(1-x^2)$
Exponential + log	75	$e^{2+\ln x}$	$e^2 x$
Trigonometric	75	$\frac{\sin x + \cos x}{\sqrt{2}}$	$\sin(x + \frac{\pi}{4})$
Roots	75	$\sqrt{x^2 - 6x + 9}$	$\lvert x - 3 \rvert$
Fractional	50	$\frac{(x^2+2)(x+1)}{x^3}$	$\frac{x^3+x^2+2x+2}{x^3}$
Mixed	100	$x \sin x + e^{\ln x}$	$x(\sin x + 1)$

errors. Figure 5 shows the performance results on mathematical answer verification based on precision. It shows that the proposed PEV algorithm has consistently out-performed Maple and Mathematica. PEV can achieve 100% performance in precision while Maple and Mathematica can only achieve about 95%. This is because PEV can avoid false negative errors and it can also reduce false positive errors to as small as possible by using many trials. Table 3 gives the performance results based on the er-ror types for Maple, Mathematica and PEV. As shown in the table, both commercial CAS systems produce false negative errors with 8.6% in Maple and 7.9% in Mathe-matica. This has shown the advantage of the proposed PEV algorithm in verifying the equivalence of two mathematical expressions numerically rather than symbolically.

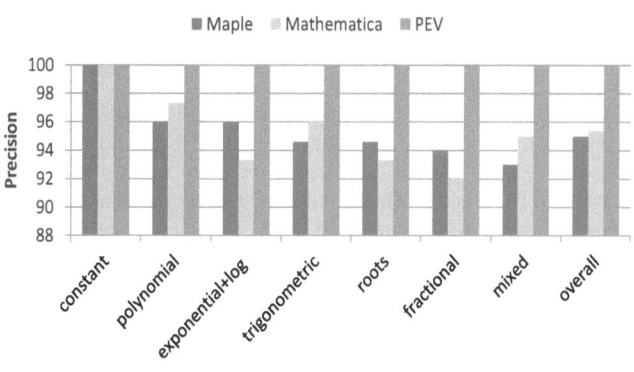

Fig. 5. Performance Results based on Precision

Table 3. Performance Results based on Error Types

	False Negative		False Positive	
	(#)	(%)	(#)	(%)
Maple	25	8.6	0	0.0
Mathematica	23	7.9	0	0.0
PEC	0	0.0	0	0.0

7   Conclusion

In this paper, we have proposed a framework for Web-based mathematics testing and assessment. The proposed framework consists of two major components on automatic test paper generation and mathematical answer verification. The proposed framework consists of an efficient constraint-based Divide-and-Conquer approach for automatic test paper generation, and an effective Probabilistic Equivalence Verification algorithm for mathematical answer verification. The performance results have shown that the DAC approach has not only achieved good quality test papers, but also satisfied the runtime requirement. In addition, the proposed PEV approach is also very effective for mathematical answer verification and assessment. For future work, we intend to investigate Web-based testing techniques for other subjects such as physics and chemistry which contain mathematical and chemical expressions as the required answer type.

References

1. Alon, N.: The probabilistic method. Wiley (2008)
2. Beckmann, N.: The r*-tree: an efficient and robust access method for points and rectangles. ACM SIGMOD Record 19(2), 322–331 (1990)
3. Conati, C.: Intelligent tutoring systems: new challenges and directions. In: Proceedings of the 14th International Conference on AI in Education (AIED), pp. 2–7 (2009)
4. Conejo, R., Guzmán, E., Millán, E., Trella, M., Pérez-De-La-Cruz, J.L., Ríos, A.: Siette: a web-based tool for adaptive testing. International Journal of Artificial Intelligence in Education 14(1), 29–61 (2004)
5. Gonzalez, T.F.: Handbook of Approximation Algorithms and Metaheuristics. Chapman & Hall/Crc (2007)
6. Guzman, E., Conejo, R.: Improving student performance using self-assessment tests. IEEE Intelligent Systems 22(4), 46–52 (2007)
7. Ho, T.F., Yin, P.Y., Hwang, G.J., Shyu, S.J.: Multi-objective parallel test-sheet composition using enhanced particle swarm optimization. Journal of Educational Technology & Society 12(4), 193–206 (2008)
8. Hu, X.M., Zhang, J., Chung, H.S.H.: An intelligent testing system embedded with an ant-colony-optimization-based test composition. IEEE Trans. on Systems, Man, and Cybernetics 39(6), 659–669 (2009)
9. Hwang, G.J.: A test-sheet-generating algorithm for multiple assessment requirements. IEEE Transactions on Education 46(3), 329–337 (2003)
10. Hwang, G.J., Lin, B., Tseng, H.H.: On the development of a computer-assisted testing system with genetic test sheet-generating approach. IEEE Trans. on Systems, Man, and Cybernetics 35(4), 590–594 (2005)
11. Hwang, G.J., Yin, P.Y., Yeh, S.H.: A tabu search approach to generating test sheets for multiple criteria. IEEE Trans. on Education 49(1), 88–97 (2006)
12. Kullback, S.: Information theory and statistics. Dover (1997)
13. Mapple: Version 15 (2011), http://www.maplesoft.com/
14. Mathematica (2011), http://www.wolfram.com/mathematica/
15. Nguyen, M.L., Hui, S.C., Fong, A.C.: An efficient multi-objective optimization approach for online test paper generation. In: IEEE Symposium on Computational Intelligence in Multi-criteria Decision-Making (MDCM), pp. 182–189 (2011)
16. Roussopoulos, N., Kelley, S.: Nearest neighbor queries. In: Proceedings of the ACM SIGMOD, pp. 71–79 (1995)

Content-Based Collaborative Filtering
for Question Difficulty Calibration

Minh Luan Nguyen[1], Siu Cheung Hui[1], and Alvis C.M. Fong[2]

[1] Nanyang Technological University, Singapore
{NGUY0093,asschui}@ntu.edu.sg
[2] Auckland University of Technology, New Zealand
acmfong@gmail.com

Abstract. Question calibration especially on difficulty degree is important for supporting Web-based testing and assessment. Currently, Item Response Theory (IRT) has traditionally applied for question difficulty calibration. However, it is tedious and time-consuming to collect sufficient historical response information manually, and computational expensive to calibrate question difficulty especially for large-scale question datasets. In this paper, we propose an effective Content-based Collaborative Filtering (CCF) approach for automatic calibration of question difficulty degree. In the proposed approach, a dataset of questions with available user responses and knowledge features is first gathered. Next, collaborative filtering is used to predict unknown user responses from known responses of questions. With all the responses, the difficulty degree of each question in the dataset is then estimated using IRT. Finally, when a new question is queried, the content-based similarity approach is used to find similar existing questions from the dataset based on the knowledge features for estimating the new question's difficulty degree. In this paper, we will present the proposed CCF approach and its performance evaluation in comparison with other techniques.

Keywords: Educational data mining, question difficulty calibration, collaborative filtering, content-based approach, Item Response Theory.

1 Introduction

With the rapid development of e-learning [14], intelligent tutoring and test generation systems have become an important component for Web-based personalized learning and self-assessment. To provide web-based testing for self-assessment, it can be done either by generating test paper automatically [9] or supporting adaptive testing [17]. For Web-based testing, the difficulty degree of each question is one of the most important question attributes that needs to be determined. To calibrate each question's difficulty degree, Item Response Theory (IRT) [4] is traditionally applied based on a large number of historical correct/incorrect response information gathered from surveys or tests. However, it is a very tedious and expensive process to collect sufficient historical response information manually.

With the recent emergence of educational data mining [1,5], question data and student response information are getting more readily available. These data may come

P. Anthony, M. Ishizuka, and D. Lukose (Eds.): PRICAI 2012, LNAI 7458, pp. 359–371, 2012.

from different sources including standardized tests combined with student demographic data and classroom interactions [2,12]. With these data, it will certainly help reduce the efforts required for calibrating the difficulty degree of questions. However, the question data and user responses gathered from online sources will not be complete due to missing user responses on certain questions. Moreover, it is also a great challenge to calibrate new question data which does not exist in the question dataset.

In this paper, we propose a Content-based Collaborative Filtering (CCF) approach for automatic calibration of difficulty degree of questions. Based on an input question query, whether it is an existing question in the dataset or a new question, the proposed CCF approach will return the difficulty degree for the input question. The rest of the paper is organized as follows. Section 2 discusses question calibration. Section 3 presents the CCF approach. Section 4 gives the performance evaluation of the CCF approach and its comparison with other techniques. Finally, Section 5 gives the conclusion.

2   Question Calibration

Item Response Theory [4] has been studied quite extensively for question calibration for the last few decades. Most of the research works focus on investigating mathematical models to predict the probability that a learner will answer correctly for a given question. Currently, several models [4] such as normal models, logistic models and Rasch model have been investigated. Such models often require model parameters which are estimated from the observed data by using typical computational algorithms such as maximum likelihood estimation, minimum χ^2 estimation and joint maximum likelihood estimation. However, in practice, obtaining a calibrated question item bank with reliable item difficulty estimates by means of IRT requires administering the item to a large sample of users in an non-adaptive manner. The recommended sample size varies between 50 and 1000 users [10]. As such, it is very expensive to gather the sample information. To overcome this problem, Proportion Correct Method (PCM) [17] was proposed to approximate question difficulty degrees. PCM computes question difficulty degree as $\beta_i = \log[\frac{1-n_i/N}{n_i/N}]$, where β_i is the question's difficulty degree, n_i is the number of users who have answered correctly out of the total N users who have answered that question. Although PCM is straightforward to compute and able to estimate the initial difficulty degree even with missing user responses, PCM still requires a large number of historical response information for accurate estimation.

Another possible way to gather question and response data is to obtain them from Web-based learning environments. It is a much more feasible and inexpensive approach. However, there is one major problem in gathering data from the learning and testing environment. The data often contain missing response information on some questions. This is because learners are free to select questions they want to practice, thereby resulting in the possibly large number of items that have not been answered by learners. Even though IRT can deal with structural incomplete datasets [4], the huge number of missing values can easily lead to non-converging estimation in the IRT model. In addition, the maximum likelihood estimation in IRT is computational expensive. To overcome the impediments of the IRT-based question calibration, other estimation methods based on small datasets have also been investigated. However, the accuracy of these methods is not comparable to IRT-based calibration under the same settings [18]. In this research, we aim to propose a feasible, efficient and accurate IRT-based question calibration for large question datasets gathered from the logging data of Web-based learning and testing environments.

3 Proposed Approach

Content-based and collaborative filtering approaches have been extensively used in recommender systems [3,16]. The content-based approach examines the properties of the items for recommendation, while collaborative filtering (CF) approach recommends items based on similarity measures between users or items from user-item interactions. The items recommended to a user are those preferred by similar users. The CF approach can also be categorized into memory-based and model-based according to user-to-user similarity and item-to-item similarity respectively. Empirically, the item-to-item CF approach has outperformed the user-to-user approach in many real-world applications. Recently, matrix factorization [13] has also been introduced for collaborative filtering. It captures the information by associating both users and items with latent profiles represented by vectors in a low dimensional space. Generally, the CF and matrix factorization approaches are used for recommending warm-start items which have already existed in the dataset. The content-based approach is used for recommending cold-start items which are new items as the CF and matrix factorization approaches are unable to function when the vector representing a new item is empty.

In this paper, we propose a hybrid Content-based Collaborative Filtering (CCF) approach for question difficulty calibration. The proposed CCF approach combines the advantages of both content-based approach and item-to-item collaborative filtering approach. Collaborative filtering is used to predict unknown responses of questions of a user (or learner) from the known responses of other users. All the gathered user responses are then used for estimating the difficulty degree of questions according to the Item Response Theory (IRT) [4]. When a user submits a new question as input query, the content-based similarities of the new question with the existing questions from the

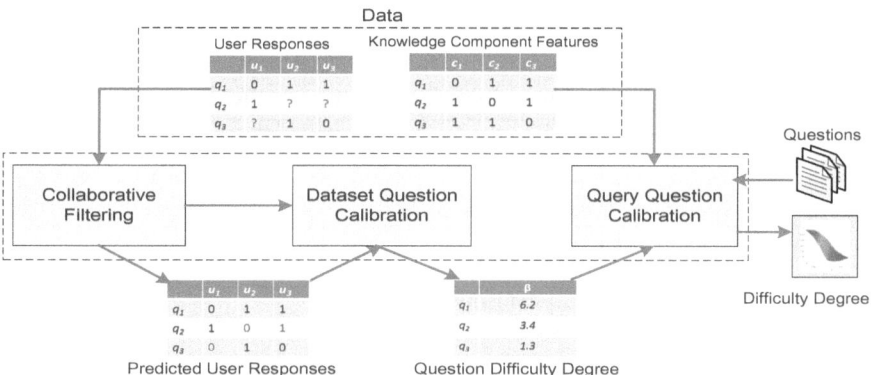

Fig. 1. The Proposed Content-based Collaborative Filtering Approach

dataset are calculated and the difficulty degree of the new question is then estimated accordingly. Figure 1 shows the proposed CCF approach which consists of 3 main processes: Collaborative Filtering, Dataset Question Calibration and Query Question Calibration.

3.1 Question Dataset

There are mainly two types of entities in the question dataset: users/learners and question items. Let $\mathcal{Q} = \{q_1, q_2, .., q_n\}$ be a dataset consisting of n questions, $\mathcal{C} = \{c_1, c_2, .., c_m\}$ be a set of m knowledge components. Each question $q_i \in \mathcal{Q}$, where $i \in \{1, 2, .., n\}$, has 5 attributes $\mathcal{A} = \{q, o, a, \beta, c\}$, where q is the question identity, o is the question content, a is the question answer, β is the difficulty degree, and c is the list of knowledge components needed to solve the question. Let $\mathcal{U} = \{u_1, u_2, .., u_N\}$ be a set of users. Each user has answered some questions with each question $q_i \in \mathcal{Q}$. And the responses to user answers are stored in a binary two-dimensional matrix $\mathcal{R}^{n \times N}$. For each response $r_{ij} \in \mathcal{R}$, its value is a 1 if the user answer is correct and 0 if the user answer is incorrect. Otherwise, r_{ij} is marked as unknown.

Table 1 shows an example Math dataset with some sample values for the question attributes, knowledge components and responses. Apart from general question information, question attributes also store the salient question content features, called Knowledge Component Features (KCF), that are highly related to the difficulty degree of the question. Specifically, a knowledge component represents the knowledge that can be used to accomplish certain tasks, that may also require the use of other knowledge components. The knowledge component feature is a generalization of terms representing the concept, principle, fact or skill, and cognitive science terms representing the schema, production rule, misconception or facet. The KCF can be labeled manually by human experts based on the contents of questions and their solutions. The Response Log stores responses based on user-question item pairs which are collaborative information such as the correct/incorrect response to a question item by a user. Such data are generally obtained from the interactions between the learners and any computer-aided tutoring systems.

Table 1. An Example Math Dataset

(a) Question

$\mathcal{Q}$_id	o	a	β	c
q_1	...	...	?	c_1, c_3
q_2	...	...	?	c_1, c_4
q_3	...	...	?	c_1
q_4	...	...	?	c_1, c_3, c_4
q_5	...	...	?	c_1, c_5
q_6	...	...	?	c_3, c_4

(b) Knowledge Component

$\mathcal{C}$_id	knowledge component
c_1	angles
c_2	triangles
c_3	polygons
c_4	congruece
c_5	similarity

(c) Response Log

$\mathcal{U}$_id	u_1	u_2	u_3	u_4	u_5	u_6
q_1	0	1	1	0	?	1
q_2	1	?	?	1	?	0
q_3	0	0	1	0	?	1
q_4	1	?	0	1	?	?

3.2   Collaborative Filtering

Collaborative filtering (CF) aims to predict unknown user responses of questions from known user responses. The probability that whether a question item can be answered correctly/incorrectly by a certain user can be predicted collaboratively based on the past user responses of all other questions. In CF, it works by gathering known responses for question items and exploiting similarities in response behavior amongst several question items to predict unknown responses. CF is a neighborhood-based approach, in which a subset of question items are chosen based on their similarities to an active question item, and a weighted combination of the responses is used to predict the response for the active question item. In this research, we use the question item-based similarity approach, which is more scalable than the user-based similarity approach. It is because the number of users could be extremely large, often up to millions, while the number of question items is much smaller. In practice, the item-based similarity approach can generally achieve faster performance and result in better predictions [15].

To predict an unknown response r_{ij} of user u_j for an active question q_i, the CF process performs the following 3 steps: assigning weights to question items, finding neighboring items and computing predicted responses.

Assigning Weights to Question Items: This step assigns a weight $w_{i,l}$ to each question item $q_l, l = 1..n$ with respect to its similarity to the active question item q_i. To achieve this, three most commonly used similarity measures, namely the Cosine Similarity, Adjusted Cosine Similarity and Pearson Correlation Similarity [6,15], could be used. Let U be the set of all users who have answered both question items q_i and q_l, let $\bar{r}_i$ be the average responses of the question item q_i among all the users in U and let $\bar{r}_v$ be the average responses of any user u_v. Table 2 gives the formulas for the three similarity measures.

Table 2. Similarity Measures

Name	Formula
Cosine Similarity	$w_{i,l} = \cos(\boldsymbol{r}_i, \boldsymbol{r}_l) = \dfrac{\boldsymbol{r}_i.\boldsymbol{r}_l}{\|\boldsymbol{r}_i\|\|\boldsymbol{r}_l\|} = \dfrac{\sum_{v=1}^{N} r_{i,v} r_{l,v}}{\sqrt{\sum_{v=1}^{N} r_{i,v}^2}\sqrt{\sum_{v=1}^{N} r_{l,v}^2}}$
Adjusted Cosine Similarity	$w_{i,l} = \dfrac{\sum_{v \in U}(r_{i,v} - \bar{r}_v)(r_{l,v} - \bar{r}_v)}{\sqrt{\sum_{v \in U}(r_{i,v} - \bar{r}_v)^2}\sqrt{\sum_{v \in U}(r_{l,v} - \bar{r}_v)^2}}$
Pearson Correlation Similarity	$w_{i,l} = \dfrac{\sum_{v \in U}(r_{i,v} - \bar{r}_i)(r_{l,v} - \bar{r}_l)}{\sqrt{\sum_{v \in U}(r_{i,v} - \bar{r}_i)^2}\sqrt{\sum_{v \in U}(r_{l,v} - \bar{r}_l)^2}}$

The Cosine Similarity measures the responses of two question items as vectors in the n-dimensional space, and computes the similarity based on the cosine of the angle between them. When computing the cosine similarity, unknown responses are treated as having the response of zero. However, there is one major drawback for the basic Cosine Similarity measure in the item-based CF approach, i.e. the difference in response scale between different users is not taken into account. The Adjusted Cosine Similarity measure overcomes this drawback by subtracting the corresponding user average from each co-response pair. The Pearson Correlation Similarity measures the extent of linear dependence between two variables. To compute the correlation, it is necessary to isolate the co-response cases, where the users have answered both questions q_i and q_l.

Finding Neighborhood Items: Based on the question item weight $w_{i,l}, l = 1..n$ obtained from the previous step, this step aims to find the top-K (a predefined value) question items, that have responses by the user u_j, having the highest similarities with the question item q_i. These items are called neighborhood items, and they are denoted as $K(q_i, u_j)$. Finding top-K question items is in fact a K-select problem [11] which can simply be done by scanning the list of items and maintaining a priority queue.

Computing Predicted Responses: This step aims to compute a predicted response r_{ij} of the active question item q_i from the weighted combination of its K selected neighborhood items' responses. Specifically, it computes the weighted average of u_j's responses on the top-K selected items weighted by similarities as the predicted response for the question item q_i by user u_j. It is computed as follows:

$$\widehat{r_{i,j}} = \frac{\sum_{l \in K(q_i, u_j)} r_{l,j} w_{i,l}}{\sum_{l \in K(q_i, u_j)} |w_{i,l}|}$$

where $w_{i,l}$ is the similarity between question items q_i and q_l.

3.3   Dataset Question Calibration

In Dataset Question Calibration, it uses the response information to calibrate the difficulty degree of each question item based on the Item Response Theory (IRT) [4]. In psychometrics, IRT models the correct/incorrect response of an examinee of given ability to each question item in a test. It is based on the idea that the probability of a correct response to an item is a statistical function of the examinee (or person) and item difficulty parameters. The person parameter is called latent trait or ability that represents the person's intelligence. Here, we use the popular one parameter logistic model (or 1-PL) [4] for discussing the calibration of difficulty degree. Calibration based on other models such as 2-PL or 3-PL could be done in a similar way.

In question calibration, user responses (i.e. r_{ij}) are used as the observed data. We need to estimate the values of the model parameters $\beta_i, i = 1..n$ and $\theta_j, j = 1..N$ such

	Users						
	u_1	u_2	...	u_j	...	u_N	Σ
q_1	r_{11}	r_{12}	...	r_{1j}	...	r_{1N}	$r_{1.} = s_1$
q_2	r_{21}	r_{22}	...	r_{2j}	...		$r_{2.} = s_2$
...	...	...	...	...	...	...	...
q_i	r_{i1}	r_{i2}	...	r_{ij}	...	r_{iN}	$r_{i.} = s_i$
...	...	...	...	...	...	...	...
q_n	r_{n1}	r_{n2}	...	r_{nj}	...	r_{nN}	$r_{n.} = s_n$
Σ	$r_{.1} = t_1$	$r_{.2} = t_2$	...	$r_{.j} = t_j$	...	$r_{.N} = t_N$	

(left margin label: Questions)

Fig. 2. Response Data Matrix

that these parameters maximize the probability of the observed data. Here, we follow the traditional approach for question calibration by using Maximum Likelihood Estimation (MLE), in which the model parameters $\beta_i, i = 1..n$ and $\theta_j, j = 1..N$ are considered as constants. The 1-PL model has an advantage that the estimation of question difficulty parameters does not require the estimation of user ability parameters.

The mathematical representation of the 1-PL model is based on a 2-dimensional data matrix obtained by the response information on the set of N users with the set of n question items. Figure 2 shows the response data matrix. The responses to the question are dichotomously scored, i.e. $r_{ij} = 0$ or 1, where $i = 1..n$ is the question index and $j = 1..N$ is the user index. For each user, there will be a column vector (r_{ij}) of item response of length n. Recall that $\mathcal{R}_{ij} = [r_{ij}]^{n \times N}$ denotes the response matrix. The vector of column marginal totals $(r_{.j})$ contains the total responses t_j of the user. The vector of row marginal totals $(r_{i.})$ contains the total responses s_i of the question item.

The probability of a correct/incorrect response in the 1-PL model is given as:

$$P(r_{ij}|\theta_j, \beta_i) = \frac{e^{r_{ij}(\theta_j - \beta_i)}}{1 + e^{(\theta_j - \beta_i)}}$$

Under the assumption that item responses are stochastically independent, the probability of user u_j yielding item response vector (r_{ij}) for the n questions is:

$$P((r_{ij})|\theta_j, (\beta_i)) = \prod_{i=1}^{n} P(r_{ij}|\theta_j, \beta_i) = \prod_{i=1}^{n} \frac{exp[r_{ij}(\theta_j - \beta_i)]}{1 + e^{(\theta_j - \beta_i)}} = \frac{exp(t_j\theta_j - \sum_{i=1}^{n} r_{ij}\beta_i))}{\prod_{i=1}^{n} 1 + e^{(\theta_j - \beta_i)}}$$

It is necessary to find the probability of a given user's total responses t_j. A user of ability θ_j can obtain a given user's total responses t_j in $\binom{n}{t_j}$ different ways and the sum of probabilities of the individual ways is the probability that $r_{.j} = t_j$. Thus,

$$P(r_{.j} = t_j|\theta_j, (\beta_i)) = \sum_{(r_{ij})}^{t_j} P((r_{ij})|\theta_j, (\beta_i)) = \frac{e^{t_j\theta_j}\gamma_{t_j}}{\prod_{i=1}^{n} 1 + e^{(\theta_j - \beta_i)}}$$

where $\sum_{(r_{ij})}^{t_j}$ represents the sum of all group combinations of $\{r_{ij}\}, i = 1..n$ such that $\sum r_{ij} = t_j$, and γ_{t_j} is the elementary symmetric function under the logarithmic transformation of the parameters:

$$\gamma_{t_j} = \sum_{(r_{ij})}^{t_j} exp(-\sum_{i=1}^{n} r_{ij}\beta_i)) = \gamma(t_j; \beta_1, \beta_2, .., \beta_n)$$

Let $d(\theta_j) = \prod_{i=1}^{n} 1 + e^{(\theta_j - \beta_i)}$. The probability of user u_j having the total responses t_j is:

$$P(r_{.j} = t_j|\theta_j, (\beta_i)) = \frac{e^{t_j\theta_j}\gamma_{t_j}}{d(\theta_j)}$$

Then, the conditional probability of the item response vector (r_{ij}), given that user u_j has the total responses t_j, is:

$$P((r_{ij})|r_{.j} = t_j) = \frac{P((r_{ij})|\theta_j, (\beta_i))}{P(r_{.j} = t_j|\theta_j, (\beta_i))} = \frac{\frac{exp(t_j\theta_j - \sum_{i=1}^{n} r_{ij}\beta_i))}{d(\theta_j)}}{\frac{e^{t_j\theta_j}\gamma_{t_j}}{d(\theta_j)}} = \frac{exp(-\sum_{i=1}^{n} r_{ij}\beta_i))}{\gamma(t_j; \beta_1, \beta_2, .., \beta_n)}$$

Note that t_j is used in place of the user's ability parameter θ_j. Hence, this probability is not a function of the user's ability level. Thus, the probability of the item response vector is conditional on the total responses and the set of item parameters (β_i). If N users have responded to the set of items and the responses from different users are assumed to be independent, then the conditional probability distribution of the matrix of item responses, given the values of raw responses $t_1, t_2,...,t_N$ of the N users, is:

$$P([r_{ij}]|t_1, t_2, .., t_N, \beta_1, \beta_2, .., \beta_n) = \frac{exp(-\sum_{j=1}^{N}\sum_{i=1}^{n} r_{ij}\beta_i))}{\prod_{j=1}^{N} \gamma(t_j; \beta_1, \beta_2, .., \beta_n)}$$

The derived equation can be used as the likelihood function for the estimation of the item parameters $\beta_2, .., \beta_n$. Here, we take advantage of the fact that the possible values of the total responses $t_j, j = 1..N$ are usually less than the number of users possessing these responses. As a result, the denominator of the equation can be rewritten as:

$$\prod_{j=1}^{N} \gamma(t_j; \beta_1, \beta_2, .., \beta_n) = \prod_{r=0}^{n} [\gamma(t_j; \beta_1, \beta_2, .., \beta_n)]^{f_r}$$

where f_r is the number of users having the total responses t. Let $s_i = \sum_{j=1}^{N} r_{ij}$ be the item response and $\beta \equiv (\beta_i) = (\beta_1, \beta_2, .., \beta_n)$, the likelihood function is:

$$l = \frac{exp(-\sum_{i=1}^{n} s_i\beta_i))}{\prod_{r=0}^{n}[\gamma(t; \beta)]^{f_r}}$$

The log likelihood is used to obtain the maximum of the likelihood function:

$$\mathcal{L} = \log l = -\sum_{i=1}^{n} s_i\beta_i - \sum_{t=1}^{n-1} f_t \log \gamma(t, \beta)$$

To obtain the maximum of the log likelihood estimates, it is necessary to take first order derivatives $\frac{\partial L}{\partial \beta_h}$ of the log likelihood function with respect to different parameters $\beta_h, h = 1..n$. Note that these first derivative values are zeros at maximum point. Hence, there are n of these equations, which need to be solved simultaneously for the estimation of β_h. Here, as n is large, these n equations can be solved approximately and iteratively by the Newton-Raphson method. To use the Newton-Raphson method, the n^2 second-order derivatives $\mathcal{L}_{hk} = \frac{\partial^2 \mathcal{L}}{\partial h \partial k}$ of the log-likelihood are also needed in order to establish the iterative approximation equation for the estimation of question difficulty degrees:

$$\begin{pmatrix} \widehat{\beta_1} \\ \widehat{\beta_2} \\ \vdots \\ \widehat{\beta_n} \end{pmatrix}_{p+1} = \begin{pmatrix} \widehat{\beta_1} \\ \widehat{\beta_2} \\ \vdots \\ \widehat{\beta_n} \end{pmatrix}_p - \begin{pmatrix} \mathcal{L}_{11} & \mathcal{L}_{12} & \cdots & \mathcal{L}_{1n} \\ \mathcal{L}_{21} & \mathcal{L}_{22} & \cdots & \mathcal{L}_{2n} \\ \vdots & \vdots & \ddots & \vdots \\ \mathcal{L}_{n1} & \mathcal{L}_{n2} & \cdots & \mathcal{L}_{nn} \end{pmatrix}_p^{-1} \times \begin{pmatrix} \widehat{\mathcal{L}_1} \\ \widehat{\mathcal{L}_2} \\ \vdots \\ \widehat{\mathcal{L}_n} \end{pmatrix}_p$$

Algorithm 1 gives the Newton-Raphson method, which solves the conditional estimates of the item difficulty parameters.

Algorithm 1. Newton_Raphson_for_Question_Calibration

Input: $\mathcal{R}_{ij} = [r_{ij}]$ - response information matrix; $t = \{t_1, t_2, .., t_N\}$ - column total
responses; $s = \{s_1, s_2, .., s_n\}$ - row total responses

Output: $\widehat{\beta_1}, \widehat{\beta_2}, .., \widehat{\beta_n}$ - question difficulty parameters

begin

1   Estimate initial item difficulty parameters: $\widehat{\beta_i}^{(0)} = \log(\frac{N-s_i}{s_i}) - \frac{\sum_{i=1}^{n} \log(\frac{N-s_i}{s_i})}{n}$;

2   **repeat**

3   Use the current set of parameters $\widehat{\beta_i}$, evaluate the symmetric functions, $\gamma(t, \beta)$,
     $\gamma(r-1, \beta(h))$, and $\gamma(r-2, \beta(h,k))$;

4   Use these values to evaluate the first and second derivatives of the log likelihood
     function: $\widehat{\mathcal{L}_h}$, $\widehat{\mathcal{L}_{hh}}$, and $\widehat{\mathcal{L}_{hk}}$;

5   Solve the Newton-Raphson Equations iteratively to obtain improved estimates of
     the n item difficulty parameters;

   until *the convergence criterion is satisfied:* $\sum_{i=1}^{n-1} \frac{(\widehat{\beta_i}^{(p+1)} - \widehat{\beta_i}^{(p)})^2}{n} < 0.001$;

6   **return** $\widehat{\beta_1}, \widehat{\beta_2}, .., \widehat{\beta_n}$;

3.4 Query Question Calibration

In the proposed approach, users can submit an input query on a question item and its difficulty degree will be returned. In the warm-start scenario, i.e. if the question has already existed in the dataset, its estimated difficulty degree will then be returned immediately. Otherwise, in the cold-start scenario, in which there is no collaborative information available for the given users or items, the proposed approach will predict the new question item's response based on the question items, which are similar in content with the new question item, the user has answered before. Intuitively, two questions are considered to be similar if they have similar content-based knowledge component features.

Let $A \in R^{|\mathcal{Q}| \times m}$ be the matrix of knowledge component attributes, where a_{ik} is a 1 iff question item q_i has the knowledge component feature k. There are totally m different knowledge component features or attributes. To measure the content-based similarity for new questions, we use the Jaccard Set Similarity [7] based on the set of knowledge component features as follows:

$$w_{i,l} = Jaccard(A_i, A_l) = \frac{|A_i \cap A_l|}{|A_i \cup A_l|}$$

where A_i and A_l are rows of matrix A w.r.t the knowledge component attributes of question items q_i and q_l.

After calculating the similarity, the remaining steps are similar to predicting the unknown responses for existing questions in the dataset. Specifically, the top-K neighborhood questions, denoted as $K(q_i)$, which are most similar to the cold-start question q_i are selected. Then, the difficulty degree β_i is computed by aggregating the weighted average of the difficulty degrees on the top-K nearest neighbors weighted by their similarities as follows:

$$\widehat{\beta}_i = \frac{\sum_{l \in K(q_i)} \beta_l w_{i,l}}{\sum_{l \in K(q_i)} |w_{i,l}|}$$

4   Performance Evaluation

Two datasets, namely *Algebra 2008-2009* and *Bridge-to-Algebra 2008-2009*, obtained from the Knowledge Discovery and Data Mining (KDD) Challenge Cup 2011 [1] are used for performance evaluation. In the experiments, we split each dataset further to form 3 test sets. Test set 1 contains questions whose number of user responses is less than 200. For the remaining questions in each dataset, 100 questions are randomly selected to form test set 2. This test set will be used for evaluating the Query Question Calibration process for cold-start question calibration. The other questions will then be used to form test set 3. Table 3 gives a summary on the statistics of the two datasets. In the experiments, test set 1 will be used for populating the data source. Test set 3 will be used for testing both the Collaborative Filtering and Dataset Question Calibration processes. To do this, we randomly hide 75% of the user response information in this test set during the experiment for evaluation. Test set 2 will be used for Query Question Calibration. The user response information are hidden in this test set during the experiment for evaluation.

Table 3. Datasets

Datasets	# Attributes	# KC Features	# Users	# Questions	# Test Sets	Questions
Algebra	23	8197	3310	5013	Test set 1	4400
					Test set 2	100
					Test set 3	513
Bridge-to-Algebra	21	7550	6043	8459	Test set 1	7400
					Test set 2	100
					Test set 3	959

We evaluate the performance based on each process of the proposed CCF approach: Collaborative Filtering, Dataset Question Calibration and Query Question Calibration. The performance of the Collaborative Filtering process can be evaluated by comparing the predictions with the actual user response information. The most commonly used metric is *Root Mean Squared Error* (RMSE) [8] which is defined as:

$$RMSE = \sqrt{\frac{\sum_{\{i,j\}} (\widehat{r_{i,j}} - r_{i,j})^2}{N}}$$

where $\widehat{r_{i,j}}$ is the predicted responses for user u_j on item q_i, $r_{i,j}$ is the actual responses, and N is the total number of responses in the test set. Similarly, we also use RMSE for evaluating the performance of the Dataset Question Calibration process of the proposed CCF approach and the Proportion Correct Method (PCM) [18] in comparison with the ground truth IRT calibration.

Figure 3 shows the performance results of the Collaborative Filtering process with different parameter settings on similarity measures and neighborhood sizes based on RMSE. In this experiment, we use the 3 different similarity measures: Cosine Similarity, Adjusted Cosine Similarity, and Pearson Correlation Similarity. We also vary the neighborhood size, ranging from 10 to 120 in an increment of 10. It can be observed that the Pearson Correlation Similarity measure has achieved significantly lower RMSE values and the performance has outperformed the two other measures. Moreover, we also observe that the Collaborative Filtering process based on the Pearson Correlation Similarity measure can achieve the best performance with the neighborhood size of about 30. Hence, we have used the Pearson Correlation Similarity measure with the neighborhood size of 30 for subsequent experiments.

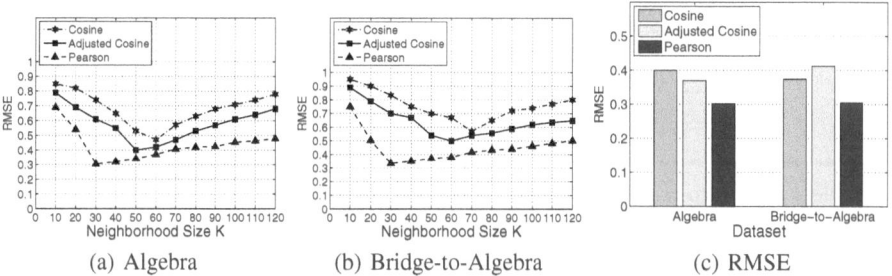

(a) Algebra   (b) Bridge-to-Algebra   (c) RMSE

Fig. 3. Performance Results of the Collaborative Filtering Process based on RMSE

To evaluate the performance of the Dataset Question Calibration process, we compare the performance of the proposed CCF approach with PCM and IRT. Figure 4(a) shows the performance results of the Dataset Question Calibration process of CCF and PCM based on RMSE. Note that IRT has achieved RMSE of 0. As can be seen, the CCF approach has consistently outperformed PCM on the 2 datasets. It has shown that the Collaborative Filtering process is effective to provide sufficient user response information for enhancing the accuracy of the Dataset Question Calibration process. Meanwhile, PCM has achieved considerably lower performance because it calibrates question difficulty with missing user response information.

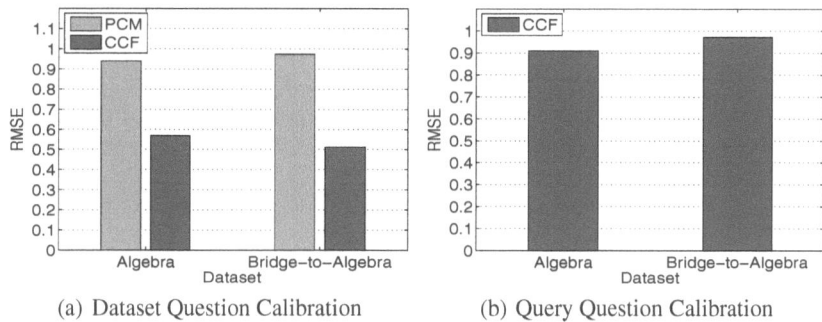

(a) Dataset Question Calibration   (b) Query Question Calibration

Fig. 4. Performance Results based on the RMSE of the 2 Question Calibration Processes

To evaluate the performance of Query Question Calibration, we only evaluate the performance of the CCF approach because both IRT and PCM cannot work with cold-start new questions. Figure 4(b) gives the performance results of the Query Question Calibration process. We observe that although the RMSE error is higher than that of the dataset questions, it is still within an acceptable range. As other approaches are unable to calibrate new questions, the calibrated question difficulty degrees are still very useful. In addtion, the performance results have also shown that the knowledge component features are effective to predict the difficulty similarity among questions.

Table 4. Performance Summary of the 3 Processes in the Proposed CCF Approach

	RMSE	
	Algebra	Bridge-to-Algebra
Collaborative Filtering	0.302	0.305
Dataset Question Calibration	0.57	0.512
Query Question Calibration	0.91	0.972

Table 4 summarizes the performance results of the 3 processes of the proposed CCF approach. Based on the evalutation criteria of RMSE error [8], the Collaborative Filtering process has achieved reliable predictions of unknown user responses. Moreover, under the situation of missing user responses, the performance of the Dataset Question Calibration process can be enhanced with predicted user responses by the Collaborative Filtering process. In addition, although cold-start question calibration has achieved about 40% higher RMSE error than that of the Dataset Question Calibration, the calibrated question difficulty degrees are still very useful to provide the initial estimation of question difficulty, which can be refined further later when user response information can be gathered more sufficiently.

5   Conclusion

In this paper, we have proposed an effective Content-based Collaborative Filtering (CCF) approach for question difficulty calibration. The proposed CCF approach consists of 3 main processes, namely Collaborative Filtering, Dataset Question Calibration and Query Question Calibration. In this paper, the proposed CCF approach and its performance evaluation have been presented. The proposed CCF approach has achieved promising results and outperformed other techniques. For future work, we intend to further enhance the runtime efficiency of the collaborative filtering and question calibration processes in the proposed CCF approach.

References

1. Kddcup 2010: Educational data mining challenge (2010),
   https://pslcdatashop.web.cmu.edu/kddcup/
2. The pslc datashop, https://pslcdatashop.web.cmu.edu/

3. Adomavicius, G., Tuzhilin, A.: Toward the next generation of recommender systems: A survey of the state-of-the-art and possible extensions. IEEE Transactions on Knowledge and Data Engineering 17(6), 734–749 (2005)
4. Baker, F.B., Kim, S.H.: Item response theory. Marcel Dekker, New York (1992)
5. Baker, R., Yacef, K.: The state of educational data mining in 2009: A review and future visions. Journal of Educational Data Mining 1(1), 3–17 (2009)
6. Deshpande, M., Karypis, G.: Item-based top-n recommendation algorithms. ACM Transactions on Information Systems (TOIS) 22(1), 143–177 (2004)
7. Geng, L., Hamilton, H.: Interestingness measures for data mining: A survey. ACM Computing Surveys (CSUR) 38(3), 9 (2006)
8. Herlocker, J., Konstan, J., Terveen, L., Riedl, J.: Evaluating collaborative filtering recommender systems. ACM Transactions on Information Systems (TOIS) 22(1), 5–53 (2004)
9. Hwang, G.J.: A test-sheet-generating algorithm for multiple assessment requirements. IEEE Transactions on Education 46(3), 329–337 (2003)
10. Kim, S.: A comparative study of irt fixed parameter calibration methods. Journal of Educational Measurement 43(4), 355–381 (2006)
11. Knuth, D.: The art of computer programming, Reading, MA, vol. 3 (1973)
12. Koedinger, K., Baker, R., Cunningham, K., Skogsholm, A., Leber, B., Stamper, J.: A data repository for the EDM community: The PSLC DataShop. CRC Press (2010)
13. Koren, Y., Bell, R., Volinsky, C.: Matrix factorization techniques for recommender systems. IEEE Computer 42(8), 30–37 (2009)
14. Li, Q.: Guest editors' introduction: Emerging internet technologies for e-learning. IEEE Internet Computing 13(4), 11–17 (2009)
15. Sarwar, B., Karypis, G., Konstan, J., Reidl, J.: Item-based collaborative filtering recommendation algorithms. In: Proceedings of the World Wide Web, pp. 285–295 (2001)
16. Su, X., Khoshgoftaar, T.: A survey of collaborative filtering techniques. In: Advances in Artificial Intelligence (2009)
17. Wainer, H.: Computerized adaptive testing. Wiley Online Library (1990)
18. Wauters, K., Desmet, P., Van Den Noortgate, W.: Acquiring item difficulty estimates: a collaborative effort of data and judgment. Education Data Mining (2011)

Natural Language Opinion Search on Blogs

Sylvester Olubolu Orimaye, Saadat M. Alhashmi, and Eu-Gene Siew

Faculty of Information Technology
Monash University, Sunway Campus, Malaysia
{sylvester.orimaye,alhashmi,siew.eu-gene}@monash.edu

Abstract. In this paper, we present natural language opinion search by unifying discourse representation structures and the subjectivity of sentences to search for relevant opinionated documents. This technique differs from existing keyword-based opinion retrieval techniques which do not consider semantic relevance of opinionated documents at discourse level. We propose a simple message model that uses the attributes of the discourse representation structures and a list of opinion words. The model compute the relevance and opinionated scores of each sentence to a given query topic. We show that the message model is able to effectively identify which entity in a sentence is directly affected by the presence of opinion words. Thus, opinionated documents containing relevant topic discourse structures are retrieved based on the instances of opinion words that directly affect the key entities in relevant sentences. In terms of MAP, experimental results show that the technique retrieves opinionated documents with better results than the standard TREC Blog 08 best run, a non-proximity technique, and a state-of-the-art proximity-based technique.

Keywords: Opinion Search, proximity, subjectivity, semantic, discourse, NLP.

1 Introduction

Retrieval of relevant and opinionated documents for a given query topic has been the focus of the research community on opinion search [1]. However, the problem with these techniques is their inability to retrieve opinionated documents in context. Opinion in context means that the retrieved opinionated documents contain sentences that are relevant to the discourse in a query topic. The sentences must also contain opinion words that "directly affect" the key entities in the discourse (i.e. *word-word dependency* between opinion words and key entities). The "key entities" in a given query are the key words that are usually the subjects of discourse. We will explain this concept as we proceed.

The opinion retrieval task mostly follows a two step approach. In the first step, traditional Information Retrieval (IR) technique (e.g. BM25 [2]) is used to retrieve documents according to their relevance to the query topic. In the second step, relevant documents are re-ranked according to their opinion scores. We focus on one additional step, and a major challenge that has been largely ignored in the opinion search task. We are aware that opinionated documents can be detected by the ordinary "presence" or

P. Anthony, M. Ishizuka, and D. Lukose (Eds.): PRICAI 2012, LNAI 7458, pp. 372–385, 2012.
© Springer-Verlag Berlin Heidelberg 2012

"proximity" of opinion words to query words in a document. However, whether such opinion words appear in context or directly affect the key entities specified in the user's query remains unsolved.

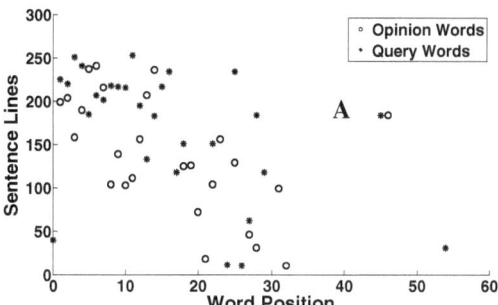

Fig. 1. Distribution of opinion and query words in a sample document retrieved by a baseline technique

The impact of this problem is that, large scale opinion search for commercial implementations (e.g. opinion analytics) can be largely biased. This may be due to unwanted opinions from the retrieved documents. Figure 1 above shows the distribution of opinion words and query words (in terms of positions) in a sample relevant document. The document was retrieved by a state-of-the-art opinion search technique. The idea is to show that using existing techniques, query words are often not directly affected by opinion words in the retrieved documents. The only position where a query word seems to directly affect an opinion word is labeled "A" in the graph. At other positions of the document, query words and opinion words either exists at different sentences or at a longer proximity. It cannot be over emphasized that the use of opinion words occurring at "varying" proximity to "any" of the query words may not be sufficient to determine the opinion relevance of documents. We show that there are two possibilities for determining opinionated sentences:

1. One or more opinion words occur at certain proximity to "any" of the query words. For example, a *proximity threshold* can be specified at which "any" opinion word may be positioned to "any" of the query words [3-4].

2. One or more opinion words directly affect the "key entities" in a sentence that is relevant to the query topic. For example, if the sentence *"The tall girl bought a beautiful iPad"* is relevant to a query topic, then the word "girl" and "iPad" are the two major entities in the sentence. The opinion words "tall" and "beautiful" directly affect the entities rather than being positioned at proximity window at which they may not necessarily affect the entities. Again, consider for example, the following adjacent sentences: *"The girl bought an iPad. What a good opportunity?"* Assuming the words "girl" and "iPad" appeared in the given query, and if we consider condition (1) above, it is more likely that the opinion word "good" would wrongly describe the entity "iPad".

In this work, our aim is to see how much improvement can be achieved by introducing condition (2) above. In order to successfully implement this idea, we propose a simple message model that is based on Discourse Representation Structures (DRS) of each sentence. The message model adopts the traditional *sender to receiver* concept. Between the *sender* and the *receiver* are other attributes of the message model which are derived from the DRS. The *sender* and the *receiver* attributes represent the *key entities* in a sentence which must be directly affected by the opinion words. A popular list of subjective adjectives is used as opinion words that must affect the entities. We summarize our contributions as follows:

- We present a novel semantic-rich natural language opinion search model that is based on DRS of sentences.
- We study the improvement made by considering opinion words which directly affect the key entities in sentences.

Our model has been evaluated against the TREC 2008 best run, a non-proximity technique, and a state-of-the-art proximity-based technique. Our results show better improvements over all the baselines.

For the rest of this paper, we discuss related work and their limitations in section 2. In section 3, we describe natural language search as applicable to the techniques to be used in our model. In section 4, we present our message model. In section 5, we discuss experiments and results. Finally, section 6 draws conclusions on our work.

2   Related Work

Early research works treated opinion retrieval as *text classification* (TC) problem thereby using Machine Leaning (ML) approach to classify text in documents that contain opinions. The TC technique enables the introduction of both traditional *supervised* and *unsupervised* ML techniques with reasonable efficiencies recorded in most research works [5]. A problem with TC technique is the identification of subjective sentences according to the domain for the purpose of training the classifiers. This is still an active area of research. For example, in the iPhone domain, a subjective sentence can read *"The iphone lacks some basic features"*, whereas in the movie domain, a subjective sentence can read *"The movie is not so interesting"*. If a classifier is trained on the iphone domain, such classifiers would have low performance on movie domain as movies cannot *lack basic features*. Thus training classifiers to retrieve opinions has been particularly successful for domain specific corpus such as *Amazon* product reviews [6], but opinion retrieval from multiple domains such as blogs is still challenging.

The *lexicon-based* approach was introduced with the aim that it can compensate for the inefficiencies of the TC techniques. Some research works have combined both lexicon-based and TC techniques for effective opinion search [7]. However, the presence of ambiguous sentences in documents that contain opinion has been one of the limitations of the lexicon-based approach. It is often challenging to understand the orientation of ambiguous words or to determine whether the words contain opinion or not. For example, there is limited success in retrieving opinion from phrases of inverted polarities, irony, idioms, and metaphor [8].

*Probabilistic approach*es have also been used to retrieve and rank opinions from documents by using statistical inferences. The idea is to automatically detect large collection of documents that contain opinions without any learning technique. Some probabilistic techniques have used proximity between query words and opinion words [3-4]. However, we believe some opinion words in proximity to any of the query words may be independent of the query words. Thus the possibility of the opinion words affecting other topics other than the query topic is high. It is better to identify words proximity in the semantic context at which opinionated information has been requested and not by the ordinary presence of query or opinion words.

With all the approaches discussed above, study has shown that it is still difficult to significantly outperform the standard TREC baselines [1]. We believe that the nature of problems that come with opinion retrieval tasks require a semantic-rich approach to detect opinion. In this paper, we will still follow the TREC opinion search definition that combines relevant score with the opinion score. The only difference is how we search for relevance and how we generate opinion scores for the relevant documents. Our technique is also proximity-based opinion search. However, we redefine proximity as one or more opinion words which "directly affect" the key entities in a given sentence.

3   Natural Language Search

Natural language search is seldom performed in opinion retrieval. One of the issues that prevent Natural Language Processing (NLP) techniques to be used in information retrieval is speed. Most NLP techniques (e.g. syntactic parsing and discourse representation structures) are rather slow to be realistically used for large datasets in IR community. However, in this work, we show that using NLP technique for large IR tasks such as opinion search is achievable within reasonable time and gives significant improvement over non-NLP techniques. We do not intend to perform real-time NLP while performing the search as most of the required NLP tasks can be performed offline. Nevertheless, we believe real-time NLP search is achievable with high performance computing. Considering the performance analysis in terms of speed and efficiency for our offline processes, the computational efficiency required would be negligible when combined with real-time search on high performance computing. For example, we performed the syntactic and semantic parse processes on a 2.66 GHz Intel Core 2 Duo CPU with 4 GB of RAM. The CPU time for a complete parse process for each blog document was 5 seconds on average. Thus for the purpose of this study, we will divide our NLP technique into two stages. First is the offline pre-processing of documents in order to make them ready for real-time NLP search by our model. The process involves pre-processing of documents and indexing for effective topic relevant search. It also involves certain syntactic and semantic processes for topic relevant documents. Second is the online opinion search process that is based on our proposed message model. This involves semantic search of subjective documents.

3.1   Syntactic Processing

Our interest in sentence-level natural language opinion search motivates the need for syntactic processing. We believe opinionated information is better shown in each relevant sentence than its adjacent as each sentence contains specific topic discourse structure. Such structure is likely to show how certain key entities are affected by one or more opinion words. Thus, an important process in our technique is to derive the DRS of sentences in each document. For this purpose, a syntactic parser is required since syntax is a preliminary to semantics.

We use a full syntactic parser named Categorial Combinatory Grammar (CCG) [9]. We are motivated by its relatively straightforward way of providing compositional semantics for the intending grammar by providing completely transparent interface between syntax and semantics [10]. However, we do not intend to discuss deep grammatical transformation of natural language sentences with CCG. This is beyond the scope of this paper as it has been discoursed in detail in existing literatures [10,9]. We use a log-linear CCG parsing model by Clark and Curran [10], supported by the result of their study that highly efficient parsing and large-scale processing is possible with linguistically motivated grammar such as CCG [11]. CCG has the advantage to recover long-range dependencies in natural language sentences with better accuracy. In terms of speed, the study in [10] shows that the syntactic parser has a parsing speed of 1.9 minutes on section 23 of the Wall Street Journal (WSJ) Penn Treebank, compared to 45 minutes by Collins parser, 28 minutes by Charniak parser, and 11 minutes by Sagae parser. The resulting CCG output parse tree for each sentence is the syntactic structure from which the required DRS will be derived.

3.2   Discourse Representation Structures

DRS is derived from Discourse Representation Theory (DRT) for the semantic analysis of natural language sentences [12]. It shows the representation of sentences using semantic and thematic roles [13]. This enables natural language applications to easily identify meaning and topic of sentences. The DRS has wide range of benefits for semantic representation with acceptable accuracy [13]. For example, appropriate predicate argument structures involving complex control or coordination of short and long-range sentences can be derived by DRS. Other benefits include better computation of conditionals and negations in sentences, better analysis of time expressions, and analysis of complex lexical phrases.

In our case, its main purpose is to identify key entities in discourse. Given any sentence, DRS differentiate the "participants of an event" by their semantic roles. For each sentence, we make the "participants of an event" to be "the key entities". Thus, we are interested in how opinion words within a sentence (if any) affect the entities in the discourse.

The understanding of the standard translation of natural language sentences to their respective DRS is beyond the scope of this paper. A detail discussion on the standard translation is provided in [13]. For the purpose of our study, we use the Boxer[1] software component for semantic analysis of syntactic structures to produce DRS [13].

3.3   Semantic Processing

The DRS output from the Boxer contains semantic roles or "attributes" that describe words in the discourse structure. In this work, we will use the term "attributes" to represents the "semantic roles" of the words in the discourse structure. Standard semantic roles are discussed in VerbNet [14], these include *patient*, *theme*, *topic*, and *event*. Our aim is to use these standard semantic roles to represent "attributes" from which the "key entities" can be derived for our message model. For example, the "topic" semantic role directly correlates and can be used to represent the "subject" attribute for our message model. We believe that words (entities) that are contained in the "subject" attribute describe what the sentence (message) is about. The idea is, by using the traditional *sender to receiver* concept, every sentence can be seen as a "message" sent by a "sender" (the initiating object) to a "receiver" (the target or subject of focus). The semantic roles adopted from the DRS include, *named*, *topic*, *theme, event*, and *patient,* respectively. We will discourse our message model and how the attributes derived from the semantic roles of the DRS can be used to compute *relevance* and *subjectivity*.

4   The Message Model

As mentioned earlier, the message model is based on the traditional *sender to receiver* concept. We believe every sentence conveys a message which is why our interest is to model every sentence as a message. The idea is that every "message" conveyed must have an "intention" and/or "opinion" on one or more entities. The DRS upon which the message model is built enables us to study the discourse-level relations between the key entities and their corresponding intentions or opinions. Thus, at discourse-level, it is straightforward to model the *relevance* of a sentence to a query topic based on the "intention" and then model the *subjectivity* of the sentence based on the "opinion" words, if any. However, it is more likely, that a sentence will have an intention than to have an opinion. In that case, the sentence is *factual*. This means the sentence is not subjective. Our interest is to retrieve documents that contain reasonable relevant and subjective sentences. Thus, both "intention" and "opinion" must be present in an opinion relevant sentence. More importantly, the intention must correlates with that of the query topic and the opinion must directly affect the key entities in the sentence. It is also possible for the opinion word to "indirectly" affect the key entities. This is somewhat challenging without enhanced *coreferencing* technique. We will only focus on opinion words that "directly affect" the key entities. We will now define the attributes used in our model.

---

[1] http://svn.ask.it.usyd.edu.au/trac/candc/wiki/boxer

4.1   Definition of Model Attributes

Sender: The sender S_d is the set of key words which act as the "initiating objects" of the discourse. The words are mostly noun or named objects. For example, in the sentence *"this team is managed by the great Declan Ryan from Clonoulty – just down the road from Ardmayle"*, the "Sender" attribute consist of the words *Declan, Ryan, Clonoulty,* and *Ardmayle.* We adopt the *named* semantic role from the DRS output to represent the "Sender" attribute as it empirically shows named entities which are most likely to be affected by opinion words.

Subject: The subject S_j is the set of key words which act as the "focus" of the discourse. Usually, this category of words can be used to summarize what a sentence is all about. For example, in the sentence *"everyone felt that Limericks hunger might prove decisive"*, the "Subject" attribute contains the word *feel (felt)* as it explains the topical focus of the sentence. We adopt the *topic* and/or *theme* semantic roles from the DRS output to represent the "Subject" attributes. The idea is that both *topic* and *theme* empirically show the focus of the discourse.

Receiver: The receiver R_v is the set of key words which act as the "subject of emphasis" in the discourse. Unlike the "Sender" attribute, these set of words may not necessarily contain only noun or named objects. It may also contain other words which have specific emphasis in the discourse. For example, in the sentence *"this team is managed by the great Declan Ryan from Clonoulty – just down the road from Ardmayle"*, the "Receiver" attribute contains the words *just* and *team.* Both words received emphasis in the discourse. For example, describing the *team* from *Clonoulty* and the location of the team (*just* down the road). We adopt the *patient* semantic role from the DRS output to represent the "Receiver" attribute as it empirically shows the subject of emphasis in the discourse.

Intention: The intention I_n is the set of words that can collectively show the relevance of a sentence in terms of meaning. Such words contain *predicates* and the *modifiers* used in a sentence. The idea is to ensure that a "discourse relationship" exists between the given query topic and a given sentence. We believe that a sentence is more likely to have the same "intention" with the query topic provided the sentence and the query topic consist of one or more words from the *subject (topic), sender (named objects),* and *receiver (subject of emphasis)* attributes. Such that $I_n = S_j \cup \{S_d \cup R_v\}$.

Key Entities: The key entities E_k is the union of the set *sender* attribute and the set *receiver* attribute such that $E_k = \{S_d \cup R_v\}$. The idea is that both *sender* and *receiver* attributes act as the significant entities in the discourse. Thus, it is more likely that *expression of feeling (opinion)* would tend towards such entities in a given sentence. This is applicable to the opinion word *"great"* in the phrase *"the great Declan Ryan"* from the earlier example sentence.

4.2   Opinion Relevance Language Model

To compute the *opinion relevance* score of a document we use the language model approach [15], which is prominent in IR tasks. Motivated by our *intention/opinion* concept, we compute the *opinion relevance* of a sentence using the probability that the given sentence contains an "intention" that correlates with that of the query topic and also contains one or more opinion words that directly affect the "key entities". Let S be a sentence and let Q_t and E_o be the query "intention" and a set of "key entities" affected by one or more opinion words in the given sentence, respectively.

$$Score(Q_t, S) \propto P(S|E_o) \propto P(E_o|S) \; P(S) \tag{1}$$

where $Score(Q_t, S)$ is the *opinion relevance score* of a given sentence given a query topic, $P(E_o|S)$ is the *opinion likelihood* of the sentence assuming entities are affected by one or more opinion words, and $P(S)$ is the *prior relevance belief* that the sentence is relevant to the query.

4.3   Prior Relevance Belief

We estimate *prior relevance belief* as the relevance of each sentence to the query topic. We assume that if a sentence is relevant to the query topic, most likely the sentence and the query share some words within the *sender, receiver,* and the *subject* attributes. Thus we use a popular relevance language model proposed by Lavrenko and Croft [16] to estimate this prior relevance belief about the sentence using the three attributes. The relevance model has shown substantial performance for effectively estimating the probabilities of words in relevant class using query alone [17]. Thus we estimate $p(\vec{w}|Q_t \cap I_n)$ in terms of joint probability of observing word vector $\vec{w}$ together with vector from intention I_n. Thus, we can make a *pairwise* independent assumption that $\vec{w}$ and terms $\vec{t}$ from $Q_t \cap I_n$ are sampled independently and identically to each other. We compute the relevance score as follows:

$$P(S) = P(\vec{w}) \prod_{i=1}^{k} \sum_{M_i \in Z} P(M_i|\vec{w}) \; P(\vec{t}_1|M_i) \tag{2}$$

where M_i is a unigram distribution from which $\vec{w}$ and $\vec{t}$ are sampled independently and identically, Z is a universal set of unigram distributions according to [16].

4.4   The Opinion Likelihood

In this section we will discuss how we use the attributes defined earlier to estimate the likelihood that a sentence generate an opinion on the key entities given in the query topic. First, we use the presence of opinion words in a sentence as the probability that a sentence is likely to be subjective if one or more opinion words exist in the sentence. A list of subjective adjectives is used as opinion words. The list is derived from a lexicon of 8000 subjective clues provided by Wilson et al [18]. The lexicon has shown to be effective in most opinion and sentiment retrieval techniques. We then estimate *opinion affect* $a(i, j)$ for opinion words w which occur at positions i and directly affect the key entities at other positions j in the sentence.

Let $a(i, j)$ be the proximity distance for an *opinion word* w from a position i (i.e. w_i) to a *key entity* position j (i.e. w_j). A straightforward way to compute *opinion affect* $a(i, j)$ is to set a fixed proximity threshold between i and j. However, using the proximity threshold alone did not give us optimal performance because of the varying short and long-range grammatical dependencies in sentences. For example, if we set the proximity threshold to 1, sentence such as *"that iPad is somewhat intelligent"* will be often ignored for the query topic *"is iPad really intelligent?"*. Thus, our aim is to better estimate $a(i, j)$.

Given $w_j \in E_k$ (Key Entities), $a(i, j)$ is a *discounting factor* $[i - j]$ which shows an *opinion word* position i that directly affects a *key entity* at position j. Thus, there are two possibilities to estimate a favorable $[i - j]$, that is, whether i directly affect j. First, i can directly affect j if i and j directly co-occur as in *"intelligent iPad"*. Second, i can directly affect j if both i and j are in the same *phrase or clause* as in *"iPad is somewhat intelligent"*. Using the first possibility, it is straight forward to compute $a(i, j)$ since the entropy between i and j would be minimum, since i directly co-occur with j. However, it is more challenging if i and j occur as part of a phrase or a clause as shown in the earlier example. To solve this problem, we use the presence of the *predicates* and the *modifiers* between i and j to measure the dependency between i and j (i.e. how i affects j). This can also be termed *word-word dependency* observed in predicate-argument structures. Thus, we estimate $E(w, i \leftrightarrow j)$ as the set of words between positions i and j. We then find the intersection between the *intention* attribute I_n (defined above) and $E(w, i \leftrightarrow j)$. The idea is that words between i and j are likely to exist as elements in I_n since I_n contains the *predicates* and the *modifiers* in the sentence. Thus, if i directly affect j, a predicate or a modifier must exist between them.

Intuitively, if $I_n \cap E(w, i \leftrightarrow j)$ is about a close relationship between i and j, then the language model is usually characterized by large probability. On the other hand, if $I_n \cap E(w, i \leftrightarrow j)$ is not about a close relationship between i and j (e.g. no *predicates* and/or *modifiers*), then the language model would be characterized by small probability. Thus, we can compute the *opinion likelihood* as follows:

$$P(E_o|S) = \sum\nolimits_{j=1}^{m} \frac{predmods_n}{E_n} \qquad (3)$$

where $P(E_o|S)$ is the *opinion likelihood* score, $predmods_n$ is the number of predicates and modifiers found in $I_n \cap E(w, i \leftrightarrow j)$, E_n is the number of words in $E(w, i \leftrightarrow j)$, and m is the number of $a(i, j)$ pairs in S. If E_n is equal to 0, this means i co-occur with j, thus we set $P(E_o|S)$ to the maximum probability of 1. If $predmods_n$ is equal to 0, this means i does not affect j, thus we set $predmods_n$ to 0.01 to avoid zero probability problem.

4.5   Opinion Relevance Score

We stated earlier that opinion search or retrieval is a two-step process. First, document must be relevant to the query topic given in natural language, and then the document must contain favorable opinion. Thus, we now define a ranking function that combines the *prior relevance belief* estimated in equation 2, with the *opinion likelihood* score estimated in equation 3.

$$Score(Q_t, S) = P(S) \, P(E_o|S),$$

$$Score(Q_t, D) = \frac{1}{N} \sum_{i=1}^{N} Score(Q_t, S)) \tag{4}$$

where $Score(Q_t, S)$ is the *opinion relevance* score for each sentence S, $Score(Q_t, D)$ is the absolute ranking score for each document, and N is the number of sentences in the document.

5   Experiment and Results

5.1   Experiment Setup

We evaluate our model against the TREC Blog 08 best run, a proximity-based technique [4], and a non-proximity technique [20]. We extracted TREC Blog08 documents by developing a heuristic rule-based algorithm (BlogTEX)[2] to retrieve blog documents from the dataset. BlogTEX retrieves only English blog documents and ignores documents written in foreign languages by using the occurrences of English function words. All markup tags and special formatting such as scripts, element attributes, comment lines, and META description tags are also removed by BlogTEX. An empirical frequency threshold is set for the number of English function words that must be found within each blog document. Blog documents that have less than the frequency threshold are considered non-English and discarded. For the purpose of this work, we set the threshold to 15 as it indeed gave appropriate English blogs compared to a lower threshold. BlogTEX also contains a highly optimized *sentence boundary* detection module which is based on LingPipe[3] *Sentence Model*. Upon extracting the English blog documents, the sentence boundary detection module is used to tokenize sentences and correctly identify boundaries (e.g. ".", "?", "!") in each document. The idea is to prepare each sentence for syntactic parsing since our model is based on sentence-level opinion search. BlogTEX is seen to have better empirical performance on TREC Blog 08 dataset compared to [19], which is why we make it available free for research purposes.

We index the retrieved blog documents using Lucene[4]. Lucene is a free Java search Application Programming Interface (API) with highly optimized indexing technique. We extracted the title and description fields of 50 TREC 2008 query topics (TREC 1001-1050) without stemming and stop words removal. The query topics are then used to retrieve topic-relevant documents. We use the popular re-ranking technique by retrieving top 1000 topic-relevant documents from the Lucene index using BM25 popular topic-relevance ranking model. Empirical parameters $k=1.2$ and $b=0.75$ are used for BM25 as they have been reported to give acceptable retrieval results [2]. We then re-rank the 1000 documents using our model.

---

[2] http://sourceforge.net/projects/blogtex
[3] http://alias-i.com/lingpipe/demos/tutorial/
sentences/read-me.html
[4] http://lucene.apache.org/java/docs/index.html

Since we are interested in searching opinion based on discourse, we transform each topic relevant document into its equivalent DRS format which was purposely designed for our message model. This process has two stages as discoursed in sections 3.1 and 3.2. First, sentences are transformed to their equivalent syntactic parse trees using the log-linear CCG parsing model [11]. Second, the syntactic output parse trees are fed into the Boxer[5] component in order to derive the equivalent DRS of sentences. We then use the output DRS of sentences to create a message model text file format for each document. The file format simply contains the transformation of sentences to their equivalent attributes-word tags (e.g. *{Sender: Ryan} {Subject: Dolphin} {Receiver: Dolphin_water}*). The log-linear CCG is available as part of the popular C&C tools[6]. As mentioned earlier, the parsing process was done on a 2.66 GHz Intel Core 2 Duo CPU with 4 GB of RAM. The CPU time for a complete transformation process (i.e. syntactic, DRS, and message model file format) for each blog document was 5 seconds on average. Note that this performance may vary on a different hardware other than what we have used for our experiments.

We transform each query topic as described in the process above. We then use our model to compute *opinion relevance* between each query topic and each document. As mentioned earlier, for detecting opinion in each sentence, we use a list of 8000 subjective adjectives from the lexicon of subjective clues provided by Wilson et al [18]. The lexicon has shown to be effective in most opinion and sentiment retrieval techniques.

The evaluation metrics used are based on Mean Average Precision (MAP), R-Precision (R-prec), and Precision at ten (P@10). More than 50% of the TREC 2008 query topics received better improvement in terms of MAP. We also perform evaluation based on negative KL-divergence by comparing the entropy between our message model and a non-proximity technique. The idea is to observe the best approximation model that has minimum cut in terms of entropy by simply measuring the uncertainty of query and opinion words shared by a given collection of blog documents.

5.2   Performance on TREC Query Set

We now show the evaluation of our model in terms of opinion search on the TREC 2008 query topics. In Table 1 below, * indicates better improvements. The results show that the proposed model has improvements over all the selected baselines.

Table 1. Results comparisons with the selected baselines

Model	MAP	R-Prec	P@10
Message Model	**0.5857** *	**0.6615** *	0.8982
TREC 08 best run	0.5610	0.6140	0.8980
Proximity-Based [4]	0.4043	0.4389	0.6660
Non-proximity [20]	0.3937	0.4231	0.6273

---

[5] http://svn.ask.it.usyd.edu.au/trac/candc/wiki/boxer
[6] http://svn.ask.it.usyd.edu.au/trac/candc/wiki

However, we recorded low performance for query topics with one or two entities. For example, topic 1050 *"What are peoples' opinions of George Clooney?"* performed poorly. We observed that *"George Clooney"* appears as the only key entity in the query topic. Therefore, the opinion relevance score is only limited to the distribution of the word *"George"* or *"Clooney"* that is favorably affected by one or more opinion words in the documents. The improvements over the non-proximity method also show that proximity-based methods are likely to retrieves better opinion relevant results that favors the key entities in a given query topic. The improvement over the TREC best run is significant and suggests the need for deep NLP opinion search.

5.3   Clarity of Opinion Proximity

To further demonstrate the effectiveness of our model, we measured the negative KL-divergence between the non-proximity technique and our message model. The purpose is to show significant relevance clarity between the opinion proximity distributions within the documents retrieved by our proximity-based model over a non-proximity model.

The negative KL-divergence method uses an information-theoretic approach that measures the distance between two distributions. A minimized negative KL-divergence indicates improved performance. We plot the graph of the measured negative KL-divergence as a function of the number of top-ranked blog documents, first from an "ideal" relevant document, and then from the retrieved documents by our model and the non-proximity method. For the "ideal" model, we selected a query topic and then manually labeled 20 opinion relevant blogs to represent the results of an ideal model. We then use our model and the non-proximity method to retrieve another 20 opinion relevant documents, respectively. The number of documents was limited to 20 because we are interested in showing the clarity of opinion distribution around the key entities at affordable browsing level.

If $P(t|Q)$ and $P(t|D)$ are the estimated query topic and document distributions respectively (assuming $P(t|D)$ has a linearly smoothed distribution). The topic relevance of document D with respect to query Q can be measured as the *negative* KL-divergence between the two models. However, since we are not only interested in the relevance but the *opinion relevance* as well, we assume each term in document D directly affect or co-occur with an opinion word. Thus we modified $P(t|D)$ to $P(i-j|D)$ for calculating the negative KL-divergence of each document as follows:

$$O_c = \sum_{w \in V} P(t|Q) \log P(i-j|D) + \left(-\sum_{w \in V} P(t|Q) \log P(t|Q)\right) \qquad (5)$$

where $P(i-j|D)$ is the newly estimated proximity distribution by using the occurrence of a query relevant term j that is favorably affected by an opinion word i.

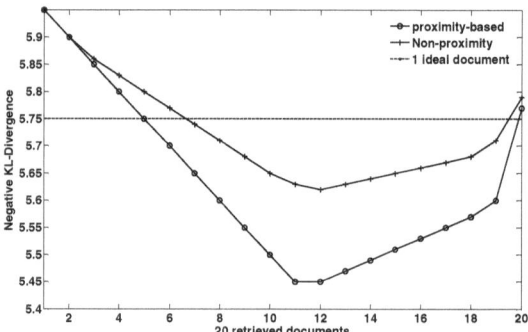

Fig. 2. Negative KL-divergence between our proximity-based model and a non-proximity technique

Figure 2 shows the clarity of opinion word distribution around the key entities for the two models on 20 retrieved documents. We observed more pronounced entropy for our model. However, out of the 20 documents, we could only observe around 7 documents which are likely to have similar proximity distributions. Since our model has more pronounced entropy, we reiterate that a proximity-based opinion search is likely to have better performance than a non-proximity search.

6   Conclusion and Future Work

We proposed natural language opinion search by using deep syntactic and semantic processes. We proposed a proximity-based model that focused on how opinion words in a document affect the key entities of a given natural language query. We evaluated non-proximity and proximity-based methods and show that the latter retrieves better opinion relevant documents that favor the key entities of a query topic. We evaluated our model on the TREC Blog 08 dataset. We observed improvements on all the selected baselines which include TREC Blog 08 best run, a state-of-the-art proximity-based technique, and a non-proximity method. Our study showed that syntactic and semantic-based opinion search using deep NLP techniques is achievable and helpful to retrieving better opinionated documents.

In future, we will investigate dependencies between opinion words in adjacent sentences. For example, an opinion word in a sentence may affect a key entity in the previous or next sentence which may not necessarily contain any opinion word of its own. We will study the contributions of these dependencies towards improving the overall opinion relevant score for each document.

References

1. Macdonald, C., Santos, R.L.T., Ounis, I., Soboroff, I.: Blog track research at TREC. In: SIGIR
2. Robertson, S., Zaragoza, H.: The Probabilistic Relevance Framework: BM25 and Beyond. Foundations and Trends in Information Retrieval 3(4), 333–389 (2009)

3. Santos, R.L.T., He, B., Macdonald, C., Ounis, I.: Integrating Proximity to Subjective Sentences for Blog Opinion Retrieval. In: Boughanem, M., Berrut, C., Mothe, J., Soule-Dupuy, C. (eds.) ECIR 2009. LNCS, vol. 5478, pp. 325–336. Springer, Heidelberg (2009)
4. Gerani, S., Carman, M.J., Crestani, F.: Proximity-Based Opinion Retrieval. In: SIGIR ACM, Geneva, Switzerland, p. 978 (2010)
5. Ounis, I., Macdonald, C., Soboroff, I.: Overview of the TREC 2008 Blog Track. In: TREC (2008)
6. Blitzer, J., Dredze, M., Pereira, F.: Biographies, Bollywood, Boom-boxes and Blenders: Domain Adaptation for Sentiment Classification. In: Proceedings of ACL (2007)
7. Zhang, M., Ye, X.: A generation model to unify topic relevance and lexicon-based sentiment for opinion retrieval. In: Proceedings of the 31st Annual International ACM SIGIR, Singapore (2008)
8. Sarmento, S., Carvalho, P., Silva, M.-J., de Oliveira, E.: Automatic creation of a reference corpus for political opinion mining in user-generated content. In: Proceedings of the 1st International CIKM Workshop on Topic-Sentiment Analysis for Mass Opinion, Hong Kong, China (2009)
9. Steedman, M.: The Syntactic Process (Language, Speech, and Communication). The MIT Press (2000)
10. Clark, S., Curran, J.R.: Wide-coverage efficient statistical parsing with ccg and log-linear models. Comput. Linguist. 33(4), 493–552 (2007)
11. Curran, J.R., Clark, S., Bos, J.: Linguistically motivated large-scale NLP with C & C and boxer. In: Proceedings of the 45th Annual Meeting of the ACL on Interactive Poster and Demonstration Sessions, Prague, Czech Republic (2007)
12. Kamp, H., Genabith, J., Reyle, U.: Discourse Representation Theory Handbook of Philosophical Logic. In: Gabbay, D.M., Guenthner, F. (eds.) Handbook of Philosophical Logic, vol. 15, pp. 125–394. Springer, Netherlands (2011)
13. Bos, J.: Wide-coverage semantic analysis with Boxer. In: Proceedings of the 2008 Conference on Semantics in Text Processing, Venice, Italy (2008)
14. Giuglea, A.-M., Moschitti, A.: Semantic role labeling via FrameNet, VerbNet and PropBank. In: Proceedings of the 21st International Conference on Computational Linguistics and the 44th Annual Meeting of the Association for Computational Linguistics, Sydney, Australia (2006)
15. Ponte, J.M., Croft, W.B.: A language modeling approach to information retrieval. In: Proceedings of the 21st Annual International ACM SIGIR, Melbourne, Australia (1998)
16. Lavrenko, V., Croft, W.B.: Relevance based language models. In: Proceedings of the 24th Annual International ACM SIGIR, New Orleans, Louisiana, United States (2001)
17. Lv, Y., Zhai, C.: A comparative study of methods for estimating query language models with pseudo feedback. In: Proceeding of the 18th ACM CIKM, Hong Kong, China (2009)
18. Wilson, T., Wiebe, J., Hoffmann, P.: Recognizing contextual polarity in phrase-level sentiment analysis. In: Proceedings of HLT/EMNLP, Vancouver, Canada (2005)
19. Nam, S.-H., Na, S.-H., Lee, Y., Lee, J.-H.: DiffPost: Filtering Non-relevant Content Based on Content Difference between Two Consecutive Blog Posts. In: Boughanem, M., Berrut, C., Mothe, J., Soule-Dupuy, C. (eds.) ECIR 2009. LNCS, vol. 5478, pp. 791–795. Springer, Heidelberg (2009)
20. Lee, Y., Na, S.H., Kim, J., Nam, S.-H., Jung, H.-Y., Lee, J.-H.: KLE at TREC 2008 Blog Track: Blog Post and Feed Retrieval. In: TREC 2008 (2008)

Buy It – Don't Buy It: Sentiment Classification on Amazon Reviews Using Sentence Polarity Shift

Sylvester Olubolu Orimaye, Saadat M. Alhashmi, and Eu-Gene Siew

Faculty of Information Technology
Monash University, Sunway Campus, Malaysia
{sylvester.orimaye,alhashmi,siew.eu-gene}@monash.edu

Abstract. In recent years, sentiment classification has been an appealing task for so many reasons. However, the subtle manner in which people write reviews has made achieving high accuracy more challenging. In this paper, we investigate the improvements on sentiment classification baselines using *sentiment polarity shift* in reviews. We focus on Amazon online reviews for different types of product. First, we use our newly-proposed Sentence Polarity Shift (SPS) algorithm on review documents, reducing the relative classification loss due to inconsistent sentiment polarities within reviews by an average of 16% over a supervised sentiment classifier. Second, we build up on a popular supervised sentiment classification baseline by adding different features which provide better improvement over the original baseline. The improvement shown by this technique suggests modeling sentiment classification systems based on polarity shift combined with sentence and document-level features.

Keywords: Sentiment classification, sentence polarity shift, reviews.

1   Introduction

Sentiment classification has attracted quite a number of research works in the past decade. The most prominent of these works is perhaps [1] which employed supervised machine learning techniques to classify positive and negative sentiments in movie reviews. The significance of this work influenced the research community and created different research directions within the field of sentiment analysis. Practical benefits also emerged as a result of automatic recommendation of movies and products by using the sentiments expressed in the related reviews. While the number of reviews has continued to grow exponentially, and with users able to express sentiments in a more subtle manner, it is important to develop more effective sentiment classification algorithms that can intuitively understand sentiments amidst natural language ambiguities.

The application of an effective sentiment classification system cannot be over-emphasized. For example, www.socialmention.com provides a system that shows the sentiment summary on popular social network and news websites. The system is able to classify the sentiments expressed in the social network contents into positive, negative, and neutral categories. In the same manner, www.tweetfeel.com is a website that

P. Anthony, M. Ishizuka, and D. Lukose (Eds.): PRICAI 2012, LNAI 7458, pp. 386–399, 2012.
© Springer-Verlag Berlin Heidelberg 2012

classifies tweets from www.twitter.com into one of positive and negative categories to form an aggregate and succinct sentiment summary. However, the performances of sentiment classification systems differ by the challenges that follow them, such as domain specific styles of expressing sentiments. Sentiment classification on reviews, for example, has been a very challenging task over the last decade with movie reviews being the most challenging of all [1-2]. In addition to the subtle manner in which people write reviews, Pang, Lee, and Vaithyanathan [1], suggested "thwarted expectation" problem as a major factor. Thwarted expectation in this regard is synonymous to inconsistent or mixed sentiment patterns in review documents. Therefore, using random sentiment polarities or ordinary bag-of-words to classify reviews into positive or negative polarity, gives many false positives and false negatives. Consider the following sentences for example:

(1) It's *hard* to imagine any director *not being at least partially* pleased with a film this good.
(2) *But* for over an hour, city of angels is well worth the time.
(3) I'm *not* saying the opening moments are *bad*, because they're pretty good.

All the three sentences above express positive sentiments. However, the presence of *negation words* and *negative adjectives* such as *not*, *but*, *hard*, *bad*, and *least* probably led to wrong classification of the sentences to the negative category. Our study shows that *negation* words are often used at almost the same proportion in both positive and negative reviews. Many negative reviews (using number of stars) also contain many positive sentences and only express negative sentiments by using just few negative sentences. Consider the following positive-labeled review:

> *"I bought myself one of these and used it minimally and was happy* <POSITIVE>
> *I am using my old 15 year old Oster* < NEGATIVE>
> *Also to my surprise it is doing a better job* <POSITIVE>
> *Just not as pretty* <NEGATIVE>
> *I have KA stand mixer, hand blender, food processors large and small...* <OBJECTIVE>
> *Will buy other KA but not this again* <NEGATIVE>*"*

In the above review, positive and negative sentiments were alternated in the earlier part but the review was concluded with negative sentiments. Thus, our proposed technique aims at reducing the error introduced into sentiment classification as a result of the problems highlighted above.

We propose a simple sentence polarity shift (SPS) algorithm that extracts sentences with consistent sentiments in a review. A key step in SPS is the identification of sentences with consistent sentiment regardless of positive and/or negative polarities. We suggest three polarity shift patterns that show consistent improvement over the baselines. We then use the extracted consistent sentences for sentiment classification by adding some newly proposed sentence-level and document-level features to a popular sentiment classification baseline provided by Pang and Lee [3]. This forms a unified sentiment classification model for reviews. We use Amazon online reviews for three different product types: *video*, *music*, and *kitchen appliances*. We say that we are able to reduce an average of 16% classification losses for two popular document classifiers, and achieve better improvement of 2% to 3%, and in some cases, equally good performance is achieved with the baseline technique.

For the rest of this paper, we discuss related work in section 2. In section 3, we present the problem formulation for this work. In section 4, we describe our SPS algorithm. Section 5 discusses the formulation of sentence-level and document level features. We perform experiments with different features in section 6. Finally, we give discussion and future work in section 7.

2   Related Work

We briefly survey previous work on sentiment classification. The most related work to ours is perhaps Pang and Lee [3]. This work proposed using subjectivity summarization based on minimum cuts to classify sentiments in movie review. The intuition is to identify and extract subjective portions of the review document using minimum cuts in graphs. The approach takes into consideration, the proximity information via graph cuts of words to be classified. The identified subjective portions resulting from the graph cuts are then classified as either negative or positive polarity. This approach showed significant improvement from 82.8% to 86.4% on the subjective portion of the documents. It also reported that equally good performance is achieved when only 60% portion of a review document is used compared to an entire review. Thus, we introduce a step further by extracting subjective sentences with consistent sentiments and then discard other subjective sentences with inconsistent sentiment polarities that may contribute noise or reduce the performance of the sentiment classifier. What distinguish our work from [3], is the ability to identify the likely subjective sentences with explicit and consistent sentiment. Our study shows some subjective sentences may not necessarily express sentiments towards the subject matter (e.g., a movie) as show in this "positive-labeled" review:

"[1]real life , however , consists of long stretches of boredom with a few dramatic moments and characters who stand around , think thoughts and do nothing , or come and go before events are resolved. [2]Spielberg gives us a visually spicy and historically accurate real life story. [3]You will like it."

Sentence 1 is a subjective sentence which does not contribute to the sentiment on the movie. Explicit sentiments are expressed in sentence 2 and 3. Discarding sentences such as 1 is likely to improve the accuracy of a sentiment classifier.

Other work include Turney and Littman [4], which use unsupervised learning of semantic orientation to classify reviews based on the number of negative and positive phrases. They achieved an accuracy of 80% over an unlabeled corpus. Similarly, Pang and Lee [1], used supervised machine learning techniques to classify movie reviews using a number of unigram and bigram features. Their unigram model reported improved performance on human classified baseline. A detailed review of sentiment classification techniques on reviews is provided in [5].

3   Problem Formulation

In [1], Pang, Lee, and Vaithyanathan discussed several challenges that are likely to affect the performance of sentiment classifiers on reviews. A common challenge is "thwarted expectation" as a result of "deliberate contrast to earlier discussion" such as shown in the example sentences discussed earlier. Thus, bag-of-words classifiers are likely to "misclassify" such review documents to an opposite sentiment polarity. The necessity for techniques which can help identify when the focus (topic) of a sentence is on the subject matter was suggested in [1]. Consequently, other off-topic sentences can be discarded. To our knowledge, detecting sentiment focus in sentences would require sophisticated "discourse analysis". A cheaper alternative is to reduce inconsistent sentiments as much as possible, such that sentiment classifier can find a clear difference between the remaining consistent classes. Thus, we focus on reducing the inconsistent sentiments polarities in subjective sentences and then add certain newly proposed features that improve the performance of a sentiment classifier.

4   Sentence Polarity Shift (SPS)

Document-level sentiment classification of reviews is challenging and sometimes gives low accuracies as most contextual information may not be directly identified by using only bag-of-words. In contrast, sentence-level information may better provide succinct summary of the overall sentiment expressed in each review. However, since sentiment per sentence in reviews is often inconsistent, a poorly modeled sentiment classifier may end up getting too many positive sentences for a negative-labeled review or too many negative sentences for a positive-labeled review. We may be able to improve sentiment classification by removing inconsistent sentiment polarities such as sentences which tend to *shift away* from prior subsequent and consistent sentiments as identified in our earlier example. We propose, as shown in Figure 1, to first employ some default techniques that have been proposed in a popular sentiment classification baseline. For example, using a *subjectivity detector* [3], which discards the objective sentences leaving only the subjective sentences. We then add another layer, *sentence polarity shift*, which identifies and extracts consistent sentiment polarity patterns from the default subjective sentences. Thus, subjective sentences with inconsistent sentiment polarities are discarded. We believe sentences with consistent polarity patterns should better represent the absolute sentiment of a review by providing a clear-cut between the polarity classes of the major consistent sentiments.

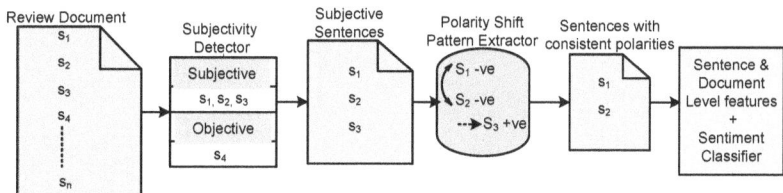

Fig. 1. Sentiment Classification using sentence polarity shift and multiple sentiment features

To our knowledge, previous work has not considered polarity shift patterns between sentences in reviews. Li et al [6], proposed *in-sentence* polarity shifting whereby the absolute polarity of a sentence is different from the sum of the polarity expressed by the content words in the sentence. For example, in the sentence *"I am not disappointed"*, the negative polarity of the word *"disappointed"* is different from the actual positive polarity expressed by the complete sentence. In contrast, we aim to capture *inter-sentence* polarity shift patterns as illustrated above. We believe this can capture the absolute sentiment of a review unlike *in-sentence* polarity shifting which may rarely occur within a review.

4.1   Polarity Shift Patterns

A major challenge in our technique is how to capture consistent polarity shift patterns that are likely to improve the performance of a sentiment classifier. Since sentiment expression at sentence-level differs greatly among review writers, one could think of a number of possible patterns which are commonly used across different review domains. However, it is non-trivial to manually capture these patterns. For example, a writer may express sentiments in a *zigzag* format, such that positive sentiment is expressed in sentence 1 and then negative sentiment in sentence 2 and then continue to alternate the sentiment polarity in this manner until the last sentence. Conversely, a writer may express sentiment in a *serialized* manner such that sentence 1 to the last sentence express the same sentiment polarity.

Considering these complexities, we adopt a hierarchical clustering technique to cluster review dataset according to the polarity shift patterns contained in each review. The dataset comprises of the three different review domains used in our experiment. We trained a language model sentence-level polarity classifier with equal number of positive and negative reviews. We then use the same dataset for testing (i.e. the training set) and classify each sentence in a review as either positive or negative. We depict positive sentence with 1 and negative sentence with 0 and then concatenate the values for all sentences in each review into a single string (e.g. #101001# for a six-sentence review). We then write the concatenated strings as the output containing the polarity shift patterns for each review in the dataset. We use *levenshtein distance*[1] to compute the similarity between the string patterns in the dataset; this gives us a symmetric distance matrix from which we formed hierarchical clusters of reviews with similar polarity shift patterns. Figure 2 shows the different clustered outputs from our combined dataset. Our SPS algorithm considers the polarity shift patterns of the three largest clusters.

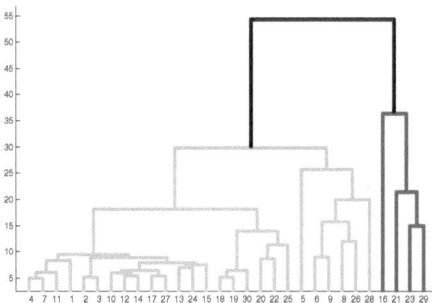

Fig. 2. Dendrogram plot of the hierarchical clusters containing three different polarity shift patterns

---

[1] It measures the edit distance between sequences of two strings.

The three clusters were observed to be *zigzag* pattern, *serialized* pattern, and *2-N* pattern, respectively. We gave an example of the *zigzag* and *serialized* patterns earlier in this section. In the 2-*N* pattern, the same consistent sentiment polarity is expressed at least twice (e.g. *<positive, positive>*), followed by another consistent *n*-number of the opposite sentiment polarity (e.g. *<negative, negative, negative>*, where $N \geq 2$. Given a review, we would therefore like our algorithm to identify the occurrence of any of the three polarity shift patterns.

A straightforward way to capture the polarity shift pattern information is to train classifiers based on the three clusters, such that an unknown review with any of the patterns is identified upon classification. However, it is very unnatural for classifiers to identify reviews which combine consistent sentiment shift patterns with inconsistent sentiments that occur intermittently, and then select only the consistent patterns from each review. For example, *<positive[1], positive[2]>. negative[1], positive[3], <negative[2], negative[3], negative[4]>*. Here, the first negative sentiment and the third positive sentiment are inconsistent with other sentiment patterns in the review. We therefore propose an alternative to overcome this problem. We use an intuitive node-based *nearest neighbor* selection technique to extract consistent sentiment polarities from a review. We are inspired by the fact that K-Nearest Neighbor algorithm has indeed been very effective in pattern classification from text [7].

4.2   Nearest Neighbor Selection

We show a worked example of the node-based *nearest neighbor* concept in Figure 3. Given a review with *n*-number of sentences $S_1, ..., S_n$, we have access to three types of information:

- *Starting polarity (polarity$_{s1}$)*: polarity of the first sentence in a review (i.e., root node) from which the consistency of the subsequent sentiment polarities can be derived;
- *Prior polarity (prior_polarity)*: polarity of the sentence where the polarity of $S_i \neq S_{i+1}$ (i.e., inconsistency is observed). This polarity consequentially becomes the prior sentiment polarity to the subsequent sentences; and
- *k-nearest neighbor (k)*: number of nearest neighbor at which consistent sentiment polarity patterns must be derived.

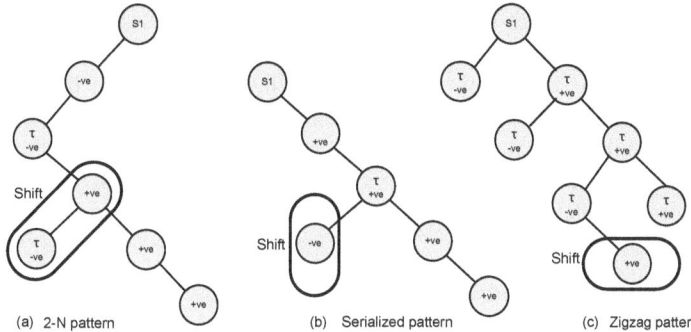

(a) 2-N pattern   (b) Serialized pattern   (c) Zigzag pattern

Fig. 3. Example of polarity shift patterns in reviews, where τ is the prior polarity as defined earlier

If we consider each review as a tree with n-number of sentence nodes, we would therefore like to formally define the node pattern selection decisions for the 2-N, *serialized,* and the *zigzag* patterns, respectively:

Definition1: a 2-N pattern is identified by $polarity_{s1} = polarity_{n,...,N} \geq 2 \leq k$, where $polarity_{n,...,N}$ are the subsequent nodes after $polarity_{s1}$, and $k \leq S_n$.

Definition2: a *serialized* pattern is identified by $polarity_{s1} = polarity_{n+1} \leq k$, where $polarity_{n+1}$ is the sentiment polarity of the subsequent nodes after the $polarity_{s1}$, and $k \leq S_n$.

Definition3: a *zigzag* pattern is identified by $polarity_{s1} \neq prior\_polarity$ and $prior\_polarity \neq polarity_{n+1} \leq k$, where $polarity_{n+1}$ is the polarity sentiment of the next node after the $prior\_polarity$, and $k \leq S_n$.

For the above scenarios, every other node outside the defined patterns is considered a *sentiment polarity shift* (i.e., inconsistent sentiments) and therefore removed from the tree. The remaining sentence nodes can thus be selected and used for sentiment classification. In our experiment, we set $k = 3$ as it empirically captures consistent polarities from our dataset.

5   Sentiment Classification

In this section, we will discuss three additional features that show improved sentiment classification performance on reviews. The default techniques have typically used bag-of-words model for sentiment classification. However, context information is often excluded by using the bag-of-words models. We propose a unifying approach that combine sentence-level and document-level models for sentiment classification. The benefits of unifying sentence-level and document-level features have recently been emphasized by [8]. Thus, the idea is to improve on the performance of ordinary bag-of-words model by compensating *"misclassification"* resulting from the bag-of-words model at document-level with fine-grained sentence-level models. Our sentence-level models combine the unique effectiveness of the Naïve Bayes (NB) classifier and the Language Model (LM) classifier. Augmenting NB with LM has also been emphasized [9], and both classifiers have shown remarkable performances in natural language problems and indeed sentiment classification. Thus, we propose a hybrid approach of using both classifiers at sentence-level such that our sentence-level models are able to capture sentiments expressed under different circumstances. The three features used in our sentiment classification include one document-level feature and two sentence-level features. These features can thus be used to train a separate classifier that forms a unified model for the overall sentiment classification.

5.1   Sentence-Level Models and Features

We propose two sentence-level features by estimating two different sentence-level models. First, we estimate a NB based sentence-level model. The NB classifier is based on conditional independence assumption, yet it has performed effectively in text classification tasks [1]. Our model assigns to a given sentence s the class $c^* = argmax_c\ P(c|s)$. Using Bayes' rule, we derive NB as follows:

$$P(c|s) = \frac{P(c)P(s|c)}{P(s)} \tag{1}$$

In Eq. 1, $P(c|s)$ plays no role in selecting the class c^*. However, estimating the term $P(s|c)$ requires NB to assume features f_i's are conditionally independent given the class of s, thus:

$$P_{\mathrm{NB}}(c|s) := \frac{P(c)(\prod_{i=1}^{m} P(f_i|c)^{n_i(s)})}{P(s)} \tag{2}$$

We implement the NB classifier using the LingPipe[2] NLP libraries and trained the sentence model with unigram features using the default LingPipe parameters. Note that in document-level model, document d is simply substituted for the sentence s.

The second sentence-level model is estimated using LM. The LM classifier performs classification using a joint probability. The probability is given by the language model for each category and a multivariate distribution of the categories. Similar to the NB sentence-level model, we assign a given sentence s the class $c^* = argmax_c\ P(c|s)$, such that:

$$P(c|s) = P(s|c) * P(c) \tag{3}$$

The LM classifier is also implemented using the LingPipe NLP libraries and the sentence model is trained using the default n-gram features (with the default smoothing parameters).

Having estimated the two sentence-level classification models, we require that we generate two different features using the *log ratios* of positive and negative sentences observed in each review. The idea is that negative sentences are often observed to outnumber positive sentences in reviews. We suggest that the reason is perhaps people tend to emphasize on the negative aspects of a product than the positive aspects. For example, we observed many positive labeled reviews to contain certain negative sentiments despite the fact that the product in review offered some satisfaction. Thus, we believe it makes sense to compensate the positive sentences in each review by estimating the *log of the ratio* of the total number of positive classified sentences to the total number of negative classified sentences. This causes the margin between the positive and negative classes to be wider since the score of a negative review would have a *decreasing function* while a positive review would have a positively *growing function*. Thus, we can estimate the document feature $f(d)$ for each sentence-level model (i.e., *log ratios* of NB sentence model and LM sentence model, respectively) as follows:

$$f(d) = \log \left(\frac{P(c^*|s_{positive}) + \theta}{P(c^*|s_{negative}) + \theta} \right) \tag{4}$$

where $s_{positive}$ and $s_{negative}$ are the total number of positive and negative classified sentences, respectively, and θ is a non-negative parameter that avoids negative and positive infinity.

---

[2] http://alias-i.com/lingpipe/

5.2   Document-Level Model and Feature

We use the LM classifier for the document-level model. The LM classifier is as described for the sentence-level LM model above. We also use *n*-gram with the default LingPipe parameters to train the classifier. One unconventional step we took was to model the classification feature for each review using the *log-probability* score computed by the document LM classifier. The *log-probability* score is the sum of the probability of the document belonging to positive and negative categories, respectively. The *log probability* is given as follows:

$$\log(P(x,y)) = -\log(e^{-x'} + e^{-y'}) \tag{5}$$

where x and y are the positive and negative categories, respectively. The idea of using the *log-probability* score as feature makes sense since the ordinary classifier decision uses the maximum probability for any of the classes at the expense of accuracy. Using *log-probability* score as the document-level feature is likely to improve the performance of the classifier.

5.3   Dataset and Baseline

We use the multi-domain sentiment dataset[3] provided by [10]. The dataset consist of Amazon online product reviews for different types of product domains. We select three different product domains: *video, music* and *kitchen appliances*. Each product domain consists of 1000 unprocessed positive reviews and 1000 unprocessed negatives reviews. We use a heuristic algorithm to extract only the review text, discarding other information such as star ratings, product name, review title, and the reviewer name. No stemming or stop words were used.

Our baseline is Pang and Lee [3], which is a popular choice in the literature. They built on their earlier work [1], to achieve an accuracy of 82.8% to 86.4% by adding *subjectivity detector* to a unigram based classifier. Although the accuracies of the baseline were achieved on the online reviews from the Internet Movie Database (IMDB), we are motivated by the fact that the movie domain has been identified as the most challenging domain for sentiment classification task [1-2]. Pang and Lee also stressed that their technique can be easily used to classify sentiments in other domains [1]. Thus, it makes sense to evaluate our technique on such baseline. However, since the selected domains in our dataset are somewhat different from the baseline, we will compare our results with the baseline lower-band of 82% for the *kitchen appliances* and *music* domains and the baseline upper-band of 86.4% for the *video* domain. The number of negative and positive reviews for each domain in our dataset is also proportional to the dataset used by the baseline (i.e. 1000).

6   Experiment and Results

We randomly select 800 positive-labeled reviews and 800 negative-labeled reviews for training each product domain. 200 positive-labeled reviews and 200 negative-labeled reviews were then used for testing each domain accordingly. Objective sentences were

---

[3] http://www.cs.jhu.edu/~mdredze/datasets/sentiment/

removed from the respective training and testing documents as per Pang and Lee [3]. Both NB and LM classifiers were trained using 5-fold cross validation.

6.1 Sentence Classifiers

The sentence classifiers require set of negative and positive sentences for training on each domain. To achieve this, we trained a NB polarity classifier[4] by supplying sentences from the 800 negative and 800 positive training documents as inputs. We then supply the same set of 800 negative and 800 positive documents to classify each sentence to one of negative or positive categories. As a result, "true negative" sentences can be pulled out from the positive training documents and "true positive" sentences can also be pulled out from the negative training documents, thereby reducing the amount of classification loss due to noise of opposite polarity. On average per the three review domains used in our experiments, this technique extracted 7.5% "true negative" sentences from the positive training documents and 7.2% "true positive" sentences from the negative training documents. We also manually verified that the sentences were indeed true negative and true positive sentences. Thus, we combine the actual positive-classified sentences from the positive training documents with the "true positive" classified sentences from the negative training documents. We also combine the actual negative-classified sentences from the negative training documents with the "true negative" classified sentences from the positive training documents. This gives us a new set of positive and negative sentences from the training documents. We refer to the new set as the "corrected" sentence polarities since it would be the actual true positives and true negatives from the sentence classifier. The original set of positive and negative sentences is simply referred to as "uncorrected" sentence polarities since it indeed contained false negatives and false positives. Table 1 shows the accuracies of NB and LM classifiers when both "corrected" and "uncorrected" set of sentences from the *kitchen appliances*[5] domain are trained and tested on the training set, respectively.

Table 1. Accuracy of sentence classifiers for the *corrected* and *uncorrected* polarity sentences

1. Corrected Set			
Classifier	**Instances**	**Features**	**Accuracy**
NB	11382	unigram	**92%**
LM	11382	n-gram = 6	**99.6%**
2. Uncorrected Set			
Classifier	**Instances**	**Features**	**Accuracy**
NB	11382	unigram	88.5%
LM	11382	n-gram = 6	98.2%

The n-gram features above (i.e. n-gram = 6) gave the best accuracy at 5-fold cross-validation. We also use *negation tagging* as suggested by Pang and Lee [1], but we recorded better improvement by tagging 1 to 3 words after a *negation* word rather

---

[4] We selected NB in order to maximize its conditional independence assumption which has indeed shown effective performance. Our baseline also use NB as the initial polarity classifier [3].

[5] For clarity, we only report for the *Kitchen and Appliances* domain. The "corrected sets" from the *video* and *music* domains also outperform their "uncorrected sets", respectively.

than all words between a *negation* word and a punctuation mark. We see that the accuracy of the classifiers for the "corrected set" of sentences is better than the "uncorrected set". Thus, we trained our two sentence-level models (i.e., NB and LM) using only the "corrected set" of positive and negative sentences derived above.

6.2   Document Classifier

Recall that each document now contains only subjective sentences as per Pang and Lee [3]. We therefore use our SPS algorithm to remove inconsistent sentiment polarities as discussed earlier (see section 4.2). Since the LM sentence classifier has a better performance as shown in Table 1 above, we use the LM sentence classifier to identify sentence polarities shift for our SPS algorithm (see section 4.1). We then train a LM document classifier using the output documents from our SPS algorithm. Again, Table 2 shows the accuracies of NB and LM classifiers on the training documents *with* and *without* our SPS algorithm. The idea is to show the likely classification loss due to inconsistent sentiment polarities in reviews. For consistency, we use only the positive and negative training documents from the *kitchen and appliances* domain.

Table 2. Accuracy of document classifiers *with* and *without* our SPS algorithm on the training set

1. With SPS algorithm			
Classifier	Instances	Features	Accuracy
NB	1600	unigram	**90.2%**
LM	1600	n-gram = 6	**92.6%**
2. Without SPS algorithm			
Classifier	Instances	Features	Accuracy
NB	1600	unigram	88.8%
LM	1600	n-gram = 6	90.8%

We see that the accuracy of the document classifiers improved by using our SPS algorithm. We are able to reduce approximately 12.5% to 19.6% classification losses of the document classifiers, respectively (i.e. improvements on the initial classifiers). Since the LM classifier performed better than the NB classifier, we therefore use the log-probability of the LM document classifier (see section 5.2) as the document-level feature.

6.3   Unifying Sentence-Level and Document-Level Features

The two sentence-level features (section 5.1) and the document-level feature (section 5.2) enable us to train a unifying classification model using any of the well performing classifiers. Our baseline presents their classification accuracies using the NB classifier and the Support Vector Machines (SVM). SVM is well established in text classification, and recently in sentiment classification [3]. We use the two classifiers by adding each of the three features one after the other. We also vary the n-gram parameters for the LM classifier in each model, but the NB sentence-level model remains unchanged since it is a unigram model. We then compare the overall performance with our baseline. SVM performed best in most cases by an average of 1.9% improvement over NB. Table 3 shows the best performance per domain by adding each feature one after the other. Again, there were *800 training* and *200 testing instances* for each domain, and the results are compared with the **82.8%** lower-band baseline and the **86.4%** upper-band baseline. Figure 4 show the performance of SVM and NB classifiers on the three domains by varying the respective *n*-gram parameters for the LM classifier at feature level.

Table 3. Best performance on testing set using *n*-gram **> 5** for LM with different feature models

Model	Kitchen	Music	Video
Sentence NB model	78.8	71.0	71.5
Sentence LM model	80.2	71.5	71.75
Document LM model	82.3	81.8	83.75
Sentence NB + Sentence LM	**84.3**	**83.8**	**87.50**
Sentence NB + Sentence LM + Document LM	**85.5**	**84.0**	**88.50**

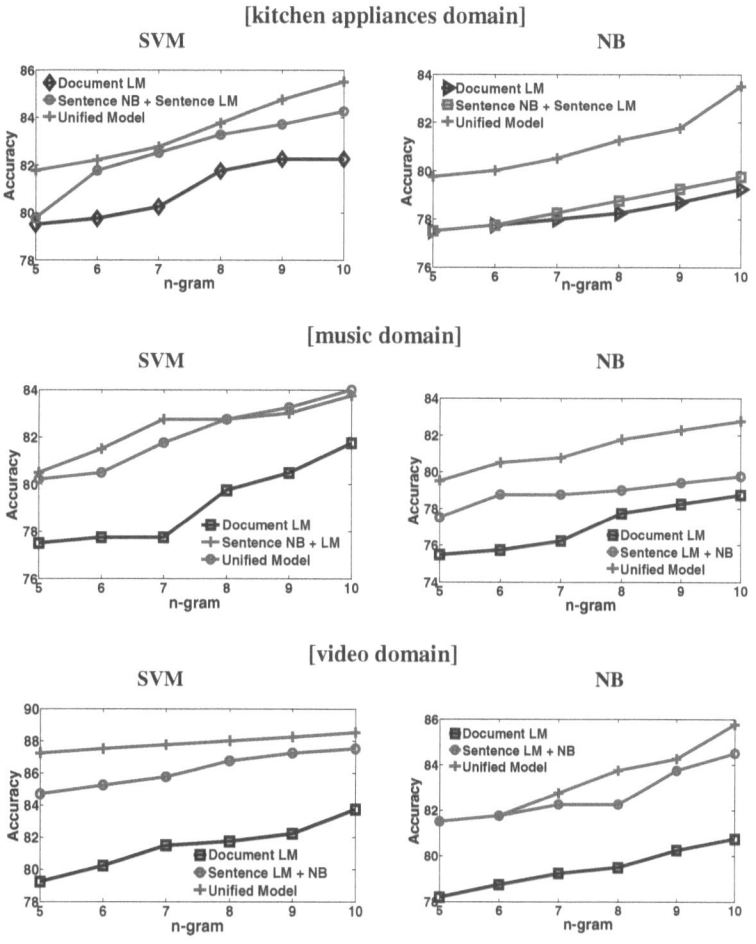

Fig. 4. Performance of SVM and NB with varying n-gram parameters for the LM classifiers

We see that the unified sentence-level and document-level model improved the lower-band baseline of 82.8% for the *kitchen appliances* and *music* domains, respectively. The unified model also improved the upper-band baseline of 86.4% for the *video* domain. However, using individual sentence-level model and individual document-level model did not improve performance, thus suggests the importance of a unified model for sentiment classification.

7   Discussion and Future Work

We proposed sentiment classification of Amazon product reviews using sentence polarity shift combined with a unified sentence-level and document level features. We see that our sentence polarity shift approach reduced the classification loss by an average of 16% over a supervised baseline. Considering the difficulty of the sentiment classification task, the unified sentence-level and document-level features show better improvement of 2% to 3%, and in some cases, equally good performance is achieved with the selected baseline. We believe that the introduction of the NB and LM classifiers at the sentence-level helped capture sentiments expressed under different circumstances and thus compensate classification errors resulting from the bag-of-words model. Figure 4 show that using LM classifiers with higher order n-gram parameters is likely to improve sentiment classification when multiple features are combined to form single unified model. This validates the study conducted by [11], emphasizing the benefits of using higher order n-grams for sentiment classification of product reviews. A unified model with SVM outperform a NB based unified model. However, the accuracy of the sentiment classification model is still less than the accuracy achieved on most traditional topic-based document classification models. We suggest effective classification models that can identify explicit sentiment features at sentence-level and then combine with other document-level features to form a more effective unified sentiment classification model.

References

1. Pang, B., Lee, L., Vaithyanathan, S.: Thumbs up?: sentiment classification using machine learning techniques. In: Proceedings of the ACL 2002 Conference on Empirical Methods in Natural Language Processing (2002)
2. Turney, P.D.: Thumbs up or thumbs down?: semantic orientation applied to unsupervised classification of reviews. In: Proceedings of the 40th Annual Meeting on Association for Computational Linguistics, Philadelphia, Pennsylvania (2002)
3. Pang, B., Lee, L.: A sentimental education: sentiment analysis using subjectivity summarization based on minimum cuts. In: Proceedings of the 42nd Annual Meeting on Association for Computational Linguistics, Barcelona, Spain (2004)
4. Turney, P., Littman, M.L.: Unsupervised Learning of Semantic Orientation from a Hundred-Billion-Word Corpus. Tech. Report EGB-1094, NRC, Canada (2002)
5. Tang, H., Tan, S., Cheng, X.: A survey on sentiment detection of reviews. Expert Systems with Applications 36(7), 10760–10773 (2009)

6. Li, S., Lee, S.Y.M., Chen, Y., Huang, C.-R., Zhou, G.: Sentiment classification and polarity shifting. In: Proceedings of the 23rd International Conference on Computational Linguistics, Beijing, China (2010)
7. Cover, T., Hart, P.: Nearest neighbor pattern classification. IEEE Transactions on Information Theory 13(1), 21–27 (1967)
8. McDonald, R., Hannan, K., Neylon, T., Wells, M., Reynar, J.: Structured Models for Fine-to-Coarse Sentiment Analysis. In: Proceedings of the 45th Annual Meeting of the Association of Computational Linguistics, pp. 432–439 (2007)
9. Peng, F., Schuurmans, D., Wang, S.: Augmenting Naive Bayes Classifiers with Statistical Language Models. Information Retrieval 7(3), 317–345 (2004)
10. Blitzer, J., Dredze, M., Pereira, F.: Biographies, Bollywood, Boom-boxes and Blenders: Domain Adaptation for Sentiment Classification. In: Proceedings of the Association of Computational Linguistics, ACL (2007)
11. Cui, H., Mittal, V., Datar, M.: Comparative Experiments on Sentiment Classification for Online Product Reviews. In: AAAI (2006)

Generation of Chord Progression Using Harmony Search Algorithm for a Constructive Adaptive User Interface

Noriko Otani[1], Katsutoshi Tadokoro[1],
Satoshi Kurihara[2], and Masayuki Numao[2]

[1] Faculty of Environmental and Information Studies, Tokyo City University
3-3-1 Ushikubo-nishi, Tsuzuki, Yokohama 224-8551, Japan
otani@tcu.ac.jp
[2] Institute of Scientific and Industrial Research, Osaka University
8-1 Mihogaoka, Ibaraki, Osaka 567-0047, Japan
{kurihara,numao}@sanken.osaka-u.ac.jp

Abstract. A series of studies have been conducted by us a *Constructive Adaptive User Interface* (CAUI). CAUI induces a personal sensibility model by using a listener's emotional impressions of music and composes music on the basis of that model. Though the experimental results show that it is possible to compose a musical piece that partially adapts to the listener's sensibility, the quality of the composed piece has not been considered thus far. In order to generate high-quality music, it is necessary to consider both the partial and the complete music structure. In this paper, we propose a method for generating chord progressions using a harmony search algorithm for CAUI. The harmony search algorithm is a music-based metaheuristic optimization algorithm that imitates the musical improvisational process. Our experimental results show that the proposed method can improve the degree of adaptability to personal sensibility.

Keywords: Music Composition, Personal Sensibility Model, Harmony Search Algorithm, CAUI.

1 Introduction

A series of works on a Constructive Adaptive User Interface (CAUI) [1] have been carried out in the past by us. CAUI induces a personal sensibility model by using a listener's emotional impressions of music and composes music on the basis of that model. Legaspi et al. [2] proposed an automatic composition system that learns the listener's sensibility model using inductive logic programming (ILP) and composes music using a genetic algorithm (GA). Though the experimental results show that it is possible to compose a musical piece that partially adapts to the listener's sensibility, the quality of the composed piece has not been considered thus far.

The personal sensibility model is represented as a rule for a partial music structure; for example, "The listener feels tender when a root of a chord progresses in the order of III, III, and VI." In order to generate high-quality music, it is necessary to consider

P. Anthony, M. Ishizuka, and D. Lukose (Eds.): PRICAI 2012, LNAI 7458, pp. 400–410, 2012.

the entire music structure. Nishikawa [3] proposed an automatic composition system that considers both partial and complete music structure. A complete music structure with high-quality music involves both "unity" and "development," which may conflict with each other. To consider both the conflicting criterions, *motif* is adopted for generating chord progression. A motif is the most basic component of a music, and it consists of two bars. A motif that adapts to the personal sensibility model is generated using GA. Subsequently, chord progression is generated on the basis of this motif by using GA with a fitness function that evaluates the development of music. The generated motif affects "unity," whereas the fitness function affects "development." However, two problems are involved in the process of chord progression. The first is that the motif may change such that it does not adapt to the model after evolution. The second is that it is impossible to generate chord progression with multiple motifs adapting to the model.

In this paper, we propose a method for generating chord progression using a harmony search (HS) algorithm for CAUI. HS is a music-based metaheuristic optimization algorithm that imitates the musical improvisational process [7]. In the HS algorithm, diversification is controlled by a pitch adjustment operation and a randomization operation, and intensification is represented by the harmony memory. Chord progression consisting motifs that adapt to the model are generated using HS.

2   Composition that Considers Both Partial and Complete Music Structure

This section explains the representation of music and the composition flow. In this study, the time signature of music is 4/4, and a musical piece is a sequence of quarter notes.

2.1   Representation of Musical Pieces

A musical piece consists of a frame structure and a chord progression. The frame structure has 10 components: *genre*, *key*, *tonic*, *tonality*, *time signature*, *tempo*, *melody instrument*, *melody instrument category*, *chord instrument*, and *chord instrument category*. The chord progression is a sequence of chords. A chord is a set of *root*, *type*, and *tension*. When the previous chord is played in succession, the chord is represented by - instead of the set. The values for each component are listed in Table 1.

A bar is a sequence of four chords, and a motif is a sequence of two bars, both of which are represented in the first-order predicate logic. Fig. 1 presents an example of predicate `music` that expresses a musical piece with eight bars. The first argument is the serial number of a music piece. The second argument `song_frame/10` is a frame structure. The third argument represents a sequence of eight bars. Each predicate `bar/4` has four chords.

A personal sensibility model involves a partial structure of music that affects a specific sensibility of a listener. Fig. 2 presents three clauses that are examples of personal sensibility models for the frame structure, motif, and chord respectively. These clauses describe the features of music that induce a feeling of tenderness in the listener. The first clause indicates that the listener feels tender upon listening to music whose tempo is andante and that is played by some kind of piano. The second clause is a feature of a motif wherein the first bar consists of four arbitrary chords and the second bar consists of an arbitrary chord, a IV add9 chord, and two beats of I major chord. The third clause is a feature of chord progression with three successive chords: minor chord, V chord, and I chord.

Table 1. Features of a frame structure and a chord

Frame structure	
Genre	pops
Key	C, Db, D, Eb, E, F, F#, Gb, G, Ab, A, B
Tonic	C, Db, D, Eb, E, F, Gb, G, Ab, A, B
Tonality	major, minor
Time signature	4/4
Tempo	larghetto, andante, moderato, allegretto, allegro, vivace, presto
Melody instrument	square lead, bass and lead, nylon guitar, chiff lead, flute, pan flute, alto sax, sawtooth lead, vibraphone, harmonica, bassoon, recorder, soprano sax, tenor sax, clarinet, strings, steel guitar, charang lead, calliope lead, english horn
Melody instrument category	lead, guitar, flute, sax, vibraphone, harmonica, bassoon, recorder, clarinet, strings, english horn
Chord instrument	synth brass, clean guitar, steel guitar, overdriven guitar, rock organ, bright piano, nylon guitar, electric grand piano, grand piano, synth voice, distortion guitar, clavi, strings, electric piano 1, halo pad
Chord instrument category	synth brass, guitar, organ, piano, synth voice, clavi, strings, pad
Chord	
Root	I, #I, bII, II, bIII, III, IV, #IV, V, #V, bVI, VI, #VI, bVII, VII
Chord type	M, 7, M7, 6, aug, aug7, m, m7, mM7, m6, m7(b5), dim7, sus4, 7sus4, add9
Tension	-, b9th, #9th, 9th, 11th, #11th, b13th, 13th

```
music(1,
  song_frame(pops,d,f,minor,four_four,allegro,
             square_lead,lead,steel_guitar,guitar),
  [bar((i,major,null),-,-,(v,minor,null)),
   bar((i,major,null),-,-,-),
   bar((i,major7,null),(i,7,null),(v,major,ninth),-),
   bar((vi,seventh_aug,null),(vi,major,null),(iv,6,null),(iv,7,null)),
   bar((i,major,null),-,(i,major7,null),(ii,minor_M7,null)),
   bar((vi,seventh_aug,null),-,(vii,minor,null),(iv,major,null)),
   bar((iiib,7,null),(iv,major7,null),(vi,minor,null),(vi,major7,null)),
   bar((vi,major7,null),(iiib,major7,null),-,(vi,major,null))]
).
```

Fig. 1. Example of music

```
frame(tender,A) :-
    tempo(A,andante),
    chord_category(A,piano).
motif(tender,A) :-
    motif(A,bar(_,_,_,_),bar(_,(iv,add9),(i,major),(i,major))).
chords(tender, A) :-
    next_to(A,B,C,_),
    has_chord(A,B,D),type(D,minor),
    has_chord(A,C,E),root(E,v),
    next_to(A,C,F,_),
    has_chord(A,F,G),root(G,i).
```

Fig. 2. Examples of personal sensibility model

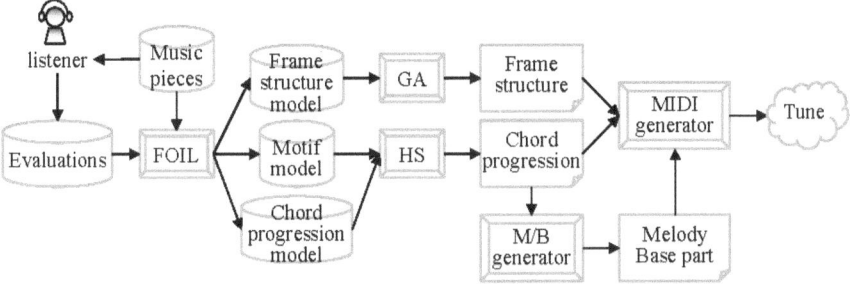

Fig. 3. Composition flow

2.2   Composition Procedure

The composition flow is illustrated in Fig. 3. The parts other than the chord progression by HS are the same as those used in Nishikawa's study [3].

The listener evaluates various musical pieces on the basis of a semantic differential method (SDM) [4], and his/her affective perceptions are collected. The listener rates a piece on a scale of 1-5 for bipolar affective adjective pairs, namely, *favorable-unfavorable*, *bright-dark*, *happy-sad*, *tender-severe*, and *tranquil-noisy*. The former adjective in each pair expresses a positive affective impression, and the latter adjective expresses a negative affective impression. The listener may give a high rating for any piece that he/she finds positive and a low rating for a piece that expresses negativity. The adjectives except for the first pair are selected from the affective value scale of music (AVSM) [5] because of their simplicity.

Personal sensibility models are learned by FOIL [6], a top-down ILP heuristic function, in which the results of the listener's evaluation are used as training examples. In the case of learning a model for positive impression, the pieces with higher rating are used as the positive examples and the remaining pieces are used as the negative examples. In the case of learning a model for negative impression, the pieces with lower rating are used as the positive examples and the remaining pieces are used as the negative examples. The threshold that determines whether a rating is high or low is respectively set to 5 and 4 for positive impressions and 1 and 2 for negative impressions. Two models with different levels are constructed according to the two thresholds. Herein, models developed using the former and latter threshold values are distinguished by the suffixes *a* and *b*, respectively.

For each of the 10 affective adjectives, three kinds of personal sensibility models are constructed: a frame structure model, a motif model, and a chord progression model.

The frame structures are generated on the basis of the frame structure model using GA. A chromosome is represented as a string of elements in a frame structure. Individuals are evolved using a two-point crossover operator and a mutation operator. The fitness value of an individual F is calculated by the fitness function $FFit(F)$ as follows.

$$FFit(F) = \begin{aligned} &Cover(F, FM_a) \times 3 + Cover(F, FM_b) \\ &-Cover(F, FM'_a) \times 3 - Cover(F, FM'_b) \end{aligned} \tag{1}$$

where FM_a and FM_b are the frame structure models for the target adjective, and FM'_a and FM'_b are the frame structure models for the antonymous adjective. The value of the function $Cover(F, FM)$ is the total number of examples covered by clauses in FM that are satisfied by F.

Chord progressions are generated on the basis of the motif model and the chord progression model using HS. The details are explained in Section 3.

The melody and base parts are generated according to the chord progression. The melody line and rhythm in the first half are generated using the first chord as the base for two bars. The melody line and rhythm in the second half are generated by randomly transforming those in the first half. The base line in each bar is a sequence of the root notes. Subsequently, all of the above results are combined to develop a tune.

3   Generation of Chord Progression Using Harmony Search

In this section, the proposed method for generating a chord progression using HS is described. A chord progression that contains a motif that does not adapt to the personal sensibility model is controlled by using both a motif model and a chord progression model. Further, generated a chord progression with two or more motifs that adapt to the model is generated.

3.1   Harmony Search Algorithm

HS imitates a musician's improvisation process. Searching for a perfect state of harmony is analogous to solving an optimization problem. When a musician is improvising, he/she plays music through any one of the following three ways.

1. Select any famous piece from his/her memory
2. Adjust the pitch of a known piece slightly
3. Compose a quite new piece

In the HS algorithm, new candidates are generated using the following three operators. Each operator corresponds to the above-mentioned ways of improvisation.

1. Choose one harmony from the harmony memory
2. Adjust a harmony in the harmony memory
3. Generate a new harmony randomly

Here harmony memory is a set of harmonies. Diversification and intensification are
necessary for an optimization algorithm. In the HS algorithm, diversification is controlled by the pitch adjustment operation and the randomization operation, and intensification is represented by the harmony memory.

The pseudo code of HS is presented in Fig. 4. G is the maximum number of iterations, R_c is a harmony-memory-considering rate, R_a is a pitch-adjusting rate, and $f(x)$
is an objective function.

```
Initialize the harmony memory;
worst := the worst harmony in the harmony memory;
worstfit := f(worst);
for i := 1 to G {
  r1 := a random number from 0.0 to 1.0;
  if(r1 < Rc) {
    new := a harmony chosen from the harmony memory randomly;
    r2 := a random number from 0.0 to 1.0;
    if(r2 < Ra) {
      new := new adjusted randomly within limits;
    }
  } else {
    new := a harmony generated randomly;
  }
  newfit := f(new);
  if(newfit > worstfit) {
    Replace worst with new;
    worst := the worst harmony in the harmony memory;
    worstfit := f(worst);
  }
}
```

Fig. 4. Pseudo code of a harmony search algorithm

3.2   Design of Harmony Search for Generating Chord Progression

In a harmony, the *root_type* note and *tension* note are alternately located in a line. Fig.
5 shows the structure of a harmony. The *root_type* note expresses the ID assigned to a
combination of *root* and *type*. The integer 0-75 is assigned as an ID to 75 (*root*, *type*)
and "-", which are contained in the existing musical pieces, respectively. The *tension*
note expresses ID 0-7 assigned to *tension*. A combination of the *root_type* note and
tension note is translated to a quarter note in phenotype. Four combinations are translated to a bar, and eight combinations are translated to a motif. A musical piece with
2N bars is represented as a sequence of N motifs.

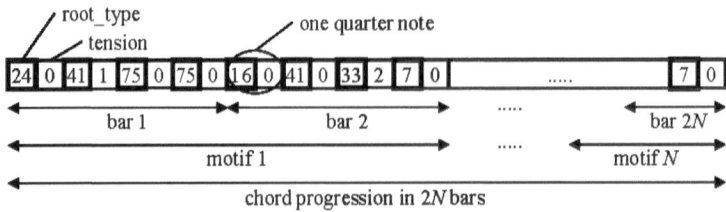

Fig. 5. Structure of a harmony

The fitness value of a harmony C is calculated by the objective function $CFit(C)$ as follows.

$$CFit(C) = \sum_{M \in C}\{Fit\_Motif(M) + Fit\_Builtin(M)\} \\ +\{Fit\_Chords(C) + Fit\_Builtin(C) + Fit\_Forms(C)\} \times N \quad (2)$$

where N is the number of motifs in a chord progression. $M \in C$ means that a motif M is contained in a chord progression C. Fit_Forms is a function that promotes the generation of music with a high development ability. $Fit\_Builtin$ is a penalty function that penalizes a chord progression when it violates the music theory. Fit_Motif and Fit_Chords are functions that indicate the degree of adaptability to the personal sensibility models. Both these functions are calculated using a $Cover$ such as $FFit$.

$$Fit\_Motif(M) = Cover(M, MM_a) \times 3 + Cover(M, MM_b) \\ -Cover(M, MM'_a) \times 3 - Cover(M, MM'_b) \quad (3)$$

$$Fit\_Chords(C) = Cover(C, CM_a) \times 3 + Cover(C, CM_b) \\ -Cover(C, CM'_a) \times 3 - Cover(C, CM'_b) \quad (4)$$

Here, MM_a and MM_b are the motif models for the target adjective, and MM'_a and MM'_b are the motif models for the antonymous adjective. CM_a and CM_b are the chord progression models for the target adjective, and CM'_a and CM'_b are the chord progression models for the antonymous adjective.

A harmony can be adjusted in any of the following four ways via the pitch adjustment operation. One of the following four ways is selected randomly.

— Change the *root_type* note to the other value randomly
— Change the *root_type* note to the other value that expresses the ID assigned to a combination of the same *type* and the different *root* randomly
— Change the *root_type* note to the other value that expresses the ID assigned to a combination of the same *root* and the different *type* randomly
— Change the *tension* note to the other value randomly

4   Experiments

Experiments were conducted with ten Japanese university students as listeners. Seven of them were male and three were female, aged between 21 and 23 years. The listeners were asked to listen to 53 well-known pieces in MIDI format and to evaluate each piece in terms of the five impression adjective pairs on a scale of 1-5. The personal sensibility models for each listener were learned on the basis of their individual evaluations, and new musical pieces were composed independently for each listener and each impression. The parameters for the generation of chord progressions are listed in Table 2. The listeners that participated in the evaluation phase of the well-known pieces were then asked to evaluate the pieces composed particularly for them in terms of the five impression adjective pairs on a scale of 1-5. They were not informed that these new pieces would induce specific impressions.

Table 2. Parameters

Parameter	Value
maximum number of iterations, G	100000
harmony-memory-considering rate, R_c	0.85
pitch-adjusting rate, R_a	0.30
size of harmony memory	500

Ten pieces were composed for each listener, i.e., one for each impression adjective. Figs. 6 and 7 show the musical pieces composed for a certain listener. Fig. 6 is a piece relating to the adjective *tranquil*, whereas Fig. 7 shows a piece relating to the adjective *noisy*. The top score numbered "1" is the melody. The middle score numbered "2" is the chord progression. The lower score numbered "3" is the base part. The melody instrument and the chord instrument for Fig. 6 are `flute` and `synth_voice`, respectively. Those for Fig. 7 are `sawtooth_lead` and `electric_piano`, respectively. The listener rated the *tranquil* piece as 5 for tranquility and the *noisy* piece as 3.

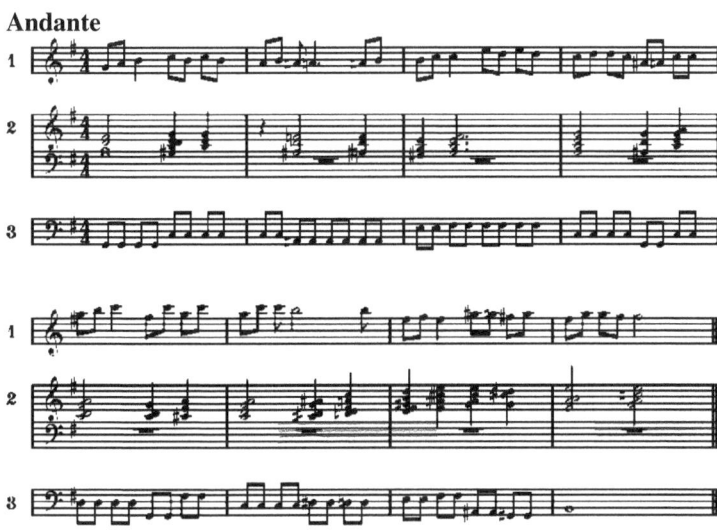

Fig. 6. A composed *tranquil* musical piece

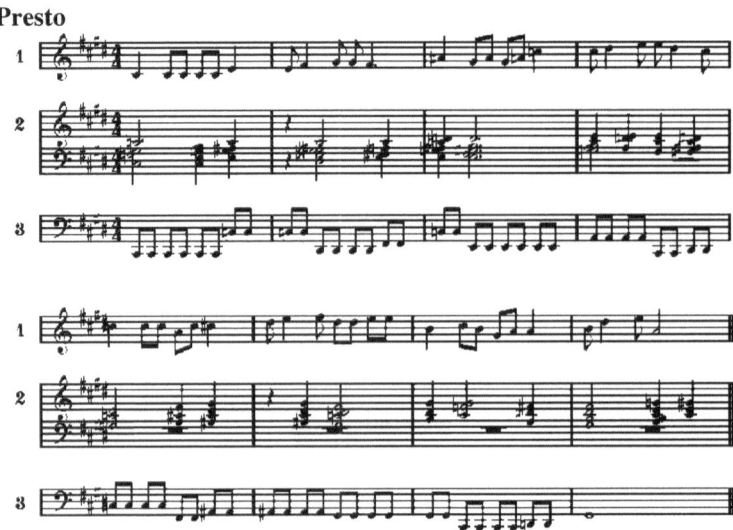

Fig. 7. A composed *noisy* musical piece

Fig. 8 shows the average of the listeners' evaluation of the impression of the composed musical pieces. Error bars show standard deviations. "Positive" shows the average result for all listeners' pieces that were composed for the target positive impression. "Negative" shows the average result for all listeners' pieces that were composed for the target negative impression. The plausibly acceptable values would be >3.0 and <3.0 for the positive and negative adjectives, respectively. The evaluation values of the positive adjective indicating brightness and tenderness were lower than 3.0. However, a large difference between the evaluation values of positive and negative adjectives was obtained for favorableness, which typically varies from person to person and may be difficult to reflect in musical pieces.

According to a Student's t-test, that is a statistical hypothesis test applied when the test statistic follows a normal distribution, the evaluation values are different for three adjective pairs at a significance level $\alpha=0.05$. They are different for one pair at a significance level $\alpha=0.01$. Fig. 9 shows the effect of melody on impressions. The evaluation values for pieces without a melody are low for all the impressions. The evaluation values for positive and negative are reversed about favorableness.

Similar to the evaluation of impressions, the listeners evaluated the pieces on a scale of 1-5 for the four quality criteria: unity, development, fun, and formation. If the listener feels that the piece has high quality about each criterion, he/she gives a high rating for the piece. Fig. 10 shows the average of all listeners' evaluation values of the quality of all composed musical pieces. Error bars show standard deviations. In all the criteria, including unity and development that tend to have the relation of a trade-off, the evaluation values exceeded not only the base value of 3.0 but also the results obtained in Nishikawa's study [3]. In addition, the results show that the pieces without melody had a higher quality. According to a Student's t-test, the evaluation values of the pieces with melody are lower than those of the pieces without melody at a significance level 0.05.

Significant difference in Student's T-test

Significance level	favorableness	brightness	happiness	tenderness	tranquility
5%	Yes	No	Yes	No	Yes
1%	No	No	No	No	Yes

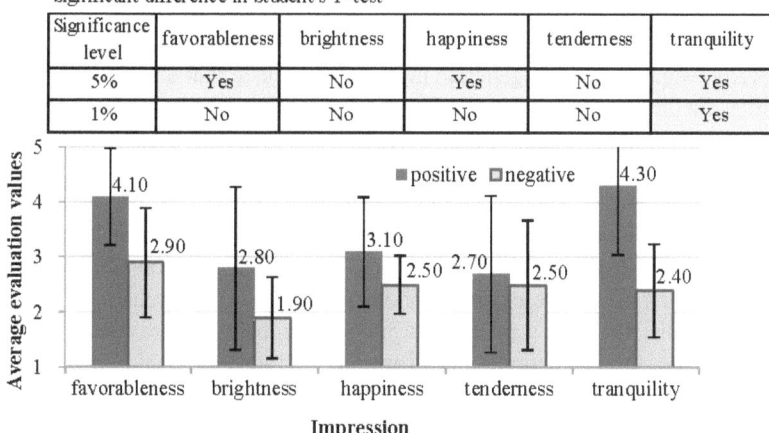

Fig. 8. Evaluation of the impression of composed musical pieces

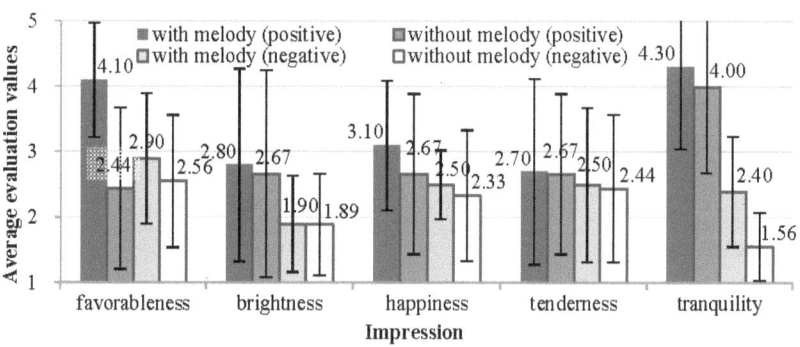

Fig. 9. Effects of melodies on impressions

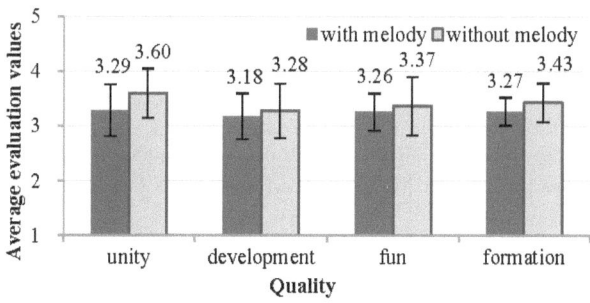

Fig. 10. Evaluation of the quality of composed musical pieces

5   Conclusion

In this study, we attempted to generate high-quality music that adapts to the listener's sensibility. The interactive evolutionary computation (IEC) can be applied for reflecting listener's sensibility to music. However, IEC needs listener's evaluation that takes time and effort, whenever a musical piece is composed. In the case of our method, once a personal sensibility model is induced, musical pieces can be composed automatically. Music composition can be repeated without listener's effort.

This study proposed a method for generating chord progressions on the basis of a HS algorithm for CAUI. The optimum chord progression is searched closely by HS, which imitates the fine-tuning process of pitches in musical improvisation. Owing to this characteristic, the evaluation values for unity and development, which apparently disagree with each other,   were found to be higher than the base value. In addition, the evaluation value for the quality criterion of fun, which is difficult to describe symbolically, was higher than the base value. Considering that the evaluation value of a musical piece changes according to the presence or absence of a melody, we can deduce that impression is considerably dependent on melody. A method for generating melody in combination with the proposed method should be investigated in order to compose relatively high-quality music, which adapts to the listener's sensibility.

Acknowledgement. This work was carried out under the Cooperative Research Program(2011233) of "Network Joint Research Center for Materials and Devices" and JSPS KAKENHI Grant Number 23300059.

References

1. Numao, M., Takagi, S., Nakamura, K.: Constructive Adaptive User Interfaces - Composing Music Based on Human Feelings. In: Proc. of the 18th National Conference on Artificial Intelligence, AAAI 2002, pp. 193–198. AAAI Press, California (2002)
2. Legaspi, R., Hashimoto, Y., Moriyama, K., Kurihara, S., Numao, M.: Music Compositional Intelligence with an Affective Flavor. In: Proc. of the 12th International Conference on Intelligent User Interfaces, pp. 216–224. ACM Press, New York (2007)
3. Nishikawa, T.: Automatic Composition System Considering both Partial and Overall Music Structure. Master's thesis, Osaka University, Japan (2009)
4. Osgood, C., Suci, G., Tannenbaum, P.: The Measurement of Meaning. University of Illinois Press, Urbana (1957)
5. Taniguchi, T.: Construction of an Affective Value Scale of Music and Examination of Relations between the Scale and a Multiple Mood Scale. The Japanese Jrl. of Psychology 65, 463–470 (1995) (in Japanese)
6. Quinlan, J.: Learning Logical Definitions from Relations. Machine Learning 5, 239–266 (1990)
7. Geem, Z., Kim, J., Loganathan, G.: A New Heuristic Optimization Algorithm: Harmony Search. Simulation 76, 60–68 (2001)

A Local Distribution Net for Data Clustering

Qiubao Ouyang, Furao Shen, and Jinxi Zhao

National Key Laboratory for Novel Software Technology,
Nanjing University, China

Abstract. In this paper, a local distribution neural network is proposed for data clustering. This competing network is designed for nonstationary and evolving environment. It represents data by means of neurons (ellipsoids) arranged on a topology map. The local distribution is stored in ellipsoids, while the global topology information is preserving in the relationship between adjacent ellipsoids. With a self-adapting threshold strategy and iteratively learning for information of local distribution, the algorithm is operated in an incremental and on-line way. During implementation, The adopted metric is an improved Mahalanobis distance which considers the local distribution and implies the anisotropy on different vector basis. Hence it can be interpreted as an incremental version of Gaussian mixture model. Experiments both on artificial data and real-world data are carried out to show the performance of the proposed method.

Keywords: self-organizing, covariance matrix, Mahalanobis distance, incremental learning, on-line learning.

1 Introduction

A self-organizing map (SOM) [1] was trained in an unsupervised learning way to produce a low-dimensional (typically two-dimensional) space. It used a neighborhood function between neurons to preserve the topological properties of the input space. However, the pre-determined neuron number and structure exposed itself to criticism. In the evolution of SOM, Teuvo Kohonen introduced another neural network named Adaptive-subspace self-organizing map (ASSOM) [2]. Different from SOM, each neuron of ASSOM was associated with a subspace. When placing a vector from data space, the projection error was obtained between subspaces (neurons) and vector, then the winning neuron was determined. After that, the extension combining the subspace and center bias was proposed. A principal components analysis self-organizing map (PCASOM) [3] was an alternative for this extension. It stored the covariance matrix, replacing the orthonormal vector basis, to represent the subspace. This was directly derived from statistic theory and has advantages in computation burden and reliability of the result. These two models contained more information in each neuron than SOM. But the critical defect in predefined size and structure was inherited. For this reason, ASSOM and PCASOM needed more priori knowledge before training procedure and could not handle the incremental learning.

P. Anthony, M. Ishizuka, and D. Lukose (Eds.): PRICAI 2012, LNAI 7458, pp. 411–422, 2012.

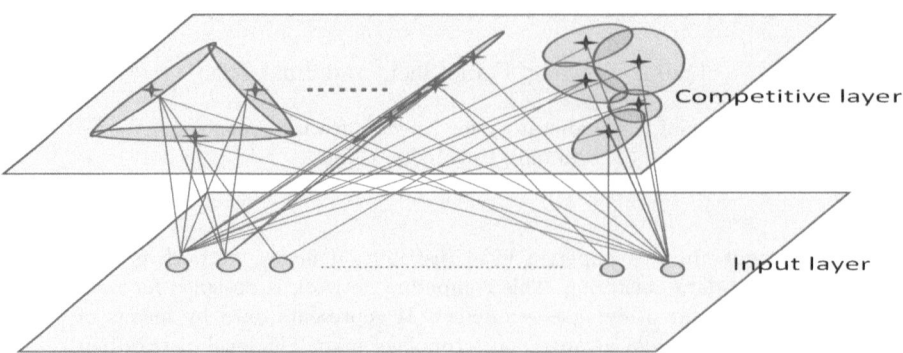

Fig. 1. Structure of Local-SOINN. The input layer accepts input data. The competitive layer dynamically generates and modifies ellipsoids (neurons) with the stimulation of the input data, and finally outputs ellipsoids to represent the initial data and cluster automatically.

A self-organizing incremental neural network (SOINN) [4] was adequate to realize the incremental learning just as its name implied. Incremental learning is a task to break away from the Stability-Plasticity Dilemma [7], which means how to adapt to new information without corrupting or forgetting previously learned information. Adopting a similarity threshold and a locally accumulated error-based insertion criterion, the system can grow incrementally and accommodate input data of non-stationary distribution. SOINN was a competition network and could deal with incremental and on-line learning task without prior knowledge such as how many classes exist.

In this paper, we propose an algorithm called local distribution learning self-organizing incremental neural network which combines the advantages of matrix learning and SOINN. Each neuron is associated with its mean vector and co-variance matrix, thus the data distribution can be represented by a collection of local PCA units [9][10] or ellipsoids. Using a self-adapting threshold strategy, the model can be implemented in an incremental learning way even priori knowledge is utterly ignorant. Then we extend the Mahalanobis distance to a more general metric and give interpretation about the relation between Local-SOINN, local PCA and Gaussian Mixture Models. Some analysis of mathematical statistics are also presented to support the stability and availability of the proposed algorithm.

2   The Local-SOINN

2.1   Structure of Local-SOINN

The structure of Local-SOINN is similar to SOM, a competition network inspired by the study of biology. As Figure 1 show, there are two layers in Local-SOINN, one is the input layer, and the other is competitive layer. In input layer, the number of neurons rest with the dimension of the input data. In competitive layer, the number of neurons is not predetermined, but automatically obtained with the training of specified data set.

The information of competitive layer is stored in two data structure: neuron set and adjacency list. Neuron set preserve the neuron information, each neuron is associated with a 3-tuple $< n, c, M >$, where n, c, M respectively represent the number, mean vector and covariance matrix of samples that assigned to this neuron. The neuron, visualized in the input space, can be described as an ellipsoid: $\{x | \sqrt{(x-c)^T M^{-1}(x-c)} < H\}$. Adjacency list save the topological relation among neurons. If two neurons (ellipsoids) overlap each other, a connection (edge) between the two neurons is created. Each connection is expressed by a 2-tuple $< id_1, id_2 >$, where id_1, id_2 respectively refer to the identity of the origin and destination neuron. These two data structure can alter dynamically with the proceeding of the proposed model, which are convenient for incremental and on-line learning.

In the next part, we detail the procedure of Local-SOINN model. The following notations are used throughout the paper except where otherwise noted. The input data set is $X = \{x_i | x_i \in R^d, i = 1, 2 \ldots t\}$, where d is the dimension of sample. The neuron set is $U = \{u^i | i = 1, 2, \ldots s\}$. Neuron u^i is associated with a 3-tuple $< n_i, c_i, M_i >$.

2.2   Algorithm of Local-SOINN

Figure 2 demonstrate the algorithm flow concisely. Initially, the neural network is empty; there is no neuron in the competitive layer. Then samples are fed into the network sequentially. When a new sample is input into the network, we obtain a set, of which the neurons contain that sample. If this set is empty, i.e., there is no neuron (ellipsoid) contain that sample, a new neuron is added in and the network get a geometrical growth. Else, acquire the winner neuron, update its information and the adjacency list, then detect whether satisfy the condition of merging with neighbors. When the result is expected to output, some post processing like denoising is implemented, and the available ellipsoids are clustered on the basis of neighbor relationship at last.

Winner Lookup and Geometrical Growth Criterion. The proposed model is based on competitive learning. The measure is an improved Mahalanobis distance. At the same time, we take a geometrical growth criterion to ensure the algorithm executed in an incremental manner. That is necessary because we can't provide the intact distribution of online-data in advance. Thus we should decide whether to create a new neuron. Assuming a new sample x access, the winner neuron should be determined. We use $D_i(x)$ to note the distance between x and neuron u^i. Set $S = \{u^i | D_i(x) < H\}$ means that sample x is located in these neurons.

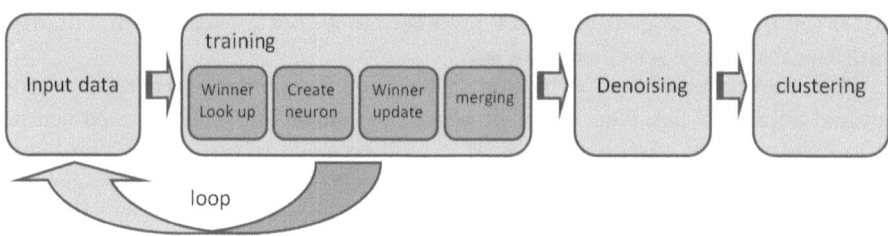

Fig. 2. Flow process of Local-SOINN

If $S = \emptyset$, a new neuron u is created and $U = U \bigcup \{u\}$. Neuron u is initialized with:

$$n = 2d, c = x, M = \frac{h^2}{d} I \tag{1}$$

where I is the identity matrix. The initialized neuron can be regarded as an ellipsoid include $2d$ samples. These samples have a h bias from x on d coordinate directions respectively. That can be represented as: $\{s|s = x \pm h\epsilon_i$ for $i = 1, 2, \ldots, d\}$ where $\epsilon_1 = (1, 0, 0, \ldots, 0), \epsilon_2 = (0, 1, 0, \ldots, 0), \cdots, \epsilon_d = (0, 0, 0, \ldots, 1)$.

Else if $S \neq \emptyset$, we select a winner neuron from set S. The winner neuron u^i is the neuron which satisfy: $u^i = \underset{u^i \in S}{\arg\min} D_i(x)$. After finding out the winner neuron, we update the relevant record of neuron set and adjacency list.

Iterative Update of Winner Neuron. Let $X = \{x_1, x_2, \ldots, x_n\}$ be a set of n samples, where $x_i \in R^d, i = 1 \ldots n$, belong to the same neuron u. Let c and M be the mean vector and covariance matrix of this data set. In case of online learning, the data are discarded after being learned. Thus, the computation of c and M must be in a recursive way. If x_{n+1} is the new sample assigned to neuron u, the relation between the old mean of dataset and the new can be present as follow:

$$c_{new} = c_{old} + (x_{n+1} - c_{old})/(n+1) \tag{2}$$

Likewise, the covariance matrix M can be written in the form of recursive relation:

$$\begin{aligned} M_{new} = M_{old} + [n(x_{n+1} - c_{old})(x_{n+1} - c_{old})^T \\ - (n+1)M_{old}]/(n+1)^2 \end{aligned} \tag{3}$$

After updating the neuron information, we handle the update of neighbourship. According to the definition of set S, we know any two neurons of S overlap each other. Thus, we add new connections $\{< u_i, u_j > |u_i \in S \land u_j \in S \land i \neq j\}$ into the adjacency list.

Merging Strategy. Since the new neurons can be automatically added into the network, there may be some number of redundant neurons. The redundant neurons are the neurons that close to each other and having some common principal components. The merging strategy for these redundant neurons should be included in the learning algorithm in order to reduce these redundant neurons. At merging stage, two neurons will merge if two conditions are satisfied: two neurons are connected by an edge; the volume of the combined neuron is not more than the sum of the two neurons that prior to the combination. Considering the combination of neuron $u_1 < n_1, c_1, M_1 >$ and $u_2 < n_2, c_2, M_2 >$, the formula is stated as follow:

$$n_{new} = n_1 + n_2$$

$$c_{new} = (n_1 * c_1 + n_2 * c_2)/n_{new}$$

$$M_{new} = \frac{n_1}{n_{new}}(M_1 + (c_{new} - c_1)(c_{new} - c_1)^T)$$

$$+ \frac{n_2}{n_{new}}(M_2 + (c_{new} - c_2)(c_{new} - c_2)^T)$$

(4)

Meanwhile, the adjacency list should be updated. We use E_1 and E_2 to signify neighbors of u_1 and u_2 respectively. Then we carry out formula $E_{new} = E_1 \bigcup E_2$ to get the neighbor set of new neuron.

Denoising Procedure. The initial dataset may involve in the pollution of noise. Many neurons are created for these noises. They occupy the network and consume a great quantity of resource. In that case, a denoiser is absolutely necessary. Since the dominated area and sample number of a neuron dominate are acquired, a measure based on density is implemented. As stated above, an ellipsoid Ω is defined as: $\{x|\sqrt{(x-c)^T M^{-1}(x-c)} < H\}$. If M is nonsingular, the volume of this ellipsoid can be figured out using formula:

$$Volume(\Omega) = 2^{\left[\frac{d+1}{2}\right]} \pi^{\left[\frac{d}{2}\right]} \left(\prod_{i=0}^{\lceil d/2 \rceil - 1} \frac{1}{d - 2 * i} \right) \sqrt{|M|} H^d$$

(5)

After obtaining the density of each ellipsoid, we can determine a threshold, such as the average density, to discriminate noise neurons from normal neuron.

Clustering Based on the Topotaxy of Neurons. The result of proposed model can ideally simulate the distribution of original data. More neurons are produced at complicated and diversified area. On the contrary, less neurons are generated to represent the simple and unitary local region. The neighbourship relation reflects the topological relationship among local areas. It gives an expression of the distribution at holistic level. With the expression of the whole distribution and the description of local information, our algorithm precisely reconstructs the original data in a condensed way.

Using the information of neighbourship relation among neurons, the neurons can be clustered into different classes. Assuming the neurons and connections are vertices and edges in a undirected graph, the task can be described as a problem of graph partition. A graph G is an ordered pair $< V, E >$ comprising a set V of vertices or nodes together with a set E of edges. A partition of V is obtained regarded to the connective relationship $\sim$ among vertices. The mathematical form can be expressed as followed:

$$V/ \sim = \{V_1, V_2, \ldots, V_k\} \tag{6}$$

where V_i is an equivalence class, for $i = 1, 2, \ldots, k$. Each equivalence class represent one cluster and the entire data contain k clusters in all.

3   Discussion and Implementation Issues

3.1   Extension of Mahalanobis Distance

Mahalanobis distance [13] is named from P. C. Mahalanobis. It is defined as follow:

$$D_M(x) = \sqrt{(x - c)^T M^{-1} (x - c)} \tag{7}$$

where M is a $d \times d$ covariance matrix and assumed to be nonsingular, c is the mean vector. Mahalanobis distance is an extension of Euclidean distance. Let the M be unit matrix, Mahalanobis distance degenerates into Euclidean distance. Further more, the Mahalanobis distance takes into account the correlation in the data, since it is calculated using the inverse of the covariance matrix of the data set.

When the investigated data are measured over a large number of variables (high dimensions), they can contain much redundant or correlated information. This so-called multicollinearity in the data leads to a singular or nearly singular covariance matrix that cannot be inverted [13]. In that case, the Mahalanobis distance is disabled and the proposed model get into trouble. So we should expand the definition of Mahalanobis distance.

As Singular Value Decomposition (SVD) formulate, a real symmetric positive semi-definite matrix M can be decomposed:

$$M = E^T \Sigma E \tag{8}$$

where E is orthogonal matrix and $\Sigma = diag(\lambda_1, \lambda_2, \ldots, \lambda_d), \lambda_1 \geq \lambda_2 \geq \ldots \geq \lambda_k > 0 = \lambda_{k+1} = \ldots = \lambda_d$. if $k < d$, the covariance matrix M is singular, the Mahalanobis distance is impracticable. Moreover, the noise and machine error may give rise to minor components which reflected in the smaller eigenvalues and corresponding eigenvectors. Similar to PCA, we can get a truncation of all eigenvalues, the minor eigenvalues are considered as noise or machine error and are cut off. Thus, predetermining a scaling factor ρ, we can obtain an approximate formula:

$$D_M^2(x) = (E(x-c))^T \underbrace{\begin{pmatrix} \frac{1}{\lambda_1} & \cdots & 0 & 0 & \cdots & 0 \\ \vdots & \ddots & \vdots & \vdots & \ddots & \vdots \\ 0 & \cdots & \frac{1}{\lambda_t} & 0 & \cdots & 0 \\ 0 & \cdots & 0 & 0 & \cdots & 0 \\ \vdots & \ddots & \vdots & \vdots & \ddots & \vdots \\ 0 & \cdots & 0 & 0 & \cdots & 0 \end{pmatrix}}_{\text{An approximation of Mahalanobis Distance}} (E(x-c)) \tag{9}$$

$$+ \underbrace{\frac{1}{\zeta} \| (x-c) - \sum_{i=1}^{t} e_i^T (x-c) e_i \|}_{Reconstruction error}$$

where $t = \underset{1 \le t \le d}{\arg\min} \sum_{i=1}^{t} \lambda_i \ge \rho \sum_{i=1}^{d} \lambda_i, \zeta = \min(\lambda_{t+1}, (1-\rho) \sum_{i=1}^{d} \lambda_i)$. It indicate that our distance measurement take two facts together, one is the Mahalanobis distance and the other is reconstruction error. Formula (9) can be showed in a more expressive form:

$$D_M^2(x) = \sum_{i=1}^{t} \frac{(e_i^T (x-c))^2}{\lambda_i} + \frac{1}{\zeta} \sum_{i=t+1}^{d} (e_i^T (x-c))^2 \tag{10}$$

this is just to make clear about the anisotropy of different directions. Bias on minor components bring large contribution to the distance. On the contrary, bias on principal components bring small contribution. That makes a lot more sense than Euclidean distance.

3.2   Statistical Interpretation

Gaussian distribution has a density function:

$$f(x) = (2\pi)^{-d/2} |M|^{-1/2} exp^{-\frac{1}{2}(x-c)^T M^{-1}(x-c)} \tag{11}$$

The central limit theorem claim that the probability density for the sum of d independent and identically distributed variables tend to a Gaussian distribution. It is one of the most important and useful distribution. In recent decades, it is widely applied to data mining to simulate the input distribution. Gaussian Mixture Models (GMM) is one implementation of these applications. Mixture models are able to approximate arbitrarily complex probability density functions. This fact make them an excellent choice for representing distribution of complex input data [14].

Take node of that, the formula of Mahalanobis distance and Gaussian density function all contain the covariance matrix M and mean vector c, which respectively refer to the unbiased estimations of expectation and variance in mathematic statistics. Each neuron contains plenty of statistical information of the original data. It stores the first-moment and second-moment in the mean vector and covariance matrix. The information is more detailed than that in SOM or SOINN, in which the neurons only be associated with mean vector. Thus our model loses less information of original data and represents the input data more ideally. In the sense of statistics theory, the common ground between Mahalanobis distance and Gaussian distribution imply Local-SOINN can be interpreted as a special GMM.

In our model, the ellipsoid (neuron) has a tendency to expand. The ellipsoid explores the density around the area it dominates. It expand itself if the density around is dense. This property is made clear below.

If M is the covariance matrix of data, we can prove that λ_i is the variance on eigenvector e_i, for $i = 1, 2, \ldots, d$. This result is a posteriori estimate of variance for the distribution. From another point of view, assumed samples uniformly distribute in a ellipsoid Ω defined as $\{x | \sqrt{(x-c)^T M^{-1}(x-c)} < H\}$, the prior estimate of variance on eigenvectors of M can be represented as:

$$
\begin{aligned}
Var(i) &= \frac{1}{Volume(\Omega)} \int_\Omega (e_i^T x)^2 \mathrm{d}\Omega \\
&= \frac{H\lambda_i}{d+2} \quad \text{for } i = 1, 2, \ldots, d
\end{aligned}
\tag{12}
$$

Comparing the two results of prior and posteriori estimate, the ellipsoid is in a dynamic balance if $\lambda_i == Var(i)$, i.e., $H == d+2$. With $H > d+2$, the newcome samples adjust the covariance matrix M and have a positive feedback on the size of ellipsoid Ω, and it's the same in reverse.

In the realization of proposed model, we set $H = (1+2*1.05^{1-n})(d+2)$. This strategy let the ellipsoid has a tendency of expansivity at preliminary stage. The ellipsoid automatically detects the density of around area, and expands itself if the periphery density is similar to the center. With the incremental of samples included in ellipsoid, H approach to $d+2$, thus the ellipsoid stay a stable state. Anyway, this parameter has a slight effect to our model. There are various alternatives as long as satisfying $H \geq d+2$ and $\lim_{n\to\infty} H = d+2$.

4   Computational Experiments

4.1   Simulation with Artificial Data

Automatically Clustering Capability Experiment. We test the experiment with a artificial data set which include three belt distributions which overlap each other. With $h^2 = 0.5$ and executed in a unsupervised and on-line model, Figure 3 illustrate local-SOINN can automatically cluster and simultaneously gain the number of classes without knowing the class number in advance. At earlier stage, Local-SOINN generates lots of ellipsoids as the same size as the initial ellipsoid. These ellipsoids just blanket the input data simply, not reflect

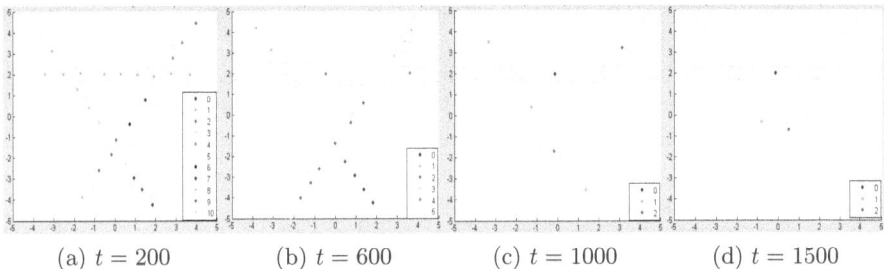

(a) $t = 200$ (b) $t = 600$ (c) $t = 1000$ (d) $t = 1500$

Fig. 3. Results at different iterations on the distribution of three overlapping belt. The near and similar neurons combine together gradually.

the correlation of data and local distribution. Then the size and boundary of ellipsoids are self-adjusting according to the input samples. The ellipsoids explore boundary and learn principal components of local area automatically. At the same time, the ellipsoids, which are close to each other and have similar principal components, may merge into a bigger ellipsoid to better display the principal components. Thus the proposed model uses as few ellipsoids as possible to describe the input data and report the categories number at last.

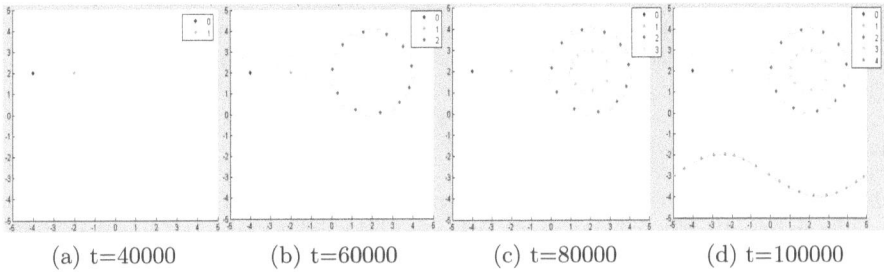

(a) t=40000   (b) t=60000   (c) t=80000   (d) t=100000

Fig. 4. In class-incremental environment, Local-SOINN can learn the new class incrementally

Incremental and On-Line Learning Ability Experiment. In this section, we adopt the artificial dataset used in [4][15]. The original data set include five clusters: two Gaussian distributions, two concentric rings and a sinusoidal curve. In addition, we add 10% noise to estimate the denoising effect. It includes two sub-experiments which are executed in two different data environments.

We first execute experiment in a class-incremental way: first, all the samples of the first Gaussian distribution are input into the net randomly, and then the samples of the second Gaussian distribution, then the big ring and small ring sequentially, the sinusoidal curve at last. Figure 4 demonstrate the incremental learning ability of Local-SOINN. It can learn the new classes without corrupting or forgetting previously learned classes.

Then, our algorithm is implemented in a more complicated environment: sample incremental and class incremental. New samples and new classes are added to the input data incrementally, the distribution of input data will change in different periods. Figure 5 illustrate that Local-SOINN can effectively simulate the data in incremental way.

Note that, experiments in this section are all run under a on-line model with $h^2 = 0.1$, the samples are discarded after learned. Three results illustrate Local-SOINN can cluster automatically under the complex non-stationary environment even the data is polluted by noise.

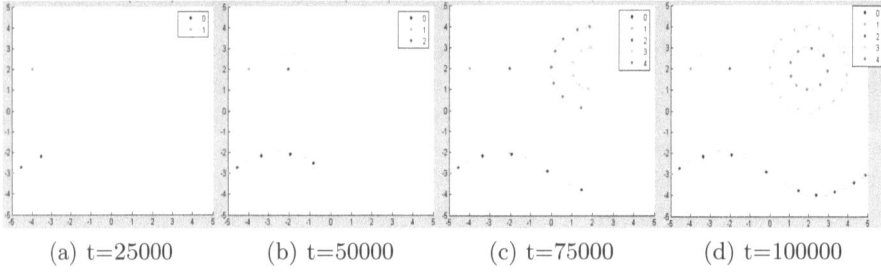

    (a) t=25000   (b) t=50000   (c) t=75000   (d) t=100000

Fig. 5. In more complicated environment (sample-incremental and class-incremental), Local-SOINN automatically adds in new neurons to simulate the input data

4.2   Simulation Using Real-World Data

Our final set of experiments is devoted to the comparison of the ability to adapt to complex multidimensional data of the ASSOM, PCASOM and Local-SOINN. The real data used are selected from the UCI Repository of Machine Learning Databases and Domain Theories.

In all these experiments, we have split the complete sample set into two disjoint subsets. The training subset has been presented to the networks in the training phase, while the test subset has been used to measure the classification performance of the trained networks. Each of the two subsets have 50% of the samples.

The experiments have been designed as follows. All the samples of the training subset have been presented to a unique network (unsupervised learning). When the training has finished, we have computed the winning neuron for all the samples of the training subset. For every neuron, we have computed its receptive field, i.e. the set of training samples for which this neuron is the winner. Each neuron has been assigned to the class with the most training samples in its receptive field. Finally, we have presented the test samples to the network, one by one. If the winning neuron corresponds with the class of the test sample, we count it as a successful classification. Otherwise, we count it as a classification failure.

Table 1. Classification performance results (the best result is in bold font)

Episodes × episode size	ASSOM		PCASOM		Local-SOINN
	10000 × 10	20000 × 20	20000 × 1	60000 × 1	1-pass-throw
BalanceScale	59.9	57.7	66.3	68.3	**79.2**
Contraceptive	44.8	46.4	42.7	42.6	**48.2**
Glass	52.4	52.4	36.2	36.2	**69.8**
Haberman	73.5	73.8	73.7	71.7	**73.9**
Ionosphere	79.4	74.9	76.1	84.2	**86.9**
PimaIndiansDiabetes	66.4	60.2	65.1	63.5	**70.4**
Yeast	39.8	39.6	38.4	45.3	**52.3**

As stated in [3], the network architecture used in both ASSOM and PCASOM is: a 4x4 rectangular lattice, with two basis vectors per neuron. The neighbourhood function has been always a Gaussian. A linear decay rate of the neighbourhood width σ has been selected in the ordering phase. In the convergence phase $\sigma = 0.04$ was fixed. The structure of Local-SOINN is not predetermined; the competitive layer is empty in initial phase. We use cross validation method to choose the best h^2 in training phase of Local-SOINN.

In the experiments with ASSOM, it is mandatory that the samples are grouped into episodes. On the other hand, the PCASOM does not need any grouping, but samples are repeatedly inputed in PCASOM to obtain fairly acceptable performance. However, the process is 1-pass-throw in Local-SOINN; samples are discarded after trained.

From Table (1), we see that the Local-SOINN model outperforms ASSOM and PCASOM in all the databases. Meanwhile, the training mode determines that the iterations of Local-SOINN are far less than ASSOM and PCASOM.

5   Conclusion

We have proposed a self-organizing incremental neural network model based on the local distribution. The neurons of this network are associated with a mean vector and a covariance matrix, which imply wealthy information of local distribution and lay a solid foundation of statistical theory for Local-SOINN. Our model uses the improved Mahalanobis distance as its measurement. This distance implicitly incorporates the reconstruction error and the anisotropy on different eigenvectors. Further more, the information storage ensure the implementation of iterative update. The denoising and automatic clustering are added in post processing at last. All these specialties make Local-SOINN realize in a incremental and on-line way. Experiments show that the new model has preferable performance both on artificial and real-word data. The future works may include extension to manifold and attributive increment learning.

Acknowledgment. This work was supported in part by the Fund of the National Natural Science Foundation of China (Grant No. 60975047, 61021062), 973 Program (2010CB327903), and Jiangsu NSF grant (BK2009080, BK2011567).

References

1. Kohonen, T.: The self-organizing map. Proceedings of the IEEE 78, 1464–1480 (1990)
2. Kohonen, T.: The adaptive-subspace SOM (ASSOM) and its use for the implementation of invariant feature detection. In: Fogelman-Soulié, F., Galniari, P. (eds.) Proceedings of the ICANN 1995, International Conference on Artificial Neural Networks, Paris: EC2 and Cie, vol. 1, pp. 3–10 (1995)
3. López-Rubio, E., Muñoz-Pérez, J., Gómez-Ruiz, J.A.: A principal components analysis self-organizing map. Neural Networks 17, 261–270 (2004)
4. Shen, F., Hasegawa, O.: An incremental network for on-line unsupervised classification and topology learning. Neural Networks 19, 90–106 (2006)
5. Arnonkijpanich, B., Hammer, B., Hasenfuss, A.: Local Matrix Adaptation in Topographic Neural Maps
6. Yin, H., Huang, W.: Adaptive nonlinear manifolds and their applications to pattern recognition. Information Sciences 180, 2649–2662 (2010)
7. Carpenter, G.A., Grossberg, S.: The ART of adaptive pattern recognition by a self-organizing neural network. IEEE Computer 21, 77–88 (1988)
8. Kambhatla, N., Leen, T.K.: Dimension reduction by local principal component analysis. Neural Computation 9(7), 1493–1516 (1997)
9. Tipping, M.E., Bishop, C.M.: Mixtures of probabilistic principal component analyzers. Neural Computation 11(2), 443–482 (1999)
10. Weingessel, A., Hornik, K.: Local PCA algorithms. IEEE Trans. Neural Networks 11(6), 1242–1250 (2000)
11. Reilly, L., Cooper, L.N., Elbaum, C.: A Neural Model for Category Learning. Biological Cybernetics 45, 35–41 (1982)
12. Wasserman, P.D.: Advanced Methods in Neural Computing. Van Nostrand Reinhold, New York (1993)
13. Maesschalck, R., Jouan-Rimbaud, D., Massart, D.L.: The Mahalanobis distance. Chemometrics and Intelligent Laboratory Systems 50, 1–18 (2000)
14. Figueiredo, M.A.T., Jain, A.K.: Unsupervised Learning of Finite Mixture Models. IEEE Transactions on Pattern Analysis and Machine Intelligence 24, 381–396 (2002)
15. Shen, F., Hasegawa, O.: An enhanced self-organizing incremental neural network for online unsupervised learning. Neural Networks 20, 893–903 (2007)

Improving Multi-label Classification Using Semi-supervised Learning and Dimensionality Reduction

Eakasit Pacharawongsakda, Cholwich Nattee, and Thanaruk Theeramunkong

School of Information, Computer, and Communication Technology
Sirindhorn International Institute of Technology
Thammasat University, Thailand
{eakasit,cholwich,thanaruk}@siit.tu.ac.th

Abstract. Multi-label classification has been increasingly recognized since it can assign multiple class labels to an object. This paper proposes a new method to solve simultaneously two major problems in multi-label classification; (1) requirement of sufficient labeled data for training and (2) high dimensionality in feature/label spaces. Towards the first issue, we extend semi-supervised learning to handle multi-label classification and then exploit unlabeled data with averagely high-confident tagged labels as additional training data. To solve the second issue, we present two alternative dimensionality-reduction approaches using Singular Value Decomposition (SVD). The first approach, namely LAbel Space Transformation for CO-training REgressor (LAST-CORE), reduces complexity in the label space while the second one namely Feature and LAbel Space Transformation for CO-training REgressor (FLAST-CORE), compress both label and feature spaces. For both approaches, the co-training regression method is used to predict the values in the lower-dimensional spaces and then the original space can be reconstructed using the orthogonal property of SVD with adaptive threshold setting. Additionally, we also introduce a method of parallel computation to fasten the co-training regression. By a set of experiments on three real world datasets, the results show that our semi-supervised learning methods gain better performance, compared to the method that uses only the labeled data. Moreover, for dimensionality reduction, the LAST-CORE approach tends to obtain better classification performance while the FLAST-CORE approach helps saving computational time.

Keywords: multi-label classification, Singular Value Decomposition, SVD, dimensionality reduction, semi-supervised learning, co-training.

1 Introduction

In the past, most traditional classification techniques usually assumed a single category for each object to be classified by means of minimum distance. However, in some tasks it is natural to assign more than one categories to an object. For examples, some news articles can be categorized into both *politic* and *crime*,

P. Anthony, M. Ishizuka, and D. Lukose (Eds.): PRICAI 2012, LNAI 7458, pp. 423–434, 2012.

or some movies can be labeled as *action* and *comedy*, simultaneously. As a special type of task, multi-label classification was initially studied by Schapire and Singer (2000) [7] in text categorization. Later many techniques in multi-label classification have been proposed for various applications. However, these methods can be grouped into two main approaches: *Algorithm Adaptation* (AA) and *Problem Transformation* (PT) as suggested in [9]. The former approach modifies existing classification methods to handle multi-label data [7]. On the other hand, the latter approach transforms a multi-label classification task into several single classification tasks and then applies traditional classification method on each task [6]. However, these two approaches might inherently have two major problems; (1) requirement of sufficient labeled data for training and (2) high dimensionality in feature/label spaces.

The first problem arises from the fact that it is a time-consuming task for human to assign suitable class labels for each object. Most real-world datasets have a few number of objects that the class labels of which are known while a large number of data available without class label information. For instance, there are million of protein sequences available in the public database but few of them are annotated with the localization or functional information. To overcome this issue, the semi-supervised multi-label learning methods have been proposed such as [2,3]. These works aimed to assign unlabeled data with suitable class labels and add the most plausible ones into the training dataset. With the use of automatically labeled data, we can improve classification performance.

The second problem comes from special characteristic of multi-label classification where each object is represented by a large number of attributes and labels. This situation may decrease classification performance due to a so-called overfitting problem. Moreover, a large number of features and labels also trigger an increment of the execution time for building classification models as well as prediction processes. Many works have been proposed to address this issue by transforming the high dimensional feature or label space to the lower-dimensional space [8,10]. Among these methods, the work of [4] proposed to reduce the complexity in both feature and label space simultaneously. However, these works focused on reducing the complexity in an input or output space but they still suffered from a limited number of available labeled data.

Recently, Qian and Davidson [5] proposed a method to incorporate a semi-supervised multi-label learning with feature space reduction. However, there are no work on coping high dimensionality in the feature as well as the label space. In this work, we study the effect of label space reduction as well as feature spaces reduction, together with semi-supervised learning. Two complimentary techniques are dimensionality reduction in the feature/label space as well as the co-training paradigm for semi-supervised multi-label classification. In addition, we also propose the parallel version of semi-supervised multi-label learning.

In the rest of this paper, Section 2 gives notations for the semi-supervised multi-label classification task and literature review to the SVD method. Section 3 presents two dimensionality-reduction approaches, namely LAbel Space Transformation for CO-training REgressor (LAST-CORE) and Feature and LAbel

Space Transformation for CO-training REgressor (FLAST-CORE). The multi-label benchmark datasets and experimental settings are described in Section 4. In Section 5, the experimental results using three datasets are given and finally Section 6 provides conclusion of this work.

2 Preliminaries

2.1 Notations

Let $\mathcal{X} = \mathbb{R}^M$ and $\mathcal{Y} = \{0,1\}^Q$ be an M-dimensional feature space and Q-dimensional binary label space, respectively, where M is the number of features and Q is a number of possible labels, i.e. classes. Let $\mathcal{L} = \{\langle \mathbf{x}_1, \mathbf{y}_1 \rangle, \langle \mathbf{x}_2, \mathbf{y}_2 \rangle,$ $..., \langle \mathbf{x}_N, \mathbf{y}_N \rangle\}$ be a set of N labeled objects (e.g., documents, images, etc.) in a training dataset, where $\mathbf{x}_i \in \mathcal{X}$ is a feature vector that represents an i-th object and $\mathbf{y}_i \in \mathcal{Y}$ is a label vector with the length of Q, $[y_{i1}, y_{i2}, ..., y_{iQ}]$. Here, y_{ij} indicates whether the i-th object belongs (1) or not (0) to the j-th class (the j-th label or not). Let $\mathcal{U} = \{\mathbf{u}_1, \mathbf{u}_2, ..., \mathbf{u}_P\}$ denote the set of P unlabeled objects whose class labels are unknown. Each object $\mathbf{u}$ is described by the feature vector in the $\mathcal{X}$ space.

For convenience, $\mathbf{X}_{N \times M} = [\mathbf{x}_1, ..., \mathbf{x}_N]^T$ denotes the *feature matrix* with N rows and M columns and $\mathbf{Y}_{N \times Q} = [\mathbf{y}_1, ..., \mathbf{y}_N]^T$ represents the *label matrix* with N rows and Q columns. Both $\mathbf{X}_{N \times M}$ and $\mathbf{Y}_{N \times Q}$ are used to represent the labeled examples. Let $\mathbf{U}_{P \times M} = [\mathbf{u}_1, ..., \mathbf{u}_p]^T$ denote the *feature matrix* of the unlabeled examples which contains P rows and M columns and $[\cdot]^T$ denotes matrix transpose.

2.2 Singular Value Decomposition (SVD)

This subsection gives a brief introduction to SVD, which was developed as a method for dimensionality reduction using a least-squared technique. The SVD transforms a feature matrix $\mathbf{X}$ to a lower-dimensional matrix $\mathbf{X}'$ such that the distance between the original matrix and a matrix in a lower-dimensional space (i.e., the 2-norm $\| \mathbf{X} - \mathbf{X}' \|_2$) are minimum.

Generally, a feature matrix $\mathbf{X}$ can be decomposed into the product of three matrices as shown in Equation (1).

$$\mathbf{X}_{N \times M} = \mathbf{T}_{N \times M} \times \mathbf{\Sigma}_{M \times M} \times \mathbf{D}_{M \times M}^T, \tag{1}$$

where N is a number of objects, M is a number of features and $M < N$. The matrices $\mathbf{T}$ and $\mathbf{D}$ are two orthogonal matrices, where $\mathbf{T}^T \times \mathbf{T} = \mathbf{I}$ and $\mathbf{D}^T \times \mathbf{D} = \mathbf{I}$. The columns in the matrix $\mathbf{T}$ are called as the *left singular vectors* while the columns in matrix $\mathbf{D}$ as the *right singular vectors*. The matrix $\mathbf{\Sigma}$ is a diagonal matrix, where $\mathbf{\Sigma}_{i,j} = 0$ for $i \neq j$, and the diagonal elements of $\mathbf{\Sigma}$ are the singular values of matrix $\mathbf{X}$. The singular values in the matrix $\mathbf{\Sigma}$ are sorted by descending order such that $\mathbf{\Sigma}_{1,1} \geq \mathbf{\Sigma}_{2,2} \geq ... \geq \mathbf{\Sigma}_{M,M}$. To discard noise, it is possible to ignore singular values less than $\Sigma_{R,R}$, where $R \ll M$. By this ignorance the three matrices are reduced to Equation (2).

$$\mathbf{X}'_{N \times M} = \mathbf{T}'_{N \times R} \times \mathbf{\Sigma}'_{R \times R} \times \mathbf{D}'^{T}_{M \times R}, \tag{2}$$

where $\mathbf{X}'_{N \times M}$ is expected to be close $\mathbf{X}_{N \times M}$, i.e. $\| \mathbf{X} - \mathbf{X}' \|_2 < \delta$, $\mathbf{T}'_{N \times R}$ is a reduced matrix of $\mathbf{T}_{N \times M}$, $\mathbf{\Sigma}'_{R \times R}$ is the reduced version of $\mathbf{\Sigma}_{M \times M}$ from M to R dimensions and $\mathbf{D}'_{M \times R}$ is a reduced matrix of $\mathbf{D}_{M \times M}$. In the next section, we show our two approaches that deploy the SVD technique to construct lower-dimensional space for both features and labels for semi-supervised multi-label classification.

3   Proposed Approaches

As mentioned earlier, most semi-supervised multi-label classification approaches either pay an attention to select some suitable unlabeled data for using as training data or to reduce dimensions in the feature space. In this work, we present two alternative approaches to improve performance of semi-supervised multi-label classification by the Singular Value Decomposition (SVD). The first approach, namely LAbel Space Transformation for CO-training REgressor (LAST-CORE), aims to reduce complexity in the label space by utilizing the SVD technique. On the other hand, the second approach which is called Feature and LAbel Space Transformation for CO-training REgressor (FLAST-CORE), considers the dependency between feature and label spaces before applying the SVD to transform both of them into the lower-dimensional spaces.

Since values in each lower-dimensional space are numeric after applying SVD, it is possible for us to apply a regression method to estimate these values. While COREG [11] is widely used as a co-training regressor, it cannot be directly applied since it is not designed to handle the multi-label data. Thus, this work extends the concept of COREG to estimate those multiple values in the reduced dimensions. Section 3.1 gives a theoretical point of view on our two alternative dimensionality reduction while section 3.2 provides an algorithmic point of view on these dimensionality reduction techniques.

3.1   Dimensionality Reductions in the Proposed Frameworks

This subsection introduces our main idea about applying SVD for dimensionality reduction to the co-training method. In multi-label classification, the sparseness problem in the label space and the correlation among labels are not much studied as pointed out in [8]. In [4], there has been a study on how to reduce both feature and label space, simultaneously. However, the work may suffer with insufficient data since labeled data are hard to find. In this work, we introduce dimensionality reduction into semi-supervised learning, or more specifically co-training. Therefore, the main idea of the LAST-CORE approach is to utilize the SVD technique to address the sparseness problem that might occurs in the label space as well as consider the correlation between labels. To reduce complexity in the label space, the LAST-CORE decomposes the label matrix $\mathbf{Y}_{N\times Q}$ into three matrices $\mathbf{T}_{N\times Q}$, $\mathbf{\Sigma}_{Q\times Q}$ and $\mathbf{D}_{Q\times Q}$. After that, to retain only significant dimensions and reduce noise, the first R ($\leq min(N,Q)$) dimensions from matrices $\mathbf{T}$ and $\mathbf{D}$ are selected as $\mathbf{T'}_{N\times R}$ and $\mathbf{D'}_{Q\times R}$. The matrix $\mathbf{D'}_{Q\times R}$ is considered as a transformation matrix that can project the higher-dimensional label matrix $\mathbf{Y}_{N\times Q}$ into a lower-dimensional label matrix $\mathbf{Y'}_{N\times R}$ as shown in Equation (3).

$$\mathbf{Y'}_{N\times R} = \mathbf{Y}_{N\times Q} \times \mathbf{D'}_{Q\times R} \qquad (3)$$

While the LAST-CORE attacks the sparseness and correlation problems in the label space, the FLAST-CORE presents an alternative approach to reduce complexity in both feature and label spaces simultaneously. It is possible to utilize the characteristic that the feature space and the label space may have some dependency with each other. Though, many methods can compute the dependencies among two sets, for example, cosine similarity, entropy or Symmetric Uncertainty (SU), we consider only the linear correlation between features and labels in this work. To reduce complexities in both spaces, it can be simultaneously compressed by performing SVD on the feature-label covariance, viewed as a dependency profile between features and labels. Equation (4) shows construction of a covariance matrix $\mathbf{S}_{M\times Q}$ to represent a dependency between feature and label spaces.

$$\mathbf{S}_{M\times Q} = E[(\mathbf{X}_{N\times M} - E[\mathbf{X}_{N\times M}])^T (\mathbf{Y}_{N\times Q} - E[\mathbf{Y}_{N\times Q}])], \qquad (4)$$

where $\mathbf{X}$ is the feature matrix, $\mathbf{Y}$ is the label matrix and $E[\cdot]$ is an expected value of the matrix. Applying SVD, the covariance matrix $\mathbf{S}_{M\times Q}$ is later decomposed to matrices $\mathbf{T}$, $\mathbf{\Sigma}$ and $\mathbf{D}$ and as described earlier, a lower-dimensional feature matrix $\mathbf{X'}$, can be created by Equation (5).

$$\mathbf{X'}_{N\times R} = \mathbf{X}_{N\times M} \times \mathbf{T'}_{M\times R} \qquad (5)$$

In the same way, a lower-dimensional label matrix $\mathbf{Y'}$ can be computed as shown in Equation (6)

$$\mathbf{Y'}_{N\times R} = \mathbf{Y}_{N\times Q} \times \mathbf{D'}_{Q\times R} \qquad (6)$$

While these two approaches use different dimensionality reduction approach, i.e., single space reduction or dual space reduction. It uses the same idea to reconstruct an original label space from a lower-dimensional label space. Using the orthogonal property of SVD which indicates that $\mathbf{T}^T \times \mathbf{T} = \mathbf{I}$ and $\mathbf{D}^T \times \mathbf{D} = \mathbf{I}$, the original label matrix $\mathbf{Y}$ can be reconstructed as shown in Equation (7).

$$\mathbf{Y}_{N\times Q} = \mathbf{Y'}_{N\times R} \times \mathbf{D'}^T_{Q\times R} \qquad (7)$$

As the result, the reconstructed label vector may include non-binary values. To interpret the values as binary decision, a threshold need to be set to map these values to either 0 or 1 for representing whether the object belongs to the class or not. As a naive approach, the fixed value of 0.5 is used to assign 0 if the value is less than 0.5, otherwise 1 [8]. For more efficiently, it is possible to apply an adaptive threshold. In this work, we propose a method to determine an optimal threshold by selecting the value that maximizes classification accuracy in the training dataset. In other words, the threshold selection is done by first sorting prediction values in each label dimension in descending order and examining performance (e.g., *macro F-measure*) for each rank position from the top to the bottom to find the point that maximizes the performance. Then, the threshold for binary decision is set at that point.

3.2   LAST-CORE and FLAST-CORE Approach

This section describes the LAST-CORE and FLAST-CORE in an algorithmic point of view. Their pseudo-codes can be simultaneously presented in Algorithm 1. Although the proposed LAST-CORE and FLAST-CORE can function with the original two-view co-training approach [1], this work slightly modifies the original co-training method to use different classifiers, instead of two different views of data. Following [1], to make the diversity among two classifiers, we use k-Nearest Neighbors (kNN) regression algorithm as the based regressor and utilize the different values of nearest neighbors (k_1 and k_2) as well as the different distance metrics (D_1 and D_2). The kNN is used since it is a lazy approach which does not build any classification model in advance. Instead it uses few nearest neighbor examples to predict the class. This concept is helpful when unlabeled data are applied.

The first step in Algorithm 1 is to assign the feature matrix $\mathbf{X}_{N \times M}$ to matrix $\mathbf{X}_1$ and $\mathbf{X}_2$. Likewise, the label matrix $\mathbf{Y}_1$ and $\mathbf{Y}_2$ are duplicated from the label matrix $\mathbf{Y}_{N \times Q}$ (line 1-2). Both matrix $\mathbf{X}_{N \times M}$ and $\mathbf{Y}_{N \times Q}$ are the labeled data and the matrix $\mathbf{X}_1$, $\mathbf{Y}_1$ and $\mathbf{X}_2$, $\mathbf{Y}_2$ are used as the training data for building the classification models. For the unlabeled data, we randomly select objects from the matrix $\mathbf{U}_{P \times M}$ to add into the matrix $\mathbf{U}'_{S \times M}$ (line 3), where S is the size of unlabeled sample set and S should be less P. With the co-training style algorithm, it iteratively selects the suitable data and appends it to the set of labeled data. The maximum number of iterations has been already defined in the variable T at line 4.

After that the SVD technique is applied to transform the higher-dimensional space to the lower-dimensional space (line 6-10). The function *labelTransformation ($\mathbf{Y}_i,R$)* is used to project the label matrix $\mathbf{Y}_i$ to the lower-dimensional label matrix with R dimensions. On the other hand, the FLAST-CORE approach uses the function *featureLabelTransformation ($\mathbf{X}_i,\mathbf{Y}_i,R$)* to transform both feature matrix $\mathbf{X}_i$ and label matrix $\mathbf{Y}_i$ to their reduced spaces. With this function, a number of dimensions in both spaces are reduced to R. The details of these two functions are described in the previous section.

Next, we utilize the concept of Binary Relevance (BR) that reduces the multi-label classification task to a set of binary classifications and then builds a classification model for each class. However, in this work, the projected label matrix $\mathbf{Y}_i$ contains numeric values rather than discrete classes. By this situation, a regression method can be applied to estimate these numeric values. While the projected label matrix $\mathbf{Y}_i$ has R dimensions, it is possible to construct a regression model for each dimension. That is, R regression models are constructed for R lower-dimensions. The function $kNN(\mathbf{X}_i, \mathbf{y}_r, k_i, D_i)$ is used as a based regressor and the return value is an average value of all k_i nearest neighbors computed using the distance metric D_i, given the data at the r-th dimension, $\mathbf{y}_r$. To take advantage of a large number of unlabeled data, each example in the matrix $\mathbf{U}'$ is fed to the kNN regressor for estimating each reduced dimension value. After that an example which is the most consistent with the labeled dataset is selected. To verify this, the mean squared error (MSE) is used as suggested in [11]. This means that the appropriate unlabeled data is the one that can reduce the overall MSE when it is appended to the training data, compared with the model that uses only the labeled data. By the high computational cost of kNN regressor, the nearest neighbors of the unlabeled example are selected to compute the MSE instead of using the entire labeled set. The function $Neighbors(\mathbf{u}_j, \mathbf{X}_i, k_i, D_i)$ returns the k_i objects which have a minimum distance from the unlabeled example $\mathbf{u}_j$. The equation at line 18 presents how to compute the MSE from the kNN regressor that uses only labeled data, denoted as $h_i(\cdot)$, and another kNN regressor, $h'_i(\cdot)$, which incorporates the unlabeled example to the training data. The value of $\delta_{u_j r}$ is the difference between those two MSE values when the unlabeled object $\mathbf{u}_j$ is applied to estimate the value in the dimension r. In addition, the larger $\delta_{u_j r}$ value, the more error is reduced. These steps are shown in the line 11-20.

After all rows in the matrix $\mathbf{U}'$ are tested, we compute the average value of δ_{u_j} for all reduced dimensions and then choose an unlabeled example that obtains the maximum value when it is added into the label dataset. In other words, the suitable unlabeled example, denoted by $\tilde{\mathbf{x}}_i$, should reduce the MSE for all dimensions (line 22). Then the corresponding values of the selected unlabeled example are estimated by the function $kNN(\mathbf{X}_i, \mathbf{y}_r, k_i, D_i)$ (line 23-25). As the result, the variable $\tilde{\mathbf{y}}_i$ stores the predicted values of all reduced dimensions. After that, to reconstruct the original label matrix, the function $labelReconstruction(\mathbf{Y}_i, \tilde{\mathbf{y}}_i)$ is used. This function multiplies a transpose of the transformation matrix to the lower-dimensional label matrix and applies an appropriate threshold to round the numeric values back to the $\{0,1\}$ (line 26). The formulation of this step has been described in the previous section. Subsequently, the suitable examples that are selected from the two regressors are appended to the training data of each others as shown in line 32. The last step in Algorithm 1 (line 32) is to append the best unlabeled example into the unlabeled matrix $\mathbf{U}'$. More concretely, the two training datasets ($\mathbf{X}_1$ & $\mathbf{X}_2$ and $\mathbf{Y}_1$ & $\mathbf{Y}_2$) will be augmented by the unlabeled data. In addition, the algorithm is terminated when the maximum number of iterations is reached or the training data is not changed. Termination, we obtained the modified labeled dataset and then we can apply any classification technique, such as Support Vector Machines (SVM), Naive Bayes or Decision Tree to build the final classification model.

Moreover, to fasten the process, it is possible for us to construct a parallel version for LAST-CORE and FAST-CORE, due to data independence. More concretely, it is possible to apply regression in parallel for each reduced dimension

Algorithm 1. The LAST-CORE and FLAST-CORE Pseudo-codes

Input: a feature matrix $\mathbf{X}_{N \times M}$, a label matrix $\mathbf{Y}_{N \times Q}$, an unlabeled matrix $\mathbf{U}_{P \times M}$, maximum number of learning iterations T, number of nearest neighbors k_1, k_2 and distance metrics D_1, D_2.

Output: two new feature matrices $\mathbf{X}_1$ and $\mathbf{X}_2$ and two new label matrices $\mathbf{Y}_1$ and $\mathbf{Y}_2$.

1: Create two new feature matrices as $\mathbf{X}_1 = \mathbf{X}_{N \times M}$ and $\mathbf{X}_2 = \mathbf{X}_{N \times M}$
2: Create two new label matrices as $\mathbf{Y}_1 = \mathbf{Y}_{N \times Q}$ and $\mathbf{Y}_2 = \mathbf{Y}_{N \times Q}$
3: Create matrix $\mathbf{U}'_{S \times M}$ by randomly selecting objects from matrix $\mathbf{U}_{P \times M}$, where $S < P$
4: **for** $t = 1$ **to** T **do**
5:    **for** $i = 1$ **to** 2 **do**
6:       **if** the method is LAST-CORE **then**
7:          $\mathbf{Y}_i \leftarrow labelTransformation(\mathbf{Y}_i, R)$
8:       **else if** the method is FLAST-CORE **then**
9:          $(\mathbf{X}_i, \mathbf{Y}_i) \leftarrow featureLabelTransformation(\mathbf{X}_i, \mathbf{Y}_i, R)$
10:      **end if**
11:      **for** $r = 0$ **to** R **do**
12:         $\mathbf{y}_r = \mathbf{Y}_i[:, r]$
13:         $h_i \leftarrow kNN(\mathbf{X}_i, \mathbf{y}_r, k_i, D_i)$
14:         **foreach** $\mathbf{u}_j \in \mathbf{U}'$ **do**
15:            $\Omega_{u_j} \leftarrow Neighbors(\mathbf{u}_j, \mathbf{X}_i, k_i, D_i)$
16:            $\hat{y}_{u_j} \leftarrow h_i(\mathbf{u}_j)$
17:            $h_i' \leftarrow kNN(\{\mathbf{X}_i \cup \mathbf{u}_j\}, \{\mathbf{y}_r \cup \hat{y}_{u_j}\}, k_i, D_i)$
18:            $\delta_{u_j r} \leftarrow \frac{1}{|\Omega_{u_j}|} \sum_{\mathbf{x}_k \in \Omega_{u_j}} ((y_k - h_i(\mathbf{x}_k))^2 - \frac{1}{|\Omega_{u_j}|} \sum_{\mathbf{x}_k \in \Omega_{u_j}} ((y_k - h_i'(\mathbf{x}_k))^2$
19:         **end for**
20:      **end for**
21:      **if** there exists a $\delta_{u_j r} > 0$ **then**
22:         $\tilde{\mathbf{x}}_i \leftarrow \arg\max_{\mathbf{u}_j \in \mathbf{U}'}(average(\delta_{u_j}))$
23:         **for** $r = 1$ **to** R **do**
24:            $\tilde{y}_{ir} \leftarrow h_i(\tilde{\mathbf{x}}_i)$
25:         **end for**
26:         $\tilde{\mathbf{y}}_i \leftarrow labelReconstruction(\mathbf{Y}_i, \tilde{\mathbf{y}}_i)$
27:         $\pi_{ix} \leftarrow \tilde{\mathbf{x}}_i; \pi_{iy} \leftarrow \tilde{\mathbf{y}}_i ; \mathbf{U}' \leftarrow \mathbf{U}' - \tilde{\mathbf{x}}_i$
28:      **else**
29:         $\pi_i \leftarrow \phi$
30:      **end if**
31:   **end for**
32:   $\mathbf{X}_1 \leftarrow \mathbf{X}_1 \cup \pi_{2x}; \mathbf{X}_2 \leftarrow \mathbf{X}_2 \cup \pi_{1x}; \mathbf{Y}_1 \leftarrow \mathbf{Y}_1 \cup \pi_{2y}; \mathbf{Y}_2 \leftarrow \mathbf{Y}_2 \cup \pi_{1y}$
33:   **if** neither of $\mathbf{X}_1$ and $\mathbf{X}_2$ changes **then**
34:      **exit**
35:   **else**
36:      Replenish $\mathbf{U}'$ to size S by randomly selecting objects from $\mathbf{U}$
37:   **end if**
38: **end for**
39: **return** new two training datasets: $\mathbf{X}_1$, $\mathbf{X}_2$, $\mathbf{Y}_1$ and $\mathbf{Y}_2$.

since they are independent with each other, as well as the nearest neighbors of each unlabeled object can be found in parallel. Therefore, the process on the labeled data and that on the unlabeled data can be separated. For the labeled data, the matrix $\mathbf{Y}_i$ is divided into R subsets which each of them represents each reduced dimension and the matrix $\mathbf{X}_i$ is duplicated to R matrices. After that, each subset is distributed to a processor (core). This means the reduced dimensions are separated to different processors (cores). In the next step, each row in the unlabeled matrix $\mathbf{U}'$ is assigned to another processor for estimating the values for each reduced dimension. Then, the prediction values computed from all processors are combined to δ_{u_j}. The parallel version of these approaches will be replaced in line 11-19. In our work, we apply the *round-robin* method to handle the load-balancing in the parallel approach.

4   Datasets and Experimental Settings

To evaluate the performance of our two proposed approaches, the benchmark multi-label datasets are downloaded from MULAN[1]. Table 1 shows the characteristics of three real world multi-label datasets. For each dataset, N, M and Q denote the total number of objects, the number of features and the number of labels, respectively. L_C represents the *label cardinality*, the average number of labels per example and L_D stands for *label density*, the normalized value of *label cardinality*, i.e., $\frac{L_C}{Q}$, as introduced by Read et al [6].

Table 1. Characteristics of the datasets used in our experiments

Dataset	Domain	N	M	Q	L_C	L_D
emotions	music	593	72	6	1.869	0.311
scene	image	2,407	294	6	1.074	0.179
yeast	biology	2,417	103	14	4.237	0.303

Since each object in the dataset can be associated with multiple labels simultaneously, the traditional evaluation metric of single-label classification could not be applied. In this work, we apply two types of well-known multi-label evaluation metrics [6]. As the label-based metric, *hamming loss* and *macro F-measure*, are used to evaluate each label separately. As the label-set-based metric, *accuracy* and *0/1 loss* are applied to assess all labels as a set.

In this work, as a baseline, we compare LAST-CORE and FLAST-CORE to a method that uses only labeled data to construct the classification model. Each dataset is randomly separated into 10% for training (labeled data), 25% for testing and the remaining 65% for unlabeled data. We repeat this setting for twenty trials and use their average. The results of the four evaluation metrics and the execution time are recorded and shown in Table 2 and 3, respectively. All semi-supervised multi-label classification methods used in this work are implemented

---

[1] http://mulan.sourceforge.net/datasets.html

in R environment [2] and Support Vector Machines (SVM) is used as the based classifier. For both approaches, we experiments with R is varied from 20% to 100% of the dimension of the original label matrix, with 20% as interval. To compare the computational time, all methods were performed on the machines that has eight AMD Opteron Quad Core 8356 1.1 GHz Processor with 512 KB of cache, 64GB RAM.

5   Experimental Results

This section provides the experimental results for investigation on the effect of dimension reduction on classification performance. The mean values of the *hamming loss*, *macro F-measure*, *accuracy* and *0/1 loss* metric of each algorithm are shown in Table 2.

Table 2. Performance comparison (mean) between traditional multi-label classification, LAST-CORE, and FLAST-CORE in terms of *hamming loss (HL)*, *macro F-measure (F1)*, *accuracy*, *0/1 loss* on the three datasets. ($\downarrow$ indicates the smaller the better; $\uparrow$ indicates the larger the better; $\bullet$/$\circ$ indicate whether the algorithm is statistically superior/inferior to the method that uses only labeled data. (two-tailed paired t-test at 5% significance level)

Dataset	Metrics	Methods	R				
			$20\% \times Q$	$40\% \times Q$	$60\% \times Q$	$80\% \times Q$	$100\% \times Q$
emotions	$HL^\downarrow$	labeled	-	-	-	-	**0.2297**
		LAST-CORE	0.5012°	0.2961°	0.2501°	0.2506°	0.2505°
		FLAST-CORE	0.4542°	0.2954°	0.2889°	0.2808°	0.2899°
	$F1^\uparrow$	labeled	-	-	-	-	0.5311
		LAST-CORE	0.5410	$0.6330^\bullet$	$0.6372^\bullet$	$0.6349^\bullet$	**$0.6390^\bullet$**
		FLAST-CORE	$0.5509^\bullet$	$0.6031^\bullet$	$0.5980^\bullet$	$0.6052^\bullet$	$0.6096^\bullet$
	$accuracy^\uparrow$	labeled	-	-	-	-	0.3998
		LAST-CORE	0.3790	$0.4735^\bullet$	**$0.5092^\bullet$**	$0.5025^\bullet$	$0.5088^\bullet$
		FLAST-CORE	0.3784	$0.4671^\bullet$	$0.4717^\bullet$	$0.4738^\bullet$	$0.4815^\bullet$
	$0/1\ loss^\downarrow$	labeled	-	-	-	-	0.8054
		LAST-CORE	0.9601°	0.8791°	**0.7868**	0.8044	0.7959
		FLAST-CORE	0.9473°	0.8537°	0.8499°	0.8348°	0.8463°
scene	$HL^\downarrow$	labeled	-	-	-	-	0.1134
		LAST-CORE	0.1269°	0.1114	$0.1088^\bullet$	$0.1054^\bullet$	$0.1063^\bullet$
		FLAST-CORE	0.1297°	0.1122	$0.1053^\bullet$	$0.1039^\bullet$	**$0.1026^\bullet$**
	$F1^\uparrow$	labeled	-	-	-	-	0.5838
		LAST-CORE	$0.6679^\bullet$	$0.6621^\bullet$	$0.6496^\bullet$	$0.6579^\bullet$	$0.6529^\bullet$
		FLAST-CORE	$0.6745^\bullet$	$0.6606^\bullet$	$0.6650^\bullet$	$0.6651^\bullet$	**$0.6698^\bullet$**
	$accuracy^\uparrow$	labeled	-	-	-	-	0.4528
		LAST-CORE	$0.5561^\bullet$	$0.5394^\bullet$	$0.5222^\bullet$	$0.5318^\bullet$	$0.5274^\bullet$
		FLAST-CORE	$0.5415^\bullet$	$0.5366^\bullet$	$0.5397^\bullet$	$0.5425^\bullet$	**$0.5512^\bullet$**
	$0/1\ loss^\uparrow$	labeled	-	-	-	-	0.5949
		LAST-CORE	$0.5478^\bullet$	$0.5224^\bullet$	$0.5202^\bullet$	$0.5066^\bullet$	$0.5101^\bullet$
		FLAST-CORE	$0.5747^\bullet$	$0.5393^\bullet$	$0.5071^\bullet$	$0.4945^\bullet$	**$0.4871^\bullet$**
yeast	$HL^\downarrow$	labeled	-	-	-	-	**0.2045**
		LAST-CORE	0.2117°	0.2096°	0.2092°	0.2088°	0.2089°
		FLAST-CORE	0.2128°	0.2075°	0.2069°	0.2067°	0.2066°
	$F1^\uparrow$	labeled	-	-	-	-	0.4131
		LAST-CORE	$0.4523^\bullet$	$0.4532^\bullet$	$0.4543^\bullet$	$0.4535^\bullet$	$0.4487^\bullet$
		FLAST-CORE	$0.4533^\bullet$	$0.4524^\bullet$	$0.4487^\bullet$	$0.4512^\bullet$	**$0.4551^\bullet$**
	$accuracy^\uparrow$	labeled	-	-	-	-	0.4705
		LAST-CORE	$0.5093^\bullet$	$0.5094^\bullet$	$0.5108^\bullet$	$0.5098^\bullet$	**$0.5111^\bullet$**
		FLAST-CORE	$0.5104^\bullet$	$0.5080^\bullet$	$0.5069^\bullet$	$0.5082^\bullet$	$0.5083^\bullet$
	$0/1\ loss^\downarrow$	labeled	-	-	-	-	0.8737
		LAST-CORE	$0.8575^\bullet$	$0.8567^\bullet$	$0.8545^\bullet$	$0.8563^\bullet$	$0.8546^\bullet$
		FLAST-CORE	$0.8607^\bullet$	$0.8545^\bullet$	**$0.8515^\bullet$**	$0.8535^\bullet$	$0.8569^\bullet$

---

[2] http://www.R-project.org/

The parameter R in this table denotes the percentages of dimension reduction from a number of dimensions in an original label matrix (Q). In the FLAST-CORE method, the percentages of dimension reduction are set to be the same for both label and feature space and the maximum number of reduced dimensions R cannot excess the minimum between the number of features and the number of labels. The best value in each dataset is emphasized by bold font. Moreover, to measure statistical significance of performance difference, a two-tailed paired t-test at 5% significance level are performed between our two proposed approaches and the traditional method uses only labeled data. Note that when the method achieves significantly better or worse performance than the traditional method on any dataset, wins and losses are marked with a marker $\bullet/\circ$ in the table. Otherwise, it means a tie or insignificant difference and they are not marked. From the table, we can make some observations as follows. To compare with multi-label classification technique that uses only labeled data, we observe that both proposed methods show better performance in terms of *macro F-measure*, *accuracy* and *0/1 loss*. On the other hand, our approaches may not be superior to the baseline in terms of *hamming loss* for the data sets with a high L_D since *hamming loss* is used to evaluate label by label and may need no consideration of label dependence. Next, the LAST-CORE approach tends to obtain better classification performance than the FLAST-CORE approach. For computational complexity, Table 3 presents execution time of the two proposed methods in serial and parallel mode. The table indicates that execution times are reduced when the data is separated to small sets and it is distributed to each processors using parallel mode. For more detail, the execution time is reduced by factor of R multiplies with a number of processors, where R is a number of reduced dimensions.

Table 3. Average execution time (in seconds) on the three datasets. Here, R is a number of reduced dimensions.

Dataset	Method	number of processors					
		1	2	4	8	16	32
emotions	LAST-CORE $(R = 20\% \times Q)$	5004.40	2691.40	1398.00	770.60	431.20	299.40
	LAST-CORE $(R = 60\% \times Q)$	20076.60	2776.40	1496.60	910.40	1037.00	976.80
	LAST-CORE $(R = 100\% \times Q)$	29919.20	2801.20	1557.00	1361.80	1267.40	1324.80
	FLAST-CORE $(R = 20\% \times Q)$	367.60	278.20	204.80	118.00	106.60	128.00
	FLAST-CORE $(R = 60\% \times Q)$	2277.80	382.40	249.00	253.40	248.00	305.60
	FLAST-CORE $(R = 100\% \times Q)$	4251.80	455.40	287.00	329.80	349.40	457.20
scene	LAST-CORE $(R = 20\% \times Q)$	56638.00	29078.20	14555.20	7655.80	4143.20	2461.00
	LAST-CORE $(R = 60\% \times Q)$	222438.40	29294.40	14904.00	8041.40	7906.00	7918.60
	LAST-CORE $(R = 100\% \times Q)$	336185.80	29557.00	15566.00	12199.60	11983.20	11884.20
	FLAST-CORE $(R = 20\% \times Q)$	405.80	269.80	187.00	132.80	113.80	122.00
	FLAST-CORE $(R = 60\% \times Q)$	2449.60	404.20	266.60	286.00	281.20	310.40
	FLAST-CORE $(R = 100\% \times Q)$	4445.60	505.40	325.40	375.20	440.40	482.00
yeast	LAST-CORE $(R = 20\% \times Q)$	23447.60	5261.80	2816.20	1695.80	1398.80	1505.60
	LAST-CORE $(R = 60\% \times Q)$	74249.80	5631.60	3484.80	4109.00	3462.40	3960.80
	LAST-CORE $(R = 100\% \times Q)$	120129.20	5463.60	5105.20	4992.60	4845.20	5749.20
	FLAST-CORE $(R = 20\% \times Q)$	1871.00	358.60	243.00	202.80	219.40	241.40
	FLAST-CORE $(R = 60\% \times Q)$	7222.60	592.40	532.20	491.80	570.20	689.40
	FLAST-CORE $(R = 100\% \times Q)$	19024.40	830.80	969.80	973.60	1162.60	1635.00

6   Conclusion

This paper presents two alternative approaches to handle the problem of high dimensionality in feature/label spaces by label and feature space compression and dependency consideration as well as the co-training approach to solve the problem of insufficient training data. The LAbel Space Transformation for CO-training REgressor (LAST-CORE) incorporates the dimensionality reduction into the co-training algorithm. The aim of this approach is to reduce complexity in the label space. On the other hand, the Feature and LAbel Space Transformation for CO-training REgressor (FLAST-CORE) considers the dependency between the feature and label spaces before transforming both spaces into a single reduced space. Experiments with a broad range of multi-label datasets show that our two proposed approaches achieve a better performance, compared to multi-label classification method that applies only labeled data.

References

1. Blum, A., Mitchell, T.: Combining labeled and unlabeled data with co-training. In: Proceedings of the Eleventh Annual Conference on Computational Learning Theory, COLT 1998, pp. 92–100. ACM, New York (1998)
2. Chen, G., Song, Y., Wang, F.: Semi-supervised multi-label learning by solving a sylvester equation. In: Proceedings of the 8th SIAM Conference on Data Mining (SDM), SDM 2008, pp. 410–419 (2008)
3. Liu, Y., Jin, R., Yang, L.: Semi-supervised multi-label learning by constrained non-negative matrix factorization. In: Proceedings of the 21st AAAI Conference on Artificial Intelligence, AAAI 2006, pp. 421–426. AAAI Press (2006)
4. Pacharawongsakda, E., Theeramunkong, T.: Towards More Efficient Multi-label Classification Using Dependent and Independent Dual Space Reduction. In: Tan, P.-N., Chawla, S., Ho, C.K., Bailey, J. (eds.) PAKDD 2012, Part II. LNCS, vol. 7302, pp. 383–394. Springer, Heidelberg (2012)
5. Qian, B., Davidson, I.: Semi-supervised dimension reduction for multi-label classification. In: Proceedings of the 24th AAAI Conference on Artificial Intelligence, AAAI 2010, pp. 569–574. AAAI Press (2010)
6. Read, J., Pfahringer, B., Holmes, G., Frank, E.: Classifier Chains for Multi-label Classification. In: Buntine, W., Grobelnik, M., Mladenić, D., Shawe-Taylor, J. (eds.) ECML PKDD 2009, Part II. LNCS, vol. 5782, pp. 254–269. Springer, Heidelberg (2009)
7. Schapire, R., Singer, Y.: Boostexter: A boosting-based system for text categorization. Machine Learning 39(2/3), 135–168 (2000)
8. Tai, F., Lin, H.-T.: Multi-label classification with principle label space transformation. In: Proceedings of the 2nd International Workshop on Learning from Multi-Label Data, MLD 2010, pp. 45–52 (2010)
9. Tsoumakas, G., Katakis, I., Vlahavas, I.: Mining Multi-label Data. In: Data Mining and Knowledge Discovery Handbook, 2nd edn. Springer (2010)
10. Zhang, Y., Zhou, Z.-H.: Multilabel dimensionality reduction via dependence maximization. ACM Transactions on Knowledge Discovery from Data (TKDD) 4(3), 1–21 (2010)
11. Zhou, Z.H., Li, M.: Semi-supervised regression with co-training style algorithms. IEEE Transactions on Knowledge and Data Engineering 19(11), 1479–1493 (2007)

Generic Multi-Document Summarization Using Topic-Oriented Information

Yulong Pei, Wenpeng Yin, and Lian'en Huang

Shenzhen Key Lab for Cloud Computing Technology and Applications
Peking University Shenzhen Graduate School
Shenzhen, Guangdong 518055, P.R. China
peiyulong@sz.pku.edu.cn, mr.yinwenpeng@gmail.com, hle@net.pku.edu.cn

Abstract. The graph-based ranking models have been widely used for multi-document summarization recently. By utilizing the correlations between sentences, the salient sentences can be extracted according to the ranking scores. However, sentences are treated in a uniform way without considering the topic-level information in traditional methods. This paper proposes the topic-oriented PageRank (ToPageRank) model, in which topic information is fully incorporated, and the topic-oriented HITS (ToHITS) model is designed to compare the influence of different graph-based algorithms. We choose the DUC2004 data set to examine the models. Experimental results demonstrate the effectiveness of ToPageRank. And the results also show that ToPageRank is more effective and robust than other models including ToHIST under different evaluation metrics.

Keywords: Multi-Document Summarization, PageRank, HITS, LDA.

1 Introduction

As a fundamental tool for understanding document data, text summarization has attracted considerable attention in recent years. Generally speaking, text summarization can be defined as the process of automatically creating a compressed version of a given text set that provides useful information for the user [3]. In the past years, two types of summarization have been explored: extractive summarization and abstractive summarization. Extractive summarization aims to generate summary directly by choosing sentences from original documents while abstractive summarization requires formulating new sentences according to the documents. Although abstractive summarization could be more concise and understandable, it usually involves heavy machinery from *natural language processing*. In this paper, we focus on extractive multi-document summarization.

In order to generate the highly comprehensive summary of a given document set, multi-document summarization can be divided into 3 steps: 1) computing the scores of sentences in the document set; 2) ranking the sentences based on the scores (or combined with some other rules); 3) choosing the proper sentences as the summary. Among the three steps, computing the scores plays the most

P. Anthony, M. Ishizuka, and D. Lukose (Eds.): PRICAI 2012, LNAI 7458, pp. 435–446, 2012.
© Springer-Verlag Berlin Heidelberg 2012

important role. Most recently, some graph-based ranking algorithms have been successfully applied for computing the sentence scores by deciding on the importance of a vertex in the graph according to the global information of the document set. Normally, a set of documents are represented as a directed or undirected graph [3] based on the relationship between sentences and then the graph-based ranking algorithm such as PageRank [12] or HITS [4] is used. According to the previous study, a concise and effective summarization should meet three requirements: diversity, coverage and balance [5]. However, these methods often employed the sentences or terms in a uniform usage [16] without considering the topic-level information, which could lead to less satisfaction of these requirements because of ignoring the information of hidden diverse topics.

To deal with the problem in previous graph-based ranking models, we propose the topic-oriented PageRank (ToPageRank) model, which is to divide traditional PageRank into multiple PageRanks with regard to various topics and then extract summaries specific to different topics. Afterwards, based on the topic distribution of all the documents the entire summaries will be obtained. Aim to explore the influence of different graph-based ranking algorithms, we compare the ToPageRank model with a modified HITS model, named topic-orientd HITS (ToHITS) accordingly. In ToHITS, topics and sentences are regarded as hubs and authorities respectively and the hub scores and authority scores are computed iteratively in a reinforcement way. Experiments on DUC2004[1] dataset have been performed and the results demonstrate the good effectiveness of the proposed ToPageRank model which outperforms all other models under various evaluation metrics. The parameters in experiments have also been investigated and the results show the robustness of the proposed model.

The rest of the paper is organized as follows. First we introduce the related work in Section 2. The basic models are discussed in Section 3. We discribe the ToHITS and ToPageRank models in Section 4. The experiments and results are represented in Section 5 and finally we conclude this paper in Section 6.

2   Related Work

Multi-document summarization aims to generate a summary by reducing documents in size while retaining the main characteristics of the original documents [18]. According to the differences of provided information, multi-document summarization can be classified into generic summarization and query-oriented summarization. In particular, generic summarization would generate the summary only based on the given documents and query-oriented summarization would form the summary on a certain question or topic. Both generic summarization and query-oriented summarization have been explored recently.

Traditional feature-based ranking methods exploited different features of sentences and terms to compute the scores and rank the sentences. One of the

---

[1] http://www-nlpir.nist.gov/projects/duc/guidelines/2004.html

most popular feature-based methods is centroid-based method [13]. MEAD as an implementation of cetroid-based summarizer applied a number of predefined features such as TF*IDF, cluster centroid and position to score the sentences. Lin and Hovy [6] used term frequency, sentence position, stigma words and simplified Maximal Marginal Relevance (MMR) to build the NeATS multi-document summarization system. You Ouyang et al. [11] studied the influence of different word positions in summarization.

Cluster information has also been explored in generating the summaries. Wang et al. [18] integrated both document-term and sentence-term matrices in a language model and utilized the mutual influence of document clusters and summaries to cluster and summary the documents simultaneously. Wan and Yang [16] clustered the documents and combined the cluster information with the Condition Markov Random model and HITS model to rank the sentences. Cai et al. [2] studied three different ranking functions and proposed a reinforcement method to integrate sentences ranking and clustering by making use of term rank distributions over clusters.

Graph-based ranking algorithms such as PageRank and HITS nowadays are applied in text processing i.e. TextRank [9] and LexPageRank [3]. Graph-based ranking algorithms take global information into consideration rather than rely only on vertex-specific information, therefore have been proved successful in multi-document summarization. Some works [17] [14] have extended the traditional graph-based models recently. In [17], a two-layer link graph was designed to represent the document set and the ranking algorithm took into account three types of relationship, i.e. relationship between sentences, relationship between words and relationship between sentences and words.

However, in the previous methods the topic-level information has seldom been studied while in our work we incorporate the topic distribution in graph-based ranking algorithms so the topic-level information can be well utilized.

3   Basic Models

3.1   Overview

The graph-based ranking algorithms stem from web link analysis. Hyperlink-Induced Topic Search (HITS) [4] algorithm and PageRank [12] algorithm are the most widely used algorithms in recent years and a number of improvements of both algorithms have been exploited recently as well. Based on PageRank, some graph-based ranking algorithms, i.e. LexPageRank and TextRank, have been introduced to text summarization. In these methods documents are represented as graph structure, more specifically, the nodes in the graph stand for the sentences and the edges stand for the relationship between a pair of sentences. The graph-based ranking algorithms are used to compute the scores of sentences and the sentences with high scores would be chosen as the summaries.

3.2 HITS

In HITS [4], Kleinberg proposed that a node had two properties, named hub and authority. A good hub node is one that points to many good authorities and a good authority node is one that is pointed to by many good hubs. In the basic HITS model, the sentences are considered as both hubs and authorities.

Given a document set D, we denote the graph as $G_H=(S, E_{SS})$ to represent the documents. $S = \{s_i | 0 \le i \le n\}$ is the set of vertices in the graph and stands for the set of sentences, and $E_{SS} = \{e_{ij} | s_i, s_j \in S, i \ne j\}$ corresponds to the relationship between each pair of sentences. Each e_{ij} is associated with a weight w_{ij} which indicates the similarity of the pair of sentences. The weight is computed by using the standard cosine measure between two sentences as follows.

$$w_{ij} = sim_{cosine}(s_i, s_j) = \frac{\vec{s_i} \cdot \vec{s_j}}{|\vec{s_i}| \times |\vec{s_j}|} \ ,$$
(1)

where $\vec{s_i}$ and $\vec{s_j}$ are the term vectors corresponding to the sentence s_i and s_j, respectively. The graph we propose to build is undirected so we have $w_{ij} = w_{ji}$ here and we follow [15] to define $w_{ii} = 0$ to avoid self transition. TF*ISF (inverse sentence frequency) value of each term is applied to describe the elements in the sentence vector. Then the weight value of sentences in the documents can be denoted as a symmetric matrix W.

The authority score $Auth^{(t+1)}(s_i)$ of sentence s_i and the hub score $Hub^{(t+1)}$ (s_j) of sentence s_j in the $(t + 1)^{th}$ iteration are computed based on the corresponding scores in the tth iteration as follows.

$$
\begin{aligned}
Auth^{(t+1)}(s_i) &= \sum_{s_j \in S, i \ne j} w_{ij} \cdot Hub^{(t)}(s_j) \\
Hub^{(t+1)}(s_j) &= \sum_{s_j \in S, i \ne j} w_{ij} \cdot Auth^{(t)}(s_i)
\end{aligned}
\ ,
$$
(2)

where $Hub^{(t+1)}(s_i)$ is equal to $Auth^{(t+1)}(s_i)$ in the $(t+1)^{th}$ iteration because the sentences are considered as both hubs and authorities; besides, the matrix W is a symmetric matrix, therefore $w_{ij} = w_{ji}$ here. To guarantee the convergence of the iterative process, $Auth(\cdot)$ (or $Hub(\cdot)$) would be normalized after each iteration as $Auth^{(t)} = Auth^{(t)}/\|Auth^{(t)}\|$. Both $Auth(s_i)$ and $Hub(s_i)$ can be initialize as $1/N$ and N is the number of sentences.

Usually the iterative process is stopped when the difference between the scores of sentences in two sequential iterations falls below a defined threshold or the number of iteration exceeds a given value. After iteration, the authority scores are used as the scores of sentences.

3.3 PageRank

PageRank is one of the most important factors for Google to rank the search results. This algorithm computes the scores of nodes by making use of the voting or recommendations between nodes.

Using the similar denotations in previous section, $G_{PR}=(S, E_{SS})$ is an undirected graph to represent the sentences and sentence correlations in a given document set D. S and E_{SS} correspond the set of sentences and the relationship between pairs of sentences respectively. The weight w_{ij} affiliated with the edge $e_{ij} \in E_{SS}$ is also computed by using the standard cosine measure represented in Equation (1). Furthermore, the definition of weight matrix W is the same as the introduction in Section 3.2.

After that, W is normalized to $\tilde{W}$ to make the sum of each row to be 1:

$$\tilde{W}_{ij} = \begin{cases} W_{ij}/\sum_{j=1}^{|S|} W_{ij}, & if \ \sum_{j=1}^{|S|} W_{ij} \neq 0 \\ 0, & otherwise \end{cases} . \tag{3}$$

Therefore, the PageRank score $Score_{PR}(s_i)$ for sentence s_i is defined based on the normalized matrix $\tilde{W}$:

$$Score_{PR}(s_i) = \lambda \cdot \sum_{j:j\neq i} Score_{PR}(s_j) \cdot \tilde{W}_{ji} + \frac{(1-\lambda)}{|S|} . \tag{4}$$

For convenience, Equation (4) can be denoted in the matrix form:

$$\vec{\omega} = \lambda \tilde{W}^T \vec{\omega} + \frac{1-\lambda}{|S|} \vec{e} , \tag{5}$$

where $\vec{\omega}$ is a $|S| \times 1$ vector made up of the scores of sentences in set S and $\vec{e}$ is a column vector with all the elements equal to 1. λ is a damping factor which ranges from 0 to 1 and $(1 - \lambda)$ indicates the probability for node s_i to jump to a random node in the graph.

4   Proposed Models

4.1   Overview

As mentioned in the Introduction, a concise summary should meet three requirements: diversity, coverage and balance. Traditional graph-based ranking algorithms vote the prestigious vertices based on the global information recursively extracted from the entire graph [9]. However, the document set normally contains a number of different topics with different importance. Computing the sentences in a uniform method would violate the coverage principle because it ignores the differences among topics.

Under the assumption that the sentences containing more important topics should be ranked higher than the sentences contain less important topics, we leverage the topic-level information. In this study, we propose the model to make use of the correlation between topics and sentences. The proposed topic-oriented PageRank (ToPageRank) model follows [8] to divide the random walk into multiple walks and compute the PageRank scores specific to individual random walk, and then combines the entire scores according to the topics distribution. Besides, we modify the basic HITS model, named topic-oriented HITS (ToHITS), to compare the influence of different graph-based algorithms on summarization incorporating the topic-level information.

4.2   Topic Detection

The goal of the topic detection is to identify the topics of a given document set automatically. In the experiments, Latent Dirichlet Allocation [1] (LDA) is utilized to detect the topics. LDA is an unsupervised machine learning technique which can identify latent topic information from a document collection. In LDA, each word w in a document d is regarded to be generated by first sampling a topic from d's topic distribution $\theta^{(d)}$, and then sampling a word from the distribution over words $\phi^{(z)}$ which characterizes the topic z. Both $\theta^{(d)}$ and $\phi^{(z)}$ have Dirichlet priors with hyper-parameters α and β, separately. Therefore, the probability of word w in the given document d and prior is represented as follows.

$$p(w|d, \alpha, \beta) = \sum_{z=1}^{T} p(w|z, \beta) \cdot p(z|d, \alpha) \ , \tag{6}$$

where T is the number of topics.

We use GibbsLDA++ toolkit[2] to detect the topics in this study. After iteration, we can get the word-topic distributions, i.e. the probability of each word w on a topic t. Since the unit used in multi-document summarization is the sentence, the probability of each sentence s on a topic t should be computed. In the experiments, we choose a heuristic method to compute the contribution degree $degree(s|t)$ of each sentence s on a topic t instead of the probability, i.e. the sum of the probability of every word occurring in the sentence on a topic. Furthermore, to keep the longer sentence from getting the larger value, we normalize the sum by dividing the length of the sentence as shown in Equation (7).

$$degree(s|t) = \frac{\sum_{w \in s} p(w|t)}{|s|} \ , \tag{7}$$

where $|s|$ is the length of the sentence s.

4.3   Topic-Oriented HITS

In ToHITS model, we build a topic-sentence bipartite graph shown in Figure 1, in which topic nodes represent the hubs and sentence nodes correspond to the authorities.

Formally, the new graph is denoted as $G_{ToH} = (A_{Sent}, H_{Topic}, E_{TS})$, where $A_{Sent} = \{s_i\}$ is the set of sentences and $H_{Topic} = \{t_i\}$ is the set of topics detected by LDA introduced in Section 4.2. $E_{TS} = \{e_{ij}|t_i \in H_{Topic}, s_j \in A_{Sent}\}$ corresponds to the correlations between a sentence and a topic. Each e_{ij} is associated with a weight w_{ij} which indicates the quantitative relationship of the topic t_i and the sentence s_j. We compute the contribution degree of the sentence s on the topic t through Equation (7) to denote the relationship. The authority score $Auth(s_i)$ of sentence s_i and the hub score $Hub(t_j)$ of topic t_j are computed by the same iteration process showed in Equation (2). After the iteration converges, the authority scores are used as the scores of sentences.

---

[2] http://gibbslda.sourceforge.net/

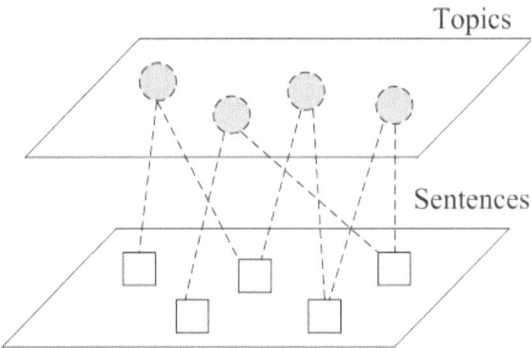

Fig. 1. The topic-sentence bipartite graph in ToHITS model

4.4 Topic-Oriented PageRank

In ToPageRank model, we leverage the topic distribution of the documents. Each node in the graph is represented by different topics rather than one [10]. In order to incorporate the topic-level information, we divide traditional PageRank into multiple PageRanks [8] based on different topics, and then the topic-specific PageRank scores is computed in each subgraph. Whereafter, the topic distribution of the entire document set is taken into account, and we can further obtain the final scores of sentences to obtain summaries that are relevant to the documents and at the same time cover the major topics in the documents set. This process is shown in Figure 2.

In PageRank model shown in Equation (4), the probability for a node to jump to a random node is set to a fixed value based on the number of sentences, however the idea in ToPageRank is to run PageRank on each individual topic. Therefore we use the Biased PageRank [8] on each topic separately. The degree value $degree(s|t)$, which is computed according to Equation (7), is assigned to each sentence s in the Biased PageRank on the specific topic. For topic t, the ToPageRank score $Score_{ToPR}^t(s_i)$ of sentence s_i is defined as follow:

$$Score_{ToPR}^t(s_i) = (1 - \lambda)degree(s|t) + \lambda \cdot \sum_{j:j\neq i} Score_{ToPR}^t(s_j) \cdot \tilde{W}_{ji} \ , \qquad (8)$$

where $\tilde{W}$ is the normalized similarity matrix introduced in Equation (3).

The final score $Score_{ToPR}(s)$ of sentence s is computed as follows:

$$Score_{ToPR}(s) = \sum_{t\in T} Score_{ToPR}^t(s) \cdot avg_p(t) \ , \qquad (9)$$

where T stands for all topics and $avg_p(t)$ represents the average value of the sum of probabilities for topic t in the document set. Instinctively every document in the dataset should be treated equally, therefore we obtain $avg_p(t)$ by computing the average value of the sum of probabilities for a topic t in every document:

$$avg_p(t) = \frac{\sum_{d \in D} p(t|d)}{|D|} \ ,\tag{10}$$

where $p(t|d)$ is the probability of topic t in the document d. D is the document set and $|D|$ is the number of documents in the set.

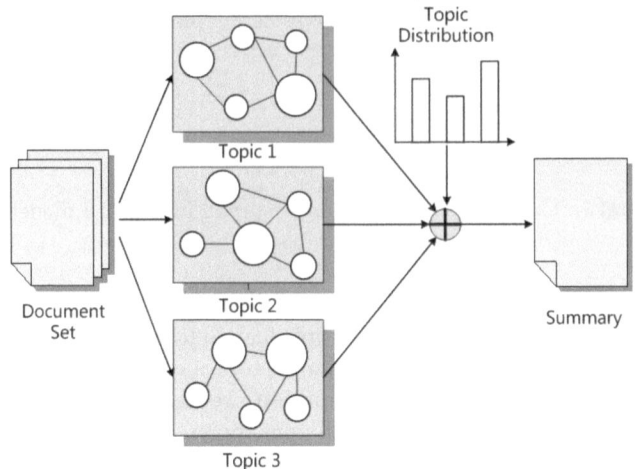

Fig. 2. The process of ToPageRank Model (reference to [8])

5   Experiments

In order to choose more informative but less redundancy sentences as the final summary, in the experiments we apply the variant version of MMR algorithm proposed in [14]. This method is to penalize the sentences that highly overlap with the sentences that have been chosen as the summary.

5.1   Data Set

To evaluate the summarization results empirically, we use DUC2004 dataset since generic multi-document summarization is one of the fundamental tasks in DUC2004. DUC2004 provided 50 document sets and every generated summary is limited to 665 bytes. Table 1 gives a brief description of the dataset.

Table 1. Description of the Data Set

	DUC2004
Data source	TDT*(Topic Detection and Tracking)
Number of collections	50
Number of documents	500
Summary length	665 bytes

* http://www.itl.nist.gov/iad/mig/tests/tdt/

5.2   Evaluation Methods

We use the ROUGE [7] toolkit[3] to evaluate these models, which has been widely applied for summarization evaluation by DUC. It evaluates the quality of a summary by counting the overlapping units between the candidate summary and model summaries. ROUGE implements multiple evaluation metrics to measure the system-generated summarization such as ROUGE-N, ROUGE-W and ROUGE-SU.

The ROUGE toolkit reported separate scores for 1, 2, 3 and 4-gram and among these different metrics, unigram-based ROUGE score (ROUGE-1) has been shown to correlate well with human judgments [7]. Besides, longest common subsequence (LCS), weighted LCS and skip-bigram co-occurrences statistics are also used in ROUGE. In this experimental results we show three of the ROUGE metrics: ROUGE-1 (unigram-based), ROUGE-2 (bigram-based), and ROUGE-SU4 (extension of ROUGE-S, which is the skip-bigram co-occurrences statistics) metrics.

5.3   Evaluation Results

The proposed ToPageRank model are compared with the ToHITS model, the basic HITS, basic PageRank, the DUC Best performing system and the lead baseline system on DUC2004. The DUC Best performing system is the system with highest ROUGE scores among all the systems submitted in DUC2004. The lead baseline takes the first sentences one by one in the last document in a document set, where documents are assumed to be ordered chronologically. Table 2 shows the comparison results on DUC2004.

It can be seen that ToPageRank model can outperform the ToHITS model, two basic models and the DUC best system. The results indicate the effectiveness of the proposed ToPageRank model. However, it is worth mentioning that the performance of ToHITS model is better than its corresponding basic HITS model but worse than basic PageRank model. It might result from that in the ToHITS model only the topic-sentence information is applied but sentence relationships are ignored, while in the PageRank model the relationships between sentences play a vital role. This comparison indicates that the sentence-level information is quite important in summarization to some extent.

To investigate how the damping factor and topic number influence the performance of these models, we compare the different combinations with λ ranges from 0 to 1 and Figure 3 shows the ROUGE-SU4 curve on the DUC2004. Correspondingly, we further compare the influence of topic number in the models and the results are shown in Figure 4 which indicate that when the topic number is relatively small (around 15) ToPageRank performs well and meanwhile ToHITS prefers relatively larger number of topics (around 20). The trends of other metrics such as ROUGE-1 and ROUGE-2 are similar.

---

[3] ROUGE version 1.5.5 is used in this study which can be found on the website
`http://www.isi.edu/licensed-sw/see/rouge/`

Table 2. Comparison results on DUC2004

Systems	ROUGE-1	ROUGE-2	ROUGE-SU4
Lead	0.33182	0.06348	0.10582
DUC Best	0.38279	0.09217	0.13349
HITS	0.36305	0.06892	0.11668
PageRank	0.38043	0.07815	0.12473
ToHITS	0.37264	0.07526	0.12095
ToPageRank	**0.40501**	**0.09555**	**0.14034**

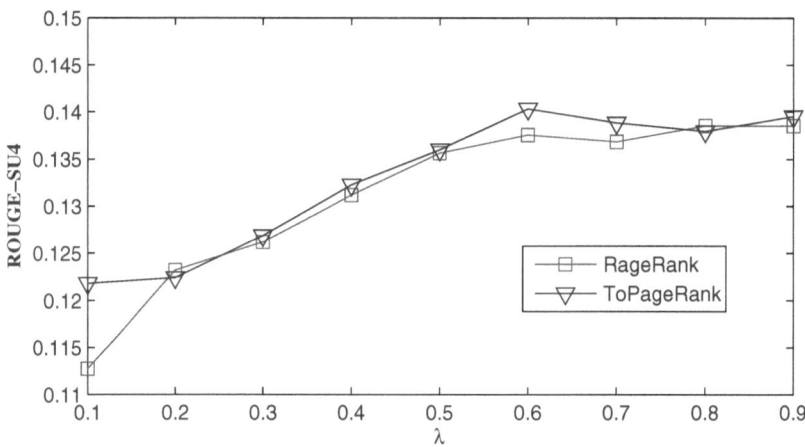

Fig. 3. ROUGE-SU4 vs. λ

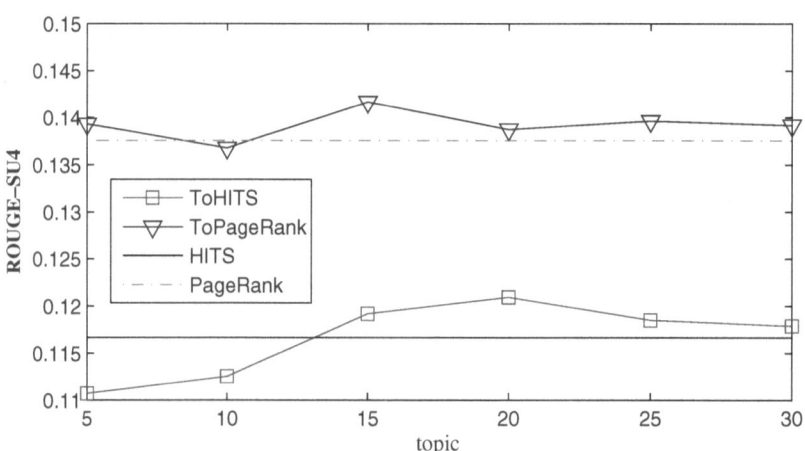

Fig. 4. ROUGE-SU4 vs. *topic*

6 Conclusion and Future Work

In this paper, we propose a novel summarization model to incorporate the topic-oriented information in the document set, named ToPageRank. To compare with different algorithms, the ToHITS model is introduced. Experimental results on DUC2004 dataset demonstrate that ToPageRank could outperform the corresponding basic models, ToHITS and the top performing systems in various evaluation metrics.

In this study, the probabilities of sentences on certain topic is computed by accumulating the probabilities of words occurring in the sentence. In future we will design some new model to compute sentence probabilities in a more meaningful method. Moreover, we will explore other correlations between sentences in graph-based ranking methods.

Acknowledgments. This research is financially supported by NSFC of China (Grant No. 60933004 and 61103027) and HGJ (Grant No. 2011ZX01042-001-001). We also thank the anonymous reviewers for their useful comments.

References

1. Blei, D.M., Ng, A.Y., Jordan, M.I.: Latent dirichlet allocation. The Journal of Machine Learning Research 3, 993–1022 (2003)
2. Cai, X., Li, W., Ouyang, Y., Yan, H.: Simultaneous ranking and clustering of sentences: a reinforcement approach to multi-document summarization. In: Proceedings of the 23rd International Conference on Computational Linguistics, pp. 134–142. Association for Computational Linguistics (2010)
3. Erkan, G., Radev, D.R.: Lexpagerank: Prestige in multi-document text summarization. In: Proceedings of EMNLP, vol. 2004, pp. 365–371 (2004)
4. Kleinberg, J.M.: Authoritative sources in a hyperlinked environment. Journal of the ACM (JACM) 46(5), 604–632 (1999)
5. Li, L., Zhou, K., Xue, G.R., Zha, H., Yu, Y.: Enhancing diversity, coverage and balance for summarization through structure learning. In: Proceedings of the 18th International Conference on World Wide Web, pp. 71–80. ACM (2009)
6. Lin, C.Y., Hovy, E.: From single to multi-document summarization: A prototype system and its evaluation. In: Proceedings of the 40th Annual Meeting on Association for Computational Linguistics, pp. 457–464. Association for Computational Linguistics (2002)
7. Lin, C.Y., Hovy, E.: Automatic evaluation of summaries using n-gram co-occurrence statistics. In: Proceedings of the 2003 Conference of the North American Chapter of the Association for Computational Linguistics on Human Language Technology, vol. 1, pp. 71–78. Association for Computational Linguistics (2003)
8. Liu, Z., Huang, W., Zheng, Y., Sun, M.: Automatic keyphrase extraction via topic decomposition. In: Proceedings of the 2010 Conference on Empirical Methods in Natural Language Processing, pp. 366–376. Association for Computational Linguistics (2010)
9. Mihalcea, R., Tarau, P.: Textrank: Bringing order into texts. In: Proceedings of EMNLP, vol. 2004, pp. 404–411. ACL, Barcelona (2004)

10. Nie, L., Davison, B., Qi, X.: Topical link analysis for web search. In: Proceedings of the 29th Annual International ACM SIGIR Conference on Research and Development in Information Retrieval, pp. 91–98. ACM (2006)
11. Ouyang, Y., Li, W., Lu, Q., Zhang, R.: A study on position information in document summarization. In: Proceedings of the 23rd International Conference on Computational Linguistics: Posters, pp. 919–927. Association for Computational Linguistics (2010)
12. Page, L., Brin, S., Motwani, R., Winograd, T.: The pagerank citation ranking: Bringing order to the web (1999)
13. Radev, D., Jing, H., Stys, M., Tam, D.: Centroid-based summarization of multiple documents. Information Processing & Management 40(6), 919–938 (2004)
14. Wan, X.: Document-Based HITS Model for Multi-document Summarization. In: Ho, T.-B., Zhou, Z.-H. (eds.) PRICAI 2008. LNCS (LNAI), vol. 5351, pp. 454–465. Springer, Heidelberg (2008)
15. Wan, X.: An exploration of document impact on graph-based multi-document summarization. In: Proceedings of the Conference on Empirical Methods in Natural Language Processing, pp. 755–762. Association for Computational Linguistics (2008)
16. Wan, X., Yang, J.: Multi-document summarization using cluster-based link analysis. In: Proceedings of the 31st Annual International ACM SIGIR Conference on Research and Development in Information Retrieval, pp. 299–306. ACM (2008)
17. Wan, X., Yang, J., Xiao, J.: Towards an iterative reinforcement approach for simultaneous document summarization and keyword extraction. In: Annual Meeting-Association for Computational Linguistics, vol. 45, p. 552 (2007)
18. Wang, D., Zhu, S., Li, T., Chi, Y., Gong, Y.: Integrating clustering and multi-document summarization to improve document understanding. In: Proceeding of the 17th ACM Conference on Information and Knowledge Management, pp. 1435–1436. ACM (2008)

Intelligent Ethical Wealth Planner:
A Multi-agent Approach

Phang Wai San, Tan Li Im, and Patricia Anthony

UMS-MIMOS Center of Excellence in Semantic Agent,
School of Engineering and Information Technology, Universiti Malaysia Sabah,
Jalan UMS, 88400, Kota Kinabalu, Sabah, Malaysia
waisanp@hotmail.com, im_87@hotmail.com,
patricia.anthony@gmail.com

Abstract. This paper presents the development of a multi-agent framework in ethical wealth management where the focus is on the design and implementation of various types of agent. This development is part of an on-going project which aims to develop a novel ethical wealth planning model based on multi-agent system. The objective of this paper is to propose a multi-agent framework that can plan, predict, assemble and recommend investment portfolio based on a set of preferences. The framework that has been addressed in this paper allows multiple agents with variable tasks to work together to achieve a certain goal.

Keywords: Intelligent agent, multi-agent, wealth planner, financial portfolio.

1   Introduction

The notion of ethical investment goes back to the attempts of some religious traditions to forbid profiting from the economic activities that violates the moral rule of the religion such as gambling and tobacco [1]. There is a strong relationship between ethical investment and Islamic principles [2]. In the Islamic context, ethical investment is the financial investment that is in accordance with Shariah rule. The companies that adopt ethical behavior, Shariah compliant, are seen as having a higher trust and corporate reputation which can be a positive screening used by Muslim or Non-Muslim investors when investing in stock market [1].

The agent concept [3] can be used to simplify the solution of large problems by distributing them to some collaborating problem solving units. This kind of strategy is called distributed problem solving. On the other hand the agent concept supports more general interactions of already distributed units, which is subject to multi-agent system [4] research. As an example, in a wealth planning setting, investment managers have to decide on a wide variety of investment options in order to ensure a profitable return to their investors. However, in order to select an investment option, there are a lot of parameters that need to be taken into account such as the track record of the companies, the revenue projection, the environmental and political conditions, the risk assessment, the nature of business etc. This generation of an investment portfolio is a complex and dynamics process as it is affected by many variables that change from time to time. Hence, a multi-agent system can be used to collaboratively solve this problem.

P. Anthony, M. Ishizuka, and D. Lukose (Eds.): PRICAI 2012, LNAI 7458, pp. 447–457, 2012.
© Springer-Verlag Berlin Heidelberg 2012

Because of the limited view of an agent and the changing environment, an agent has to be adaptive. New information will restrict or relax the possible actions of an agent. When usual constraint solvers face relaxations they re-compute the entire system again. Thereby, the efficiency of the computation gets worse as well as the stability of a solution. Incremental approaches [5] overcome this by updating only the affected parts, rather than re-computing the whole system. As this problem plays an important role in various fields, a lot of work on incremental approaches has been done such as the Dynamic Constraint Satisfaction Problem [6]. There is also a need for dynamic adaptation during the planning phase. As the optimal plan length is not known in advance, the size of the constraint system might change during the search for a solution. Examples of constraint-based planning systems which perform a problem expansion during the search are Graphplan and Descartes [6].

In order to get information about the constraints and parameters for the changing environment, an agent will be required to mine the web for additional resources in order to analyze the status of the current environment as well as to predict the current state of the environment in the future. In a financial setting, companies would usually publish information relating to their financial performance to the public. It is possible to build a web mining agent that is capable of extracting this semi-structured information, merge the information and create investment ontology. By having this investment ontology, the planning agent can then make full use of this information to generate a more sound and reliable investment portfolio.

Another advantage of utilizing agents in a financial setting is in its ability to forecast or predict estimated revenue for the future based on the current available information. For an investment manager to recommend a portfolio, he needs to select potential companies to invest, and predict the likely revenue that will be generated by this particular company within a given period of time. Obviously, companies that are expected to guarantee higher revenue will be selected to populate the investment portfolio. However, predicting the possible profit that can be gained by a particular company is not a simple process. The investment manager needs to take into account the financial status of the company, the type of investment, the environmental variables that might affect the operation of the company as well as the company's past financial performances. In the past, utilization of prediction methods such as time series, genetic algorithm and artificial neural network [7][8][9] have been widely adopted in forecasting future data. Potentially the predictor agent can use any of these techniques or its combination. However, the accuracy of these prediction techniques is influenced by the nature of the problem and its dynamicity or complexity.

The final stage of the investment portfolio generation is the planning process. By drawing the values obtained from the mined information and the predicted values, the agent now needs to determine an investment plan. This can be done by establishing different combinations of investment paths that can be explored by the agent with the objectives of selecting the path that will maximize on the investment profit. One such novel technique is the constraint based semantic planning which allows for the generation of the optimal path by combining the investment and constraint based planning technique.

Based on the setting described above, this research project will develop a semantic multi-agent framework that utilize semantic and agent technology to generate a plan for a given problem. To demonstrate the effectiveness of this framework, we will implement the framework with the ethical wealth management setting in order to develop an intelligent investment planner that will generate a profitable investment portfolio for its investors based on certain parameters. The rest of the paper is organized as follows. Section 2 describes related work on web based wealth planner. Section 3 elaborates on the architecture and agent components in the multi-agent framework. Section 4 evaluates the performance of the ethical wealth planner. Finally, Section 5 concludes and elaborates on the future work.

2   Related Works

Over the last few years, a number of conventional web based wealth planners have been published. Among them, the most famous financial services provider system is Financial Engines which was founded by William F. Sharpe [10]. The Financial Engines system provides subscribers with specific buy-and-sell recommendations for managing tax-deferred and taxable investments. Besides that, there are some well-known finance institutions that provide their own on-line system for financial and investment advice such as HSBC [11], Citibank [12], Maybank [13], etc.

However, most of the wealth planners focus too narrowly on investment strategy and risk management services, and they lack flexible and autonomous problem solving behavior [14][15]. Due to these imperfections, Gao proposed a novel web-services-intelligent-agent-based family wealth management system [14][15]. They proposed that a group of logical, interacting and autonomous agents are applied to deal with the complex, dynamic and distributed processes in wealth management system. In addition, Web-services technology is integrated into the agent architecture for the utilization of interaction among agents. Wealth planner is actually a kind of complex, dynamic and distributed process that requires degree of cooperative problem solving capabilities system [16]. Therefore, multi-agent based approach is appropriate to the problem where a group of agents with variable tasks work together to achieve a common goal.

Kiekintveld et al. [17] presented an agent called Deep Maize, to forecast market prices in the Trading Agent Competition Supply Chain Management Game. As a guiding principle they exploited as many sources of available information as possible to inform predictions. Since information comes in several different forms, they integrated well-known methods into a novel way to make predictions. The core of the predictor is a nearest neighbour machine learning algorithm that identifies historical instances with similar economic indicators. They also augmented this with online learning procedures that transform predictions by optimizing the scoring rules. This allowed them to select more relevant historical contexts using additional information that are made available during individual games. They explored the advantages of two different representations for predicting price distributions.

Another work done by Ricardo Antonello and Ricardo Azambuja Silveira [18], proposed a multi-agent systems to reduce the uncertainty about investment in stock index prediction. The proposed system has a group of several agents that will monitor and predict the movement direction for each asset in Bovespa Index. The econometric models are adapted as multiple linear regression in the decision mechanism of the agents. The neural network is used in the decision engine of the agents to allow the agents to learn from the historical stock prices and then make decision. The authors claimed that the use of population of agents in predicting the stock movement is more efficient compared to the prediction based only on data in the index time series.

Tamer et al. [19] proposed the Stock Trading Multi-Agent System (STMAS) that utilized agent technology. STMAS is a web role-based multi-agent system that was implemented in the field of business intelligence. The system helps investors in buying and selling of stocks through the Egyptian Exchanges. In STMAS, roles are assigned to the agent dynamically by INRIN System. The INRIN System will assign the Role-Interface and Role-Implementation to a suitable agent to perform the required task. There were a total of five agents that were created in this system, Trading Information Agent, System Manager Agent, Role-Interface Generator Agent, Trading Expert Agent and Broker Agent.

3   The Multi-agent Ethical Wealth Planner

This work focuses on the development of a semantic multi-agent framework that can plan, predict, assemble and recommend investment portfolio based on a set of preferences. This framework will allow multiple agents with variable tasks to work together to achieve a certain goal. In order to investigate the effectiveness of the framework, an Ethical Wealth Planner (EWP) model is developed using the multi-agent framework. In this work, three types of agent were developed which are the Crawler Agent, the Wealth Forecasting Agent and the Wealth Planning Agent. The implementation is aimed to develop an intelligent investment planner that will generate a profitable investment portfolio for its users.

3.1   Service-Oriented Multi-agent Architecture

The Ethical Wealth Planner consists of a front end EWP visualizer and back end components that are made up of three agents. These agents are the Crawler Agent, the Wealth Forecasting Agent and the Wealth Planning Agent. In EWP, this multi-agent framework is deployed in a Service Oriented Architecture (SOA) [20] where each of the agents is wrapped as a standalone service provider. These services are accessible via web service protocol. Among the advantages of using SOA are standardized Service Contract, Service Loose Coupling, Service Autonomy, Service Abstraction, Service Reusability, and Service Statelessness. Above all, the service discovery, composition and invocation can be autonomously done by the intelligent agents.

The Agent Lookup service is responsible to handle the interaction among the agents. Agent Lookup plays the role of "Yellow Pages", in which every agent must register itself to the Agent Lookup service on startup. The Agent Lookup will keep track of the location of the agents. When one component or agent wishes to communicate with another agent, it gets the location of the agent through Agent Lookup before performing the communication. In the case of EWP, the Simple Object Access Protocol (SOAP) [21] is used for the inter-component communication.

Figure 1 illustrates the architecture of the system and how each agent interacts with each other. Firstly, the Crawler Agent will mine and collect company profile from the web. The data crawled will be stored in the knowledge base that resides in the Alle-groGraph [22] server. AllegroGraph is a graph database that uses efficient memory utilization with disk-based storage. It supports SPARQL from numerous client applications [22]. At the same time, the Crawler Agent will also collect the daily stock prices and stores them into a relational database. The information will then be used by the Wealth Forecasting Agent in predicting the next day's stock closing price. This agent has learning skills due to the use of MLP neural network which consist of one input layer, one hidden layer and one output layer. The nodes in the network are connected in feed forward manner. The network is then trained by using back propagation algorithm (learning with sigmoid function). The Wealth Planning Agent is responsible to perform decision making based on the prediction results and other constraints. The EWP visualizer is a web-based user interface which is responsible for the user's visualization and navigation. It interacts with the agent components through SOAP in order to retrieve the wealth planner.

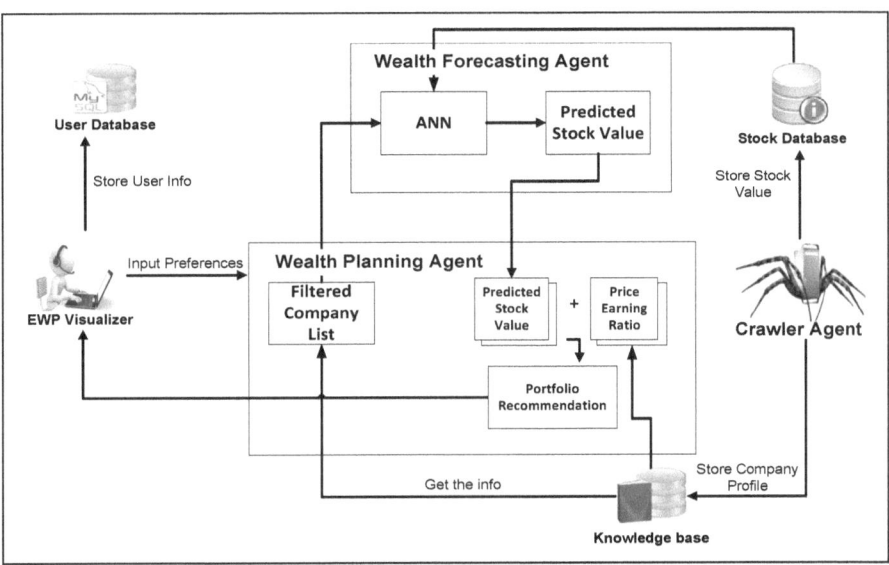

Fig. 1. Ethical Wealth Planner Multi-Agent Architecture

3.2   Agents' Roles and Responsibilities

The Crawler Agent. This agent mines the web to collect data relating to company profile and information such as the company's name, business description, director and officers, and news related to its company. This data is crawled from the Wall Street Journal [23]. We created a company's ontology to capture the information relating to company. The extracted semi-structured information is merged and added to the knowledge base. The Crawler Agent is triggered to run on a monthly basis to keep the company profiles up-to-date. Besides, the Crawler Agent also collects the historical stock price of the company. This historical stock price is to be used as an input for predicting the stock movement. In this research, fifteen stocks from Kuala Lumpur Stock Exchange (KLSE) are monitored. The agent crawls and updates the stock prices on daily basis.

The Wealth Forecasting Agent. The role of the Wealth Forecasting Agent is to learn the trend of stock price from the historical data sets for a particular stock in a given period. This agent is responsible to predict the next day's closing price of the stocks. The forecasting technique that is used for the closing price prediction is artificial neural network (ANN). A set of historical stock prices of 15 companies that are listed on KLSE gathered from the World Wide Web is used for the training and testing of the neural network. The set of data was downloaded for the period of 1st June 2006 until 31st December of 2011. This historical data set from 1st June 2006 until 31st May 2011 are set apart for learning purpose while the remaining are reserved for testing. The Wealth Forecasting Agent will return a prediction result to the Wealth Planning Agent for further action.

The Wealth Planning Agent. The Wealth Planning Agent accepts input from user and filters out a list of potential companies. This set of companies will be passed to the Wealth Forecasting Agent for stock price prediction. The Wealth Planning Agent will use the prediction results together with the Price Earnings Ratio (PE) [24] to determine the best returned result among all the companies and recommends it to the investor. The decision making is completed in a rule-based manner.

3.3   Agents Interaction

Agent interaction is the most important process in EWP. All agents interact with each other to perform the task of generating the investment portfolio to its investors. The investment portfolio generation process is as follow:

1. The Crawler Agent is triggered to insert/update the company profile and stock prices in the knowledge base and the relational database.
2. The EWP visualizer accepts the user investment preferences (such as amount of investment, type of company, duration of investment, and etc.) and passes it to the Wealth Planning Agent.
3. The Wealth Planning Agent accepts the input and filters out the potentials stocks that conform to the investor's preferences. The list of potential stocks is then passed to the Wealth Forecasting Agent.
4. The Wealth Forecasting Agent accepts the list of stocks passed by the Wealth Forecasting Agent. The Wealth Forecasting Agent executes the prediction algorithms (ANN) to predict the future movement of each of the stocks.
5. The Wealth Forecasting Agent returns the predicted results to the Wealth Planning Agent.
6. The Wealth Planning Agent receives the predicted results from the Wealth Forecasting Agent.
7. The Wealth Planning Agent gets the Price Earnings ratio value from knowledge base which was collected by the Crawler Agent previously.

8. The Wealth Planning Agent makes decision based on the predicted results and Price Earnings ratio to select the most profitable investment portfolio.
9. The Wealth Planning Agent generates the investment portfolio and passes the results to the EWP visualizer.
10. The EWP visualizer receives the recommended investment portfolio and displays it to the investor.

At the time of writing, this research work is still ongoing. However, the multi-agent framework proposed has been developed. The EWP has been partially implemented using this multi-agent framework.

4   Experimental Evaluation

Experiments were conducted to test the EWP performance. The first experiment is carried out in order to test the prediction's accuracy of the neural network utilized by the Wealth Forecasting Agent. Fifteen stocks that comprised of Shariah and non-Shariah compliance stocks from KLSE are chosen to represent the virtual stock market. Five years historical data set are used in this experiment. The data set is divided into two parts, the training set and testing set. The training set is used as input to train the network. After the network is trained, the testing set is used to predict the movement of the stock price. These predicted values are then compared with the historical values to calculate the accuracy of the prediction. The percentage of the predicted value will be calculated using the Residual Error (RE) for each single stock. After that, the Average Residual Error (ARE) will be calculated to evaluate the overall performance of the ANN. Table 1 shows the accuracy and RE of the stocks prediction results in percentage.

Table 1. Accuracy and Residual Error of the Stocks Prediction based on 5 years historical data

Stock	Accuracy (%)	RE (%)
BAT_KLS	98.31	1.09
BPURI_KLS (S)	97.30	2.20
COMMERZ_KLS	98.22	1.39
DIGI_KLS (S)	96.77	3.18
DRBHCOM_KLS	97.54	2.30
FAJAR_KLS	98.41	1.40
JETSON_KLS (S)	94.80	5.75
KEN_KLS (S)	97.53	2.17
MAXIS_KLS (S)	97.84	1.28
MAYBANK_KLS	98.80	1.06
NESTLE_KLS (S)	98.70	1.12
NOTION_KLS (S)	97.86	2.20
PBBANK_KLS	98.33	1.43
SPRITZR_KLS (S)	97.47	2.09
TAKAFUL_KLS (S)	97.15	2.38

Based on the result obtained, the prediction value of MAYBANK_KLS stock recorded the highest accuracy rate of 98.80% and obtained the lowest RE of 1.06%. JETSON_KLS recorded the lowest accuracy of 94.80% with RE of 5.75%. Using these results, we obtained the average prediction accuracy of 97.67%. The average residual error obtained for this experiment is 2.07%.

The second experiment is conducted to test the performance of the Wealth Planning Agent. Out of fifteen stocks, nine of the Shariah compliance stocks (indicated as (S)) are tested. The settings of the investment are fixed and the duration of the investment is between one and six months. Table 2 and Table 3 show the results of the Wealth Planning Agent in that suggests the most profitable stocks over the given duration.

Table 2. Predicted profit margin (in RM per share) for each stock from one to six months

Duration (Month/s) Stock	1	2	3	4	5	6
DIGI_KLS (S)	0.0396	0.1082	0.1393	0.235	0.2594	0.4013
MAXIS_KLS (S)	0.0293	0.0058	0.021	0.0115	0.0886	0.3012
NOTION_KLS (S)	0.000	0.000	0.000	0.0166	0.000	0.0135
TAKAFUL_KLS (S)	0.2097	0.4394	0.3835	0.388	0.3545	0.3599
NESTLE_KLS (S)	0.2669	0.2053	0.235	2.6887	2.2486	3.6166
SPRITZR_KLS (S)	0.0261	0.0276	0.0247	0.0214	0.0299	0.1706
BPURI_KLS (S)	0.0114	0.1402	0.1469	0.1847	0.1742	0.1382
JETSON_KLS (S)	0.000	0.000	0.000	0.000	0.000	0.4602
KEN_KLS (S)	0.0515	0.0356	0.0459	0.0393	0.0727	0.0272

Table 3. Results of the Stock Chosen from one to six months

Duration (Month/s)	Stock	Predicted Buy Date	Predicted Sell Date	Predicted Buy Price (RM Per Share)	Predicted Sell Price (RM Per Share)	Profit Margin (RM Per Share)
1	NESTLE _KLS (S)	1/6/2011	22/6/2011	46.67	49.94	0.27
2	TAKAFUL _KLS (S)	1/6/2011	13/7/2011	1.68	2.12	0.44
3	TAKAFUL _KLS (S)	1/6/2011	12/7/2011	1.69	2.07	0.38
4	NESTLE _KLS (S)	13/7/2011	19/9/2011	49.13	1.82	2.69
5	NESTLE _KLS (S)	12/7/2011	19/9/2011	48.14	50.39	2.25
6	NESTLE _KLS (S)	12/7/2011	29/11/2011	49.70	53.31	3.62

In Table 2, the profit margin per share for each stock are calculated based on the predicted closing prices, by calculating the difference of the highest price and the lowest price within the specified time frame. Some of the profit margin for some stocks is 0, this is because the highest closing price is screened at the beginning, and at the same time the price movement of the stock is predicted to decrease along the selected duration. In Table 3, the stocks that gains highest profit margin for each of the duration were collected. The final output of the Ethical Wealth Planner is the investment portfolio which includes the predicted prices, company's profile and the predicted profit that could be earned by the investor. Figure 2 shows the screen shot of investment portfolio that was generated during the experiment. Inside the investment portfolio, the company's description is showed. A link is provided to view the details of the company which were stored in knowledge base. A result showing the recommended buy and sell time is displayed in table form together with the graph that compared the prediction and actual stock price over the time period.

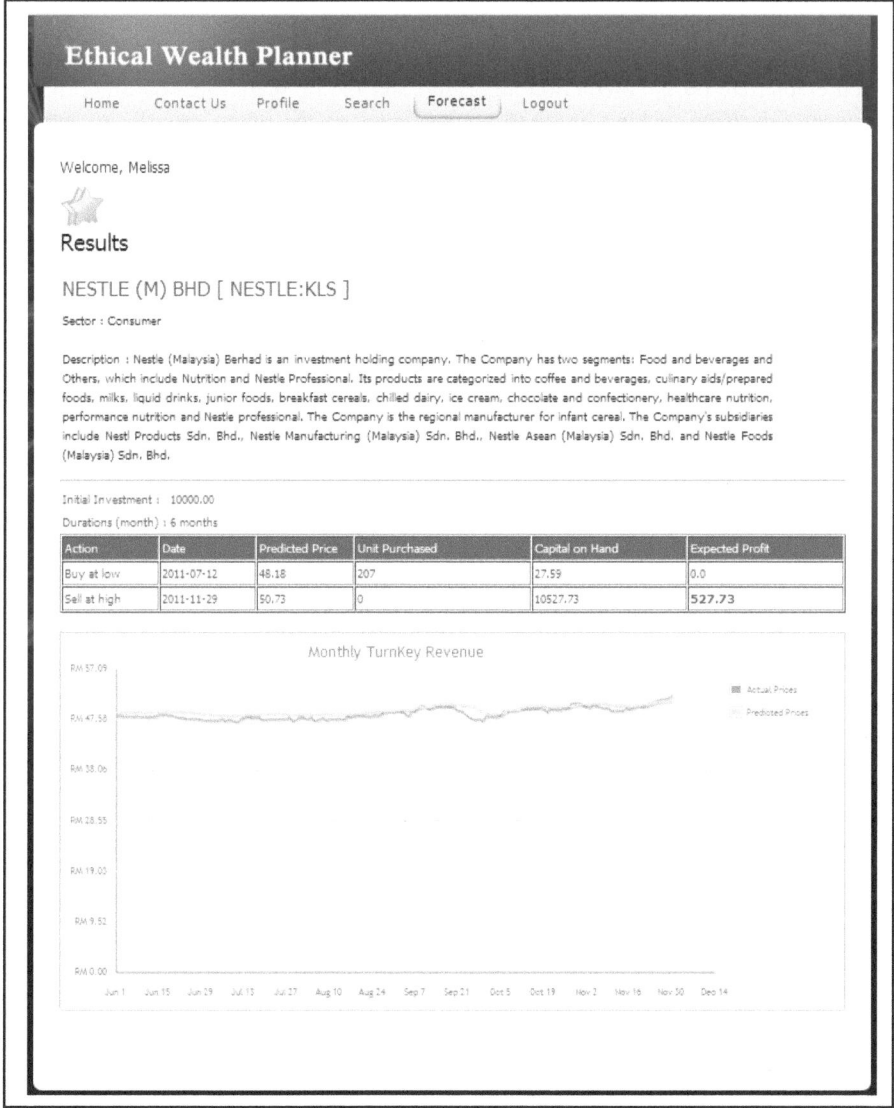

Fig. 2. Screen shot of the generated investment portfolio

5   Conclusion and Future Work

This paper presented the development of Ethical Wealth Planner. The system is developed to conform to a multi-agent framework. It consists of three types of agent, the Crawler Agent, the Wealth Planning Agent and the Wealth Forecasting Agent. Each of these agents is assigned a distinct task. SOAP is adopted as an interaction model between these agents. Furthermore, the interaction between agents in order to achieve their goals in this framework is described. Several tests have been conducted to evaluate the efficiency of the multi-agent framework. Currently, the prediction algorithm used by the Wealth Forecasting Agent is Feed Forward Neural Network with back propagation which recorded 97.67% of prediction accuracy. However, more research which emphasize on the prediction algorithm and the optimal selection algorithm in the agent's decision making needs to be conducted in the future in order to achieve higher accuracy in this wealth planner. In this paper, the multi-agent framework is proved to be workable where each of the agents is able to work together in generating an investment portfolio, and the decision of the agent is performed in rules basis. In the future work, these rules should be captured in an investment ontology which will be utilized by the agent in making decision. The utilization of investment ontology should allow the agent to reasoning semantically and generate a more meaningful portfolio.

References

1. Hussein, A.K.: Ethical Investment: Empirical Evidence from FTSE Islamic Index. Islamic Economic Studies 12(1), 21–40 (2004)
2. Elmelki, A., Ben Arab, M.: Ethical Investment and the Social Resposibilties of the Islamic Banks. International Business Research 2, 123–130 (2009)
3. Jennings, N.R., Wooldridge, M.: Applications of Intelligent Agents. In: Jennings, N.R., Wooldridge, M.J. (eds.) Agent Technology Foundations, Applications, and Markets (1998)
4. Wooldridge, M.: Intelligent agents. In: Weiß, G. (ed.) Multiagent Systems: A Modern Approach to Distributed Artificial Intelligence, pp. 27–77. MIT Press, Cambridge (1999)
5. Ramalingarn, G., Reps, T.: A Categorized Bibliography on Incremental Computation. In: Proceedings of the 20th Annual ACM Symposium on Principles of Programming Languages, pp. 502–510 (1993)
6. Joslin, D., Pollack, M.E.: Is "early commitment" in plan generation ever a good idea? In: Proceedings of the Thirteenth National Conference on Artificial Intelligence, AAAI 1996, pp. 1188–1193 (1996)
7. Keith, C.C.C., Foo, K.T.: Enhancing Technical Analysis in the Forex Market Using Neural Network. In: IEEE International Conference on Neural Network (1995)
8. Baba, N., Inoue, N., Yan, Y.J.: Utilization of Soft Computing Techniques for Constructing Reliable Decision Support Systems for Dealing Stocks. In: IJCNN 2002: Proceedings of The 2002 International Joint Conference on Neural Network (2002)
9. Dhar, S., Mukherjee, T., Ghoshal, A.K.: Proceedings of the International Conference on Communication and Computational Intelligence-2010, pp. 597–602 (2010)
10. Financial Engines (2011) from Financial Engines,
    http://www.financialengines.com (retrieved 2011)

11. HSBC (2011) from HSBC, `http://www.hsbc.com/` (retrieved 2011)
12. Citibank (2011) from Citibank, `http://www.citibank.com/` (retrieved 2011)
13. maybank2u.com (2011) from maybank2u.com, `http://www.maybank2u.com.my/` (retrieved 2011)
14. Gao, S.J., Wang, H.Q., Wang, Y.F., Shen, W.Q., Yeung, S.B.: Web-service-agents-based Family Wealth Management System. Expert Systems with Applications 29, 219–228 (2005)
15. Gao, S.J., Wang, H.Q., Xu, D.M., Wang, Y.F.: An intelligent Agent-assisted Decision Support System for Family Financial Planning. Decision Support Systems 44, 60–78 (2007)
16. Dugdale, J.: A Cooperative Problem-solver for Investment Management. International Journal of Information Management 16(2), 133–147 (1996)
17. Kiekintveld, C., Miller, J., Jordan, P.R., Wellman, M.P.: Forecasting Market Prices in a Supply Chain Game. In: Sixth International Joint Conference on Autonomous Agents and Multi-Agent Systems, pp. 1–8 (2007)
18. Antonello, R., Silveira, R.A.: Multiagent Systems in Stock Index Prediction. In: Demazeau, Y., Dignum, F., Corchado, J.M., Bajo, J., Corchuelo, R., Corchado, E., Fernández-Riverola, F., Julián, V.J., Pawlewski, P., Campbell, A. (eds.) Trends in PAAMS. AISC, vol. 71, pp. 563–571. Springer, Heidelberg (2010)
19. Mabrouk, T.F.: A Multi-Agent Role-Based System for Business Intelligence. In: Innovations and Advances in Computer Science and Engineering, pp. 203–208. Springer Science+Business Media B.V. (2010)
20. SOA (2011) from SOA Principles, `http://www.soaprinciples.com/` (retrieved 2011)
21. SOAP (2007) from SOAP Specification, W3C Recommendation, 2nd edn., `http://www.w3.org/TR/soap/` (retrieved 2011)
22. AllegroGraphRDFStore 4.2 (2011) from Franz, Franz Inc., `http://www.franz.com/agraph/allegrograph/` (retrieved 2011)
23. Market Data Center (2011) fromThe Wall Street Journal, `http://online.wsj.com/mdc/public/page/marketsdata_asia.html` (retrieved 2011)
24. Understanding the P/E Ratio (2011) from Investopedia, `http://www.investopedia.com/university/peratio/` (retrieved 2011)

Scalable Text Classification with Sparse Generative Modeling

Antti Puurula

Department of Computer Science, The University of Waikato, Private Bag 3105,
Hamilton 3240, New Zealand

Abstract. Machine learning technology faces challenges in handling
"Big Data": vast volumes of online data such as web pages, news stories and articles. A dominant solution has been parallelization, but this
does not make the tasks less challenging. An alternative solution is using
sparse computation methods to fundamentally change the complexity of
the processing tasks themselves. This can be done by using both the sparsity found in natural data and sparsified models. In this paper we show
that sparse representations can be used to reduce the time complexity
of generative classifiers to build fundamentally more scalable classifiers.
We reduce the time complexity of Multinomial Naive Bayes classification
with sparsity and show how to extend these findings into three multi-label
extensions: Binary Relevance, Label Powerset and Multi-label Mixture
Models. To provide competitive performance we provide the methods
with smoothing and pruning modifications and optimize model meta-
parameters using direct search optimization. We report on classification
experiments on 5 publicly available datasets for large-scale multi-label
classification. All three methods scale easily to the largest available tasks,
with training times measured in seconds and classification times in milliseconds, even with millions of training documents, features and classes.
The presented sparse modeling techniques should be applicable to many
other classifiers, providing the same types of fundamental complexity
reductions when applied to large scale tasks.

Keywords: sparse modeling, multi-label mixture model, generative
classifiers, Multinomial Naive Bayes, sparse representation, scalable computing, big data.

1 Introduction

Machine learning systems are operating on increasingly larger amounts of data,
or "Big Data". A dominant idea has been to tackle these challenges by paral-
lelizing algorithms with cluster computing and more recently cloud computing.
An alternative approach to the scalability problem is to change the algorithms
to more scalable ones. Most types of web data are naturally sparse, including
graph and text data. The models can be made sparse as well. Sparse computing
methods offer the possibility of solving the scalability problem by reducing the

P. Anthony, M. Ishizuka, and D. Lukose (Eds.): PRICAI 2012, LNAI 7458, pp. 458–469, 2012.
© Springer-Verlag Berlin Heidelberg 2012

computational complexity of the algorithms themselves, offering fundamentally more efficient solutions.

Sparse computing works by representing data and models using sparse matrix representations. For example, a vector of word counts of a document $\boldsymbol{w}$ can be represented by two smaller vectors of indexes and non-zero counts. Alternatively a hash table can be used for this, for constant time lookups and additions. In both cases the complexity of storing sparse information is reduced from full $|\boldsymbol{w}|$ to sparse complexity $s(\boldsymbol{w})$, where $s(\boldsymbol{w})$ is the number of non-zero counts. Fundamental reductions in computing complexities can be gained by choosing the correct sparse representation.

In this paper we show that by using the correct sparse representations the time complexity of generative classifiers can be reduced. We propose a sparse time complexity algorithm for MNB classification. We then demonstrate sparse generative classification with three multi-label extensions of Multinomial Naive Bayes (MNB), representing baseline approaches to multi-label classification. Binary Relevance (BR) method extends MNB by considering each label in multi-label classification as a separate problem, performing binary-class classifications. Label Powerset (PS) converts each labelset to a label, performing multi-class classification. Finally, Multi-label Mixture Modeling (MLMM) decomposes labelsets into mixture components, performing full multi-label classification. For each method a couple of meta-parameters are optimized using a direct search algorithm to provide realistic performance on the datasets. The direct search optimizations are done using a parallelized random search algorithm, to optimize the microaveraged F-score of development sets for each method.

Five freely available large-scale multi-label datasets are used for the experiments, using reported preprocessing for comparison of results. It is demonstrated that the use of sparse computing results in training times measured in seconds and classification times in milliseconds, even on the largest datasets with millions of documents, features and classes.

The paper continues as follows. Section 2 proposes sparse computation with the MNB model. Section 3 proposes three extensions of sparse MNB to multi-label classification. Section 4 presents experimental results on the five datasets and Section 5 completes the paper with a discussion.

2   Sparse Computation with Multinomial Naive Bayes

2.1   Multinomial Naive Bayes

Naive Bayes (NB) models [1, 2, 3] are generative graphical models, models of the joint probability distribution of features and classes. In text classification the joint distribution $p(\boldsymbol{w}, m)$ is that of word count vectors $\boldsymbol{w} = [w_1, ..., w_N]$ and label variables $m : 1 \leq m \leq M$, where N is the number of possible words and M the number of possible labels. Bayes classifiers use the Bayes theorem to factorize the joint distribution into *label prior* $p(m)$ and *label conditional* $p_m(\boldsymbol{w})$ models with separate parameters, so that $p(\boldsymbol{w}, m) = p(m)p_m(\boldsymbol{w})$. NB uses the additional assumption that the label conditional probabilities are independent, so that $p_m(\boldsymbol{w}) = \prod_n p_m(w_n, n)$. Multinomial Naive Bayes (MNB) parameterizes the label conditional probabilities with a Multinomial distribution, so that $p_m(w_n, n) \propto p_m(n)^{w_n}$. In summary, MNB takes the form:

$$p(\boldsymbol{w}, m) = p_m(\boldsymbol{w})p(m) \propto p(m) \prod_{n=1}^{N} p_m(n)^{w_n}, \tag{1}$$

where $p(m)$ is Categorical and $p_m(n)^{w_n}$ Multinomial.

2.2   Feature Normalization

Modern implementations of MNB use feature normalizations such as TF-IDF [4]. Surprisingly, this method developed for improving information retrieval performance has been shown to correct many of the incorrect data assumptions that the MNB makes [3]. The version of TF-IDF we use here takes the form:

$$w_n = \frac{\log[1 + w_n^u]}{s(\boldsymbol{w}^u)} \log[\max(1, \frac{D}{D_n} - 1)], \tag{2}$$

where w_n^u is the original word count, $s(\boldsymbol{w}^u)$ the number of non-zero counts in the word vector, D_n the number of training documents word n occurs in, and D the number of training documents.

The first factor (TF) in the function performs unique length normalization [5] and word frequency log transform. Unique length normalization is used, as it has been shown to be consistent across different types of text data [5]. The log transform corrects for the "burstiness effect" of word occurrences in documents, not captured by a Multinomial. As shown in [3], performing a simple log-transform corrects this relatively well. The second factor (IDF) performs an unsmoothed Croft-Harper IDF transform. This downweights words that occur in many documents and gives more importance to rare words. The Croft-Harper IDF downweights the common more words severely, actually setting the weight of words occurring in more than half the documents to 0. This induces sparsity and can be useful when scaling to large-scale tasks.

2.3   Sparse Representatation and Classification with Generative Models

It is a common practice in text classification to represent word count vectors $\boldsymbol{w}$ sparsely using two smaller vectors, one for indices of non-zero counts $\boldsymbol{v}$ and one for the counts $\boldsymbol{c}$. A mapping from dense to sparse vector representation can be defined $k(\boldsymbol{w}) = [\boldsymbol{v}, \boldsymbol{c}]$. Less commonly, the Multinomial models can be represented sparsely in the same way, using a vector for non-zero probability indices and one for the probabilities. By using the right type of sparse representations we can reduce both the space and time complexity of generative models much further.

Instead of having either a dense or sparse vector for each Multinomial, all Multinomial counts can be represented together in the same hashtable, using tuples of indices $\{m, n\}$ as the key and the log-probability $\log(p_m(n))$ as the stored value. This is known as the *dictionary of keys* representation for sparse matrices. Like dense vectors, the counts can be updated and queried in constant time. Like sparse vectors, the space use of storing the Multinomial is $s(\boldsymbol{w})$. But unlike sparse vectors, there is no need for allocating vectors when a resize is needed, or when a new label is encountered in training data.

The dictionary of keys is an efficient representation for model training, but for classification an even more efficient sparse representation exists. The *inverted index* forms the core technique of modern information retrieval, but surprisingly it has been proposed for classification use only recently [6]. An inverted index can be used to access the multinomials, consisting of a vector $\boldsymbol{\kappa}$ of label lists called *postings lists* $\boldsymbol{\kappa}_n$. In this paper it is shown that the inverted index can be used to reduce the time complexity of inference with generative models.

A naive algorithm for MNB classification computes the probability of the document for each label and keeps the label maximizing this probability. Taking the data sparsity $s(\boldsymbol{w})$ into account, this has the time complexity $O(s(\boldsymbol{w})M)$. With the inverted index, we can substantially reduce this complexity. Given a word vector, only the labels occurring in the postings lists can be considered for evaluation. To avoid a classification error in abnormal cases, the probability for the apriori most likely label needs to be precomputed. The *evaluation list* $\boldsymbol{\varrho}$ of labels can be computed by taking a union of the occurring labels, $\boldsymbol{\varrho} = \cup_{n:w_n>0}\boldsymbol{\kappa}_n$. Replacing the full set of labels with the evaluation list results in sparse $O(s(\boldsymbol{w})|\boldsymbol{\varrho}|)$ time complexity.

When conventional smoothing methods such as Dirichlet prior or interpolation are used, the smoothing probabilities can be precomputed and only updated for each label. Using this with generative modeling, we get another time complexity reduction. When constructing the evaluation list, the matching words between the word vector and unsmoothed multinomial can be saved as *update lists* $\boldsymbol{\nu}_m$ for each label. This reduces the time complexity to $O(s(\boldsymbol{w})+\sum_{m\in\boldsymbol{\varrho}}|\boldsymbol{\nu}_m|))$ where $|\boldsymbol{\nu}_m|$ are the update list sizes, at worst $\min(s(\boldsymbol{w}), s(p_m^u))$. Algorithm 1 gives a pseudocode description of sparse MNB classification.

3   Multi-label Extensions of Naive Bayes

Multi-label classification deals with the extension of single-label classification from a single label to set of labels, or equivalently a binary labelvector of label occurrences $\boldsymbol{l} = [l_1, ..., l_M]$. In supervised multi-label tasks the training dataset is labeled with the labelsets. Evaluation is done using a variety of metrics, most commonly the micro-averaged and macro-averaged F-scores [7]. We optimize and evaluate using micro-averaged F-score as the measure, as this been most commonly used with the text datasets we are experimenting on.

We evaluate three scalable extensions to MNB for sparse multi-label classification. The first two are the problem transformation methods of Binary Relevance(BR) [8] and Label Powerset(PS) [9], that are commonly used as baselines in multi-label classification [10, 11]. The third one is a multi-label mixture model

Algorithm 1. Sparse MNB Classification

1: $log\_smooth = 0$
2: **for all** $n \in k(\boldsymbol{w})_1$ **do** ▷ Iterate document words $k(\boldsymbol{w})_1$
3:   $log\_smooth+ = \log(p^s(n)) * w_n$
4:   **for all** $m \in \boldsymbol{\kappa}_n$ **do** ▷ Iterate postings list $\boldsymbol{\kappa}_n$
5:     $\boldsymbol{\nu}_m = \cup(\boldsymbol{\nu}_m, (n))$
6: $m^{max} = m^{apriori}$
7: $p^{max} = \log(p(m^{apriori})) + log\_smooth$
8: **for all** $m \in \boldsymbol{\varrho}$ **do** ▷ Iterate evaluation list $\boldsymbol{\varrho}$
9:   $p^{new} = \log(p(m)) + log\_smooth$
10:   **for all** $n \in \boldsymbol{\nu}_m$ **do** ▷ Iterate update list $\boldsymbol{\nu}_m$
11:     $p^{new}+ = (\log(p_m(n)) - \log(p^s(n))) * w_n$
12:     **if** $p^{new} > p^{max}$ **then**
13:       $p^{max} = p^{new}$
14:       $m^{max} = m$

(MLMM) that we have developed, that uses mixture modeling to decompose label combinations.

3.1 Binary Relevance

The binary relevance method [8] considers each label in the labelvector independently, by performing a binary classification for each label. The advantages of BR are that it is very efficient and easy to implement with any classifier capable of binary-label classification. The disadvantage is that it totally ignores label correlations, in the worst cases classifying all labels as positive or none. A relevance thresholding scheme is commonly used to improve the results, by adjusting the relevance decision boundary to maximize the evaluation score on a held-out development set. Extending the MNB classifier for BR is straightforward, with a positive Multinomial and a corresponding negative Multinomial for each label. Since we work with very large numbers of labels, we can approximate the negative Multinomial with a background distribution $p(n)$ with little loss in accuracy.

3.2 Label Powerset

The label powerset method [9] is a straightforward way to perform multi-label classification with single-label classifiers. Each labelset in training is converted to a single label identifier and converted back to labelsets after classification. Both operations can be done using hash table lookups, externally to the classifier. The main disadvantage is the increased space and time complexity of classification. Instead of models for at most M labels, PS constructs models for each labelset configuration l occurring in the training data. It neither takes into account similarities between the labelsets, and can only classify labelsets that are seen in the training data. Despite the theoretical problems, PS forms the basis of some of the most successful multi-label classification algorithms.

3.3   Multi-label Mixture Model

Multi-label mixture models [12, 13, 14] attempt generalization of MNB to multi-label data by decomposing labelset-conditional Multinomials into mixtures of label-conditional Multinomials. Here we propose a simple multi-label mixture model that is closely related to the original Multi-label Mixture Model [12] and Parametric Mixture Model [13], taking the form:

$$p(\boldsymbol{w}, \boldsymbol{l}) \propto p(\boldsymbol{l}) \prod_{n=1}^{N} [\sum_{m=1}^{M} \frac{l_m}{s(\boldsymbol{l})} p_m(n)]^{w_n} \tag{3}$$

By constraining the labelvector to a single label $s(\boldsymbol{l}) = 1$, the model reduces to MNB. A number of choices exist for modeling $p(\boldsymbol{l})$, one being a Categorical distribution over the labelvectors [12]. We use a fixed mixture of a Categorical with a smoothing distribution:

$$p(\boldsymbol{l}) = 0.5p^u(\boldsymbol{l}) + 0.5p^s(s(\boldsymbol{l})), \tag{4}$$

where p^u is the unsmoothed Categorical, p^s is a Categorical over labelcounts $s(\boldsymbol{l})$ corrected to $\boldsymbol{l}$ event space.

Classification with multi-label mixture models requires heuristics in practice, as a naive algorithm would perform a 2^M enumeration of all the possible labelsets. A common greedy algorithm [12, 13] starts with a labelvector of zeros and iteratively sets to 1 the label m that improves $p(\boldsymbol{w}, \boldsymbol{l})$ the most. The iteration ends if no label improves $p(\boldsymbol{w}, \boldsymbol{l})$. The labelcount prior p^s constrains what would be M^2M to a maximum of Mq^2 evaluations, where $q = \max_{\boldsymbol{l}:p(\boldsymbol{l})>0}(s(\boldsymbol{l}))$ is the largest labelcount in training data, resulting in $O(q^2 M s(\boldsymbol{w}))$ complexity.

This algorithm can be improved by taking into account the sparse improvements we've proposed for MNB and some additional heuristics. We can add caching of probabilities, so that in each iteration the probabilities $p(w_n)$ can be updated instead of recomputed. This reduces the time complexity to $O(Mqs(\boldsymbol{w}))$. We can also remove all non-improving labels in each iteration from the evaluation list as a weak heuristic. Finally, we can combine these with the sparse classification done in Algorithm 1, to get the worst time complexity of $O(q(s(\boldsymbol{w}) + \sum_{m \in \varrho} |\boldsymbol{\nu}_m|))$, or simply q times the sparse MNB complexity.

3.4   Model Modifications

For competitive performance and to deal with realistic data some modifications are required in generative modeling. An example of a mandatory modification is smoothing for the conditional Multinomials. In this paper we use four meta-parameters $\boldsymbol{a}$ to produce realistic performance for each compared method, except for the additional label thresholding meta-parameter a_5 used with BR.

The first modification is the *conditional smoothing*. We use Jelinek-Mercer interpolation, or linear interpolation with a background model. For each method, we estimate a background Multinomial concurrently to the Multinomials and interpolate this with the conditionals, so that $p_m(n) = (1-a_1)p_m^u(n) + a_1 p^s(n)$.

The second modification is to enable training with limited memory to very large datasets. We first constrain the hashtable for the conditionals to a maximum of 8 million counts, so that adding keys above that is not allowed. In addition we use pruning with the IDF weights. When a count is incremented, we compare its IDF-weighted value to an *insertion pruning* threshold a_2. If the value is under the threshold, the count is removed from the hashtable. When stream training is used, the IDF-values can be approximated with running estimates, giving gradually more accurate pruning.

A third modification is to scale the priors, replacing $p(l)$ by $p(l)^{a_3}$. This is commonly done in speech recognition, where language models are scaled. We've added this modification as it has a very considerable effect, especially in cases where the prior is less usable for a dataset and the generative model still weights labels according to the prior.

As a fourth modification we add a pruning criteria to the classification algorithm. We add a sorting to the evaluation lists, by scoring each label as the sum of matching TF-IDF weighted word counts $q_m = \sum_n w_n \mathbf{1}_{p_m^u(n)>0}$, and sorting the evaluation list by the scores q_m. We can then use the ranked evaluation list, by stopping the classification once the mean log probability of the evaluated labels deviates too far from the maximum log probability found, $mean\_logprob - a_4 < max\_logprob$. The scoring adds a $O(|\varrho|\log(|\varrho|))$ term to the time complexity, but this does not increase the worst time complexity in typical cases.

3.5   Meta-parameter Optimization

We use a direct search [15] approach for optimizing the meta-parameters. The function value $f(a)$ we optimize is the micro-averaged F-score of held-out development data. The type of direct search we use is a random search algorithm [16, 17], an approach best suited for the low-dimensional, noisy and multimodal function we are dealing with.

In random search the current best point a is improved by generating a new point $d \leftarrow a + \Delta$ with step Δ and replacing a with d if the new point is as good or better, if $f(a) \leq f(d)$, $a \leftarrow d$. The iterations are then continued to a maximum number of iterations I. A number of heuristics are commonly used to improve basic random search. We use a couple common ones, along a novel Bernoulli-Lognormal function for step generation.

Instead of generating a single point, we generate a population of J points in order to fully use parallel computing. Instead of having a single point for generation, we keep many, so that if any point $d_j \leftarrow a + \Delta_j$ improves or equals $f(a)$, we replace the current set of points by the Z best points sharing the best value. Subsequent points are then generated evenly from the current set of best points $a_{1+j\%Z}$.

We generate steps with a Bernoulli-Lognormal function, so that for each a_t we first generate a step direction with a uniform Bernoulli $b_{jt} \in (-1, 1)$ and then multiply the stepsize by a Lognormal $e^{\mathcal{N}(0,2)}$. We combine this with a global adaptive stepsize decrease, so that we start stepsizes at half range $c_t = 0.5 * (max_t - min_t)$ and multiply them by 1.2 after an improving iteration and by 0.8 otherwise. This gives the step generation process as $\Delta_{jt} \leftarrow b_{jt} \, c_t \, e^{\mathcal{N}(0,2)}$. In addition we improve variance among the generated points by generating each alternate step in a mirrored direction: if $j\%2 = 0$, $\boldsymbol{b}_j \leftarrow -\boldsymbol{b}_{(j-1)}$. A heuristic starting point $\boldsymbol{a}$ is used and to reduce the search space each meta-parameter is constrained to a suitable range $d_{jt} \leftarrow min(max(a_{1+j\%Z} + \Delta_{jt}, min_t), max_t)$ found in model development.

4 Experiments

4.1 Experiment Setup

The experiment software was implemented using Java and executed on single thread on a 3.40GHz Intel i7-2600 processor using 4GB of RAM. Five recent large-scale multi-label datasets were used for the experiments, all freely available for download. Table 1 shows the dataset statistics. The numbers of distinct labelsets are omitted from the table, but in the worst case this is 1468718 classes for the WikipL dataset, close to a million and a half. The datasets were preprocessed by lowercasing, stopwording and stemming, and stored in LIBSVM sparse format.

Table 1. Statistics for the five training datasets. Documents D in training, unique labels M, unique words N, mean of unique labels per document $e(s(\boldsymbol{l}))$ and mean of unique words per document $e(s(\boldsymbol{w}))$.

Dataset	Train D	Labels M	Words N	$e(s(\boldsymbol{l}))$	$e(s(\boldsymbol{w}))$	Task description
RCV1-v2	343117	350	161218	1.595	63.169	News articles
Eurlex	17381	3870	179540	5.319	**270.346**	Legal documents
Ohsu-trec	197555	14379	291299	**12.389**	81.225	Medical abstracts
DMOZ	392756	27874	593769	1.028	174.124	Web pages
WikipL	**2363436**	**325014**	**1617125**	3.262	42.534	Wikipedia articles

The datasets consist of WikipL, DMOZ, Ohsu-trec, Eurlex and RCV1-v2, each split to a training set, held-out development set and evaluation set. WikipL and DMOZ are the larger datasets from LSHTC2[1] evaluation of multi-label classification. For Eurlex[2] [18] the first 16381 documents of eurlex_tokenstring_CV1-10_train.arff were used as the training set, the last 1000 as the development set and eurlex_tokenstring_CV1-10_test.arff as the evaluation set. Ohsu-trec[3] used

---

[1] http://lshtc.iit.demokritos.gr/
[2] http://www.ke.tu-darmstadt.de/resources/eurlex
[3] http://trec.nist.gov/data/t9_filtering.html

both the title and abstract as document text, preprocessed by lowercasing, removing <3 letter words and using the Porter stemmer. The MeSH terms were
used as labels with additional specifiers to terms discarded. Ohsumed.88-91 is
used as the training dataset, the first 1000 documents of ohsumed.87 as the development set and the rest as the evaluation set. The files lyrl2004_tokens_test_pt*.
dat for RCV1-v2[4] [19] were used as the training set, the first 1000 documents
of lyrl2004_tokens_train.dat were used as the development set and the last 8644
as the evaluation set. The categorization rcv1-v2.industries.qrels was used for
labeling.

4.2   Experiment Results

The three evaluated methods were optimized for micro-averaged F-score on each
development set using a 50x30 (50 iterations, 30 points) random search. Figure
1 compiles the results from the runs, showing training times in seconds, development set classification times in milliseconds and evaluation set micro-averaged
F-scores. The time estimates were computed as the median of 8 runs. Table 2
shows the same numbers in a table form in addition to the development set
F-scores.

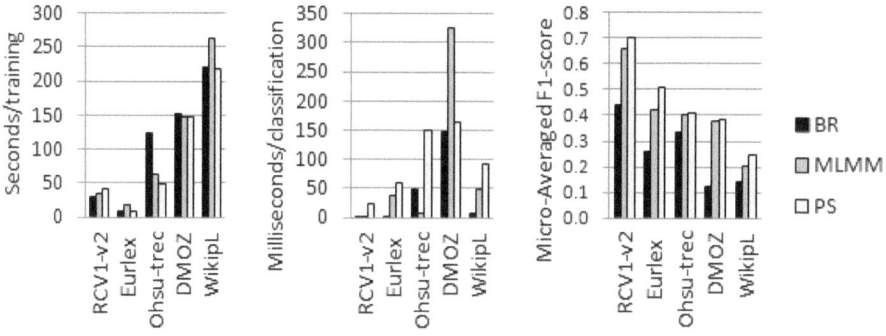

Fig. 1. Median training times, median classification times and micro-averaged F-scores

One-tailed paired t-tests were used to test the differences in F-scores and times,
verified by Wilcoxon signed rank tests with $p < 0.05$. Significances from t-tests $p <$
0.05 are shown in parenthesis. BR is outperformed in accuracy by both MLMM
($p < 0.006$) and PS ($p < 0.004$). In addition the effect size is very large, with
BR falling behind by almost half the F-score on average. The difference between
MLMM and PS accuracy is not significant, although on average PS is over 3%
F-score better than MLMM. In terms of training set times all models perform
similarly, with no significant differences and very similar mean times. In terms of

---

[4] http://www.daviddlewis.com/resources/testcollections/rcv1/

classification times the large variance causes only the difference between BR and PS to be significant ($p < 0.013$), although the mean times suggest BR is twice as fast as both MLMM and PS, and MLMM is somewhat faster than PS.

Table 2. Median training times, median classification times and micro-averaged F-scores

(a) Training times in seconds

	BR	MLMM	PS
RCV1-v2	30.73	35.53	42.50
Eurlex	9.56	18.60	9.01
Ohsu-trec	123.51	63.43	48.32
DMOZ	152.95	148.29	147.14
WikipL	219.82	261.92	218.01

(b) Classification times in ms

	BR	MLMM	PS
RCV1-v2	0.25	2.18	24.73
Eurlex	1.71	38.47	59.70
Ohsu-trec	47.72	5.90	150.85
DMOZ	148.55	324.79	164.98
WikipL	5.96	49.18	91.21

(c) Development set micro-averaged F-scores

	BR	MLMM	PS
RCV1-v2	0.439	0.660	**0.702**
Eurlex	0.259	0.422	**0.508**
Ohsu-trec	0.332	0.405	**0.407**
DMOZ	0.121	0.381	**0.383**
WikipL	0.143	0.206	**0.249**

(d) Evaluation set micro-averaged F-scores

	BR	MLMM	PS
RCV1-v2	0.434	0.665	**0.705**
Eurlex	0.242	0.408	**0.498**
Ohsu-trec	0.318	**0.402**	0.401
DMOZ	0.101	**0.362**	0.358
WikipL	0.111	0.190	**0.228**

5 Discussion

This paper showed how sparse matrix representations can be applied to reduce the complexity requirements of generative models. Although sparse representations such as the inverted index are fundamental in fields such as information retrieval, we have found no prior work on explicitly applying sparse representations for reducing space and time complexities for probabilistic models. It is likely that sparse representations are used in practical implementations of existing models, but the connections to complexity theory have been so far omitted.

We demonstrated how generative classifiers such as MNB can utilize sparse representations for reducing the time complexity of classification. We then used these representations with three extensions of MNB on the largest publicly available multi-label classification datasets. To get representable performance, a couple of parameterized modifications were used. The meta-parameters required by these were optimized regarding the micro-averaged F-score of development sets with a direct search algorithm. All 3 classifiers could be trained in some minutes on a single processor, with millions of documents, features and classes. Although not optimized with classification speed in mind, the 3 classifiers performed classifications in times ranging from microseconds to some hundreds of milliseconds, even when using over a million classes.

In the experiments presented here no comparisons to dense classifiers were made, as it would be tedious to compare dense classification speeds with the large datasets. In preliminary work we attempted several toolkits, but did not find ones that could scale to the databases discussed here. For future research it will be interesting to see what other classifiers can benefit from sparse representations. This can potentially change what classifiers are preferred in large scale tasks. It is expectable that the use of sparse representations becomes a mainstay of machine learning with the processing of web-scale tasks.

References

[1] Maron, M.E.: Automatic indexing: An experimental inquiry. J. ACM 8, 404–417 (1961)

[2] McCallum, A., Nigam, K.: A comparison of event models for Naive Bayes text classification. In: AAAI 1998 Workshop on Learning for Text Categorization, pp. 41–48. AAAI Press (1998)

[3] Rennie, J.D., Shih, L., Teevan, J., Karger, D.R.: Tackling the poor assumptions of naive bayes text classifiers. In: ICML 2003, pp. 616–623 (2003)

[4] Jones, K.S.: A Statistical Interpretation of Term Specificity and its Application in Retrieval. Journal of Documentation 28(1), 11–21 (1972)

[5] Singhal, A., Buckley, C., Mitra, M.: Pivoted document length normalization. In: Proceedings of the 19th Annual International ACM SIGIR Conference on Research and Development in Information Retrieval, SIGIR 1996, pp. 21–29. ACM, New York (1996)

[6] Shanks, V.R., Williams, H.E., Cannane, A.: Indexing for fast categorisation. In: Proceedings of the 26th Australasian Computer Science Conference, ACSC 2003, vol. 16, pp. 119–127. Australian Computer Society, Inc., Darlinghurst (2003)

[7] Tsoumakas, G., Katakis, I., Vlahavas, I.P.: Mining multi-label data. In: Maimon, O., Rokach, L. (eds.) Data Mining and Knowledge Discovery Handbook, pp. 667–685. Springer (2010)

[8] Godbole, S., Sarawagi, S.: Discriminative methods for multi-labeled classification, pp. 22–30 (2004)

[9] Boutell, M.R., Luo, J., Shen, X., Brown, C.M.: Learning multi-label scene classification. Pattern Recognition 37(9), 1757 (2004)

[10] Tsoumakas, G., Katakis, I., Vlahavas, I.: A Review of Multi-Label Classification Methods. In: Proceedings of the 2nd ADBIS Workshop on Data Mining and Knowledge Discovery, ADMKD 2006, pp. 99–109 (2006)

[11] Read, J., Pfahringer, B., Holmes, G., Frank, E.: Classifier Chains for Multi-label Classification. In: Buntine, W., Grobelnik, M., Mladenić, D., Shawe-Taylor, J. (eds.) ECML PKDD 2009, Part II. LNCS, vol. 5782, pp. 254–269. Springer, Heidelberg (2009)

[12] McCallum, A.: Multi-label text classification with a mixture model trained by EM. In: Proceedings of the AAAI 1999 Workshop on Text Learning (1999)

[13] Ueda, N., Saito, K.: Parametric mixture models for multi-labeled text. In: Advances in Neural Information Processing Systems, vol. 15, pp. 721–728. MIT Press (2002)

[14] Wang, H., Huang, M., Zhu, X.: A generative probabilistic model for multi-label classification. In: Proceedings of the 2008 Eighth IEEE International Conference on Data Mining, pp. 628–637. IEEE Computer Society, Washington, DC (2008)

[15] Powell, M.J.D.: Direct search algorithms for optimization calculations. Acta Numerica 7, 287–336 (1998)

[16] Favreau, R.R., Franks, R.G.: Statistical optimization. In: Proceedings Second International Analog Computer Conference (1958)
[17] Brunato, M., Battiti, R.: Rash: A self-adaptive random search method. In: Cotta, C., Sevaux, M., Sörensen, K. (eds.) Adaptive and Multilevel Metaheuristics. SCI, vol. 136, pp. 95–117. Springer (2008)
[18] Loza Mencía, E., Fürnkranz, J.: Efficient Multilabel Classification Algorithms for Large-Scale Problems in the Legal Domain. In: Francesconi, E., Montemagni, S., Peters, W., Tiscornia, D. (eds.) Semantic Processing of Legal Texts. LNCS, vol. 6036, pp. 192–215. Springer, Heidelberg (2010)
[19] Lewis, D.D., Yang, Y., Rose, T.G., Li, F.: RCV1: A New Benchmark Collection for Text Categorization Research. J. Mach. Learn. Res. 5, 361–397 (2004)

UPM-3D Facial Expression Recognition Database(UPM-3DFE)

Rabiu Habibu*, Mashohor Syamsiah, Marhaban Mohammad Hamiruce,
and Saripan M. Iqbal

Multimedia System Laboratory,
Department of Computer and Communication Systems,
Universiti Putra Malaysia, Serdang Selangor 43400, Malaysia
http://www.upm.edu.my

Abstract. Facial expression studies have now become the central topic among computer vision community; this can be attributed to application it finds in security, human computer interaction, entertainment industries, etc. Using the state of art equipment, We Built a 3D facial expression database named UPM-3DFE database. This database contained 350 face images of 50 persons, with each posing the six universally accepted facial expressions ie; happy, sad, angry, fear, disgust and surprise. The participants are drawn from different ancestral/ethnic background. The database was evaluated using both subjective and objective analysis. We further investigated the relationship between the machine expression recognition and the human effort required to mimic the expression. The result shows a negative correlation.

Keywords: Facial expression recognition, 3D face Database, Expression mimicking.

1 Introduction

Research in the area of human behavior via information displayed on the human face is becoming more passionate among computer vision community. This can be connected with the growing application it finds in areas such as; human computer interaction, security, psychological studies, pain detection, robotics, facial animation, etc. Due to the fact that two dimensional (2D) databases have been available for a very long time, the problem of facial expression recognition was initially dominated by 2D methods [1–3]. Illumination changes and pose variations are paired issues that remain a challenge to a fully successful 2D system [4]. The recent technological advancement in three dimensional (3D) acquisition devices makes the modality more accurate and affordable. Data acquired in 3D is invariant to both illumination and pose changes. Moreover, it provides far richer information than its 2D counterpart [5]. Although, the appearance of the six basic expressions namely; happy, sad, angry, fear, disgust and surprise are

---

* Corresponding author.

P. Anthony, M. Ishizuka, and D. Lukose (Eds.): PRICAI 2012, LNAI 7458, pp. 470–479, 2012.

universal across all ethnic groups and culture [11]. However, factors such as face shape, appearance, and facial hair vary with sex and ethnic background. Therefore, expression recognition accuracy increases where the subjects are drawn from a single ethnic group, this is due to the homogeneity of face appearances and shapes within same group [8, 14]. To foster research in 3D facial expression recognition (FER), we developed a multi expression 3D facial expression database called UPM-3DFE containing 50 subjects of different ancestral/ethnic background, detail of the database's ethnic distribution is as shown in table 1. Albeit there are few devotedly 3D facial expression database available [6, 7]. UPM-3DFE is unique in terms of its ethnic contents and distribution, which is crucial as argued by Hewahi et al [8], it is also more realistic in terms of variability in subject pose and outfit.

Table 1. Comparison of some 3D facial expression databases

	Bosphorus	BU-3DFE	UPM-3DFE
No of subjects	81	100	50
male-female ration	51 by 30	44 by 56	30 by 20
age variation	25 to 35	18 to 70	20 to 60
basic emotion	√	√	√
White	about 90%	51.0%	0.0%
Blacks/Africans	N/A	9.0%	18.0%
Chinese/east-Asian	N/A	24.0%	20.0%
Indians	N/A	6.0%	8.0%
Middle-east Asian	N/A	2.0%	26.0%
Latino-Hispanic	N/A	8.0%	0.0%
south-east Asia	N/A	0.0%	28.0%

N/A means not available.   .

1.1   3D Facial Expression Acquisition Process

1.2   Studio Setup

We made use of 3D Flexscan (V2.6) system in acquiring our 3D facial images. The system consists of two high vision cameras placed at a distance of 30 inches apart with a projector mounted in-between them. The projector projects different binary patterns onto the subject's face, while the two cameras captures the pattern as deformed by the subject's face components.

Using the stereo photometry technique, the system automatically determines correspondences between the images captured by the two cameras and merge them into a single 3D face model, with a resolution of 25K to 35K polygons per model. The whole exercise is controlled and coordinated by a computer system. Fig 1 depicted the overall system setup.

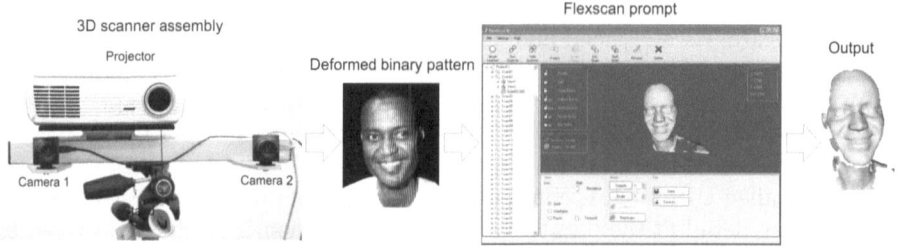

Fig. 1. Facial expression acquisition setup

1.3   Facial Expression Capturing

The currently available 3D capturing devices are mostly use under controlled studio condition, where the distance between the scanner and subject to be imaged is calibrated and fixed. Obtaining a spontaneous expression under such condition is a very difficult task, as the subject will be in a full picture of the whole set-up. Further more, it is unethical to image human being without their prior consent. For these reasons we resort to develop a deliberate expression based database. Before the capturing process start, each participant is initiated through an introductory session, where he or she is briefed on what he is about to do. In addition, sets of photographs of each of the six basic expressions as displayed by expert actors/actress are shown to him. A mirror is as well made available to him so that he can see him self mimicking the expressions. After the introductory session, the subject is asked to sit at a distance of about 1 meter from the scanner. He is then instructed to perform the six universally accepted expressions ie; happy, sad, angry, fear, disgust and surprise. An additional neutral expression was also considered. Each expression was displayed for a very brief time enough to allow the scanner capture the guise. Participants were allowed to have caps on, head veils, hair falling on the face region, necklaces and ear rings, while the only constrain was the spectacles. Similarly, the participants were allowed to move their head within $\pm 30^o$ in the left-right directions. At the end of the exercise we have collected 350 face meshes from 50 subjects, 30 males and 20 females; whose ages span between 20 to 60 years. The participants were drawn from a different ancestral background, which includes Arabs, Malays, Africans, Chinese, Indians and Persians. This is a very important factor as ethnicity variation influences expression recognition [8].

2   Mesh Preprocessing

The acquired 3D facial models were preprocessed to remove the extraneous data that are not part of the face region, for example, the head veils, neck and other parts of the subject's body. We further apply median filtering on the z-axis to remove spikes; finally cubic interpolation was used in filling holes from the mesh.

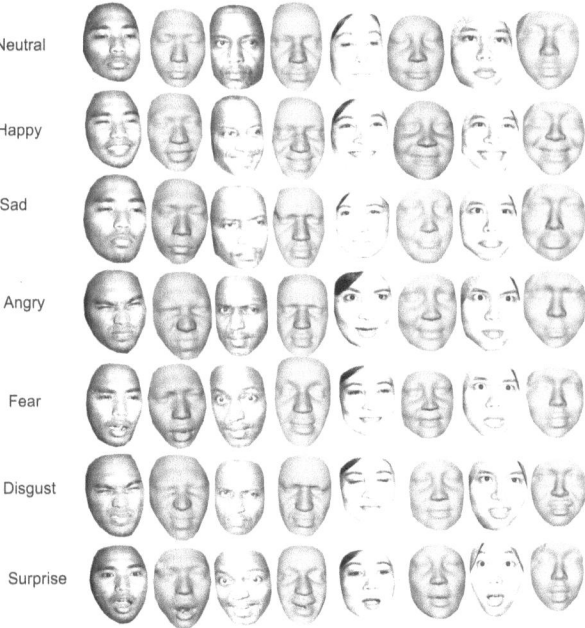

Fig. 2. Samples facial shape models and their corresponding intensity images for four subjects displaying the seven basic expressions

2.1 Fiducial Points

We manually identified and annotated 32 expressive sensitive points on each face mesh in the database. The selection of these points was guided by MPEG4 and FACS [9, 10]. To have a more reliable land mark localization, the database was annotated by two different persons independently. The final land mark points are determined by averaging the two manually labelled face meshes. These feature points are intended to be used as ground-truth reference for researchers.

3 Database Validation

We applied three different measurement techniques to evaluate the performance of the database. This is important so that the machine interpretation can be compared with human assessment (ground truth) and see how closely related these measures are:

1) Performer assessment: at the end of the capturing session, each participant was given a set of photograph depicting his expression poses to identify the seven expressions and to fill-in the space provided underneath the expression posed. He was as well requested to state how much effort is needed to mimic each of seven expressions by ticking one out of four boxes provided in the forced-choice

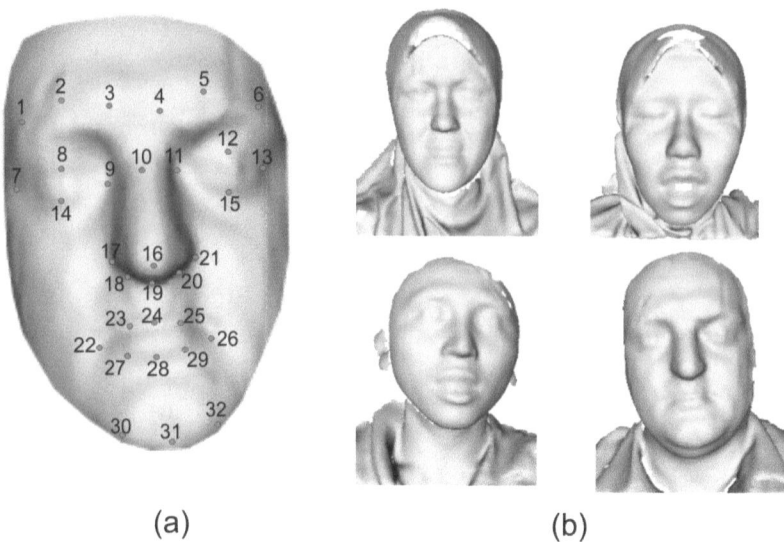

(a)   (b)

Fig. 3. (a) 32 manually annotated feature points (b) samples of unprocessed meshes

type likert scale. These boxes were marked as very difficult (4), difficult (3), easy (2) and very easy (1). Table 2 depicts the result of participants scores for effort exerted against the expression posed.

Table 2. Percentage votes for effort required vs expression pose

	Very easy (1)	Easy (2)	Difficult (3)	Very difficult (4)
Neutral	75.0%	25.0%	0.0 %	0.0%
Happy	50.0%	50.0%	0.0%	0.0%
Sad	0.0%	50.0%	50.0%	0.0%
Angry	12.5%	50.0%	25%	12.5%
Fear	0.0%	50.0%	37.5%	12.5%
Disgust	0.0%	75.0%	12.5%	12.5%
Surprise	25.0%	75.0%	0.0%	0.0%

2) Expert assessment: since human understand images more clearer in 2D format than in 3D, which appear ghost-like to them. A complete expression data set of all subjects in the database in 2D format (intensity image) was sent to four experts in psychology department. They were requested to identify the expression expressed by each subject in the database. To get a consolidated human

Table 3. Average confusion matrix for human evaluation

	Neutral	Happy	Sad	Angry	Fear	Disgust	Surprise
Neutral	98.8%	0.0%	0.5%	0.7%	0.0%	0.0%	0.0%
Happy	0.0%	99.7%	0.0%	0.0%	0.3%	0.0%	0.0%
Sad	2.1%	0.0%	95.4%	1.8%	0.4%	0.3%	0.0%
Angry	1.2%	0.0%	0.7%	95.1%	1.1%	1.9%	0.0%
Fear	0.6%	0.3%	1.5%	0.2%	97.1%	0.3%	0.0%
Disgust	0.4%	0.0%	0.6%	0.8%	0.4%	97.8%	0.0%
Surprise	0.0%	0.5%	0.0%	0.0%	0.2%	0.0%	99.3%

interpretation the performer assessment was merged with the expert assessment to evaluate the average human interpretation. The mean confusion matrix of human interpretation is shown in table 3.

3) Machine assessment: we carried out an experiment of facial expression recognition on the UPM-3DFE database. Given the 32 feature points we constructed two sets of geometrically feature vectors, the distance feature vectors consisting of 27 distances and angle feature vectors consisting of 29 angle vectors. Fig 4a and 4b depict the distances and angles vectors respectively. These features are further fused together to form an extended feature vector. A maximum relevance minimum redundancy algorithm (mRMR) was invoked to select the most discriminating features among them [12, 13]. The candidate features were then randomly partitioned into two sets, with training set containing 45 subjects while the remaining 5 subjects were used as test set. We then constructed one-against-one multi-class SVM classifier, which is a successfully supervised learning scheme for pattern classification [15]. For N-class classification, the training feature vector set is use in training $\frac{1}{2}N(N-1)$ unique pair SVM classifiers. The test feature vector on the other hand was tested against all the trained SVM models to produce the classification results. The majority voting scheme was finally employed to predict the class of ensemble outputs. To ensure person in-dependency, the intersection between the training and testing sets is always equal to zero, meaning that any subject belonging to the test set will not appear in the training set. The classification result is shown in table 4.

3.1 Discussion

To measure the participants' response to the question of how much effort is required to mimic a particular expression. We evaluated the participants' percentage votes for each expression (table 2) and determined to which side of the scale it shifted. The result was then compared with the machine recognition accuracy of table 4. It is from this comparison that we inferred the relationship between these quantities.

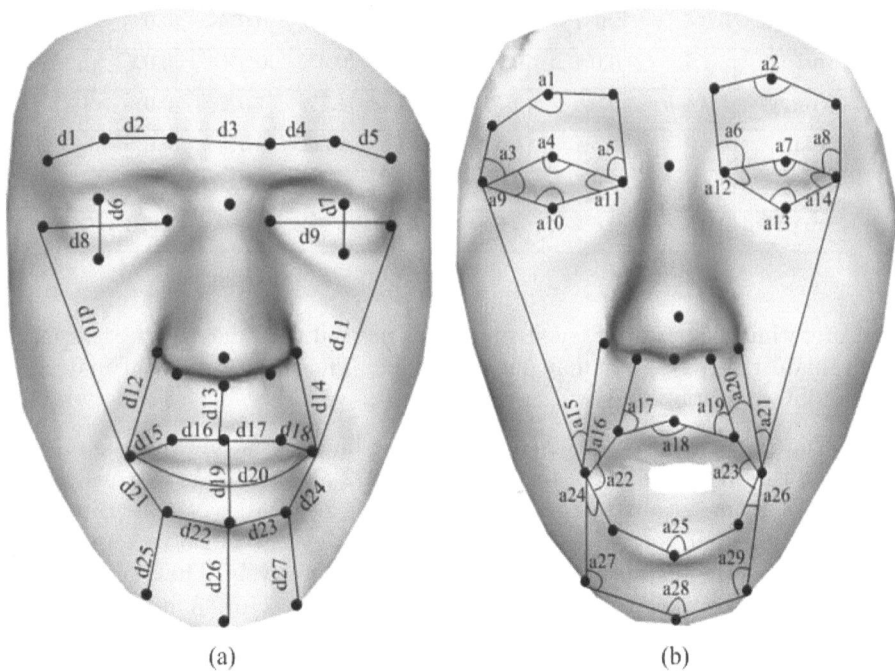

(a)   (b)

Fig. 4. (a) 27 selected distance vectors (d1 to d27), (b) 29 selected angles (a1 to a29) drawn across a face sample

Table 4. Average confusion matrix for machine evaluation

	Neutral	Happy	Sad	Angry	Fear	Disgust	Surprise
Neutral	**90.3%**	0.0%	6.1%	3.6%	0.0%	0.0%	0.0%
Happy	0.0%	**97.1%**	0.0%	0.2%	2.1%	0.6%	0.0%
Sad	6.4%	0.0%	**86.8%**	5.9%	0.9%	0.0%	0.0%
Angry	6.2%	0.0%	3.8%	**86.3%**	1.2%	2.5%	0.0%
Fear	4.6%	1.5%	3.2%	1.8%	**88.9%**	0.0%	0.0%
Disgust	1.9%	0.0%	3.0%	0.0%	0.4%	**94.7%**	0.0%
Surprise	0.0%	0.4%	0.0%	0.0%	0.0%	1.7%	**97.9%**

Fig. 5. Subject posing of fear expression, (but showing combination of fear & happy expression)

1) Neutral: from table 2 the participants' score for this expression is 100% towards the easy side of the scale. This implied that the expression requires less effort to portray. The corresponding machine recognition rate for the expression from table 3 is 90.3%

2) Happy: the percentage score for this expression clearly shifted towards the easy side of the scale, implying that all the participants agreed that happy expression requires less effort to mimic. The corresponding machine recognition rate for the expression is 97.1%

3)Sad: from table 2, the expression recorded 50% on each side of the scale, implying that sad expression is not easy to mimic as compared to neutral or happy expressions. Its machine recognition rate stands at 86.8%

4)Angry: percentage score for this expression is 62.5% towards the easy side of the scale and 37.5% towards the difficult side. Although the weight is more to the easy side, but is not as clear as that of happy or neutral. The machine recognition rate score is 86.3%

5) Fear: the participants' percentage score for fear expression is given as 50% on each side of the scale. The machine recognition rate recorded for the expression is 88.9%

6) Disgust: participant score for disgust is 75% towards the easy side and 25% towards the difficult side. The machine recognition rate is 94.7%

7) Surprise: the participant score for this expression is 100% towards the easy side of the scale; its machine recognition rate reach up to 97.9%

From the above discussion, it can be seen that all expressions with participants' score of 75 % and above toward the easy side of the scale have also recorded a machine recognition rate of 90% and above. This inferred that the lesser the effort required to mimic an expression, the higher is its machine recognition accuracy. On the other hand, expressions that are centred around the mid scale or shifted towards the difficult side of the scale, have their machine recognition rate lower than 90%. This implies that the more the effort required to mimic an expression, the less likely the intended expression is shown correctly. For example the subject in fig. 5 intends to pose the fear expression, but the bulging of his checks and up lift of the mouth corners indicated a blend of happy expression into the intended fear expression. Such failure of mimicking the intended expression correctly due to its difficulty leads to lower machine recognition accuracy.

4   Conclusion

We presented the UPM-3DFE database containing 3D face images of 50 subjects, including 30 males and 20 females. Each subject displayed the six basic expressions, plus the neutral expression. The subjects were drawn from a different ancestral/ethnic background. The database was evaluated using both subjective and objective means. The study also investigated the correlation between machine expression recognition accuracy and effort needed to mimic the expression by human. The result shows negative correlation between expression recognition and effort required to mimic it. It is hope that this database will be helpful in developing researches in human computer interaction, security, psychological studies and biomedical studies.

References

1. Wang, J., Yin, L.: Static topographic modeling for facial expression recognition and analysis. Comput. Vis. Image Und. 108, 19–34 (2007)
2. Xie, X., Lam, K.M.: Facial expression recognition based on shape and texture. Pattern Recogn. 42, 1003–1011 (2009)
3. Kotsia, I., Buciu, I., Pitas, I.: An analysis of facial expression recognition under partial facial image occlusion. Image Vision Comput. 26, 1052–1067 (2008)
4. Li, S.Z., Jain, A.K.C.: The Grid: Handbook of face recognition. Springer, New York (2011)
5. Wang, J., Yin, L., Wei, X., Sun, Y.: 3D facial expression recognition based on primitive surface feature distribution. In: IEEE Computer Society Conference on Computer Vision and Pattern Recognition, pp. 1399–1406. IEEE Press, New York (2006)
6. Savran, A., Alyz, N., Dibekliolu, H., Eliktutan, O., Gkberk, B., Sankur, B., Akarun, L.: Bosphorus database for 3D face analysis. Biometrics and Identity Management, 47–56 (2008)
7. Yin, L., Wei, X., Sun, Y., Wang, J., Rosato, M.J.: A 3D facial expression database for facial behavior research. In: 7th IEEE International Conference on Automatic Face and Gesture Recognition, pp. 211–216. IEEE Press, New York (2006)
8. Hewahi, N.M., Baraka, A.R.M.: Impact of Ethnic Group on Human Emotion Recognition Using Backpropagation Neural Network. BRAIN, Broad Research in Artificial Intelligence and Neuroscience 2, 20 (2012)

9. Ostermann, J.: Face Animation in MPEG-4. In: MPEG-4 Facial Animation, pp. 17–55 (2002)
10. Ekman, P., Friesen, W.V.: Facial action coding system (1977)
11. Darwin, C., Ekman, P., Prodger, P.: The expression of the emotions in man and animals. Oxford University Press, USA (2002)
12. Peng, H., Long, F., Ding, C.: Feature selection based on mutual information criteria of max-dependency, max-relevance, and min-redundancy. In: IEEE Transactions on Pattern Analysis and Machine Intelligence, pp. 1226–1238. IEEE Press, New York (2005)
13. Bonev, B., Escolano, F., Cazorla, M.: Feature selection, mutual information, and the classification of high-dimensional patterns. Pattern Anal. Appl. 11, 309–319 (2008)
14. Farkas, L.G., Munro, I.R.: Anthropometric Facial Proportions in Medicine. Charles C. Thomas, Springfield (1987)
15. Joutsijoki, H., Juhola, M.: Kernel selection in multi-class support vector machines and its consequence to the number of ties in majority voting method. Artif. Intell. Rev., 1–18 (2011)

Probabilistic Reasoning in DL-Lite

Raghav Ramachandran[1], Guilin Qi[2,3], Kewen Wang[1],
Junhu Wang[1], and John Thornton[1]

[1] School of Information and Communication Technology, Griffith University, Australia
[2] School of Computer Science and Engineering, Southeast University, China
[3] State Key Lab for Novel Software Technology, Nanjing University, P.R. China

Abstract. The problem of extending description logics with uncertainty has received significant attention in recent years. In this paper, we investigate a probabilistic extension of DL-Lite, a family of tractable description logics. We first present a new probabilistic semantics for terminological knowledge bases based on the notion of *types*. The semantics proposed is not capable of handling assertional knowledge. In order to reason with both terminological and assertional probabilistic knowledge, we propose a probabilistic semantics based on a finite semantics for DL-Lite called features. This approach enables us to infer new information from the existing knowledge base by drawing on the inherent relation between a *probabilistic TBox* and a *probabilistic ABox*.

Keywords: Description logics, Probabilistic reasoning, Nonmonotonic reasoning, DL-Lite.

1 Introduction

Description logics (DLs) have proved to be a successful formalism for representing and reasoning about knowledge [2]. They provide logical underpinning for Web Ontology Language (OWL). A knowledge base in description logics typically consists of two parts: a *TBox* which consists of terminological knowledge; and an *ABox* which stores instance-level information about the state of the world. A knowledge base in DLs models a domain of interest in terms of concepts and roles, which represent classes of individuals and binary relations on classes of individuals, respectively. There are numerous applications which need working with uncertain knowledge in ontologies [5, 6, 14], and probabilistic extensions of description logics have been developed to represent and reason with uncertainty in terminological knowledge [8–10, 17–21]. These studies are based on different description logics such as $\mathcal{ALC}$, $\mathcal{SHIF}$ and $\mathcal{SHOIN}$ [8, 9, 18].

Probabilistic description logic knowledge bases include *terminological probabilistic knowledge* and *assertional probabilistic knowledge*. The former incorporates probabilistic knowledge about concepts and roles and the latter holds probabilistic knowledge about instances of the same [5, 6]. Let us consider the following example which is presented in [18].

Example 1. Consider the information *'generally a pacemaker patient is a male with probability of at least 0.4'*. This is an example of terminological probabilistic knowledge and can be represented by

P. Anthony, M. Ishizuka, and D. Lukose (Eds.): PRICAI 2012, LNAI 7458, pp. 480–491, 2012.

$$(male|pacemaker)[0.4, 1].$$

An instance of assertional probabilistic knowledge is, *'Tom is a pacemaker patient with a probability of at least 0.6'*. It can be represented by

$$pacemaker(Tom)[0.6, 1].$$

The existing different accounts of probabilistic description logics are generally based on the one proposed in [9]. Some of the existing accounts of probabilistic description logics only focus on terminological knowledge [8, 9] and not assertional knowledge. Probabilistic extensions of assertional knowledge have also been considered [17, 18]. However, existing semantics for probabilistic knowledge bases is not intuitive to represent assertional knowledge. For example, the framework proposed in [17, 18] does not take in to account the relation between an *ABox* and the *TBox* and cannot represent the probability of an assertion of the form $R(a, b)$. Furthermore, existing probabilistic description logics are often based on expressive description logics, thus are not tailored to deal with large scale uncertain data.

In this work, we begin by presenting a simple generalization of the probabilistic framework proposed in [17, 18] in DL-Lite, which is a family of tractable description logics. We present a new probabilistic semantics for the terminological knowledge base based on the notion of *types* [11, 22], where a type is a set of basic concepts. We show that the results obtained in [17, 18] can be obtained in our framework as well, when restricted to knowledge bases with just terminological knowledge bases. We then extend this initial probabilistic framework to present an extended framework that can handle the complete knowledge base, including assertional probabilistic knowledge. We base our study on the alternative semantics given in [22] defined by features. We show that our framework enables us to infer new assertional knowledge from a given probabilistic knowledge base.

The organization of the paper is as follows: In the following section, we present the preliminaries and the syntax of DL-Lite$_{bool}^N$ language. In Section 3, we recall the notions of types and the notion of features. In Section 4, we present a comparison of our initial probabilistic framework with the one proposed in [17, 18]. Next, we present an extended framework based on features in Section 5, followed by a brief comparison with the existing frameworks in Section 6.

2   The DL-Lite Family

We begin by presenting the preliminaries of the DL-Lite$_{bool}^N$ language [1, 3, 4]. A *signature* is a finite set $\mathcal{S} = \mathcal{S}_C \cup \mathcal{S}_R \cup \mathcal{S}_I \cup \mathcal{S}_N$ where $\mathcal{S}_C$ is the set of atomic concepts, $\mathcal{S}_R$ is the set of atomic roles, $\mathcal{S}_I$ is the set of individual names and $\mathcal{S}_N$ is the set of natural numbers in $\mathcal{S}$. $\top$ and $\bot$ will not be considered as atomic concepts or atomic roles. Formally, given a signature $\mathcal{S}$, a DL-lite$_{bool}^N$ language has the following syntax:

$$R \leftarrow P|P^-,$$
$$B \leftarrow \top|\bot|A| \geq nR,$$
$$C \leftarrow B|\neg C|C_1 \sqcap C_2,$$

where $n \in \mathcal{S}_N$, $A \in \mathcal{S}_C$ and $P \in \mathcal{S}_R$.

A *TBox* $\mathcal{T}$ is a finite set of concept subsumptions of the form $C \sqsubseteq D$, where C and D are general concepts. An *ABox* $\mathcal{A}$ is a finite set of membership assertions of the form $C(a)$, $R(a,b)$, where a, b are individual names. We call $C(a)$ a membership assertion and $R(a,b)$ a role assertion.

The semantics of a DL-Lite$_{bool}^{N}$ is given by interpretations. An interpretation I is a pair $(\Delta^I, \cdot^I)$, where Δ^I is a non-empty set called the domain, and $\cdot^I$ is an interpretation function that associates each individual name a with an element $a^I \in \Delta^I$, each atomic role P with a subset of $\Delta^I \times \Delta^I$ and each atomic concept A with a subset of Δ^I. The interpretation function $\cdot^I$ can be extended to general concept and role descriptions, P^-, $\geq nR$ and $\neg C$.

An interpretation I satisfies

- an inclusion axiom $C \sqsubseteq D$ if $C^I \subseteq D^I$,
- a membership assertion $C(a)$ if $a^I \in C^I$,
- a role assertion $R(a,b)$ if $(a^I, b^I) \in R^I$.

The interpretation I satisfies a *TBox* $\mathcal{T}$ if it satisfies every subsumption axiom in $\mathcal{T}$ and I satisfies an *ABox* $\mathcal{A}$ if it satisfies every membership assertion and every role assertion in $\mathcal{A}$. Interpretation I satisfies a classical knowledge base $\mathcal{K} = (\mathcal{T}, \mathcal{A})$, if it satisfies both $\mathcal{T}$ and $\mathcal{A}$. Such an interpretation is termed a model of the knowledge base.

3 Background

There are certain drawbacks when working with interpretations for a description logic knowledge base. One of them being that a description logic knowledge base can have infinitely many models. Another being that given a set of interpretations there may not exist any knowledge base whose set of models is exactly the given set of interpretations. In order to avoid these drawbacks, alternative semantics have been proposed for a description logic knowledge base based on *features* [22]. A feature can be seen a simplification of a standard interpretation and at the same time it is a Herbrand interpretation equipped with limited structure. In this section, we recall the notion of features. Features for DL-Lite$_{bool}^{N}$ are in turn based on the notion of types, defined in [11].

Given a signature $\mathcal{S}$, a type is defined as follows:

Definition 1. *An $\mathcal{S}$-type τ is a set of basic concepts over $\mathcal{S}$, such that*

1. $\top \in \tau$,
2. $\bot \notin \tau$, and
3. *for any $m, n \in \mathcal{S_N}$ with $m < n$, $\geq nR \in \tau$ implies $\geq mR \in \tau$.*

The set of all types, denoted by Θ, is finite, since the signature $\mathcal{S}$ is finite.

Such a definition of types can handle the semantics of the terminological knowledge. A type τ is said to satisfy a basic concept B iff $B \in \tau$, and is denoted by $\tau \models B$. This is extended to all terminological knowledge as follows:

- $\tau \models \neg C$ iff $\tau \not\models C$,
- $\tau \models C \sqcap D$ iff $\tau \models C$ and $\tau \models D$, and
- $\tau \models C \sqsubseteq D$ iff $\tau \models \neg C$ or $\tau \models D$.

A type τ is said to satisfy a set of concept subsumptions $\mathcal{C}$ if it satisfies each element in $\mathcal{C}$. Given a *TBox* $\mathcal{T}$, a type satisfies $\mathcal{T}$ when it satisfies every element in $\mathcal{T}$. Thus while types provide a semantic characterization for DL-Lite$_{bool}^N$ *TBox*, they cannot handle assertional knowledge. Types are extended to the notion of features in order to provide an alternative semantic characterization for the complete DL-Lite$_{bool}^N$ knowledge base.

A feature is a collection of a set of types and a Herbrand set of assertional knowledge. A Herbrand set of assertional knowledge is defined as follows [22]:

Definition 2. *A Herbrand set $\mathcal{H}$ is a finite set of assertions of the form $B(a)$ or $P(a,b)$ where $a, b \in \mathcal{S}_I$, $P \in \mathcal{S}_R$ and B is a basic concept over $\mathcal{S}$, satisfying the following conditions:*

1. *For each $a \in \mathcal{S}_I$, $\top(a) \in \mathcal{H}$, $\bot(a) \notin \mathcal{H}$, and $\geq nR(a) \in \mathcal{H}$ implies $\geq mR(a) \in \mathcal{H}$ for $m, n \in \mathcal{S}_N$ with $m < n$.*
2. *For each $P \in \mathcal{S}_R$, $P(a, b_i) \in \mathcal{H}$ $(i = 1, \ldots, n)$ implies $\geq mP(a) \in \mathcal{H}$ for any $m \in \mathcal{S}_N$ such that $m \leq n$.*
3. *For each $P \in \mathcal{S}_R$, $P(b_i, a) \in \mathcal{H}$ $(i = 1, \ldots, n)$ implies $\geq mP^-(a) \in \mathcal{H}$ for any $m \in \mathcal{S}_N$ such that $m \leq n$.*

The intuition behind the above conditions is quite clear. For further discussions on these conditions, the readers are referred to [22]. Given a Herbrand set $\mathcal{H}$, let the set of all concept assertions of a in $\mathcal{H}$ be $\{B_1(a), \ldots, B_k(a)\}$. Then the set of concepts $\{B_1, \ldots, B_k\}$ is a type and is denoted by τ_a. It is termed as *type of a in $\mathcal{H}$*.

Definition 3. *Given a signature $\mathcal{S}$, a feature is defined as a pair $\mathcal{F} = \langle \varXi, \mathcal{H} \rangle$, where $\varXi$ is a non-empty set of types and $\mathcal{H}$ a Herbrand set, satisfying the following conditions:*

1. *$\exists P \in \bigcup \varXi$ iff $\exists P^- \in \bigcup \varXi$, for each $P \in \mathcal{S}_R$.*
2. *$\tau_a \in \varXi$, for each $a \in \mathcal{S}_I$.*

A Herbrand set $\mathcal{H}$ is said to satisfy a concept assertion $C(a)$ if the type τ_a satisfies the concept C. We say the $\mathcal{H}$ satisfies a role assertion $P(a, b)$ or $P^-(b, a)$ if $P(a, b)$ is in $\mathcal{H}$. We extend this to say a Herbrand set $\mathcal{H}$ satisfies an *ABox* $\mathcal{A}$ if $\mathcal{H}$ satisfies every assertion in $\mathcal{A}$. A feature $\mathcal{F} = \langle \varXi, \mathcal{H} \rangle$ is said to satisfy a membership assertion $C(a)$, if and only if the Herbrand set $\mathcal{H}$ satisfies $C(a)$. Similarly, we say $\mathcal{F}$ satisfies a role assertion $P(a, b)$ if and only if $\mathcal{H}$ satisfies $P(a, b)$. Also, we say $\mathcal{F}$ satisfies an inclusion axiom $C \sqsubseteq D$, if and only if every type in $\varXi$ satisfies the inclusion axiom.

A feature $\mathcal{F}$ is termed a model or a model feature of the classical knowledge base if and only if it satisfies every inclusion axiom in the *TBox* and every assertion in the *ABox*. The set of all model features of a membership assertion $C(a)$ is denoted by $MF(C(a))$, model features of an inclusion axiom $C \sqsubseteq D$ is denoted by $MF(C \sqsubseteq D)$ and the model features of a knowledge base $\mathcal{K}$ is denoted by $MF(\mathcal{K})$.

4 A Probabilistic Framework Based on Types

We make use of the medical knowledge base given in [18] to illustrate a probabilistic knowledge base.

Example 2. *Consider medical records of patients related to cardiological illnesses. The knowledge base distinguishes between heart patients, patients with pacemaker, by the patient's gender, illnesses, illness levels and symptoms. The patients whose records are present in the knowledge base are Tom, Maria and John. Tom is a heart patient, while John and Maria are male and female pacemaker patients.*

A subsumption axiom is of the form "generally heart patients suffer from high blood pressure" and "generally pacemaker patients do not suffer from high blood pressure". These are represented as

- *$heartpatient \sqsubseteq highbloodpressure$ and*
- *$pacemaker \sqsubseteq \neg highbloodpressure$.*

The probabilistic terminological knowledge is of the form, "generally a pacemaker patient is a male with probability of at least 0.4", represented as $(male|pacemaker)$ $[0.4, 1]$. Assertional probabilistic knowledge is of the form, "Tom is a pacemaker patient with a probability of at least 0.6", represented as $pacemaker(Tom)[0.6, 1]$.

Formally, a probabilistic terminological box $\mathcal{PT}$ is a finite set of conditional constraints of the form $(C|D)[l, u]$ where C and D are general concepts and $0 \leq l \leq u \leq 1$. Probabilistic assertional knowledge is given by formulae of the form $C(a)[l, u]$ and $R(a, b)[l, u]$ with $0 \leq l \leq u \leq 1$. Assertional knowledge $C(a)$ is represented in terms of probabilities as $C(a)[1, 1]$ and similarly $R(a, b)$ is represented as $R(a, b)[1, 1]$. The set of all probabilistic assertional knowledge is termed probabilistic assertional box, $\mathcal{PA}$. A probabilistic knowledge base $\mathcal{K}$ is a triple $\mathcal{K} = (\mathcal{T}, \mathcal{PT}, \mathcal{PA})$. In the following discussion, we only consider the terminological probabilistic knowledge, i.e., the knowledge base is a pair $(\mathcal{T}, \mathcal{PT})$.

Let the set of all types be denoted by Θ. A probabilistic interpretation Pr is a probability function on Θ, i.e., $Pr : \Theta \longrightarrow [0, 1]$ such that $Pr(\tau)$ for all $\tau \in \Theta$ sums up to 1. The probability of a concept C under the interpretation Pr is the sum of the probabilities of all types that satisfy C, i.e.,

$$Pr(C) = \sum_{\tau \in \Theta, \, \tau \models C} Pr(\tau).$$

For concepts C, D, with $Pr(D) > 0$, we define $Pr(C|D)$ as $Pr(C \sqcap D)/Pr(D)$. A probabilistic interpretation Pr is said to satisfy a conditional constraint $(C|D)[l, u]$ iff $Pr(D) = 0$ or $Pr(C|D) \in [l, u]$. We say a probabilistic interpretation Pr satisfies a set of conditional constraints $\mathcal{R}$ iff $Pr \models F$ for all $F \in \mathcal{R}$. We say a probabilistic interpretation Pr satisfies a subsumption axiom $C \sqsubseteq D$ if and only if the subsumption axiom is satisfied by every type with non-zero probability, denoted by $Pr \models C \sqsubseteq D$. In other words, Pr satisfies $C \sqsubseteq D$ when every type τ with $Pr(\tau) > 0$ satisfies the same. A probabilistic interpretation is said to satisfy a set of concept subsumptions when it satisfies every element in the set. We denote Pr satisfies X, by $Pr \models X$, where X could be a concept subsumption or a conditional constraint or a set of either. Such an interpretation Pr is said to be a model of X. A *TBox* $\mathcal{T}$ is said to be *satisfiable* when there exists a model for $\mathcal{T}$.

Consistency in a probabilistic knowledge base has also been defined based on consistency in probabilistic default reasoning [15, 16]. This is based on the assumption that the axioms in the deterministic *TBox* ($\mathcal{T}$) and probabilistic *TBox* ($\mathcal{PT}$) are assigned different status, in that, the axioms in $\mathcal{T}$ need necessarily be satisfied by a model of the knowledge base, while it is not necessary for the axioms in $\mathcal{PT}$ to be satisfied by a model of the knowledge base. The elements in $\mathcal{PT}$ are considered as default knowledge while the elements in $\mathcal{T}$ are taken as strict knowledge. Default reasoning from conditional knowledge bases employs the notions of verification, falsification and toleration [7], which in probabilistic description logic can be defined as follows: We say a probabilistic interpretation Pr *verifies* a conditional constraint $(C|D)[l, u]$ iff $Pr(D) = 1$ and $Pr(C) \in [l, u]$. A probabilistic interpretation Pr *falsifies* a conditional constraint $(C|D)[l, u]$ iff $Pr(D) = 1$ and $Pr(C) \notin [l, u]$. Given a set of conditional constraints $\mathcal{R}$, $\mathcal{R}$ is said to *tolerate* a conditional constraint $(C|D)[l, u]$ under $\mathcal{T}$ iff $\mathcal{T} \cup \mathcal{R}$ has a model which verifies $(C|D)[l, u]$.

A knowledge base $(\mathcal{T}, \mathcal{PT})$ is *consistent* iff (i) $\mathcal{T}$ is satisfiable, and (ii) there exists an ordered partition $(P_0, \ldots, P_k)$ of $\mathcal{PT}$ such that each P_i with $i \in \{0, \ldots, k\}$ is the set of all conditional constraints in $P \backslash (P_0 \cup \ldots \cup P_{i-1})$ that are tolerated under $\mathcal{T}$ by $P \backslash (P_0 \cup \ldots \cup P_{i-1})$.

The definition of *consistency* is aimed at resolving inconsistencies that could arise within $\mathcal{PT}$ as a whole. The possible inconsistencies could only occur between two different partitions. As it is an ordered partition, the inconsistency can be resolved by preferring the highly ranked partition over the others. This partitioning of $\mathcal{PT}$ also presents a tool to build an ordering among the set of all possible probabilistic interpretations Pr.

Lexicographic entailment [15, 16], which is based on Lehmann's lexicographic entailment [13], is the reasoning formalism employed in [18]. Given a partition of $\mathcal{PT}$ it is possible to order the set of all possible probabilistic interpretations as follows [17, 18]: A probabilistic interpretation Pr is *more preferable* (or *lex-preferable* or *lexicographically preferable*) to an interpretation Pr' iff some $i \in \{0, \ldots, k\}$ exists such that the number of elements in P_i satisfied by Pr is more than the number satisfied by Pr' and for every $i < j \leq k$ both Pr and Pr' satisfy same number of elements in P_j.

An interpretation Pr that satisfies $\mathcal{T}$ and a set of conditional constraints $\mathcal{R}$ is a *lexicographically minimal model* of $\mathcal{T} \cup \mathcal{R}$ iff no model of $\mathcal{T} \cup \mathcal{R}$ is lex-preferable to Pr. A sentence $(C|D)[l, u]$ is said to be a lexicographic consequence of $\mathcal{T} \cup \mathcal{R}$ iff it is satisfied by every lexicographically minimal model of $\mathcal{T} \cup \mathcal{R}$.

The following result shows that lexicographic entailment satisfies the general nonmonotonic properties, such as the postulates of *Right weakening*, *Reflexivity*, *Left logical equivalence*, *Cut* and *Cautious monotonicity* as studied in [12].

Theorem 1. *Let the knowledge base* $(\mathcal{T}, \mathcal{PT})$ *be a p-consistent knowledge base,* C, D, C', D' *be concepts and* $l, u, l', u' \in [0, 1]$. *Then,*

RW If $(C|\top)[l, u] \Rightarrow (C'|\top)[l', u']$ *is logically valid and* $(\mathcal{T}, \mathcal{PT}) \vdash_{lex} (C|D)[l, u]$, *then* $(\mathcal{T}, \mathcal{PT}) \vdash_{lex} (C'|D)[l', u']$.

Ref $(\mathcal{T}, \mathcal{PT}) \vdash_{lex} (C|C)[1, 1]$.

LLE If $C \Leftrightarrow C'$ *is logically valid, then* $(\mathcal{T}, \mathcal{PT}) \vdash_{lex} (D|C)[l, u]$ *iff* $(\mathcal{T}, \mathcal{PT}) \vdash_{lex} (D|C')[l, u]$.

Cut If $(\mathcal{T}, \mathcal{PT}) \vdash_{lex} (C|C')[1, 1]$ *and* $(\mathcal{T}, \mathcal{PT}) \vdash_{lex} (D|C \wedge C')[l, u]$, *then* $(\mathcal{T}, \mathcal{PT}) \vdash_{lex} (D|C')[l, u]$.

CM If $(\mathcal{T}, \mathcal{PT}) \vdash_{lex} (C|C')[1, 1]$ *and* $(\mathcal{T}, \mathcal{PT}) \vdash_{lex} (D|C')[l, u]$, *then* $(\mathcal{T}, \mathcal{PT}) \vdash_{lex} (D|C \wedge C')[l, u]$.

Proof: Majority of this proof follows similar to the one provided in [17, 18].

RW Assume that $(\mathcal{T}, \mathcal{PT}) \vdash_{lex} (C|D)[l, u]$. Since $(C|\top)[l, u] \Rightarrow (C'|\top)[l', u']$ is logically valid, we have $Pr \models (C'|D)[l', u']$ for every lexicographically minimal model of $(\mathcal{T}, \mathcal{PT})$. Hence $(\mathcal{T}, \mathcal{PT}) \vdash_{lex} (C'|D)[l', u']$.

Ref Given an interpretation Pr, we have $Pr \models (C|C)[1, 1]$. Hence $(C|C)[1, 1]$ is satisfied by every lexicographically minimal model of $(\mathcal{T}, \mathcal{PT})$.

LLE Suppose that $C \Leftrightarrow C'$ is logically valid. Assume that $(\mathcal{T}, \mathcal{PT}) \vdash_{lex} (D|C)[l, u]$. Since $C \Leftrightarrow C'$ is logically valid, we have $Pr \models (D|C')[l, u]$ for every lexicographically minimal model of $(\mathcal{T}, \mathcal{PT})$. Hence, $(\mathcal{T}, \mathcal{PT}) \vdash_{lex} (D|C')[l, u]$. Similarly the converse can be proven.

Cut Let $(\mathcal{T}, \mathcal{PT}) \vdash_{lex} (C|C')[1, 1]$ and $(\mathcal{T}, \mathcal{PT}) \vdash_{lex} (D|C \wedge C')[l, u]$. Therefore, for every lexicographically minimal model Pr of $(\mathcal{T}, \mathcal{PT})$ we have $Pr \models (D|C')[l, u]$, that is, $(\mathcal{T}, \mathcal{PT}) \vdash_{lex} (D|C')[l, u]$.

CM Let $(\mathcal{T}, \mathcal{PT}) \vdash_{lex} (C|C')[1, 1]$ and $(\mathcal{T}, \mathcal{PT}) \vdash_{lex} (D|C')[l, u]$. Hence, for every lexicographically minimal model Pr of $(\mathcal{T}, \mathcal{PT})$, $Pr \models (D|C \wedge C')[l, u]$. Therefore, we have $(\mathcal{T}, \mathcal{PT}) \vdash_{lex} (D|C \wedge C')[l, u]$.

∎

It must be noted that the probabilistic framework in [18] is based on probability functions on the set of all possible worlds, where a possible world is defined as a set of *basic c-concepts*. Thus a possible world is a special case of types in DL-Lite. Thus our framework which is based on probability functions on the set of all types, that are sets of concepts not restricted to just *basic c-concepts*, generalizes the approach in [18]. However, a framework purely based on types cannot take in to account the assertional probabilistic knowledge. In order to study a more complete probabilistic knowledge base in DL-Lite$_{bool}^{N}$, i.e., one that includes assertional probabilistic knowledge, we extend the probabilistic framework given above to one based on features instead of types.

5 Extended Framework

Features, as given in Section 3, are pairs of sets of types and a Herbrand set of membership and role assertions. Let $\mathcal{F}$ be a feature, then $\mathcal{F} = \langle \varXi, \mathcal{H} \rangle$, where $\varXi$ is a set of types and $\mathcal{H}$ is a Herbrand set. The relation between $\mathcal{H}$ and $\varXi$ is as given in Definition 3.

We define the new semantics for a probabilistic description logic based on features as follows: When the signature $\mathcal{S}$ is finite, the set of all features is also finite. A probabilistic interpretation Pr is a probability function on the set of all features such that $Pr(\mathcal{F})$ is non-negative for every $\mathcal{F}$ and sums up to 1. The probability associated with any concept C is given by the sum of the probabilities associated its model features. For concepts C, D with $Pr(D) > 0$, we define $Pr(C|D)$ as $Pr(C \cup D)/Pr(D)$. Probabilities associated with any membership assertion $Pr(C(a))$ or role assertion $Pr(P(a, b))$ is given by the sum of the probabilities associated with the model features of $C(a)$ and $P(a, b)$, respectively.

We say that a probabilistic interpretation Pr satisfies an inclusion axiom $C \sqsubseteq D$ if and only if every feature $\mathcal{F}$ with non-zero probability satisfies $C \sqsubseteq D$. A probabilistic interpretation is said to satisfy a formula of the form $(C|D)[l, u]$ when $Pr(D) = 0$ or $Pr(C|D) \in [l, u]$. Similarly Pr is said to satisfy an assertion of the form $C(a)[l, u]$ or $P(a, b)[l', u']$ when probabilities of $C(a)$ and $P(a, b)$ under Pr lie in the intervals $[l, u]$ and $[l', u']$, respectively. When a probabilistic interpretation Pr satisfies any formula X, it is denoted by $Pr \models X$. By extending these satisfiability properties, we have that a probabilistic interpretation satisfies a knowledge base $\langle \mathcal{T}, \mathcal{PT}, \mathcal{A} \rangle$ when it satisfies all the elements in $\mathcal{T}, \mathcal{PT}$ and $\mathcal{A}$. Such a probabilistic interpretation Pr is termed a *model* of the knowledge base $\mathcal{K}$.

Given a knowledge base, one of the main tasks is to infer knowledge from it. However, it is necessary for a knowledge base to be consistent in order to infer reasonable information from it. Since when the knowledge base is inconsistent it is possible to infer unreasonable or at times inconsistent information from the same. Therefore it is necessary to define the notion of consistency of a knowledge base. A knowledge base $\mathcal{K}$ is said to be consistent if and only if it has at least one model. The knowledge base is said to be inconsistent otherwise. A formula X is said to be *entailed* by a given knowledge base $\mathcal{K}$ when every model of $\mathcal{K}$ satisfies X.

Given a consistent knowledge base, the following results reflect the inferences that can be made in our framework. By definition of a feature, there is an intuitive relation between the set of types $\varXi$ and the assertional knowledge in a feature. Our framework ensures that this relation is maintained when incorporating probabilities. This relation aids us in our reasoning tasks. Let C, D be concepts such that C is subsumed by D. Suppose that $C(a)$ is a membership assertion related to C for any individual a, then by understanding of subsumption, it is expected that $D(a)$ is entailed by the knowledge base. When the probabilistic assertional knowledge $C(a)[l, u]$ is present in the knowledge base, then Theorem 2 shows that the corresponding probabilistic assertional knowledge that should be entailed is $D(a)[l, 1]$.

Theorem 2. *Let* $\mathcal{K} = \langle \mathcal{T}, \mathcal{PT}, \mathcal{A} \rangle$ *be a consistent knowledge base such that* $C \sqsubseteq D \in \mathcal{T}$ *and* $C(a)[l, u] \in \mathcal{A}$. *Then* $\mathcal{K} \models D(a)[l, 1]$.

Proof: Let Pr be an arbitrary model of the knowledge base $\mathcal{K}$. Then $Pr \models C \sqsubseteq D$ and $Pr \models C(a)[l, u]$. Let $Pr(C(a)) = x$. Therefore, there exists a feature $\mathcal{F} = \langle \varXi, \mathcal{H} \rangle$ such that $C(a) \in \mathcal{H}$ and $Pr(\mathcal{F}) > 0$. From the definition, we know that the sum of probabilities associated with such $\mathcal{F}$ equals to x. Since $C(a) \in \mathcal{H}$, we have $C \in \tau$, where τ is the *type of* a in $\mathcal{H}$ and therefore $\tau \in \varXi$. From $Pr \models C \sqsubseteq D$, we have $\mathcal{F}' \models C \sqsubseteq D$ for every $\mathcal{F}'$ such that $Pr(\mathcal{F}') > 0$. Therefore, $\mathcal{F} \models C \sqsubseteq D$ which implies that $\tau' \models C \sqsubseteq D$ for every $\tau' \in \varXi$. Hence we have $\tau \models C \sqsubseteq D$, which implies that $D \in \tau$, and therefore $D(a) \in \mathcal{H}$. Thus $\mathcal{F} \models D(a)$, whenever $\mathcal{F} \models C(a)$. Therefore $Pr(D(a)) \geq x$. ∎

Example 3. *Consider the medical ontology given in Example 1. We have that "generally heart patients have high blood pressure". Maria's illness symptoms suggest that she is a heart patient with a probability of at least 0.4, but not more than 0.8. Thus we have the following:*

- *heartpatient* $\sqsubseteq$ *highbloodpressure and*
- *heartpatient*$(Maria)[0.4, 0.8]$.

Then, from Theorem 2, we can infer that Maria has high blood pressure with a probability of at least 0.4, i.e., $highbloodpressure(Maria)[0.4, 1]$.

Let $C(a)$ be a membership assertion in the knowledge base. When given a terminological probabilistic knowledge with respect to concepts C and D, i.e., $(D|C)[l, u]$, then one should expect to infer some information regarding the probabilistic assertional knowledge relating to $D(a)$. This is given by the following theorem.

Theorem 3. *Let $\mathcal{K} = \langle \mathcal{T}, \mathcal{PT}, \mathcal{A} \rangle$ be a consistent knowledge base such that $C(a)[1, 1] \in \mathcal{A}$ and $(D|C)[l, u] \in \mathcal{PT}$. Suppose $\mathcal{K} \models (C|\top)[r, s]$ then $\mathcal{K} \models D(a)[l', u']$ where $l' = r \cdot l$ and $u' = u \cdot s$.*

Proof: Let $\mathcal{K} = (T, PT, A)$ be a knowledge base such that $C(a)[1, 1] \in A$ and $(D|C)[l, u] \in PT$. Suppose Pr be a model of $\mathcal{K}$, hence $Pr \models (D|C)[l, u]$. We assume that $Pr(C \wedge D)/Pr(C) = x$. That is, there exists a feature $\mathcal{F} = \langle \Xi, \mathcal{H} \rangle$ with $Pr(\mathcal{F}) > 0$ such that $\mathcal{F} \models C \wedge D$. Since $Pr(C(a)) = 1$, we have $\mathcal{F} \models C(a)$. Therefore $C \in \tau$, where τ is the *type of* a in $\mathcal{H}$. Since $\mathcal{F} \models C \wedge D$, we have $\tau \models C \wedge D$ and $D \in \tau$. Therefore $D(a) \in \mathcal{H}$, which implies $\mathcal{F} \models D(a)$. Thus we have $Pr(D(a)) = Pr(C \wedge D) = x \cdot Pr(C)$. Hence $\mathcal{K} \models D(a)[l', u$, where $l' = r \cdot l$ and $u' = u \cdot s$. ∎

Example 4. *Continuing our medical ontology example, John is a pacemaker patient. At least 40% of the patients in the database are pacemaker patients. The following probabilistic knowledge is also present in the knowledge base, " Pacemaker patients have a private health insurance with a probability of at least 0.9". Thus we have:*

- *$pacemakerJohn[1, 1]$,*
- *$(pacemaker|\top)[0.4, 1]$, and*
- *$(\exists HasHealthInsurance.PrivateHealthInsurance|pacemaker)[0.9, 1]$.*

Then, from Theorem 3, we can infer that John has a private health insurance with a probability of at least 0.36.

The above two results, help is to infer new assertional probabilistic knowledge from the knowledge base. The following result presents inference of new terminological probabilistic knowledge. Let C, D and E be three concepts. Given that D subsumes C and the probabilistic knowledge $(C|E)[l, u]$, from the following theorem, we can infer about $(D|E)$.

Theorem 4. *Given a consistent knowledge base $\mathcal{K} = \langle \mathcal{T}, \mathcal{PT}, \mathcal{A} \rangle$ with $C \sqsubseteq D \in \mathcal{T}$ and $(C|E)[l, u] \in \mathcal{PT}$. Then $\mathcal{K} \models (D|E)[l, 1]$.*

Proof: Let $\mathcal{K} = (T, PT, A)$ be a knowledge base. Assume that $C \sqsubseteq D \in T$ and $(C|E)[l, u] \in PT$. Suppose Pr is a model of the knowledge base and $Pr(C|E) = x \in [l, u]$. That is, $Pr(C \wedge E) = xPr(E)$. Let $\mathcal{F}$ be an arbitrary feature with $Pr(\mathcal{F}) > 0$, such that $\mathcal{F} \models C \wedge E$. From $Pr \models C \sqsubseteq D$, we have $\mathcal{F} \models C \sqsubseteq D$. Since $\mathcal{F} \models C \wedge E$ implies $\mathcal{F} \models C$ and hence $\mathcal{F} \models D$. Therefore, every feature with positive probability that satisfies $C \wedge E$, also satisfies $D \wedge E$. Hence $Pr(D \wedge E) \geq Pr(C \wedge E)$ for every model of $\mathcal{K}$. Thus we have $\mathcal{K} \models (D|E)[l, 1]$. ∎

Example 5. *(Medical ontology continued.) All the male pacemaker patients are pace-maker patients. The knowledge base also contains the information that heart patients are male with a probability of at least 0.4 and less than 0.8. Thus we have,*

- *malepacemakerpatient ⊑ pacemaker, and*
- *(malepacemakerpatient|heartpatient)[0.4, 0.8].*

Then, from Theorem 4, we can infer that heart patients are pacemaker patients with a probability of at least 0.4, i.e., (pacemaker|heartpatient)[0.4, 1].

These are but a few instances of the inferences that can be made from a given proba-bilistic description logic knowledge base. These results show that by using probabilities based on the set of all features our framework captures the intuitive relation between ter-minological and assertional knowledge. This enables us to infer new information from a given knowledge base.

6   Comparison with the Existing Frameworks

In the preceding sections, we have proposed a fresh look at probabilistic description logics based on features. Our framework has certain positives when compared with some of the existing frameworks [8–10, 17–20].

Heinsohn's framework for probabilistic description logic [8] is based on a fairly fundamental understanding of probability. The probability associated with each concept is defined in terms of the cardinality of the set of individuals that belong to the concept. While in the framework proposed by Jaeger [9], the probability distribution is over a set of concept descriptions. Moreover, Jaeger's framework assumes an asymmetry between the deterministic terminological axioms and deterministic assertional axioms. While an assertional axiom of the form $C(a)$ is replaced by a probabilistic assertion $P(a \in C) = 1$, the axiom $C \sqsubseteq D$ is not considered equivalent to $P(D|C) = 1$. It is evident that there is no such asymmetry in our framework. In our framework, it turns out that whenever a deterministic axiom $C \sqsubseteq D$ belongs to the *Tbox*, we can infer $P(D|C) = 1$ from the knowledge base.

In [18] it is assumed that the set of individuals, or objects, can be partitioned into two disjoint subsets (i) probabilistic individuals and (ii) classical individuals. Probabilistic individuals are those for which there are probabilistic assertional information stored in the knowledge base. However such an assumption is not required for our framework. There are a few similarities between the two frameworks as well. For instance, in [18] the probability distribution is over a sets of basic c-concepts and in our framework the distribution is considered to be over set of all types. Hence our work is a generalization of the framework in [18] when considering only the terminological knowledge.

Also the treatment of assertional probabilistic knowledge in the existing frameworks is not very intuitive. They assume a separate probability function with respect to each individual [9, 18]. Thus a probabilistic interpretation function is a family of probability functions corresponding to the set of individuals and one for the terminological knowl-edge. The relation between these probability functions is also not very clear in these frameworks. Our framework offers a simpler and elegant alternative to this situation. This enables us to make new inferences from the knowledge base.

7   Conclusion

In this work, we began by proposing a new semantics for probabilistic terminological logics based on probability functions over *types*. Such a framework generalizes the framework presented in [17, 18], when restricted to discussion only about terminological knowledge in the framework of DL-Lite. We have proposed another semantics for a complete probabilistic DL-Lite knowledge base based on *features*. The semantics based on features empowers us to make interesting inferences from the existing knowledge base. It must be noted that we do not present all possible inference results and that more results can be obtained in this framework. We aim to study such results in the future, along with investigating the computational complexity of these reasoning tasks.

Acknowledgements. This work was supported by the Australia Research Council (ARC) Discovery grants DP1093652 and DP110101042. Guilin Qi is supported by Excellent Youth Scholars Program of Southeast University and Doctoral Discipline Foundation for Young Teachers in the Higher Education Institutions of Ministry of Education (20100092120029).

References

1. Artale, A., Calvanese, D., Kontchakov, R., Zakharyaschev, M.: DL-lite in the light of first-order logic. In: Proceedings of the 22nd AAAI Conference on Artificial Intelligence, AAAI 2007, pp. 361–366 (2007)
2. Baader, F., Calvanese, D., McGuinness, D.L., Nardi, D., Patel-Schneider, P.F. (eds.): The Description Logic Handbook: Theory, Implementation, and Applications. Cambridge University Press (2003)
3. Calvanese, D., De Giacomo, G., Lembo, D., Lenzerini, M., Rosati, R.: DL-lite: Tractable description logics for ontologies. In: Proceedings of the 20th AAAI Conference on Artificial Intelligence, AAAI 2005, pp. 602–607 (2005)
4. Calvanese, D., De Giacomo, G., Lembo, D., Lenzerini, M., Rosati, R.: Tractable reasoning and efficient query answering in description logics: The DL-lite family. Journal of Automated Reasoning 39(3), 385–429 (2007)
5. Paulo, C.G.: Costa. Bayesian Semantics for the Semantic Web. PhD Thesis, George Mason University, Fairfax, VA, USA (2005)
6. Costa, P.C.G., Laskey, K.B.: Pr-owl: A framework for probabilistic ontologies. In: Proceedings of the 4th Conference on Formal Ontologies in Information Systems, FOIS 2006, pp. 237–249 (2006)
7. Goldszmidt, M., Pearl, J.: On the consistency of defeasible databases. Artificial Intelligence 52, 121–149 (1991)
8. Heinsohn, J.: Probabilistic description logics. In: Proceedings of the 10th Conference on Uncertainty in Artificial Intelligence, UAI 1994, pp. 311–318 (1994)
9. Jaeger, M.: Probabilistic reasoning in terminological logics. In: Proceedings of the 5th International Conference on Principles of Knowledge Representation and Reasoning, KR 1994, pp. 305–316 (1994)
10. Koller, D., Levy, A., Pfeffer, A.: P-classic: A tractable probabilistic description logic. In: Proceedings of the 14th AAAI Conference on Artificial Intelligence, AAAI 1997, pp. 390–397 (1997)

11. Kontchakov, R., Wolter, F., Zakharyaschev, M.: Can you tell the difference between DL-lite ontologies? In: Proceedings of the 11th International Conference on Principles of Knowledge Representation and Reasoningm, KR 2008, pp. 285–295 (2008)
12. Kraus, S., Lehmann, D., Magidor, M.: Nonmonotonic reasoning, preferential models and cumulative logics. Artificial Intelligence 44(1-2), 167–207 (1990)
13. Lehmann, D.J.: Another perspective on default reasoning. Annals of Mathematics and Artificial Intelligence 15(1), 61–82 (1995)
14. Liu, H., Lutz, C., Miličić, M., Wolter, F.: Foundations of instance level updates in expressive description logics. Artificial Intelligence 175(18), 2170–2197 (2011)
15. Lukasiewicz, T.: Probabilistic logic programming under inheritance with overriding. In: Proceedings of the 17th Conference on Uncertainty in Artificial Intelligence, UAI 2001, pp. 329–336 (2001)
16. Lukasiewicz, T.: Probabilistic default reasoning with conditional constraints. Annals of Mathematics and Artificial Intelligence 34(1-3), 35–88 (2002)
17. Lukasiewicz, T.: Probabilistic Lexicographic Entailment under Variable-Strength Inheritance with Overriding. In: Nielsen, T.D., Zhang, N.L. (eds.) ECSQARU 2003. LNCS (LNAI), vol. 2711, pp. 576–587. Springer, Heidelberg (2003)
18. Lukasiewicz, T.: Expressive probabilistic description logics. Artificial Intelligence 172(6-7), 852–883 (2008)
19. Lutz, C., Schröder, L.: Probabilistic description logics for subjective uncertainty. In: Proceedings of the 12th International Conference on Principles of Knowledge Representation and Reasoning, KR 2010. AAAI Press (2010)
20. Sebastiani, F.: A probabilistic terminological logic for modelling information retrieval. In: Proceedings of the 17th Annual International ACM-SIGIR Conference on Research and Development in Information Retrieval, SIGIR 1994, pp. 122–130 (1994)
21. Udrea, O., Yu, D., Hung, E., Subrahmanian, V.S.: Probabilistic Ontologies and Relational Databases. In: Meersman, R. (ed.) OTM 2005. LNCS, vol. 3760, pp. 1–17. Springer, Heidelberg (2005)
22. Wang, Z., Wang, K., Topor, R.W.: A new approach to knowledge base revision in DL-lite. In: Proceedings of the 24th AAAI Conference on Artificial Intelligence, AAAI 2010, pp. 369–374 (2010)

A Semantics Driven User Interface
for Virtual Saarlouis

Deborah Richards[1] and Stefan Warwas[2]

[1] Department of Computing Faculty of Science,
Macquarie University Sydney 2109, Australia
`deborah.richards@mq.edu.au`
[2] German Research Center for Artificial Intelligence,
Stuhlsatzenhausweg 3, 66123 Saarbrücken, Germany
`stefan.warwas@dfki.de`

Abstract. Virtual reality can support edutainment applications which
seek to provide an engaging experience with virtual objects and spaces.
However, these environments often contain scripted avatars and activities that lack the ability to respond or adapt to the user or situation;
intelligent agent technology and a semantically annotated 3D world can
address these problems. This paper shows how the use of agents and semantic annotations in the ISReal platform can be applied to the virtual
heritage application involving the historic fortified town of Saarlouis.

Keywords: Semantic Virtual Worlds, Agents, Virtual Saarlouis,
ISReal.

1 Introduction

While virtual reality is often associated with fantasy and entertainment, the
application of virtual reality technology to travel and heritage applications for
(re)creating actual and/or historical events and places is clear. Beside photo realistic graphics running in real-time, intelligent and flexible behavior of avatars
and non player characters (NPCs) are required to produce a highly realistic
virtual world. In the case of massively multi-player online role-playing games
like World of Warcraft, avatar intelligence is primarily achieved by using real
(i.e. human) intelligences who create their own (shared) meaning. System (artificial) intelligence is usually limited to the use of production-rule type triggers
or hard-coded scripts to create the illusion of intelligent behavior in NPCs. The
intelligent agent paradigm offers a clean, intuitive, and powerful way of modeling
the behavior of intelligent entities in virtual worlds which tend to be dynamic,
partially observable and non-deterministic. To address these challenges, agents
can be represented in virtual worlds through avatars (their virtual bodies), interact with their environment through sensors and actuators, reason about their
beliefs and plan their actions in order to achieve a given goal. In order to enable agents to interact with their virtual environment they need, beside purely
geometric data, additional semantic information about their environment.

P. Anthony, M. Ishizuka, and D. Lukose (Eds.): PRICAI 2012, LNAI 7458, pp. 492–503, 2012.
© Springer-Verlag Berlin Heidelberg 2012

The ISReal platform offers such features by combining technologies from various research areas such as computer graphics and the semantic web in a unique way [12]. In particular, highly realistic 3D simulation of scene objects are semantically annotated in a way that allows agents to reason upon them for their deliberative action planning in the semantic virtual world. Novelly semantics are not used purely to allow the software agent to reason; in the application reported in this paper involving the historic German town of Saarlouis, the object and environment semantics are used to allow the human user to intelligently navigate around and interact with the virtual world.

This paper first introduces the application domain of Saarlouis in Section 2 followed by consideration of the tourism and museum guide domain and approaches in Section 3. We then provide an overview of the ISReal platform in Section 4 looking in more detail at the agent component in Section 5. A semantics-driven user interface to Saarlouis is presented in Section 6. Section 7 concludes the paper and considers some future work.

2   Introducing Historical Saarlouis

The historical fortified town of Saarlouis was founded by Louis XIV in 1680 and built according to plans drawn up by the renowned fortress architect Vauban. The virtual environments we have created concern two distinct time periods for Saarlouis: 1700 and and 1870 covering the French (1680-1815) and Prussian (1815-1918) periods, respectively. Such time transitions are not possible in the real world, however, virtual worlds are not restricted by time or space.

Numerous avatars in period costume walk the streets of Saarlouis, including soliders and craftmen such as wigmakers, bakers and butchers. These characters have been created using the Xaitment[1] game engine and programmed to avoid collisions, detect the location of the user, to remain within defined walking areas and to move between specified locations. To act as intelligent agents, we have some specific semantically-annotated characters, e.g.Ludwig (Louis) XIV, founder and namesake of Saarlouis; Sebastian le Prestre de Vauban, city and fortress architect; Thomas de Choisy, the first governer and fortress engineer; Kiefer a wig maker; and Lacroix, the *forgotten soldier*. The annotated buildings or historical sites may be connected with the above people such as the Kommandantur, Lock Bridge, Vauban Island and Kaserne 1 Barracks. As with *people* objects, these *places* can provide services such as being open.

In addition to the user being able to meet a person or visit a place, the user is able to choose an *activity* (for example, the baker baking bread or the wigmaker making a wig). A particularly interesting activity in this application, is the fortification and protection of Saarlouis which is a complex process requiring many services that would need to be planned and multiple agents which would need to be coordinated. Here we only consider a simplified single agent version. We explore the fortification of Saarlouis further in Sections 4, 5 and 6.

---

[1] http://www.xaitment.com/

3   Virtual Guides for Objects of Interest

To assist us in the design of a user interface for Saarlouis, we considered applications and approaches to museum and tour guides as well as product and exhibit guides. All of these applications are concerned with presentation of objects of interest, rather than provision of information, to the user. The object of interest may be many different things such as a building, work of art, artefact, person, event, activity (e.g. cultural activity such as a folk dance). The object of interest may exist physically, conceptually, historically and/or virtually. The type of interaction and the devices used will depend on the nature of the object and whether the object and/or user are physically and/or virtually present in the environment. We are mostly interested in the situation where the user is telepresent. Some applications support both situations (e.g. the MINERVA [18] robot at the Smithsonian's National Museum of American History). We are particularly interested in smart applications which use Semantic Web and/or agent-based technologies.

The use of semantic annotations to enrich museum content has been considered within the PEACH project ([15], [16]). The Multimedia Data Base (MMDB) stores semantic annotation of audio, texts and images to enable multiple applications to browse the content repository via XML transactions. The annotations are "shallow representations of its meaning" where "*meaning* refers to a set of key words that represent the topics of each paragraph and image detail and that are organized in a classification." (page 4). The topics are keywords which represent entities (characters or animals) or processes (e.g. hunting or leisure activities) and can be assigned to a class using the *member-of* relation. By annotating the multimedia content, [16] are able to determine what story beats [10] should be provided and to support a number of "communicative strategies". One strategy proposes further links to other texts based on links in the current text. A second strategy uses the theme(s)/class(es) of the current text to suggest other links with the same theme or class. A third strategy offers specific examples related to the current text. We use a similar semantics-driven strategy (see Section 6), but navigation of a multimedia library and virtual Saarlouis are different experiences. A semantically annotated virtual world might be relevant in geographical simulation work. For example, the Crowdmags: Multi-Agent Geo-Simulation to manage the interactions of crowds and control forces [11], could semantically enhance the virtual environment (VE) to guide the agents in their decision making.

Kleinermann et al. [9] use stored navigation paths and semantics to reduce time and provide guidance on how to interact with an object in a VE. Kleinermann et al. seeks to address the difficulty of acquiring annotations. To maximise reuse of these annotations they have a tool to allow easy creation and modification of multiple sets of annotations to support different domains, tasks, purposes or groups of people. To address this issue we present a tool to assist with semantic annotation of our agents (see Section 5.3). However, our key focus is on how the annotations can be machine-processed by agents to guide user interactions (see Section 6).

4   An Overview of ISReal

This section provides an overview of the ISReal platform and its components [12]. The ISReal platform consists of components from different technology domains required for intelligent 3D simulation (see Figure 1).

Graphics. The central components regarding the computer graphics domain are XML3D[2] for managing the scene graph at run-time and the rendering engine which is responsible for visualizing the scene graph. The graphics components offer 3D animation services and virtual 3D sensors which can be used by agents to sense semantic objects. Furthermore, the platform can be extended by a virtual reality (VR) system for immersive user interaction in a given 3D scene.

Semantics and Services. The *Global Semantic Environment* (GSE) of the IS-Real platform manages a global domain ontology describing the semantics of 3D objects, and a repository of semantic services for 3D object animations executed by the graphics environment. The ISReal platform consists of two semantic management systems (i) one for maintaining and querying the global ontology, and (ii) one for semantic service registration and execution handling. The ontology management system (OMS) allows to plugin different RDF triple stores of which one is uniquely selected for object queries in SPARQL, and different semantic reasoners like Pellet[3] (with internal Jena) and STAR[7] for complementary concept and relational object queries. The implementation uses the LarKC architecture [4]. The OMS is implemented as a LarKC Decider consisting of different LarKC reasoner plug-ins. As processing plug-ins the OWLIM triple store system [8] and Pellet are used.

Semantic World Model. The basic semantic grounding of the ISReal platform is called *semantic world model*. It is the set of semantically annotated scene graph objects with references to the global domain and task ontology, services, and hybrid automata. The semantics of each XML3D scene object is described by its RDFa[4] linkage to (i) an appropriate ontological concept and object definition in OWL, (ii) animation services in OWL-S, and (iii) hybrid automata for its behavior over continuous time.

Other Components. *Agent technology* is used to model the behavior of intelligent entities in a 3D scene. Due to the important role played in supporting user interactions and queries, Section 5 of this paper provides further information on the agent architecture and query handling for the ISReal platform. The *Verification* component manages and composes hybrid automata each describing temporal and spatial properties of 3D objects and their interaction in a scene, and verifies them against safety requirements given by the user at design time. The ISReal platform can run as a full fledged VR with immersive user interaction but it is also possible to connect to a scene with a web browser like Firefox or Chrome. User interaction involves a selection of choices and presentation of answers (in audio and text form) based on the semantics of the scene.

---

[2] http://graphics.cs.uni-sb.de/489/
[3] Pellet: http://clarkparsia.com/pellet
[4] http://www.w3.org/TR/xhtml-rdfa-primer/

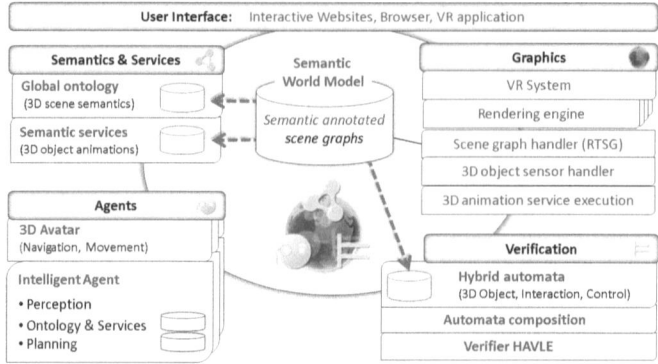

Fig. 1. The ISReal platform components

Fig. 2. Simulation of historic Saarlouis

Related Work. In different contexts, others have considered semantic annotation of 3D environments e.g. in [13]. Kalman et al. [6] proposed the concept of *smart objects* which is a geometrical object enriched with meta information about how to interact with it (e.g. grasp points). Abaci et al. [1] extended smart objects with PDDL data in order to plan with them. There are numerous examples of game engines being connected to BDI agents (e.g. [3]) and virtual environments using multi-agent systems technology (e.g. [2]). ISReal differs from this body of work as it integrates virtual worlds, semantic web and agent technology into one coherent platform for semantically-enabled 3D simulation of realities. Work using classical planners aim on improving the look-ahead behavior of BDI agents and on explicit invocation of a classical planner for sub-goals, assuming plan templates to be static at run-time. Our agents discover new services at run-time.

Saarlouis Context. As an example of how ISReal and Semantic Web technologies are used in the Saarlouis application, see Figure 2. The webservices shown allow a novice soldier agent to learn how the fortifications, particularly the lock bridge, can be used to protect Saarlouis from attack.

5   The ISReal Agent

Intelligent agents that inhabit virtual worlds require a flexible architecture in order to reason and plan in real-time with the information they perceive from their environment. Section 5.1 provides on overview of the general ISReal agent architecture. Section 5.2 introduces the ISReal query taxonomy. Finally Section 5.3 explains how to develop ISReal agents.

5.1   Architecture

ISReal agents are based on the *belief-desire-intention* (BDI) agent architecture which is well suited for dynamic and real-time environments [14]. The ISReal agent controls an avatar (its virtual body) which is situated in a 3D scene. We extended the agent with an individual sensor-based object perception facility (its interface to the virtual world), and a component for handling semantic web data, called *Local Semantic Environment* (LSE). In ISReal we follow the paradigm *"What you perceive is what you know"*, meaning an agent only has knowledge about object instances and services it already perceived during run-time through its sensor(s) and the components that manage the global semantic world model (GSE and RTSG) are strictly passive. This paradigm enables us to simulate realistic information gathering. Figure 3 shows an overview of the ISReal agent architecture. The agent's BDI plan library implements (i) the basic interaction with the ISReal platform and its LSE (e.g. initialization and perception handling), (ii) the agent's domain independent behavior (e.g. basic movement) and (iii) domain specific behavior patterns (customized behaviors regarding a certain domain or application). The domain independent behavior patterns allow us to deploy an agent into a virtual world it is not explicitly designed for and it is still able to perform basic tasks.Section 6 provides an example.

The LocalSE contains the agent's knowledge base. It is important to note that the agent's local knowledge is probably incomplete and can be inconsistent to the real world (e.g. due to outdated knowledge). The LocalSE is only updated if the agent perceives something. The LocalSE consists of (i) an *Ontology Management System* (OMS) to a) govern the semantic description of the world that is known to the agent, b) allow queries in SPARQL[5], a query language to primarily query RDF graphs, and c) semantic reasoning, and (ii) a service registry to maintain semantic services for the interaction with the objects in the 3D environment (object services). In addition it also provides tools for semantic service handling like a *service composition planner* (SCP) and a matchmaker.

5.2   Query Handling

Users need an interface for querying their user agents and assigning goals to them. To achieve a given goal, agents need powerful reasoners to access the knowledge in their LocalSE. For this purpose we created a query taxonomy. We use a plug-in architecture for query types to be open for new technologies

---

[5] W3C: http://www.w3.org/TR/rdf-sparql-query

and tools. We distinguish between (i) informational queries for querying the agent's local knowledge base, and (ii) transactional queries that cause the agent to perform actions in order to reach an assigned goal. Query classes include:

Spatial Query. To determine the spatial relationship between objects. For example, *"Is object A inside object B?"*.

Object Query. To retrieve information about instances (A-Box knowledge), such as *"Is passage1 open?"*. To handle these queries (expressed in SPARQL[6]) we use the RDF triple store system OWLIM [8].

Concept Query. Concept queries, such as *"Is Passage a subclass of Open?"*, retrieve information about concepts (T-Box knowledge). We use the OWL2 semantic reasoner Pellet to answer such queries.

Object Relation Query. To find non-obvious relations between different objects, i.e. a set of entities $\{e_1, \ldots, e_n\}$. The OMS can find the smallest tree of the RDF graph representing the KB, such that it contains all entities $\{e_1, \ldots, e_n\}$ [7]. A simple query is *"How are objects passage1, passage2, passage3, and bridgeA related?"*.

Funcional Relational Query. To answer questions like *"How to close passage X?"*. The query is purely informational and does not cause the agent to execute the found plan.

Temporal Query such as *"When did you see object X the last time?"*.

Matchmaking Query. The user can ask the agent whether it knows a certain type of service. The query is answered by using a semantic matchmaker.

Declarative Goal Query. A declarative goal specifies a target state the agent should bring about (e.g. $at(agent1, bridgeA)$). The agent has to do means-end reasoning and execute the found plan (if there is one).

Perform Goal Query. Perform goals are purely procedural goals that do not contain declarative information about the goal state. Perform goals trigger agent plans that are able to achieve the desired behavior (e.g. turn around).

5.3   Developing and Semantically Annotating ISReal Agents

To provide a complete agent configuration (see Figure 4) a graphical designer has to develop the body geometry, movement animations, and position the sensor in XML3D, using a state of the art 3D modeling tool. From the agent perspective we assume that the domain ontology already exists, so that the avatar can be semantically annotated using our annotation tool. OWL-S service descriptions have to be developed and implemented that describe the agent's basic actions (OWLS-ED). The initial agent knowledge base (A-Box) can be created using an ontology tool like Protege[7]. The modeling of the actual agent and its behavior is done using our model-driven development environment for multiagent systems *DSML4MAS Development Environment* (DDE) [19].

---

[6] SPARQL: `http://www.w3.org/TR/rdf-sparql-query/`
[7] Protege: `http://protege.stanford.edu/`

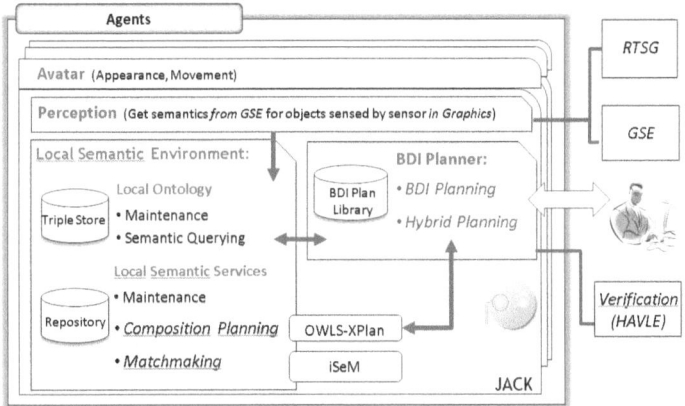

Fig. 3. ISReal agent architecture

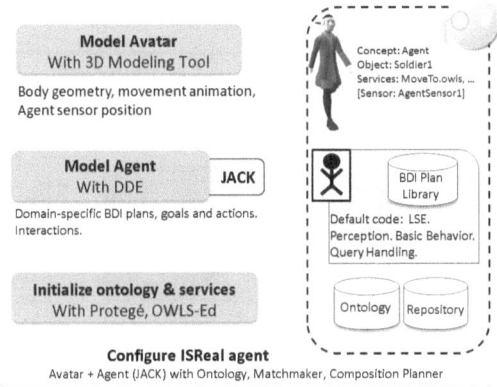

Fig. 4. ISReal agent configuration

6   Semantics and Agent-Driven User Interface

While the ISReal architecture supports a very large number of possible queries, rather than require the user to form the query or our system to interpret the request, we use the object semantics and agent reasoning and planning capabilities to determine and present what queries are available. Basically, in the approach the user is given the opportunity to select an object of interest and later to follow links related to that object (see how objects can be related to one another in Table 1), return to the original object or select a new object of interest. To create an engaging experience which will allow more of Saarlouis to be visited, satisfy user preferences and achieve teacher/curator learning goals, the ability of the agent to perceive the objects in the environment and to plan will be used to add elements of serendepity or dramatic effect "on the way" to the user's chosen object.

Table 1. Partial meta-model for three *Person* objects

Property	Example 1	Example 2	Example 3
Name:string	Ludwig XIV	Sebastian de Vauban	Thomas de Choisy
Date of birth:date	5-11-1638	??-5-1633	1632
Place of birth:Place	St Germain en Laye	St-Leger Moigneville	
Date of death: date	1-11-1715	30-3-1707	20-2-1710
Place of death: Place	Versailles	Paris	Saarlouis
Lived In:Place	Versailles		Kommandantur
Commissioned:Person	Vauban	Choisy	
Role:string	King of France	Architect of Saarlouis	The first governor
services		introduceSelf	introduceSelf explainFortifications

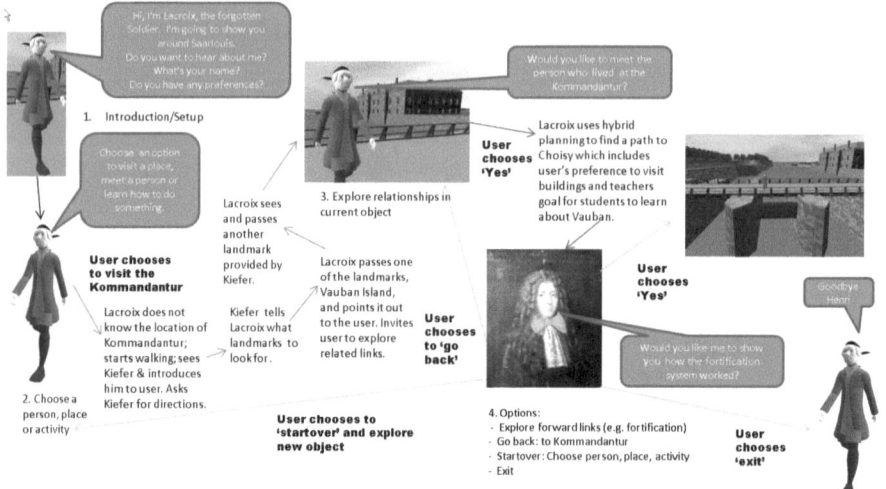

Fig. 5. Possible interaction between user, agent and environment

Example. A possible interaction is depicted in Figure 5 where we see one complete session starting from setup to exit. Not shown in the diagram, when the user starts up the virtual environment, the user (who may be a tourist, history buff or school child) is able to select their personal guide, consisting of the avatar and its role (here novice). The intelligent guide agent is attached to the avatar and initialized. As shown in step 1 the agent introduces themself and requests some information from the user (name and preferences).

As a basic interaction strategy, the user is asked (see step 2) to choose if they want to meet a person, visit a place or learn about an activity/task. Objects can be classified as either a place or person and they may know how to perform an activity (e.g. introduceSelf, explainFortifications). As this knowledge is part of the object meta-model, when the user selects a person, place or activity, the user-interface can query the objects in the world/scene to produce an appropriate list for the user to choose from. This list will take into account what the user has already seen, either hiding already visited objects from the list or placing them at the end of the list. User and teacher preferences could also be taken into account in the ordering provided. Once a user chooses the object of interest, the agent is assigned the goal to find an "interesting" pathway from the current location in the virtual world to the location of the chosen object.

Three simplified *tours* are depicted, each in a different shade (tour 1-blue, tour 2-red, tour 3-green). Four alternative green tours are shown and reflect the choices that were available to the user. The end node of each tour depicts the choice taken by the user. Alternative paths from one user-chosen end node to another are possible. Tours 1 and 2 show one possible path that might have occured. To determine the path from the current location to the object chosen by the user or to determine if an object passed along a pathway should be brought to the attention of the user, the agent uses a number of filters and strategies. These filters are based on the user's preferences for certain types of objects or themes; the existence of a pedagogical model which identifies any objects or activities which must be encountered (these are not necessarily added to the first tour) or the sequence in which certain objects must be visited; the possible inclusion of dramatic elements such as a visit to the house of a colourful character or building with mysterious past. The agents access to services and ability to reason and plan about the objects and environment allow factors such, as if the place has already been visited, to be taken into account. Adding further to the element of surprise, a strategy recommended by [16] to improve attention and memory retention that needs to be balanced with meeting user's expectations , and drawing on the ability of the agent to perceive its world to discover new objects and services, we also allow the agent to interrupt its current plan to, for instance, introduce a place, person or activity discovered on the way. See, for example, the disovery of Kiefer and Vauban Island in Tour 1(blue). The relationships between objects, and objects and services, are used to drive possible further interactions. Similar to the communicative strategies used in the PEACH project (Section 3), in step 3 the user is able to follow a link related to the Kommandantur, i.e. to visit Choisy who lived in this building (see Table 1). This allows the user to traverse between objects in a forward fashion. Note that other links and options were also possible for the user but not shown for clarity in the example: step 4 depicts a more complete set of options. As well as following links forward, the user can choose to **go back** to access options related to the previously chosen object, **start over** with a top level search for place, person or activity or **exit** to quit the program. At any point, the user is able to take these three options.

Query Processing. As introduced in Section 5.2, the agent is capable of handling several types of queries. These queries are primarily used by the agent to satisfy its own informational or transactional needs, rather than for the end user as we believe they would have some difficulty identifying which type of query was most appropriate and how to form the query. For the reasons given above, for the Saarlouis application we do not allow the user to enter natural language requests. For the Saarlouis application we may allow the user to select certain queries in order, for instance, to satisfy their own particular interests or a set of learning goals set by a teacher.

In the context of Saarlouis and in line with Silveira et al. [17] user information needs, our approach provides information which addresses: *Choice:* what places can I choose to visit?; who can I choose to meet?; *Procedural:* how do I do $< activity >$; where can I find $< person >$?; how do I get to $< place >$; in what order should I visit Saarlouis? *Informative:* what/who can I do/see/visit/meet? *Guidance:* who should I talk to next?; where should I visit next?; who should I ask? What should I do now? *History:* who have I already met?; what have I already visited?; what have I already done? *Descriptive:* what does this person do?; who is this?; why are they famous?; who lived here? what can you tell me about $< person, place, activity >$?; how is this person related to $< person, place >$?

Silvieira's user information needs which are not supported relate more to explanations and system motivations and include: *Interpretive:* why are you telling me about $< object >$?; where are we going?; why are we going to see $< object >$; *Investigative:* did I miss anything?; what happened before that time?; what do I still need to do?; and *Navigational:* where am I? In line with the findings of [5] there are some additional status-related questions that could be considered to allow greater transparency and minimize user frustration: What are you doing now? Why did you do that? When will you be finished? What sources did you use? The status, interpretive, investigative and navigational questions we intend to offer in an interrupt mode allowing the user at any time to pause the current explanation, choose the question of interest to them and then return to the current interaction. These questions may be affected by which time period was chosen and other user preferences.

In ISReal the user is able to select and right-click on an object to receive the set of services it offers and a description. These services can be selected and added to a stack for execution by the agent. We anticipate that in addition we will offer the user a set of relevant queries for that object based on the query-type modules available in ISReal and in line with the types of questions posed above. Being able to click on an object in the scene and gain information about it and execute its services provides an alternative and complementary interaction method and would better suit a user or application where being taken on a tour is less appropriate or appealing.

7   Conclusion

We have presented the use of the ISReal platform in the Saarlouis virtual heritage application. Rather than providing a virtual world for free exploration, our intelligent agent acts as a tour guide that escorts the user around Saarlouis, providing explanations and allowing choices at certain points in the tour. The agent uses the semantics embedded in the world to dynamically plan suitable tours that take into account user preferences and history, pedagogical goals and potentially incorporate dramatic elements to make the experience more enjoyable and memorable.

References

1. Abaci, T., Ciger, J.: Planning with Smart Objects. In: WSCG SHORT Papers Proceedings, pp. 25–28. UNION Agency - Science Press (2005)
2. Anastassakis, G., Ritchings, T., Panayiotopoulos, T.: Multi-agent Systems as Intelligent Virtual Environments. In: Baader, F., Brewka, G., Eiter, T. (eds.) KI 2001. LNCS (LNAI), vol. 2174, pp. 381–395. Springer, Heidelberg (2001)
3. Davies, N.P., Mehdi, Q.: BDI for Intelligent Agents in Computer Games. In: Proc. 8th Int. Conf. on Comp. Games, CGAMES 2006. Uni. of Wolverhampton (2006)
4. Fensel, D., et al.: Towards LarKC: a Platform for Web-scale Reasoning. IEEE Computer Society Press, Los Alamitos (2008)
5. Glass, A., McGuinness, D.L., Wolverton, M.: Toward establishing trust in adaptive agents. In: Proc. 13th IUI 2008, Gran Canaria, Spain, pp. 227–236. ACM, NY (2008)
6. Kallmann, M.: Object Interaction in Real-Time Virtual Environments. PhD thesis, École Polytechnique Fédérale de Lausanne (2001)
7. Kasneci, G., Ramanath, M., Sozio, M., Suchanek, F.M., Weikum: STAR: Steiner Tree Approximation in Relationship-Graphs. In: Proc. of 25th IEEE Intl. Conf. on Data Engineering, ICDE (2009)
8. Kiryakov, A., Ognyanov, D., Manov, D.: OWLIM - a Pragmatic Semantic Repository for OWL. In: WISE Workshops, pp. 182-192 (2005)
9. Kleinermann, F., De Troyer, O., Creelle, C., Pellens, B.: Adding Semantic Annotations, Navigation Paths and Tour Guides to Existing Virtual Environments. In: Wyeld, T.G., Kenderdine, S., Docherty, M. (eds.) VSMM 2007. LNCS, vol. 4820, pp. 100–111. Springer, Heidelberg (2008)
10. Mateas, M., Stern, A.: Towards Integrating Plot and Character for Interactive Drama. In: Socially Intelligent Agents, pp. 221–228 (2002)
11. Moulin, B., Larochelle, B.: Crowdmags: Multi-Agent Geo-Simulation of Crowd and Control Forces Interactions. In: Mohamed, A. (ed.) Modelling, Simulation and Identification. InTech (2010),
    http://www.intechopen.com/books/modelling--simulation-
    and-identification/title-crowdmags-multi-agent-geo-
    simulation-of-crowd-and-control-forces-interactions-
12. Nesbigall, S., Warwas, S., Kapahnke, P., Schubotz, R., Klusch, M., Fischer, K., Slusallek, P.: Intelligent Agents for Semantic Simulated Realities - The ISReal Platform. In: Proc. 2nd Conf. on Agents & AI, ICAART 2010, pp. 72–79 (2010)
13. Pittarello, F., De Faveri, A.: Semantic Description of 3D Environments: A Proposal Based on Web Standarts. In: Proc. of Intl. Web3D Conf. ACM Press (2006)
14. Rao, A.S., Georgeff, M.P.: BDI Agents: From Theory to Practice. In: Proc. of the 1st Intl. Conf. on Multi-Agent Systems, pp. 312–319. AAAI Press (1995)
15. Rocchi, C., Stock, O., Zancanaro, M.: Semantic-based Multimedia Representations for the Museum Experience. In: Proc. HCI in Mobile Guides Wkshp, Mobile HCI 2003 (2003)
16. Rocchi, C., Stock, O., Zancanaro, M., Kruppa, M., Krueger, A.: The Museum Visit: Generating Seamless Personalized Presentations on Multiple Devices. In: Proc. Intelligent User Interfaces IUI 2004, pp. 316–318. ACM Press (2004)
17. Silveira, M., de Souza, C., Barbosa, S.: Semiotic Engineering Contributions for Designing Online Help Systems. In: SIGDOC 2001. ACM (2001)
18. Thrun, S., et al.: MINERVA: A second generation mobile tour-guide robot. In: Proc. Int. Conf. on Robotics & Automation, ICRA 1999, Detroit, Mich, May 10-15 (1999)
19. Warwas, S., et al.: The DSML4MAS Development Environment. In: Proc. of 8th Intl. Conf. on Autonomous Agents and Multiagent Systems, AAMAS 2009 (2009)

Fuzzified Game Tree Search – Precision vs Speed

Dmitrijs Rutko

Faculty of Computing, University of Latvia,
19 Raina blvd., LV-1586, Riga, Latvia
dim_rut@inbox.lv

Abstract. Most game tree search algorithms consider finding the optimal move. That is, given an evaluation function they guarantee that selected move will be the best according to it. However, in practice most evaluation functions are themselves approximations and cannot be considered "optimal". Besides, we might be satisfied with nearly optimal solution if it gives us a considerable performance improvement.

In this paper we present the approximation based implementations of the fuzzified game tree search algorithm. The paradigm of the algorithm allows us to efficiently find nearly optimal solutions so we can choose the "target quality" of the search with arbitrary precision – either it is 100% (providing the optimal move), or selecting a move which is superior to 95% of the solution space, or any other specified value.

Our results show that in games this kind of approximation could be an acceptable tradeoff. For example, while keeping error rate below 2%, the algorithm achieved over 30% speed improvement, which potentially gives us the possibility to search deeper over the same time period and therefore make our search smarter. Experiments also demonstrated 15% speed improvement without significantly affecting the overall playing strength of the algorithm.

Keywords: game tree search, alpha-beta pruning, fuzzified search algorithm, performance.

1   Introduction

Game tree search remains one of the challenging problems in artificial intelligence and an area of active research. While classical results have been achieved based on Alpha-Beta pruning there are a lot of further enhancements and improvements including NegaScout, NegaC*, PVS, SSS* / Dual*, MTD(f) and others [1-6].

Alternative approaches also exist. For instance, Berliner's algorithm B*, which uses interval bounds, allows to select the optimal move without needing to compute the exact game value [19].

While these algorithms are based on optimal search (always returning an optimal solution) recent approaches incorporate approximation ideas. Mostly, they are based on probabilistic forward pruning techniques where we are trying to prune less optimistic nodes and sub-trees based on heuristics or shallow search.

P. Anthony, M. Ishizuka, and D. Lukose (Eds.): PRICAI 2012, LNAI 7458, pp. 504–515, 2012.
© Springer-Verlag Berlin Heidelberg 2012

In this paper we present a new approach for nearly optimal solutions based on a fuzzified tree search algorithm. The main idea is that exact game tree evaluation is not required to find the best node. Thus, we can search less nodes and therefore improve the speed. It has been proven to be more efficient compared to other algorithms in specific games [12]. We propose the notion of quality of search, so this way we can adjust both target quality and performance accordingly to suit our needs. Our experiments show that by applying this technique it is possible to improve algorithm performance significantly while keeping the error rate very low, which, in turn, does not affect overall playing strength of the program.

The paper is organized as follows: the current status of approximation patterns in game tree search is discussed; then a brief overview of fuzzified tree search is given. Following this, a new enhancement to this algorithm based on quality of search is proposed. Thereafter, the experimental setup and empirical results on search quality and performance obtained in a real game are shown. The paper is concluded with future research directions.

2 Approximation and Near Optimal Search

Approximation search paradigms have become popular recently. It may be due to the following reasons. Firstly, classical search algorithms, which try to find optimal solution, are close to their theoretical limits and it is difficult to improve them further. Secondly, these algorithms rely on the quality of the evaluation functions which are not optimal by their nature – they are an approximation of utility value of board position which we are not aware of. So, if our search is based on approximate position estimations, there is no reason we cannot change the nature or tree search and make it also approximate.

In fact, it is possible and a lot of research has been done in this direction:

- *ProbCut* is selective search modification of Alpha-Beta. It assumes that estimation of a node could be done based on a shallow search. It excludes sub-trees form the search, which are probably irrelevant to the main line of play. This approach has been successfully used in his Othello program Logistello significantly improving the playing strength [13].
- *Multi-ProbCut* is a further improvement of Prob-Cut which uses additional parameters and pruning thresholds for different stages of the game. Further experiments demonstrated that this approach could be efficiently used in chess [14, 16].
- *Multi-cut* - Speculative pruning technique which takes into account not only probability of cutting off relevant lines of play, but also chances of such wrong decision affecting move selection at the root of the search tree [15].
- *RankCut* - a domain-independent forward pruning technique that exploits move ordering and prunes once no better move is likely to be available. It has been implemented in an open source chess program, CRAFTY, where RankCut reduced the game-tree size by approximately 10%-40% for search depths 8-12 while retaining tactical reliability [17].

- *Game Tree Search with Adaptive Resolution* – an approach, where value returned by the modified algorithm, called Negascout-withresolution, differs from that of the original version by at most R. The experiment results demonstrate that Negascout-with-resolution yields a significant performance improvement over the original algorithm on the domains of random trees and real game trees in Chinese chess [18].

3   Fuzzified Tree Search

Fuzzified tree search, which is used in our research, is based on the idea that the exact position evaluation is not required to find the best move, as we are in fact interested in the node which guarantees the highest outcome [7, 8].

Most game tree search approaches are based on Alpha-Beta pruning Fig. 1. We could instead look at our game tree from a relative perspective e.g. if this move is better or worse than some value X (Fig. 2). At each level, we identify if a sub-tree satisfies "greater or equal" criteria. For instance, passing the search algorithm argument "5", we can obtain the information that the left branch has a value less than 5 and the right branch has a value greater or equal than 5. We do not know exact sub-tree evaluation, but we have found the move that leads to the best result.

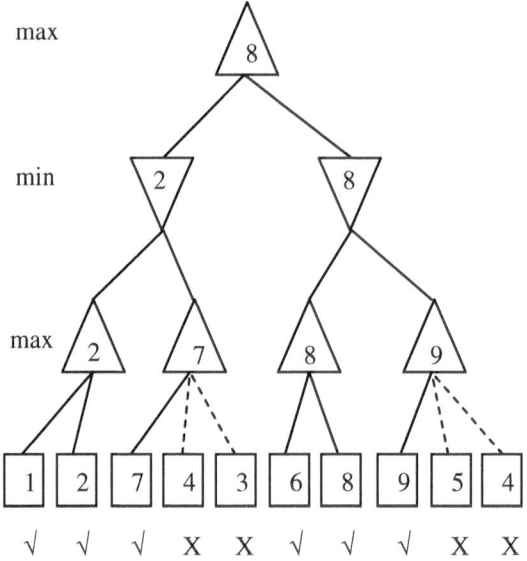

Fig. 1. The Alpha-Beta approach

In this case, different cut-offs are possible:

- at max level, if the evaluation is greater (or equal) than the search value;
- at min level, if the evaluation is less than the search value.

In the given example, the reduced nodes are shown with dashed line. Comparing to Fig. 1 it can be seen that not only more cut-offs are possible, but also pruning occurs at a higher level which results in better performance.

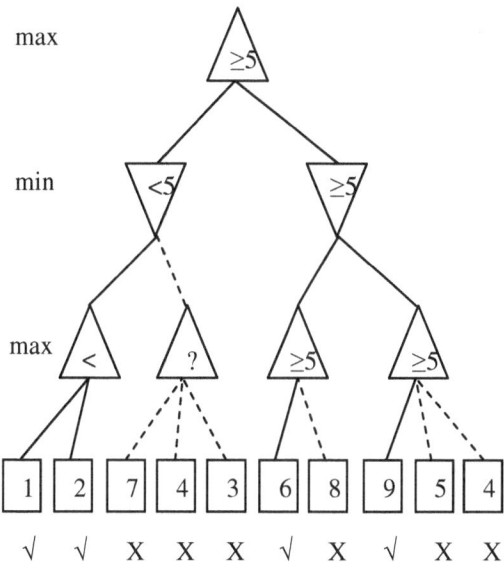

Fig. 2. Fuzzy best node approach

4   Fuzzified Search Algorithm

Best Node Search (BNS) is a game tree search algorithm based on the idea described in the previous section. The main difference between the classical approach and BNS is that it does not require knowledge of the exact game tree minimax value to select a move. We only need to know which sub-tree has the higher estimation. By iteratively performing search attempts the algorithm can obtain information about which branch has a higher estimation without knowing the exact value. So less information is required and, as a result, the best move can be found faster – the total number of searched nodes is smaller and total algorithm execution time is reduced comparing to the algorithms based on the exact game tree evaluation [12].

From technical point of view, BNS-based approach is similar to MTD(f) and the difference is that BNS stops searching without computing the accurate minimax value, while MTD(f) guarantees that the minimax value at the root node is computed.

```
function BNS(node, α, β)
    do
        test := NextGuess(node, α, β)
        betterCount := 0
        foreach child of node
            bestVal := -AlphaBeta(child, -test, -(test - 1))
            if bestVal ≥ test
                betterCount := betterCount + 1
                bestNode := child
        update alpha-beta range
    while not((β - α < 2) or (betterCount = 1))
    return bestNode
```

Fig. 3. The BNS algorithm

BNS uses a standard call of Alpha-Beta search with 'zero window'. While scanning a game tree, the algorithm checks all sub-trees and returns a node which leads to the best result. In general, BNS is expected to be more efficient comparing to the classical algorithms in terms of number of nodes checked as it does not obtain additional information which is not required in many cases – the exact game tree minimax value.

The BNS algorithm is given in Fig. 3 which makes use of the following functions:

1. NextGuess() – returns the next separation value to be tested by the algorithm;
2. AlphaBeta() – alpha-beta search with Zero Window (Null Window) performs a boolean test whether a move produces a worse or a better score than the passed value.

BNS algorithm finds essentially the same move as alpha-beta (or its variants) - the move which has highest evaluation, so they will play identically (taking into account differences in move ordering). But BNS will perform faster by searching fewer nodes on average.

5 Quality of the Search and Nearly Optimal Solutions

Evaluation functions are approximations and might be imprecise, otherwise there would be no reason to perform deeper search. They are not optimal by their nature, but it is important how close it is to real utility of board position and how it is related to the chance of winning.

Besides, in many game specific situations very often the program is required to respond in a limited time. Thereby, it becomes more important to find solutions fast and the optimality of the solution moves to a secondary role. So we are interested in techniques which would allow us to control the quality of our search while focusing on performance and overall search speed.

Fuzzified Game Tree Search algorithm fits this idea very well. That is given an evaluation function we can extend our search by introducing an additional parameter - quality of search. Or, in other words, this is a guaranteed probability of finding the move, which is in the top N% of possible moves. For example, the traditional implementation uses maximum quality of the search, finding the best move with 100 % probability. However, if we were satisfied with a move in the top 10%, we could easily perform this search.

This target quality logically means that if we choose some level of confidence, for example 95% this means that the move that is returned by search algorithm would be better (not worse) than all other 95% of possible moves. On the contrary, if we choose the highest level of confidence (100%), the move found would be the best over all possible moves.

The most important part of this algorithm is that it is designed to allow choosing target quality of the search (your level of confidence) arbitrarily. It may depend on level of playing strength, time remaining for search and could always be updated dynamically as all information about already checked moves is stored in memory. So, if you choose additional move tuning so only new unexplored parts of the tree would be researched with no loss of time (double search).

So, we propose additional improvement to the existing algorithm to adhere to the aforementioned observations.

Let us explore the existing algorithm and note additional changes required to make nearly optimal search or arbitrary search of quality available.

The following is the same search algorithm listed in Fig. 3 except for one enhancement – the addition of the search quality parameter.

```
function BNS(node, α, β, quality)
    do
        test := NextGuess(node, α, β)
        betterCount := 0
        foreach child of node
            bestVal := -AlphaBeta(child, -test, -(test - 1))
            if bestVal ≥ test
                betterCount := betterCount + 1
                bestNode := child
                if expectedQuality ≥ quality
                    return bestNode
        update alpha-beta range
    while not((β - α < 2) or (betterCount = 1))
    return bestNode
```

Fig. 4. BNS with quality

Expected quality of search is measured as a probability of randomly choosing a move over all possible moves which satisfy our search criteria (Fig. 5).

It depends on the number of nodes checked so far and the total number of nodes. Initially we use the ratio of "better nodes" amongst checked nodes to calculate the expected number of "better nodes" amongst all children. And then, overall expected quality is measured as probability of having found the best node.

So the main idea is to stop our search and return the best move found so far as soon as we are confident about the quality of our search results. This version of the algorithm is used in the following experiments.

$$ExpectedQuality = \frac{1}{\frac{betterCount}{number\ of\ nodes\ already\ checked} * total\ number\ of\ nodes}$$

Fig. 5. Estimation of expected quality

6   Game Setup

For our research experiments we have chosen the same game "Hey! That's My Fish!", in which fuzzified search algorithm was already tested and previously demonstrated better efficiency compared to other existing search algorithms [12].

"Hey! That's My Fish!" is a 2-4 player board game. The aim is to collect as many fish as you can with your penguins.

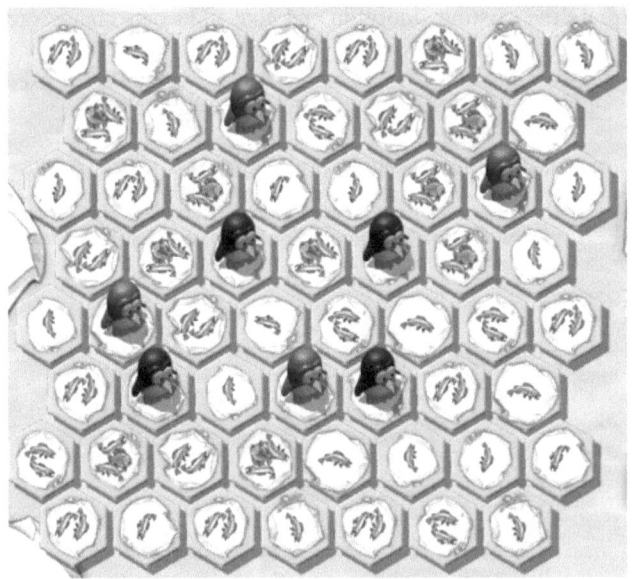

Fig. 6. "Hey! That's My Fish!" game board

Setup: 60 hexes are randomly laid out in 8 rows, alternating between 7 and 8 hexes each - Fig. 6. They are all face up so that you can clearly see where the fish clusters are. Each hex has either 1, 2, or 3 fish on it. There are 30 "1" fish hexes, 20 "2" fish hexes and 10 "3" fish hexes, which results in maximum 100 points (fish) which is possible to get in a game. Each player then places 2-4 penguins (depending on the number of players). They place these one at a time and they must be placed on "1" fish hexes.

- Two Player game: each person has 4 penguins
- Three Player game: each player has 3 penguins
- Four Player game: each player has 2 penguins

Play: On his turn a player moves one of his penguins. He must move it in a straight line, and may move it as little as 1 hex and as many hexes as is legal. Penguins must stop before they reach: the edge of the board; a break in the hexes; or another penguin. After moving his penguin the player then picks up the hex which his penguin started on.

Ending the game: a player leaves the game when he can't move any more of his penguins, claiming the last hexes his penguins are standing on. When all the players are done moving, the game is over. (Practically, the game actually ends when each player can see that his penguins are each on their own "islands" of ice. Each player then picks up any hexes on these islands which he could reach.) The players then count up all their fish, and the player with the most wins.

This game is simple, but involves subtle strategy and a fair amount of tactical content [9]. The moves are not entirely obvious. You can be very aggressive in the game, making clever blocking moves and carefully analyzing the vector-based movement [10].

The main strategic element of gameplay is working to isolate other players' penguins, trapping them in small areas; when this happens, and an area is isolated that contains only one penguin, the owner scores all tiles in the isolated area, and the penguin is removed from the play. Maximizing score by getting high-fish hexes is a secondary but important strategic consideration [11].

7   Experimental Results

The described algorithm BNS with quality was implemented and tested in this game. Now, we present additional details and experimental results obtained in our study.

To perform game tree search a straight-forward evaluation function was implemented. At each step, we count the number of fish consumed so far, so overall evaluation function is the amount of fish obtained by the current player minus the amount of fish obtained by the opponent limited by search depth.

Evaluation function = Fish Amount (player) – Fish Amount (opponent)

The main advantages of this approach are the following: it is fast, simple, reasonable, and it converges to correct estimation towards the game end. Downsides of this function are that it does not address specific aspects of the game like strategic area isolation and opponent blocking, but we are mainly focused on search algorithm comparison, which gives us a wide area for research. Additional features like iterative deepening were also used in these experiments [12].

We conduct experiments with different values of expected confidence ranging from 100% to 0%. However, 100% means that we are looking for the best move without compromise, but on the other hand 0% does not mean that we are acting randomly. Instead we search for the first move that satisfy selection criteria, and only then return it. So in practical experiments error rate is measured as the ratio of non-optimal to optimal solutions selected during the entire game and the actual quality is calculated as 1 - error rate.

We also measure performance improvement as total number of nodes searched during the entire game divided by number of nodes required to search for the highest level of confidence (100% quality).

As it could be seen from Fig. 7, the actual quality of the search remains 100% for first iterations (for expected quality 100%, 90% and 80% respectively). That means that for this particular case, we receive 20% of performance improvement while algorithm is still able to find the best moves all the time. Then actual quality goes down performance gets increased considerably. For example, with retaining 98% quality we achieve 30% speed improvement, and with 95% quality we achieve 40% speed improvement. Our diagram terminates with having around 75% of performance improvement while retaining 75% search quality.

An additional parameter we are tracking is the average error per wrong move found (in game points) – the difference between evaluation of the best move and the move returned by our algorithm. Initially there is no error, as the algorithm shows 100% actual quality, then for two cases (70% and 60% expected quality) it stays within 1 point per move error, and for remaining experiments it converges to around 1.2 points of error per move. This is a really important indicator, showing that even when there is an error in algorithm search, the error is small and this move is very close to the optimal one.

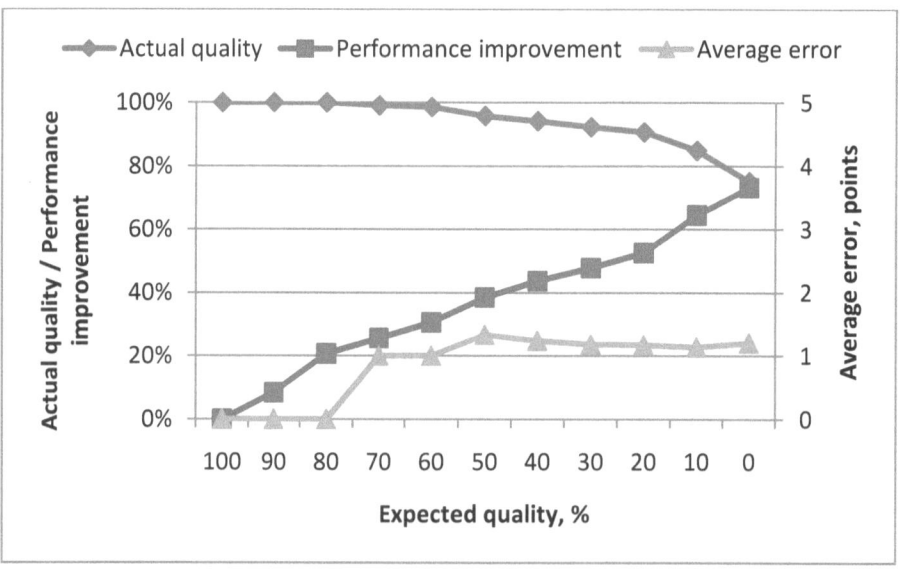

Fig. 7. Actual quality vs. Performance improvement. Average error (points)

It is important to note that, currently we operate with minimal guaranteed (expected) confidence level, so actual quality is much higher than the expected values. This is mainly because when searching we are not acting randomly, we are looking for at least one move satisfying our search criteria to guarantee our minimal confidence level.

To conclude, our research with more objective results additional experiments have been performed based on a series of games.

As "Hey! That's my Fish" starts with a random setup and the outcome of the game largely depends on beneficial location of valuable tiles we measure algorithm performance based on a series of matches. Program plays with itself 100 matches calculating winning probability (number of games won by each algorithm). As player order (who starts first) is also important, we switch players during these series of games, so player who started first, starts second. It results in a total series of 200 games for each configuration. Results are displayed on Fig. 8.

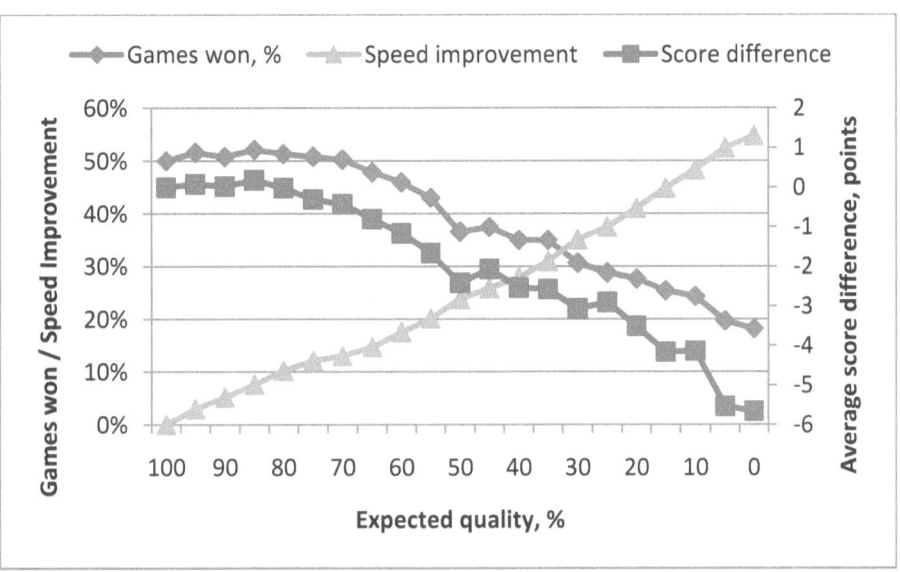

Fig. 8. Number of games won (%) vs Speed Improvement. Average points difference.

The following configuration has been used: search depth 6, expected quality of search varying from 100% to 0%. As a reference algorithm (opponent against whom we play) we use the same algorithm but with 100% (maximum) expected search quality.

Performance improvement is measured as difference between total number of nodes searched by each algorithm during entire series of games. Score difference represents an average variance in points obtained by each player.

As it could be seen from the chart, algorithms start equally with having winning probability 50% and score difference 0. As long as expected quality decreases, the speed improvement grows linearly. It is important to note that winning probability remains close to 50% (with fluctuations in 1-2% range) while decreasing the expected quality up to 70%. It gives 15% speed improvement without significantly affecting the overall playing strength of the algorithm. However, further decreasing of expected search quality results in considerable performance degradation.

8   Conclusions and Future Work

The main goal of this paper was to show that fuzzified tree search algorithm could be easily extended and efficiently used for finding nearly optimal solutions. The experiments demonstrate a 30% speed improvement over the standard approach while retaining an error rate below 2%. Moreover, in case of error the selected move is still very close to the optimal solution. Further experiments show 15% speed improvement without significantly affecting the overall playing strength of the algorithm. It could be concluded that the proposed approximation search paradigm could be used in real domain games with high a level of confidence.

However, additional research might be needed in the following areas:

- improve estimation precision of expected quality vs. actual quality achieved;
- apply different heuristic based evaluation functions.

The future experiments should also consider analyzing algorithm performance and efficiency in other games, but we believe that the proposed approach could be successfully applied to any type of game.

Acknowledgements. This research is supported by the European Social Fund project No. 2009/0138/1DP/1.1.2.1.2/09/IPIA/VIAA/004.

References

1. Pearl, J.: The solution for the branching factor of the alpha-beta pruning algorithm and its optimality. Communications of the ACM (1982)
2. Reinefeld, A.: An Improvement to the Scout Tree-Search Algorithm. ICCA Journal 6(4), 4–14 (1983)
3. Weill, J.C.: The NegaC* search. ICCA Journal (March 1992)
4. Plaat, A., Schaeffer, J., Pijls, W., de Bruin, A.: Best-First Fixed-Depth Minimax Algorithms. Artificial Intelligence 87 (1996)
5. Björnsson, Y.: Selective Depth-First Game-Tree Search. Ph.D. thesis, University of Alberta (2002)
6. Russell, S.J., Norvig, P.: Artificial Intelligence: A Modern Approach, 3rd edn. Pearson Education, Inc., Upper Saddle River (2010)
7. Rutko, D.: Fuzzified Algorithm for Game Tree Search. In: Second Brazilian Workshop of the Game Theory Society, BWGT (2010)
8. Rutko, D.: Fuzzified Algorithm for Game Tree Search with Statistical and Analytical Evaluation. Scientific Papers, vol. 770, pp. 90–111. University of Latvia (2011)
9. Hey That's My Fish - German Board Games,
   http://www.mrbass.org/boardgames/heythatsmyfish/
10. Review of Hey! That's My Fish! – RPGnet,
    http://www.rpg.net/reviews/archive/11/11936.phtml

11. Hey! That's My Fish! Remarkable Depth for Minimal Rules,
    http://playthisthing.com/hey-thats-my-fish
12. Rutko, D.: Fuzzified Tree Search in Real Domain Games. In: Batyrshin, I., Sidorov, G. (eds.) MICAI 2011, Part I. LNCS (LNAI), vol. 7094, pp. 149–161. Springer, Heidelberg (2011)
13. Buro, M.: ProbCut: An effective selective extension of the alpha–beta algorithm. ICCA Journal 18(2), 71–76 (1995)
14. Buro, M.: Experiments with Multi–ProbCut and a new high–quality evaluation function for Othello. In: Workshop on Game–Tree Search, NECI (1997)
15. Björnsson, Y., Anthony Marsland, T.: Multi-cut alpha-beta-pruning in game-tree search. Theoretical Computer Science 252(1-2), 177–196 (2001)
16. Jiang, A.X., Buro, M.: First Experimental Results of ProbCut Applied to Chess. In: Proceedings of the Advances in Computer Games Conference, Graz, vol. 10 (2003)
17. Lim, Y.J., Lee, W.S.: RankCut - A Domain Independent Forward Pruning Method for Games. In: AAAI (2006)
18. Chang, H.-J., Tsai, M.-T., Hsu, T.-S.: Game Tree Search with Adaptive Resolution. In: Proceedings of the 13th Advances in Computer Games Conference, ACG 2011, Tilburg, Netherlands (2011)
19. Berliner, H.J.: The B* Tree Search Algorithm: A Best-First Proof Procedure. Artificial Intelligence 12(1), 23–40 (1979)

Processing Incomplete Temporal Information in Controlled Natural Language

Rolf Schwitter

Centre for Language Technology
Macquarie University
Sydney, Australia
Rolf.Schwitter@mq.edu.au

Abstract. If not all temporal information is available in a text, humans usually use additional background knowledge to answer questions about the text. In this paper, we first investigate how humans answer this kind of questions and then suggest two general strategies that we can use to answer these questions automatically in a controlled natural language context. We show how background knowledge about events and their effects can be made explicit in a controlled natural language and how this additional information can be translated together with the textual information into a formal notation for automated reasoning. For this purpose, we introduce an Answer Set Programming based version of the Event Calculus and use this reasoning framework as a starting point for answering questions over the formalised textual information.

1 Introduction

This work is situated at the intersection of controlled natural language processing, automated reasoning and question answering. Recently, a number of controlled natural languages have been developed and used as high-level knowledge acquisition and specification languages [2,4,22]. These controlled natural languages are engineered subsets of full natural languages whose grammar and vocabulary have been restricted in a systematic way to reduce both ambiguity and complexity of full natural languages [17]. Controlled natural languages can significantly improve the knowledge acquisition process and the understandability of specifications compared to specifications written in formal languages, in particular if the writing process of these specifications is supported by a predictive authoring tool [11].

Since most of these controlled natural languages can be translated unambiguously into formal notations, automated reasoning and question answering becomes immediately possible. We show in this paper that controlled natural languages can also serve as an interesting test field for exploring new ideas of combining textual information with background knowledge and automated reasoning techniques. In particular, we investigate how complex question answering can be performed in a controlled natural language context where not all relevant

P. Anthony, M. Ishizuka, and D. Lukose (Eds.): PRICAI 2012, LNAI 7458, pp. 516–527, 2012.
© Springer-Verlag Berlin Heidelberg 2012

temporal information is available in a text using a novel Answer Set Programming [12] based version of the Event Calculus [10,15].

In order to present our research results in a compact way, we use the following short text (1) that has been written in the controlled natural language PENG Light [22]. This text consists of a sequence of temporally ordered events and two explicit temporal event modifiers (*at 11:00* and *at 12:10*):

1. John enters the university library at 11:00 and leaves the library with three books. John gets on the city bus and gets off the bus in the city centre at 12:10. John gives the books to Gaby at Starbucks.

At first glance, this text is easy to understand for a human reader, and a human reader usually does not have any problems to answer the following questions about this text:

2. When does John enter the university library?
3. When does John get off the bus?

All relevant information for answering these two questions can be extracted directly from the textual information. However, the situation gets a bit more difficult if the text does not provide the relevant temporal information as it is the case for the following three questions:

4. Where is John at 11:30?
5. When does John get on the bus?
6. When does John give the books to Gaby?

As we will see in Section 2, human subjects do normally not simply answer with *I don't know* but use additional background knowledge to answer these questions in a cooperative way [8]. It is interesting to investigate what kind of background knowledge and strategies humans use to answer these questions and to see if it is possible to make the relevant background knowledge that they use explicit in a controlled natural language. If this is indeed the case, then we can combine the background knowledge with the textual information and translate the resulting information into a formal notation for automated reasoning and question answering.

The rest of this paper is organised as follows: In Section 2, we look at how humans answer the questions (4-6), hypothesize what kind of background knowledge they use and suggest two strategies that a machine can use to answer these questions. In Section 3, we present the controlled language processor that we use to process the text and the background knowledge so that the resulting notation can be used by an automated reasoner. In Section 4, we give a brief introduction to Answer Set Programming (ASP) since we use this relatively new programming paradigm later in Section 5 to implement a modified version of the Event Calculus. In Section 6, we evaluate our approach by sketching how question answering is implemented on top of the ASP-based Event Calculus and by showing how the questions are answered. Finally, in Section 7, we conclude.

2   An Experiment

In order to better understand how humans answer the questions (4-6), we gave the text (1) in its written form to ten human subjects. All of them are students or members of a computational linguistics department. We asked these human subjects without imposing any time limit to provide written answers to these questions (repeated below as Q4-Q6).

Question Q4 is a *wh*-question and is asking for the location of a person at a specific point in time. Since there is no explicit timepoint available in the text, the answers that the human subjects provide show some interesting variations:

Q4. Where is John at 11:30?
4.1 *either in the library, on the city bus or somewhere in between*
4.2 *in the bus or in the library*
4.3 *library, bus or on his way to the bus station*
4.4 *in the library*
4.5 *we don't know*
4.6 *maybe still at the library, maybe already on the city bus*
4.7 *in the library or in the city bus*
4.8 *library?*
4.9 *library / waiting for the bus / in the bus*
4.10 *waiting for the bus OR in the bus*

Four subjects (4.2, 4.6, 4.7, and 4.10) distinguish two elements in their answers; these elements (e.g., *in the library*) denote states that are derived from the events in the text. Three subjects (4.1, 4.3, and 4.9) distinguish three elements in their answers whereas in each answer one of these elements (e.g., *somewhere in between*) is derived in an indirect way from the events in the text. Two subjects (4.4 and 4.8) provide only one element as answer, and finally one subject (4.5) seems to take the text as the only source of information and answers with *we don't know*.

Question Q5 is asking for a timepoint or a period of time but there is no explicit temporal marker available in the text, apart from the two temporal markers (*at 11:00* and *at 12:10*) that modify different events than the one under investigation:

Q5. When does John get on the bus?
5.1 *between 11.00 and 12.10*
5.2 *unknown, between 11:00 & around 12:00*
5.3 *between 11.10 - 12*
5.4 *before 12:00*
5.5 *we don't know*
5.6 *after leaving the library*
5.7 *between approx. 11:05 and 12:05*
5.8 *maybe at 12.00?*
5.9 *[11:01 - 12:09]*
5.10 *between 11:01 and 12:09*

It is interesting to see that six subjects (5.1, 5.2, 5.3, 5.7, 5.9 and 5.10) take the two temporal markers in the text into consideration and provide an interval as answer. However, five of these subjects (5.2, 5.3, 5.7, 5.9 and 5.10) modify these temporal markers slightly in their answers. One subject (5.4) takes the second temporal marker as a rough starting point and provides an open interval together with a temporal operator as answer. One subject (5.5) suggests the answer *we don't know*, and another subject (5.2) proposes a similar answer as part of the entire answer. Another subject (5.8) offers a modified form of the second temporal marker as possible answer. And finally, one subject (5.6) uses a rather different strategy, it takes the temporal operator *after* and the previous event *leaving the library* to construct the answer *after leaving the library*.

The form of question Q6 is similar to Q5 but in contrast to Q5, the timepoint of the event that is here under investigation does not occur between two explicit temporal markers in the text (1):

Q6. When does John give the books to Gaby?
  6.1 *after 12:10*
  6.2 *after 11:00*
  6.3 *we don't know*
  6.4 *after meeting her at Starbucks*
  6.5 *we don't know*
  6.6 *when he sees her at Starbucks*
  6.7 *after 12:10*
  6.8 *at 12.15?*
  6.9 *[12:11 - ...]*
 6.10 *after 12:11*

Three subjects (6.1, 6.7, and 6.10) take the closest temporal marker (*12:10*) that is available in the text as starting point and use the temporal operator *after* to construct an answer. One subject (6.2) choses a similar strategy but takes the temporal marker (*11:00*) that is farther away as reference point for answering the question. An other subject (6.8) suggests a possible timepoint (*12:15*) derived from the closest temporal marker as answer. One subject (6.9) takes again the closest temporal marker as a rough starting point but constructs an open interval as answer. Two subjects (6.4 and 6.6) refer indirectly to the event under investigation by constructing a related event with a temporal operator as answer. And finally, two subjects (6.3 and 6.5) focus entirely on the textual information and answer the question with: *we don't know*.

Against the backdrop of the variations in these answers, we would like to suggest two general strategies to answer these questions automatically: strategy I for questions like Q4 and strategy II for questions like Q5 and Q6.

2.1   Strategy I

If there is no temporal marker available in the text, then return *I don't know* as answer but return the last location (e.g., *at Starbucks*) as answer if the explicit timepoint (e.g., *11:30*) is dropped from the question. If there is only one temporal marker available in the text (e.g., *11:00*) that occurs before the timepoint in the question (e.g., *11:30*), then return all locations after that temporal marker in form of a disjunction as answer (e.g., *in the university library or in the bus or at Starbucks*). If there are two temporal markers in the text (e.g., *11:00* and *12:10*) and the timepoint in the question (e.g., *11:30*) falls between these temporal markers, then return all relevant locations in form of a disjunction as answer (e.g., *in the university library or in the bus*).

2.2 Strategy II

If there is no temporal marker in the text, then return the previous event together with a temporal operator as answer (e.g., *after leaving the library*). If there exists only one temporal marker in the text (e.g., *12:10*) that modifies an event that occurs before the event in the question, then return the temporal marker together with a temporal operator as answer (e.g., *after 12:10*). If there are two temporal markers in the text (e.g., *11:00* and *12:10*), then return the corresponding interval together with temporal operators as answer (e.g., *after 11:00 and before 12:10*).

3   Controlled Natural Language Processing

In order to implement these two strategies, we need to be able to process the textual information and the relevant background knowledge automatically. For this purpose, we rely on the controlled natural language processor PENG Light [22] that uses a unification-based grammar together with a chart parser. The writing process of this controlled natural language is supported by a predictive authoring tool [18]. The language processor resolves anaphoric references and translates the text (1) via discourse representation structures [9] into a number of predefined predicates and represents them as a set of facts (f(.)). For example, the translation of the first sentence of the text (1) results in the following representation:

7. f(named(x1,john)). f(event(e1,entering,x1,x2)).
   f(object(x2,university_library)). f(time(e1,at,x3)).
   f(timex(x3,1100)). f(event(e2,leaving,x1,x2)). f(theme(e2,with,x4)).
   f(cardinality(x4,eq,3)). f(object(x4,book)). f(time(e2,at,x5)).
   f(timex(x5,0)). f(before(e1,e2)).

This representation reifies events (e.g., e1) and states (not shown here), and uses a flat notation for atoms together with a small number of predefined predicates for events, states, complements and modifiers. Note that the order of events is recorded with the help of the **before**-predicate and that the temporal modifiers of events are represented with the help of **time**- and **timex**-predicates. Note also that if no explicit temporal modifier can be derived from the text for an event, then the constant 0 is used instead of a concrete temporal expression.

As we will see in more detail in Section 5, we specify also the relevant background knowledge in controlled natural language; for example, domain-dependent effect axioms such as (8) and (9) and supporting statements such as (10-12), and then translate this background knowledge automatically into the formal target notation:

8. If an agent enters a building then the agent will be in that building.
9. If an agent leaves a building then the agent will no longer be in that building.
10. John is an agent.
11. Every university library is a building.
12. Every building is an artefact.

Before we do that we will give a brief introduction to Answer Set Programming in the next section, since we will use this rather novel programming paradigm to implement a modified version of the Event Calculus that is then used to process the textual information together with the background knowledge.

4   Answer Set Programming (ASP)

Answer Set Programming (ASP) is a form of declarative problem solving that has its roots in logic-based knowledge representation and non-monotonic reasoning [1,7,12]. ASP programs look similar to Prolog programs [3] but use an entirely different computational mechanism. Instead of deriving a solution from a program specification via SLDNF resolution [13] like in Prolog, finding a solution in ASP corresponds to computing stable models (= answer sets) with the help of an ASP solver [5,21]. Compared to Prolog, standard ASP programs have the following main characteristics: (a) the order of program rules and conditions within rules do not matter, (b) termination is (often) not an issue, and (c) answer sets and not proofs represent solutions. An ASP program may have zero, one or multiple answer sets as solution.

ASP programs use a Prolog-style rule notation of the form: `<head> :- <body>`. If the body of a rule is empty, then the symbol ':-' can be dropped and the rule is called a fact. The following is a variable-free ASP program with negation as failure (not) in the body of the second rule:

13. `p :- q. q :- not r.`

This program has exactly one answer set {p,q} as solution consisting of the two atoms p and q. We can extend this program by adding a rule with a disjunction | in its head:

14. `p :- q. q :- not r. s | t :- p.`

This program has two answer sets: {p,q,s} and {p,q,t}. Instead of using a disjunctive rule in (14), we can try to replace it by a choice rule of the form:

15. `· {s,t} :- p.`

However, this rule generates four stable models: {p,q}, {p,q,t}, {p,q,s}, {p,q,s,t}. This happens because we use the choice rule without numerical bounds. In order to achieve the same effect as with a disjunctive rule, we have to write:

16.   `1 {s,t} 1 :- p.`

This modified rule specifies a cardinality constraint and says that if the atom `p` is included in an answer set, then include exactly one of the atoms `s` or `t` into that answer set.

In ASP, we can exclude answer sets from a program by adding integrity constraints to the program, for example:

17.   `:- s, not t.`

If we add this constraint to the program (14), then only the answer set $\{p,q,t\}$ is generated, since the constraint in (17) prevents adding the atom `s` to an answer set if `t` is not generated.

ASP programs that contain variables are first transformed into variable-free programs by a grounding component (= ASP grounder) before they can be processed by an ASP solver. For example, the ASP program:

18. `p(a,b). q(X) :- p(X,Y), not r(Y).`

results after grounding in the subsequent grounded program:

19. `p(a,b). q(a) :- p(a,a), not r(a). q(a) :- p(a,b), not r(b).`
    `q(b) :- p(b,a), not r(a). q(b) :- p(b,b), not r(b).`

An ASP solver can then find the following answer set $\{p(a,b),q(a)\}$ for this grounded program. In general, checking whether S is an answer set of a program P follows the following idea: Generate the S-reduct P^S of P by first deleting any rule in P that has a condition *not C* in its body where $C \in S$, and then by deleting every condition of the form *not C* in the bodies of the remaining rules. That means the resulting reduct P^S is a logic program consisting of a set of definite clauses without negation as failure. We say that S is a *stable model* of P iff it satisfies the equation $S = M(P^S)$ where $M(P^S)$ denotes the minimal Herbrand model [13] of the definite clause program P^S. Let us illustrate this by an example, if we take $S = \{p(a,b),q(a)\}$ and the grounded program (19) as starting point, then the reduct P^S is:

20. `p(a,b). q(a) :- p(a,b). q(b) :- p(b,b).`

An interesting aspect of ASP is that it distinguishes two kinds of negation: negation as failure (`not`) and classical negation (`-`). By combining these two kinds of negation, it is possible to express, for example, closed world assumption (21) and that something holds by default (22):

21. `-p :- not p.`
22. `p :- not -p.`

In Section 6, we will see that these two kinds of negation can also be used to sort atoms in an elegant way. Modern ASP tools provide additional functionality, for example for computing aggregates and optimal answer sets; also ASP tools exist that combine the grounder and the solver in one component (see [6] for details).

5   Towards an ASP-Based Event Calculus

The Event Calculus is a general logic-based formalism for representing and reasoning about events and their effects [10,14,15,20]. It takes the notion of events, fluents, timepoints and time-intervals as ontological primitives and provides a notation that allows us to specify which events happen at what timepoints and which fluents (= time-varying properties) are initiated or terminated by these events. This approach is very useful in our context since we need to be able to represent the relevant background knowledge of the text and have an inference mechanism at hand that makes it possible to compute the expected answers.

The Event Calculus uses domain-dependent axioms to specify the effects of events and a fixed number of domain-independent core axioms that implement the inference engine and take the domain-dependent axioms as input. These domain-dependent axioms process the facts derived from the textual information.

In our case a number of domain-dependent axioms are used to specify which events initiate or terminate which fluents. These axioms are context-sensitive and can be expressed directly as conditional statements in controlled language:

23. If an agent enters a building then the agent will be in that building.
24. If an agent leaves a building then the agent will no longer be in that building.
25. If an agent leaves a building with an object then the agent will be owning that object.
26. If an agent gets on a vehicle then the agent will be in that vehicle.
27. If an agent gets off a vehicle then the agent will no longer be in that vehicle.
28. If an agent A gives an object to an agent B then the agent A will no longer be owning that object and the agent B will be owning that object.
29. If an agent A gives an object to an agent B at a location then the agent A will be at that location and the agent B will be at that location.

These conditional statements are then automatically translated into effect axioms by the language processor; for example, (23) into (30) and (24) into (31):

30. `initiates(E1,fluent(s(E1),location(in),X1,X2)) :-`
    `    object(X1,agent), event(E1,entering,X1,X2), object(X2,building).`

31. `terminates(E2,fluent(s(E1),location(in),X1,X2)) :-`
    `    object(X1,agent), event(E2,leaving,X1,X2), object(X2,building),`
    `    before(E1,E2).`

Axiom (30) states that an event `E1` initiates a fluent `s(E1)` of a certain type if `X1` is an `agent`, `E1` is an `entering`-event, and `X2` is a `building`. Note that the axiom (30) uses the term `s(E1)` as identifier for the fluent and relates this fluent to the initiating event `E1`.

Axiom (31) terminates the fluent `s(E1)` of the same type, if `E2` is a `leaving`-event, `X1` is an `agent`, `X2` is a `building`, and `E1` occurs before `E2`. This axiom uses a similar structure as (30) but the variable for the terminating event `E2` is different to the variable used in the identifier `s(E1)` of the fluent. In ASP, variables that appear in the head of a rule must also occur in the body of the same rule in order to make the rule *safe*; in (31), for example, the `before`-predicate is used to guarantee rule safety for `E1`.

The predicates used in domain-dependent axioms such as (30) and (31) do not directly access the facts in (7) derived from the text since these predicates rely on additional interface predicates derived from supporting statements such as (10-12), for example:

```
32. object(X,building) :-
      f(object(X,university_library)).
```

Domain-dependent axioms such as (30) and (31) are processed by a small number of **core** axioms. These core axioms do not depend on the domain but on the task; that means similar tasks use similar core axioms. For our task, we only need four core axioms that have some similarities to the core axioms used by the Simplified Event Calculus [16,20], but we use a richer representation that deals with intervals instead of timepoints.

The first core axiom (33) specifies that an interval (`intvl`) holds if `E1` is an event that `happens` at timepoint `T1` and `initiates` a fluent `F`, and `E2` is another event that `happens` at `T2` and `terminates` the fluent `F`, and `E1` occurs `before E2` and the fluent `F` is `not broken` between `E1` and `E2`:

```
33. holds(intvl(E1,T1,F,E2,T2)) :-
      happens(E1,T1), initiates(E1,F), happens(E2,T2), terminates(E2,F),
      before(E1,E2), not broken(E1,F,E2).
```

The second core axiom (34) is similar to (33) but deals with open intervals where an event `E1 happens` at a timepoint `T1` but the terminating event and its timepoint are both unknown (0):

```
34. holds(intvl(E1,T1,F,0,0)) :-
      happens(E1,T1), initiates(E1,F), not broken(E1,F).
```

The remaining core axioms (35) and (36) below guarantee default persistence in combination with the negation-as-failure operator (`not`) used in (33) and (34). These two axioms specify under which conditions a fluent `F` is `broken`:

```
35. broken(E1,F,E2) :-
      happens(E,T), before(E1,E), before(E,E2), terminates(E,F).
36. broken(E1,F) :-
      happens(E,T), before(E1,E), terminates(E,F).
```

Note that the `happens`-predicates rely on an additional interface predicate that accesses a suitable `event`-predicate with its temporal modifier; the arity of this `event`-predicate depends on the number of its complements:

```
37. happens(E,T) :-
      f(event(E,N,X1)) : f(event(E,N,X1,X2)) : f(event(E,N,X1,X2,X3)),
      f(time(E,P,X4)), f(timex(X4,T)).
```

Note also that we exploit in the core axioms (33, 35, 36) the transitive closure of the information defined by means of the `before`-predicate (38) that accesses the temporal ordering of the events derived from the textual information:

38. `before(E1,E2)  :- f(before(E1,E2)).`
    `before(E1,E3)  :- f(before(E1,E2)), before(E2,E3).`

Given facts such as in (7), domain-dependent axioms like (30) and (31) plus interface and additional predicates like (32, 37, 38), we can use the core axioms of the Event Calculus to compute the following intervals (39) for the fluents:

39. `holds(intvl(e1,1100,fluent(s(e1),location(in),x1,x2),e2,0))`
    `holds(intvl(e2,0,fluent(s(e2),owning,x1,x4),e5,0))`
    `holds(intvl(e3,0,fluent(s(e3),location(in),x1,x6),e4,1210))`
    `holds(intvl(e5,0,fluent(s(e5),location(at),x9,x10),0,0))`
    `holds(intvl(e5,0,fluent(s(e5),location(at),x1,x10),0,0))`
    `holds(intvl(e5,0,fluent(s(e5),owning,x9,x4),0,0))`

These intervals build the starting point for the question answering process that can be implemented on **top** of the ASP-based Event Calculus.

6   Evaluation

In order to be answered automatically, the questions Q4-Q6 are first translated by the controlled natural language processor into ASP rules and then added to the formalised knowledge. Each question results in a number of ASP rules that implement the two strategies introduced in Section 2. The ASP tool *glingo* [5] is then used to compute the expected answers.

Because of the limited space in this paper, we discuss here only one ASP rule in detail, namely the one for question Q4: *Where is John at 11:30?* The rule (40) below deals with the case where the temporal marker in the question does not occur in the text (there exists another rule that deals with the case where the temporal marker actually occurs in the text):

40. `answer(P,O,D)  :-`
    `    holds(intv(E1,TL,fluent(s(E),location(P),X1,X2),E2,TR,D)),`
    `    TL < 1130, 1130 < TR, f(named(X1,john)), f(object(X2,O)).`

This rule specifies that if a particular fluent holds between timepoints `TL` and `TR`, and `1130` lies somewhere between these two timepoints, and there exists a named entity `X1` who is `john` and an object `X2`, then the answer consists of the name `P` of the prepositional modifier, the name `O` of the object, plus a value `D` for the distance between the event at timepoint `TL` and the event that triggered the fluent. This gives us a way to order the answers. Given the rule in (40) together with the following integrity constraint in controlled language:

41. It is not the case that an agent is in two artefacts between the timepoint T1 and the timepoint T2.

that has been translated automatically beforehand into (42):

42. :- 2 { holds(intvl(E1,T1,fluent(s(E),location(P),X,Y),E2,T2,D))
         : object(Y,artefact) }, object(X,agent).

An answer set solver like *glingo* will generate **two** stable models that contain exactly one **answer**-predicate because of the above integrity constraint:

43. **Answer 1**: answer(in,university_library,0), ...

    Answer 2: answer(in,city_bus,2), ...

The two choice rules in (44) illustrate that the holds-predicate used in (40) to extract the fluent between the two timepoints TR and TL operates over the intervals (intvl) that have been generated by the Event Calculus in (39):

44. { holds(intvl(E1,TL,F1,E3,TR,0)) } :-
       prior_to(TL,TR), holds(intvl(E1,TL,F1,E2,0)),
       holds(intvl(E3,TR,F2,E4,T4)).

    { holds(intvl(E1,TL,F2,E6,TR,D)) } :-
       prior_to(TL,TR), holds(intvl(E1,TL,F1,E2,T2)),
       holds(intvl(E3,0,F2,E4,0)), holds(intvl(E5,T5,F3,E6,TR)),
       before(E1,E3,D), before(E3,E5), before(E4,E6).

The prior_to-predicate in (44) calculates the timepoint TL that immediately precedes the timepoint TR; we conduct this search by combining classical negation (-) with negation-as-failure (**not**) and do this in the following way:

45. prior_to(TL,TR) :-
       timepoint(TL), timepoint(TR), TL < TR, not -prior_to(TL,TR).

    -prior_to(TL,TR) :-
       timepoint(TL), timepoint(TR), timepoint(T), TL < T, T < TR.

A final point to note is that the first of the three before-predicates in (44) uses three arguments; this predicate returns the distance D between the event at timepoint TL and the event that triggered the fluent. The value for D is then used to order the answers.

7   Conclusion

In this paper we investigated what human subjects do when they answer *wh*-questions over a sequence of events where not all relevant information is explicitly available in the text. We identified two general strategies that they use to answer these questions and showed what kind of additional domain-dependent background knowledge is required to answer these questions automatically by a machine. We argued that this background knowledge can be expressed in controlled natural language and translated together with the text into a formal target notation for automated reasoning. In particular, we introduced and used an ASP-based version of the Event Calculus as inference engine. This inference engine derives open and closed intervals during which fluents hold that have been initiated by events. We showed how the inference engine can be extended in an elegant way for question answering. The presented strategies and the implemented solution are very general and can be used for other application scenarios.

References

1. Baral, C.: Knowledge Representation, Reasoning and Declarative Problem Solving. Cambridge University Press (2003)
2. Clark, P., Harrison, P., Jenkins, T., Thompson, J., Wojcik, R.: Acquiring and Using World Knowledge using a Restricted Subset of English. In: The 18th International FLAIRS Conference (FLAIRS 2005), pp. 506–511 (2005)
3. Clocksin, W.F., Mellish, C.S.: Programming in Prolog: Using the ISO Standard, 5th edn. Springer, Heidelberg (2003)
4. Fuchs, N.E., Kaljurand, K., Kuhn, T.: Attempto Controlled English for Knowledge Representation. In: Baroglio, C., Bonatti, P.A., Małuszyński, J., Marchiori, M., Polleres, A., Schaffert, S. (eds.) Reasoning Web 2008. LNCS, vol. 5224, pp. 104–124. Springer, Heidelberg (2008)
5. Gebser, M., Kaufmann, B., Neumann, A., Schaub, T.: Conflict-driven answer set solving. In: Proceedings of IJCAI, pp. 386–392 (2007)
6. Gebser, M., Kaufmann, B., Kaminski, R., Ostrowski, M., Schaub, T.: Potassco: The Potsdam Answer Set Solving Collection. AI Communictions 24(2), 107–124 (2011)
7. Gelfond, M., Lifschitz, V.: The stable model semantics for logic programming. In: Proceedings of the 5th ICLP, pp. 1070–1080 (1988)
8. Grice, P.: Logic and conversation. In: Cole, P., Morgan, J. (eds.) Syntax and Semantics, vol. 3. Academic Press, New York (1975)
9. Kamp, H., Reyle, U.: From Discourse to Logic: Introduction to Model-theoretic Semantics of Natural Language, Formal Logic and Discourse Representation Theory. Kluwer, Dordrecht (1993)
10. Kowalski, R., Sergot, M.: A Logic-based Calculus of Events. New Generation Computing 4, 67–95 (1986)
11. Kuhn, T.: Controlled English for Knowledge Representation. Doctoral Thesis. Faculty of Economics, Business Administration and Information Technology of the University of Zurich (2010)
12. Lifschitz, V.: What Is Answer Set Programming? In: Proceedings of AAAI 2008, vol. 3, pp. 1594–1597 (2008)
13. Lloyd, J.: Foundations of Logic Programming. Springer (1987)
14. Miller, R., Shanahan, M.: Some Alternative Formulations of the Event Calculus. In: Kakas, A.C., Sadri, F. (eds.) Computat. Logic (Kowalski Festschrift). LNCS (LNAI), vol. 2408, pp. 452–490. Springer, Heidelberg (2002)
15. Mueller, E.T.: Commonsense Reasoning. Morgan Kaufmann Publishers (2006)
16. Sadri, F., Kowalski, B.: Variants of the Event Calculus. In: Proceedings of the International Conference on Logic Programming, pp. 67–81 (1995)
17. Schwitter, R.: Controlled Natural Language for Knowledge Representation. In: Proceedings of COLING 2010, pp. 1113–1121 (2010)
18. Schwitter, R., Ljungberg, A., Hood, D.: ECOLE - A Look-ahead Editor for a Controlled Language. In: Proceedings of EAMT-CLAW 2003, pp. 141–150 (2003)
19. Shanahan, M.: Solving the Frame Problem: A Mathematical Investigation of the Common Sense Law of Inertia. MIT Press, Cambridge (1997)
20. Shanahan, M.: The Event Calculus Explained. In: Veloso, M.M., Wooldridge, M.J. (eds.) Artificial Intelligence Today. LNCS (LNAI), vol. 1600, pp. 409–430. Springer, Heidelberg (1999)
21. Simons, P., Niemelä, I., Soininen, T.: Extending and implementing the stable model semantics. Artificial Intelligence 138(1-2), 181–234 (2002)
22. White, C., Schwitter, R.: An Update on PENG Light. In: Pizzato, L., Schwitter, R. (eds.) Proceedings of ALTA 2009, pp. 80–88 (2009)

Opinion Target Extraction for Short Comments

Lin Shang[1], Haipeng Wang[1,2], Xinyu Dai[1], and Mengjie Zhang[3]

[1] State Key Laboratory of Novel Software Technology, Nanjing University
Nanjing 210046, China
[2] Trend Micro China Develop Center, Nanjing 210000
[3] School of Engineering and Computer Science,Victoria University of Wellington
P.O. Box 600, Wellington 6140, New Zealand

Abstract. Target extraction is an important task in sentiment analysis. Many existing methods have worked well in news and blogs. However, they are not effective for short product comments. In this paper, we firstly prove that a well-known method, Ku's method, cannot obtain good results for short comments. Then we propose a new method to extract opinion targets by developing a two-dimensional vector representation for words and a back propagation neural network for classification. The proposed method is examined and compared with two well-known opinion extraction methods (Ku's and LDA methods) on an crawled network mobile phone corpus from "Zhongguancun online" with 14408 comments. The strict evaluation and the lenient evaluation are used in the experiments to determine the goodness of the extracted opinion targets. Experimental results show that under the strict evaluation, the proposed method can achieve better precision by 8.33% improvement over Ku's and 16.67 % improvement over LDA. Under the lenient evaluation, the proposed method can achieve a 28.33% improvement in precision over Ku's and 33.33% over LDA. In addition, the opinion targets extracted by our method are much closer to the true topics and much more meaningful than those extracted by the other two methods.

Keywords: sentiment analysis, target extraction, neural networks.

1 Introduction

With the development of internet, more and more subjective texts are coming. The analysis and extraction of the sentiment information have been a hot topic in natural language processing and related areas. Among these, the sentiment analysis for text mining has attracted an increasing attention [1], especially in the product reviews [2, 3]. Sentiment analysis has many tasks, such as extraction of sentimental information, classification (polarity or extent), retrieval and induction [4, 5]. Early work in this area includes some machine learning methods to detect the products or movie reviews polarity [4, 6].

In opinion analysis for products, a classic sentiment application would be tracking what bloggers or reviewers are saying about a product like new iPad. The subjective text needs to be extracted from comments for products in different aspects. It is important to choose the words that can exactly represent the main concepts of a relevant document set derived from the product comments. The work of this paper aims to address the opinion target(topic) extraction problem, which is one of the two fundamental problems in opinion analysis [7, 8]. Opinion targets usually refer to product features, which may be defined as product components or attributes [7]. Contributions have been proposed in the targets extraction for product review mining [9--12].

P. Anthony, M. Ishizuka, and D. Lukose (Eds.): PRICAI 2012, LNAI 7458, pp. 528–539, 2012.

The most straightforward method to extract opinion targets is the manual construction of topics [5]. However, it may bring lots of work and has less portability. In [9] the method based on the relations between features and topics can obtain better topic features for articles and web pages, but it has never been applied to short comments. In recent work, target extraction was treated as a topic coreference resolution problem by Stoyanov et al. [13], but it may ignore sentiment information in words. Some topic modeling methods such as LDA(Latent Dirichlet Allocation) and PLSA(Probabilistic Latent Semantic Analysis) modeled the documents generation and mined the implied targets. However, they did not perform well when applied to very short documents [14]. For Chinese news comments, Ma and Wan [15] proposed an approach to extracting not only explicit but also implicit opinion topics. Also, in [16, 17], Ku's method achieved significant results in news and blog corpora. However, due to lack of professional knowledge for most customers, the comments on the products from the customers are only confined to their feelings and experience. The comments on the product would only relate to certain aspects. This is because comments are generally short, which are different from news and blogs with long reviews.

The overall goal of this paper is to develop a new approach to extracting opinion targets for short comments with the expectation of achieving meaningful opinion targets with high precision and close to topic. To achieve this goal, we will develop a new algorithm to automatically learn and extract opinion targets in short comments. The proposed algorithm will be examined and compared with two well-known target extraction methods on a large corpus of mobile online short comments. Specifically, we will:

-- theoretically prove whether the existing well-known opinion extraction approaches in long comments can be used for extracting opinion targets in short comments, and
-- propose a new opinion extraction approach for short comments by developing a novel words representation and investigate whether the proposed algorithm can successfully extract opinion targets with better precision, closer to true topic and more meaningful than opinion targets obtained by the well-known methods.

2   Analysis of Ku's Method

2.1   Ku's Method

In opinion target extraction, it is a typical approach to choosing meaningful words that can clearly cover the concepts of a product [18, 16]. Such words form the major topic of the corpus. Ku et al. [17] proposed a method to detect major topics, which achieved good results on news and blogs. In [17], the weights were assigned to each word at both the document level and the paragraph level. In the following formulas, W denotes the weights, and S denotes the document level while P is the paragraph level. TF is term frequency, and N is word count.

$$W_{s_i t} = TF_{s_i t} * \log \frac{N}{N_{s_i}} \tag{1}$$

$$W_{p_j t} = TF_{p_j t} * \log \frac{N}{N_{p_j}} \tag{2}$$

where symbol i is the document index, symbol j is the paragraph index, and symbol t is the word index.

Based on the weights assigned to each word, the following formulas were used to judge whether the term is thought as representative.

$$Disp_{st} = \sqrt{\frac{\sum_{i=1}^{m} (W_{s_i t} - mean)^2}{m}} * TH \tag{3}$$

$$Dev_{s_i t} = \frac{w_{s_i t} - mean}{Disp_{st}} \tag{4}$$

$$Disp_{pt} = \sqrt{\frac{\sum_{j=1}^{n} (W_{p_j t} - mean)^2}{n}} \tag{5}$$

$$Dev_{p_j t} = \frac{w_{p_j t} - mean}{Disp_{pt}} \tag{6}$$

Formula (1) gives the TF*IDF (Term Frequency, Inverse Document Frequency) value of the words in the article level, while formula (2) in the paragraph level. Formula (3) gives the word frequency between the documents, where TH is a threshold to constrain the extraction number of topics and m is the document number in the corpus. Formula (4) gives the words frequency within an article. Formula (5) gives the frequency of words between paragraphs. Formula (6) gives word frequency within a paragraph. In all the equations, parameter $mean$ is the average value.

If a word is a topic of the corpus, then either of the two conditions should be satisfied:

Condition One: The word appears in many documents, but in a few paragraphs.

$$Disp_{st} \leq Disp_{pt} \Longrightarrow \exists S_i, \forall P_j \in S_i Dev_{(s_i t)} \leq Dev_{(p_j t)} \tag{7}$$

Condition Two: The word appears in a few documents, but in many paragraphs.

$$Disp_{st} > Disp_{pt} \Longrightarrow \exists S_i, \forall P_j \in S_i Dev_{(s_i t)} > Dev_{(p_j t)} \tag{8}$$

According to the definitions described above, the score of a term, defined as $|Dev_{s_i t} - Dev_{p_j t}|$ could measure how significant the word was for a relevant document set. From the topical set, opinion-oriented sentences could be extracted. Then the tendencies could be determined.

Ku's method has been applied to many opinion mining tasks and achieved good results particularly on news and blogs[16, 17]. However, it could not achieve good results on short comments. In the next subsection we will prove why it cannot work well for short comments.

2.2   Proof of Ku's Method not Being Effective for Short Comments

Property1. In Ku's method, if the word is only in one comment and the review has only one paragraph, then whether the word is the opinion target depends only on the number of paragraphs in the comment, the number of documents in the corpus and the TH values.

Proof: Assume that a word only appears in one document, which has only one paragraph, then $N_{(s_i)} = N_{(p_j)}$.

According to formulas (1) and (2), we can conclude that the word weights in the paragraph level and in the document level are equal, and assume that the value is w:

$$W_{s_i t} = W_{p_j t} = w \tag{9}$$

The following are the average values of the word weights in the document level and the paragraph level:

$$mean_{st} = \frac{w}{m} \qquad mean_{pt} = \frac{w}{n} \tag{10}$$

where m is the number of documents in the corpus, n is the number of paragraphs in the comment.

$$Disp_{st} = \sqrt{\frac{\sum_{i=1}^{m}(W_{s_i t} - mean)^2}{m}} * TH = \sqrt{\frac{(m-1)(\frac{w}{m})^2 + w - \frac{w}{m}^2}{m}} * TH$$

$$= \sqrt{\frac{w^2 * (\frac{m-1}{m^2} + \frac{(m-1)^2}{m^2})}{m}} * TH = \frac{\sqrt{m-1}}{m} * w * TH \tag{11}$$

Then in the documents, we get:

$$Dev_{s_i t} = \frac{w_{s_i t} - mean_{st}}{Disp_{st}} = \frac{w - \frac{w}{m}}{\frac{\sqrt{m-1}}{m} * w * TH} = \frac{\sqrt{m-1}}{TH} \tag{12}$$

Corresponding to paragraphs, we get:

$$Disp_{pt} = \sqrt{\frac{\sum_{j=1}^{n}(W_{p_j t} - mean)^2}{n}} = \sqrt{\frac{(n-1)(\frac{w}{n})^2 + (w - \frac{w}{n})^2}{n}}$$

$$= w * \sqrt{\frac{\frac{n-1}{n^2} + (\frac{n-1}{n})^2}{n}} = w * \frac{\sqrt{n-1}}{n} \tag{13}$$

while the $Dev_{p_j t}$ is:

$$Dev_{p_j t} = \frac{W_{p_j t} - mean}{Disp_{pt}} = \frac{w - \frac{w}{n}}{w * \frac{\sqrt{n-1}}{n}} = \sqrt{n-1} \tag{14}$$

Based on the knowledge above, we can see that the $Disp_{st}, Disp_{pt}, Dev_{s_i t}, Dev_{p_i t}$ values are only related to the number of documents in the corpus m, the paragraphs number n and the threshold value TH, but not related to the frequency of the words in the documents.

We know that in short product reviews, the comments are usually in one paragraph. Concluded from **Property1**, we can obtain that if we use the Ku's method, the opinion targets extraction would have no relevance to the context and the words counts. However, in practice, the context and the word count do represent the product topics. For example, reviewers may comment a laptop product using word "screen" several times and the same conditions of the website name like "zhongguancun online". In this case, the extraction method should clearly extract word "screen", but not "zhongguancun online", which Ku's method cannot achieve. It would be important to consider the words context and word count. Based on the analysis of the products comments, in this paper we propose new definitions of word representation, and present a new method to extract opinion targets, which will be described in the next section.

3   The New Method

3.1   New Representation of Subjective Words: $word_{wf}$ and $word_{occur}$

Product comments significantly differ from those in news and long comments. We can find the following properties for short product reviews in the word level.

1. **Opinion targets and the emotional words appear together at a high proportion.** The opinion targets are mainly used for evaluation and description. In the semantic unit of the consumers' emotional expression, the opinion targets and sentimental words appear together at a high proportion. For example, in the mobile phone comment 'screen is big' the topic 'screen' and the sentiment word 'big' appear together. That is common for short reviews. However, in the long review the topic 'screen' may be followed by some descriptions like '...laptop screen updated with...'. In expression of a semantic unit, two factors are generally regarded to give the view of some products, which are sentimental words and opinion targets. Therefore, in the same semantic unit it is a higher proportion together with opinion targets and sentimental words.

2. **Low frequency words may also be sentimental words.** In long news or comments, low frequency word may be neglected. In papers [2, 3, 9], we can see that it is the problem during the frequent items mining for opinion mining to extract the low frequency opinion target directly. To manage that, some algorithms have been implemented to get low-frequency opinion targets. However, the results are not good. Based on the priori knowledge, we know that whether a word is an opinion target not only relates to the category of itself, but also the frequency accompanying with sentimental words.

In this paper, we propose a two-dimensional vector $word_{wf}$ and $word_{occur}$ to represent a word, where $word_{wf}$ is the proportion of the word accompanying with sentimental words, and $word_{occur}$ is the normalized value of the word frequency. By vector calculating, we can decide whether the word is an opinion target.

Proportion with Sentiment Words:

$$word_{occur} = \log \frac{N_{si}}{S} \tag{15}$$

where S is the number of the sentiment words in one comment, N_{si} is the number of word i accompanying with sentiment words.

Word Frequency:

$$word_{wf} = \log \frac{1}{N} \tag{16}$$

where N is the number of words in one comment (document). With the new word representation, we can construct the words input vector. The computing algorithm will be described in the following section.

3.2   The Method

We propose a new extracting method based on the neural network classifier (Fig 1).

There are three main steps in our method: preprocessing, computing the word vector representation $word_{wf}$ and $word_{occur}$, and classifying them using a neural network. In the first step, we input the corpus, use the segmentation and tagging to obtain the words and derive sentiment words from HOWNET [19]. In the second step, we compute the words values $word_{wf}$ and $word_{occur}$ according to the new definitions in section 3.1. Then we use a back propagation neural network to train and classify the words. We will give the detailed descriptions in the rest of this section.

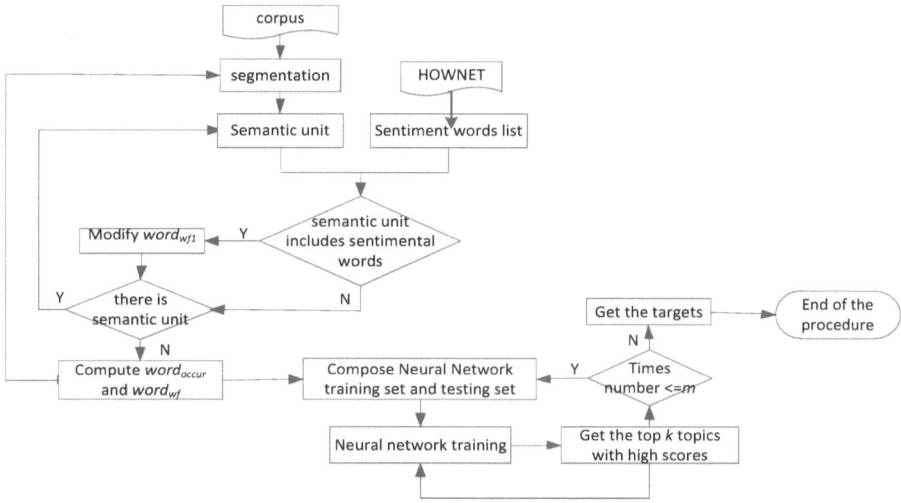

Fig. 1. Flowchart to get the opinion target

Preprocessing. When corpus is input, we firstly segment words and obtain noun words list. Text is divided into different semantic units according to the punctuation set. Sentimental words are derived from HOWNET. The details are described in Algorithm 1.

Algorithm 1. Processing for the Opinion Targets Extraction

Input: Documents in the corpus
Output: The nouns words list and the sentimental words list
 1: Segment words and compose the proper nouns list
 2: For each document in the corpus, construct semantic units separated by punctuations
 3: From HOWNET download the subjective documents and compose a set of sentiment words
 4: Return words list $PreWord$ and sentiment words set $OpinionWord$.

Computing the Words Values $word_{wf}$ and $word_{occur}$. The calculation of the two word representations $word_{wf}$ and $word_{occur}$ is given in Algorithm 2.

Training and Classifying. We use a neural network to extract topics. Firstly we generate a training set, and set initial opinion targets in the training set with score 1, non-opinion targets with score 0. Then, a three-layer back propagation neural network is adopted with the hyperbolic tangent function as the transfer function. The neural network contains one hidden layer, one input layer and one output layer. The input layer has three neurons, which are $word_{wf}$, $word_{occur}$, and constants input. The hidden layer has two neurons. The output layer has one neuron. The initial weights between the input layer and hidden layer are random numbers between [-0.2, 0.2], the initial weights between the hidden layer and output layer are random numbers between [-2.0, 2.0]. These parameter values were determined through empirical search via experiments to obtain good results. The details of the main parts are presented in Algorithm 3.

Algorithm 2. Computing the Words Values: $word_{wf}$ and $word_{occur}$.

Input: words list$PreWord$ and sentiment words set $OpinionWord$ output by **Algorithm 1**.
Output: The words values: $word_{wf}$ and $word_{occur}$.
1: for each document doc_i
2:    for each sentiment unit $sentence_{i,j}$
3:        for each word $word_t$ in list $PreWord$
4:            if $word_i$ is in $OpinionWord$
5:                $word_{wf}$ for the word i increases by one
6:                $word_{occur}$ increases by sentences count
7: for each word in $Preword$
8:    Calculate $word_{wf}$ according to the definitions
9:    Normalize($word_{occur}$)
10: **return** $word_{wf}$, $word_{occre}$.

Algorithm 3. Opinion Target Extraction Algorithm Using Neural Network

---

Input: Opinion target list $list_0$ and non-opinion target list $list_1$.
Output: Opinion target list.
1: Construct the words input vector . The words vector values are $word_{wf}$ and $word_{occur}$ calculated by Algorithm 2.
2: Train the neural network.
3: $value$ $n.test$. Compute and save the words and the words' scores for output.
4: Give 0.5 as the cut-off point between opinion targets and non-opinion targets. The words of scores greater than or equal to 0.5 are divided into opinion targets, less than 0.5 into non-opinion targets.
5: Select k number of the highest scores in the opinion targets, and add into $list_1$ and same number of the lowest scores in the non-opinion targets to add into $list_0$. (Here, in our experiment, we set $k = 6$)
6: Continue the training and classifying for m times (here in our experiment, we set $m =10$), compute new opinion targets and add in.
7: **return** Opinion target list n_{list}.

---

4   Experiments and Results

4.1   Data Set

We obtain online comments about mobile phones from "Zhongguancun online" website by web crawler with 14408 comments. According to the description of section 3.2, we initialize the targets to 1 and the non-targets to 0. The 37 opinion targets have been manually selected from the nouns and subjective documents in the corpus, which are manually labelled. Words in the training set are randomly selected from the segmentation words list $PreWord$, with proportion of the targets covering around 50%.

4.2   Evaluation Setup

To evaluate the new approach, we compare it with existing two popular methods: Ku's method [16, 17] and the LDA method [14]. Ku's method [16, 17] was used in opinion extraction for news and blog corpora both in English and in Chinese. It performed well with high F-measure for verbs and nouns. However, in section 2.2 we have proven that Ku's method can not work well for short comments. In the experiment, we compare the results for short comments to check whether the experimental results match the proof or not. For Ku's method, three different values for parameter TH were set and extraction is tested three times. The LDA-based approaches are the most popular topic modeling methods to extract the opinion targets. Titov's LDA method [14] had significant improvements in performance. In the experiment, we apply Titov's LDA method to the crawled mobile reviews to show how the methods work in short comments.

We use precision, recall and F-measure to evaluate the results. The metric scores for precision, recall and F-measure are computed over the top 20 extracted targets. Both the strict and lenient standards are applied for evaluation, where a strict standard means the same annotation must be achieved by all three annotators while a majority agreement of the annotators is required for a lenient evaluation [8]. The extraction program is running on Ubuntu version 10.10, and MySQL database MySQLdb package is from [20]. The Python language is mainly implemented. We use the bpnn package [21] for computing the values of words in opinion target extraction process, the mmseg package [22] for the words segmentation, and program [ICTCALS50_Linux_RHAS_32_c] for tagging [23].

A Chinese term is expressed as (A, B, "C") or A(B, "C"), in which A stands for Chinese word, B for the same word in English and C for the Chinese Pinyin.

4.3   Results and Comparison

Table 1 shows the precision, recall and F-measure results of our method and the other two methods (Ku's and LDA). The Para. in the second column of Table 1 stands for different parameters in the three methods, which is TH in Ku's method, topic number in LDA and times in our method.

Table 1. The Precision, Recall and F-measure Results Comparing with Ku's and LDA Methods

Methods	Para.	Strict			Lenient		
		P	R	F-measure	P	R	F-measures
Ku's	0.5	0.4000	0.2286	0.2909	0.3500	0.2000	0.2545
	2	0.3500	0.2000	0.2545	0.3500	0.2000	0.2545
	8	0.3500	0.2000	0.2545	0.3500	0.2000	0.2545
	Average	**0.3667**	**0.2095**	**0.2667**	**0.3500**	**0.2000**	**0.2545**
LDA	10	0.2500	0.1429	0.1819	0.2500	0.1429	0.1819
	100	0.3000	0.1714	0.2182	0.3000	0.1714	0.2182
	800	0.3000	0.1714	0.2182	0.3500	0.2000	0.2545
	Average	**0.2833**	**0.1619**	**0.2060**	**0.3**	**0.1714**	**0.2182**
Ours	1	0.5000	0.2857	0.3636	0.6500	0.3714	0.4727
	2	0.4500	0.2517	0.3228	0.6000	0.3429	0.4364
	3	0.4000	0.2286	0.2909	0.6500	0.3714	0.4727
	Average	**0.4500**	**0.2463**	**0.3184**	**0.6333**	**0.3619**	**0.4606**

According to Table 1, our new method consistently achieved much better performances than the Ku's method and LDA in terms of precision, recall and F-measure under both strict and lenient evaluations. This is reflected by the average results and also individual experiment runs. There is no single experiment run of our new method performed worse than the other two methods and different experiment runs of our new method achieved very similar performances, suggesting that the proposed new method is quite stable although it includes a stochastic component.

In terms of the strict evaluation, our approach achieved an average precision of 45%, which represents an 8.33% improvement over Ku's method and 16.67 % over LDA. At the same time, the new method also achieved much better recall and F-measure than the other two methods. For the lenient evaluation, the new method achieved an average precision of 63.33%, an average recall of 36.19%, and an average F-measure of 0.4606, which are all significantly better than the Ku's method and the LDA methods.

Comparing the two evaluations with the new method, we noticed that the use of the Lenient evaluation resulted in a much better precision, recall and F-measure than the strict evaluation (63.33% vs 45.00% in average precision, 36.19% vs 24.63% in average recall, and 0.4606 vs 0.3184 in average F-measure). This suggest that the lenient evaluation is more effective than the strict evaluation in the proposed new method. Inspection of the detailed results reveals that this is because there are many word abbreviations in the data set, and the lenient evaluation scheme can much better cope that situation than the strict evaluation scheme. For example, some topics such as "屏 (screen,'ping')", and "屏幕 (screen, 'pingmu')", really refer to the same topic ("screen"), but the strict evaluation scheme incorrectly treats them differently and assigned a negative value for matching while the lenient evaluation correctly classified them as the same topic and assigned a positive match output. Clearly, the new method with the lenient evaluation scheme can better deal with such cases than the strict evaluation scheme. We also noticed that this characteristic was not so clear for the Ku's and the LDA methods, where the two evaluation schemes achieved very similar results. This is mainly because we introduced a learning/adaptive process in the new method, which can automatically learn such features from existing instances.

4.4 Further Analysis

To illustrate why our new method works well on the opinion target extraction and how good the results are, we list the extracted top 20 topics by our method, Ku's method and the LDA method respectively in Tables 2, 3 and 4.

Table 2. The Top 20 Targets with Different TH Values by Ku's Method

NO.	TH		
	0.5	2	8
1	是, yes, "shi"	是 , yes, "shi"	是, yes, "shi"
2	很, very, "hen"	很, very, "hen"	很, very, "hen"
3	手机, mobile phone, "shouji"	手机, mobile phone, "shouji"	手机, mobile phone, "shouji"
4	也, too, "ye"	不, not, "bu"	功能, function, "gongneng"
5	功能, function, "gongneng"	我, I, "wo"	我, I, "wo"
6	不, not, "bu"	功能, function, "gongneng"	屏幕, screen, "pingmu"
7	屏幕, screen, "pingmu"	就, as soon as, "jiu"	不, not, "bu"
8	我, I, "wo"	和, and, "he"	就, as soon as, "jiu"
9	有, have, "you"	在, at, "zai"	不错, not bad, "bucuo"
10	在, at, "zai"	机, machine, "ji"	和, and, "he"
11	)	不错, not bad, "bucuo"	也, also, "ye"
12	屏, screen, "ping"	用, use, "yong"	在, at, "zai"
13	就, as soon as, "jiu"	屏幕, screen, "pingmu"	有, have, "you"
14	机, machine, "ji"	不错, not bad, "bucuo"	屏, screen, "ping"
15	系统, system, "xitong"	软件,, software, "ruanjian"	机, machine, "ji"
16	不错, not bad, "bucuo"	屏, screen, "ping"	软件, software, "ruanjian"
17	软件, software, "ruanjian"	电池, battery, "dianchi"	电池, battery, "dianchi"
18	电池, battery, "dianchi"	比, ratio, "bi"	也, also, "ye"
19	用, use, "yong"	电池, battery, "dianchi"	用, use, "yong"
20	比, ratio, "bi"	可以, can, "keyi"	比, ratio, "bi"

As can be seen from Table 2, the Ku's method incorrectly extracted many topics such as '在 (at "zai")', '是(yes "shi")', '很(very "very")', '也(too "ye")', '不(not "bu")', '我(I, "wo")', '用(use, "yong")', '不错(not bad, "bucuo")', '就(as soon as "jiu")' and '可以(can, "keyi")' . These words/topics have little relevance to the mobile phone product review. Even for the different TH values, many of these words/topics such as the topic '在(at "zai")', '是(yes "shi")', '很(very "very")', '也(too "ye")', '不(not "bu")" have been extracted every time. For the three different TH values of 0.5, 2 and 8, the number of irrelevant words/topics extracted by the Ku's method is 11/20, 12/20 and 12/20, respectively. This suggests that the Ku's method consistently extracted more than half irrelevant topics no matter what TH value was used.

Table 3. The Top 20 Targets with Different Topic Numbers by LDA

NO.	topic numbers		
	10	100	800
1	手机, mobile phone, "shouji"	价格,price, "jiage"	价格,price, "jiage"
2	是, yes, "shi"	一般, usual, "yiban"	屏, screen, "ping"
3	东西,goods,"dongxi"	说了, say,"shuole"	大, big, "da"
4	点, dot, "dian"	不 , not, "bu"	有点, some, "youdian"
5	就,as soon as, "jiu"	很, too, "hen"	电池, battery, "dianchi"
6	电池, battery, "dianchi"	是, yes, "shi"	一般, usual, "一般"
7	屏幕, screen, "pingmu"	还是, still, "haishi"	功能, function, "gongneng"
8	我, I, "wo"	手感,hands feel,"shougan"	什么, what, "shenme"
9	功能, function, "gongneng"	触摸, touch, "chumo"	外, outside, "wai"
10	减少, reduce, "jianshao"	好看,good-looking,"haokan"	GPS
11	苹果,apple,"pingguo"	功能, function, "gongneng"	好看,good-looking,"haokan"
12	8310	什么, what, "shenme"	没, no, "mei"
13	屏, screen, "ping"	很好, very good, "henhao"	很, very, "hen"
14	乔布, JOB, "qiaobu"	和, and, "he"	系统, system, "xitong"
15	很, very, "hen"	点, dot, "dian"	点, dot, "dian"
16	太贵, too exprensive, "taigui"	可能 ,maybe, "keneng"	都,also, "dou"
17	朴素, simple, "朴素"	锁, lock, "duo"	真的,real, "zhende"
18	易, easy, "yi"	外观,appearence, "waiguan"	给,give, "gei"
19	指纹,fingerprint , "zhiwen"	起来, up, "qilai"	丰富,plenty, "fengfu"
20	不错, good, "bucuo"	手机, mobile phone, "shouji"	待机,standby, "daiji"

The top 20 topics extracted by the LDA method are shown in Table 3 when the topic number varies (10, 100 and 800). Similarly to the Ku's method, many topics such as '东西(goods "dongxi")', '是(yes "shi")', '很(very "very")', '就(as soon as "jiu")', '什么(what, "shenme")' and '点(dot, "dian")' extracted by this method do not really have relevance to the mobile phone product review. The numbers of irrelevant topics extracted by this method under different topic numbers (10, 100 and 800) are 11/20, 13/20 and 12/20, which are even slightly more than those by the Ku's method. This is consistent with the performance shown in the previous subsection.

Table 4 shows the 20 top topics extracted by our new method. In Table 4, we can see that the words like '东西(goods "dongxi")' , '点(dot, "dian")' , '在(at "zai")' and '可以(can, "keyi")' are not listed in Table 4. Also some topics, such as '容量,capacity,"rongliang"', '后台,back-end,"houtai"', '分辨率,resolution"fenbianlv"', and '闪光灯, flash"shanguangdeng"' can give user useful information about the mobile phone product. In fact, almost all the topics extracted by our new methods are relevant to the mobile product review. This is due mainly to the introduction of the new representation of the subjective words $word_{wf}$ and $word_{occur}$, which allows the new method to successfully filter most of the irrelevant topics.

For the product review, the sentiment polarity and degree mainly depends on the product profile, such as color, appearance and power, etc. in mobile phones. These descriptive words are more important than the general words, which should be firstly ex-

Table 4. The Extraction Opinion Objects of Our Methods

NO	1	2	3
1	软件,software,"ruanjian"	机子,machine,"jizi"	人,person,"ren"
2	系统,system,"xitong"	电,electric,"dian"	网,web,"wang"
3	价格,price,"jiage"	缺点,shortcoming,"quedian"	歌,song,"ge"
4	机,machine,"ji"	电话,telephone,"dianhua"	能力,ability,"nengli"
5	感觉,feel,"ganjue"	时,time,"shi"	个人,single,"geren"
6	卡,card,"ka"	优点,advantage,"youdian"	电脑,computer,"diannao"
7	效果,effect,"xiaoguo"	机身,body,"jishen"	版,version,"ban"
8	时间,time,"shijian"	诺,No,"nuo"	线,line,"xian"
9	问题,problem,"wenti"	按键,press key,"anjian"	音,sound,"yin"
10	头,head,"tou"	时尚,fashion,"shishang"	闪光灯,flash,"shanguangdeng"
11	键盘,keyboard,"jianpan"	外形,shape,"waixing"	东西,goods,"dongxi"
12	速度,speed,"sudu"	键,keyboard,"jian"	牙,tooth,"牙"
13	智能,intelligence,"zhineng"	基,Ki,"ji"	版本,version,"banben"
14	音乐,music,"yinyue"	话,talk,"hua"	电容,capacitance,"dianrong"
15	款,style,"kuan"	天,sky,"tian"	容量,capacity,"rongliang"
16	游戏,game,"youxi"	质量,quality,"zhiliang"	苹果,apple,"pingguo"
17	点,dot,"dian"	做工,work,"zuogong"	主题,topic,"zhuti"
18	耳机,earphone,"erji"	界面,interface,"jiemian"	后台,back-end,"houtai"
19	分辨率,resolution,"fenbianlv"	机器,machine,"jiqi"	网络,web,"wangluo"
20	视频,video,"shipin"	程序,program,"chengxu"	手,hand,"shou"

tracted. Using our method with the new representation and the learning/adaptive process, we can successfully extract more descriptive words than the other two methods.

5 Conclusions

The goal of this paper was to investigate whether the existing commonly used opinion mining approaches in long review comments can be used for extracting opinion targets in short product reviews and develop a new approach to successfully extracting opinion targets from short comments. This goal has been successfully achieved by theoretically proving Ku's method cannot successfully extract opinion targets from short comments, introducing a new representation of subjective words, and proposing a three step approach to automatically learning and extracting opinion targets from short comment reviews. The new approach was examined and compared with the Ku's method and another common method LDA for opinion mining on a large corpus of mobile phone network (short) comment review from "Zhongguancun online". The experimental results show that the Ku's method could not achieve good performance on extracting opinion targets from these short comments (which is consistent with the theoretical results) and that the proposed method significantly outperformed the Ku's and LDA method on this data set. Further inspection of the results of the three methods reveals that more than half of the top 20 targets extracted by the two existing methods are irrelevant to the mobile phone products and that almost all of the top 20 opinion targets extracted by the new methods are relevant. These results in turn show that the new representations and learning/adaptive process introduced to the new approach are effective and the new method can obtain much more meaningful topics than the two existing, commonly used methods in the literature.

Although the proposed method achieve better performance than the two popular existing methods, it still has space for improvement as the precisions and recalls are still not sufficiently high. For future work, we will consider new ways for improving the system performance, particularly investigating new methods for better segmentation of words, and considering approaches to better using prior knowledge for word representation and automatically learning the weights of opinion targets.

Acknowledgements. This work is supported in part by the National Science Foundation of China (NSFC No. 61170180, NSFC No. 61035003). We would like thank Bing Xue for her comments and help for improving the paper.

References

1. Morinaga, S., Yamanishi, K., Tateishi, K., Fukushima, T.: Mining product reputations on the web. In: KDD, pp. 341–349 (2002)
2. Popescu, A.-M., Etzioni, O.: Extracting product features and opinions from reviews. In: HLT/EMNLP (2005)
3. Hu, M., Liu, B.: Mining opinion features in customer reviews. In: AAAI, pp. 755–760 (2004)
4. Pang, B., Lee, L., Vaithyanathan, S.: Sentiment classification using machine learning techniques. CoRR, cs.CL/0205070 (2002)
5. Yao, T., Uszkoreit, H.: Building a lexical sports ontology for chinese ie using reusable strategy. In: FSKD, vol. (2), pp. 668–672 (2007)
6. Turney, P.D.: Thumbs up or thumbs down? semantic orientation applied to unsupervised classification of reviews. CoRR, cs.LG/0212032 (2002)
7. Liu, B.: Web Data Mining: Exploring Hyperlinks, Contents, and Usage Data. In: Data-Centric Systems and Applications. Springer (2007)
8. Pang, B., Lee, L.: Opinion mining and sentiment analysis. Foundations and Trends in Information Retrieval 2(1-2), 1–135 (2007)
9. Yi, J., Nasukawa, T., Bunescu, R.C., Niblack, W.: Sentiment analyzer: Extracting sentiments about a given topic using natural language processing techniques. In: ICDM, pp. 427–434 (2003)
10. Blei, D.M., Lafferty, J.D.: Correlated topic models. In: NIPS (2005)
11. Qiu, G., Liu, B., Bu, J., Chen, C.: Opinion word expansion and target extraction through double propagation. Computational Linguistics 37(1), 9–27 (2011)
12. Zhang, L., Liu, B.: Identifying noun product features that imply opinions. In: ACL (Short Papers), pp. 575–580 (2011)
13. Stoyanov, V., Cardie, C.: Topic identification for fine-grained opinion analysis. In: COLING, pp. 817–824 (2008)
14. Titov, I., McDonald, R.T.: Modeling online reviews with multi-grain topic models. In: WWW, pp. 111–120 (2008)
15. Ma, T., Wan, X.: Opinion target extraction in chinese news comments. In: COLING (Posters), pp. 782–790 (2010)
16. Ku, L.-W., Lee, L.-Y., Wu, T.-H., Chen, H.-H.: Major topic detection and its application to opinion summarization. In: SIGIR, pp. 627–628 (2005)
17. Ku, L.-W., Liang, Y.-T., Chen, H.-H.: Opinion extraction, summarization and tracking in news and blog corpora. In: AAAI Spring Symposium: Computational Approaches to Analyzing Weblogs, pp. 100–107 (2006)
18. Fukumoto, F., Suzuki, Y.: Event tracking based on domain dependency. In: SIGIR, pp. 57–64 (2000)
19. Dong, Z., Dong, Q., Hao, C.: Hownet and its computation of meaning. In: COLING (Demos), pp. 53–56 (2010)
20. http://mysql-python.sourceforge.net/mysqldb.html
21. http://dir.filewatcher.com/d/other/src/development/languages/bpnn-0.2.0-2.src.rpm.4066.html
22. http://www.coreseek.com/opensource/mmseg/
23. http://ictclas.org/ictclas_download.aspx

Managing the Complexity of Large-Scale Agent-Based Social Diffusion Models with Different Network Topologies

Alexei Sharpanskykh

VU University Amsterdam, Department of Artificial Intelligence
De Boelelaan 1081, 1081 HV Amsterdam, The Netherlands

Abstract. Social diffusion models have been extensively applied to study biological and social processes on a large scale. Previously two issues with these models were identified: understanding emerging dynamic properties of complex systems, and a high computational complexity of large-scale social simulations. Both these issues were tackled by abstraction techniques developed previously for social diffusion models with underlying random networks. In the paper it is shown that these techniques perform poorly on scale-free networks. To address this limitation, new model abstraction methods are proposed and evaluated for three network types: scale-free, regular and random. These methods are inspired by node centrality measures from the social networks area. The proposed methods increase the computational efficiency of the original model significantly (up to 40 times for regular networks).

Keywords: group dynamics, model abstraction, social diffusion, large-scale agent-based simulation.

1 Introduction

Social diffusion models have been extensively applied to study diverse biological and social processes on a large scale, such as spread of innovation [10], dynamics of epidemics, formation and spread of opinions [4-7, 9, 12, 13]. These models describe a gradual spread of states (e.g., information) of agents in a network. Previously two issues with large-scale social diffusion models were identified [12, 13]: understanding emerging dynamic properties of systems, and a high computational complexity of large-scale social simulations. To address these issues, model abstraction techniques were proposed for social diffusion models with underlying random networks [12, 13]. The main idea behind these techniques was to identify dynamic clusters of agents with similar states, and to replace them by super-agents that were used in simulation as aggregate objects. The dynamic properties of these super-agents were inferred from properties of the agents of which they comprised. When some of the agents forming a super-agent changed their states, dynamic re-grouping of the super-agent took place.

The existing abstraction techniques for social diffusion models [12, 13] and model reduction methods [1, 2] are based on direct calculation of the equilibrium state of a network. However, as shown in the paper, when a network does not converge to

P. Anthony, M. Ishizuka, and D. Lukose (Eds.): PRICAI 2012, LNAI 7458, pp. 540–551, 2012.
© Springer-Verlag Berlin Heidelberg 2012

equilibrium quickly and/or the equilibrium state is disturbed frequently, these abstraction methods perform poorly. To address this limitation, three methods for approximating the dynamics of a large-scale social diffusion model are proposed, which do not rely on the system reaching equilibrium. For the illustration of the methods a social decision making model based on social diffusion is used in the paper. The methods proposed are based on determining a relative contribution of each agent from a group to the joint group's opinion. The agent's contribution depends on the agent's degree of influence in the network. The abstraction methods proposed differ only in the way how the agent's degree of influence is defined. In contrast to the existing model abstraction techniques, the abstraction methods proposed are inspired by node centrality measures used in social network analysis [3, 8].

The second contribution of the paper is that the proposed abstraction methods, as well as the most promising existing abstraction technique from [12], are evaluated on three network types, which often occur in many fields [3]: scale-free, regular and random. A few works exist that investigate effects of network topologies on the dynamics of social diffusion models [6]. To our best knowledge, analysis of abstraction techniques for social diffusion models was only attempted for random networks [12, 13]. Our analysis showed that the network type is essential for the applicability of the model abstraction techniques: whereas equilibrium-based techniques are well-suited for random and regular networks, these techniques perform poorly on scale-free networks. For the latter networks, the abstraction methods proposed in this paper show the best approximation results. Furthermore, the developed abstraction methods increase the computational efficiency of the original model significantly. The acceleration factor is the highest for regular networks (~40).

In many applications the size of dynamic groups, which could be numerous, is (much) smaller than the total number of agents. Previously, abstraction techniques were applied in a large-scale crowd evacuation study (~10000 agents) [11]. Although the number of agents was significant, the maximal size of emergent dynamic groups was 174. The maximal size of groups considered in this paper is 1000 agents.

The paper is organized as follows. An agent-based social decision making model based on social diffusion is described in Section 2. The proposed model abstraction methods are explained in Section 3 and evaluated in Section 4. Section 5 concludes the paper.

2   A Social Decision Making Model Based on Social Diffusion

In this section first a model description is provided in Section 2.1. Then, in Section 2.2 model analysis is given.

2.1   Model Description

The model describes decision making by agents in a group as a process of social diffusion. Two decision options – $s1$ and $s2$ - are considered by the agents. The opinion $q_{s,i}$ of an agent i for a decision option $s \in \{s1, s2\}$ is expressed by a real number in the range $[0, 1]$, reflecting the degree of the agent's support for the option. For each option each agent communicates its opinion to other agents, and thus influences their opinions. Agents communicate only with those agents to which they are connected in a social network representing a group. In this study we consider three network topologies (see Fig.1):

- *Scale-free network*: a connected graph with the property that the number of links originating from a given node representing an agent has a power law distribution. In such networks the majority of the agents have one or two links, but a few agents have a large number of links.
- *Regular network*: a connected graph with the property that most nodes have approximately the same number of links (i.e., a homogeneous network). In this study each node has the number of links equal to the half of the number of agents.
- *Random network*: a graph, in which links between nodes occur at random, with equal probability $p=0.5$. Only connected graphs are considered in this study.

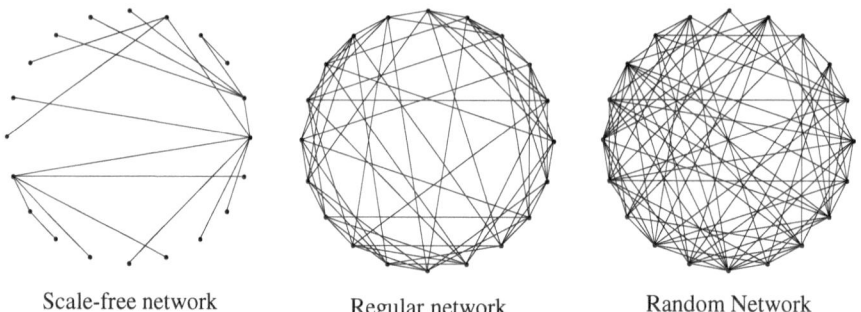

Scale-free network   Regular network   Random Network

Fig. 1. Examples of different network topologies with 20 agents

It is assumed that the agents are able to both communicate and receive opinions to/from the agents, to which they are connected (i.e., the links between agents are bidirectional). Furthermore, a weight $\gamma_{i,j} \in [0,1]$, indicating the degree of influence of agent i on agent j, is associated with each link for each direction of interaction. This weight determines to which extent the opinion of agent i influences the opinion of agent j for each option.

In the literature on social diffusion two modes of interaction of agents have been repeatedly considered:

- *synchronous* or *parallel mode*, in which all agents interact with each other at the same time (synchronously), as in [9];
- *asynchronous* or *sequential mode*, in which at every time point, only one pair of connected agents chosen randomly interacts, as in [4].

Both these modes are considered in this paper. In the parallel mode, the opinion states of the agents are updated at the same time as follows:

$$q_{s,i}(t+\Delta t) = q_{s,i}(t) + \eta_i\, \delta_{s,i}(t)\Delta t \qquad (1)$$

Here η_i is an agent-dependent parameter within the range $[0,1]$, which determines how fast the agent adjusts to the opinion of other agents, $\delta_{s,i}(t)$ is the amount of change of the agent i's opinion due to the influence of other agents defined as:

$$\delta_{s,i}(t) = \sum_{j\neq i} \gamma_{j,i}(q_{s,j}(t) - q_{s,i}(t))/\sum_{j\neq i} \gamma_{j,i} \qquad \text{when } \sum_{k\neq i} \gamma_{k,i} \neq 0$$

and

$$\delta_{s,i}(t) = 0 \qquad \text{when } \sum_{k\neq i} \gamma_{k,i} = 0$$

In the sequential mode two agents interact at each time point. The opinion states of interacting agents i and j are updated similarly to (1):

$$q_{s,i}(t+\Delta t) = q_{s,i}(t) + \eta_i \gamma_{j,i}/\sum_{j\neq i} \gamma_{j,i} (q_{s,j}(t) - q_{s,i}(t))\Delta t \qquad (2)$$

$$q_{s,j}(t+\Delta t) = q_{s,j}(t) + \eta_i \gamma_{i,j}/\sum_{i\neq j} \gamma_{i,j}(q_{s,i}(t) - q_{s,j}(t))\Delta t \qquad (3)$$

The opinion states of the agents not interacting remain the same:

$$q_{s,k}(t+\Delta t) = q_{s,k}(t) \qquad \text{when } k\neq i, k\neq j$$

In the real world, opinions of human agents are influenced not only by their peers with whom they have mutual connections, but also by external information sources, such as media. To represent such sources in the model, a special agent type is introduced called *information source agent*. Such agents can influence normal agents, but cannot be influenced by any of the agents, i.e., $\gamma_{i,a}=0$ for an information source a and all agents i. It is assumed that interaction between normal agents and information source agents lasts shortly (1 time point in the simulation). For the time point, when agent i interacts with information source a, the following formula is used to update the agent i's opinion state instead of the formulae (1)-(3) in both parallel and sequential modes:

$$q_{s,i}(t+\Delta t) = q_{s,i}(t) + \eta_i \gamma_{a,i}(q_{s,a}(t) - q_{s,i}(t))\Delta t \qquad (4)$$

Different interaction frequencies of agents with information sources are examined by simulation in Sections 2.2 and 4.

2.2   Model Analysis

The model abstraction methods developed previously (e.g., [12, 13, 2]) are based on approximation of the behaviour of a model by the model's equilibrium state, calculated explicitly. However, if a model does not reach the equilibrium state quickly and/or the equilibrium state is disturbed frequently, the precision of the equilibrium-based abstraction methods may decline drastically (more precise results are given in Section 4). In this section we examine by simulation how the speed of convergence of the agent-based social diffusion model from Section 2.1 depends on the topology of the underlying network and on the number of agents in the network.

It was shown in [12] that every static network, in which each agent is influenced by at least one another agent, reaches an equilibrium state. Thus, since only connected networks are considered in this study, every model in this study reaches an equilibrium state. Also, previously it was proved for social diffusion models that the agents are in an equilibrium state if and only if their opinions are the same [12]. Based on this result, the degree of convergence cd of the model is evaluated in this study by the sum of the variances of the opinions of the agents for each option:

$$cd(t) = \sum_{s=\{s1,s2\}} var(q_{s,1..n}(t)),$$

where n is the number of agents.

To evaluate the degree of convergence for the model, 18 simulation settings were introduced. Each setting is characterized by the mode (parallel, sequential), network type (scale-free, regular, random) and the number of agents (100, 500, 1000). Then, each setting was simulated 1000 times and *cd* was calculated and averaged over all simulations for every 100th time point. The results are given in Figures 2-5.

As can be seen from Fig.2, in the parallel mode, during the first 100 time points the variance of the agents' opinions decreases very quickly to a very low value, and then remains low and decreases slowly. The speed of convergence of the model to equilibrium is higher for the random and regular networks in both sequential and parallel modes. The results for the regular networks, not shown in the figures, are very close to the ones for the random networks. This explains why equilibrium-based abstraction methods developed previously (as in [12, 13]) perform well on random networks. The rate of convergence of the model decreases with an increase in the number of agents in all simulation settings. In the sequential mode (see Fig.3), the model stabilizes much slower than in the parallel mode, especially when the number of agents is high.

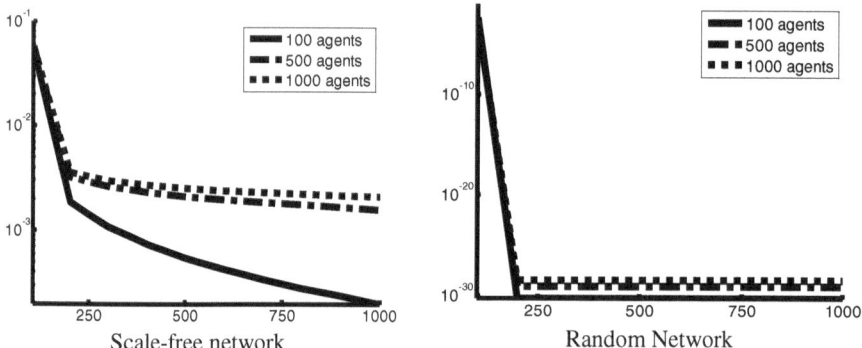

Fig. 2. Change of the variance of the opinion states of the agents over time in the parallel mode; the horizontal axis is time, the vertical axis is variance (log scale)

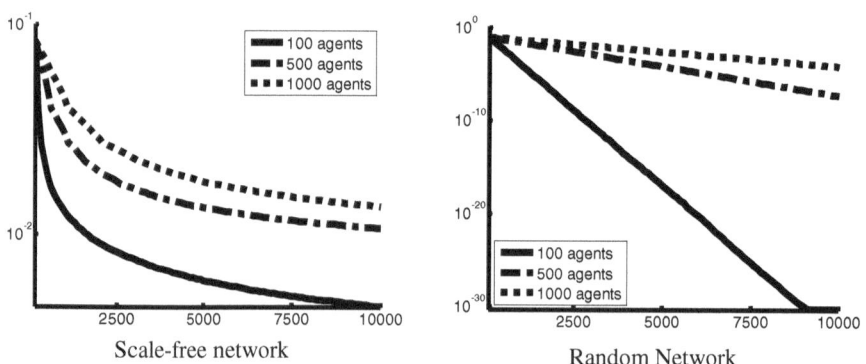

Fig. 3. Change of the variance of the opinion states of the agents over time in the sequential mode; the horizontal axis is time, the vertical axis is variance (log scale)

When information source agents send messages to other agents, they may disturb the convergence of the network. In Fig. 4 and 5 it is shown how the degree of convergence of the model is influenced by information source agents. All simulations were performed for the model with 500 normal agents and 50 information source agents. The average time between messages (ATBM) was varied between 10 and 5000. As can be seen in Fig. 4, in the parallel mode, the model stabilizes quickly after each interaction with information source agents. Thus, the number of messages does not influence the degree of convergence of the model in the parallel mode greatly. In the sequential mode (see Fig. 5), the number of messages has a greater effect on the degree of convergence than in the parallel mode.

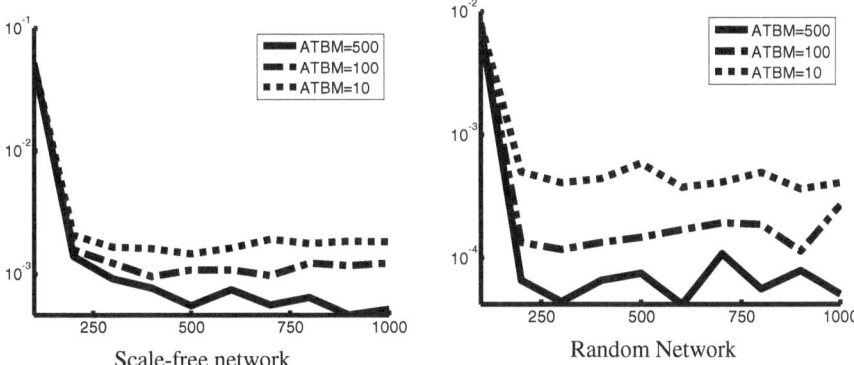

Fig. 4. Change of the variance of the opinion states of 500 agents over time in the parallel mode with the average time between messages (ATBM) > 0; the horizontal axis is time, the vertical axis is the variance of the opinion states (log scale)

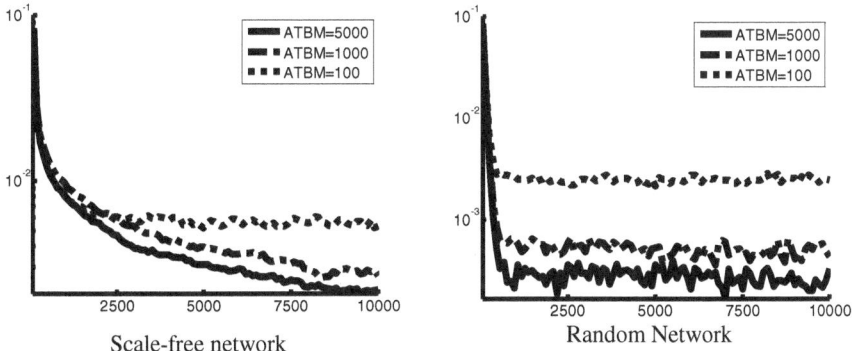

Fig. 5. Change of the variance of the opinion states of 500 agents over time in the sequential mode with the average time between messages (ATBM) > 0; the horizontal axis is time, the vertical axis is the variance of the opinion states (log scale)

In conclusion, whereas equilibrium-based abstraction methods may be effective for approximating the model with random and regular networks in the parallel mode, other abstraction methods need to be developed for models with scale-free networks. Three such methods are proposed in the next section.

3   Model Abstraction Methods

In this section three new model abstraction methods are proposed. The methods approximate the averaged opinion states of a model with network G with n agents: $q_{s,G}(t)=\Sigma_{i \in G}\ q_{s,i}(t)/n$. The approximated states $q^*_{s,G}$ are determined at the beginning of the simulation by weighted aggregation of the agents' initial opinions:

$$q^*_{s,G}(1)=\Sigma_{i \in G}\ sv_i\ q_{s,i}(1)/\ \Sigma_{i \in G}\ sv_i \tag{5}$$

The weight sv_i of an agent i (significance value) is an indication of the agent's amount of influence in the network. The abstraction methods proposed in this section differ only in the ways how the significance values are calculated.

The approximated states are updated after every interaction of an agent j with an information source agent:

$$q^*_{s,G}(t)= (q^*_{s,G}(t-\Delta t)\Sigma_{i \in Gj}\ sv_i + sv_j\ q_{s,j}(t))/\Sigma_{i \in G}\ sv_i \tag{6}$$

where $q_{s,j}(t)$ is calculated by (4), and the agent j's opinion states before the interaction are equal to the approximated states $q^*_{s,G}(t-\Delta t)$.

For the rest of the time the approximated states do not change: $q^*_{s,G}(t)= q^*_{s,G}(t-\Delta t)$.

In the first abstraction method $M1$, significance value sv_i of agent i is determined by the sum of the degrees of the agent's influence on the other agents in the network:

$$sv^{M1}_i = \Sigma_{j \in G}\ \gamma_{i,j} \tag{7}$$

This measure is similar to the degree centrality measure often used in social networks [3]. In [8] a centrality measure for weighted networks is proposed, in which both weights of links and the number of links are taken into account. Preliminary experiments with this measure produced results worse than the ones of method $M1$. Instead of this measure, the sum of relative degrees of influence $\gamma_{i,j}/\Sigma_{k \in G}\ \gamma_{k,j}$ was used as the basis for method $M2$:

$$sv^{M2}_i = \Sigma_{j \in G}\ \gamma_{i,j}*, \tag{8}$$

where $\gamma_{i,j}* = \gamma_{i,j}/\Sigma_{k \in G}\ \gamma_{k,j}$, if $\Sigma_{k \in G}\ \gamma_{k,j} > \gamma_{i,j}$, and $\gamma_{i,j}* = \gamma_{i,j}$ otherwise.

This measure stems from the model itself, and thus reflects better its dynamics than the measures from both $M1$ and [8].

Both measures sv^{M1}_i and sv^{M2}_i are calculated for each agent locally, by taking into account direct influences only. In the third abstraction method $M3$, also indirect influences of agents in the network are considered. In $M3$ the significance values of the agents are propagated through the network. When the value is propagated, it is multiplied by the strength of the link. Thus, agents, which have a high influence on influential agents, have also a high significance value. The significance value sv^{M3}_i is calculated by an iterative algorithm provided below.

Initially, the significance values of the agents are set to sv^{M2}_i. Then, at every iteration the agent j's significance value from the previous iteration multiplied by the relative degree of influence of agent i on agent j (i.e., $sv^{M3}_j(iter-1)\gamma_{i,j}\ /\Sigma_{k \in G}\ \gamma_{k,j}$) is propagated to agent i. The new significance value of each agent is the sum of the received (propagated) significance values. Thus, with every iteration, indirect influences of more and more distant agents in the network are taken into account in the calculation of the significance values.

Algorithm 1. Calculating significance values sv^{M3}_i

1: for all agents i: $sv^{M3}_i(1) \Leftarrow sv^{M2}_i$

2: **for** $iter=2$ to $<number\text{-}iterations>$ **do**

3:    For all agents i: $sv^{M3}_i(iter)= \sum_{j\in G} sv^{M3}_j(iter\text{-}1)\, \gamma_{i,j}{}^*$,

   $\gamma_{i,j}{}^* = \gamma_{i,j}/\sum_{k\in G} \gamma_{k,j}$, if $\sum_{k\in G} \gamma_{k,j} > \gamma_{i,j}$, and $\gamma_{i,j}{}^* = \gamma_{i,j}$ otherwise.

4: **end for**

The precision of the abstraction methods is evaluated by calculating root-mean-square error (RMSE) defined as

$$(\textstyle\sum_{t=1..end\_time} ((q^*_{1,G}(t)- q_{1,G}(t))^2 +(q^*_{2,G}(t)- q_{2,G}(t))^2)/end\_time)^{1/2}$$

In Algorithm 1 the precision of abstraction method $M3$ depends on the number of iterations. In the scale-free networks the decrease of the error with an increase of the number of iterations is gradual (see Fig.6, left), whereas in the regular and random networks the error drops only during a few first iterations and remains almost constant afterwards.

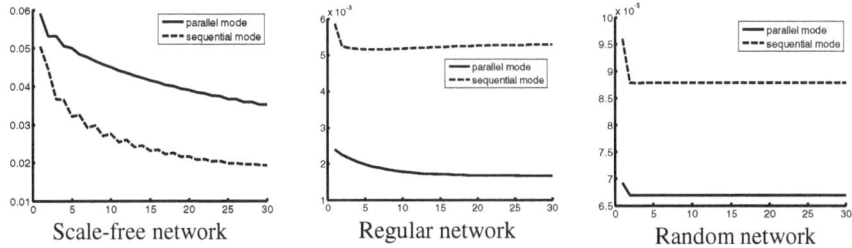

Fig. 6. Change of the root-mean-square error during the first 30 iterations (averaged over 1000 simulation trials) in networks with 100 agents

This may be explained by that regular and random networks are more densely connected than scale-free networks. Thus, in the former networks indirect influence of agents reaches every agent in a fewer number of iterations than in the latter networks.

In the next section the abstraction methods proposed will be evaluated by measuring RMSE and simulation time for the network topologies under consideration.

4   Evaluation of the Model Abstraction Methods

To evaluate the model abstraction methods described in Section 3, 72 simulation settings were introduced. Each setting is characterized by the mode (parallel, sequential), network topology type (scale-free, regular, random), the number of agents (100, 500, 1000), and the average time between messages (1, 2, 5, 10 for the parallel mode and 10, 50, 100, 500 for the sequential mode). In addition, the number of information source agents was 10 times less than the number of normal agents. The initial opinion states of the agents $q_{s,i}(1)$ were uniformly distributed in the interval [0,1]. The parameters $\gamma_{i,j}$ and η_i were taken from the uniform distribution in the interval]0,1]. Moreover, $\Delta t = 1$. The simulation time was 1000 for the parallel mode and 10000 for the sequential mode. For method *M3* in Algoritm1 30 iterations were performed for the scale-free networks and 10 iterations were performed for the regular and random networks. Each simulation setting was simulated 1000 times for each abstraction method, and RMSE and simulation time was determined and averaged over all simulations runs for each setting. The results obtained for three abstraction methods proposed were also compared to the results of the invariant-based abstraction method, which showed the best performance in [12]. The RMSE obtained for the four abstraction methods for networks with 500 and 1000 agents are provided in Figures 7 and 8. In the following the results obtained are discussed.

It was established in Section 2.2 that the speed of convergence of the model to equilibrium is higher for random and regular networks than for scale-free networks in both sequential and parallel modes. Because of a relatively slow convergence of scale-free networks, the invariant-based abstraction method, which relies on a quick convergence to an equilibrium state, performed poorly in the parallel mode in comparison to the methods proposed in this paper (Fig.7). Also, in the sequential mode, *M3* outperformed the invariant-based method (Fig.8).

From the three methods proposed *M3* provides the best approximation results for the scale-free network topology. In the parallel interaction mode *M1* slightly outperforms *M2*, again because *M2* approximates the equilibrium state of the network slightly better than *M1*.

In the sequential mode, the scale-free network has more time to reach an equilibrium state, due to longer periods between subsequent messages. Thus, the averaged opinion states of the network are closer to the equilibrium states. Therefore, the invariant-based method outperforms methods *M1* and *M2*, however, still worse than *M3*. As *M2* produces approximations that are closer to the equilibrium states than approximations obtained by *M1*, *M2* shows better results than *M1* for the scale-free networks in the sequential mode.

For the regular network, in the parallel mode the invariant-based method outperforms greatly all other methods. This can be partially explained by a relatively fast convergence of regular networks to equilibrium in the parallel mode. In sequential mode, all the methods show similar performance for both regular and random networks (Fig.8).

For the random networks in the parallel mode the invariant-based method and *M2* show the best approximation results. Partially this can be explained by fast convergence of random networks to equilibrium. As both the invariant-based and *M2* methods provide a close approximation of an equilibrium state, these methods perform the best for random networks.

Note that for all methods the error decreases with the increase of the number of agents. This result was also observed in [12]. In many cases the error grows with the increase of the average time between messages from information source agents. This is because the update of the approximated states occurs only after interaction with information source agents. The longer periods between updates are, the more error accumulates. However, frequent messages disturb the process of convergence and, thus, also contribute to the error growth. The errors for the regular and random networks are lower than the errors for the scale-free networks. Furthermore, the errors in the sequential mode on average lower than the errors in the parallel mode.

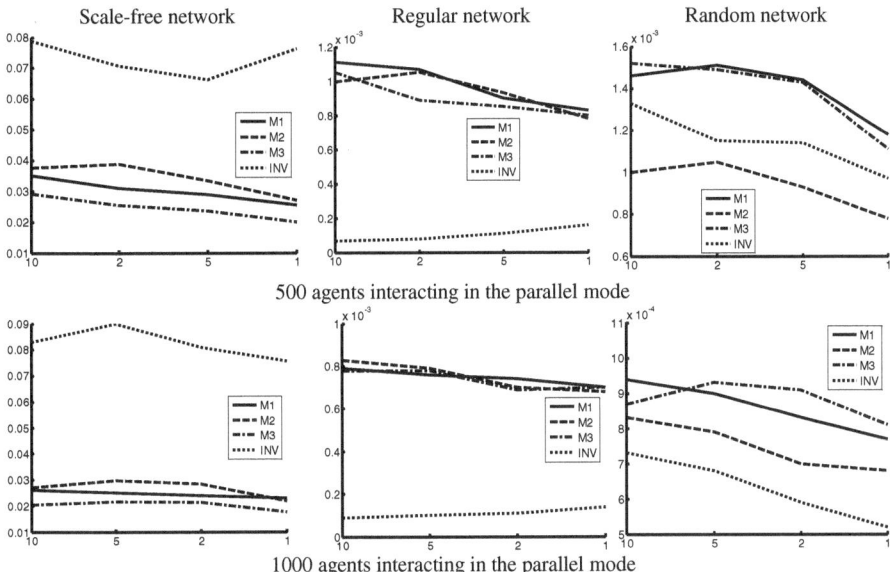

Fig. 7. Mean RMSE for the proposed abstraction methods *M1-M3* and the invariant-based method (*INV*) in the parallel mode; the horizontal axis is the average time between messages

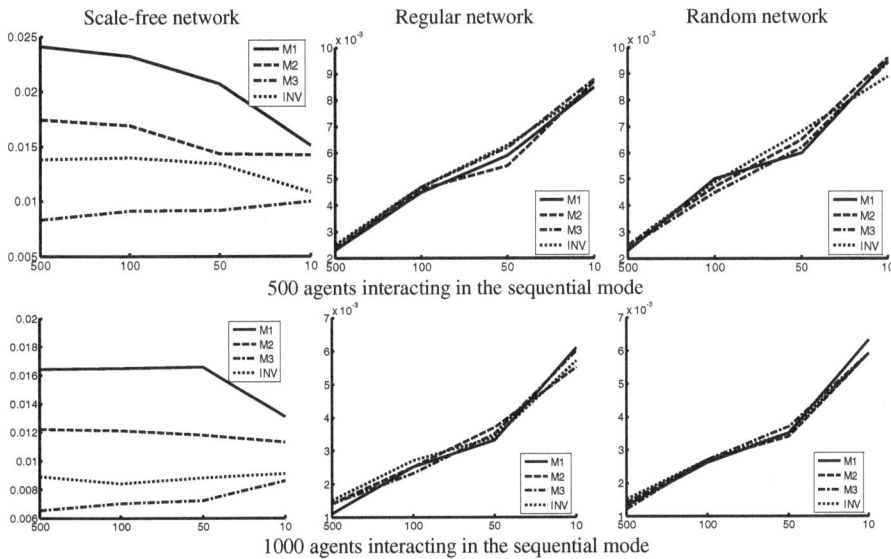

Fig. 8. Mean RMSE errors for the proposed abstraction methods *M1-M3* and the invariant-based method (*INV*) in the sequential mode; the horizontal axis is the average time between messages

The mean time complexity for the original model from Section 2 and for the proposed abstraction methods is provided in Table 1. The variances of these results are very low (of the order of 10^{-5}).

The developed abstraction methods increase the computational efficiency of the original model in the parallel mode significantly. The fastest simulation models are obtained by abstraction method *M1*, whereas abstraction by method *M3* is the slowest. However, the time difference between *M1* and *M3* is small (less than 1.5 in the worst case). The fastest simulation model is obtained for the scale-free networks, the slowest – for the random networks. Thus, the simulation time depends on the density of a network. The largest acceleration factor due to abstraction is obtained for the regular networks (up to 40); for scale-free networks – up to 25, and for random networks – up to 30. With the increase of the number of messages, the simulation time increases. This is because of the update of the approximated states after each interaction with information source agents.

Note that the simulation time of the model in the sequential mode is low, and depends weakly on the number of agents. This is because the states of only one pair of agents are updated at each time point.

Table 1. Mean simulation time in seconds for the original and abstracted models; the average time between messages is 1 for the parallel mode and 10 for the sequential mode

Network type	Scale-free			Regular			Random		
# of agents	100	500	1000	100	500	1000	100	500	1000
Parallel mode									
Original model	0.51	9.9	41	0.73	15.9	54.2	1	25.1	99.48
Method *M1*	0.02	0.05	2.7	0.03	0.74	4.42	0.05	0.92	4.62
Method *M2*	0.02	0.05	2.7	0.03	0.75	4.51	0.05	0.97	4.83
Method *M3*	0.03	0.07	3.6	0.03	0.93	5.3	0.05	0.99	5.8
Sequential mode									
Original model	0.56	3.3	9.4	0.55	3.5	10.9	0.59	3.8	11.7
Method *M1*	0.31	2.1	6.8	0.3	2.2	8.3	0.33	2.4	8.6
Method *M2*	0.31	2.1	6.8	0.3	2.2	8.3	0.33	2.4	8.6
Method *M3*	0.31	2.3	7.8	0.31	2.4	9.2	0.36	2.7	9.5

5   Conclusions and Discussion

Model abstraction methods, as proposed in the paper, provide information about global dynamics of a system by approximating global system states. Furthermore, these methods allow tackling computational efficiency problems of large-scale simulations. As demonstrated in the paper, the precision and efficiency of model abstraction methods depends greatly on the network type underlying the model. The equilibrium-based methods are well-suited for random and regular networks, which allow a fast convergence of the model to equilibrium. At the same time, the methods proposed in this paper are much better than the equilibrium-based methods for models with scale-free networks. In contrast to the existing model abstraction techniques, our methods do not rely on the system reaching equilibrium, which is a desirable property for slow converging scale-free networks.

As demonstrated in the paper, the developed abstraction methods increase the computational efficiency of the original model in the parallel mode significantly (up to 40 times for the regular topology).

Currently several techniques for abstraction of models based on hybrid automata [1] and differential equations [2] exist. However, such approaches can be applied efficiently only for systems described by sparse matrixes. Social diffusion models represent tightly connected systems, which do not allow a significant reduction of the state space using such techniques. In particular, a previous study showed that common model reduction techniques such as balanced truncation [2] do not allow decreasing the rank of the matrix describing the model from Section 2.

Acknowledgement. This research is supported by the Dutch Technology Foundation STW, which is the applied science division of NWO, and the Technology Programme of the Ministry of Economic Affairs.

References

1. Alur, R., Henzinger, T.A., Lafferriere, G., Pappas, G.J.: Discrete abstraction of hybrid systems. Proceedings of IEEE 88(7), 971–984 (2000)
2. Antoulas, A.C., Sorensen, D.C.: Approximation of large-scale dynamical systems: An overview. International Journal of Appl. Math. Comp. Sci. 11, 1093–1121 (2001)
3. Barabási, A.-L., Albert, R.: Emergence of Scaling in Random Networks. Science 286, 509–512 (1999)
4. Deffuant, G., Neau, D., Amblard, F., Weisbuch, G.: Mixing beliefs among interacting agents. Advances in Complex Systems 3, 87–98 (2001)
5. Hegselmann, R., Krause, U.: Opinion Dynamics and Bounded Confidence Models, Analysis and Simulation. Journal of Artificial Societies and Social Simulation 5(3) (2002)
6. Long, G., Yun-feng, C., Xu, C.: The evolution of opinions on scale-free networks. Frontiers of Physics in China 1(4), 506–509 (2006)
7. Macy, M., Kitts, J.A., Flache, A.: Polarization in Dynamic Networks: A Hopfield Model of Emergent Structure. In: Dynamic Social Network Modeling and Analysis, pp. 162–173. National Academies Press, Washington, DC (2003)
8. Opsahl, T., Agneessens, F., Skvoretz, J.: Node centrality in weighted networks: Generalizing degree and shortest paths. Social Networks 32(3), 245–251 (2010)
9. Parunak, H.V.D., Belding, T.C., Hilscher, R., Brueckner, S.: Modeling and Managing Collective Cognitive Convergence. In: Proceedings of AAMAS 2008, pp. 1505–1508. ACM Press (2008)
10. Rogers, E.M.: Diffusion of innovations. Free Press, New York (2003)
11. Sharpanskykh, A., Zia, K.: Grouping Behaviour in AmI-Enabled Crowd Evacuation. In: Novais, P., Preuveneers, D., Corchado, J.M. (eds.) ISAmI 2011. AISC, vol. 92, pp. 233–240. Springer, Heidelberg (2011)
12. Sharpanskykh, A., Treur, J.: Group Abstraction for Large-Scale Agent-Based Social Diffusion Models. In: Zhan, J., et al. (eds.) Proceedings of the 3rd International Conference on Social Computing, SocialCom 2011, pp. 830–837. IEEE Computer Society Press (2011)
13. Sharpanskykh, A., Treur, J.: Group Abstraction for Large-Scale Agent-Based Social Diffusion Models with Unaffected Agents. In: Kinny, D., Hsu, J.Y.-j., Governatori, G., Ghose, A.K. (eds.) PRIMA 2011. LNCS (LNAI), vol. 7047, pp. 129–142. Springer, Heidelberg (2011)

Building Detection with Loosely-Coupled Hybrid Feature Descriptors

Sieow Yeek Tan, Chin Wei Bong, and Dickson Lukose

Artificial Intelligence Center, Mimos Berhad,
Technology Park Malaysia, Bukit Jalil, Kuala Lumpur, Malaysia
{tan.sy,cw.bong,dickson.lukose}@mimos.my

Abstract. The paper presents a hybrid approach that ultilizes multiple low-level feature descriptors for performing building detection in 2D images. The proposed method is a symbiosis of two feature descriptors, namely Color and Edge Directivity Descriptor (CEDD) and Fuzzy Color and Texture Histrogram (FCTH) . The use of edge detection, texture and color combined features using fuzzy technique in encoding low-level visual information from images are embedded in the hybridization. First, multiple locations from a target image are chosen in the feature extraction process. Then, a hybridized vector index is proposed for measuring the low-level visual features distance between the target natural images with the training images, allowing a building content to be detected. Size and resolution of the source of images are not restricted in the proposed model and thus it can enhance the computational effectiveness. The empirical assessment, in term of the accuracy in detecting building objects in a set of images, validates the feasibility and potentiality of the proposed techniques.

Keywords: Building Detection, Image Understanding, Multiple Feature Descriptors, Loosely-coupled features.

1   Introduction

Buildings detection in digital images plays an important role in image understanding and information retrieval process. For instance, a tourist obtains information of her current location by taking environmental pictures containing important buildings using a handheld device for a communication center [1]. The main objective of the tourist is to recognize her current location. The first important process in this image understanding example is to detect the buildings in the image in order to relate to her location. This common process is usually conducted by matching the user's image with the images pre-indexed in an image bank. Subsequently, a building is recognized by comparing to the indexed information to obtain the highest closeness with the pre-indexed image. Furthermore, the building detection research and technology is crucial for image indexing, image retrieval, surveillance, robotic navigation, virtual reality, automatic navigation and etc [1].

P. Anthony, M. Ishizuka, and D. Lukose (Eds.): PRICAI 2012, LNAI 7458, pp. 552–563, 2012.
© Springer-Verlag Berlin Heidelberg 2012

Various methodologies are still being investigated to achieve effective and robust object recognition such as building in images. Indeed, building detection problem appeared in aerial images and natural images. Studies in [3 - 7] explored methods in solving the building detection problem in aerial images by using low-level image information such as edges, lines, road junctions, illumination, shadow, height and so on. The main concern of those buildings detection in aerial images is on detecting the roof presence as shown in [8]. Unlike the method proposed in studies [9] and [10], where natural images are used, their research concern is to differentiate building from other objects in a scene or its background. Thus, researchers in this area start exploiting several salient low-level visual information of building for performing the tasks. For instance, study in [2] has used the line-structure as low-level features captured as the statistic properties of buildings and use it together with other generic statistic properties for building detection purpose.

This paper explains the detection of building content in natural images using single and multiple visual descriptors. Hence, our concern is on the choice of features in representing such structural object. Indeed, a standard set of feature descriptors have been proposed by MPEG-7 standard [11] such as colour Layout (CLD), Colour Structure (CSD), Dominant Colour (DCD) and Edge Histrogram (EHD) [11-12] to efficiently encode colour and texture pattern. They are widely used in Content Based Image Retrieval (CBIR) system, such as QBIC[13], SIMPLIcity[14] and MIRROR[15] for image indexing and retrieval purpose. Several research studies have tried to apply local visual features such as SIFT [18] and Gabor [19], but they are limited to building detection in satelite images. There are also researchers that showed a few methods in recognizing specific objects by combining multiple feature descriptors, yet their focus is on image classification and retrievel [16-17].

The research problem of this paper is specific to the pattern, colour and structure of the building appearance in a natural image. Consequently, the contribution of this paper is related to the use of multiple visual descriptors for detecting buildings in 2D images, located at medium and long distances from camera. We proposed a hybrid model of two descriptors, namely CEDD: Color and Edge Directivity Descriptor [20] and FCTH: Fuzzy Color And Texture Histrogram [21]. A loosely-coupled mechanism is used in fusing these feature descriptors, and use it for making a decision of building detection. The proposed feature descriptors adopts a hybrid scheme in encoding low-level visual content that includes colour, texture and direction of edges via a fuzzy manner.

The paper is organized as follows: Section 2 describes the background and related works in building detection. Literature studies about existing works on employing visual descriptors in building detection are discussed. In addition, the background of CEDD and FCTH is presented together with information related to visual descriptors and how multiple features is used for various objects detection. Next, Section 3 elaborates the details of our proposed loosely-coupled algorithm that hybridizes the aforementioned feature descriptors. We then explain the feature vectors collection, calculation and comparison processes on how we calculate the hybridized index and how we set the threshold values. Section 4 presents the dataset used and experiment setup details. The results are finally discussed in Section 5, followed by conclusion and further works in last section.

2   Literature Study and Related Works

The idea of using low-level features in detecting building in digital image is an important method and it has been explored in [1, 9-10]. In [2], the author proposed a novel building detection algorithm by using structure features such as Haar-like features. A set of generic low-level building features are extracted and classified by machine learning algorithm, namely Support Vector Machine (SVM) to detect building content in images. On the other hand, study [9] incorporates a Bayesian framework with a set of perceptual grouping rules for the retrieval of building images with a large number of significant edges, junction, parallel lines, in comparison with image predominantly non-building objects. These structural visual patterns exist because of the presence of corners, windows, doors and boundaries of the building. By exploiting such primitive features, study [9] claimed an effective way for detecting building in an image. Study [10] presents the usage of multiscale feature vectors to capture general statistical properties of man-made structures. A set of feature vectors is extracted from different scales of source images to characterize straight lines and edges, encoded in term of its gradient magnitude and orientation. These feature vectors are then used for comparing the features extracted from the real-world image. When it yields a high value, building structure is considered detected.

To the best of our knowledge, the use of multiple features is proved to be one of the most effective object detection methods in digital image. In [16] and [17] studies, the authors used various feature descriptors proposed by MPEG-7. The authors tried to detect building object in their experiment and concluded the use of multiple features with a better result as compared to the use of one feature only. However, using more visual descriptors in encoding such low-level visual information will lead to higher demand of computational power. Especially when dealing with large image dataset, computational efficiency becomes a very crucial factor.

In recent year, the CEDD and FCTH were developed as low-level feature descriptors. They are mainly used for image indexing and retrieval purpose. The size of CEDD descriptors is limited to 54 bytes vector. It focuses in encoding the color and edge information. On the other hand, FCTH size is limited to 72 bytes vector, where this descriptor combines the color and texture information using fuzzy systems. This descriptor was found to be appropriate for accurately retrieving images even in distortion cases such as deformations, noise and smoothing [21]. Due to the small size of these descriptors, they required low computation power in the extraction and comparison process [20]. In contrary, MPEG-7 descriptors might require higher computation power as compared to CEDD or FCTH because both of them are considered as compact descriptors. Combining these two descriptors allow us to represent the low-level visual content of building in term of color, edge and texture effectively. The following subsections discuss the details of the two descriptors before we describe our hybridization model.

2.1   CEDD: Color and Edge Directivity Descriptor

CEDD is a visual descriptor that maps the visual content of an image, in term of color and direction of edges, into a new feature space, where it has to be discriminative enough for a description of an object [20]. In order to construct CEDD feature vector, the image is required to break into a pre-set number of blocks. Coordinate Logic Filters (CLF) [27] is used as the edge detection techniques to extract the detail of edge information. It is also incorporated with H (Hue), S (Saturation) and V (Value) color channel into histograms. A set of fuzzy rules, which undertake the extraction of a Fuzzy-Linking histogram proposed in [22], stems the color information of the image into HSV color space. A set of 20 TSK-like (Takagi-Sugeno-Kang) rules [23] are applied to a 3-input fuzzy system to generate a 10-bin quantized histogram, where each histogram represents a pre-set color selected based on works presented in [24]. Then, four extra TSK-like rules are applied to a 2-input fuzzy system in order to change these 10-bins histogram into 24-bins histogram, importing additional information related to the hue of each color. Five digital filters that were proposed in the MPEG-7 Edge Histogram Descriptors [25] are used to export the information   related to the texture of the image. Six texture regions are defined, thus shaping the original number of histogram bins to become 144. Finally, Gustafson Kessel fuzzy classifier [26] is used in order to quantize the 144 CEDD factors in the interval of 0 to 7, limiting the length of the descriptors in 432 bits, which is 54 bytes.

2.2   FCTH: Fuzzy Color and Texture Histogram

FCTH is another visual descriptor which encodes the color histogram and texture information via a combination of 3 fuzzy units. The first fuzzy unit and the second fuzzy unit are similar to what CEDD descriptor comprises. It is also constituted by a set of fuzzy rules for constructing a Fuzzy-Linking histogram stems from HSV color space [22], and applying 20 TSK-like rules [23] to a 3-input fuzzy system to generate a 10-bin quantized histogram. Similarly, four extra TSK-like rules are applied to a 2-input fuzzy system to relate hue information of each color into 24-bins histogram. A difference to CEDD is the third fuzzy unit process. The texture elements of each image block are extracted via Haar Wavelet transform. The motivation of using this transformation is the moments of wavelet coefficients in various frequency bands that have been proven effective for discerning texture. These details served as the input for a 3-input fuzzy system, with 8 rules applied [21]. For the purpose of incorporating additional texture information from the image to the original 24-bins histogram, it was converted into a 192-bins histogram. Finally, these 192 FCTH factors are quantized into the interval range from 0 to 7, limiting the length of the descriptor to 576 bits per image.

3   Loosely-Coupled Hybrid Features Descriptors

3.1   The Hybridization Model

We have modeled a loosely-coupled hybrid feature descriptors machanism, which incorporate CEDD and FCTH for detecting building in 2D images. Fig. 1 shows an overview for the suggested model. Given a set of experiment images, each image will be cropped multiple times to get a number of image blocks, N, from various random location within the image. Each image block will then go through a process to obtain the CEDD and FCTH feature vectors, denoted as $F_c{}^i$ and $F_f{}^i$ respectively, where $i =$ 1,...,N. Both $F_c{}^i$ and $F_f{}^i$ belonged to the feature vectors obtained from the ith image block.

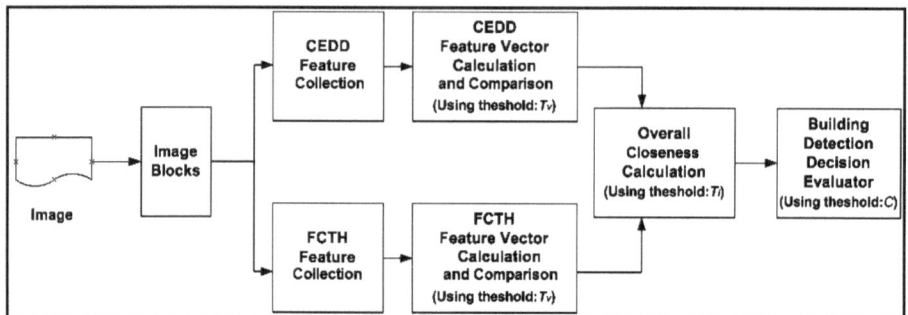

Fig. 1. Loosely-coupled hybrid feature descriptors model

The process is followed by performing the hybridized vector calculation, comparing the vectors obtained from the experiment image blocks from a set of training images. The training image dataset is a set of images that contains predominantly building content. Similarly, we obtain the CEDD and FCTH vectors from each of the image in the training set, denoted as $V_c{}^j$ and $V_f{}^j$ respectively, where $j = 1,..., M$. M is the size of the training dataset.

Each image blocks will draw two sets of hybridized vector value after the comparison process. One of the value set is from the CEDD vectors comparison while another set is from FCTH vectors comparison. If the CEDD hybridized index value exceed a pre-set threshold, T_v, where $T_v \in [0, 1]$, when comparing with the jth training vector, we considered the experiment image block is matched for the CEDD feature with respect to the jth training image. On the other hand, if the hybridized value is less than the mentioned threshold, it is considered unmatched. Similar mechanism applied exactly for FCTH vectors comparison. The detail of hybridized vector calculation is illustrated in the next section.

After the feature vectors obtained from an experiment image block have been compared with all the feature vectors from the training image set, we perform an overall hybridized index calculation. We normalize the hybridized index value to a range of 0 and 1, so that it is feasible to compare with another pre-set threshold, denoted as T_i, where $T_i \in [0, 1]$. If the overall value exceeds the threshold, the image block is marked with a positive count. Hence, if the total count of image blocks has exceed a satisfy threshold value, C, where $C \in [0, 1]$, the decision of building content has been detected is set to positive.

3.2   Hybridized Vector Calculation and Threshold Definition

The hybridized index calculation, $I(F, V)$, is essential in our methodology, in order to measure the distance between the training feature vector set, denoted as set $V = \{ V^j \mid j \in [1, M]\}$ and the feature vectors collected from an experiment image, denoted as set $F = \{ F^i \mid i \in [1, N]\}$. In this case, higher vector index indicates there are a lot of visual closeness between the training images and the experiment image. Thus, it helps to conclude whether the object in training images is available in the experiment image.

We proposed Cosine Similarity to measure the closeness between two feature vectors. Cosine Similarity is a measure of closeness of two vectors by measuring the cosine of the angle between them. Assumed we are interested in acquiring the closeness betwen the CEDD vector collected from the ith image block in the experiment image and the trained CEDD vector collected from the jth image in the training set, the closeness distance value is given as:

$$S_c^{i,j} = I(F_c^i, V_c^j) = \frac{(F_c^i \cdot V_c^j)}{||F_c^i|| \times || V_c^j||}$$

Similarly, for the FCTH vector, the equation is given as:

$$S_f^{i,j} = I(F_f^i, V_f^j) = \frac{(F_f^i \cdot V_f^j)}{||F_f^i|| \times || V_f^j||}$$

As mentioned in previous section, an overall hybridized index calculation has peformed after all the index values between i^{th} block and the training set have obtained. The evaluation is to determine whether the ith image block matched in comparison process. We considered the matching signal is triggered when

$$\left(\sum_{j=1}^{M} d(j) \right) \Big/ M \geq T_i$$

where

$$d(j) = 1, S^{i,j} \geq T_v$$

and

$$d(j) = 0, S^{i,j} < T_v$$

Let $B(i) = 1$ when the matching signal is triggered for the ith image block, and a building detection decision is considered positive when

$$\left(\sum_{i=1}^{N} B(i) \right) \Big/ N \geq C$$

where N is the total number of image blocks and C is the pre-set threshold value mentioned.

4   Experiment Environment Setup

4.1   Experiment and Training Image Dataset

Our experiment dataset contains of 100 colored, natural images, generally downloaded from Internet. 50 of them are not only containing building as the major content, but also containing other objects such as sky, sea, trees, and etc. Fig. 2 shows some of the sample images available from the dataset. Another 50 of them do not contain any buildings. These are images of wheat field, forest, galaxy, wave and etc; partly shown in Fig. 3. The size of the width and height of the images can be varying in pixel, averagely ranging between 100 and 3000, or even more. Also, the resolution of the image is not a vital parameter of concern.

Fig. 2. Samples of experiment images which containing building as the primary object

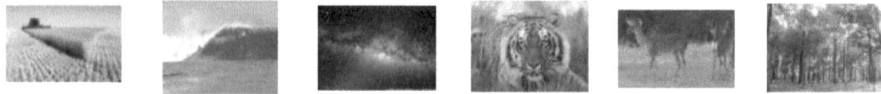

Fig. 3. Samples of experiment images which do not contain any concept of building

We prepared another set of images, which only contains building as the major object, served as training image dataset. We manually crop the training images into the area of just containing building object, in the same time removing other objects. This is because low-level visual content from other objects will create unwanted features to be encoded in feature descriptors. This has performed to minimize the feature descriptor noise. In other words, such pre-processing steps hold the purpose to enable the feature descriptors to focus in encoding building feature.

Type of image	Sample Images				
Original training images					
Pre-processed training images					

Fig. 4. Samples of pre-processed training images for feature extraction during training process

In this case, we have prepared 20 (M = 20) training images and extracted the CEDD and FCTH feature descriptors from them. Fig. 4 shows some samples of the training images, which have gone through the abovementioned pre-processing steps.

4.2   Experiment Environment Setup

We executed an empirical study through multiple experiments by using both CEDD and FCTH visual descriptors individually and hybrid. For this experiment, we set N = 15, which mean that there will be 15 regions from different locations of an experiment image will be chosen for feature extraction. The width and height of the regions will be varying from one and others, ranging from 50 to 150 pixels. Fig. 5 shows a sample of regions cropped from an experiment image.

Fig. 5. Sample of images blocks extracted from multiple regions randomly located in experiment image

For the threshold value, we pre-set T_v to different cases, ranging from 0.5 to 0.75. We proposed the value of 0.5 for T_i. In this paper, we performed our experiment by assuming two values of C, where one is 0, and another one is 0.5. We also carry out two set of experiments, using only CEDD and FCTH descriptor respectively. These two set of experiment have the abovementioned setting unchanged. The results are elaborated further in the following section.

5   Discussion of Experiment Results

Our main focus is on evaluating the accuracy and recall of our proposed method. Given the experiment image dataset that contains building object, denoted as A, the calculation of accuracy value normally defined as:

$$Accuracy = \frac{|D|}{|A|}$$

where D is the subset of the experiment images detected with building content proposed by the method. On the other hand, the recall calculation is defined as:

$$Recall = \frac{|D'|}{|A'|}$$

where A' is the experiment image set that do not contain any building object and D' is the image set determined by the proposed method that do not detect any building concept. Here, the higher the recall's value, the better the proposed method performed.

We presented 3 types of results: building detection by using only CEDD and FCTH respectively, and the third one using hybridization of CEDD and FCTH. For each type, we presented the accuracy and recall values, which normalized into percentage. All experiment settings, including the number of image blocks as well as the threshold values remained same. However, several set of experiments were carried out with the aim to examine the accuracy and recall with respect to different threshold values.

Fig. 6 and Fig. 7 below show the graphs of the experiment results with the accuracy values. Fig. 6 shows the experiment result with the C value equals to 0, meanwhile the C value is equaled to 0.5 in Fig. 7. Fig. 8 and Fig. 9 show the experiment results with the recall values, using the similar setting of both C values shown in Fig. 6 and Fig. 7.

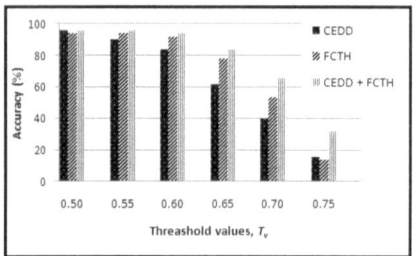

Fig. 6. Accuracy result for building detection with the setting of $T_i = C = 0$

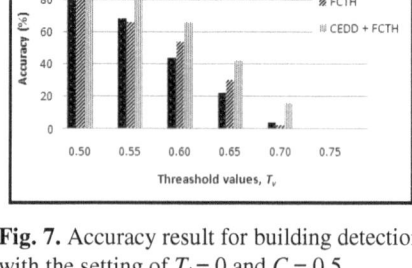

Fig. 7. Accuracy result for building detection with the setting of $T_i = 0$ and $C = 0.5$

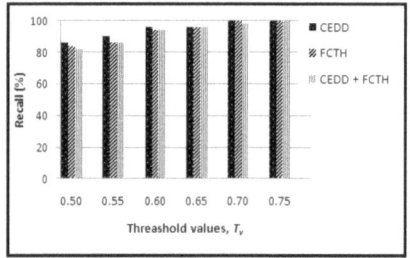

Fig. 8. Recall result for building detection with the setting of $T_i = C = 0$

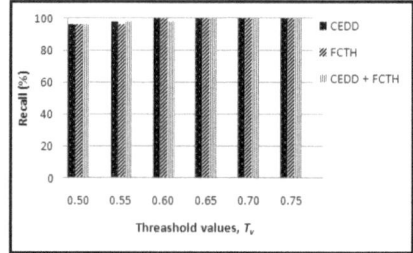

Fig. 9. Recall result for building detection with the setting of $T_i = 0$ and $C = 0.5$

We obtained nearly 100% of accuracy when the thresholds were set in the range of 0.5 to 0.6. We noticed that by using CEDD feature descriptor alone, we are able to achieve quite high accuracy in detecting building content. This shows the color and edge information are contributing factors in recognizing buildings. Also, the result marked relatively quite high accuracy when we use FCTH feature descriptor alone. This demonstrates that texture feature can be considered as another vital factor. On the other hand, the accuracy of using hybridization of two descriptors has obviously outperformed the use of single descriptor in every other case of experiment setting. This has proved that the hybridization can fully utilize the color, edge and texture information in representing building object and thus optimize the recognition process.

On the other hand, we obtained a high rank of recall values, ranging from 85% to 100%, for all type of results. The high percentage of recall percentage reveals the fact that CEDD and FCTH visual descriptors are considerably feasible used for building detection operation. One of the possible reasons of some non-building images are still detected with building concepts is the color and texture patterns are too similar as the building objects perceived. We are also convinced that the hybridized descriptors maintained a good performance in building detection operation. This has shown from the result as the hybridized features captured a 100% recall percentage for threshold values $T_v \geq 0.65$ and $C = 0.5$.

From the result, we observed that when we increased the value of threshold T_v, from 0.5 to 0.75 with the interval of 0.05, the accuracy dropped gradually. This is because a higher closeness between the trained features and the extracted feature are required for concluding a matching decision. Similarly, with incrementing value of C from 0 to 0.5, we observed the same effect. We conjecture that this behavior may be due to the noise and a variety of buildings' feature patterns available between the training images and the experiment images. It is suggested that in the scenario of different objects detection, it will has differing range of threshold settings in order to maintain high accuracy result.

6   Conclusion and Future Work

A technique using loosely-coupled hybridized low-level feature descriptors to perform a building detection task in a 2D natural image is presented. Experiment results show that the proposed technique with fusion of multiple visual descriptors for building detection is feasible. We show that the loosely coupled mechanism in combining multiple feature descriptors can to be one of the best way that significantly reduce the computation power while preserving it's representation power in describing objects.

A more comprehensive evaluation of the proposed technique and additional improvements are being undertaken. Immediate work includes using other type of visual descriptors, especially those proposed by MPEG-7 standard, to perform object detection using the similar methodology. The effectiveness and robustness in using such multiple low-level feature will be experimented and evaluated. We envision to detect not only the building object, but also other common object available in natural images, such as tree, sea, sky, rock, car, human, animal, road, sky, moutain and etc.

References

1. Wei, Z.Q., Han, L.R., Yang, M., Ji, X.P., Yin, B., Chu, J.: Building Extraction Based on Hue Cluster Analysis in Complex Scene. In: CISP, pp. 1–5 (2009)
2. Yanyun, Q., Nanning, Z., Cuihua, L., et al.: Salient building detection based on SVM. Journal of Computer Research and Development 44(1), 141–147 (2007)

3. McKeown, D.M.: Toward Automatic Cartographic Feature Extraction. In: Pau, L.F. (ed.) Mapping and Spatial Modelling for Navigation. NATO ASI series, vol. F65, pp. 149–180 (1990)

4. Irvin, R.B., McKeown, D.M.: Methods for Exploiting the Relationship Between Buildings and Their Shadows in Aerial Imagery. IEEE Trans. Systems, Man, and Cybernetics 19(6), 1,564–1,575 (1989)

5. McGlone, J.A., Shufelt, J.C.: Projective and Object Space Geometry for Monocular Building Extraction. In: Proc. IEEE Conf. Computer Vision and Pattern Recognition, pp. 54–61 (1994)

6. Shufelt, J.A.: Exploiting Photogrammetric Methods for Building Extraction in Aerial Images. Int'l Archives of Photogrammetry and Remote Sensing XXXI(B6/S), 74–79 (1996)

7. Shufelt, J.A.: Projective Geometry and Photometry for Object Detection and Delineation. PhD thesis, Computer Science Dept., Carnegie Mellon Univ., available as Technical Report CMU-CS-96-164 (1996)

8. Mayer, H.: Automatic object extraction from aerial imagery - a survey focusing on buildings. Computer Vision and Image Understanding 74(2), 138–149 (1999)

9. Iqbal, A., Aggarwal, J.K.: Applying perceptual grouping to content-based image retrieval:Builing images. In: Proc. IEEE Int. Conf. CVPR, vol. 1, pp. 42–48 (1999)

10. Kumar, S., Hebert, S.: Man-made Structure Detection in Natural Images using a Causal Multiscale Random Field. In: Proc. IEEE Int. Conf. on CVPR, vol. 1, pp. 119–126 (2003)

11. Martinez, J.M.: Mpeg-7 Overview,
    `http://www.chiariglione.org/mpeg/standards/mpeg-7/mpeg-7.htm`

12. Manjunath, B.S., Ohm, J.-R., Vasudevan, V.V., Yamada, A.: Color and Texture Descriptors. IEEE Transactions on Circuits and Systems for Video Technology 11(6), 703–715 (2001)

13. Flickner, M., Sawhney, H., Niblack, W., Ashley, J., Huang, Q., Dom, B., Gorkani, M., Hafner, J., Lee, D., Petkovic, D., Steele, D.: Flickner: Query by image and video content: the QBIC system. Computer 28(9), 23–32 (1995)

14. James, Z.W., Jia, L., Gio, W.: SIMPLIcity: Semantics-Sensitive Integrate, Matching for Picture Libraries. IEEE Transactions on Pattern Analysis and Machine Intelligence 23(9) (2001)

15. Ka-Man, W., Kwok-Wai, C., Lai-Man, P.: MIRROR: an interactive content based image retrieval system. In: Proceedings of IEEE International Symposium on Circuit and Systems 2005, Japan, vol. 2, pp. 1541–1544 (2005)

16. Zhang, Q., Izquierdo, E.: A Multi-feature Optimization Approach to Object-Based Image Classification. In: Sundaram, H., Naphade, M., Smith, J.R., Rui, Y. (eds.) CIVR 2006. LNCS, vol. 4071, pp. 310–319. Springer, Heidelberg (2006)

17. Zhang, Q., Izquierdo, E.: Combining low-level features for semantic extraction in image retrieval. EURASIP Journal on Advances in Signal Processing 2007(4), 1–12 (2007)

18. Sırmaçek, B., Ünsalan, C.: Urban Area and Building Detection Using SIFT Keypoints and Graph Theory. IEEE Transactions on Geoscience and Remote Sensing 47(4), 1156–1167 (2009)

19. Sırmaçek, B., Ünsalan, C.: Building detection using local Gabor features in very high resolution satellite images. In: Proceedings of RAST 2009, Istanbul, Turkey (2009)

20. Savvas, A.C., Yiannis, S.B.: CEDD: color and edge directivity descriptor: a compact descriptor for image indexing and retrieval. In: Proceedings of the 6th International Conference on Computer Vision Systems, Santorini, Greece (2008)

21. Savvas, A.C., Yiannis, S.B.: FCTH: Fuzzy Color and Texture Histogram - A Low Level Feature for Accurate Image Retrieval. In: Proceedings of the Ninth International Workshop on Image Analysis for Multimedia Interactive Services, pp. 191–196 (2008)
22. Chatzichristofis, S., Boutalis, Y.: A Hybrid Scheme for Fast and Accurate Image Retrieval Based on Color Descriptors. In: De Mallorca, P. (ed.) IASTED International Conference on Artificial Intelligence and Soft Computing (ASC 2007), Spain (2007)
23. Zimmerman, H.J.: Fuzzy Sets, Decision Making and Expert Systems. Kluwer Academic Publ., Boston (1987)
24. Konstantinidis, K., Gasteratos, A., Andreadis, I.: Image Retrieval Based on Fuzzy Color Histogram Processing. Optics Communications 248(4-6), 15, 375–386 (2005)
25. Won, C.S., Park, D.K., Park, S.-J.: Efficient Use of MPEG-7 Edge Histogram Descriptor. ETRI Journal 24 (2002)
26. Gustafson, E.E., Kessel, W.C.: Fuzzy Clustering with a Fuzzy Covariance Matrix. In: IEEE CDC, San Diego, California, pp. 761–766 (1979)
27. Mertzios, B., Tsirikolias, K.: Coordinate Logic Filters: Theory and Applications Nonlinear Image Processing. In: Mitra, S., Sicuranza, G. (eds.) ch. 11. Academic Press, London (2004) ISBN: 0125004516

Towards Optimal Cooperative Path Planning
in Hard Setups through Satisfiability Solving[*]

Pavel Surynek[1,2]

[1] Charles University in Prague, Faculty of Mathematics and Physics
Malostranské náměstí 2/25, 118 00 Praha 1, Czech Republic
[2] Kobe University, Graduate School of Maritime Sciences, Intelligent Informatics Laboratory
5-1-1 Fukae-minamimachi, Higashinada-ku, Kobe 658-0022, Japan
pavel.surynek@mff.cuni.cz

Abstract. A novel approach to cooperative path-planning is presented. A SAT solver is used not to solve the whole instance but for optimizing the makespan of a sub-optimal solution. This approach is trying to exploit the ability of state-of-the-art SAT solvers to give a solution to relatively small instance quickly. A sub-optimal solution to the instance is obtained by some existent method first. It is then submitted to the optimization process which decomposes it into small subsequences for which optimal solutions are found by a SAT solver. The new shorter solution is subsequently obtained as concatenation of optimal sub-solutions. The process is iterated until a fixed point is reached. This is the first method to produce near optimal solutions for densely populated environments; it can be also applied to domain-independent planning supposed that sub-optimal planner is available.

1   Introduction and Context

Cooperative path-planning recently attracted considerable interest of the AI community. This interest is motivated by the broad range of areas where cooperative path-planning can be applied (robotics, computer entertainment, traffic optimization, etc.) as well as by challenging aspects which it offers. The task consists in finding spatial temporal paths for agents which want to reach certain destinations without colliding with each other. One of the most important breakthrough in solving the task is represented by the WHCA* algorithm [8] which decouples the search for cooperative plan into searches for plans for individual agents. Recently, an optimal decoupled method appeared [10]. The common drawback of decoupled approach is that it is applicable only on instances with small occupancy of the environment by agents.

The opposite of the spectrum of solving algorithms is represented by complete sub-optimal methods [4, 11]. These algorithms are able to provide solution irrespectively of the portion of space occupied by agents. Especially good performance is reported for highly occupied instances. On the other side, too long solutions are usually generated for sparsely populated environments. Other methods are trying to exploit the

[*] This work is supported by the by the Japan Society for the Promotion of Science (contract no. P11743) and by the Czech Science Foundation (contract no. GAP103/10/1287).

P. Anthony, M. Ishizuka, and D. Lukose (Eds.): PRICAI 2012, LNAI 7458, pp. 564–576, 2012.

structure of environment [7] or the structure of current arrangement of agents [12]. Again these methods require relatively unoccupied environment and in some cases they are not complete.

Here we are trying to contribute to a not yet addressed case with high occupancy and the requirement on solution to have short makespan. Our approach is basically complete. We use SAT solving technology in a novel and unique way to address this case. First a sub-optimal solution is generated by some of the existent fast algorithms. The sub-optimal solution is then decomposed into small sub-sequences that are replaced by optimal sub-solutions generated by a SAT solver. The process is iterated until the makespan converges. This decomposition of the original problem allowed us to exploit the strongest aspect of SAT solvers – that is, their ability to satisfy relatively small yet complex enough SAT instance very quickly.

The rest of the paper describes cooperative path planning formally first. Then our special domain dependent SAT encoding and the optimization methods are introduced. An experimental comparison with several existent techniques is presented finally.

2   Cooperative Path Planning Formally

Arbitrary **undirected graph** can be used to model the environment. Let $G = (V, E)$ be such a graph where V is a finite set of vertices and $E \subseteq \binom{V}{2}$ is a set of edges.

The placement of agents in the environment is modeled by assigning them vertices of the graph. Let $A = \{a_1, a_2, \ldots, a_n\}$ be a finite set of *agents*. Then, an arrangement of agents in vertices of graph G will be fully described by a *location* function $\alpha: A \longrightarrow V$; the interpretation is that an agent $a \in A$ is located in a vertex $\alpha(a)$. At most **one agent** can be located in each vertex; that is α is uniquely invertible. A generalized inverse of α denoted as $\alpha^{-1}: V \longrightarrow A \cup \{\perp\}$ will provide us an agent located in a given vertex or $\perp$ if the vertex is empty.

Definition 1 (COOPERATIVE PATH PLANNING). An instance of *cooperative path-planning* problem is a quadruple $\Sigma = [G = (V, E), A, \alpha_0, \alpha_+]$ where location functions α_0 and α_+ define the initial and the goal arrangement of a set of agents A in G respectively.   □

The dynamicity of the model supposes a discrete time divided into time steps. An arrangement α_i at the i-th time step can be transformed by a transition action which instantaneously moves agents in the non-colliding way to form a new arrangement α_{i+1}. The resulting arrangement α_{i+1} must satisfy the following *validity conditions*:

(i) $\forall a \in A$ either $\alpha_i(a) = \alpha_{i+1}(a)$ or $\{\alpha_i(a), \alpha_{i+1}(a)\} \in E$ holds
       (agents move along edges or not move at all),
(ii) $\forall a \in A \ \alpha_i(a) \neq \alpha_{i+1}(a) \Rightarrow \alpha_i^{-1}(a) = \perp$
       (agents move to vacant vertices only), and
(iii) $\forall a, b \in A \ a \neq b \Rightarrow \alpha_{i+1}(a) \neq \alpha_{i+1}(b)$
       (no two agents enter the same target/unique invertibility of resulting arrangement).

The task in cooperative path planning is to transform α_0 using above valid transitions to α_+.

Definition 2 (SOLUTION, MAKESPAN). A *solution* of a *makespan* m to a cooperative path planning instance $\Sigma = [G, A, \alpha_0, \alpha_+]$ is a sequence of arrangements $\vec{s} = [\alpha_0, \alpha_1, \alpha_2, ..., \alpha_m]$ where $\alpha_m = \alpha_+$ and α_{i+1} is a result of valid transformation of α_i for every $= 1, 2, ..., m - 1$. □

If it is a question whether there is a solution of Σ of the makespan at most a given bound we are speaking about the *bounded variant*. Notice that due to no-ops introduced in valid transitions it is equivalent to finding a solution of the makespan equal to the given bound.

3   SAT Encoding of the Bounded Variant

Our goal was to devise a SAT encoding of bounded cooperative path planning suitable for relatively densely populated environments. At the same time we needed to keep the encoding compact. We followed the classical *Graphplan* inspired encodings [2, 3] as for we also encode each time step.

3.1   Using Multi-valued State Variables

We were primarily inspired by SATPLAN [3] and SASE [2] encodings in our design. But unlike these generic encodings we were working with the specific domain so we could facilitate the domain knowledge in the design of the instance encoding. We a priori know what the candidates for *multi-valued state variables* are in our domain – basically, these are represented by location function and its inverse. Using techniques proposed by Rintanen [6] each state variable can be encoded by logarithmic number of propositional variables with respect to the number its values. Another considerable aspect is how to encode transition actions together with validity conditions.

Representing arrangement of agents by **inverse locations** (that is, there is a state variable for each vertex) allowed us to encode transitions efficiently. There are two *primitive actions* for each edge adjacent to the given vertex plus one no-op action. Half of the primitive actions corresponding to a vertex are for incoming agents while the other half is for outgoing agents. If the outgoing primitive action is selected it is necessary to propagate the selection as corresponding selection of incoming primitive action in the target vertex. Representing the selection of the primitive action as a multi-values state variable automatically ensures that conditions (i) and (iii) are encoded. Moreover, we do not need any *mutex* constraints in the encoding. Notice also, that the degree of vertices in G is typically low for real-life environments, thus action selection in the vertex can be captured by few propositional variables.

Let $\Sigma = [G = (V, E), A, \alpha_0, \alpha_+]$ be a cooperative instance and $k \in \mathbb{N}$ be a makespan bound. Our encoding has layers numbered $0, 1, ..., k$. Suppose that neighboring vertices of a given vertex are ordered in the fixed order. That is, $\forall v \in V$ we have function $\sigma_v \colon \{u | \{v, u\} \in E\} \longrightarrow \{1, 2, ..., \mathrm{dg}_G(v)\}$ and its inverse σ_v^{-1}.

Definition 3 (LAYER ENCODING). The i-th regular layer consists of the following integer interval **state variables**:

- $\mathcal{A}_i^v \in \{0,1,2,\ldots,n\}$ for all $v \in V$ such that
  $\mathcal{A}_i^v = j$ iff $\alpha_i(a_j) = v$
- $\mathcal{T}_i^v \in \{0,1,2,\ldots,2\deg_G(v)\}$ for all $v \in V$ such that
  $\mathcal{T}_i^v = 0$ iff no-op was selected in v;
  $\mathcal{T}_i^v = \sigma_v(u)$ iff an outgoing primitive action with
                  the target $u \in V$ was selected in v;
  $\mathcal{T}_i^v = \deg_G(v) + \sigma_v(u)$ iff an incoming primitive action with $u \in V$ as the
  source was selected in v.

and **constraints**:

- $\mathcal{T}_i^v = 0 \Rightarrow \mathcal{A}_{i+1}^v = \mathcal{A}_i^v$ for all $v \in V$ (**no-op** case);
- $0 < \mathcal{T}_i^v \leq \deg_G(v) \Rightarrow \mathcal{A}_i^u = 0 \wedge \mathcal{A}_{i+1}^u = \mathcal{A}_i^v \wedge \mathcal{T}_i^u = \sigma_u(v) + \deg_G(u)$
  where $u = o_v^{-1}(\mathcal{T}_i^v)$ for all $v \in V$ (**outgoing** agent case);
- $\deg_G(v) < \mathcal{T}_i^v \leq 2\deg_G(v) \Rightarrow \mathcal{T}_i^u = \sigma_u(v)$
  where $u = \sigma_v^{-1}(\mathcal{T}_i^v - \deg_G(v))$ for all $v \in V$ (**incoming** agent case).   □

State variables $\mathcal{A}_i^v$ for $v \in V$ represent inverse location function at the time step i. Analogically, state variables $\mathcal{T}_i^v$ for $v \in V$ represent transition actions selected in vertices at time step i. Constraints merely encode the validity conditions.

The last encoding layer is irregular as it has location state variables only. To finish the encoding of bounded cooperative instance we need to encode the initial and the goal arrangement straightforwardly as follows:

$$\mathcal{A}_0^v = j \quad \text{iff } \alpha_0^{-1}(v) = a_j,$$
$$\mathcal{A}_0^v = 0 \quad \text{iff } \alpha_0^{-1}(v) = \perp,$$
$$\mathcal{A}_k^v = j \quad \text{iff } \alpha_+^{-1}(v) = a_j,$$
$$\mathcal{A}_k^v = 0 \quad \text{iff } \alpha_+^{-1}(v) = \perp.$$

Transformation of the encoding from the above integer representation to the propositional one is also straightforward. To reduce size of clauses we should use standard Tseitin's hierarchical encoding with auxiliary variables.

Proposition 1 (ENCODING SIZE). *A regular layer of the propositional encoding of the bounded cooperative instance requires*

$$|V|\lceil\log_2|A|\rceil + \sum_{v \in V}\lceil\log_2(2\deg_G(v)+1)\rceil \tag{1}$$

*propositional **variables** for representing state variables,*

$$2|E|\lceil\log_2|A|\rceil + |V|\lceil\log_2|A|\rceil \tag{2}$$

*auxiliary propositional **variables** from Tseitin's translation*

$$\sum_{v \in V}\deg_G(v)\,(2\lceil\log_2(2\deg_G(v)+1)\rceil + 7\lceil\log_2|A|\rceil) + |V|\lceil\log_2|A|\rceil \tag{3}$$

***clauses** for representing constraints, and*

$$|V|(2^{\lceil\log_2|A|\rceil} - |A|), \tag{4}$$

$$\sum_{v \in V} 2^{\lceil \log_2(2\,dg_G(v)+1) \rceil} - 2 \sum_{v \in V} dg_G(v) - |V| \qquad (5)$$

clauses *for excluding unused location and transition action states respectively.* ■

Proof. We just need to observe that $2\,dg_G(v) + 1$ cases (preconditions) need to be distinguished for each vertex $v \in V$ and for each of these cases a corresponding effect needs to be enforced. The cases with **outgoing** agent need each $\lceil \log_2|A| \rceil$ auxiliary propositional variables which come from Tseitin's encoding and $\lceil \log_2(2\,dg_G(v) + 1) \rceil + 2\lceil \log_2|A| \rceil + 5\lceil \log_2|A| \rceil$ clauses (1 equality between transition action and a constant + 2 equalities between inverse location and a constant, and 1 equality between two inverse locations). The cases with **incoming** agent do not require any auxiliary variable and only $\lceil \log_2(2\,dg_G(v) + 1) \rceil$ clauses are needed (1 equality between transition action and a constant). Finally, the single **no-op** case requires $\lceil \log_2|A| \rceil$ auxiliary variables and $5\lceil \log_2|A| \rceil$ clauses (1 equality between two inverse locations). From this overview expressions (1)-(5) are straightforward. ■

Most of clauses generated in our encoding have arity of $\lceil \log_2(2\,dg_G(v) + 1) \rceil + 1$ for some $v \in V$ or $\lceil \log_2|A| \rceil + 1$. The comparison with the graph-plan based encoding used in SATPLAN is shown in Table 1. Our domain-specific encoding is clearly smaller while the difference is growing as the number of agents increases.

Table 1. Comparison *of encoding sizes.* The smallest number of layers for which SATPLAN was unable to detect unreachability of the goal using mutex reasoning is indicated as *goal level* – it is used as the makespan bound.

|Agents| in 4-connected grid 8x8	Goal level	SATPLAN encoding		Our **domain specific** encoding	
		|Variables|	|Clauses|	|Variables|	|Clauses|
4	8	5864	55330	9432	55008
8	8	10022	165660	11968	70400
12	8	14471	356410	11968	68352
16	10	30157	1169198	18490	112580
24	10	43451	2473813	18490	107360
32	14	99398	8530312	32116	200768

4   COBOPT: A New Approach to (Path) Planning

Our novel cooperative path planning technique called COBOPT exploits SAT solving technology [1] not to produce a solution but to optimize it with respect to the makespan. To be able to use SAT solvers in this way we need to obtain some (sub-optimal) solution to the cooperative instance first. Let this initial solution be called base solution. As we mentioned, many solving techniques for cooperative path planning are available at the present time [7], [8] (WHCA*); [11] (BIBOX); [4] (PUSH-SWAP); [10] - OD+ID; [12] (MAPP). Any of them can be used to produce base solution within our framework. Our approach is completely generic in this sense. Notice however, that particular solving technique is always designed for a specific class of the problem while outside this class it may provide worse performance. The typical weakness is for example that decoupled techniques (WHCA*, MAPP) admit that not all the agents need to reach their destination [8, 12].

In our initial experiments, we found that it is becoming dramatically more difficult for SAT solvers to solve bounded cooperative instance as the bound is growing. To be more concrete, a SAT solver usually struggles with the instance consisting of the graph containing 100 vertices, 30 agents, and the bound of 10 for several minutes if the presented SAT encoding is used. In case of the SATPLAN encoding the situation is even worse – the solver even struggles with generating the formula for minutes. This finding renders possibility of using SAT solvers to solve a cooperative instance of considerable size in the SATPLAN style [2, 3] as infeasible at the current state-of-the-art since it may require hundreds of time steps. But using a SAT solver in the SATPLAN style has one undisputable advantage if we manage to get a solution from it – it is makespan optimal.

After producing a base solution, this is submitted to a SAT based optimization process. A maximum bound k^+ for encoding cooperative instances is specified. Then sub-sequences in the base solution are replaced with computed optimal sub-solution. Suppose that we are currently optimizing at time step t. It is computed what is the largest $t^+ > t$ such that the time step t^+ can be reached from the time step t with no more than k^+ steps. Then sub-solution of the base solution from the time step t to t^+ is replaced by the optimal one obtained from the SAT solver. The process then continues with optimization at time step t^+ until the whole base solution is processed.

Algorithm 1. COBOPT: *SAT-based cooperative path planning solution optimization – basic scheme based on binary search.*

---

function *COBOPT-Optimize-Cooperative-Plan* $(\Sigma, \vec{s}, k^+)$: **solution**
1: $\vec{s}_+ \leftarrow \vec{s}$
2: **do**
3: $\quad \vec{s}_- \leftarrow \vec{s}_+$
4: $\quad$ **let** $\vec{s}_- = [\alpha_0, \alpha_1, \alpha_2, \dots, \alpha_m]$
5: $\quad t \leftarrow 0; \vec{s}_+ \leftarrow []$
6: $\quad$ **while** $t < m$ **do**
7: $\quad\quad t^+ \leftarrow$ *Find-Last-Reachable-Arrangement*$(\Sigma, \alpha_t, \vec{s}_-, k^+)$
8: $\quad\quad \vec{s}_+ \leftarrow \vec{s}_+.$*Compute-Optimal-Solution*$(\Sigma, \alpha_t, \alpha_{t^+})$
9: $\quad\quad t \leftarrow t^+$
10: **while** $|\vec{s}_-| > |\vec{s}_+|$
11: **return** $\vec{s}_+$

function *Find-Last-Reachable-Arrangement* $(\Sigma, \alpha_t, \vec{s}, k^+)$: **integer**
1: **let** $\vec{s} = [\alpha_0, \alpha_1, \alpha_2, \dots, \alpha_m]$
2: $l \leftarrow t; u \leftarrow m + 1$
3: **while** $u - l > 1$ **do**
4: $\quad r \leftarrow (u + l)/2$
5: $\quad k \leftarrow \min(m - t, k^+)$
6: $\quad$ **if** *Check-Reachability*$(\Sigma, \alpha_t, \alpha_r, k)$ **then**
7: $\quad\quad \Xi \leftarrow$*Encode*$(\Sigma, \alpha_t, \alpha_r, k)$
8: $\quad\quad$ **if** *Solve-SAT* (Ξ) **then** $l \leftarrow r$
9: $\quad\quad$ **else** $u \leftarrow r$
10: $\quad$ **else**
11: $\quad\quad u \leftarrow r$
12: **return** l

function *Check-Reachability* $(\Sigma, \alpha_t, \alpha_r, k)$: **boolean**
1: **let** $\Sigma = [G, A, \alpha_0, \alpha_+]$
2: **for each** $a \in A$ **do**
3: $\quad$**if** $\text{dist}_G(\alpha_t(a), \alpha_r(a)) > k$ **then return** *FALSE*
4: **return** *TRUE*

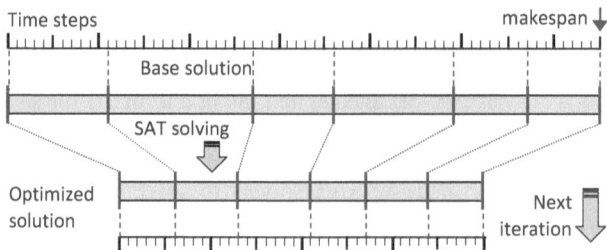

Fig. 1. Illustration *of the optimization process.* A single iteration is shown – these are repeated until a fixed point is reached.

The optimization process can be iterated by taking new solution as the base one until a fixed point is reached. The binary search is exploited to find t^+ and the optimal sub-solution in order to reduce the number of SAT solver invocations – see Algorithm 1 which summarizes basic CoBopt optimization method formally. Notice that some extra care is needed to obtain optimal sub-solution at the end of the base solution sequence.

The process of optimization is illustrated in Figure 1. Notice that separation points in the base solution are selected on the greedy basis – optimization always continues on the first not yet processed time step. We also considered optimizing placement of separation point by dynamic programming techniques. This approach generates slightly better base solution decomposition. However it is at the great expense in overall runtime as many more invocation of the SAT solver are necessary.

5   Experimental Evaluation

We implemented the proposed CoBopt optimization method in C++ to conduct an experimental evaluation. A competitive comparison against 3 existent methods was made – WHCA*, SatPlan, and Bibox. WHCA* was chosen as reference method as it is considered to be standard decoupled method for cooperative path planning and its properties and performance are well known.

Table 2. Optimal *solutions obtained by* SatPlan. No more agents can be solved by SatPlan within the time limit of 7200s.

|Agents|	4-connected grid 8x8		4-connected grid 16x16	
	Optimal makespan	Runtime (s)	Optimal makespan	Runtime (s)
1	5	0.0	4	0.68
4	6	0.15	21	195.5
8	8	19.85	15	1396.07

As no implementation of WHCA* was available we re-implemented it in C++ by ourselves. SATPLAN is the most similar method to our approach and very importantly it produces optimal solutions – we used implementation provided by the authors. Finally, BIBOX was selected as major method for producing base solutions in hard setups.

Our choice was not discouraged by the wrong statement of Standley and Korf [10] who consider it together with the method of Ryan [7] to have memory and time requirements that limit their applicability. According to our findings, these algorithms have important theoretical guarantees and good practical performance. Particularly, BIBOX has polynomial time complexity (solutions to all the benchmarks presented here were generated within less than 0.1 seconds) and generates good quality suboptimal solutions irrespectively how many agents are contained in the instance – together with the algorithm PUSH-SWAP by Luna and Berkis [4] it is the only algorithm able to generate base solution for hard setups. Authors provide working implementation of BIBOX which we exploited within our experiments. COBOPT using BIBOX as a base solver will be referred to as COBOPT(BIBOX). As a SAT solver within our method, MINISAT 2.2 [1] was used.

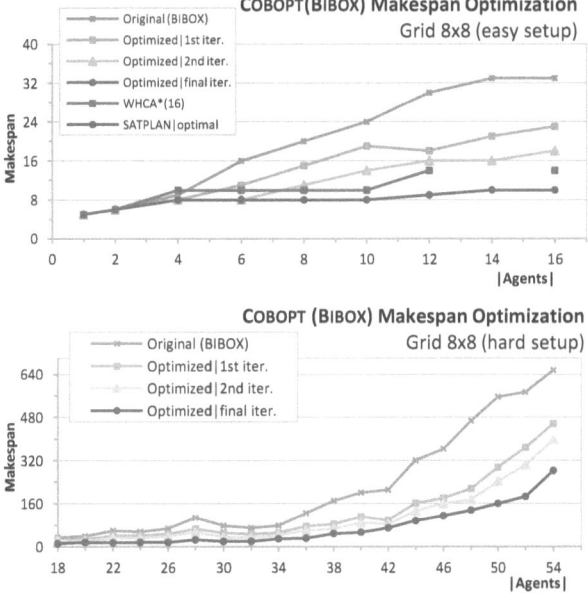

Fig. 2. Makespan *optimization in the 4-connected grid 8×8*. A comparison with the optimal SATPLAN and near optimal WHCA* is shown.

Standard benchmark setups for cooperative path planning which consists of a 4-connected grid graph and randomly arranged initial and goal locations for agents were used. Various parameters of the COBOPT(BIBOX) and other methods were observed in the dependence of the increasing number of agents in the instance. Two setups were used: grids of size 8×8 and 16×16 with number of agents ranging from 1 to 54 and 1 to 128 respectively. The timeout of 240s and 120s per SAT solver invocation was used for these setups respectively. Makespan bounds of 8 and 6 were used respectively. Additionally there was an overall timeout of 7200s (2 hours) after which the optimization process was terminated.

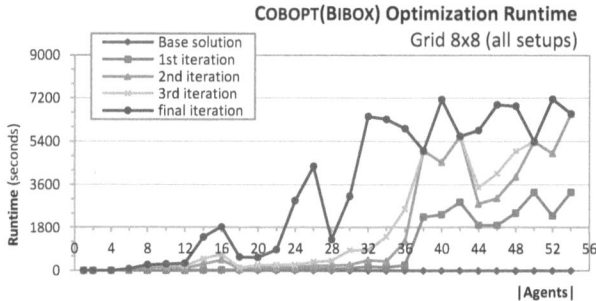

Fig. 3. Runtime *measurements per optimization iteration in 8×8 grid.* The base solution can be produced in less than 0.1s.

Due to the bigger size of SAT encodings for the 16×16 grid the optimization method uses less aggressive setup with respect to the SAT solver. Using WHCA* we observed that setups with up to approximately 20% of occupied vertices are in fact easy as only very limited cooperation among agents is necessary. This observation ruled out from our consideration the method OD+ID as it is reported to be efficient only in the setups with less than 10% of occupied vertices. Here we are interested primarily in setups with occupancy in the range 20% - 50% which is increasingly harder as cooperation between agents gradually increases.

Table 3. MINISAT *statistics (8×8 grid).* Each invocation of MINISAT within COBOPT optimization has the timeout of 240s.

|Agents| in 4-connected grid 8x8	Number of MINISAT results in final iteration		
	SAT instances	**UNSAT** instances	**INDET** instances
4	13	2	0
8	44	4	0
12	79	5	0
16	96	15	0
24	253	28	0
32	194	27	2

To learn what the optimal makespan for tested instances is we tried SATPLAN (Table 2). Unfortunately SATPLAN was able to generate solution only to instances with small number of agents. The reason is primarily inefficiency of domain-independent SAT encoding (Table 1).

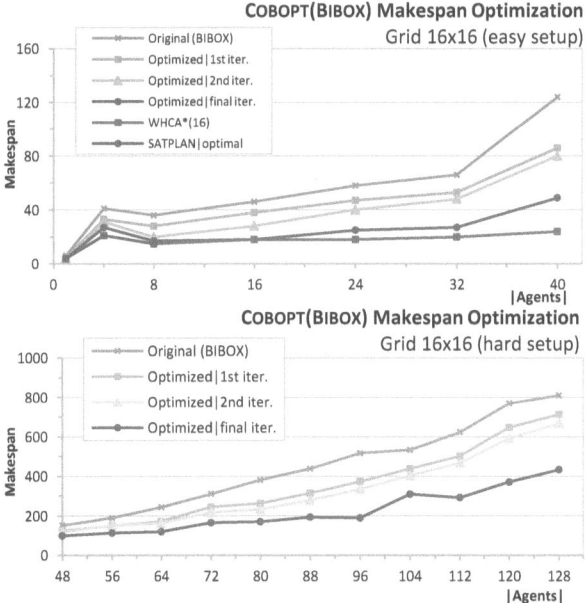

Fig. 4. Makespan optimization in the 4-connected grid 16×16

In the following experiments we exploited the decoupled WHCA* method. Expectably it is able to generate near optimal solutions (Figure 2, Figure 4) since near optimal path is tried to be found for each agent separately. However, this method is principally unable to solve instances where non-trivial cooperation among agents is necessary. WHCA* was used to classify instances on **easy** and **hard** – the easy ones are those solvable by WHCA*.

Contrary to SATPLAN and WHCA* COBOPT is more successful; it is able to provide solution to every instance to which base solving method can do so – in case of the BIBOX algorithm these are all the instances in our test suite.

In case of the 8×8 grid COBOPT(BIBOX) generates very near optimal solutions for easy setups (same as SATPLAN; same as or better than WHCA*) - Figure 2. Nevertheless, the most interesting behavior is exhibited in the hard region where compression up to the ratio of $\frac{1}{3}$ with respect to the makespan of base solution can be achieved. Although it is not known if optimum was actually reached, this is a big qualitative leap from the base solution and it demonstrates efficiency of the COBOPT optimization process.

Supposed that certain simplification is accepted then we can calculate expected lower bound for the optimal makespan in the 4-connected grid environment. Let us suppose that if an agent is blocked on its path by another agent, it will either wait or go into unblocked neighborhood all with the same probability of $\frac{1}{4}$. This behavior deflects the agent from its original path and some extra steps are then necessary to continue in the right direction. It is supposed that original path continues to vertices in

Expected number of extra steps to

$\frac{1}{3}0 + \frac{1}{3}2 + \frac{1}{3}2 = \frac{4}{3}$

$\frac{1}{3}2 + \frac{1}{3}2 + \frac{1}{3}2 = 2$

$\frac{1}{3}0 + \frac{1}{3}2 + \frac{1}{3}2 = \frac{4}{3}$ $\frac{1}{3}1 + \frac{1}{3}1 + \frac{1}{3}1 = 1$

the neighborhood of blocked vertex with the same probability of $\frac{1}{3}$. Under these assumptions we obtain that the expected number of extra steps is $\frac{1}{4}(2\frac{4}{3} + 2 + 1) = \frac{19}{12}$ per two original steps. Simply it means that two original steps require almost two extra steps. We will adopt quite strong assumption and round it up to exactly two extra steps which consequently implies that the agent actually does not reduce its distance from the destination.

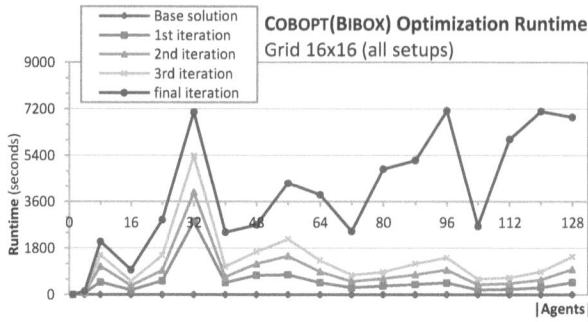

Fig. 5. Runtime measurements in the 16×16 grid

Proposition 2 (EXPECTED MAKESPAN). *The expected make-span required to travel distance $d \in \mathbb{N}$ in a 4-connected grid with occupancy ratio c under our assumptions is:*

$$\mu(d) = \frac{c(19-7d)-12d}{12(c-1)}.$$ ∎

Proof. The following recurrence holds under assumptions stated above: $\mu(d) \geq c(\mu(d) + \frac{19}{12}) + (1 - c)(\mu(d - 1) + 1)$ where we can put $\mu(1) = 1$. From this we quickly obtain the required explicit form.   ∎

According to above calculations COBOPT(BIBOX) generates near optimal solutions that differs from the expected optimum by less than 25% in setups with up to occupancy of 60% in the grid 8×8. Expectably in the grid 16×16 the situation is not so optimistic, solutions differ here from the expected optimum by factor of 3.0 to 6.0 in hard setups with occupancy up to 50%. This worse performance is mainly because of the size of the grid which prevented us from using more aggressive optimization.

The number of iterations until the fixed point was reached ranged from 1 to 20 with median of 7 in case of 8×8 grid and from 2 to 31 with median of 11 for the grid 16×16. The number of SAT solver invocations is reported in Table 3. It is clear that in our approach the SAT solver is invoked many times with relatively easy instances.

Runtime[1] is reported in Figure 3 and Figure 5. Despite hundreds of SAT solver invocations the overall runtime is kept in acceptable bounds. Fortunately, the COBOPT method is very friendly to multithreaded implementation. Hence the scalability of the method is extremely good (provided that computational resources are available).

---

[1] All the runtime measurements were done on a machine with the 6 core CPU Intel Xeon 2.0GHz and 12GiB RAM under Linux kernel 2.6.24-19. All the 6 cores of the CPU were exploited in parallel.

Moreover, if the method for producing base solutions is fast enough then COBOPT represents the anytime method in fact – at any time step the solving process can be terminated and feasible (sub-optimal) solution is returned.

To get insight what happen when a solver is used for optimization we investigated distribution of the number of actions executed in parallel – Figure 6. Base solutions seem to suffer from locked agents which are forced to wait until their path is freed. In optimized solutions, as many as possible agents are actively moving towards goals – it is possible to observe that agents utilize almost all the available unoccupied space.

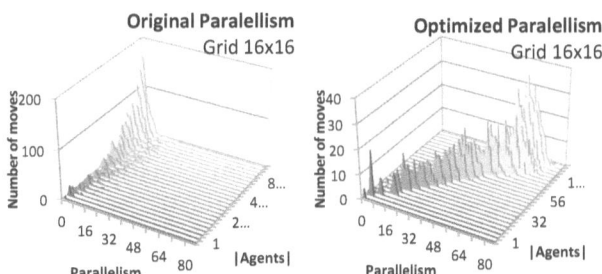

Fig. 6. Distribution *of parallelism in the grid 16×16.* Almost all the free space is used for moving in the optimized solution.

6   Discussion, Conclusions, and Future Works

The new **SAT based** solving method for cooperative path planning called COBOPT has been presented. To be able to use a SAT solver for cooperative path-planning we also developed a new SAT **encoding** for cooperative instances. The encoding utilizes properties of cooperative planning in order to reduce its size and increase efficiency.

The COBOPT method was shown that it is able to generate **near optimal** or **good quality** solutions in setups with high occupancy of the environment by agents. It is the **first** method capable of doing so. In our experiments we solved 4-connected grid instances of size up to 16×16 with up to 50% space occupied by agents with high quality makespans. One of the positive aspects of the new approach is also the fact that it can be easily parallelized for multi-core architectures which supports better scalability.

The COBOPT method has also quite strong implications for classical planning. Provided that efficient makespan sub-optimal planner is available, COBOPT can be immediately used to optimize its output (SASE and SATPLAN encodings are ready). A possible future improvement is to reduce the size of the domain dependent encoding for sparsely populated instances. The application of binary search for solvable instance may be also revised as other types of search may be more efficient.

References

1. Eén, N., Sörensson, N.: An Extensible SAT-solver. In: Giunchiglia, E., Tacchella, A. (eds.) SAT 2003. LNCS, vol. 2919, pp. 502–518. Springer, Heidelberg (2004)
2. Huang, R., Chen, Y., Zhang, W.: A Novel Transition Based Encoding Scheme for Planning as Satisfiability. In: Proceedings AAAI 2010. AAAI Press (2010)

3. Kautz, H., Selman, B.: Unifying SAT-based and Graph-based Planning. In: Proceedings of the 16th International Joint Conference on Artificial Intelligence (IJCAI 1999), pp. 318–325. Morgan Kaufmann (1999)
4. Luna, R., Berkis, K.E.: Push-and-Swap: Fast Cooperative Path-Finding with Completeness Guarantees. In: Proceedings of the 22nd International Joint Conference on Artificial Intelligence (IJCAI 2011), pp. 294–300. IJCAI/AAAI Press (2011)
5. Ratner, D., Warmuth, M.K.: Finding a Shortest Solution for the N × N Extension of the 15-PUZZLE Is Intractable. In: Proceedings of AAAI 1986, pp. 168–172. Morgan Kaufmann (1986)
6. Rintanen, J.: Compact Representation of Sets of Binary Constraints. In: Proceedings of the 17th European Conference on Artificial Intelligence (ECAI 2006), pp. 143–147. IOS Press (2006)
7. Ryan, M.R.K.: Exploiting Subgraph Structure in Multi-Robot Path Planning. Journal of Artificial Intelligence Research (JAIR) 31, 497–542 (2008)
8. Silver, D.: Cooperative Pathfinding. In: Proceedings of the 1st Artificial Intelligence and Interactive Digital Entertainment Conference (AIIDE 2005), pp. 117–122. AAAI Press (2005)
9. Standley, T.S.: Finding Optimal Solutions to Cooperative Pathfinding Problems. In: Proceedings of the 24th Conference on Artificial Intelligence (AAAI 2010). AAAI Press (2010)
10. Standley, T.S., Korf, R.E.: Complete Algorithms for Cooperative Pathfinding Problems. In: Proceedings of IJCAI 2011, pp. 668–673. IJCAI/AAAI Press (2011)
11. Surynek, P.: A Novel Approach to Path Planning for Multiple Robots in Bi-connected Graphs. In: Proceedings of the International Conference on Robotics and Automation (ICRA 2009), pp. 3613–3619. IEEE Press (2009)
12. Wang, K.C., Botea, A.: MAPP: a Scalable Multi-Agent Path Planning Algorithm with Tractability and Completeness Guarantees. JAIR 42, 55–90 (2011)

A New Algorithm for Multilevel Optimization Problems Using Evolutionary Strategy, Inspired by Natural Adaptation

Surafel Luleseged Tilahun[1], Semu Mitiku Kassa[2], and Hong Choon Ong[1]

[1] Universiti Sains Malaysia, School of Mathematical Sciences, 11800, Penang, Malaysia
`surafelaau@yahoo.com, hcong@usm.cs.my`
[2] Addis Ababa University, Department of Mathematics, 1176, Science faculty, A.A., Ethiopia
`smtk@math.aau.edu.et`

Abstract. Multilevel optimization problems deals with mathematical programming problems whose feasible set is implicitly determined by a sequence of nested optimization problems. These kind of problems are common in different applications where there is a hierarchy of decision makers exists. Solving such problems has been a challenge especially when they are non linear and non convex. In this paper we introduce a new algorithm, inspired by natural adaptation, using (1+1)-evolutionary strategy iteratively. Suppose there are k level optimization problem. First, the leader's level will be solved alone for all the variables under all the constraint set. Then that solution will adapt itself according to the objective function in each level going through all the levels down. When a particular level's optimization problem is solved the solution will be adapted the level's variable while the other variables remain being a fixed parameter. This updating process of the solution continues until a stopping criterion is met. Bilevel and trilevel optimization problems are used to show how the algorithm works. From the simulation result on the two problems, it is shown that it is promising to uses the proposed metaheuristic algorithm in solving multilevel optimization problems.

Keywords: Multilevel optimization, (1+1)-Evolutionary strategy, metaheuristic algorithms, Natural adaptation.

1   Introduction

Many resource allocation or planning problems require compromises among the objectives of several interacting individuals or agencies; most of the time, arranged in hierarchical administrative structure and can have independent even sometimes conflicting objectives. A planner at one level of the hierarchy may have its objective function determined partly by variables controlled at other levels. Assuming that the decision process has a preemptive nature and having k levels of hierarchy, we consider the decision maker at level 1 to be the *leader* and those at lower levels to be *followers*. These kind of problems can be modeled as a nested optimization problem, referred to as multilevel programming [1]. Mathematical programming models to solve problems of these kind has been studied since 1960s, [2].

P. Anthony, M. Ishizuka, and D. Lukose (Eds.): PRICAI 2012, LNAI 7458, pp. 577–588, 2012.

Multilevel optimization analysis becomes more and more applicable in different fields. Its role in agricultural economics has been studied by Candler et. al [3] . Kocara and Outrata studied its use in engineering design [4]. Its application to transport network was also studied in [5]. Generally whenever there is a hierarchy of decision maker in such a way that each decision maker controls part of the decision variable, multilevel optimization problem model is the one suitable for the situation [6].

Due to its many applications, multilevel programming, in particular bilevel programming, has evolved significantly [7, 8]. In the late nineties Bahatia and Biegler proposed an approach with periodic property [9]. Stochastic programming like method was also proposed by Acevedo and Pistikopoulos [10]. Furthermore Pistikopoulos et. al. and other researchers proposed a new algorithm based on parametric programming theory [8, 11, 12, 13, 14]. Most of the solution methods proposed are mainly for bilevel and trilevel optimization problems with linear or convex property. The search for solution methods still continues, especially methods which are not affected by behavior of the objective functions. Perhaps metaheuristic algorithms are suitable for such purpose. That is why some recent solution methods involve metaheuristic algorithms. Among many metaheuristic algorithms evolution algorithm [15, 16, 17] and particle swarm optimization [18, 19, 20] are used in many researches and application

This paper introduces a new algorithm inspired by natural adaptation and based on $(1+1)$-evolutionary strategy. The format of the paper is as follows; in the next section basic concepts will be discussed followed by a discussion on the introduced algorithm in section 3. The algorithm will be tested using a bilevel and a trilevel optimization problem in section 4. At last a conclusion will be given in section 5.

2   Preliminaries

2.1   Multilevel Optimization Problem (MLOP)

Optimization problems of the following form, as in equation 1, are called k-level optimization problems.

$$\underset{x_1}{Optimize} \quad f_1(x_1, x_2, ..., x_k)$$

$$s.t. \quad (x_1, x_2, ..., x_k) \in S_1 \subseteq \Re^n \text{ and } x_2 \text{ solves}$$

$$\underset{x_2}{Optimize} \quad f_2(x_1, x_2, ..., x_k)$$

$$s.t \quad (x_1, x_2, ..., x_k) \in S_2 \subseteq \Re^n \text{ and } x_3 \text{ solves}$$

$$\cdot \qquad\qquad\qquad (1)$$

$$\cdot$$

$$\cdot$$

$$\underset{x_k}{Optimize} \quad f_k(x_1, x_2, ..., x_k)$$

$$s.t. \quad (x_1, x_2, ..., x_k) \in S_k \subseteq \Re^n$$

$$(x_1, x_2, ..., x_k) \in S \subseteq \Re^n \quad \text{common constraint}$$

where $x_i \in \Re^{n_i}$ $\forall i \in \{1,2,...,k\}$, $n = \sum_{j=1}^{k} n_j$ and "Optimize" can be either maximize or minimize.

Generally, multilevel optimization problems (MLOP) are optimization problems which have a subset of their variables constrained to be optimal solutions of other optimization problems parameterized by the remaining variables. Depending on the number of optimization problems in the constraint set a level will be assigned. k-level optimization problem is an optimization problem which has k-1 optimization problems in the constraints.

The first optimization problem, f_1, is called leader's or level one problem and the others are followers' with level number increasing when going down. The decision maker at level j controls only x_j, whereas the other parts of the variables are controlled by the decision makers in other levels.

A point $x^* = (x_1^*, x_2^*,..., x_k^*)$ is said to be an optimal solution for the multilevel optimization problem if x* is an optimal solution for the leader's problem, satisfying lower level problems as a constraint set. Since we have different levels and different optimization problems in each level there usually will be a conflict of objectives, hence the concept of compromise optimality needs to be defined. A compromise optimal solution is a member of the feasible set for which there doesn't exist another feasible point which does the same in all objectives and better at least in one objective function. For a given multilevel optimization problems it is possible to have many solutions depending on the decision power of the decision maker in each level. Furthermore, unlike single level optimization problems, convexity doesn't guarantee the existence of an optimal solution, and generally it is a non convex problem even when the involved functions are linear. These behaviors make multilevel optimization problems challenging compared to single level optimization problems.

2.2   Evolutionary Strategy

Evolutionary strategy is a metaheuristic algorithm which is inspired by natural evolution. It has an operator which corresponds to the mutation operator in genetic algorithm. Depending on the number of children each solution member gives, we have many kind of evolutionary strategy. In this paper we consider a (1+1)-evolutionary strategy. (1+1)-evolutionary strategy is an evolutionary strategy in which a parent gives a birth to one child [21]. (1+1)-evolutionary strategy has the following main steps:

1. Generate a random set of solutions, $\{x_1, x_2, ..., x_m\}$
2. Move each solution member x_i, in a randomly generated direction d, $x_i' = x_i + d$. d is from a normal distribution $N(0,\delta)$, where δ is algorithm parameter.
3. Compare the performance of x_i and x_i' according to the objective function; and take the one which does better.
4. If termination criterion is not met go to step (2).

3   Metaheuristic Algorithm for MLOP

A metaheuristic algorithm is an algorithm with randomness property which tries to find a solution for optimization problems by improving the solution set iteratively. Most of these algorithms are inspired by a certain natural phenomenon. Perhaps, it is a good idea to face the challenge of multilevel optimization using metaheuristic solution methods. In this paper we introduce a new metaheuristic algorithm. The algorithm proposed in this paper uses the concept of evolutionary strategy and is inspired by natural adaptation. The leader's problem is solved for all the variables satisfying all constraint sets, as if it controls all the variables. Then that solution will adapt itself according to the objective functions in each level while going through all the levels.

Consider a k-level optimization problem shown below:

$$\min_{x_1} \quad f_1(x_1, x_2, ..., x_k)$$

$$s.t. \quad (x_1, x_2, ..., x_k) \in S_1 \subseteq \mathfrak{R}^n \text{ where } x_2 \text{ solves}$$

$$\min_{x_2} \quad f_2(x_1, x_2, ..., x_k)$$

$$s.t \quad (x_1, x_2, ..., x_k) \in S_2 \subseteq \mathfrak{R}^n \text{ where } x_3 \text{ solves}$$

$$\cdot \tag{2}$$

$$\cdot$$

$$\cdot$$

$$\min_{x_k} \quad f_k(x_1, x_2, ..., x_k)$$

$$s.t. \quad (x_1, x_2, ..., x_k) \in S_k \subseteq \mathfrak{R}^n$$

$$(x_1, x_2, ..., x_k) \in S \subseteq \mathfrak{R}^n \quad \text{common constraint}$$

In the algorithm the leader's problem will be solved for $(x_1^0, x_2^0, ..., x_k^0)$ with all the constraints in all the levels including the common constraint, given as:

$$\min_{x_1, x_2, ..., x_k} \quad f_1(x_1, x_2, ..., x_k)$$

$$s.t. \quad (x_1, x_2, ..., x_k) \in \overline{S_1} \tag{3}$$

$$\text{where } \overline{S_1} = S \cap S_1 \cap S_2 \cap ... \cap S_k$$

This implies that $f_1(x_1^0, x_3^0, ..., x_k^0) \leq f_1(x_1, x_2, ..., x_k) \quad \forall (x_1, x_2, ..., x_k) \in \overline{S_1}$. Then by fixing x_1^0 we solve the second level problem and update $(x_2^0, x_3^0, ..., x_k^0)$ as shown in equation (4).

$$\min_{x_2, x_3, ..., x_k} \quad f_1(x_1^0, x_2, ..., x_k)$$

$$s.t. \quad (x_1^0, x_2, ..., x_k) \in \overline{S_2} \tag{4}$$

$$\text{where } \overline{S_2} = S \cap S_2 \cap S_3 \cap ... \cap S_k$$

Generally for any level i we will solve the corresponding problem using evolutionary strategy, as shown in (5).

$$\min_{x_i, x_{i+1},...,x_k} f_1(x_1^0, x_2^0,...,x_{i=1}^0, x_i, x_{i+1},..., x_k)$$

$$s.t. \quad (x_1^0, x_2^0,..., x_{i=1}^0, x_i, x_{i+1},..., x_k) \in \overline{S}_i \tag{5}$$

$$where \overline{S}_i = S \cap S_i \cap S_{i+1} \cap ... \cap S_k$$

Once the kth level problem is solved, then $(x_2^0, x_3^0,..., x_k^0)$ will be used as parameters to solve the problem given in equation (6) for x_1^1.

$$\min_{x_1} f_1(x_1, x_2^0,..., x_k^0)$$

$$s.t. \quad (x_1, x_2^0,..., x_k^0) \in S_1 \tag{6}$$

$$(x_1, x_2^0,..., x_k^0) \in S$$

Similarly, to solve the second level for x_2^1 we fix $(x_1^1, x_3^0, x_4^0,..., x_k^0)$ and use evolutionary strategy. Once x_2^1 is computed we fix $(x_1, x_2, x_4,..., x_k)$ as $(x_1^1, x_2^1, x_4^0,..., x_k^0)$ and solve for $x_3 = x_3^1$. By continuing in similar way for all the levels down, at last $(x_2^1, x_3^1,,..., x_k^1)$ will be fixed to solve for the first level problem for $x_1 = x_1^2$. This process will continue until a termination criterion is fulfilled. It means at j^{th} iteration and optimizing i^{th} level we will have the following optimization problem:

$$\min_{x_i} f_i(x_1^j, x_2^j,..., x_{i-1}^j, x_i, x_{i+1}^{j-1},..., x_k^{j-1})$$

$$s.t. \quad (x_1^j, x_2^j,..., x_{i-1}^j, x_i, x_{i+1}^{j-1},..., x_k^{j-1}) \in S'_i \tag{7}$$

$$(x_1^j, x_2^j,..., x_{i-1}^j, x_i, x_{i+1}^{j-1},..., x_k^{j-1}) \in S^i$$

where S'_i is the ith level constraint set with all the other variables are fixed and S^i is also the common constraint with all the variables, except variable i, are fixed.

At each step (1+1)-evolutionary algorithm will be used with a property of passing the previous solution. It means, suppose we are solving the ith level problem at a particular iteration j. $(x_1^j, x_2^j,..., x_{i-1}^j, x_{i+1}^{j-1},..., x_k^{j-1})$ is fixed then when evolutionary strategy is used x_i^{j-1} will be taken as a member among the randomly generated initial solution set. This will help the algorithm not to move away from a good solution because of the conflict of objective functions.

The algorithm is summarized in the following tables:

Table 1. The algorithm

Input: $f_i(x_1, x_2,...,x_k), S, S_i \ \forall i \in \{1,2,...,k\}$

for p=1:k

$$(x_1^0, x_2^0, ..., x_k^0) == \text{EvolutionaryStrategy } (f_p(x_1^0, x_2^0, ... x_{p-1}^0, x_p, ..., x_k), S_p, S, n_p)$$

$$\text{where } S_p = S_p \cap S_{p+1} \cap ... \cap S_k$$

end

for j=1:MaxGen

for i=1:k

if $(i > 1)$

$$x_i^j == \text{EvolutionaryStrategy } (f_i(x_1^j, x_2^j, ..., x_{i-1}^j, x_i, x_{i+1}^{j-1} ..., x_k^{j-1}), S_i, S, n_i, x_i^{j-1})$$

else

$$x_1^j == \text{EvolutionaryStrategy } (f_1(x_1, x_2^{j-1}, ..., x_k^{j-1}), S_1, S, n_1, x_1^{j-1})$$

end if

end for

$$(x_1^*, x_2^*, ..., x_k^*) = (x_1^j, x_2^j, ..., x_k^j)$$

terminate if termination criteria is fulfilled

end for

Output: $(x_1^*, x_2^*, ..., x_k^*)$

Table 2. Evolutionary strategy with passing a solution,
EvolutionaryStrategy $(f(x), S_1, S, n, x')$

Input: $f(x)$, S_1, S, n, x'

Algorithm Parameter: δ

Do:

Randomly generate m-1 solutions for x from the feasible region, say $x_1, x_2,$
$..., x_{m-1}$.

Put $x_m = x'$ (*if x' is given, else generate x_m also randomly*)

for i=1:m

Generate d from $N(0, \delta)$

$x_i^* = x_i + d$

Check feasibility

if $(f(x_i^*) \leq f(x_i))$

$x_i = x_i^*$

end if

end for

Repeat until termination criteria is met

$x^* = x_j$, such that $f(x_j) \leq f(x_i) \ \forall i \in \{1,2,...,m\}$

Output: x^*

4   Simulation Examples

To demonstrate the algorithm we use a bilevel and a trilevel optimization problems.

a) Bilevel Example

The bilevel problem is taken from a book chapter [22]. It is as given in equation (8). After solving the problem using the algorithm, the solution is compared to the solution given in the book. According to the book the solution is (x', y') = (0.609, 0.391, 0, 0, 1.828), with values 0.6429 and 1.6708 for f_1 and f_2, respectively.

$$\min_{x} f_1(x, y) = y_1^2 + y_3^2 - y_1 y_3 - 4y_2 - 7x_1 + 4x_2$$

$$s.t \quad x_1 + x_2 \leq 1$$

$$\min_{y} f_2(x, y) = y_1^2 + \frac{1}{2} y_2^2 + \frac{1}{2} y_3^2 + y_1 y_2 + (1 - 3x_1)y_1 + (1 + x_2)y_2 \quad (8)$$

$$s.t \quad 2y_1 + y_2 - y_3 + x_1 - 2x_2 + 2 \leq 0$$

$$x \geq 0$$

$$y \geq 0$$

The algorithm parameter δ was set to be 1 and number of initial solutions, m, was set to be 50. Furthermore the number of iteration was set to be 30 for the evolutionary strategy and 50 for the algorithm.

First the leader's problem is solved with all the constraints as shown in equation (9)

$$\min_{x,y} f_1(x, y) = y_1^2 + y_3^2 - y_1 y_3 - 4y_2 - 7x_1 + 4x_2$$

$$s.t \quad x_1 + x_2 \leq 1$$

$$2y_1 + y_2 - y_3 + x_1 - 2x_2 + 2 \leq 0 \quad (9)$$

$$x \geq 0$$

$$y \geq 0$$

Using (1+1)-evolutionary strategy, the solution for the problem in equation (9) is found to be $x^0 = (0.2756 \; 0.7117)$ and $y^0 = (0.0167 \; 1.1095 \; 2.0073)$ with $f_1(x^0, y^0) = 0.4763$.

Then by fixing x^0 the second level problem is solved.

$$\min_{y} f_2(x^0, y) = y_1^2 + \frac{1}{2} y_2^2 + \frac{1}{2} y_3^2 + y_1 y_2 + (1 - 3x_1^0)y_1 + (1 + x_2^0)y_2$$

$$s.t \quad 2y_1 + y_2 - y_3 + x_1^0 - 2x_2^0 + 2 \leq 0 \quad (10)$$

$$x_1^0, x_2^0 \geq 0$$

$$y \geq 0$$

Then equation (10) is solved using evolutionary strategy for y', in such a way that y^0 will be taken as one of the initial solutions in the evolutionary strategy. After

y^1 is computed, it will be fixed to solve the leader's problem for x^1, again using evolutionary strategy with passing the previous best, x^0, as one of the initial population member.

$$\min_x f_1(x, y^1) = (y_1^1)^2 + (y_3^1)^2 - y_1^1 y_3^1 - 4 y_2^1 - 7 x_1 + 4 x_2$$

$$s.t \quad x_1 + x_2 \leq 1 \tag{11}$$

After that the second problem will be solved for y^2 by fixing x^1 and passing y^1 as a member of initial solutions in the evolutionary strategy. This pattern repeats itself until a preset iteration number, which in our case is 50, is reached.

After running the program using matlab the optimal solution was found to be (x^*, y^*) = (0.5307 0.4683 0.0012 0.0002 1.5989), with 0.7122 and 1.2778 for f_1 and f_2, respectively. The performance of the algorithm compared to the solution in the book is presented in Figure (1). Furthermore, it compares the performance in terms of the sum of the functional values, $f_1(x',y')+f_2(x',y')$ and $f_1(x^*,y^*)+f_2(x^*,y^*)$.

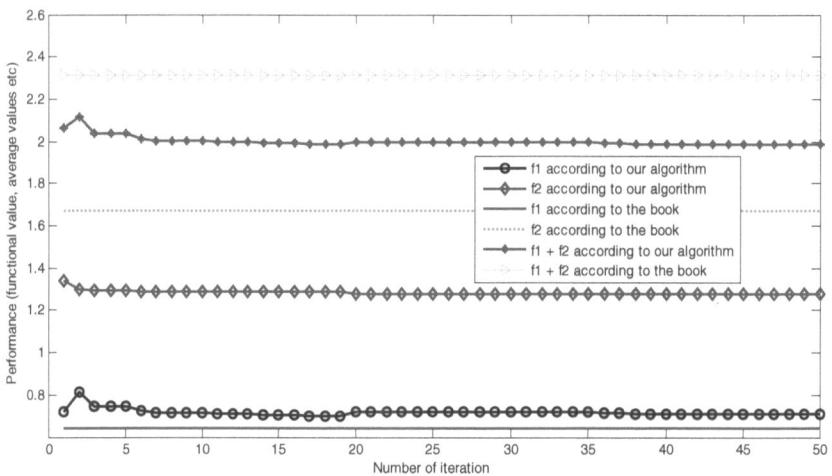

Fig. 1. Performance of the algorithm in the bilevel problem

From the result it is clear that if equal weight is given to the objective functions the algorithm performs better because $f_1(x',y')+f_2(x',y')$ = 2.3134 and $f_1(x^*,y^*)+f_2(x^*,y^*)$ = 1.9900, where x' is the solution from the book and x* is the solution after running the algorithm.

b) Trilevel Example
The second test problem is a trilevel optimization problem taken from a thesis done by Molla [23]. It is given in equation (12)

$$\min_{x} f_1(x, y, z) = -x + 4y$$

s.t. $x + y \leq 1$ and y solves

$$\min_{y} f_2(x, y, z) = 2y + z$$

s.t. $-2x + y \leq -z$ and z solves   (12)

$$\min_{z} f_3(x, y, z) = -z^2 + y$$

s.t. $z \leq x$

$$\left.\begin{array}{l} 0 \leq x \leq 0.5 \\ 0 \leq y, z \leq 1 \end{array}\right\} \text{is common constraint}$$

The algorithm parameter and number of iterations are the same in the previous case. According to the algorithm first the leader's problem, as shown in equation (13), is solved.

$$\min_{x,y,z} f_1(x, y, z) = -x + 4y$$

s.t. $x + y \leq 1$

$-2x + y \leq -z$ (13)

$z \leq x$

$0 \leq x \leq 0.5$

$0 \leq y, z \leq 1$

After running the code the solution is found to be $(x^0,\ y^0,\ z^0)$ = (0.4999 0.0003 0.1687). Now by fixing x^0 we solve for the problem in equation (14).

$$\min_{y,z} f_2(x^0, y, z) = 2y + z$$

s.t. $-2x^0 + y \leq -z$

$z \leq x^0$ (14)

$0 \leq x^0 \leq 0.5$

$0 \leq y, z \leq 1$

The solution of equation (14) is $10^{-3}(0.1097\ 0.2957)$, with $f_2(0.4999,\ (10^{-3})(0.1097),\ (10^{-3})(0.2957))=10^{-4}(5.1513)$. Hence $(x^0,\ y^0,\ z^0)$ is updated to $(0.4999,\ (10^{-3})(0.1097),\ (10^{-3})(0.2957))$.

Afterwards we fix (x^0, y^0) and solve the third level problem for z^0.

$$\min_{z} f_3(x^0, y^0, z) = (-z)^2 + y^0$$

s.t. $z \leq x^0$ (15)

$0 \leq x^0 \leq 0.5$

$0 \leq y^0, z \leq 1$

z^0 is updated to be 0.4978. Hence (x^0, y^0, z^0) is computed, then y^0 and z^0 will be fixed using evolutionary strategy to solve for x^1, with x^0 as one of the solution candidate for the evolutionary strategy algorithm. After running the matlab code the result becomes $(x^*, y^*, z^*) = (0.5, 0, 0.0095)$. And $f_1(x^*, y^*, z^*) = -0.5$, $f_2(x^*, y^*, z^*) = 0.0127$ and $f_3(x^*, y^*, z^*) = -0.2499$. The performance of the functional values as a function of iteration number is shown in the graph below:

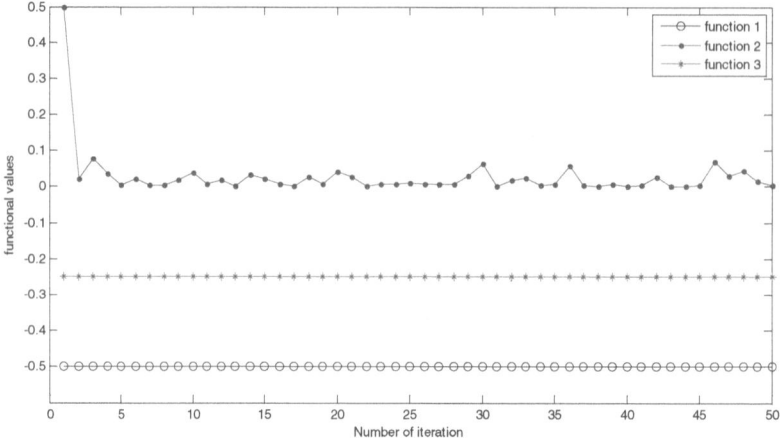

Fig. 2. Performance of the algorithm in the trilevel problem

Our result is better than the result reported by Molla [17], which is $(x_m, y_m, z_m)=(0.5, 1, 1)$. This implies that $f_1(x_m, y_m, z_m)=3.5$, $f_2(x_m, y_m, z_m)=3$ and $f_3(x_m, y_m, z_m)=0$, which are worse in sense of minimization when compared to our result not only in average but also for each individual objective functions.

5   Conclusion

In this paper a new metaheuristic algorithm for multilevel optimization problem which mimic the concept of natural adaptation is introduced. The algorithm uses (1+1)-evolutionary strategy to solve one of the level's optimization problem at once. The leader's problem will be solved for all the variables satisfying all the constraint sets in all the levels. That solution will go through each level iteratively by adapting itself with the corresponding objective function and constraint set of each level. In each iteration the solution will respond to the change in the parameters and tries to update itself compared to the previous solution and the new parameters. (1+1)-evolutionary strategy is used in the updating process with a property of considering the previous best as a candidate solution for the current stage. These updating will continue iteratively until termination criterion is met. From the simulation results on a bilevel and trilevel optimization problem, it is shown that the algorithm gives a promising result for multilevel optimization problems.

Acknowledgements. This work is in part supported by Universiti Sains Malaysia short term grant number 304/PMATHS/6311126. The first author would like to acknowledge a support from TWAS-USM fellowship.

References

1. Bard, J.F., Falk, J.E.: An explicit solution to the multilevel programming problem. Computers and Operations Research 9(1), 77–100 (1982)
2. Dantzig, G.B., Wolfe, P.: Decomposition Principle for Linear Programs. Operations Research 8(1), 101–111 (1960)
3. Candler, W., Fortuny-Amat, J., McCarl, B.: The potential role of multilevel programming in agricultural economics. American Journal of Agricultural Economics 63, 521–531 (1981)
4. Vincent, N.L., Calamai, H.P.: Bilevel and multilevel programming: A bibliography review. Journal of Global Optimization 5, 1–9 (1994)
5. Marcotte, P.: Network design problem with congestion effects: A case of bilevel programming. Mathematical Programming 43, 142–162 (1986)
6. Rao, S.S.: Engineering Optimization: theory and practice, 4th edn. John Wiley & Sons Inc. (2009)
7. Faisca, P.N., Dua, V., Rustem, B., Saraiva, M.P., Pistikopoulos, N.E.: Parametric global optimization for bilevel programming. Journal of Global Optimization 38, 609–623 (2007)
8. Pistikopoulos, N.E., Georgiads, C.M., Dua, V.: Multiparametric Programming: theory, algorithm and application. Wiely-Vich Verlag Gmbh and Co. KGaA (2007)
9. Bahatia, T.K., Biegler, L.T.: Multiperiod design and planning with interior point method. Computers and Chemical Engineering 23(14), 919–932 (1999)
10. Acevedo, J., Pistikopoulos, E.N.: Stochastic optimization based algorithms for process synthesis under uncertainty. Computer and Chemical Engineering 22, 647–671 (1998)
11. Dua, V., Pistikopoulos, E.N.: An algorithm for the solution of parametric mixed integer linear programming problems. Annals of Operations Research 99(3), 123–139 (2000)
12. Dua, V., Bozinis, N.A., Pistikopoulos, E.N.: A multiparametric programming approach for mixed-integer quadratic engineering problem. Computers and Chemical Engineering 26(4/5), 715–733 (2002)
13. Pistikopoulos, E.N., Dua, V., Ryu, J.H.: Global optimization of bilevel pro-gramming problems via parametric programming. Computational Management Science 2(3), 181–212 (2003)
14. Li, Z., Lerapetritou, M.G.: A New Methodology for the General Multi-parametric Mixed-Integer Linear Programming (MILP) Problems. Industrial & Engineering Chemistry Research 46(15), 5141–5151 (2007)
15. Wang, Y., Jiao, Y.C., Li, H.: An evolutionary algorithm for solving nonlinear bilevel programming based on a new constraint-handling scheme. IEEE Transactions on Systems, Man and Cybernetics Part C 35(2), 221–232 (2005)
16. Deb, K., Sinha, A.: Solving Bilevel Multi-Objective Optimization Problems Using Evolutionary Algorithms. In: Ehrgott, M., Fonseca, C.M., Gandibleux, X., Hao, J.-K., Sevaux, M. (eds.) EMO 2009. LNCS, vol. 5467, pp. 110–124. Springer, Heidelberg (2009)
17. Sinha, A.: Bilevel Multi-objective Optimization Problem Solving Using Progressively Interactive EMO. In: Takahashi, R.H.C., Deb, K., Wanner, E.F., Greco, S. (eds.) EMO 2011. LNCS, vol. 6576, pp. 269–284. Springer, Heidelberg (2011)

18. Kuo, R.J., Huang, C.C.: Application of particle swarm optimization algorithm for solving bi-level linear programming problem. Computers & Mathematics with Applications 58(4), 678–685 (2009)
19. Gao, Y., Zhang, G., Lu, J., Wee, H.M.: Particle swarm optimization for bi-level pricing problems in supply chains. Journal of Global Optimization 51, 245–254 (2011)
20. Zhang, T., Hu, T., Zheng, Y., Guo, X.: An Improved Particle Swarm Optimiza-tion for Solving Bilevel Multiobjective Programming Problem. Journal of Applied Mathematics 2012, Article ID 626717 (2012), doi:10.1155/2012/626717
21. Negnevitsky, M.: Artificial Intelligence: A Guide to Intelligent System. Henry Ling Limited, Harlow (2005)
22. Faisca, N.P., Rustem, B., Dua, V.: Bilevel and multilevel programming, Multi-Parametric Programming. WILEY-VCH Verlag GmbH & Co. KGaA, Weinheim (2007)
23. Molla, A.: A multiparametric programming approach for multilevel optimization, A thesis submitted to the department of mathematics, Addis Ababa University (2011)

Hamming Selection Pruned Sets (HSPS)
for Efficient Multi-label Video Classification

Tiong Yew Tang, Saadat M. Alhashmi, and Mohamed Hisham Jaward

Monash University Sunway Campus, Jalan Lagoon Selatan, 46150, Selangor, Malaysia
{tang.tiong.yew,alhashmi,mohamed.hisham}@monash.edu

Abstract. Videos have become an integral part of our life, from watching movies online to the use of videos in classroom teaching. Existing machine learning techniques are constrained with this scaled up activity because of this huge upsurge in online activity. A lot of research is now focused on reducing the time and accuracy of video classification. Content-Based Video Information Retrieval CBVIR implementation (E.g. Columbia374) is one such approach. We propose a fast Hamming Selection Pruned Sets (HSPS) algorithm that efficiently transforms multi-label video dataset into single-label representation. Thus, multi-label relationship between the labels can be retained for later single label classifier learning stage. Hamming distance (HD) is used to detect similarity between label-sets. HSPS captures new potential label-set relationships that were previously undetected by baseline approach. Experiments show a significant 22.9% dataset building time reduction and consistent accuracy improvement over the baseline method. HSPS also works on general multi-label dataset.

Keywords: Index Terms— Semantic Concept, Classification, Videos, Label Combinations, Multi-Label, Pruned Sets, TRECVID, Index and Retrieval.

1 Introduction

Multimedia information indexing and retrieval is an integral part of modern video search engines to describe, store and organize multimedia information and to assist people in finding multimedia information conveniently. Many research papers have been published on the area of *Content-Based Video Information Retrieval* (CBVIR) [1, 2]. CBVIR is an information retrieval research problem which involves machine learning over the video content (e.g. colors, shapes, textures, texts, motions and etc.).

In this paper, our focus is on *Video Semantic Concept Classification* (VSCC) which is a sub-theme of CBVIR for detecting the correct semantic concept of a video shot. For example, let us consider a video shot containing keyframes of a car moving on a road. The detected semantic concepts for this video shot are car, weather and road. In fact this example shows that the video semantic concept classification naturally is a multi-label research problem.[1] Relationship between these labels can be exploited for enhancing classification accuracy as demonstrated by these research

---

[1] Throughout this paper, we use multi-label instead of multi-concept for our research context.

P. Anthony, M. Ishizuka, and D. Lukose (Eds.): PRICAI 2012, LNAI 7458, pp. 589–600, 2012.
© Springer-Verlag Berlin Heidelberg 2012

papers [3-5]. Video semantic concept classification is an important indexing step for later video search and retrieval tasks.

The existing TRECVID[2] pre-processed multi-label video corpus such as Columbia 374 [6], Vireo 374 [7], MediaMill 101 [8] and combination of both Columbia and Vireo [9] are freely available baseline systems for VSCC. In these datasets, the human annotated concepts are available in multi-label format. Our experiments in this paper are based on MediaMill 101 [8] video corpus. Multi-label relationship information in the video datasets can be utilized to enhance video semantic concept classification accuracy by applying Problem Transformation (PT). PT is the process whereby a multi-label problem is transformed into one or more single-label problems [10]. After PT, any single label machine learning methods can be applied to the transformed datasets and can train a multi-label model for classification.

In this paper, an efficient Hamming distance based Pruned Sets Selection algorithm is proposed. This algorithm (abbreviated as HSPS) transforms a multi-label dataset into single-label representation. Hamming distance [11] is commonly used in channel coding in Telecommunication but we used it to find similarity between label-sets. This approach improves the performance obtained from the state of the art multi-label classification algorithm proposed in [12] and the approach in [12] is considered as our baseline approach for performance comparison. HSPS can detect new potential label-set relationships that were previously undetected by baseline approach [12]. Our contributions in this paper are in two folds: we introduce an intuitive way of capturing undetected new label sets and thus improve multi-label classification accuracy; our proposed method reduces the dataset build time and improves learning efficiency.

For the formal description in this paper, finite set of labels in a multi-label video set is denoted as $\mathcal{L} = (\lambda_j : j = 1 \dots m)$. We use $D = ((\boldsymbol{x}_i, \boldsymbol{y}_i), i = 1 \dots n)$ to denote a set of multi-label training instances (n denotes total number of available data instances). Here m denotes the total number of possible labels and is equal to the length of label vector, $\boldsymbol{y}_i$. The label set $\boldsymbol{y}_i \subseteq \mathcal{L}$ is the set of labels of the i-th example and $\boldsymbol{x}_i$ is the corresponding feature vector. We denote the i-th data instance as $\boldsymbol{d}_i = \{\boldsymbol{x}_i, \boldsymbol{y}_i\}$ and $|\boldsymbol{y}|_i$ is the sum of all labels in $\boldsymbol{y}_i$ such that $|\boldsymbol{y}|_i = \sum_{j=0}^{m} y_{ij}$.

This paper is organized as follows. Section 2 provides an overview of the related work. The baseline method is explained in section 3. Our novel HSPS algorithm is then proposed in section 4. Section 5 discusses our evaluation methods and our results are presented in section 6. Finally, conclusions and future research directions are presented in section 7.

2   Related Work

Many different problem transformation methods have been proposed to solve multi-label problem. We can categorize them into four different groups. First is the *Binary Relevance* method (BR) or one versus all. BR method is the most popular PT approach for video data classification [6, 9]. BR approach transforms any multi-label

---

[2] http://trecvid.nist.gov/

problem into m binary problems, and then a binary classifier is assigned for predicting the association of a single label for each binary problem. Although BR is simple and relatively fast, it has been criticized for not explicitly modeling label correlations [13, 14]. This is also known as label independent problem where label correlations are not captured and thus labels are independent from each other.

The second method is *Pair-Wise* classification method (PW) or one versus one. PW is a paradigm where one classifier learner is assigned with each pair of labels [15, 16]. However, PW methods are known for not handling well with overlapping labels and establish disjoint assignments in the multi-label context. PW also suffers from time complexity because the training time is quadratic with respect to the number of labels [17, 18].

Third is *Label Power Set* method (LP) which handles all label sets as combinations of labels to form a single-label representation. In this single-label representation, the set of single labels represents all distinct label sets in the multi-label training data [19, 20]. *Pruned Set* (PS) algorithm proposed by Read [12] is a complexity simplified version of LP method. The benefit of LP is to take account of label's relationship directly, but LP require many class labels in the single-label transformation as there are $min(m, 2^m - 1)$ number of distinct classes in the worst case. Another issue with LP is that it can over fit the training dataset because it can only classify new examples with classes that already exist in the training set [20, 21].

Fourth is *Context-Based Framework* method (CBF) which is a framework based on constructing context spaces of concepts, such that the contextual correlations are used to improve the performance of concept detector [3-5]. Most of the VSCC task used these methods for enhancing classification accuracy. CBF uses a two-layer structure context framework built on top of BR method. Accuracy of CBF depends heavily on BR base concept detectors. Any noise introduced in the first BR layer can scale up in the second layer because the two-layers are not closely integrated compared to LP. TRECVID video corpus is naturally biased towards negative examples [1] and thus can deteriorate the performance of the second layer. In our approach, we avoid this problem by not using the context-based framework's two layer design.

As our baseline method, we have used the PS approach [12] and this approach provides the following advantages compared to other approaches. (1) PS algorithm achieves the best multi-label classification performance compared to other available methods. (2) It offers reasonable trade-off between building time and multi-label relationship information preservation where other PT methods tend to be biased toward different sides.

3   Pruned Sets (PS)

Label power set method overcomes label independence problem by capturing relationship information of all possible combination of label sets. For example, let us consider a multi-label instance y_i with three labels $\lambda_1, \lambda_2, \lambda_3$ are all equal to 1. The feature vector x_i associated with this label set $y_i = (1,1,1)$ will be transformed to label power set of y_{i0} (100), y_{i1} (010), y_{i2} (001), y_{i3} (110), y_{i4} (011), y_{i5} (101),

$y_{i6}(111)$ and each of these label set is associated to the same x_i feature vector. However this problem of transforming method suffers from infrequent label sets after the transformation. Large number of infrequent label set instances will create huge dataset building time overhead. Furthermore, these infrequent label set instances does not improve the classification performance. To resolve these issues, we use an algorithm named Pruned Set (PS) proposed by Read [12]. Pruned Set reduces the relationship complexity between the labels and captures only the frequently occurring relationship in the dataset. It also "prunes" away infrequent label sets to improve learning time. Pruned Set Algorithm introduces two parameters p and n to control the total number of relationship for the dataset without sacrificing the dataset building time and prediction performance.

There are two major steps in Pruned Set algorithm. Step one is to prune away infrequent occurring label set combinations according to parameter p. Parameter p is the label set count threshold parameter. It is used to capture the frequently occurring label set instances by counting the label sets where their count of occurrence frequency, c is more than p.

In second step, the infrequent data instances D_p with an occurrence count c is less than p are processed with a sub-sampling step to retain information from the first pruned process. Sub-sampling is a process to extract subset of label sets $(y_i, c) \in L', p > c$ from the infrequent data instances D_p and then assigning the feature vector x_i to all label sets. Parameter n is to determine how many subsets should be extracted from the infrequent data instances and transform it to dataset D. For example, Fig. 1 (b) shows a Pruned Set algorithm with parameter $p = 2$, $n = 2$. Label set $y_a = 00011000000$ and $y_b = 00001010000$ occur 3 times ($c = 3$) and they are considered as frequently occurring label sets. Now let us consider an infrequent label set example with label set $y_h = 00011110110$ at Fig. 1 (a). Pruned Set algorithm will assign two new label sets y_a and y_b to this data instance (as $n = 2$). This sub-sampling provides $n = 2$ number of label sets such that y_a and y_b are the subset of y_h ($y_a \subseteq y_h$ and $y_b \subseteq y_h$). Note that the label set 00010000000 is also a subset of y_h. However, label set 00010000000 is not selected by the sub-sampling process as its occurrence count is less than p. If parameter $n = 1$ then either of y_a or y_b will be selected.

If we observe the first three data instances in Fig. 1 (a), you can see that they only differ in one bit position. It is likely that these three instances are caused by similar data instances with same concepts. By utilizing these data instances, it is expected to provide better classification accuracy and this is our motivation for proposed algorithm. In our work, we capture these unutilized data instances by using a novel Hamming distance based pruning approach.

4   Hamming Selection Pruned Sets (HSPS)

In this paper, we propose a LP method that is called Hamming Selection Pruned Sets (HSPS) for VSCC tasks. HSPS is an enhanced version of Pruned Sets (PS) algorithm [12] where it can identify new concepts of label sets that cannot be detected by PS. HSPS algorithm measures the similarity between the potential candidates for new concepts by using Hamming Distance calculation. We denote Hamming Distance calculation operator as $\otimes$. We define k as label set's index for pair comparison with label set y_i. Hamming distance calculates the difference between two labels such that $y_i \otimes y_h$ is the symmetrical difference between these two label sets. The potential candidates in the new class label sets will then be transformed to a common label sets by using binary AND operator (denoted as $\oplus$) as explained below.

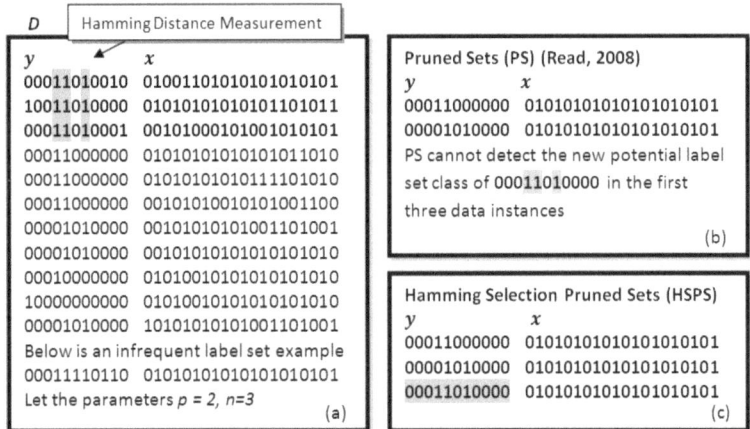

Fig. 1. This figure illustrates the HSPS algorithm successfully captures the undetected label set class relationship from baseline PS algorithm

Fig. 1 (c) illustrates the HSPS algorithm. The example shows that the highlighted labels are the label set that are not detected by PS algorithm. Thus this non-frequent label sets class of 00011110110 will not be added in the PS sub sampling step. On the other hand, our novel HSPS algorithm will capture the label sets relationship by measuring Hamming distance between the potential new label sets classes (00011010010, 10011010000, 00011010001) using Hamming distance. After successfully capturing the new label set classes, HSPS will merge the new class 00011010000 to the existing pruned subset candidates. Fig. 2 is the overall flow chart of HSPS algorithm. Our original contribution is included in step 3 of Fig. 2. Fig. 3 explains our original approach for Hamming selection algorithm. Our proposed algorithm in Fig. 2 consists of the following steps:

Step 1. This step is to search for each data instance d_i with label set y_i from dataset D. If found with similar label set with y_i then occurrence frequency count c_i will be incremented by 1.

Step 2. In this step we consider the p pruning parameter. The data instance d_i with its label set y_i with an occurrence frequency, $c_i < p_i$ will be excluded from set D'. The excluded data instances D_p will be passed to Step 3 and Step 4. D' is the transformed dataset for single label learning.

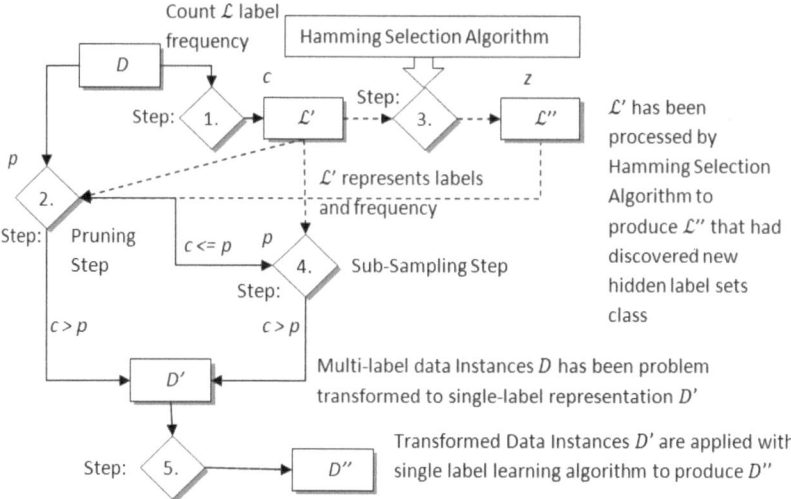

Fig. 2. This is the flow chart of HSPS algorithm

Step 3. Fig. 3 explains the Hamming Selection algorithm in detail. For all the pruned data instances D_p is sorted according to $|\mathbf{y}|$ with a higher value on top of the list. Then, for each available data instances $\mathbf{d}_i$ will be compared with its label set to each following label set using Hamming Distance calculation and it gives out a value denoted as h. A threshold parameter for our Hamming Distance value is denoted as z. This z parameter determines the likelihood of the compared label sets. For each available data instances d_i, if label set $\mathbf{y}_i$ is found to be more than p times $h < z$, then we select the data instances for common label set representation. For common label set representation, each selected label set is transformed with AND operation $\mathbf{y}_{i1} = \mathbf{y}_{i0} \oplus \mathbf{y}_{i1}$, $\mathbf{y}_{i2} = \mathbf{y}_{i1} \oplus \mathbf{y}_{i2}, \ldots \mathbf{y}_{in} = \mathbf{y}_{in-1} \oplus \mathbf{y}_{in}$ to represent them. Label comparison operations will continue until there are no more available label sets for comparison for each available data instances $\mathbf{d}_i$, where label $\mathbf{y}_i$ is found to be more than p times of $h < z$. If there is no more available label sets to be compared, all the duplicated common representation label sets in the last stage will be removed. These duplicated label sets was created by AND operation.

Step 4. The excluded data instances D_p from Step 2 will be considered in this step. The data instances label set $\mathbf{y}_i$ will be decomposed to individual $\mathbf{y}_{i0}, \mathbf{y}_{i1}, \ldots \mathbf{y}_{in}$ where $(\mathbf{y}_{in}, c) \in L'$ such that $c > p$. Then, these decomposed new examples $\mathbf{d}_{i0}(\mathbf{y}_{i0}, \mathbf{x}_i), \mathbf{d}_{i1}(\mathbf{y}_{i1}, \mathbf{x}_i), \ldots \mathbf{d}_{in}(\mathbf{y}_{in}, \mathbf{x}_i)$ will be added to D'. Note that all instances are assigned with the same $\mathbf{x}_i$ feature vector after the data transformation.

Step 5. In this last step, a single-label representation is created from D' and use a single-label learning method to train the dataset. In this way, the relationship between the label sets will be preserved.

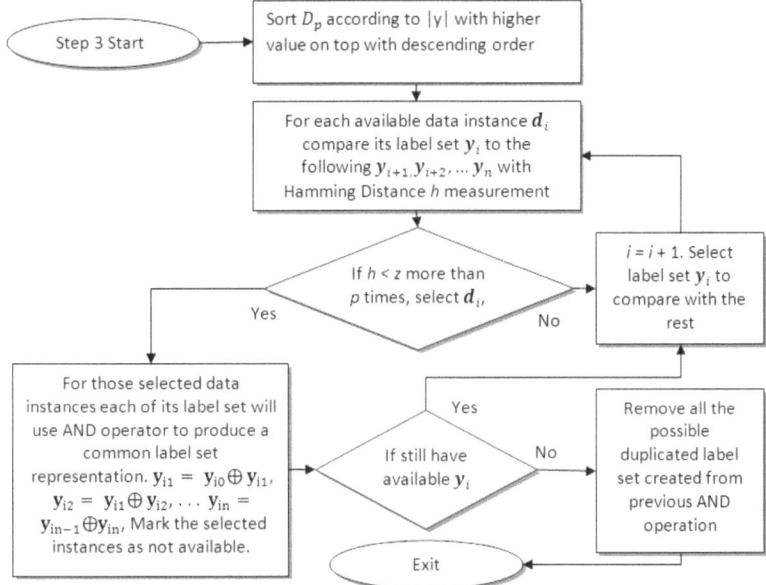

Fig. 3. This is a detailed Hamming Selection Algorithm flow chart

4.1 Ensembles of Hamming Selection Pruned Sets (EHSPS)

HSPS method can detect the potential new label sets that PS method cannot detect. Moreover, HSPS can be introduced as a standalone learning method for the training datasets. However, as a standalone method, HSPS still has the limitation of not being able to create new label sets which have not been seen in the training data. This can create an issue when working with datasets where label set is not frequent and complex. A method presented by Read [12] called *Ensemble of Pruned Sets (EPS)* can create new label sets combinations using probability distributions of the single-label classifier. As a result, combined learning of several classifiers forms an ensemble of HSPS. Ensemble method is similar to bagging approach but with one exception, its training set is without replacement. In our work, we use similar ensemble scheme such as *Ensembles of Hamming Selection Pruned Sets (EHSPS)* to compare with EPS method directly to reduce the complexity of experimental parameter settings.

5   Experimental Evaluation

In our experiment settings, we empirically set Hamming Distance parameter to $z = 2$ for all EHSPS experiments. WEKA[3] is a collection of machine learning algorithms and tools for data mining tasks. SMO is a WEKA implementation of John C. Platt's Sequential Minimal Optimization algorithm for training a support vector classifier using RBF kernels [22]. The single-label classifier SMO is used to train the transformed dataset. We extended our HSPS experiment development to the baseline

---

[3] http://www.cs.waikato.ac.nz/ml/weka/

system called MEKA[4] by Read [12]. Default setting from MEKA ensemble setting is used for the each experiment with EPS and EHSPS. Default cross validation setting values in MEKA tool are used for our experiments.

5.1 Evaluation Measures and Methodology

Evaluation measure of multi-label classification is different to that used in single-label classification. A multi-label classification will create a label subset $y_i \subseteq \mathcal{L}$ as a classification for an instance d_i. The label set ground truth is denoted as $s_i \subseteq \mathcal{L}$ for evaluation. Accuracy is defined as:

$$Accuracy(D) = \frac{1}{|D|} \sum_{i=1}^{|D|} \frac{|s_i \cap y_i|}{|s_i \cup y_i|} \tag{1}$$

In our evaluation we also considered the $F_{1\ macro}$ evaluation. $F_{1\ macro}$ is used as a standard evaluation metric by information retrieval tasks. Let's denote D as classified multi-label dataset. If p_i is the precision and r_i is the recall of the predicted labels y_i from the ground truth s_i for each data instance d_i, then $F_{1\ macro}$ is expressed as:

$$F_{1\ macro} = \frac{1}{|D|} \sum_{i=1}^{|D|} \frac{2\,p_i\,r_i}{p_i + r_i} \tag{2}$$

5.2 Datasets

MediaMill [8] dataset is created from the 2005 NIST TRECVID challenge[5] dataset which included user annotated video data. The annotated dataset is organized into 101 video semantic concepts such as Aircraft, Face, Explosion, Prisoner, and etc. The feature vector x_i are visual features extracted from [23].

Another dataset is Enron (is a sub set of the Enron e-mail corpus),[6] which has labels with a hierarchical relationship of categories created by the UC Berkeley Enron Email Analysis Project.[7] We had included email corpus experiment's results to demonstrate general usage of our algorithm for multi-label classification tasks.

6   Results

In our experiment, our parameter settings on EPS and EHSPS are divided into seven different settings, A (p=2, n=1), B (p=2, n=3), C (p=3, n=2), D (p=3, n=3), E (p=3, n=4), F (p=4, n=3), G (p=4, n=4). Table 1 shows experimental results on MediaMill and Table 2 is our experiment results on Enron email dataset. Symbol + indicates positive improvement and symbol – indicates decrease of performance over baseline.

In our experiment with the same p and n parameter settings for EPS and EHSPS, we discover that there is a significant increase up to 22.9% efficiency performance gains over the baseline in term of dataset building time with MediaMill dataset. We can observe similar improvement with Enron email dataset with 31.2% of efficiency gains. The efficiency performance gains are acquired by the common label

---

[4] http://meka.sourceforge.net/

[5] http://www.science.uva.nl/research/mediamill/challenge/

[6] http://www.cs.cmu.edu/~enron/

[7] http://bailando.sims.berkeley.edu/enron_email.html

representation step which combines much of the label sets together and thus minimized the total data instances count. Hence, the reduced data instances during the subsampling step will contribute to the overall dataset building time reduction. Fig. 4(a) is the efficiency gains on MediaMill dataset. Our proposed method demonstrates general application of our method to different type of datasets with consistent building time improvements over these datasets.

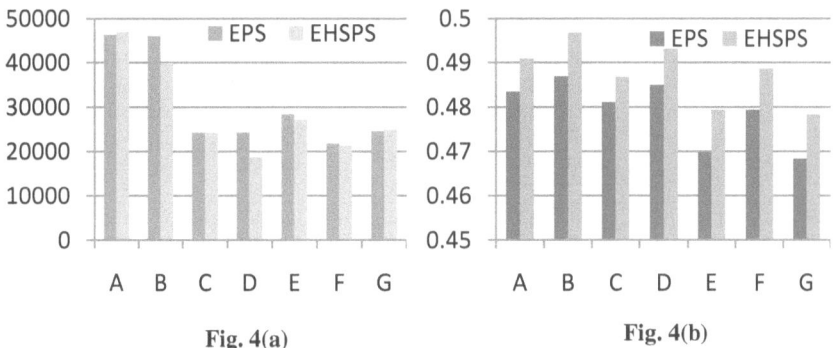

Fig. 4(a)   Fig. 4(b)

Fig. 4. Bar charts above are experiment analysis for both EPS and EHSPS algorithm. Fig. 4(a) is the comparison with MediaMill dataset building time and Fig. 4(b) is the comparison of $F_{1\,macro}$ with Enron dataset.

$F_{1\,macro}$ measurement indicates all the parameter settings for Enron dataset have a consistent positive gains over the baseline. On the other hand, the best experiment's accuracy results is $F_{1\,macro}$ = 0.52 for MediaMill dataset and $F_{1\,macro}$ = 0.4968 for Enron dataset compared over the baseline. However, there are some minor precision value drops in MediaMill dataset measurement. The precision value drop is an expected outcome for our results. The reason is that for recall measurement can recognizes a more general label sets representation rather than specific label recognition featured by precision measurement. Hence the common label set representation produced by EHSPS algorithm will predict a better result in a general metric such as recall measurement. Moreover, $F_{1\,macro}$ is a combination for both precision and recall measurement, thus the minor precision value drop do not decrease the overall $F_{1\,macro}$ results. In Fig. 4 bar chart the maximum $F_{1\,macro}$ increment is up to 1.256% for MediaMill and 2.129% for Enron which are as expected. This is because the undiscovered label set relationship is a small subset over the whole datasets. Therefore, the small accuracy gain is an expected outcome for the small subset relationship discovery.

For our experiment design, we have limited options to compare our EHSPS algorithm with other multi-label classification baseline methods (E.g. BR, PW and etc). The reason behind is that we cannot compare with those approaches with different parameter settings. Moreover, PS is having best performance against others multi-label approaches as described by Read [12]. Therefore, we only use PS to compare with EHSPS because the parameter settings are the same with one additional of z parameter that is specific to HSPS. Same parameters settings are used to prove our method provides performance gains over our baseline on a consistent basis. Furthermore, we had set $z = 2$ to all experiments to ensure integrity in our results.

We demonstrated that our novel HSPS algorithm has a better efficiency and accuracy performance gain over the baseline implementation. Our method can consistently outperform baseline system in various parameter settings.

Table 1. The EHSPS experiment using Mediamill dataset comparing with state-of-the-art Ensemble Pruned Sets algorithm

Algorithm	Parameters Settings	Building Time	Accuracy	Precision	Recall	F1 macro
EPS	p=2,n=1	46345.909	0.3773	0.6357	0.4077	0.4967
EPS	p=2,n=3	46094.245	0.3815	0.649	0.4124	0.5043
EPS	p=3,n=2	24267.738	0.3818	0.6404	0.4165	0.0594
EPS	p=3,n=3	24253.719	0.3841	0.6544	0.4163	0.5089
EPS	p=3,n=4	28372.544	0.3806	0.6648	0.4113	0.5082
EPS	p=4,n=3	21813.294	0.3921	0.6519	0.4323	0.5198
EPS	p=4,n=4	24578.62	0.3831	0.6686	0.418	0.5144
EHSPS	p=2,n=1	46941.013- 1.284%	0.3767- 0.1716%	0.6317- 0.6304%	0.4081+ 0.1182%	0.4959- 0.1756%
EHSPS	p=2,n=3	39917.204+ 13.4009%	0.3811- 0.1097%	0.6432- 0.8959%	0.4141+ 0.4213%	0.5038- 0.0946%
EHSPS	p=3,n=2	24178.752+ 0.3667%	0.3813- 0.1196%	0.6359- 0.6921%	0.4177+ 0.274%	0.0602+ **1.256%**
EHSPS	p=3,n=3	18700.063+ **22.8982%**	0.3842+ 0.0188%	0.6479- 0.988%	0.4194+ 0.7249%	0.5092+ 0.0519%
EHSPS	p=3,n=4	27186.6+ 4.1799%	0.3804- 0.0593%	0.6597- 0.7674%	0.4131+ 0.4224%	0.508- 0.0358%
EHSPS	p=4,n=3	21317.416+ 2.2733%	0.3921- 0.0032%	0.6464- 0.8384%	0.4349+ 0.6161%	**0.52+** 0.031%
EHSPS	p=4,n=4	24953.479- 1.5251%	0.3827- 0.1055%	0.6627- 0.8701%	0.4197+ 0.4004%	0.5139- 0.0922%

Table 2. EHSPS algorithm with Enron email dataset and comparing with EPS algorithm

Algorithm	Parameters Settings	Building Time	Accuracy	Precision	Recall	F1 macro
EPS	p=2,n=1	273.8820	0.3476	0.4850	0.4819	0.4835
EPS	p=2,n=3	213.6550	0.3483	0.4911	0.4829	0.4869
EPS	p=3,n=2	89.8590	0.3472	0.4904	0.4721	0.4811
EPS	p=3,n=3	139.9930	0.3478	0.4895	0.4804	0.4850
EPS	p=3,n=4	112.5450	0.3365	0.4866	0.4541	0.4698
EPS	p=4,n=3	67.1900	0.3435	0.4920	0.4673	0.4793
EPS	p=4,n=4	70.6090	0.3367	0.4922	0.4467	0.4684
EHSPS	p=2,n=1	188.488+ **31.1791%**	0.3555+ 2.2715%	0.4886+ 0.7469%	0.4932+ 2.3327%	0.4909+ 1.5361%
EHSPS	p=2,n=3	164.113+ 23.1878%	0.3567+ 2.4014%	0.4951+ 0.8333%	0.4985+ 3.2389%	**0.4968+** 2.032%
EHSPS	p=3,n=2	92.299- 2.7154%	0.3521+ 1.4099%	0.4937+ 0.6852%	0.48+ 1.6563%	0.4867+ 1.1776%
EHSPS	p=3,n=3	98.631+ 29.5458%	0.3551+ 2.0961%	0.4922+ 0.5451%	0.4941+ 2.8484%	0.4932+ 1.6945%
EHSPS	p=3,n=4	86.323+ 23.2991%	0.3437+ 2.1339%	0.4888+ 0.4477%	0.4702+ 3.5522%	0.4793+ 2.0301%
EHSPS	p=4,n=3	**47.355+** 29.5208%	0.3514+ 2.2871%	0.4935+ 0.3042%	0.4839+ 3.5565%	0.4886+ 1.9464%
EHSPS	p=4,n=4	49.49+ 29.9098%	0.3437+ 2.1024%	0.4938+ 0.3179%	0.4638+ 3.8293%	0.4783+ **2.1285%**

7   Conclusion and Future Work

In this paper, we had reviewed many current multi-label VSCC tasks in our literature review studies. Based on these literature findings, we had addressed these baseline approaches limitations. Therefore, we presented our novel HSPS algorithm which is has a better efficiency and accuracy performance gain over the baseline implementation. Our method is able to run on two real-world datasets in different domains such as MediaMill from multimedia and Enron from email and produced positive results. Our results show that HSPS algorithm will perform efficiently and accurately in many different parameter setting scenarios.

There are many possible ways to further improve our algorithm. We are considering to implement a streaming experiment settings environment for our HSPS algorithm because of our efficiency improvement over the baseline system. This is to show that HSPS algorithm can be implemented to streaming video dataset instead of keyframes based batch approach for VSCC.

References

1. Snoek, C.G.M., Worring, M.: Concept-Based Video Retrieval. Found. Trends Inf. Retr. 2, 215–322 (2009)
2. Bhatt, C., Kankanhalli, M.: Multimedia data mining: state of the art and challenges. Multimedia Tools and Applications 51, 35–76 (2011)
3. Yu-Gang, J., Jun, W., Shih-Fu, C., Chong-Wah, N.: Domain adaptive semantic diffusion for large scale context-based video annotation. In: 2009 IEEE 12th International Conference on Computer Vision, pp. 1420–1427 (2009)
4. Kennedy, L.S., Chang, S.-F.: A reranking approach for context-based concept fusion in video indexing and retrieval. In: Proceedings of the 6th ACM International Conference on Image and Video Retrieval, pp. 333–340. ACM, Amsterdam (2007)
5. Wei, X.-Y., Jiang, Y.-G., Ngo, C.-W.: Exploring inter-concept relationship with context space for semantic video indexing. In: Proceedings of the ACM International Conference on Image and Video Retrieval, pp. 1–8. ACM, Santorini (2009)
6. Yanagawa, A., Chang, S.F., Kennedy, L., Hsu, W.: Columbia university's baseline detectors for 374 lscom semantic visual concepts. Columbia University ADVENT technical report 222-2006 (2007)
7. Yu-Gang, J., Jun, Y., Chong-Wah, N., Hauptmann, A.G.: Representations of Keypoint-Based Semantic Concept Detection: A Comprehensive Study. IEEE Transactions on Multimedia 12, 42–53 (2010)
8. Snoek, C.G.M., Worring, M., van Gemert, J.C., Geusebroek, J.-M., Smeulders, A.W.M.: The challenge problem for automated detection of 101 semantic concepts in multimedia. In: Proceedings of the 14th Annual ACM International Conference on Multimedia, pp. 421–430. ACM, Santa Barbara (2006)
9. Jiang, Y.G., Yanagawa, A., Chang, S.F., Ngo, C.W.: CU-VIREO374: fusing Columbia374 and VIREO374 for large scale semantic concept detection 223, 1 (2008)
10. Tsoumakas, G., Katakis, I., Vlahavas, I.: Mining Multi-label Data. In: Maimon, O., Rokach, L. (eds.) Data Mining and Knowledge Discovery Handbook, pp. 667–685. Springer, US (2010)

11. Hamming, R.W.: Error detecting and error correcting codes. Bell System Technical Journal 29, 147–160 (1950)
12. Read, J.: A pruned problem transformation method for multi-label classification, pp. 143–150 (2008)
13. Park, S.H., Fürnkranz, J.: Multi-label classification with label constraints, pp. 157–171 (2008)
14. Elisseeff, A., Weston, J.: A kernel method for multi-labelled classification. In: Advances in Neural Information Processing Systems, vol. 14, pp. 681–687 (2001)
15. Xun, Y., Wei, L., Tao, M., Xian-Sheng, H., Xiu-Qing, W., Shipeng, L.: Automatic Video Genre Categorization using Hierarchical SVM. In: 2006 IEEE International Conference on Image Processing, pp. 2905–2908 (2006)
16. Nowak, E., Juries, F.: Vehicle categorization: parts for speed and accuracy. In: 2nd Joint IEEE International Workshop on Visual Surveillance and Performance Evaluation of Tracking and Surveillance, pp. 277–283 (2005)
17. Petrovskiy, M.: Paired Comparisons Method for Solving Multi-Label Learning Problem. In: Sixth International Conference on Hybrid Intelligent Systems, HIS 2006, pp. 42–42 (2006)
18. Ráez, A.M., López, L.A.U., Steinberger, R.: Adaptive Selection of Base Classifiers in One-Against-All Learning for Large Multi-labeled Collections. In: Vicedo, J.L., Martínez-Barco, P., Muñoz, R., Saiz Noeda, M. (eds.) EsTAL 2004. LNCS (LNAI), vol. 3230, pp. 1–12. Springer, Heidelberg (2004)
19. Read, J., Bifet, A., Holmes, G., Pfahringer, B.: Efficient multi-label classification for evolving data streams (2010)
20. Tsoumakas, G., Vlahavas, I.P.: Random k-Labelsets: An Ensemble Method for Multilabel Classification. In: Kok, J.N., Koronacki, J., Lopez de Mantaras, R., Matwin, S., Mladenič, D., Skowron, A. (eds.) ECML 2007. LNCS (LNAI), vol. 4701, pp. 406–417. Springer, Heidelberg (2007)
21. Cheng, W., Hüllermeier, E.: Combining instance-based learning and logistic regression for multilabel classification. Machine Learning 76, 211–225 (2009)
22. Platt, J.C.: 12 Fast Training of Support Vector Machines using Sequential Minimal Optimization (1998)
23. Van Gemert, J.C., Geusebroek, J., Veenman, C.J., Snoek, C.G.M., Smeulders, A.W.M.: Robust Scene Categorization by Learning Image Statistics in Context. In: Conference on Computer Vision and Pattern Recognition Workshop, CVPRW 2006, pp. 105–105 (2006)

Hierarchical Workflow Management in Wireless Sensor Network[*]

Endong Tong[1], Wenjia Niu[1], Gang Li[2],
Hui Tang[1], Ding Tang[1], and Song Ci[1,3]

[1] High Performance Network Laboratory
Institute of Acoustics, Chinese Academy of Science
Beijing, China, 100190
{Tonged,niuwj,tangh,tangd,sci}@hpnl.ac.cn
[2] School of Information Technology, Deakin University
221 Burwood Highway, Vic 3125, Australia
ligang@ieee.org
[3] University of Nebraska-Lincoln, Omaha, NE 68182 USA

Abstract. To build the service-oriented applications in a wireless sensor network (*WSN*), the workflow can be utilized to compose a set of atomic services and execute the corresponding pre-designed processes. In general, *WSN* applications rely closely on the sensor data which are usually inaccurate or even incomplete in the resource-constrained *WSN*. Then, the erroneous sensor data will affect the execution of atomic services and furthermore the workflows, which form an important part in the bottom-to-up dynamics of *WSN* applications. In order to alleviate this issue, it is necessary to manage the workflow hierarchically. However, the hierarchical workflow management remains an open and challenging problem. In this paper, by adopting the *Bloom filter* as an effective connection between the *sensor node layer* and the *upper application layer*, a hierarchical workflow management approach is proposed to ensure the *QoS* of workflow-based *WSN* application . The case study and experimental evaluations demonstrate the capability of the proposed approach.

Keywords: Wireless Sensor Network, Workflow Management, QoS, Bloom Filter.

1 Introduction

In the past decade, *WSNs* have been successfully used in various applications, ranging from *environment sensing, surveillance, smart home* to *precision agriculture* etc. Consisting of a large number of sensors, Wireless Sensor Networks

[*] This research is supported by the National Natural Science Foundation of China (No. 61103158), the National S&T Major Project (No. 2010ZX03004-002-01), the Deakin CRGS 2011 and the Securing CyberSpaces Research Cluster of Deakin University, the Sino-Finnish International S&T Cooperation and Exchange Program (NO. 2010DFB10570), the Strategic Pilot Project of Chinese Academy of Sciences (No.XDA06010302), and the Ultra-realistic Acoustic Interactive Communication on Next-Generation Internet (No. 11110036).

P. Anthony, M. Ishizuka, and D. Lukose (Eds.): PRICAI 2012, LNAI 7458, pp. 601–612, 2012.

(*WSNs*) [1] provide the capability of monitoring physical or environmental conditions as well as sending the collected data to a base station. Recently, many attempts have been devoted to utilize *service-oriented architecture (SOA)* [2] to effectively support the loose-coupled *WSN* applications[3] [4]. The current *WSN* paradigm mainly contains isolated networks that are often self-organized and self-managed to provide services, while *SOA* exhibits its potentials in facilitating collaborations between distributed autonomous services in an open dynamic environment, and realizing the interoperability between heterogenous *WSNs*.

Due to the strong data gathering capability and data-centric characteristic of *WSN*, the number of atomic services is inevitably large. A simple adoption of *SOA* into *WSN* may bring issues such as the *waste of resources* and the *less efficient service management*. In order to alleviate this problem, one pioneering work was proposed by Tong et al. [5], who designed a *Reasoning-based Context-aware Workflow* management model named *Recow*, in which a rule-based reasoning module is responsible for extracting semantic information so that the lower level sensor data will have a loose-coupled connection with the upper level logic process. This model can significantly reduce the number of atomic services and simplify the atomic service management. However, as the implementation of workflow in *WSN* applications is hierarchical, any fault in atomic services and sensor nodes will affect the execution of workflow. Hence, the effective workflow management in *WSN* should consider both the atomic service aspect and the sensor node aspect. This calls for an effective and efficient hierarchical workflow management method. As a natural extension of Tong el al.'s work [5], this paper will focus on the hierarchical management of dynamic workflows in WSN More specifically, the following two challenges will be addressed in this paper:

- As *WSNs* are constrained in energy and computation resources, the data gathered by sensors may be inaccurate because of energy exhaustion, device fault or environment obstruction. However, as *WSN* applications closely rely on sensor data, the inaccuracy in the lower level sensed data will affect the correct execution of upper level atomic services, and further lead to *QoS* deterioration in corresponding workflows. This is usually referred to as the bottom-to-up dynamical characteristic [6] of *WSN* applications.
- In the hierarchical workflow management, the implementation details on the lower level is transparent to the upper level. While most *WSN* applications require low latency: Once the *QoS* deteriorates to an unacceptable threshold, the management system should promptly locate and fix the problem.

For these challenges, we should develop a fast-responsing dynamic interaction mechanism for workflow management in *WSN* in order to ensure the application *QoS*. In this paper, we will introduce the Bloom filter into our workflow management. Bloom filter is an efficient approach to represent a set of elements in order to support the membership queries. Compared with other methods to represent a set, such as the linked list method, Bloom filter is time and space efficient, and this makes it an excellent choice to store the set of atomic services in current running workflows and the set of sensor nodes in *WSN*. Through the flexible interaction between hierarchical Bloom filters at each level of workflow management, an efficient hierarchical workflow management in *WSN* can be achieved.

The rest of the paper is organized as follows. Sec. 2 discusses the related work, followed by the proposed approach in Section 3. Section 4 presents experiment results that illustrate the benefits of the proposed scheme. Section 5 concludes the paper.

2   Preliminaries and Related Work

2.1   Preliminaries

Bloom Filter. Bloom filter (*BF*) [7], is a hashing-based data structure that succinctly represents a set of elements to support membership queries. Due to its temporal and spatial efficiency, Bloom filter can be utilized in our framework. We will exploit two types of Bloom filter, i.e. the *standard Bloom filter* for the membership query on the node level to save space cost, and the *counting Bloom filter* for the membership query and manipulation on the service level to guarantee the accuracy.

Standard Bloom Filter. It initializes a bit array of m bits with 0 and uses k independent hash functions to hash an element into k of m array positions. In this paper, we regard each sensor node's attribute (represented as A) as an element to be inserted into the bit array. A_i will be hashed by k hash functions and get array positions $h_1(A_i)$, $h_2(A_i)$ $\cdots$, $h_k(A_i)$. Then the corresponding positions in the bit array will be set to 1. Here, one position can be set to 1 multiple times, but only the first change takes effect. To check if A_i' is in the bit array, we can check if all the positions $h_1(A_i')$, $h_2(A_i')$ $\cdots$, $h_k(A_i')$ are set to *1*. If yes, then A_i' belongs to the bit array.

Counting Bloom Filter. One key problem in *BF* is that we cannot perform the deletion operation which reverses the insertion process. If we set corresponding bits to *0*, we may be setting a location to 0 that is hashed to by other elements in the set. In this case, the Bloom filter can no longer correctly reflect all elements in the set. To avoid this problem, Fan et al. [8] introduced the counting Bloom filter (*CBF*), whose entry is a counter rather than a bit. When an item is inserted, the corresponding counter increases by 1; when an item is deleted, the corresponding counter decreases by 1.

Hierarchical Workflow. In *WSN*, sensor data are usually the input of the atomic services. Based on these atomic services, workflows can be built to implement *WSN* applications according to requirements. Tong et al. [5] developed a hierarchical framework for workflow management in *WSN*, which contains three levels: the *workflow level*, the *semantic service level* and the *node level*.

- *Workflow level* is the direct level to accomplish concrete *WSN* applications. It is responsible for monitoring the workflows' state and ensuring the correct execution of running workflows.

- *Semantic service level* is the middle level which is responsible for getting sensor data from the *base station* and further as components to build workflows. The *QoS* of atomic services can be changing in real world, which makes it hard to guarantee the workflow *QoS*. Hence, some managements should be done on this level through re-composing or replacing abnormal services.
- *Node level* contains a large set of sensor nodes for sensing the surrounding environment and transmitting data to the *base station*. Due to *WSN* applications closely rely on the quality of sensor data, management on the node level also plays an important role in workflow management.

In order to implement the effective execution of workflows, the management on all these three levels is necessary, and this calls for a mechanism to realize the interaction among the three levels, which is proposed in Sec. 3.

2.2   Related Work

Workflow technique has been successfully applied to automate complex business processes. In the execution process of workflow, there exist some problems, which can be roughly generalized as the functional problem (e.g. the execution exceptions) and the non-functional problem (e.g. the *QoS* deterioration), which has attracted significant research efforts. Kammer et al. [9] argued that workflow should be dynamically adaptive to the changing contexts so that it will not affect the ongoing execution of underlying workflow. Ardissano et al. [10] proposed the abstract workflow which is composed by abstract services, which is associated with a set of concrete atomic services which are different in *QoS* attributes. When workflow is running, one concrete atomic service will be chosen in order to meet required characteristics. Patel et al. [11] proposed a *QoS*-oriented framework, called *WebQ*, to perform the refinement of existing services. It utilized a monitoring module to dynamically monitor the associated *QoS* parameters, and *QoS* deteriorating will trigger some pre-defined rules that force the de-selection of problematic services and re-selection of new set of best services to meet the *QoS* requirements. Cardoso et al. [12] utilized simulation analysis to tune the *QoS* metrics of workflows. When the need to adapt a workflow is detected, the possible effects of adaptation can be explored with simulation.

One common observation in existing work is that the performance of *QoS*-oriented dynamic workflow solutions closely relies on their special application characteristics. Hence, follow the dynamic workflow research [9][10][11], our work will focus on the adaptation of workflow which aims to resolve the *QoS* deterioration in *WSN*. In addition, existing work mainly focus on service-level adaption to realize the dynamic workflow management. However, in *WSN*, the workflow *QoS* deterioration may also come from sensor node level, i.e. the physic level. Several work has argued that the node level should be considered in the workflow management. Nan Hua et al. [13] presented an active *QoS* infrastructure of *WSN*, named *QISM*, which is based on service-oriented middleware. Through the feedback and negotiation between application and sensor node network, the *QoS* can be guaranteed in a better way. Anastasi et al. [14] also presented a service-oriented, flexible and adaptable middleware (*SensorMW*). Through the co-operation among different components, *SensorMW* allows applications to configure the *WSN* according to the *QoS* requirements. Hence, a bottom-to-up workflow management becomes natural and necessary in this context.

Our work further extend the bottom-to-up hierarchical workflow management [14]. Existing work only implement the workflow management at the atomic service level [5], but it fails to resolve the problem fundamentally as the problem may coming from the *node level*. Hence, we should manage the workflow hierarchically. However, how to make effective interaction among different levels for dynamic workflow management in *WSN* remains an open and challenging issue, which is also the research objective of this paper. More specifically, we will firstly exploit the Bloom filter technique to build an efficient interaction mechanism for *QoS* guarantee on different levels. Then based on this interaction mechanism, the Bloom filter-based hierarchical workflow management approach in *WSN* is developed.

3 Bloom Filter-Based Workflow Management Approach

3.1 Hierarchical Workflow Model

In the work [5], we adopted a reasoning-based approach to construct semantic services. Each network entity (see Fig. 1 (a)) is defined and implemented in the three levels described in Sec. 2. On the node level, the *base station* is assumed to be with enough energy and computation capability, so it can be used as the *sink node* to connect with external network. In addition, common nodes are grouped into clusters and regions. Region covers several clusters while cluster covers some sensor nodes that are only one-hop away from the *cluster head*. Normally, the *cluster head*s can be determined by *LEACH-C* [15], while the *cluster head* with the highest residual energy can be selected as the *region head*.

In this paper, we utilize Bloom filter to build a hierarchical workflow management model, as shown in Fig. 1 (b). Our model mainly consists of the *Node-level BF*, the *Service-level CBF*, the *QoS Monitoring*, the *GAI* and the *Interaction Module*. The right part out of the dashed box in Fig. 1 (b) represents the basic workflow implementation process. When the *workflow engine* constructs a workflow, all the atomic services involved in the workflow will be inserted into the *service-level CBF*. As one atomic service may be involved in more than one workflow, so repeated insertion is supported by increasing the corresponding counter upon each insertion. On the *node level* of *WSN*, sensor nodes with the same sensing task in the same sensing region will be activated in turn in order to minimize the energy consumption. Hence, when a sensor node is activated, it will take turns be inserted into the *Bloom filter arrays* stored by its cluster head and region head. We use $\mathcal{R}_{BF}^{i}$ to represent the *i*-th *region head*'s *Bloom filter array*, and $\mathcal{C}_{BF}^{i,j}$ to represent the *i*-th *region head*'s *j*-th *cluster head*'s *Bloom filter array*. Then, the *node-level BF* is hierarchical as shown in Fig. 2.

When the deterioration of workflow *QoS* is monitored, the *interaction module* will be invoked. Through the interaction among the *interaction module*, the *QoS*

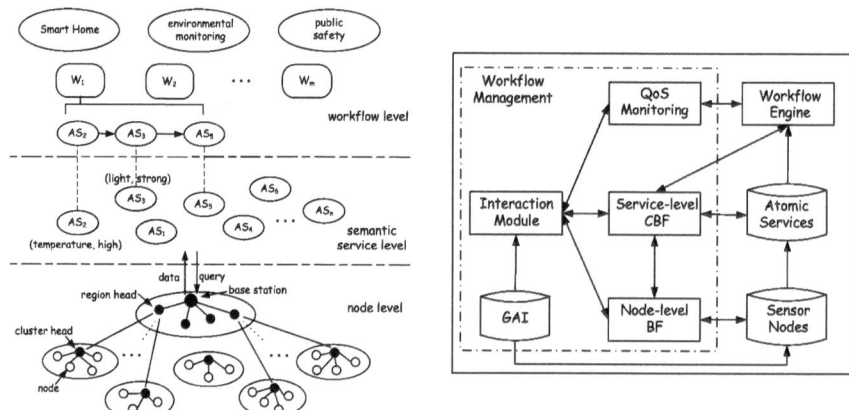

Fig. 1. (a)Network Entities in Three Levels; (b)Bloom filter-based Hierarchical Workflow Model

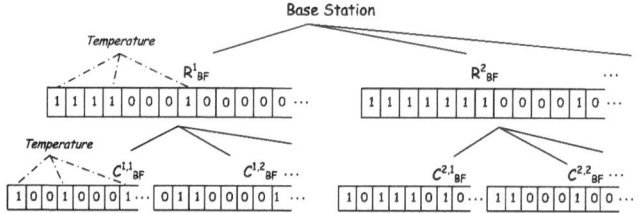

Fig. 2. Hierarchical Node-level BF

monitoring, the *service-level CBF* and the *node-level BF*, an efficient workflow management can be achieved.

3.2   The Interaction Mechanism

In this section, we will describe how the *interaction module* will cooperate with other modules to realize efficient workflow management.

Input/Output. As mentioned above, the *service-level CBF* maintains the set of current atomic services. Due to the implementation of *CBF* is based on the *hash function*, we assign each atomic service with a unique *ID AS.id*, which can be hashed and inserted into the *service-level CBF*.

Moreover, the *node-level BF* aims to manage the set of activated sensor nodes in *WSN*. While most sensor nodes could be integrated with more than one sensing units and can sense several different types of data. For a sensor node which senses *temperature*, *humidity* and *lighting* data, we can represent it as $Dev_i = \{attr_1, attr_2, attr_3\}$, in which $Dev_i.attr_1 = $ "*Temperature*", $Dev_i.attr_2 = $ "*Humidity*", $Dev_i.attr_3 = $ "*Lighting*". The individual attribute will be represented in type *string* and can be inserted into the *node-level BF*.

The *QoS monitoring* module is responsible for receiving the trigger command from the *interaction module*, as well as activate the monitoring process to find all workflows with low *QoS*. The output of the *QoS monitoring* module is a workflow list, denoted as $W_{list} = \{W_a, W_b, \cdots W_n\}$, in which $W_a = \{AS_a, AS_b, AS_c, \cdots\}$, $W_b = \{AS_b, AS_c, AS_f, \cdots\}$, $\cdots$. As we can see, the same element can be contained in more than one workflow. For example, the element AS_b and AS_c belong to both W_a and W_b.

Operation. There are four types of operations in the workflow management model as follows.

- $\mathcal{I}nsert(CBF, AS_i.id)$: insert $AS_i.id$ into the *service-level CBF*.
- $\mathcal{D}elete(CBF, AS_i.id)$: delete $AS_i.id$ from the *service-level CBF*.
- $\mathcal{Q}uery(CBF, AS_i.id)$: check whether $AS_i.id$ is the member of *service-level CBF*.
- $\mathcal{T}rigger(QoS)$: a command to trigger the *QoS monitoring* module.

In the *node-level BF*, we define two types of operations, i.e. $\mathcal{I}nsert(BF, Dev_i.attr_j)$ and $\mathcal{Q}uery(BF, Dev_i.attr_j)$, where $Dev_i.attr_j$ means the j-th attribute of Dev_i. In addition, the *return value* for $\mathcal{I}nsert$, $\mathcal{D}elete$ and $\mathcal{Q}uery$ are from the set of $\{0, 1\}$, where 0 means that $\mathcal{I}nsert$ and $\mathcal{D}elete$ are successful and $\mathcal{Q}uery$ is true. The *return value* for $\mathcal{T}rigger$ will be a workflow list as mentioned above.

GAI. In order to realize the interaction between the *service-level CBF* and the *node-level BF*, we need a unified scheme. As the output from the *service-level CBF* is a set of atomic services, while the input of *node-level BF* is the attribute of sensor node. Hence, we will add an attribute label for each atomic service, denoted as $AS_i.attr_j$, which means the j-th attribute of the atomic service AS_i. This attribute is related with node's attribute. Hence, we can do $\mathcal{Q}uery(BF, AS_i.attr_j)$ on the *node-level BF*. Obviously, we must utilize a *GAI* (Global Attribute Information) to define the unified representation between *Dev.attr* and *AS.attr*.

Process. The *interaction module* is responsible for the interaction process among all the modules to guarantee workflow *QoS*, as shown in Fig. 3.

Here, *QoS* attributes mainly come from the *QoS monitoring module* periodically [12] or users' feedback. When a workflow *QoS* deterioration occurs, there must be some services in this workflow have low *QoS*. While, these low *QoS* services may be also included in other workflows. Hence, *interaction module* sends $\mathcal{T}rigger(QoS)$ to the *QoS monitoring* module to collect all the workflows with low *QoS*, represented as W_{list}. Due to the *service-level CBF* maintains only the set of activated services involved in running workflows, *interaction module* will do $\mathcal{D}elete(CBF, AS_a.id)$, $\mathcal{D}elete(CBF, AS_b.id)$, $\cdots$ to the *service-level CBF* to delete all the atomic services in W_{list} from the *service-level CBF*. Next, based on W_{list}, the *interaction module* can get the union of all the required workflows as: $PPS_{list} = W_a \cap W_b \cap \cdots W_n = \{AS_b, AS_c, \cdots\}$.

After this process, there may be only one element in PPS_{list}. In this case, we will go through to *node-level BF* process directly; if $PPS_{list} = \emptyset$, all the atomic services consisted in W_{list} will be inserted into PPS_{list}, and the *interaction module* will do $Query(CBF,\ PPS_1.id)$, $Query(CBF,\ PPS_2.id)$, $\cdots$ to the *service-level CBF* to find if it is the member of the *service-level CBF*. If not, then abnormal services found, represent as PS_{list}.

Then the attributes of abnormal services will be retrieved and as the input elements to do $Query(BF,\ PS_i.attr_j)$ in *node-level BF*. Due to the hierarchical characteristic of *node-level BF*, the membership query will be executed in a stepwise manner. If the *returned value* from a certain region head is 0, we believe that the data is gathered from this region. Furthermore, we do membership query to each cluster head consisted in this region. Similarly, if the *returned value* from a certain cluster is 0, we believe that the data is gathered from this cluster. Due to the cluster head is one-hop to its member sensor nodes, we finally locate the specific malfunctional sensor nodes. Finally, this problem will be reported to *WSN manager* to trigger the corresponding repairing processes to provide *QoS*-ensured workflows again.

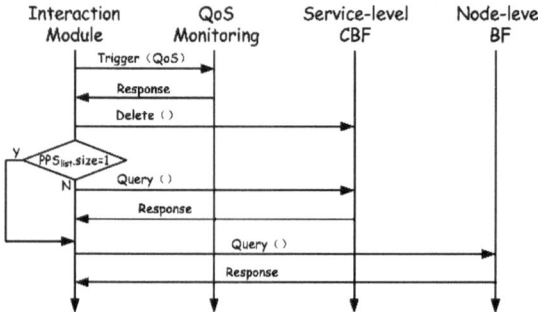

Fig. 3. Interaction Process

4   Experimental Evaluation

In this section, we will illustrate the process of our approach through a case study. Furthermore, an experiment will be carried out to show the time and space efficiency of the proposed approach.

4.1   Case Study

Suppose the *node-level BF* is given, as shown in Fig. 2. We first assign an array $\mathcal{CBF}[m]$, with an appropriate size $m = n \cdot lg(1/E) \cdot lge$, in which n means the total number of atomic services that will be inserted into $\mathcal{CBF}[m]$, and e is the natural number. In order to support *WSN* applications, four activated workflows: W_1 $(AS_2 \rightarrow AS_3 \rightarrow AS_5)$, W_2 $(AS_1 \rightarrow AS_3)$, W_3 $(AS_2 \rightarrow AS_4 \rightarrow AS_5)$ and W_4 $(AS_1 \rightarrow AS_2 \rightarrow AS_4)$ are built. Initially, the atomic services involved in these four workflows will be inserted into $\mathcal{CBF}[m]$, as shown in Fig. 4 (a).

After a period of time, deterioration in the workflow QoS may be from uses' feedback. Then, a QoS monitoring process should be invoked to get the abnormal workflow list: $W_{list} = \{W_3, W_4\}$. As the *service-level CBF* maintain the set of current activated atomic services, the atomic services in W_3 and W_4 (i.e. AS_1, AS_2, AS_4 *and* AS_5) will be deleted from $\mathcal{C}BF[m]$ as shown in Fig. 4 (b). Next, we will union W_3 and W_4 and get possible abnormal services $PPS_{list} = W_3 \cap W_4 = \{AS_2, AS_4\}$. As the size of PPS_{list} is 2, we continue to do membership query on the *service-level CBF*. As shown in Fig. 4 (b), we can find that AS_3 hashed positions are $\mathcal{C}BF[0] = 2, \mathcal{C}BF[3] = 2$ and $\mathcal{C}BF[8] = 3$ while AS_4 hashed positions are $\mathcal{C}BF[5] = 1, \mathcal{C}BF[10] = 0$ and $\mathcal{C}BF[12] = 0$. Obviously, AS_4 is not the member of *service-level CBF*. Hence, we believe that AS_4 will have low QoS. Furthermore, we get the $AS_4.attr = $ "*temperature*" and do membership query to *node-level BF* in a stepwise manner. As we can see in Fig. 2, "*temperature*" is the member of *region head 1* and *cluster head 1* respectively. Therefore, we finally get the results that there exists some problem in *cluster 1*.

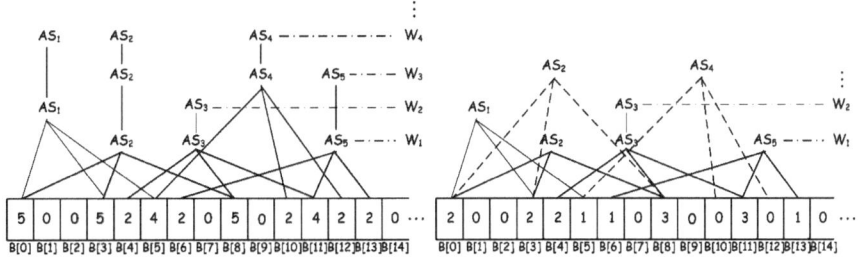

Fig. 4. (a)Initial CBF-based Storage; (b)CBF-based Storage After Deletion

4.2   Performance Evaluation

For the interaction between the *workflow level* and the *service level*, one alternative method is through the *linked list*. Hence, we carry out experiments to investigate the performance of our approach compared with *linked list*-implemented approach, and the following notations are used:

- k: the number of *hash functions* in *Bloom filter*,
- n_w: the number of current activated workflows in *WSN*,
- n_i: the number of atomic services in i-th workflows,
- n_{pps}: the number of possible abnormal service list PPS_{list},
- n_s: the total number of services,
- m: the array size of *CBF*,
- E: the *false positive probability*.

Cost Analysis. We assume that the PPS_{list} is given. In the *linked list*-implemented approach, we could use these n_{pps} services to do membership query, and the query operation numbers in the worst case is $O\{n_{pps} \cdot \sum_{i=1}^{n_w} n_i\}$. While in our approach, we only need do k hashes for each service in PPS_{list}. So, the number of operations of our approach is $O\{n_{pps} \cdot k\}$, which is more efficient than the *linked list*-implemented approach.

For the space complexity, we suppose that there are $1,000$ activated atomic services. In the *linked list*-implemented approach, if we represent the atomic service *ID* as an *int* integer, we should assign *16* bits for each *ID*. Assuming that each atomic service is reused on average at *6* times, then we can calculate the total space usage is $1,000 \cdot 6 \cdot 16 bits = 96,000 bits$. In our approach, we expect the false positive E is *0.01*, so we should assign a Bloom filter array with the number is $1,000 \cdot lg(1/E) \cdot lge \doteq 9,600$. In *CBF*, we should also maintain a counter at each entry. Analysis from [8] reveals that *4* bits per counter should be adequate for most applications. Considering of the reuse rate of services is *6*, so we assign each array position *5* bits. Therefore, the total space usage is $9,600 \cdot 5 = 48,000 bits$. We can see that our approach can save up to 50% in space cost. The detail comparison is shown in Fig. 5 (a). As we can see, the space cost in the linked list-implemented approach is larger than in our approach. To be noted that in the linked list-implemented approach, when the number of services is smaller than *30,000*, each service *ID* can be assigned only with *16* bits; while when the number of services is larger than *35,000*, each service *ID* should be assigned with *32* bits.

Empirical Results. Before the experiments, we should determine the parameters. Bloom filter is a probabilistic data structure for membership query. When querying an element, the element should be hashed and get k array positions. Sometimes, all the k array positions may have by chance been set to *1* during the insertion of other elements. So, false positives are possible.

Assume that a hash function will select each array position with equal probability, then after all n_s elements are inserted into the Bloom filter, the probability that a specific bit is *0* is $p = (1 - 1/m)^{kn_s}$. The probability of a *false positive* for an element not in the set is $E = (1 - (1 - 1/m)^{kn_s})^k \doteq (1 - e^{-kn_s/m})^k$. Through mathematical analysis, we can find that E is minimum when $k = (m/n_s) \ln 2$. We replace k with $(m/n_s) \ln 2$ and get $E = (1 - e^{-((m/n_s) \ln 2)n_s/m})^{(m/n_s) \ln 2}$. Then we have $m = n_s lg(1/E) \cdot lge$, where e is the natural number.

Here, we assume that there are totally $100,000$ atomic services in our workflow management system. In the *linked list*-implemented approach, we can use *long int* type to represent a service *ID*. While, in our approach, we require that E is *0.01* and set $m = nlg(1/E) \cdot lge \doteq 960,000$, $k = (m/n) \ln 2 \doteq 7$. When constructing workflows, we assume it will follows the *Zipf* [16] distribution. Because one service may exist in more than one workflow, we assume that the maximum reuse rate is *6*. We construct $10,000, 20,000, \cdots, 50,000$ workflows respectively, compute the PPS_{list} by assuming a service is in problem and then analyze the time consumed to find the abnormal services. The experiments repeat five times with different PPS_{list} and the results are as follows. As we can see in Fig. 5 (b), our approach improves the time consumption by two orders of magnitude than the linked list-implemented approach. Furthermore, the time cost in the *linked list*-implemented approach increases with the workflow scale while the proposed approach is almost remaining unvaried.

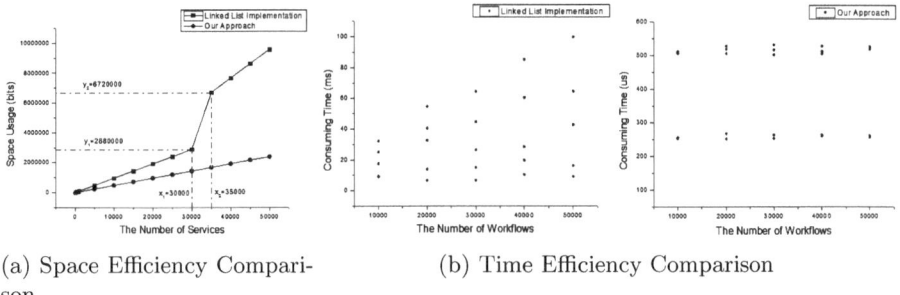

(a) Space Efficiency Compari-   (b) Time Efficiency Comparison
son

Fig. 5. Time and Space Efficiency Comparison

5   Conclusions

In this paper, we propose a Bloom filter-based workflow management approach, in which both the *CBF* and the *BF* are utilized on the *service level* and the *node level* respectively. Through a series of *Insert* or *Delete* operations, *CBF* and *BF* maintains the set of correctly running atomic services and activated sensor nodes. Then, we develop a workflow management model consisted of the *Service-level CBF*, the *Node-level BF*, the *QoS Monitoring*, the *GAI* and the *Interaction Module*. Through the interaction among above modules, more effective and efficient workflow management in *WSN* can be achieved. Compared with the simple linked list-implemented approach, our approach is much more efficient in space and time. Furthermore, our future work will continue to focus on the node-level management, namely after finding the abnormal sensor nodes, identifying which action should be done to guarantee the workflow *QoS* again.

References

1. Yick, J., Mukherjee, B., Ghosal, D.: Wireless sensor network survey. J. Computer Networks 52(12), 2292–2330 (2008)
2. Erl, T.: Service-oriented architecture: concepts, technology, and design. Prentice Hall PTR (2005)
3. Leguay, J., Lopez-Ramos, M., Jean-Marie, K., Conan, V.: An efficient service oriented architecture for heterogeneous and dynamic wireless sensor networks. In: 33rd IEEE Conference on Local Computer Networks, pp. 740–747. IEEE Press, Quebec (2008)
4. Caete, E., Chen, J., Diaz, M., Llopis, L., Rubio, B.: A service-oriented middleware for wireless sensor and actor networks. In: 6th IEEE Conference on Information Technology: New Generations, pp. 575–580. IEEE Press, Las Vegas (2009)
5. Tong, E., Niu, W., Tang, H., Li, G., Zhao, Z.: Reasoning-based Context-aware Workflow Management in Wireless Sensor Network. In: 9th IEEE Conference on Service Oriented Computing, pp. 270–282. IEEE Press, Paphos (2011)

6. Chen, D., Varshney, P.K.: QoS support in wireless sensor networks: A survey. In: International Conference on Wireless Networks, pp. 227–233. CSREA Press, Las Vegas (2004)

7. Bloom, B.H.: Space/time trade-offs in hash coding with allowable errors. J. Communications of the ACM 13(7), 422–426 (1970)

8. Fan, L., Cao, P., Almeida, J., Broder, A.Z.: Summary Cache: A Scalable Wide-Area Web Cache Sharing Protocol. J. IEEE/ACM Transactions on Networking 8(3), 281–293 (2000)

9. Kammer, P.J., Bolcer, G.A., Taylor, R.N., Hitomi, A.S., Bergman, M.: Techniques for Supporting Dynamic and Adaptive Workflow. J. Computer Supported Cooperative Work 9(3), 269–292 (2000)

10. Ardissono, L., Furnari, R., Goy, A., Petrone, G., Segnan, M.: A framework for the management of context-aware workflow systems. In: 3rd International Conference on Web Information Systems and Technologies, pp. 1–9. IEEE Press, Florian Lautenbacher (2007)

11. Patel, C., Supekar, K., Lee, Y.: A QoS Oriented Framework for Adaptive Management of Web Service Based Workflows. In: Mařík, V., Štěpánková, O., Retschitzegger, W. (eds.) DEXA 2003. LNCS, vol. 2736, pp. 826–835. Springer, Heidelberg (2003)

12. Cardoso, J., Miller, J., Sheth, A., Arnold, J., Kochut, K.: Quality of service for workflows and web service processes. J. Web Semantics: Science, Services and Agents on the World Wide Web 1(3), 281–308 (2004)

13. Hua, N., Yu, N., Guo, Y.: Research on Service oriented and Middleware based active QoS Infrastructure of Wireless Sensor Networks. In: 10th International Symposium on Pervasive Systems, Algorithms, and Networks, pp. 208–213. IEEE Press, Kaohsiung (2009)

14. Anastasi, G.F., Bini, E., Romano, A., Lipari, G.: A Service-Oriented Architecture for QoS Configuration and Management of Wireless Sensor Networks. In: 15th International Conference on Emerging Technologies and Factory Automation, pp. 1–8. IEEE Press, Bilbao (2010)

15. Heinzelman, W.B., Chandrakasan, A.P., Balakrishnan, H.: An application-specific protocol architecture for wireless microsensor networks. J. IEEE Transactions on Wireless Communications 1(4), 660–670 (2002)

16. Zipf, G.K.: Human behavior and the principle of least effort. J. Cambridge, Mass., 386–409 (1949)

Multimodal Biometric Person Authentication Using Fingerprint, Face Features

Tran Binh Long[1], Le Hoang Thai[2], and Tran Hanh[1]

[1] Department of Computer Science, University of Lac Hong 10 Huynh Van Nghe,
DongNai 71000, Viet Nam
tblong@lhu.edu.vn
[2] Department of Computer Science, Ho Chi Minh City University of Science
227 Nguyen Van Cu, HoChiMinh 70000, Viet Nam
lhthai@fit.hcmus.edu.vn

Abstract. In this paper, the authors present a multimodal biometric system using face and fingerprint features with the incorporation of Zernike Moment (ZM) and Radial Basis Function (RBF) Neural Network for personal authentication. It has been proven that face authentication is fast but not reliable while fingerprint authentication is reliable but inefficient in database retrieval. With regard to this fact, our proposed system has been developed in such a way that it can overcome the limitations of those uni-modal biometric systems and can tolerate local variations in the face or fingerprint image of an individual. The experimental results demonstrate that our proposed method can assure a higher level of forge resistance in comparison to that of the systems with single biometric traits.

Keywords: Biometrics, Personal Authentication, Fingerprint, Face, Zernike Moment, Radial Basis Function.

1   Introduction

Biometrics refers to automatic identification of a person based on his physiological or behavioral characteristics [1],[2]. Thus, it is inherently more reliable and more capable in differentiating between an authorized person and a fraudulent imposter [3]. Biometric-based personal authentication systems have gained intensive research interest for the fact that compared to the traditional systems using passwords, pin numbers, key cards and smart cards [4] they are considered more secure and convenient since they can't be borrowed, stolen or even forgotten. Currently, there are different biometric techniques that are either widely-used or under development, including face, facial thermo-grams, fingerprint, hand geometry, hand vein, iris, retinal pattern, signature, and voice-print (figure.1) [3], [5]. Each of these biometric techniques has its own advantages and disadvantages and hence is admissible, depending on the application domain. However, a proper biometric system to be used in a particular application should possess the following distinguishing traits: uniqueness, stability, collectability, performance, acceptability and forge resistance [6].

P. Anthony, M. Ishizuka, and D. Lukose (Eds.): PRICAI 2012, LNAI 7458, pp. 613–624, 2012.
© Springer-Verlag Berlin Heidelberg 2012

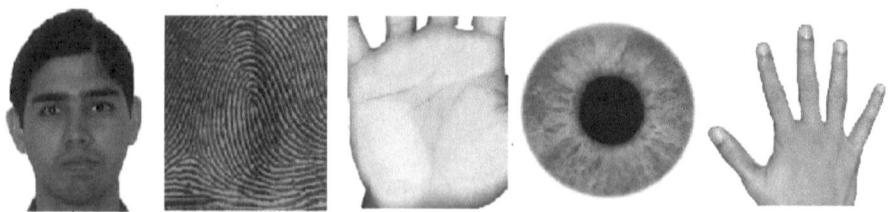

Fig. 1. Examples of biometric characteristic

Most biometric systems that are currently in use employ a single biometric trait; such systems are called uni-biometric systems. Despite their considerable advances in recent years, there are still challenges that negatively influence their resulting performance, such as noisy data, restricted degree of freedom, intra-class variability, non-universality, spoof attack and unacceptable error rates. Some of these restrictions can be lifted by multi-biometric systems [7] which utilize more than one physiological or behavioral characteristic for enrollment and verification/ identification, such as different sensors, multiple samples of the same biometrics, different feature representations, or multi-modalities. These systems can remove some of the drawbacks of the uni-biometric systems by grouping the multiple sources of information [8]. In this paper, multi-modalities are focused.

Multimodal biometrics systems are gaining acceptance among designers and practitioners due to (i) their performance superiority over uni-modal systems, and (ii) the admissible and satisfactory improvement of their system speed. Hence, it is hypothesized that our employment of multiple modalities (face and fingerprint) can conquer the limitations of the single modality- based techniques. Under some hypotheses, the combination scheme has proven to be superior in terms of accuracy; nevertheless, practically some precautions need to be taken as Ross and Jain [7] put that Multimodal Biometrics has various levels of fusion, namely sensor level, feature level, matching score level and decision level, among which [8] fusion at the feature level is usually difficult. One of the reasons for it is that different biometrics, especially in the multi-modality case, would have different feature representations and different similarity measures. In this paper, we proposed a method using face and fingerprint traits with feature level fusion. Our work aims at investigating how to combine the features extracted from different modalities, and constructing templates from the combined features. To achieve these aims, Zernike Moment (ZM)[9] was used to extract both face and fingerprint features. First, the basis functions of Zernike moment (ZM) were defined on a unit circle. Namely, the moment was computed in a circular domain. This moment is widely-used because its magnitudes are invariant to image rotation, scaling and noise, thus making the feature level fusion of face and fingerprints possible. Then, the authentication was carried out by Radial Basis Function (RBF) network, based on the fused features.

The remainder of the paper is organized as follows: section 2 describes the methodology; section 3 reports and discusses the experimental results, and section 4 presents the conclusion.

2 Methodology

Our face and fingerprint authentication system is composed of two phases which are enrollment and verification. Both phases consist of preprocessing for face and fingerprint images, extracting the feature vectors invariant with ZMI, fusing at feature level, and classifying with RBF. (Figure 2)

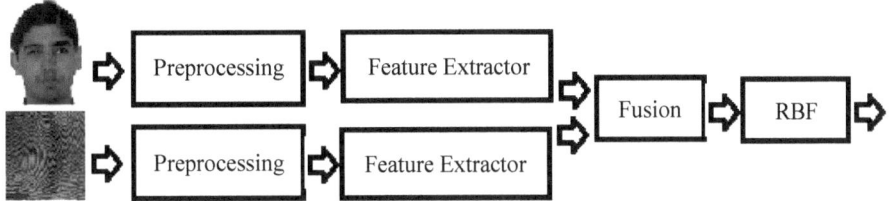

Fig. 2. The chart for face and fingerprint authentication system

2.1 Preprocessing

The purpose of the pre-processing module is to reduce or to eliminate some of the image variations for illumination. In this stage, the image had been preprocessed before the feature extraction. Our multimodal authentication system used histogram equalization, wavelet transform [10] to preprocess the image normalization, noise elimination, illumination normalization etc., and different features were extracted from the derived image normalization (feature domain) in parallel structure with the use of Zernike Moment (ZM).

Wavelet transform [10] is a representation of a signal in terms of a set of basic functions, obtained by dilation and translation of a basis wavelet. Since wavelets are short-time oscillatory functions with finite support length (limited duration both in time and frequency), they are localized in both time (spatial) and frequency domains. The joint spatial-frequency resolution obtained by wavelet transform makes it a good candidate for the extraction of details as well as approximations of images. In the two-band multi-resolution wavelet transform, signals can be expressed by wavelet and scaling basis functions at different scale, in a hierarchical manner. (Figure.3)

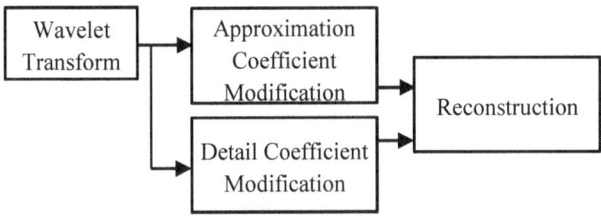

Fig. 3. Block diagram of normalization

$$f(x) = \sum_k a_{0,k} \emptyset_{0,k}(x) + \sum_j \sum_k d_{j,k} \psi_{j,k}(x) \tag{1}$$

$\emptyset_{j,k}$ are scaling functions at scale j and $\psi_{j,k}$ are wavelet functions at scale j. $a_{j,k}$, $d_{j,k}$ are scaling coefficients and wavelet coefficients.

After the application of wavelet transform, the derived image was decomposed into several frequency components in multi-resolution. Using different wavelet filter sets and/or different number of transform-levels will bring about different decomposition results. Since selecting wavelets is not the focus of this paper, we randomly chose 1-level db10 wavelets in our experiments. However, any wavelet-filters can be used in our proposed method.

2.2   Feature Extraction with Zernike Moment

The purpose of feature extraction is to extract the feature vectors or information which represents the image. To do it, Zernike Moment (ZM) was used. Zernike moment (ZM) used for face and fingerprint recognition in our work is based on the global information. This approach, also known as statistical method [11], or moment- and model- based approach [12][13], extracts the relevant information in an image. In order to design a good face and fingerprint authentication system, the choice of feature extractor is very crucial. The chosen feature vectors should contain the most pertinent information about the face and the fingerprint to be recognized. In our system, different feature domains were extracted from the derived images in parallel structure. In this way, more characteristics of face and fingerprint images for authentication were obtained. Among them, two different feature domains- ZM for Face and ZM for fingerprint - were selected.

Given a 2D image function f(x, y), it can be transformed from Cartesian coordinate to polar coordinate f(r, θ), where r and θ denote radius and azimuth respectively. The following formulae transform from Cartesian coordinate to polar coordinate,

$$r = \sqrt{x^2 + y^2}, \tag{2}$$

and

$$\theta = \arctan\left(\frac{y}{x}\right) \tag{3}$$

Image is defined on the unit circle that $r \leq 1$, and can be expanded with respect to the basic functions $V_{nm}(r, \theta)$.

For an image $f(x, y)$, it is first transformed into the polar coordinates and denoted by $f(r, \theta)$. The Zernike moment with order n and repetition m is defined as

$$M_{nm} = \frac{n+1}{\pi} \int_0^{2\pi} \int_0^1 [V_{nm}(r, \theta)]^* f(r, \theta) r dr d\theta \tag{4}$$

Where * denotes complex conjugate, n = 0, 1, 2. . . ∞, m is an integer subject to the constraint that n - |m| is nonnegative and even. $V_{nm}(r, \theta)$ is the Zernike polynomial, and it is defined over the unit disk as follows

$$V_{nm}(r, \theta) = R_{nm}(r)e^{im\theta} \tag{5}$$

With the radial polynomial $R_{nm}(r)$ defined as

$$R_{nm}(r) = \sum_{s=0}^{\frac{(n-|m|)}{2}} \frac{(-1)^s(n-s)! r^{n-2s}}{s!\left(\frac{n+|m|}{2}-s\right)!\left(\frac{n-|m|}{2}-s\right)!} \quad (6)$$

The kernels of ZMs are orthogonal so that any image can be represented in terms of the complex ZMs. Given all ZMs of an image, it can be reconstructed as follows.

$$f(r,\theta) = \sum_n \sum_{(All\ m's)} M_{nm} V_{nm}(r,\theta) \quad (7)$$

The defined features of Zernike moments themselves are only invariant to rotation. To achieve scale and translation invariance, the image needs to be normalized first by using the regular Zernike moments.

The translation invariance is achieved by translating the original image $f(x,y)$ to $f(x+\bar{x}, y+\bar{y})$, where $\bar{x} = m_{10}/m_{00}$ and $\bar{y} = m_{01}/m_{00}$.

In other words, the original image's center is moved to the centroid before the Zernike moment's calculation. Scale invariance is achieved by enlarging or reducing each shape so that the image's 0th regular moment m'_{00} equals to a predetermined value β. For a binary image, m_{00} equals to the total number of shape pixels in the image, for a scaled image $f(\alpha x, \alpha y)$, its regular moments $m'_{pq} = \alpha^{p+q+2} m_{pq}$, m_{pq} is the regular moments of $f(x,y)$.

Since the objective is to make $m'_{00} = \beta$, we can let $= \sqrt{\beta/m_{00}}$. By substituting $= \sqrt{\beta/m_{00}}$ into m'_{00}, we can obtain $m'_{00} = \alpha^2 m_{00} = \beta$.

The fundamental feature of the Zernike moments is their rotational invariance. If $f(x,y)$ is rotated by an angle α, then we can obtain that the Zernike moment Z_{nm} of the rotated image is given by

$$Z'_{nm} = Z_{nm} e(-jm\alpha) \quad (8)$$

Thus, the magnitudes of the Zernike moments can be used as rotationally invariant image features.

2.3 ZM-Based Features

It is known from the experiments that ZM performs better than other moments (e.g. Tchebichef moment [14], Krawtchouk moment [15]) do. In practice, when the orders of ZM exceed a certain value, the quality of the reconstructed image degrades quickly

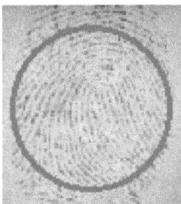

Fig. 4. Example of ZM for feature extraction with face and fingerprint

Table 1. The first 10 order Zernike moments

Order	Dimensionality	Zernike moments
0	1	M_{00}
1	2	M_{11}
2	4	M_{20}, M_{22}
3	6	M_{31}, M_{33}
4	9	M_{40}, M_{42}, M_{44}
5	12	M_{51}, M_{53}, M_{55}
6	16	M_{60}, M_{62}, M_{64}, M_{66}
7	20	M_{71}, M_{73}, M_{75}, M_{77}
8	25	M_{80}, M_{82}, M_{84}, M_{86}, M_{88}
9	30	M_{91}, M_{93}, M_{95}, M_{97}, M_{99}
10	36	$M_{10\,0}$, $M_{10\,2}$, $M_{10\,4}$, $M_{10\,6}$, $M_{10\,8}$, $M_{10\,10}$

because of the numerical instability problem inherent with ZM. From the noted problem, we decided to choose the first 10 orders of ZM with 36 feature vector elements. In this way, ZM can perform better. (Table 1)

Fingerprint Feature Extraction
In the paper, the fingerprint image was first enhanced by means of histogram equalization, wavelet transform, and then features were extracted by Zernike Moments invariant (ZMI) that was used as feature descriptor so that each feature vector extracted from each image normalization can represent the fingerprint. And to obtain a feature vector, that is, $F^{(1)} = (z_1, \ldots, z_k)$, where z_k is feature vector elements $1 \leq k \leq 36$, let the feature for the i-th user be $F_i^{(1)} = (z_1, ..., z_k)$. (Figure.4)

Face Feature Extraction
To generate feature vector of size n, first the given face image was normalized by histogram equalization, wavelet transform, and then computed by the Zernike moment. Let the result be the vector $F^{(2)} = (v_1, \ldots v_n)$. Similar to the extraction of fingerprint features, where v_n is feature vector elements $1 \leq n \leq 36$, let the feature for the i-th user is $F_i^{(2)} = (v_1, ..., v_n)$.(Figure.4)

Feature Combination
After the generation of the features from both fingerprint and the face image of the same person (say, the i-th user), it is possible to combine the two vectors $F_i^{(1)}$ and $F_i^{(2)}$ into one, with the total number of n+k component. That is, the feature vector for the i-th user is $F_i = (u_1, \ldots, u_{n+k})$, where feature vector elements $1 \leq n+k \leq 72$ are combined.

2.4   Classification

In this paper, an RBF neural network was used as a classifier in a face and fingerprint recognition system in which the inputs to the neural network are the feature vectors derived from the proposed feature extraction technique described in the previous section.

RBF Neural Network Description
RBF neural network (RBFNN)[16][17] is a universal approximator that is of the best approximation property and has very fast learning speed thanks to locally- tuned neurons (Park and Wsandberg, 1991; Girosi and Poggio, 1990; Huang, 1999a; Huang, 1999b). Hence, RBFNNs have been widely used for function approximation and pattern recognition.

A RBFNN can be considered as a mapping: $\Re^r \rightarrow \Re^s$. Let $P \in \Re^r$ be the input vector, and $C_i \in \Re^r$ ($1 \ll i \ll u$) be the prototype of the input vectors, then the output of each RBF unit can be written as:

$$R_i(P) = R_i(\|P - C_i\|) \quad i = 1, \dots, u \tag{9}$$

where $\| . \|$ indicates the Euclidean norm on the input space. Usually, the Gaussian function is preferred among all possible radial basis function due to the fact that it is factorable. Hence,

$$R_i(P) = \exp\left(-\frac{\|P - C_i\|^2}{\sigma_i^2}\right) \tag{10}$$

where σ_i is the width of the ith RBF unit. The jth output $y_i(P)$ of a RBFNN is

$$y_i(P) = \sum_{i=1}^{u} R_i(P) \times w(j, i) \tag{11}$$

where w(j,i) is the weight of the jth receptive field to the jth output.

In our experiments, the weight w(j,i), the hidden center Ci and the shape parameter of Gaussian kernel function σ_i were all adjusted in accordance with a hybrid learning algorithm combining the gradient paradigm with the linear least square (LLS)[18] paradigm.

System Architecture of the Proposed RBFNN
In order to design a classifier based on RBF neural network, we set a fixed number of input nodes in the input layer of the network. This number is equal to that of the combined feature vector elements. Also, the number of nodes in the output layer was set to be equal to that of the image classes, equivalent to 8 combined fingerprint and facial images. The RBF units were selected equal to the set number of the input nodes in the input layer.

For a neural network: feature vector elements of ZM, equal to 72, correspond to 72 input nodes of input layer. Our chosen number of RBF units of hidden layer is 72, and the number of nodes in the output layer is 8.

3   Experimental Results

3.1   Database of the Experiment

Our experiment was conducted on the public domain fingerprint images dataset DB4 FVC2004 [19], ORL face database [20].

Fig. 5. Captured sample fingerprint images from FVC 2004 database

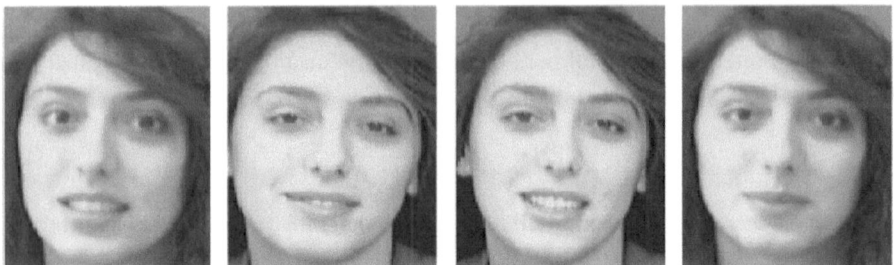

Fig. 6. Sample face images from ORL face database

In DB4 FVC2004 database, the size of each fingerprint image is 288x384 pixels, and its resolution is 500 dpi. FVC2004 DB4 has 800 fingerprints of 100 fingers (8 images of each finger). Some sample fingerprint images used in the experimentation were depicted by Figure.5.

ORL face database is comprised of 400 images of 40 persons with variations in facial expressions (e.g. open/close eyes, smiling/non-smiling), and facial details (e.g. with wearing glasses/without wearing glasses). All the images were taken on a dark background with a 92 x 112 pixels resolution. Figure.6 shows an individual's sample images from the ORL database.

With the assumption that certain face images in ORL and fingerprint images in FVC belong to an individual, in our experiment, we used 320 face images (8 images from each of 40 individuals) in ORL face database, and 320 fingerprint images (8 images from each of 40 individuals) in FVC fingerprint database. Combining those images in pairs, we had our own database of 320 double images from 40 different individual, 8 images from each one that we named ORL-FVC database.

3.2   Evaluation

In this section, the capabilities of the proposed ZM-RBFN approach in multimodal authentication are demonstrated. A sample of the proposed system with two different feature domains and of the RBF neural network was developed. In this example, concerning the ZM, all moments from the first 10 orders were considered as feature vectors, and the number of combined feature vector elements for these domains is 72. The proposed method was evaluated in terms of its recognition performance with the use of ORL-FVC database. Five images of each of 40 individuals in the database were randomly selected as training samples while the remaining samples without overlapping were used as test data. Consequently, we have 200 training images and 120 testing images for RBF neural network for each trial. Since the number of the ORL-FVC database is limited, we had performed the trial over 3 times to get the average authentication rate. Our achieved authentication rate is 96.55% (Table 2).

Table 2. Recognition rate of our proposed method

Test	Rate
1	97.25%
2	95.98%
3	96.43%
Mean	96.55%

In our paper, the effectiveness of the proposed method was compared with that of the mono-modal traits, typically human face recognition systems [21], and fingerprint recognition systems [22], of which the ZM has first 10 orders with 36 feature elements. From the comparative results of MMT shown in Table 3, it can be seen that the recognition rate of our multimodal system is much better than that of any other individual recognition, and that although the output of individual recognition may agree or conflict with each other, our system still searches for a maximum degree of agreement between the conflicting supports of the face pattern.

Table 3. The FAR,FRR and Accuracy values obtained from the monomodal traits

Trait	FRR(%)	FAR(%)	Accuracy
Face[21]	13.47	11.52	73.20
Fingerprint[22]	7.151	7.108	92.892

Also in our work, we conducted separated experiments on the technique of face, fingerprint, fusion at matching score and feature level. The comparison between the achieved accuracy of our proposed technique with that of each mentioned technique has indicated its striking usefulness and utility. (See in figure.7).

For the recognition performance evaluation, a False Acceptance Rate (FAR) and a False Rejection Rate (FRR) test were performed. These two measurements yield another performance measure, namely Total Success Rate (TSR):

$$\text{TSR} = \left(1 - \frac{\text{FAR+FRR}}{\text{total number of accesses}}\right) \times 100\% \qquad (12)$$

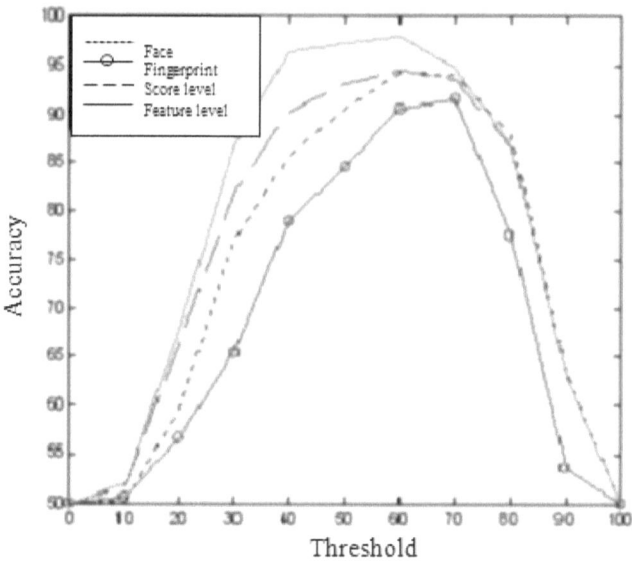

Fig. 7. The Accuracy curve of face, fingerprint, fusion at score and feature level

The system performance was evaluated by Equal Error Rate (EER) where FAR=FRR. A threshold value was obtained, based on Equal Error Rate criteria where FAR=FRR. Threshold value of 0.2954 was gained for ZM-RBF as a measure of dissimilarity.

Table 4 shows the testing results of verification rate with the first 10 order moments for ZM, based on their defined threshold value.

The results demonstrate that the application of ZM as feature extractors can best perform the recognition.

Table 4. Testing result of authentication rate of Multimodal

Method	Thres	FAR(%)	FRR(%)	TSR(%)
Proposed method	0.2954	4.95	1.12	96.55

4   Conclusion

This paper has outlined the possibility to augment the verification accuracy by integrating multiple biometric traits. In the paper, the authors have presented a novel approach in which both fingerprint and face images are processed with Zernike Moment- Radial Basis Functions to obtain comparable features. The reported experimental results have demonstrated a remarkable improvement in the accuracy achieved from the proper fusion of feature sets. It is also noted that fusing information from independent/ uncorrelated sources (face and fingerprint) at the feature level fusion enables better authentication than doing it at score level. This preliminary achievement does not constitute an end in itself, but suggests an attempt of a multimodal data fusion as early as possible in parallel processing. However, the real feasibility of this approach, in a real application scenario, may heavily depend on the physical nature of the acquired signal; thus, it is assumed that further experiments on "standard" multimodal databases will allow better validation of the overall system performances.

References

1. Campbell Jr., J., Alyea, L., Dunn, J.: Biometric security: Government application and operations (1996), `http://www.vitro.bloomington.in.us:8080/~BC/`
2. Davies, S.G.: Touching big brother: How biometric technology will fuse flesh and machine. Information Technology @ People 7(4), 60–69 (1994)
3. Newham, E.: The Biometric Report. SJB Services, New York (1995)
4. Jain, K., Hong, L., Pankanti, S.: Biometrics: Promising Frontiers for Emerging Identification Market. Comm. ACM, 91–98 (February 2000)
5. Clarke, R.: Human identification in information systems: Management challenges and public policy issues. Information Technology @ People 7(4), 6–37 (1994)
6. Ross, A., Nandakumar, D., Jain, A.K.: Handbook of Multibiometrics. Springer, Heidelberg (2006)
7. Ross, A., Jain, A.K.: Information Fusion in Biometrics. Pattern Recognition Letters 24(13), 2115–2125 (2003)
8. Jain, A.K., Ross, A.: Multibiometric systems. Communications of the ACM 47(1), 34–40 (2004)
9. Zernike, F.: Physica (1934)
10. Du, S., Ward, R.: Wavelet based illumination normalization for face recognition. Department of Electrical and Computer Engineering. The University of British Coloumbia, Vancouver, BC, Canada; IEEE Transactions on Pattern Analysis and Machine Intelligence, 0-7803-9134-9 (2005)
11. Belhumeur, P.N., Hespanha, J.P., Kriegman, D.J.: Eigenfaces va. Fisherfaces: Recognition using class specific linear projection. IEEE Transactions on Pattern Analysis and Machine Intelligence 19(7), 711–720 (1997)
12. Cootes, T., Taylor, C., Cooper, D., Graham, J.: Active shape models-their training and applications. Computer Vision and Image Understanding 61(1), 38–59 (1995)
13. Cootes, T., Edwards, G., Taylor, C.: Active appearance models. IEEE Transactions on Pattern Analysis and Machine Intelligence 23(6), 681–685 (2001)
14. Mukundan, R., Ong, S.H., Lee, P.A.: Image analysis by Tchebichef moments. IEEE Transactions on Image Processing 10(9), 1357–1364 (2001)
15. Yap, P.T., Paramesran, R., Ong, S.H.: Image analysis by Krawtchouk moments. IEEE Transactions on Image Processing 12(11), 1367–1377 (2003)
16. Haddadnia, J., Faez, K.: Human face recognition using radial basis function neural network. In: Proc. 3rd International Conference on Human and Computer, HC 2000, Aizu, Japan, pp. 137–142 (September 2000)
17. Haddadnia, J., Ahmadi, M., Faez, K.: A hybrid learning RBF neural network for human face recognition with pseudo Zernike moment invariant. In: IEEE International Joint Conference on Neural Network, IJCNN 2002, Honolulu, Hawaii, USA, pp. 11–16 (May 2002)
18. Jang, J.-S.R.: ANFIS: Adaptive-Network-Based Fuzzy Inference System. IEEE Trans. Syst. Man. Cybern. 23(3), 665–684 (1993)

19. FVC. Finger print verification contest (2004),
    http://bias.csr.unibo.it/fvc2004.html
20. ORL. The ORL face database at the AT&T (Olivetti) Research Laboratory (1992),
    http://www.uk.research.att.com/facedatabase.html
21. Lajevardi, S.M., Hussain, Z.M.: Zernike Moments for Facial Expression Recognition. In:
    Internation Conference on Communication, Computer and Power, ICCCP 2009, Muscat,
    February 15-18, pp. 378–381 (2009)
22. Qader, H.A., Ramli, A.R., Al-Haddad, S.: Fingerprint Recognition Using Zernike Moments. The Internation Arab Journal of Information Technology 4(4) (October 2007)

Explaining Subgroups through Ontologies

Anže Vavpetič[1], Vid Podpečan[1], Stijn Meganck[2], and Nada Lavrač[1,3]

[1] Department of Knowledge Technologies, Jožef Stefan Institute, Ljubljana, Slovenia
[2] Computational Modeling Lab, Vrije Universiteit Brussel, Brussel, Belgium
[3] University of Nova Gorica, Nova Gorica, Slovenia
anze.vavpetic@ijs.si

Abstract. Subgroup discovery (SD) methods can be used to find interesting subsets of objects of a given class. Subgroup descriptions (rules) are themselves good explanations of the subgroups. Domain ontologies provide additional descriptions to data and can provide alternative explanations of discovered rules; such explanations in terms of higher level ontology concepts have the potential of providing new insights into the domain of investigation. We show that this additional explanatory power can be ensured by using recently developed semantic SD methods. We present the new approach to explaining subgroups through ontologies and demonstrate its utility on a gene expression profiling use case where groups of patients, identified through SD in terms of gene expression, are further explained through concepts from the Gene Ontology and KEGG orthology.

Keywords: data mining, subgroup discovery, ontologies, microarray data.

1 Introduction

The paper addresses the task of subgroup discovery, first defined by Klösgen [10] and Wrobel [27], which is at the intersection of classification and association discovery. The goal is to find subgroups of individuals that are statistically important according to some property of interest from a given population of individuals. For example, a subgroup should be as large as possible and exhibit the most unusual distribution of the target class compared to the rest of the population.

Subgroup discovery methods can be used to find descriptions called of interesting subsets of objects of a given class in binary as well as in multi-class problems. Subgroup descriptions, formed as rules with a class label in the rule conclusion and a conjunction of attribute values in the rule condition, typically provide sufficiently informative explanations of the discovered subgroups. However, with the expansion of the Semantic Web and the availability of numerous domain ontologies which provide domain background knowledge and semantic descriptors to the data, we are faced with a challenge of using this publicly available information also to provide explanations of rules initially discovered by standard symbolic data mining and machine learning algorithms. Approaches which would enhance symbolic rule learning with the capability of providing explanations of the rules also in terms of higher-level concepts than those used in

P. Anthony, M. Ishizuka, and D. Lukose (Eds.): PRICAI 2012, LNAI 7458, pp. 625–636, 2012.
© Springer-Verlag Berlin Heidelberg 2012

rule descriptors, have a potential of providing new insights into the domain of investigation.

In this paper we show that an additional explanatory step can be performed by using recently developed semantic subgroup discovery approaches [12, 17]. The new methodology is show-cased on a gene expression profiling use case, where groups of patients of a selected grade of breast cancer, identified through subgroup discovery in terms of gene expression, are further explained through terms from the Gene Ontology and KEGG. The motivation for the use case in breast cancer patient analysis comes from the experts' assumption that there are several subtypes of breast cancer. Hence, in addition to distinguish between patients with breast cancer (the positive cases) and healthy patients, the challenge is first to identify breast cancer subtypes by finding subgroups of patients followed by inducing explanations in terms of the same biological functions, processes and pathways of genes, characterizing different molecular subtypes of breast cancer.

While the presented use case is specific to the systems biology domain, the proposed approach is general and can be applied in any application area, provided the existence of domain ontologies.

The paper is structured as follows. Section 2 discusses the related work. The new methodology is presented in detail in Section 3 and illustrated in Section 4 on the breast cancer gene expression (microarray) data. Section 5 concludes the paper and provides plans for further work.

2   Related Work

This section discusses the work related to the steps of the presented methodology. As a complex multi-step approach the presented work relates to contrast data mining, subgroup discovery and semantic data mining. Although the latter is also related to semantic web technologies, the related work on these topics is out of the scope of this overview.

Mining of contrasts in data has been recognized as one of the the fundamental tasks in data mining [26]. The general idea is to discover and understand contrasts (differences) between objects of different classes. One of the first algorithms which has explicitly addressed the task of mining contrast sets is the STUCCO algorithm, developed by Bay and Pazzani [2]. It searches for conjunctions of attributes and values (contrast sets) which exhibit different levels of support in mutually exclusive groups. Mining for contrasting sets is also related to exception rule mining as defined by Suzuki [21, 22] where the goal is to discover rare deviating patterns which complement strong base rules to form rule pairs. Suzuki [22] defines an exception as something different from most of the rest of the data which can be also seen as a contrast to given data and/or existing domain knowledge. However, while being efficient at discovering statistically significant contrasting and exceptional patterns, STUCCO, PEDRE [21] and similar systems do not address the representation and explanation of contrasts using available related background knowledge and ontologies.

The general problem of subgroup discovery was defined by Klösgen [10] and Wrobel [27] as search for population subgroups which are statistically interesting and which exhibit unusual distributional characteristics with respect to the property of interest. Similarly to contrasting patterns, subgroup descriptions are conjunctions of attributes and values which characterize selected class of individuals. Furthermore, Kralj Novak et al. [11] have shown that contrast set mining, emerging pattern mining [4] as well as subgroup discovery can be viewed as variants of rule learning by providing appropriate definitions of compatibility.

Several algorithms were developed for mining interesting subgroups using exhaustive search or using heuristic approaches: Explora [10], APRIORI-SD [9], SD-Map [1], SD [7], CN2-SD [13]. These algorithms employ different heuristics to asses the interestingness of the discovered rules, which is usually defined in terms of rule unusualness and size.

While subgroup descriptions in the form of rules are relatively good descriptions of subgroups there is also abundance of background knowledge in the form of taxonomies and ontologies readily available to be incorporated to provide better high-level descriptions and explanations of discovered subgroups. Especially in the domain of systems biology the GO ontology[1] and KEGG orthology[2] are good examples of structured domain knowledge.

The challenge of incorporating domain ontologies in data mining was addressed in recent work on semantic data mining [12]. The SDM-SEGS system developed by Vavpetič et. al. [25] is an extension of the earlier domain-specific algorithm SEGS [24] which allows for semantic subgroup discovery in gene expression data. SEGS constructs gene sets as combinations of GO ontology terms, KEGG orthology terms, and terms describing gene-gene interactions obtained from the Entrez [14] database. SDM-SEGS extends and generalizes this approach by allowing the user to input any set of ontologies in the OWL format and an empirical data collection which is annotated by domain ontology terms.

In the presented methodology semantic subgroup discovery approaches such as SEGS, SDM-SEGS and SDM-Aleph[3] serve as explanatory subsystems which semantically describe and explain contrasting groups in input data.

3   Methodology

This section presents the steps of the proposed methodology. The first step involves finding relevant sets of examples (relevant to the user) by using subgroup discovery, thus creating a new labeling for the examples pertaining to the selected sets. The second step deals with ranking the attributes according to their ability to distinguish between the sets. The third step of the methodology induces symbolic explanations of a selected target example set (subgroup detected in the first step) by using ontological concepts.

---

[1] http://www.geneontology.org/
[2] http://www.genome.jp/kegg/
[3] http://kt.ijs.si/software/SDM/

3.1   Obtaining Sets and Creating a New Labeling

To find a potentially interesting set of examples, the user can choose from a number of data mining algorithms. Data mining platforms such as Orange [3] and Weka [8] offer various clustering, classification and visualization techniques. A potentially interesting set of examples can be a cluster of examples, examples in a node of a decision tree, a set of examples revealed by a visualization method, a set of examples covered by a subgroup description, and others; in this work we concentrate on subgroup discovery.

First, some basic notation needs to be established. Let $D = \{e_1, e_2, \ldots, e_n\}$ be a dataset with attributes $a_1, a_2, \ldots, a_m$ and a continuous or discrete target variable y (note that unsupervised methods do not require a target variable). Let v_{ij} denote the value of attribute a_j for example e_i. If a_j is continuous, let Min_j and Max_j be the minimum and maximum values of the attribute in the dataset.

In the following, subgroups and clusters are represented as sets of examples. Let S_A and S_B denote two sets of examples ($S_A \cap S_B = \emptyset$) that are of interest to the user who wants to determine which groups of attributes (expressed as ontological concepts) differentiate S_A from S_B.

Regardless of how S_A and S_B are obtained, the new re-labeled dataset D' is constructed as follows. The target variable y is replaced by a binary target variable y' and for each example e_i the new label c' is defined as:

$$c' = \begin{cases} 1, \text{ if } e_i \in S_A \\ 0, \text{ otherwise} \end{cases}$$

Note that if D is unlabeled, the new target variable y' is added to the domain. We shall now illustrate how to determine S_A and S_B using subgroup discovery (SD). SD algorithms induce symbolic subgroup descriptions of the form

$$(y = c) \leftarrow t_1 \wedge t_2 \wedge \ldots t_l$$

where t_j is a conjunct of the form $(a_i = v_{ij})$. If a_i is continuous and the selected subgroup discovery algorithm can deal with continuous attributes, t_i can also be defined as an interval such that $(a_i \geq v_{ij})$ or $(a_i \leq v_{ij})$. An example of a subgroup description on the well known UCI lenses dataset is:

$$(lenses = hard) \leftarrow (prescr. = myope) \wedge (astigm. = yes) \wedge (tear\_rate = normal)$$

A subgroup description R can be also viewed as a set of constraints (conjuncts t_i) on the dataset, and the corresponding subgroup as the set of examples $cov(R)$ which satisfy the constraints, i.e., examples covered by rule R.

Suppose the user is presented with a set of subgroup descriptions $R = \{R_1, R_2, \ldots, R_k\}$. Then the set of examples S_A can either be defined as a single subgroup $S_A = cov(R_i)$ or a union of subgroups. S_B can represent all other examples $S_B = D \setminus S_A$, a single subgroup or a union of subgroups - depending on the user's preference.

3.2   Ranking of Attributes

Once the relabeled dataset D' is available, the attributes are assigned ranks according to their ability to distinguish between the two sets of examples S_A and S_B. To calculate the ranks, any attribute quality measure can be used, but in practice attribute ranking using the ReliefF [18] algorithm has proven to yield reliable scores for this methodology to work.

In contrast to myopic measures (e.g., Gain Ratio), ReliefF takes into account the context of other attributes when evaluating an attribute. This is an important benefit when applying this methodology to datasets such as microarray data since it is known that there are dependencies among many genes.

The ReliefF algorithm works as follows. A random subset of examples of size $m \leq n$ is chosen. Each attribute starts with a ReliefF score of 0. For each randomly selected example e_i and each class c, k nearest examples are selected. The algorithm then goes through each attribute a_l and nearest neighbor e_j $(i \neq j)$, and updates the score of the attribute as follows:

- if e_i and e_j belong to the same class and at the same time have different values of a_l, then the attribute's score is decreased;
- if the examples have different attribute values and belong to different classes, then the attribute's score is increased.

This step of the methodology results in a list of ReliefF attribute scores such that $L = [(a_1, r_1), \ldots, (a_m, r_m)]$ where r_i is the ReliefF score representing the ability of attribute a_i to distinguish between sets S_A and S_B.

3.3   Inducing Explanations Using Ontologies

At this stage of the methodology a semantic subgroup discovery algorithm [12] is applied to generate explanations using the list of ranked attributes L. Each subgroup description (rule) induced by a semantic subgroup discovery algorithm represents one explanation, and each explanation is a conjunction of ontological concepts. The assumption here is that a domain ontology O (one or more) is available or, more specifically, that a mapping between the attributes and ontological concepts exists. For example, in the case of microarray data, an attribute (gene) IDH1 is mapped to (annotated by) the ontological concept *Isocitrate metabolic process* from the Gene Ontology, indicating that this gene takes part in this particular biological process. Thus, each ontological concept, as well as each explanation, describes a set of attributes.

Annotations enable the explanations to have strictly defined semantics, and from a data mining perspective, this information enables the algorithm to generalize better than by using attribute values alone. The explanations can be made even richer if additional relations that are available among the attributes (or ontological concepts) are included in the explanations. Using the microarray example, genes are known to *interact*, and this information can be directly used to form explanations.

Currently, there are three publicly available SDM systems that can be used for this purpose of inducing explanations: SEGS [24], a domain specific system for analyzing microarray data using the Gene Ontology and KEGG, SDM-SEGS [12], the general purpose version of SEGS, that enables the use of OWL ontologies, but is limited to a maximum of three ontologies, and SDM-Aleph[4], a general purpose SDM system based on the ILP system Aleph, that can use any number of OWL ontologies.

All three systems focus on inducing explanations in the form of rules whose conjuncts correspond to ontological concepts. To illustrate how explanations are induced, consider that SEGS is selected to be used on a microarray domain. Note that in the following description, genes can be thought of as instances/examples, since the algorithm is not limited only to genes (this fact is exploited in SDM-SEGS). The idea behind SEGS as well as SDM-SEGS, illustrated on the problem of finding explanations for top-ranked genes, is as follows.

Top-down bounded exhaustive search is used to construct the set of explanations/subgroup descriptions (rules) according to the user-defined constraints (e.g., minimum support), and the algorithm considers all explanations that can be formed by taking one concept from each ontology as a conjunct. The input list L of ranked genes is first split into two classes. The set of genes above a selected threshold value is the set of differentially expressed genes for which a set of rules is constructed (these rules describe sets of genes which distinguish set S_A from set S_B).

The construction procedure starts with a default rule $top\_genes(X) \leftarrow$, with an empty set of conjuncts in the rule condition, which covers all the genes. Next, the algorithm tries to conjunctively add the top concept of the first ontology (yielding e.g., $top\_genes(X) \leftarrow biological\_process(X)$) and if the new rule satisfies all of the size constraints, it adds it to the rule set and recursively tries to add the top concept of the next ontology (e.g., $top\_genes(X) \leftarrow biological\_process(X) \wedge molecular\_function(X)$). In the next step all the child concepts of the current conjunct/concept are considered by recursively calling the procedure. Due to the transitivity of the *subClassOf* relation between concepts in the ontologies, the algorithm can employ an efficient pruning strategy. If the currently evaluated rule does not satisfy the size constraints, the algorithm can prune all rules which would be generated if this rule were further specialized. Additionally, the user can specify gene interaction data by specifying the *interacts* relation. In this case, for each concept which the algorithm tries to conjunctively add to the rule, it also tries to add its interacting counterpart.

In SEGS, the constructed explanations are assigned scores using several well-known methods (e.g., GSEA [20]) and the significance of the explanations is evaluated using permutation testing.

In our setting, the resulting descriptions correspond to subgroups of attributes (e.g., genes) which enable distinguishing between the sets S_A and S_B.

---

[4] kt.ijs.si/software/SDM/

3.4   Implementation

The described methodology is implemented and integrated within Orange4WS
[16], a service-oriented environment which extends the Orange [3] data mining
platform by adding support for knowledge discovery workflow construction using distributed web services. As a result, researchers and end-users are able to
achieve repeatability of experiments, integration of distributed and heterogeneous data and knowledge sources, and simple sharing of workflows and implemented solutions. The methodology can be modified and extended by composing
the already implemented visual programming components (widgets) or by adding
new components, either by providing local implementations or by including web
services. This flexibility is especially useful for the evaluation of different algorithms performing certain steps of the methodology, such as ranking of attributes
or clustering.

4   Biomedical Use Case

This section presents and discusses the application of the presented methodology
on gene expression data. More specifically, we evaluate the methodology on the
breast cancer dataset using our implementation of the methodology as a workflow
in the Orange4WS environment.

The gene expression dataset used in our analysis is the dataset published by
Sotiriou et al. [19] (GEO series GSE2990). It is a merge of the KJX64 and KJ125
datasets and contains expression values of 12,718 genes from 189 patients with
primary operable invasive breast cancer. It also provides 22 metadata attributes
such as age, grade, tumor size and survival time. We have used the expert-curated
re-normalized binarized version of the dataset from the InSilico database [23].
Within the InSilico framework, the raw data was renormalized using fRMA[15]
and a genetic barcode (0/1) was generated based on whether the expression of
a gene was significantly higher (K standard deviations) than the no expression
level estimated on a reference of approx. 800 samples. In this setting $g_i = 1$ means
that gene g_i is over-expressed and $g_i = 0$ means that it is not. The ultimate goal
of the experiment was to induce meaningful high-level semantic descriptions of
subgroups found in the data which could provide important information in the
clinical decision making process.

Our main motivation for developing the presented methodology is to descriptively characterize various breast cancer subtypes, while, in the experiments
presented here we focus on describing breast cancer grades, which enables us to
focus only on the evaluation of the methodology.

The conducted experiment on the presented dataset in the Orange4WS environment employs processing components (widgets) from Orange and Orange4WS
in a complex data analysis workflow[5] which is shown in Figure 1.

In the first step, the *InSilico database search* widget is used to download the
breast cancer patient data, i.e., a binarized version of the gene expression data

---

[5] The software, workflow and the results are available at `http://orange4ws.ijs.si`.

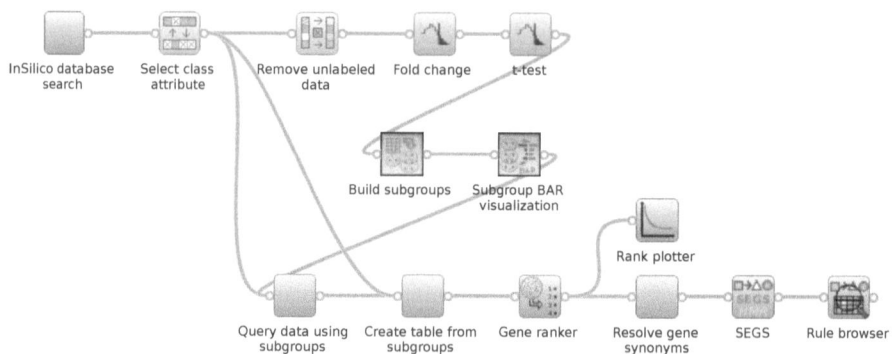

Fig. 1. A sample workflow implementing the proposed methodology

(note that the frozen robust multiarray analysis (fRMA) normalization [15] is also available). As the GSE2990 dataset does not have pre-specified classes we have selected the *Grade* attribute as the target attribute. According to Elston [5] and Galea [6], histologic grade of breast carcinomas provides clinically important prognostic information. Approximately one half of all breast cancers are assigned histologic grade 1 or 3 status (low or high risk of recurrence) but a substantial percentage of tumors (30% - 60%) are classified as histologic grade 2 (intermediate risk of recurrence) which is not informative for clinical decision making [19]. Obviously, to increase the prognostic value of tumor grading, further refinement of histologic grade 2 status is necessary [19].

The second step of the workflow is to use the *Remove unlabeled data* widget to remove 17 unclassified examples for which the histologic grade is unknown. Although these examples may contain important information, this would require using unsupervised methods (e.g. clustering) instead of supervised subgroup discovery algorithms used in our experiments (note, however, that subgroup discovery in the presented workflow can easily be replaced by clustering or some other unsupervised method).

Next, attribute (gene) selection is performed using a gene selection component which allows to filter the genes according to various scoring methods such as fold change, t-test, ANOVA, signal-to-noise-ratio and others. Removal of genes considered to be unimportant by the selected scoring is needed to reduce the search space of subgroup discovery methods. In our approach we have selected the genes in two stages: first, only the genes with a fold change of > 1 are selected, and second, only the genes with p-value $< 0,01$ given by the t-test are selected. This yields a total of 399 genes to be used in the subgroup discovery process.

The *Build subgroups* widget implements SD [7], APRIORI-SD [9] and CN2-SD [13] subgroup discovery algorithms while the *Subgroup BAR visualization* component provides a facility of bar chart visualization and selection of subgroups. The selected subgroups are used to query the original data to obtain the covered set of examples which are then merged with the rest of the data.

Table 1. The best-scoring subgroups found using CN2-SD with default parameters for the Grade 3 patients. TP and FP are the true positive and false positive rates, respectively.

#	Subgroup Description	TP	FP
1	Grade = 3 ← DDX39A = 1 ∧ DDX47 = 1 ∧ RACGAP1 = 1 ∧ ZWINT = 1 ∧ PITPNB = 1	43	5
2	Grade = 3 ← TPX2 = 1 ∧ DDX47 = 1 ∧ PITPNB = 1 ∧ HN1 = 1	26	0

As a result it is possible to rank the genes in the re-constructed dataset according to their ability to differentiate between the discovered subgroups and the rest of the data. The ranking of genes is performed by the *Gene ranker* widget implementing the ReliefF algorithm.

Finally, the computed ranking is sent to the SEGS semantic subgroup discovery algorithm (SDM-SEGS and SDM-Aleph can also be used). As the SEGS algorithm has large time and space requirements it is implemented as a web service which allows it to run on a powerful server. SEGS induces rules providing explanations of the top ranked attributes by building conjunctions of ontology terms from the GO ontology, KEGG orthology, and interacting terms using the Entrez gene-gene interactions database as described in Section 3.3. In our experiments we have used the latest updates of the ontologies and annotations provided by NCBI[6] and the Gene Ontology project.

The subgroup discovery analysis yielded two large subgroups (Table 1) of Grade 3 patients. Using the GeneCards[7] on-line tool, we have confirmed that all of the genes from the subgroup descriptions are typically differentially expressed (up-regulated) in breast cancer tissue when compared with normal tissue.

In the rest of this section we focus on the larger subgroup #1, for which we have generated explanations (Table 2). A total of 90 explanations with p-value < 0.05 (estimated using permutation testing) were found. Due to space restrictions we display only the top 10 explanations generated by SEGS (the complete list is available at `http://orange4ws.ijs.si`). For example, Explanation #1 describes genes which are annotated by GO/KEGG terms: chromosome and cell cycle.

In the study by Sotiriou et al. [19] where the the expression profiles of Grade 3 and Grade 1 patients were compared, the genes that are associated with histologic grade were shown to be mainly involved in cell cycle regulation and proliferation (uncontrollable division of cells is one of the hallmarks of cancer). The explanations of Subgroup #1 of Grade 3 patients in Table 2 agree with their findings. In general, the explanations describe genes that take part in cell cycle regulation (Explanations #1-#10), cell division (Explanation #3) and other components that indirectly affect cell division (e.g., Explanations #4 and #5: microtubules are structures that pull the cell apart when it divides).

---

[6] `http://www.ncbi.nlm.nih.gov/gene`
[7] `http://www.genecards.org`

Table 2. The explanations for the patients from subgroup #1 from Figure 2. We omit the variables from the rules for better readability. Note that since the p-values are estimations, some can also have a value of 0.

#	Explanation	p-value
1	chromosome ∧ cell cycle	0.000
2	cellular macromolecule metabolic process ∧ intracellular non-membrane-bounded organelle ∧ cell cycle	0.000
3	cell division ∧ nucleus ∧ cell cycle	0.000
4	regulation of mitotic cell cycle ∧ cytoskeletal part	0.000
5	regulation of mitotic cell cycle ∧ microtubule cytoskeleton	0.000
6	regulation of G2/M transition of mitotic cell cycle	0.000
7	regulation of cell cycle process ∧ chromosomal part	0.000
8	regulation of cell cycle process ∧ spindle	0.000
9	enzyme binding ∧ regulation of cell cycle process ∧ intracellular non-membrane-bounded organelle	0.000
10	ATP binding ∧ mitotic cell cycle ∧ nucleus	0.005

Our study shows that by using our methodology one can automatically reproduce the observations noted in the earlier work by Sotiriou et al. This can encourage the researchers to apply the presented methodology in similar exploratory analytics tasks. Given the public availability of the software the methodology can be simply reused in other domains.

5 Conclusions

In this paper we presented a methodology for explaining subgroups of examples using higher-level ontological concepts. First, a subgroup of examples is identified (e.g., using subgroup discovery), which is then characterized using ontological concepts thus providing insight into the main differences between the given subgroup and the remaining data.

The proposed approach is general, and can be employed in any application area, provided the existence of available domain ontologies and annotated data to be analyzed. Note, however, that the experiments presented in this work are limited to gene expression data, more specifically, to explaining subgroups of breast cancer patients in terms of gene expression data.

As the experts assume that there are several molecular subtypes of breast cancer, our main research interest is to employ the presented methodology to descriptively characterize the hypothesized cancer subtypes. The approach presented in this paper has the potential of discovering groups of patients which correspond to the subtypes while explaining them using ontology terms describing gene functions, processes and pathways; but in this paper, we applied the methodology only to describe breast cancer grades with the aim of evaluation.

Using subgroup discovery we have identified two main subgroups that characterize Grade 3 breast cancer patients. These were then additionally explained using Gene Ontology concepts and KEGG pathways and the explanations (rules or subgroup descriptions of gene sets) agree with previous findings characterizing grades using microarray profiling.

The results of the conducted experiments are encouraging and show the capabilities of the presented approach. In further work we will employ the methodology to detecting and characterizing subtypes of breast cancer while the results will be evaluated by medical experts. Furthermore, we wish to apply this methodology to other domains, as well as advance the level of exploitation of domain ontologies for providing explanations of the results of data mining.

References

[1] Atzmüller, M., Puppe, F.: SD-Map – A Fast Algorithm for Exhaustive Subgroup Discovery. In: Fürnkranz, J., Scheffer, T., Spiliopoulou, M. (eds.) PKDD 2006. LNCS (LNAI), vol. 4213, pp. 6–17. Springer, Heidelberg (2006)

[2] Bay, S.D., Pazzani, M.J.: Detecting group differences: Mining contrast sets. Data Mining and Knowledge Discovery 5(3), 213–246 (2001)

[3] Demšar, J., Zupan, B., Leban, G.: From experimental machine learning to interactive data mining, white paper. Faculty of Computer and Information Science. University of Ljubljana (2004), http://www.ailab.si/orange

[4] Dong, G., Li, J.: Efficient mining of emerging patterns: Discovering trends and differences. In: Proceedings of the 5th ACM SIGKDD International Conference on Knowledge Discovery and Data Mining, KDD 1999, pp. 43–52 (1999)

[5] Elston, C.W., Ellis, I.O.: Pathological prognostic factors in breast cancer. I. The value of histological grade in breast cancer: experience from a large study with long-term follow-up. Histopathology 19(5), 403–410 (1991)

[6] Galea, M., Blamey, R., Elston, C., Ellis, I.: The Nottingham prognostic index in primary breast cancer. Breast Cancer Research and Treatment 22, 207–219 (1992)

[7] Gamberger, D., Lavrač, N.: Expert-guided subgroup discovery: Methodology and application. Journal of Artificial Intelligence Research 17, 501–527 (2002)

[8] Hall, M., Frank, E., Holmes, G., Pfahringer, B., Reutemann, P., Witten, I.H.: The WEKA data mining software: an update. SIGKDD Explor. Newsl. 11, 10–18 (2009)

[9] Kavšek, B., Lavrač, N.: APRIORI-SD: Adapting association rule learning to subgroup discovery. Applied Artificial Intelligence 20(7), 543–583 (2006)

[10] Klösgen, W.: Explora: a multipattern and multistrategy discovery assistant. In: Advances in Knowledge Discovery and Data Mining, pp. 249–271. American Association for Artificial Intelligence, Menlo Park (1996)

[11] Kralj Novak, P., Lavrač, N., Webb, G.I.: Supervised descriptive rule discovery: A unifying survey of contrast set, emerging pattern and subgroup mining. Journal of Machine Learning Research 10, 377–403 (2009)

[12] Lavrač, N., Vavpetič, A., Soldatova, L., Trajkovski, I., Novak, P.K.: Using Ontologies in Semantic Data Mining with SEGS and g-SEGS. In: Elomaa, T., Hollmén, J., Mannila, H. (eds.) DS 2011. LNCS, vol. 6926, pp. 165–178. Springer, Heidelberg (2011)

[13] Lavrač, N., Kavšek, B., Flach, P.A., Todorovski, L.: Subgroup discovery with CN2-SD. Journal of Machine Learning Research 5, 153–188 (2004)

[14] Maglott, D., Ostell, J., Pruitt, K.D., Tatusova, T.: Entrez gene: gene-centered information at NCBI. Nucleic Acids Research 33(Database issue) (2005)

[15] McCall, M.N., Bolstad, B.M., Irizarry, R.A.: Frozen robust multiarray analysis (fRMA). Biostatistics 11(2), 242–253 (2010)

[16] Podpečan, V., Zemenova, M., Lavrač, N.: Orange4WS environment for service-oriented data mining. The Computer Journal Online Access (2011); advanced Access Published August 7, 2011: 10.1093/comjnl/bxr077

[17] Podpečan, V., Lavrač, N., Mozetič, I., Novak, P.K., Trajkovski, I., Langohr, L., Kulovesi, K., Toivonen, H., Petek, M., Motaln, H., Gruden, K.: SegMine workflows for semantic microarray data analysis in Orange4WS. BMC Bioinformatics 12, 416 (2011)

[18] Robnik-Šikonja, M., Kononenko, I.: Theoretical and empirical analysis of ReliefF and RReliefF. Machine Learning 53, 23–69 (2003)

[19] Sotiriou, C., Wirapati, P., Loi, S., Harris, A., Fox, S., Smeds, J., Nordgren, H., Farmer, P., Praz, V., Haibe-Kains, B., Desmedt, C., Larsimont, D., Cardoso, F., Peterse, H., Nuyten, D., Buyse, M., Van de Vijver, M.J., Bergh, J., Piccart, M., Delorenzi, M.: Gene expression profiling in breast cancer: understanding the molecular basis of histologic grade to improve prognosis 98(4), 262–272 (2006)

[20] Subramanian, A., Tamayo, P., Mootha, V.K., Mukherjee, S., Ebert, B.L., Gillette, M.A., Paulovich, A., Pomeroy, S.L., Golub, T.R., Lander, E.S., Mesirov, J.P.: Gene set enrichment analysis: A knowledge-based approach for interpreting genome-wide expression profiles. Proceedings of the National Academy of Sciences of the United States of America 102(43), 15545–15550 (2005)

[21] Suzuki, E.: Autonomous discovery of reliable exception rules. In: Proceedings of the Third International Conference on Knowledge Discovery and Data Mining, pp. 259–262 (1997)

[22] Suzuki, E.: Data mining methods for discovering interesting exceptions from an unsupervised table. Journal of Universal Computer Science 12(6), 627–653 (2006)

[23] Taminau, J., Steenhoff, D., Coletta, A., Meganck, S., Lazar, C., de Schaetzen, V., Duque, R., Molter, C., Bersini, H., Nowé, A., Weiss Solís, D.Y.: InSilicoDB: an R/Bioconductor package for accessing human Affymetrix expert-curated datasets from GEO. Bioinformatics (2011)

[24] Trajkovski, I., Lavrač, N., Tolar, J.: SEGS: Search for enriched gene sets in microarray data. Journal of Biomedical Informatics 41(4), 588–601 (2008)

[25] Vavpetič, A., Lavrač, N.: Semantic data mining system g-SEGS. In: Proceedings of the Workshop on Planning to Learn and Service-Oriented Knowledge Discovery, PlanSoKD 2011, ECML PKDD Conference, Athens, Greece, September 5-9, pp. 17–29 (2011)

[26] Webb, G.I., Butler, S.M., Newlands, D.: On detecting differences between groups. In: Proceedings of the 9th ACM SIGKDD International Conference on Knowledge Discovery and Data Mining, KDD 2003, pp. 256–265 (2003)

[27] Wrobel, S.: An Algorithm for Multi-relational Discovery of Subgroups. In: Komorowski, J., Żytkow, J.M. (eds.) PKDD 1997. LNCS, vol. 1263, pp. 78–87. Springer, Heidelberg (1997)

Automatic Keyword Extraction
from Single-Sentence Natural Language Queries

David X. Wang, Xiaoying Gao, and Peter Andreae

School of Engineering and Computer Science,
Victoria University of Wellington, P.O. Box 600, Wellington, New Zealand
{david.wang,xgao,peter.andreae}@ecs.vuw.ac.nz

Abstract. This paper presents a novel algorithm of extracting keywords from single-sentence natural language queries in English. The process involves applying a series of rules to a parsed query in order to pick out potential keywords based on part-of-speech and the surrounding phrase structure. A supervised machine learning method is also explored in order to find suitable rules, which has shown promising results when cross-validated with various training sets.

1 Introduction

Queries expressed as questions in natural language are easy for humans to construct and to understand, and it would be desirable to have natural language user interfaces for search engines and other computer based query interfaces that could handle such queries. However, traditional search engines are keyword based, and require the user to express their query as a list of keywords. Natural language queries typically contain many non-keywords within the query which will confuse the search engine and lead to poor performance. One approach to solving this mismatch is to preprocess an incoming natural language query to extract the searchable keywords in the query, before feeding them into a search engine. This would allow the development of a natural language user interface (LUI) which is backed by a traditional search engine. The potential uses of such a LUI can also include natural language question answering and voice-enabled software such as Siri.

The idea of keyword extraction is not a new concept; however, the focus is on extracting keywords from documents in order to summarise them or to provide meta-information regarding them. For example, Anette Hulth [1–3] has published a range of papers describing many ways to achieve this. Extracting keywords from documents allows the use of frequency analysis to identify words that have high importance within a document, which is a crucial step in existing keyword extraction techniques. Frequency analysis cannot be applied to queries because queries are typically very much shorter than documents, and therefore frequency based extraction techniques do not work.

This paper proposes a solution that uses natural language processing to recognise patterns in a single-sentence query (generally in question form). The

P. Anthony, M. Ishizuka, and D. Lukose (Eds.): PRICAI 2012, LNAI 7458, pp. 637–648, 2012.

basic idea is that keywords can be identified by looking at a word's part-of-speech(PoS), the type of phrase in which it resides, and its neighbouring phrases. The key techniques used are representing the knowledge for keyword extraction as a set of selection rules and a genetic algorithm method for learning this knowledge automatically.

The rest of this paper is organised as follows. Section 2 summarises the related work and Section 3 details our algorithm and the rule representation of heuristics using a working example. Sections 4 explains the supervised learning method and Section 5 describes how the training data is automatically generated and presents the results. Section 6 concludes this paper and points out future work.

2   Related Work

Three of the most closely related research projects are summarised in this section.

Hulth [2] describes experiments on the automatic extraction of keywords from documents and abstracts using a supervised machine learning algorithm. Some of the features used for training included term frequency, collection frequency and position of first occurrence. While these features are suitable for sizable blocks of text, they would not produce desirable results when used on very short documents such as single-sentence queries.

Hulth also presented various methods for extracting candidate terms which are applicable to very short documents, and we have built on these methods. One is to extract noun phrases from the text. We have expanded this to extract all phrases, not just noun phrases (See Section 3.3). Another method selects terms when they match a set of part-of-speech tag sequences, which is the basis of the selection rules approach in Sections 3.4 and 3.5. However, while Hulth defined 56 tag patterns by hand, this paper describes a supervised machine learning solution to find suitable selection rules.

AlchemyAPI [4] is a text analysis and mining tool used to transform text into knowledge. One of the tools it provides is a keyword extraction tool, which works on webpages, documents and short queries. AlchemyAPI requires a large database of indexed and tagged content (webpages). A natural language query is analysed statistically to extract the the most relevant topics related to the query. A disadvantages to this method is that it requires a large database and the current implementation requires the keyword extraction process to be performed remotely. Second, the relevance ranking of topics is constrained by the content available in the underlying database. For example, the query "What is the tallest mountain in New Zealand?" should ideally be converted to "tallest mountain New Zealand". But AlchemyAPI's keyword extractor only produces "tallest mountain". The result is that the keyword extraction is inconsistent and unpredictable.

Despite its limitations, we use this keyword extraction tool to produce training data in order to test the effectiveness of our machine learning implementation (See Section 4).

QuestionBank[5] is a corpus of 4000 parse annotated English queries and questions produced by John Judge in 2006 and updated in 2010. We have used this corpus to create large training sets, with the help of AlchemyAPI (See Section 5), in order to test the performance of our keyword extraction algorithm.

3   Our Rule-Based Approach

We first give a very high level description of the algorithm, followed by details for each step using a working example.

3.1   Overview

The algorithm takes a natural single-sentence query as input and generates a set of keywords labeled with corresponding PoS tags.

In the first step, the query is partitioned into a sequence of non-overlapping phrases/chunks either by performing a shallow parse (a.k.a. chunking) or by performing a full parse and identifying the lowest phrase-level nodes. Each word is labeled with the phrase it is part of.

In the second step, a set of rules is then applied to each word to determine whether it should be extracted as a keyword or not. Later subsections present two representations of such rules, one working at the phrase-level, and the other working at the word-level.

3.2   Parsing

The implementation of the first step uses Java libraries such as OpenNLP[6] and Stanford Parser[7]. The nodes of the parse tree are tagged using the Penn Treebank notation[8] so that any path from the root of the tree to a leaf passes through a clause-level, a phrase-level, and finally a word-level(PoS) node in that order (see Figure 1 for an example).

3.3   Chunking

Given the complete parse tree, the original query must be split into non-overlapping chunks. The easiest way to extract chunks from a parse tree is to simply split the tree into subtrees rooted at phrase-level nodes. Table 1 lists the phrase-level nodes we used — a subset of the most relevant phrase-level tags taken from the Penn Treebank guidelines[8].

The algorithm performs an in-order traversal of the parse tree. For each leaf, it finds the closest ancestor which is a valid phrase-level node. These nodes will then be put into an ordered set. For example, given the parse tree in Figure 1, the algorithm would produce the set shown below in Figure 2.

At this point, the chunks which have been extracted may overlap, which occurs whenever one of the chunks is a descendant of another. For example, in Figure 2, NP_3 is a descendant of PP. The algorithm therefore searches for pairs of chunks that overlap, and removes the lower level chunk from the higher level chunk. The

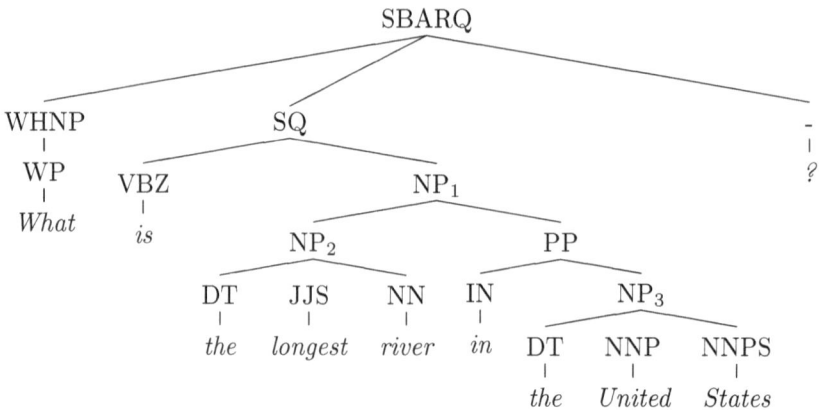

Fig. 1. A parse of the sentence "What is the longest river in the United States?"

Table 1. List of Penn Treebank phrase-level tags

ADJP	Adjective Phrase	ADVP	Adverb Phrase
CONJP	Conjuction Phrase	NP	Noun Phrase
PP	Prepositional Phrase	VP	Verb Phrase
WHADJP	Wh-adjective Phrase	WHADVP	Wh-adverb Phrase
WHNP	Wh-noun Phrase	WHPP	Wh-prepositional Phrase

output will be the set of remaining chunks that contain at least one leaf (*i.e.*, word), as in Figure 3.

Note: If an input query is not in the form of a question (*i.e.*, does not end with a question mark), then the parsing and chunking steps above can be condensed into a single step by performing only a shallow parse on the input. The Illinois Chunker[9] and Apache OpenNLP[6] library both perform this well. However, these tools do not perform well on queries in the form of a question because they often fail to identify Wh-phrases, so the WHNP phrase in the above example would have been identified as simply being a noun phrase, and the full-parse tools are needed in these cases.

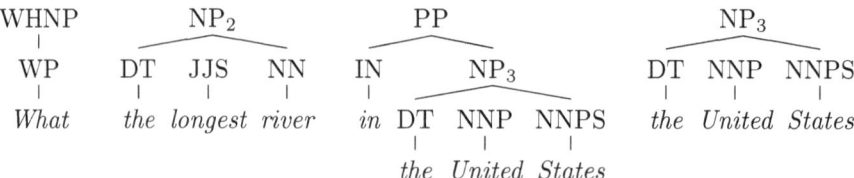

Fig. 2. Raw chunks with overlapping

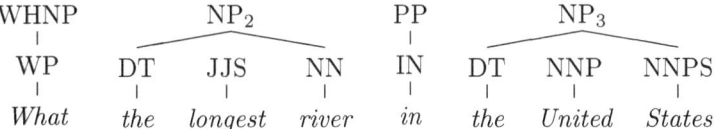

Fig. 3. Non-overlapping chunks

3.4   Rule-Based Chunk Selection

Chunk selection is the core part of the algorithm and has to select the most relevant of the phrases from the previous stage and discard all the other phrases. The algorithm assumes that the immediate context of a phrase/chunk contains important syntactic cues about the semantic value of a phrase.

To perform the selection, a list of selection rules are applied to the extracted chunks/phrases. Each rule consists of a series of up to three phrase-level tags where each tag is marked for selection or rejection.

Table 2. An example of a simple selection rule

NP	PP	NP
✓	✗	✓

Table 2 is an example selection rule, which asserts that if a prepositional phrase is between two noun phrases, then both the noun phrases should be retained, and the prepositional phrase should be discarded. If this rule is applied to the previous example (Figure 3), just two phrases will remains as shown in Figure 4.

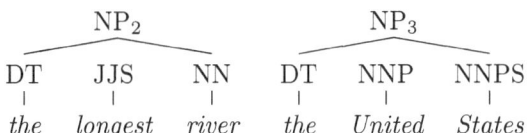

Fig. 4. Selected chunks after applying selection rule

Wildcard tags (which may be matched to any phrase) are also allowed as part of a selection rule to provide them with greater versatility. The rule shown in Table 3a will select any phrase between two verb phrases while the rule shown in Table 3b will select any phrase preceding an adverb phrase.

Given a list of these selection rules (either hand-constructed or machine-learnt), each one is applied to the extracted phrases in order. All the extracted phrases are initially flagged to be discarded. When a sequence of phrases match the pattern in a rule, the flag of each phrase is changed to be selected or discarded as specified by that rule. The rules are listed in increasing priority and

Table 3. More examples of simple selection rule(with wildcards)

(a)

VP	*	VP
✗	✓	✗

(b)

*	ADVP
✓	✗

will override any decisions made by previous rules so far if the same phrase is matched multiple times. The intention is that early rules (low priority) should take care of generic cases, while later rules (high priority) should take care of any special cases that remain.

3.5 Rule-Based Part-of-Speech Selection

Selecting specific chunks/phrases using the simple selection rules in the previous section works for simple cases, but does not take into account any information about the contents of the phrases, and is obviously inadequate by itself. The second part of the selection process uses rules that consider the PoS of each word within phrases. Table 4 lists the PoS tags that were used. These "advanced selection rules" use a different rule representation that extends the expressiveness of simple selection rules.

Table 4. List of Penn Treebank part-of-speech tags

CC	CD	DT	EX	FW	IN	JJ	JJR	JJS
LS	MD	NN	NNS	NNP	NNPS	PDT	POS	PRP
PRP$	RB	RBR	RBS	RP	SYM	TO	UH	VB
VBD	VBG	VBN	VBP	VBZ	WDT	WP	WP$	WRB

Advanced selection rules also contain up to 3 phrase-tags, just like their simple counterparts, but each tag has a subrule attached to it consisting of a set of PoS tags which are individually marked for selection or discarding. (Note that the order of the PoS tags within a subrule is not significant.) Table 5 shows an example of an advanced selection rule.

Table 5. An example of an advanced selection rule

NP			VP	ADVP		
✓			✗	✓		
NNS	NNPS	DT		RB	RBR	RBS
✓	✓	✗	-	✓	✗	✗

The sequence of phrase-tags in the rule is matched to the list of input phrases in the same manner as a simple selection rule – noun phrase, followed by a verb and adverb phrase — but when a match is found, the subrules are then applied to the words in each matched phrase, which are then marked for selection or deletion. For example, if the starting noun phrase in Table 5 is matched to a chunk, then all the proper nouns (singular and plural) in the words of the chunk are selected while all determiners are discarded. For the ending adverb phrase, standard adverbs are selected while comparative and superlative adverbs are discarded. No rules are applied to the middle verb phrase. After the entire rule and its subrules have been applied, every word in each phrase should be flagged to be either selected or discarded. As for the simple selection process, all words are initially flagged to be discarded.

Please note that an unselected phrase is not discarded; it simply has no sub-rules applied to it. After a list of advanced selection rules have been applied, the words which are flagged for selection are extracted as keywords.

Also, most PoS tags can only be found as part of one or two different types of phrases (e.g. an adjective should not be found within a verb phrase). Advanced selection rules could be made more efficient by limiting the types of PoS tags in a subrule to the ones which are allowed to occur in the phrase its attached to.

3.6   Trimming Stopwords

A different version of our algorithm is to use stopwords trimming instead of PoS selection (Section 3.5). Stopwords need to be trimmed if only chunk selection is used (Section 3.4); If PoS selection (Section 3.5) is used instead, there is no need to trim stopwords. Both versions are tested and the comparison results are shown in Section 5.

Even though only relevant phrases are left after chunking, there may still be stopwords within those phrases which provide little meaning to the overall query. Each word within a phrase is compared to a list of stopwords. All words that appear in the list are removed.

Choosing a list of stopwords must be done carefully, since having the wrong stopwords or missing the right stopwords may reduce the overall quality of the extracted keywords. We believe that is is better to have a moderately strict set of stopwords as long as it does not include many prepositions, since these tend to have greater relevance in natural language queries/questions than in normal speech.

We used a list of approximately 150 stopwords, which contained very few prepositional stopwords (e.g. after, near, etc..). Applying this step to the ongoing example (Figure 4), the determiner ("the") at the start of each phrase would be removed.

$$JJS(longest) \ NN(river) \ NNP(United) \ NNPS(States)$$

Removing stopwords in this way will work most of the time, however, there are some exceptions. For example, with the query "What papers were published by Havigli and Crisafulli in 2010", the final extracted phrases will likely be [what papers], [published], [Havigli and Crisafulli] and [2010]. Removing stopwords will leave "papers published Havigli Crisafulli", but removing the stopword "and" between the two author names is inappropriate in this particular case. To correct this, stopwords should only be removed from the edges of the chunks, in other words, trimmed from either end of each phrase instead of simply removing all of them.

Note that the step is not necessary if the keywords are to be used in an online search engine such as Google, since the search engine will automatically remove stopwords.

In addition, some search engines (such as Google) provide the option to search for exact matches for any keywords enclosed within double quotes. To utilise this feature, double quotes could be added to the final extracted phrases to produce more relevant results. In this case, the running example would produce: "longest river" "United States".

4   Learning Selection Rules

There are a huge number of possible rules even with a limit of three phrases per rule – over 1,000 patterns of phrases and almost 10,000 if each pattern can have multiple boolean mappings (even more for advanced selection rules). So it can be rather difficulty to construct a set of rules by hand. We used genetic algorithms (a type of machine learning) to enable the computer to come up with a suitable set of rules on its own.

Listing 1.1. Example of a training element

```
(SBARQ (WHNP (WP What))(SQ (VBZ is )(NP (DT a)(NN prism )))(.?))
prism
```

First, a training set has to be created. The training file format is as follows – each training element consists of a parsed query and a set of relevant keywords which should be extracted from it (See Listing 1.1). Using pre-parsed queries removes the need to perform parsing during the training process, thus improving performance. In practise, having a training set size of a few hundred is enough to produce suitable selection rules.

Traditionally in genetic algorithms, a chromosome is represented as a linear bit-string. However, there was no easy way to represent selection rules this way, so instead, each chromosome consists of an array of selection rule objects. In this manner, a chromosome is essentially the list of selection rules described in Section 3.4. We used the same method for learning the advanced rules described in Section 3.5. The learning performance for two kinds of rules are presented in Section 5.

4.1   Fitness Function

When evaluating the performance of a chromosome, it is converted into a list of selection rules and then applied to each training query in order; with a later rule overriding a previous in the event of a conflict. The selected keywords are then compared to the expected (or ideal) keywords provided by each training element. If simple selection rules are used, stopwords will have to be trimmed (See Section 3.6) before the fitness assessment.

The accuracy of the extracted keywords is assessed using a weighted F-score. Equation 3 shows how this is calculated. The F-score is weighted towards recall because it is more important for the relevant keywords to be extracted than for irrelevant words to be discarded. To give an overall fitness value to the chromosome, the F-score of each training element is averaged over the entire training set.

$$precision = \frac{|\{idealKeywords\} \cap \{extractedKeywords\}|}{|\{extractedKeywords\}|} \tag{1}$$

$$recall = \frac{|\{idealKeywords\} \cap \{extractedKeywords\}|}{|\{idealKeywords\}|} \tag{2}$$

$$F_2 = 5 \cdot \frac{precision \cdot recall}{(4 \cdot precision) + recall} \tag{3}$$

4.2 Overfitting

On smaller training sets (*i.e.*, around 100), a solution will converge quite rapidly and will likely be overfitted to the training data. To combat this, the training set was divided into smaller groups. Each group is then used in separate training runs to produce different optimal solutions. The top solutions were taken from each pool of solutions created by each group, and combined into a final set of selection rule lists.

When using a set of selection rule lists instead of just a single list, chunk selection is done slightly differently. Each selection rule list is applied to the training set individually, as before, and the subset of phrases selected is then recorded. At the end, instead of each phrase being marked as either selected or discarded, a weighting will be attached corresponding to how often it was chosen to be selected – 0 if it was never selected and 1 if it was always selected. This weighting can then be compared to some threshold to determine whether it should be discarded or kept. The weighting threshold used in our experiments is 0.5, so the final selection is made by the majority vote of each selection rule list.

5 Evaluation

We created two training sets in order to test the effectiveness of our supervised machine learning method at finding suitable selection rules. The first training set contained 150 random queries which were labelled by hand with their ideal keywords. The second training set was created using 4000 queries chosen from QuestionBank [5], in which 3600 queries which were labelled using AlchemyAPI[4], and the other 400 queries did not return any results from AlchemyAPI and so were excluded.

When testing with the smaller training set, ten-fold cross-validation was used to determine the accuracy. When testing with the larger training set, the first

800 queries were used for training while the rest was used for testing. The same experiment was done again using the next 800 queries for training and the rest for testing. After four iterations of this, the accuracy is determined by averaging the performance on each of the four test sets.

Using simple and advanced selection rules worked well when tested on the smaller training set – achieving an average F-score of 0.86 and 0.92 respectively. However, when tested with the larger training set, using simple selection rules performed very poorly – achieving an average F-score of 0.65; while using advanced selection rules performed OK – achieving an average F-score of 0.84. This is most likely due to the inconsistent nature of AlchemyAPI's keyword extractor when used in this manner (as discussed in Section 2); working at a phrase-level simply wasn't enough to make the fine selections needed for this training set. Furthermore, the list of stop words used in conjunction with the simple selection rules were tailored for the smaller training set which we labelled ourselves, thus further impeding performance. Advanced selection rules could bypass both of these problems – by making fine selections at a word-level without the need of a tailored list of stop words – and therefore performed much better.

An question for discussion is why not use only advanced selection rules if they are better. The reason is that a much longer list of rules are needed for them to work and a much larger population is also needed for training. The result is a noticeable slowdown in the overall keyword extraction process and a large increase in training time. Therefore, simple selection rules should be used if the queries are relatively simple and a suitable list of stop words can be found. Table 6 shows the recommended training parameters for both simple and advanced selection rules, and their respective training speeds.

Table 6. Recommended training parameters

Learning Parameter	Simple Rules	Advanced Rules
Chromosome/Rule List Length	10-15	100-200
Population Size	50-100	500-1000
Reproduction	1-pt crossover	1 or 2-pt crossover
Relative Training Speed	1x	100x

Figure 5 shows the effect of increasing the number of rules and the population size for learning the advanced selection rules. Our further investigation shows that increasing the chromosome length beyond 200 will allow more and more edge cases to be covered, but the overall improvement in the F-score is so small that it isn't worth the trade-off against speed.

Splitting the training set into smaller groups did manage to reduce overfitting (See Section 4.2) on the smaller training set – achieving on average a 10% improvement when using 5 groups compared to just 1; further increasing the number of groups showed little improvement. On the larger training set, overfitting was not much of an issue and splitting the training set into smaller groups made little to no improvement to the overall performance; all it did was add additional overhead to the training time, so it isn't recommended for larger training sets.

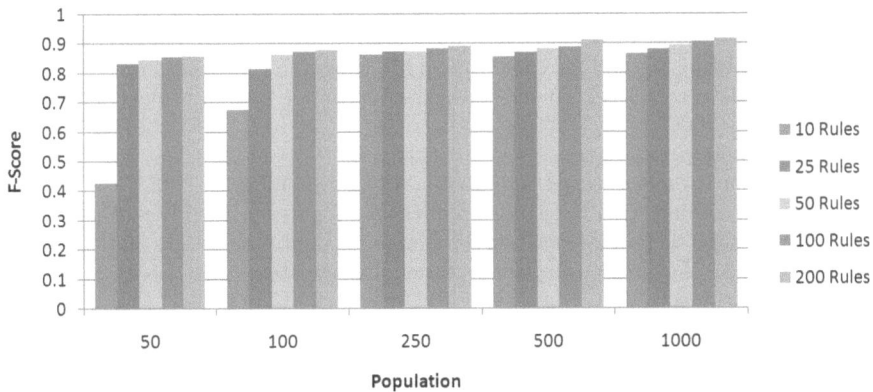

Fig. 5. The effect of increasing the number of rules and population size

6   Conclusions and Future Work

This paper introduces a novel algorithm for keyword extraction from single-sentence natural language queries. We developed a rule-based representation for extraction patterns and successfully applied a genetic algorithm to learn the rules automatically. The keyword extraction algorithm proposed in this paper has shown promising results and the supervised machine learning implementation performed well when cross-validated with various training sets.

Much still needs to be done in order to refine the method. The optimisation suggested in Section 3.5 for advanced selection rules has not yet been implemented; doing so could drastically reduce the number of rules and population needed when training with advanced selection rules. If this is the case, then simple selection rules are likely to become inferior in comparison.

In addition, the paper presented just two representations of selection rules. Further work needs to be done to expand on this as there are potentially better rule representations to be explored and tested.

References

1. Hulth, A., Karlgren, J., Jonsson, A., Boström, H., Asker, L.: Automatic Keyword Extraction Using Domain Knowledge. In: Gelbukh, A. (ed.) CICLing 2001. LNCS, vol. 2004, pp. 472–482. Springer, Heidelberg (2001)
2. Hulth, A.: Improved automatic keyword extraction given more linguistic knowledge. In: Proceedings of the 2003 Conference on Empirical Methods in Natural Language Processing, EMNLP 2003, pp. 216–223. Association for Computational Linguistics, Stroudsburg (2003)
3. Hulth, A.: Enhancing linguistically oriented automatic keyword extraction. In: Proceedings of HLT-NAACL 2004: Short Papers. HLT-NAACL-Short 2004, pp. 17–20. Association for Computational Linguistics, Stroudsburg (2004)

4. Orchestr8: Alchemyapi,
   http://www.alchemyapi.com (last accessed February 10, 2012)
5. Judge, J., Cahill, A., van Genabith, J.: Questionbank: creating a corpus of parse-annotated questions. In: Proceedings of the 21st International Conference on Computational Linguistics and the 44th Annual Meeting of the Association for Computational Linguistics, ACL-44, pp. 497–504. Association for Computational Linguistics, Stroudsburg (2006)
6. Apache: Opennlp,
   http://opennlp.sourceforge.net (last accessed December 21, 2011)
7. Klein, D., Manning, C.D.: Accurate unlexicalized parsing. In: Proceedings of the 41st Annual Meeting on Association for Computational Linguistics, ACL 2003, vol. 1, pp. 423–430. Association for Computational Linguistics, Stroudsburg (2003)
8. Bies, A., Ferguson, M., Katz, K., MacIntyre, R., Tredinnick, V., Kim, G., Marcinkiewicz, M.A., Schasberger, B.: Bracketing guidelines for treebank ii style penn treebank project. Technical report, University of Pennsylvania (1995)
9. C.C.G., Illinois chunker,
   http://cogcomp.cs.illinois.edu/page/software_view/Chunker
   (last accessed January 17, 2012)

Model Combination for Support Vector Regression via Regularization Path

Mei Wang[1,2] and Shizhong Liao[1]

[1] School of Computer Science and Technology,
Tianjin University,Tianjin, China
[2] School of Computer and Information Technology,
Northeast Petroleum University, Daqing, China
szliao@tju.edu.cn

Abstract. In order to improve the generalization performance of support vector regression (SVR), we propose a novel model combination method for SVR on regularization path. First, we construct the initial candidate model set using the regularization path, whose inherent piecewise linearity makes the construction easy and effective. Then, we elaborately select the models for combination from the initial model set through the improved Occam's Window method and the input-dependent strategy. Finally, we carry out the combination on the selected models using the Bayesian model averaging. Experimental results on benchmark data sets show that our combination method has significant advantage over the model selection methods based on generalized cross validation (GCV) and Bayesian information criterion (BIC). The results also verify that the improved Occam's Window method and the input-dependent strategy can enhance the predictive performance of the combination model.

Keywords: Model combination, Support vector regression, Regularization path, Occam's Window.

1 Introduction

Support vector regression (SVR) [1] is an extension of the support vector method to regression problem, which maintains all the main characteristics of the maximal margin algorithm. The generalization performance of SVM depends on the parameters of regularization and kernels. Various algorithms [2,3] have been developed for choosing the best parameters. Regularization path algorithm is another important algorithm to address the SVR model selection problem [4,5], which can fit the entire path of SVR solutions for every value of the regularization parameter. Gunter and Zhu [4] proposed an unbiased estimate for the degrees of freedom of the SVR model, then applied the generalized cross validation (GCV) criterion [6] to select the optimal model. However, single model only has limited information and usually exists uncertainty [7,8]. Model combination is an alternative way to overcome the limitations of model selection,

P. Anthony, M. Ishizuka, and D. Lukose (Eds.): PRICAI 2012, LNAI 7458, pp. 649–660, 2012.
© Springer-Verlag Berlin Heidelberg 2012

which can integrate all useful information from the candidate models into the final hypothesis to improve generalization performance. There are a lot of experimental works showing that combining learning machines often leads to improved generalization performance [9,10,11,12,13].

In this paper, we study the model combination for SVR on regularization path. First, the initial candidate model set is obtained according to the regularization path, whose inherent piecewise linearity makes the construction easy and effective. All possible models are involved in the initial model set, including good performance ones and bad performance ones. Then, a subset from all available individual SVR models is selected by the improved Occam's Window method and the input-dependent strategy. The improved Occam's Window method can eliminate the model with poor performance and select the sparse model. The input-dependent strategy can determine the combination model set according to the estimation of the generalization error of the input. Finally, The combination on the selected models is carried out using the Bayesian model averaging, in which the model posterior probability is estimated by Bayesian information criterion (BIC) approximation.

2 ϵ-SVR Regularization Path

In this section, we briefly introduce the ϵ-SVR regularization path algorithm and refer readers to [4] for a detailed tutorial. The training data set has been taken as $T = \{(\boldsymbol{x}_1, y_1), ..., (\boldsymbol{x}_n, y_n)\} \subset \mathbb{R}^p \times \mathbb{R}$, where the input $\boldsymbol{x}_i$ is a vector with p predictor variables, and the output y_i denotes the response. In ϵ-SVR, our goal is to find a function

$$f(\boldsymbol{x}) = \beta_0 + \langle \boldsymbol{\beta}, \boldsymbol{x} \rangle, \text{ with } \boldsymbol{\beta} \in \mathbb{R}^p, \ \beta_0 \in \mathbb{R},$$

that has at most ϵ deviation from the actually obtained targets y for all the training data, and at the same time is as flat as possible. In practice, one often maps $\boldsymbol{x}$ onto a high dimensional reproducing kernel Hilbert space (RKHS), and fits a nonlinear kernel SVR model. Using the following ϵ-insensitive loss function

$$|y - f(\boldsymbol{x})|_\epsilon = \begin{cases} 0, & \text{if } |y - f(\boldsymbol{x})| < \epsilon, \\ |y - f(\boldsymbol{x})| - \epsilon, & \text{otherwise,} \end{cases}$$

the standard *loss + penalty* criterion of the ϵ-SVR model may be written as

$$\min_{f \in \mathcal{H}_K} \sum_{i=1}^n |y_i - f(\boldsymbol{x}_i)|_\epsilon + \frac{\lambda}{2} \|f\|_{\mathcal{H}_K}^2, \tag{1}$$

where λ is the regularization parameter, and $\mathcal{H}_K$ is a structured RKHS generated by a positive definite kernel $K(\boldsymbol{x}, \boldsymbol{x}')$. Using the representer theorem [14], the solution to equation (1) has a finite form

$$f(\boldsymbol{x}) = \beta_0 + \frac{1}{\lambda} \sum_{i=1}^n \theta_i K(\boldsymbol{x}, \boldsymbol{x}_i), \text{ with } \theta_i \in [-1, +1], \ i = 1, \ldots, n. \tag{2}$$

In this paper, we use $\boldsymbol{\theta}$ to denote the coefficient vector $\boldsymbol{\theta} = (\theta_1, \ldots, \theta_n)^\top$.

According to the piecewise of the ϵ-insensitive loss function and the Karush-Kuhn-Tucker conditions, the training data set is partitioned into the following five disjoint sets:

$$
\begin{aligned}
\mathcal{R} &= \{i : y_i - f(\boldsymbol{x}_i) > \epsilon,\ \theta_i = 1\}, \\
\mathcal{E}_\mathcal{R} &= \{i : y_i - f(\boldsymbol{x}_i) = \epsilon,\ 0 \leq \theta_i \leq 1\}, \\
\mathcal{C} &= \{i : -\epsilon < y_i - f(\boldsymbol{x}_i) < \epsilon,\ \theta_i = 0\}, \\
\mathcal{E}_\mathcal{L} &= \{i : y_i - f(\boldsymbol{x}_i) = -\epsilon, -1 \leq \theta_i \leq 0\}, \\
\mathcal{L} &= \{i : y_i - f(\boldsymbol{x}_i) < -\epsilon,\ \theta_i = -1\}.
\end{aligned}
$$

The ϵ-SVR regularization path algorithm keeps track of the five sets, and examines these sets until one or both of them change. For example, a point from $\mathcal{C}$ enters $\mathcal{E}_\mathcal{R}$. Once there is a change in the elements of the sets, we will say an event has occurred and a breakpoint will appear on the regularization path. As a data point passes through $\mathcal{E}_\mathcal{R}$ or $\mathcal{E}_\mathcal{L}$, its respective θ_i must change from 1 to 0 or -1 to 0 or vice versa. The algorithm begins with $\lambda^0 = \infty$ and the initial sets $\mathcal{E}_\mathcal{R}$ and $\mathcal{E}_\mathcal{L}$ have at most one point combined. The initial solution is obtained by solving a linear programming problem. Then the algorithm recursively computes λ^l ($l \in \mathbb{N}$). Each λ^l corresponds to the value of λ when an event occurs. The λ^{l+1} will be the largest λ less than λ^l such that either θ_i ($i \in \mathcal{E}_\mathcal{R}^l$) reaches 0 or 1, or θ_j ($j \in \mathcal{E}_\mathcal{L}^l$) reaches 0 or -1, or one of the points in $\mathcal{R}$, $\mathcal{L}$, or $\mathcal{C}$ reaches an elbow. When λ^{l+1} is known, the index sets $\mathcal{R}$, $\mathcal{E}_\mathcal{R}$, $\mathcal{C}$, $\mathcal{E}_\mathcal{L}$, $\mathcal{L}$ and $\boldsymbol{\theta}$ are updated according to the nature of the transition that had taken place to yield $\mathcal{R}^{l+1}$, $\mathcal{E}_\mathcal{R}^{l+1}$, $\mathcal{C}^{l+1}$, $\mathcal{E}_\mathcal{L}^{l+1}$, $\mathcal{L}^{l+1}$ and $\boldsymbol{\theta}^{l+1}$. This main phase proceeds repeatedly in increasing value of l and decreasing value of λ^l starting from λ^0 until termination. It is worth noting that the θ_is ($i \in \mathcal{L} \cup \mathcal{R}$) do not change in value when no new event happens. The algorithm will be terminated either when the sets $\mathcal{R}$ and $\mathcal{L}$ become empty or when λ has become sufficiently close to zero.

The whole solution path $\boldsymbol{\theta}(\lambda)$ is piecewise linear. As long as the break points can be establish, all values in between can be found by simple linear interpolation. Figure 1 shows the paths of all the $\{\theta_i(\lambda) \mid 0 < \lambda < \infty\}$ for data set $pyrim$ with $n = 7$.

3   Model Combination for SVR

In this section, we will present how to construct the candidate model set according to the ϵ-SVR regularization path and how to combine the models.

3.1   Initial Model Set Based on Regularization Path

As we have stated in the former section, the regularization path algorithm can compute the exact entire regularization path, which can facilitate the selection of a model. The path $\{\boldsymbol{\theta}(\lambda), 0 \leq \lambda \leq \infty\}$ ranges from the least regularized model to the most regularized model. We adopt the notation $f(\boldsymbol{x}; \boldsymbol{\theta}, \lambda)$ for a model with

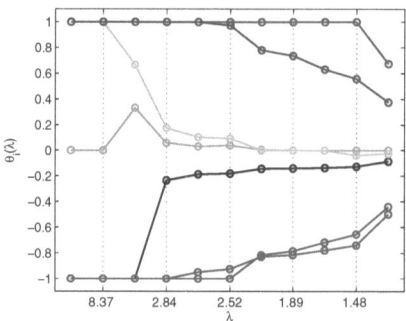

Fig. 1. The entire collection of piecewise liner paths $\theta_i(\lambda)$, $i = 1, \ldots, n$ for the data set *pyrim*

parameter $\boldsymbol{\theta}$ and regularization parameter λ. It should be understood that the different models may be parameterized differently. Hence by $f(\boldsymbol{x}; \boldsymbol{\theta}, \lambda)$ we really mean $f(\boldsymbol{x}; \boldsymbol{\theta}(\lambda), \lambda)$ or $f_\lambda(\boldsymbol{x}; \boldsymbol{\theta})$, and we use the notation f_λ for simplicity.

The solution $\boldsymbol{\theta}(\lambda)$ is piecewise linear as a function of λ. We let the sequence $\infty > \lambda_1 > \cdots > \lambda_k > 0$ denote the corresponding break points on the path. In the interior of any interval $(\lambda_{l+1}, \lambda_l)$, $0 < l < k$, the set $\mathcal{L}, \mathcal{E}_\mathcal{L}, \mathcal{C}, \mathcal{E}_\mathcal{R}, \mathcal{R}$ are constant with respect to λ, such that the support vectors (i.e. the points with $\theta_i \neq 0$) remain unchanged. Therefore, all the regularization parameter in $(\lambda_{l+1}, \lambda_l)$ can lead to models with the same complexity. So we select $\boldsymbol{\theta}$s and λs on the break points to obtain the initial candidate model set $\mathcal{M}_{init} = \{f_{\lambda_1}, \ldots, f_{\lambda_k}\}$. The total number k of break points is $c \times n$, where n is the size of the training set and c is some small number around 1-6.

From the initial candidate ϵ-SVR model set, we can not directly perform the model combination over it, because most of the models in $\mathcal{M}_{init}$ are trivial in the sense that they predict the data far less well than the best models. So, we perform the model combination over a subset of parsimonious, data-supported models. In the latter subsections, we propose two simple and efficient ways of selecting models to guarantee the good performance of model combination.

3.2   Improved Occam's Window

All possible models are involved in the initial model set $\mathcal{M}_{init}$, including the good performance ones and the poor performance ones. Madigan and Raftery [15] used the Occam's Window method for graphical model and showed combination on the selected models provided better inference performance than basing inference on a single model in each of the examples they considered. In this paper, we apply an improved Occam's Window method to eliminate the poor performance models. In the proposed method, posterior model probabilities are used as a metric to guide model selection. There are two basic principles underlying this approach.

First, if a model predicts the data far less well than the model which provides the best prediction, then it has been discredited and should no longer be considered. Thus the model not belonging to should be excluded from the combination candidate model set, where posterior probability ratio W is chosen by the data analyst and $\max_l\{\Pr(f_{\lambda_l}\,|\,T)\}$ denotes the model in initial candidate model set $\mathcal{M}_{init}$ with the highest posterior model probability.

$$\mathcal{M}' = \left\{ f_{\lambda_j} : \frac{\max_l\{\Pr(f_{\lambda_l}\,|\,T)\}}{\Pr(f_{\lambda_j}\,|\,T)} \leq W \right\}$$

Secondly, appealing to Occam's razor, we exclude models which receive less support from the data than any of their simpler submodels. Here we give the definition of submodel. If f_{λ_i} is a submodel of f_{λ_j}, we mean that all the support vectors involved in f_{λ_i} are also in f_{λ_j}. Thus we also exclude from models belonging to then we obtain the model set $\mathcal{M}_{ao} = \mathcal{M}' \backslash \mathcal{M}'' \subseteq \mathcal{M}_{init}$.

$$\mathcal{M}'' = \left\{ f_{\lambda_j} : \exists f_{\lambda_l} \in \mathcal{M}_{init}, f_{\lambda_l} \subset f_{\lambda_j}, \frac{\Pr(f_{\lambda_l}\,|\,T)}{\Pr(f_{\lambda_j}\,|\,T)} > 1 \right\},$$

The posterior probability ratio W is usually a constant as in [15]. However, the statistical results show that only few of the ϵ-SVR models in $\mathcal{M}_{init}$ have strongly peak posterior probabilities, as shown in Figure 2. So, we apply a query-dependent method to determine the ratio W. Starting from $W = k/20$, we double it for every iteration and examine the number of models in set $\mathcal{M}''$. Once the number changes dramatically, we terminate the iteration process and use the last W value. In the experiments of the next section, this would be the case with the model set size $|\mathcal{M}'|$ increasing more than 4. Here, if the model set size enlarges dramatically, it means that many models with low posterior probabilities enter the model set $|\mathcal{M}'|$.

The improved Occam's Window algorithm, as shown in the Algorithm 1, can greatly reduce the number of models in the candidate model set. Typically, in our experience, the number of the candidate model set is reduced to fewer than $k/20$.

3.3   Input-Dependent Strategy

Though the improved Occam's Window method, we have obtained a credible candidate model set on the training data. Further, in order to perform good prediction on new input, we need a more credible input-dependent subset for model combination.

The generalization performance of the model combination can be evaluated by prediction error on the new input x. For regression, we apply quadratic error $(f_{bma}(x) - y)^2$ to calculate the prediction error of the model combination, where f_{bma} denotes the combined model.

Since the probability distribution according to which the data generated is unknown, it is impossible for us to compute the expectation of the combination error. In this paper, we use the nearest neighbor method to estimate the combination expected error on the input x. Specifically, we adopt the search strategy

---

Algorithm 1. The improved Occam's Window algorithm.

Input: $\mathcal{M}_{init} = \{f_{\lambda_1}, \ldots, f_{\lambda_k}\}$, $\mathcal{P} = \{\mathrm{Pr}(f_{\lambda_1}), \ldots, \mathrm{Pr}(f_{\lambda_k})\}$, k, s
Output: $\mathcal{M}_{ao}$
$MP \leftarrow \max(\mathcal{P})$;
$\mathcal{M}_{ao} \leftarrow \emptyset$;
$\mathcal{M}_{tmp} \leftarrow \emptyset$;
$W \leftarrow k/20$;
while $\mathcal{M}_{init} \neq \emptyset$ **do**
    for $f \in \mathcal{M}_{init}$ **do**
        if $MP/\mathrm{Pr}(f) \leq W$ **then**
            $\mathcal{M}_{tmp} \leftarrow \mathcal{M}_{tmp} \cup \{f\}$;
            $\mathcal{M}_{init} \leftarrow \mathcal{M}_{init} \backslash \{f\}$
        end
    end
    if $|\mathcal{M}_{tmp}| - |\mathcal{M}_{ao}| \leq s$ **then**
        $\mathcal{M}_{ao} \leftarrow \mathcal{M}_{tmp}$;
        $W \leftarrow W * 2$;
    end
    else
        $\mathcal{M}_{ao} \leftarrow \mathcal{M}_{tmp}$;
        break
    end
end
for $f \in \mathcal{M}_{ao}$ **do**
    for $f_1 \in \mathcal{M}_{ao} \backslash \{f\}$ **do**
        if $(\mathrm{Pr}(f_1) > \mathrm{Pr}(f))$ *and* $(f_1 \subset f)$ **then**
            $\mathcal{M}_{ao} \leftarrow \mathcal{M}_{ao} \backslash \{f\}$;
            break;
        end
    end
end

---

which computes the Euclidean distances between the input x and each point in the training set and then selects the one with smallest distance.

Suppose the input x's nearest neighbor we find is x_e, $e \in [1, n]$ and its output is denoted by y_e. For each ϵ-SVR model $f_{\lambda_j} \in \mathcal{M}_{ao}$, we compute the prediction error $(f_{\lambda_j}(x_e) - y_e)^2$, and sort them by ascending order. We add the model with current smallest error in $\mathcal{M}_{ao}$ to candidate model set, denoted by $\mathcal{M}_{aa}$, meanwhile remove it from $\mathcal{M}_{ao}$. Then we perform the model combination over $\mathcal{M}_{aa}$, and then compute the combination error $(f_{bma}(x_e) - y_e)^2$. Until the combination error no longer declines, the model selection process will be terminated.

Since the whole combination process is dynamic, once a model is added to the model set $\mathcal{M}_{aa}$, we should update the posterior probabilities for each model. The model selection process is shown in Algorithm 2.

Algorithm 2. Input-dependant model selection algorithm.

Input: $\mathcal{M}_{ao}$, $T = \{(\boldsymbol{x}_1, y_1), \ldots, (\boldsymbol{x}_n, y_n)\}$, $\boldsymbol{x}$
Output: $\mathcal{M}_{aa}$
Find the nearest neighbor point $\boldsymbol{x}_e$ of $\boldsymbol{x}$ from T;
Compute the prediction error for each model on $\boldsymbol{x}_e$;
for $\mathcal{M}_{ao} \neq \emptyset$ **do**

$\quad f_m \leftarrow f \in \mathcal{M}_{ao}$ with lowest prediction error;
$\quad \mathcal{M}_{aa} \leftarrow \mathcal{M}_{aa} \cup \{f_m\}$;
$\quad \mathcal{M}_{ao} \leftarrow \mathcal{M}_{ao} \setminus \{f_m\}$;
$\quad$ Update posterior probabilities of models in $\mathcal{M}_{aa}$;
$\quad$ Compute $f_{bma}(\boldsymbol{x}_e)$;
$\quad$ **if** $(f_{bma}(\boldsymbol{x}_e) - y_e)^2$ *is greater than last time* **then**
$\quad \quad$ break;
$\quad$ **end**

end

3.4   Bayesian Model Averaging

For ϵ-SVR, we apply the model combination method—Bayesian model averaging. Suppose we have a ϵ-SVR candidate model set $\mathcal{M} = \{f_1, \ldots, f_m\}$. The Bayesian model averaging over $\mathcal{M}$ has the form

$$f_{bma}(\boldsymbol{x}) = \sum_{j=1}^{m} f_j(\boldsymbol{x}) \, \Pr(f_j \,|\, T), \tag{3}$$

where $\Pr(f_j \,|\, T)$ is the posterior probability of model f_j, $j = 1, \ldots, m$. Then this is the process of estimating the prediction under each model f_j and then averaging the estimates according to how likely each model is.

We can perform Bayesian model averaging over any ϵ-SVR model set, such as $\mathcal{M}_{init}$, $\mathcal{M}_{ao}$ and $\mathcal{M}_{aa}$, while for new input we should use the selected model set $\mathcal{M}_{aa}$.

In general, the posterior probability of model f_j in equation (3) is given by

$$\Pr(f_j \,|\, T) \propto \Pr(T \,|\, f_j) \Pr(f_j), \tag{4}$$

where $\Pr(T \,|\, f_j)$ is the marginal likelihood of model f_j and $\Pr(f_j)$ is the prior probability that f_j is the true model. In this paper, we will propose a simple and efficient method to estimate the model posterior probability for fixed regularization parameter λ, which depends on the ϵ-SVR regularization path algorithm. We estimate the posterior probability of each model f_j as [16]

$$\widehat{\Pr}(f_j \,|\, T) = \frac{e^{-\frac{1}{2} \cdot \mathrm{BIC}_j}}{\sum_{f_m \in \mathcal{M}} e^{-\frac{1}{2} \cdot \mathrm{BIC}_m}}, \tag{5}$$

and for each model f_j in model set $\mathcal{M}$, its BIC value can be calculated as

$$\mathrm{BIC}(f_j) = \frac{\|\boldsymbol{y} - \boldsymbol{f}_j\|^2}{n\sigma^2} + \frac{\log(n)}{n} \mathrm{df}(f_j), \tag{6}$$

where $\boldsymbol{y} = (y_1, \ldots, y_n)^\top$, $\boldsymbol{f}_j = (f_j(\boldsymbol{x}_1), \ldots, f_j(\boldsymbol{x}_n))^\top$, $j = 1, \ldots, k$.

3.5 Computational Complexity

The main computational burden of the model combination for ϵ-SVR centers on building the ϵ-SVR regularization path, proceeding improved Occam's Window procedure to exclude models to obtain the model set $\mathcal{M}_{ao}$, and selecting models using the input-dependant strategy to obtain the model set $\mathcal{M}_{aa}$.

The approximate computational complexity of the ϵ-SVR regularization path algorithm is $O(cn^2m + nm^2)$ [4], where n is the size of the training data and m is the average size of $\mathcal{E}_R \cup \mathcal{E}_L$, and c is some small number as previously mentioned.

The search strategy in the improved Occam's Window method is to identify the models in $\mathcal{M}_{ao}$. First part of the method involves $O(dk)$ operations, including finding the largest posterior probability and excluding the models with low posterior probability. Here, $k = c \times n$ is the size of the initial candidate model set, and d is the iteration for adjusting W, our experience so far suggests that d is around $3 - 8$. Second part of the method involves $O(k^2m)$ operations, including determining subset relationship and comparing the posterior probabilities between each pair of the models, and the m is the average size of $\mathcal{E}_R \cup \mathcal{E}_L$. So the approximate computational complexity of the improved Occam's Window is $O(cn^2m)$.

The approximate computational complexity of the input-dependant strategy is $O(cn)$, including finding the nearest neighbor data point from the training data and determining the final candidate model set for combination.

So, the total computational complexity of the model combination for ϵ-SVR on regularization path is $O(cn^2m + nm^2)$.

4   Experiments

In this section, we investigate the performance of our model combination with GCV-based and BIC-based model selection on seven benchmark data sets used in [2] (available online at http://www.csie.ntu.edu.tw/~cjlin/libsvmtools/datasets/), and we consider Gaussian radial basis kernel $K(\boldsymbol{x}, \boldsymbol{x}') = \exp(-\gamma \|\boldsymbol{x} - \boldsymbol{x}'\|^2)$, where γ is the prespecified kernel parameter. We use the same values for γ and ϵ as specified in [2] shown in Table 1, where n denotes the size of the data set, and p denotes the dimension of the input $\boldsymbol{x}$.

For data set *abalone* we randomly sample 1000 examples from the 4177 examples; for *cpusmall* we randomly sample 1000 examples from the 8292 examples; for *spacega* we randomly sample 1000 examples from the 3107 examples.

Since the usual goal of regression analysis is to minimize the predicted squared-error loss, the prediction error is defined as

$$\mathrm{PredE} = \frac{1}{m} \sum_{i=1}^{m} (y_i - f(\boldsymbol{x}_i))^2,$$

where m is the number of the test data.

Table 1. Summary of the Seven Benchmark Data Sets

DataSet	n	p	γ	ϵ	DataSet	n	p	γ	ϵ
pyrim	74	27	0.0167	0.00136	cpusmall	1000	12	0.0913	0.12246
triazines	186	60	0.0092	0.00910	spacega	1000	6	0.1664	0.01005
mpg	392	7	0.3352	0.18268	abalone	1000	8	0.1506	0.12246
housing	566	13	0.1233	0.18268					

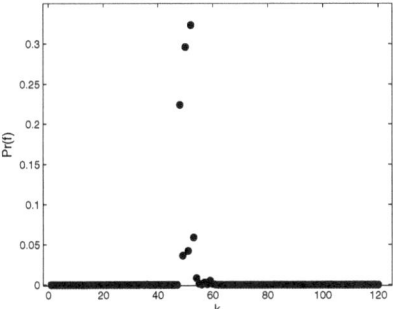

Fig. 2. The posterior probability distribution of the models according to the regularization path

4.1 Model Posterior Probability Distribution

First, we verify the model posterior probabilities according to the regularization path. We randomly sample on data set *pyrim*, and then build each regularization path using the algorithm proposed in [4]. We compute the model posterior probability as described in Section 3. The posterior probability of models from most regularized to least regularized is shown in Figure 2.

From the figure we find that only few models have higher posterior probabilities and most of the other models have very small posterior probabilities around zero. Therefore, applying the improved Occam's Window method we can discard most of the models in the initial candidate model set. We record an example of adjusting procedure on posterior probability ratio W in Table 2, where the ATs is the iteration on the W. We observe that once the ratio is large enough, many models enter the candidate model set. From the last line of the table, we can conclude that poor performance models can decrease the performance of model combination.

4.2 Performance Comparison

In this subsection, we first compare the prediction performance of the model combination over $\mathcal{M}_{aa}$ with the model selection methods based GCV [4] and

Table 2. Adjusting procedure on posterior probability ratio W with data set *pyrim*

Ats	W	$|\mathcal{M}'|$	PredE
1	6	2	0.00593
2	12	5	0.00601
3	24	7	0.00596
4	48	9	0.00570
5	96	22	0.00713

BIC. We randomly split the data into training and test sets, with the training set comprising 80% of the data. We repeat this process 30 times and compute the average prediction errors and their corresponding standard errors. We calculate the prediction error with each test data for each method. The results are summarized in Table 3. From the tables we find that the model combination has the lowest prediction error and standard error. In a sense, this experiment shows that model combination has the property of "many can be better than one".

Table 3. Comparisons of the prediction error on real data for model selection and model combination

	GCV	BIC	BMC
pyrim	0.0055 (0.0026)	0.0052 (0.0027)	0.0049 (0.0023)
triazines	0.0242 (0.0081)	0.0239 (0.0080)	0.0237 (0.0078)
mpg	7.32 (2.35)	7.25 (2.32)	7.13 (2.12)
housing	10.82 (3.65)	10.77 (3.60)	10.04 (3.49)
cpusmall	27.48 (10.25)	27.32 (10.25)	27.10 (10.23)
spacega	0.0125 (0.0015)	0.0122 (0.0015)	0.0120 (0.0015)
abalone	4.31 (1.05)	4.29 (1.05)	4.27 (1.04)

In the second part of the experiment, we compare the prediction performance of model combination over the model set $\mathcal{M}_{init}$, $\mathcal{M}_{ao}$ and $\mathcal{M}_{aa}$. We compute the prediction error for the model combination with and without the model selection strategy. The results are summarized in Table 4, where BMC_i denotes the model combination over the initial candidate model set $\mathcal{M}_{init}$; BMC_o denotes combination over the model set $\mathcal{M}_{ao}$, and BMC_p denotes combination over the final model set $\mathcal{M}_{aa}$. From the tables we find that the model combination over the selected candidate model set has lower prediction error than over the initial model set. In a sense, this experiment shows that the model combination has the property of "many can be better than all".

Table 4. Comparisons of the prediction error on real data for model combination with and without model selection strategy

	BMC$_i$	BMC$_o$	BMC$_p$
pyrim	0.0067 (0.0037)	0.0051 (0.0027)	0.0049 (0.0023)
triazines	0.0317 (0.0092)	0.0277 (0.0083)	0.0237 (0.0078)
mpg	7.98 (2.68)	7.28 (2.32)	7.13 (2.12)
housing	11.73 (3.98)	10.73 (3.56)	10.04 (3.49)
cpusmall	29.49 (11.08)	28.01 (10.78)	27.10 (10.23)
abalone	4.83 (1.27)	4.36 (1.09)	4.27 (1.04)
spacega	0.0204 (0.0019)	0.0128 (0.0015)	0.0120 (0.0015)

5   Conclusion

In this paper, we propose a new model combination framework for ϵ-SVR. We can obtain all possible models according to the regularization path. Applying the improved Occam's Window method and the input-dependant strategy, we greatly reduce the number of candidate models and improve the model combination prediction performance on test data. The model combination on regularization path can reduce the risk of single model selection, and improve the prediction performance. The experimental results on real data show that with some pre-processing of model set the combination prediction accuracy significantly exceeds that of a single model.

Our model combination for ϵ-SVR on regularization path provide a common framework for model combination, which can be extended to support vector machines (SVMs)[17] and other regularized models.

Acknowledgment. The authors thank the three referees for their helpful comments. This work is supported by Natural Science Foundation of China under Grant No. 61170019 and Natural Science Foundation of Tianjin under Grant No. 11JCY BJC00700.

References

1. Smola, A.J., Schölkopf, B.: A tutorial on support vector regression. Statistics and Computing 14(3), 199–222 (2004)
2. Chang, M.W., Lin, C.J.: Leave-one-out bounds for support vector regression model selection. Neural Computation 17(5), 1188–1222 (2005)
3. Wang, G., Yeung, D., Lochovsky, F.: Two-dimensional solution path for support vector regression. In: Proceedings of the 23th International Conference on Machine Learning, pp. 993–1000 (2006)
4. Gunter, L., Zhu, J.: Efficient computation and model selection for the support vector regression. Neural Computation 19(6), 1633–1655 (2007)

5. Wang, G., Yeung, D.Y., Lochovsky, F.H.: A new solution path algorithm in support vector regression. IEEE Transactions on Neural Networks 19(10), 1753–1767 (2008)
6. Craven, P., Wahba, G.: Smoothing noisy data with spline functions. Numerische Mathematik 31(4), 377–403 (1978)
7. Petridis, V., Kehagias, A., Petrou, L., Bakirtzis, A., Kiartzis, S., Panagiotou, H., Maslaris, N.: A bayesian multiple models combination method for time series prediction. Journal of Intelligent and Robotic Systems 31(1), 69–89 (2001)
8. Freund, Y., Mansour, Y., Schapire, R.E.: Why averaging classifiers can protect against overfitting. In: Proceedings of the Eighth International Workshop on Artificial Intelligence and Statistics, vol. 304. Citeseer (2001)
9. Ji, C., Ma, S.: Combinations of weak classifiers. IEEE Transactions on Neural Networks 8(1), 32–42 (1997)
10. Raftery, A.E., Madigan, D., Hoeting, J.A.: Bayesian model averaging for linear regression models. Journal of the American Statistical Association 92, 179–191 (1997)
11. Kittler, J.: Combining classifiers: A theoretical framework. Pattern Analysis and Applications 1, 18–27 (1998)
12. Evgeniou, T., Pontil, M., Elisseeff, A.: Leave one out error, stability, and generalization of voting combinations of classifiers. Machine Learning 55(1), 71–97 (2004)
13. Bagui, S.C.: Combining pattern classifiers: methods and algorithms. Technometrics 47(4), 517–518 (2005)
14. Kimeldorf, G., Wahba, G.: Some results on Tchebycheffian spline functions. Journal of Mathematical Analysis and Applications 33(1), 82–95 (1971)
15. Madigan, D., Raftery, A.E.: Model selection and accounting for model uncertainty in graphical models using Occam's window. Journal of the American Statistical Association 89(428), 1535–1546 (1994)
16. Hastie, T., Tibshirani, R., Friedman, J.: The Elements of Statistical Learning: Data Mining, Inference, and Prediction. Springer (2008)
17. Zhao, N., Zhao, Z., Liao, S.: Probabilistic Model Combination for Support Vector Machine Using Positive-Definite Kernel-Based Regularization Path. In: Wang, Y., Li, T. (eds.) ISKE2011. AISC, vol. 122, pp. 201–206. Springer, Heidelberg (2011)

Subspace Regularized Linear Discriminant Analysis for Small Sample Size Problems

Zhidong Wang[1] and Wuyi Yang[2],[*]

[1] School of Computer Science and Engineering, Beihang University,
Beijing, 100191, China
[2] Key Laboratory of Underwater Acoustic Communication and Marine Information
Technology of the Ministry of Education, Xiamen University,
Xiamen, 361005, China
wyyang@xmu.edu.cn

Abstract. Linear discriminant analysis (LDA) can extract features that preserve class separability. For small sample size (SSS) problems, the number of data samples is smaller than the dimension of data space, and the within-class scatter matrix of data samples is singular. LDA cannot be directly applied to SSS problems, since LDA requires the within-class scatter matrix to be non-singular. Regularized linear discriminant analysis (RLDA) is a common way to deal with singularity problems. In this paper, a subspace RLDA (SRLDA) algorithm is presented for SSS problems, which is performed in a subspace containing the range space of the total scatter matrix. The use of different parameter values in the regularization of the within-class matrix connects SRLDA to several extensions of LDA including discriminative common vectors (DCV), complete LDA (CLDA) and pseudo-inverse LDA (PLDA). An efficient algorithm to perform SRLDA based on the QR decomposition is developed, which makes the algorithm feasible and efficient for SSS problems. Face recognition is a well-known SSS problem and extensive experiments on various datasets of face images are conducted to evaluate the proposed algorithm and compare SRLDA with other extended LDA algorithms. Experimental results show the proposed algorithm can fuse discriminative information in the range and the null space of the within-class scatter matrix to find optimal discriminant vectors.

Keywords: Regularized linear discriminant analysis, QR decomposition, small sample size problem.

1 Introduction

Linear discriminant analysis (LDA) is a very popular method for dimensionality reduction because of its relative simplicity and effectiveness. LDA seeks for a linear transformation that maximizes class separability in the projection subspace. So, the criteria of LDA for dimension reduction are formulated to maximize

---

[*] Corresponding author

P. Anthony, M. Ishizuka, and D. Lukose (Eds.): PRICAI 2012, LNAI 7458, pp. 661–672, 2012.
© Springer-Verlag Berlin Heidelberg 2012

the ratio of the between-class scatter to the within-class scatter in the reduced dimensional space.

Assume that there are n data points $x_1, \cdots, x_n \in \mathbb{R}^m$ belonging to c classes. The number of data points that belong to the i-th class is n_i, and $n = \sum_{i=1}^{c} n_i$. LDA finds a linear transformation $A = [a_1, \cdots, a_d] \in \mathbb{R}^{m \times d}$ that maps data point x_j in the m-dimensional space to a vector y_j in the d-dimensional space:

$$A : x_j \in \mathbb{R}^m \rightarrow y_j = A^T x_j \in \mathbb{R}^d (d < m)$$

Let N_i be the index set of data items in the i-th class, i.e. x_j, for $j \in N_i$, belongs to the i-th class. The between-class scatter matrix S_b, the within-class scatter matrix S_w, and the total scatter matrix S_t are defined as

$$S_b = \sum_{i=1}^{c} n_i (\mu_i - \mu)(\mu_i - \mu)^T,$$

$$S_w = \sum_{i=1}^{c} \sum_{j \in N_i} (x_j - \mu_i)(x_j - \mu_i)^T,$$

$$S_t = S_b + S_w = \sum_{j=1}^{n} (x_j - \mu)(x_j - \mu)^T,$$

where $\mu_i = \frac{1}{n_i} \sum_{j \in N_i} x_j$ is the centroid of the i-th class, and $\mu = \frac{1}{n} \sum_{j=1}^{n} x_j$ is the global centroid. An optimal transformation $A = [a_1, \cdots, a_d]$ would maximize the objective function:

$$J(A) = \arg\max_{A} \frac{\text{trace}(A^T S_b A)}{\text{trace}(A^T S_w A)}. \tag{1}$$

When S_w is non-singular, the transformation matrix A can be obtained by finding the generalized eigenvectors corresponding to the d largest eigenvalues of $S_b a = \lambda S_w a$. The solution can also be obtained by applying an eigendecomposition on the matrix $S_w^{-1} S_b$. Since the rank of the matrix S_b is bounded from above by $c-1$, there are at most $c-1$ eigenvectors corresponding to nonzero eigenvalues.

In many interesting machine learning and data mining problems, the dimension of the data points in general is higher than the number of data points, i.e. $m > n$. These problems are known as "small sample size" (SSS) problems. Since LDA requires the within-class scatter matrix to be non-singular and the within-class scatter matrix can be singular in SSS problems, LDA cannot be directly applied to these problems. In order to deal with SSS problems, many extensions of LDA, including Pseudo-inverse LDA (PLDA) [1], Regularized LDA (RLDA)[2,3], discriminative common vectors (DCV) [4, 5], Complete LDA (CLDA) [6, 7], Direct LDA (DLDA) [8], etc, were proposed.

RLDA regularizes S_w by adding a multiple of identity matrix to S_w, as $S_w + \alpha I_m$, for $\alpha > 0$, where I_m is a $m \times m$ identity matrix. $S_w + \alpha I_m$ is non-singular, and the optimal transformation A would maximize the objective function:

$$J(A) = \arg\max_A \frac{\text{trace}(A^T S_b A)}{\text{trace}(A^T (S_w + \alpha I_m) A)}. \tag{2}$$

In this paper, a subspace RLDA (SRLDA) algorithm using two different parameter values in the regularization of the within-class matrix is proposed to solve SSS problems. The rest of the paper is organized as follows. SRLDA is presented in Section 2. In section 3, the relationship among SRLDA and other extended LDA methods is discussed. In section 4, an efficient algorithm to perform SRLDA based on the QR decomposition is developed. Experimental results are reported in section 5. Last, a conclusion is drawn in section 6.

2   Subspace Regularized Linear Discriminant Analysis

Liu et al. [9] proved that RLDA can be performed in the principal component analysis transformed space. For linear methods, one can first go to a subspace that contains all the data and then optimize the objective function in this subspace[10]. So, RLDA can be performed in a subspace that contains the range space of the total scatter matrix, and all optimal discriminant vectors can be derived from the subspace without any loss of the discriminatory information.

The proposed subspace regularized linear discriminant analysis (SRLDA) select a subspace which contains the range space of the total scatter matrix. Assume that Ψ is a selected subspace and $\{w_1, w_2, \cdots, w_l\}$ is an orthonormal basis for Ψ, where l is the dimension of Ψ. Then, data points are projected into this subspace. In the subspace, the between-class scatter matrix is $\hat{S}_b \in \mathbb{R}^{l \times l}$, and the within-class scatter matrix is $\hat{S}_w \in \mathbb{R}^{l \times l}$. We denote $\hat{S}_w$ as $\hat{S}_w = Q\Sigma Q^T$, where Q is orthogonal, $\Sigma = \text{diag}(\Sigma_w, 0)$, $\Sigma_w \in \mathbb{R}^{s \times s}$ is diagonal, and $s = \text{rank}(\hat{S}_w)$. The diagonal elements of Σ are eigenvalues of $\hat{S}_w$ ordered in non-increasing order, and the columns of Q are their corresponding eigenvectors.

Two parameter values are used in the regularization of the within-class matrix. SRLDA regularizes $\hat{S}_w$ by adding a diagonal matrix as $\hat{S}_w + Q(\Sigma_r)Q^T$, where $\Sigma_r = \text{diag}(\sigma I_s, \alpha I_{l-s})$, $\sigma > 0$, $\alpha > 0$, I_s is a $s \times s$ identity matrix, and I_{l-s} is a $(l-s) \times (l-s)$ identity matrix. In the subspace Ψ, the optimization criteria is formulated as the maximization problem of the objective function:

$$J(B) = \arg\max_B \frac{\text{trace}(B^T \hat{S}_b B)}{\text{trace}(B^T (\hat{S}_w + Q\Sigma_r Q^T) B)} \tag{3}$$

An optimal transformation matrix B can be obtained by finding the generalized eigenvectors corresponding to the d largest eigenvalues of $\hat{S}_b b = \lambda(\hat{S}_w + Q\Sigma_r Q^T)b$. Suppose $b_1, b_2, \cdots, b_d$ are the generalized discriminant vectors corresponding to the d largest nonzero eigenvalues. We get the transformation matrix $A = WB$, where $W = (w_1, w_2, \cdots, w_l)$ and $B = (b_1, b_2, \cdots, b_d)$.

3   Connections to Extended LDA Methods

In this section, the relationship among SRLDA and some extended LDA methods, including DCV, CLDA, and PLDA, is presented.

Let $S_t = U\Sigma_1 U^T$ be the spectral decomposition of S_t, where U is orthogonal, $\Sigma_1 = \mathrm{diag}(\Sigma_t, 0)$, $\Sigma_t \in \mathbb{R}^{t \times t}$ is diagonal, and $t = \mathrm{rank}(S_t)$. The diagonal elements of Σ_1 are the eigenvalues of S_t ordered in non-increasing order, and the columns of U are their corresponding eigenvectors. Let $U = (U_1, U_2)$ be a partition of U, such that $U_1 \in \mathbb{R}^{m \times t}$ and $U_2 \in \mathbb{R}^{m \times (m-t)}$. Since $S_t = S_b + S_w$, we have

$$U^T S_t U = \begin{pmatrix} \Sigma_t & 0 \\ 0 & 0 \end{pmatrix} = \begin{pmatrix} U_1^T(S_b + S_w)U_1 & U_1^T(S_b + S_w)U_2 \\ U_2^T(S_b + S_w)U_1 & U_2^T(S_b + S_w)U_2 \end{pmatrix} \qquad (4)$$

It follows that $U_2^T(S_b + S_w)U_2 = 0$. S_b and S_w both are positive semidefinite, and therefore $U_2^T S_b U_2 = 0$ and $U_2^T S_w U_2 = 0$. Thus, $S_w U_2 = 0$, and $U_1^T S_w U_2 = 0$. So, we have

$$U^T S_w U = \begin{pmatrix} U_1^T S_w U_1 & 0 \\ 0 & 0 \end{pmatrix}$$

Suppose $u_1, u_2, \cdots, u_t$ are the columns of U_1. Define a subspace $\Psi = R(U_1)$ spanned by the columns of U_1. Obviously, Ψ is the range space of S_t. SRLDA can find optimal discriminant vectors in the subspace Ψ. Then, the within-class scatter matrix of the mapped data in Ψ is $\hat{S}_w = W^T S_w W \in \mathbb{R}^{l \times l}$, where $W = U_1$ and $l = t$. Let $\hat{S}_w = U_1^T S_w U_1 = V\Sigma_2 V^T$ be the spectral decomposition of $\hat{S}_w$, where V is orthogonal, $\Sigma_2 = \mathrm{diag}(\Sigma_w, 0)$, $\Sigma_w \in \mathbb{R}^{s \times s}$ is diagonal, $s = \mathrm{rank}(\hat{S}_w)$. The diagonal elements of Σ_2 are eigenvalues of $\hat{S}_w$ ordered in non-increasing order, and the columns of V are their corresponding eigenvectors. So, we have

$$\begin{pmatrix} V & 0 \\ 0 & I \end{pmatrix}^T U^T S_w U \begin{pmatrix} V & 0 \\ 0 & I \end{pmatrix} = \begin{pmatrix} \Sigma_2 & 0 \\ 0 & 0 \end{pmatrix}$$

Let $V = (V_1, V_2)$ be a partition of V, such that $V_1 \in \mathbb{R}^{t \times s}$ and $V_2 \in \mathbb{R}^{t \times (t-s)}$. Define $P = (P_1, P_2, P_3)$, $P_1 = U_1 V_1$, $P_2 = U_1 V_2$, $P_3 = U_2$, and $\Sigma_3 = \mathrm{diag}(\Sigma_w, 0, 0) \in \mathbb{R}^{m \times m}$. Then we have

$$S_w = P\Sigma_3 P^T. \qquad (5)$$

So, the diagonal elements of Σ_3 are eigenvalues of S_w ordered in non-increasing order, and the columns of P are their corresponding eigenvectors.

The regularized within-class scatter matrix is defined as $\hat{S}_w + Q\Sigma_r Q^T$, where $Q = V$, $\Sigma_r = \mathrm{diag}(\sigma I_s, \alpha I_{t-s})$, $\sigma > 0$, $\alpha > 0$, I_s is a $s \times s$ identity matrix, and I_{t-s} is a $(t-s) \times (t-s)$ identity matrix. An optimal transformation matrix B in (3) can be obtained by finding the generalized eigenvectors corresponding to the d largest eigenvalues of $\hat{S}_b b = \lambda(\hat{S}_w + V\Sigma_r V^T)b$.

3.1 Connections to DCV and CLDA

DCV [4, 5] finds discriminant vectors in the null space of the within-class scatter matrix and overlooks discriminant information contained in the range space of the within-class scatter matrix. CLDA [6, 7] finds "irregular" discriminant vectors in the null space of the within-class scatter matrix and "regular" discriminant vectors in the range space of the within-class scatter matrix. Two kinds of discriminant vectors are fused in the classification level.

Suppose b is a generalized eigenvector of $\hat{S}_b$ and $\hat{S}_w + V\Sigma_r V^T$ corresponding to the nonzero eigenvalue λ, i.e., $\hat{S}_b b = \lambda(\hat{S}_w + V\Sigma_r V^T)b$. Recall that $W = U_1$, $\hat{S}_b = W^T S_b W = U_1^T S_b U_1$. $\hat{S}_w + V\Sigma_r V^T = V\hat{\Sigma}V^T$, where $\hat{\Sigma} = \Sigma_2 + \Sigma_r$. It follows that $U_1^T S_b U_1 b = \lambda V\hat{\Sigma}V^T b$. $V\hat{\Sigma}V^T$ can be expressed as

$$V\hat{\Sigma}V^T = (V_1, V_2)\begin{pmatrix} \Sigma_w + \sigma I_s & 0 \\ 0 & \alpha I_{t-s} \end{pmatrix}\begin{pmatrix} V_1^T \\ V_2^T \end{pmatrix} = V_1(\Sigma_w + \sigma I_s)V_1^T + \alpha V_2 V_2^T.$$

Since $b \in \mathbb{R}^{t \times 1}$ and $V = (V_1, V_2) \in \mathbb{R}^{t \times t}$ is orthogonal, b can be expressed as $b = V_1 b_1 + V_2 b_2$, where $b_1 \in \mathbb{R}^{s \times 1}$ and $b_1 \in \mathbb{R}^{(t-s) \times 1}$. It follows that

$$U_1^T S_b U_1 V_1 b_1 + U_1^T S_b U_1 V_2 b_2 = \lambda V_1(\Sigma_w + \sigma I_s)b_1 + \alpha \lambda V_2 b_2.$$

Since $V_2^T V_1 = 0$, we have

$$V_1^T U_1^T S_b U_1 V_1 b_1 + V_1^T U_1^T S_b U_1 V_2 b_2 = \lambda(\Sigma_w + \sigma I_s)b_1$$

and

$$V_2^T U_1^T S_b U_1 V_1 b_1 + V_2^T U_1^T S_b U_1 V_2 b_2 = \alpha \lambda b_2.$$

Since $\Sigma_w + \sigma I_s$ is positive definite and $\alpha > 0$, we have

$$(\Sigma_w + \sigma I_s)^{-1}(V_1^T U_1^T S_b U_1 V_1 b_1 + V_1^T U_1^T S_b U_1 V_2 b_2) = \lambda b_1$$

and

$$\alpha^{-1}(V_2^T U_1^T S_b U_1 V_1 b_1 + V_2^T U_1^T S_b U_1 V_2 b_2) = \lambda b_2.$$

When $\alpha = 1$ and σ tends to ∞, $b_1 = 0$ and

$$V_2^T U_1^T S_b U_1 V_2 b_2 = \lambda b_2.$$

Since $V_2^T U_1^T S_w U_1 V_2 = 0$, the column vectors of $U_1 V_2$ forms an orthonormal basis for $R(S_t) \cap N(S_w)$, where $R(S_t)$ denotes the range space of S_t, and $N(S_w)$ denotes the null space of S_w. Suppose $b_2^1, \cdots, b_2^d$ are the eigenvectors of matrix $V_2^T U_1^T S_b U_1 V_2$ corresponding to the d largest nonzero eigenvlaues. Then, $U_1 V_2 b_2^1, \cdots, U_1 V_2 b_2^d$ are the discriminant vectors in the null space of the within-class scatter matrix, which are found by DCV. These vectors also are the "irregular" discriminant vectors found in CLDA.

When $\sigma = 0$ and α tends to ∞ , $b_2 = 0$ and

$$(\Sigma_w)^{-1} V_1^T U_1^T S_b U_1 V_1 b_1 = \lambda b_1.$$

Since $V_1^T U_1^T S_w U_1 V_1 = \Sigma_w$, the space spanned by the column vectors of $U_1 V_1$ is the range space of the within-class scatter matrix. Suppose $b_1^1, \cdots, b_1^d$ are the generalized eigenvectors of matrices $V_1^T U_1^T S_b U_1 V_1$ and Σ_w corresponding to the d largest nonzero eigenvlaues. Then, $U_1 V_1 b_1^1, \cdots, U_1 V_1 b_1^d$ are the "regular" discriminant vectors found in CLDA.

3.2   Connections to Pseudoinverse LDA

Pseudo-inverse is a common way to deal with singularity problems. When a matrix is singular, the inverse of the matrix does not exist. However, the pseudo-inverse of any matrix is well defined. Moreover, the pseudo-inverse of a matrix coincides with its inverse when it is invertible. Pseudoinverse LDA (PLDA) applys the eigen-decomposition to the matrix $S_w^+ S_b$.

Suppose b is a generalized eigenvector of $\hat{S}_b$ and $\hat{S}_w + V \Sigma_r V^T$ corresponding to the nonzero eigenvalue λ, i.e. $\hat{S}_b b = \lambda (\hat{S}_w + V \Sigma_r V^T) b$. Since $\hat{S}_w + V \Sigma_r V^T$ is positive definite, b is also the eigenvector of $(\hat{S}_w + V \Sigma_r V^T)^{-1} \hat{S}_b$ corresponding to the nonzero eigenvalue λ, i.e. $(\hat{S}_w + V \Sigma_r V^T)^{-1} \hat{S}_b b = \lambda b$. Recall that $\hat{S}_b = U_1^T S_b U_1$ and $\hat{S}_w + V \Sigma_r V^T = V \hat{\Sigma} V^T$. It follows that

$$U_1 (V \hat{\Sigma} V^T)^{-1} U_1^T S_b U_1 b = \lambda U_1 b.$$

In the original space, the corresponding discriminant vector of b is $\varphi = U_1 b$. Hence,

$$U_1 (V \hat{\Sigma} V^T)^{-1} U_1^T S_b \varphi = \lambda \varphi, \tag{6}$$

which implies that φ is an eigenvector of $U_1 (V \hat{\Sigma} V^T)^{-1} U_1^T S_b$. Since V is orthogonal, we have

$$(V \hat{\Sigma} V^T)^{-1} = V(\hat{\Sigma})^{-1} V^T = V \begin{pmatrix} (\Sigma_w + \sigma I_s)^{-1} & 0 \\ 0 & \alpha^{-1} I_{t-s} \end{pmatrix} V^T. \tag{7}$$

When $\sigma = 0$ and α tends to ∞ , (7) becomes

$$(V \hat{\Sigma} V^T)^{-1} = V(\hat{\Sigma})^{-1} V^T = V \begin{pmatrix} \Sigma_w^{-1} & 0 \\ 0 & 0 \end{pmatrix} V^T.$$

Recall that $V = (V1, V2)$. It follows that

$$U_1 (V \hat{\Sigma} V^T)^{-1} U_1^T = (U_1 V_1) \Sigma_w^{-1} (U_1 V_1)^T. \tag{8}$$

Recall that $S_w = P \Sigma_3 P^T$, $P = (P_1, P_2, P_3)$, $\Sigma_3 = \text{diag}(\Sigma_w, 0, 0)$ and $P_1 = U_1 V_1$. We have

$$(S_w)^+ = (P_1, P_2, P_3) \begin{pmatrix} \Sigma_w^{-1} & 0 & 0 \\ 0 & 0 & 0 \\ 0 & 0 & 0 \end{pmatrix} (P_1, P_2, P_3)^T = (U_1 V_1)(\Sigma_w)^{-1}(U_1 V_1)^T. \quad (9)$$

From (6), (8) and (9), we can find that φ is an eigenvector of $(S_w)^+ S_b$ corresponding to the nonzero eigenvalue λ, i.e. $S_w^+ S_b \varphi = \lambda \varphi$. When $\sigma = 0$ and α tends to ∞, φ is also a "regular" discriminant vector in the range space of the within-class scatter matrix. So, PLDA finds discriminative vectors in the range space of the within-class scatter matrix, and ignores discriminative information in the null space of the within-class scatter matrix.

4   Efficient Algorithm for SRLDA via QR Decomposition

The subspace $R(X)$ spanned by the columns of X contains the range space of the total scatter matrix, where $X = (x_1, \cdots, x_n) \in \mathbb{R}^{m \times n}$ is the training data set. All optimal discriminant vectors can be derived from $R(X)$. For SSS problems, the number of dimensions m is usually much larger than the total number of points n, i.e. $m >> n$, and training data points usually are linear independent. So, the dimension of the space spanned by the columns of X usually is n. An orthonormal basis of $R(X)$ can be obtained through the QR decomposition.

Let $X = WR$ be the QR decomposition of X, where $W = (w_1, w_2, \cdots, w_n) \in \mathbb{R}^{m \times n}$ has orthonormal columns, and $R \in \mathbb{R}^{n \times n}$ is an upper triangular. Then, the vectors $w_1, w_2, \cdots, w_n$ form an orthonormal basis of $R(X)$.

The steps for the SRLDA algorithm are as follows:

1. Calculate the QR decomposition of X: $X = WR$;
2. Map training data set: $Z = W^T X$;
3. Calculate the between-class scatter matrix $\hat{S}_b$ and the within-class scatter matrix $\hat{S}_w$;
4. Regularize the within-class scatter matrix: $\hat{S}_w = \hat{S}_w + V\Sigma_r V^T$;
5. Work out d generalized eigenvectors $b_1, b_2, \cdots, b_d$ of matrices $\hat{S}_b$ and $\hat{S}_w$ corresponding to the d largest nonzero generalized eigenvlaues;
6. Calculate the transformation matrix $A = WB$, where $B = (b_1, \cdots, b_d)$.

The rank of the between-class matrix S_b is at most $c - 1$. In practice, c centroids in the data set are usually linearly independent. In this case, the number of retained dimensions is $d = c - 1$.

The time complexity of the algorithm can be analyzed as follows: Step 1 takes $O(mn^2)$ time for the QR decomposition. Step 2 takes $O(mn^2)$ time for multiplication of two matrices. Step 3 takes $O(n^3)$ time to calculate $\hat{S}_w$ and $\hat{S}_b$. The time complexity of Step 4 is $O(n^3)$, since the algorithm has to compute the eigen-decomposition of a n by n matrix and calculate the multiplication of three matrices. Step 5 calculates d nonzero generalized eigenvalues and their corresponding generalized eigenvectors of a n by n matrix, hence the time complexity of Step 5 is $O(dn^2)$. Finally, in Step 6, it takes $O(mnd)$ time for matrix multiplication. Recall that the number of dimensions m is usually much larger than the total number of points n. Therefore, the total time complexity of the algorithm is $O(mn^2)$.

5 Experiments

In this section, experiments are carried out to compare SRLDA with DCV, CLDA, and PLDA, in terms of classification accuracy. The data sets for experiments are presented in section 5.1. Experimental setting is described in section 5.2. In section 5.3, experiment results are reported.

5.1 Datasets

Face recognition is a well-known SSS problem and experiments are carried out on three face databases: the CMU PIE face database, the Yale face database, and the ORL face database.

The CMU PIE database has more than 40,000 face images of 68 people. Under varying pose, illumination, and expression, the face images were captured. In the experiments, we randomly select 20 persons and 80 images for each person to compare the performance of different algorithms.

The Yale Center for Computational Vision and Control constructed the Yale face database. The database has 165 face images of 15 people. The images have different variations including lighting condition, facial expression.

The ORL face database contains 40 people and there are ten face images for each people. All the images were taken against a dark homogeneous background with the subjects in an upright, frontal position (with tolerance for some side movement). Some images were captured at different times and demonstrate variations in expression and facial details.

Original images were normalized (in scale and orientation) such that the two eyes were aligned at the same position. The size of each face image is 32×32 pixels.

5.2 Experimental Setting

We partition the image sets into training and testing set with different numbers. To simplify representation, Gm/Pn is used to denotes that we randomly select m images per person for training and the remaining n images are for testing. For each Gm/Pn, we average the classification results over 50 random splits and report the mean as well as the standard deviation.

The Euclidean distance is used. Since the size of each face image is 32×32 pixels, each image is concatenated into a 1024-dimensional vector. Testing images are mapped to a low-dimensional space via the optimal transformation learned from training image. Then, the testing images are classified by the nearest neighbor criterion.

For CLDA, we search for the best coefficient to fuse regular and irregular discriminant feature vectors for classification and report the best classification accuracies. For SRLDA, we let $\sigma = 2^r$ and $\alpha = 2^g$, where r and g vary from -15 to 15 incremented by 1. We report the best classification performance of SRLDA.

5.3   Experimental Results

Table 1 shows the classification accuracies and the standard deviations of different algorithms on the three face image data sets: PIE, Yale, and ORL. Furthermore, we also plot the classification performance of SRLDA under different values of the regularization parameters on the PIE database for G15/P65 in Figs. 1, 2, and 3.

From the results in Table 1, we observe that:

1) DCV always achieved better performance than PLDA. The reason may be that in SSS problems, the discriminative information in the null space of the within-class scatter matrix is more important than the discriminative information in the range space of the within-class scatter matrix for classification. With the increasing number of training samples, the difference in classification accuracy between DCV and PLDA decreased, which implies that the discriminative information in the range space of the within-class scatter matrix increases with the increasing number of training samples.

Table 1. Recognition accuracies and standard deviations (%)

Method	PIE				
	G5/P75	G8/P72	G10/P70	G12/P68	G15/P65
SRLDA	81.27(5.37)	**88.30(3.11)**	**90.37(3.71)**	**92.95(2.11)**	**94.35(1.89)**
DCV	80.99(5.49)	87.59(3.26)	89.13(4.28)	91.62(2.50)	92.89(2.13)
PLDA	67.49(6.34)	77.74(4.63)	81.76(4.88)	86.23(3.12)	89.20(2.44)
CLDA	**81.40(5.42)**	87.98(3.23)	89.49(4.11)	91.94(2.36)	93.15(2.07)

Method	Yale				
	G3/P8	G4/P7	G5/P6	G6/P5	G7/P4
SRLDA	**71.43(3.60)**	**78.42(4.00)**	**82.56(3.30)**	84.99(3.87)	86.77(3.39)
DCV	**71.43(3.60)**	78.23(4.10)	82.44(3.28)	84.85(4.11)	86.77(3.39)
PLDA	38.05(4.12)	50.69(5.31)	61.87(4.48)	68.85(5.34)	74.67(4.97)
CLDA	69.98(3.67)	77.30(4.37)	81.93(3.41)	**85.01(3.58)**	**86.93(3.25)**

Method	ORL				
	G3/P7	G4/P6	G5/P5	G6/P4	G7/P3
SRLDA	89.24(1.97)	**93.69(1.58)**	**96.37(1.24)**	**97.29(1.57)**	**97.93(1.37)**
DCV	88.84(1.99)	93.09(1.70)	95.76(1.35)	96.52(1.92)	97.23(1.70)
PLDA	79.02(2.54)	88.79(2.60)	93.04(1.92)	94.67(1.60)	96.00(1.74)
CLDA	**89.29(1.91)**	93.53(1.59)	96.07(1.41)	96.80(1.72)	97.43(1.65)

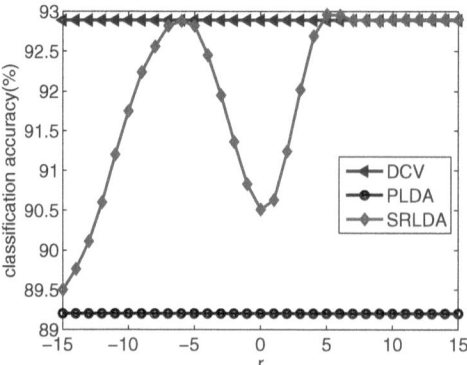

Fig. 1. Classification accuracies of SRLDA under different σ's, where $\sigma = 2^r$ and $\alpha = 1$

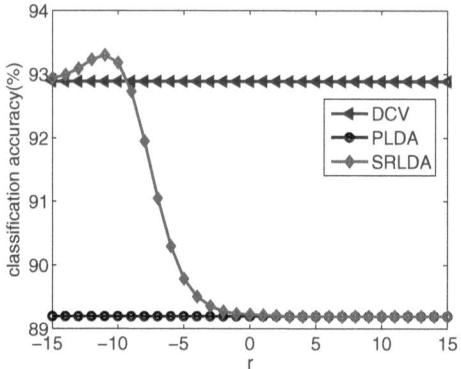

Fig. 2. Classification accuracies of SRLDA under different α's, where $\alpha = 2^r$ and $\sigma = 0$

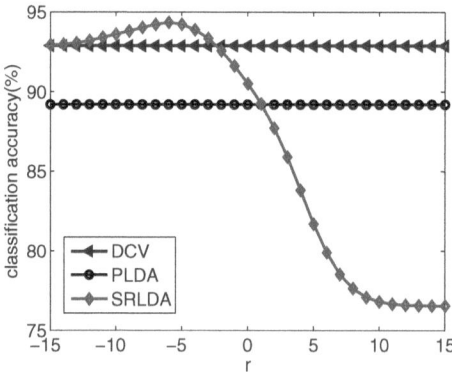

Fig. 3. Classification accuracies of SRLDA under different σ's and α's, where $\alpha = \sigma = 2^r$

2) On the dataset PIE, SRLDA achieved poorer performance than CLDA for G5/P75 and outperformed CLDA for other cases. On the dataset ORL, SRLDA achieved poorer performance than CLDA for G3/P8 and outperformed CLDA for other cases. On the Yale dataset, SRLDA achieved poorer performance than CLDA for G6/P5 and G7/P4, and outperformed CLDA for other cases. Generally speaking, SRLDA achieved better performance than CLDA.

3) On the dataset PIE and the dataset ORL, the classification accuracy differences between DCV and PLDA were small, and SRLDA and CLDA achieved better performance than DCV and PLDA. However, on the Yale dataset, the differences in classification accuracy between DCV and PLDA were large for G3/P8, G4/P7, and G5/P6, and CLDA achieved poorer performance than DCV. The reason may be that when the difference in classification accuracy between DCV and PLDA is large, it is difficult for CLDA to fuse irregular discriminative vectors and regular discriminative vectors for classification. However, SRLDA can more efficiently fuse two kinds of discriminative information in the range and the null space of the within-class scatter matrix and achieved better performance than CLDA or similar performance as CLDA.

4) From the standard deviation of accuracies shown in the table, we find that SRLDA is more stable than other algorithms.

From the results in Figs. 1, 2, and 3., we observe that:

1) When $\alpha = 1$, and $\sigma = 2^r (r \geq 7)$, the performance of SRLDA was the same as DCV. The underlying reason is that when $\alpha = 1$, with the increasing value of σ, SRLDA reduces to DCV and finds discriminant vectors in the null space of the within-class scatter matrix.

2) When $\sigma = 0$, and $\alpha = 2^r (r \geq 3)$, the performance of SRLDA was the same as PLDA. The underlying reason is that when $\sigma = 0$, with the increasing value of α, SRLDA reduces to PLDA and finds discriminant vectors in the range space of the within-class scatter matrix.

3) When $\sigma = \alpha = 2^r (r \leq -3)$, SRLDA fused two kinds of discriminant information in the null and the range space of the within-class scatter matrix to find discriminative vectors for classification and achieved better performance than DCV and PLDA. In order to fuse two kinds of discriminative information for classification, small values should be set to α and σ.

6   Conclusion

In this paper, SRLDA was proposed to solve SSS problems. By using different parameter values in the regularization of the within-class matrix, SRLDA are closely related to several extended LDA methods, including DCV, CLDA, and PLDA. An efficient algorithm to perform SRLDA was developed, which performs RLDA in the space spanned by the data samples and finds an orthonormal basis of the space through the QR decomposition. Experimental results showed that SRLDA can efficiently fuse discriminative information in the range and the null space of the within-class scatter matrix and finds optimal discriminant vectors for classification.

Acknowledgments. This study is supported by the Fundamental Research Funds for the Central Universities under the grant (No. 2011121010).

References

1. Fukunaga, K.: Introduction to Statistical Pattern Recognition, 2nd edn. Academic Press, New York (1990)
2. Friedman, J.H.: Regularized discriminant analysis. J. Am. Stat. Assoc. 84(405), 165–175 (1989)
3. Dai, D.Q., Yuen, P.C.: Regularized discriminant analysis and its application to face recognition. Pattern Recognition 36, 845–847 (2003)
4. Chen, L.F., Liao, H.Y.M., Lin, J.C., Kao, M.D., Yu, G.J.: A New LDA-Based Face Recognition System which Can Solve the Small Sample Size Problem. Pattern Recognition 33(10), 1713–1726 (2000)
5. Cevikalp H., Neamtu M., Wilkes M., Barkana A.: Discriminative common vectors for face recognition. IEEE Trans. Pattern Anal. Mach. Intell. 27 (1), 4C13 (2005)
6. Yang, J., Yang, J.Y.: Why Can LDA Be Performed in PCA Transformed Space? Pattern Recognition 36(2), 563–566 (2003)
7. Yang, J., Frangi, A.F., Yang, J.Y., Zhang, D., Jin, Z.: KPCA plus LDA: A Complete Kernel Fisher Discriminant Framework for Feature Extraction and Recognition. IEEE Trans. Pattern Analysis and Machine Intelligence 27(2), 230–244 (2005)
8. Yu, H., Yang, J.: A Direct LDA Algorithm for High-Dimensional Data-With Application to Face Recognition. Pattern Recognition 34(10), 2067–2070 (2001)
9. Liu, J., Chen, S.C., Tan, X.Y.: A study on three linear discriminant analysis based methods in small sample size problem. Pattern Recognition 41, 102–116 (2008)
10. Schölkopf, B., Herbrich, R., Smola, A.J.: A Generalized Representer Theorem. In: Helmbold, D.P., Williamson, B. (eds.) COLT 2001 and EuroCOLT 2001. LNCS (LNAI), vol. 2111, pp. 416–426. Springer, Heidelberg (2001)

A Particle Swarm Optimisation Based Multi-objective Filter Approach to Feature Selection for Classification

Bing Xue[1,2], Liam Cervante[1], Lin Shang[2], and Mengjie Zhang[1]

[1] Victoria University of Wellington, P.O. Box 600, Wellington 6140, New Zealand
{Bing.Xue,Liam.Cervante,Mengjie.Zhang}@ecs.vuw.ac.nz
[2] State Key Laboratory of Novel Software Technology,
Nanjing University, Nanjing 210046, China
shanglin@nju.edu.cn

Abstract. Feature selection (FS) has two main objectives of minimising the number of features and maximising the classification performance. Based on binary particle swarm optimisation (BPSO), we develop a multi-objective FS framework for classification, which is *NSBPSO* based on multi-objective BPSO using the idea of non-dominated sorting. Two multi-objective FS algorithms are then developed by applying mutual information and entropy as two different filter evaluation criteria in the proposed framework. The two proposed multi-objective algorithms are examined and compared with two single objective FS methods on six benchmark datasets. A decision tree is employed to evaluate the classification accuracy. Experimental results show that the proposed multi-objective algorithms can automatically evolve a set of non-dominated solutions to reduce the number of features and improve the classification performance. Regardless of the evaluation criteria, NSBPSO achieves higher classification performance than the single objective algorithms. NSBPSO with entropy achieves better results than all other methods. This work represents the first study on multi-objective BPSO for filter FS in classification problems.

Keywords: Feature Selection, Particle Swarm Optimisation, Multi-Objective Optimisation, Filter Approaches.

1 Introduction

Feature selection (FS) is an important pre-processing technique for effective data analysis in many areas such as classification. In classification, without prior knowledge, relevant features are usually difficult to determine. Therefore, a large number of features are often involved, but irrelevant and redundant features may even reduce the classification performance due to the unnecessarily large search space. FS can address this problem by selecting only relevant features for classification. By eliminating/reducing irrelevant and redundant features, FS could reduce the number of features, shorten the training time, simplify the learned classifiers, and/or improve the classification performance [1].

FS algorithms explore the search space of different feature combinations to reduce the number of features and optimise the classification performance. They have two key factors: the evaluation criterion and the search strategy. Based on the evaluation criterion, existing FS approaches can be broadly classified into two categories: wrapper

P. Anthony, M. Ishizuka, and D. Lukose (Eds.): PRICAI 2012, LNAI 7458, pp. 673–685, 2012.
© Springer-Verlag Berlin Heidelberg 2012

approaches and filter approaches. In wrapper approaches, a learning/classification algorithm is used as part of the evaluation function to determine the goodness of the selected feature subset. Wrappers can usually achieve better results than filters approaches, but the main drawbacks are their computational deficiency and loss of generality [2]. Filter approaches use statistical characteristics of the data for evaluation and the FS search process is independent of a learning/classification algorithm. Compared with wrappers, filter approaches are computationally less expensive and more general [1].

The search strategy is a key factor in FS because of the large search space (2^n for n features). In most situations, it is impractical to conduct an exhaustive search [2]. A variety of search strategies have been applied to FS. However, existing FS methods still suffer from different problems such as stagnation in local optima and high computational cost [3, 4]. Therefore, an efficient global search technique is needed to better address FS problems. Particle swarm optimisation (PSO) [5, 6] is one of the relatively recent evolutionary computation techniques, which are well-known for their global search ability. Compared with genetic algorithms (GAs) and genetic programming (GP), PSO is computationally less expensive and can converge more quickly. Therefore, PSO has been used as an effective technique in many fields, including FS in recent years [3, 4, 7].

Generally, FS has two main objectives of minimising both the classification error rate and the number of features. These two objectives are usually conflicting and the optimal decision needs to be made in the presence of a trade-off between them. However, most existing FS approaches are single objective algorithms and belong to wrapper approaches, which are less general and computationally more expensive than filter approaches. There has been no work conducted to use PSO to develop a multi-objective filter FS approach to date.

The overall goal of this paper is to develop a new PSO based multi-objective filter approach to FS for classification for finding a set of non-dominated solutions, which contain a small number of features and achieve similar or even better classification performance than using all features. To achieve this goal, we will develop a multi-objective binary PSO framework, *NSBPSO*, and apply two information measurements (mutual information and entropy) to the proposed framework. These proposed FS algorithms will be examined on six benchmark tasks/problems of varying difficulty. Specifically, we will investigate

- whether using single objective BPSO and the two information measurements can select a small number of features and improve classification performance over using all features;
- whether NSBPSO with mutual information can evolve a set of non-dominated solutions, which can outperform all features and the single objective BPSO with mutual information; and
- whether NSBPSO with entropy can outperform all other methods above.

2   Background

2.1   Particle Swarm Optimisation (PSO)

PSO is an evolutionary computation technique proposed by Kennedy and Eberhart in 1995 [5, 6]. Candidate solutions in PSO are encoded as particles. Particles move in the search space to search for the best solution by updating their positions according to the experience of a particle itself and its neighbours. $x_i = (x_{i1}, x_{i2}, ..., x_{iD})$ and $v_i = (v_{i1}, v_{i2}, ..., v_{iD})$ represent the position and velocity of particle i, where D is the dimensionality of the search space. *pbest* represents the best previous position of a particle and *gbest* represents the best position obtained by the swarm so far. PSO starts with random initialisations of a population of particles and searches for the optimal solution by updating the velocity and the position of each particle according to the following equations:

$$x_{id}^{t+1} = x_{id}^t + v_{id}^{t+1} \tag{1}$$

$$v_{id}^{t+1} = w * v_{id}^t + c_1 * r_{1i} * (p_{id} - x_{id}^t) + c_2 * r_{2i} * (p_{gd} - x_{id}^t) \tag{2}$$

where t denotes the tth iteration. d denotes the dth dimension. w is inertia weight. c_1 and c_2 are acceleration constants. r_{1i} and r_{2i} are random values uniformly distributed in [0, 1]. p_{id} and p_{gd} represent the elements of *pbest* and *gbest*. v_{id}^t is limited by a predefined maximum velocity, v_{max} and $v_{id}^t \in [-v_{max}, v_{max}]$.

PSO was originally proposed to address continuous problems [5]. Later, Kennedy and Eberhart [8] developed a binary PSO (BPSO) to solve discrete problems. In BPSO, x_{id}, p_{id} and p_{gd} are restricted to 1 or 0. v_{id}^t in BPSO indicates the probability of the corresponding element in the position vector taking value 1. A sigmoid function is used to transform v_{id} to the range of (0, 1). BPSO updates the position of each particle according to the following formula:

$$x_{id} = \begin{cases} 1, & \text{if } rand() < \frac{1}{1+e^{-v_{id}}} \\ 0, & otherwise \end{cases} \tag{3}$$

where $rand()$ is a random number chosen from a uniform distribution in [0,1].

2.2   Entropy and Mutual Information

Information theory developed by Shannon [9] provides a way to measure the information of random variables with entropy and mutual information. The entropy is a measure of the uncertainty of random variables. Let X be a random variable with discrete values, its uncertainty can be measured by entropy $H(X)$:

$$H(X) = -\sum_{x \in \mathcal{X}} p(x) \log_2 p(x) \tag{4}$$

where $p(x) = Pr(X = x)$ is the probability density function of X.

For two discrete random variables X and Y with their probability density function $p(x, y)$, the joint entropy $H(X, Y)$ is defined as

$$H(X, Y) = -\sum_{x \in \mathcal{X}, y \in \mathcal{Y}} p(x, y) \log_2 p(x, y) \tag{5}$$

When a variable is known and others are unknown, the remaining uncertainty is measured by the conditional entropy. Given Y, the conditional entropy $H(X|Y)$ of X with respect to Y is

$$H(X|Y) = - \sum_{x \in \mathcal{X}, y \in \mathcal{Y}} p(x,y) \log_2 p(x|y) \tag{6}$$

where $p(x|y)$ is the posterior probabilities of X given Y. If X completely depends on Y, then $H(X|Y)$ is zero, which means that no more other information is required to describe X when Y is known. On the other hand, $H(X|Y) = H(X)$ denotes that knowing Y will do nothing to observe X.

The information shared between two random variables is defined as mutual information. Given variable X, mutual information $I(X;Y)$ is how much information one can gain about variable Y.

$$
\begin{aligned}
I(X;Y) &= H(X) - H(X|Y) \\
&= - \sum_{x \in \mathcal{X}, y \in \mathcal{Y}} p(x,y) \log_2 \frac{p(x,y)}{p(x)p(y)}
\end{aligned} \tag{7}
$$

According to Equation 7, the mutual information $I(X;Y)$ will be large if X and Y are closely related. $I(X;Y) = 0$ if X and Y are totally unrelated.

2.3 Multi-objective Optimisation

Multi-objective optimisation involves minimising or maximising multiple conflicting objective functions. The formulae of a k-objective minimisation problem with multiple objective functions can be written as follows:

$$minimise \ \ F(x) = [f_1(x), f_2(x), \ ... \ , f_k(x)] \tag{8}$$

subject to:

$$g_i(x) \leqslant 0, \ (i = 1, 2, \ ... \ m) \quad and \quad h_i(x) = 0, \ (i = 1, 2, \ ... \ l) \tag{9}$$

where x is the vector of decision variables, $f_i(x)$ is a function of x, $g_i(x)$ and $h_i(x)$ are the constraint functions of the problem.

In multi-objective optimisation, the quality of a solution is explained in terms of trade-offs between the conflicting objectives. Let y and z be two solutions of the above k-objective minimisation problem. If the following conditions are met, one can say y dominates z:

$$\forall i : f_i(y) \leqslant f_i(z) \quad and \quad \exists j : f_j(y) < f_j(z) \tag{10}$$

where $i, j \in \{1, 2, ..., k\}$. When y is not dominated by any other solutions, y is referred as a Pareto-optimal solution. The set of all Pareto-optimal solutions forms the trade-off surface in the search space, the Pareto front. A multi-objective algorithm is designed to search for a set of non-dominated solutions.

2.4   Related Work on FS

A number of FS algorithms have been recently proposed [1] and typical FS algorithms are reviewed in this section.

Traditional FS Approaches. The Relief algorithm [10] is a classical filter FS algorithm. Relief assigns a weight to each feature to denote the relevance of the feature to the target concept. However, Relief does not deal with redundant features, because it attempts to find all relevant features regardless of the redundancy between them. The FOCUS algorithm [11] exhaustively examines all possible feature subsets, then selects the smallest feature subset. However, it is computationally inefficient because of the exhaustive search.

Two commonly used wrapper FS methods are sequential forward selection (SFS) [12] and sequential backward selection (SBS) [13]. SFS (SBS) starts with no features (all features), then candidate features are sequentially added to (removed from) the initial feature subset until the further addition (removal) does not increase the classification performance. The limitation of SFS and SBS is that once a feature is selected (eliminated) it cannot be eliminated (selected) later, which is so-called nesting effect. Stearns addressed this limitation by proposing the "plus-l-take away-r" to perform l times forward selection followed by r times backward elimination [14]. However, the optimal values of (l, r) are difficult to determine.

Evolutionary Computation Algorithms (Non-PSO) for FS. Recently, evolutionary computation techniques have been applied to FS problems. Based on GAs, Chakraborty [15] proposes a FS algorithm using a fuzzy-set based fitness function. However, BPSO with the same fitness function achieves better performance than this GA based algorithm. Hamdani et al. [16] develop a multi-objective FS algorithm using non-dominated sorting based multi-objective genetic algorithm II (NSGAII), but its performance has not been compared with any other FS algorithm.

Muni et al. [17] develop a multi-tree GP algorithm for FS (GPmtfs) to simultaneously select a feature subset and design a classifier using the selected features. For a c-class problem, each classifier in GPmtfs has c trees. Comparisons suggest GPmtfs achieves better results than SFS, SBS and other methods. However, the number of features selected increases when there are noisy features. Kourosh and Zhang [18] propose a GP relevance measure (GPRM) to evaluate and rank subsets of features in binary classification tasks, and GPRM is also efficient in terms of FS.

PSO Based FS Approaches. PSO has recently gained more attention for solving FS problems. Wang et al. [19] propose a filter FS algorithm based on an improved binary PSO and rough sets. Each particle is evaluated by the dependency degree between class labels and selected features, which is measured by rough sets. This work also shows that the computation of the rough sets consumes most of the running time, which is a drawback of using rough sets in FS problems.

Azevedo et al. [20] propose a FS algorithm using PSO and support vector machines (SVM) for personal identification in a keystroke dynamic system. However, the proposed algorithm obtains a relatively high false acceptance rate, which should be low in most identification systems. Mohemmed et al. [7] propose a FS method (PSOAdaBoost) based on PSO and an AdaBoost framework. PSOAdaBoost simultaneously searches for the best feature subset and determines the decision thresholds of AdaBoost. Liu et al. [4] introduce multi-swarm PSO to search for the optimal feature subset and optimise the parameters of SVM simultaneously. The proposed FS method achieved better classification accuracy than grid search, standard PSO and GA. However, it is computationally more expensive than the other three methods because of the large population size and complicated communication rules between different subswarms.

A variety of FS approaches have been proposed, but most of them treat FS as a single objective problem. Although Hamdani et al. [16] develop a NSGAII based multi-objective algorithm, there is no comparison to test its performance. Studies have shown that PSO is an efficient technique for FS, but the use of PSO for multi-objective FS has never been investigated. Moreover, most existing approaches are wrappers, which are computationally expensive and less general than filter approaches. Therefore, investigation of a PSO based multi-objective filter FS approach is still an open issue and we make an effort in this paper.

3   Proposed Multi-objective FS Algorithms

Two filter measurements based on mutual information and entropy [21] are firstly described. Then we propose a new multi-objective BPSO framework, which forms two new algorithms to address FS problems.

3.1   Mutual Information and Entropy for FS

Mutual information can be used in FS to evaluate the relevance between a feature and the class labels and the redundancy between two features. In [21], we proposed a BPSO based filter FS algorithm (PSOfsMI) using mutual information to evaluate the relevance and redundancy in the fitness function (Equation 11). The objectives are to maximise the relevance between features and class labels to improve the classification performance, and to minimise the redundancy among features to reduce the number of features.

$$Fitness_1 = D_1 - R_1 \tag{11}$$

where

$$D_1 = \sum_{x \in X} I(x; c), \quad and \quad R_1 = \sum_{x_i, x_j \in X} I(x_i, x_j).$$

where X is the set of selected features and c is the class labels. Each selected feature and the class labels are treated as discrete random variables. D_1 calculates the mutual information between each feature and the class labels, which determine the relevance of the selected feature subset to the class labels. R_1 evaluates the mutual information shared by each pair of selected features, which indicates the redundancy contained in the selected feature subset.

Mutual information can find the two-way relevance and redundancy in FS, but could not handle multi-way complex feature interaction, which is one of the challenges in FS. Therefore, a group evaluation using entropy was proposed in [21] to discover multi-way relevance and redundancy among features. A single objective filter FS algorithm

(PSOfsE) [21] was then developed based on the group evaluation and BPSO, where Equation 12 was used as the fitness function.

$$Fitness_2 = D_2 - R_2 \qquad (12)$$

where

$$D_2 = IG(c|X) \quad and \quad R_2 = \frac{1}{|X|} \sum_{x \in X} IG(x|\{X/x\})$$

where X and c have the same meaning as in Equation 11. D_2 evaluates the information gain in c given information of the features in X, which show the relevance between the selected feature subset and the class labels. R_2 evaluates the joint entropy of all the features in X, which indicates the redundancy in the selected feature subset. Detailed calculation of D_2 and R_2 is given in [21].

The representation of a particle in PSOfsMI and PSOfsE is a n-bit binary string, where n is the number of available features in the dataset and also the dimensionality of the search space. In the binary string, "1" represents that the feature is selected and "0" otherwise.

3.2   New Algorithms: NSfsMI and NSfsE

PSOfsMI and PSOfsE [21] have shown that mutual information or entropy can be an effective measurement for filter FS. Therefore, we develop a multi-objective filter FS approach based on BPSO and mutual information (or entropy) with the objectives of minimising the number of features and maximising the relevance between features and class labels. Standard PSO could not be directly used to address multi-objective problems because it was originally proposed for single objective optimisation. In order to use PSO to develop a multi-objective FS algorithm, one of the most important tasks is to determine a good leader ($gbest$) for each particle from a set of potential non-dominated solutions. NSGAII is one of the most popular evolutionary multi-objective techniques [22]. Li [23] introduces the idea of NSGAII into PSO to develop a multi-objective PSO algorithm and achieves promising results on several benchmark functions.

In this study, we develop a binary multi-objective PSO framework (NSBPSO) for filter FS based on the idea of non-dominated sorting. Two filter multi-objective FS algorithms are then developed based on NSBPSO, which are NSfsMI using D_1 to evaluate the relevance between features and class labels, and NSfsE using D_2 to measure the relevance. Algorithm 1 shows the pseudo-code of NSfsMI and NSfsE. The main idea is to use non-dominated sorting to select a $gbest$ for each particle and update the swarm in the evolutionary process. As shown in Algorithm 1, in each iteration, the algorithm firstly identifies the non-dominated solutions in the swarm and calculates the crowding distance, then all the non-dominated solutions are sorted according to the crowding distance. For each particle, a $gbest$ is randomly selected from the highest ranked part of the sorted non-dominated solutions, which are the least crowded solutions. After determining the $gbest$ and $pbest$ for each particle, the new velocity and the new position of each particle are calculated according to the equations. The old positions (solutions) and the new positions of all particles are combined into one union. The non-dominated solutions in the union are called the first non-dominated front, which are excluded from the union. Then the non-dominated solutions in the new union are called the second

---

Algorithm 1. Pseudo-Code of NSfsMI and NSfsE

---

begin
   divide $Dataset$ into a Training set and a Test set, initialise the swarm ($Swarm$);
   while $Maximum\ Iterations$ *is not met* **do**
      evaluate two objective values of each particle;   /* number of features
      and the relevance (D_1 in NSfsMI and D_2 in NSfsE) on
      the Training set */
      identify the particles ($nonDomS$) (non-dominated solutions in $Swarm$);
      calculate crowding distance particles in $nonDomS$ and then sort them;
      for $i=1$ **to** *Population Size (P)* **do**
         update the *pbest* of particle i;
         randomly select a *gbest* for particle i from the highest ranked solutions in
         $nonDomS$;
         update the velocity and the position of particle i;
      add the original particles $Swarm$ and the updated particles to $Union$;
      identify different levels of non-dominated fronts $F = (F_1, F_2, F_3, ...)$ in
      $Union$;
      empty the $Swarm$ for the next iteration;
      $i = 1$;
      while $|Swarm| < P$ **do**
         if $(|Swarm| + |F_i| \leq P)$ **then**
            calculate crowding distance of each particle in F_i;
            add F_i to $Swarm$;
            $i = i + 1$;
         if $(|Swarm| + |F_i| > P)$ **then**
            calculate crowding distance of each particle in F_i;
            sort particles in F_i;
            add the $(P - |Swarm|)$ least crowded particles to $Swarm$;
   calculate the classification error rate of the solutions (feature subsets) in the F_1 on the
   test set;
   return the solutions in F_1 and their testing classification error rates;

---

non-dominated front. The following levels of non-dominated fronts are identified by repeating this procedure. For the next iteration, solutions (particles) are selected from the top levels of the non-dominated fronts, starting from the first front.

4   Experimental Design

Table 1 shows the six datasets used in the experiments, which are chosen from the UCI machine learning repository [24]. The six datasets were selected to have different numbers of features, classes and instances and they are used as representative samples of the problems that the proposed algorithms will address. In the experiments, the instances in each dataset are randomly divided into two sets: 70% as the training set and 30% as the test set. All FS algorithms firstly run on the training set to select feature subsets and then the classification performance of the selected features will be calculated on the test set by a learning algorithm. There are many learning algorithms that can be used here,

Table 1. Datasets

Dataset	Type of the Data	#Features	#Classes	#Instances
Lymphography (Lymph)	Categorical	18	4	148
Mushroom	Categorical	22	2	5644
Spect	Categorical	22	2	267
Leddisplay	Categorical	24	10	1000
Soybean Large	Categorical	35	19	307
Connect4	Categorical	42	3	44473

such as K-nearest neighbour (KNN), NB, and DT. A DT learning algorithm is selected in this study to calculate the classification accuracy.

In all FS algorithms, the fully connected topology is used, $v_{max} = 6.0$, the population size is 30 and the maximum iteration is 500. $w = 0.7298$, $c_1 = c_2 = 1.49618$. These values are chosen based on the common settings in the literature [6]. Each algorithm has been conducted for 40 independent runs on each dataset.

For each dataset, PSOfsMI and PSOfsE obtain a single solution in each of the 40 runs. NSfsMI and NSfsE obtain a set of non-dominated solutions in each run. In order to compare these two kinds of results, 40 solutions in PSOfsMI and PSOfsE are presented in the next section. 40 sets of feature subsets achieved by each multi-objective algorithm are firstly combined into one union set. In the union set, for the feature subsets including the same number of features (e.g. m), their classification error rates are averaged. Therefore, a set of average solutions is obtained by using the average classification error rates and the corresponding number of features (e.g. m). The set of average solutions is called the *average* Pareto front and presented in the next section. Besides the average Pareto front, the non-dominated solutions in the union set are also presented in the next section.

5   Results and Discussions

Figures 1 and 2 show the results of NSfsMI and PSOfsMI, NSfsE and PSOfsE. On the top of each chart, the numbers in the brackets show the number of available features and the classification error rate using all features. In each chart, the horizontal axis shows the number of features selected and the vertical axis shows the classification error rate. In the figures, "-A" stands for the average Pareto front and "-B" represents the non-dominated solutions resulted from NSfsMI and NSfsE in the 40 independent runs. "PSOfsMI" and "PSOfsE" show the 40 solutions achieved by PSOfsMI and PSOfsE.

In some datasets, PSOfsMI or PSOfsE may evolve the same feature subset in different runs and they are shown in the same point in the chart. Therefore, although 40 results are presented, there may be fewer than 40 distinct points shown in a chart. For "-B", each of these non-dominated solution sets may also have duplicate feature subsets, which are shown in the same point in a chart.

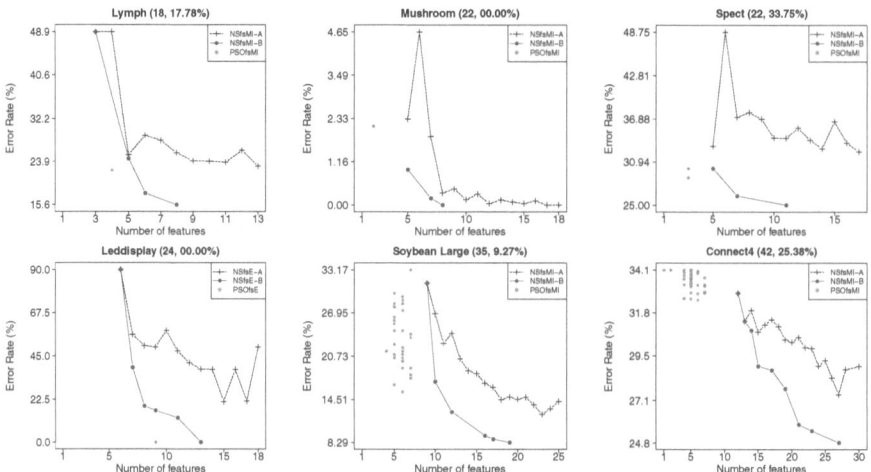

Fig. 1. Experimental Results of PSOfsMI and NSfsMI

5.1 Results of NSfsMI

According to Figures 1, it can be seen that on average, PSOfsMI reduced around 75% of the available features in most cases although the classification error rates are slightly higher than using all features in some datasets.

Figure 1 shows that in three datasets, the average Pareto fronts of NSfsMI (NSfsMI-A) include two or more solutions, which selected a smaller number of features and achieved a lower classification error rate than using all features. For the same number of features, there are a variety of combinations of features with different classification performances. The feature subsets obtained in different runs may include the same number of features but different classification error rates. Therefore, although the solutions obtained in each run are non-dominated, some solutions in the average Pareto front may dominate others. This also happens in NSfsE. In almost all datasets, the non-dominated solutions (NSfsMI-B) include one or more feature subsets, which selected less than 50% of the available features and achieved better classification performance than using all features. For example, in the Spect dataset, one non-dominated solution selected 11 features from 22 available features and the classification error rate was decreased from 33.75% to 25.00%. The results suggests that NSfsMI as a multi-objective algorithm can automatically evolve a set of feature subsets to reduce the number of features and improve the classification performance.

Comparing NSfsMI with PSOfsMI, it can be seen that in most cases, NSfsMI achieved better classification performance than PSOfsMI although the number of features are slightly larger. Comparisons show that with *mutual information* as the evaluation criterion, the proposed multi-objective FS algorithms, NSfsMI can outperform single objective FS algorithm (PSOfsMI).

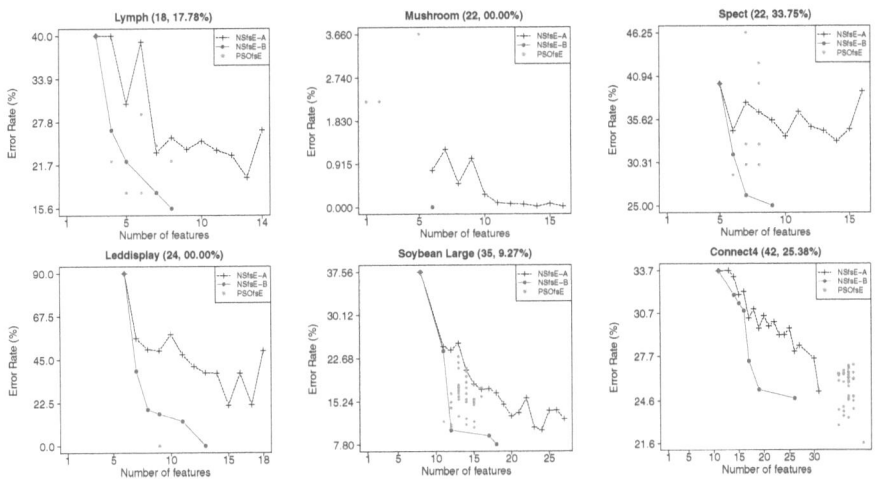

Fig. 2. Experimental Results of PSOfsE and NSfsE.

5.2   Results of NSfsE

Figure 2 shows that PSOfsE selected around half of the available features and achieved similar or even better classification performance than using all features in most cases.

Figure 2 shows that in most cases, NSfsE-A contains more than one solution that selected a smaller number of features and achieved better classification performance than using all features. In almost all datasets, NSfsMI-B reduced the classification error rate by only selecting around half of available features. Take the Spect dataset as an example, NSfsE reduced the classification error rate from 33.75% to 25.00% by selecting only 9 features from the 22 available features. The results suggest that the proposed NSfsE with *entropy* as the evaluation criterion can evolve a set of feature subsets to simultaneously improve the classification performance and reduce the number of features.

Comparing NSfsE with PSOfsE, it can be observed that NSfsE outperformed PSOfsE because NSfsE achieved better classification performance than PSOfsE in all datasets although NSfsE selected slightly more features than PSOfsE in most cases. Comparisons show that with *entropy* as the evaluation criterion, the proposed multi-objective FS algorithms (NSfsE) can achieve better solutions than single objective FS algorithm (PSOfsE).

5.3   Further Comparisons

Comparing *mutual information* and *entropy*, Figures 1 and 2 show that PSOfsE and NSfsE using entropy usually achieved better classification performance than PSOfsMI and NSfsMI using mutual information. PSOfsMI using mutual information usually selected a smaller number of features than PSOfsE using entropy. The proposed multi-objective algorithms, NSfsE usually evolved a smaller number of features and achieved better classification performance than NSfsMI. The comparisons suggest that the algorithms

with entropy as the evaluation criterion can discover the multiple-way relevancy and redundancy among a group of features to further increase the classification performance.
Because the evaluation is based on a group of features (instead of a pair of features), the
number of features involved is usually larger in PSOfsE than PSOfsMI. However, the
number of features in the proposed multi-objective algorithms is always small because
they can explore the search space more effectively to minimise the number of features.
Moreover, NSfsE can utilise the discovered multiple-way relevancy to simultaneously
increase the classification performance.

6   Conclusions

This paper aimed to propose a filter multi-objective FS approach based on BPSO to
search for a small number of features and achieve high classification performance. The
goal was successfully achieved by developing two multi-objective FS algorithms (NS-
fsMI and NSfsE) based on two multi-objective BPSO (NSBPSO) and two information
evaluation criteria (mutual information and entropy). The proposed algorithms were
examined and compared with two BPSO based single objective FS algorithms, namely
PSOfsMI and PSOfsE, based on mutual information and entropy on six benchmark
datasets. The experimental results show that in almost all cases, the proposed multi-
objective algorithms are able to automatically evolve a Pareto front of feature subsets,
which included a small number of features and achieved better classification performance than using all features. NSfsMI and NSfsE achieved better classification performance than BPSOfsMI and BPSOfsE in most cases.

The proposed multi-objective FS algorithms can achieve a set of good feature subsets, but it is unknown whether the achieved Pareto fronts can be improved or not. In
the future, we will further investigate the multi-objective PSO based filter FS approach
to better address FS problems.

Acknowledgments. This work is supported in part by the National Science Foundation
of China (NSFC No. 61170180) and the Marsden Fund of New Zealand (VUW0806).

References

[1] Dash, M., Liu, H.: Feature selection for classification. Intelligent Data Analysis 1(4), 131–
    156 (1997)
[2] Kohavi, R., John, G.H.: Wrappers for feature subset selection. Artificial Intelligence 97,
    273–324 (1997)
[3] Unler, A., Murat, A.: A discrete particle swarm optimization method for feature selection in
    binary classification problems. European Journal of Operational Research 206(3), 528–539
    (2010)
[4] Liu, Y., Wang, G., Chen, H., Dong, H.: An improved particle swarm optimization for feature
    selection. Journal of Bionic Engineering 8(2), 191–200 (2011)
[5] Kennedy, J., Eberhart, R.: Particle swarm optimization. In: IEEE International Conference
    on Neural Networks, vol. 4, pp. 1942–1948 (1995)
[6] Shi, Y., Eberhart, R.: A modified particle swarm optimizer. In: IEEE International Confer-
    ence on Evolutionary Computation, CEC 1998, pp. 69–73 (1998)

[7] Mohemmed, A., Zhang, M., Johnston, M.: Particle swarm optimization based adaboost for face detection. In: IEEE Congress on Evolutionary Computation, CEC 2009, pp. 2494–2501 (2009)

[8] Kennedy, J., Eberhart, R.: A discrete binary version of the particle swarm algorithm. In: IEEE International Conference on Systems, Man, and Cybernetics. Computational Cybernetics and Simulation, vol. 5, pp. 4104–4108 (1997)

[9] Shannon, C., Weaver, W.: The Mathematical Theory of Communication. The University of Illinois Press, Urbana (1949)

[10] Kira, K., Rendell, L.A.: A practical approach to feature selection. In: Assorted Conferences and Workshops, pp. 249–256 (1992)

[11] Almuallim, H., Dietterich, T.G.: Learning boolean concepts in the presence of many irrelevant features. Artificial Intelligence 69, 279–305 (1994)

[12] Whitney, A.: A direct method of nonparametric measurement selection. IEEE Transactions on Computers C-20(9), 1100–1103 (1971)

[13] Marill, T., Green, D.: On the effectiveness of receptors in recognition systems. IEEE Transactions on Information Theory 9(1), 11–17 (1963)

[14] Stearns, S.: On selecting features for pattern classifier. In: Proceedings of the 3rd International Conference on Pattern Recognition, Coronado, CA, pp. 71–75 (1976)

[15] Chakraborty, B.: Genetic algorithm with fuzzy fitness function for feature selection. In: IEEE International Symposium on Industrial Electronics, ISIE 2002, vol. 1, pp. 315–319 (2002)

[16] Hamdani, T.M., Won, J.-M., Alimi, M.A.M., Karray, F.: Multi-objective Feature Selection with NSGA II. In: Beliczynski, B., Dzielinski, A., Iwanowski, M., Ribeiro, B. (eds.) ICANNGA 2007, Part I. LNCS, vol. 4431, pp. 240–247. Springer, Heidelberg (2007)

[17] Muni, D., Pal, N., Das, J.: Genetic programming for simultaneous feature selection and classifier design. IEEE Transactions on Systems, Man, and Cybernetics, Part B: Cybernetics 36(1), 106–117 (2006)

[18] Neshatian, K., Zhang, M.: Genetic Programming for Feature Subset Ranking in Binary Classification Problems. In: Vanneschi, L., Gustafson, S., Moraglio, A., De Falco, I., Ebner, M. (eds.) EuroGP 2009. LNCS, vol. 5481, pp. 121–132. Springer, Heidelberg (2009)

[19] Wang, X., Yang, J., Teng, X., Xia, W., Jensen, R.: Feature selection based on rough sets and particle swarm optimization. Pattern Recognition Letters 28(4), 459–471 (2007)

[20] Azevedo, G., Cavalcanti, G., Filho, E.: An approach to feature selection for keystroke dynamics systems based on PSO and feature weighting. In: IEEE Congress on Evolutionary Computation, CEC 2007, pp. 3577–3584 (2007)

[21] Cervante, L., Xue, B., Zhang, M., Lin, S.: Binary particle swarm optimisation for feature selection: A filter based approach. In: IEEE Congress on Evolutionary Computation, CEC 2012, pp. 889–896 (2012)

[22] Deb, K., Pratap, A., Agarwal, S., Meyarivan, T.: A fast and elitist multiobjective genetic algorithm: NSGA-II. IEEE Transactions on Evolutionary Computation 6(2), 182–197 (2002)

[23] Li, X.: A Non-dominated Sorting Particle Swarm Optimizer for Multiobjective Optimization. In: Cantú-Paz, E., Foster, J.A., Deb, K., Davis, L., Roy, R., O'Reilly, U.-M., Beyer, H.-G., Kendall, G., Wilson, S.W., Harman, M., Wegener, J., Dasgupta, D., Potter, M.A., Schultz, A., Dowsland, K.A., Jonoska, N., Miller, J., Standish, R.K. (eds.) GECCO 2003, Part I. LNCS, vol. 2723, pp. 37–48. Springer, Heidelberg (2003)

[24] Frank, A., Asuncion, A.: UCI machine learning repository (2010)

Exploiting Independent Relationships in Multiagent Systems for Coordinated Learning

Chao Yu, Minjie Zhang, and Fenghui Ren

School of Computer Science and Software Engineering,
University of Wollongong, Wollongong, 2522, NSW, Australia
cy496@uowmail.edu.au, {minjie,fren}@uow.edu.au

Abstract. Creating coordinated multiagent policies in an environment with uncertainties is a challenging issue in the research of multiagent learning. In this paper, a coordinated learning approach is proposed to enable agents to learn both individual policies and coordinated behaviors by exploiting independent relationships inherent in many multiagent systems. We illustrate how this approach is employed to solve coordination problems in robot navigation domains. Experimental results of different scales of domains prove the effectiveness of our learning approach.

Keywords: Multiagent Learning, Coordinated Learning, Independence, Decentralized Markov Decision Processes.

1 Introduction

Multiagent Learning (MAL) is an important issue in the research of Multiagent Systems (MASs), finding increasing applications in a variety of domains ranging from robotics, distributed traffic control, resource management to automated trading, etc. [6]. In current literature, many MAL problems are modeled as Decentralized Markov Decision Processes (Dec-MDPs), which extend single-agent Markov Decision Processes (MDPs) to model distributed decision making processes of multiple agents. However, due to the joint transition/reward functions shared among all the agents and agents' partial observability of the environment as well as limited communication capability, constructing an efficient coordinated learning policy in MASs is recognized as a challenging problem.

There have been a profusion of approaches proposed to capture the structure features in Dec-MDPs for MAL with different assumptions and emphasis. One such approach is based on Multiagent MDPs (MMDPs) [12], in which each agent receives the same reward and has a full joint-state observability. MMDPs represent fully cooperative multiagent tasks and the assumption of a full joint-state observability implies that the learning process in MMDPs can be conducted in a centralized way. Other approaches include hierarchical MAL approaches which exploit Multi-agent Semi-MDP (MSMDP) model to decompose the task into subtasks hierarchically [8], and Coordination Graph (CG) based approaches that capture the local interactions among agents in Dec-MDPs so that a jointly optimal action can be determined without considering every possible action in

P. Anthony, M. Ishizuka, and D. Lukose (Eds.): PRICAI 2012, LNAI 7458, pp. 686–697, 2012.
© Springer-Verlag Berlin Heidelberg 2012

an exponentially large joint action space [11,9]. All these approaches are focused on agents' learning of a coordinated policy under an assumption that an explicit representation of coordination necessities, either through a predefined task decomposition or a fixed interaction structure, is given beforehand. However, learning the necessities of coordination is not addressed in these approaches.

In this paper, we propose a MAL approach to solve coordination problems in robot navigation domains. Our approach is based on the claim that many real-world multiagent domains exhibit a large amount of context-specific independence, thus can be represented compactly. Learning of coordination in theses domains can be solved efficiently by exploiting independent relationships among agents combined with factored representations [7]. To reflect this feature, we firstly introduce the uncertainties of transition and reward independence in factored Dec-MDPs, with the aim to decompose the decision making process more efficiently than that in general Dec-MDPs. We then propose a learning approach to enable agents to learn when and how to coordinate with each other by exploiting the independent relationships among them. Unlike the aforementioned approaches, we do not assume prior knowledge on the structure of the problem or each agent's global observability of the joint state information of all agents, so as to make our approach more suitable for practical applications. We give an illustration of how this approach is applied to solve the coordination problems in robot navigation domains. Experimental results show that our approach can achieve a near-optimal performance with low computational complexity based only on each agent's local observability of the environment.

The rest of this paper is structured as follows. Section 2 formally presents the concept of independence in factored Dec-MDPs. Section 3 introduces our learning approach in detail and illustrates how to apply it in robot navigation domains. The experimental results are presented in Section 4. Section 5 compares our approach with some related approaches and Section 6 concludes the paper.

2   Independence in Factored Dec-MDPs

A Dec-MDPs model can be defined by a tuple $N = (\{\alpha_i\}, S, \{A_i\}, P, R)$, where $\alpha_i(i \in [1, n])$ represents an agent; S is a finite set of joint states of all agents; $A_i(i \in [1, n])$ is a finite set of actions available to agent α_i; $P(s, a, s')$ represents the transition probability from state s to state s' when the joint action $a = \langle a_1, ..., a_n \rangle (a_i \in A_i)$ is taken and $R(s, a)$ represents the expected reward received by all agents for taking the joint action a in state s. We call a n-agent Dec-MDPs a factored Dec-MDPs when the system state can be factored into $n + 1$ distinct components so that $S = S_0 \times S_1 \times ... \times S_n$, where S_0 denotes an agent-independent component of the state, $S_i(i \in [1, n])$ is the state space of agent α_i.

A factored, n-agent Dec-MDPs model is *transition independent* [4] if the overall transition function P can be separated into $n + 1$ distinct functions $P_0, ..., P_n$ as $P = P_0 \prod_{i=1}^{n} P_i$, where each $P_i(i \in [1, n])$ stands for the individual transition function of agent α_i, P_0 is an initial transition component. Let $\hat{s}_i = \langle s_0, s_i \rangle$ $(s_i \in S_i)$ represent the *local state* of agent α_i and $s_{-i} = \langle s_0, s_1, ...s_{(i-1)}, s_{(i+1)}..., s_n \rangle$

represent the sequence of state components of all agents except agent α_i. For any next state $s'_i \in S_i$ of agent α_i, P_i is given by Equation 1.

$$P_i(\langle s_0, ..., s_n \rangle, \langle a_1, ..., a_n \rangle, \langle s'_i, s'_{-i} \rangle) = \begin{cases} P_0(s_0, s'_0), & \text{if i=0;} \\ P_i(\hat{s}_i, a_i, \hat{s'_i}), & \text{else.} \end{cases} \quad (1)$$

Similarly, a factored Dec-MDPs model is *reward independent* [4] if the overall reward function R can be represented as a function of $R_1, ..., R_n$ by Equation 2, where each R_i $i \in [1, n]$ represents the individual reward function of agent α_i.

$$R(\langle s_0, ..., s_n \rangle, \langle a_1, ..., a_n \rangle) = f(R_1(\hat{s}_1, a_1), ..., R_n(\hat{s}_n, a_n)) \quad (2)$$

As can be seen from the factorizations given by Equation 1 and Equation 2, general *transition* and *reward independent* Dec-MDPs decompose the overall functions P, R to individual functions P_i, R_i that depend only on each agent's local information, i.e., local state $\hat{s}_i$ and individual action a_i. However, an agent's local information usually cannot determine the individual functions fully as each agent's individual functions are potentially affected by other agents. As such, in [5], the authors parameterized an agent's individual functions on the overall state and action of all the agents, and decomposed an individual function to a local individual component based on the agent's local information $(\hat{s}_i, a_i)$ and an interaction component based on all the agents' information (s, a), where $s = \langle s_i, s_{-i} \rangle, a = \langle a_i, a_{-i} \rangle$. The factorization in [5], however, assumed that the states can be divided in a black-and-white way, i.e., the agents are either completely independent or completely dependent on each other in a state. This handling of independence is always intractable due to the complexity of Dec-MDPs and lack of prior knowledge of the domain structure. To better reflect the uncertainties of independent relationships among agents, we introduce transition/reward independent degree in the factorization of the individual functions as given by Equation 3 and Equation 4.

$$P_k(\langle s_k, s_{-k} \rangle, \langle a_k, a_{-k} \rangle, \langle s'_k, s'_{-k} \rangle) = \mu P_k(\hat{s}_k, a_k, \hat{s'_k}) +$$
$$(1 - \mu)P^I(\langle s_k, s_{-k} \rangle, \langle a_k, a_{-k} \rangle, \langle s'_k, s'_{-k} \rangle) \quad (3)$$

$$R_k(\langle s_k, s_{-k} \rangle, \langle a_k, a_{-k} \rangle) = \nu(R_k(\hat{s}_k, a_k)) + (1 - \nu)R^I(\langle s_k, s_{-k} \rangle, \langle a_k, a_{-k} \rangle) \quad (4)$$

where the individual transition(reward) function $P_k(R_k)$ is decomposed to a local individual function $P_k(\hat{s}_k, a_k, \hat{s'_k})(R_k(\hat{s}_k, a_k))$ and an interaction function $P^I(R^I)$, combined with a transition(reward) independent degree $\mu(\nu) \in [0, 1]$ signifying the uncertainties of the transition(reward) independent relationships between agent k and the remaining agents.

The overall independent degree, which is signified by the value functions, thus can be formally defined by Definition 1.

Definition 1 (Independent Degree). *Given a factored Dec-MDPs model that satisfies Equation 3 and Equation 4, agent α_k is independent of the remaining agents by an independent degree $\xi \in [0, 1]$ in state $s_k \in S_k$ if the individual value* *function can be decomposed according to Equation 5, where ξ is confined by the values of μ and ν, i.e., $(\xi - \mu)(\xi - \nu) \leq 0$.*

$$V_k(\langle s_k, s_{-k}\rangle, \langle a_k, a_{-k}\rangle) = \xi V_k(\hat{s}_k, a_k) + (1 - \xi)V^I(\langle s_k, s_{-k}\rangle, \langle a_k, a_{-k}\rangle) \quad (5)$$

An independent degree is a value signifying the extent that an agent is independent of other agents. The larger this value, the more independent this agent is of other agents. V_k (V^I) is the expected local (interaction) value function, which reflects the real independent level of agent α_k regarding other agents.

It is noted that, for all agents, if $\xi = 1$ holds in every state, namely, both the transition and reward independent are completely achieved for these agents, a Dec-MDPs problem reduces to a set of independent MDP problems, each of which can be solved separately, reducing the complexity of this class of problems to P-complete of standard MDPs. In other situations, the Dec-MDPs problem becomes a NEXP-complete puzzle, even for the simplest scenario involving only two agents [10]. To bring down the complexity, a straightforward way is to provide the agents with sufficient information to overcome the uncertainties caused by the non-stationary environment. This means allowing the agents to observe other agents' information for decision making, either through a global observability of the environment or an unlimited communication capability (as in Multiagent MDP (MMDP) [12]). However, keeping all the agents' information during learning soon becomes intractable as the search space grows exponentially in the number of agents and communication/observability are always restricted in real-life applications. A solution to dodge this dilemma is to let agents learn to use other agents' information for coordination only when it is necessary. The independent degrees that capture different levels of independence in Dec-MDPs signify the extent of such a necessity, therefore can be exploited for more efficient decision making by an agent without considering all the information of other agents, most of which is redundant as indicated by the context-specific independence. In the following section, a learning approach is proposed to solve coordination problems in robot navigation domains through dynamically approximating the independent degrees during a learning process.

3   Solving Robot Navigation Problems

Consider three simple domains in which two robots R_1, R_2 are navigating in an environment, each trying to reach its respective goal G_1 and G_2 in Figure 1. This kind of domains can be modeled as Dec-MDPs, in which each robot has to make a decision to optimize a global reward function through its local observation. It is showed that even if the states, where full dependent relationships exist, are known beforehand to the robots, learning to achieve an efficient coordination policy is a challenging task [1]. Intuitively, robots have different independent degrees in these domains. Each robot can choose its action individually when the independent degrees are high. However in some specific situations when robots have low independent degrees, they will mutually affect each other's transitions and rewards, therefore the other robot's information should be taken into account when the robot makes its own decision.

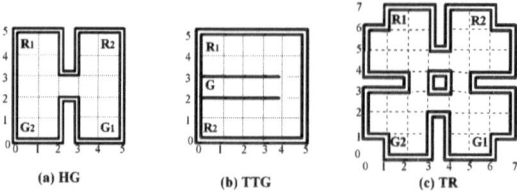

Fig. 1. Three small robot navigation domains

3.1 The Principle of the Learning Approach

We propose a MAL approach to estimate the optimal policy when the model is not well specified. We say a model is not well specified if the transition and reward functions as well as the independent degrees are unknown to the agents. When agent α_i has an independent degree of ξ_i^k in state s_i^k, it will apply independent learning with a probability of ξ_i^k and otherwise coordinated learning to combine other agents' information for decision making with a probability of $1 - \xi_i^k$. Here, we assume that the coordinated learning process imposes no cost to the agents, i.e., whether or not to coordinate is completely determined by the independent degrees. An agent starts with an initial belief that there is no dependencies between other agents, and then learn to approximate the independent degrees through interactions with others to achieve efficient coordinated policies.

Algorithm 1. General learning approach for agent α_i

---
1 Initialize learning information, for all state $s_i^k \in S_i$, $\xi_i^k \leftarrow 1$;
2 **if** *agent α_i is in state s_i^k* **then**
3   with probability of ξ_i^k: *independent learning*;
4   with probability of $1 - \xi_i^k$: *coordinated learning*;
5   Receive penalized reward r_i and search for causes of the reward;
6   **for** *each state $s_i^j \in S_i$* **do**
7   Calculate similarity $\zeta(s_i^k, s_i^j)$ and update eligibility trace ε_i^j;
8   Adjust ξ_i^j according to $\zeta(s_i^k, s_i^j)$ and ε_i^k based on the spreading function $f_{s_i^k}^{r_i}(s_i^j)$;
9   **end**
10 **end**

The sketch of our learning approach is given by Algorithm 1. The basic idea is to adjust the independent degrees according to an immediate penalized reward received by an agent. A penalized reward caused by a conflict signifies that coordination is required in the corresponding state, thus the independent degree should decrease. However, a reward received in a state s is contributed implicitly by all other states in the domain, with the extent of affect determined by the similarity between those states and state s. Let $\bar{s}$ be the attribute vector that describes a state s. The similarity $\zeta_{\langle s^i, s^j \rangle}$ between two local states s^i and s^j can be given by $\zeta_{\langle s^i, s^j \rangle} = \|\bar{s^i} - \bar{s^j}\|$. In robot navigation problems, similarity between two local states is reflected by the Euclidean distance between these two states.

Based on the similarity of states, the cause of a penalized reward can be spread to all the states through a *spreading function* defined as below.

Definition 2 (Spreading Function). *A spreading function of reward r received in s^*, $f_{s^*}^r(s) \in [0,1]$, is a function where the following conditions hold:*

- *Condition 1.* $\exists s_1^*, \forall s^i \in S, s^i \neq s_1^* : f_{s^*}^r(s_1^*) > f_{s^*}^r(s^i)$ *and* $\exists s_2^*, \forall s^i \in S, s^i \neq s_2^* : f_{s^*}^r(s_2^*) > f_{s^*}^r(s^i) \Rightarrow s_1^* = s_2^* = s^*$;
- *Condition 2.* $\forall s^i, s^j : \zeta_{\langle s^i, s^* \rangle} > \zeta_{\langle s^j, s^* \rangle} \Rightarrow f_{s^*}^r(s^i) > f_{s^*}^r(s^j)$.
- *Condition 3.* $\forall s^i, s^j : \zeta_{\langle s^i, s^* \rangle} = \zeta_{\langle s^j, s^* \rangle} \Rightarrow f_{s^*}^r(s^i) = f_{s^*}^r(s^j)$;

Spreading function is a valid representation that enables to reflect the contribution each state s in the domain makes to a penalized reward received in conflicting state s^*. When a reward r is achieved in state s^*, $f_{s^*}^r(s)$ has the highest value signifying that reward r is mainly caused by state s^* (Condition 1). As the similarity between s and s^* decreases, state s has a lower effect on causing reward r, which is reflexed by the lower value of $f_{s^*}^r(s)$ (Condition 2).

The spreading function gives an overall estimation of the cause of a reward when agents are facing uncertainties, assuming that states having the same similarities with the conflicting state s^* play the same role of causing the reward (Condition 3). However, in many cases, the role of these states can be different. Agents should assign credit or penalty to those states which are *eligible* for the resulting rewards. As such, *eligibility trace (ET)* is introduced to make a temporary record of the occurrence of an event that a penalized reward is received. Let S^c be the state trajectory that causes an event, ε_i^j be the ET value of agent α_i in state s_i^j, then for each state $s_i^k \in S_i$, the ET value can be updated as Equation 6, where $\gamma \in [0,1)$ is a discount rate and $\lambda \in [0,1]$ is the trace-decay parameter.

$$\varepsilon_i^k(t+1) = \begin{cases} \gamma\lambda\varepsilon_i^k(t) + 1, & s_i^k \in S^c; \\ \gamma\lambda\varepsilon_i^k(t), & else. \end{cases} \tag{6}$$

Based on the spreading function and eligibility trace, a penalized reward $r(t)$ that signifies the necessity of coordination can be spread among all the states that are potentially eligible for this penalized outcome. The independent degree then can be adjusted according to $\xi_i^k(t+1) = \daleth(\psi_i^k(t+1))$, where $\psi_i^k(t+1)$ is a value signifying the necessity of coordination and can be updated by $\psi_i^k(t+1) = \psi_i^k(t) + \varepsilon_i^k(t)f_{s^*(t)}^{r(t)}(s_i^k)$, where $s^*(t)$ is the state resulting in the penalized reward $r(t)$ and $\daleth(x)$ is a normalization function that maps the value x to interval $[0,1]$, with lower value of x corresponding higher $\daleth(x)$.

3.2   An Explicit Learning Algorithm

A concrete learning algorithm derived from the general approach is proposed in Algorithm 2 in the perspective of robot i, where $Q_i(s_i, a_i)$ is the single-state Q-value table, $Q_c(js_i, a_i)$ is the joint-state Q-value table, ξ_i^k and ε_i^k ($k = 1, ..., m$) are the independent degree and eligibility trace value when robot i is in state s_i^k, respectively, and *trajectory_list* is a list to store the trajectory of the robot.

Algorithm 2. Coordinated learning algorithm for robot i

---

1 Initialize $Q_i(s_i, a_i)$; $Q_c(js_i, a_i) \leftarrow \emptyset$; $\xi_i^k(t) \leftarrow 1$; $\varepsilon_i^k(t) \leftarrow 0$; $trajectory\_list \leftarrow \emptyset$;
2 **for** *each episode n (n=1,...,E)* **do**
3   **for** *each step t (t=1,...,T)* **do**
4   Generate a random number $\tau, \tau \in [0, 1]$;
5   **if** $\xi(s_i) \leq \tau$ **then**
6   **if** *agent j is in vision* **then**
7                  $PerceptionFlag$ = True and read s_j, $js_i \leftarrow \langle s_i, s_j \rangle$;
8   **if** js_i not in table $Q_c(js_i, a_i)$ **then**
9   add js_i to table $Q_c(js_k, a_k)$
10   **end**
11   select $a_i(t)$ from $Q_c(js_i, a_i)$;
12   **else** select $a_i(t)$ from $Q_i(s_i, a_i)$;
13   **else** select $a_i(t)$ from $Q_i(s_i, a_i)$;
14         $tracjecotry\_list.add(s_i)$, transit to state s_i' and receive $r_i(t)$;
15   **if** $\xi(s_i) \geq \tau$ and PerceptionFlag = $True$ **then**
16   Update $Q_c(js_i, a_i) \leftarrow Q_c(js_i, a_i) + \alpha[r_i(t) + \gamma \max_{a_i'} Q_i(s_i', a_i') - Q_c(js_i, a_i)]$;
17   **else**
18   Update $Q_i(s_i, a_i) \leftarrow Q_i(s_i, a_i) + \alpha[r_i(t) + \gamma \max_{a_i'} Q_i(s_i', a_i') - Q_i(js_i, a_i)]$;
19   **end**
20   Call Algorithm 3 to adjust $\xi_i^k(t)$, $s_i \leftarrow s_i'$;
21   **end**
22 **end**

Algorithm 3. Adjusting the independent degree $\xi_i^k(t)$

---

1 Input: time step t, $\xi_i^k(t-1)$, eligible trace value $\varepsilon_i^k(t-1)$, $(k = 1, ..., m)$, $trajectory\_list$;
2 **if** *A collision occurs in state $s_i(t)$* **then**
3      $max_{temp} \leftarrow 0$;
4   **for** *each $s_i^k \in S_i$* **do**
5   **if** $s_i^k \in tracjecotry\_list$ **then** $\varepsilon_i^k(t) \leftarrow \gamma^\lambda \varepsilon_i^k(t-1) + 1$;
6   **else** $\varepsilon_i^k(t) \leftarrow \gamma^\lambda \varepsilon_i^k(t-1)$;
7          $\psi_i^k(t) \leftarrow \psi_i^k(t-1) + \frac{\varepsilon_i^k(t)}{\sqrt{2\pi}} e^{-\frac{1}{2}[\frac{(x_k - x_u)^2}{\delta_1^2} + \frac{(y_k - y_u)^2}{\delta_2^2}]}$;
8   **if** $\psi_i^k(t) \geq max_{temp}$ **then** $max_{temp} = \psi_i^k(t)$;
9   **end**
10   **for** *each state $s_i^k \in S_i$* **do**
11         $\xi_i^k(t) = 1 - \frac{\psi_i^k(t)}{max_{temp}}$;
12   **end**
13 **end**

Learning robot i chooses its action to decide whether to coordinate with other robots based on the independent degree. If the robot chooses action *coordinate*, it will activate its perception process to determine the local state information of other robots. After the successful perception process, robot i makes use of the local state information from other robots to choose its action based on the joint-state Q-value table $Q_c(js_i, a_i)$. Otherwise, it makes its decision only based on single-state Q-value table $Q_i(s_i, a_i)$. After each transition, robot i updates its independent degrees based on the updating process given by Algorithm 3.

In Algorithm 3, $s_i^k(x_k, y_k)$ is a local state with coordinate (x_k, y_k), (x_u, y_u) is the coordinate of the conflicting state $s_i(t)$, and ψ is a value to signify the necessity of coordination. As the similarity between s_i^k and $s_i(t)$ decreases (i.e., the Euclidean distance increases), the value of spreading function decreases. To simplify expression, we choose a two-dimensional normal distribution function as the spreading function. As for the normalization function $\daleth(x)$, we simply choose the maximum ψ after each episode, which is denoted as max_{temp}, and let the values of ψ in other states be compared with max_{temp} to confine ψ to $[0, 1]$.

4   Experiment

In this section, we test the performance of our learning approach, which is denoted as *IDL* (*Independent Degree Learning*), and compare it with other two approaches. The first approach is to let each agent learn its policy independently, which is denoted as *IL* (*Independent Learning*). The second approach is *JSAL* (*Joint State Action Learning*), in which agents either communicate unlimitedly with a central controller or have a full observability of the environment to receive the joint-state-action information of all agents to guide the learning process.

We firstly test these approaches in three small grid-world domains given by Figure 1. These approaches are then applied to several larger domains given by Figure 3 in [1]. Robots navigate with 4 actions, i.e., Move "East", "South", "West" and "North". Each action moves the robot to the corresponding direction deterministically. When robots collide into a wall, they will rebound back. If they collide into each other, both are transferred back to their original states. The exploration policy we adopt is the fixed $\varepsilon - greedy$ policy with $\varepsilon = 0.1$. The learning rate $\alpha = 0.05$, discount factor $\gamma = 0.95$, $\lambda = 0.8$. We let the length of a grid in the domain be 1, the scanning distance of robots be 2 and $\delta_1^2 = \delta_2^2 = 1$. Rewards are given as follows: $+20$ for reaching the goal state, -1 for colliding into a wall and -10 for colliding into the other robot. We run all approaches for 10,000 episodes and average the last 2000 episodes to compute the overall performance. All results are then averaged over 25 runs.

Figure 2 (a) - (c) depict the learning processes in small domains. We can see that *JSAL* can obtain an optimal reward because robots learn their policies based on joint-state-action information. On the contrary, *IL* obtains a very low reward due to the lack of coordination. Our approach *IDL* learns independently as the *IL* does at the beginning of learning process and then converges to the optimal value gradually as the robots learn the different independent degrees between them so that coordinated behaviors can be achieved when a low independent degree indicates that coordination is required in a corresponding state.

Table 1 gives the overall results in small domains. We can see that our approach reduces the computational complexity significantly compared with *JSAL* in terms of state and action space. This reduction is more desirable in larger domains where the computational complexity of *JSAL* is too high to be implemented, which can be verified by the results in larger domains given later. The results show that our learning approach can achieve a good performance, which is quite close to the optimal performance of *JSAL*. As for the steps to goals, robots in *IL* always find the shortest paths to their goals, which, in turn, causes the high percentage of collision because they do not coordinate with each

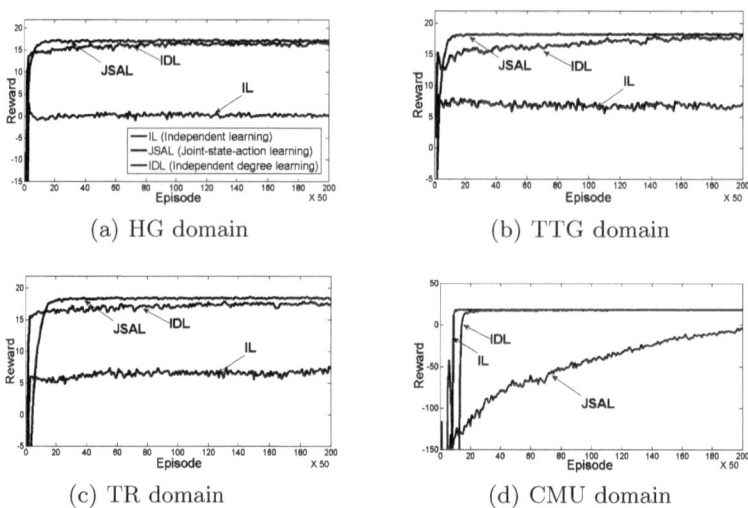

(a) HG domain   (b) TTG domain

(c) TR domain   (d) CMU domain

Fig. 2. Learning processes of different learning approaches

Table 1. Learning performances of different approaches in small scale domains

Domain	Approach	States	Actions	Reward	Collision(%)	Steps
HG	IL	21	4	**0.16**±*0.19*	**0.65**±*0.01*	**12.50**±*0.17*
	IDL	25.40	4	**16.38**±*0.19*	**0.09**±*0.01*	**17.13**±*0.55*
	JSAL	441	16	**17.10**±*0.49*	**0.05**±*0.02*	**21.66**±*2.79*
TTG	IL	25	4	**6.77**±*0.21*	**0.42**±*0.01*	**12.53**±*0.04*
	IDL	29.27	4	**17.66**±*0.10*	**0.04**±*0.00*	**15.78**±*0.22*
	JSAL	625	16	**18.22**±*0.29*	**0.00**±*0.00*	**22.42**±*3.12*
TR	IL	36	4	**6.71**±*2.29*	**0.43**±*0.08*	**13.66**±*0.53*
	IDL	41.01	4	**17.47**±*0.43*	**0.06**±*0.01*	**16.55**±*1.01*
	JSAL	1296	16	**18.19**±*0.46*	**0.01**±*0.01*	**29.94**±*8.63*

other when coming to dangerous states with low independent degrees. On the contrary, *JSAL* can learn a safe detour strategy to reduce the likelihood of collision, which accordingly increases the steps to the goals. The 95% confidence intervals of step number to goals also imply that the learnt policy in *JSAL* is not deterministic and can be affected by the stochastic learning process. However, our approach combines merits of both *IL* and *JSAL*, allowing robots to find the shortest path with a higher certainty while only make small detours in the states where coordination is most needed.

Figure 3 shows the values of $1 - \xi$ of robot R_1 in three small domains. As expected, the values in the conflicting states (i.e., areas around the entrance or doorway where robots are more inclined to have collisions) are much higher than the values of those "safe" states where robot R_1 can generally choose its actions disregarding robot R_2. It is also interesting to notice that the dependent degrees

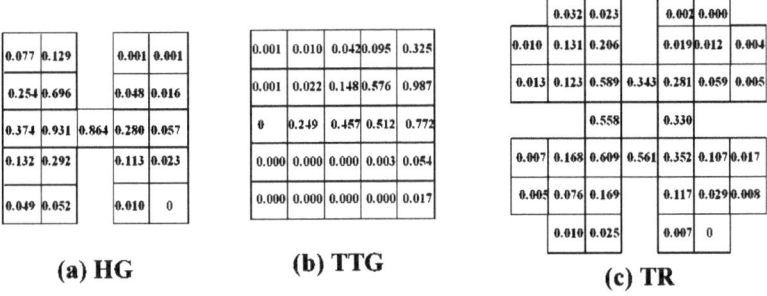

(a) HG   **(b) TTG**   **(c) TR**

Fig. 3. Dependent degrees $(1 - \xi)$ of robot R_1 in small scale domains

Table 2. Learning performances of different approaches in large scale domains

Domain	Approach	State	Action	Reward	Collision(%)	Step
ISR	IL	43	4	**10.11**±*2.61*	**0.32**±*0.09*	**6.28**±*0.26*
	IDL	45.75	4	**15.74**±*0.25*	**0.10**±*0.00*	**10.67**±*0.45*
	JSAL	1849	16	**16.86**±*0.30*	**0.06**±*0.01*	**14.56**±*5.10*
MIT	IL	49	4	**16.84**±*1.23*	**0.08**±*0.02*	**23.15**±*0.98*
	IDL	54.76	4	**17.20**±*0.72*	**0.05**±*0.02*	**25.11**±*1.17*
	JSAL	2401	16	**16.49**±*0.38*	**0.02**±*0.01*	**44.02**±*4.57*
PTG	IL	52	4	**12.18**±*3.75*	**0.24**±*0.13*	**10.04**±*1.06*
	IDL	55.66	4	**17.12**±*0.43*	**0.06**±*0.01*	**11.27**±*0.40*
	JSAL	2704	16	**17.76**±*0.43*	**0.04**±*0.01*	**14.32**±*4.30*
CIT	IL	70	4	**15.10**±*2.98*	**0.12**±*0.10*	**20.87**±*1.80*
	IDL	76.00	4	**16.02**±*1.94*	**0.08**±*0.06*	**22.04**±*1.42*
	JSAL	4900	16	**16.81**±*0.45*	**0.03**±*0.01*	**22.42**±*1.92*
SUNY	IL	74	4	**19.10**±*0.42*	**0.01**±*0.01*	**12.17**±*0.14*
	IDL	79.86	4	**18.80**±*0.64*	**0.02**±*0.02*	**12.94**±*1.00*
	JSAL	5476	16	**17.60**±*0.32*	**0.02**±*0.01*	**26.47**±*3.98*
CMU	IL	133	4	**17.03**±*0.86*	**0.03**±*0.02*	**42.65**±*1.81*
	IDL	141.74	4	**17.69**±*0.69*	**0.02**±*0.02*	**44.04**±*2.15*
	JSAL	17689	16	**-9.96**±*1.41*	**0.05**±*0.01*	**236.50**±*9.62*

in domain TR are comparatively lower than those in the other two domains. This can be easily explained by that robots in domain TR are not so tightly coupled as there are more than one route for the robots to reach their own goals. This is opposite to domain HG and TTG, where only one route is available so that the robots heavily depend on each other for decision making, especially in the conflicting states, causing high values of $1 - \xi$ around these states.

Table 2 gives the learning results in large domains. In these domains, robots have high independent degrees in most states such that coordination actions are not heavily required. This explains why in some domains (e.g., MIT, CIT), *IL* already receives a very good performance. Even in these domains, our learning approach can still improve the performance further with the exception in SUNY, where coordination is not necessary at all. Another interesting finding is that as the state and action sizes grow, the performance of *JSAL* decreases. Figure 2 (d) plots the learning process in CMU domain. We can see that *JSAL* approach converges too slow to reach an optimal value. *IDL* only considers the joint state information when the independent degrees signify this consideration necessary. In this way, the search space is reduced substantially compared with *JSAL*.

In general, our learning approach can learn an efficient coordination policy with a low computational complexity during approximating the independent degrees. Our approach outperforms the uncoordinated approach by considering coordination only when it is necessary and enables robots to learn a shorter path with a higher certainty to the goal than the centralized approach.

5   Related Work

Much attention has been paid to the problem of coordinated learning in recent MAL research. In [8], a hierarchical multi-agent RL algorithm called Cooperative HRL was proposed for agents to learn coordinated skills at the level of subtasks instead of primitive actions. However, each agent is assumed to be given an initial hierarchical decomposition of the overall task. In [11], an coordinated reinforcement learning approach was developed for agents to coordinate their behaviors based on parameterized, structured representation of a value function. However, this approach requires an independent relationship, represented as a Coordination Graph (CG), to be known to all agents. In [9], an algorithm, called Utile Coordination, was introduced for agents to learn where and how to coordinate their behaviors by learning the dependencies among these agents represented by a CG. However, agents need a predefined period of time to collect statistical information to determine the dependent relationships among agents, which means the learning is conducted in an off-line way. The authors in [2] also used an off-line learning approach to determine in which states coordination is most needed, and then agents learn how to coordinate their behaviors by using joint-state-action information of all the agents. Unlike these approaches, our approach does not assume prior knowledge of the domain structure and the independent relationships among agents are learnt by agents during the learning process, which means agents learn where and how to coordinate their actions concurrently.

In [1], a two-layer extension of Q-learning approach was proposed to exploit the sparse interactions in MASs for an efficient coordinated learning. However, the outcome of this approach is strongly dependent on the value of the penalty that the pseudo-coordination action is chosen. In [3], an algorithm, called CQ-learning, was proposed to enable agents to adapt the state representation for efficient coordination. CQ-learning approach differs from our approach in that it depends on the assumption that each agent has already learnt an optimal individual policy such that the agent has a model of its expected rewards to detect the states where coordination is most beneficial by collecting statistics information during learning.

6   Conclusion

In this paper, we proposed a MAL approach to solve coordination problems in MASs. Our approach exploits independent relationships among agents exhibited in many real-world multiagent domains to bring down the overall computational complexity in factored Dec-MDPs. We gave an illustration of how to apply the approach to robot navigation problems. Experimental results showed that robots could learn a near-optimal performance in different scales of domains with a low computational complexity. Although we only tested our learning approach in robot navigation problems in this paper, the proposed learning approach can be applied with small adaptations to more general problems, in which certain extent of independent relationships exit among agents. Currently, we are working on applying the approach to predator-prey and robot soccer domains.

References

1. Melo, F.S., Veloso, M.: Decentralized MDPs with sparse interactions. Artif. Intel. 175, 1757–1789 (2011)
2. Yu, C., Zhang, M., Ren, F.: Coordinated Learning for Loosely Coupled Agents with Sparse Interactions. In: Wang, D., Reynolds, M. (eds.) AI 2011. LNCS, vol. 7106, pp. 392–401. Springer, Heidelberg (2011)
3. De Hauwere, Y.M., Vrancx, P., Nowé, A.: Learning multi-agent state space representations. In: AAMAS 2010, pp. 715–722. IFAAMAS, Richland (2010)
4. Allen, M., Zilberstein, S.: Complexity of decentralized control: Special cases. In: Adv. Neural Inform. Proc. Systems, vol. 22, pp. 19–27 (2009)
5. Spaan, M., Melo, F.S.: Interaction-driven Markov games for decentralized multiagent planning under uncertainty. In: AAMAS 2008, pp. 525–532. IFAAMAS, Richland (2008)
6. Busoniu, L., Babuska, R., De Schutter, B.: A comprehensive survey of multiagent reinforcement learning. IEEE Trans. Syst. Man Cybern. C. Appl. Re. 38(2), 156–172 (2008)
7. Roth, M., Simmons, R., Veloso, M.: Exploiting factored representations for decentralized execution in multiagent teams. In: AAMAS 2007, pp. 469–475. ACM Press, New York (2007)
8. Ghavamzadeh, M., Mahadevan, S., Makar, R.: Hierarchical multi-agent reinforcement learning. In: AAMAS, vol. 13(2), pp. 197–229. Springer, Heidelberg (2006)
9. Kok, J.R., Hoen, P., Bakker, B., Vlassis, N.: Utile coordination: Learning interdependencies among cooperative agents. In: CIG 2005, pp. 29–36. IEEE Press, New York (2005)
10. Bernstein, D.S., Givan, R., Immerman, N., Zilberstein, S.: The complexity of decentralized control of Markov decision processes. Math. Oper. Re. 27(4), 819–840 (2002)
11. Guestrin, C., Lagoudakis, M., Parr, R.: Coordinated reinforcement learning. In: ICML, pp. 227–234 (2002)
12. Boutilier, C.: Planning, learning and coordination in multiagent decision processes. In: TARK, pp. 195–210 (1996)

A Model of Intention with (Un)Conditional Commitments*

Dongmo Zhang

Intelligent Systems Laboratory,
University of Western Sydney, Australia
d.zhang@uws.edu.au

Abstract. This paper proposes a model of intention with conditional/ unconditional commitments based on Cohen and Levesque's (C&L's for short) framework of intention. We first examine C&L's framework with a well-known philosophical puzzle, the Toxin Puzzle, and point out its insufficiency in modelling conditional and unconditional commitments. We then propose a model theory of a specific modal logic with modalities representing the typical mental attributes as well as action feasibility and realisibility. Instead of defining intention as a persistent goal, we define an intention as a conditional/unconditional commitment made for achieving a goal that is believed to be achievable. Finally we check our framework with the Toxin Puzzle and show our solution to the puzzle.

1 Introduction

With the motivation of modelling autonomous agents and multi-agent systems, a great number of logical frameworks, just to name a few, Cohen and Levesque's formalism of intention [1], Rao and Georgeff's BDI logic [2], and Meyer et al.s KARO logic [3], were proposed in the past two decades, attempting to capture the rationality of human agency and imitate rational behaviour of human being for implementation of software agents.

Most of the early works on agent modelling in the AI literature were stimulated by philosophical investigations on human mental states, most significantly Bratman's theory of intention [4]. It has been widely accepted that human rational behaviour is highly effected, if not determined, by their mental attitudes, such as knowledge, belief, desire and intention. However, formalising each of the mental attitudes and their relationships have been proved to be hard and tend to be complicated. This is because any separation of these attitudes leads to insufficiency of explanation in human rational behaviour. The central of these attitudes is human intention, which is in general intertwined with other attitudes, such as belief, desires and commitments, involving reasoning about actions and time [1, 4]. Any formal analysis of intention must explicate the relationships among those facets of mental state. This explains why the existing logics of intention are highly complicated [1, 2, 5, 3, 6–10]. Even worse, there is a theoretical boundary that prevents us from putting all the facets into a single logic system. Schmidt and Tishkovsky showed that if we combine propositional dynamic logic, used for specifying actions and time,

---

* This research was partially supported by the Australian Research Council through Discovery Project DP0988750.

P. Anthony, M. Ishizuka, and D. Lukose (Eds.): PRICAI 2012, LNAI 7458, pp. 698–709, 2012.

with doxastic modal logics, used for specifying knowledge and beliefs, the outcome logic system collapses to propositional dynamic logic if we admit substitution (a fundamental rule for using axiom schemata) [11]. Certain compromise or isolation of concepts has to be made in order to get a logic of intention with manageable complexity.

Another tricky part of modelling intention is its intimate connection to the concept of commitment. On the one hand, an agent intending to do an action would mean that the agent is committing to the action to be done now or in the future. On the other hand, an agent may continuously weigh his competing goals and choose one that is most desirable to achieve, which means that the agent does not have to bond to his commitments [4]. To deal with such a dilemma, Cohen and Levesque model intention as a kind of persistent goal: *an agent intending to do an action will form a persistent goal with which the agent acts until either the action is done or is believed that it can never be done* [1]. However, as we will illustrate in the next section, such a definition of intention is insufficient to capture the concept of intention with conditional/unconditional commitments, which often occur in the real world.

In this paper, we propose a model of intention based on C&L's framework but reformulate the concept of intention and commitment. Instead of defining intention as a persistent goal, we define an intention as a conditional/unconditional commitment made for achieving a goal that is believed to be achievable. We check our framework by providing a solution to a well-known philosophical problem, the Toxin Puzzle [12].

The paper is organised as the following. In the next section, we briefly recall the basic concepts of C&L's framework and using their language to describe the Toxin Puzzle. Section 3 presents our model of intention and discuss its properties. Section 4 gives our solution to the Toxin Puzzle before we conclude the paper in Section 5. Due to space limit, we omit all the proofs and some technique lemmas.

2   C&L's Model of Intention

C&L's model combines doxastic logics and segments of dynamic logic in model-theoretic fashion. Agent's beliefs and goals are represented by modalities, BEL and GOAL, and are specified via possible world semantics. It is assumed that BEL satisfies KD45 (thus its accessibility relation is Euclidean, transitive and serial) and that GOAL satisfies KD (its accessibility relation is then serial). Actions are represented by action expressions which compound primitive events via program connectives: ; (*sequential*), — (*nondeterministic choice*), ? (*test*) and * (*iteration*). Time is modeled in linear structure with infinite future and infinite past. Each course of event represents a time unit. Therefore a time period is identical to a sequence of events that happen during the period. A possible world is then an infinite sequence of events with two open ends, representing the events happened in the past and the events happen now or in the future.

Two temporal operators HAPPENS and DONE are defined to indicate the actions that are about to happen and the actions that have just been done, respectively. Other temporal operators can be defined accordingly:

$\Diamond\varphi =_{def} \exists e$ HAPPENS $e; \varphi?$
$\Box\varphi =_{bef} \neg\Diamond\neg\varphi$

LATER $\varphi =_{def} \neg\varphi \wedge \Diamond\varphi$

BEFORE $\varphi \ \psi =_{def} \forall e(\text{HAPPENS } e; \psi? \rightarrow \exists e'(e' \leq e \wedge \text{HAPPENS } e'; \varphi?))$

where $e \leq e'$ means that e' happens before e.

Based on the concepts, they introduce the concept of persistent goals to capture "one grade" of commitments. Intuitively, a goal of an agent is a *persistent goal* (P-GOAL) if the goal will not give up once it is established unless the agent believes that it has been achieved or it can never be achievable.

$$\text{P-GOAL } \varphi =_{def} (\text{GOAL LATER } \varphi) \wedge (\text{BEL } \neg\varphi) \wedge \\ (\text{BEFORE } (\text{BEL } \varphi \vee \text{BEL } \Box\neg\varphi) \ \neg(\text{GOAL LATER } \varphi)) \tag{1}$$

With the concept of commitment, intention can be simply defined as a persistent goal to perform an action:

$$\text{INTEND } \alpha =_{def} \text{P-GOAL } (\text{DONE } ([\text{BEL } (\text{HAPPENS } \alpha)]?; \alpha)) \tag{2}$$

where α is an action expression. Intuitively, an agent intending to do an action is a commitment (persistent goal) to have done this action if he believes the action is executable.

The INTEND operator can be overloaded to take a proposition as argument:

$$\text{INTEND } \varphi =_{def} \text{P-GOAL } \exists e(\text{DONE } [(\text{BEL } \exists e'(\text{HAPPENS } e'; \varphi?)) \wedge \\ \neg(\text{GOAL}\neg\text{HAPPENS}(e; \varphi?))]?; e; \varphi?) \tag{3}$$

where φ is a proposition. It reads that, to intend to bring about φ, an agent sets up a persistent goal that is to find a sequence of events after doing which himself, φ holds. For a full understanding of C&L's theory of intention, including comments, criticisms and follow-ups, the reader is referred to Cohen and Levesque (1990) [1], Singh (1992) [13], Rao and Georgeff [14], Hoek and Wooldridge [6], and etc.

2.1 The Toxin Puzzle

Gregory Kavka challenges theories of intention by inventing the following thought experiment, known as *the Toxin Puzzle* [12]:

> *Suppose that a billionaire offers you on Monday a million dollars if on Tuesday you intend to drink a certain toxin on Wednesday. It is common knowledge that drinking this toxin will make you very sick for a day but will not have a lasting effect. If you do so intend on Tuesday, the million dollars will be irreversibly transferred to your bank account; this will happen whether or not you actually go ahead and drink the toxin on Wednesday.* [1]

Although the puzzle sounds a bit unrealistic, we can find large amount of similar scenarios in the real-world, such as government funded research grants, scholarships, and high-risk investments.

The Toxin Puzzle has induced tremendous debates in the philosophical literature [15–18]. It challenges the theories of intentions from two aspects: *the nature of intentions* and *the rationality of intentions*. On the one hand, if intentions were inner

---

[1] This simplified version of the Toxin Puzzle is cited from [4] p.101.

performances or self-directed commands, an agent would have no trouble to form an intention. On the other hand, when reasons for intending and reasons for acting diverge, as they do in the puzzle, confusion often reigns: either giving way to rational action, rational intention, or aspects of the agent's own rationality [12].

2.2   Formalising the Toxin Puzzle in C&L's Language

We describe the Toxin Puzzle in C&L's language. Let $wait\_a\_day$ denote the action "wait for one day" and $drink$ the action "drink the toxin". We use $millionaire$ to denote the fact "given a million dollars". Then the billionaire's offer, which was made on Monday, can be expressed as:

$$(\text{INTEND } wait\_a\_day; drink) \to millionaire \qquad (4)$$

Assume that an agent had a goal to be a millionaire thus he formed an intention INTEND $wait\_a\_day; drink$ on Tuesday. As the billionaire had promised, the agent was given one million dollars and became a millionaire on the same day. The question is: *Whether the agent will drink the toxin on the next day as the billionaire expected?*

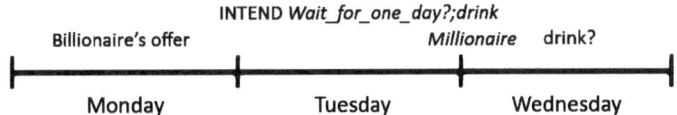

According to C&L's definition (Equation 2), INTEND $wait\_a\_day; drink$ represents the following persistent goal:

$$\text{P-GOAL}[\text{DONE}((\text{BEL}(\text{HAPPENS } wait\_a\_day; drink))?; wait\_a\_day; drink)] \qquad (5)$$

which says that the agent has a persistent goal to drink the toxin on the next day as long as he believes it can happen in due course. Obviously the agent does not have to believe on Tuesday that the action will happen because an agent is allowed to withdraw his commitment if there is other competing goals and he does not know whether there is any competing goals on next day. Therefore (BEL(HAPPENS $wait\_a\_day; drink$))? fails on Tuesday. Then the persistence goal is carried over to Wednesday. However, execution of the action $wait\_a\_day; drink$ does not make sense. No matter it is executed or not, the outcome is not what is expected. In fact, since a persistent goal does not require when the goal is achieved, any time-restricted intention cannot be rightly characterised using persistent goals.

As the matter of fact, there is no standard answer to the Toxin Puzzle. The puzzle is used to illustrate the divergence of reasons for intending and reasons for acting [12]. The reason for the agent to form the intention is to be a millionaire but the reasons to act, if it happens, is due to his commitment. Both sides (the billionaire and the agent) agree on the first reason but diverge on the second reason if the agent does not drink the toxin. The billionaire expects the agent to form a *unconditional commitment* while the agent might have formed a *conditional commitment*. Nevertheless, none of conditional or unconditional commitments can be properly specified in C&L's model.

3   The Model of Intention

In this section we propose a model of intention with condition/unconditional commitments. The model is built our on C&L's framework. We do so because C&L's framework is well-known in AI community and relatively simple. We will introduce a formal language that consists of two doxastic modalities BEL (for belief) and DES (for desire), and three action modalities ATH, FEASIBLE and COMMITTED. ATH stands for "*About To Happen*". Similar to C&L's framework, we shall present our model in possible-words semantics with linear-time model. Each world represents a linear sequence of events. However, our time is future-directed without referring to the past. Another important difference from the C&L model is that, for this paper, we assume that any events that happens are due to the actions performed by a single agent. The beliefs, desires, intentions are the mental states of the same agent. Commitments are meant self-commitment. Any changes of worlds are resulted from the agent's actions. We admit that such a simplification would lead to the insufficiency of modeling social behaviours of rational agency. However, such a sacrifice of completeness can be worth if a simple theory of intention can help us to understand the concept deeper.

3.1   Syntax

Consider a propositional language with modalities BEL, DES, ATH, FEASIBLE, COMMITTED, and regular action expressions. Let Φ be a countable set of atomic formulas, denoted by $\phi, \phi_1, \phi_2, \cdots$, and Act_P be a countable set of primitive events or actions, denoted by $a, b, \cdots$. The set of all formulas generated from Φ, Act_P, and the modalities, are denoted by Fma. Its members are denoted by $\varphi, \psi, \varphi_1, \cdots$, etc. Act is the set of all compound actions through action connectives $;, \cup, *$, with typical members denoted by $\alpha, \beta, \cdots$. We use $\top$ and $\bot$ to denote the propositional constants "**true**" and "**false**", respectively.

Formally, formulas ($\varphi \in Fma$) and actions ($\alpha \in Act$) are defined by the following BNF rules:

$$\varphi ::= \phi \mid \neg\varphi \mid \varphi_1 \to \varphi_2 \mid \text{BEL}\varphi \mid \text{DES}\varphi \mid \text{ATH } \alpha \mid \text{FEASIBLE } \alpha \mid \text{COMMITTED } \alpha$$
$$\alpha ::= a \mid \alpha_1; \alpha_2 \mid \alpha_1 \cup \alpha_2 \mid \alpha^* \mid \varphi?$$

where $\phi \in \Phi$ and $a \in Act_P$.

For any formula $\varphi \in Fma$ and action $\alpha \in Act$, we write α_φ to represent another action that is yielded from α by replacing each primitive action a in α with $\varphi?; a$. It is easy to give an inductive definition of α_φ based on α's structure so omitted.

Given an action α, we define all the computation sequences of α by induction on the structure of α as follows (see [19]):

$CS(a) =_{def} \{a\}$, where $a \in Act_P$, an atomic program
$CS(\varphi?) =_{def} \{\varphi?\}$
$CS(\alpha; \beta) =_{def} \{\sigma'; \sigma'' : \sigma' \in CS(\alpha), \sigma'' \in CS(\beta)\}$
$CS(\alpha \cup \beta) =_{def} CS(\alpha) \cup CS(\beta)$
$CS(\alpha^*) =_{def} \bigcup_{n \geq 0} CS(\alpha^n)$.
$CS = \bigcup_{\alpha \in Act} CS(\alpha)$.
where $\alpha^0 = \top?$ and $\alpha^{n+1} = \alpha; \alpha^n$.

We remark that with non-deterministic choice and iteration, we can express "very rough" plans. Suppose that we have only finite number of primitive actions $a_1, a_2, \cdots ,$ a_n. Then the action **any** $= (a_1 \cup \cdots \cup a_n)^*$ can be any action the agent can perform except test actions. Therefore, whenever we talk about an action description, it actual mean a "rough plan" or a "partial plan" in Bratman's terminology [4] because the execution of such a auction relies on a high-level planner to find an executable computation sequence of primitive events that implements the action.

3.2   Semantic Model

A model M is a structure $(E, W, \bar{\sigma}, B, D, C, V)$, where E is a set of primitive event types[2]. $W \subseteq [\mathbb{N} \to E]$ is a set of possible courses of events or worlds specified as a function from the natural numbers, representing time points, to elements of E, $\bar{\sigma} \in W$ is a special world (the real world) representing the actual events that have happened or will happen, $B \subseteq W \times \mathbb{N} \times W$ is the accessibility relation for belief modality, $D \subseteq W \times \mathbb{N} \times W$ is the accessibility relation for desire modality, $C : W \times \mathbb{N} \to \wp(CS)$ is a function that assign a set of computation sequences to each world at each time[3], and $V : W \times \mathbb{N} \to \wp(\Phi)$ is the valuation function of the atomic formulae.

Definition 1. Given a model M, let $\sigma \in W$ be any possible world and $n \in \mathbb{N}$ a natural number. We define the satisfaction relation $M, \sigma, n \models \varphi$ as follows:

1. $M, \sigma, n \models p$ iff $p \in V(\sigma, n)$, where $p \in \Phi$.
2. $M, \sigma, n \models \neg\varphi$ iff $M, \sigma, n \not\models \varphi$.
3. $M, \sigma, n \models \varphi \to \psi$ iff $M, \sigma, n \models \varphi$ implies $M, \sigma, n \models \psi$.
4. $M, \sigma, n \models \mathtt{BEL}\ \varphi$ iff for all σ' such that $(\sigma, n, \sigma') \in B$, $M, \sigma', n \models \varphi$.
5. $M, \sigma, n \models \mathtt{DES}\ \varphi$ iff for all σ' such that $(\sigma, n, \sigma') \in D$, $M, \sigma', n \models \varphi$.
6. $M, \sigma, n \models \mathtt{ATH}\ \alpha$ iff for all $i \leq n$, $\sigma(i) = \bar{\sigma}(i)$ and $\exists m, m \geq n$ such that $M, \bar{\sigma}, n[\![\alpha]\!]m$.
7. $M, \sigma, n \models \mathtt{FEASIBLE}\ \alpha$ iff there exists σ' such that for all $i \leq n$, $\sigma(i) = \sigma'(i)$ and $\exists m, m \geq n$ such that $M, \sigma', n[\![\alpha]\!]m$.
8. $M, \sigma, n \models \mathtt{COMMITTED}\ \alpha$ iff $CS(\alpha) \subseteq C(\sigma, n)$.
9. $M, \sigma, n[\![a]\!]n + 1$ iff $a = \sigma(n + 1)$, where a is a primitive action.
10. $M, \sigma, n[\![\alpha \cup \beta]\!]m$ iff $M, \sigma, n[\![\alpha]\!]m$ or $M, \sigma, n[\![\beta]\!]m$.
11. $M, \sigma, n[\![\alpha; \beta]\!]m$ iff $\exists k, n \leq k \leq m$, such that $M, \sigma, n[\![\alpha]\!]k$ and $M, \sigma, k[\![\beta]\!]m$.
12. $M, \sigma, n[\![\varphi?]\!]m$ iff $M, \sigma, n \models \varphi$.
13. $M, \sigma, n[\![\alpha^*]\!]m$ iff $\exists n_1, \cdots, n_k$, where $n_1 = n$ and $n_k = m$ and for every i such that $1 \leq i \leq k$, $M, \sigma, n_i[\![\alpha]\!]n_{i+1}$.

A formula φ is satisfiable if there is at least one model M, world σ and index n such that $M, \sigma, n \models \varphi$. A formula φ is valid iff for every model M, world σ, and index n, $M, \sigma, n \models \varphi$.

It is not hard to find that our semantic model is a variation of C&L's model. The semantics for $\mathtt{BEL}$ is exactly the same as C&L's. The one for $\mathtt{DES}$ are similar to C&L's

---

[2] For simplicity, we will identify each primitive event symbol with its type. Equivalently, we assume that there is a unique one to one mapping from Act_P to E.

[3] $C(\sigma, n)$ can be interpreted as the searching space of the event sequences for planning purposes.

semantics for GOAL operator but with less constraints (see more details in next subsection). We shall introduce GOAL as a composite concept defined by BEL and DES.

ATH α represents the statement that action α is about to happen in the real world ($\bar{\sigma}$). Note that it is different from C&L's HAPPENS for its truth value relies on the real world $\bar{\sigma}$, which is unique to a model.

FEASIBLE α represents the feasibility of action α performed by the agent under consideration. That α is feasible in a world σ at time n means that there is a possible world σ' which has the same history as σ but could branch at time n such that α is executable in the world σ'.

The operator COMMITTED captures the key feature of agent commitments. $CS(\alpha)$ lists all the alternative computation paths of action α. The set $C(\sigma, n)$ collects all the computation sequences the agent has committed to at each time n and each world σ. $CS(\alpha) \subseteq C(\sigma, n)$ means that no matter which path the agent takes to execute α, the path has been committed.

3.3   Constraints and Properties

In this subsection, we introduce a few constraints on the semantic conditions for BEL, DES and COMMITTED. Similar to C&L's model, we assume that B is transitive, serial and Euclidean. Therefore BEL is a doxastic modality that satisfies KD45.

We assume that DES is a modal operator that satisfies K and define goal as a composite concept from belief and desire:

Definition 2. GOAL $\varphi =_{def}$ DES $\varphi \wedge \neg$BEL φ.

In words, φ is a goal of an agent if the agent desires or wishes φ to be true and does not believe it is true now. Assume that you have a goal to be a millionaire. It means that you desire to be a millionaire but you believe that currently you are not a millionaire. The idea is inspirited by Meyer et al.'s KARO logic[4].

The semantic condition for COMMITTED is much more complicated. Intuitively, the accessibility relation $C(\sigma, n)$ collects all the computation sequences the agent committed to at time n in world σ. The agent will choose a computation sequence from this collection to execute. We assume that C satisfies the following conditions:

(c1) *If $\delta \in C(\sigma, n)$, then for any initial subsequence δ' of δ, $\delta' \in C(\sigma, n)$.*
  In other words, $C(\sigma, n)$ is closed under initial subsequences.
(c2) *If φ?; $\delta \in C(\sigma, n)$, then $\delta \in C(\sigma, n)$.*
  Similar to C&L's model, we assume that a test action does not take time. Thus tests have to be carried out in conjunction with one primitive event.
(c3) *For any sequences $\delta \in C(\sigma, n)$ and any world $\sigma' \in W$, if for all $i \leq n$ $\sigma(i) = \sigma'(i)$ and $\delta = \sigma(n); \delta'$, then $\delta' \in C(\sigma', n+1)$.*
  In other words, if one committed to performing a sequence of actions, then he should carry over the commitment (the remaining computation sequences) to the next state of any possible world he might move to once he fulfils the initial primitive action in the sequence. This conditions captures the phenomenon of commitment persistence.

---

[4] In [3], Meyer et al. recognise beliefs and wishes as the primitive mental attributes instead of belief-desire. We shall not discuss the philosophical differences between wish and desire.

Example 1. *Given an action* $\alpha = (a\cup b); c; (p?; d \cup (\neg p); e)$. *Then* α *has the following computation sequences:* $a; c; p?; d$, $a; c; (\neg p)?; e$, $b; c; p?; d$, *and* $b; c; (\neg p)?; e$. *Assume that an agent has committed to performing* α *at time 0 and makes no other commitments further on. Firstly, the agent takes action* a. *Thus he has to carry over the committed sequences* $c; p?; d$ *and* $c; (\neg p)?; e$ *but drop the other two. After done* c, *if* p *is true, he shall continues with his committed action* d; *otherwise, he shall do* e. *Assume the latter case holds, the actual execution sequence will be* $a; c; (\neg p)?; e$, *which fulfils his commitment.*

According to our semantics, the following property is easy to verify,

Proposition 1. $\models$ ATH $\alpha \rightarrow$ FEASIBLE α.

It shows that any action that is about to happen must be feasible for the agent to execute.

Proposition 2. *Properties of* COMMITTED *operator:*

1. $\models$ COMMITTED$(\alpha; \beta) \rightarrow$ [COMMITTED $\alpha \wedge$ (ATH $\alpha \rightarrow$ ATH $(\alpha;$ (COMMITTED $\beta)?)$)].
2. $\models$ COMMITTED$(\alpha \cup \beta) \rightarrow$ COMMITTED$(\alpha) \wedge$ COMMITTED(β).
3. $\models$ COMMITTED$(\alpha^*) \rightarrow$ [COMMITTED$(\alpha) \wedge$ (ATH $\alpha \rightarrow$ ATH $(\alpha;$ (COMMITTED $\alpha^*)?)$)].

These statements shows that a commitment is carried over along the path of an execution.

3.4   Implementation Assumption and Conditional Commitments

Suppose that an agent has committed to performing an action α. Doing this action is also feasible for him. Then we should expect that the action is about to happen. This ideas can be described as the following assumption.

Implementation Assumption (IA) $\models$ (FEASIBLE $\alpha \wedge$ COMMITTED $\alpha) \rightarrow$ ATH α.

Figure 1 illustrates the relation between commitment, feasibility and actual execution. Assume that an agent has committed to do action α which has five possible computation sequences a, b, c, d, e. The current state is at node d1. Among all the computation sequences, the paths a, c, e are not feasible (marked in red). The agent is actual taking

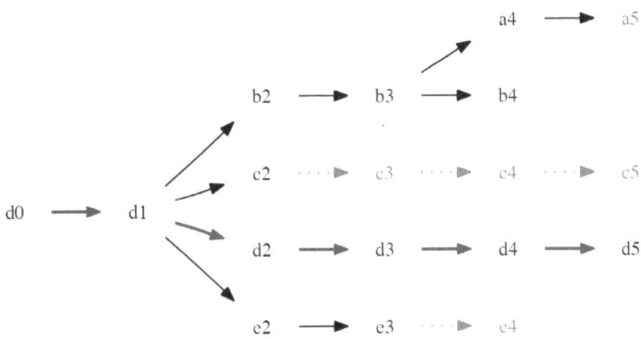

Fig. 1. Promise, feasibility and execution

path d (marked in bold and blue). Note that Implementation Assumption assumes that an agent has the ability of choosing a committed path from all feasible computation paths and execute the path.

This assumption is similar to C&L's assumption on persistent goal: *if a goal is not achieved, the agent will keep the goal until he believes that it is not achievable*. However, for action commitments, the assumption sounds a bit strong because if the commitment of performing an action was to achieve a goal and the goal has achieved or is believed not achivable, the commitment for performing the action might be dropped if the commitment is conditional. For example, assume that I intend to go to the airport tomorrow to pick my friend up from the airport. So I commit myself to going the airport tomorrow. However, if the friend is not coming tomorrow, the commitment to going the airport should be dropped. We call such a commitment a *conditional commitment*. I committed to going airport is because I wanted to pick my friend up. Without such a demand, I would not make the commitment. We define conditional commitment as follows:

Definition 3. *For any formula φ and action α,*
    $\text{COMMITTED}_\varphi\ \alpha =_{def} \text{COMMITTED}\ \alpha_\varphi$

$\text{COMMITTED}_\varphi\ \alpha$ means "commit to α under condition φ". Recall that α_φ is generated from α by replacing all primitive action a with $\varphi?; a$. The condition φ is checked all the way during the execution of α. Note that we can view $\text{COMMITTED}\ \alpha$ as $\text{COMMITTED}_\top\ \alpha$. Therefore the original COMMITTED operator can be treated as unconditional commitment. In this sense, **IA** is not strong.

3.5   Definition of Intentions

C&L consider intending an action and intending a proposition as separate concepts. As many philosophers argue, "[i]t is (logically) impossible to perform an intentional action without some appropriate reason." ([20] p264). Meanwhile we also think it is not a valid intention of achieving a goal without refer to a (rough) plan to achieve the goal. In this paper, we define an intention as a binary operator with the arguments: the intended action and the reason for doing the action. Similar idea also appears in [5, 3, 7]. Different from C&L's definition again, we treat conditional intentions and unconditional intentions as two different objects. We will give separate definition for each of them: *unconditional intention* and *conditional intention*. We start with the simply one first.

Definition 4. (Unconditional Intention)
    $\text{INTEND}_1\ \alpha\ \varphi =_{def} \text{GOAL}\ \varphi\ \wedge\ \text{BEL}(\text{FEASIBLE}\ \alpha; \varphi?) \wedge \text{COMMITTED}\ \alpha.$

Intuitively, I intend to do α to achieve a goal φ iff the following conditions are satisfied:

– φ is one of my goals;
– I believe that it is feasible to bring about φ via conducting α.
– I commit to performing α.

Note that for unconditional intention, any committed action has to been fulfilled no matter whether the goal has achieved, in which case the intention no longer exists. This is the intention the billionaire understood.

Definition 5. (Conditional Intention)
INTEND$_2$ α φ $=_{def}$ GOAL φ $\wedge$ BEL(FEASIBLE α; φ?) $\wedge$ COMMITTED$_{DEC\varphi}$ α.

The only difference between unconditional intention and conditional intention is the form of commitments. A conditional intention commits to perform the intended action as long as the goal is still desirable.

Example 2. Consider the example of chopping a tree (see Cohen and Levesque [1]). The intention of knocking down a tree can be expressed as follows:

$$\text{INTEND}_2 \ (chop)^* \ (tree\_down)$$

To form the intention, it is required:

1. Knocking down the tree is my goal;
2. I believes that if I chop the tree repeatedly, the tree can be knocked down eventually (I may not know how many chops are actually needed).
3. I commit to doing the action as long as I still desire to knock down the tree (I may stop without knocking down the tree).

Note that our representation of this example using the concept of conditional intention is much clearer and simpler than C&L's version, thanks the specific definition of α_φ (Section 3.1).

4   Solution to the Toxin Puzzle

Now we are ready to present our solution to the Toxin Puzzle. Assume that the agent forms a unconditional intention on Tuesday as follows:

$$\text{INTEND}_1 \ (wait\_a\_day; drink) \ millionaire \tag{6}$$

which means

1. The agent has a goal to be a millionaire, i.e., GOAL $millionaire$.
2. The agent believes that $wait\_a\_day$; $drink$ can bring about the goal, i.e.,
   BEL[FEASIBLE($wait\_a\_day$; $drink$; $millionaire$?)].
3. The agent commits to doing the action $wait\_a\_day$; $drink$ unconditionally, i.e.,
   COMMITTED($wait\_a\_day$; $drink$).

The billionaire's offer can be represented by the following statement:

$$[\text{INTEND}_1 \ (wait\_a\_day; drink) \ \varphi] \rightarrow millionaire \tag{7}$$

where φ can be any reason as long as the billionaire can accept. Since the intention has been formed, the agent will receive a payment on Tuesday. The agent's goal is then achieved and the intention no longer exists. However, he has commit to drinking the toxin on next day, i.e., COMMITTED($wait\_a\_day$; $drink$). By Proposition 2, after done the action $wait\_a\_day$, the agent needs to keep the rest of commitment, i.e., COMMITTED($drink$). According to **IA**, as long as drinking the toxin is feasible to him, he has to make it happen on Wednesday, i.e., **ATH**$(drink)$. This explains why the billionaire does not care too much about what is the goal of the agent to form the intention.

Assume instead that the agent forms the following conditional intention on Tuesday:

$$\text{INTEND}_2 \ (wait\_a\_day; drink) \ millionaire \tag{8}$$

and, assume (but not too sure!) that the billionaire agrees on the intention and pay the agent one million dollars. The agent then have no goal to be a millionaire. However, whether the agent keeps the commitment of drinking the toxin will depend on whether he still keep the desire to be a millionaire. Either way can be rational. As main philosophers have pointed out, the reason for an agent to do an action, here is a million, can be different from the reason for the agent to drink. Whether the agent should keep the commitment does not necessarily reply on whether the agent has been a millionaire but on the agent's desire.

5   Conclusion

We have proposed a formal model of intention based on Cohen and Levesque's framework. We have shown that C&L's account of intention as persistent goal is inadequate to capture the key features of commitments of rational agency. We have argued that commitment plays a crucial role in intentional reasoning. With a unconditional commitment, the agent picks up a feasible computation path from a partial plan (specified by the intended action) to achieve his goal. The agent executes the actions alone the path and carries over unfulfilled actions to next state, which creates reason-giving force on the agent to execute his intended actions. With a conditional commitment, the condition is checked before executing any primitive actions. Both the accounts of commitment provide rational explanations to the Toxin Puzzle.

Conditional intention has been a traditional topic in philosophy for a long history even though it has not been well-investigated in the AI literature [21]. It has been widely accepted that a conditional intention and an unconditional intention refer to different objects and should be viewed as different intentions. In this paper, we treat the condition of a conditional intention as the way to determine whether the committed action should be carry out or stop while we treat the unconditional intention as an extreme case of conditional intention (blind commitment). More importantly, for a conditional intention, we do not assume that the commitment of the intention is automatically dropped if the intended goal has achieved. Whether an agent drops his commitments or not is determined by his desires.

A few issues are excluded from the current work. First, this work deals with single agent intentional reasoning. Commitment is meant self-commitment. However, the consideration of intention in multia-gent environment will give rise of issues on social commitments. In such a case, game-theoretic approach might be needed to model the social behaviour of rational agents. Secondly, we consider belief, desire and commitment as the primacy of intention. Changes of these mental states are left unspecified. The investigation of the dynamics of belief, desire and commitment will certainly provide a better understanding of human intention. Finally, the current work is based on propositional logic. Extending the framework to first-order language will allow us to define more temporal operator, such as $\Diamond$, $\Box$, $achievable$ and etc.

References

1. Cohen, P.R., Levesque, H.J.: Intention is choice with commitment. Artif. Intell. 42(2-3), 213–261 (1990)
2. Rao, A.S., Georgeff, M.P.: Modelling rational agents within BDI architecture. In: Proceedings of the 2nd International Conference on Principles of Knowledge Representation and Reasoning, KR 1991, pp. 473–484 (1991)
3. Meyer, J.J.C., van der Hoek, W., van Linder, B.: A logical approach to the dynamics of commitments. Artif. Intell. 113(1-2), 1–40 (1999)
4. Bratman, M.E.: Intentions, Plans, and Practical Reason. Harvard University Press (1987)
5. van Linder, B., van der Hoek, W., Meyer, J.J.C.: Formalising abilities and opportunities of agents. Fundamenta Informaticae 34(1-2), 53–101 (1998)
6. van der Hoek, W., Wooldridge, M.: Towards a logic of rational agency. Logic Journal of the IGPL 11(2), 135–159 (2003)
7. Schmdit, R., Tishkovsky, D., Hustadt, U.: Interactions between knowledge, action and commitment within agent dynamic logic. Studia Logica 78, 381–415 (2004)
8. Herzig, A., Longin, D.: C&l intention revisited. In: KR, pp. 527–535 (2004)
9. Lorini, E., Herzig, A.: A logic of intention and attempt. Synthese 163(1), 45–77 (2008)
10. Lorini, E., van Ditmarsch, H.P., Lima, T.D., Lima, T.D.: A logical model of intention and plan dynamics. In: ECAI, pp. 1075–1076 (2010)
11. Schmidt, R.A., Tishkovsky, D.: On combinations of propositional dynamic logic and doxastic modal logics. Journal of Logic, Language and Information 17(1), 109–129 (2008)
12. Kavka, G.S.: The toxin puzzle. Analysis 43(1), 33–36 (1983)
13. Singh, M.P.: A critical examination of the cohen-levesque theory of intentions. In: Proceedings of the 10th European Conference on Artificial intelligence, ECAI 1992, pp. 364–368. John Wiley & Sons, Inc. (1992)
14. Rao, A.S., Georgeff, M.P.: Decision procedures for bdi logics. J. Log. Comput. 8(3), 293–342 (1998)
15. Andreou, C.: The newxin puzzle. Philosophical Studies, 1–9 (July 2007)
16. Bratman, M.E.: Toxin, temptation, and the stability of intention. In: Coleman, J.L., Morris, C.W. (eds.) Rational Commitment and Social Justice: Essays for Gregory Kavka, pp. 59–83. Cambridge University (1998)
17. Gauthier, D.: Rethinking the toxin puzzle. In: Coleman, J.L., Morris, C.W. (eds.) Rational Commitment and Social Justice: Essays for Gregory Kavka, pp. 47–58. Cambridge University Press (1998)
18. Harman, G.: The toxin puzzle. In: Coleman, J.L., Morris, C.W. (eds.) Rational Commitment and Social Justice: Essays for Gregory Kavka, pp. 84–89. Cambridge University (1998)
19. Harel, D., Kozen, D., Tiuryn, J.: Dynamic Logic. The MIT Press (2000)
20. Davidson, D.: Essays on Actions and Events. Clarendon Press, Oxford (1980)
21. Meiland, J.W.: The Nature of Intention. Methuen & Co Ltd. (1970)

Group Polarization: Connecting, Influence and Balance, a Simulation Study Based on Hopfield Modeling

Zhenpeng Li and Xijin Tang

Institute of Systems Science, Academy of Mathematics and Systems Science, Chinese
Academy of Sciences, Beijing 100190, P.R. China
{lizhenpeng,xjtang}@amss.ac.cn

Abstract. In this paper, we address the general question of how negative social
influence determines global voting patterns for a group-based population, when
individuals face binary decisions. The intrinsic relation between global patterns
and local structure motifs distribution is investigated based on Hopfield model.
By simulating results with the model, we examine the group opinions polarization processes, and find that the global pattern of group opinions polarization is
closely linked with local level structure balance. This computing result is well
agreed with the classic structure balance theory of social psychology.

Keywords: global polarization, structure balance, Hopfield model, social
influence.

1   Introduction

As one of the important collective actions, global polarization patterns are widely
observed in human society. Examples of these situations include culture, social
norms, political voting and on-line public hotspots debates. From decision-making
perspective, the global pattern of collective dynamics is rooted from individuals' micro-level decision making processes.

The individual decision-making is always as a result of several complex and dynamic social psychological and economic behavior processes.   Examples include
"herding effect [1]", "wisdom of crowds[2]", "information cascade[3]" etc. Modeling
of social system dynamics inevitably involves underlining social psychological
processes which have a close connection with real world phenomena. From a bottom-
up point of view, in modeling social processes, individuals' locally cumulative interacting behaviors would evolve into different global patterns.   The emergence of
broad global features of social structure is often from the local interconnecting mechanism, e.g., short average path and high clustering coefficients contribute to small
world mechanism [4]. However, it is difficult to infer how the global patterns are
generated from certain simple local aggregated social processes. In such cases, agent-
based computational techniques are necessary because analytic solutions are simply
not available.

P. Anthony, M. Ishizuka, and D. Lukose (Eds.): PRICAI 2012, LNAI 7458, pp. 710–721, 2012.
© Springer-Verlag Berlin Heidelberg 2012

In this study, through agent-based Hopfield network simulation, we investigate the underlying relationship between macroscopic group polarization patterns and local dynamic structure balance, which is based on three classic social psychological processes-----social influence, social identity and structure balance.   Our simulation shows that global polarization of collective voting behavior has the implicit close relation with local dyadic and triadic motifs dynamic structure variation processes. This computing result is well agreed with theories of Heider's cognitive balance [5], Cartwright and Harary generalized structure balance [6].

The rest of the paper is organized as follows: in Section 2, according to social identity theory, we discuss social influence implication and classify the social influences into three types, positive, negative and neutral. Section 3 addresses the classic structure balance theory, and 16 types of triadic classification. Then we define influence structures on the basis of the 16 types of triads. In Section 4, we extend Hopfield model by adding triad social structure. The triad structure implications are also discussed. Section 5 is our model computing analysis. We examine the relationship between negative social influence and polarization, and then focus on the connection between global polarization pattern and dyad/triad balance at the micro-level. Section 6 is our conclusion remarks.

2   Social Influence

Social influence refers to the way people are affected by the thoughts, feelings, and behaviors of others. It is a well-studied and core topic in social psychology. It studies the change in behavior that one affects another, intentionally or unintentionally, as a result of the way the changed person perceives themselves in relationship to the influencer, other people and society in general [7].

In many circumstances, instead of independent rational choice, individual decisons are influenced by their perceptions, observations, or expectations of decisions made by others, and then herding effect might appear. Many theories and studies account for this collective phenomenon. Individuals, for example, may be susceptible to social influence out of a desire to identify with the certain social groups, or to classify oneself from the groups. In order to avoid unexpected sanctions or risks, they may resort to group behavior, or response to influential authorities, as a way of reducing one's decision making difficulty[8].

Recently, an increasing number of empirical studies showed that social networks, such as MySpace, Twitter and Facebook are showing unexpected power, affecting every aspects of our social life, and significantly impacting on individual options, opinions or attitudes as society becomes more inter-connected.   The recent "Arab Spring" on Twitter is one of the best annotations [9].

Social influence, therefore, is related both to an individual's cognition of the social world, and to the dynamics of group patterns. Next we classify social influence into three concrete types based on the theory of social identity.

2.1   A Classification of Social Influences

As one of the important theoretical basis in social simulation, social influence mechanism was widely studied. Most social simulation literatures consider the principle of homogeneous influence (homogenous attraction and ingroup impact), i.e. similarity leads to interaction and interaction leads to more similarity. So the more similar an actor is to a neighbor, the more likely the actor will adopt his/her neighbor's opinion. Individuals blindly follow majority behaviors or options, then the homogeneous influence may create "herding effect", which means individuals don not consider their own subgroup identity (do as most people do).   For example, based on the single interaction principle of homogeneous influence, Axelord observed a local convergence and global multiple polarization pattern[10]. And the voter model shows one-polarization domination results with any initial binary opinions percentages [11].

However, the homogeneous impact is not the only social factor to glue individuals together and form different groups. Nation's split, religious conflicts, culture diversity and radical segregation, etc. account for the repulsive attitudes deeply rooted in human collective dynamics. Herding effects only partially explain one aspect of collective behaviors, individuals in a group act together without planned direction, based on indefinite individuals' group recognition. Next, we seek explanation from social identity theory.

Social identity as a basic theory is a powerful tool to predict certain intergroup behaviors on the basis of the perceived status of the intergroup environment. It states that social behavior often vary along a mutual influence processes between interpersonal and intergroup behaviors [12].

Individuals are likely to display favoritism among ingroup and disapproval among outgroup. In other words, individuals usually display positive attitudes toward ingroup members, while negative toward outgroup ones. For example in the case of voting and debating about distribution of national income, different classes may have different interests and political tendencies, individuals often favor ingroup and against outgroup opinions or stands.

In many real situations, negative repulsive impact among social groups is an important ingredient while it has been barely focused together with positive attractive influence behavior in studies. Here, we concern about what the pattern will be if consider both the positive attractive and negative repulsive impact. Next we make a detailed classification for general social influence from social identity perspective.

Positive influence which refers to homogeneous impact among ingroup numbers. Individuals within the same group, share the same tagged consensus, such as beliefs, interests, education or other similar social attributes. During group decision making, homogenous positive influence will play vital role for achieving the group consensus.

The second one is negative social impact which may block the formation of consensus among different outgroups. Individuals within different groups find it difficult to gain the agreement during group decision making even under the pre-condition that they share the same initial opinions. Since different groups have different social group unified interests, emotions, behaviors and value orientations, they act differently. Such kind of impact for individuals' opinions selection is regarded as heterogeneous repulsion. The use of both positive and negative interactions in social systems has been previously introduced to coalitions study among a set of countries [13].

Recently, many studies suggest online social networks in which relationships can be either positive (indicating relations such as trust, friendship) or negative (indicating relations such as opposition, distrust or antagonism). Such a mix of positive and negative links arises in a variety of online settings, e.g., Epinions, Slashdot and Wikipedia[14].

Apart from positive and negative influence, we also observe individuals who seldom care about others and share no commons. As a type of special individuals' attitudes, the individuals might not belong to any labeled subgroup, members in the group have no common social identity, no firm stand about some opinions and are in a state of something else. Then we introduce the third one, unsocial phenomena as a type of special individuals' attitude, in which the individuals do not belong to any labeled subgroup. Members in this group have no common social identity, no firm position about some social opinions and in a state of neither fish nor fowl. According to the above analysis, Figure 1 presents the influence relations among ingroup and outgroup individuals.

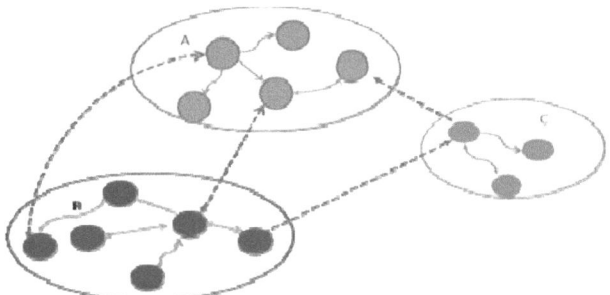

Fig. 1. Three types of social influence in a networked group

Illustration in Fig.1, A, B, C represent three different group. No connection among nodes (individuals) means that they have neutral influence. ◄───► stands for positive mutual influence among ingroup members, ───► stands for unilateral positive impact. While ◄- -► represents mutually negative repulsive among outgroup members, - - - -► stands for unilateral negative against outgroup individuals.

2.2   Structure Balance

Heider's balance theory is one of the cognitive consistency theories, which addresses the balance on the relationship between three things: the perceiver, another person, and an object[5]. Based on one of Heider's propositions stating that an individual tends to choose balance state in her interpersonal relation, and avoid tension or imbalance state in his/her interpersonal relations. This enforces someone to change her sentiment relation toward balance formation or to lesser force/tension. Cartwright and Harary generalized Heider's cognitive balance to structure balance[6].

Structure balance suggests that some social relationships are more usual and stable than others. It focuses on triadic relationships such as friendship and antagonistic, e.g., graphs whose signed edges represent friendship/hostile relationship among individuals. Structure balance theory affirms that signed social networks tend to be organized so as to avoid tense or nervous situations, based on the common principles that "the friend of my friend is my friend, the enemy of my friend is my enemy, the friend of my enemy is my enemy", this balanced and unbalanced triadic relationship are illustrated in Fig.2. In the illustration, T1 and T2 are balanced triad since the algebraic multiplication of edges signs has a positive value, while T3, T4 are unbalanced.

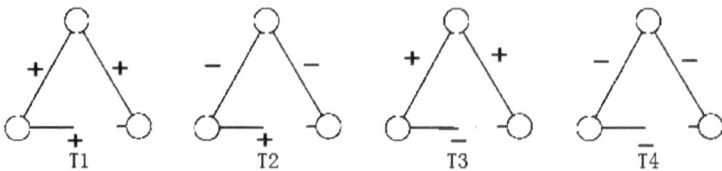

Fig. 2. Balanced and unbalanced triadic relationship ("+" denotes friendship, " − "denotes enemy relation)

Holland and Leinhardt addressed that classic balance theory offers a set of simple local rules for relational change and classified local triadic motifs into 16 types, according to mutual reciprocity, asymmetry relation, non-relationship etc[15]. We can see Code 300 triad relation corresponding to structure balance under the condition of the triad product signs satisfies "+",   as illustrated in Figure 3.

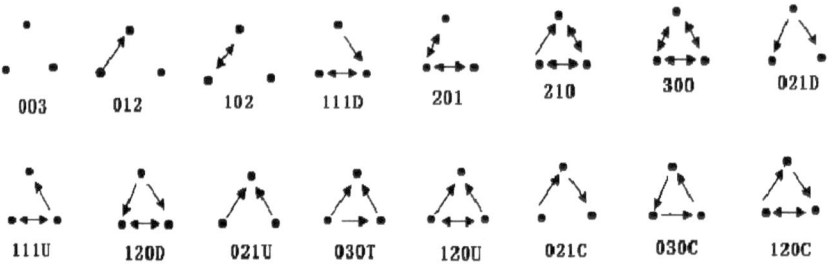

Fig. 3. 16 types of triad distributions in classic structure balance theory

The triad as one of important interpersonal relation has been excessively investigated by sociologists, e.g., Watt and Strogatz suggested that clustering coefficient is one of the important small world local structure features [16]. Wasserman and Faust defined a global clustering coefficient to indicate the clustering in the whole network[17]. In the next section, we discuss the influence balance on triad and dyad.

2.3   Influence Balance on Dyad and Triad

For any individual in a group, his/her options adoption comes from the influence around him/her and the corresponding cumulative social pressure. For example, if one's main friends have bought the same brand mobile phone, one might have high possibility to buy a mobile phone with this brand.

Here we focus on triadic relation influence balance. In Fig.3, we use directed edge represents directed influence relation. For example, the interpersonal influence on triad 111D includes three dependent dyadic social influence structure, in detail $i \xrightarrow{+} k$ stands for i has positive influence on k, $j \longleftrightarrow k$ stands for j, k have mutual negative influence and no connection represents i, j have   neutral influence as illustrated in Fig. 4.

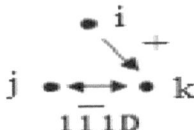

Fig. 4. The influence relation on code 111 D

There exist two basic structures that could satisfy influence balance in 16 motifs. For dyadic influence relation, Code 120 stands for mutual positive or negative influence. For triadic influence relation Code 300 is balanced if and only if product of any mutual influence relation signs satisfies +.

Next we examine Hopfield network model based on the aforementioned three types of social influence mechanism. We focus on two aspects, one is negative impact on group voting stable pattern, and the other is the relation between global pattern and local dyad, triad distribution. This model is described in the following section.

3   Hopfield Network Model

Macy et al. presented a Hopfield model to describe group polarization problems, with continuous connection weights and principles of homogeneous attraction and heterogeneous repulsion [18]. Their study found that group can display consensus, bipolarization and pluralistic alignments under different social pressures and exterior interventions. Their model assumed that each individual face binary opinions and has N-1 undirected ties to others. These ties measured by weights, which determine the strength and valance of connection between agents. Formally, social pressure P_{is} on individual i to adopt a binary state $s_i = \pm 1$ is the sum of the states of all other individuals j, conditioned by the weights I_{ij} (-1< I_{ij} <1) of dyadic tie between i and j:

$$P_{is} = \frac{\sum_{j=1}^{N} I_{ij} s_j}{N-1}, \ j \neq i,$$

(1)

If consider the external intervention, i.e., the influence for individuals opinion comes from other out-group impact, then we can replace Equ.(1) with Equ. (2) and obtain the logistic form

$$\tau_{is} = \frac{v_s}{1 + e^{-KP_{is}}} + (1 - v_s) X_i.$$

(2)

where v_s is used to trade off the internal and external group influence for individual i opinion, K is the size of opinions dimension. Given a randomly selected threshold $\pi_{thresh} = 0.5 + \varepsilon \chi$, if $\tau_{is} \geq \pi_{threshold}$, individual i chooses +1 (support), else chooses -1 (oppose), where ε is Harsanyi smooth responding parameter χ is subject to uniform distribution between -0.5 and $+0.5$. Equ. (3) describes the update of influence processes of individual j to i ($j \neq i$).

$$I_{ij}(t+1) = I_{ij}(t)(1-\lambda) + \frac{\lambda}{K}\sum_{k=1}^{K} s_{jk}(t)s_{ik}(t), j \neq i, \tag{3}$$

where t is the time step, λ is an adjustable parameter between 0 and 1. Comparing with [18], we extend the original Hopfield model from two aspects.

Firstly, with the motivation of investigating the relationship between non-positive social influence and group opinions polarization, instead the assigning of continuous values between -1 and $+1$ to I_{ij}, we assign three discrete values $-1, +1, \ 0$ to I_{ij} to indicate the three types of social influence. Individuals (agents) are influenced by others and also influence others, as conditioned by the valence of the social identity tie I_{ij}, where $I_{ij} = \{+1, 0, -1\}$ listed and explained as follows:

1) "+1" denotes the positive homogeneous social influence,
2) "−1" stands for xenophobia, antagonistic, negative social influence,
3) "0" represents neutral influence.

Secondly, Equ.(1) shows that individual i cumulative social pressure is from dyadic structure, and the triadic influence relation and corresponding cumulative social pressure is not included in the Hopfield model. Here, we add triadic influence into the model. Equ.(1) is evolved into Equ.(4) as following form,

$$P_{is} = \frac{\sum_{j=1}^{N} I_{ij} s_j}{N-1} + \frac{2\sum_{j \neq m \neq i = 1}^{N} I_{ijm} s_j s_m}{(N-1)(N-2)}, j \neq i. \tag{4}$$

where the second term represents the cumulative social pressure that any two individuals j, m impose on i in triadic structure, simultaneously. $I_{ijm} = \{+1, 0, -1\}$ stands for the social influence that j, m impose on i. With triad influence structures included in the model, it can better describe the real world interpersonal dynamic processes and opinions formation. For example, if we only consider dyadic influence, the cumulative social pressure of any individual i just depends on his/her in-degree d_i. However, with triadic influence included, one part of the social pressure will come from local clustering coefficient C_i. Since we put triad influence structure included in Equ.(4), the corresponding triadic social influence weight update processes is described with Equ.(5),

$$I_{ijm}(t+1) = I_{ijm}(t)(1-\lambda) + \frac{\lambda}{K}\sum_{k=1}^{K} s_{jk}(t)s_{ik}(t)s_{mk}(t), j \neq i. \tag{5}$$

As in Equ.(3), λ is an adjustable parameter between 0 and 1. Here we name λ is the social influence evolutionary parameter, which is used to adjust social influence strength variation.

Next, we will focus on the intrinsic relation between group polarization and social influence from both dyads and triads based on the extended Hopfield model simulation. The pseudo code is listed as follows.

4   Simulation and Results Analysis

In this section, firstly we use Matlab computing platform to simulate group voting polarization processes, then detect local dyad, triad distribution by R package sna[19].

4.1   Simulating Procedure

The pseudo codes of implementing Hopfield network social influence processes are listed as below:

```
Step 1:Let t = 0 , given v_s , λ , ε
```
Initialize each voter k-th dimension options
$s_{ik}(0) = \pm 1$, $k = 1,...,K$; $i, j, m = 1,...,N$, Randomly generate each pair of voters' dyadic and triadic social influence I_{ij}, I_{ijm} , respectively.

```
Step 2:t=t+1, compute (4),(2) for each agent i
```
Randomly generate $\chi \sim U(-0.5,0.5)$, then compute $\pi_{thresh} = 0.5 + \varepsilon\chi$ and logistic social pressure τ_{is}

```
       if  τ_is > π_thresh
       s_ik(t) = 1
       else
       s_ik(t) = -1
```

```
Step 3:For a given small positive real number δ, compute
(3),(5)
```
$$\text{If}\quad \max(|I_{ij}(t) - I_{ij}(t-1)| \cup |I_{ijm}(t) - I_{ijm}(t)|) < \delta$$

```
       stop
       else
       go to Step 2.
```

4.2   Negative Social Influence Promotes Group Bi-polarization

We take the test by setting $N = 100$, $K = 5$, $\varepsilon = 0.01$ and $\lambda = 0.5$. To illustrate the group voting pattern, we generate $N \times K$ matrix which N denotes for group size, K for the number of options. We run 100 times for average. Fig.5 shows the group initial random opinions states when each agent i faces K-dimension options (before group polarization). Fig.6 illustrates the group bi-polarization state under the condition of no imposing external influence ($v_s = 1$) and with three types of influence. We can observe that two patterns appear after group polarization, i.e., one pattern is $(-1, -1, -1, -1, -1)$, i.e., (black, black, black, black, black) (marked by V_1), the other is $(+1, +1, +1, +1, +1)$, i.e., (white, white, white, white, white) (represented by V_2). The ratio of the 2-pattern size is approximate to 1:1.

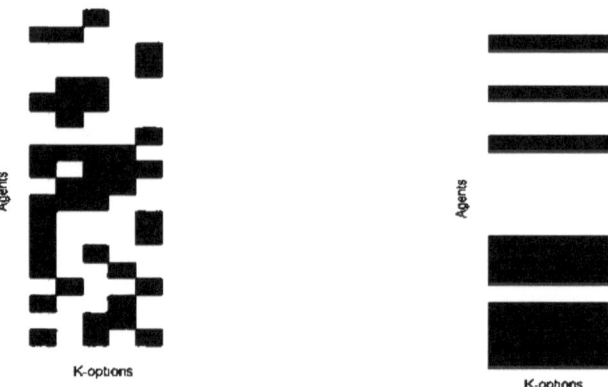

Fig. 5. Initialization of group options

Fig. 6. Bi-polarization pattern of group options

The relationship between exogenous intervention parameter v_s to group polarization is shown in Figure 7. We can see that when $v_s = 1$ (no external intervention to the group interaction processes), the ratio of $V_1 / (V_1 + V_2)$ is approximate to 0.5. However, the fifty to fifty well matched equilibrium will be destroyed with a little cut off v_s. In other words, external intervention will lead to majority pattern appeared. For example, when $v_s = 0.9$ we observe the group consensus appears, i.e., $V_1 / (V_1 + V_2)$ is approximate to 1, the pattern V_2 nearly disappears. It is worth pointing out that the parameters ε, λ variation also impacts the group polarization patterns; however, exogenous intervention is the dominant factor that leads to group polarization.

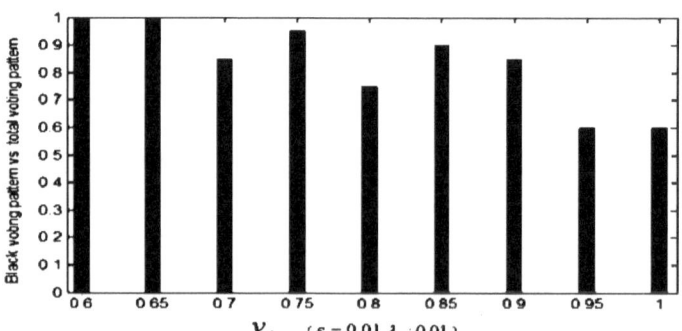

Fig. 7. v_s impact on group opinions polarization

4.3   Dyad and Triad Distribution Before and After Bi-polarization

Furthermore, we also investigate the triadic relation motifs distribution before and after bi-polarization by using R package sna. We find that the overwhelming structure balance motifs emerge and concurrently with the polarization process.

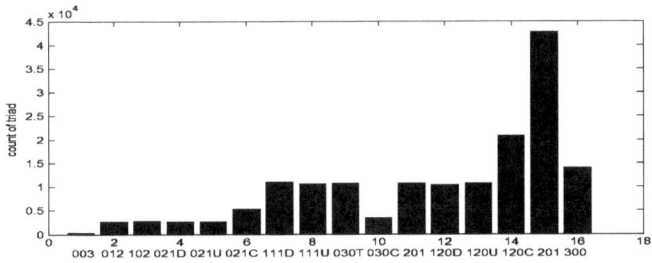

Fig. 8. Initial triad distribution before bi-polarization of group opinions

Fig.8 shows the initial local triads distribution according to randomly generated social influence matrix ($t=0$). We can observe that all 16 types of triads exist in the initial triadic relationships. With the social influence matrix updating and individuals' option changes, group voting bi-polarization emergences at step $t = 30$. Corresponding triad distribution as illustrated in Fig. 9, we observe that other triads disappear and only triad Code 300 remains. We also find the algebraic multiplication of influence signs in all Code 300 relation has a positive value. This result suggests social influence balance emerges from interpersonal negative or positive influence relations among agents. Simultaneously, neutral influence relation disappears.

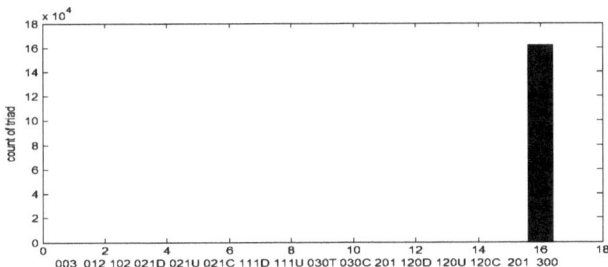

Fig. 9. The triad distribution after bi-polarization of group opinions (code 300 corresponding to the structure balance triad)

We also investigate the dyadic influence relations distribution before and after group bi-polarization. Fig.10 shows the initial local dyad distribution according to randomly generated social influence matrix ($t=0$). We can observe that three types of dyads exist in the initial triadic relationships. Asymmetrical dyadic influence relation is dominant among the three types.

When the global voting pattern of bi-polarization reaches, the corresponding dyad distribution is illustrated in Fig. 11. We find that asymmetrical and neutral influence relations between each individual disappear.   Only mutual influence relation (mutual positive or negative impact) remains.

According to above computing results analysis, we observe that dyadic and triadic influence balance among agents has inherent relation with global bi-polarization pattern. This is similar to the macro-micro linkage: sentiment relations among agents (localized as triad and dyad) lead to the collective balance of the group.   In other words, we observe the micro foundation (at dyadic/triadic level) of the collective global pattern.

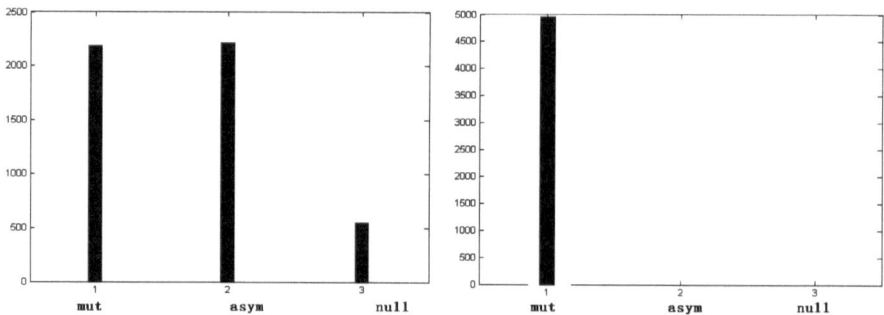

Fig. 10. Initial dyad distribution before bi-polarization of group opinions

Fig. 11. Final dyad distribution after bi-polarization of group opinions

5   Conclusion

In this paper, we address the implications of three types of social influence based on social identity theory. We investigate the non-positive social impact on group polarization based on Hopfield network model with both dyad and triad influence considered. By simulation we find that bi-polarization pattern tends to emerge with no imposing external intervention, and consensus may occur among group members if the non-positive influence is neglected.

Most literatures suggest that the homogeneous social influence will bring the global stability of social homogeneity, where convergence to one leading polarization is almost irresistible in a closely interconnected or interrelated population. However, in this paper the simulation based on Hopfield network model demonstrates that social homogeneous stable state is highly brittle if "influence ties" are either to be negative or zero. The result argues that bi-polarization may also be attributed to in-group/out-group differentiation and rejection antagonism, which conclusion is consistent with our former study [20].

It is worth pointing out that individuals can evolve into local dyadic and triadic balanced states after group opinion polarization. This conclusion shows that opinions polarization in a group is coexisted with local level structure balance, which reveals some interesting internal connection between global collective pattern and local social structure stability.

Acknowledgments. This research was supported by National Basic Research Program of China under Grant No. 2010CB731405, National Natural Science Foundation of China under Grant No.71171187.

References

1. Raafat, R.M., Chater, N., Frith, C.: Herding in humans. Trends in Cognitive Sciences 13, 420–428 (2009)
2. Mannes, A.E.: Are we wise about the wisdom of crowds? the use of group judgments in belief revision. Management Science 55(8), 1267–1279 (2009)

3. Bikhchandani, S., Hirshleifer, D., Welch, I.: Learning from the Behavior of others: Conformity, Fads, and Informational Cascades. Journal of Economic Perspectives 12, 151–170 (1998)
4. Watts, D.J., Strogatz, S.: Collective dynamics of "small-world" networks. Nature 393(6684), 440–442 (1998)
5. Heider, F.: Attitudes and Cognitive Organization. Journal of Psychology 21, 107–112 (1946)
6. Cartwright, D., Harary, F.: Structural balance: A generalization of Heider's theory. Psychological Review 63, 277–292 (1956)
7. Asch, S.E.: Effects of group pressure upon the modification and distortion of judgment. In: Guetzkow, H. (ed.) Groups, Leadership and Men. Carnegie Press, Pittsburgh (1951)
8. Dunia, L.P., Watts, D.J.: Social Influence, Binary Decisions and Collective Dynamics. Rationality and Society 20, 399–443 (2008)
9. Rane, H., Sumra, S.: Social media, social movements and the diffusion of ideas in the Arab uprisings. Journal of International Communication 18(1), 97–111 (2012)
10. Axelrod, R.: The dissemination of culture: A model with local convergence and global polarization. Journal of Conflict Resolution 41(2), 203–226 (1997)
11. Castellano, C., Vilone, D., Vespignani, A.: Incomplete Ordering of the voter model on Small-World Networks. EuroPhysics Letters 63(1), 153–158 (2003)
12. Tajfel, H.: Social identity and intergroup relations. Cambridge University Press, Cambridge (1982)
13. Galam, S.: Fragmentation versus stability in bimodal coalitions. Physica A 230, 174–188 (1996)
14. Leskovec, J., Huttenlocher, D., Kleinberg, J.: Predicting Positive and Negative Links in Online Social Networks. In: ACM WWW International Conference on World Wide Web (2010)
15. Holland, P.W., Leinhardt, S.: A method for detecting structure in sociometric data. American Journal of Sociology 70, 492–513 (1970)
16. Watts, D.J., Strogatz, S.: Collective dynamics of "small-world" networks. Nature 393(6684), 440–442 (1998)
17. Wasserman, S., Faust, K.: Social Network Analysis: Methods and Applications. Cambridge University Press, Cambridge (1994)
18. Macy, M.W., Kitts, J.A., Flache, A.: Polarization in Dynamic Networks A Hopfield Model of Emergent Structure. In: Breiger, R., Carley, K., Pattison, P. (eds.) Dynamic Social Network Modeling and Analysis: Workshop Summary and Papers, pp. 162–173. The National Academies Press, Washington DC (2003)
19. Butts, C.T.: Social Network Analysis with sna. Journal of Statistical Software 24(6), 1–51 (2008)
20. Li, Z., Tang, X.: Group Polarization and Non-positive Social Influence: A Revised Voter Model Study. In: Hu, B., Liu, J., Chen, L., Zhong, N. (eds.) BI 2011. LNCS, vol. 6889, pp. 295–303. Springer, Heidelberg (2011)

Combined Optimal Wavelet Filters with Morphological Watershed Transform for the Segmentation of Dermoscopic Skin Lesions

Alaa Ahmed Abbas, Wooi-Haw Tan, and Xiao-Ning Guo

Faculty of Engineering, Multimedia University, 63100 Cyberjaya, Selangor, Malaysia
alaa.abayechi@gmail.com,
{twhaw,guo.xiaoning}@mmu.edu.my

Abstract. In this paper, a technique is proposed to segment skin lesions from dermoscopic images through a combination of watershed transform and wavelet filters. In our technique, eight types of wavelet filters such as Daubechies and bi-orthogonal filters were applied before watershed transform. The resulting image was then classified into two classes: background and foreground. As watershed transform generated many spurious regions on the background, morphological post-processing was conducted. The post-processing split and merged spurious regions depending on a set of predefined criteria. As a result, a binary image was obtained and a boundary around the lesion was drawn. Next, the automatic boundary was compared with the manually delineated boundary by medical experts on 70 images with different types of skin lesions. We have obtained the highest accuracy of 94.61% using watershed transform with level 2 bi-orthogonal 3.3 wavelet filter. Thus, the proposed method has effectively achieved segmentation of the skin lesions, as shown in this paper.

Keywords: Segmentation, dermoscopic images, watersheds transform, wavelets transform, region merging.

1   Introduction

For the last two decades, automated systems have been designed to help physicians make decisions on complex diseases, such as dermatological diseases. Melanoma is one of the most dangerous dermatological diseases in humans; some types of this disease involve malignant or benign skin tumours. It penetrates rapidly and deeply into the skin and thus has increasing mortality rates. The main symptom of melanoma is a change in the colour, shape, and texture of spots on the skin. Physicians should detect melanoma in its early stage before it spreads.

In this paper, we propose a technique to classify melanoma with the use of dermoscopic images. The first stage in our technique is image segmentation. The aim of segmentation is to change the representation of an image into objects for easier analysis.

P. Anthony, M. Ishizuka, and D. Lukose (Eds.): PRICAI 2012, LNAI 7458, pp. 722–727, 2012.
© Springer-Verlag Berlin Heidelberg 2012

2   Related Works

Although watershed transformation is a common technique for image segmentation, its application in automatic medical image segmentation is plagued by over-segmentation and sensitivity to noise. Random walk is a probabilistic approach which is also used to solve the problem in the watershed algorithm by reducing the processing time and by comparing the image directly with the original image with few watershed regions that did not merge [1]. The combination of watershed transform with a hierarchical merging process is proposed to reduce noise and to preserve edges through application on two-dimensional/three-dimensional magnetic resonance images [2]. Moreover, watershed algorithm based on connected components is proposed to improve watershed efficiency through adaptation of the connected component operator. However, the algorithm does not modify the principle of watershed segmentation and does not build watershed lines [3]. Partial differential equations for image de-noising are combined with watershed segmentation to enhance the merging of edges and regions through application of the equations on both MR and natural images [4].

Gaussian scale mixtures in wavelet domain represent a method to remove noise from digital images; they are based on a statistical model that involves the coefficients of an over-complete multiscale-oriented basis and have been applied on images such as a boat, Lena, Barbara, and pepper [5]. In addition, robust watershed segmentation with the use of wavelets has been proposed to obtain a good gradient image. However, some small regions were not removed or merged based on the specified criteria [6]. K-means, watershed segmentation, and difference in strength map are combined to solve the problem of watershed in brain images, but this combination is too sensitive in the selection of the correct threshold values and depending on K-means results [7].

This paper presents a technique that uses wavelet transform to decompose the image prior to watershed transformation to remove over-segmentation and enhance the image. In addition, the segments are post-processed by merging and removing spurious segments based on predefined criteria as to produce more discernible results.

3   Methodology

In order to prepare the image for the subsequent processes, a 3 by 3 median filter is applied to remove white noise and small or thin hairs from the image. This would enhance the quality of watershed segmentation and to obtain images with well-defined boundaries as well as to remove small noise and enhance image contrast.

Wavelet transform (WT) is a mathematical tool that can be used to decompose the original image and to produce perfect reconstructions. WT is classified into two categories: discrete wavelet transform (DWT) and continuous WT. WT is better than Fourier transform because the former obtains not only frequency information but also temporal information. DWT decomposes an original image into sub-images with three details and one approximation of a low–high filter associated with the mother wavelet and the downsampling.

Wavelets come in different forms, such as Haar, Daubechies, Symlets, Coiflets, Bi-orthogonal and Meyer, among others. Bi-orthogonal WT is used by numerous studies that proposed discrete image wavelet decomposition for two detailed images in the horizontal and the vertical directions in the same mother wavelet. The goal is to obtain smooth images at each scale 2^j for $j = 1, \ldots, 4$ for gradient estimation. At scale 2^j, good approximations of the detail images along horizontal and vertical direction and edge magnitudes in the same scale can be calculated as follows [8-9]:

$$M_{2^j} = \sqrt{\left(W^1 f\right)^2 + \left(W^2 f\right)^2} \ . \tag{1}$$

where $W^1 f$ and $W^2 f$ are the wavelet coefficients

$$W_\psi^i (i, m, n) = \frac{1}{\sqrt{MN}} \sum_{x=0}^{M-1} \sum_{y=0}^{N-1} f(x, y) \psi_{j,m,n}^i (x, y), i = \{H, V, D\} \ . \tag{2}$$

where ψ is a measure of variation along horizontal edges, vertical edges and diagonal.

Inverse WT is applied to enhance edges. Two requirements must be met to obtain good performance for the denoising image. Firstly, signal information should be extracted from noisy wavelet coefficients with the use of inter scale models. Secondly, a high degree of agreement must exist between wavelet coefficient distribution and Gaussian distribution.

Watershed transform is one of the morphological approaches for the segmentation of objects within the gradient images. It generates serious problems such as over or under-segmentation that results from badly contrasted images when the algorithm is used directly on gradient data for low-contrast images. The idea of watershed as the input for a 2D greyscale image involves bright regions as peak and dark parts as valley. If the valley is filled with water, it can meet one or more of the neighbouring valleys. If we want to prevent flooding water from one valley to another, then a dam of one pixel width must be secured to avoid the merging of catchment basins that separate individuals by a watershed line. These dams should also be high enough to prevent the water from spilling at any point of the process. This process is repeated until the water rises to the height of the highest peak in the image [10].

Researchers have achieved good image segmentation with the use of several approaches to merge watershed regions. For example, Weicker used the contrast difference between adjacent regions as merging criterion [4]. Some researchers also used the minimum size of a region and the minimum edge intensity to separate adjacent regions. Their idea for merging regions include weak or wide borders or borders that lie near segmented areas through the use of two threshold values to determine the border [11].

In this work, we determined two of the largest regions that have a lesion smaller in size than a healthy skin region. In addition, we also computed the distance between the lesion and the other regions in order to determine those regions that are outside the lesion. Thus, we merged the regions close to or inside the maximum area. Conversely, those regions that have small areas and have diameters that are the same as lines or points are removed to reduce the number of segmented regions. This would produce an image with two areas, which are the lesion and the skin. The methodology of the proposed technique is summarized in the flowchart shown in Fig. 1.

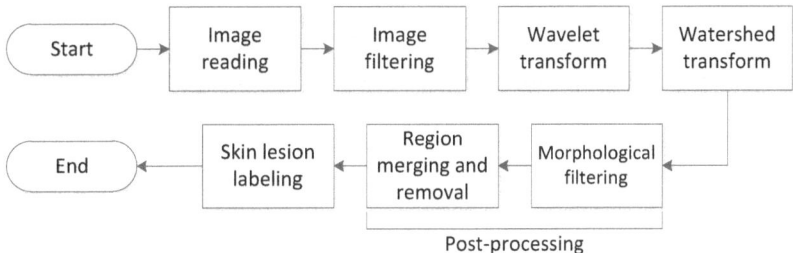

Fig. 1. The flowchart of the proposed technique

4   Experimental Results and Evaluation

In this paper, we used 70 dermoscopic images, each with a size of 200 × 200 pixels. The goal was to select the optimal wavelet filters among the following eight types of wavelets: bior1.1, bior1.3, bior2.2, bior2.4, bior3.3, db2, db3, and db4. Before we applied the wavelet filters, we first converted the images into gray level. In the experiment, the ground truth is a manual border drawn by dermatologists to compare the manual border with the automatic one. Three performance metrics, sensitivity, specificity and accuracy, are used to assess the efficiency of each type of wavelet filter by comparing automatic borders with manual borders.

In the experiment, the true positive (TP) is the number of overlapping pixels in the lesion; the true negative (TN) is the number of overlapping pixels outside the lesion; the false positive (FP) is the number of overlapping pixels between the automatically labelled lesion and those outside the manually labelled lesion; the false negative (FN) is the number of overlapping pixels between the manually labelled lesion and those outside the automatically labelled lesion.

Another metric known as similarity is used to select the best wavelet filter in level 2^j by comparing automatic border and manual border drawn by expert dermatologists. The similarity is computed based on the following expression [12]:

$$Similarity = \frac{2 \times TP}{2 \times TP + FN + FP}. \tag{3}$$

5   Experimental Results and Discussion

We applied eight types of wavelet filters with four levels and determined the border of each image after the combination of watershed with WT. We then determined the edge of the lesion and compared it with the manual border. Moreover, the proposed technique was compared with thresholding, watershed and both combined.

The performance of each type of wavelet filter was computed by comparing the automatic border with the manual border. The best one or the one with the highest accuracy, specificity and similarity is bior3.3 in level 2 ($j = 2$). Level 2 in all types of wavelet filters obtained higher accuracy than level 3 and 4 ($j = 3, 4$). Moreover, the accuracy of bi-orthogonal WT is higher than those of the Daubechies (db2, db3, db4) after WT is combined with watershed. These results are shown in Table 1. Our method also obtained better results when compared to other methods as shown in Table 2. Nevertheless, 2% of the images have exhibited poor results. Some results are shown in Fig. 2 to illustrate the performance of the proposed technique.

Table 1. The results of applying eight types of wavelet for $j = 2$ before watershed transform

Methods	Sensitivity	Specificity	Accuracy	Similarity
Bi-orthogonal 1.1 (bior1.1)	87.75	97.83	94.23	91.71
Bi-orthogonal 1.3 (bior1.3)	89.62	96.65	94.19	91.97
Bi-orthogonal 2.2 (bior2.2)	88.54	97.90	94.36	92.21
Bi-orthogonal 2.4 (bior2.4)	89.27	97.30	94.53	92.40
Bi-orthogonal 3.3 (bior3.3)	88.60	98.21	**94.61**	92.45
Daubechies 2 (db2)	89.51	97.04	94.46	92.42
Daubechies 3 (db3)	89.04	97.35	94.44	92.32
Daubechies 4 (db4)	88.98	97.19	94.18	92.11

Table 2. Comparisons with three standard methods

Methods	Sensitivity	Specificity	Accuracy	Similarity
Thresholding	85.48	94.82	92.02	89.01
Watershed RGB	86.55	97.38	92.61	89.86
Watershed and thresholding	84.91	98.35	92.50	89.41

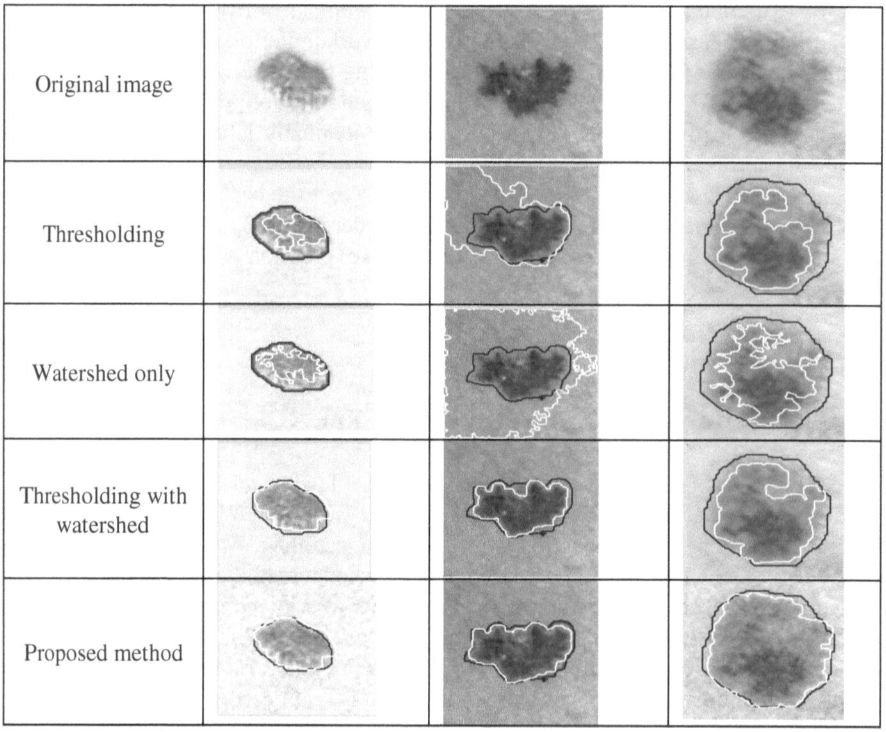

Fig. 2. Comparisons between automatic border (white) and manual border (blue)

6   Conclusion and Future Work

In this paper, we propose the segmentation of skin lesion from the dermoscopy images by combining watershed transform with wavelet filters. The highest accuracy was achieved with level 2 bior3.3 wavelet filter combined with watershed transform. Our method effectively reduced the problems of over-segmentation in watershed transform. Moreover, our proposed method facilitated good object recognition, as was demonstrated in this paper. In the future, we plan on refining the watershed method with the use of other approaches.

Acknowledgement. The authors would like to thank Dr. Joaquim M. da Cunha Viana, for his kindness in providing the dermoscopic images for use in this research work.

References

1. Malik, S.H., Aihab, K., Amna, B.: Modified Watershed Algorithm for Segmentation of 2D Images. Issues in Informing Science and Information Technology 6, 877–886 (2009)
2. Kostas, H., Serafim, N., Efstratiadis, N.M., Aggelos, K.K.: Hybrid Image Segmentation Using Watersheds and Fast Region Merging. IEEE Transactions on Image Processing 7, 1684–1699 (1998)
3. Bieniek, A., Moga, A.: An Efficient Watershed Algorithm Based on Connected Components. Pattern Recognition 33, 907–916 (2000)
4. Weickert, J.: Efficient Image Segmentation Using Partial Differential Equations and Morphology. Pattern Recognition 34, 1813–1824 (2000)
5. Javier, P., Vasily, S., Martin, J.W., Eero, P.S.: Image Denoising Using Scale Mixtures of Gaussians in the Wavelet Domain. IEEE Transactions on Image Processing 12, 1338–1350 (2003)
6. Scharcanski, J., Jung, C.R., Clarke, R.T.: Adaptive Image Denoising Using Scale and Space Consistency. IEEE Transactions on Image Processing 11, 1092–1101 (2002)
7. Naser, S.: Image Segmentation Based on Watershed and Edge Detection Techniques. The International Arab Journal of Information Technology 3, 104–110 (2002)
8. Stephane, M., Sifen, Z.: Characterization of Signals from Multiscale Edges. IEEE Transactions on Pattern Analysis and Machine Intelligence 14, 710–732 (1992)
9. Rafael, C.G., Richard, E.W.: Digital Image Processing. Prentice-Hall, New Jersey (2010)
10. Jung, C.R., Scharcanski, J.: Robust Watershed Segmentation Using Wavelets. Image and Vision Computing 23, 661–669 (2005)
11. Manuela, P., Mario, F.: Biomedical Diagnostics and Clinical Technologies: Applying High-performance Cluster and Grid Computing. IGI Global, Hershey (2011)
12. Rahil, G., Mohammad, A., Celebi, M.E., Alauddin, B., Constantinos, D., George, V.: Automatic Segmentation of Dermoscopy Image Using Histogram Thresholding on Optimal Colour Channels. International Journal of Medical Sciences 1, 125–134 (2010)

Predicting Academic Emotions Based on Brainwaves, Mouse Behaviour and Personality Profile

Judith Azcarraga and Merlin Teodosia Suarez

Center for Empathic Human-Computer Interactions,
De La Salle Universtiy, Manila, Philippines
{jay.azcarraga,merlin.suarez}@delasalle.ph

Abstract. This research presents how the level of academic emotions such as confidence, excitement, frustration and interest, may be predicted based on brainwaves and mouse behaviour, while taking into account the student's personality. Twenty five (25) college students of different personalities were asked to use the Aplusix® algebra learning software while an EEG sensor was attached to their head to capture their brainwaves. Brainwaves were carefully synchronized with the mouse behaviour and the assigned student activity. The collected brainwaves were then filtered, pre-processed and transformed to different frequency bands (alpha, beta, gamma). A number of classifiers were then built using different combinations of frequencies and mouse information which were used to predict the intensity level (low, average, high) of each emotion.

Keywords: Academic Emotions, Brainwaves, Mouse Behavior, Affective Computing, Learning Systems.

1 Affective Computing

Since learning is greatly affected by the emotional state of a student, a learning system that is aware of a student's emotion may be able to provide more appropriate and timely feedback, thus improving the over-all efficiency of the system. This research presents how the level of academic emotions such as confidence, excitement, frustration and interest [10], may be predicted based on brainwaves and mouse behaviour, while taking into account student's personality and tutorial event.

A student may experience a wide range of emotions during a learning process which can eventually influence his/her cognitive performance. Positive emotions can enhance memory processes such as recall and problem solving [10]. Negative emotions, on the other hand, such as frustration, anger and confusion, may lead to loss of interest in learning, self-destructive thinking or giving up of the task [7].

Emotions can be expressed in many ways and most current researches on emotion recognition consider external features such as facial expression, voice, posture and gestures [1] as well as internal features such as the brainwaves. For studies that recognize emotion based on the brainwaves, the electroencephalogram (EEG) is a common sensor used to capture electro-magnetic signals of brainwaves.

P. Anthony, M. Ishizuka, and D. Lukose (Eds.): PRICAI 2012, LNAI 7458, pp. 728–733, 2012.
© Springer-Verlag Berlin Heidelberg 2012

In [4], brainwaves are combined with other physiological signals from skin conductance, respiration, blood volume pulse and finger temperature in order to assess emotions under different emotional stimulations. In a previous work by the authors [3], brainwaves are combined with mouse behavior information in predicting academic emotions, while in [2], the user personality profile is also taken into account.

The use of the mouse in assessing the emotional state of its user had been explored in [13], where a special biometric mouse is used to assess the user's emotional state and labor productivity. Similarly, in [12], with the use of a standard input mouse, the mouse clicks of users while engaged in a frustrating task was investigated.

As to the learners' personality profile, some researches have explored its influence in predicting emotions, infer student goals, interaction patterns, student actions, emotional reaction to a tutor event and even performance [2][5]. In [14], it is suggested that personality traits (i.e. accommodation) can affect the thinking style of a person.

2   Experimental Set-Up and Methodology

Twenty five (25) undergraduate students were taken as subjects for this study. The participants were asked to solve problems using the Aplusix® [9] algebra learning software while an EEG sensor with 14 channels was attached to their head. Four algebra problems were purposely designed to be of varying difficulty levels in order to induce the four emotions (hopefully in different intensities).

Prior to the learning session, the participants were asked to take a personality test, and were asked to relax, before they were given instructions on how to use the software. With the EEG sensor attached to the head, the relaxation period lasts for about 3 minutes, with eyes closed .

During the learning session, which last for about 15 minutes, brainwaves and mouse behavior information are automatically captured by a special software module. Every two minutes, the student is asked to report his/her level of confidence, excitement, frustration and interest as well as his/her assessment of the difficulty of the task. The design of the self-reporting window is designed to as unobstrusive as possible – with just 5 sliding bars - one for each of the four emotions and the fifth for the perceived difficulty of the task. The students did not find the sliding bars distracting and they would use very little time in moving the bars left/right.

3   Data Pre-processing and Data Preparation

To find which frequency band(s) would yield better performance in predicting the intensity of a particular emotion, twenty datasets (20) were formed according to the different combinations of the four (4) emotions, brainwaves frequency bands, and mouse information (alpha, beta, gamma, all brainwaves, all brainwaves & mouse). Each dataset were further divided into training and test sets. From the 25 participants, 20 were selected for the training set (11 male and 9 female) and 5 were randomly selected for the test set (3 male and 2 female).

For all the datasets, each instance is tagged with the intensity level of the emotion associated with the dataset. Each instance is composed of synchronized brainwaves, mouse behavior information, assigned student activity, reported difficulty level and user personality score. Each instance covers a 2-second epoch of brainwaves and mouse information, implemented with 1-second overlap between instances.

Raw brainwaves are transformed into different frequency bands in order to remove noise, to filter them and to extract useful information for data analysis [11]. In this study, raw brainwave data from the 14-channel EEG sensor were transformed using Fast Fourier Transform (FFT). For each band, peak magnitude, peak power and mean spectral power were computed using functions of GNU Octave®. Moreover, filtering techniques such as BandPass filter and Moving Average are applied. BandPass filter allows the signal to be segmented into Alpha (8-12 Hz), Beta with sub-bands Beta High (20-30 Hz) and Beta Low (12-20 Hz), and Gamma (30 Hz and above) bands.

Brainwaves that were captured during the relaxed period were used to compute the baseline of each participant. The average of each brainwave feature serves as the baseline value of that particular feature and of that particular participant. The transformed frequency value taken during the tutorial session is subtracted from its corresponding baseline value in order to determine how far it deviates from its baseline.

The reported score for confidence, excitement, frustration, interest task difficulty are converted to their corresponding intensity level (*low, average, high*). This is done by computing the mean and standard deviation of that particular emotion value of all the subjects. The label *low* was given to instances with emotion value 1 standard deviation below the mean. The label *high* was given to the instances with emotion value 1 standard deviation above the mean and the label *average* was given to all other instances near the mean. Such intensity level is used to tag each 2-second segment instance that occurs for the last 2-minute time period of the corresponding emotion. Moreover, no segments were included in the dataset during the annotation period.

Based on the mouse information, mouse behavior features were computed, such as the Euclidean distance (spatial displacement) of mouse positions, the total number of clicks, and the average click duration over a 2-second time period.

The log file produced by the algebra software is used to determine which activity a student is engaged in - *answering, thinking* or *hinting*. "Answering" occurs when the participant has typed an answer. The participant is said to be "thinking" when the participant has not typed anything. "Hinting" occurs when the participant asked for a hint or received a warning from the software.

Each participant was made to take an online Big 5 personality test (http://www.similarminds.com/big5.html) which provides percentile scores for five personality factors such as extroversion, orderliness, emotional stability, accommodation and inquisitiveness [14].

4   Results and Discussion

The accuracy of predicting the level of intensity for each emotion (low, average, high) is evaluated using several machine learning and computational intelligence techniques of WEKA [8] such as Support Vector Machines (SVM), k-Nearest Neighbor (kNN) and C4.5 Decision Trees. For the kNN, the value of k was set to 5. Table 1 provides the performance of the classifiers based on f-measure.

One of the goals of this research is to investigate which frequency bands, if any, are relevant in predicting the intensity of each emotion. The results, indeed, show that certain frequency bands are more significant than the others in predicting some emotional intensities. Based on the weighted average in Table 1, the level of intensity for *confidence* can be predicted with an average of 82% for SVM when alpha waves features are used. High prediction rate (74%) was also achieved based on beta waves

Table 1. F-Measure values in classifying each emotion intensity

Classifier		Alpha (α)	Beta (β)	Gamma (γ)	α+β+ γ	α+β+υ+ Mouse	Alpha (α)	Beta (β)	Gamma (γ)	α+β+ γ	α+β+γ+ Mouse
		Confidence					Frustration				
C4.5	Low	0.64	0.55	0.42	0.04	0.08	0.00	0.00	0.00	0.00	0.00
	Ave	0.60	0.79	0.52	0.58	0.60	0.83	0.54	0.78	0.78	0.78
	High	0.41	0.54	0.37	0.39	0.38	0.49	0.23	0.45	0.54	0.38
	WtdAve	0.57	**0.74**	0.49	0.53	0.54	**0.74**	0.47	0.70	0.71	0.69
kNN k=5	Low	0.14	0.03	0.18	0.17	0.20	0.09	0.02	0.14	0.05	0.09
	Ave	0.49	0.48	0.60	0.69	0.63	0.50	0.71	0.73	0.78	0.72
	High	0.52	0.41	0.44	0.38	0.40	0.32	0.24	0.23	0.16	0.14
	WtdAve	0.48	0.45	0.56	0.62	0.58	0.45	0.61	0.62	0.65	0.61
SVM	Low	0.86	0.74	0.85	0.82	0.84	0.01	0.00	0.00	0.08	0.10
	Ave	0.86	0.60	0.57	0.84	0.83	0.24	0.58	0.72	0.53	0.62
	High	0.61	0.39	0.36	0.55	0.55	0.37	0.43	0.44	0.41	0.40
	Wtd Ave	**0.82**	0.58	0.55	0.79	0.79	0.25	0.53	0.65	0.49	0.56
		Excitement					Interest				
C4.5	Low	0.00	0.00	0.00	0.00	0.00	0.00	0.00	0.00	0.00	0.00
	Ave	0.80	0.63	0.80	0.82	0.82	0.59	0.65	0.49	0.72	0.72
	High	0.18	0.34	0.18	0.24	0.28	0.00	0.00	0.00	0.00	0.00
	Wtd Ave	0.67	0.56	0.67	**0.70**	**0.70**	0.48	0.52	0.39	**0.58**	**0.58**
kNN k=5	Low	0.12	0.00	0.11	0.04	0.07	0.00	0.00	0.00	0.00	0.00
	Ave	0.61	0.54	0.58	0.74	0.62	0.54	0.56	0.51	0.69	0.70
	High	0.36	0.29	0.24	0.13	0.11	0.00	0.00	0.00	0.04	0.07
	Wtd Ave	0.55	0.47	0.51	0.62	0.51	0.43	0.45	0.41	0.56	0.57
SVM	Low	0.00	0.00	0.00	0.00	0.00	0.00	0.00	0.00	0.00	0.00
	Ave	0.53	0.54	0.66	0.47	0.48	0.44	0.49	0.46	0.50	0.50
	High	0.36	0.14	0.16	0.03	0.02	0.00	0.00	0.00	0.00	0.00
	Wtd Ave	0.49	0.46	0.55	0.38	0.39	0.35	0.39	0.37	0.40	0.40

using C4.5. For *excitement*, using alpha or gamma waves, the average accuracy yielded 67%. However, the accuracy is higher, reaching up to 70%, when all the brainwaves features are used. Good results were likewise achieved for *frustration* reachingup to 74% by using just the alpha waves features. On the other hand, all the brainwaves features have to be combined to achieve at most 58% accuracy rate in predicting level of intensity for *interest*.

Based on the results, alpha waves are significant in predicting intensity levels of various emotions such as *confidence, excitement* and *frustration*. Alpha waves are also explored in [6], where the difference between upper and lower alpha frequencies were used to study the brainwaves of students as they solve complex problems.

Table 1 shows high accuracy mostly in predicting *average* intensity. This may be attributed to the imbalanced datasets since the majority class is *average*. However, the results in predicting *low* and *high* classes, although poor, may be considered significant and the features that contribute to such intensities need further investigation. A tutoring system is considered effective if it intervenes only when a student is highly frustrated or low in confidence. It would be very important for such system to detect when these events start to occur in order to provide a timely intervention. Thus, we would like to focus not so much on the average but on the low and high intensities because these are the critical emotion states for such systems. Table 1 clearly also shows that the level of *confidence* can be well predicted using the SVM approach. The accuracy of predicting low level of *confidence* has reached up to 86% whereas predicting a high level of confidence has reached 61% accuracy. In the case of *frustration*, a high level can only be predicted with 54% accuracy, and worse, a low level frustration was almost always incorrectly classified (almost 0%). Incorrect classification also occurred in predicting low *excitement* and for both low and high *interest*. High intensity level for *excitement* can only be predicted up to 36%. This may imply that the level of interest is difficult to predict by just using the brainwave features. Other features, maybe from other physiological devices, may be necessary.

5   Conclusion and Future Work

In designing effective tutoring systems, it is important for the system to monitor whether a learner is experiencing some negative emotions that may affect his/her learning performance. Awareness of such emotions and their intensity level is necessary in order for the system to have timely intervention and scaffolding.

This research attempts to predict the intensity level of academic emotions such as *confidence, excitement, frustration* and *interest* based on brainwaves, mouse behavior, personality profile and tutorial activity. Different frequency bands were investigated to be able to determine which band is significant in predicting different emotion intensities. Alpha waves were found to be significant in predicting intensity levels of *confidence, excitement* and *frustration*. Low level of *confidence* can be predicted up to 86% while high level of *confidence* can be predicted with up to 61% accuracy, using alpha frequency features only. High *frustration* can only be predicted correctly up to 54% using all the frequencies. Low performance in predicting intensity of *frustration, interest* and *excitement* would suggest that other feature selection and data mining techniques need to be considered. Further pre-processing of the data may also be necessary. Moreover, classification might improve if different classification models are created for different personality types.

Acknowledgements. We thank all the subjects who participated in this experiment and the Philippine Council for Industry, Energy and Emerging Technology Research and Development (PCIEERD) of the Department of Science and Technology (DOST) of the Philippine government for all the support in conducting this research.

References

1. Arroyo, I., Cooper, D.G., Burleson, W., Woolf, B.P., Muldner, K., Christopherson, R.M.: Emotion Sensors Go To School. In: Dimitrova, V., Mizoguchi, R., Du Boulay, B., Graesser, A. (eds.) AIED, pp. 17–24. IOS Press (2009)
2. Azcarraga, J.J., Ibanez, J.F., Lim, I.R., Lumanas, N.: Use of Personality Profile in Predicting Academic Emotion Based on Brainwaves and Mouse Behavior. In: 2011 Third International Conference on Knowledge and Systems Engineering, pp. 239–244 (2011)
3. Azcarraga, J., Ibanez, J.R.: J.F., Lim, I.R., Lumanas JR, N., Trogo, R., Suarez, M.T.: Predicting Academic Emotion based on Brainwaves Signals and Mouse Click Behavior. In: Hirashima, T., et al. (eds.) 19th International Conference on Computers in Education (ICCE), pp. 42–49. NECTEC, Thailand (2011)
4. Chanel, G.: Emotion assessment for affective computing based on brain and peripheral signals. University of Geneva (2009)
5. Conati, C., Maclaren, H.: Empirically building and evaluating a probabilistic model of user affect. User Modeling and User Adapted Interaction 19, 267–303 (2009)
6. Jausovec, N., Jausovec, K.: EEG activity during the performance of complex mental problems. International Journal of Psychophysiology 36, 73–88 (2000)
7. Kort, B., Reilly, R., Picard, R.: An Affective Model of interplay between emotions and learning: Reengineering Educational Pedagogy-Building a Learning Companion. In: International Conference on Advanced Learning Technologies (2001)
8. Hall, M., Frank, E., Holmes, G., Pfahringer, B., Reutemann, P., Witten, I.H.: The WEKA Data Mining Software. SIGKDD Explorations 11 (2009)
9. Nicaud, J.-F., Bouhineau, D., Huguet, T.: The Aplusix-Editor: A New Kind of Software for the Learning of Algebra. In: Cerri, S.A., Gouardéres, G., Paraguaçu, F. (eds.) ITS 2002. LNCS, vol. 2363, pp. 178–187. Springer, Heidelberg (2002)
10. Pekrun, R., Goetz, T., Titz, W., Perry, R.P.: Academic Emotions in Students' Self-Regulated Learning and Achievement: A Program of Qualitative and Quantitative Research. Routledge (2002)
11. Sanei, S., Chambers, J.A.: EEG signal processing. Wiley-Interscience (2007)
12. Scheirer, J., Fernandez, R., Klein, J., Picard, R.W.: Frustrating the user on purpose: a step toward building an affective computer. Interacting with Computers 14, 93–118 (2001)
13. Zavadskas, E.K., Kaklauskas, A., Seniut, M., Dzemyda, G., Ivanikovas, S., Stankevic, V., Simkevicius, C., Jarusevicius, A.: Web-based biometric mouse intelligent system for analysis of emotional state and labour productivity. In: The 25th International Symposium on Automation and Robotics in Construction ISARC 2008, pp. 429–434 (2008)
14. Zhang, L.F.: Thinking styles and the big five personality traits revisited. Personality and Individual Differences, 11 (2006)

Extraction of Chinese Multiword Expressions Based on Artificial Neural Network with Feedbacks

Yiwen Fu[*], Naisheng Ge, Zhongguang Zheng, Shu Zhang,
Yao Meng, and Hao Yu

Fujitsu R&D Center Co.,Ltd.
Beijing P.R. China 100025
{art.yiwenfu,genaisheng,zhengzhg,zhangshu,
mengyao,yu}@cn.fujitsu.com

Abstract. Multiword Expressions present idiosyncratic features in the application of Natural Language Processing. This paper focuses on Multiword Expressions extraction from bilingual corpus with alignment information constructed by Statistical Machine Translation (SMT) and word alignment method. A pattern based extraction system and an Artificial Neural Network (ANN) with feedback are applied for extracting MWEs. The results show that both of these approaches can achieve satisfying performance.

1   Introduction

In many literatures, multiword expressions (MWEs) are interpreted as a sequence of words which has fixed structure or combination and carries a certain meaning. According to [1], MWEs are interpreted as "idiosyncratic interpretations that cross word boundaries (or spaces) ". Researchers put their focus on some specific groups of MWEs such as phrasal verbs, light verbs, idioms, adjectives and nouns [2-4]. In recent decades, many studies are conducted on extraction of MWEs of various languages based on lexical information and grammatical information. These languages include English, Greek, French and so on [5]. There are mainly three ways to achieve desired MWEs. First, extract MWEs based on lexical information (contingency table[6] and Loglike measure[7] for example). Second, extract MWEs based on grammatical data which refers to syntactic and morphological information. The last approach refers to a hybrid way which applies both lexical and grammatical information.

This paper focuses on Chinese MWEs which refer to those idiomatic phrases and proper nouns which are built up by more than one Chinese segmentations and have specific part of speech (POS) in the context, and are commonly used in a certain domain. Since POS tagging is not reliable for Chinese in many cases, Statistical Machine Translation (SMT) is introduced to assist in analyzing Chinese MWEs through corresponding English translation. This paper is based on the hypothesis that SMT is capable of making the potential MWEs more clear.

---

[*] Corresponding author.

P. Anthony, M. Ishizuka, and D. Lukose (Eds.): PRICAI 2012, LNAI 7458, pp. 734–739, 2012.

2   Multiword Expressions Extraction

In this section, pattern based extraction system and ANN are introduced. Pattern based extraction system extract those Chinese MWEs whose corresponding English chunk matches with defined patterns. Generated MWEs by pattern based system are applied to ANN as training set. The Statistical Machine Translation (SMT) implemented in this paper is based on the hierarchical phrase-based translation model (HPB)[8]. Word alignment refers to finding the translation correspondences between target language and source language. We use Giza++[9] which is a widely used to obtain word alignment automatically.

2.1   Data Set

We apply several patents in Chinese as the data set. SMT is applied to translate Chinese patent into English and provides alignment information from Chinese to English. After translation, we get the data as shown in Fig. 1.

---

the/DT hybrid/JJ fuel/NN control/NN system/NN preferably/RB includes/VBZ a/DT plurality/NN of/IN for/IN sensing/VBG the/DT diesel/NN fuel/NN blend/VB and/or/JJ additional/JJ sensor/NN operating/VBG parameters/NNS of/IN the/DT system/NN ./.

该 混合 燃料 控制 系统 优选 包括 多 个 用于 感应 柴油机 和 / 或 附加 混合 燃料 系统 的 工作 参数 的 传感器 。

1:1 2:2 3:3 4:4 5:5 6:6 7:7 9:8 9:9 11:10 12:11 14:12 15:18 16:17 17:13 17:14 17:15 18:16 19:24 20:21 21:22 22:20 24:19 25:25

Fig. 1. Example of Source Data Set

As shown in Fig. 1, the first block contains English sentence translated by SMT. In addition, Stanford POS tagger[10] is applied to English translation and the POS tagging information is maintained. The second block enclose original Chinese sentence which is separated into several segmentations. Alignment information is presented in the third block in which there are many number pairs (num1:num2). The first number in the number pair is the index of the word in English sentence and the second number is the corresponding Chinese segmentation in Chinese sentence. In some cases, Chinese segmentation or English words does not have corresponding English chunk/word or Chinese segmentations and this phenomenon is called "empty alignment".

2.2   Pattern Based MWEs Extraction System

By analyzing the source data set, our linguistic expert find out a fairly simple way to extract MWEs from bilingual corpus constructed by SMT. We realize that most of the

Chinese MWEs sharing limited combination of POS patterns of corresponding English. In this case, linguistic expert designs plenty of POS patterns for extracting MWEs. If a match between English chunk's POS pattern and designed pattern is detected, the English chunks and corresponding Chinese is extracted.

The patterns we used are listed in Table 1: mark "+" means corresponding tag should be present at least once; sign "?" means appear once or do not appear. Pattern 5 and 6 are extracted with additional restrictions that English word must be mapped into more than one Chinese segmentations.

Table 1. Examples for MWEs Extraction Patterns

Index	POS tagging Pattern	Index	POS tagging Pattern	Index	POS tagging Pattern
1	(/DT)(/NN)+	4	(/DT)?(/JJ)+(/NN)+	7	((/IN)(/:)(/VV))
2	(/VV)+(/NN)+	5	(/JJ)	8	(/NN)+(/CC)(/NN)+
3	(/NN)(/NN)+	6	(/NN)	10	.etc

2.3   Artificial Neural Network with Feedbacks

In this paper, ANN is applied to extract MWEs due to its capability of modeling complex patterns. A simple structure of ANN is shown in Fig. 2 on the left side. Dots are neurons and arrows show the propagation of signal flows. Black lines indicate the connections between neurons. On the right side of Fig. 2, the structure of ANN in this paper is presented. In each iteration, the classifying result of previous instance is fed into ANN to classify current instances. In this paper, there are 154 neurons in the input layer and two neurons in the output layer. 78 neurons are settled in the hidden layer.

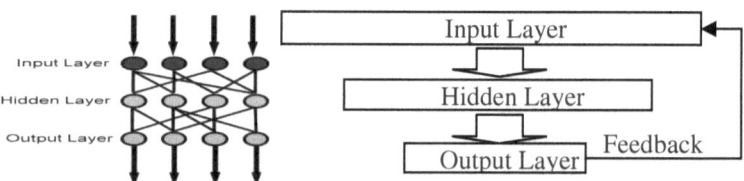

Fig. 2. Structure of A Simple ANN(Left)[11] Structure of ANN in This Paper(Right)

Supervised learning is applied as learning paradigm in this paper. Regarding to the 154 input factors, examples of input values are presented in Table 2 listed in column attribute1 to attribute 154. Instances are fed into ANN according to the order of appearance in Chinese sentence in terms of three grams. The three grams are constructed by central segmentation and two segmentations which are consecutive to central segmentation. For the first and the last instances, they do not have first element and the last element in the three grams respectively. POS of the three grams are applied as feature value. Statistical information of the Chinese three grams is fed into

ANN. Most importantly, the results of previous three grams are kept and reused in the next iteration of training and calculation. For the instance whose index is 1, feedback is always set as negative. Back-Propagation [12] is implemented to train the ANN.

Table 2. Data Structure Applied to ANN

Chinese	English(SMT)	Attribute1	Attribute2	Attribute3	...	Attribute154	Labels
最初	initial/JJ	1	2	23	...	False	False
施用	apply-ing/VBG	2	2	6	...	Feedback	False
引	pri-mer/NN	3	1	14	...	Feedback	True
物	pri-mer/NN	4	1	14	...	Feedback	True
的	of/IN	5	1	5	...	Feedback	False
步骤	step/NN	6	2	23	...	Feedback	False
将	will/MD	7	1	5	...	Feedback	False
...	...	..	..	...	...	...	...

3   Experiments

In this part, the steps of how experiments are conducted will be drawn. We believe that those patterns kept in SMT system can be applied to make MWEs clear. Following, we present two approaches to extract MWEs. In total, there are 1984 sentences 76,057 Chinese segmentations included in corpus.

For constructing ANN, pattern based MWEs extraction system is applied to generate training and testing data set. Extracted MWEs are check by human. Data set is saved in CSV format as shown in Table 2. One of third of the data set is applied as training data for ANN and two of third of data set is applied for testing.

Two experiments are conducted in Section 4. ANN with feedback and ANN without feedback are compared. In addition, ANN is compared with pattern based extraction system. ANN returns "True" or "False" to each Chinese segmentation. More than one consecutive "True" indicates that corresponding Chinese segmentations are one integral MWE.

4   Evaluation

Following, the results of pattern based and ANN MWEs extraction systems are depicted in terms of quantity of extracted MWEs, accuracy and average length.

In the beginning, we apply the same training data to ANNs with and without feedback structure. The accuracy of extracted MWEs by these two kinds of ANN is presented in Table 3. According to the results, it is obvious that ANN with feedback is much better than the ANN without feedback.

Table 3. Comparing the Accuracy of ANN with Feedback and without Feedback

Type of ANN(Neurons in Input, Hidden and Output Layer)	Accuracy
ANN without feedback(153,78,2)	75.0%
ANN with feedback(154,78,2)	81.0%

Following, the comparison between ANN with feedback and pattern based extraction system are conducted. Duplicated MWEs has been removed. Results are shown in Table 4 . Pattern based system and ANN have the similar performance.

Table 4. Comparison of MWEs between Pattern Based System and ANN

Method	Quantity of Extracted MWEs	Quantity of Complete MWEs	Accuracy	Average Length of MWEs
Pattern Based System	4106	3408	83%	7
ANN	4064	3292	82.4%	6

Although SMT can provide information for helping to extract MWEs, it still makes mistakes. Fig. 3 presents a counterexample of SMT translated Chinese sentence. The Chinese means that studying the influence of fish oil fat emulsion on the monocyte-macrophage cell function of cancer patients with intestinal obstruction. Incorrect translation leads to the failure extraction of "癌 性 肠梗阻 患者" by pattern based extraction system. However these wrong translation would not influence the extraction of "癌 性 肠梗阻 患者" by ANN. Alignment algorithm will also introduce mistakes which lead to incomplete MWEs extracted by both of the approaches. In addition, empty alignment lies in the head and the tail of MWEs will cause incomplete MWEs extracted by pattern based system. However, empty alignment nearly has no impact on ANN.

patients/NNS study/VBP fish/NN oil/NN fat/JJ emulsion/NN to/TO for/IN ileus/NN cancer/NN …
研究 鱼油 脂肪 乳剂 对 癌 性 肠梗阻 患者 单 核 巨 噬 细胞 功能 的 影响
1:9 2:1 3:2 4:2 5:3 6:4 7:5 9:7 9:8 10:6 11:17 13:10 13:11 15:12 15:13 15:14 16:15

Fig. 3. Example of Mistake made by SMT

Besides the recalled MWEs by ANN, ANN successfully identifies new MWEs which take 22% of the entire set of MWEs extracted by ANN. Both of the pattern based system and ANN have pros and cons. For pattern based system, designing a new pattern requires linguistic expert goes through entire data set. Fortunately, in this experiment, data set is relatively small. Improving the performance of pattern based system on a large data set is a tough work. The advantage of pattern based system is that results are completely and only dependent on designed patterns so it is easy to extend the range of extracted MWEs. However, the result of ANN is unpredictable through its weights. The merit of ANN is that ANN is not heavily dependent on SMT and word alignment results. We have successfully substitute SMT by Chinese POS tagger and achieve an accuracy which is slightly lower than 82.4%. At the same time, we save nearly 30 minutes cost by SMT and word alignment algorithm. Due to the page limit, it will not be presented here.

5   Conclusions

As the results presented in Table 4 in Section 4, pattern based system and ANN with feedback achieve almost the same performance in quantity, accuracy and average length. However, ANN is more robust and more efficient than pattern based system. This paper just throws some light on how to utilize available resources to alleviate Chinese NLP problems. In the future, we will introduce available POS patterns and morphology patterns into ANN.

References

1. Sag, I.A., Baldwin, T., Bond, F., Copestake, A., Flickinger, D.: Multiword Expressions: A Pain in the Neck for NLP. In: Gelbukh, A. (ed.) CICLing 2002. LNCS, vol. 2276, pp. 1–15. Springer, Heidelberg (2002)
2. Ren, Z., et al.: Improving statistical machine translation using domain bilingual multiword expressions. In: Proceedings of the Workshop on Multiword Expressions: Identification, Interpretation, Disambiguation and Applications, pp. 47–54. Association for Computational Linguistics, Suntec (2009)
3. Vincze, V., István Nagy T., Berend, G.: Multiword expressions and named entities in the Wiki50 corpus. In: Proceedings of RANLP 2011, Hissar, Bulgaria (2011)
4. Attia, M., Toral, A., Tounsi, L., Pecina, P., van Genabith, J.: Automatic extraction of Arabic multiword expressions. In: 7th Conference on Language Resources and Evaluation (LREC 2010), Valletta, Malta (2010)
5. Fung, P.: Extracting Key Terms from Chinese and Japanese texts. In: Computer Processing of Oriental Languages, pp. 99–121 (1998)
6. Church, K.W.G., William, A., Hanks, P., Hindle, D.: Using Statistical Linguistics in Lexical Analysis. In: Lexical Acquisition: Using On-line Resourceto Build a Lexicon. Lawrence Erlbaum, Hilldale (1991)
7. Da Silva, J.F., Lopes, G.P.: A Local Maxima method and a Fair Dispersion Normalization for extracting multi-word units from corpora. World Trade, 369–381 (1999)
8. Chiang, D.: A hierarchical phrase-based model for statistical machine translation. In: Proceedings of the 43rd Annual Meeting on Association for Computational Linguistics, pp. 263–270. Association for Computational Linguistics, Ann Arbor (2005)
9. Och, F.J., Ney, H.: A systematic comparison of various statistical alignment models. Comput. Linguist. 29(1), 19–51 (2003)
10. Toutanova, K., Klein, D., Manning, C., Singer, Y.: Feature-Rich Part-of-Speech Tagging with a Cyclic Dependency Network. In: Proceedings of HLT-NAACL 2003 (2003)
11. Wikipedia. Feedforward neural network, http://en.wikipedia.org/wiki/Feedforward_neural_network (cited March 30, 2012)
12. Haikin, S.: Neural Networks: A Comprehensive Foundation. Pearson Education (1998)

Intelligent Situation Awareness on the EYECANE

Jihye Hwang, Yeounggwang Ji, and Eun Yi Kim

Visual Information Processing Labratory, Department of Advanced Technology Fusion,
Konkuk University, Hwayang-dong, Gwangjin-gu, Seoul, South Korea
{hjh881120,ji861020,eykim04}@gmail.com

Abstract. In this paper, the navigating system with camera embedded white-cane, which is called EYECANE, is presented to help the visually impaired in safely traveling on unfamiliar environments. It provides three environmental information to the user: 1) the current situation (place type) where a user stands on, 2) the positions and sizes of obstacles around a user, and 3) the viable paths to prevent the collisions of such obstacles. The experimental results demonstrated the effectiveness of the proposed method.

Keywords: EYECANE, Situation recognition, obstacle detection, the visually impaired people, assistive device.

1   Introduction

Urban intersections are the most dangerous parts of a blind person's travel. Practically, the average rate of 22% among total accidents is occurred on traffic intersections. Accordingly, it is essential to discriminate the place type, where a user is standing on, as intersections and sidewalks.

So far various solutions to safely cross intersections have been proposed and implemented. The accessible pedestrian signals (APS) [1] and Talking Signs [2] have been developed to help the visually impaired in knowing when to cross intersections. However, although their adoption is spreading, they are still only available in very few places, and some additional devices should be involved. As alternative to these approaches, the vision-based methods such as "Crosswatch" [3] and "Bionic Eyeglasses" [4] have been recently proposed. The Crosswatch is a hand-held vision system for locating crosswalk and for finding its orientation, and Bionic Eyeglasses is used to localize the crosswalk. However, while they are working well on localizing the crosswalk at the intersection, they are unable to recognize if the user is standing on intersection.

This paper describes a new development in the EYECANE, where we have added the outdoor situation awareness to discriminate the user's current place as intersection and sidewalk. Thus, it can automatically recognize the following environmental information: 1) user's current situation, 2) the position and size of obstacles to be placed around a user, and 3) the viable paths to prevent the collisions of obstacles.

P. Anthony, M. Ishizuka, and D. Lukose (Eds.): PRICAI 2012, LNAI 7458, pp. 740–745, 2012.
© Springer-Verlag Berlin Heidelberg 2012

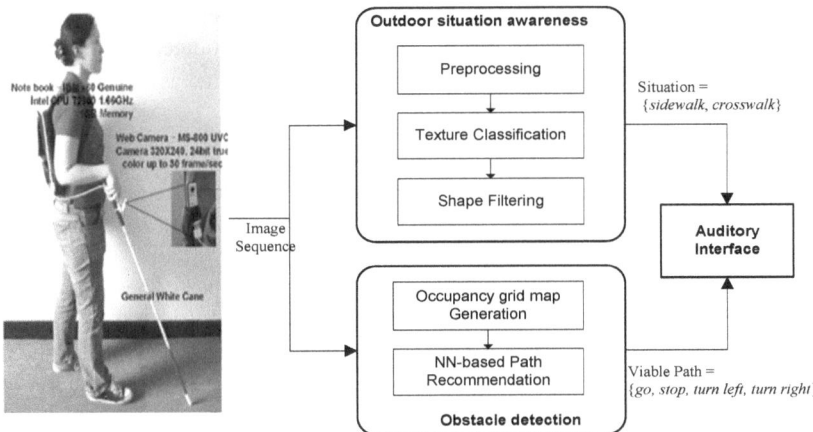

Fig. 1. The prototype of our EYECANE

Fig. 1 shows our EYECANE, which is composed of three main modules: outdoor situation awareness, obstacle detection and auditory interface. In the outdoor situation awareness, the place type where a user is standing on is determined based on textural and shape properties. And, in obstacle detection module, various obstacles are first localized using online background model, then viable paths to avoid them are determined by neural network-based classifier. Finally, the recognized results are verbally notified to the user through an auditory interface.

2   Outdoor Situation Awareness

In this work, *a situation means the type of place the user is located*, which is categorized into sidewalk and intersection. For recognizing outdoor situation, texture classification and shape filtering were performed on the input image.

Before texture classification and shape filtering, we apply Gaussian filter and histogram equalization to the input image in turn. Then, for further processing, the input image sized at 640×480 is divided into 768 sub-regions sized at 20×20.

2.1   Texture Classification

The easiest way to extract such boundaries is to apply the edge operator on the input image. Fig. 2(b) shows the edge images for Fig. 2(a), where they includes many lines that occur as a natural part of a scene, as well as the boundaries between sidewalks and roadways (those belong to the boundary class).

To discriminate the boundaries between sidewalks and roadways from other lines, the texture properties of sub-regions are investigated.

In this work, to characterize the variability in a texture pattern, both HOG (Histogram of Oriented Gradient) and color information are used.

The HOG is the feature descriptor to count the occurrences of gradient orientation in the sub-regions of an image, which is many used for object detection. For all 20×20 sized sub-regions, the HOGs are calculated, which is identified as

$$\text{HOG}_R = \left\{ \text{HOG}_R(i) = \frac{1}{\text{Area}(R)} \times \sum_{j \in R} \text{magnitude}(j), \text{if orientation}(j) = i \, (1 \leq i \leq 6) \right\}.$$

In addition, the average value of pixels' saturation within a sub-region is used to describe the color information, as the pixels corresponding to the roadway have the distinctive saturation distribution.

Based on these properties, a rule-based classification is performed on every sub-region. A sub-region is classified as the boundary class if both of the following conditions are satisfied: 1) HOG_R has the larger variance than a predefined threshold θ_H; 2) S_R is smaller than a threshold θ_S.

Fig. 2 shows the process of the texture classification. For all sub-regions, the Sobel operator is first performed, which is shown in Fig. 2(b). Then, the sub-regions are filtered by their textural properties. Firstly, the sub-regions with uniformly distributed HOGs are filtered, which is shown in Fig. 2(c), where most of sub-regions corresponding to the boundary class are preserved and others are eliminated, however, it still includes some misclassified sub-regions. Those false alarms are filtered again by color information, then the results are shown in Fig. 2(d). As can be seen in Fig. 2(d), the classification results include most of the sub-regions with correct boundary class.

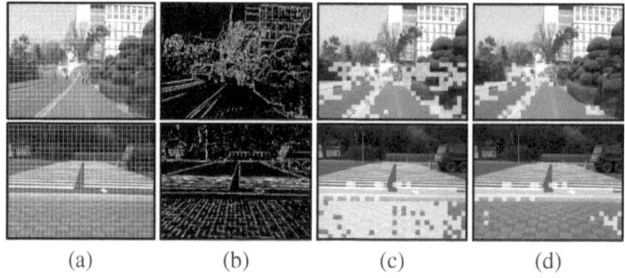

(a)	(b)	(c)	(d)

Fig. 2. Texture classification results (a) Input images (b) edge images by Sobel operator (c) filtered images by HOG distribution (d) filtered images by color information

2.2   Shape Filtering

In this step, we determine the situation based on the orientations of the boundaries: if they are horizontally aligned, the intersection is assigned to the current image; if they are aligned close to the vertical, the current image is labeled as sidewalk.

To eliminate the affects by misclassified sub-regions and determine correct situation, the profile analysis is performed on the classified images.

Accordingly, a classified image is projected along a y-axis and x-axis and two histograms are computed: horizontal histogram and vertical one. Thereafter, the following three heuristics are applied in turn, to determine the current situation: (1) An *intersection* is assigned to the current image in which some horizontal histogram values are more than a threshold; (2) An *intersection* is assigned to the input image in which the vertical histogram is uniformly distributed; (3) A *sidewalk* is assigned in which the vertical histogram has the larger variance than a threshold σ.

Figs. 3(c) and (d) illustrate how the situation is determined. Fig. 3(c) is a y-axis projection profile of a classified image, and Fig. 3(d) is an x-axis projection profile of a classified image. As you can see in Fig. 3(c), the top image has the vertical histogram with larger variance, thus its situation is considered as sidewalk. On the other hand, the bottom image has some horizontal histogram values larger than a threshold, thus its situation is determined as intersection.

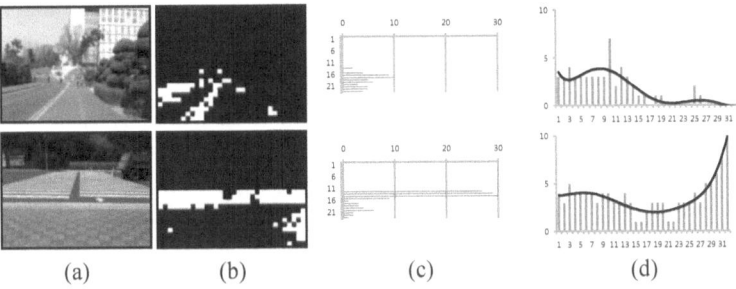

(a)   (b)   (c)   (d)

Fig. 3. Shape filtering results (a) Input image (b) classified image by texture (c) horizontal projection profile (c) vertical projection profile

3   Obstacle Detection

The method for detecting obstacles around user and for determining viable paths was proposed in our previous work [5]. Briefly speaking, the obstacle detection is performed by two steps: occupancy map generation and NN-based path recommendation.

Firstly, the obstacles are extracted using online background estimation, thereafter occupancy grid maps (OGMs) are made where each cells is allocated at a walking area and it has the different gray levels according the occupancy of obstacles. These OGMs are given to path recommendation which estimates possible paths at current position. Here, for robustness to illuminations, weather condition and complex environments, the machine learning using neural network (NN) is used. The details are described in our previous work [5].

4   Experimental Results

To assess the effectiveness of the proposed recognition method, experiments were performed on the images obtained from outdoors. For the practical use as mobility aids of the visually impaired, the EYECANE should be robust to environmental factors, such as different place types and lightening conditions. Therefore, 80,000 indoor and outdoor images, including official buildings, department stores, and underground areas, were collected over one year at different times.

4.1   Situation Awareness Results

To assess the effectiveness of the proposed method, experiments were performed on the images obtained from outdoors. A total of 2243 images were collected, which were then categorized into 8 datasets, according to their environmental complexity. Among them, 174 images were used as training data for finding optimal parameter set (θ_H, θ_S, σ), which were used for texture classification and shape filtering. And the other images were used for testing.

Table 1. Accuracy of outdoor situation awareness (%)

Illumination type	Direct sunlight with little shadow				Direct sunlight with complex shadow				
Type of ground pattern	Non-textured ground		Highly textured ground		Non-textured ground		Highly textured ground		
DB	DB1	DB2	DB3	DB4	DB5	DB6	DB7	DB8	Total
Accuracy	91	95.3	100	100	94.8	97.4	53.5	69	87.6

Table 1 summarizes the performance of the situation recognition under various outdoor environments. The average accuracy was about 87.6%. For the DB1 to DB4, the proposed method showed the accuracy of above 96%. And, it showed still good performance on the illumination type with complex shadows, except for DB7 and DB8. On both DB7 and DB8, the proposed method showed the lowest accuracies.

However, these errors can be easily solved if we use the history of the situations in between certain time slot.

4.2   Obstacle Detection Results

In this section, we investigated the performance of obstacle detection with a huge data. Unlike the outdoor situation awareness, this module was evaluated using images obtained from both indoors and outdoors.

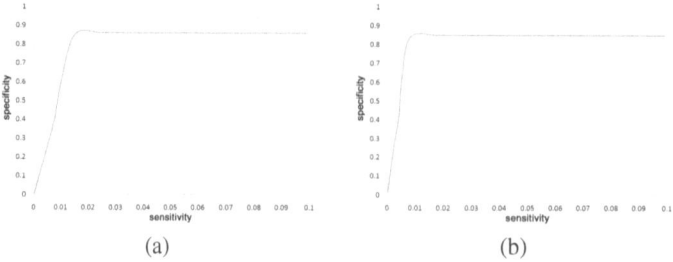

(a)   (b)

Fig. 4. Performance summarization of obstacle detection (a) the accuracy on indoors (b) the accuracy on outdoors

Fig. 4 shows the performance summarization of obstacle detection under indoors and outdoors, where it showed accuracies of 86% and 91.5%, respectively. In particular, it showed the better performance on outdoors.

The main purpose of the EYECANE is to help the safe mobility of the visually impaired and to prevent some dangerous collisions with vehicles or obstacles. For its practical use as an assistive device, the real time processing should be supported.

The average frame processing times for the outdoor situation awareness and for obstacle detection were about 228.54ms and 5.03ms, respectively. As such, the proposed method can process more than 4 frames per second on low-performance computer.

Consequently, the experiments proved that the proposed method produced the superior accuracy for situation awareness and safe path prediction, thereby assisting safer navigation for the visually impaired person in real-time.

5   Conclusion

In this study, an assistive device, EYECANE was presented to help the safe mobility of the visually impaired person. The main goal of the EYECANE is to automatically recognize some dangerous situations such as intersection, vehicle and obstacle and prevent the collisions of them. To assess the validity of the EYECANE, we have conducted experiments with 80,000 images, then the results showed that can recognize the outdoor situation with an accuracy of 87.6% and produce the viable paths with an accuracy of 91.5% on outdoors.

Acknowledgement. "This research was supported by Basic Science Research Program through the National Research Foundation of Korea(NRF) funded by the Ministry of Education, Science and Technology(grant number)" (20110002565) and (20110005900).

References

1. Barlow, J.M., Bentzen, B.L., Tabor, L.: Accessible pedestrian signals: Synthesis and guide to best practice. National Cooperative Highway Research Program (2003)
2. Brabyn, J., Crandall, W., Gerrey, W.: Talking signs: a remote signage, solution for the blind, visually impaired and reading disabled. Engineering in Medicine and Biology Society (1993)
3. Ivanchenko, V., Coughlan, J.M., Shen, H.: Crosswatch: A Camera Phone System for Orienting Visually Impaired Pedestrians at Traffic Intersections. In: Miesenberger, K., Klaus, J., Zagler, W.L., Karshmer, A.I. (eds.) ICCHP 2008. LNCS, vol. 5105, pp. 1122–1128. Springer, Heidelberg (2008)
4. Karacs, K., Lazar, A., Wagner, R., Balya, D., Roska, T., Szuhaj, M.: Bionic Eyeglass: an Audio Guide for Visually Impaired. In: Biomedical Circuits and Systems, BioCAS, pp. 190–193 (2006)
5. Ju, J., Ko, E., Kim, E.: EYECane: Navigating with camera embedded white cane for visually impaired person. In: ACM ASSETS 2009, pp. 237–238 (2009)

The Use of Sound Symbolism in Sentiment Classification

Takuma Igarashi, Ryohei Sasano, Hiroya Takamura, and Manabu Okumura

Tokyo Institute of Technology
{igarashi,sasano,takamura,oku}@lr.pi.titech.ac.jp

Abstract. In this paper, we present a method for estimating the sentiment polarity of Japanese sentences including onomatopoeic words. Onomatopoeic words imitate the sounds they represent and can help us understand the sentiment of the sentence. Although there are many onomatopoeic words in Japanese, conventional sentiment classification methods have not taken them into consideration. The sentiment polarity of onomatopoeic words can be estimated using the sound symbolism derived from their vocal sounds. Our experimental results show that the proposed method with sound symbolism can significantly outperform the baseline method that is not with sound symbolism.

1 Introduction

Köhler [7] reported that there is a close relationship between vocal sounds and visual impressions of shape. After experimentation, Ramachandran and Hubberd [8] named this phenomenon the **Bouba/Kiki effect**.A similar phenomenon is referred to as **sound symbolism** [2], which is the idea that the vocal sounds of words bring impressions of the thing indicated by the words. Analysis of these impressions is likely to be beneficial to several natural language processing (NLP) applications. In this paper, we focus on sentiment classification to show the benefit of sound symbolism.

Sentiment classification is the task of classifying words, sentences, and documents according to sentiment polarity. The polarity is "positive" when the text suggests good sentiment, "negative" when the text suggests bad sentiment, and "neutral" when the text suggests neither good nor bad sentiment. The most common approaches to sentiment classification use adjectives (such as "delicious") in context [6]. In these approaches, researchers prepare a lexicon that contains adjectives and their polarities in advance and then use the lexicon to estimate the polarity of the target text. This adjective-based approach, however, cannot estimate the polarity of sentences that include no adjectives.To resolve this problem, some researchers have expanded the adjective polarity lexicon to other parts of speech by using co-occurrence information in corpus [9].

However, such approaches cannot deal with unknown words. Japanese onomatopoeia, in particular, are used at a fairly high frequency, and new onomatopoeic words are being created all the time. This results in many unknown onomatopoeic words in Japanese text. Consider the following example sentence:

(1) *Te-ga **bettori/becho-becho** suru.*
 hands   sticky/humid   are
 My hands are sticky/humid.

P. Anthony, M. Ishizuka, and D. Lukose (Eds.): PRICAI 2012, LNAI 7458, pp. 746–752, 2012.

"*Bettori*" and "*becho-becho*" are onomatopoeic words in Japanese. The sentence has a negative polarity. While "*Bettori*" is commonly used and usually included in a typical lexicon, "*becho-becho*" is not. The problem is that lexicon-based approaches cannot deal with such unknown onomatopoeic words.

On the other hand, onomatopoeia reflects sound symbolism to a large extent [4]. We can understand the nuance of onomatopoeic words from the impression of sound symbolism, even if we do not know the meaning of the words. Hence, modeling sound symbolism can help estimate the polarity of unknown onomatopoeic words. We present a method for modeling sound symbolism and demonstrate its effectiveness in sentiment classification.

2 Background

Onomatopoeic words either imitate the sounds they represent or suggest the relevant situation or emotion.

(2) *Kare-ha **niko-niko** waratta.*
     He   smiled.

"*Niko-niko*" is an onomatopoeic word indicating the bland smile. "*Waratta*" is a generic verb indicating smile, laugh, chuckle, simper, or guffaw. In Japanese, such situational or emotional information is represented by onomatopoeic words that are separate from verbs, while in English verbs tend to include such information.

Onomatopoeic words have phonological and morphological systematicity. For example, "*niko-niko*," "*nikotto*," and "*nikkori*" suggest similar situations. "*Niko*" is the **base** form of these words. Morphemes are made up from base forms through a kind of conjugation.

The main aim of the proposed system is to interpret unknown onomatopoeic words.

(3) *Kare-ha **gehi-gehi** waratta.*
     He   vulgarly   laughed.

Sentence (3) has negative polarity. This polarity is derived from the onomatopoeic word "*gehi-gehi*," i.e., the sentence "*Kare ha waratta* (He laughed)." does not have negative polarity. When an onomatopoeic word has polarity, the polarity usually relates to its sound symbolic impression. Hamano [4] studied the sound symbolism of Japanese onomatopoeic words in detail and found that voiced consonants often give a negative impression in Japanese. In other words, we can interpret the word as negative from the impression we derive from sound symbolism, even if we have never seen the word before.

3 Proposed Method

Our method estimates the polarity label of a sentence that includes onomatopoeic words. The system receives as input a sentence that includes onomatopoeic words, classifies it into one of three classes (positive, negative, and neutral) using contextual and sound symbolism features, and finally outputs the polarity label. The proposed system is based on the support vector machine (SVM) [1]. The bag-of-words features and the context polarity feature are baseline features that are not related to sound symbolism; these are described in detail in Section 3.1. The vocal sound features and the onomatopoeia dictionary feature are sound symbolism features; we describe them in Sections 3.2 and 3.3. The sound symbolism model, which is trained from an onomatopoeia dictionary using vocal sound features, takes an onomatopoeic word as input and outputs a binary value: 1 if the inputted word is estimated to be positive, and 0 otherwise. The main SVM classifier uses this value as an onomatopoeia dictionary feature.

3.1 Baseline Features

Context information is often used for sentiment classification. We, therefore, model context information with the bag-of-words model.

We also use a sentiment lexicon created by Takamura et al. [9]. They assessed the polarity probability of words in WordNet [3] with a **spin model**, assuming that words next to each other in context tend to have the same polarity. We use these probabilities as the basis of contextual polarity in addition to the bag-of-words features.

3.2 Vocal Sound Features

We consider three forms in the representation of Japanese vocal sounds:

1. **Kana characters**
   Kana is a type of Japanese character. Each kana consists of a consonant onset and vowel nucleus (CV): e.g., " (ha)" or only a vowel (V): e.g., " (i)."
2. **Phoneme symbols**
   A phoneme is a representation to describe vocal sounds recognized by native speakers in a language. The phoneme /k/ in words such as **k**it or s**k**ill is recognized to be the same sound, although indeed they are different sounds; [k^h] (aspirated) for **k**it and [k] (unaspirated) for s**k**ill.
3. **Phonetic symbols**
   Phonetics is the objective and strict representation of all vocal sounds. Unlike phoneme, vocal sounds such as [k^h] and [k] are distinguished in phonetic symbols. We use the International Phonetic Alphabet (IPA)[1] for this representation.

These representations are used to make sound symbolic features. We convert onomatopoeic words to each of the three forms and create N-gram features from them. For example, an onomatopoeic word " (*shito-shito*, drizzlingly)" is converted to [ʃitoʃito] with phonetic symbols. N-gram features are derived from the phonetic representation: [ʃ], [i], [t], [o], [ʃi], [it], [to], and [oʃ] (uni-gram and bi-gram).

These features denote what sounds exist and what connections of the sounds exist in the word, but do not consider relations among them. For example, [m] and [p] are simply regarded as different sounds, although both are bilabial consonants. We, therefore, also use consonant categories in the IPA to take such relations into consideration (see

---

[1] http://www.langsci.ucl.ac.uk/ipa/

Table 1). We use these categories as binary features for the first and second consonants of an onomatopoeic word on the basis of Hamano's study [4], which states that the **base** of an onomatopoeic word generally includes the first and second consonants, and that consonants in the base play a key role in sound symbolism.

Table 1. Consonant categories in the IPA. We considered only sounds that appear in Japanese.

1. plosive (e.g. [p], [k])	2. fricative (e.g. [ɸ], [s])	3. affricate (e.g. [ʦ], [ʣ])
4. nasal (e.g. [m], [n])	5. flap or tap (e.g. [ɾ], [ɾʲ])	6. approximant (e.g. [j], [w])
7. bilabial (e.g. [ɸ], [p])	8. alveolar (e.g. [t], [n])	9. palatal (e.g. [ç], [j])
10. velar (e.g. [k], [w])	11. coronal (e.g. [z], [ʃ])	12. dorsal (e.g. [ç], [ɴ])
13. voiced (e.g. [g], [z])		

3.3 Which Representation Is Best?

Our hypothesis is that sound symbolic impressions of onomatopoeic words are composed of the vocal sounds of the words. We think that a more detailed representation is better for modeling sound symbolism: in other words, phoneme symbols are better than kana, and that phonetic symbols are better than phoneme symbols. We conducted experiments to test this hypothesis.

First, we trained an SVM classifier[2]. When the classifier takes an onomatopoeic word, it classifies the word as either positive or negative. For this experiment, we used a Japanese onomatopoeic dictionary [5] that consists of 1064 onomatopoeic words along with a description of the senses and impressions of each. The impression of each sense of each word takes one of five grades: "positive," "a little positive," "neither positive nor negative," "a little negative," and "negative." We picked out 845 onomatopoeic words that did not contain any contradictory impressions (such as "positive" and "a little negative" in the same word) and then classified 225 onomatopoeic words that were "positive" or "a little positive" into a positive class and 620 onomatopoeic words that were "negative" or "a little negative" into a negative class. We evaluated the performance with accuracy measure by performing ten-fold cross-validation.

The following are the models we compared. *Baseline* is the model that classifies test instances into the most frequent class in training data, i.e., in this case, it always classifies test data as "negative". *Kana*, *Phoneme*, and *IPA* use only N-gram features (uni, bi, and tri-gram) with each vocal sound representation detailed in Section 3.2: kana, phoneme, and phonetic (IPA). *IPACat* is the model with consonant category features in addition to *IPA*, and *IPACat2* is the model with consonant bi-gram features in addition to *IPACat*. Consonant bi-gram features are created from sequences of only those consonants in the IPA that are obtained to eliminate vowels from the phonetic representation of onomatopoeic words. For example, the onomatopoeic word " (*shito-shito*)" is converted to a phonetic representation [ʃitoʃito] and then further converted to [ʃtʃt] by eliminating the vowels. In this case, the consonant bi-gram features are [ʃt] and [tʃ]. We considered consonant bi-gram features in the experiment because we felt that consonant connections can affect the overall impression of sound symbolism.

---

[2] We used SVM*light* (http://svmlight.joachims.org/).

The experimental result is shown in Fig. 1. The horizontal axis indicates the classification accuracy[3]. This result suggests that the hypothesis in Section 3.2 is valid; IPA phonetic representation is the best representation for modeling sound symbolism among the kana, phoneme and phonetic representations. This is why we use the *IPACat2* model as the sound symbolism model in the proposed system.

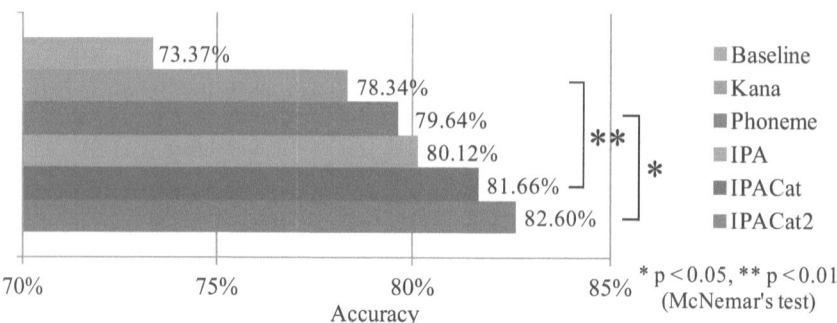

Fig. 1. Experimental result of estimating polarity of onomatopoeic words

4   Experiments

The proposed system estimates the polarity of a sentence that includes onomatopoeic words. Here, we present experiments to demonstrate that sound symbolism features are useful in sentiment classification. When the system takes as input a sentence that includes an onomatopoeic word, it classifies the sentence into one of three classes (positive, negative, or neutral) and outputs the class label of the sentence.

4.1   Experimental Setup

In the experiment, we used documents collected from Tabelog[4], which is a review site for restaurants in Japan. Food and drink reviews include more onomatopoeic words than other domains since onomatopoeic words are subjective and convenient for describing tastes of food and drink. We extracted sentences that include onomatopoeic words with a Japanese POS tagger, JUMAN Ver.7.0[5], that can automatically recognize onomatopoeic words even if they are not contained in the JUMAN dictionary by using morphologic patterns. We manually selected 1000 sentences for 50 onomatopoeic words (20 sentences per one onomatopoeic word). Since we wanted to focus on the sound symbolism of onomatopoeia, we excluded sentences that included negation words that would affect the sentence polarity.

The gold standard labels were tagged by two annotators. Any discrepancies that arose were discussed by the annotators in order to decide on one label. The labels

---

[3] We empirically choose the best C parameter for SVM from {0.01, 0.1, 1, 10, 100, 1000} for each model.

[4] http://tabelog.com/

[5] http://nlp.ist.i.kyoto-u.ac.jp/index.php?JUMAN

can be grouped into positive (label-1, 2), negative (label-3, 4), neutral (label-5), and other (label-6). We removed the label-6 data from the dataset because they could not be identified even by human annotators and are therefore considered noise. Positive and negative classes were split into two on the basis of polarity dependence on an onomatopoeic word so that we can investigate whether sound symbolism is more useful in a limited dataset (label-1, 3, 5) in which the polarity depends on onomatopoeic words than in a full dataset (label-1 to 5). We call the full dataset Dataset1 and the limited dataset Dataset2.

In this experiment, we evaluated the performance of classifying sentences that include unknown onomatopoeic words; that is, onomatopoeic words that appear in the test set but do not appear in the training set. We split the dataset into 10 parts and used one part for testing, another for development, and the rest for training. We used a linear SVM with a one-versus-rest method and evaluated the performance with accuracy by performing ten-fold cross-validation[6].

We tested the following five settings:

M_B BOW, *Context Polarity* (Baseline)
M_{BP} BOW, *Context Polarity, Phonetic*
M_{BPD} BOW, *Context Polarity, Phonetic, Onomatopoeia Dictionary*
M_{BD} BOW, *Context Polarity, Onomatopoeia Dictionary*
M_D *Onomatopoeia Dictionary*

The baseline method M_B uses a bag-of-words model (BOW) and context polarity (as described in Section 3.1). This model does not take sound symbolism into consideration, while the other models do. The *Phonetic* features are the same features as *IPACat2* in Section 3.3, which are the N-gram features of phonetic symbols (uni-gram, bi-gram, and tri-gram), consonant categories in the IPA, and consonant bi-gram. The *Onomatopoeia Dictionary* feature is a binary feature that is the output of the sound symbolism model.This model is the same as *IPACat2* and was trained on a dataset that contains 820 onomatopoeic words[7].

4.2   Results and Discussion

The experimental results are shown in Fig. 2. In Dataset1, the baseline had a 47.27% accuracy, while M_{BD}, at 62.94%, had the strongest performance. There was a significant difference between M_{BD} and the baseline with McNemar's test (P < 0.01). The *Onomatopoeia Dictionary* feature was effective because the sound symbolism model learned sound symbolic impressions from many onomatopoeic words.

In Dataset2, the accuracy of the baseline was just 44.85% and lower than its accuracy in Dataset1. This decline makes sense because Dataset2 contains only limited data where the polarity of a sentence is dependent on an onomatopoeic word. M_{BD} again achieved the best performance, with an accuracy of 62.40%.

---

[6] The C parameter of SVM is automatically decided on the best value in {0.1, 1, 10, 100} by testing with the development set.
[7] No onomatopoeic words in this dataset occur in Tabelog data.

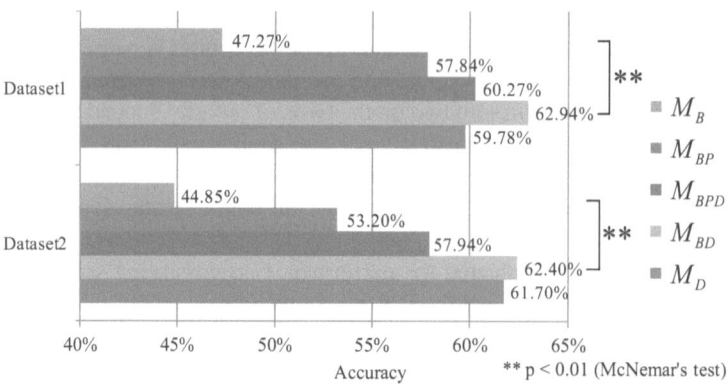

Fig. 2. Result of experiment to estimate the polarity of sentences that include onomatopoeic words

5   Conclusion

In this paper, we introduced a sentiment classification method that uses sound symbolism, which conventional approaches have not taken into consideration. We modeled the sound symbolism with vocal sound representations of onomatopoeic words and showed that the polarity of unknown onomatopoeic words can be estimated by using the N-gram features of a phonetic representation and consonant category features of phonetic symbols.

The experimental results clearly showed that the methods using sound symbolism significantly outperformed the baseline model, which indicates that sound symbolism is useful in the sentiment classification task.

References

1. Cortes, C., Vapnik, V.: Support-vector networks. Machine Learning 20(3), 273–297 (1995)
2. David, C.: A Dictionary of Linguistics and Phonetics. Wiley-Blackwell (2008)
3. Fellbaum, C.: WordNet: An Electronic Lexical Database. MIT Press (1998)
4. Hamano, S.: The Sound-Symbolic System of Japanese. CSLI Publications (1998)
5. Hida, Y., Asada, H.: Gendai Giongo Gitaigo Yoho Jiten. Tokyo-do (2002) (in Japanese) ISBN: 9784490106107
6. Kamps, J., Marx, M., Mokken, R.J., de Rijke, M.: Using Wordnet to Measure Semantic Orientations of Adjectives. In: Proceedings of the 4th International Conference on Language Resources and Evaluation (LREC 2004), pp. 1115–1118 (2004)
7. Köhler, W.: Gestalt Psychology. Liveright, New York (1929)
8. Ramachandran, V.S., Hubbard, E.M.: Synaesthesia: A window into perception, thought and language. Journal of Consciousness Studies, 3–34 (2001)
9. Takamura, H., Inui, T., Okumura, M.: Extracting semantic orientations of words using spin model. In: Proceedings of the 43rd Annual Meeting on Association for Computational Linguistics (ACL 2005), pp. 133–140 (2005)

Enhancing Malay Stemming Algorithm
with Background Knowledge

Leow Ching Leong, Surayaini Basri, and Rayner Alfred

School of Engineering and Information Technology,Universiti Malaysia Sabah, Jalan UMS,
88400,Kota Kinabalu, Sabah, Malaysia
dragon_july14@hotmail.com, surayaini_basri@yahoo.com,
ralfred@ums.edu.my

Abstract. Stemming is a process of reducing the inflected words to their root form. Stemming algorithm for Malay language is very important especially in building an effective information retrieval system. Although there are many existing Malay stemmers such as Othman's and Fatimah's algorithms, they are not complete stemmers because their algorithms fail to stem all the Malay words as there is still a room for improvement. It is difficult to implement a perfect stemmer for Malay language due to the complexity of words morphology in Malay language. This paper presents a new approach to stem Malay word with higher percentage of correctly stemmed words. In the proposed approach, additional background knowledge is provided in order to increase the accuracy of stemming words in Malay language. This new approach is called a Malay stemmer with background knowledge. Besides having reference to a dictionary that contains all root words, a second reference to a dictionary is added that contains all affixed words. These two files are considered as the background knowledge that will serve as references for the stemming process. A Rule Frequency Order (RFO) is applied as the basis stemming algorithm due to its high accuracy of correctly stemming Malay words. Based on the results obtained, it is proven that the proposed stemmer with background knowledge produces less error in comparison to previously published stemmers that do not apply any background knowledge in stemming Malay words.

Keywords: Malay stemming, affixes, Rule Frequency Order, Background Knowledge, Rule-based Affix Elimination.

1   Introduction

Information retrieval (IR) is an area of study that focuses on retrieving information based on a query submitted by user. It is used in searching for documents, information within documents, and for metadata about documents. Some of IR applications include text search, machine translation, document summarization, and text classification. Since Malay language is among the most widely spoken languages in the world [16], Information Retrieval in Malay language is also important. A well-developed Information Retrieval application should be established in order to process documents

P. Anthony, M. Ishizuka, and D. Lukose (Eds.): PRICAI 2012, LNAI 7458, pp. 753–758, 2012.
© Springer-Verlag Berlin Heidelberg 2012

written in Malay language as there is still much information can be extracted from these documents.

An IR process begins by sending a query to the IR application. In order to find the information that is most relevant to the query, the IR application is required to search for all relevant information related to the query and process all these information obtained by going through several processes. One of the processes in processing the information is a process called stemming.

Stemming is a process of reducing affixed word into its root word and it often influences the quality of results obtained in retrieving relevant documents or categorizing bilingual corpus [18]. The stemming process often uses dictionary as part of its process. The dictionary containing root words will be used to check if the root word of the affixed word exists in the dictionary or not. During the checking, a traditional algorithm will scan for all words stored inside the dictionary regardless whether the word has affixes or not and this may consume a lot of time. Besides that, if the dictionary that contains root words that do not have any affixes, the stemmer may incorrectly return the stemmed word. For instance, a word with affixes, e.g., *"katakan"*, will be stemmed to become *"katak"* if the suffix *"an"* is removed. This is because the root word *"katak"* exists in the dictionary although it doesn't have affixes. As a result, the affixed word *"katakan"* will be stemmed to become *"katak"*.

This problem can be solved if we have additional dictionary for reference that stores root words that has affixes only. For instance, when the stemmer removes the suffix *"an"* from the word *"katakan"*, the root word obtained would be *"katak"* and this root word exists in the first dictionary. However, since the word *"katak"* does not exists in the second dictionary that contains root words that has affixes only, then the word *"katak"* cannot be the root word. As a result, the Malay stemmer would move on to another rule to stem the word. In this case, it will check for the suffix *"kan"*. Then, if the stemmer removes the suffix *"kan"* from the word *"katakan"*, then the root word would be *"kata"*. Then, the stemmer will refer again the second dictionary that contains root words that have affixes only and find that the root word "kata" exists in the second dictionary. Since the root word *"kata"* exists in the second dictionary, then the final result of stemmed word for the unstemmed word *"katakan"* would be *"kata"*, which is a correctly stemmed word. Therefore, this paper proposes a stemmer that applies two types of dictionary. The first dictionary stores all root words obtained from Kamus Dewan Fourth Edition [17] and this is called root words dictionary. The latter dictionary contains words that only have affixes which are also obtained from Kamus Dewan Fourth Edition [17]. The second dictionary is called affixed words background knowledge dictionary.

Since stemming can be used to reduce the derived word into its root word, the number of words to be processed by the IR system can also be reduced because the derived words may be formed from the same root word. Thus it can increase the performance of the IR application since the number of words that are required to be processed is less. Therefore, a well-developed stemmer is essentially required in order to support better Information Retrieval system.

There are several researches that had been conducted related to Malay language stemmers [1, 9, 11]. The number of rules recommended by researchers keeps increasing and thus the framework of a Malay stemmer keeps changing over time. Some researches discussed about affixes rules that are derived from morphological rules in Malay. There are also researches that propose algorithms for stemming Malay words. Some of the algorithms are derived from the implementation of English stemming algorithm such as Porter Stemming. Porter stemming is a well-known algorithm that is used to stem English language text [12].There are a lot of English stemming algorithms such as Lovins' algorithm [13], Paice/Husk's algorithm [14] and Dawson's algorithm [15]. However, there is no complete stemming algorithm existed that has been really built in order to resolve the problem mentioned earlier. Each algorithm may cause errors that include over-stemming or under-stemming. Hence, there are still some improvements that are required in order to assist the task of effectively and efficiently retrieving information for documents written in Malay language.

This paper is organized as follows. Section 2 discusses about some of the related works for Malay stemming algorithms. Section 3 describes the proposed algorithm for the Malay stemming approach with affixed words background knowledge. Section 4 shows the results obtained and finally this paper is concluded in Section 5.

2   Related Works

Most languages use affixes to carry out another meaning of the words. Malay language uses four types of affixes in the word which are *prefix, suffix, infix* and *circumfix* [1]. *Prefix* is the additional word located at the left of the base word. Handling prefixes in Malay language is more complicated because the usage of the prefix depends on the initial character of the root word [3, 5]. Some of the prefixes might cause changes of the initial character of the root word (e.g., the word "*penyapu*" changes to "*sapu*") while some are not (e.g., the word "*pekerja*" changes to "*kerja*").

Suffix is the additional word located at the right side of the base word. Suffix in Malay do not affect the root word. *Infix* is the additional word that is inserted after the first consonant of the base and the usage of infix in Malay corpus is not much. The combination of prefix and suffix is known as *circumfix*. Currently there are 418 rules and spelling variations rules for stemming purposes that have been developed [2].

There are several types of errors that may occur in stemming words that include under-stemming, over-stemming, spelling exceptions and unchanged. Under-stemming will occur when words, that should be grouped together, are grouped in another group due to various different stems [10]. On the other hand, over-stemming will occur when root words, that should not be grouped together, are grouped in the same group. The effectiveness of recall in IR application will decrease due to the existence of under-stemming while over-stemming will reduce the precision of IR.

The first Malay stemming algorithm was developed by Othman in 1993, in which there were 121 morphological rules introduced which are derived from Kamus Dewan (1991) [8]. One of the weaknesses of this stemming process is that this algorithm consumes a lot of time with the adoption of rule-based approach. The stemmer introduced does not have any references to any dictionaries before any rules are applied. Hence some root words that fulfill the rule of the affixes might be incorrectly stemmed and produced ambiguous meaning of the root word. For example, the word *"tempatan"* will be stemmed to *"tempat"* although the root word *"tempatan"* has its own meaning. Sembok's algorithm is designed by adopting Othman's algorithm and adding two set of morphological rules [11]. The first set contains 432 rules of affixes which cater Quran words and the second set contains 561 rules of affixes which cater modern Malay words. In addition to that, there is additional step implemented in Sembok's algorithm based on Othman's algorithm. This step includes checking the input word against a dictionary before applying any rules of the affixes to avoid stemming root words which fulfill the affixes rules.

Besides that, Porter-based stemmer [6] is also used in stemming words in Malay language. However, Porter-based stemmer cannot be applied directly for stemming words in Malay language. This is because there are some differences between word structures in Malay and English which include the combination of affixes attached to a word, the differences in syllables used to construct a word and finally the presence of infixes in Malay.

There are some works conducted that are related to the order of rules applied when stemming words in Malay language [7]. Abdullah *et al.* investigated the effectiveness of a Malay stemmer based on the arrangement of the rules applied to the words in Malay language. Abdullah's algorithm sorts the stemming rules based on the frequency of their usages from the previous execution [7].

Besides that, Abdullah *et al.* have also conducted a series of testing process with Rules Application Order (RAO), Rules Application Order 2 (RAO2), New Rules Application Order (NRAO) and Rules Frequency Order (RFO). Abdullah *et al.* have shown that the best order of the affixes rules depends on the documents collection and the best order of the affixes for their collection is *prefix*, *suffix*, *prefix-suffix* and *infix*. Based on Abdullah *et al.*, RFO algorithm produces the best result among the rest of the algorithms. Although RFO algorithm produces the best result, there are still some errors produced by the algorithm. For example the word *"bacakan"*, the stemmed word produces by the RFO algorithm is *"bacak"* and the actual root word should be *"baca"* since the word *"bacak"* exists in the dictionary that contains root words without affixes. Hence, this paper proposes a new approach that ignores certain root words that do not contain any affixes when checking for affixes rules.

3   Malay Stemmer Algorithm Based on Background Knowledge

In the proposed Malay stemmer based on background knowledge, RFO will be used as the basis algorithm. In this algorithm, checking for prefixes will be the first process in the affixes rule checking, followed by suffixes, prefixes-suffixes and infix. Instead of using an online dictionary to check for root words, this paper proposes two types of dictionaries as references in order to check for root words. The first dictionary contains all root words and the second dictionary contains only the root words that have affixes which is called background knowledge. The background knowledge is proposed because there are some root words that do not need to be checked when running the affixes rules because those words do not consist of any affixes. The proposed algorithm can be described as follows:

Step 1: Get the next word
Step 2: Check the word against the first dictionary. If the word exists, go to step 1.
Step 3: Apply the rule on the word to get a stem;
Step 4: Recode for prefix spelling exceptions and check the second dictionary.
    If exist, the stem is a root word and go to step 1 else continue.
Step 5: Check the stem for spelling variations and check the second dictionary.
    If exist, the stem is a root word and go to step 1 else continue.
Step 6: Recode for suffix spelling exceptions and check the second dictionary.
    If exist, the stem is a root word and go to step 1 else go to step 3.

4   Experiment Results and Discussion

Two experiments have been conducted in this study. The first experiment is conducted by using only a single dictionary root words text file. The second experiment is conducted with two dictionaries included as references which are the root words dictionary and the affixed words only background knowledge dictionary. The online news articles are collected and used as the input word and we managed to obtain a total of 5319 words. The percentage of error is 0.21% when stemming Malay words without using the background knowledge dictionary. On the other hand, the percentage of error is 0.09% when stemming Malay words using the background knowledge dictionary. It is also found that stemming process without background knowledge produces more under stemmed errors. However, a new resolution is required to solve the errors produced by the proposed stemming process with background knowledge.

Based on the result obtained, it is proven that the stemming method with additional background knowledge produces lesser error compared to the stemming algorithm without background knowledge. It is because there are words that do not need to be checked for the affixes rules checking because those words do not have any affixes. Hence, by removing those words from being checked during the affixes removal process, the stemming error can be avoided. In addition to that, the number of words used in the background knowledge dictionary is less than the first dictionary. This also helps in reducing the time required to run the stemming algorithm. However, there is also a disadvantage of this stemming. The dictionary used must be updated from time to time manually. Hence, the dictionary needs to be updated if a new word is found that does not have any affixes in Malay language.

5   Conclusion

In this paper, we have formally presented a new approach to stem Malay words with higher percentage of correctly stemmed words. In the proposed approach, additional background knowledge is provided in order to increase the accuracy of stemming

Malay words. Based on the results obtained from experiments, it can be concluded that the Malay stemming approach equipped with background knowledge produces a better stemming result compared to the stemming approach without background knowledge. The proposed approach can be improved by studying the pattern of words that are stemmed incorrectly by the proposed algorithm. For future works, a better algorithm is needed to solve this problem. In addition to that, there is still weakness in this study as both dictionaries are keyed in manually. It is hoped that an automated mechanism should be studied that will automatically update the dictionary that contains all affixed words only.

References

1. Bali, R.M.: Computational Analysis of Affixed Words in Malay Language. Unit Terjemahan Melalui Komputer School of Computer Sciences. Universiti Sains Malaysia (2004)
2. Ahmad, F.: A Malay Language Document Retrieval System: An Experiment Approach and Analysis. Universiti Kebangsaan Malaysia, Bangi (1995)
3. Yee, L.T.: A Minimally-Supervised Malay Affix Learner. In: Proceedings of the Class of 2003 Senior Conference, Computer Science Department, Swarthmore College, pp. 55–62 (2003)
4. Ahmad, F., Yusoff, M., Sembok, T.M.T.: Experiments with a Stemming Algorithm for Malay Words. Journal of the American Society for Information Science 47(12), 909–918 (1996)
5. Idris, N., Syed Mustapha, S.M.F.D.: Stemming for Term Conflation in Malay Texts (2001)
6. Mangalam, S.: Malay-Language Stemmer. Sunway Academix Journal (2006)
7. Abdullah, M.T., Ahmad, F., Mahmod, R., Sembok, M.T.: Rules Frequency Order Stemmer for Malay Language. IJCSNS International Journal of Computer Science and Network Security 9(2) (2009)
8. Othman, A.: Pengakar Perkataan Melayu untuk Sistem Capaian Dokumen. Unpublished master's thesis. UniversitiKebangsaan Malaysia, Bangi, Malaysia (1993)
9. Sembok, T.M.T., Bakar, Z.A.: Effectiveness of Stemming and n-grams String Similarity Matching on Malay Documents. International Journal of Applied Mathematics and Informatics 5(3) (2011)
10. Al-Shammari, E.T.: Towards an Error-Free Stemming. In: IADIS European Conference Data Mining (2008)
11. Sembok, T.M.T., Bakar, Z.A., Ahmad, F.: Experiments in Malay Information Retrieval. In: International Conference on Electrical Engineering and Informatics (2011)
12. Porter, M.F.: An algorithm for suffix stripping. Program (1980)
13. Porter, M.F.: An Algorithm for Suffix Stripping. Program 14, 130–137 (1980)
14. Paice, C.D.: Another Stemmer. ACM SIGIR Forum 24(3) (1990)
15. Dawson, J.: Suffix Removal and Word Conflation. ALLC Bulletin 2 (1974)
16. James, N.S.: The Indonesian Language: Its History and Role in Modern Society, p. 14. UNSW Press (2004)
17. Noresah, B., Ibrahim, A.: Kamus Dewan Edisi Keempat, Dewan Bahasadan Pustaka, Kuala Lumpur (2010)
18. Alfred, R., Paskaleva, E., Kazakov, D., Bartlett, M.: Hierarchical Agglomerative Clustering of English-Bulgarian Parallel Corpora. In: RANLP 2011, pp. 24–29 (2007)

Using Tagged and Untagged Corpora to Improve Thai Morphological Analysis with Unknown Word Boundary Detections

Wimvipa Luangpiensamut, Kanako Komiya, and Yoshiyuki Kotani

Tokyo University of Agriculture and Thechnology, 2-24-16 Naka-cho, Koganei,
Tokyo, 184-8588 Japan
wimvipa@gmail.com, {kkomiya,kotani}@cc.tuat.ac.jp
http://www.tuat.ac.jp/~kotani/

Abstract. Word boundary ambiguity is a major problem for the Thai morphological analysis since the Thai words are written consecutively with no word delimiters. However the part of speech (POS) tagged corpus which has been used is constructed from the academic papers and there are no researches that worked on the documents written in the informal language. This paper presents Thai morphological analysis with unknown word boundary detection using both POS tagged and untagged corpora. Viterbi algorithm and Maximum Entropy (ME) - Viterbi algorithm are employed separately to evaluate our methods. The unknown word problem is handled by making use of string's length in order to estimate word boundaries. The experiments are performed on documents written in formal language and documents written in informal language. The experiments show that the method we proposed to use untagged corpus in addition to tagged corpus is efficient for the text written in informal language.

Keywords: Natural Language Processing, Morphological Analysis, Word Segmentation.

1 Introduction

Word segmentation is a challenging problem for the Thai language because the Thai words are not tokenized by whitespaces. There are many researches focused on the Thai morphological analysis. However, the resource that is frequently used so far is the Orchid corpus which was constructed from academic paper. The performance of using Orchid corpus for processing texts written in informal language is still unknown.

In this paper, we propose two morphological analysis methods which are used to experiment on two test sets (formal language and informal language). Along with the Orchid corpus, we use a large text corpus. We assume that the large text corpus will provide up-to-dates words and could be more flexibly used than dictionaries. We also focus on solving the unknown word problem. Each unknown word handling feature is created for each method specifically. The inputs to the system are strings. The system will output word segmented strings tagged by POS tags.

P. Anthony, M. Ishizuka, and D. Lukose (Eds.): PRICAI 2012, LNAI 7458, pp. 759–764, 2012.
© Springer-Verlag Berlin Heidelberg 2012

2   Corpora

In this research, three corpora are used. The first one is a POS tagged corpus called Orchid Corpus [1-2]. The second one is BEST 2009 corpus [3] which is a text corpus containing words from articles, encyclopedias, news archives and novels. Although the BEST corpus is not POS tagged, there are <NE> tags indicating name entities. The Orchid corpus and the BEST corpus are used as training data. Our systems are designed to output results in the Orchid corpus' standard.

The third corpus is a five thousand word POS tagged corpus called Web Document. This corpus was privately constructed to be used as a test data in our research. The collected articles are written in formal and informal language. The articles are manually segmented and POS tagged using the Orchid corpus' standard.

3   System and Algorithms

3.1   Preprocessing

Firstly, we filtered out output candidates (substring) that are not capable of forming syllable using the following rules:

1. A substring that starts with postposed vowel such as ะ, ำ and ๅ is non-syllabic.
2. A substring that starts with subscript vowel such as ุ and ู is non-syllabic.
3. A substring that starts with superscript vowel such as ั, ิ, ี, ึ, ื and ็ is non-syllabic.
4. A substring that starts with any of the four tonal marks is non-syllabic.
5. A substring that starts with Mai Yamok (ๆ) or Pai-Yan (ฯ) is non-syllabic.
6. A substring that ends with preposed vowel such as เ, แ, โ, ใ and ไ is non-syllabic.
7. A substring that ends with ฿ or ฿ is non-syllabic.
8. A two-alphabet long substring that ends with voiceless mark is non-syllabic.

3.2   Unknown Word Handling Feature for the Viterbi Algorithm

The first method is Viterbi algorithm [4]. Our unknown word handling feature is applied when the system calculates emission probabilities. An emission probability E of a word w being tagged with a POS p can be calculated by the following equation. "w_p" means the amount of the word w found in the Orchid corpus as the POS p. "ALLW$_{op}$" means the amount of any words found in the Orchid corpus as the POS p.

$$E = \frac{w_P}{ALLW_{OP}} \tag{1}$$

The emission probabilities of words that are found only in BEST corpus or unknown strings cannot be calculated by using the equation 1 (E1) since their POSs are unknown. The equations for calculating the emission probabilities of words found only in BEST corpus (E2) and unknown strings (E3) are designed specifically. The emission probabilities are estimated according to in which corpus a word is found or not found.

The system will treat words that POS is unknown as a proper noun, a common noun or a verb in order that the system can calculate transition probabilities as normally.

We assume that the proportions of each type of word classified by POS in the BEST corpus is similar to that of the Orchid corpus. An E2 is calculated by the following equation. The w_b means the amount of the words w found in BEST corpus. ALLW$_{bp}$ refers to the amount of words assigned to POS in BEST corpus, assumedly calculated.

$$E2 = \frac{w_b}{ALLW_{bp}} \tag{2}$$

An E3 is calculated by the following equation. Only the length of the unknown string w_{uk} is employed. However, we need to limit the unknown words' lengths to 20 characters. If not, the system will estimate the entire input as one word.

$$E3 = 10^{-2 \times w_{uk}.length} \tag{3}$$

The E3 works well for detecting hidden unknown words that contain several short rarely-used known words. If an unknown string is consisted of rarely used short known words, their E2s must be low and splitting into short words leads to the occurrences of transition probabilities. We found that Emission probabilities of rarely used known words are normally around $10^{-7} \sim 10^{-5}$. That means, the E3 of the unknown words which are 2.5-3.5 (2-4) characters long are approximately equal to the E1-2 of rarely used known words. Thus, a sequence of several short known words and an unknown word composed of these several words will give almost equivalent emission probabilities value. However, with concerning of the transition probabilities, the score of the path containing only one state will win the score of the path containing several states.

3.3   Maximum Entropy – Viterbi Algorithm

Word Segmentation Using ME. The ME is an algorithm that attempts to characterize a sequence of alphabets as the most likely set of strings. It offers more freedom in grouping alphabets into strings using a set of features which are designed similarly to patterns in the Orchid corpus. We use N-gram features, Jump Bigram feature, Punctuation feature, Latin Scripts Word feature, Numeral feature, Non-Syllabic Containing Path feature, Too Short Known Word feature and Too Long Unknown Word feature.

The N-gram (Unigram, Bigram, and Trigram) features are effective for known words. Given a string, if the function can find the string as an n-gram in any corpora, the function will assess to the string a value. This value depends on the string's length. Longer words will give more extra value than short words to make sure that the system will choose longer known word over splitting it into short known words.

The Jump Bigram feature is designed to detect an unknown word between two known words. For example, the trigram ร้าน-สุกีโคคา-สาขา Ran-Sukikoka-Saka 'shop – Sukikoka – branch' is found in the BEST corpus. Given an input string ร้านไดโซสาขา Ran-Daiso-Saka 'shop – Daiso – branch', this function will match the first and the last token of the trigram to the left and the right substrings of the input. The middle substring ไดโซ which is not found in any corpora will be considered as an unknown word.

The punctuation, Latin scripts word and numeral features group the adjoining characters which are similar in scripts' classes. The last three features give negative values to any substrings that are not capable to be syllables.

After the word segmentation procedure, the segmented words will be tagged by a set of POS tag candidates depending on types of words such as Latin scripts written words, numerals, punctuation, etc.

POS Tagging Using MEMM's Viterbi. Once an input string is segmented into words, the system performs POS tagging using the MEMM's Viterbi algorithm [5]. The probability of a word w being a POS s given the amount of words w found in corpora is calculated from the following equation.

$$p(s|w) = \frac{P(s \cap w)}{P(w)} \tag{4}$$

For unknown words or words found only in the BEST corpus, the P(s|w) depend on how many POS classes they can be assigned to. The P(s|w) can be calculated as the following equation, where N_s refers to the amount of POS classes the word w can be assigned to. If the word w can be tagged with two POS tags, the P(s|w) will be 0.5.

$$p(s|w) = \frac{1}{N_s} \tag{5}$$

4   Experiment

The Orchid corpus' articles which published in the years 1989-1990 and the entire BEST corpus are used as the training data. Two sets of test data are employed. One is extracted from the Orchid corpus randomly (no training data included). Another is the Web Document corpus. The amount and proportion of words in the training and the test set are provided in Table 1 and Table 2. Table 3 and Table 4 show the performance of our system in word segmentation and POS tagging. Our baselines are the methods that do not apply BEST corpus.

Table 1. Amount of words in training and test data

Training Data	
Corpus	Word
Orchid Corpus	138045
BEST Corpus	5022672
Test Data	
Orchid Corpus	7802
Web Documents	5061

Table 2. Proportion of Known and Unknown Words in Test Data

Test Data	known in Orchid	Unknown in Orchid	unknown in Orchid but exist in BEST	unknown in both corpora
Orchid	86.7%	13.3%	4.85%	8.45%
Web Doc	74.2%	25.8%	16.7%	9.1%

Table 3. Results of Word Segmentation

		Test Data: Orchid Corpus			Test Data: Web Documents		
		Precision	Recall	F_1	Precision	Recall	F_1
W/O BEST	Viterbi	0.823	0.843	0.833	0.731	0.767	0.748
	ME-V	**0.844**	**0.859**	**0.851**	0.732	0.735	0.734
W/ BEST	Viterbi	0.8	0.83	0.81	**0.813**	**0.874**	**0.843**
	ME-V	0.799	0.848	0.823	0.712	0.818	0.76

Table 4. Results of WS & POS tagging

		Test Data: Orchid Corpus			Test Data: Web Document		
		Precision	Recall	F_1	Precision	Recall	F_1
W/O BEST	Viterbi	0.748	0.767	0.757	0.654	0.687	0.67
	ME-V	**0.771**	**0.785**	**0.778**	0.649	0.652	0.651
W/ BEST	Viterbi	0.722	0.749	0.735	**0.701**	**0.753**	**0.726**
	ME-V	0.729	0.774	0.751	0.61	0.7	0.651

5   Discussion

5.1   Using the Web Document corpus as Test Data

The results show that when the test data is the Web Document corpus, the methods that employ the BEST corpus gave the better results. According to the Table 2, using the BEST corpus also decreases the amount of unknown words by 16.7%.

Moreover, since the result of Viterbi method is better than ME-Viterbi method, we checked the output and discovered that the ME-Viterbi method tends to separate compound words into shorter words. The cause of this problem is that the ME-Viterbi method does not use the frequency of word appearance. On the other hand, Viterbi method which depends on frequency of word appearance and word' length tends to group words more successfully. It also works well on transliterated words such as personal names. However, the Viterbi method tends to group any first name and last name into one word including the first name, a white space and the last name which is not exact to the Orchid corpus' standard.

The accuracy of the POS tagging using the Viterbi algorithm is low because the system usually assigned the unknown proper nouns to the common noun class. The POS tagging ambiguities that exist in the Orchid corpus also causes problems.

5.2   Using the Orchid Corpus as Test Data

When the Orchid corpus is used as the test data, the Orchid corpus alone is adequate. The results of the experiments done without the BEST corpus are even slightly better. We think it is because the different word segmentation style between the

Orchid corpus and the BEST corpus may distract the system from choosing the right path especially when the system processes partially hidden unknown words which some substrings are found only in the BEST corpus.

In addition, we believe that the word segmentation using the ME and specified features is more efficient than depending only on the probabilities if the training data and the test data are composed in the same standard and the standard is observed. The features handling Latin scripts words and numerals are very helpful since the unknown words in the test data extracted from the Orchid corpus are mostly English words and numerals.

6   Conclusion

This paper described our research on the Thai morphological analysis using POS tagged and untagged corpora. We proposed two methods that are Viterbi-algorithm and ME - Viterbi algorithm with unknown word detection features. We used two test data sets. One is quite similar to the benchmark. Another is written in various writing styles. The result proved that the Orchid corpus alone is not adequate for processing all kinds of documents. The text corpus is very useful as a word list because it contains a great amount of proper nouns and commonly used words.

Our research also shows that each method has a strong point. The Viterbi method gives a promising result for unknown word problem solving. The ME-Viterbi works well if training and test data have similar writing style and vocabulary. The strong points of the both methods we have discovered from this research will be useful in the future as we will try to adapt each method's strong point to more efficient feature.

References

1. Charoenporn, T., Sornlertlamvanich, V., Isahara, H.: Building A Large Thai Text Corpus - Part-Of-Speech Tagged Corpus ORCHID. In: Proceedings of the NLPRS 1997, pp. 509–512 (1997)
2. Sornlertlamvanich, V., Charoenporn, T., Isahara, H.: ORCHID: Thai Part-Of-Speech Tagged Corpus Technical Report TRNECTEC-1997-001, NECTEC (1997)
3. Boriboon, M., Kriengket, K., Chootrakool, P., Phaholphinyo, S., Purodakananda, S., Thanakulwarapas, T., Kosawat, K.: BEST Corpus Development and Analysis. In: Proceedings of the 2009 International Conference on Asian Language Processing, pp. 322–327 (2009)
4. Viterbi, A.J.: Viterbi, Error bounds for convolutional codes and anasymptotically optimum decoding algorithm. IEEE Transactions on Information Theory 13(2), 260–269 (1967)
5. McCallum, A., Freitag, D., Pereira, F.: Maximum Entropy Markov Models for Information Extraction and Segmentation. In: The 17th International Conference on Machine Learning (2000)

Image Retrieval Using Most Similar Highest Priority Principle Based on Fusion of Colour and Texture Features

Fatin A. Mahdi, Mohammad Faizal Ahmad Fauzi, and Nurul Nadia Ahmad

Multimedia University, Malaysia
fatinmahdi2000@yahoo.com,
{faizal1,nurulnadia.ahmad}@mmu.edu.my

Abstract. We propose Content-Based Image Retrieval (CBIR) system using local RGB colour and texture features. Firstly, the image is divided into sub-blocks, and then the local features are extracted. Colour is represented by Colour Histogram (CH) and Colour Moment (CM). Texture is obtained by using Gabor filter (Gab) and Local Binary Pattern (LBP). An integrated matching scheme based on Most Similar Highest Priority (MSHP) principle is used to compare the blocks of query and database image. Since each feature extracted from images just characterizes certain aspect of image content, features fusion are necessary to increase the retrieval performance. We present a novel fusion method based on fusing the distance value for each feature instead of the feature itself to avoid the curse of dimensionality. Experimental results in terms of the precision/recall estimates demonstrate that the performance of the proposed fusion method gives better performance than that when either method is used alone.

Keywords: Colour feature, Texture feature, Image sub-block, Feature fusion, MSHP principle.

1 Introduction

Multimedia has an important place in many applications area such as trademark, medical imaging, etc. Generally, CBIR can be defined as any system that helps to retrieve and organize digital image archives by their visual content [1]. Most of CBIR automatically extract set of low-level features such as colour, texture and shape, and then stored these features in the database as feature vector (FV) for similarity measurement between query image and database images by comparing the FV differences [2]. Commonly, image retrieval systems which are based on only one feature do not achieve good retrieval accuracy. To yield better retrieval performance, some retrieval systems were developed based on combination of appropriate relevant features [3]. It is also observed that the local features based image retrieval play a significant role in determining similarity of images [3, 4]. In this paper we extract colour and texture features from each image sub-block as local feature. To investigate for a better retrieval results we further fused the colour and texture descriptors using fusion distance

P. Anthony, M. Ishizuka, and D. Lukose (Eds.): PRICAI 2012, LNAI 7458, pp. 765–770, 2012.

method. This paper is organized as follow: Section 2 covers the system overview and the proposed fusion method. Section 3 deals with the experimental results. Finally, Section 4 presents the conclusions and future work.

2   System Overview and the Proposed Method

Our proposed system is based on fusion of colour and texture features of image sub-blocks, and using similarity measure based on Most Similarity Highest Priority principle for matching process. Figure 1 illustrates the proposed image retrieval system.

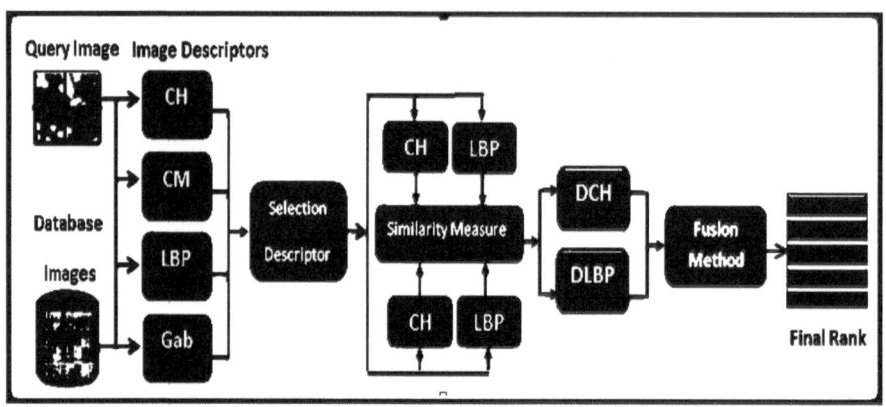

Fig. 1. Block diagram of the proposed image retrieval system

2.1   Low Level Features Extraction

Firstly, the image is partitioned into 2x2 and 3x3 non-overlap equal size sub-blocks and 5 overlap sub-blocks. Then the local low level features (Colour and Texture) are extracted from each sub-block of the database to distinguishable image content. Basically, the key to accomplishing an effective retrieval system is to select the accurate features that represent the image as strong as possible [1].

2.1.1   Colour Feature Extraction
Colour feature is commonly used in CBIR systems because of its effectiveness in searching and its simplicity in implementation. In our system we considered CH and CM. CH is a representation technique of colour distribution [1]. It combined probabilistic properties of the various colour channels (R=Red, G=Green, and B=Blue channels), by capturing the number of pixels having particular properties. CM has also been successfully used in many retrieval systems [5]. In our system we extract the first three central moments (mean, standard deviation and variance) of each channel as that proposed in [6]. In this case, only 9 values are used to represent the colour feature of each sub-block.

2.1.2   Texture Feature Extraction

Texture has been one of the most important features used to recognize the objects in the image. It is a repeated pattern of pixels over a spatial domain.

Local Binary Pattern (LBP)

LBP was introduced in [7]; it is a computationally simple and very powerful method of analyzing image textures. LBP is based on a binary code created by dividing the image into several blocks. The eight neighbors are labeled using a binary code {0, 1} obtained by comparing the values of eight neighbors to the central pixel value. If the center pixel's value is greater than the neighbor, then it is labeled 0, otherwise it is labeled 1. This gives an 8-digit binary number which is multiplied by weights 2^{i-1} given to the corresponding pixels in order to convert to a decimal value that will represent the new center value. We then compute the histogram of these numbers. Lastly, the histograms are concatenated to represent the LBP FV of all image blocks.

Gabor Transform Filters

Our implementation to compute the Gabor filters is based on [8]. The feature is computed by filtering the image with a bank of orientation and scale sensitive filters and then the mean of the output is computed in the frequency domain.

2.2   Integrated Image Matching

In this paper we proposed the integrated image matching procedure similar to the one used in [3, 6, 7]. In our method, a sub-block from query image is matched to all sub-blocks in the target image. A bipartite graph of sub-blocks for the query image and the target image is built as shown in Figure 2 using 2×2 sub-blocks. The labeled edges of the bipartite graph refer to the distances between each sub-block, and then we will save all 16 distances in adjacency matrix of the bipartite graph. The first row in this matrix represents the distance between the top left sub-block of the query image with all sub-blocks of target image and so on. A minimum cost matching based on MSHP principle is done for this graph. The minimum distance d_{ij} of this matrix is found between sub-blocks i of query image and j of target image. The distance value is stored and the row corresponding to sub-block i and column corresponding to sub-block j, are ignored (i.e. replaced by some high value, say 999). This will prevent sub-block i of query image and sub-block j of target image from participating in the next matching process. A particular sub-block is to participate in the matching process only once. This process is repeated until every sub-block finds a matching, then we collect the 4 distance values to get the new distance between the query and target images using equation (1). Figure 3 illustrates this process. The complexity of the matching procedure is reduced from $O(n^2)$ to $O(n)$, where n is the number of sub-blocks involved.

$$NewD_{qt} = \sum_{i=1}^{n} \sum_{j=1}^{n} d_{ij} \dots\dots\dots\dots\dots \quad \dots (1)$$

where i = 1, 2,..., n, j = 1, 2, ..., n, and d_{ij} is the best-match distance between sub-block i of query image q and sub-block j of target image t.

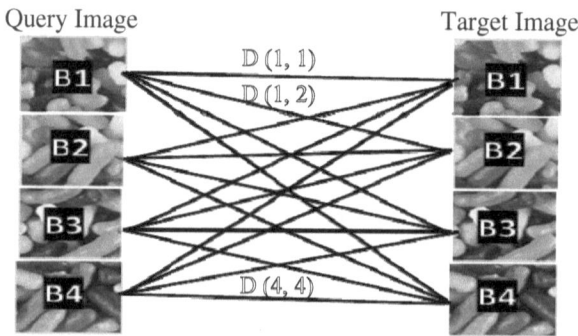

Fig. 2. Bipartite graph using 2 x 2 = 4 sub-blocks of query and target image

54.7322	55.1945	62.2743	60.5692
73.8617	77.7636	79.4756	81.6315
58.4133	60.7407	66.2487	66.1349
121.4567	123.6607	130.4524	129.9424

(a)

54.7322	999	999	999
999	77.7636	79.4756	81.6315
999	60.7407	66.2487	66.1349
999	123.6607	130.4524	129.9424

(b)

54.7322	999	999	999
999	999	79.4756	81.6315
999	60.7407	999	999
999	999	130.4524	129.9424

(c)

54.7322	999	999	999
999	999	79.4756	999
999	60.7407	999	999
999	999	999	129.9424

(d)

NewDqt = D(QB1,TB1) + D(QB3,TB2) + D(QB2,TB3) + D(QB4,TB4)
New (D) = 54.7322 + 60.7407 + 79.4756 + 129.9424= 324.8909

Fig. 3. Image similarity computation based on MSHP principle, (a) first pair of matched sub-blocks i=1, j=1 (b) second pair of matched sub-blocks i=3, j=2 (c) third pair of matched sub-blocks i=2, j=3 (d) fourth pair of matched sub-blocks i=4, j=4, yielding the integrated minimum cost match distance 324.8909.

2.3   Fusion of Colour and Texture Features

Information fusion is define as "an information process that associates, correlates and combines data and information from single or multiple sources to achieve refined estimates of parameters, characteristics, events and behaviors" [9]. In this paper we present a novel fusion method based on fusing MSHP distance value for each feature instead of FV fusion to avoid the curse of dimensionality.

2.4   Fusion Algorithm

We performed features fusion based on the following steps:

1. Extract the colour and texture feature for each 2×2 sub-block of target image and query image, and then compute the similarity distance for each feature.

2. Fuse the distance values of colour and texture features by summing them to obtain 16 fusion distance values, and store these values in a 4-by-4 matrix.
3. Apply MSHP principle onto the matrix of distance values to get the new distance value between query and database images.
4. Repeat step (1-3) for all database images.
5. Sort the new distance values in ascending order to retrieve most similar images.

3   Experimental Results

The dataset used in our experiment consist of 864 heterogeneous images from original Vision texture image. In order to evaluate the effectiveness of proposed systems, all the images in the database will be used as query image. To evaluate the efficiency of the system two well-known metrics are used, i.e. precision (the ratio of the relevant images that are retrieved to the total number of retrieved images), and recall (the ratio of the relevant images that are retrieved to the total number of relevant images in the database) [1]. These metrics give an estimate of retrieval effectiveness in the range from 0 (worst case) to 1 (perfect retrieval). The experimental results for evaluating the retrieval performance are done by comparing between the individual descriptor (CH, CM, LBP, and Gab) with features fusion, as shown in Figure 4 and also between the 2×2, 3x3 non-overlapping sub-blocks and 5-overlapping sub-blocks based on the best fusion (CH & LBP – from Figure 4) as shown in Figure 5.

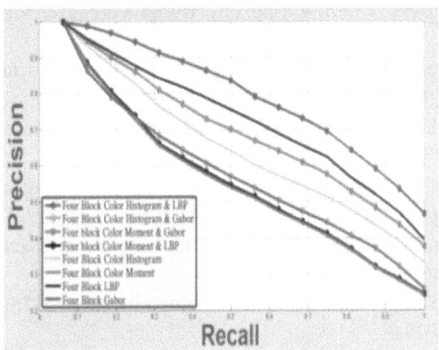
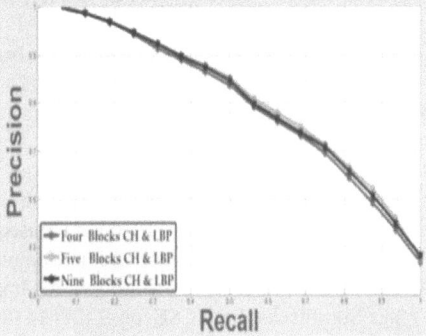

Fig. 4. The Recall-precision graph of individual and feature fusion using MSHP principle

Fig. 5. The Recall-precision graph of different sub-blocks based feature fusion using MSHP principle

Table 1 shows the comparison of average precision obtained by image retrieval. It can clearly be seen that fusion CH with LBP is significantly (Pre % = 0.7893) more effective than individual CH (Pre % = 0.6444) and LBP (Pre % = 0.7249).

Table 1. Comparison in term of precision percentage (Pre %) between individual feature descriptors and feature fusion using MSHP distance

CH (Pre %)	LBP (Pre %)	CH &LBP (Pre %)	CM (Pre %)	Gab (Pre %)	CM&Gab (Pre %)	CH&Gab (Pre %)	CM&LBP (Pre %)
0.6444	0.7249	0.7893	0.5519	0.5761	0.6904	0.5759	0.5578

4   Conclusions and Future Work

The main contribution of this work is a comprehensive comparison between single features, features fusion, and between different sub-blocks image. In particular, the fusion of CH and LBP, and CM and Gab resulted to better precision-recall than that obtained for the corresponding individual feature. We also found that the performance of fusion method for different image sub-blocks are equally similar with the 5-blocks recorded insignificantly higher performance between 50% to 90% recall. For our future work, we will explore other methods for fusing colour and texture features, apart from investigating other colour and texture descriptors.

References

1. Chatzichristofisand, S.A., Boutalis, Y.S.: Compact Composite Descriptors for Content Based Image Retrieval: Basics, Concepts, Tools. VDM Verlag Dr. Muller GmbH & Co. KG, book (2011)
2. Datta, R., Joshi, D., Li, J., Wang, J.Z.: Image retrieval: ideas, influences, and trends of the new age. ACM Computing Surveys 40(2), 1–60 (2008)
3. Saad, M.: Content Based Image Retrieval Literature Survey. Multi-Dimensional Digital Signal Processing (March 18, 2008)
4. Liu, P., Jia, K., Wang, Z., Lv, Z.: A New and Effective Image Retrieval Method Based on Combined Features. In: 4th Int. Conf. on Image and Graphics, P. S. (2007)
5. Hiremath, P.S., Pujarii, J.: Content Based Image Retrieval based on Colour, Texture and Shape Features using Image and its Complement. IJCSS 1(4) (2007)
6. Kavitha, C., Rao, B.P., Govardhan, A.: Image Retrieval based on Combined Features of Image Sub-blocks. In: IJCSE, pp. 1429–1438 (2011)
7. Li, J., Wang, J.Z., Wiederhold, G.: IRM: Integrated Region Matching for Image Retrieval. In: Proc. of the 8th ACM Int. Conf. on Multimedia, pp. 147–156 (2000)
8. Howarth, P., Ruger, S.: Robust texture features for still-image retrieval. In: IEE Proceedings of Visual Image Signal Processing, vol. 152(6) (2005)
9. Mangai, U.G., Samanta, S., Das, S., Chowdhury, P.R.: A Survey of Decision Fusion and Feature Fusion Strategies for Pattern Classification. IETE Technical 27(4), 293–307 (2010)

Computational Intelligence for Human Interactive Communication of Robot Partners

Naoki Masuyama[1], Chee Seng Chan[1], Naoyuki Kuobota[2], and Jinseok Woo[2]

[1] University of Malaya
Lembah Pantai, 50603 Kuala Lumpur, Malaysia
[2] Tokyo Metropolitan University
6-6 Asahigaoka, Hino, Tokyo 191-0065, Japan
naoki.masuyama17@gmail.com, cs.chan@um.edu.my,
kubota@tmu.ac.jp, woo-jinseok@sd.tmu.ac.jp

Abstract. This paper proposes a multi-modal communication method for human-friendly robot partners based on various types of sensors. We explain informationally structured space to extend the cognitive capabilities of robot partners based on environmental systems. We propose an integration method for estimating human behaviors using sound source angle information, and gesture recognition by the multi-layered spiking neural network with the time series of human hand positions. Finally, we show several experimental results of the proposed method, and discuss the future direction on this research.

Keywords: Intelligent Robots, Computational Intelligence, Human robot interaction.

1 Introduction

Recently, the rate of elderly people rises in the super-aging society. For example, the rate is estimated to reach 25.2% in Tokyo in 2015. In general, the mental and physical care are very important in order to restrain the progress of dementia of elderly people living home alone. Therefore, it is essential to use human-friendly robots to support the mental and physical care for elderly people and to assist the care of caregivers to elderly people. However, it is difficult for a robot to converse appropriately with a person even if many contents of the conversation are designed in advance because the performance of voice recognition is not enough in the daily conversation. Furthermore, in addition to verbal communication, the robot should understand non-verbal communication e.g. facial expressions, emotional gestures and pointing gestures.

According to the relevance theory [1], each person has his or her own cognitive environment, and the communication between people is restricted by their cognitive environments. The shared cognitive environment is called a mutual cognitive environment. A human-friendly robot partner also should have such a cognitive environment, and the robot should keep updating the cognitive environment according to current perception through the interaction with a person in order to realize the natural communication.

P. Anthony, M. Ishizuka, and D. Lukose (Eds.): PRICAI 2012, LNAI 7458, pp. 771–776, 2012.
© Springer-Verlag Berlin Heidelberg 2012

We have proposed an information support system using human-friendly conversation to elderly people by integrating robot technology, network technology, information technology, and intelligence technology based on the concept of mutual cognitive environments [2,3]. In order to share their cognitive environments between a person and robot partner, the environment surrounding people and robot partners should have a structured platform for gathering, storing, transforming, and providing information. Such an environment is called informationally structured space [4,5]. If the robot can easily share the environmental information with people, the social communication with people might become very smooth and natural. Therefore, in this paper, we propose a method of estimating human behaviors using various types of sensors for multi-modal communication between a person and robot partner.

2   Robot Partners and Multi-modal Communication

We have used various types of robot partners such as MOBiMac, Hubot, Apri Poco, Palro, Miuro, and others for the support to elderly people, rehabilitation support, and robot edutainment [2-4, 6-8].

In this paper, the robot partner needs to do not only measurement of human position but also gesture recognition. However, It is difficult to recognize the human gesture by using own sensor. Therefore, we use Microsoft Kinect sensor instead of own sensor in order to do these tasks. Furthermore, the robot itself is enough to be equipped with only cheap sensors.

To natural verbal communication between robot partner and human, it's necessary for robot partner to understand the multi cognitive environment with human. Kinect sensor can help robot partner to understand positions of human and non-verbal information like gesture, they will make natural communication between robot partner and human.

3   Human Interactive Communication System

Fig.1 shows a system configuration for the human interactive communication. The robot is placed on the desk and Kinect is placed behind the robot to compensate for perception ability of the robot. In Kinect, the measurement angle of the sound source is wider than the measurement angle of the view. Therefore, it enables for the robot to recognize a sound if there aren't people inside scope of the camera image.

4   Gesture Recognition

The positions of human face and hands are extracted by OpenCV and OpenNI. The relative position of hands is calculated where the facial position is defined as the origin for gesture recognition. In order to extract human gesture, we apply a multilayered spiking neural network.

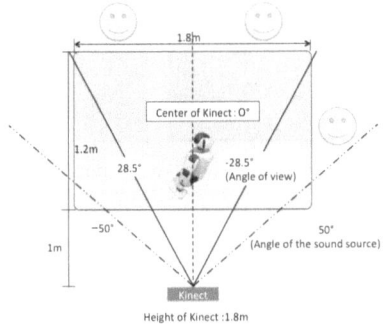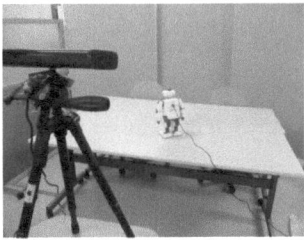

Fig. 1. Overview of the system

Various types of artificial neural networks have been proposed to realize clustering, classification, nonlinear mapping, and control [9]. We use a simple spike response model to reduce the computational cost.

The sequence of human hand positions is used for recognizing spatio-temporal pattern as a human gesture by spiking neural network (SNN). SNN is often called a pulsed neural network, and is one of the artificial NNs imitating the dynamics introduced the ignition phenomenon of a cell, and the propagation mechanism of the pulse between cells [10,11]. In this paper, we use a simple spike response model to reduce the computational cost.

First of all, the internal state $h_{1,i}(t)$ of a spiking neuron in the first layer is calculated as follows;

$$h_{1,i}(t) = \tanh\left(h_{1,i}^{syn}(t) + h_{1,i}^{ext}(t) + h_{1,i}^{ref}(t)\right) \tag{1}$$

The internal state $h_{2,i}(t)$ of the ith spiking neuron in the second layer is calculated as follows;

$$h_{2,i}(t) = \tanh\left(\gamma_2^{sys} \cdot h_{2,i}(t-1) + h_{2,i}^{ext}(t) + h_{2,i}^{ref}(t)\right) \tag{2}$$

The internal state $h_{3,i}(t)$ of the ith spiking neuron in the third layer is calculated as follows;

$$h_{3,i}(t) = \tanh\left(\gamma_3^{sys} \cdot h_{3,i}(t-1) + h_{3,i}^{ext}(t) + h_{3,i}^{ref}(t)\right) \tag{3}$$

The robot recognizes the ith gesture when the ith spiking neuron is fired. Figures 2 and 3 show a preliminary experimental result of gesture recognition based on human hand motions where the number of recognizable gestures is 3; (a) Bye-bye A (left – right - left), (b) Bye-bye B (upperleft – lowerright - upperleft), (c) Clock-wise square.

5   Experimental Results

Fig. 4 shows experimental result of verbal communication. We divided the area where can be estimated sound source angle by Kinect into five regions. We placed the user on area 2 and area 4, and placed PALRO on area 3. The values of vertical axis in Fig. 4 are defined as Table 2. Using the sound source angle, it can identified User1, User 2 and PALRO. After the statement of the User, PALRO performs a remark. Therefore area 3 will be detection after other areas.

(a) Bye-bye A   (c) Clock-wise square

Fig. 2. Image processing for human gesture recognition ((a)and(c))

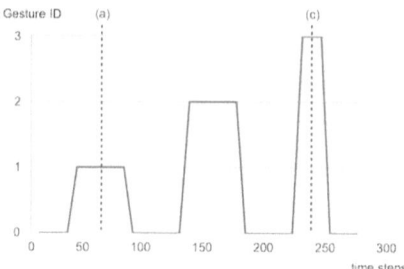

Fig. 3. The history of Gesture ID in a preliminary experimental result; (0) No Gesture, (a) Bye-bye A (left – right - left), (b) Bye-bye B (upperleft – lowerright - upperleft), (c) Clock-wise square

Table 1. Separation angle of estimation the sound source area

Area ID	Area [deg]
5	50 - 30
4	30 - 10
3	10 - -10
2	-10 - -30
1	-30 - -50
0	Silence

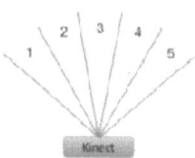

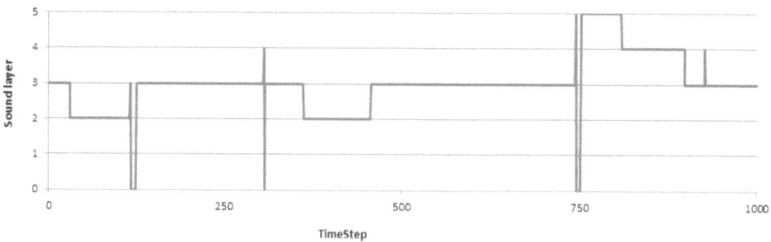

Fig. 4. The history of sound source estimation

Fig. 5 and Table 2 show experimental flow of interaction between the users and robot partner. At first, User 2 is at out range of Kinect's camera so PALRO can't recognize. Therefore, User 1 needs to use non-verbal information by pointing User 2 to PALRO. With this, PALRO can recognize User 2 who is outside of perception.

This has shown that it is possible to transmit clear information about direction using gesture of pointing.

After recognized pointing, PALRO performs remark to check and induce active communication with User 2. (It's assume that User 2 stays place where can be estimate sound source angle.)

(a) Pointing to outside of the view

(b) Recognition of person by using sound

(c) Human Detection source of camera image

Fig. 5. Experimental flow of interaction between human and robot

Table 2. Example of interaction between human and robot

User1	: Pointing Gesture (Fig.5 (a))
PALRO	: "Is there the person to the left side?" (Fig.5 (b))
User 2	: "Yes, there is."
PALRO	: "Please come here."
User 2	: "Hello."
PALRO	: "Hello, how are you?" (Fig.5 (c))

Fig. 6 shows experimental result using verbal and non-verbal communication together. It is performed by the recognition of user identification and user positions using a sound source angle and a gesture. Therefore, this result shows robot partner and human can natural communication using verbal and non-verbal information.

6   Summary

In this paper, we proposed a method of multi-modal communication combining various types of sensors based on the interaction range. First, we explained the concept of informationally structured space, and robot partners developed in this paper. Next, we propose an estimation method of human behaviors using Kinect. In the experimental results, the proposed method can estimate human behaviors based on sound source angle information, and gesture recognition by the multi-layered spiking neural network using the time series of human hand positions. We will incorporate the interactive learning algorithm of gesture recognition to realize on-line acquisition of gestures for natural communication. As another future work, we will flexibly and adaptively integrate these robot partners with the informationally structured space to realize natural communication and human-friendly interaction.

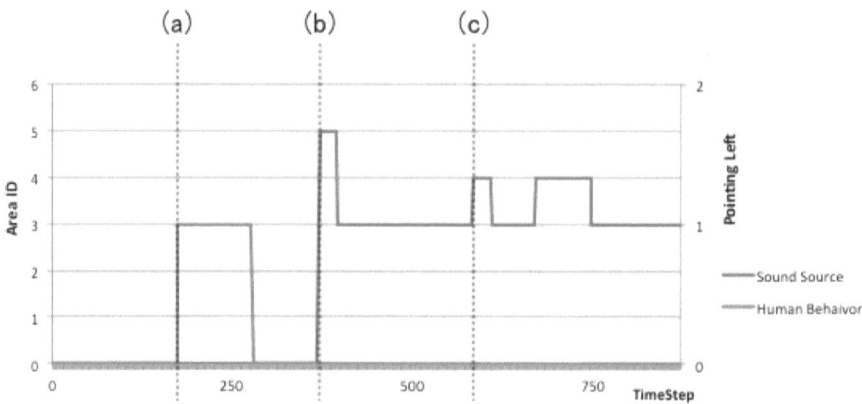

Fig. 6. The history of interaction between human and robot

References

1. Sperber, D., Wilson, D.: Relevance - Communication and Cognition. Blackwell Publishing Ltd (1995)
2. Kubota, N., Yorita, A.: Structured Learning for Partner Robots based on Natural Communication. In: Proc (CD-ROM) of 2008 IEEE Conference on Soft Computing in Industrial Applications, pp. 303–308 (2008)
3. Kimura, H., Kubota, N., Cao, J.: Natural Communication for Robot Partners Based on Computational Intelligence for Edutainment. In: Mechatronics 2010, pp. 610–615 (2010)
4. Kubota, N., Yorita, A.: Topological Environment Reconstruction in Informationally Structured Space for Pocket Robot Partners. In: 2009 IEEE International Symposium on Computational Intelligence in Robotics and Automation, pp. 165–170 (2009)
5. Obo, T., Kubota, N., Lee, B.H.: Localization of Human in Informationally Structured Space Based on Sensor Networks. In: Proc. of IEEE WCC 2010, pp. 2215–2221 (2010)
6. Mirza, N.A., Nehaniv, C.L., Dautenhahn, K., te Boekhorst, R.: Using temporal information distance to locate sensorimotor experience in a metric space. In: Proc. of IEEE CEC 2005, pp. 150–157 (2005)
7. Kubota, N., Hisajima, D., Kojima, F., Fukuda, T.: Fuzzy and Neural Computing for Communication of A Partner Robot. Journal of Multiple-Valued Logic and Soft-Computing 9(2), 221–239 (2003)
8. Kubota, N., Ito, Y., Abe, M., Kojima, H.: Computational Intelligence for Natural Communication of Partner Robots. In: ASAEM 2005, Technical Section B-11 (2005)
9. Obo, T., Kubota, N., Taniguchi, K., Sawaya, T.: Human Localization Based on Spiking Neural Network in Intelligence Sensor Networks. In: SSCI 2011, pp. 125–130 (2011)
10. Maass, W., Bishop, C.M.: Pulsed Neural Networks. MIT Press (1999)
11. Gerstner, A.W.: Spiking Neurons. In: Maass, W., Bishop, C.M. (eds.) Pulsed Neural Networks, ch. 1, pp. 3–53. MIT Press (1999)

Healthy or Harmful? Polarity Analysis Applied to Biomedical Entity Relationships

Qingliang Miao[*], Shu Zhang, Yao Meng, Yiwen Fu, and Hao Yu

Fujitsu R&D Center Co., LTD.
No.56 Dong Si Huan Zhong Rd, Chaoyang District, Beijing P.R. China
{qingliang.miao,zhangshu,mengyao,
art.yiwenfu,yu}@cn.fujitsu.com

Abstract. In this paper, we investigate how to automatically identify the polarity of relationships between food and disease in biomedical text. In particular, we first analyze the characteristic and challenging of relation polarity analysis, and then propose a general approach, which utilizes background knowledge in terms of word-class association, and refines this information by using domain-specific training data. In addition, we propose several novel learning features. Experimental results on real world datasets show that the proposed approach is effective.

1 Introduction

To discover the underlying knowledge hidden in biomedical data, much work has been focusing on biomedical relationship extraction. However, there is little work conducted on polarity analysis of these relationships [5]. For instance, in Figure 1, Sentence 1 describes two positive relationships, while Sentence 2 indicates a negative relationship.

Sentence 1: "Genistein was shown to arrest the growth of malignant melanoma in vitro and to inhibit ultraviolet (UV) light-induced oxidative DNA damage."
Sentence 2: "In addition, eating soybean pastes was associated with the increased risk of gastric cancer in individuals with the GSTM1."

Fig. 1. Example of Relation-bearing Sentence

In biomedical literature, different studies may report different findings of the same biological relationship. For instant, some studies suggest "green tea" have inhibitory effects against some tumor diseases, while another study shows that they are irrelevant. Consequently, it is essential to identify the polarity of biomedical relationships.

Recently, there has been much research on sentiment analysis in subjective text [3]. However, the characteristics of relation-bearing sentences vary from subjective ones.

---

[*] Corresponding author.

P. Anthony, M. Ishizuka, and D. Lukose (Eds.): PRICAI 2012, LNAI 7458, pp. 777–782, 2012.

Firstly, subjective sentences express people's opinions, while relation-bearing sentences declare objective information, although both of them contain polarity information. Secondly, in subjective sentences, polarity information is mostly included in opinion words, which are usually adjectives. However, in relation-bearing sentences, polarity information is often embodied in verbs, verb or prepositional phrases. Finally, opinion words in subjective sentences are usually stative adjectives, such as "perfect", "expensive", while dynamic phrases are usually used to express polarity information, such as "reduce mortality" and "increase risk".

Due to the characteristics of relation-bearing sentences, traditional methods in sentiment analysis field could not be adopted to solve relation polarity analysis effectively. In this paper, we propose an integrated approach to identify polarity of relationships. Specifically, this paper makes the following contributions. First, we analyze the challenges of relation polarity analysis and point out its difference from traditional sentiment analysis issue. Second, we propose a general approach which combines background knowledge with supervised learning. Third, we develop several learning features, which could capture semantic and dynamic change information.

2   Terminology Definition

In order to facilitate the explanation, we first define the basic terminologies of relation polarity analysis.

Definition 1: Multiple Relation-bearing Sentence

Multiple relation-bearing sentences (*MRS*) contain more than two entities and their mutual relationships. Sentence 1 in Figure 1 is an example, in which there are three entities and two positive relationships.   Generally speaking, MRS could be represented by the following patterns, where *M-M, O-M* and *M-O* represent many-to-many, one-to-many and many-to-one relationships respectively. Table 1 show the multiple relation patterns, where *e* represent entity, *r* represent relation words.

Table 1. Multiple relation patterns

Pattern Name	Multiple relation patterns
M-M	$\{e_1, e_2, ..., e_m, r, e_1^{'}, e_2^{'}, ..., e_n^{'}\}$
	$\{e_1, e_2, ..., e_m, (r_1), e_1^{'}, (r_2), e_2^{'}, ..., (r_n), e_n^{'}\}$
O-M	$\{e, r, e_1^{'}, e_2^{'}, ..., e_n^{'}\}$
	$\{e, (r_1), e_1^{'}, (r_2), e_2^{'}, ..., (r_n), e_n^{'}\}$
M-O	$\{e_1, e_2, ..., e_m, r, e^{'}\}$

Definition 2: Single Relation-bearing Sentence

Single relation-bearing sentences (*SRS*) contain two entities and one relationship (referring to Sentence 2 in Figure 1). In this study, we utilize predefined rules to simplify *MRS* to *SRS*, and then identify the relation polarity in SRS ultimately.

Definition 3: Relation Polarity Category

In this paper, we focus on relation polarity analysis of SRS and classify relation polarity into three classes, namely, positive, negative and neutral class. Table 2 shows some example sentences.

Table 2. Relation polarity and example sentences, symbols "+", "-" and "=" represent positive, negative and neutral class, respectively

Polarity class	Example sentence
Positive (+)	Soy isoflavones <u>protect against</u> colorectal cancer.
Negative (-)	Soybean pastes <u>increased risk of</u> gastric cancer.
Neutral (=)	HDL cholesterol was <u>associated with</u> cancer.

3   The Proposed Approach

In particular, the proposed approach is based on a sentiment lexicon of relation words and a classification model trained on labeled instances. The probability distributions from these two models are then adaptively pooled to create a composite multinomial classifier that captures both sources of information. Next, we will introduce the overall algorithm for relation polarity classification and learning feature design, respectively.

3.1   Relation Polarity Classification

Inspired by the work in [1], we adopt a linear opinion pool to integrate information from training data and background knowledge. Specifically, the aggregate probability:

$$P(c_i|r_j)= \sum_{e=1}^{E} \lambda_e P_e(c_i|r_j) \tag{1}$$

where E is the number of experts; $P_e(c_i|r_j)$ represents the probability assigned by expert e to relation r_j belongs to class c_i; and the weights λ_e sum to one. We use the same weighting scheme as [1]:

$$\lambda_e = \log\frac{1-err_e}{err_e} \tag{2}$$

In this paper, we consider two experts: one is learned based on training data, $P_t(c_i|r_j)$, and the other one is a generative model that explains the background knowledge $P_k(c_i|r_j)$. Probability distribution $P_t(c_i|r_j)$ could be estimated from training data based on Bayes' theorem as follows:

$$P_t(c_i|r_j)=\frac{P(c_i)\prod_n P(w_n|c_i)}{\prod_n P(w_n)} \tag{3}$$

where w_n represents nth word in relation r_j.

We use polarity words in proportion to relation r_j to estimate $P_k(c_i|r_j)$ according to the knowledge base:

$$P_k(c_i|r_j) = \frac{|W_j \cap K_i|}{|W_j|} \qquad (4)$$

where W_j represents the word set in relation r_j; K_i represents knowledge base of polarity class c_i such as positive, negative and neutral.

3.2   Learning Feature Design

Besides the traditional word features such as unigram and bigram features, we consider following properties of relation polarity. First, relation polarity information is mostly represented in verbs, prepositional and verb phrases. Second, polarity information is usually embedded in dynamic change. With these properties in mind, we develop the following learning features:

Part of Speech

As relation words are mainly verbs, prepositional and verb phrases, part of speech might also play an important role in encouraging relation polarity identification. In particular, we adopt Stanford tagger [6] to produce POS features.

Entity Category

Entity category information can alleviate the data sparseness problem in the learning process. All mentions of specific food and disease in the biomedical text are generated to the food and disease category by replacing them with the tag "*FOOD*" and "*DISEASE*".

Entity Order

In this paper, we use entity order to represent sequential features, specifically, we use four binary functions "*isFoodLeft()*", "*isFoodRight()*", "*isDiseaseLeft()*" and "*isDiseaseRight()*" to capture entity order information. For example, in sentence "Green tea decreases cancer risk.", features *isFoodLeft()=1, isFoodRight()=0, isDiseaseLeft()=0* and *isDiseaseRight()=1*.

Dynamic Feature

In relation-bearing sentences, polarity information usually involves dynamic changes. Thus the polarity of a relation is often determined by how change happens: if a bad thing was decreased then it is a positive relation, and vice versa. In particular, we manually collect four classes of words: those indicating more, such as "increase" and "enhance", those indicating less, such as "reduce" and "decrease", those indicating good, such as "benefit" and "improvement" and those indicating bad, such as "mortality" and "risk". We use four binary functions "*isMore()*", "*isLess()*", "*isGood()*" and "*isBad()*" to capture dynamic change information. For example, in the above sentence features *isMore()=0, isLess()=1, isGood()=0* and *isBad()=1*.

Negation

At last, we consider two kinds of negative words "not"-related words such as "does not", and word "no"-related words such as "no" and "without". These negative words usually suggest a neutral polarity. We use *"NEG"* to represent negation.

4   Experiments

In the experiments, we use the datasets in [5]. We first extract the relation-bearing sentences and then simplify the MRS to SRS. The statistic of the dataset is shown in Table 3. We evaluate our approach from two perspectives: (1) performance of the proposed approach; (2) effectiveness of the learning features. We perform 10-fold cross validation throughout our evaluation. *Precision*, *recall* and *F-measure* are primarily used to measure the performance.

Table 3. The statistics of the dataset

#Positive	#Negative	#Neutral
748	341	162

4.1   Evaluation of Proposed Approach

In order to evaluate the effectiveness of our approach, we use a supervised learning based method, specifically, Naïve Bayes classifier algorithm [4] and a lexicon-based method [2] as baselines. In the proposed approach, the weights of P_t and P_k are 0.6 and 0.4 respectively.

Table 4. Polarity classification results by using different methods. LB: lexicon-base method; NB: Naïve Bayes and IM: integrated method.

Method	Ave Precision	Ave Recall	Ave F-measure
LB	0.516	0.522	0.420
NB	0.727	0.708	0.714
IM	**0.736**	**0.726**	**0.727**

Table 4 shows the relation polarity classification results. From Table 4, we can see that the proposed approach outperforms the two baselines. In order to evaluate whether the improvement is significant, we conduct pairwise *t*-test on *F-measure* and the *p-value* is 0.043<0.05, which indicates the improvement is significant. One main reason might be that background knowledge could provide complementary information that training data cannot capture.

4.2   Evaluation of Learning Features

In this experiment, unigram features are used as a baseline and other features are incrementally added to the training process. We chose unigram features as baseline since these features alone could achieve competitive results. Table 5 shows the experiment results, from which we find that bigram, dynamic and negation features are more predictive than part of speech features.

Table 5. Polarity classification results by using Naïve Bayes classifier and different feature sets. U: unigram features, P: part of speech features, B: bigram features, D: dynamic features and N: negation features.

Feature Set	*Precision*	*Recall*	*F-measure*
U	0.676	0.641	0.645
U,P	0.680	0.631	0.645
U,P,B	0.705	0.642	0.659
U,P,B,D	0.708	0.672	0.685
U,P,B,D,N	**0.727**	**0.708**	**0.714**

5   Conclusion and Future Work

In this study, we analyze the challenges of relation polarity analysis and point out its difference from traditional sentiment analysis issue. We propose an integrated approach which combines background knowledge with domain-specific training data. Several novel learning features are proposed to capture polarity information in relation-bearing sentences. We conduct separated experiments to evaluate the effectiveness of the proposed approach and learning features as well. In our future work, we will expand the annotated corpus to facilitate further validation of the findings. Secondly, we want to estimate the strength of the relationships.

References

1. Melville, P., Gryc, W., Lawrence, R.: Sentiment Analysis of Blogs by Combining Lexical Knowledge with Text Classification. In: Proceedings of Knowledge Discovery and Data Mining, pp. 1275–1284 (2009)
2. Miao, Q.L., Li, Q.D., Zeng, D.: Fine-grained Opinion Mining by Integrating Multiple Review Sources. Journal of the American Society for Information Science and Technology 61(11), 2288–2299 (2010)
3. Pang, B., Lee, L.: Opinion Mining and Sentiment Analysis. Foundations and Trends in Information Retrieval 2(1-2), 1–135 (2008)
4. Pang, B., Lee, L., Vaithyanathan, S.: Thumbs up? Sentiment Classification using Machine Learning Techniques. In: Proceedings of the ACL 2002 Conference on Empirical Methods in Natural Language Processing, pp. 79–86 (2002)
5. Swaminathan, R., Sharma, Yang, H.: Opinion Mining for Biomedical Text Data: Feature Space Design and Feature Selection. In: Proceedings of the 9th International Workshop on Data Mining in Bioinformatics (2010)
6. Toutanova, K., Klein, D., Manning, C., Singer, Y.: Feature-Rich Part-of-Speech Tagging with a Cyclic Dependency Network. In: Proceedings of HLT-NAACL, pp. 252–259 (2003)

Semi-automatic Segmentation of 3D Point Clouds Skeleton without Explicit Computation for Critical Points

Kok-Why Ng[1], Abdullah Junaidi[1], and Sew-Lai Ng[2]

[1] Faculty of Computing and Informatics, Multimedia University
[2] Foundation Studies and Extension Education, Multimedia University
Jalan Multimedia, 63100, Cyberjaya, Selangor, Malaysia
{kwng,junaidi.abdullah,slng}@mmu.edu.my

Abstract. Segmentation of 3D point clouds is vigorously discussed in recent years. Many existing techniques pre-process the data to identify critical points for meaningful features. Very often, the critical points are predicted at curvature objects and this does not always generate promising outcome. This paper proposes to segment the point clouds based on its skeleton as this robustly reflects the global shape of an object. The skeleton is constructed via Laplacian-based contraction method. Spherical approach is applied along the skeleton to segment the point clouds. The entire process is automatic. Only moderate user-input is applied to align the skeleton to the object. The output of the proposed method is to be compared with another popular segmentation method with critical points input. The result shows that the proposed method generates more accurate segmented features.

Keywords: Point Cloud, Skeleton Extraction, Segmentation, Laplacian, Polygonal Model.

1 Introduction

Segmentation is a crucial process in many research fields to separate a collection of data into different subsets, attribute to a specific property or to human visualization of a particular description. The segmented features can be applied for collision detection, morphing, mesh editing and etc. This paper will investigate and segment the model based on skeleton.

2 Related Work

To convert 3D model into 1D skeleton, distance field [1] measures the global maximum Euclidean distance from the boundary as the point source and a wave front evolved over time to find the centered-point for skeleton. Geometric method such as Reeb Graph [2] obtains an approximate medial axis by retrieving the internal edges and faces of a Voronoi diagram. Edge-contraction method [3] collapses edges into separated line structure before joining them to be a complete connected skeleton.

P. Anthony, M. Ishizuka, and D. Lukose (Eds.): PRICAI 2012, LNAI 7458, pp. 783–788, 2012.
© Springer-Verlag Berlin Heidelberg 2012

For segmentation, a fine to coarse approach in [4] is applied to combine the segmented tiny parts into meaningful features. The approach is fast but the transition
process is a waste of computation power and is not straight-forward. Edge contraction
and space sweeping in [3] is applied to segment an object automatically, however, the
collapsed edges are not stable and can produce weird result. Hierarchical shape
segmentation in [5] identifies the critical points on curve skeleton and uses geodesic
distant to segment the model based on the critical points. The method produces satisfactory segmentation with the user input critical points. However, it depends very
much on the accuracy of the geodesic approach. It is also doubted of the segmented
result if the critical points are to be computed automatically. Their work will be compared with our proposed method since both involve partially human input for accurate
segmentation.

3   Proposed Algorithm

The proposed algorithm is divided into three main processes: voxelization, skeletonization and segmentation. The input 3D point clouds will first be voxelized to reduce
the complexity of the data for accelerating the skeletonization and segmentation
process. Implicit surface reconstruction is computed to obtain the detail volume of the
model for Laplacian smoothing process. This process aims for constructing the skeleton. Figure 1 shows the proposed algorithm of this work.

It begins with primary input of point clouds data. Each point is voxelized using
Octree data structure to reduce the density and redundancy of points for accelerating
the post-processing. This step indirectly bridges any moderately missing data and
reduces the noisy data. The voxelized point clouds will then be triangulated to obtain
the detail volume for skeleton extraction through Laplacian smoothing process.

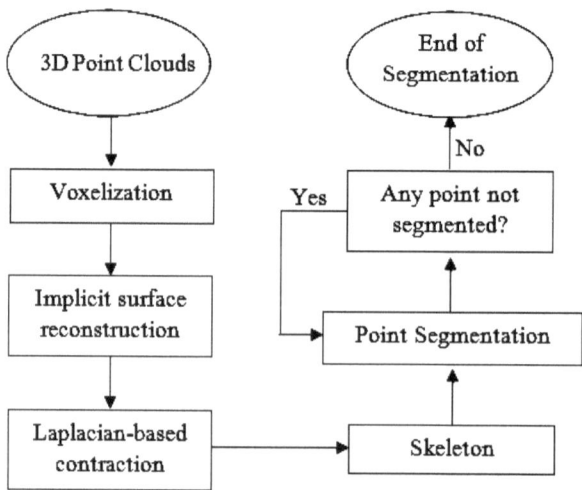

Fig. 1. The proposed algorithm

The crucial points which the edge-contraction attaches to, will be determined by Laplacian smoothing process. This paper borrows the idea from [6] to shrink the model. The Laplacian takes a sum of second partial differential of function with respect to Euclidean space to measure the curvature of a model. While iteratively shrinking the edges, cotangent weights are applied to multiply with Laplacian so that the constructed skeleton will be within the internal region of the original model shape. This process is to be solved in linear system below:

$$\begin{bmatrix} W_L L \\ W_H \end{bmatrix} P' = \begin{bmatrix} 0 \\ W_H P \end{bmatrix}$$

(1)

where P is the input data and P' is the contracted data. L is an n x n Laplacian matrix. W_L and W_H are the cotangent weights to balance the contraction and attraction constraints. At the end, edge-collapsed approach is applied to obtain thin 1D skeleton.

Figure 2(a) shows the original hand model with many small pieces of skeleton (various colors) constructed along the curve. Each skeleton joint will be computed for the number of skeleton edges connected to it (see Figure 2(b)). The end joint will be counted as 1's as it is only connected by one edge. The middle joints will take either 2's or more of the edges connection. Any joint with more than two connected edges will form a junction. The entire links from one end joint to the junction will be merged and is considered a feature. In Figure 2(c), dihedral angle (ɵ) and the length (l) of each skeleton edge will be computed to distinguish the need to further separate a feature into more features. This is important especially at the curvature shape in order to preserve meaningful feature and to reduce unnecessary short skeletons.

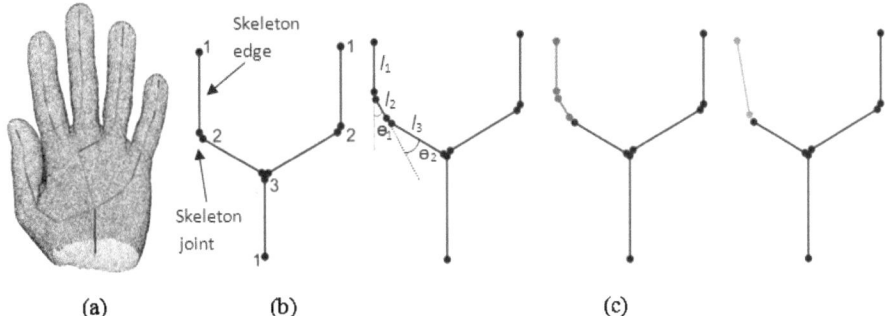

(a)   (b)   (c)

Fig. 2. (a) Original hand with many small pieces of skeletons constructed along the curve. (b, c) Computation for number of skeleton edges connected to each skeleton joint.

With the skeleton features, a simple spherical distance check is proposed to trace the surrounding point clouds of the model. The spherical size is determined by taking the average distant of two consecutive skeleton edges. Spherical size, $s = 0.5\ (e_i + e_{i+1})$ where e is the skeleton edge and i is the skeleton point. Note that a threshold is set for any outlier skeleton edge to ensure none extreme spherical is generated. Figure 3 shows all the point clouds that have been segmented according to the skeleton edges respectively.

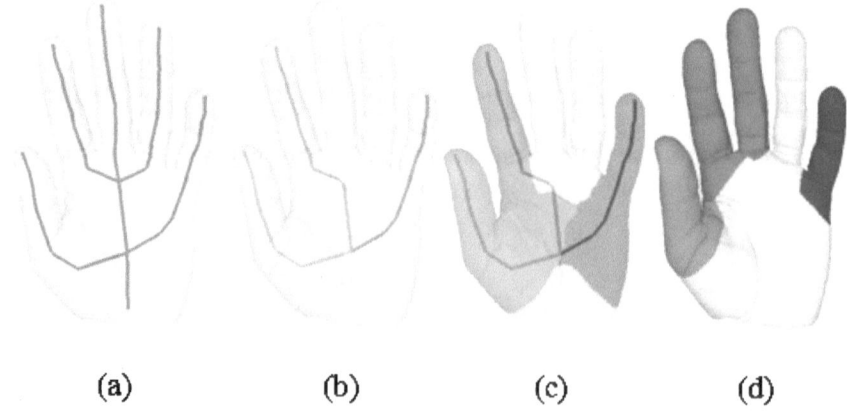

$$(a) \qquad (b) \qquad (c) \qquad (d)$$

Fig. 3. (a) Adjusted skeleton point at the junction. (b) Segmented skeleton edges from (a). (c) Segmented point clouds from (b). (d) Result of skeletons nearby the junction is removed.

4   Experimental Result

The proposed algorithm runs on 3.07GHz Intel® Core™ i3 CPU Windows 7 Home Premium with 4GB RAM. Various 3D models are tested and shown in Figure 4 below.

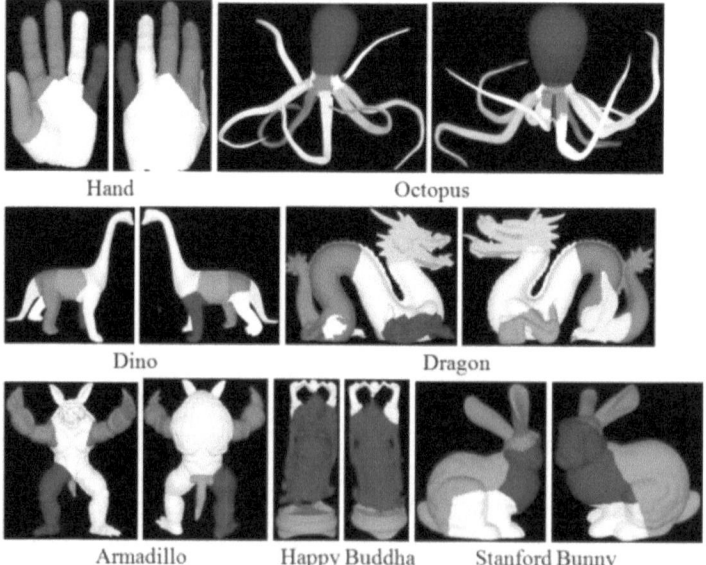

Fig. 4. Result of the 3D models to be tested using the proposed algorithm

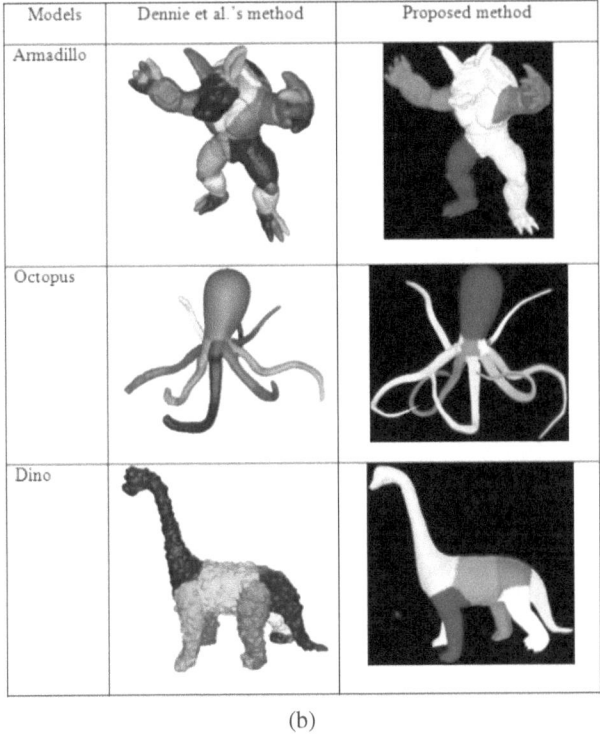

(b)

Fig. 5. The three segmented models on the left column are from Dennie et al.'s paper [5] and the segmented models on the right column are generated by the proposed method in this paper

Of the hand model, it is noted that the segmented result does not occur exactly at the contour (the texture) of the model. This is because the constructed skeleton does not stop at the texture but to continue forming a smooth curve along the model. The rest of the models are well-segmented.

5   Comparison and Analysis

This paper compares the proposed method with skeleton-based hierarchical segmentation algorithm by Dennie et al. [5]. They manually identify critical points on the curve skeleton for segmenting the models. Figure 5 shows the comparison for three segmented models of Armadillo, Octopus and Dino. The segmented Armadillo's body by Dennie et al. is a mixture of red, pink and yellow colors. This does not reflect a complete feature of a body. But the segmented Armadillo's body by the proposed method shows more complete (entire cyan color) feature. Dennie et al.'s segmented octopus looks more solid than the proposed method. The center part of the segmented octopus by the proposed method shows extra green color. The segmented Dino model of the proposed method is finer than Dennie et al.'s method as there is extra segmented color (green) at the butt of the Dino.

6   Conclusion

This paper proposes segmenting the 3D point clouds skeleton without explicitly computation of critical points for meaningful features. The skeleton constructed via Laplacian-based contraction is precise though little user-input is required in order to produce more accurate result. In near future, this work can further be improved to completely run on automatic mode.

References

1. Cornea, N.D., Min, P., Silver, D.: Curve-skeleton properties, applications, and algorithms. IEEE Transactions on Visualization and Computer Graphics 13(3), 530–548 (2007)
2. Pascucci, V., Scorzelli, G., Bremer, P.-T., Mascarenhas, A.: Robust on-line computation of Reeb graphs: simplicity and speed. ACM Trans. Graph 26(3), Article No.58 (2007)
3. Li, X., Woon, T.-W., Tan, T.-S., Huang, Z.: Decomposing polygon meshes for interactive applications. In: Symposium on Interactive 3D Graphics, pp. 35–42 (2001)
4. Tierny, J., Vandeborre, J.P., Daoudi, M.: Topology driven 3D mesh hierarchical segmentation. In: IEEE International Conference on Shape Modeling and Applications (SMI 2007), pp. 215–220 (2007)
5. Dennie, R., Telea, A.: Skeleton-based Hierarchical Shape Segmentation. In: IEEE International Conference on Shape Modeling and Applications (SMI 2007), pp. 179–188 (2007)
6. Junjie, C., Tagliasacchi, A., Olson, M., Zhang, H., Su, Z.: Point Cloud Skeletons via Laplacian-Based Contraction. In: IEEE International Conference on Shape Modeling and Applications (SMI 2010), pp. 187–197 (2010)

Implementation of Automatic Nutrient Calculation System for Cooking Recipes Based on Text Analysis

Jun Takahashi, Tsuguya Ueda, Chika Nishikawa,
Takayuki Ito, and Akihiko Nagai

Nagoya Institute of Technology, Gokiso-cho, Showa-ku, Nagoya, Aichi, Japan
{takahashi.jun,nishikawa.chika}@itolab.nitech.jp
{ito.takayuki,nagai.akihiko}@nitech.ac.jp

Abstract. This paper proposes a system that automatically calculates the nutrients in recipes on the Web for users with little technical knowledge of nutrition. This system collects cooking information from web sites, identified ingredients, converted food quantities into grams and calculated the nutrients in each recipe. In addition, the system analyzed cooking procedures, such as boiling or roasting, and calculated the nutrients taking those into account. With this system, consumer who does not have knowledge about nutrients can learn that information and easily can understand nutrients of recipes.

Keywords: nutrient automatic calculation system.

1 Introduction

With the spread of the Internet and improvement of IT literacy, people are increasingly searching for recipes online when they cook. In a recent survey[1], 44.4% of homemakers said they use recipes routinely; thus, it is anticipated that use of recipe web sites will increase. "COOKPAD" [3] is the largest recipe site in Japan, in which users can search for and contribute recipes. The site contains more than 1 million recipes so far. But,these recipes sites have problem that they cannot get detailed nutrition information. Those that do, such as "Bob & Angie"[2], which displays the quantity of nutrients such as protein and fat, tend to have a limited number of recipes. While there is a lot of software nowadays for calculating recipe nutrition, users must input the quantity of food in grams and select foods from lists, which is a big burden.

Therefore, in this paper we provide a system that automatically calculates the nutrients in recipes on web sites, and consumer users can understand easily nutrients of recipes.

2 An Automatic Nutrient Calculation System

2.1 System Overview

This paper explain the system flow in Figure 1.

P. Anthony, M. Ishizuka, and D. Lukose (Eds.): PRICAI 2012, LNAI 7458, pp. 789–794, 2012.

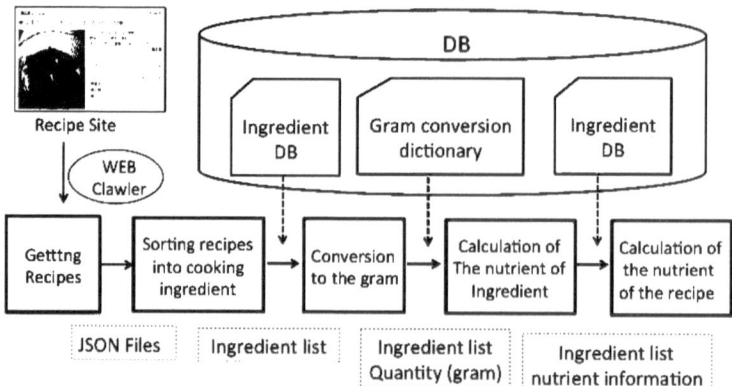

Fig. 1. System Overview

Table 1. Pantothenic acid in 100 grams of certain ingredients

Ingredient name	Content	Unit
Chicken, Liver - Raw	10.1	mg
Shiitake mushrooms - Dried	7.93	mg
Pork, Smoked Liver	7.28	mg
Pork, Liver - Raw	7.19	mg
Meat, Liver - Raw	6.4	mg

2.2   Databases Used in System

Here, ingredients refer to the food materials such as meat, dairy products, and spices that make up a dish. This database contains information on the amount of nutrients in an ingredient. We were able to collect such data for 1,861 ingredients using a crawler that we developed. Each ingredient contains each nutrient in different proportions and amounts. Table 1 shows the amount of pantothenic acid in 100 grams of certain ingredients.

When the system converts the quantity of the ingredients into a gram, the system refers to the gram conversion dictionary that we built. We built the gram conversion dictionary using Digest version 5 correct food composition table[4] and the 2001 national survey on nutrition food number list [5]. The gram conversion dictionary now contains 1,930 cases. Table 2 shows a part of the gram conversion dictionary. There are ingredient names and standard quantities in this dictionary.

We can build the recipe database including nutrient information by implementing this system. In the database, recipe name, ingredient name, quantity (gram unit), and the quantity of the nutrient that the system calculated are registered. Information about approximately 1 million recipes is registered now.

Table 2. Part of Gram conversion dictionary

Ingredient name	quantity
Burdock root - Raw	150g
Komatsuna leaves - Raw	255g
Radish roots, with skin - Raw	900g
Spinach leaves - Raw	270g
Eggplant fruit - Raw	72g

2.3   Concrete Method of Nutrient Calculation

The step after the acquisition of the recipe is identifying the ingredient name. The system identifies an ingredient name by matching an ingredient name used in a recipe and an ingredient registered in the ingredient nutrient database. The system analyzes recipes by MeCab (the morphological analyzer software)[7].

The nutrient content greatly changes according to the cooking processes. In the ingredient nutrient database, the quantity of each nutrient is registered separately when the ingredient is raw, boiled or broiled. The ingredient names registered with the ingredient nutrient database include information on cooking processing method. However, the existing system [8] does not consider cooking processing when the system calculates the nutrients of a recipe. Therefore, this paper analyze the cooking method in the recipe.

Table 3 shows examples of "broccoli", "sponge gourd" and "spinach" registered in the ingredient nutrient database. For example, "spinach" is registered as "spinach, leaf - raw" and "spinach, leaf -boiled" in the ingredient nutrient database. "Raw" is not cooked by heat processing.

This means that the quantity of nutrients for "spinach" is different for each cooking condition.

Table 3. Part of registration for cooking processing about ingredients

ingredientID	ingredient name
6263	broccoli, inflorescence - raw
6264	broccoli, inflorescence - boiled
6265	sponge gourd, fruit - raw
6266	sponge gourd, fruit - boiled
6267	spinach, leaf - raw
6268	spinach, leaf - boiled

Although the recipes describe quantities of ingredients in different units, This system changed all of them into grams. Words of "half" in the recipeh converted it to 1/2 of a gram (per foods) registered with a gram conversion dictionary. We also calculated numbers such as 2 or 3 and 1-2 to the average between those numbers. Our system assumes that a tablespoon equals 15 grams, a teaspoon is equal to 5 grams, and the words a "little" and an "appropriate amount" correspond to 10 grams of the ingredient material.

3   Performance Evaluation of System

3.1   Performance Evaluation of Identifying Ingredient Name

Our system confirms precision of identifying an ingredient name(Figure 1) in the nutrient calculation process. Evaluation was done on 100 recipes, which are registered in the recipe database. 628 ingredients were contained in 100 recipes. Our system compared the results of our system to the results we checked manually. Our system was able to identify 596 of them correctly. Our system was not able to identify 32 of them correctly. Even though this system could not identify some of the ingredients, it could have been the result of typographical errors. The other ingredients were simply not registered ingredient nutrient database to system.

3.2   Performance Evaluation of Reflection of Cooking Processing

Our system confirm whether the system reflects the processing method of cooking such as hbroilh, hstewh, h boilh and hrawh. Evaluation was done on 20 ingredients for each cooking processing. Ingredients that determined by this system compared by a dietitian. the results is the following table 4.

Table 4. System succeeded in identifying

Cooking process	Number of Identify	%
Boil	15	75
broil	20	100
stew	18g	90
raw	65	65

This system was able to reflect the cooking process at a high rate. But, there were some cases that were incorrectly determined as "raw". System sometimes erroneously determined the processing method.

3.3   Performance Evaluation of Conversion to Grams of the Quantity of Ingredient

Our system confirmed the precision of conversion to grams of the quantity of the ingredient. System checked the result of changing into grams by using it to analyze the 596 ingredients shown above. System was able to identify 512 of them correctly. System succeeded in identifying 86%. Our system was able to convert into grams the amount of ingredient at a relatively high rate. However, system was not able to convert units correctly in some cases. These are cases in which the unit of the recipe is not registered in the gram conversion dictionary, a quantity is not noted, etc. In the future, this system must expand the gram conversion dictionary.

3.4   Performance Evaluation of Calculation of the Nutrients of the Recipe

This paper evaluated the effectiveness of this system when it calculated the amount of nutrients in a dish (recipe). The system first had to identify the ingredients, next calculate the amount of each ingredient, and then use the ingredient nutrient table to calculate the amount of nutrients in each ingredient. Evaluation was done on 100 dishes. This paper compared the results of our system to the results of the existing system [8]. We used data on nutrients calculated on 100 recipes by experts. Figure 2 shows the result that compare recipe and nutrient with on respectively 100 recipes.

(1) Error of calculation result of nutrients by experts and calculation results of this system

In Figures 2, we can see that the gray bar graph rises generally. This means that our system can calculate a nutrient with higher precision than the existing system [8].

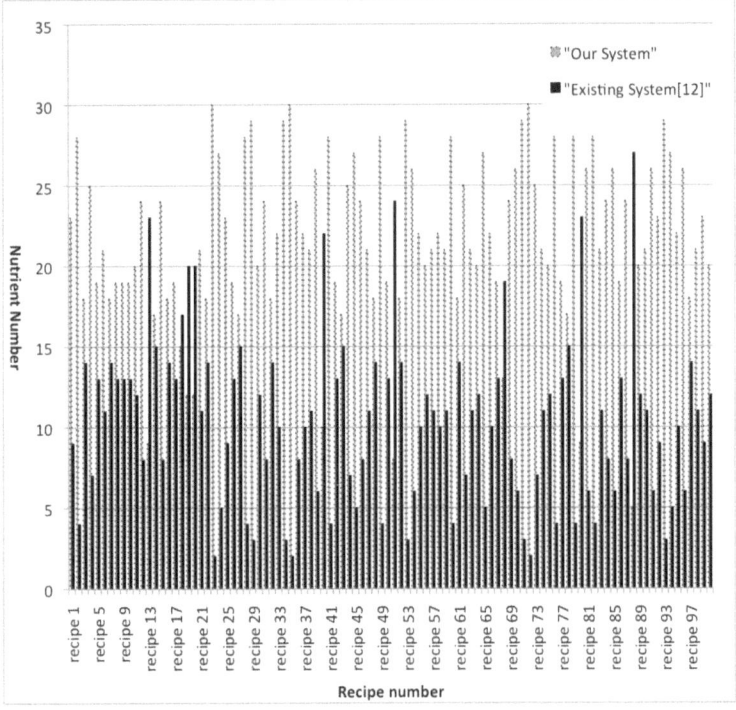

Fig. 2. Comparison result for each recipe

4   Conclusion and Future Work

In this paper, we described a system that automatically calculates the nutrients in recipes on web sites. For the system to do so, it first collects information on the recipe from the web site, identifies ingredients, converts quantities into grams, and calculates the nutrients of each ingredient. In addition, the system analyzes cooking procedure and calculates the nutrients of the ingredients according to such methods as "boil" or "bake". Our system can calculate the nutrients included in a large quantity of recipes on the Web automatically.

In the future,this system need to increasing the scope of the ingredient nutrient database and gram conversion dictionary for supporting various ingredients and quantities in recipe.

References

1. MOBILE MARKETING DATA LABO., Site Usage Survey housewife recipes
2. Bob and Angie recipe site, http://www.bob-an.com/recipe/daily/daily.asp
3. COOCPAD, http://cookpad.com/, http://msrsoft.com/hm/432/index.htm
4. Kagawa, Y.: Digest version 5 correct food composition table, Kagawa Nutrition University
5. 2001 national survey on nutrition Food number list, The study group on standardization of diet surveys, http://www.nih.go.jp/eiken/nns/system/bangohyo.pdf
6. Food Composition Database, http://fooddb.jp/
7. MeCab:Yet Another Part-of-Speech and Morphological Analyzer, http://mecab.sourceforge.net/
8. Iwakami, M., Ando, S., Ito, T., Tanaka, M.: An Implementation of Goal-Oriented Recipe Recommendation System with Nutritious Information. In: The 73rd National Convention of IPSJ (2011)

A Feature Space Alignment Learning Algorithm

Chao Tan and Jihong Guan

Department of Computer Science and Technology,
Tongji University, Shanghai 201804, China
tanchao222@gmail.com, jhguan@tongji.edu.cn

Abstract. Manifold learning algorithms have been proved to be effective in pattern recognition, image analysis and data mining etc. The local tangent space alignment (LTSA) algorithm is a representative manifold learning method for dimension reduction. However, datasets cannot preserve their original features very well after dimension reduction by LTSA, due to the deficiency of local subspace construction in LTSA, especially when data dimensionality is large. To solve this problem, a novel subspace manifold learning algorithm called *feature space alignment learning* (FSAL in short) is proposed in this paper. In this algorithm, we employ candid covariance-free independent principal component analysis (CCIPCA) as a preprocessing step. Experiments over artificial and real datasets validate the effectiveness of the proposed method.

Keywords: Manifold learning, dimension reduction, incremental tangent space.

1 Introduction

In many intelligent data analysis areas such as pattern recognition, machine learning and data mining etc., feature reduction is a crucial preprocessing step that transforms a dataset from its original dimensional space to a lower dimensional space, while trying to retain the dataset's major characteristics. Principal component analysis (PCA) [1] is a widely used feature reduction method, which includes non-incremental PCA and incremental PCA (IPCA). Non-incremental PCA needs to compute covariance matrix by eigenvalue decomposition (EVD) to get the eigenvectors. When data dimensionality is large, the computation of covariance matrix and eigenvalue decomposition is very time-consuming. The candid covariance-free IPCA (CCIPCA) algorithm [6] was proposed as an incremental PCA method, which obtains the principal components without computing the covariance matrix. So it is very efficient. However, as a linear dimension reduction method, CCIPCA cannot detect the nonlinear structures in datasets.

Manifold learning is a type of effective dimension reduction method, which has been investigated extensively in the fields of machine learning, pattern recognition abd computer vision etc. Typical manifold learning algorithms include isometric feature mapping (ISOMAP) [2], locally linear embedding (LLE) [3], Laplacian eigenmaps (LE) [4] and local tangent space alignment (LTSA) [5].

P. Anthony, M. Ishizuka, and D. Lukose (Eds.): PRICAI 2012, LNAI 7458, pp. 795–800, 2012.
© Springer-Verlag Berlin Heidelberg 2012

Though these algorithms above have been widely applied, they have various inherent deficiencies. Each algorithm can preserve only certain characteristics of the underlying dataset. For example, the LTSA algorithm constructs a local tangent space for each data point and obtains the global low-dimensional embedding coordinates through affine transformation of the local tangent spaces. LTSA uses PCA to construct local subspace, which has some problems. On the one hand, when being applied to pattern recognition, the recognition accuracy decreases rapidly as data dimensionality increases. On the other hand, the order of the matrix used for eigen-decomposition is equal to the samples' number, which makes it sensitive to nearest neighborhood choice. These problems degrade its performance in many applications. Min et al. [7] proved that the local tangent space can be represented by the eigenvectors of samples' covariance matrix. Therefore, the problem of constructing local tangent space can be transformed to a local principal component analysis problem. This inspires us to find the local principal components of a dataset using the idea of CCIPCA.

This paper proposes a new manifold learning algorithm to improve the LTSA algorithm, which is called *feature space alignment learning*, FSAL in short. The new algorithm need not compute the local covariance matrices of data points and thus solves the large-scale matrix eigen-decomposition problem in LTSA. The basic idea is first to use incremental principal components analysis as a preprocessing step, and then to re-construct the local feature space that is composed of the computed principal eigenvectors, through eigen-decomposition and singular value decomposition. Experiments on face datasets show that the new method performs well when sample datasets are large and data dimensionality is high, and it is not impacted by the choice of neighbors.

2   Feature Space Alignment Learning

The proposed feature space alignment learning (FSAL) method exploits the idea of CCIPCA to avoid constructing and re-computing the covariance matrix of a dataset to obtain the coordinates of points in the neighborhood constructed in the local feature space. Here, we present the details of the FSAL method.

2.1   Preprocessing

For N data points $x_1, x_2, ..., x_N \subset \mathbb{R}^m$, we want to compute the dominated eigenvectors. We first construct the $m \times m$ covariance matrix $A = \frac{1}{N} \sum_{i=1}^{N} x_i x_i^T$.

Let v be the eigenvector of A. If we define $w_j = x_i x_i^T v_j$, then $Av = \frac{1}{N} \sum_{i=1}^{N} x_i x_i^T v_j$ can be regarded as the mean of w_j, where v_j is the j-th estimation of v.

According to the statistical efficiency theory mentioned in [6], the mean of w_j, $\bar{w} = \frac{1}{N} \sum_{j=1}^{N} w_j$ is the efficient estimation of $x_i x_i^T v_j$ with a known standard deviation. Based on the stochastic gradient ascent (SGA) algorithm proposed by Oja & Karhunen [8,9], we can obtain the estimate of w_j in each step.

Let u_j denote the j−th estimate of Av. Then we can lead to an incremental formula of u_j, $u_j = \frac{1}{N} \sum_{i=1}^{N} x_i x_i^T \frac{u_{j-1}}{||u_{j-1}||}$.

According to SGA, u_j can be written as the following recursive form:

$$u_j = w_1 u_{j-1} + w_2 x_i x_i^T \frac{u_{j-1}}{||u_{j-1}||} \tag{1}$$

Where w_1 and w_2 are the weights of the last estimate and the new data point respectively. While w_2 is the learning rate which plays a part in the gradient method mentioned in SGA, suggested to be $\frac{1}{N}$ in [6], which can accomplish the statistical efficiency. So w_1 can be set as $1 - w_2 = \frac{N-1}{N}$. Because the manual tuning of the learning rate is impractical for on-line learning condition, an improvement using an amnesic factor l for the rate is mentioned in [6], to make w_1 smaller. Some "noise" arise in early estimation stage can be kept off and the estimation of (1) can be sped up. Then we get the improved estimate procedure as follows,

$$u_j = \frac{N - 1 - l}{N} u_{j-1} + \frac{1 + l}{N} x_i x_i^T \frac{u_{j-1}}{||u_{j-1}||} \tag{2}$$

To evaluate the next order eigenvector, Weng et al subtract the first component's projection from the data before computing the next component, which is similar with GHA (generalized Hebbian algorithm) [10]. GHA is a model of Linear feedforward neural network belonging to unsupervised learning algorithm which is mainly used in PCA. Combining Oja's theory and GHA, there forms the learning rule,

$$x_i^{j+1} = x_i^j - (x_i^j)^T \frac{u_j}{||u_j||} \frac{u_j}{||u_j||} \tag{3}$$

So the computation of the next order eigenvector is in the complementary space of u_j, for the eigenvectors are orthogonal to each other. This is also the reason we use the preprocess step to construct the feature space of sample points consist of principal eigenvectors.

2.2   Constructing Alignment Matrix Using Dominated Eigenvectors

Now we consider how to construct the low dimensional($d, d < m$) coordinates in the feature space obtained above. For $j = 1, ..., k$, determine k nearest neighbors u_{i_j} of each eigenvectors u_i after preprocessing step. Let $U_i = [u_{i_1}, \cdots, u_{i_k}]$ be the matrix composed by its k nearest neighbors including itself.

The dataset of principal eigenvectors are assumed to be sampled from a d dimensional affine subspace, the orthonormal basis of the affine subspace compose matrix can be written as $U = WA + E$ in matrix form.

By minimizing the cost function: $\min ||E|| = \min_{W} ||U - WA||_F$, we can get orthonormal basis W to learn the manifold of the affine subspace, where $|| \cdot ||_F$ indicates the matrixs Frobenius norm. Implement eigen-decomposition of matrix $U_i^T U_i$ and the first largest d eigenvectors construct the optimal solution.

Now we construct matrix $B = V_i V_i^T$ and extract its feature, where V_i is d eigenvectors corresponding to d largest eigenvalues of $U_i^T U_i$. The feature extraction problem can be solved by singular value decomposition (SVD) on B. Compute d smallest unit eigenvectors $q_1, ..., q_d$ of matrix B.

2.3   Calculating Low Dimensional Coordinates

Motivated by Zhang and Zha's work in LTSA [5], we suppose Y_i to compose orthonormal basis forming the feature space spanned by the principal eigenvectors q_i. To construct the low dimensional geometry in the feature space, we minimize the reconstruction error E_i as follows, where $T_i = [t_{i_1}, ..., t_{i_k}]$ is optimal low dimensional coordinates the matrix based on the obtained coordinates Q_i in the feature space spanned by the principal components of B.

$$\min_{T_i} ||E_i||_F^2 = \min_{T_i} ||T_i(I - \frac{1}{k}ee^T) - Y_i Q_i||_F^2 = Tr(T_i R_i R_i^T T_i^T) \qquad (4)$$

Where $R_i = (I - \frac{1}{k}ee^T)(I - Q_i^T Q_i)$. On the basis of Zhang's theory [5], the eigenvector matrix $[t_2, ..., t_{d+1}]$ composed by the d eigenvectors of $R_i R_i^T$ corresponding to the 2nd to $d + 1^{st}$ smallest eigenvalues corresponds to the optimal low dimensional coordinates to minimize $||E_i||_F^2$.

3   Performance Evaluation

We use prevalent real-life datasets to test the effectiveness of the proposed approach in dimension reduction, classification and clustering, and compare it with some existing related algorithms.

3.1   Results on the ORL Face Dataset

The ORL face dataset [11] contains 400 face images of 40 persons with different poses and expressions. Each person has 10 images with 64*64 pixels. We first use three different algorithms FSAL, LE and LTSA to get low dimensional results(here we set 2 dimension) of ORL face image dataset. Then classify them using k nearest neighbors algorithm (kNN) as the classifier to compare their classification accuracy respectively.

10 images from one individual are selected for testing while the rest ones are training set. The individuals are randomly selected for 10 times and the number of the nearest neighbors k is set to 8. When methods like LE and LLE

use all the samples' adjacent information to extract features, LTSA uses only part of samples' adjacent information like local tangent space to get probable distribution of original manifold. While FSAL extracts principal features of the space of original manifold firstly, then utilizes alignment method in k nearest neighborhood of each principal eigenvectors to construct the low dimensional coordinates. Results in Fig. 1 denote that our algorithm's classification accuracy is better than other algorithms on all 10 testing sets.

Now we display clustering effect on the same dimensional reduction results. Here we choose K-means as clustering method and set K=6. The algorithms' clustering result curves along the changing of neighborhood parameter k are shown in Fig. 1. The algorithm proposed has better performance of clustering effect than other algorithms for different k.

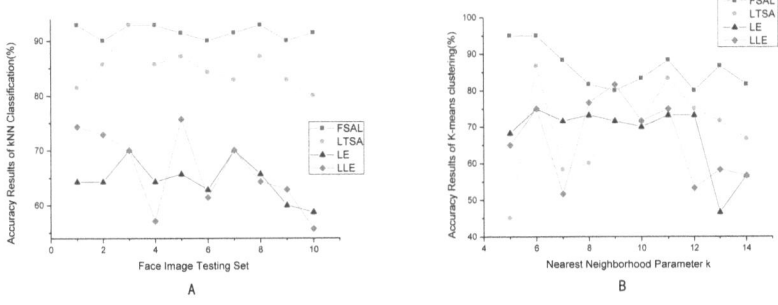

Fig. 1. A. Accuracy results of kNN classification after dimensional reduction on 10 testing sets;B. Accuracy results of K-means clustering after dimensional reduction on different neighborhood parameter k

3.2   Results on the Frey Face Dataset

In this part, we choose Frey Face image dataset [11] as dataset, which contains large quantity of images of one individual with different poses and expressions. Each image is a picture of 560 pixels. We take six hundred images as the training samples and randomly select ten images from remaining part for testing in this scheme.

Fig. 2 shows the classification accuracy rates of various algorithms on 1st to 10th testing sets, representing that FSAL's performance is generally better than other algorithms on first 10 testing sets. Without loss of generality, we perform a second experiment of classification on the other group of testing sets. From the graph we can find that FSAL has better performance on classification effect in a large scale.

As mentioned previously, the good dimension reduction result will be the key factor of following operations to obtain good results. So the proposed algorithm FSAL can play an important role in many practical applications such as Data Mining, Image Processing, Intelligent Information Processing and so on.

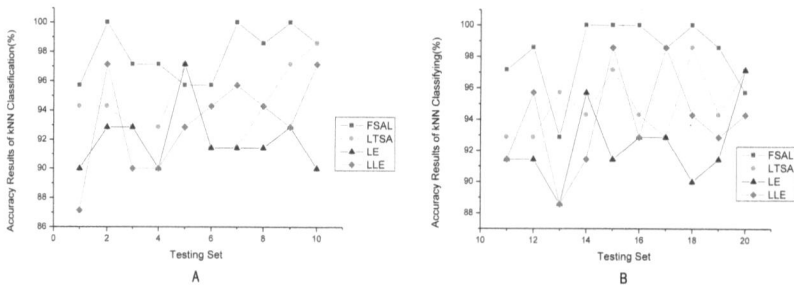

Fig. 2. A. Accuracy results of kNN Classification after dimensional reduction on first 10 testing sets;B. Accuracy results on next 10 testing sets

4   Conclusion

This paper presents a new manifold learning algorithm named FSAL. The experiments show the improved recognition accuracy and dimension reduction ability after dimension reduction. It is noticeable that FSAL's properties will not be affected by the number of the nearest neighbors, k. Meanwhile when the number and dimension of dataset are high our algorithm still preserve the geometric property of the nonlinear manifold exactly. It should be indicated that FSAL provides higher accuracies with a reasonable computational cost.

References

1. Turk, M., Pentland, A.P.: Face recognition using eigenfaces. In: IEEE Conf. on Computer Vision and Pattern Recognition (1991)
2. Tenenbaum, J.B., de Silva, V., Langford, J.C.: A global geometric framework for nonlinear dimensionality reduction. Science 290, 2319–2323 (2000)
3. Roweis, S.T., Saul, L.K.: Nonlinear dimensionality reduction by locally linear embedding. Science 290, 2323–2326 (2000)
4. Belkin, M., Niyogi, P.: Laplacian eigenmaps for dimensionality reduction and data representation. Neural Computation 15(6), 1373–1396 (2003)
5. Zhang, Z.Y., Zha, H.Y.: Principal manifolds and nonlinear dimensionality reduction via tangent space alignment. SIAM Journal of Scientific Computing 26, 313–338 (2004)
6. Weng, J.Y., Zhang, Y.L., Hwang, W.S.: Candid Covariance-free Incremental Principal Component Analysis. IEEE Trans. on Pattern Analysis and Machine Intelligence 25 (2003)
7. Min, W.L., Lu, L., He, X.F.: Locality pursuit embedding. Pattern Recognition 37(4), 781–788 (2004)
8. Oja, E.: Subspace Methods of Pattern Recognition. Research Studies Press, Letchworth (1983)
9. Oja, E., Karhunen, J.: On stochastic approximation of the eigenvectors and eigenvalues of the expectation of a random matrix. Journal of Mathematical Analysis and Application 106, 69–84 (1985)
10. Kreyszig, E.: Advanced engineering mathematics. Wiley, New York (1988)
11. http://www.cs.toronto.edu/~roweis/data.html

An Evolutionary Multi-objective Optimization Approach to Computer Go Controller Synthesis

Kar Bin Tan[1], Jason Teo[1], Kim On Chin[1], and Patricia Anthony[2]

[1] Evolutionary Computing Laboratory, School of Engineering and Information Technology,
Universiti Malaysia Sabah, Kota Kinabalu, Sabah, Malaysia
tankarbin85@gmail.com, {jtwteo,kimonchin}@ums.edu.my
[2] Center of Excellence in Semantic Agents, School of Engineering and Information
Technology, Universiti Malaysia Sabah, Kota Kinabalu, Sabah, Malaysia
panthony@ums.edu.my

Abstract. Evolutionary multi-objective optimization (EMO) has gained popularity and it has been successfully applied in several research areas. Based on the literature review conducted, EMO approach has not been applied in any Go game application. In this study, artificial neural networks (ANNs) are evolved with an EMO algorithm, Pareto Archived Evolution Strategies (PAES) for computer player to learn and play the 7x7 board Go game against GNU Go. In this study, two conflicting objectives are investigated: first, maximize the ability of neural player to play the Go game and second, minimize the complexity of the ANN by reducing the hidden units. Several comparative empirical experiments were conducted that showed EMO which optimize two distinct and conflicting objectives outperformed the single-objective (SO) optimization which only optimized the first objective with no pressure selection on the second objective.

Keywords: artificial intelligence, evolutionary multi-objective optimization, single-objective optimization, artificial neural networks, computer go.

1   Introduction

Evolutionary algorithms (EAs) become a popular optimization approach to handle real life (multiple conflicting objectives) problems because of its ability of population-based stochastic search methods and apply the principle of survival of the fittest to produce a set of Pareto optimal solutions [2-4],[14]. This type of algorithm is called multi-objective evolutionary algorithms (MOEAs). Since the research area of MOEAs are continue to advance and a tremendous amount of researches have been done, now it known as EMO [4].

Board games have been the focus as the test-bed for new approaches of artificial intelligence (AI) in recent year [5],[6],[8-13],[15-17]. Many board games like 8x8 Hex board game, 6x6 Lines of Actions (LOA), Chess, Backgammon, Checkers, Othello, Tic-Tac-Toc have been solved or even beat the human best player [1],[9],[15],[16]. Unfortunately, Go game is different. Go is ancient, hard, complex

P. Anthony, M. Ishizuka, and D. Lukose (Eds.): PRICAI 2012, LNAI 7458, pp. 801–806, 2012.

board game and difficult for computer to master it. Currently, Go human expert can consistently beat the strongest computer Go program without any handicap [12],[15-17]. AI researchers faced many challenges in building strong Go game playing computer program. These because of the obscure and dynamic changes of its blocks form by the stones in playing games and the branching factors (number of possible moves) also larger than others board games. Instead, current Go programming systems still difficult to evaluate the Go game board position and hence cause its progress to remain elusive [1],[9-11],[17].

Most board games have been solved using search-based approaches which included 5x5 board Go game [15],[16]. Van der Werf (2009) stated that the largest square Go board which has computer proof is 5x5 board and for other larger rectangular board like 4x7, 3x8, 5x6, only the human solutions are published. Variety machine learning approaches which involve supervised learning and Temporal Difference Learning (TDL) incorporate with ANN to teach and train the computer player to play the Go game have been reported and shown have some degree of success [1],[5],[6],[8],[10-13],[17]. Besides, many of them start the learning process from small board before go deep into the advanced strategies in larger board size [6],[15],[16]. However, practically all of the reported studies rely on SO EAs to conduct this artificial evolution process; multi-objective optimization has not been applied yet in any Go application. Therefore in this study, the main motivation is to fill some of these gaps in the literature as well as to explore previously untested areas of application for Go game. This study used EMO as an approach and evolved with ANNs which act as a Go game player to play 7x7 board Go game and compare the performance between EMO approach and SO optimization approach.

2   Fitness Functions

The fitness function (1) is used by Stanley (2004). It selected and used in this study because GNU Go game engine is used and it able to estimate the score for every move of the game and this makes the fitness calculation more accurate and hence, the quality of play throughout the game is thereby considered. In (1), e_i is the estimated score on move i, n is the number of moves before the final move, and e_f is the final score. This fitness equation weights the mean of the estimated score twice as much as the final score, which emphasizes the performance over the course of the whole game over the final position. Since GNU Go returns positive scores for White, they must be subtracted from 100 to correlate higher fitness with greater success for Black. The means of the estimated score as well as the final score will be negative for Black since the positive score is for White. Besides that, this experimental setting fixed that at the end of each game, at least one quarter of the game board must be filled by the stones of both players and these will cause larger values of the mean of the estimated score to be calculated. Therefore if Black wins the game, the fitness scores in (1) will exceed 110. For (2), h_i is the number of hidden units in the ANN. It will be evolved and minimized during evolution.

$$f_1 = 100 - \left(\frac{2\sum_{i=1}^{n} e_i}{n} - e_f\right) \qquad (1)$$

$$f_2 = \sum_{i=1}^{100} h_i \qquad (2)$$

3 Experimental Setup

GNU Go version 3.6's game engine is used and acted as the opponent of the neural players. GNU Go is playing White while the neural players are playing Black where komi 6.5 is given to White. The Japanese scoring system is used. Level 10 is used in GNU Go and the seed values will be change in every new game. This will cause the GNU Go to become partially stochastic and give the implication that the neural players always learn to play Go game against different opponents. All the experiments are run in ACSII mode in order to reduce the computational cost.

A simple three layered feed-forward ANN is used and all games are conducted on a 7x7 board. The structure of the ANN is composed of 49 inputs which are the current board positions on a 7x7 board. The board situation is encoded to an integer for each intersection, empty = 0, white = 1 and black = 2, which will fed into the input layer of ANN. The structure of the ANN is composed of 50 output neurons which will decide the next move for Black, representing every possible playing position on a 7x7 board plus a pass move. The highest output activation value will indicate the position on the board that the Black will choose to move on. If the highest output activation value indicates an invalid position on the board, the second highest output activation value will be selected instead, and so on. All neurons employ the sigmoid activation function. The maximum number of generations is 5000.

There are three investigated experimental setups in this study. First, using fixed number of hidden units (set to 100) in the ANN and evolved with (1+1) Evolution Strategies (ES) and optimize the Go fitness score in (1) only (SO). Second, using variable hidden units in the ANN and evolved with (1+1) ES and optimize the Go fitness score in (1) only (SO). Initially, the numbers of hidden units is set to 100. The binary number for the hidden units represents a switch to turn a hidden unit on or off. Hidden units will be evolved using a bit-wise mutator from 1 to 100. Third, optimize the fitness score in (1) as well as minimize the hidden units in the ANN (multi-objective). ANN is evolved with the (1+1) PAES and the setting is same as second one but the hidden units will be minimized during evolution using (2). ES is used in the first and second experimental setups because they involve SO optimization of the fitness score only. For the third experimental setting, PAES is used because an archive is needed in ES, which is used to store all the Pareto solutions since it involves multi-objective optimization that results in more than one optimal solution.

There are two types of metrics used for the performance evaluation due to two different experimental analysis involved in this study: (1) evolution analysis, and (2) testing evaluation. The evolution analysis involves best fitness score, average fitness scores, number of hidden units in ANN and standard deviation. To obtain more accuracy and consistency results in this research, ten runs for each mutation rate ranging from 0.1 to 0.9 will be carried out. Testing evaluation should be conducted if the environment setting in the evolution process is noisy. In this study, the playing Go board is a noisy environment because of the position of the stones on the board will be change frequently. For SO optimization setting, the network which obtained highest fitness score will be selected and tested. For multi-objective optimization setting, the networks selected for testing is the solution obtained highest fitness score and consists of the least hidden units in the ANN. All the selected networks had been tested for ten times in this study.

4   Results and Discussion

The evolution analysis and the testing evaluation using SO and EMO approaches respectively have been shown in Fig. 1.

The training results obtained using fixed number of hidden units in ANN.

Mutation Rate	Best Fitness Score	Hidden Units	Average Fitness Score	Standard Deviation
0.1	76	100	58.9	11.19
0.2	87	100	76.3	8.03
0.3	**202**	100	77.8	44.79
0.4	80	100	66.4	7.78
0.5	89	100	69.7	8.59
0.6	71	100	64.5	5.58
0.7	86	100	64.4	8.73
0.8	84	100	64.6	8.69
0.9	78	100	65.7	7.70

The training results obtained using variable number of hidden units in ANN.

Mutation Rate	Best Fitness Score	Hidden Units	Average Fitness Score	Standard Deviation
0.1	98	57	67.2	12.73
0.2	90	46	62.1	14.09
0.3	72	42	59.5	10.74
0.4	92	56	62.1	13.38
0.5	64	57	56.7	6.98
0.6	65	52	56.9	5.70
0.7	63	54	51.6	8.41
0.8	70	54	50.7	12.20
0.9	69	42	53.4	7.24

The training results obtained using EMO approach.

Mutation Rate	Best Fitness Score	Hidden Units	Average Fitness Score	Standard Deviation
0.1	62	51	48.0	6.65
0.2	70	37	51.8	8.87
0.3	60	44	47.2	10.73
0.4	64	56	49.5	8.83
0.5	60	45	42.3	11.21
0.6	66	50	41.9	10.31
0.7	58	43	38.0	13.82
0.8	**172**	**53**	42.1	46.96
0.9	**164**	**60**	50.7	40.93

The testing results of highest fitness score network using fixed number of hidden units, variable number of hidden units and EMO approaches.

Mutation Rate	Best Fitness Score	Hidden Units	Successful Runs (%)	Final Game Score
0.3 (Fixed HN)	202	100	10	B+32.5
0.1 (Variable HN)	98	57	0	W+7.5
0.8 (EMO)	**172**	**53**	**100**	B+38.5
0.9 (EMO)	164	60	100	B+22.5

Fig. 1. Evolution analysis and testing evaluation

Successful runs mean the evolved networks can beat the GNU Go 3.6 in 7x7 board Go game. The first character in final game score means the winner, B=Black and W=White and followed by score which is the total territories obtained plus the komi value. The experimental results shows that the percentage of the neural player either using SO optimization or EMO approaches to beat the GNU Go 3.6 in 7x7 board size is very low. The highest fitness score obtained among all the repeated runs is 202 and was obtained from the fixed number of hidden unit architecture with 0.3 mutation rate used in ES. This architecture can obtain high fitness score because of it only concerned single objective which maximize the playing fitness score only. Furthermore, it only can beat the GNU Go once among and the hidden units used in ANN are 100.

For variable number of hidden units architecture, no single run can beat the GNU Go in 7x7 board. The highest fitness score obtained is only 98 and the hidden units used are 57 with 0.1 mutation rate used in ES. Second and third highest fitness scores among all the repeated runs obtained in EMO approach system. The fitness scores obtained are 172 and 164 and the number of hidden units of the ANNs was also reduced from 100 to 53 and 60 respectively. The experimental results also significantly show that using large mutation rates in EMO algorithm can generate good Go playing strategies and even beat the GNU Go in 7x7 board game. The complexity of the ANNs was successfully and significantly minimized from 100 hidden units to only 53 hidden units and still provided very good playing abilities.

Only 10% of the successful run obtained by the evolved network which using 0.3 mutation rate in ES and using SO optimization approach in testing evaluation. Black able to win the GNU Go 32.5 point in 7x7 board game. On the other hand, the percentage of both networks using EMO approach to beat GNU Go in 7x7 board games is 100%. 6 final game points different between the networks using SO optimization which using 0.3 mutation rate in ES and the network using EMO approach and obtained 38.5 final game points. This performance is surprisingly good which mean that using 0.8 mutation rate in (1+1) PAES which consists of 53 hidden units is able to generate good Go game strategies and frequently win the level 10 GNU Go in 7x7 board.

This study shows that ANNs evolved with an EMO algorithm in the form of the PAES is able to generate good Go playing strategies and can beat GNU Go on 7x7 Go board. The testing results have proven that EMO that utilizes two conflicting fitness function that are maximizing the fitness score and minimizing the complexity of ANN outperforms the single-objective optimization systems and illustrates using multi-objective optimization is better than SO optimization approaches in machine learning. As for the future works, co-evolution approach and incremental evolution concept may perform well in larger Go board game using the further refine fitness functions. Besides, the neural player should play the Go game with other competitive computer Go to learn new Go playing strategies in various board sizes as well.

References

1. Bouzy, B., Cazenave, T.: Computer Go: an AI Oriented Survey. Artificial Intelligence Journal 132, 39–103 (2001)
2. Thu Bui, L., Alam, S.: Multi-Objective Optimization in Computational Intelligence: Theory and Practice. Information Science Reference, Hershey (2008)

3. Coello Coello, C.A., Pulido, G.T., Montes, E.M.: Current and Future Research Trends in Evolutionary Multiobjective Optimization. In: Grana, M., Duro, R., d'Anjou, A., Wang, P.P. (eds.) Information Processing with Evolutionary Algorithms: From Industrial Applications to Academic Speculations, pp. 213–231. Springer, London (2005)
4. Corne, D.W., Knowles, J.D.: Techniques for Highly Multiobjective Optimisation: Some Nondominated Points are Better than Others. Computing Research Repository (CoRR) abs/0908.3025 (2009)
5. Enzenberger, M.: Evaluation in Go by a Neural Network Using Soft Segmentation. In: 10th Advances in Computer Games Conference (ACG-10), Graz, Styria, Austria, pp. 97–108 (2003)
6. Gauci, J., Stanley, K.O.: Indirect Encoding of Neural Networks for Scalable Go. In: Schaefer, R., Cotta, C., Kołodziej, J., Rudolph, G. (eds.) PPSN XI, Part I. LNCS, vol. 6238, pp. 354–363. Springer, Heidelberg (2010)
7. GnuGo Documentation, http://www.gnu.org/software/gnugo/gnugo_toc.html (access on January 2012)
8. Lubberts, A., Miikkulainen, R.: Co-Evolving a Go-Playing Neural Network. In: Coevolution: Turning Adaptive Algorithms upon Themselves, Birds-of-a-Feather Workshop, Genetic and Evolutionary Computation Conference (GECCO), San Francisco, MA, USA (2001)
9. Lucas, S.M.: Computational Intelligence and Games: Challenges and Opportunities. International Journal of Automation and Computing 5(1), 45–57 (2008)
10. Mayer, H.A.: Board Representations for Neural Go Players Learning by Temporal Difference. In: IEEE Symposium on Computational Intelligence and Games (CIG), Honolulu, Hawaii, USA, pp. 183–188 (2007)
11. Mayer, H.A., Maier, P.: Coevolution of Neural Go Players in a Cultural Environment. In: Congress on Evolutionary Computation (CEC), Edinburgh, Scotland, pp. 1017–1024 (2005)
12. Runarsson, T.P., Lucas, S.M.: Coevolution Versus Self-Play Temporal Difference Learning for Acquiring Position Evaluation in Small-Board Go. IEEE Transactions on Evolutionary Computation 9(6), 628–640 (2005)
13. Stanley, K.O., Miikkulainen, R.: Evolving a Roving Eye for Go. In: Deb, K., Tari, Z. (eds.) GECCO 2004. LNCS, vol. 3103, pp. 1226–1238. Springer, Heidelberg (2004)
14. Teo, J., Abbass, H.A.: Elucidating the Benefits of A Self-Adaptive Pareto EMO Approach for Evolving Legged Locomotion in Artificial Creatures. In: IEEE Conference on Evolutionary Computation (CEC), vol. 2, pp. 755–762. IEEE Press, Canberra (2003)
15. Van der Werf, E.C.D., Winands, M.H.M.: Solving Go for Rectangular Boards. ICGA Journal 32(2), 77–88 (2009)
16. Van der Werf, E.C.D., Van den Herik, H.J., Uiterwijk, J.W.H.M.: Solving Go on Small Boards. International Computer Games Association (ICGA) Journal 26(2), 92–107 (2003)
17. Wu, L., Baldi, P.: Learning to Play Go Using Recursive Neural Networks. Neural Network 21(9), 1392–1400 (2008)

Mining Frequent Itemsets with Dualistic Constraints

Anh Tran[1], Hai Duong[1], Tin Truong[1], and Bac Le[2]

[1] Department of Mathematics and Computer Science, University of Dalat, Dalat, Vietnam
{anhtn,haidc,tintc}@dlu.edu.vn
[2] University of Natural Science Ho Chi Minh, Ho Chi Minh, Vietnam
lhbac@fit.hcmus.edu.vn

Abstract. Mining frequent itemsets can often generate a large number of frequent itemsets. Recent studies proposed mining itemset with the different types of constraint. The paper is to mine *frequent itemsets*, where a one: *does not contain any item of C_0 or contains at least one item of C_0.* The set of all those ones is partitioned into equivalence classes. Without loss of generality, we only investigate each class independently. One class is represented by a frequent closed set L and splits into two disjoint sub-classes. The first contains frequent itemsets that do not contain any item of C_0. It is generated from the corresponding generators. The second includes in two subsets of the frequent itemsets coming from the generators containing in C_0, and the ones obtained by connecting each non-empty subset of $L \cap C_0$ with each element of the first.

Keywords: Closed itemsets, frequent itemsets, dualistic, constraints, generators.

1   Introduction

First introduced and researched by Agrawal et al. [1] in 1993, mining frequent itemsets has been become one of the important problems in data mining. As usual, users are only interested in frequent itemsets that satisfy given constraints. The problem has been receiving attentions of many researchers [3, 4, 5, 7]. Let us consider searching documents in the Internet. The databases for obtaining them are usually saved into the tables. Each row in a table contains keywords appeared in a document. Assume that, a new user U wants to touch in the document D. Thus, U needs to know some keywords according to D. It is difficult to U, in the meaning that, U can get the set C of all keywords related to the subject that D belongs to, but he can not to know the keywords containing in D. Hence, the important task is to determine keyword sets containing in C. Those sets help the users to touch in documents quickly. For a transaction database T included in the set A of all items, let A^F be the set of all frequent ones. The paper focuses on the following problem: *Given a constraint C ($C \subseteq A^F$), mine frequent itemsets (keyword sets) whose items are in C (Cons1) ?".* They do not contain any item of the complement set C_0 of C ($C_0 := A \setminus C$). On the extension, we consider its dualistic problem: *Generate frequent itemsets L' such that L' contains at least one item of C_0 (Cons2).* For example, users need to know frequent keyword sets based on a few given keywords. They can lead users quickly to the desired documents.

P. Anthony, M. Ishizuka, and D. Lukose (Eds.): PRICAI 2012, LNAI 7458, pp. 807–813, 2012.

Solving these two problems by the algorithms of mining directly frequent itemsets such as Eclat [10], Apriori [1], etc is not suitable because minimum support and constraint often change (see [2] for details). Recently, in [2], we proposed the suitable model for mining frequent itemsets restricted on constraint. It can be applied to mine frequent itemsets with above dualistic constraints. Let us consider the class represented by frequent closed itemset L. For mining itemsets with *Cons2*, we split it into two parts. The first contains the ones generated from the generators that each of them contains at least one item of C_0. The other includes in the ones created by connecting each frequent itemset that does not contain any item of C_0 and each non-empty subset of $L \cap C_0$. The paper is organized as follows. Section 2 recalls some concepts of frequent itemset mining. Section 3 proposes the ways to generate non-repeatedly all frequent itemsets with dualistic constraints. Experimental results and the conclusion are shown in Sections 4 and 5.

2   Preliminaries

Given non-empty set $\mathcal{O}$ containing transactions. Let A be the set of items that are in transactions and $\mathcal{R}$ a binary relation on $\mathcal{O} \times A$. Consider two functions: λ: $2^{\mathcal{O}} \rightarrow 2^A$, ρ: $2^A \rightarrow 2^{\mathcal{O}}$ defined as follows: $\forall A \subseteq A$, $O \subseteq \mathcal{O}$: $\lambda(O) = \{a \in A \mid (o,a) \in \mathcal{R}, \forall o \in O\}$, $\rho(A) = \{o \in \mathcal{O} \mid (o, a) \in R, \forall a \in A\}$. Assign that $h = \lambda_o \rho$, h(A) are called the closure of A. A is called a closed set [12] if $h(A) = A$. A set of items containing at least one transaction is called an *itemset*. For itemset S, $supp(S):=|\rho(S)| / |\mathcal{O}|$ is called the support of S. Let $s_0 \in (0; 1]$ be minimum support, S is frequent iff $supp(S) \geq s_0$ [1]. Let FS and FCS be respectively the classes of all frequent itemsets and all frequent closed itemsets. For G, A: $\emptyset \neq G \subseteq A \subseteq A$, G is called a generator [6] of A if $h(G)=h(A)$ and $(\forall G': \emptyset \neq G' \subset G \Rightarrow h(G') \subset h(G))$. Since the cardinality of the class of all generators of A is finite, they can be numbered as follows: $G(A) = \{A_i, i \in \{1, 2, ..., |G(A)|\}\}$.

3   Mining Frequent Itemsets with Dualistic Constraints

Theorem 1 [8] (A partition of FS). $$FS = \sum_{L \in FCS} [L].$$

Each class contains frequent itemsets of the same closure L. Without loss of generality, *it only needs to exploit independently mining frequent itemsets with the dualistic constraints in each class.* Afterwards, we write L $\in FCS$ simply L.

Definition 1. The set of the elements in [L] "*containing in C* (*Cons1*) and the set of the ones "*involved with C_0*" (*Cons2*) are defined as follows:

$$FS_C(L) = \{L' \in [L] \mid L' \cap C_0 = \emptyset\}, \quad FS_{\cap C0}(L) = \{L' \in [L] \mid L' \cap C_0 \neq \emptyset\}.$$

Theorem 2 (Structure of Each Equivalence Class). Let us denote "+" as the union of two disjoint sets, we have:

$$[L] = FS_C(L) + FS_{\cap c0}(L).$$

Definition 2. The set of generators L_i containing in C and its complement on $G(L)$ containing the generators involved with C_0 are defined in the following:

$$G_C(L) = \{G \in G(L) \mid G \cap C_0 = \varnothing\}, \quad G_{\neg C}(L) = \{G \in G(L) \mid G \cap C_0 \neq \varnothing\}.$$

Let N be the cardinality of $G(L)$. All n generators of L in $G_C(L)$ are numbered as L_1, L_2, .., L_n. The ones in $G_{\neg C}(L)$ are L_{n+1}, L_{n+2}, .., L_N.

3.1   Generating Non-repeatedly Frequent Itemsets Containing in C

Using GEN-ITEMSETS [2], we derive the class [L]. For each $L' \in [L]$, we test "$L' \cap C_0 = \varnothing$?". The ones passed are in $FS_C(L)$. This way is simple (MFS-CC-SIMPLE is the corresponding algorithm). However, when the cardinality of [L] is big, and the one of $FS_C(L)$ is small, it runs slowly. How to generate directly the elements of $FS_C(L)$? Based on propositions 2 and 3 in [2], we can do it quickly.

```
MFS-CC (L, G(L)):
1.  NewGL = CLASSIFY-GENERATORS (L, G(L), out n, out N)
2.  return Sub-MFS-CC (L, NewGL, n)  // FS_C (L)


Sub-MFS-CC (L, NewGL, n):
3.  FS_C(L) = Ø and X_U = ∪_{L_i ∈ G_C(L)} L_i and X_ = (L∩C)\X_U

4.  for (i=1; i <= n; i++) do
5.      X_{U,i} = X_U\L_i; // L_i ∈ G_C(L)
6.      for all X'_i ⊆ X_{U,i} do
7.          IsDuplicate = false
8.          for (k=1; k<i; k++) do
9.              if X_k ⊂ X_i+X'_i then  IsDuplicate = true and break
10.         if not(IsDuplicate) then
11.             for all X˜ ⊆ X_ do  FS_C(L).add(X_i+X'_i+X˜)
12. return FS_C(L)
```

Fig. 1. The algorithm MFS-CC for mining frequent itemsets containing in C

Table 1. Database T

Trans	Items	Trans	Items
1	aceg	5	aceg
2	acfh	6	bceg
3	adfh	7	acfh
4	bceg		

Example 1. Consider database T in Table 1. Fix now $s_0 = 2/7$, using *Charm-L* [11] and *MinimalGenerators* [9], we have the frequent closed itemset L=aceg together $G(L) = \{ae, ag\}$. With C_0 = cfh. Then, $C = A^f \setminus C_0$ = abcefgh \ cfh = abeg. From definition 2, $G_C(L) = \{G_{C,1}=ae, G_{C,2}=ag\}$. Using MFS-CC, X_U = aeg, $X_{U,1}$ = g, $X_{U,2}$ = e, $X_\_ = (L \cap abeg) \setminus X_U = \varnothing$. For $G_{C,1}$, we have: ae+$\varnothing$, ae+g $\in$ FS_C(aceg). For $G_{C,2}$: ag+$\varnothing$ $\in$ FS_C(aceg) (ag+e does not appear again). Thus, FS_C(aceg) = {ae, aeg, ag}.

3.2   Mining Frequent Itemsets Involved with C_0

Based on directly definition 1, we can obtain the algorithm MFS-IC-SIMPLE (by replacing "L'$\cap C_0 = \varnothing$" in MFS-CC-SIMPLE by "L'$\cap C_0 \neq \varnothing$") for mining the class of frequent itemsets involved with C_0. However, it works slowly. So, how to generate quickly elements of $FS_{\cap c0}(L)$? We could assume that $L \cap C_0 \neq \varnothing$ (conversely, $FS_{\cap c0}(L) = \varnothing$). We split $FS_{\cap c0}(L)$ into two parts. The first one $FS^+_C(L)$ contains frequent itemsets created by adding each non-empty subset of $L \cap C_0$ into each frequent itemset containing in C. We have:

$$FS^+_C(L) = FS_C(L) \oplus (2^{L \cap C_0} \setminus \{\varnothing\}),$$

where: the "sum" operator $\oplus$ of X and Y (X, Y $\subseteq 2^A \setminus \{\varnothing\}$) is defined: $X \oplus Y = \{A+B: \varnothing \neq A \in X, \varnothing \neq B \in Y\}$. Using the generators involved with C_0, we generate the frequent itemsets involved with it. For L $\in$ *FCS*, $L_U = \bigcup_{L_i \in G(L)} L_i$, $L_k \in G_{\neg c}(L)$, $L_{U,k} = L_U \setminus L_k$, $L_\_ = L \setminus L_U$, let assign

$$FS_{\neg c}(L) = \{L_k + L'_k + L\sim: L_k \in G_{\neg c}(L), L'_k \subseteq L_{U,k}, L\sim \subseteq L_\_$$
$$and\ (L_j \not\subset L_k + L'_k, L_j \in G(L), \forall j: 1 \leq j < k)\}.$$

Theorem 3 (Structure of the Set of Frequent Itemsets Involved with C_0). For L $\in$

$$FCS:\ FS_{\cap c0}(L) = FS^+_C(L) + FS_{\neg c}(L).$$

Proof: It is easy to see that FS^+_C (L) and $FS_{\neg c}(L)$ are two disjoint subsets of $FS_{\cap c0}(L)$. Then, let us prove that: $FS_{\cap c0}(L) \subseteq FS^+_C$ (L) + $FS_{\neg c}(L)$. Denoted that $L_U = \bigcup_{L_i \in G_c(L)} L_i$, $L_{U,C,i} := L_{U,C} \setminus L_i$, $L_{\neg,c} := (L \cap C) \setminus L_{U,C}$. $\forall L' = L_k + L'_k + L\sim \in FS_{\cap c0}(L)$, $L_k \in G(L)$, $L'_k \subseteq L_{U,k}$, $L\sim \subseteq L_\_$ and L'$\cap C_0 \neq \varnothing$, consider two cases: [*Case 1*] If $L_k \in G_C(L)$, then $(L'_k + L\sim) \cap C_0 \neq \varnothing$. Let us call $L''_k = L'_k \cap L_U$, $T \subseteq L_{U,C,k}$, $L'''_k = (L'_k \setminus L_{U,C}) \cap T \subseteq L_{\neg,c}$, $L''''_k = (L'_k \setminus L_{U,C}) \setminus T \subseteq L \setminus T$, $L\sim_T = L\sim \cap T \subseteq L_{\neg,c}$, $L\sim_{\neg T} = L\sim \setminus T \subseteq L \setminus T$, so $(L'''k + L\sim_T) \subseteq L_{\neg,c}$ and $\varnothing \neq (L''''_k + L\sim_{\neg T}) \subseteq L \setminus T$. Indeed, if $(L''''_k + L\sim_{\neg T}) = \varnothing$, $L''''_k = L\sim_{\neg T} = \varnothing$. Then, $L \sim C_0$, $L'_k \setminus L_{U,C} \subseteq C_0$ and $\varnothing \neq (L'_k + L\sim) \cap C_0 = L'_k + L\sim = L''''_k + L\sim_{\neg T} = \varnothing$: contradiction! Hence, $L' = [L_k + L''_k + (L'''_k + L\sim_T)] + (L''''_k + L\sim_{\neg T}) \in FS^+_C(L)$. [*Case 2*] If $L_k \in G_{\neg c}(L)$, $L' \in FS_{\neg c}(L)$.

Example 2. Consider L=aceg, $\mathcal{G}(L)=\{ae, ag\}$ and C_0=e. So $\mathcal{C}$=abcfgh. Thus, $\mathcal{G}_C(L)=\{ag\}$. Since $\mathcal{FS}_C$(aceg)=$\{ag, agc\}$, $L \cap C_0$=e, so $\mathcal{FS}^+_C(L)=\{ag+e, agc+e\}$. Moreover, $\mathcal{G}_{\neg C}(L) = \{ae\}$, $L_U = aeg$, $L_{U,1} = g$ and $L_\_= c$. Thus, $2^{L_\_} = \{0, c\}$. Then $\mathcal{FS}_{\neg C}(L) = \{ae, ae+c\}$. Hence, $\mathcal{FS}_{\neg C0}(L) = \mathcal{FS}^+_C(L) + \mathcal{FS}_{\neg C}(L) = \{ag+e, agc+e, ae, ae+c\}$.

MFS-IC $(L, \mathcal{G}L))$:
1. NewGL = CLASSIFY-GENERATORS (L, $\mathcal{G}(L)$, out n, out N)
2. $\mathcal{FS}^+_C(L) = \varnothing$ and LC = $L \cap C_0$
3. **if** LC $\neq \varnothing$ **then**
4. $\mathcal{FS}_C(L)$ = Sub-MFS-CC (L, $\mathcal{G}(L)$) and LC_Class = $2^{LC} \backslash \{\varnothing\}$
6.   **for all** L' $\in \mathcal{FS}_C(L)$ **do**
7.   **for all** L''' $\in \mathcal{FS}_C(L)$ **do** $\mathcal{FS}^+_C(L)$.add (L'+L''')
8. $\mathcal{FS}_{\neg C}(L) = \varnothing$ and $L_U = \bigcup\limits_{L_i \in \mathcal{G}(L)} L_i$ and $L_\_ = L \backslash L_U$;

9. **for** (k=n+1; k <= N; k++) **do**
10. $L_{U,k} = L_U \backslash L_k$ // $L_k \in \mathcal{G}_{\neg C}(L)$
11.   **for all** L'$_k \subseteq L_{U,k}$ **do**
12.   IsDuplicate = false
13.   **for** (j=1; j<k; j++) **do**
14.   **if** $L_j \subset L_k+L'_k$ **then** *IsDuplicate* = true and *break*
15.   **if** *not(IsDuplicate)* **then**
16.   **for all** L$^-\subseteq L_\_$ **do** $\mathcal{FS}_{\neg C}(L)$.add($L_k+L'_k+L^-$)
17. **return** $\mathcal{FS}^+_C(L) + \mathcal{FS}_{\neg C}(L)$

Fig. 2. The algorithm MFS-IC to mine frequent itemsets involved with C_0

4   Experimental Results

The following experiments were performed on i5-2400 CPU, 3.10 GHz @ 3.09 GHz, 3.16 GB RAM, running Windows. The algorithms were coded in C#. Four databases in FIMDR (http://fimi.cs.helsinki.fi/data/) are used during these experiments: Pumsb contains 49046 transactions, 7117 items (P, 49046, 7117); Mushroom (M), 8124, 119; Connect (C), 67557, 129; and Pumsb* (P*), 49046, 7117.

We compare the running times of MFS-CC-SIMPLE with MFS-CC when mining frequent itemsets containing in C. For each pair of database (DB) and minimum support (MS), we consider the lengths of C ranging from 20% to 70% of |A^F| (step 2%). For each one, 15 constraints are considered. We test 234 cases for each algorithm. Experiments showed that in almost cases, MINE-CC runs quickly than MFS-CC-SIMPLE. We can save the amounts of the running time ranging from *88.7% to 95.5%*.

Next, we compare the running time of MFS-IC-SIMPLE (T_ICS) with the one of MFS-IC (T_IC) when mining frequent itemset involved with C_0,. For each (DB, MS), the lengths of C_0 are ranged from 1 to at most 20% of $|A^f|$ with step 2. Table 2 shows the comparison, where: NCons is the number of the selected constraints, NLess is the number of constraints such that T_IC < T_ICS. The percent ratio of NLess to NCons and the average number of the percent ratios of T_IC to T_ICS are shown in columns *RNLess* and R_IC (%). In almost cases, MFS-IC runs more quickly than MFS-IC-SIMPLE. The time can be reduced into a value of *62.3% to 19.1%*.

Table 2. The reductions in the time for mining frequent itemsets involved with C_0

(DB, MS)	NCons	RNLess (%)	R_IC (%)	(DB, MS)	NCons	RNLess (%)	R_IC (%)
M, 0.15	135	*100.0*	27.6	P, 0.75	135	99.3	*62.3*
M, 0.1	150	100.0	22.4	P, 0.7	135	98.5	49.0
M, 0.05	150	100.0	*19.1*	P, 0.65	135	100.0	43.3
Co, 0.65	135	100.0	26.3	P*, 0.35	150	99.3	36.9
Co, 0.6	135	100.0	23.2	P*, 0.3	150	100.0	34.3
Co, 0.55	135	100.0	26.2	P*, 0.25	150	100.0	30.2

5   Conclusions

We presented the efficient algorithms MFS-CC and MFS-IC for mining frequent itemsets with the dualistic constraints. Those algorithms are built based on the explicit structure of frequent itemset class. The class is split into two sub-classes. Each sub-class is found by applied the efficient representation of itemsets to the suitable generators. The tests on four benchmark databases showed the efficiency of our approach.

References

1. Agrawal, R., Srikant, R.: Fast algorithms for mining association rules. In: Proceeding of the 20th International Conference on Very Large Data Bases, pp. 478–499 (1994)
2. Anh, T., Hai, D., Tin, T., Bac, L.: Efficient Algorithms for Mining Frequent Itemsets with Constraint. In: Proceedings of the Third International Conference on Knowledge and Systems Engineering, pp. 19–25 (2011)
3. Bayardo, R.J., Agrawal, R., Gunopulos, D.: Constraint-Based Rule Mining in Large, Dense Databases. Data Mining and Knowledge Discovery 4(2/3), 217–240 (2000)
4. Cong, G., Liu, B.: Speed-up Iterative Frequent Itemset Mining with Constraint Changes. In: ICDM, pp. 107–114 (2002)
5. Nguyen, R.T., Lakshmanan, V.S., Han, J., Pang, A.: Exploratory Mining and Pruning Optimizations of Constrained Association Rules. In: Proceedings of the 1998 ACM-SIG-MOD Int'l Conf. on the Management of Data, pp. 13–24 (1998)

6. Pasquier, N., Taouil, R., Bastide, Y., Stumme, G., Lakhal, L.: Generating a condensed representation for association rules. J. of Intelligent Information Systems 24(1), 29–60 (2005)
7. Srikant, R., Vu, Q., Agrawal, R.: Mining association rules with item constraints. In: Proceeding KDD 1997, pp. 67–73 (1997)
8. Truong, T.C., Tran, A.N.: Structure of Set of Association Rules Based on Concept Lattice. In: Nguyen, N.T., Katarzyniak, R., Chen, S.-M. (eds.) Advances in Intelligent Information and Database Systems. SCI, vol. 283, pp. 217–227. Springer, Heidelberg (2010)
9. Zaki, M.J.: Mining non-redundant association rules. Data Mining and Knowledge Discovery (9), 223–248 (2004)
10. Zaki, M.J., Parthasarathy, S., Ogihara, M., Li, W.: New algorithms for fast discovery of association rules. In: Proc. 3rd Int. Conf. on Knowledge Discovery and Data Mining (KDD 1997), pp. 283–296 (1997)
11. Zaki, M.J., Hsiao, C.J.: Efficient algorithms for mining closed itemsets and their lattice structure. IEEE Trans. Knowledge and Data Engineering 17(4), 462–478 (2005)
12. Wille, R.: Concept lattices and conceptual knowledge systems. Computers and Math. with App. 23, 493–515 (1992)

Extraction of Semantic Relation Based on Feature Vector from Wikipedia

Duc-Thuan Vo and Cheol-Young Ock

Natural Language Processing Lab, School of Computer Engineering
and Information Technology, University of Ulsan, Korea
thuanvd@gmail.com, okcy@ulsan.ac.kr

Abstract. In this paper, we propose a feature vector to extract semantic relations using dependency tree and parse tree. We exploit relation descriptions from infoboxes on Wikipedia documents. The features include part-of-speech, phrase label in dependency tree, and grammatical structure of phrase label, path of phrase label inherent in parse tree. In our experiments, support vector machine and k-nearest neighbor are applied to extract relations from Wikipedia documents.

Keywords: Relation extraction, feature-based, parse tree, dependency tree, support vector machine and k-nearest neighbor.

1   Introduction

With the rapid increase of amount of text documents, extraction of relation between entities has become important. The goal of relation extraction is often to discover the relevant segments of information in a data stream that will be useful for structuring data. A large amount of researches have been conducted to extract relations using seed instances. [1,2,9,10] employed bootstrapping based on pattern matching approaches. The approaches exploited information redundancy of Web instances to be extracted will tend to appear in uniform contexts. However, information redundancy in Web cannot be guaranteed in Wikipedia.

In kernel-based approaches, [3] performed word contexts around entities to indicate a relation. This approach used only the shortest path between entities in a dependency tree, and each word in the shortest path is emerged with POS, entity type. Thus, the structure of dependency path is encoded. With tree kernels-based, [14] used parse tree to construct each node which contains Entity-Role, Named entity, Chunking tag with semantic information corresponding to entities and relations for extraction on organization-location, person-affiliation detection. [4] examined the utility of different features such as WordNet, parts of speech, and entity types, and found that the dependency tree kernel achieves.

In feature-based approaches [7,15], the researchers focused into sets of feature definitions and measuring similarities. [7] used textual analyses such as POS, Parsing, NER to define features which include entities, type of entities (person, location), number of entities between the two entities whether both entities belong to same chunk, number of

P. Anthony, M. Ishizuka, and D. Lukose (Eds.): PRICAI 2012, LNAI 7458, pp. 814–819, 2012.

words separating the two entities, and paths between the two entities in a parse tree. The authors [15] presented supervised approach, which employed lexical, syntactic and semantic knowledge in feature-based relation extraction, for applying ACE RDC task. From these studies, syntactic of full parsing and semantic information is being one of effective extraction relation information tools in WordNet.

In Wikipedia, YAGO [12] built an ontology by extracting relations from Wikipedia categories using WordNet and heuristic rules. The limitation in exploring free text is the main source of relations. [11] described an extraction pattern-based method for extracting is-a and part-whole relations from Wikipedia text to enrich WordNet. [13] exploited various features in Wikipedia to enhance extraction of relations from Wikipedia text. However, these methods require manual tuning of the similarity thresholds for each pattern. Therefore, extracting semantic relations on Wikipedia corpus is challenging.

In this paper we employ vector model as feature-based approach to relation extraction that combines diverse lexical, syntactic features. We define features by determining the dependencies of part-of-speech, phrase label in dependency tree and grammatical structure of phrase label, path of phrase label inherent parse tree. We use information obtained from Wikipedia infoboxes to describe relations as a pre-filter to identify relevant sentences in Wikipedia articles for extracting semantic relations. Finally, we apply classification algorithms to extract semantic relations.

The rest of paper is organized as follows. Section 2 describes features in vector model, and presents methods used for classification. Section 3 shows experimental results applying in Wikipedia documents and our discussion. The last Section ends with conclusion and future work.

2   Relation Extraction from Wikipedia

We conduct two categories for features based on [15]. The first includes information about part-of-speech and phrase label, which are dependent on dependency tree. The second is the phrase label, path of phrase label be inherent from parse tree. Both of them employ information from the phrase labels of entities, relations, and paths from entities to relations in sentence, are used to collect features for relation extraction. [5] defined relations and the problem of relation extraction follows.

- Input:
    - Set of D documents
    - Set of entities E=$\{e_i\}$ i=$\overline{1,n}$ in D documents
    - Set of relations R= $\{Rj\}$ j=$\overline{1,m}$
- Output:
    - Set of relations Rj (e_{i1}, e_{i2}) with $1 \leq i \leq n$, $1 \leq j \leq m$

Following the problem to extract relation R_j (e_{i1}, e_{i2}) from D documents. In Wikipedia, we find pairs of entities (e_{i1}, e_{i2}) that satisfy relations $R_j \in R$. Assuming "One considered relation can be presented in one sentence", we exploit information of infobox on each Wikipedia document to identify pair of entities (e_{i1}, e_{i2}) related with R_j. Our approach for relation extraction from Wikipedia documents describes as.

— Identify pair of entities $R_j(e_{i1}, e_{i2})$ by exploiting information from Wikipedia info-boxes.
— Collect sentences which contain pair of entities R_j (e_{i1}, e_{i2}), then we convert them into feature vectors. The mentioned features include word dependencies, part-of-speech labels, grammatical phrase labels, path of phrase labels using dependency tree and parse tree analysis.
— Classify feature vectors to extract relations using classification algorithms.

2.1   Feature Vector for Relation Extraction

For a given sentence "Harvard University, established in 1636 by the Massachusetts Legislature, is located in Cambridge, Massachusetts, United States.", we concentrate analyzing the relation $R(e_1, e_2)$ as *Located("Harvard university", "Massachusetts, United States")*. We consider a set of features including word dependencies, part-of-speech, grammatical structure of phrase label and path of phrase label of sentence, and then construct into a feature vector $V(vector\text{-}label; f_1; f_2; f_3; f_4; f_5)$. The vector shows positive (+) or negative (-) in relation R_j; and components of $f_1, f_2, f_3, f_4,$ and f_5 describe features' characteristics presenting relation in a sentence.

Vector Label. We identify vector label using dependency tree [9] that exploit features of part-of-speech and grammatical structure of phrase label in sentence. It shows the grammatical dependencies of relation *Located("Harvard University", "Massachusetts, United States")* on dependency tree analyzing as follows.

```
nn(University-2, Harvard-1)
nsubjpass(located-11, University,-2)
prep_in(located-11,States-16)
nn(States-16, Massachusetts-14)
nn(States-16, United-15)
```

The relationships of words <"University-2"-"Harvard-1">, <"States-16"-"United-15"> and <"States-16"-"Massachusetts-14"> describe noun compound modifiers of an NP. It can be any noun that serves to modify the head noun. The relationship of <"located-11"-"University-2"> describes a passive nominal subject as a noun phrase which is the syntactic subject of a passive clause. Relationship <"located-11"-"States-16"> is a prepositional modifier of a verb or a noun that serves to modify the meaning of verb "Located" and noun "States".

Features of Vector. In the feature vector $V(vector\text{-}label; f_1; f_2; f_3; f_4; f_5)$, the mentioned components of features in grammatical structure of phrase labels describe entities, relation, and paths of phrase labels from entities to relation. They are determined using parse tree as:

f_1: phrase label feature presenting entity e_1
f_2: phrase label feature presenting relation R_j
f_3: phrase label feature presenting entity e_2
f_4: path the phrase label feature from e_1 to R_j
f_5: path the phrase label feature from e_2 to R_j

To evaluate f_1, f_2, f_3, f_4, f_5 we transformed the structure of parse tree in Figure 1 into the structure presenting the relation as shown in Figure 2. We exploited grammatical phrase labels presenting entities (e_1, e_2) and relation R_j. Values of f_1, f_2, f_3, f_4, f_5 are computed as follows.

$$f_{1,2,3} = \begin{cases} 0 \text{ if node label contains negative words \{no, not, etc\}} \\ \text{\# related nodes / \# nodes on mentioned phrase} \end{cases}$$

$f_{4,5}$ = # path links from entities to relation

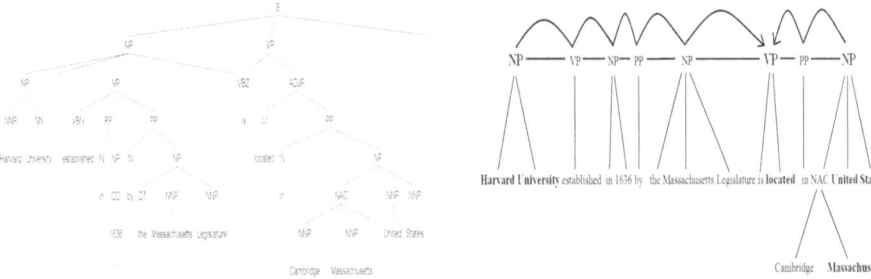

Fig. 1. Parse tree analysis

Fig. 2. Paths from entities to relation

Thus, the NP phrase contains two nodes that are related to "*Harvard University*", f_1=2/2, and the VP phrase where label "*located*" belongs has 2 nodes, f_2=1/2 and f_3=3/4. In Figure 2, value f_4 is computed from the path of the NP phrase containing entity "*Harvard university*" to the VP phrase containing relation "*located*", thus f_4=5. Value f_5 is the path of the NP phrase containing entity "*Massachusetts, United States*" to the VP phrase containing relation "*located*", f_5=2. Finally, the feature vector for considered relation obtains $V(vector\text{-}label=+1;f_1:1;f_2:1/2;f_3:3/4;f_4:5;f_5:2)$.

Feature vector is a model representing relation between two entities in the sentence, the vector label indicates positive (+1) or negative (-1) of relation. The values of f_1, f_2 and f_3 in [0, 1] will show ability to present entities and relation in sentence. Higher values indicate stronger representations of entities and their relation in the sentence. The values of f_4, f_5 are relevance values between entities and relation by phrase distances in the sentence. Lower values indicate greater relevance of entities and their relation; and contrariwise, the weaker one shows. With the converted feature vectors, we employ classification methods as support vector machine and k-nearest neighbor to extract semantic relations.

3   Experimental Results and Discussions

We used 9820 Wikipedia documents in "Education" category to extract information from Wikipedia infoboxes. We use Wikipedia API for searching and limit within 25 searching results, and then extract sentences related to each relation for data setting as in Table 1. These sentences are converted into feature vectors by using Stanford Parser for analyzing dependency tree and parse tree. Table 2 shows the result which is applied classification methods on feature vectors using support vector machine (SVM) and k-nearest neighbor (kNN) in four types of relations. The evaluation follows 10-fold cross validation schema. The highest value is achieved 91.51% and 87.67% in F-measure of SVM and kNN in *"Located"* relation. The *"Type"* relation is obtained low values 74.10% and 76.16% in F-measure of SVM and kNN due to high error rate of feature vectors.

Table 1. Data setting for four types of relations

Type	#pairs of entities	#feature vectors		#error vectors	Error ratio(%)
		#positive	#negative		
Established/Founded	2128	1009	2843	243	6.30%
Person_president	512	432	2571	112	3.72%
Type	1398	1030	1484	345	13.72%
Located	1418	1176	2134	212	6.40%

Errors of feature vector are detected in the list of relations. Their ratios are shown in Table 1. They reflect corresponding F-measure of classified instances in Table 2. In fact, high error's ratio in *"Type"* relation (13.72%) leads to low value with SVM (74.10%) and with kNN (76.16%). While the error's ratio of *"Person_president"* is the lowest (3.72%) but still creates low result after trained, 62.28% with SVM and 63.74% with kNN. The unbalance of #positive and #negative in feature vectors causes this problem, see on Table 1. Thus, there are very few training examples as relations, the classifier is less likely to identify a testing instance as a relation. And Table 3 shows classification in type of relations which are similar in grammatical structure or semantic relation between "Located" relation of our approach and related work [14].

Table 2. The Precision, Recall, and F-measure on feature vectors classification

Table 3. Comparison between "Located" relation and similar relation in related work

Type	Precision(%)		Recall(%)		F-measure (%)	
	SVM	kNN	SVM	kNN	SVM	kNN
Established/Founded	85.65	83.50	96.39	91.36	90.59	87.25
Person_president	81.17	79.63	50.52	53.13	62.28	63.74
Type	74.32	76.98	73.88	74.51	74.10	76.16
Located	93.36	86.64	89.74	88.74	91.51	87.67

Type	P(%)	R(%)	F(%)
Our: SVM(Located)	93.36	89.74	91.51
Our: kNN(Located	86.64	88.74	87.67
Zelenko(2003): SVM(org-location)	76.33	91.78	83.30

We think that effective performance in our approach is due to two reasons. First, detecting entities and relation in sentence that that exploit Wikipedia characteristics as infoboxes is important. Second, most of relations have two mentions being close to each other. While short-distance relations dominate and can be solved by syntactic features in dependency tree. And the further parse tree features can only take effect in remaining long-distance relations. But full parsing can take much difficult in long-distance relations. Transformation from full parsing into tree presenting relation focused on syntactic phrases to get more effectiveness in long-distance relations in our approach.

4   Conclusion and Future Work

We have presented an approach to extract exact relation between entities from Wikipedia. We have employed infoboxes as relation detection. So we do not have to annotate manually. Our approach is integration parse tree and dependency tree to analysis features for constructing feature vector. The approach can be considered as a step towards with vector model on N-grams in relations by using DBPedia [6] as Wikipedia characteristics as hierarchical categories, word linking, etc to identify entities and relations. Moreover, we continue to consider our model of feature vector in unsupervised approach and make approximate comparison.

References

1. Agichtein, E., Gravano, L.: Snowball: Extracting Relation from Large Plaintext Collections. In: Proceedings ACM DL 2000, pp. 85–94 (2000)
2. Brin, S.: Extracting Patterns and Relations from the World Wide Web. In: Atzeni, P., Mendelzon, A.O., Mecca, G. (eds.) WebDB 1998. LNCS, vol. 1590, pp. 172–183. Springer, Heidelberg (1999)
3. Bunescu, R., Mooney, R.J.: A Shortest Path Dependency Kernel for Relation Extraction. In: HLT/EMNLP, pp. 724–731 (2005)
4. Culotta, A., Sorensen, J.: Dependency Tree Kernels for Relation Extraction. In: Proceedings of ACL 2004, Barcelona, Spain, July 21-26, pp. 423–429 (2004)
5. Girju, C.R.: Text Mining for Semantic Relations. PhD. Thesis. The University of Texas at Dallas (2002)
6. Jentzsch, Z.: DBpedia – Extracting structured data from Wikipedia. Presentation at Semantic Web In Bibliotheken (SWIB 2009), Cologne, Germany (2009)
7. Kambhatla, N.: Combining Lexical, Syntactic and Semantic Features with Maximum Entropy Models for Extracting Relations. In: Proceedings of ACL 2004, pp. 178–181 (2004)
8. Marneffe, M.C., Manning, C.D.: The Stanford Typed Dependencies Representation. In: Proceedings of CrossParser 2008 COLING 2008, pp. 1–8 (2008)
9. Pantel, P., Pennacchiotti, M.: Espresso: Leveraging Generic Patterns for Automatically Harvesting Semantic Relations. In: Proceedings of COLING/ACL 2006, pp. 113–120 (2006)
10. Ravichandran, D., Hovy, E.: Learning Surface Text Patterns for a Question Answering System. In: Proceedings of ACL 2002, pp. 41–47 (2002)
11. Ruiz-Casado, M., Alfonseca, E., Castells, P.: Automatic Extraction of Semantic Relationships for WordNet by Means of Pattern Learning from Wikipedia. In: Montoyo, A., Muñoz, R., Métais, E. (eds.) NLDB 2005. LNCS, vol. 3513, pp. 67–79. Springer, Heidelberg (2005)
12. Suchanek, F.M., Kasneci, G., Weikum, G.: YAGO: A Core of Semantic Knowledge Unifying WordNet and Wikipedia. In: WWW (2007)
13. Wang, G., Zhang, H., Wang, H., Yu, Y.: Enhancing Relation Extraction by Eliciting Selectional Constraint Features from Wikipedia. In: Kedad, Z., Lammari, N., Métais, E., Meziane, F., Rezgui, Y. (eds.) NLDB 2007. LNCS, vol. 4592, pp. 329–340. Springer, Heidelberg (2007)
14. Zelenko, D., Aone, C., Richardella, A.: Kernel Methods for Relation Extraction. Journal of Machine Learning Research 3, 1083–1106 (2003)
15. Zhou, G., Zhang, M.: Extracting Relation Information from Text Documents by Exploring Various Types of Knowledge. Information Processing and Management 43, 969–982 (2007)

τε2asp : Implementing $\mathcal{TE}$ via Answer Set Programming

Hai Wan[1], Yu Ma[2], Zhanhao Xiao[2], and Yuping Shen[3]

[1] Software School, Sun Yat-Sen University, Guangzhou, China, 510275
[2] School of Information Science and Technology, Sun Yat-Sen University,
Guangzhou, China, 510275
[3] Institution of Logic and Cognition, Sun Yat-Sen University, Guangzhou,
China, 510275

Abstract. This paper studies computational issues related to the problem of reasoning about action and change in timed domains by translating it into answer set programming paradigm. Based on this idea, we implement a new action and change reasoning solver - τε2asp with a polynomial translation without increasing its complexity for checking satisfiability and entailment in $\mathcal{TE}$ and report some experimental results.

1 Introduction

In order to extend the study of reasoning about action and change [1–3] to handle *timed domains*, the so-called timed action language $\mathcal{A}_\mathcal{T}$ [4], narrative-based action logic AL^2_{TC} [5], timed action language $\mathcal{TE}$ [6] are proposed. This paper focuses on providing a computational realization of solutions to reasoning about problems described by $\mathcal{TE}$, regarded as an extension of action language $\mathcal{E}$ [7]. Reasoning about action and change in *timed domains*, is not only considered as toy problems, e.g., *timed Yale Shooting*, but also studied as real-world applications, e.g., *timed automaton*[8], *assembly plant*[8], and *rail road crossing control*[9]. Compared with $\mathcal{A}_\mathcal{T}$ and AL^2_{TC}, $\mathcal{TE}$ overcomes semantics defect of $\mathcal{A}_\mathcal{T}$ and can be implemented by *satisfiability modulo theory*(SMT) [10] solvers.

This paper studies a link between timed action language $\mathcal{TE}$ and declarative logic programming approach of *answer set programming* (ASP) [11], intending to implement reasoning about action and change in *timed domains* by translating $\mathcal{TE}$ into ASP [12], a promising approach implemented by a number of sophisticated solvers [13–16]. A study on encodings of reasoning problems in language $\mathcal{E}$ was developed by exploiting the relation between $\mathcal{E}$ and ASP[7].

This translation is not only theoretically interesting but also of practical relevance. Based on the translation, we implement a new action and change reasoning solver, called τε2asp. We report some experimental results, which demonstrate that the performance of τε2asp is rather satisfactory.

The paper is organized as follows. Section 2 recalls some basic notions and definitions in timed action language $\mathcal{TE}$. Section 3 presents the translation from $\mathcal{TE}$ to answer set programming. Section 4 explains the implementation of τε2asp and reports some experiments. Finally, Section 5 concludes the paper.

P. Anthony, M. Ishizuka, and D. Lukose (Eds.): PRICAI 2012, LNAI 7458, pp. 820–825, 2012.
© Springer-Verlag Berlin Heidelberg 2012

2   Preliminaries

We consider Shen's *timed action language* $\mathcal{TE}$[6], in which a $\mathcal{TE}$ theory should be composed of a 5-tuple and propositions in $\mathcal{TE}$ are defined as follows:

- C-proposition is either of the form: A **initiates** F **resets** λ **if** C **when** Ψ, or A **terminates** F **resets** λ **if** C **when** Ψ, means action A can make fluent F hold (or not hold) and reset a set of clocks C;
- H-proposition: A **happens-at** T means action A happens at time T;
- T-proposition: L **holds-at** T means fluent L does (or doesn't) hold at T.

C-propositions introduce a new kind of action effect: clock resetting, and action precondition: clock constraint. Resetting a clock is to start counting its ticks and execute an action satisfying related clock constraints besides fluent preconditions. H- and t-propositions remain the same w.r.t. $\mathcal{E}$ [17]. Model of $\mathcal{TE}$ is defined by Definition 2 in [6]. The following is an example [4–6]:

Example 1. **Timed Yale Shooting in** $\mathcal{TE}$
Let $\mathcal{TE} = \langle \mathbf{N}, \{Load, Shoot\}, \{Loaded, Alive\}, \{x\}, \mathcal{B}(\{x\})\rangle$,a $\mathcal{TE}$ theory D_{TYS} of timed Yale Shooting consists of the following propositions:

- *Shoot* **terminates** *Alive* **if** $\{Loaded\}$ **when** $\{x < 5\}$
- *Load* **initiates** *Loaded* **resets** $\{x\}$
- *Alive* **holds-at** 0, $\neg Loaded$ **holds-at** 0
- *Load* **happens-at** 1, *Shoot* **happens-at** 3

Based on [6], we implement the translation from $\mathcal{TE}$ to SMT and can reason by using SMT solver $Z3$ [1], which tops solvers of kind.

3   From $\mathcal{TE}$ to Answer Set Programming

3.1   Translation Algorithm

In order to construct translation from $\mathcal{TE}$ to ASP, we define a $\mathcal{TE}$ theory D with five sets, $\alpha, \beta, Hp, Tp, Cp$, and a N-time sequence, where α and β is a set composed of all fluent and clocks in D respectively and Hp, Tp, Cp is a set composed of all h-, t-, or c-propositions in D respectively.

The translation from $\mathcal{TE}$ to ASP can be accomplished by algorithm 1. We can get the facts corresponding to this ASP program, named as Δ. To specify the time sequence, a series of propositions denoting time(*i.e.* $time(0), time(1), \ldots,$ $time(N - 1)$) should be taken into Δ. Note that we add $time(N - 1)$ except $time(N)$ into Δ to terminate the reasoning at time N. For every fluent F in D would be expressed in the form of proposition like "$fluent(F)$." in Δ_α. Similarly, for every clock C in D would be translated to "$clock(C)$." in Δ_β.

Assuming that h-propositions $Hp_1, Hp_2, ..., Hp_p$ comprise of set Hp and an h-proposition Hp_i is in the form like "a_i **happens-at** t_i". When translated into

---

[1] $Z3$-3.2 at http://research.microsoft.com/en-us/um/redmond/projects/z3/

Algorithm 1. Translating $\mathcal{TE}$ to answer set programming

---

input : A $\mathcal{TE}$ theory $D = \{\alpha, \beta, Hp, Tp, Cp\}$ with a N-time sequence
output: An ASP program Δ

```
 1  Δ ← time(0..N − 1).
 2  forall the 1 ≤ i ≤ l do                                              // fluent
 3  └   Δα ← Δα ∪ {fluent(αᵢ).}
 4  forall the 1 ≤ i ≤ m do                                             // clocks
 5  └   Δβ ← Δβ ∪ {clock(βᵢ).}
 6  forall the 1 ≤ i ≤ p do                                    // H-propositions
 7  └   ΔH ← ΔH ∪ {aᵢ(tᵢ).}
 8  forall the 1 ≤ i ≤ q do                                    // T-propositions
 9  │   if fᵢ = αⱼ then
10  │   └   ΔT ← ΔT ∪ {fluent(fᵢ, tᵢ).}
11  │   if fᵢ = ¬αⱼ then
12  │   └   ΔT ← ΔT ∪ {−fluent(fᵢ, tᵢ).}
13  forall the 1 ≤ i ≤ r do                                    // C-propositions
14  │   if initiates flᵢ then
15  │   └   ΔC ← ΔC ∪ {acᵢ(T), Literal(Cᵢ), Number(Ψᵢ) → ini(flᵢ, T + 1).}
16  │   if terminates flᵢ then
17  │   └   ΔC ← ΔC ∪ {acᵢ(T), Literal(Cᵢ), Number(Ψᵢ) → tmn(flᵢ, T + 1).}
18  │   if resets λᵢ then
19  │   │   forall the 1 ≤ j ≤ s do
20  │   │   └   ΔC ← ΔC ∪ {acᵢ(T), Literal(Cᵢ), Number(Ψᵢ) → rst(λᵢⱼ, T).}
21  return Δ ← Δ ∪ Δα ∪ Δβ ∪ ΔH ∪ ΔT ∪ ΔC
```

ASP, Hp_i is switched to a proposition of "$a_i(t_i)$." standing for an action occurrence a_i at time t_i. A t-proposition like "F **holds-at** T" means that the positive literal of F at time T is true, while "$fluent(F,T)$." should be included by the translated ASP program. Correspondingly, "$\neg F$ **holds-at** T" means the negative one is true and "$-fluent(F,T)$" should be included. Like set Hp, set Tp is made up of t-propositions $Tp_1, Tp_2, ..., Tp_q$ and a t-proposition Tp_i is presented as "f_i **holds-at** t_i". The translation can generate according to the proposition with the different literal of f_i, positive and negative literal. In order for translation, we introduce two binary functions in syntax: $Literal$ and $Number$. $Literal$ translates a set of fluent literals into a series of atoms and $Number$ translates a set of temporal constraints into atoms of body in Δ_C.

3.2 Reasoning Rules

Besides Δ obtained from Algorithm 1, we need 6 reasoning rules, denoted by Γ. Because rules in Γ construct the basic rules of reasoning about $\mathcal{TE}$ via ASP, Γ can't change with the change of $\mathcal{TE}$ theory.

Rule 1: generating initial complete knowledge.

$$1\{fluent(F,0), -fluent(F,0)\}1 \leftarrow fluent(F). \tag{1}$$

Rule 2: specifying initial clock value.

$$clock(C, 0, 0) \leftarrow clock(C). \tag{2}$$

Rule 3: eliminating inconsistent fluent literals.

$$\leftarrow fluent(F, T), -fluent(F, T). \tag{3}$$

Rule 4: specifying fluent persistent.

$$fluent(F, T + 1) \leftarrow fluent(F, T), -tmn(F, T + 1), time(T). \tag{4}$$

$$- fluent(F, T + 1) \leftarrow -fluent(F, T), -ini(F, T + 1), time(T). \tag{5}$$

$$- ini(F, T + 1) \leftarrow not\ ini(F, T + 1), time(T), fluent(F). \tag{6}$$

Rule 5: describing initiation and termination effects.

$$fluent(F, T + 1) \leftarrow ini(F, T + 1). \tag{7}$$

$$- fluent(F, T + 1) \leftarrow tmn(F, T + 1). \tag{8}$$

Rule 6: describing resetting effects.

$$clock(C, XX, T + 1) \leftarrow XX := X + 1, clock(C, X, T), -rst(C, T), time(T). \tag{9}$$

$$clock(C, 1, T + 1) \leftarrow rst(F, T), time(T). \tag{10}$$

3.3   Temporal Constraints

In addition, an ASP program of rules for defining temporal constraints is necessary, which is called by Λ.

$$eq(C, n, T) \leftarrow X == n, clock(C, X, T). \tag{11}$$

$$eq(C1, C2, n, T) \leftarrow X1 - X2 == n, clock(C1, X1, T), clock(C2, X2, T). \tag{12}$$

Because temporal constraints are different with the change of theory, Λ change accordingly. To reduce the reasoning space, all temporal constraints in theory D are substituted into Λ and those not in D should be removed from Λ.

So far the translation from a $\mathcal{TE}$ theory to ASP is accomplished and we can obtain ASP program Σ consisting of Δ, Λ, Γ. Observe that Δ, Λ and Δ can be translated in polynomial time, so the process is polynomial.

Theorem 1. *Let D be a $\mathcal{TE}$ theory whose time is complete and Δ be its translated program by Algorithm 1, Λ be its program of temporal constraints rules, Γ be the ASP program of reasoning rules and an ASP program Σ consisting of Δ, Λ, Γ. Then D has a model $< H, K >$ if and only if Σ has an answer set M s.t. every fluent F, every clock C and time T in D,*

1. *$H(F, T) = \top$ if and only if $fluent(F, T) \in M$;*
2. *$H(F, T) = \bot$ if and only if $-fluent(F, T) \in M$;*
3. *$K(T)(x) = x_T$ if and only if $clock(x, x_T, T) \in M$.*

4   Implementation and Application

4.1   Implementation

A new solver is implemented for $\mathcal{TE}$, called $\tau\varepsilon$2asp (Fig.1). An input $\mathcal{TE}$ theory is firstly translated to an ASP program by the *translator* in $\tau\varepsilon$2asp. Then, an ASP solver $Clasp^2$ is called to compute answer sets, which will be interpreted to the original $\mathcal{TE}$ theory by a *convertor*. $\tau\varepsilon$2asp is written in C, running on a machine with 2 processors($Intel(R)\ Core(TM)2\ Duo\ CPU\ T7600$) under Ubuntu 10.04.

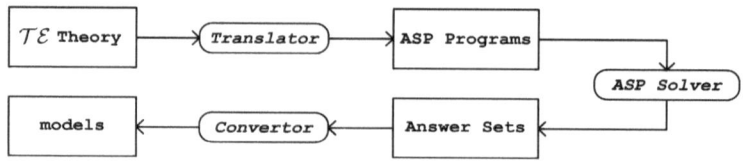

Fig. 1. Outline of $\tau\varepsilon$2asp

4.2   Experimental Results

Table 1 presents some experimental results of $\tau\varepsilon$2asp and compare it with SMT solver $Z3$ and timed automaton tool $HyTech^3$, which record computing time of series timed Yale Shooting problem in seconds, taking the average of 50 runs.

Table 1. Experimental results of timed Yale Shooting problem

Load	Shoot	Bound	$\tau\varepsilon$2asp	Z3	HyTech
1	3	8	0.00538	0.03348	0.02668
5	10	15	0.00716	0.04372	0.02142
20	24	30	0.00822	0.05134	0.02822
60	65	80	0.01590	0.11776	0.02108
120	123	150	0.03472	0.23290	0.02732
250	255	300	0.09556	3.85892	0.02076

Because *HyTech* solver can only find traces, a comparator is necessary to confirm whether the trace is consistent with the action sequence. Therefore the data in column *HyTech* in Table 1 is the time spent in finding traces to each state, omitting comparison time which is not a polynomial. Because of different computation mechanism, *HyTech* has little change with the increasing scale of the problem, while time via $\tau\varepsilon$2asp and SMT solver $Z3$ increases radically with the expansion of problem's scale. To sum up, considering the comparison time of Hytech, $\tau\varepsilon$2asp solver is the best.

5   Conclusions and Future Work

This paper contributes the study of reasoning about action and change in timed domains $\mathcal{TE}$. Theoretically, $\mathcal{TE}$ can be translated into ASP via a polynomial algorithm and a series of rules without increasing its computation complexity.

---

² Clasp-2.0.5 at http://www.cs.uni-potsdam.de/clasp/
³ *HyTech*-1.0.4 at http://embedded.eecs.berkeley.edu/research/hytech/

Practically, a new solver $\tau\varepsilon$2asp is developed. In addition, a series of experiments about the comparison among $\tau\varepsilon$2asp, SMT solver ($Z3$) and a timed automaton tool ($HyTech$) have been done. It is a pleasure that $\tau\varepsilon$2asp performs much better than $Z3$ and $HyTech$. For future work, it is an important task to extend $\mathcal{TE}$ with ramification and qualification and improve $\tau\varepsilon$2asp. Last but not least, to explore more applications of $\mathcal{TE}$ by using $\tau\varepsilon$2asp is a job of significance.

Acknowledgments. This research has been partially supported by *doctoral program foundation of institutions of higher education of China* under Grant 20110171120041 and under Grant GD10YZX03, WYM10114.

References

1. Shoham, Y.: Reasoning about action and change. MIT Press (1987)
2. Harmelen, F.V., Lifschitz, V., Porter, B.: Handbook of knowledge representation. Elsevier Science (2008)
3. Reiter, R.: Knowledge in action: logical foundations for specifying and implementing dynamical systems. MIT Press (2001)
4. Simon, L., Mallya, A., Gupta, G.: Design and Implementation of A_T: a Real-Time Action Description Language. In: Hill, P.M. (ed.) LOPSTR 2005. LNCS, vol. 3901, pp. 44–60. Springer, Heidelberg (2006)
5. Cabalar, P., Otero, R.P., Pose, S.G.: Temporal constraint networks in action. In: ECAI, pp. 543–547 (2000)
6. Shen, Y.P., Dang, G.R., Zhao, X.S.: Reasoning about action and change in timed domains. In: NMR (2010)
7. Dimopoulos, Y., Kakas, A.C., Michael, L.: Reasoning About Actions and Change in Answer Set Programming. In: Lifschitz, V., Niemelä, I. (eds.) LPNMR 2004. LNCS (LNAI), vol. 2923, pp. 61–73. Springer, Heidelberg (2003)
8. Clarke, E.M., Grumberg, O., Peled, D.A.: Model checking. MIT Press (2000)
9. Alur, R.: Timed Automata. In: Halbwachs, N., Peled, D.A. (eds.) CAV 1999. LNCS, vol. 1633, pp. 8–22. Springer, Heidelberg (1999)
10. Nieuwenhuis, R.: SAT Modulo Theories: Enhancing SAT with Special-Purpose Algorithms. In: Kullmann, O. (ed.) SAT 2009. LNCS, vol. 5584, p. 1. Springer, Heidelberg (2009)
11. Lifschitz, V.: Action languages, answer sets and planning. The Logic Programming Paradigm: a 25 Year Perspective 25, 357–373 (1999)
12. Gelfond, M., Lifschitz, V.: The stable model semantics for logic programming. In: ICLP/SLP, pp. 1070–1080 (1988)
13. Gebser, M., Kaufmann, B., et al.: Conflict-driven answer set solving. In: IJCAI, pp. 386–392 (2007)
14. Niemelä, I., Simons, P.: Smodels Implementation of the Stable Model and Well-Founded Semantics for Normal Logic Programs. In: Fuhrbach, U., Dix, J., Nerode, A. (eds.) LPNMR 1997. LNCS, vol. 1265, pp. 421–430. Springer, Heidelberg (1997)
15. Leone, N., Pfeifer, G., et al.: The dlv system for knowledge representation and reasoning. ACM Trans. Comput. Log. 7(3), 499–562 (2006)
16. Lierler, Y., Maratea, M.: Cmodels-2: SAT-based Answer Set Solver Enhanced to Non-tight Programs. In: Lifschitz, V., Niemelä, I. (eds.) LPNMR 2004. LNCS (LNAI), vol. 2923, pp. 346–350. Springer, Heidelberg (2003)
17. Kakas, A.C., Miller, R.: A simple declarative language for describing narratives with actions. Journal of Logic Programming 31(1-3), 157–200 (1997)

Investigating Individual Decision Making Patterns in Games Using Growing Self Organizing Maps

Manjusri Wickramasinghe[1], Jayantha Rajapakse[1], and Damminda Alahakoon[2]

[1] School of Information Technology - Sunway Campus
{manjusri,jayantha.rajapakse}@monash.edu
[2] Clayton School of Information Technology
Monash University
damminda.alahakoon@monash.edu

Abstract. Each player has an unique and a distinct way of interacting with a computer game due to the preconceived notion and the experience gained through playing the game title. During the game play a player adapts to the challenges posed by the game and a pattern of interaction emerge corresponding to factors such as tackling opponents, movement strategies and even decision making at certain game environments. Understanding decision making patterns provide valuable information about the players which could be exploited to enhance the total game play experience. This paper investigates the possibility of understanding the decision making patterns of a player whilst playing the 2D arcade game Pac-Man using an unsupervised approach known as the growing self organized map (GSOM). Results of this study motivated us to conjecture that player decision making patterns could be identified and explained via unsupervised learning.

Keywords: GSOM, Pattern Recognition, Player Profiling.

1 Introduction

In most computer game genres available today the way artificially controlled opponents behave in-game seems to be the most pivotal point of immersion for all the players involved. Computer game opponents come in diverse forms such as a character/unit controlled by another human player in a multi-player gaming environment; a bot which is controlled by the computer using AI techniques also known as an NPC (Non Player Class Character). It's a known fact that game play with human controlled opponents is challenging and competitive. Many multi-player games such as World of WarCraft[1], StarCraft[1], Unreal Tournament[2] and Quake[3] gained immense popularity due to this aspect. The same

---

[1] http://sea.blizzard.com/en-sg/
[2] http://epicgames.com/
[3] http://www.idsoftware.com/games/quake/

P. Anthony, M. Ishizuka, and D. Lukose (Eds.): PRICAI 2012, LNAI 7458, pp. 826–831, 2012.
© Springer-Verlag Berlin Heidelberg 2012

level of enthusiasm cannot be observed about the opponents controlled by the computer using AI techniques. This is primarily due to the fact that game AI (only concerned on NPC's behavior) does not account for the level of expertise the human player portrays as a human opponent would depending on the context of the game. The main reason for this observation is that games are supposed to cater to a large audience and therefore it is impossible to make game AI adapt to a specific player by means of a scripted NPC behavior. However, game AI could be made personalized to a player by using adaptive game AI techniques such as unsupervised learning and genetic algorithms.

Although adaptive game AI is coined as an elegant solution for the mundane and overly predictable behavior of the NPC's, its still at it's nascent and is mainly confined to research laboratories. One viable approach for adaptive gaming is to initially identify the possible attributes of opponents (end user/gamer/player) in-game behavior such as strategy and preference. The best possible way to capture such information would be via unsupervised learning techniques such as the growing self organizing map(GSOM) [1]. This paper presents such an attempt to identify the end user according to decisions made whilst playing the 2D arcade game Pac-man[4] by Namco.

2 Related Work: Adaptive Gaming

In order to build adaptive opponents with minimized predictability we need techniques which are able to anticipate, predict, plan and adapt to the changes in the game environment caused by the human antagonist. Learning or analyzing patterns from human players were done by Drachen et al. [3] using the game Tomb Raider: Underworld. This analysis was conducted using the emergent self organizing map (ESOM) and was successful in identifying four player types using six statistical measures. Another study in player modeling was conducted by Tychsen et al. [2] using the game Hit-man: Blood Money which used game metrics to decompose player controlled characters. Few research efforts have been carried out using the arcade game Pac-man on various game play aspects. In game AI, Pac-man based research comes in two segments that is to create AI controllers for the Pac-man to avoid ghosts and the other is to create AI algorithms to control ghosts which are more challenging and fun to play with. An evolutionary, rule based [4] Pac-man controller was proposed by Gallagher et al. where the agent adaptively learns through a population based incremental approach. Two neuro-evolution controllers were proposed by Gallagher et al. [5] and Yannakakis et al. [6] where the former is a controller for the Pac-man using minimal on-screen information and the latter is a controller for the ghosts using a computer-guided fixed strategy opponent. In literature apart from a very few articles there is a void in using unsupervised learning methodologies[7] to model human players. This study focuses on one such approach to identify viable patterns in the player decision making during game play.

---

[4] http://en.wikipedia.org/wiki/Pac-Man/

3   Why an Unsupervised Approach?

Behavior depicted by players when playing through a game title is as diverse as their respective personalities and ethnicities. Therefore customizing game AI to make game-play interesting for each and every player is an impossible task using conventional AI techniques such as scripting. Hence to customize or rather adapt AI to a player, one should identify the traits portrayed by an individual whilst playing the game. This task can only be achieved by a methodology which is able to learn from each action the player performs in game without the help of any targets or templates. Therefore an unsupervised learning methodology seems the logical choice in order to identify viable traits portrayed by the player during game play.

4   Experiment Setup

The experiment was set up using the timeless arcade game Pac-man designed by Toru Iwatani in it's classical version. Since this study is concerned with identifying decision making patterns of players it was needed to identify where in the Pac-man game a player's decision could be monitored. Therefore, it was decided to track player movements (direction taken) at each junction point available in the Pac-man maze. Hence, the game was modified to track the player's movement and the position's of the opponents. A sample of ninety four players with diverse backgrounds on gaming was selected and given to play the game in walk-through mode and the game mode. The walk-through mode consists of no opponents and the Pac-man was allowed to move around the maze unchallenged and complete the game level. Five such game walk-throughs of the same player were recorded. Then each player was given the game with the opponents to play until the completion of five levels or until all the lives are exhausted. The experiment described in the next section was performed on three randomly chosen samples as an preliminary study.

5   Experimental Results

This section presents the results obtained by applying the GSOM algorithm to the Pac-man data set. The raw data collected as specified in section 4 was preprocessed to fit the GSOM input format. There is a total of thirteen attributes (one attribute each for nine junctions and four attributes for the four possible decisions) which make up a single decision or a move by the player at a given junction in the Pac-man game. Two separate experiments were conducted for the walk-through data (walk-through mode) and for the game data (game mode). However, in both experiments, the network parameter and initialization weights were kept the same. The reason for setting initialization weights as the same throughout both experiments is to make the maps more comparable between different players. The comparison is required since the authors are of the view that the structure of the network is unique to each individual player and game play pattern.

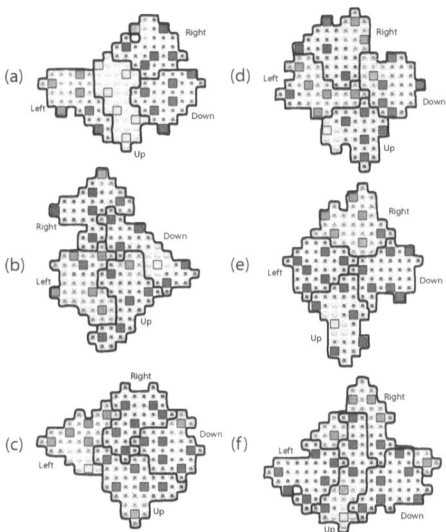

Fig. 1. GSOM generated for three individual players. (a)-(c) Map generated by using the maze walk-through as input data. (d)-(f) Map generated by using the real game as input data.The larger squares represent a winning neuron in at least one occasion in the training process and the smaller squares represents a neuron where no input was mapped to. Neurons with the same colors on a map belongs to a single cluster.

5.1   Analysis of Data

Before interpreting the results obtained on Figure 1 it is worth mentioning that the data each GSOM is trained on belongs to a single player. Therefore the clusters represent salient properties of decisions made by the player during the two experiments. According to Figure 1(a) - (c) it can be observed that the silhouette of walk-through maps is unique to each player. The reason for this observation is that the map is generated upon the order of the decisions presented to the network. If the order of the inputs are changed the corresponding winners of the network would also be different which includes the adapting neighborhoods. Therefore the different neurons in the network would be grown depending on the error accumulated which would result in a different silhouette. Hence no two maps would be the same in structure unless decision making patterns are identical in other players. This is also verified further by Figure 1(d) - (f). As expected the maps generated for the game data and the walk-through data for the same player is different due to the decisions which was forced upon by the movement of ghosts.

On further analyzing the maps obtained in Figure 1 it was observed that each of the six maps presented are hierarchically organized in to four clusters (The basic four clusters are depicted in each of the maps in Figure 1 by the black outline). These clusters map directly to the four decisions available to the player when playing the game. Furthermore, inside each of these four high level clusters there exist low level clusters which belongs to the junction types available in the maze (These clusters are depicted in color in each map in Figure 1). The low level clusters can be categorized into two distinct types (Let's denote them as Type 1 and Type 2 clusters). Type 1 clusters uniquely identifies a single decision and a junction type pair where as the Type 2 clusters identify the decision uniquely. For example in Table 1 cells highlighted in yellow represents Type 1 clusters. Each cluster number assigned to these cells are unique to that cell in the entirety of that experiment mode. All other uncolored cells represent Type 2 clusters. The thing to note here is that for a single decision column in Table 1 same cluster number is assigned to different junction type cells. Furthermore, it can be observed from Table 1 that the formation of Type 1 clusters when considering both experiment modes presents an interesting pattern which is explained next. This pattern has two basic observable aspects. That is, it can be

Table 1. Cluster analysis for a single player (Figure 1(c)&(f)) for both the experiments. The yellow colored bold font cell represents a type 1 cluster. The gray cells represent an impossible move from the given junction and the uncolored cells represent type 2 clusters.

Junction Type	Walk-through				Game			
	Right	Left	Up	Down	Right	Left	Up	Down
1	0			2	10			7
2		**10**		**5**		**6**		7
3	0		**9**		10		2	
4		1	6			0	2	
5	**3**	**4**		2	**5**	**3**		7
6	0	1	6		**1**	0	2	
7	0		6	2	10		**9**	7
8		1	**7**	2		0	**4**	7
9	**8**	1	6	2	**8**	0	2	7

observed that seven Type 1 clusters (not required to be present in the same number on both modes) are available in each experiment mode and out of these seven Type 1 clusters, five appears to have formed in the same location (with respect to junction type, decision pair) for both modes of the experiment. For example junction type 5's left and the right decision are represented by Type 1 clusters in both occasions. This type of an observation could only be made if there is a significant amount of input data which represents these junction type, decision pairs which belongs to these Type 1 clusters. Therefore, it could be derived using Table 1 that the player who's data is presented here is less likely to move down when encountered with a junction of type 5. This derivation was indeed verified by using the raw data collected where only 8 out of the 65 (Right:46.15%, Left:41.54% & Down:12.31%) recorded moves pertaining to junction type 5 resulted in opting to move down in game mode data. This means that the individual decision making patterns of a player remains consistent and intact and therefore could be identified and exploited using the GSOM technique.

6   Conclusions and Future Work

Each player of a computer game interacts with the game in a manner which they are accustomed to. Even with different control options, different play strategies, a pattern will emerge with time. A player would adopt the same pattern if successful, unless it is identified by the opponent and acted upon which will pave the way for an absorbing game play experience. Therefore as a first step in moving towards player centric games, we presented an experimental study on identifying player decision making patterns using an unsupervised learning technique. Also it is important to note that our explanations on some aspects of our work was forcibly made brief by space constraints.

Nevertheless, we believe this pattern identification via unsupervised learning algorithms can be further improved to the level of player profiling whilst playing the game in real time. Therefore as an immediate future work we propose to analyze the importance of type 2 clusters in generated maps and also to integrate the unsupervised learning classifier in to the game and recognize patterns in real-time. Finally, we conjecture that the player decision making patterns in predator/pray games such as Pac-man are unique and that unsupervised learning classifiers can be used to recognize them.

References

1. Alahakoon, D., Halgamuge, S.K., Srinivasan, B.: Dynamic self-organizing maps with controlled growth for knowledge discovery. IEEE Transactions on Neural Networks 11, 601–614 (2000)
2. Tychsen, A., Canossa, A.: Defining personas in games using metrics. In: Proceedings of the 2008 Conference on Future Play: Research, Play, Share, pp. 73–80. ACM (2008)
3. Drachen, A., Canossa, A., Yannakakis, G.N.: Player modeling using self-organization in tomb raider: underworld. In: IEEE Symposium on Computational Intelligence and Games (CIG), pp. 1–8 (2009)
4. Gallagher, M., Ryan, A.: Learning to play Pac-Man: An evolutionary, rule-based approach. In: The 2003 Congress on Evolutionary Computation (CEC), vol. 4, pp. 2462–2469. IEEE (2003)
5. Gallagher, M., Ledwich, M.: Evolving pac-man players: Can we learn from raw input? In: IEEE Symposium on Computational Intelligence and Games (CIG), pp. 282–287 (2007)
6. Yannakakis, G.N., Hallam, J.: Evolving opponents for interesting interactive computer games. In: From Animals to Animats, vol. 8, pp. 499–508 (2004)
7. Lopes, R., Bidarra, R.: Adaptivity Challenges in Games and Simulations: A Survey. IEEE Transactions on Computational Intelligence and AI in Games 3, 85–99 (2011)

Automatic Multi-document Summarization Based on New Sentence Similarity Measures

Wenpeng Yin, Yulong Pei, and Lian'en Huang

Shenzhen Key Lab for Cloud Computing Technology and Applications
Peking University Shenzhen Graduate School
Shenzhen, Guangdong 518055, P.R. China
mr.yinwenpeng@gmail.com, peiyulong@sz.pku.edu.cn, hle@net.pku.edu.cn

Abstract. The acquiring of sentence similarity has become a crucial step in graph-based multi-document summarization algorithms which have been intensively studied during the past decade. Previous algorithms generally considered sentence-level structure information and semantic similarity separately, which, consequently, had no access to grab similarity information comprehensively. In this paper, we present a general framework to exemplify how to combine the two factors above together so as to derive a corpus-oriented and more discriminative sentence similarity. Experimental results on the DUC2004 dataset demonstrate that our approaches could improve the multi-document summarization performance to a considerable extent.

Keywords: graph-based multi-document summarization, sentence similarity, LDA.

1 Introduction

Sentence-based extractive summarization is a typical category of automatic document summarization and it is commonly on the basis of graph-based ranking algorithms, such as TextRank [7]. Usually, such ranking approaches use some kinds of similarity metrics to rank sentences for inclusion in the summary. The similarity of sentences can be determined by many means which can be roughly comprehended in two levels: word space based level and semantic space based level. However, the former is somewhat strict and inflexible because it depends on hard matching of terms, in which case, synonyms, hypernyms and hyponyms are treated thoroughly differently even though a term is supposed to share some similar treatments with its relatives. The other extreme is that the semantic level places too much emphasis on semantic relationship between sentences, which results in losing sentence-level structure similarity that could play as an important indicator in differentiating sentences while measuring similarity.

In order to improve the quality of summary produced via graph-based summarization algorithm, we present a framework combining sentence-level structure similarity and semantic similarity together to address the limitations of existing approaches in deriving sentence similarity. Lin et al. [5] described three methods

P. Anthony, M. Ishizuka, and D. Lukose (Eds.): PRICAI 2012, LNAI 7458, pp. 832–837, 2012.
© Springer-Verlag Berlin Heidelberg 2012

to measure sentence similarity based on term order information: longest common subsequence (LCS), weighted longest common subsequence (WLCS) and skip-bigram co-occurrence statistics, all of which could reveal the sentence-level structure similarity very well. When it turns to semantic aspect, Latent Dirichlet Allocation (LDA) [2], a latent topic model, is an appropriate tool to measure word similarity because it could capture the patterns of word usage by analyzing its context. Thus, we combine LDA with LCS, WLCS and skip-bigram respectively to design three soft matching algorithms to illuminate our intention. One advantage of our approaches is that they consider lexical order information as well as semantic relationship. The other advantage lies in the ability to identify the different senses of words with respect to their co-occurring context and consequently acquire the similarity variably. Experiments on the DUC2004 corpus demonstrate the good effectiveness of the proposed algorithms in promoting multi-document summarization performance.

2   Related Work

Since this work focuses on proposing new sentence similarity measures for graph-based summarization algorithm so as to improve the system performance, we briefly introduce some summarization methods relevant to sentence similarity and some representative approaches measuring sentence relatedness.

Famous graph-based ranking algorithms TextRank [7] and LexPageRank [3] have been successfully applied to document summarization domain, they conduct PageRank algorithm on a weighted graph, where the vertices are sentences and the weighted edge indicates the relevance of two sentences, which is acquired by using cosine measure. The task in [1] presented a method to measure dissimilarity between sentences using the normalized google distance, then performed sentence clustering for automatic text summarization.

Zhang et al. [9] indicated that sentence similarities based on word set and word order have better performance than other sentence similarities. Sentence similarity based on TF-IDF has lower precision rate, recall rate and F-measure. The work in [4] is similar to ours. It presented an algorithm that took account of semantic information and word order information implied in the sentences. The semantic similarity of two sentences was calculated using information from a structured lexical database and from corpus statistics. Word order similarity was determined by the normalized difference of word order.

3   A New Word Similarity Algorithm Based on LDA

The ability capturing semantic relations between words of Latent Dirichlet Allocation (LDA) [2] is achieved by exploiting word co-occurrence: words which co-occur in the same contexts are projected onto the same latent topic, and words that occur in different contexts are projected onto different latent topics. That's to say, words with same latent topic are supposed to possess a certain degree of similarity in semantic respect. In this study, we propose that terms assigned same latent topic have a similarity value ranging from 0 to 1 and the concrete value could be determined by calculating the Kullback-Leibler(KL) Divergence of their distributions over latent topics. According to the Bayes rule, the probability of a specific topic z_k given a word w_v in the documents D is:

$$P(z_k|w_v, D) = \frac{P(w_v|z_k) \cdot P(z_k|D)}{P(w_v|D)} \quad . \tag{1}$$

Then the divergence of two terms w_i (probability distribution P_{w_i}) and w_j (probability distribution P_{w_j}) is determined as follows:

$$D\left(w_i, w_j\right) = KL\left(P_{w_i}, P_{w_j}\right) + KL\left(P_{w_j}, P_{w_i}\right) \quad . \tag{2}$$

Since KL divergence is asymmetric, we apply the above KL divergence-based symmetric measure. The divergence is transformed into similarity measure [6]:

$$Simi\left(w_i, w_j\right) = 10^{-\delta D(w_i, w_j)} \quad . \tag{3}$$

In experiments, we use the GibbsLDA++[1], a C/C++ implementation of LDA using Gibbs Sampling.

4   Sentence Similarity Measures

4.1   LCS_LDA

In our modified LCS (hereafter LCS_LDA), given the following original sentences:

$$S1 : boy_1 \ enjoy_2 \ happy_3 \ holiday_7 \quad S2 : boy_1 \ enjoy_2 \ happy_3 \ vacation_7$$

The subscript denotes the topic index assigned to the corresponding word. We could easily derive that the traditional LCS of S1 and S2 is 3 (hereafter, we use LCS on behalf of the length of LCS directly, such principle also applies to WLCS and Skip-Bigram cases). Instead, in our new scenario, we consider their topic sequences firstly. Consequently, the LCS of S1 and S2 seems to be 4. However, the rationale of our method lies in that although the topic indexes of two words in two different sentences are the same, the similarity value of the two word strings depends on LDA. For instance, assume that in Equation 3, "holiday" and "vacation" own a similarity value 0.9, in other words, the LCS_LDA of S1 and S2 has changed to be 3.9 in our proposed algorithm. Undoubtedly, 3.9 could reflect the length of *longest approximate subsequence* between S1 and S2 more exactly than 3 obtained in traditional LCS algorithm.

   In general, the LCS_LDA score of two sentences could be computed using a analogous algorithm with LCS in [5], the key difference lies in that the variation of score in each step during the entire computing process is more likely a decimal based on Equation 3 rather than an integer 1. Therefore, inspired by [5], given the LCS_LDA score (LL for convenience) of two sentences X of length m and Y

---

[1] GibbsLDA++: http://gibbslda.sourceforge.net

of length n, we could derive the their similarity $Simi_{LL}(X,Y)$ using the following equations:

$$R_{LL} = \frac{LL(X,Y)}{m} \quad P_{LL} = \frac{LL(X,Y)}{n}$$

$$Simi_{LL}(X,Y) = \frac{(1+\beta^2)R_{LL} \cdot P_{LL}}{R_{LL} + \beta^2 P_{LL}} \quad , \tag{4}$$

where $\beta = P_{LL}/R_{LL}$.

4.2 WLCS_LDA

As [5] indicated, while LCS has many good properties, it does not differentiate LCSes of different spatial relations within their embedding sequences. To improve the basic LCS method, $f(\cdot)$, a function of consecutive matches, is adopted to assign different credits to consecutive in-sequence matches, which is called Weighted LCS (WLCS). Similarly, we integrate LDA with WLCS based on the similar principle in LCS_LDA. Given the WLCS_LDA score (WL for convenience) of two sentences X of length m and Y of length n, their similarity $Simi_{WL}(X,Y)$ could be derived using the following equations:

$$R_{WL} = f^{-1}\left(\frac{WL(X,Y)}{f(m)}\right) \quad P_{WL} = f^{-1}\left(\frac{WL(X,Y)}{f(n)}\right)$$

$$Simi_{WL}(X,Y) = \frac{(1+\beta^2)R_{WL} \cdot P_{WL}}{R_{WL} + \beta^2 P_{WL}} \quad , \tag{5}$$

where $\beta = P_{WL}/R_{WL}$ and $f(k) = k^2$.

4.3 Skip-Bigram_LDA

In this section, we firstly redefine skip-bigram match as a soft one (SM) rather than a strict co-occurrence as follow:

$$SM[(t_1,t_2),(t_3,t_4)] = \begin{cases} \frac{Simi(w_1,w_3)+Simi(w_2,w_4)}{2} & \text{if } t_1 = t_3 \text{ and } t_2 = t_4 \\ 0 & \text{otherwise} \end{cases} \quad , \tag{6}$$

where (t_1,t_2) and (t_3,t_4) are two topic skip-bigrams. w_1, w_2, w_3, and w_4 are word strings to which t_1, t_2, t_3 and t_4 correspond, respectively. Word similarity values $Simi(w_1,w_3)$ and $Simi(w_2,w_4)$ are computed using Equation 3.

Consider the example in Section 4.1 again, topic skip-bigram (1, 7) exists in both $S1_{topic}$ and $S2_{topic}$, $SKIP2(S1,S2)$ should increase 1 according to traditional Skip-Bigram co-ocurrence, whereas, in our soft match algorithm, the match degree between $(1,7)_{S1}$ and $(1,7)_{S2}$ is the average value between $Simi(employee, employee)$ and $Simi(holiday, vacation)$. Therefore, say $Simi(employee, employee) = 1$ and $Simi(holiday, vacation) = 0.96$, then $SM((1,7)_{S1}, (1,7)_{S2})=0.98$. Consequently, $SKIP2(S1,S2)$ should be merely added to 0.98. Hereafter, we use $SKIP2_{LDA}(S1, S2)$ to represent the sum of the SM values which always result from the optimal matching between topic pairs of S1 and S2. Actually, any topic pair in a sentence is likely to match more than one topic pair of the other sentence, in such case, it bears close analysis to take the optimal pair match into account.

Given two sentences X of length m and Y of length n, the Skip-Bigram_LDA (SBL) similarity $Simi_{SBL}(X, Y)$ can be derived using the following equations:

$$R_{SBL} = \frac{SKIP2_{LDA}(X, Y)}{C(m, 2)} \quad P_{SBL} = \frac{SKIP2_{LDA}(X, Y)}{C(n, 2)}$$

$$Simi_{SBL}(X, Y) = \frac{(1 + \beta^2)R_{SBL} \cdot P_{SBL}}{R_{SBL} + \beta^2 P_{SBL}} \quad , \tag{7}$$

where $\beta = P_{SBL}/R_{SBL}$ and $C(\cdot, \cdot)$ represents the combination calculation.

5   Experiments

5.1   Data Set and Evaluation Metric

We conduct experiments on DUC2004[2] benchmark dataset. It provides 50 document sets. According to the task definitions, systems are required to produce a concise summary for each document set and the length of summaries is limited to 665 bytes. We use the ROUGE 1.5.5[3] toolkit for evaluation, which is officially adopted by DUC for evaluating automatic generated summaries.

Documents are pre-processed by segmenting sentences and splitting words. Stop words are removed and the remaining words are stemmed using Porter stemmer[4]. Then, we utilize sentence similarity discussed in Section 4 to construct undirected weighted graphs based on the algorithm proposed in [3] for scoring and ranking all the sentences. A modified version of the MMR algorithm [8] is used to remove redundancy and choose both informative and novel sentences into the summary. In experiments we set the parameters empirically. The damping factor λ in graph algorithm is set to 0.85. The penalty degree factor ω in the modified MMR is set to 0.4. Besides, we set the parameter δ in Equation 3 to 1 and the topic number in LDA is 50.

5.2   Performance Evaluation and Comparison

In experiments we compare our improved measures with three basic methods (LCS, WLCS and Skip-Bigram) and LexPageRank which is a PageRank-based summarization algorithm on the basis of cosine similarity measure taking into account only the term co-occurrence rather than the order of words. Table 1 shows the comparison results. Seen from Table 1, LCS, WLCS and Skip-Bigram

---

[2] Refer to http://www-nlpir.nist.gov/projects/duc/data.html for a detailed description of the dataset.
[3] http://www.isi.edu/licensed-sw/see/rouge/
[4] Porter stemmer:http://tartarus.org/martin/PorterStemmer/

Table 1. Comparison results on DUC2004

Systems	ROUGE-1	ROUGE-2	ROUGE-SU4
LexPageRank	0.37875	0.08354	0.12770
LCS	0.35404	0.06521	0.11019
WLCS	0.35332	0.06987	0.11175
Skip-Bigram	0.36540	0.07764	0.11950
LCS_LDA	0.38101	0.08466	0.12989
WLCS_LDA	0.38161	0.08858	0.12993
Skip-Bigram_LDA	**0.38523**	**0.09109**	**0.13123**

all have poor performances compared with LexPagaRank, which might result from that although LexPageRank ignores the order information, LCS, WLCS and Skip-Bigram neglect some words that co-occur in two sentences while not in the common subsequence. Nevertheless, their modified versions could considerably improve the evaluation results over all three metrics, which demonstrates that combining word semantic similarity and sentence structure information does benefit the calculating of sentence semantic similarity.

Acknowledgments. This research is financially supported by NSFC of China (Grant No. 60933004 and 61103027) and HGJ (Grant No. 2011ZX01042-001-001). We thank the anonymous reviewers for their useful comments.

References

1. Aliguliyev, R.M.: A new sentence similarity measure and sentence based extractive technique for automatic text summarization. Expert Systems with Applications 36(4), 7764–7772 (2009)
2. Blei, D.M., Ng, A.Y., Jordan, M.I.: Latent dirichlet allocation. The Journal of Machine Learning Research 3, 993–1022 (2003)
3. Erkan, G., Radev, D.R.: Lexpagerank: Prestige in multi-document text summarization. In: Proceedings of EMNLP, pp. 365–371 (2004)
4. Li, Y., McLean, D., Bandar, Z.A., O'Shea, J.D., Crockett, K.: Sentence similarity based on semantic nets and corpus statistics. IEEE Transactions on Knowledge and Data Engineering 18(8), 1138–1150 (2006)
5. Lin, C., Och, F.: Automatic evaluation of machine translation quality using longest common subsequence and skip-bigram statistics. In: Proceedings of the 42nd Annual Meeting on Association for Computational Linguistics, p. 605. Association for Computational Linguistics (2004)
6. Manning, C., Schutze, H.: Foundations of statistical natural language processing. In: Enhancing Semantic Distances With Context Awareness (1999)
7. Mihalcea, R., Tarau, P.: Textrank: Bringing order into texts. In: Proceedings of EMNLP, pp. 404–411. ACL, Barcelona (2004)
8. Wan, X.: Document-Based HITS Model for Multi-document Summarization. In: Ho, T.-B., Zhou, Z.-H. (eds.) PRICAI 2008. LNCS (LNAI), vol. 5351, pp. 454–465. Springer, Heidelberg (2008)
9. Zhang, J., Sun, Y., Wang, H., He, Y.: Calculating statistical similarity between sentences. Journal of Convergence Information Technology 6(2) (2011)

Using Common-Sense Knowledge in Generating Stories

Sherie Yu and Ethel Ong

Center for Language Technologies, De La Salle University, Manila, Philippines
sherie_yu@yahoo.com, ethel.ong@delasalle.ph

Abstract. A problem with most story generation systems is the lack of an adequately-sized body of knowledge to generate stories from. This paper presents
an approach that focuses on providing a large amount of common-sense
knowledge to automatic story generators while keeping extensive manual
handcrafting of knowledge to a minimum. It does so by combining manually-
created resources with freely-available common-sense knowledge in machine-
readable format for the generation of stories.

Keywords: Automatic Story Generation, Storytelling Knowledge, Knowledge
Representation, Common-sense Reasoning.

1 Introduction

Story generation systems have been used for endowing computational creativity to
computers [1], as a tool for writing [2], and as an educational tool [3]. However,
computers seem to have a hard time at generating stories that make sense [1]. This is
attributed to the lack of an adequately-sized body of knowledge to generate stories
from. Systems such as Mexica [4], Picture Books [5] and SUMO Stories [6] make use
of manually built resources that contain domain-dependent information, which make
them not work well for unexpected inputs. They do not have the same basic general or
common-sense knowledge that humans have to reason about everyday life [9].

This paper presents an approach to providing a large amount of common-sense
knowledge to story generation systems by combining manually-created resources with
freely-available common-sense knowledge in machine-readable format. Such knowledge is often referred to as storytelling knowledge. The next section identifies the
types of knowledge needed by our story generator. This is followed by a description
of the architecture that utilized the storytelling knowledge. Preliminary results are
provided, ending with a discussion of issues and recommendations for future work.

2 Storytelling Knowledge

Our storytelling knowledge adapts Swartjes' [10] two-layer ontology for representing
storytelling knowledge. The upper story world ontology contains existing common-
sense knowledge resources. These include ConceptNet [7], a semantic network of
common-sense concepts classified into thematic categories such as events, causal and

P. Anthony, M. Ishizuka, and D. Lukose (Eds.): PRICAI 2012, LNAI 7458, pp. 838–843, 2012.
© Springer-Verlag Berlin Heidelberg 2012

affective, which fit with some of the characteristics inherent in stories; VerbNet [14], a semantic lexicon of verbs classified according to abstract classes with thematic roles and frames that are suitable for use by natural language generation systems, the broader area where story generation falls under; and WordNet [15].

The domain-specific world ontology, on the other hand, models elements that are typical to the target domain, in this case, children's stories. The elements include themes, story characters, and events.

Story themes for children usually center on everyday life experiences (such as *going to camp*) and behavior development (such as *honesty* and *bravery*). Characters are given names to identify them from other characters. The interpersonal relationships that often exist between characters in children's stories such as "*ParentOf*" and "*TeacherOf*" are also modeled. Most critical to the representation of characters are the set of roles and traits that influence their goals and ultimately determine the actions they perform. For instance, a character with the role '*student*' and a trait of "*lazy*" may fail to fulfill the goal to "*have good grade*" by choosing to "*play*" rather than to "*study*". On the other hand, a similar character with a trait of "*dishonest*" rather "*lazy*" may choose to "*cheat*" instead.

Events represent the atomic units of a story. They represent actions and states. When represented in the context of a story, each event is associated with a *timepoint*. States are represented as primitives or predicates that signify a particular meaning as illustrated in Table 1 while actions are represented with their respective VerbNet thematic roles as shown in Table 2.

Table 1. Predicates

Predicate	Meaning
HasPossession (?character, ?object)	character owns object
RelationshipChange (?character1, ?character2, -10)	character2 holds ill feelings towards character1

Table 2. Representation of Events

Event	*steal*	*break*	*scold*
Agent	?character1	?object	?character1
Source	?character2	---	---
Patient	---	---	?character2
Theme	?object	---	---

Themes revolve around the behavioral development of the main character based on a moral or virtue. They are represented as rules in the ontology and are patterned after the classical plot structure for children's stories as presented in [16].

3   System Architecture

The architecture shown in Figure 1 is based on a model of story writing that identifies a balance between the plotting of the characters and the author [17]. The prompt is a user input which specifies the basic elements (characters, objects and optional location) that should be present in the story.

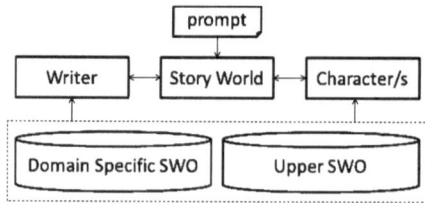

Fig. 1. System Architecture

The writer and the character agents interact through the story world that represents the story being created. It contains a history of the events that have occurred so far with their corresponding *timepoints*. In order to move the story forward, both agents make queries to the storytelling knowledge and store the results to the story world. However, whereas the writer has full access to all the information in the storytelling knowledge, characters have a limited access (i.e., cannot see the themes).

3.1   ConceptNet Disambiguation

Disambiguating between the senses of a concept enables the agent to make better choices. When a '*child*' character agent makes a query for the list of candidate objects a "*child*" could perform with the action "*play*", the upper ontology should return "*toy*" rather than "*piano*" since it is more likely that the sense of "*play*" being referred to is "*being engaged in playful activity*".

Whenever a query is made, the concepts are disambiguated into their proper senses through a modified implementation of the algorithm described by [12]. A *Word Sense Profile* (WSP) or list of terms is constructed for each sense of the ambiguous concept based on WordNet. Then, the relatedness between the terms is measured based on *Normalized Google Distance* (NGD). Lastly, the noisy terms that would decrease the performance of the algorithm are filtered out.

In our story generator, AnalogySpace [13] is used in place of NGD to measure the relatedness between terms. Filtering of noisy terms it not performed. Furthermore, the part-of-speech (POS) tags of each concept in the assertion were identified prior to creating the WSP to optimize the performance of the disambiguation.

Once disambiguated, the concepts are now associated with their WordNet keys to supplement the current ConceptNet labels. Furthermore, verbs get a VerbNet *class id* information since the verbs in VerbNet contain mappings to WordNet.

3.2   Character Goal Formulation

To simulate the plotting of the characters' individual goals and plans, a mechanism for formulating character goals is followed. Every main character agent in the story queries the upper story world ontology for a goal to pursue. For each role assigned to a character, a pool of candidate goals is obtained from ConceptNet through the query:

```
Desires (?role, goal)
```

A random goal for a character is selected after factoring in the scores of each "*Desires*" assertion in the goals pool. For instance, if the query, "*Desires ('child', goal)*" returns the following assertions and their respective scores:

Desires (child, learn) +2, Desires (child, play) +5

Then "*Desires (child, play)*" would have a better chance of being selected.

3.3   Character Goal Completion

Each character goal can be completed only by the character that formulated the goal. A higher priority is given to actions from the domain-specific ontology.

Querying the domain-specific ontology, events at the previous *timepoint* must satisfy the preconditions of an action whose post-condition matches with the goal. A '*dishonest*' character might achieve the goal '*good grade*' with the action '*cheat*' if it had '*dishonest*' and '*good grade*' as its pre- and post-conditions, respectively.

Querying the upper story world ontology is a bit more complicated. First, AnalogySpace [13] is used to obtain the top *n* similar concepts. A goal is selected from this list of concepts with the highest probability given to the original concept. For each goal, the events that achieve the goal are selected through the following queries:

MotivatedByGoal(?event,goal), HasPrerequisite(goal,?event)
Causes(?event,goal), UsedFor(?event,goal)

The results are stored disjointly so that any single action can be performed in order to achieve the goal. For the "*good grade*" goal, some of the possible actions that could lead to its achievement are "*attend class, study hard*". These actions are represented in the way actions are described in the domain-specific ontology since they have also been disambiguated and therefore assigned their particular WordNet senses.

3.4   Plot Progression

At every *timepoint*, the writer inspects the story world to determine if the intended story theme has been realized. Since the themes concern behavior development from a negative trait to a positive one (e.g. a "*dishonest character*" learning to be "*honest*"), the rules are generalized as "*when a main character performs an action based on a negative trait that leads to a negative consequence, the character realizes his/her mistake and does the right thing in the future*". Performing an action based on a trait is already embedded into the mechanism by which a character agent chooses an action. The link between the negative action and the consequence that follows is provided by profluence. The creation of a similar circumstance for allowing the character to do the right thing in the future is the task of the writer.

Table 3. Queries for Consequential Progression

Query	Meaning
Causes(current,?event)	What does the current event cause to happen?
HasSubevent (current,?event)	What normally happens when current occurs?

3.5   Story Profluence

Profluence is the logical progression of events. This is manifested through the completion of character goals; and the consequential and motivational progressions as shown in Tables 3 and 4 respectively. The current goal is labeled as *current*.

Table 4. Queries for Motivational Progression

Query	Meaning
MotivatedByGoal (current, ?event)	What events motivate current to happen?
CausesDesire (?event, current)	What events can possibly motivate current?
	(special case: current must be a goal)

4   Preliminary Results

ConceptNet 4 is used in this system. Only assertions with scores of at least two were considered. The number of concepts, n, has been limited to 1.

Listing 1 shows a sample trace of character goals, story world state and character actions. The theme selected is for that of a *greedy* character. The prompt specified only the characters that should be included in the story: *Danny who is greedy* and *Hannah who is stingy*. The two plot points that completed the plot for "*greedy*" were "*steal*" and "*punch*" in which *Hannah punched Danny for stealing her candy.*

Listing 1. Sample Output

```
Timepoint 0:  goal:HasPossession(Danny,candy)
              state:HasPossession(Hannah,candy)
              state:!HasPossession(Danny,candy,)
Timepoint 1:  action:steal(Danny,Hannah,candy,Danny)
Timepoint 2:  state:HasPossession(Danny,candy)
              state:RelationshipChange(Hannah,Danny,-10)
              state:EmotionalState(Danny,happy)
Timepoint 3:  state:EmotionalState(Hannah,anger,Danny)
              action:dance(Danny)
Timepoint 4:  action:punch(Hannah,Danny)
```

5   Conclusion

The paper presented a preliminary work on providing a story generator with existing common-sense knowledge. It describes the representation of the storytelling knowledge as well as how the story generator would make use of this knowledge.

An evaluation system must be devised in close communication with an evaluator (possibly a story writer) to identify an evaluation scheme and the most appropriate story representation to use. The representation must also contain the appropriate contextual information to allow for conversion into natural language text in the future.

ConceptNet was built on user-supplied knowledge through the Open Mind Common Sense (OMCS) project [7] in which common-sense was acquired from the general public online. Despite obtaining a high evaluation score regarding the accuracy of information contained in this resource, the generated output may be inappropriate for children and may also be inaccurate. Some form of automatic validation of an assertion or concept before usage into the system would be beneficial to address the inaccuracy concern.

References

1. Lonneker, B., Meister, J.C.: Dream On: Designing the Ideal Story Generator Algorithm. In: ACH and the ALLC, pp. 1–3. University of Victoria, Victoria (2005)
2. Riedl, M.O.: Narrative Generation: Balancing Plot and Character. Ph.D. Dissertation. North Carolina State University (2004)
3. Rowe, J.P., Mcquiggan, S.W., Lester, J.C.: Narrative Presence in Intelligent Learning Environments. In: AAAI Symposium on Intelligent Narrative Technologies, Arlington, Virginia, pp. 126–133 (2007)
4. Péréz y Péréz, R., Sharples, M.: MEXICA: A Computer Model of a Cognitive Account of Creative Writing. JETAI 13(2), 119–139 (2001)
5. Ong, E.: A Commonsense Knowledge Base for Generating Children's Stories. In: AAAI 2010 Fall Symposium Series on Common Sense Knowledge, pp. 82–87. AAAI, Virginia (2010)
6. Cua, J., Ong, E., Manurung, R., Pease, A.: Representing Story Plans in SUMO. In: NAACL Human Language Technology 2010 Second Workshop on Computational Approaches to Linguistic Creativity, pp. 40–48. ACL, Stroudsburg (2010)
7. Liu, H., Singh, P.: Commonsense Reasoning in and over Natural Language. In: 8th International Conference on Knowledge-Based Intelligent Information and Engineering Systems. Springer, Berlin (2004)
8. Niles, I., Pease, A.: Towards a Standard Upper Ontology. In: FOIS, Ogunquit, Maine, October 17-19, pp. 2–9 (2001)
9. Lieberman, H., Liu, H., Singh, P., Barry, B.: Beating Common Sense into Interactive Applications. AI Magazine 25, 63–76 (2004)
10. Swartjes, I.: The Plot Thickens: Bringing Structure and Meaning into Automated Story Generation. Master's Thesis. University of Twente, The Netherlands (2006)
11. Ang, K., Yu, S., Ong, E.: Theme-Based Cause-Effect Planning for Multiple-Scene Story Generation. In: 2nd ICCC, Mexico City, Mexico, April 27-29, pp. 48–53 (2011)
12. Chen, J., Liu, J.: Disambiguating ConceptNet and WordNet for Word Sense Disambiguation. In: 5th IJCP 2011, Chiang Mai, Thailand, pp. 686–694 (2011)
13. Speer, R., Havasi, C., Lieberman, H.: AnalogySpace: Reducing the Dimensionality of Common Sense Knowledge. In: Cohn, A. (ed.) 23rd National Conference on Artificial Intelligence vol. 1, pp. 548–553. AAAI Press (2008)
14. Kipper, K.S.: VerbNet: A Broad-Coverage, Comprehensive Verb Lexicon. Ph.D. Dissertation. University of Pennsylvania, Philadelphia (2005)
15. Miller, G.A.: WordNet: A Lexical Database for English. Communications of the ACM 38(11), 39–41 (1995)
16. Machado, J.: Storytelling. In: Early Childhood Experiences in Language Arts: Emerging Literacy. Thomson/Delmar Learning, New York (2003)
17. Ryan, M.: Cheap Plot Trick, Plot Holes, and Narrative Design. Narrative 17(1), 56–75 (2009)

Context-Based Interaction Design for Smart Homes

Zuraini Zainol and Keiichi Nakata

Informatics Research Centre, University of Reading RG6 6UD Reading, United Kingdom
z.zuraini@pgr.reading.ac.uk, k.nakata@henley.reading.ac.uk

Abstract. This paper presents the notion of Context-based Activity Design (CoBAD) that represents context with its dynamic changes and normative activities in an interactive system design. The development of CoBAD requires an appropriate context ontology model and inference mechanisms. The incorporation of norms and information field theory into Context State Transition Model, and the implementation of new conflict resolution strategies based on the specific situation are discussed. A demonstration of CoBAD using a human agent scenario in a smart home is also presented. Finally, a method of treating conflicting norms in multiple information fields is proposed.

Keywords: Context-based Activity Design, Context Reasoning, State Transition Rule.

1 Introduction

The emergence of pervasive, context-aware and mobile computing requires system design to be adaptive and responsive to aspects of setting in which the tasks are performed, including other users, devices and environment [1]. An important aspect in the emergence of these technologies is context. Although numerous studies on context have been conducted, the meaning of context is not well-defined. Mostly, context refers to fixed elements: person, locations, and surrounding elements involved in an interaction. However, not much attention has been paid on the context of the interaction itself [1]. Work by Dourish [2] provides different viewpoints towards context, i.e., i) context arises from the activity itself and both entities: context and activity cannot be separated; ii) context is dynamic depending on the activities carried out; iii) context is not fixed information; and finally iv) context is a relational property that holds between objects or activities. Accordingly, we argue that the concept of context is relevant and influential to the interaction in an activity system. Hence, context should be incorporated in design process as it becomes more important in pervasive computing. This paper presents a context-based interactive system design that captures dynamic changes in context and normative activities. The paper is organised as follows: first we provide an overview of CoBAD, then a discussion on the concept of information field and how they can be captured and represented in CoBAD. An introduction of conflict resolution strategies in relation to information fields is made. Section 4 discusses the analysis and reasoning about conflicting norms. Finally, the paper concludes with a summary of the contributions made and future work.

P. Anthony, M. Ishizuka, and D. Lukose (Eds.): PRICAI 2012, LNAI 7458, pp. 844–849, 2012.

2   Context-Based Activity Design (CoBAD)

Zainol and Nakata [3] proposed a context ontology model, COM, which can be applied in CoBAD. The important feature of COM is, it categorises context identifiers into three types: extrinsic, interface, and intrinsic. In particular, interface context contains activities as it captures the context at the interface between intrinsic and extrinsic contexts. CoBAD is based on context reasoning (CR) where various types of contexts are employed to infer new context. CR is a condition-action rule that specifies the activity reasoning of human agent. A CR rule can be expressed as follows:

[ExtrinsicContext ∧ InterfaceContext ∧ IntrinsicContext] → [newContext]

The condition on the left hand side (LHS) of the CR rule refers to the situational conditions and it consists of context identifiers, while the right hand side (RHS) consists of new context of any category. A special type of CR rule is state transition (ST). In ST rules, the Interface Context in LHS is not empty and the RHS is a new Interface Context, i.e., they specify the possible activity states that human agents can perform in the next activity state. A ST rule can be expressed as follows:

[ExtrinsicContext ∧ InterfaceContext ∧ IntrinsicContext] → [newInterfaceContext]

The LHS is for the current activity state in that particular situation, whereas the RHS is for the updates of the current activity state, which yields the next state.

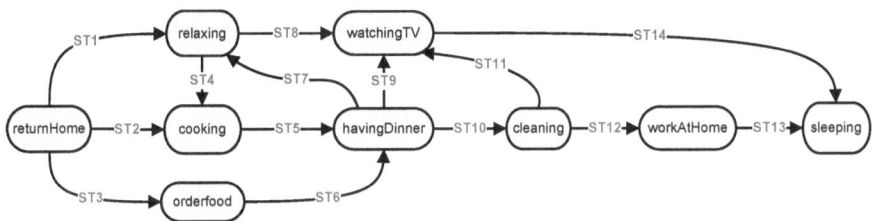

Fig. 1. An example of CSTM with ST rules

Table 1. The example of ST rules using FOL predicates where T indicates 'true'

Rule	Activity Reasoning
ST1	[T ∧ activity(returnHome) ∧ state(tired)] → [activity(relaxing)]
ST2	[T ∧ activity(returnHome) ∧ desire(hunger)] → [activity(cooking)]
ST3	[T ∧ activity(returnHome) ∧ desire(hunger) ∧ state(tired)] → [activity(orderFood)]
ST4	[T ∧ activity(relaxing) ∧ desire(hunger)] → [activity(cooking)]
ST5	[T ∧ activity(cooking) ∧ desire(hunger)] → [activity(havingDinner)]
ST6	[T ∧ activity(orderFood) ∧ desire(hunger)] → [activity(havingDinner)]
ST7	[T ∧ activity(havingDinner) ∧ T] → [activity(relaxing)]
ST8	[T ∧ activity(relaxing) ∧ T] → [activity(watchingTV)]
ST9	[T ∧ activity(havingDinner) ∧ T] → [activity(watchingTV)]
ST10	[T ∧ activity(havingDinner) ∧ T] → [activity(cleaning)]
ST11	[T ∧ activity(cleaning) ∧ T] → [activity(watchingTV)]
ST12	[T ∧ activity(cleaning) ∧ T] → [activity(workAtHome)]
ST13	[T ∧ activity(workAtHome) ∧ desire(sleepy)] → [activity(sleeping)]
ST14	[T ∧ activity(watchingTV) ∧ desire(sleepy)] → [activity(sleeping)]

To design the transitions of human agent activities in a smart home, we apply the method of state space representation (see Fig. 1). We present the Context State Transition Model (CSTM) that contains a set of activity states and ST rules (see Table 1). Depending on a particular activity state a human agent is in, different ST rules that are available and can be triggered. As in any state transition model, this model is characterised by multiple possible transitions in some states. Here we assume that activities (Interface Context) are mutually exclusive, i.e., an agent cannot be engaged in more than one activity at any time. Consider a case where a resident returned home and he/she is tired and hungry. This satisfies the LHS of multiple rules (i.e., ST1, ST2 and ST3), which in turn introduces a conflict among them. To resolve the conflict we incorporate into CSTM, concepts from organisational semiotics [4]: norms and information field. Information field (IF) is a set of shared norms that governs the behaviour of a group of member in an organised fashion, in which norms is a field of force that requires the members of a community to behave or think in a particular way [4]. Humans are seen as agents and their actions are influenced by the forces that present in IFs, and these forces originate from norms that are shared in a community [5]. Generally, these norms can be categorised into: perceptual, evaluative, cognitive and behavioural. Based on this, the following mappings of these norms can be considered: i) perceptual norms are concerned with how human agent acts in accordance with his/her perception based on facts – hence they can be represented as conditions; ii) cognitive norms represent the aspect of human agent's belief about actions that can be represented as activities; iii) behavioural norms determine how human agent should behave and define what he/she is expected to do under a given situation, and they can be represented by ST rules; and iv) evaluative norms represent the aspect of choices or preferences of a human agent to choose his/her next action based on the available context information – hence they can be represented by a set of activities or conditions. What we can observe here is that the intersections of multiple IFs in the model may introduce conflicts among norms. To resolve a conflict, we define the precedence of norms among the shared norms. By setting this, a human agent is expected to be able comply with an appropriate norm based on the situation [3].

3   Conflict Resolution Strategies for CSTM

In resolving a conflict, a conflict resolution strategy (CRS) is required to decide which rule to fire. The most common CRSs used in the production system are recency, specificity, and refractoriness. However, none of them support the preference setting of norms. Therefore, we formulate new CSRs and define them into three main situations:

- *Situation1*: when the IF 1 is more dominant than the IF 2, i.e., $[d(IF_1) > d(IF_2)]$, where two IFs, IF 1 and IF 2, overlap,   then apply dominant IF.

— DominantIF: choose the rule from dominant IF;
— DominantPreferredOutcome: choose the rule that result in preferred outcome specified by evaluative norms in dominant IF;
— DominantPreferredCondition: choose the rule that contains preferred condition specified by evaluative norms in dominant IF;
— DominantRuleCondition: choose the rule that contains condition in dominant IF.

- *Situation 2*: If no dominant IF specified, i.e., $[d(IF_1) = d(IF_2)]$ then apply any IF.
- — PreferredOutcome: choose the rule that result in preferred outcome in any IF;
- — PreferredCondition: choose the rule that result in preferred condition in any IF.
- *Situation 3:* If the strategies specify in both situation 1 and 2 failed then apply the standard strategies: Random, Recency, Specificity, and Refractoriness.

4   The Analysis and Reasoning of Conflicting Norms in CSTM

Based on the example in Fig. 1, we first analyse the base information field (IF) called IF England since it assumes the activities to be taking place in the norms of England. To perform reasoning, we applied a production system (PS). In this study, the ordering of strategies is set as follows: DominantIF, DominantPreferredOutcome, DominantPreferredCondition, DominantRuleCondition, PreferredOutcome, PreferredCondition, Random, Recency, Specificity, and Refractoriness. The sequences of rule firing and actions are presented in Table 2 and 4.

Table 2. Trace of a PS for CSTM for base IF (IF England)

Cycle	Working Memory(WM)		Conflict	Conflict	Rule
	IF England	new/derived facts	Set	resolution	fired
0	desire(hunger)	activity(returnHome)	ST2	NULL	ST2
1	desire(hunger)	activity(cooking)	ST5	NULL	ST5
2	desire(hunger)	activity(havingDinner)	ST7, ST9,ST10	Random	ST9
3	desire(hunger)	activity(watchingTV)	NULL	NULL	HALT

Assume that when we begin, the rules and facts have been loaded into production rules and working memory (WM), respectively. The inference begins from the top of the rule ST1 and goes on downward until the first true condition is found. Given that activity(returnHome) represents the starting state, in the cycle 0, the recognise-act-cycle (RAC) matches the current state: desire(hunger) and activity(returnHome) against the ST rules. At this stage, the ST2 matches the facts. Therefore, ST2 is fired and its action, cooking is asserted in the WM; indicating the transition from activity state returnHome to cooking. The cycle 1 uses the facts: desire(hunger) and activity(cooking) would match with ST5. The ST5 is fired and the activity state havingDinner is added in the WM; indicating a transition from activity state, cooking to havingDinner. In the cycle 2, the RAC again matches the facts: desire(hunger) and activity(havingDinner) against the ST rules. At this stage, ST7, ST9 and ST10 are enabled for firing. To resolve the conflict, a strategy, random is used to fire the ST9. Therefore, ST9 is fired and its action, watchingTV is then asserted in WM, which in turn moves from the activity state, havingDinner to watchingTV. Finally, the execution halts in cycle 3 as no more rules to fire.

Next, we analyse the overlapping of dominant IF (IF Family) that captures the norms of a specific family, which is superimposed onto the IF England. In this example, a human agent is belonging to a family, and at the same time she is following the England's culture where she lives (see Fig. 2). The overlapping of these IFs will bring different set of norms into the IF England (refer Table 3).

Table 3. A set of norms added from the IF Family into IF England

Type of norms	Norms added from the IF Family
Behavioural	STF1: activity(returnHome) → activity(cooking)
Evaluative	pref(cleaning) > $pref$(watchingTV) > $pref$(relaxing),
	pref(watchingTV) > $pref$(workAtHome)
Cognitive	activity(sleeping)
Perceptual	desire(sleepy), state(tired)

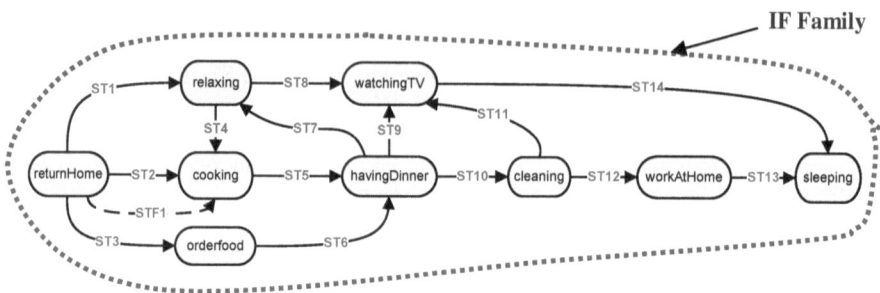

Fig. 2. The overlapping of IF Family to IF England in the CSTM

Again, given that activity(returnHome) represents the starting state, the inference begins with ST1 (see Table 4). In the cycle 0, the RAC matches the current state: desire(hunger), desire(sleepy), state(tired) and activity(returnHome) in the WM against the ST rules. At this stage, ST1, ST2, ST3 and STF1 are enabled for firing. To resolve the conflict, the strategy DominantRuleIF is applied to select a rule from the dominant IF. Therefore, STF1 is fired and its action, cooking is placed in the WM, which in turn moves the activity state from returnHome to cooking. The cycle 1 uses the updated facts: desire(hunger), desire(sleepy), state(tired) and activity(cooking) would match with ST5. The ST5 is fired and the activity state, havingDinner is then added in the WM; indicating a transition from activity state, cooking to havingDinner. In the cycle 2, the RAC again matches the facts: desire(hunger), desire(sleepy), state(tired) and activity(havingDinner) in the WM against the ST rules. At this stage, ST7, ST9 and ST10 are enabled for firing. To resolve the conflict, the strategy DominantPreferredOutcome is used to fire the ST10 based on the following preferences: [pref(cleaning)> pref(watchingTV) > pref(relaxing)]. Therefore, ST10 is fired and its action, activity(cleaning) is asserted in the WM, which in turn moves the activity state from havingDinner to cleaning. In the cycle 3, the RAC again matches the facts: desire(hunger), desire(sleepy), state(tired) and activity(cleaning) in the WM against the ST rules. Two rules: ST11 and ST12 are matched with the facts. To resolve the conflict, the strategy, DominantPreferredOutcome is applied to choose ST11 based on the following preferences: [pref(watchingTV) > $pref$(workAtHome)]. The transition of activity state is now changed from the activity state, cleaning to watching TV. In the cycle 4, the RAC again matches the facts: desire(hunger), desire(sleepy), state(tired) and activity(watchingTV) in WM against the ST rules. At this stage, only ST14 matches with the facts. Hence, ST14 is fired and its action, activity(sleeping) is then added in the WM; indicating a transition from activity state, watchingTV to sleeping. Finally, in cycle 5, the execution halts as no more rules to fire.

Table 4. Trace of a PS for CSTM based on the overlapping IF Family and IF England

Cycle	Working Memory(WM)			Conflict set	Conflict resolution	Rule fired
	IF England	**IF Family**	**new/derive facts**			
0	desire(hunger)	desire(sleepy) state(tired)	activi-ty(returnHome)	ST1, ST2, ST3, STF1	DominantRuleIF	STF1
1	desire(hunger)	desire(sleepy) state(tired)	activity(cooking)	ST5	NULL	ST5
2	desire(hunger)	desire(sleepy) state(tired)	activi-ty(havingDinner)	ST7,ST9, ST10	DominantPreferrd Outcome	ST10
3	desire(hunger)	desire(sleepy) state(tired)	activity(cleaning)	ST11, ST12	DominantPreferd Outcome	ST11
4	desire(hunger)	desire(sleepy) state(tired)	activi-ty(watchingTV)	ST14	NULL	ST14
5	desire(hunger)	desire(sleepy) state(tired)	activity(sleeping)	NULL	NULL	HALT

5   Conclusions and Future Work

In this paper we have developed a notion of CoBAD that represents context with its dynamic changes and normative activities in the interactive system design. We also discussed the integration of semiotics theory in the CSTM that represent the aspect of default and dynamic social norms that governs the behaviour of human agent in a specific situation. However, the overlapping of information fields may introduce conflicts among them. To resolve a conflict, we proposed new conflict resolution strategies based on organisational semiotics perspective. In the future we will continue our work on the development of multiple scenarios of human agent activities in a smart home, and then followed by the implementation based on a rule-based expert system.

References

1. Chen, Y., Atwood, M.E.: Context-Centered Design: Bridging the Gap Between Understanding and Designing. In: Jacko, J.A. (ed.) HCI 2007, Part I. LNCS, vol. 4550, pp. 40–48. Springer, Heidelberg (2007)
2. Dourish, P.: What we talk about when we talk about context. Personal Ubiquitous Computing 8(1), 19–30 (2004)
3. Zainol, Z., Nakata, K.: Information Fields in Context-based Activity Design. In: 13th International Conference on Informatics and Semiotics in Organisations: Problems and Possibilities of Computational Humanities, Leeuwarden, The Netherlands, pp. 103–110 (2011)
4. Stamper, R., Liu, K., Sun, L., Tan, S., Shah, H., Sharp, B., Dong, D.: Semiotic Methods for Enterprise Design and IT Applications. In: Proceedings of the 7th International Workshop on Organisational Semiotics, pp. 190–213 (2004)
5. Gazendam.: Organizational Semiotics: A State of the Art Report, vol. (1) (2004)

A Visual Approach for Classification Based on Data Projection

Ke-Bing Zhang[1], Mehmet A. Orgun[1], Rajan Shankaran[1], and Du Zhang[2]

[1] Department of Computing, Macquarie University, Sydney, NSW 2109, Australia
{kebing.zhang,mehmet.orgun,rajan.shankaran}@mq.edu.au
[2] Department of Computer Science, California State University, Sacramento,
CA 95819-6021, USA
zhangd@ecs.csus.edu

Abstract. In this paper we present a visual approach for classification in data mining, based on the enhanced separation feature of a visual technique, called Hypothesis-Oriented Verification and Validation by Visualization (HOV^3). In this approach, the user first projects a labeled dataset by HOV^3 with a statistical measurement of the dataset on a *2d* space, where data points with the same class label are well separated into groups. Then each well separated group and its measure vector are employed as a visual classifier to classify unlabeled data points by projecting and grouping them together with the overlapping labeled data points. The experiments demonstrate that our approach is effective to assist the user on classification of data by visualization.

Keywords: Classification, Data Mining, Visualization, Data Projection.

1   Introduction

Classification is a supervised learning process by which objects are partitioned into several predefined groups. The predefined groups are built up from a training dataset where each object in the predefined groups is assigned a class label. The method of partitioning the objects of the training dataset is called a classifier, which is used to classify unknown objects further with particular class labels based on their similarity to the predefined groups.

There have been many algorithmic methods developed to deal with the real world applications of classification [7]. However, with the increasing amount and complexity of electronic data produced in the real world, the user cannot rely on fully automatic techniques for data analysis, because algorithmic classification approaches need "human-in-the-loop" to adjust parameters for the classification process in order to enable the users to interactively refine their predictions based on their modeling assumptions that are impossible for a fully automated approach to reach on its own.

Graphical displays help people think better, since a visual display is often the best way of transmitting complex data to the human brain. Therefore, visualization and interaction with algorithmic classification methods may serve a central role in the classification process [3]. Several visual techniques, such as PBC [1], StarClass [5] and Parallel Coordinates [3], have been developed to facilitate classification in data

P. Anthony, M. Ishizuka, and D. Lukose (Eds.): PRICAI 2012, LNAI 7458, pp. 850–856, 2012.
© Springer-Verlag Berlin Heidelberg 2012

mining. However, those techniques have certain issues on classifying very large datasets. For example, Parallel Coordinates cannot easily deal with very high dimensional data in classification. The classification results of PBC and StarClass heavily depend on the experience of individual users for visually classifying the unlabeled datasets. The interactive classification function of Weka [6] is not powerful enough to represent the overview of data projection of high dimensional data. The fuzzy-based visual technique proposed in [4] is not suited for classifying very large datasets, due to its high computation complexity.

Motivated by the limitations of these approaches to classification by visualization, we propose an approach to classification based on a visual technique called Hypothesis-Oriented Verification and Validation by Visualization (HOV^3) [8, 9]. A more detailed explanation of our approach and its demonstration on several benchmark classification datasets are provided in the following sections.

2   Classification by HOV^3

2.1   The HOV^3 Technique and Its Enhanced Separation Feature

The aim of projecting high dimensional data onto a $2d$ or $3d$ space is to present a more intuitive, visual perception of the data to the user. The HOV^3 technique [8] adopts the Polar Coordinates representation to project high-dimensional data onto a $2d$ surface. First, a $2d$ Polar Coordinates plane is divided into n equal size sectors with n coordinate axes, where each axis represents a dimension and all axes share the center of a circle on the $2d$ space. Then the values of all axes are respectively mapped to the orthogonal x and y coordinates which share the initial point on the $2d$ space. Thus data with any dimensions can be transformed to the orthogonal coordinates x and y, and represented by a point on the x-y $2d$ plane. The projection of HOV^3 can be viewed as a mapping from high-dimensional real space to a $2d$ complex number space ($R^n \rightarrow C^2$) by employing a measure vector m. The real part and the imaginary part of data are mapped to the x axis and y axis respectively.

HOV^3 provides a quantitative mechanism to the user on analysis and visualization of high dimensional data. This study is based on the enhanced separation feature of HOV^3 [9]. Based on the quantitative measurement feature of HOV^3, it is proved that, applying a measure vector m multiple times to a dataset by HOV^3, would make some data points to be more tightly packed in geometry [8]. In practice, the number of times of applying m to the studied data is usually decided by the user based on his/her observation of the data projection. This feature is significant for data classification by HOV^3, since a clear geometrical separation of data groups provides the user with an intuitive insight on whether a data point belongs to a specific class.

2.2   The Classification Procedure by HOV^3

The process of classification by HOV^3 features three phases described as follows:

- **Establishing a Visual Classifier:** The user first projects the predefined groups (training dataset) by HOV^3, and then each group is well separated by the user based on the enhanced separation feature of HOV^3. A well separated predefined group and its corresponding separation vector are recorded as a *visual classifier*.

- **Classifying Unlabelled Data:** The user mixes the test dataset (unlabelled data objects) with the predefined group data of a visual classifier. Then the mixed dataset is projected by HOV^3 with the separation vector of the visual classifier. Based on a given threshold, the user can select the unlabelled data points which overlap the predefined group objects in the data projection. The selected data points then are assigned the corresponding class label and removed from the test dataset. This process is performed until all visual classifiers are applied.
- **Analysis of Classification Accuracy:** Based on the result in phase two above, the user can obtain the classification accuracy by comparing the correctly classified objects with the total sample dataset.

3   The Experiments

This section discusses the classification experiments by HOV^3 on three classification benchmark datasets of *Shuttle*, *Segment* and *Satimagine*, which are available from the machine learning website http://archive.ics.uci.edu/ml/machine-learning-databases/. The experiments are implemented in MATLAB.

3.1   Establishing a Visual Classifier

We demonstrate how to separate predefined groups and obtain visual classifiers by our approach on the Shuttle training dataset. The Shuttle dataset has 9 dimensions and 58,000 instances where 45,300 instances are used for training and 14,500 for the test.

The Shuttle training data are labeled with numbers from 1 to 7. We first plot the Shuttle training dataset D_l by HOV^3 with the vector $m = [1, 1, 1, 1, 1, 1, 1, 1, 1]$, i.e., no measurement case. The original data distribution of the Shuttle training data is plotted in Fig. 1, where the class data points are highly overlapping. We choose two times of the 7th row of correlation coefficients of the Shuttle training data as a measure vector m to project D_l by HOV^3. The data distribution is shown in Fig.2, where the data points of Class 5 in the Shuttle training data are clearly separated by the projection. Then we record (color) the dataset of Class 5 as δ_l, and the measure vector $m_l=m$ as a visual classifier $C_l = (\delta_l, m_l)$. Then Class 5 is removed from the Shuttle training dataset $(D_l=D_l-\delta_l)$. The remaining data points are plotted in Fig. 3.

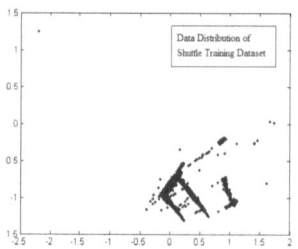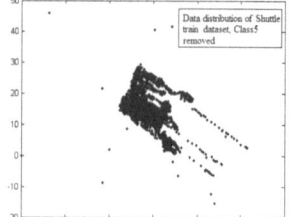

Fig. 1. The original data distribution of the Shuttle training dataset projected by HOV^3

Fig. 2. The separated and marked data points of Class 5 of the Shuttle training dataset

Fig. 3. The data distribution of the Shuttle training dataset, with the data points in Class 5 removed

Next we select the second row of the correlation coefficients of the new D_l (with Class 5 removed) as the measure vector m and apply m to D_l by HOV^3. The projected data distribution of D_l is plotted in Fig.4, where the data points of Class6 δ_2 are marked (red colored) and separated from the data points labeled with different class numbers. As mentioned above, the visual classifier $C_2 = (\delta_2, m_2)$ is recorded and Class 6 δ_2 is removed from D_l. We repeat the process to separate the remaining classes individually. Fig. 5 presents the separated and marked/colored data points of Class 7 in the Shuttle training set, where there may be overlapping data points. But when we zoom in the diagram by MATLAB, and find that all red points clearly separated from other blue items. This situation is same as in Fig. 7 and Fig. 8.

Then we select the second row of the correlation coefficients of the new D_l (with Class 5 removed) as the measure vector m and apply m to D_l by HOV^3. The projected data distribution of D_l is plotted in Fig.4, where the data points of Class6 δ_2 are marked (red colored) and separated from the data points labeled with different class numbers. As mentioned above, the visual classifier $C_2 = (\delta_2, m_2)$ is recorded and Class6 δ_2 is removed from D_l. We repeat the process to separate the remaining classes individually. Fig. 5 presents the separated and marked/colored data points of Class 7 in the Shuttle training set, where there may be overlapping data points. But when we zoom in the diagram by MATLAB, and find that all red points clearly separated from other blue items. This situation is same as in Fig. 7 and Fig. 8.

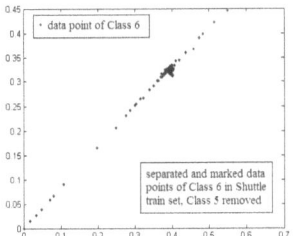 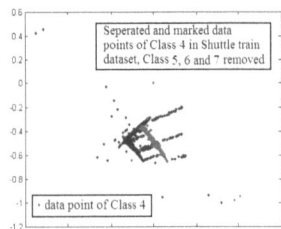

Fig. 4. The separated and marked Class 6 data points of the Shuttle training set, with Class 5 removed

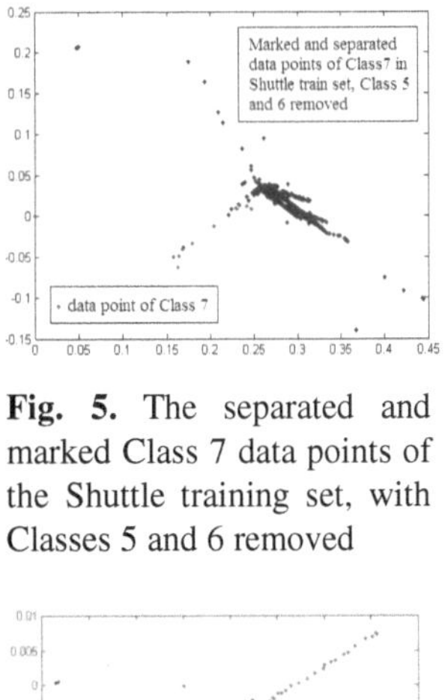

Fig. 5. The separated and marked Class 7 data points of the Shuttle training set, with Classes 5 and 6 removed

Fig. 6. The separated and marked Class 4 data points of the Shuttle training set, with Classes 5, 6 and 7 removed

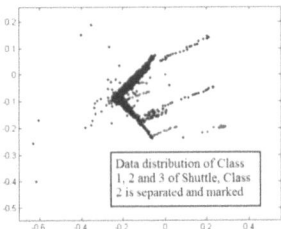 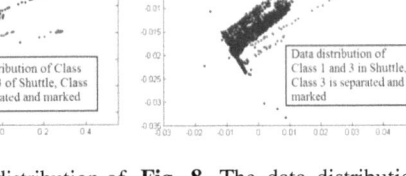 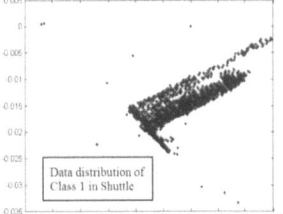

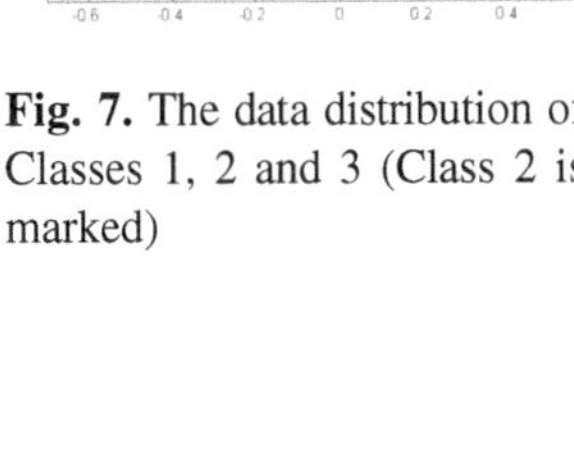

Fig. 7. The data distribution of Classes 1, 2 and 3 (Class 2 is marked)

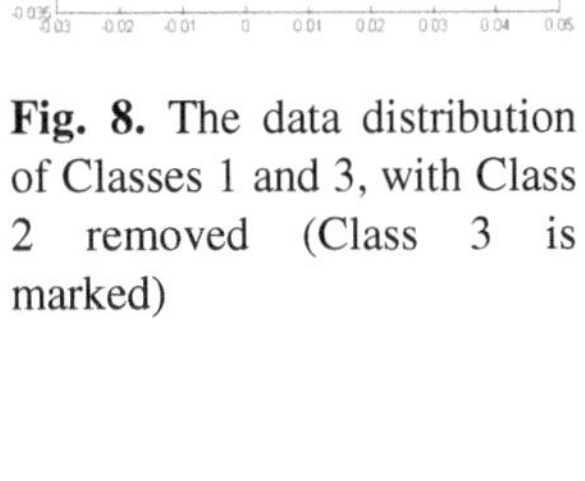

Fig. 8. The data distribution of Classes 1 and 3, with Class 2 removed (Class 3 is marked)

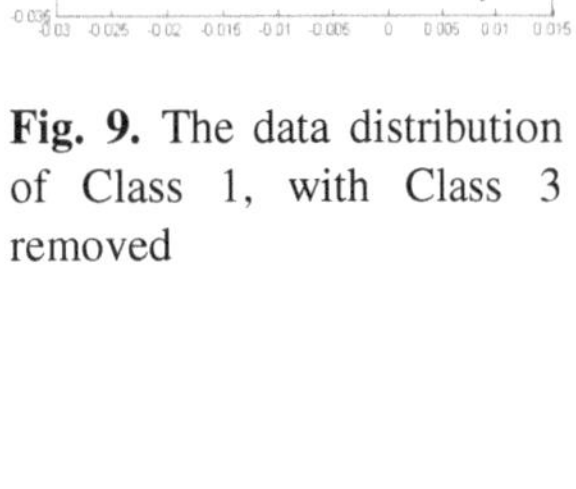

Fig. 9. The data distribution of Class 1, with Class 3 removed

In Fig. 9, we can find that there exist 7 overlapping data points between Class 3 (red color) and Class 1 (blue ones). It is hard to separate these two classes well. But compared to the total size of Class 1 (34,108) and Class 3 (132), as shown in Table 1, this minor overlapping rate 0.02044% is acceptable.

3.2   Classifying Unlabelled Data and Analysis of Classification Accuracy

We next apply the set of visual classifiers generated above to the test dataset of Shuttle for selecting the overlapping data points by their corresponding predefined groups. First, we mix the labeled 2,458 data points of Class 5 of the Shuttle training set and the unlabeled test dataset D_0 (14,500 instances). Their original data distribution projected by HOV^3 is plotted in Fig.10. Then we apply the $C1.m_1 =$ [-0.4343, 3.5231e-4, 0.0795, 4.7315e-05, 0.0108, -6.2927e-07, 1.0, -1.4896e-4, -0.1711] to plot the mixed data $D_0+C_1.\delta_1$. Their data distribution is shown in Fig.11, where data points of $C_1.\delta_1$ are red-colored.

We have also applied our approach to the *Segment* and *Satimage* datasets. Fig.13 and Fig.14 show the data projections of the Segment and Satimage training datasets respectively, where the data points of Class 2 in each dataset are colored as red dots. The process of classifying the unlabeled data of *Segment* and *Satimage* is performed as in the case for *Shuttle*. The experimental results of using HOV^3 to classify unlabeled data of *Shuttle, Segment* and *Satimage* are summarized in Table 2.

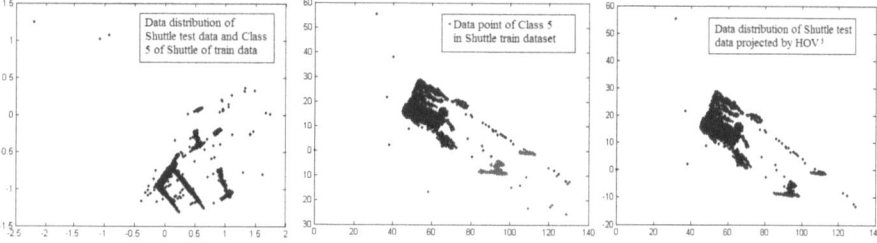

Fig. 10. The data distribution of the Shuttle test data and Class 5 projected by HOV^3 without any measurement

Fig. 11. The data distribution of the Shuttle test data and Class 5 projected by HOV^3 with the separation vector

Fig. 12. The data distribution of the Shuttle test data, with the data points in Class 5 removed

We may find that the similarity of data distributions in Fig.2 and Fig. 11, and Fig.3 and Fig.12 are high. Based on the zoom in function by MATLAB, we carefully collect the data items overlapping the red colored data points as δ_{c1}. Then δ_{c1} is removed from D_0 ($D_0=D_0-\delta_{ci}$). This process is performed iteratively by each visual classifier to D_0 until all the visual classifiers are applied. The collected δ_{ci} are listed in Table 1. We compare the labels we marked on the data items in each δ_{ci} (i=1..7) to their original labels. The contrast of their correctness/incorrectness is also listed in Table 1. We then calculate the accuracy of the classified data items by HOV^3as (correct collection) / (total observation) = (14,500-20)/14,500x100% =99.86%.

Table 1. Experimental Results on Classification of the Shuttle Dataset by HOV^3

Class	Training set	Test set	Overlapped test items	Correct items	Wrong items
1	34,108	11478	11480	11475	5
2	37	13	11	11	0
3	132	39	30	30	0
4	6,748	2155	2160	2154	6
5	2,458	809	811	806	5
6	6	4	4	2	2
7	11	2	4	2	2

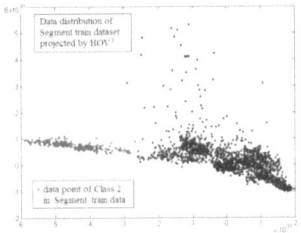 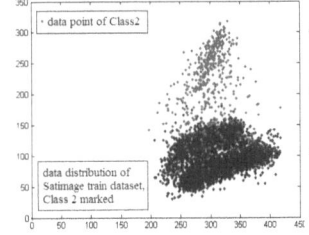

Fig. 13. The data distribution of the *Segment* training dataset (with marked Class 2 data points) projected by HOV^3 with the measure vector m_{v2}

Fig. 14. The data distribution of the *Satimage* training dataset projected by HOV^3 with twice the standard deviation of *Satimage* as the statistical prediction

Fig. 15. The data distribution of the *Shuttle* training dataset (with class labels) projected by HOV^3 with two times of standard dariation of D as the measure vector

3.3   Discussion

We have compared the accuracy of our approach with those of other classification experiments on the same datasets from the paper [5]. We summarize those experimental results and ours in Table 2. By analyzing the results in Table 2, we observe that the accuracy of classification is related to the dimensionality of data. That is, the higher the dimensionality of data, the lower the accuracy of the classification result. This is a typical example of the "curse of dimensionality" as Bellman noted [2]. The experiments also showed that, the accuracy of classification highly depends on the distinction of classes in the training dataset. For example, we have also included the class indices of the Shuttle training data, i.e., the 10th dimension of Shuttle as a special case in our experiments. We adopt the two times of

Table 2. The contrast of accuracy between HOV^3 and other classification methods (including the analysis of the classification results given by Teoh S. T. and Ma K-L [5])

Datasets				Algorithmic Approaches to Classification				Visual Approaches		
	Size	Dims	Class	CART	C4	SPRINT	CLOUDS	PBC	StarClass	HOV^3
Shuttle	43500	9	7	99.9	99.9	99.9	99.9	99.9	99.9	99.9
Segment	2130	19	7	84.9	95.9	94.6	94.7	94.8	95.2	94.8
Satimage	4435	36	6	85.3	85.2	86.3	85.9	83.5	85.3	83.5

the standard deviation of the Shuttle training data (with 10 dimensions) as a measure vector to project it by HOV^3. Its data projection is presented in Fig. 15, where we colored the 7 clearly separated classes. That is why all approaches in Table 2 have a very high accuracy on the classification of the Shuttle dataset.

4   Conclusions

We have presented a visual approach for classification based on the enhanced feature of HOV^3. In this approach, the user first separates the predefined groups in a training dataset by applying the statistical predictions of the dataset in HOV^3. Then each separated predefined group and its separation vector are used as a visual classification model to classify an unlabeled dataset. The effectiveness of this approach has been demonstrated by the experiments reported in the paper. Furthermore, our approach is not sensitive to the order of the dimensions of data, whereas some other approaches such as PBC [1] and StarClass [5] are, which may cause the final visualized result to vary.

References

1. Ankerst, M., Ester, M., Kriegel, H.-P.: Towards an effective cooperation of the user and the computer for classification. In: Proceedings of KDD 2000, pp. 179–188. ACM press (2000)
2. Bellman, R.E.: Dynamic programming. Princeton University Press (1957)
3. Inselberg, A.: Parallel Coordinates: Visualization, Exploration and Classification of High-Dimensional Data. In: Inselberg, A. (ed.) Handbook of Data Visualization, Springer Handbooks of Computational Statistics, vol. III, pp. 643–680 (2008)
4. Rehm, F., Klawonn, F., Kruse, R.: Rule Classification Visualization of High-Dimensional Data. In: Proceedings of IPMU 2006, pp. 1944–1948 (2006)
5. Teoh, S.T., Ma, K.-L.: StarClass: Interactive Visual Classification Using Star Coordinates. In: Proceedings of SDM 2003, pp. 178–185 (2003)
6. http://www.cs.waikato.ac.nz/ml/weka/index.html
7. Witten, L.H., Frank, E., Hall, M.A.: Data Mining Practical Machine Learning Tools and Technologies. Morgan Kaufmann Publishers (2011)
8. Zhang, K.-B., Orgun, M.A., Zhang, K.: A Prediction-Based Visual Approach for Cluster Exploration and Cluster Validation by HOV^3. In: Kok, J.N., Koronacki, J., Lopez de Mantaras, R., Matwin, S., Mladenič, D., Skowron, A. (eds.) PKDD 2007. LNCS (LNAI), vol. 4702, pp. 336–349. Springer, Heidelberg (2007)
9. Zhang, K.-B., Orgun, M.A., Zhang, K.: Enhanced Visual Separation of Clusters by M-Mapping to Facilitate Cluster Analysis. In: Qiu, G., Leung, C., Xue, X.-Y., Laurini, R. (eds.) VISUAL 2007. LNCS, vol. 4781, pp. 285–297. Springer, Heidelberg (2007)

On Modelling Emotional Responses
to Rhythm Features

Jocelynn Cu[1], Rafael Cabredo[2],
Roberto Legaspi[2], and Merlin Teodosia Suarez[1]

[1] Center for Empathic Human-Computer Interactions,
De La Salle University, Philippines
{jiji.cu,merlin.suarez}@delasalle.ph
[2] Institute of Scientific and Industrial Research, Osaka University, Japan
{cabredo,roberto}@ai.sanken.osaka-u.ac.jp

Abstract. Rhythm is one of the most essential elements of music that can easily capture the attention of the listener. In this study, we explored various rhythm features and used them to build emotion models. The emotion labels used are based on Thayers Model of Mood, which includes contentment, exuberance, anxiety, and depression. Empirical results identify 11 low-level rhythmic features to classify music emotion. We also determined that KStar can be used to build user-specific emotion models with a precision value of 0.476, recall of 0.480, and F-measure of 0.475.

Keywords: Rhythm features, emotion models, classification.

1 Introduction

Music induces emotional and physical responses from people. According to [4], music allows the listener to re-live a personal experience and remember other events of heightened emotional moments. The past decade has seen a lot of studies that characterizes music according to emotion, for example, that of [1,3]. Studies on music and its correlation to emotion have led to applications of music in health and well-being, education and marketing among others.

According to [7], among all elements of music, rhythm is the most fundamental, essential, structural and organizational music feature because it captures the attention of an individual. It is also used in all temporal aspects of a musical work whether represented in a score, measured from a performance, or existing only in the perception of the listener [2].

Since rhythm is closely related to a person's emotional response while listening to music [5], our objective then is to determine specific rhythm features that are useful in classifying emotions induced by music.

2 Methodology

Our approach is to build a music emotion database and then extract rhythm features for modelling and classification. The music emotion corpus is built by

P. Anthony, M. Ishizuka, and D. Lukose (Eds.): PRICAI 2012, LNAI 7458, pp. 857–860, 2012.
© Springer-Verlag Berlin Heidelberg 2012

Table 1. Song segment distribution based on tempo markings

Tempo marking	Beats per minutes (BPM)	Number of songs
Adagio	66–76	1
Andante	76–108	4
Moderato	108–120	47
Allegro	120–168	35
Presto	168–200	13

using 100 instrumental pieces with mixed genres. For each song, a 30-second segment having consistent tempo were chosen. Table 1 shows the distribution of song segments based on tempo markings. Segments were sampled at 44.1 KHz.

Six participants (three male and three female), aged 17–23 years old and with various musical background, were engaged to listen to the music segments in a controlled environment and label these with one emotion which the music induced in them. Their choices were limited to the four labels described in Thayers Model of Mood: Contentment, Exuberance, Anxiety, and Depression. Contentment refers to music that gives a sense of relaxation and positive feelings to the listener. Exuberance refers to music that is energetic and gives a heightened sense of happiness in the listener. Anxiety refers to music that is energetic but influences negative feelings in the listener. Depression refers to music that is calm yet gives the feeling of negativity to the listener.

Music Analysis, Retrieval and Synthesis for Audio Signals (MARSYAS) [6] was used to extract 18 rhythm features from each music segment. These include the beat histogram; the amplitudes of the first, second, and third highest peaks in the beat histogram; the BPM of the first, second, and third highest peaks in the beat histogram; the ratio between the low BPM and high BPM; the mean autocorrelation value; the fundamental frequency, spectral flatness, standard deviation, spectral centroids, spectral spread of the beat histogram; the number of maxima; and the tempo. Using the rhythm features, 100 instances were produced for the dataset. WEKA[1] was used to build the emotion model for each participant. The following classifiers were used in the experiments: C4.5, Naïve Bayes, Support Vector Machines, KStar, and JRip. These classifiers were compared against each other using true positive (TP) rate, false positive (FP) rate, precision, recall and F-measure.

3   Results and Analysis

We observe the dataset is lopsided and a music segment is labelled differently by the participants. Only 4% of the music segments were labelled consistently by all participants with Contentment. Most of the music segments were labeled with Contentment. Music that was not labeled as such (14%), were equally labelled with Exuberant (high-energy, low-stress) or Anxiety (high-energy, high-stress).

---

[1] http://www.cs.waikato.ac.nz/ml/weka/

Table 2. Comparison of various classifiers for user-specific modelling

Classifier	TP rate	FP rate	Precision	Recall	F-Measure
C4.5	0.460	0.216	0.454	0.460	0.454
Nave Bayes	0.352	**0.194**	0.402	0.352	0.347
SVM	0.435	0.248	0.347	0.435	0.381
KStar	**0.480**	0.203	**0.476**	**0.480**	**0.475**
JRip	0.430	0.287	0.447	0.430	0.374

Table 3. Rhythm features that are significant for classifying music emotions

Exhuberance	Anxiety	Contentment	Depression
LowPeakAmp	LowPeakAmp	MidBPM	LowPeakAmp
MidPeakAmp	MidPeakAmp	MeanACR	MidBPM
HighPeakAmp	HighPeakAmp	Flatness	HighBPM
MidBPM	Spread2	StdDev	Flatness
HighBPM	Tempo	NumMax	Spread2
MeanACR			NumMax
Flatness			Tempo
StdDev			
NumMax			

Only 16 music segments were equally labelled with Contentment or Depression. This indicates that participants have difficulty differentiating the stress level (i.e., high-stress level or low-stress level). Only 2 music segments were given the label Depression, which may indicate two things: (1) that Depression is not easily recognizable by the listener, or (2) the dataset does not contain songs that are depressing.

From these observations, the dataset could not be used to build a general emotion model. A user-specific music emotion model was built instead. Table 2 shows the results of the various ML approaches for one participant. In this example, KStar returned the highest values in TP rate, precision, recall, and F-measure; and, Naïve Bayes returned the lowest value in FP rate. Therefore, for this participant, the music affect model was built using KStar classifier.

We used the 3-sigma rule to determine the best set of rhythm features per emotion label. We computed the mean and standard deviation of each rhythm feature and created a normal distribution for each. Those features that lie within one standard deviation of the mean were taken as the best rhythm feature set. For this same example, the important rhythm features for each type of emotion is listed in Table 3. Results indicate that:

- LowPeakAmp is significant in differentiating Exuberance, Anxiety, and Depression;
- MidBPM, flatness, and NumMax are significant features for Exuberance, Contentment, and Depression;
- MidPeakAmp and HighPeakAmp are significant to differentiate between Exuberance and Anxiety;
- MeanACR and StdDev are significant to differentiate between Exuberance and Contentment;

- Spread2 and Tempo are significant to differentiate between Anxiety and Depression; and
- HighBPM is significant to different between Exuberance and Depression.

Results show that there are no features that differentiate Contentment from Depression. This could be the reason why it was difficult for a listener to differentiate the stress level in low-energy music. The MidPeakAmp and HighPeakAmp values for Exuberance and Anxiety overlaps, this could be the reason why it is difficult to differentiate the stress level even in high-energy music.

4   Concluding Remarks

This study focused on identifying relevant rhythm features for detecting music emotions. Preliminary results have shown that only 11 of the original 18 rhythm features are actually significant in differentiating emotions. Based on analysis of empirical data from one participant, it is indeed possible to build user-specific music emotion models. Among the various classifiers tested, KStar yielded the best results in music emotion modelling based on TP rate, precision, recall, and F-measure values. However, further experiments and analysis should be carried out to make more general conclusions.

Acknowledgments. This research is supported in part by the Philippine Council for Industry, Energy and Emerging Technology Research and Development of the Department of Science and Technology, the Management Expenses Grants for National Universities Corporations through the Ministry of Education, Culture, Sports, Science and Technology (MEXT) of Japan, by the Global COE (Centers of Excellence) Program of MEXT, and by KAKENHI 23300059.

References

1. Avisado, H.G., Cocjin, J.V., Gaverza, J.A., Cabredo, R., Cu, J., Suarez, M.: Analysis of music timbre features for the construction of user-specific affect model. In: WCTP 2011, Quezon City, Philippines (2011)
2. Gouyon, F., Dixon, S.: A review of automatic rhythm description system. Computer Music Journal 29(1) (2004)
3. Han, B.J., Ho, S.M., Dannenberg, R.B., Hwang, E.J.: SMERS: music emotion recognition using support vector regression. In: ISMIR (2009)
4. Juslin, P.N., Sloboda, J.A. (eds.): Music and emotion: theory and practice. Oxford University Press, Oxford (2001)
5. Liu, D., Lu, L., Zhang, H.: Automatic mood detection from acoustic music data. In: ISMIR (2003)
6. Tzanetakis, G., Jones, R., Castillo, C., Martins, L.G., Teixeira, L.F., Lagrange, M.: Interoperability and the Marsyas 0.2 Runtime. In: Proc. Int. Comptuer Music Conference, Belfast, UK (2008)
7. Murrock, C.J.: Music and mood. In: Clard, A.V. (ed.) Psychology of Mood. Hauppage Nova Science Publishers (2006)

Harnessing Wikipedia Semantics for Computing Contextual Relatedness

Shahida Jabeen, Xiaoying Gao, and Peter Andreae

School of Engineering and Computer Science
Victoria University of Wellington, P.O. Box 600, Wellington, New Zealand
{shahidarao,xgao,peter.andreae}@ecs.vuw.ac.nz

Abstract. This paper proposes a new method of automatically measuring semantic relatedness by exploiting Wikipedia as an external knowledge source. The main contribution of our research is to propose a relatedness measure based on Wikipedia senses and hyperlink structure for computing contextual relatedness of any two terms. We have evaluated the effectiveness of our approach using three datasets and have shown that our approach competes well with other well known existing methods.

Keywords: contextual relatedness, Wikipedia hyperlinks, relatedness measures, semantics.

1 Introduction

The task of measuring relatedness of two concepts requires the understanding of implicit relations between concepts based on deeper level of world knowledge about the entities in consideration. For instance, to correctly assess that "Forensic Science" and "Teeth" are related, lot of background knowledge of "Forensic Odontology" is required but this knowledge can not be found in phrases themselves. The limiting factor for computing semantic relatedness is undoubtedly the requirement of background knowledge. Fortunately, Wikipedia stores a great deal of information not only about the concepts themselves but also about various aspects of the relations between concepts.

The rest of the paper is organized as follows: Section 2 discusses three broad categories of semantic measures proposed in literature. Section 3 presents our new approach. Section 4 analyzes the performance of our proposed methodology by comparing it to other existing strategies using three well known datasets. Section 5 presents conclusions and discusses future research directions.

2 Related Work

Many researchers have explored the use of Wikipedia components to extract semantics in a variety of ways. WikiRelates! [1] used Wikipedia category network to identify semantics. Eric Yeh et al. [2] constructed a Wikipedia graph with

P. Anthony, M. Ishizuka, and D. Lukose (Eds.): PRICAI 2012, LNAI 7458, pp. 861–865, 2012.
© Springer-Verlag Berlin Heidelberg 2012

Wikipedia articles as nodes and the link between them as the edges in the graph and used it as an effective source of computing semantic relatedness. Explicit Semantic Analysis(ESA) [3] compared two text fragments or words by projecting them into the space of Wikipedia articles and comparing their resultant concept vectors like the traditional Vector Space model. WLM [4] used Wikipedia links for calculating semantic relatedness and proposed a relatedness measure based on weighted average of *tfidf* and Normalized Google Distance (NGD) inspired measures.

3   WikiSim - A New Measure of Semantic Relatedness

"WikiSim" is our approach for computing Wikipedia based contextual relatedness of two terms. WikiSim has two phases. It first extracts candidate senses from Wikipedia's disambiguation pages for both input words, then it computes the semantic relatedness of candidate senses using a Wikipedia hyperlinks based relatedness measure.

3.1   Candidate Senses Extraction

To compute the relatedness of two terms, each of which may have several different senses , it is important to identify the appropriate sense of each word. "Crane" and "implement" are unrelated terms if we use their senses "the bird" and "the action of implementing", but are closely related in the senses "the machine" and "the tool".

   Wikipedia containing information about different senses of words, represents them in three ways. A term followed by the context in parenthesis e.g. Crane (machine); an alternative word, usually a synonym, hyponym or hypernym e.g. sense "tool" for the term "implement"; or a phrase e.g. "writing implements". WikiSim does not use the phrasal senses possibly because most of the phrasal senses are extremely specific e.g. ""Forest Township, Missaukee County, Michigan".

3.2   Contextual Relatedness Computation

Then, all unique in-links (all articles referring to the input word article) and out-links (all articles referred by the input word article) of each candidate sense of the input word w_a are extracted and compared with that of every sense of the input word w_b. Each sense pair is assigned a relatedness weight based on link sharing using the following formula:

$$w(s_i, s_j) = \left\lceil \frac{|S|}{|T|} \right\rceil \ if S \neq \{\o\} \tag{1}$$

In the above formula, s_i and s_j are senses of input words w_a and w_b respectively. $|S|$ is the set of all the links shared by a sense pair and $|T|$ represents total number of distinct in links and out-links of both senses. In other words, the weight of a sense pair is the probability of links shared by them, or 0 if S is an empty set. For the sense pairs which do not share any links, we calculated relatedness scores by computing weighted average of a link based Vector Space Model(VSM) inspired measure and the Normalized Google Distance(NGD) inspired measure based on the work of [4]. The highest weight is picked up as a measure of contextual relatedness with the associated senses representing disambiguated contexts of input words.

4 Evaluation

To compare our results with other existing approaches, we evaluated our work on three well known benchmark datasets namely Miller and Charles (M&C) [5], Rubenstein and Goodenough (R&G) [6] and WordSimilarity-353 [7] using Spearman's correlation coefficient.

Table 1. Correlation based Performance of WikiSim on two subsets of WordSim-353

Subset	Correlation (individual words)	Correlation (Phrase)	Matched Articles
Set-1	0.59	0.63	57
Set-2	0.55	0.67	81
Averaged Correlation	0.57	0.65	

The averaged correlation value of wikiSim on both subsets of WordSimilarity-353 turned out to be much low as represented by correlation with individual words in Table 2. On analyzing the dataset, we observed that most of word pairs in this dataset are those that make a new context such as "secret weapons" when occur together but are quite distinctive otherwise. For handling such word pairs, we concatenated both words in every word pair and searched for a corresponding Wikipedia article. Such word pair is assigned weight based on following formula:

$$w(Art_{ab}) = \left\lceil \frac{|I|}{|T|} \right\rceil \tag{2}$$

where, $|I|$ represents the total number of distinct in-links of the article Art_{ab} and $|T|$ represents the sum of both distinct in-links and out-links. This formula represents the extent of popularity of the phrase based article. Popularity refers to the number of references made to a target article. The higher the popularity of an article, the higher will be its significance in the Wikipedia corpus. The number of matched Wikipedia articles for all datasets shown in table 2 explain well the reason for shortfall in the correlation values of wikiSim on wordSim353. By considering word pair popularity, we improved the average correlation on WordSim-353 from 0.57 to 0.65. The correlation value of WikiSim, averaged on all three dataset is significantly better than Wikirelates and WLM but still falls behind ESA. One possible reason for that is, WikiSim maps input words to their corresponding Wikipdia articles which is a knowledge acquisition bottleneck of

Table 2. Correlation based Performance Comparison of WikiSim with other Wikipedia based approaches on Three datasets

Dataset	WikiRelate	ESA	WLM	WikiSim
M&C	0.45	0.73	0.70	0.74
R&G	0.52	0.82	0.64	0.75
WordSim-353	0.49	0.75	0.69	0.65
Average	0.48	0.76	0.67	0.71

our system just like WikiRelates and WLM approaches. Another reason is that WikiSim is based on the Wikipedia senses which are manually annotated so existence of insufficient or no senses corresponding to an input word could again be a knowledge acquisition bottleneck. Our approach is cheaper than ESA which preprocesses the whole Wikipedia dump. The difference between our approach and WLM [4] is that they have used manually disambiguated datasets for evaluation and have followed a different methodology for computing the contextual relatedness. Our approach is different from WikiRelates in that WikiRelates combined Wikipedia's category structure and some lexical measures to compute semantic relatedness.

5   Conclusions

In this paper we proposed and evaluated a novel approach for computing semantic relatedness using Wikipedia senses and hyperlink structure. Our approach presented a measure for computing semantic relatedness. We have shown that with an average Spearman's correlation coefficient of 0.71 on three datasets, our approach performs consistently better than many other well known Wikipedia based approaches. In future, we plan to optimize our relatedness measure by considering other semantic elements of Wikipedia.

References

1. Strube, M., Ponzetto, S.P.: Wikirelate! computing semantic relatedness using wikipedia. In: Proceedings of Association for the Advancement of Artificial Intelligence, AAAI (2006)
2. Yeh, E., Ramage, D., Manning, C.D., Agirre, E., Soroa, A.: Wikiwalk: random walks on wikipedia for semantic relatedness. In: 2009 Workshop on Graph-based Methods for Natural Language Processing, TextGraphs-4, pp. 41–49. Association for Computational Linguistics, Suntec, Singapore (2009)
3. Gabrilovich, E., Markovitch, S.: Computing semantic relatedness using wikipedia-based explicit semantic analysis. In: Proceedings of the 20th International Joint Conference on Artificial Intelligence, pp. 1606–1611 (2007)

4. Milne, D., Witten, I.H.: An effective, low-cost measure of semantic relatedness obtained from wikipedia links. In: Proceeding of AAAI Workshop on Wikipedia and Artificial Intelligence: an Evolving Synergy, pp. 25–30. AAAI Press (2008)
5. Miller, G.A., Charles, W.G.: Contextual correlates of semantic similarity. Language and Cognitive Processes 6(1), 1–28 (1991)
6. Rubenstein, H., Goodenough, J.B.: Contextual correlates of synonymy. Commun. ACM 8, 627–633 (1965)
7. Finkelstein, L., Gabrilovich, E., Matias, Y., Rivlin, E., Solan, Z., Wolfman, G., Ruppin, E.: Placing search in context: the concept revisited. ACM Trans. Inf. Syst. 20(1), 116–131 (2002)

Chinese Morphological Analysis
Using Morpheme and Character Features

Kanako Komiya[1], Haixia Hou[1], Kazutomo Shibahara[1,2], Koji Fujimoto[1,2],
and Yoshiyuki Kotani[1]

[1] Tokyo University of Agriculture and Thechnology, 2-24-16 Naka-cho, Koganei,
Tokyo, 184-8588 Japan
{kkomiya,kotani}@cc.tuat.ac.jp, hhxlovehhx@hotmail.com
http://www.tuat.ac.jp/~kotani/
[2] Tensor Consulting Co.Ltd, 2-10-1-306 Koujimachi Chiyoda-Ku, Tokyo,102-0083 Japan
{kazutomo.shibahara,koji.fujimoto}@tensor.co.jp
http://www.tensor.co.jp/

Abstract. One of the most important problems of the morphological analysis is
processing of unknown words. This paper proposes to use morpheme and character features to relieve the problem of the unknown words without decreasing
of the precision for the known words. We used the maximum entropy method
which is flexible to the information of the morphemes and the characters. The
experiments revealed that both the morpheme and character features are effective for Chinese morphological analysis and the character features are useful for
the processing of the unknown words.

Keywords: Chinese morphological analysis, morpheme features, character features, maximum entropy method.

1 Introduction

The morphological analysis is an important technique of the natural language
processing. It is particularly important in Chinese because it has no word boundaries.
The important problem of most of the morphological analyzers which use only morpheme features such as in [5] and [4] is processing of unknown words. In contrast, [3]
used character features to relieve this problem because Chinese characters are ideo-
grams but showed that the precision for known words tends to be low only with character features. Hence, we propose to use morpheme and character features to solve
these two problems simultaneously. The experiments revealed that both the morpheme and character features are effective for Chinese morphological analysis and the
character features are useful for the processing of the unknown words. The closest
work to ours is that by [2] who developed the word segmentation system that uses the
morpheme and character features, which focused on only the word segmentation, and
the morphological analysis is not accomplished so far.

P. Anthony, M. Ishizuka, and D. Lukose (Eds.): PRICAI 2012, LNAI 7458, pp. 866–869, 2012.
© Springer-Verlag Berlin Heidelberg 2012

2   The Morphological Analysis Using the Morpheme and Character Features

The morphological analysis consists of the following five steps:

1. Make a dictionary which lists words and their POSs from the training corpus,
2. The costs of the feature functions are learned,
3. When the sentence was input into the system, prepare a node set, i.e., the set of the candidate morphemes,
4. Link the candidate morphemes and construct a lattice. The candidates are the nodes in the lattice. Fig. 1 shows the lattice when the input is "我是学生。(I am a student.) ". Some nodes are omitted in this figure,
5. The optical path whose probability is the greatest in the lattice is determined using the ME model.

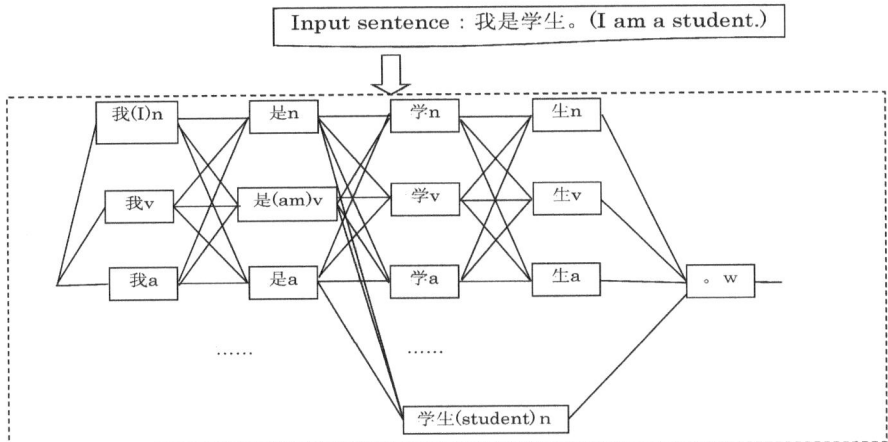

Fig. 1. The lattice when the input is "我是学生。 (I am a student.)"

3   The Morpheme and Character Features

We used the maximum entropy (ME) model of [1]. In the model, P(y|x), the probability that the POS of a morpheme x is y, is calculated.   The features are used as arguments of the feature functions, which are defined as follows:

$$f_{ijk}(x, y) = f(x) = \begin{cases} 1, & \text{if x have information } g_{ij} \text{ and y=k,} \\ 0, & \text{otherwise,} \end{cases} \quad (1)$$

where, x: a candidate morpheme (a character string), y: a POS of x, g_{ij}: the feature which is formation j of type i, i: an index of a type of features, j: an index of a feature of type i, and k: an index of a label. In this paper, two types of morpheme and one type of character feature are introduced. The index i represents these three types. Note that one feature will make 46 feature functions for our POS tag set includes 46 POSs.

A. The POS bigram of morphemes (47 features): The first type of the features is the POS of the previous morpheme of the morpheme which is focused on. We have 47 features of this type, because we have 46 POSs and one more feature: the beginning of a sentence. The combinations of the POSs and the morpheme which are not in the training corpus were also considered in this type. Index j represents the POS of the previous morpheme.

B. The bag-of-morphemes (57,760 features): We have 57,760 features because we have 57,760 morphemes (words) in our corpus. Index j represents the morpheme itself.

C. The position of a character in a candidate morpheme (12,977 features): The notation of [2] is used for this feature: S, B, I, and E for a character which constitutes a morpheme on its own, is positioned at the beginning of the morpheme, is positioned at the intermediate of the morpheme, and is positioned at the end of the morpheme, respectively. Index j represents the combination of the positions and the characters. We used only the correct combinations for the reason of the memory limitation.

4   Experiment

We carried out experiments using the PFR (Peita-Fujitsu-Renmin Ribao) People's. Daily POS Tagged Chinese Corpus. We randomly selected 1541 sentences from the corpus for the experiments. The data set includes 46,251 sentences (44,710 for the training and 1,541 for the test) and 1,083,411 morphemes (1,048,121 for the training and 35,290 for the test). To address the unknown word problem, our system consider not only the character strings in a dictionary but also the character strings which are not. The system did not consider the morphemes whose length is more than five characters except those which are already in the dictionary because only 0.62% of the training corpus were these morphemes. Marks like periods were treated as a morpheme to speed up the system. The open and closed tests were carried out three times. The 300 sentences were randomly selected and used for the test. The averaged number of the morphemes of the test data of the closed and open tests are 7,025 and 6,807, respectively.

5   Results and Discussion

Table 1 lists the average precisions and recalls of the closed and open tests. These results revealed that the character features are not effective in the closed tests, which have no unknown words, but are obviously effective in both precision and recall in the open tests, which have many unknown words. These results indicate that both the morpheme and character features are effective for Chinese morphological analysis and the character features are useful for the processing of the unknown words.

The following example shows that the character features are useful for the processing of the unknown words. The experiments also revealed that some characters tended to be positioned in the fixed position. We think these features are useful.

Table 1. The average precisions and recalls of the closed and open tests

Features	The closed test		The open test	
	Precision	Recall	Precision	Recall
Morphemes	96.10%	95.40%	83.71%	89.20%
Morphemes and characters	**96.10%**	**95.90%**	**90.31%**	**93.20%**
Deference	**0.0**	**0.5**	**6.6**	**4.0**

Adjectives (A) 细致 (careful) and (B) 圆润 (mellow) were correctly processed only in our system because the system could use the two features of the position of a character in a candidate morpheme for each: (1A) 细 (thin) and (1B) 圆 (round) at the beginning of a sentence and (2A) 致 (detailed) and (2B) 润 (smooth) at the end of a sentence. This is because although these morpheme were not in a dictionary nor training data but other adjectives which consist of the same characters such as 细心(careful), 细微 (slight), 细嫩(tender), 别致(unique), 雅致(elegant) , 圆满 (perfect), 圆浑 (mellow), 滋润 (moist), 红润 (rosy), and 湿润 (moist) were there.

6   Conclusion and Future Work

This paper proposed to use morpheme and character features to relieve the problem of the unknown words without decreasing of the performance of the known words. We used the ME method using the features about the bigram of POSs, the bag-of-morphemes, and the position of the characters in a candidate morpheme. We got the performances that exceed the baseline system which uses only the morpheme features especially in the open tests and these results revealed that both the morpheme and character features are effective for Chinese morphological analysis and the character features are useful for the processing of the unknown words. We think that the categories of the character can be used as the character features in the future.

References

1. Berger, A.L., Pietra, V.J.D., Pietra, S.A.D.: A maximum entropy approach to natural language processing. Computational Linguistics 22(1), 39–71 (1996)
2. Nakagawa, T.: Chinese and japanese word segmentation using word-level and character-level information. In: Proceedings of the 20th International Conference on Computational Linguistics (COLING 2004), pp. 466–472 (2004)
3. Oda, H., Mori, S., Kita, K.: A japanese word segmenter by a character class model. Journal of Natural Language Processing 6(7), 93–108 (1999)
4. Takeuchi, K., Matsumoto, Y.: Hmm parameter learning for japanese morphological analyzer. In: Proceedings of the 10th Pacific Asia Conference on Language Information and Computation (PACLIC 10), pp. 163–172 (1995)
5. Uchimoto, K., Sekine, S., Isahara, H.: The unknown word problem: a morphological analysis of japanese using maximum entropy aided by a dictionary. In: Proceedings of the 2001 Conference on Empirical Methods in Natural Language Processing, pp. 91–99 (2001)

An Improved Recommender Based on Hidden Markov Model

Jialing Li, Li Li*, Yuheng Wu, and Shangxiong Chen

Institute of Logic and Intelligence
Faculty of Computer and Information Science
Southwest University, Chongqing 400715, China

Abstract. In reality, users rarely dedicate excessive interest into only one topic over a long time. We propose a topic-based hidden Markov model to analyze temporal dynamics of users' preference. Experiments show that given observations of a new entrant, the proposed model is able to recommend a specific user group he/she can be classified into and also can anticipate what topic he/she will be mostly interested in.

1 Introduction

The key point of recommendation is to capture users' preference as precisely as possible [1]. Xu et al. [2] model user preferences on various topics in a topic oriented graph, and devise a topic-oriented recommendation system by preference propagation. Experiments manifest that their approach outperforms several collaborative filtering methods. It, however, does not take users' time information and relevant user information into account. Daud's work [3] clearly shows that users' interests and relationships are changing over times. Experiments show the effectiveness of the proposed approach. Based on observations and the existing works, we thus make an assumption that a user's preference is influenced by users who have the same tastes, the topic of resources and time information.

The contributions of the paper are: 1) Similar users are grouped by K-means and topic-based classification; 2) Temporal dynamics are considered when dealing with users' preference. Topic-based hidden Markov model is introduced to learn user's preference with changing circumstances.

2 Problem Definition

Topics of resources are $C = \{c_1, c_2, ..., c_k\}$. The topic sequence with time information is denoted as $c_{t_1}{}^{t_1} c_{t_2}{}^{t_2} \cdots c_{t_n}{}^{t_n}$. Firstly users are classified into different classes for two reasons: 1) The possibility of the sparsity of data can be eliminated; 2) Grouping users sharing the same rating patterns into the same class can simplify the task. Secondly, the collection of topic sequences of classes of users are used to train hidden Markov model for each class of user.

---

* Corresponding author.

P. Anthony, M. Ishizuka, and D. Lukose (Eds.): PRICAI 2012, LNAI 7458, pp. 870–874, 2012.

Usually people are mostly like to change their focuses among the general topic categories of the resources periodically for reasons like varying social trends, different thoughts, and a diversity of sentiment and moods, but they are concealed from us and can only be disclosed by sophisticated techniques, say HMM. Thus we assume hidden factors affecting users' moods are $H = (h_1, h_2, ..., h_6)$.

3   Topic-Based Hidden Markov Model

Topic-based HMM is also defined as $\lambda = (Q, O, \pi, A, B)$, where $Q = (q_1, q_2, ..., q_T)$ stands for the finite set of states, where each finite state represents a corresponding hidden state in H; $O = (o_1, o_2, ..., o_T)$ represents the finite set of observations, where each observation state represents a corresponding topic in C; π is the vector of initial probability distribution on Q; A is a stochastic transition matrix, where $A : Q \times Q \to [0, 1]$, $A_{ij} = P(q_{t+1} = j | q_t = i)$ and $\sum_{j \in Q} A_{ij} = 1$; and B is the confusion matrix which describes the transition between states and observations, $B : Q \times O \to [0, 1]$, with $B_{ik} = P(O_t = k | q_t = i)$, $\sum_{k \in O} B_{ik} = 1$. The state sequence $q = q_1 q_2 ... q_t$ satisfies the Markov property. The topic-based HMM can be used to solve 3 kinds of problem:

- Estimation Problem. The probability of a sequence $o = o_1 o_2 ... o_n$ given a HMM λ is given by the Eq. (1) which is calculated by the forward-backward algorithm.

$$p(o|\lambda) = \sum_{q_1, ..., q_n \in Q} \pi(q_1) \cdot B_{q_1 o_1} \cdot A_{q_1 q_2} \cdot B_{q_2 o_2} \cdots A_{q_{n-1} q_n} \cdot B_{q_n o_n} \qquad (1)$$

- Decoding Problem. The problem is defined in Eq. (2) as follows:

$$\delta_t(i) = p(q_1 q_2 ... q_{t-1}, q_t = i, o_1 o_2 \cdots o_t | \lambda) \qquad (2)$$

Given a sequence $o = o_1 o_2 ... o_T$ and $\lambda = (\pi, A, B)$, how to find the state sequence $q = q_1 q_2 ... q_N$ which is represented by $\delta_T(i)$ at time T and satisfy the max likelihood which is calculated by Viterbi and represented as $\arg\max_Q P(Q|O)$.

- Parameter Learning Problem. The problem of HMMs to find the model λ which maximises the probability in Eq. (2) is solved by Baum-Welch algorithm.

4   Experiments and Evaluations

The HMM toolbox we use was written by Kevin Murphy. Our proposed method is evaluated in two datasets, one is MovieLens[1] (M for short), topics in MovieLens are initialized into 18 kinds. Another one is Jester Joker (J for short) [5].

Firstly, we use K-means to cluster users in each dataset into 3 classes, respectively. Then we split the data in each class for two datasets into two parts: the training data and test data with a proportion of 80% and 20%.

---

[1] http://movielens.umn.edu

The first evaluation is depicted as follows. A HMM will be trained for each user group. Namely, there will be three HMMs (refer as HMM1, HMM2, and HMM3) for each dataset. After training, the results expressed as Eq. (1) are shown in "log" form in Fig. 1. The first three figures in Fig. 1 (Fig.1(a)-(c)) present log likelihood comparison among three trained HMMs in MovieLens, while the rest of figures (Fig.1(d)-(f)) present comparisons in Jester Joker.

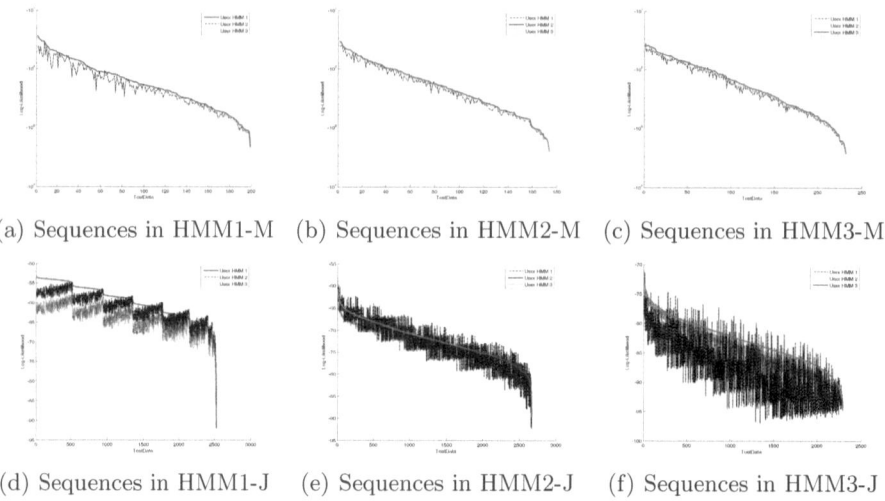

(a) Sequences in HMM1-M   (b) Sequences in HMM2-M   (c) Sequences in HMM3-M

(d) Sequences in HMM1-J   (e) Sequences in HMM2-J   (f) Sequences in HMM3-J

Fig. 1. Log Likelihoods Comparison of Different HMMs

In Fig. 1, the horizontal axis presents each test sequence of one user group, the vertical axis presents log likelihood values. In each figure, the red solid legend works as baseline, and denotes the results of a given test sequence in the corresponding HMM. The other two black line legends denote likelihood value of the given test sequence from other two HMMs.

In Fig. 1, results in each figure are analyzed against the value of the red solid line in descending order. The bigger the log values the more likely this particular user belonging to the specific (user group) HMM . It is clear that the red solid lines are higher than the other two black lines in most cases. This manifests that the proposed model can identify which user class these sequences come from. Namely, given the observation sequence of the user, the trained HMM can be used to identify which category the user belongs to. Furthermore, useful recommendations can also be obtained from the results.

Secondly, we use the proposed model to predict what will be the most interesting topic. Firstly, the last topic word within each test sequence of each user group is taken away intentionally. Then the pre-defined 18 topics are used to replace the removed topic word one by one, which incurs 17 more repetition in total. In doing so, we want to find out the interesting or suitable topic within the given HMM. Finally we calculate the log likelihood of all the replacements in

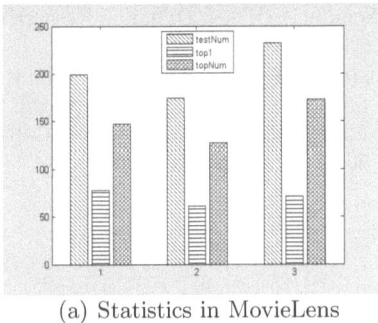

(a) Statistics in MovieLens

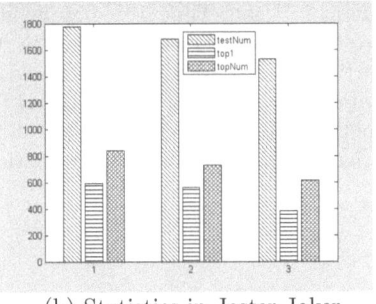

(b) Statistics in Jester Joker

Fig. 2. Log Likelihood Sequences Comparison of 18 Topics

the corresponding HMMs. We then count how many times that the biggest log likelihood sequences are in accordance with the real sequence we have observed.

In Fig. 2(a), the horizontal axis presents three HMMs with MovieLens dataset (Fig. 2(a)), the vertical axis presents the counter. The meanings of the legends are as follows: "testNum" denotes the total number of test sequences in the corresponding model before substitution; "top1" is the number of events happened that the sequence we really observed gained the biggest log likelihood; "topNum" is the number of times that the sequence we really observed has achieved beyond third biggest log likelihood value. Fig. 2(b) is obtained from Jester Joker dataset.

In Fig. 2, when the real observed sequence achieved the biggest log likelihood, it means that the trained HMM can generate exactly the same topic sequence as observed. Namely, the proposed model is able to recommend the user with the topic generated by the corresponding user group HMM. A particular user's next most likely interesting topic can be foretold with our method to some degree.

5   Conclusion

We propose a new collaborative filtering method using hidden Markov models which can be used to learn an appointed user's rating behavior pattern and predict a user's preference trends with certain probability as well.

Acknowledgements. This work was partially supported by the Research Fund (09SKB34), National Natural Science Foundations of China(61170192), and Natural Science Foundations of CQ(2007BB2372).

References

1. Burke, R.: Hybrid Recommender Systems: Survey and Experiments. User Modeling and User-Adapted Interaction 12, 331–370 (2002)
2. Xu, G., Gu, Y., Zhang, Y., Yang, Z., Kitsuregawa, M.: TOAST: A Topic-Oriented Tag-Based Recommender System. In: Bouguettaya, A., Hauswirth, M., Liu, L. (eds.) WISE 2011. LNCS, vol. 6997, pp. 158–171. Springer, Heidelberg (2011)

3. Daud, A.: Using Time Topic Modeling for Semantics-Based Dynamic Research Interest Finding. Knowledge Based Systems 26 (2010)
4. Hartigan, J.A., Wong, M.A.: A K-means Clustering Algorithm. Journal of the Royal Statistical Society 28(1), 100–108 (1979)
5. Goldberg, K., Roeder, T., Gupta, D., et al.: Eigentaste: A Constant Time Collaborative Filtering Algorithm. Information Retrieval 4(2), 133–151 (2001)

Event Recognition in Parking Lot Surveillance System

Lih Lin Ng and Hong Siang Chua[*]

School of Engineering, Computing and Science,
Swinburne University of Technology (Sarawak Campus), Malaysia
{lng,hschua}@swinburne.edu.my

Abstract. This paper presents a novel event recognition framework in video surveillance system, particularly for parking lot environment. An event is represented by feature vector that contains dynamic information and the contextual information of the motion trajectory is incorporated into the recognition process. Experimental results have demonstrated great accuracy of the proposed event recognition algorithm.

Keywords: Video surveillance, event recognition, anomaly detection.

1 Introduction

Automatic detection of event is becoming increasingly important for video surveillance applications. In this paper, a novel event recognition and classification technique in parking lot surveillance system is presented by trajectory learning and analysis. The proposed system is adapted from the semi-supervised method where a set of 'normal' events are trained from the known data set that are extracted from the object motion trajectories such as the time and location when the object enters/exits the scene, and the average size of the object. The remaining sections of this paper are organized as follows. Section 2 describes the system architecture of the proposed event recognition and anomaly detection algorithms and the experimental results from the real video sequences taken from the parking lot are presented in section 3.

2 System Architecture

The proposed video surveillance system has four main processes: object tracking, event representation, training and event recognition as is shown in Fig. 1. The object tracking process detects the moving object and collects the spatial-temporal information of the tracked object. Event representation process constructs the feature vector from the motion trajectory. In training process, feature vectors of the known events are extracted and low-dimensional representation of the known events are computed. Lastly, the event recognition process identifies the type of event.

---

[*] Corresponding author.

P. Anthony, M. Ishizuka, and D. Lukose (Eds.): PRICAI 2012, LNAI 7458, pp. 875–878, 2012.
© Springer-Verlag Berlin Heidelberg 2012

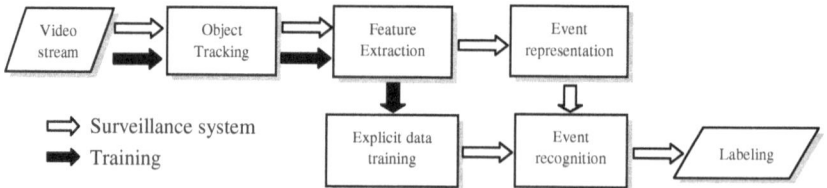

Fig. 1. Event classification block diagram

2.1   Object Tracking and Feature Extraction

Adaptive Gaussians Mixture Model (GMM) is employed in our object tracking process [1]. Each pixel is modeled by a mixture of five weighted Gaussian distributions. The advantage of this approach is the robustness to lighting changes and unstructured motion such as vegetation waving. Also, it can handle local and global change of the background appearance. Median filtering and morphological operations are employed to improve the accuracy of the detected object trajectory.

An event e is defined over a temporal interval $[t_s, t_f]$ where t_s is the time when object is detected and t_f is the time when the object exits the scene. The event e over a sequence of N consecutive frames is represented by:

$$e(t_s, t_f) = \{f_t | f_t \in O_i, t \in [1 \dots N]\} \tag{1}$$

where $f_t = \{x_t, y_t, v_t, A_t\}$, which v_t and A_t are the velocity and area profiles of the moving object. As a consequence of time-varying tracks of different event, trajectories are often unequal in dimension. Thus, normalization and scaling operations are applied to obtain a standardized representation of the event e.

2.2   Event Representation and Explicit Data Training

Numerous efforts have been devoted to trajectory analysis by learning semantic scene [2-3].The type of event can be estimated from the scene information of the trajectory such as the starting and ending points. For example, a vehicle passes by the scene will produce a trajectory with a starting point at the entry region and an ending point at the exit region. Also, the type of the object that initiates the event (vehicle or human) can be determined by the area information of the object. Therefore, a logical table of these events can be formed as shown in Table 1. The spatial information was also extracted from the object trajectory and an event tag $T_e = \{X_s, X_f, \bar{A}, e\}$ is formed, where Xs and X_f are the starting and ending centroid location, $\bar{A}$ is the mean object area.

Explicit definitions of 'normal' events are trained with labelled video samples. To obtain a compact and low-dimensional representation of these events, principal component analysis (PCA) [4] is performed on the extracted event vector e. Fig.2 shows the trained feature vector computed from the event *vehicle passes by the scene* and their principal component representation of the feature vector.

Table 1. Logical scene information for different events

Event	Description	Starting	Ending	Area
1	Vehicle passes by the scene	entrance	exit	Large
2	Vehicle enters the scene and occupies a parking space	entrance	internal	Large
3	Vehicle moves out of parking space and exits the scene	internal	exit	Large
4	A person walks through the scene	entrance	exit	Small
5	A person enters the scene and picks up a vehicle	entrance	internal	Small
6	A person drops off from vehicle and exits the scene	internal	exit	Small

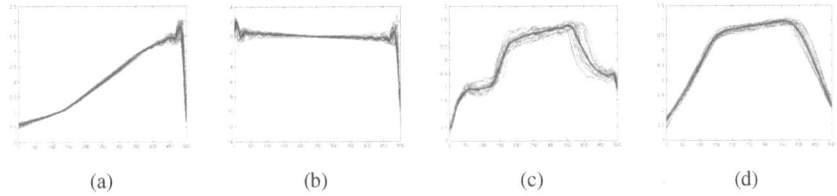

| (a) | (b) | (c) | (d) |

Fig. 2. Principal component representations of the (a) x axis projection, (b) y axis projection, (c) velocity, and (d) area of the trajectory for event *'vehicle passes by the scene'*

2.3   Event Recognition

The event recognition is accomplished by evaluating the event tag T_e. Feature matching and contextual information analysis are incorporated to improve accuracy of the recognition outcome. The explicit definitions are used as a map to classify the incoming event. Fig. 3 shows the decision tree of the event recognition process. Firstly, the type of object is identified by computing the mean area $\bar{A}$ of the object. Next, entry and exit points of the trajectory are analyzed and the possible type of event is estimated. The final decision is made by evaluating the similarity of the feature vector f_t of the incoming event to the explicit definition of the same event.

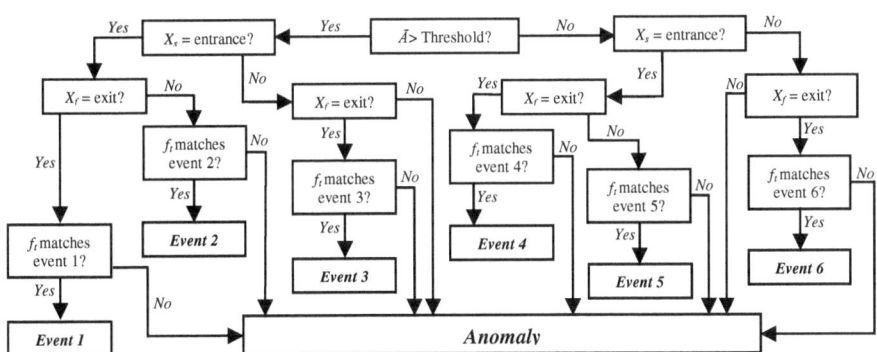

Fig. 3. Decision tree of the event recognition process

3   Experimental Results

The test videos were captured by using a static camera that monitoring the outdoor parking lot of Swinburne University. A sample video of more than 1000 frames and contains 4 events is illustrated in Fig. 4 and the experiment outcome of the proposed event recognition system is tabulated in Table 2.

Fig. 4. Object tracking result of the test video. From left to right, car passed by the scene, people walked by the scene, people picked up a car, and car moved out of the parking space.

Table 2. Results of event recognition from the test video

Event no	Start Frame	End Frame	X_s	X_f	$\bar{A}$	Event e	Corr.
#1	53	82	entrance	exit	car	1	0.9894
#2	124	246	entrance	exit	people	4	0.9038
#3	372	563	entrance	internal	people	5	0.9616
#4	679	824	internal	exit	car	3	0.7650

4   Conclusions

An event recognition system specifically in parking lot environment is proposed based on the spatial and dynamic information of the trajectory. Contextual informa-tion of the trajectory was incorporated to enhance the accuracy of the outcome. The experiments show the events were recognized accurately using the datasets collected from parking lot surveillance scenarios.

References

1. Stauffer, C., Grimson, W.E.L.: Adaptive background mixture models for real-time tracking. In: Computer Vision and Pattern Recognition (CVPR 1999), Colordo Springs (June 1999)
2. Morris, B.T., Trivedi, M.M.: Learning and Classification of Trajectories in Dynamic Scenes: A General Framework for Live Video Analysis. In: IEEE Fifth International Confe-rence on Advanced Video and Signal Based Surveillance
3. Makris, D., Ellis, T.: Learning semantic scene models from observing activity in visual sur-veillance. IEEE Transactions on Systems, Man, and Cybernetics, Part B: Cybernetics 35(3), 397–408 (2005)
4. Bashir, F.I., Khokhar, A.A., Schonfeld, D.: Object Trajectory-Based Activity Classification and Recognition Using Hidden Markov Models. IEEE Transactions on Image Processing 16(7), 1912–1919 (2007)

Heuristics- and Statistics-Based Wikification

Hien T. Nguyen[1], Tru H. Cao[2], Trong T. Nguyen[2], and Thuy-Linh Vo-Thi[2]

[1] Ton Duc Thang University, Vietnam
`hien@tdt.edu.vn`
[2] Ho Chi Minh City University of Technology and John von Neumann Institute, Vietnam
`tru@cse.hcmut.edu.vn`

Abstract. With the wide usage of Wikipedia in research and applications, disambiguation of concepts and entities to Wikipedia is an essential component in natural language processing. This paper addresses the task of identifying and linking specific words or phrases in a text to their referents described by Wikipedia articles. In this work, we propose a method that combines some heuristics with a statistical model for disambiguation. The method exploits disambiguated entities to disambiguate the others in an incremental process. Experiments are conducted to evaluate and show the advantages of the proposed method.

1 Introduction

Recently, precisely identifying and linking specific words or phrases (*mention*s or *surface form*s henceforth) in a text to their referent entities in a knowledge base (KB) have attracted much research attention. This task was well-known as entity linking [1] or entity disambiguation (ED) [5]. When the used KB is Wikipedia, the task is also known as wikification [3]. ED is a crucial task in natural language processing (NLP). It annotates/maps mentions of entities or concepts in a text with/to identifiers (IDs) from a given KB. In this paper, we propose a method for ED using Wikipedia.

For the past five years, many approaches have been proposed for ED [2], [3], etc. In addition, since 2009, entity linking (EL) shared task held at Text Analysis Conference [1] has attracted more and more attentions in linking entity mentions to KB entries. In EL task, given a query containing a named entity (NE) and a background document including that named entity, the system is required to provide the ID of the KB entry to which the name refers; or NIL if there is no such KB entry [1]; the used KB is Wikipedia. Even though those approaches to EL exploited diverse features and employed many learning models [1], [4], [5], a hybrid approach that combines heuristics and statistics has not been explored.

In this paper, we propose a disambiguation approach that extends the one proposed in [9] to deal with both types of entity, namely, NEs and general concepts. The approach combines heuristics and statistics for entity disambiguation in an incremental process. For statistical ranking of candidate entities, we investigate various features proposed in [8] and adapt them for both named entity and general concept mentions, and explore more information of disambiguated entities to disambiguate the others.

P. Anthony, M. Ishizuka, and D. Lukose (Eds.): PRICAI 2012, LNAI 7458, pp. 879–882, 2012.

2   Proposed Method

Firstly, we built a controlled vocabulary based on Wikipedia. For identifying entity mentions, we employed a NE recognition tagger to identify NEs and a noun phrase chunker to identify base noun phrases, each of which is considered as an entity mention. The entity mentions were then used to retrieve their candidates based on the controlled vocabulary. To date, there are two popular ways used in previous work to generate candidates of mentions. The first one is based on outgoing links listed in disambiguation pages of Wikipedia; and the second one makes use of labels of outgoing links in Wikipedia articles and destinations of these links. However, they both produce incorrect results. To reduce the number of incorrect candidates of a mention, we employ both of them and get the intersection of the result sets as the final one.

Next, heuristics were applied to disambiguate mentions. We propose two main heuristics, named H_1 and H_2. These heuristics exploit title-hints of articles in Wikipedia. Note that each article in Wikipedia has a title, belongs to some categories, and may have several *redirect* pages whose titles are alternative names of the described entity or concept in the article; and, different articles have different titles. If an entity has the same surface form with others, the title of the article describing it may contain further information that we call *title-hint* to distinguish the described entity from the others. The title-hint is separated from the surface form by parentheses, e.g. "John McCarthy (computer scientist)", or a comma, e.g. "Columbia, South Carolina". Let m be the mention to be disambiguated. These two heuristics are stated as follows:

- H_1: among candidate entities of m, the ones whose title-hints occur around m in a context window are chosen. For instance, given the sentence "A state of emergency has been declared in the US state of Georgia after two people died in storms, a day after a tornado hit the city of Atlanta," for the mention "Atlanta", the candidate entity having the title "Atlanta, Georgia" is chosen because its title-hint "Georgia" occurs around the mention; and for the mention "Georgia" the candidate entity having the title "Georgia (U.S state)" is chosen because its title-hint "US state" occurs around it.
- H_2: if m is a title-hint of an already disambiguated entity around it, the chosen candidate entities are the ones that have outgoing links to the disambiguated entity or this disambiguated entity has outgoing links to these candidates. For instance, given the phrase "Atlanta, Georgia", after applying H_1, the mention "Atlanta" is annotated with the title "Atlanta, Georgia" in Wikipedia; for the mention "Georgia", after applying H_2, it is annotated with the title "Georgia (U.S state)" in Wikipedia because both Wikipedia articles "Atlanta, Georgia" and "Georgia (U.S state)" have reciprocal links to each other.

Finally, we employed the vector space model where entities are represented by feature vectors and cosine similarity is used to rank candidates. Features are divided into two groups: text features (representing mentions in a text) and Wikipedia features (representing Wikipedia articles). For each article, the Wikipedia features contain the entity title (ET), titles of its redirect pages (RT), its category labels (CAT), its labels of outgoing links (OL).

The text features contain the following types: (i) all *entity mentions* (EM); (ii) *local words* (LW), that are all the words, not including special tokens such as $, #, ?, etc., found inside a specified context window around the mention to be disambiguated and are not part of mentions occurring in the window context to avoid duplicate features; (iii) *coreferential words* (CW), used exclusively to represent NEs, that are all local words of the mentions co-referent with the mention to be disambiguated; (iv) all identifiers of the entities whose mentions have already been annotated (*IDs*); and (v) *extended text* (ExT): let *m* be a mention to be disambiguated, U_1 be a set of titles corresponding to already disambiguated entities, and U_2 be a set of destinations of outgoing links in a certain candidate, *Ext* is a set (of titles) that is the intersection of U_1 and U_2.

After extracting features for a mention in a text or an entity described in Wikipedia, we put them into a bag-of-words. Then we normalize the bag of words by removing special characters, punctuation marks, and special tokens; removing stop words such as *a*, *an*, *the*, etc.; and stemming words using Porter stemming algorithm. Then each bag-of-words is converted into a feature vector.

3   Evaluation

For evaluating the performance of our disambiguation method, we have built two corpora in which entities are manually annotated with Wikipedia titles using the English version of Wikipedia downloaded on 22 July 2011. The first corpus D_1 contains 2038 mentions that are NEs and the second corpus D_2 contains 3898 mentions of both general concepts and NEs. We evaluate our methods against the methods of Milne and Witten [6] (denoted as M&W) and Ratinov *et al.* [3] (denoted as Wikifier) using a previously-employed "bag of titles" (BOT) evaluation presented in [3]. To this end, we run the disambiguation algorithms of M&W[1] and Wikifier[2] on our datasets respectively and utilize BOT evaluation implemented in Wikifier.

Table 1. The overall results on datasets after running M&W, Wikifier, ours Method 1&2

Method	Datasets	Precision	Recall	F-Measure
M&W	D_1	86.74%	27.95%	42.28%
	D_2	86.75%	36.65%	51.53%
Wikifier	D_1	83.72%	28.08%	42.06%
	D_2	83.91%	36.02%	50.41%
Method 1	D_1	84.21%	74.88%	79.27%
	D_2	64.35%	57.02%	60.47%
Method 2	D_1	73.28%	68.79%	70.97%
	D_2	58.54%	60.46%	59.48%

We combine the features by two ways to form two methods. The first one (henceforth *Method 1*) uses the features EM, LW, CW (only for NEs), and IDs of disambiguated entities to represent mentions. The second one (henceforth *Method 2*) uses the features EM in windows whose sizes are 55 tokens around the mention to be

---

[1] Available at http://wikipedia-miner.sourceforge.net/
[2] Available at http://cogcomp.cs.illinois.edu/page/download_view/Wikifier

disambiguated, LW, CW (only for NEs), IDs of disambiguated entities, and Ext to represent mentions. Both methods represent entities described in Wikipedia using the features presented above.

Table 1 shows the results on datasets D_1 and D_2 after running the disambiguation algorithm of M&W, the disambiguation algorithm of Wikifier, and our Methods 1&2. The results on the table show that our methods achieve lower precision but higher recall as compared to M&W and Wikifier. However, they have a balance of the precision and recall measures, and achieve the F-measure higher than those of M&W and Wikifier on the same datasets.

4   Conclusion

In this paper, we propose a new method for Wikification of both named entities and general concepts. Our method combines some heuristics and a statistical model for Wikification in an incremental process where already disambiguated entities are exploited to disambiguate the remaining ones. We have explored diverse features from local to global ones in a text and Wikipedia for entity disambiguation. The experimental results show that the proposed method achieves comparable performance, with balanced precision and recall, to the state-of-the-art methods.

References

[1] Ji, H., Grishman, R., Dang, H.T.: An Overview of the TAC 2011 Knowledge Base Population Track. In: Proc. of Text Analysis Conference (2011)
[2] Han, X., Sun, L., Zhao, J.: Collective Entity Linking in Web Text: A Graph-Based Method. In: Proc. of SIGIR 2011, pp. 765–774 (2011)
[3] Ratinov, L., Roth, D., Downey, D., Anderson, M.: Local and Global Algorithms for Disambiguation to Wikipedia. In: Proc. of ACL-HLT 2011 (2011)
[4] Zhang, W., Su, J., Tan, C.-L., Wang, W.: Entity Linking Leveraging Automatically Generated Annotation. In: Proc. of COLING 2012 (2010)
[5] Dredze, M., McNamee, P., Rao, D., Gerber, A., Finin, T.: Entity Disambiguation for Knowledge Base Population. In: Proc. of COLING 2010 (2010)
[6] Milne, D., Witten, I.H.: Learning to Link with Wikipedia. In: Proc. of the 17th ACM CIKM, pp. 509–518 (2008)
[7] Barker, K., Cornacchia, N.: Using noun phrase heads to extract document keyphrases. In: Proc. of the 13th Biennial Conf. of the Canadian Society on Computational Studies of Intelligence, pp. 40–52 (2000)
[8] Nguyen, H.T., Cao, T.H.: Exploring Wikipedia and Text Features for Named Entity Disambiguation. In: Nguyen, N.T., Le, M.T., Świątek, J. (eds.) ACIIDS 2010, Part II. LNCS (LNAI), vol. 5991, pp. 11–20. Springer, Heidelberg (2010)
[9] Nguyen, H.T., Cao, T.H.: Named Entity Disambiguation: A Hybrid Statistical and Rule-Based Incremental Approach. In: Domingue, J., Anutariya, C. (eds.) ASWC 2008. LNCS, vol. 5367, pp. 420–433. Springer, Heidelberg (2008)

Identification and Visualisation of Pattern Migrations in Big Network Data

Puteri N.E. Nohuddin[1,4], Frans Coenen[1], Rob Christley[2], and Wataru Sunayama[3]

[1] Department of Computer Science, University of Liverpool, UK
{puteri,frans}@liverpool.ac.uk
[2] School of Veterinary Science, University of Liverpool and National Centre for Zoonosis Research, Leahurst, Neston, UK
robc@liverpool.ac.uk
[3] Graduate School of Information Sciences, Hiroshima City University, Japan
sunayama@sys.info.hiroshima-cu.ac.jp
[4] Universiti Pertahanan Nasional Malaysia, Kuala Lumpur, Malaysia
puteri@upnm.edu.my

Abstract. In this paper, we described a technique for identifying and presenting frequent pattern migrations in temporal network data. The migrations are identified using the concept of a Migration Matrix and presented using a visualisation tool. The technique has been built into the Pattern Migration Identification and Visualisation (PMIV) framework which is designed to operate using trend clusters which have been extracted from "big" network data using a Self Organising Map technique.

Keywords: Frequent Patterns, Trend Analysis, Pattern Clustering, Visualisation, Self Organising Maps.

1   Introduction

A Pattern Migration Identification and Visualisation (PMIV) framework is proposed directed at detecting and visualising changes in frequent pattern trends in social network data. Given a set of frequent pattern trends collected over a sequence of temporal episodes we can group the global set of trends into a set of clusters. In this paper, the trends are clustered using a Self Organising Map (SOM). Each cluster (SOM node) will contain frequent patterns that have similar trends associated with them. We refer to such clusters as *trend clusters*. Furthermore, we can detect changes in temporal trend clusters. More specifically the changes in which particular frequent patterns associated with one type of trend may "migrate" to a different type of trend over a sequence of episodes. In addition we can identify communities of trend clusters that are connected with one another in terms of pattern migrations. To aid the proposed trend migration analysis a visualisation mechanism is also supported. The rest of the paper is organised as follows: Section 2 presents an overview of the proposed PMIV framework, Section 3 presents a demonstration of the operation of PMIV and the paper is concluded in Section 4.

P. Anthony, M. Ishizuka, and D. Lukose (Eds.): PRICAI 2012, LNAI 7458, pp. 883–886, 2012.
© Springer-Verlag Berlin Heidelberg 2012

2   The Pattern Migration Identification and Visualisation (PMIV) Framework

The PMIV framework is directed at finding interesting pattern migrations between trend clusters and trend changes in social network data. In the context of this paper, trends are trend lines representing frequency counts associated with binary valued frequent patterns discovered using an appropriate frequent pattern trend mining algorithm (for example the TM-TFP algorithm [1, 2]). A trend is considered to be interesting if its "shape" changes significantly between episodes. The application of the proposed PMIV framework requires as input, a SOM map [3, 4], where each node represents a trend cluster.

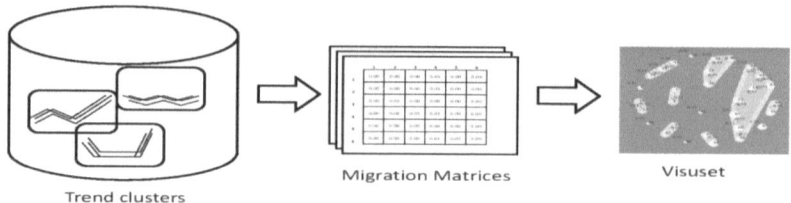

Trend clusters   Migration Matrices   Visuset

Fig. 1. Schematic illustrating the operation of the proposed framework

Figure 1 gives a schematic of the PMIV framework. The input to the PMIV framework is a set of e SOM maps describing a sequence of Trend Cluster Configurations (TC) such that $TC = \{SOM_1, SOM_2, ..., SOM_e\}$ where e is the number of episodes. Each SOM then comprises m nodes describing m different trend clusters, $T_i = \{\tau_1, \tau_2, ..., \tau_m\}$ ($0 < i \leq e$). Each trend cluster comprises the identifiers for zero, one or more frequent patterns. Note the number of trends clusters (nodes), m, in each SOM is constant across the episodes, this is why it may be the case that for some applications a trend cluster (SOM node) τ_k is empty.

The first stage in the PMIV process is to detect pattern migrations in the set of trend cluster configurations, TC, between sequential pairs of episodes E_j and E_{j+1}. This is achieved by generating a sequence of migration matrices for each SOM pair. A Migration Matrix (MM) is a two dimensional grid with $m \times m$ rows and columns (recall that m is the number of nodes, trend clusters, in a SOM map, and that the number of nodes is constant over an entire sequence of episodes). The rows represent the trend clusters associated with some episode E_j (SOM$_j$) and the columns the trend clusters associated with the next episode in the sequence E_{j+1} (SOM$_{j+1}$). Each cell in the MM holds a count of the number of frequent pattern trends that have moved from trend cluster τ_r (represented by row r) in episode E_j to trend cluster τ_c (represented by column c) in episode E_{j+1}. The leading diagonal therefore represents the number of frequent pattern trends that have stayed in the same cluster. The resulting migration matrix determines the number of migrations from one type of trend (trend cluster) to another, this information is then used in the visualization process.

The second stage is to illustrate the identified pattern migrations using Visuset [5]. The aim of the visualisation of pattern migrations is to produce a "pattern migration maps" using the Visuset software [5], which used the concept of a *Spring Model* [6]. Using this model the map to be depicted is conceptualised in terms of a physical system where the links represent springs and the nodes objects connected by the springs (using spring values). In the context of the PMIV framework, network nodes are represented by the trend clusters in SOM E and the spring value defined in terms of a *correlation coefficient (C)*. Visuset is also used to identify "communities" within networks using the Newman method [7]. The process starts with a number of clusters equivalent to the number of nodes. Iteratively, the trend clusters (nodes) with the greatest "similarity" are then combined to form a number of merged clusters. Best similarity is defined in terms of the *Q-value*. The C and Q-values are determined using information derived in MMs. More information can be found in [8].

3   Analysis of the PMIV Framework Using a Social Network Dataset

The work described in this paper has been evaluated using the Cattle Tracing System (CTS) database in operation in Great Britain (GB). The CTS database can be interpreted as a social network, where each node represents a geographical location and the links the number of cattle moved between locations. The network comprises a set of n time stamped datasets $D = \{d_1, d_2, ..., d_n\}$ partitioned into m episodes. Each data set d_i comprises a set of records such that each record describes a social network node paring indicating "traffic" between the two nodes. Each record description then consists of some subset of a global set of attributes A that describes the network elements. The process for identifying the frequent patterns of interest is outside the scope of this paper. However, a pattern is considered frequent if its support count is above some support threshold σ. The trend line for a particular frequent pattern is then defined by the sequence of support counts obtained for the datasets making up an episode.

Once the SOM maps had been generated the maps were processed so as to generate the associated migration matrices. A fragment of the migration matrix generated from the CTS dataset is given in Table 1. The table shows the number of CTS patterns that have migrated between episodes 2003 and 2004. For example in the case of node n_1, 71 trends have remained in the same node (self links), however 13 trends have migrated from node n_1 to node n_2, and so on. The values held in the migration matrices are also used to determine the C and Q-values for the visualization process.

Table 1. Migration Matrix for CTS network Pattern Migrations from E_{2003} to E_{2004}

	n_1	n_2	n_3	n_4	n_5	...	n_{100}
n_1	71	13	11	0	0	...	0
n_2	10	6	13	0	1	...	1
n_3	8	0	21	2	0	...	0
n_4	0	0	0	3	0	...	0
n_5	0	0	0	0	14	...	0
...	...	...	...	...	...	...	0
n_{100}	0	0	0	0	0	...	7

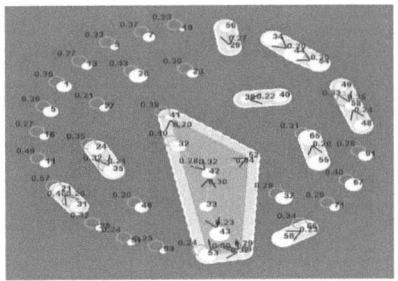

Fig. 2. CTS network Pattern Migrations for episodes 2003 and 2004

Figure 2 shows the CTS network pattern migration map for episodes 2003 and 2004. 45 nodes, out of a total of 100, had a C-value greater than 0.2 and were therefore displayed. The islands indicate communities of pattern migrations. The nodes are labeled with the node number from SOM_j, and the links with their C-value. The link direction shows the direction of migration from SOM_j to SOM_{j+1}. From the map we can identify that there are a 30 nodes with self-links. However, we can also deduce that (for example) patterns are migrating from node 34 to node 44, and from node 44 to 54 (thus indicating a trend change). The size of the nodes indicates how many patterns there are in each trend cluster, the bigger the node the large the number of patterns.

4   Conclusions

In this paper, the authors have described the PMIV framework for detecting changes in trend clusters within social network data. The PMIV framework then supports the analysis of trend clusters for the purpose of identifying pattern migrations between pairs of trend clusters found in a SOM_j and a SOM_{j+1}. The pattern migrations are detected using the concept of Migration Matrices and illustrated using an extension of Visuset.

References

1. Nohuddin, P.N.E., Christley, R., Coenen, F., Setzkorn, C.: Trend Mining in Social Networks: A Study Using a Large Cattle Movement Database. In: Perner, P. (ed.) ICDM 2010. LNCS (LNAI), vol. 6171, pp. 464–475. Springer, Heidelberg (2010)
2. Nohuddin, P.N.E., Coenen, F., Christley, R., Setzkorn, C.: Detecting Temporal Pattern and Cluster Changes in Social Networks: A Study Focusing UK Cattle Movement Database. In: Shi, Z., Vadera, S., Aamodt, A., Leake, D. (eds.) IIP 2010. IFIP AICT, vol. 340, pp. 163–172. Springer, Heidelberg (2010)
3. Kohonen, T.: The Self Organizing Maps. Series in Information Sciences, vol. 30. Springer, Heidelberg (1995)
4. Kohonen, T.: The Self Organizing Maps. Neurocomputing Elsevier Science 21, 1–6 (1998)
5. Sugiyama, K., Misue, K.: Graph Drawing by the Magnetic Spring Model. Journal of Visual Languages and Computing 6(3), 217–231 (1995)
6. Hido, S., Idé, T., Kashima, H., Kubo, H., Matsuzawa, H.: Unsupervised Change Analysis Using Supervised Learning. In: Washio, T., Suzuki, E., Ting, K.M., Inokuchi, A. (eds.) PAKDD 2008. LNCS (LNAI), vol. 5012, pp. 148–159. Springer, Heidelberg (2008)
7. Newman, M.E.J.: Fast Algorithms for Detecting Community Structure in Networks. Journal of Physical Review E, 1-5 (2004)
8. Nohuddin, P.N.E., Sunayama, W., Christley, R., Coenen, F., Setzkorn, C.: Trend Mining and Visualisation in Social Networks. In: Proc. of the 31st SGAI International Conference on Artificial Intelligence (2012)

Improving Top-N Recommendations with User Consuming Profiles

Yongli Ren, Gang Li, and Wanlei Zhou

School of Information Technology, Deakin University,
221 Burwood Highway, Vic 3125, Australia
{yongli,gang.li,wanlei}@deakin.edu.au

Abstract. In this work, we observe that user consuming styles tend to change regularly following some profiles. Therefore, we propose a *consuming profile* model to capture the user *consuming styles*, then apply it to improve the Top-N recommendation. The basic idea is to model user *consuming styles* by constructing a representative subspace. Then, a set of candidate items can be estimated by measuring its reconstruction error from its projection on the representative subspace. The experiment results show that the proposed model can improve the accuracy of *Top-N recommendation*s much better than the state-of-the-art algorithms.

1 Introduction

The Top-N recommendations are generally produced by two steps: firstly, the system predicts ratings on all candidate items; secondly, it will sort those candidate items according to their predicted ratings, and recommend the Top-N ranked items to the active user. Recently, Cremonesi et. al. argued that, for the task of *Top-N recommendation*, it is not necessary to predict the exact ratings. They proposed two new Top-N recommendation algorithms, *PureSVD* and *NNcosNgbr*, that are designed to provide correct ranking of items to the active user [3]. Their experiment results show that the *PureSVD* algorithm can always perform better on the Top-N recommendation task than those latent factor models based on rating prediction, including the well-known *SVD++* model that played a key role in the winning of the *Netflix* progress award [5]. However, the research on Top-N recommendation is still underdeveloped, especially compared with the rating prediction task.

In this paper, we argue that each user has a *consuming profile* that is different from his/her rating profile that is pursued by most rating prediction methods. In this work, we propose a *consuming profile* model to capture users' consuming styles. The idea is to construct a representative subspace for user *consuming profile*, then estimate how well a candidate item matches the user's consuming style by measuring its reconstruction error from its projection on the representative subspace. Due to the ability to model *user consuming styles* and their temporal dynamics, the *consuming profile* model can generate Top-N recommendations that can satisfy users' various needs better, and provide more accurate

P. Anthony, M. Ishizuka, and D. Lukose (Eds.): PRICAI 2012, LNAI 7458, pp. 887–890, 2012.
© Springer-Verlag Berlin Heidelberg 2012

recommendations than other *Top-N recommendation* methods, including those blending/ensemble techniques.

The rest of the paper is organized as follows. we present the *consuming profile* model in section 2, evaluate it in section 3, and summarize this paper in section 4.

2   The Consuming Profile

Let $\mathcal{U} = \{u_1, \cdots, u_x, \cdots, u_m\}$ denote a set of m users, $\mathcal{T} = \{t_1, \cdots, t_l, \cdots, t_n\}$ denote a set of n items, and $\mathcal{C} = \{c_1, \cdots, c_j, \cdots, c_q\}$ denote a set of q categories. $T_N(u_x)$ represents the list of Top-N recommendations to user u_x. $\mathcal{R}$ represents the $m \times n$ *rating matrix*, and r_{xl} denote the rating on item t_l by user u_x. $\mathcal{P} = \{\mathbf{p}_1, \cdots, \mathbf{p}_x, \cdots, \mathbf{p}_m\}$ denotes the *consuming profiles* for m users, and $\mathbf{p}_x$ denotes the *consuming profile* for user u_x. $\mathcal{Z} = \{\mathbf{z}_1, \cdots, \mathbf{z}_x, \cdots, \mathbf{z}_m\}$ denotes the projection of $\mathcal{P}$, and $\mathbf{z}_x$ denotes the projection of $\mathbf{p}_x$. The active user is the user for whom the recommendations are generated.

2.1   The Consuming Profile Model

We define that a *consuming profile* is a sequence of personal consuming styles. By representing user consuming styles in this way, users' various needs and their dynamics are naturally integrated. Each consuming profile is created for one user. The consuming style at a certain time is captured by the corresponding consuming profile vector.

However, to make use of this consuming profile for Top-N recommendations, there are two challenges: 1) the processing algorithm must have the ability to deal with incomplete data, since there are lots of missing values. 2) in the recommendation step, the algorithm must be able to measure how well a set of candidate items fit the user's consuming profile, then to decide whether this item will be recommended or not. In the following section, we will address these two challenges by constructing a *consuming profile subspace*.

In this paper, we build up a representative subspace for *consuming profiles* by information theory. In this work, we use *Principal Component Analysis* (PCA) to construct a subspace to capture the main variance of the *incomplete* consuming profiles in the data set. The projection from the subspace is given by:

$$\mathbf{z}_x = \mathcal{W}^T \cdot (\mathbf{p}_x - \mu), \tag{1}$$

where $\mu = \frac{1}{m} \sum_{x=1}^{m} \mathbf{p}_x$, $\mathbf{p}_x \in \mathcal{P}$ is the consuming profile for user u_x, $\mathbf{z}_x$ is the projection of $\mathbf{p}_x$ on the subspace, and $\mathcal{W}$ is the orthogonal eigenvectors of the covariance matrix of $\mathcal{P}$. Due to the high incompleteness of consuming profiles, we introduce an EM-like algorithm to learn the representative subspace from the observed values in the consuming profiles $\mathcal{P}$.

- E-Step: the missing values are replaced with corresponding values in μ, then apply the standard PCA to calculate $\mathcal{W}$. After that, $\mathbf{z}_x$ can be estimated by using the least squares solution with $\mathcal{W}_j^k$ and $[\mathbf{p}_x]^a$, where $[\mathbf{p}_x]^a$ denotes

the $\mathbf{p}_x$ in the current iteration step, but only have values on the positions that correspond to $\mathbf{p}_x^a$, and $\mathcal{W}_j^k$ denotes the first k principal component coefficients in the j-th iteration.
- M-Step: the reconstruction $\hat{\mathbf{p}}_x$ can be calculated as:

$$\hat{\mathbf{p}}_x = \mathcal{W} \cdot \mathbf{z}_x + \mu, \tag{2}$$

where $\mathbf{z}_x$ is the projection of $\mathbf{p}_x$ on the subspace.

2.2   The *CPS* Algorithm

When recommending a list of items to a user, it is actually trying to construct the user's consuming profile with minimum errors. Then, we apply the reconstruction error to estimate how much of a set of items match a user's consuming profile.

Given a set of items $S(u_x)$ to user u_x at time i, all items can be temporarily absorbed into u_x's consuming profile vector p_{xi}, then a tentatively changed consuming profile vector $\tilde{p}_{xi}$ is available. The *reconstruction error* for $S(u_x)$ is defined by using the squared distance between this changed consuming profile and the original one:

$$\| \tilde{\mathbf{p}}_x^a - \mu^a - [\mathcal{W}^k \cdot \mathbf{z}_x]^a \|^2, \tag{3}$$

The proposed *CPS* Algorithm can used implemented in two steps: 1) since there is no item recommended to the active user, his/her corresponding consuming profile is empty. We calculate the reconstruction error for each candidate item with Eq. 3, then initialize the empty consuming profile by absorbing the item with the minimum reconstruction error. 2) the algorithm will iteratively estimate each remaining candidate item. Specifically, it tentatively absorbs each remaining item into the current consuming profile, and take this one into the recommendation list if the reconstruction error is greater than it was before absorbing.

3   Experiment and Analysis

The data set used in this experiment is the *MovieLens* data set, which includes around 1 million ratings collected from $6,040$ users on $3,900$ movies. The data set is split into the *training set* and the *test set*. The *test set* only contains 5-star ratings. Following literature [2,3], we reasonably assume that the 5-star rated items are relevant to the active user, and use a similar methodology to conduct experiments. The quality of Top-N recommendations is measured by the *Recall*, which is defined based on the overall number of *hits*. For the active user, if the test item is in the Top-N recommendation list, we call this a *hit*.

To fully examine the performance of the proposed model, we compare the proposed *CPS* algorithm with other 6 state-of-the-art Top-N recommendation algorithms, including the *Logistic* Regression [1,6] and the *Mean* strategy, *PureSVD* and *NNcosNgbr*, Top Popular (*TopPop*) and Movie Average (*MovieAvg*). For these blending algorithms *CPS*, *Logistic* and *Mean*, they all blend the results

Table 1. Recall at N on *All-But-One* data set

Items	Algorithm	$N = 1$	$N = 5$	$N = 10$	$N = 15$	$N = 20$
All Items	CPS	**0.3843**	**0.5715**	**0.6321**	**0.6825**	**0.7260**
	Logistic	0.2194	0.4697	0.5866	0.6556	0.7022
	Mean	0.2010	0.4093	0.5298	0.6007	0.6543
	PureSVD	0.2111	0.4593	0.5781	0.6460	0.6881
	NNcosNgbr	0.1377	0.3240	0.4242	0.4886	0.5399
	TopPop	0.0740	0.2123	0.3030	0.3568	0.3969
	MovieAvg	0.0000	0.0217	0.0800	0.1306	0.1821

from *PureSVD*, *NNcosNgbr*, *TopPop* and *MovieAvg* for Top-N recommendations. For our *CPS*, the dimension of *consuming profile subspace* is set to 50, the max iteration of training is set to 20, and the error threshold is set to 10^{-5}.

Table 1 shows the performance of the algorithms on the *All-But-One* data set, when N equals to 1, 5, 10, 15, and 20, respectively. It is clear that the proposed *CPS* algorithm achieves significant better results than all the other algorithms. So, the proposed *CPS* algorithm can discover much more items than any other algorithms based on user rating estimation.

4   Conclusions

In this paper, a novel user *consuming profile* model is proposed to capture user consuming styles. Based on this model, we introduce a *CPS* algorithm for *Top-N recommendations*. Experiment results show that *CPS* performs significantly better than other blending methods in the field of Top-N recommendations.

References

1. Chapelle, O.: Yahoo! Learning to Rank Challenge Overview. Journal of Machine Learning Research - Proceedings Track 14, 1–24 (2011)
2. Cremonesi, P., Garzotto, F., Negro, S., Papadopoulos, A.: Comparative Evaluation of Recommender System Quality. In: CHI Extended Abstracts, pp. 1927–1932 (2011)
3. Cremonesi, P., Koren, Y., Turrin, R.: Performance of Recommender Algorithms on Top-N Recommendation Tasks. In: Recsys 2010, pp. 39–46. ACM (2010)
4. Herlocker, J.L., Konstan, J.A., Terveen, L.G., Riedl, J.T.: Evaluating Collaborative Filtering Recommender Systems. ACM Transactions on Information Systems 22(1), 5–53 (2004)
5. Koren, Y.: Factorization meets the neighborhood: a multifaceted collaborative filtering model. In: SIGKDD 2008, pp. 426–434. ACM (2008)
6. Sueiras, J., Salafranca, A., Florez, J.L.: A classical predictive modeling approach for Task: Who rated what? of the KDD CUP 2007. SIGKDD Explorations 9(2), 57–61 (2007)

Towards Providing Music for Academic and Leisurely Activities of Computer Users

Roman Joseph Aquino*, Joshua Rafael Battad, Charlene Frances Ngo,
Gemilene Uy, Rhia Trogo, Roberto Legaspi, and Merlin Teodosia Suarez

De La Salle University Manila,
2401 Taft Avenue Manila 1004
{rhia.trogo,merlin.suarez}@delasalle.ph

Abstract. This paper uses brainwaves to recognize the computer activity of the user and provides music recommendation. Twenty-three (23) hours of data collection was performed by asking the computer user to wear a device that collects electroencephalogram (EEG) signals from his brain as he performed whatever tasks he wanted to perform while listening to music. The features of the preferred song given the activity of the user is used to provide songs for the user automatically. Activities were classified as either academic or leisure. The music provision model was able to predict the music features preferred by the user with accuracy of 76%.

Keywords: support provision, brain-computer interface, intention recognition, behavior recognition.

1 Introduction

This presents a music provision system that automatically provides music for a computer user given the computer activity of the person (i.e., academic or leisurely). The music played while the computer user is engaged in an activity helps his well-being however, the user may not have the time to preselect songs that he or she would like to listen to. Thus, automating the music selection for the user based on the activities performed offloads the user the task of having to select his songs that matches his given activity. This research assumes that the song features are indicators of the type of songs the user prefers to listen to while performing an activity.

A more straightforward solution in providing music support for the computer user is by looking at the applications that are being run. However, this is only plausible if there is a complete listing of all the possible applications and websites that can be used by the user and there is a corresponding annotation for each that indicates which type of music is preferred given these applications. Hence, this paper classifies the activity using the brainwaves of the computer user.

---

* This research would was funded by DOST-PCASTRD.

P. Anthony, M. Ishizuka, and D. Lukose (Eds.): PRICAI 2012, LNAI 7458, pp. 891–894, 2012.
© Springer-Verlag Berlin Heidelberg 2012

2   Data Gathering Sessions

The researchers interviewed a group of students without revealing that the interview is to find out whether they listen to music as they perform their activities. Among the students interviewed, only one subject was selected based on how the user plays music as he performs his computer activities.

The music database from [2] was used alongside the activity-music tracker program in order to gather data. The researchers elicited data from the subjects for a total of forty-six (46) sessions where each session lasted for thirty (30) minutes. Twenty-three (23) hours of data gathering was performed.

The subject chose to listen to one hundred ninety-two (192) songs. The subject was seated in front of a computer while wearing the Emotiv EPOC Neuroheadset EEG. The sessions were held in an isolated room where the subject is accompanied only by one researcher.

The subject was given the free rein to choose whichever activity he wants to perform. The Music Tracker software logged all the applications used as well as the sites visited and music listened to by the user. Music that was skipped by the user, meaning, the user did not finish the song (i.e., song is played for less than 20 seconds), are considered to be inappropriate songs played given the current context (i.e., activity type and specific activity of the user). This approach was also applied in the works of [3] where it is assumed that if the user cancels an action, the action is not preferred by the user. The music that is not skipped by the user is assumed to be music that the user deems acceptable or tolerable. After 30 minutes of data gathering, the subject was asked to annotate the activities that were performed as academic or leisurely.

3   Music Provision System

The inputs for the Music Provision System are the EEG signals of the computer user. The EEG signals of the computer user are segmented, and then bandpass filtering and FFT are applied and feature extraction ensues. After the features are extracted, these are then classified as academic activity or leisure activity brainwaves based on the Brainwaves-Activity Model [1].

C4.5, the classifier used, generated a tree that makes use of the music features to identify the activity type. The tree describes the type of music that is deemed to be something that the user does not find to be distracting (i.e., music with features that is similar to music that was not skipped the user before). The tree generated gives the rules for music (i.e., features of music associated to the type of activity) that can be played for a certain activity. Hence, the activity type can now be used to find the matching features that are used for an activity type (song features for activity type). The rules that were generated from the Music Features-Activity Type Model[1] are then used to provide the music. There are however, many rules that can be derived for a general activity, hence, the specific activity is also used to extract rules (i.e., song features for a specific activity) from the Music Features-Specific Activity[1]. This takes into account the specific activity of the user. From this specific activity, music rules are derived and these rules serve as basis for the music that is provided. Each set of song features for general activity will be assessed based on its similarity to the set of song features for the specific activity. Similarity is measured using the Manhattan Distance.

A distance score is computed for each set of rule of the general activity. The rule with the least value will be used as the best rule and its set of song features will be used to get the suggested song. The distance score is computed as the summation of the results of the Manhattan Distance discussed earlier between the set of song features of the general activity rule and the sets of song features of the specific activity rules.

After acquiring the best rule among the general activity rules, its set of song features will be compared to every set of song features of every song in the music repository to get the closest song to the rule. Another distance score is assigned to each song in the music repository. Manhattan Distance Algorithm will be used to compute the distance score of each song. The song with the least distance score will be suggested to the user.

3.1 Static Rule Learning

In order for the system to know that a song was previously rejected by the user, the distance score discussed previously is modified in a way that past events will be considered in the calculation. The Distance Score will be updated so that the new Distance Score will also consider the past feedback of the user. The Learning Score ranges from 0 to 1, the Maximum Learning Score therefore is 1. This Learning Score will be used in calculating the new Distance Score of a given rule. The new computation for the Distance Score is shown in Equation 1. The function uses Learning Scores assigned to each rule of the General Activity rules.

$$\text{Distance Score} = \alpha\,(D\text{\_}Score) - (1 - \alpha)\,L\text{\_}Score \qquad (1)$$

where, $D\text{\_}Score$ is the Distance Score and the $L\text{\_}Score$ is the Learning Score. α ranges from $0 < 1$ inclusive. If α approaches 1, the Distance Score or the basis of the song decision will be solely dependent on the Distance Score, therefore there is no learning involved. If $\alpha = 0$, the new Distance Score will be solely dependent on the Learning Score, therefore this considers only the full learning. When a song is rejected by the user, the system deducts 0.2 from the Learning Score of its rule. Clicking the Next button in the system is interpreted as rejection of the song provided for the user [3]. As seen from the Equation 1, having a lower Learning Score lowers the chances of it being suggested. Initial value of Learning Score is 1.

Tests were conducted where 4 exact copies of brainwaves were used as an input to the system. Following the algorithm, the songs should vary if the user continues to click next for all the four instances. The result of the tests shows that if $\alpha > 0.6$, the songs are not varying, therefore the system is not learning. Learning here is the ability of the system to adapt. The system uses $\alpha = 0.6$.

This gives the Distance Score priority, so that the system will be using the right rule given the situation, but is also closely considering the Learning Score which covers the past feedbacks of the user.

3.2   Test Results for Static Rule Learning

To test the accuracy of the model, the model was evaluated according to the correctness of its output compared to the actual song of the given instance in the test set (2-hour unseen data). If the Manhattan Distance of the songs is below the threshold 1, then the predicted song is close to the actual. If the distance approaches 0, then the song features of the songs being compared become closer. Refer to Equation 2.

$$Accuracy = \frac{Accepted\ Comparisons}{Total\ Instances} * 100\% \tag{2}$$

The Brainwave-Activity model is 54% accurate. Given that the accuracy of the Brainwave-Activity model is such, the accuracy of the Static Rule Learning is 66%. If the Brainwave-Activity model is omitted and the activity is just manually provided hence, this is 100% accurate, the accuracy of Static Rule Learning is 76%.

4   Conclusion

This paper presents the results for music provision using the user's brainwaves, specific activities and activity types. This paper shows that the music preference of the user given the user's activities may be provided automatically using the features used in [2]. The experiment conducted shows that the model was able to accurately provide the music preference of the user with the accuracy of 76% given that the brainwaves were able to characterize the type of the activity correctly.

References

1. Aquino, R.J., Battad, J., Ngo, C.F., Uy, G., Trogo, R., Suarez, M.: Towards Empathic Support Provision for Computer Users. In: Nishizaki, S., et al. (eds.) WCTP 2011. PICT, vol. 5, pp. 15–27. Springer, Japan (2012)
2. Azcarraga, A., Manalili, S.: Design of a Structured 3D SOM as a Music Archive. In: Laaksonen, J., Honkela, T. (eds.) WSOM 2011. LNCS, vol. 6731, pp. 188–197. Springer, Heidelberg (2011)
3. Mozer, M.C., Miller, D.: Parsing the Stream of Time: The Value of Event-Based Segmentation in a Complex Real-World Control Problem. In: Giles, C.L., Gori, M. (eds.) IIASS-EMFCSC-School 1997. LNCS (LNAI), vol. 1387, pp. 370–388. Springer, Heidelberg (1998)

Overlapping Community Discovery via Weighted Line Graphs of Networks

Tetsuya Yoshida

Graduate School of Information Science and Technology,
Hokkaido University
N-14 W-9, Sapporo 060-0814, Japan
yoshida@meme.hokudai.ac.jp

Abstract. We propose an approach for overlapping community discovery via weighted line graphs of networks. For undirected connected networks without self-loops, we generalize previous weighted line graphs by: 1) defining weights of a line graph based on the weights in the original network, and 2) removing self-loops in weighted line graphs, while sustaining their properties. By applying some off-the-shelf node partitioning method to the weighted line graph, a node in the original network can be assigned to more than one community based on the community labels of its adjacent links. Various properties of the proposed weighted line graphs are clarified. Furthermore, we propose a generalized quality measure for soft assignment of nodes in overlapping communities.

1 Introduction

We propose an approach for overlapping community discovery via weighted line graphs of networks. For undirected connected networks without self-loops, we generalize previous weighted line graphs [3] by: 1) defining weights of a line graph based on the weights in the original network, and 2) removing self-loops in weighted line graphs, while sustaining their properties. After transforming the original network into a weighted line graph, community discovery is conducted by applying some off-the-shelf node partitioning method to the transformed graph. The node in the original network can be assigned to more than one community based on the community labels of its adjacent links.

2 Weighted Line Graphs of Networks

2.1 Previous Weighted Line Graphs

Definition 1 (line graph [2]). *For a simple graph $G=(V, E)$, the line graph $L(G)$ of G is the graph on E in which $x, y \in E$ are adjacent as vertices if and only if they are adjacent as edges in G.*

P. Anthony, M. Ishizuka, and D. Lukose (Eds.): PRICAI 2012, LNAI 7458, pp. 895–898, 2012.
© Springer-Verlag Berlin Heidelberg 2012

For a simple graph G, the adjacency matrix $\mathbf{C}$ of the corresponding line graph $L(G)$ can be represented in terms of the incidence matrix $\mathbf{B}$ of G as:

$$\mathbf{C} = \mathbf{B}^T\mathbf{B} - 2\mathbf{I}_m \qquad (1)$$

where $\mathbf{I}_m$ stands for the square identity matrix of size m [1].

Several kinds of weighted line graphs were proposed [3]. Among them, the following weighted adjacency matrices were proposed as the recommended representation matrices for their weighted line graphs:

$$\mathbf{E} = \mathbf{B}^T\mathbf{D}^{-1}\mathbf{B} \qquad (2)$$
$$\mathbf{E}_1 = \mathbf{B}^T\mathbf{D}^{-1}\mathbf{A}\mathbf{D}^{-1}\mathbf{B} \qquad (3)$$

where $\mathbf{D}$ is a diagonal matrix with $\mathbf{D}_{ii} = k_i$ [2]. The diagonal matrix $\mathbf{D}$ is also represented as $diag(\boldsymbol{k})$ in this paper.

2.2   Weighted Line Graphs for Weighted Networks

Representation Matrices for Weighted Networks. Let $\tilde{\mathbf{A}}$ stands for a weighted adjacency matrix of a network G where $\tilde{\mathbf{A}}_{ij} = w_{ij}$. Based on $\tilde{\mathbf{A}}$, we define the following vector and matrix:

$$\tilde{\boldsymbol{k}} = \tilde{\mathbf{A}}\mathbf{1}_n \qquad (4)$$
$$\tilde{\mathbf{D}} = diag(\tilde{\boldsymbol{k}}) \qquad (5)$$

The vector $\tilde{\boldsymbol{k}}$ in eq.(4) represents the sum of weights on links which are adjacent to each node in G. The matrix $\tilde{\mathbf{D}}$ in eq.(5) corresponds to the weighted counterpart of the diagonal matrix $\mathbf{D}$ in eq.(2) and eq.(3).

Finally, we define a weighted incidence matrix $\tilde{\mathbf{B}}$ based on the weights in G. For a link $\alpha=(i, j)$ which is adjacent to node i and node j in G, $\tilde{\mathbf{B}}_{i\alpha}$ and $\tilde{\mathbf{B}}_{j\alpha}$ are set to the weight w_{ij} in G; other elements in the α-th column of $\tilde{\mathbf{B}}$ are set to zeros. Thus, as in the generalization from $\mathbf{A}$ to $\tilde{\mathbf{A}}$, the standard 0-1 incidence matrix $\mathbf{B}$ is generalized to $\tilde{\mathbf{B}}$ based on the weights in G.

Weighted Line Graphs for Weighted Networks. Based on the above matrices, we define the following representation matrices for weighted line graphs, which are the generalization of the standard line graph as well as the ones in [3]:

$$\tilde{\mathbf{C}} = \tilde{\mathbf{B}}^T\tilde{\mathbf{B}} - 2diag(\boldsymbol{w}') \qquad (6)$$
$$\tilde{\mathbf{E}} = \tilde{\mathbf{B}}^T\tilde{\mathbf{D}}^{-1}\tilde{\mathbf{B}} \qquad (7)$$
$$\tilde{\mathbf{E}}_1 = \tilde{\mathbf{B}}^T\tilde{\mathbf{D}}^{-1}\tilde{\mathbf{A}}\tilde{\mathbf{D}}^{-1}\tilde{\mathbf{B}} \qquad (8)$$

where $\boldsymbol{w}' = \boldsymbol{w} \odot \boldsymbol{w}$. Here, $\odot$ stands for the Hadamard product.

---

[1] We represent the number of nodes as n, the number of links as m.
[2] The i-th diagonal element is set to the i-th element of $\boldsymbol{k}$ in $\mathbf{D}$.

Theorem 1. *Following properties hold for the previous weighted line graphs and the proposed weighted line graphs:*

$$1_m^T \mathbf{C} = (\mathbf{k} - 1_n)^T \mathbf{B} \qquad 1_m^T \tilde{\mathbf{C}} = (\tilde{\mathbf{k}} - 1_n)^T \tilde{\mathbf{B}} \qquad (9)$$

$$1_m^T \mathbf{E} = 21_m^T \qquad 1_m^T \tilde{\mathbf{E}} = 2\mathbf{w}^T \qquad (10)$$

$$1_m^T \mathbf{E}_1 = 21_m^T \qquad 1_m^T \tilde{\mathbf{E}}_1 = 2\mathbf{w}^T \qquad (11)$$

Proofs of other properties are omitted due to space shortage.

2.3   Weighted Line Graphs without Self-loops

According to Definition 1, for a simple network, the corresponding line graph should also be simple network. However, previous representation matrices $\mathbf{E}$ and $\mathbf{E}_1$ contain self-loops (i.e., diagonal elements are not zeros). Thus, the structural correspondence of the original network and its transformed graph does not hold in the previous approach.

Suppose $\mathbf{M} \in \mathbb{R}_+^{\ell \times \ell}$ is a (weighted) adjacency matrix for a network with non-zero diagonal elements (i.e., with self-loops). The proposed method distributes the diagonal elements of $\mathbf{M}$ into off-diagonal elements while sustaining some properties of $\mathbf{M}$. To realize this, we define the following matrices and vectors:

$$\mathbf{m} = diag(\mathbf{M}) \qquad (12)$$

$$\mathbf{D}_M = diag(\mathbf{m}) \qquad (13)$$

$$\mathbf{M}_{wo} = \mathbf{M} - \mathbf{D}_M \qquad (14)$$

$$\mathbf{m}_{wo} = \mathbf{M}_{wo} 1_\ell \qquad (15)$$

Based on the above matrices and vectors, our method transforms the matrix $\mathbf{M} \in \mathbb{R}_+^{\ell \times \ell}$ into the following matrix $\mathbf{N} \in \mathbb{R}_+^{\ell \times \ell}$ as:

$$\mathbf{N} = \mathbf{M}_{wo} + \mathbf{D}_M^{1/2} \, diag(\mathbf{m}_{wo})^{-1/2} \, \mathbf{M}_{wo} \, diag(\mathbf{m}_{wo})^{-1/2} \, \mathbf{D}_M^{1/2} \qquad (16)$$

where $diag(\mathbf{m}_{wo})$ is a diagonal matrix with $\mathbf{m}_{wo}$ (as $\mathbf{D}_M$ in eq.(13)).

The following properties hold for the transformed matrix $\mathbf{N}$:

Theorem 2. *For a symmetric square matrix $\mathbf{M}$ with non-negative real values, the following properties for the matrix $\mathbf{N}$ with eq.(16):*

$$diag(\mathbf{N}) = \mathbf{0}_\ell \qquad (17)$$

$$\mathbf{N}^T = \mathbf{N} \qquad (18)$$

$$\mathbf{N}1_\ell = \mathbf{M}1_\ell \qquad (19)$$

Proofs are omitted due to space shortage.

2.4   A Generalized Modularity for Overlapping Communities

The standard modularity for node partitioning [1] is based on the "hard" assignment of each node into a community, since it is based on node partitioning.

We propose a generalized modulariy based on the "soft" assignment of nodes for overlapping communities.

For overlapping communities, we generalize the 0-1 indicator matrix $\mathbf{S} \in \{0,1\}^{n \times c}$ to $\tilde{\mathbf{S}} \in \mathbb{R}_+^{n \times c}$. The matrix $\tilde{\mathbf{S}}$ represents a soft assignment of nodes into communities, and needs to satisfy the following:

$$\tilde{\mathbf{S}}_{ij} \geq 0, \quad \forall i,j \tag{20}$$

$$\tilde{\mathbf{S}}\mathbf{1}_c = \mathbf{1}_n \tag{21}$$

With $\tilde{\mathbf{S}}$, we propose the following soft modularity Q_s, we propose the following function as a generalized modularity for overlapping communities:

$$Q_s = \frac{1}{\mathbf{1}_n^T \mathbf{A} \mathbf{1}_n} \operatorname{tr}(\tilde{\mathbf{S}}^T (\mathbf{A} - \mathbf{P}) \tilde{\mathbf{S}}) \tag{22}$$

2.5   Soft Indicator Matrix for Link Partitioning

In link partitioning, each link is assigned to a community. Based on the community labels of links, we define a matrix $\tilde{\mathbf{H}} \in R_+^{m \times c}$ as:

$$\tilde{\mathbf{H}}_{\alpha k} = \begin{cases} w_\alpha & if \;\; link \; \alpha \text{ (with weight } w_\alpha) \text{ is assigned to community } k \\ 0 & \text{otherwise} \end{cases} \tag{23}$$

Based on $\tilde{\mathbf{H}}$ in eq.(23), we construct a soft indicator matrix $\tilde{\mathbf{S}}$ as:

$$\tilde{\mathbf{S}}_{ik} = \frac{\sum_\alpha \text{ adjacent to } _i \tilde{\mathbf{H}}_{\alpha k}}{\sum_{k'} \sum_\alpha \text{ adjacent to } _i \tilde{\mathbf{H}}_{\alpha k'}} \tag{24}$$

3   Concluding Remarks

In this paper we proposed weighted line graphs of networks for overlapping community discovery. Various properties of the proposed weighted line graphs are clarified, and the properties indicate that the proposed weighted line graphs are natural extensions of previous ones. Furthermore, we propose a generalized quality measure for soft assignment of nodes in overlapping community discovery. Evaluations of the proposed approach are left for future work.

Acknowledgments. This work is partially supported by the grant-in-aid for scientific research (No. 24300049) funded by MEXT, Japan, Casio Science Promotion Foundation and Toyota Physical & Chemical Research Institute.

References

1. Clauset, A., Newman, M.E.J., Moore, C.: Finding community structure in very large networks. Physical Review E 70(6), 066111 (2004)
2. Diestel, R.: Graph Theory. Springer (2006)
3. Evans, T., Lambiotte, R.: Line graphs, link partitions, and overlapping communities. Physical Review E 80(1), 016105:1–016105:8 (2009)

Community Structure Based Node Scores
for Network Immunization

Tetsuya Yoshida and Yuu Yamada

Graduate School of Information Science and Technology,
Hokkaido University
N-14 W-9, Sapporo 060-0814, Japan
{yoshida,yamayuu}@meme.hokudai.ac.jp

Abstract. We propose community structure based node scores for network immunization. Since epidemics (e.g, virus) are propagated among groups of nodes (communities) in a network, network immunization has often been conducted by removing nodes with large score (e.g., centrality) so that the major part of the network can be protected from the contamination. Since communities are often interwoven through intermediating nodes, we propose to identify such nodes based on the community structure of a network. By regarding the community structure in terms of nodes, we construct a vector representation of each node based on a quality measure of communities for node partitioning. Two types of node score are proposed based on the direction and the norm of the constructed node vectors.

1 Introduction

When a node is contaminated with some disease or virus, it is necessary to remove (or, vaccinate) it in order to prevent the propagation of contamination over the whole network; otherwise, contamination is propagated to other groups of nodes (communities). However, compared with the whole size of a network, it is often the case that the amount of available doze is limited. Thus, it is necessary to selectively utilize the available doze for removing nodes. Since communities are often interwoven through intermediating nodes, it is important to identify such nodes for network immunization.

We propose community structure based node scores for network immunization. Contrary to other approaches based on the graph cut of links in a network [3], we consider the community structure in terms of nodes in a network. Based on a quality measure of communities for node partitioning [1], a vector representation of nodes in a network is constructed, and the community structure in terms of the distribution of node vectors is utilized for calculating node scores. Two types of node score are proposed based on the direction and the norm of the constructed node vectors.

P. Anthony, M. Ishizuka, and D. Lukose (Eds.): PRICAI 2012, LNAI 7458, pp. 899–902, 2012.

2   Network Immunization

Epidemics (e.g, virus) are often propagated through the interaction between nodes (e.g., individuals, computers) in a network. If a contaminated node interacts with other nodes, contamination can spread over the whole network. In order to protect the nodes in the network as much as possible, it is necessary to disconnect (or, remove) the contaminated node so that the major part of the network, such as the largest connected component (LCC), can be prevented from the contamination.

The standard approach for network immunization is to remove nodes which play the major role in message propagation. Based on the assumption that such nodes are in some sense "central" nodes in networks, removal of nodes with large centrality has been widely used as a heuristic immunization strategy.

2.1   Node Centrality

Various notions of "node centrality" have been studied in social network analysis [2,4]. Since nodes with many links can be considered as a hub in a network, the degree (number of links) of a node is called degree centrality. On the other hand, betweenness centrality focuses on the shortest path along which information is propagated over a network. By enumerating the shortest paths between each pair of nodes, betweenness centrality of a node is defined as the number of shortest paths which go through the node.

Similar to the famous Page Rank, eigenvector centrality utilizes the leading eigenvector of the adjacency matrix $\mathbf{A}$ of a network, and each element (value) of the eigenvector is considered as the score of the corresponding node. Based on the approximate calculation of eigenvector centrality via perturbation analysis, another centrality (called dynamical importance) was also proposed in [5].

3   Community Structure Based Node Scores

3.1   Node Vectors

It was shown that the maximization of modularity, which has been widely utilized as a quality measure of communities for node partitioning [1], can be sought by finding the leading eigenvector of the following matrix [3]:

$$\mathbf{B} = \mathbf{A} - \mathbf{P} \qquad (1)$$

By utilizing several eigenvectors of $\mathbf{B}$ in eq.(1) with the largest positive eigenvalues, the modularity matrix $\mathbf{B}$ can be approximately decomposed as:

$$\mathbf{B} \simeq \mathbf{U}\mathbf{\Lambda}\mathbf{U}^T \qquad (2)$$

where $\mathbf{U}=[\boldsymbol{u}_1,\cdots,\boldsymbol{u}_q]$ are the eigenvectors of $\mathbf{B}$ with the descending order of eigenvalues, and $\mathbf{\Lambda}$ is the diagonal matrix with the corresponding eigenvalues. Based on eq.(2), the following data representation was proposed [3]:

$$\mathbf{X} = \mathbf{U}\mathbf{\Lambda}^{1/2} \qquad (3)$$

Each row of $\mathbf{X}$ corresponds to the representation of a node in $\mathbb{R}^q$ (q is the number of eigenvectors). Hereafter, the i-th node in a network is represented as a vector $\boldsymbol{x}_i \in \mathbb{R}^q$ (i-th row of $\mathbf{X}$ in eq.(3)), and is called a node vector.

With this vector representation, community centrality was defined as [3] [1]:

$$cc(\boldsymbol{x}_i) = \boldsymbol{x}_i^T \boldsymbol{x}_i \tag{4}$$

3.2 Inverse Vector Density

We propose a node score in terms of the mutual angle between node vectors in eq.(3). The number of "near-by" node vectors for each node is utilized for identifying border nodes (which act as bridges between communities). To realize this, we calculate angle θ_{ij} between each pair of nodes (i, j), and count the number of vectors which are inside the cone with angle θ for each node.

The above idea is formalized as follows:

$$\mathbf{D} = diag(\|\boldsymbol{x}_1\|, \ldots, \|\boldsymbol{x}_n\|) \tag{5}$$

$$\mathbf{X}_1 = \mathbf{D}^{-1}\mathbf{X} \tag{6}$$

$$\boldsymbol{\Theta} = \cos^{-1}(\mathbf{X}_1 \mathbf{X}_1{}^T) \tag{7}$$

$$f(\boldsymbol{\Theta}_{ij}, \theta) = \begin{cases} 1 & (\boldsymbol{\Theta}_{ij} < \theta) \\ 0 & (otherwise) \end{cases} \tag{8}$$

$$ivd(\boldsymbol{x}_i) = \frac{1}{\sum_j^n f(\boldsymbol{\Theta}_{ij}, \theta)} \tag{9}$$

where $\mathbf{D}$ in eq.(5) is a diagonal matrix with elements $\|\boldsymbol{x}_1\|, \ldots, \|\boldsymbol{x}_n\|$, and θ is a threshold. The value of $ivd(\cdot)$ in eq.(9) corresponds to the node score.

By scaling the matrix $\mathbf{X}$ in eq.(3) with $\mathbf{D}^{-1}$, each row of $\mathbf{X}_1$ is normalized as one in L2 norm. Thus, the angle of each pair of node vectors can be calculated with respect to the normalized $\mathbf{X}_1$ in eq.(7). The function f in eq.(8) is an indicator function and checks if the mutual angle θ_{ij} is less than the threshold θ. Finally, since border nodes have relatively small number of near-by node vectors (as in Section ??), the node score is calculated by taking the inverse of the number of vectors inside the cone with angle θ. We call this node score as IVD (inverse vector density), since it is based on the inverse of the number of node vectors over the base of a cone with radius one and angle θ.

3.3 Community Centrality Based Inverse Vector Density

As in other centrality based immunization methods, removal of hub nodes, which act as mediators of information diffusion over the network, also seems effective for network immunization. For instance, hub nodes have large centrality in terms of degree centrality and eigenvector centrality in Section 2.1. However, since node

---

[1] Both the square sum ($\boldsymbol{x}_i^T \boldsymbol{x}_i$) and the absolute sum (norm) ($\|\boldsymbol{x}_i\|$) were called community centrality in [3].

vectors tend to concentrate around community centers, the node score of a hub node gets rather small with IVD in Section 3.2, since it is based on the inverse of the number of near-by node vectors.

One of the reasons is that, the direction of each node vector in eq.(3) is utilized in IVD, but another property, namely its norm, is not utilized yet. The norm (or, square norm) of a node vector was regarded to what extent the node is central to a community [3], and was named as community centrality in eq.(4). We reflect the community centrality on IVD and define another node score as:

$$ccivd(\boldsymbol{x}_i) = cc(\boldsymbol{x}_i) \times ivd(\boldsymbol{x}_i) \tag{10}$$

We call this node score as CCIVD (Community Centrality based Inverse Vector Density).

4   Concluding Remarks

We proposed community structure based node scores for network immunization. A vector representation of each node is constructed based on a quality measure of communities, and the community structure in terms of the distribution of node vectors is utilized for calculating node scores. Two types of node score were proposed based on the direction and the norm of the constructed node vectors. Evaluations of the proposed node scores are left for future work.

Acknowledgments. This work is partially supported by the grant-in-aid for scientific research (No. 24300049) funded by MEXT, Japan, Casio Science Promotion Foundation and Toyota Physical & Chemical Research Institute.

References

1. Clauset, A., Newman, M.E.J., Moore, C.: Finding community structure in very large networks. Physical Review E 70(6), 066111 (2004)
2. Mika, P.: Social Networks and the Semantic Web. Springer (2007)
3. Newman, M.: Finding community structure using the eigenvectors of matrices. Physical Review E 76(3), 036104 (2006)
4. Newman, M.: Networks: An Introduction. Oxford University Press (2010)
5. Restrepo, J.G., Ott, E., Hunt, B.R.: Characterizing the dynamical importance of network nodes and links. Physical Review Letters 97, 094102 (2006)

Author Index